Forms Involving $\sqrt{u^2 - a^2}$

39 $\displaystyle\int \sqrt{u^2 - a^2}\, du = \frac{u}{2}\sqrt{u^2 - a^2} - \frac{a^2}{2}\ln|u + \sqrt{u^2 - a^2}| + C$

40 $\displaystyle\int u^2\sqrt{u^2 - a^2}\, du = \frac{u}{8}(2u^2 - a^2)\sqrt{u^2 - a^2} - \frac{a^4}{8}\ln|u + \sqrt{u^2 - a^2}| + C$

41 $\displaystyle\int \frac{\sqrt{u^2 - a^2}}{u}\, du = \sqrt{u^2 - a^2} - a\cos^{-1}\frac{a}{u} + C$

42 $\displaystyle\int \frac{\sqrt{u^2 - a^2}}{u^2}\, du = -\frac{\sqrt{u^2 - a^2}}{u} + \ln|u + \sqrt{u^2 - a^2}| + C$

43 $\displaystyle\int \frac{du}{\sqrt{u^2 - a^2}} = \ln|u + \sqrt{u^2 - a^2}| + C$

44 $\displaystyle\int \frac{u^2\, du}{\sqrt{u^2 - a^2}} = \frac{u}{2}\sqrt{u^2 - a^2} + \frac{a^2}{2}\ln|u + \sqrt{u^2 - a^2}| + C$

45 $\displaystyle\int \frac{du}{u^2\sqrt{u^2 - a^2}} = \frac{\sqrt{u^2 - a^2}}{a^2 u} + C$

46 $\displaystyle\int \frac{du}{(u^2 - a^2)^{3/2}} = -\frac{u}{a^2\sqrt{u^2 - a^2}} + C$

Forms Involving $a + bu$

47 $\displaystyle\int \frac{u\, du}{a + bu} = \frac{1}{b^2}(a + bu - a\ln|a + bu|) + C$

48 $\displaystyle\int \frac{u^2\, du}{a + bu} = \frac{1}{2b^3}[(a + bu)^2 - 4a(a + bu) + 2a^2\ln|a + bu|] + C$

49 $\displaystyle\int \frac{du}{u(a + bu)} = \frac{1}{a}\ln\left|\frac{u}{a + bu}\right| + C$

50 $\displaystyle\int \frac{du}{u^2(a + bu)} = -\frac{1}{au} + \frac{b}{a^2}\ln\left|\frac{a + bu}{u}\right| + C$

51 $\displaystyle\int \frac{u\, du}{(a + bu)^2} = \frac{a}{b^2(a + bu)} + \frac{1}{b^2}\ln|a + bu| + C$

52 $\displaystyle\int \frac{du}{u(a + bu)^2} = \frac{1}{a(a + bu)} - \frac{1}{a^2}\ln\left|\frac{a + bu}{u}\right| + C$

53 $\displaystyle\int \frac{u^2\, du}{(a + bu)^2} = \frac{1}{b^3}\left(a + bu - \frac{a^2}{a + bu} - 2a\ln|a + bu|\right) + C$

54 $\displaystyle\int u\sqrt{a + bu}\, du = \frac{2}{15b^2}(3bu - 2a)(a + bu)^{3/2} + C$

55 $\displaystyle\int \frac{u\, du}{\sqrt{a + bu}} = \frac{2}{3b^2}(bu - 2a)\sqrt{a + bu}$

56 $\displaystyle\int \frac{u^2\, du}{\sqrt{a + bu}} = \frac{2}{15b^3}(8a^2 + 3b^2u^2 - 4abu)\sqrt{a + bu}$

57 $\displaystyle\int \frac{du}{u\sqrt{a + bu}} = \frac{1}{\sqrt{a}}\ln\left|\frac{\sqrt{a + bu} - \sqrt{a}}{\sqrt{a + bu} + \sqrt{a}}\right| + C,$ if $a > 0$
$\displaystyle\qquad\qquad\qquad = \frac{2}{\sqrt{-a}}\tan^{-1}\sqrt{\frac{a + bu}{-a}} + C,$ if $a < 0$

58 $\displaystyle\int \frac{\sqrt{a + bu}}{u}\, du = 2\sqrt{a + bu} + a\int \frac{du}{u\sqrt{a + bu}}$

59 $\displaystyle\int \frac{\sqrt{a + bu}}{u^2}\, du = -\frac{\sqrt{a + bu}}{u} + \frac{b}{2}\int \frac{du}{u\sqrt{a + bu}}$

60 $\displaystyle\int u^n\sqrt{a + bu}\, du = \frac{2}{b(2n + 3)}\left[u^n(a + bu)^{3/2} - na\int u^{n-1}\sqrt{a + bu}\, du\right]$

61 $\displaystyle\int \frac{u^n\, du}{\sqrt{a + bu}} = \frac{2u^n\sqrt{a + bu}}{b(2n + 1)} - \frac{2na}{b(2n + 1)}\int \frac{u^{n-1}\, du}{\sqrt{a + bu}}$

62 $\displaystyle\int \frac{du}{u^n\sqrt{a + bu}} = -\frac{\sqrt{a + bu}}{a(n - 1)u^{n-1}} - \frac{b(2n - 3)}{2a(n - 1)}\int \frac{du}{u^{n-1}\sqrt{a + bu}}$

(Continued inside back cover)

Calculus

Calculus

James Stewart
McMaster University

Brooks/Cole Publishing Company
Pacific Grove, California

Brooks/Cole Publishing Company
A Division of Wadsworth, Inc.

Printed in the United States of America

10 9 8 7 6 5 4 3

Library of Congress Cataloging-in-Publication Data

Stewart, James [date]
  Calculus.

  Includes index.
  1. Calculus. 2. Geometry, Analytic. I. Title.
QA303.S8825 1986   515'.15   86-6078

ISBN 0-534-06690-9

Sponsoring Editor: Jeremy Hayhurst
Editorial Assistant: Amy Mayfield
Production Services Coordinator: Joan Marsh
Production: Cece Munson, The Cooper Company
Manuscript Editor: Carol Reitz
Interior Design: Detta Penna
Interior Illustration: Vantage Art, Inc.
Typesetting: Syntax International
Cover Design: Jamie Sue Brooks
Cover Photograph: Richard Bellomy
Violin Owned by: Ann Silva Busch
Violin Courtesy of: Van Butler
Cover Printing: Phoenix Color Corporation
Printing and Binding: Rand McNally & Company

Endpapers: From *Calculus with Analytic Geometry* by E. W. Swokowski.
Copyright © 1983 by PWS Publishers. Reprinted by permission.

To my mother
and to the memory of my father

Preface

A successful course in calculus must provide a foundation for future work in mathematics, although we recognize that the course can also be valuable training in analytical thinking and comprise an important part of a general education. Our teaching responsibilities are to instill a certain technical competence in our students, which stresses when and where to apply calculus, and we teach those skills within a tight budget of time and opportunity. Within those constraints we must try to convey a sense of the power of the discipline while we progress through its fundamentals and mechanics. A good textbook, like a good teacher, should accomplish both aims.

What motivates someone to complete a writing project of this magnitude is the belief that a good textbook makes a difference. The challenge in writing one is in striving for clarity, with enough development and design pedagogy for the student who will read but is uncomfortable with mathematics, yet also enough interest and integrity for the student who is comfortable and motivated and will go on in mathematics. My motivation for writing this book was the challenge of making this material comprehensible and usable for the student while doing justice to the intrinsic beauty of the subject.

I have always been guided in my teaching by the conviction that "the idea is the thing"—that mathematical formalism and much of the interesting minutiae of the discipline must be secondary to the goal of making this material understandable for the student. I have concentrated on being detailed and thorough in my presentation of the essentials of the subject. An author of an introductory text must resist the temptation to tell "all he knows," yet I have included a historical flavor and details of applications that many students find highly motivational.

In this book you will find a crisp, mathematically precise presentation that will allow you to teach your own course. The emphasis is on understanding. Enough mathematical detail is presented so that the treatment is precise, but without allowing formalism to become intrusive. The instructor can follow an appropriate course between intuition and rigor by choosing to include or exclude optional sections and proofs. Section 2.3, for example, on the precise definition of the limit is an optional section. Although a majority of theorems are proved in the text, some of the more difficult proofs are given in Appendix C.

Problem Solving Emphasis

My educational philosophy was strongly influenced by attending the lectures of George Polya and Gabor Szego when I was a student at Stanford University. Both Polya and Szego consistently introduced a topic by relating it to something concrete and familiar. I have emulated them by carefully motivating new ideas within an established conceptual framework. Wherever practical I have introduced topics with an intuitive geometrical or physical description and consistently attempted to tie mathematical concepts to the students' experience.

Polya's lectures led me to think about developing skills that could be used to approach problems in flexible, incisive ways. I was greatly impressed by his lectures on strategies of problem solving. His books *How to Solve It*, *Mathematical Discovery*, and *Mathematics and Plausible Reasoning* have become the core text material for a scientific problem solving course that I instituted and teach at McMaster University. I have adapted these problem solving strategies to the study of calculus both explicitly, by outlining strategies, and implicitly, by illustration and example.

Students have difficulties in situations where there is no single well-defined procedure for obtaining the answer, such as in related rate problems, maximum and minimum problems, integration, testing series, and solving differential equations. In these and other situations I have set out some strategies in the spirit of Polya's principles of problem solving. In particular there are three separate special sections devoted to problem solving: 7.6 (Strategy for Integration), 10.7 (Strategy for Testing Series), and 15.5 (Strategy for Solving First Order Differential Equations).

Early Trigonometric Functions and "Early Transcendentals" Option

For the most part the order of topics presented is fairly traditional. The trend toward the early introduction of trigonometric functions is reflected in the placement of the differentiation of all six trigonometric functions in Chapter 3 before the Chain Rule. However, many instructors would also like to be able to use the other transcendental functions prior to the coverage of the definite integral. I have encouraged them to do so by making the introduction of the transcendental functions as flexible as possible. In Chapter 6 the exponential function is defined first, followed by the logarithmic function as its inverse. (Students have seen these functions introduced this way since high school.) Later I present the more elegant definition of the logarithm as an integral. This presentation allows the coverage of much of Chapter 6 before Chapter 5, if desired. To accommodate this choice of presentation there are specially identified problems involving integrals of exponential and logarithmic functions at the end of the appropriate sections of Chapter 5. This order of presentation will allow a faster-paced course to teach the transcendental functions and the definite integral in the first semester of the course.

Curve Sketching

Curve sketching is such a valuable general mathematical skill that I have encouraged its use by giving it a lot of emphasis in this text. There is a wealth of examples and exercises on both the usual rational functions and the transcendental functions.

Exercises and Examples

There is nothing in any text that can compensate for a deficiency of good problems. I have selected my exercises from those used in over 15 years of calculus classes and have expressly chosen examples for their instructional value. There are over 6500 exercises ranging from the essential routine ones to those that will challenge your best students. I've made special effort to include some more unusual problems at both extremes of the spectrum of difficulty. The first 18 problems in Exercises 7.4, for example, ask only for the form of the partial fraction decomposition, whereas most exercise sets include challenging problems selected because they are theoretical or applied or because they are outside the students' experience and therefore thought provoking. Applied problems have been selected from the social, physical, and engineering sciences, as well as from mathematics itself. I have tried to emphasize the essential mathematical similarities in these apparently varied situations. The unifying role of mathematics is especially stressed in Section 3.7.

Some substantial applications appear in the exercises rather than in the text itself. For example, in Exercises 11.9 I ask students to derive Kepler's Laws after having supplied them with the necessary background and numerous hints. The number of exercises significantly exceeds what you will need in a single course and your students can trust the accuracy of the answers and solutions. Problem checking and proofing, while never absolutely perfect, have been exhaustive.

Margin Symbols

The calculator symbol in the margin indicates an exercise, or group of exercises, that requires the use of calculator or computer. Although most of these exercises are designed to illustrate the power of algorithmic computation, some of them are there to show its limitations. See for instance Exercise 48 in Section 2.2. Appendix D, "Lies My Calculator Told Me," alerts the student to many of the situations in which calculators give unreliable answers.

From repeatedly correcting the same mistakes I have come to recognize consistent pitfalls that trap many students. I believe that it is best to alert students to them and in many cases have done so with the "caution" symbol. For example, on page 387, the symbol is used to warn the student against a common misuse of l'Hospital's Rule and on page 584 it is used to emphasize that if $\lim_{n \to \infty} a_n = 0$, then we cannot conclude that $\sum a_n$ is convergent.

Introduction and Review Material

The introduction consists of a Preview of Calculus, which could be left for the student to discover or assigned as supplementary reading for motivational reasons as well as an overview of the course. There is a review of precalculus topics in Chapter 1, which can be largely omitted or covered at a rapid pace. Most students, however, will profit from a look at Section 1.7, which lists the common functions, their graphs, and rules for transforming graphs.

Appendixes

I have also included an algebra review appendix, because so many of the errors made by students writing calculus exams are not errors in calculus itself, or even in precalculus material, but rather in very basic algebra. Appendix A contains a substantial review of elementary algebra and many drill exercises. Similarly, a review of trigonometry appears in Appendix B, because a review thorough enough for students who really need it would be intrusive in the text per se.

Acknowledgments

Newton said that he stood on the shoulders of giants and he was not an especially modest man. We all have more to be modest about and no one could look back on a project of this magnitude without gratefully acknowledging the efforts of others who contributed so much of their time and the benefit of their scholarship. I sincerely thank the following reviewers: John Alberghini, Manchester Community College; Daniel Anderson, University of Iowa; David Berman, University of New Orleans; Richard Biggs, University of Western Ontario; the late Stephen Brown, Auburn University; David Buchthal, University of Akron; James Choike, Oklahoma State University; Carl Cowen, Purdue University; Daniel Cyphert, Armstrong State College; Robert Dahlin, University of Wisconsin; Daniel DiMaria, Suffolk Community College; Daniel Drucker, Wayne State University; Dennis Dunninger, Michigan State University; Bruce Edwards, University of Florida; Garrett Etgen, University of Houston; Frederick Gass, Miami University of Ohio; Bruce Gilligan, University of Regina; Stuart Goldenberg, California Polytechnic State University; Michael Gregory, University of North Dakota; Charles Groetsch, University of Cincinnati; D. W. Hall, Michigan State University; Allen Hesse, Rochester Community College; Matt Kaufmann; David Leeming, University of Victoria; Mark Pinsky, Northwestern University; Lothar Redlin, Pennsylvania State University, Ogontz Campus; Eric Schreiner, Western Michigan University; Wayne Skrapek, University of Saskatchewan; William Smith, University of North Carolina; Richard St. Andre, Central Michigan University; and Steven Willard, University of Alberta.

In addition, I thank the following colleagues at McMaster who helped me class-test this material over a period of seven years and made useful suggestions: Joe Csima, Carl Riehm, Evelyn Nelson, Zdislav Kovarik, Gunter Bruns, Hans Heinig, Eric Sawyer, Alex Rosa, Ian Hambleton, Bruno Mueller, Claude Billigheimer, John Chadam. In particular, the ideas of Appendix D are due to Zdislav Kovarik.

I also owe thanks to Marlene Mirza, Deane Maynard, and Patsy Chan, who typed the manuscript; to Dan Anderson, Dan Drucker, Barbara Frank, Eric Bosch, and Ron Donaberger, who worked the exercises; and to Ron Scoins, Terry Tiballi, and Clifton Whyburn who read the page proofs and checked for mathematical accuracy.

I am grateful for the splendid job done by the staff of Brooks/Cole Publishing Company and, in particular, the production editor, Cece Munson; the production coordinator, Joan Marsh; the designer, Detta Penna; the art director, Jamie Sue Brooks; and the editorial associate, Amy Mayfield. My most heartfelt thanks are reserved for the editors who have guided this project to its present state: Ron Munro, Harry Campbell, Craig Barth, and Jeremy Hayhurst.

JAMES STEWART

To the Student

Reading a calculus textbook is different from reading a newspaper or a novel, or even a physics textbook. Don't be discouraged if you have to read a passage more than once in order to understand it. You should have pencil and paper at hand to make a calculation or sketch a diagram.

Some students start by trying their homework problems and only read the text if they get stuck on an exercise. I suggest that a far better plan is to read and understand a section of the text before attempting the exercises. In particular, you should study the definitions to see the exact meanings of the terms.

The way to master calculus is to solve problems—lots of problems. Each exercise set starts out with drill problems—straightforward applications of formulas. Then the calculations become more involved or the reasoning becomes less straightforward. Finally, the exercises at the very end become quite challenging and require more thought.

Part of the aim of this course is to train you to think logically. Learn to write the solutions of the exercises in a connected step-by-step fashion with explanatory words or symbols—not just a string of disconnected equations or formulas.

The answers to the odd-numbered exercises appear at the back of the book, in Appendix E. There are often several different forms in which to express an answer, so if your answer differs from mine, don't immediately assume that you are wrong. There may be an algebraic or trigonometric identity that connects the answers. For example, if the answer given in the back of the book is $\sqrt{2} - 1$ and you obtain $1/(1 + \sqrt{2})$, then you are right and rationalizing the denominator will show that the expressions are equivalent.

The symbol ▦ means that a calculator (or computer) is required to do a calculation in an example or exercise. The symbol ● indicates the end of a proof or an exercise. You will also encounter the symbol ⊘, which warns you against committing an error. I have placed this symbol in the margin in situations where I have observed that a large proportion of my students tend to make the same mistake.

Calculus is an exciting subject; I hope you find it both useful and interesting in its own right. If you would like to see what lies ahead in this course, have a look at "A Preview of Calculus" which follows on page xxi.

A Note on Logic

In understanding the theorems it is important to know the meaning of certain logical terms and symbols. If P and Q are mathematical statements, then $P \Rightarrow Q$ is read as "P implies Q" and means the same as "If P is true, then Q is true." The *converse* of a theorem of the form $P \Rightarrow Q$ is the statement $Q \Rightarrow P$. (The converse of a theorem may or may not be true. For example, the converse of the statement "If it rains, then I take my umbrella" is "If I take my umbrella, then it rains.") The symbol $\Leftrightarrow$ indicates that two statements are equivalent. Thus $P \Leftrightarrow Q$ means that both $P \Rightarrow Q$ and $Q \Rightarrow P$. The phrase "if and only if" is also used in this situation. Thus "P is true if and only if Q is true" means the same as $P \Leftrightarrow Q$.

Contents

A Preview of Calculus   xxi

CHAPTER 1   **Functions and Graphs**   1

1.1   Numbers, Inequalities, and Absolute Values   2
1.2   The Cartesian Plane   10
1.3   Lines   14
1.4   Second-Degree Equations   20
1.5   Functions and Their Graphs   27
1.6   Combinations of Functions   36
1.7   Types of Functions and Transformed Functions   40
        Review of Chapter One   43

CHAPTER 2   **Limits and Rates of Change**   48

2.1   Tangents and Velocities   50
2.2   The Limit of a Function   59
2.3   The Precise Definition of a Limit   67
2.4   Properties of Limits   74
2.5   Continuity   82
        Review of Chapter Two   92

CHAPTER 3   **Derivatives**   94

3.1   Derivatives   96
3.2   Differentiation Formulas   106
3.3   Derivatives of Trigonometric Functions   117
3.4   The Chain Rule   124
3.5   Implicit Differentiation; Angles between Curves   133
3.6   Higher Derivatives   141
3.7   Rates of Change in the Natural and Social Sciences   145
3.8   Related Rates   155
3.9   Differentials   160
3.10 Newton's Method   164
        Review of Chapter Three   168

CHAPTER 4   **The Mean Value Theorem and Its Applications 172**

4.1 Maximum and Minimum Values   174
4.2 The Mean Value Theorem   182
4.3 Monotonic Functions and the First Derivative Test   187
4.4 Concavity and Points of Inflection   194
4.5 Limits at Infinity: Horizontal Asymptotes   199
4.6 Infinite Limits: Vertical Asymptotes   211
4.7 Curve Sketching   221
4.8 Applied Maximum and Minimum Problems   231
4.9 Applications to Economics   239
4.10 Antiderivatives   243
       Review of Chapter Four   249

CHAPTER 5   **Integrals 252**

5.1 Sigma Notation   254
5.2 Area   259
5.3 The Definite Integral   268
5.4 Properties of the Definite Integral   274
5.5 The Fundamental Theorem of Calculus   279
5.6 The Substitution Rule   289
5.7 Areas between Curves   297
5.8 Applications to Science   303
5.9 Average Value of a Function   307
       Review of Chapter Five   311

CHAPTER 6   **Inverse Functions: Exponential, Logarithmic, and Inverse Trigonometric Functions 314**

6.1 Exponential Functions   316
6.2 Inverse Functions   319
6.3 Logarithmic Functions   327
6.4 Derivatives of Logarithmic Functions   335
6.5 Derivatives of Exponential Functions   343
6.6 An Alternative Approach to Logarithmic and Exponential Functions   351
6.7 Exponential Growth and Decay   359
6.8 Inverse Trigonometric Functions   365
6.9 Hyperbolic Functions   376
6.10 Indeterminate Forms and L'Hospital's Rule   383
       Review of Chapter Six   394

CHAPTER 7   **Techniques of Integration 398**

7.1 Integration by Parts   400
7.2 Trigonometric Integrals   406
7.3 Trigonometric Substitution   412

7.4   Integration of Rational Functions by Partial Fractions   419

7.5   Rationalizing Substitutions   430

7.6   Strategy for Integration   433

7.7   Approximations: The Trapezoidal Rule and Simpson's Rule   440

7.8   Improper Integrals   448

7.9   Using Tables of Integrals   457

     Review of Chapter Seven   459

CHAPTER 8   **Further Applications of Integration**   462

8.1   Volume   464

8.2   Volumes by Cylindrical Shells   474

8.3   Arc Length   480

8.4   Area of a Surface of Revolution   486

8.5   Work   493

8.6   Hydrostatic Pressure   497

8.7   Moments and Centers of Mass   501

8.8   Differential Equations   508

     Review of Chapter Eight   516

CHAPTER 9   **Parametric Equations and Polar Coordinates**   520

9.1   Curves Defined by Parametric Equations   522

9.2   Tangents and Areas   526

9.3   Arc Length and Surface Area   530

9.4   Polar Coordinates   535

9.5   Areas and Lengths in Polar Coordinates   543

9.6   Conic Sections   549

9.7   Conic Sections in Polar Coordinates   557

9.8   Rotation of Axes   562

     Review of Chapter Nine   566

CHAPTER 10   **Infinite Sequences and Series**   568

10.1   Sequences   570

10.2   Series   579

10.3   The Integral Test   588

10.4   The Comparison Tests   592

10.5   Alternating Series   596

10.6   Absolute Convergence and the Ratio and Root Tests   601

10.7   Strategy for Testing Series   608

10.8   Power Series   610

10.9   Taylor and Maclaurin Series   615
10.10 The Binomial Series   625
10.11 Approximation by Taylor Polynomials   629
         Review of Chapter Ten   638

CHAPTER 11   **Three-Dimensional Analytic Geometry and Vectors   642**

11.1   Three Dimensional Coordinate Systems   644
11.2   Vectors   648
11.3   The Dot Product   655
11.4   The Cross Product   662
11.5   Equations of Lines and Planes   669
11.6   Quadric Surfaces   677
11.7   Vector Functions and Space Curves   683
11.8   Arc Length and Curvature   690
11.9   Motion in Space: Velocity and Acceleration   697
11.10 Cylindrical and Spherical Coordinates   704
         Review of Chapter Eleven   708

CHAPTER 12   **Partial Derivatives   712**

12.1   Functions of Several Variables   714
12.2   Limits and Continuity   724
12.3   Partial Derivatives   733
12.4   Tangent Planes and Differentials   741
12.5   The Chain Rule   749
12.6   Directional Derivatives and the Gradient Vector   758
12.7   Maximum and Minimum Values   768
12.8   Lagrange Multipliers   776
         Review of Chapter Twelve   782

CHAPTER 13   **Multiple Integrals   786**

13.1   Double Integrals over Rectangles   788
13.2   Iterated Integrals   793
13.3   Double Integrals over General Regions   799
13.4   Double Integrals in Polar Coordinates   809
13.5   Applications of Double Integrals   814
13.6   Triple Integrals   821
13.7   Triple Integrals in Cylindrical and Spherical Coordinates   831
13.8   Change of Variables in Multiple Integrals   837
         Review of Chapter Thirteen   846

CHAPTER 14 **Vector Calculus** 850

14.1 Vector Fields 852
14.2 Line Integrals 855
14.3 The Fundamental Theorem for Line Integrals 866
14.4 Green's Theorem 875
14.5 Curl and Divergence 883
14.6 The Area of a Surface 890
14.7 Surface Integrals 898
14.8 Stokes' Theorem 910
14.9 The Divergence Theorem 916
14.10 Summary 922
Review of Chapter Fourteen 923

CHAPTER 15 **Differential Equations** 926

15.1 Basic Concepts 928
15.2 Homogeneous Equations 932
15.3 First Order Linear Equations 934
15.4 Exact Equations 940
15.5 Strategy for Solving First Order Equations 944
15.6 Second Order Linear Equations 947
15.7 Complex Roots 953
15.8 Nonhomogeneous Linear Equations 956
15.9 Applications of Second Order Differential Equations 963
15.10 Series Solutions 971
Review of Chapter Fifteen 976

APPENDIX A **Review of Algebra** A1

APPENDIX B **Review of Trigonometry** A15

APPENDIX C **Proofs of Theorems** A27

APPENDIX D **Lies My Calculator Told Me** A39

APPENDIX E **Answers to Odd-Numbered Exercises** A43

INDEX A101

A Preview of Calculus

Calculus is fundamentally different from the mathematics that you have previously studied. Calculus is less static and more dynamic. It is concerned with change and motion; it deals with quantities that approach other quantities.

For that reason it may be useful to have an overview of the subject before beginning its intensive study. In what follows, we give a glimpse of some of the main ideas of calculus by showing how limits arise when we attempt to solve various problems.

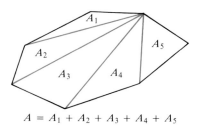

$$A = A_1 + A_2 + A_3 + A_4 + A_5$$

Figure 1

The Area Problem

The origins of calculus go back at least 2500 years to the ancient Greeks, who found areas using the "method of exhaustion." They knew how to find the area A of any polygon by dividing it into triangles as in Figure 1 and adding the areas of these triangles.

It is a much more difficult problem to find the area of a curved figure. Their method of exhaustion was to inscribe polygons in the figure and circumscribe polygons about the figure and then let the number of sides of the polygons increase. Figure 2 illustrates this process for the special case of a circle with inscribed regular polygons.

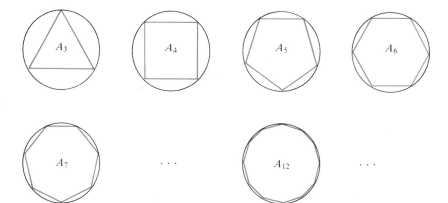

Figure 2

Let A_n be the area of the inscribed polygon with n sides. As n increases, it appears that A_n becomes closer and closer to the area of the circle. We say that the area of the circle is the *limit* of the areas of the inscribed polygons, and we write

$$A = \lim_{n \to \infty} A_n$$

The Greeks themselves did not explicitly use limits. However, by indirect reasoning, Eudoxus (fifth century B.C.) used exhaustion to prove the familiar formula for the area of a circle: $A = \pi r^2$.

We shall use a similar idea in Chapter 5 to find areas of regions of the type shown in Figure 3. We shall approximate the desired area A by areas of rectangles (as in Figure 4), let the width of the rectangles decrease, and then calculate A as the limit of these sums of areas of rectangles.

The area problem is the central problem in the branch of calculus called *integral calculus*. The techniques that we shall develop in Chapter 5 for finding areas will also enable us to compute the volume of a solid, the length of a curve, the force against a dam, the mass and center of gravity of a rod, and the work done in pumping water out of a tank.

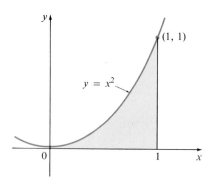

Figure 3

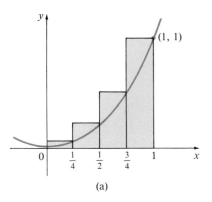
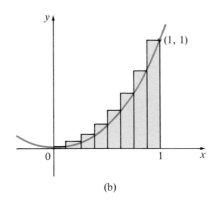
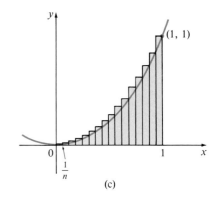

(a)   (b)   (c)

Figure 4

The Tangent Problem

Consider the problem of trying to find the equation of the tangent line t to a curve with equation $y = f(x)$ at a given point P. (We shall give a precise definition of a tangent line in Sections 2.1 and 3.1. For now you can think of it as a line that touches the curve at P as in Figure 5.) Since we know that the point P lies on the tangent line, we can find the equation of t if we know its slope m. The problem is that we need two points to compute the slope and we know only one point, P, on t. To get around this problem we first find an approximation to m by taking a nearby point Q on the curve and computing the slope m_{PQ} of the secant line PQ. From Figure 6 we see that

(1)
$$m_{PQ} = \frac{f(x) - f(a)}{x - a}$$

Now imagine that Q moves along the curve toward P as in Figure 7. You can see that the secant line rotates and approaches the tangent line as

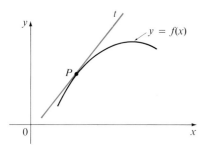

Figure 5
THE TANGENT LINE AT P

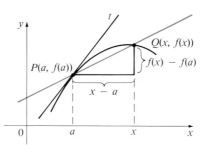

Figure 6
THE SECANT LINE PQ

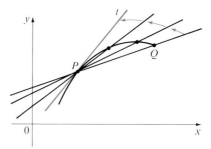

Figure 7

its limiting position. This means that the slope m_{PQ} of the secant line becomes closer and closer to the slope m of the tangent line. We write

$$m = \lim_{Q \to P} m_{PQ}$$

and we say that m is the limit of m_{PQ} as Q approaches P along the curve. Since x approaches a as Q approaches P, we could also use Equation 1 to write

(2)
$$m = \lim_{x \to a} \frac{f(x) - f(a)}{x - a}$$

Specific examples of this procedure will be given in Sections 2.1 and 3.1.

The tangent problem has given rise to the branch of calculus called *differential calculus*, which was not invented until more than 2000 years after integral calculus. The main ideas behind differential calculus are due to the French mathematician Pierre Fermat (1601–1665) and were developed by the English mathematicians John Wallis (1616–1703), Isaac Barrow (1630–1677), and Isaac Newton (1642–1723) and the German mathematician Gottfried Leibniz (1648–1716).

The two branches of calculus and their chief problems, the area problem and the tangent problem, appear to be very different, but it turns out that there is a very close connection between them. The tangent problem and the area problem are inverse problems in a sense that will be described in Chapter 5.

Velocity

When we look at the speedometer of a car and read that the car is traveling at 85 km/h, what does that indicate to us? We know that if our velocity remains constant, then after an hour we will have traveled 85 km. But if the velocity of the car varies, what does it mean to say that the velocity at a given instant is 85 km/h?

In order to analyze this question, let us analyze the motion of a car that travels along a straight road and assume that we can measure the distance traveled by the car (in kilometers) at 1-minute intervals as in the following chart:

t = time elapsed (min)	0	1	2	3	4	5
d = distance (km)	0	0.559	1.647	3.307	5.223	7.158

As a first step toward finding the velocity after 2 minutes have elapsed, let us find the average velocity during the time interval $2 \leqslant t \leqslant 4$:

$$\text{average velocity} = \frac{\text{distance traveled}}{\text{time elapsed}}$$

$$= \frac{5.223 - 1.647}{4 - 2}$$

$$= 1.788 \text{ km/min}$$

Similarly the average velocity in the time interval $2 \leqslant t \leqslant 3$ is

$$\text{average velocity} = \frac{3.307 - 1.647}{3 - 2} = 1.660 \text{ km/min}$$

We have the feeling that the velocity at the instant $t = 2$ cannot be much different from the average velocity during a short time interval starting at $t = 2$. So let us imagine that the distance traveled has been measured at 6-second (0.1 min) time intervals as in the following chart:

t	2.0	2.1	2.2	2.3	2.4	2.5	2.6	2.7	2.8	2.9	3.0
d	1.647	1.759	1.874	1.993	2.119	2.264	2.442	2.634	2.845	3.067	3.307

Then we can compute, for instance, the average velocity over the time interval $[2, 2.5]$:

$$\text{average velocity} = \frac{2.264 - 1.647}{2.5 - 2} = 1.234 \text{ km/min}$$

The results of such calculations are shown in the following chart:

time interval	$[2, 3]$	$[2, 2.7]$	$[2, 2.5]$	$[2, 2.4]$	$[2, 2.3]$	$[2, 2.2]$	$[2, 2.1]$
average velocity (km/min)	1.660	1.410	1.234	1.180	1.153	1.135	1.120

The average velocities over successively smaller intervals appear to be getting closer to a number near 1.1, and so we expect that the velocity at exactly $t = 2$ is about 1.1 km/min. (This corresponds to $1.1 \times 60 = 66$ km/h.) In Section 2.1 we shall define the instantaneous velocity of a moving object as the limiting value of the average velocities over smaller and smaller time intervals.

In Figure 8 we have shown a graphical representation of the motion of the car by plotting the distance traveled as a function of time. If we write

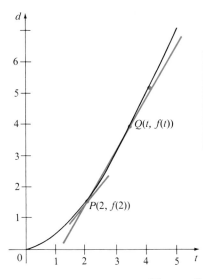

Figure 8

$d = f(t)$, then $f(t)$ is the number of kilometers traveled after t minutes. The average velocity in the time interval $[2, t]$ is

$$\text{average velocity} = \frac{\text{distance traveled}}{\text{time elapsed}} = \frac{f(t) - f(2)}{t - 2}$$

which is the same as the slope of the secant line PQ in Figure 8. The velocity v when $t = 2$ is the limiting value of this average velocity as t approaches 2; that is,

$$v = \lim_{t \to 2} \frac{f(t) - f(2)}{t - 2}$$

and we recognize from Equation 2 that this is the same as the slope of the tangent curve at P.

Thus when we solve the tangent problem in differential calculus, we will also be solving problems concerning velocities. The same techniques will also enable us to solve problems involving rates of change in all of the natural and social sciences.

The Limit of a Sequence

In the fifth century B.C. the Greek philosopher Zeno of Elea posed four problems, now known as *Zeno's paradoxes*, that were intended to challenge some of the ideas concerning space and time that were held in his day. Zeno's second paradox concerns a race between the Greek hero Achilles and a tortoise that has been given a head start. Zeno argued as follows that Achilles could never pass the tortoise: Suppose that Achilles starts at position a_1 and the tortoise starts at position t_1 (see Figure 9). When Achilles reaches the point $a_2 = t_1$, the tortoise is further ahead at position t_2. When Achilles reaches $a_3 = t_2$, the tortoise is at t_3. This process continues indefinitely and so it appears that the tortoise will always be ahead! But this defies common sense.

Figure 9

One way of explaining this paradox is with the idea of a *sequence*. The successive positions of Achilles $(a_1, a_2, a_3, \ldots)$ or the successive positions of the tortoise $(t_1, t_2, t_3, \ldots)$ form what is known as a sequence.

In general, a sequence $\{a_n\}$ is a set of numbers written in a definite order. For instance, the sequence

$$\left\{ 1, \frac{1}{2}, \frac{1}{3}, \frac{1}{4}, \frac{1}{5}, \ldots \right\}$$

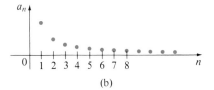

(a)

(b)

Figure 10

can be described by giving the following formula for the nth term:

$$a_n = \frac{1}{n}$$

We can visualize this sequence by plotting its terms on a number line as in Figure 10(a) or by drawing its graph as in Figure 10(b). Observe from either picture that the terms of the sequence $a_n = 1/n$ are becoming closer and closer to 0 as n increases. In fact we can make the terms as small as we please by making n large enough. We say that the limit of the sequence is 0, and we indicate this by writing

$$\lim_{n \to \infty} \frac{1}{n} = 0$$

In general, the notation

$$\lim_{n \to \infty} a_n = L$$

is used if the terms a_n approach the number L as n becomes large. This means that the numbers a_n can be made as close as we like to the number L by taking n sufficiently large.

The concept of the limit of a sequence occurs whenever we use the decimal representation of a real number. For instance, if

$$a_1 = 3.1$$
$$a_2 = 3.14$$
$$a_3 = 3.141$$
$$a_4 = 3.1415$$
$$a_5 = 3.14159$$
$$a_6 = 3.141592$$
$$a_7 = 3.1415926$$
$$\vdots$$

then
$$\lim_{n \to \infty} a_n = \pi$$

The terms in this sequence are rational approximations to π.

Let us return to Zeno's paradox. The successive positions of Achilles and the tortoise form sequences $\{a_n\}$ and $\{t_n\}$, where $a_n < t_n$ for all n. But using the more precise treatment of sequences given in Chapter 10, it can be shown that both sequences have the same limit:

$$\lim_{n \to \infty} a_n = p = \lim_{n \to \infty} t_n$$

It is precisely at this point p that Achilles overtakes the tortoise.

The Sum of a Series

Another of Zeno's paradoxes, as passed on to us by Aristotle, is the following: "A man standing in a room cannot walk to the wall. In order to do so, he would first have to go half the distance, then half the remaining distance, and then again half of what still remains. This process can always be continued and can never be ended." (See Figure 11.)

Figure 11

Of course we know that the man can actually reach the wall, so this suggests that perhaps the total distance can be expressed as the sum of infinitely many smaller distances as follows:

(3)
$$1 = \frac{1}{2} + \frac{1}{4} + \frac{1}{8} + \frac{1}{16} + \cdots + \frac{1}{2^n} + \cdots$$

Zeno was arguing that it does not make sense to add infinitely many numbers together. But there are other situations in which we implicitly use infinite sums. For instance, in our decimal notation, the symbol $0.\bar{3} = 0.33333\ldots$ means

$$\frac{3}{10} + \frac{3}{100} + \frac{3}{1000} + \frac{3}{10,000} + \cdots$$

and so, in some sense, it must be true that

$$\frac{3}{10} + \frac{3}{100} + \frac{3}{1000} + \frac{3}{10,000} + \cdots = \frac{1}{3}$$

More generally, if d_n denotes the nth digit in the decimal representation of a number, then

$$0.d_1 d_2 d_3 d_4 \ldots = \frac{d_1}{10} + \frac{d_2}{10^2} + \frac{d_3}{10^3} + \cdots + \frac{d_n}{10^n} + \cdots$$

Therefore some infinite sums, or infinite series as they are called, have a meaning. But we must define carefully what the sum of an infinite series is.

Returning to the series in Equation 3, we denote by s_n the sum of the first n terms of the series. Thus,

$$s_1 = \tfrac{1}{2} = 0.5$$

$$s_2 = \tfrac{1}{2} + \tfrac{1}{4} = 0.75$$

$$s_3 = \tfrac{1}{2} + \tfrac{1}{4} + \tfrac{1}{8} = 0.875$$

$$s_4 = \tfrac{1}{2} + \tfrac{1}{4} + \tfrac{1}{8} + \tfrac{1}{16} = 0.9375$$

$$s_5 = \tfrac{1}{2} + \tfrac{1}{4} + \tfrac{1}{8} + \tfrac{1}{16} + \tfrac{1}{32} = 0.96875$$

$$s_6 = \tfrac{1}{2} + \tfrac{1}{4} + \tfrac{1}{8} + \tfrac{1}{16} + \tfrac{1}{32} + \tfrac{1}{64} = 0.984375$$

$$s_7 = \tfrac{1}{2} + \tfrac{1}{4} + \tfrac{1}{8} + \tfrac{1}{16} + \tfrac{1}{32} + \tfrac{1}{64} + \tfrac{1}{128} = 0.9921875$$

$$\vdots$$

$$s_{10} = \tfrac{1}{2} + \tfrac{1}{4} + \cdots + \tfrac{1}{1024} \approx 0.99902344$$

$$\vdots$$

$$s_{16} = \frac{1}{2} + \frac{1}{4} + \cdots + \frac{1}{2^{16}} \approx 0.99998474$$

Observe that as we add more and more terms, the partial sums become closer and closer to 1. In fact, it can be shown that by taking n large enough (that is, by adding sufficiently many terms of the series) we can make the partial sum s_n as close as we please to the number 1. It therefore seems reasonable to say that the sum of the infinite series is 1 and to write

$$\frac{1}{2} + \frac{1}{4} + \frac{1}{8} + \cdots + \frac{1}{2^n} + \cdots = 1$$

In other words, the reason the sum of the series is 1 is that

$$\lim_{n \to \infty} s_n = 1$$

In Chapter 10 we shall make these ideas precise and we shall use Newton's idea of combining infinite series with differential and integral calculus.

Summary

We have seen that the concept of a limit arises in trying to find the area of a region, the slope of a tangent to a curve, the velocity of a car, or the sum of an infinite series. In each case the common theme is the calculation of a quantity as the limit of other, easily calculated, quantities. It is this basic idea of a limit that sets calculus apart from other areas of mathematics. In fact, we could define calculus as the part of mathematics that deals with limits.

Sir Isaac Newton invented his version of calculus in order to explain the motion of the planets around the sun. Today calculus is used not only in calculating the orbits of satellites and spacecraft and in the study of astronomy, nuclear physics, electricity, thermodynamics, acoustics, design of machines, chemical reactions, growth of organisms, weather prediction, and the calculation of life insurance premiums, but also in such everyday concerns as fencing off a field so as to enclose the maximum area or computing the most economical speed for driving a car. Some of these uses of calculus will be explored throughout this book.

1 Functions and Graphs

In most sciences one generation tears down what another has built and what one has established another undoes. In Mathematics alone each generation builds a new story to the old structure.

Hermann Hankel

In this first chapter we review the mathematical topics—algebra, analytic geometry, and functions—that are needed for the study of calculus.

Numbers, Inequalities, and Absolute Values

Calculus is based on the real number system. We start with the **integers:**

$$\ldots, -3, -2, -1, 0, 1, 2, 3, 4, \ldots$$

Then we construct the **rational numbers,** which are ratios of integers. Thus any rational number r can be expressed as

$$r = \frac{m}{n} \qquad \text{where } m \text{ and } n \text{ are integers and } n \neq 0$$

Examples are

$$\tfrac{1}{2} \qquad -\tfrac{3}{7} \qquad 46 = \tfrac{46}{1} \qquad 0.17 = \tfrac{17}{100}$$

(Recall that division by 0 is always ruled out, so expressions like $\frac{3}{0}$ and $\frac{0}{0}$ are undefined.) There are also real numbers, such as $\sqrt{2}$, that cannot be expressed as a ratio of integers and are therefore called **irrational numbers.** It can be shown, with varying degrees of difficulty, that the following are also examples of irrational numbers:

$$\sqrt{3} \qquad \sqrt{5} \qquad \sqrt[3]{2} \qquad \pi \qquad \sin 1^\circ \qquad \log_{10} 2$$

The set of all real numbers is usually denoted by the symbol R. When we use the word "number" without qualification, we shall mean "real number."

Every number has a decimal representation. If the number is rational, then the corresponding decimal will be repeating. For example,

$$\tfrac{1}{2} = 0.5000\ldots = 0.5\bar{0} \qquad\qquad \tfrac{2}{3} = 0.66666\ldots = 0.\bar{6}$$

$$\tfrac{157}{495} = 0.3171717\ldots = 0.3\overline{17} \qquad \tfrac{9}{7} = 1.285714285714\ldots = 1.\overline{285714}$$

(The bar indicates that the sequence of digits repeats forever.) On the other hand, if the number is irrational, the decimal will be nonrepeating:

$$\sqrt{2} = 1.414213562373095\ldots \qquad \pi = 3.141592653589793\ldots$$

If we stop the decimal expansion of any number at a certain place, we get an approximation to the number. For instance, we can write

$$\pi \approx 3.14159265$$

where the symbol $\approx$ is read "is approximately equal to." The more decimal places we retain, the better the approximation we get.

The real numbers can be represented by points on a line as in Figure 1.1. The positive direction (to the right) is indicated by an arrow. We choose an arbitrary reference point O, called the **origin,** which corresponds to the real number 0. Given any convenient unit of measurement, each positive number x is represented by the point on the line a distance of x units to the right of the origin, and each negative number $-x$ is represented by the

Figure 1.1

point x units to the left of the origin. Thus every real number is represented by a point on the line, and every point P on the line corresponds to exactly one real number. The number associated with the point P is called the **coordinate** of P and the line is then called a **coordinate line,** or a **real number line,** or simply a **real line.** Often we identify the point with its coordinate and think of a number as being a point on the real line.

The real numbers are ordered. We say a *is less than* b and write $a < b$ if $b - a$ is a positive number. Geometrically this means that a lies to the left of b on the number line. (Equivalently, we say b *is greater than* a and write $b > a$.) The symbol $a \leqslant b$ (or $b \geqslant a$) means that either $a < b$ or $a = b$ and is read "a is less than or equal to b." For instance, the following are true inequalities:

$$7 < 7.4 < 7.5 \qquad -3 > -\pi \qquad \sqrt{2} < 2 \qquad \sqrt{2} \leqslant 2 \qquad 2 \leqslant 2$$

In what follows we shall need to use set notation. A **set** is a collection of objects, and these objects are called the **elements** of the set. If S is a set, the notation $a \in S$ means that a is an element of S, and $a \notin S$ means that a is not an element of S. For example, if Z represents the set of integers, then $-3 \in Z$ but $\pi \notin Z$. If S and T are sets, then their **union** $S \cup T$ is the set consisting of all elements that are in S or T (or in both S and T). The **intersection** of S and T is the set $S \cap T$ consisting of all elements that are in both S and T. In other words, $S \cap T$ is the common part of S and T. The empty set, denoted by $\varnothing$, is the set that contains no elements.

Some sets can be described by listing their elements between braces. For instance, the set A consisting of all positive integers less than 7 can be written as

$$A = \{1, 2, 3, 4, 5, 6\}$$

We could also write A in set-builder notation as

$$A = \{x \mid x \text{ is an integer and } 0 < x < 7\}$$

which is read "A is the set of x such that x is an integer and $0 < x < 7$."

There are certain sets of real numbers, called **intervals,** that occur frequently in calculus and correspond geometrically to line segments. For example, if $a < b$, the **open interval** from a to b consists of all numbers between a and b and is denoted by the symbol (a, b). Using set-builder notation, we can write

$$(a, b) = \{x \mid a < x < b\}$$

Notice that the endpoints of the interval—namely, a and b—are excluded. This is indicated by the round brackets ( ) and by the open dots in Figure 1.2. The **closed interval** from a to b is the set

$$[a, b] = \{x \mid a \leqslant x \leqslant b\}$$

Figure 1.2

OPEN INTERVAL (a, b)

Figure 1.3

CLOSED INTERVAL [*a*, *b*]

Here the endpoints of the interval are included. This is indicated by the square brackets [ ] and by the solid dots in Figure 1.3. It is also possible to include only one endpoint in an interval, as shown in Table 1.1.

We shall also need to consider infinite intervals such as

$$(a, \infty) = \{x \mid x > a\}$$

This does not mean that ∞ ("infinity") is a number. The notation (a, ∞) stands for the set of all numbers which are greater than a, so the symbol ∞ simply indicates that the interval extends indefinitely far in the positive direction.

Table 1.1 gives a list of the nine possible types of intervals. When these intervals are discussed, it will always be assumed that $a < b$.

Table of Intervals (1.1)

Notation	Set Description	Picture
(a, b)	$\{x \mid a < x < b\}$	
$[a, b]$	$\{x \mid a \le x \le b\}$	
$[a, b)$	$\{x \mid a \le x < b\}$	
$(a, b]$	$\{x \mid a < x \le b\}$	
(a, ∞)	$\{x \mid x > a\}$	
$[a, \infty)$	$\{x \mid x \ge a\}$	
$(-\infty, b)$	$\{x \mid x < b\}$	
$(-\infty, b]$	$\{x \mid x \le b\}$	
$(-\infty, \infty)$	R (set of all real numbers)	

When working with inequalities, the following rules should be noted:

Rules for Inequalities (1.2)

(a)   If $a < b$, then $a + c < b + c$.
(b)   If $a < b$ and $c < d$, then $a + c < b + d$.
(c)   If $a < b$ and $c > 0$, then $ac < bc$.
(d)   If $a < b$ and $c < 0$, then $ac > bc$.
(e)   If $0 < a < b$, then $1/a > 1/b$.

Rule (a) says that we can add any number to both sides of an inequality, and Rule (b) says that two inequalities can be added. However, we have to be careful with multiplication. Rule (c) says that we can multiply both sides of an inequality by a *positive* number, but Rule (d) says that *if we multiply both sides of an inequality by a negative number, then we reverse the direction of the inequality.* For example, if we take the inequality $3 < 5$ and multiply by 2, we get $6 < 10$, but if we multiply by -2, we get $-6 > -10$. Finally,

Rule (e) says that if we take reciprocals, then we reverse the direction of an inequality (provided the numbers are positive).

● **Example 1**   Solve the inequality $1 + x < 7x + 5$.

Solution   The given inequality is satisfied by some values of x but not by others. To solve an inequality means to determine the set of numbers x for which the inequality is true. This is called the *solution set*.

First we subtract 1 from each side of the inequality [using Rule (a) with $c = -1$]:

$$x < 7x + 4$$

Then we subtract $7x$ from both sides [Rule (a) with $c = -7x$]:

$$-6x < 4$$

Now we divide both sides by -6 [Rule (d) with $c = -\frac{1}{6}$]:

$$x > -\tfrac{4}{6} = -\tfrac{2}{3}$$

These steps can all be reversed, so the solution set consists of all numbers greater than $-\frac{2}{3}$. In other words, the solution of the inequality is given by the interval $(-\frac{2}{3}, \infty)$.   ●

● **Example 2**   Solve the inequalities $4 \leqslant 3x - 2 < 13$.

Solution   Here the solution set consists of all values of x that satisfy both inequalities. Using the rules given in (1.2), we see that the following inequalities are equivalent:

$$4 \leqslant 3x - 2 < 13$$

$$6 \leqslant 3x < 15 \qquad (add\ 2)$$

$$2 \leqslant x < 5 \qquad (divide\ by\ 3)$$

Therefore the solution set is $[2, 5)$.   ●

● **Example 3**   Solve $2x + 1 \leqslant 4x - 3 \leqslant x + 7$.

Solution   This time we first solve the inequalities separately:

$$2x + 1 \leqslant 4x - 3 \qquad\qquad 4x - 3 \leqslant x + 7$$

$$4 \leqslant 2x \qquad\qquad\qquad 3x \leqslant 10$$

$$2 \leqslant x \qquad\qquad\qquad x \leqslant \tfrac{10}{3}$$

Since x must satisfy both inequalities, we have

$$2 \leqslant x \leqslant \tfrac{10}{3}$$

Thus the solution set is the closed interval $[2, \frac{10}{3}]$.   ●

● **Example 4**   Solve $\dfrac{1 + x}{1 - x} > 1$.

Solution   We would like to multiply both sides by $1 - x$, but in view of Rules (c) and (d) we must consider separately the cases where $1 - x$ is positive or negative.

CASE I:

If $1 - x > 0$, that is, $x < 1$, then multiplying the given inequality by $1 - x$ gives

$$1 + x > 1 - x$$

which becomes $2x > 0$, that is, $x > 0$. So we have

$$0 < x < 1$$

CASE II:

If $1 - x < 0$, that is, $x > 1$, then multiplying the given inequality by $1 - x$ gives

$$1 + x < 1 - x$$

which becomes $2x < 0$, that is, $x < 0$. But the conditions $x > 1$ and $x < 0$ are incompatible, so there is no solution in Case II.

Therefore the solution set is the open interval $(0, 1)$.   ●

● **Example 5**   Solve $x^2 + 3x > 4$.

Solution 1   First we write the inequality in the form

$$x^2 + 3x - 4 > 0 \qquad \text{or} \qquad (x - 1)(x + 4) > 0$$

The product of two numbers is positive when both factors are positive or when both factors are negative.

CASE I:

When $x - 1 > 0$ and $x + 4 > 0$, we have $x > 1$ and $x > -4$. But if $x > 1$, then $x > -4$ holds automatically. So the solution in this case is given by $x > 1$.

CASE II:

When $x - 1 < 0$ and $x + 4 < 0$, we have $x < 1$ and $x < -4$. But if $x < -4$, then $x < 1$ holds automatically. So the solution in this case is given by $x < -4$.

Combining both cases, we see that the solution set is

$$\{x \mid x < -4 \text{ or } x > 1\}$$

which can also be written as a union of intervals:

$$(-\infty, -4) \cup (1, \infty)$$

Recall that this notation means the solution set consists of the numbers that are either in the interval $(-\infty, -4)$ or in the interval $(1, \infty)$. The solution set is pictured in Figure 1.4.

Figure 1.4

$-4 \qquad\qquad 0 \quad 1$

Solution 2   Another method is to argue that the product $(x - 1)(x + 4)$ is 0 when $x = 1$ or -4. These numbers divide the real line into three intervals $(-\infty, -4)$, $(-4, 1)$, and $(1, \infty)$, on each of which the product keeps a constant sign as shown in the following chart:

Interval	$x - 1$	$x + 4$	$(x - 1)(x + 4)$
$x < -4$	$-$	$-$	$+$
$-4 < x < 1$	$-$	$+$	$-$
$x > 1$	$+$	$+$	$+$

Then we can simply read from the chart that the solution set is

$$\{x \mid x < -4 \text{ or } x > 1\} = (-\infty, -4) \cup (1, \infty) \qquad \bullet$$

The method given in Solution 2 of Example 5 is particularly useful when there are three or more factors to keep track of.

Absolute Value

The **absolute value** of a number a, denoted by $|a|$, is the distance from a to 0 on the real number line. Distances are always positive or 0, so we have

$$|a| \geqslant 0 \qquad \text{for every number } a$$

For example,

$$|3| = 3 \qquad |-3| = 3 \qquad |0| = 0 \qquad |\sqrt{2} - 1| = \sqrt{2} - 1 \qquad |3 - \pi| = \pi - 3$$

In general, we have

(1.3)

$$\begin{aligned} |a| &= a \qquad \text{if } a \geqslant 0 \\ |a| &= -a \quad\;\; \text{if } a < 0 \end{aligned}$$

(Remember that if a is negative, then $-a$ is positive.)

 Let us recall that the symbol $\sqrt{}$ means "the positive square root of." Thus $\sqrt{r} = s$ means $s^2 = r$ and $s \geqslant 0$. Therefore the equation $\sqrt{a^2} = a$ is not always true. It is true only when $a \geqslant 0$. If $a < 0$, then $-a > 0$, so we have

$\sqrt{a^2} = -a$. In view of (1.3), we then have the equation

(1.4)
$$\sqrt{a^2} = |a|$$

which is true for all values of a.

Hints for the proofs of the following properties are given in Exercises 1.1.

Properties of Absolute Values (1.5)

Suppose a and b are any real numbers and n is an integer. Then

(a) $|ab| = |a||b|$

(b) $\left|\dfrac{a}{b}\right| = \dfrac{|a|}{|b|}$ $b \neq 0$

(c) $|a^n| = |a|^n$

(d) $|a + b| \leq |a| + |b|$ *(the Triangle Inequality)*

When solving equations or inequalities involving absolute values, it is often very helpful to use the following statements:

(1.6)

Suppose $a > 0$. Then

(a) $|x| = a$ if and only if $x = \pm a$

(b) $|x| < a$ if and only if $-a < x < a$

(c) $|x| > a$ if and only if $x > a$ or $x < -a$

For instance, the inequality $|x| < a$ says that the distance from x to the origin is less than a, and you can see from Figure 1.5 that this is true if and only if x lies between $-a$ and a.

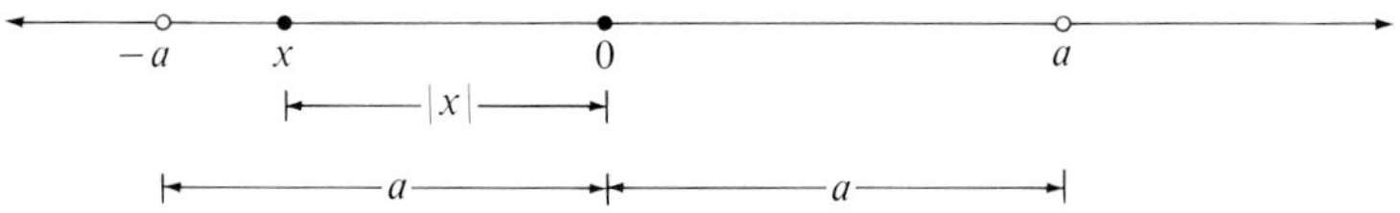

Figure 1.5

If a and b are any real numbers, then the distance between a and b is the absolute value of the difference, namely, $|a - b|$, which is also equal to $|b - a|$ (see Figure 1.6).

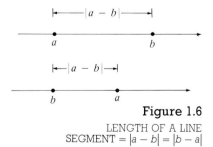

Figure 1.6

LENGTH OF A LINE
SEGMENT $= |a - b| = |b - a|$

● **Example 6**   Solve $|2x - 5| = 3$.

Solution   By statement (a) of (1.6), $|2x - 5| = 3$ is equivalent to

$$2x - 5 = 3 \qquad \text{or} \qquad 2x - 5 = -3$$

So $2x = 8$ or $2x = 2$. Thus $x = 4$ or $x = 1$.   ●

● **Example 7**   Solve $|x - 5| < 2$.

Solution 1   By Statement (b) of (1.6), $|x - 5| < 2$ is equivalent to

$$-2 < x - 5 < 2$$

Therefore, adding 5 to each side, we have

$$3 < x < 7$$

and the solution set is the open interval $(3, 7)$.

Solution 2   Geometrically, the solution set consists of all numbers x whose distance from 5 is less than 2. From Figure 1.7 we see that this is the interval $(3, 7)$.

Figure 1.7

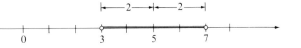

● **Example 8**   Solve $|3x + 2| \geqslant 4$.

Solution   By statements (a) and (c) of (1.6), $|3x + 2| \geqslant 4$ is equivalent to

$$3x + 2 \geqslant 4 \qquad \text{or} \qquad 3x + 2 \leqslant -4$$

In the first case $3x \geqslant 2$, which gives $x \geqslant \frac{2}{3}$. In the second case $3x \leqslant -6$, which gives $x \leqslant -2$. So the solution set is

$$\{x \mid x \leqslant -2 \text{ or } x \geqslant \frac{2}{3}\} = (-\infty, -2] \cup [\tfrac{2}{3}, \infty)$$

EXERCISES 1.1

In Exercises 1–32 solve the given inequalities in terms of intervals and illustrate the solution sets on the real number line.

1. $2x + 7 > 3$

2. $3x - 11 < 4$

3. $1 - x \leqslant 2$

4. $4 - 3x \geqslant 6$

5. $2x + 1 < 5x - 8$

6. $1 + 5x > 5 - 3x$

7. $-1 < 2x - 5 < 7$

8. $1 < 3x + 4 \leqslant 16$

9. $0 \leqslant 1 - x < 1$

10. $-5 \leqslant 3 - 2x \leqslant 9$

11. $4x < 2x + 1 \leqslant 3x + 2$

12. $2x + 3 < x + 4 < 3x - 2$

13. $1 - x \geqslant 3 - 2x \geqslant x - 6$

14. $x > 1 - x \geqslant 3 + 2x$

15. $(x - 1)(x - 2) > 0$

16. $(2x + 3)(x - 1) \geqslant 0$

17. $2x^2 + x \leqslant 1$

18. $x^2 < 2x + 8$

19. $x^2 + x + 1 > 0$

20. $x^2 + x > 1$

21. $x^2 < 3$

22. $x^2 \geqslant 5$

23. $x^3 > x$

24. $x^3 + 3x < 4x^2$

25. $\dfrac{1}{x} < 4$

26. $-3 < \dfrac{1}{x} \leqslant 1$

27. $\dfrac{4}{x} < x$

28. $\dfrac{x}{x+1} > 3$

29. $\dfrac{2x+1}{x-5} < 3$

30. $\dfrac{2+x}{3-x} \leqslant 1$

31. $\dfrac{x^2-1}{x^2+1} \geqslant 0$

32. $\dfrac{x^2-2x}{x^2-2} > 0$

In Exercises 33–36 solve the given equation for x.

33. $|2x| = 3$

34. $|3x+5| = 1$

35. $|x+3| = |2x+1|$

36. $\left|\dfrac{2x-1}{x+1}\right| = 3$

In Exercises 37–52 solve the given inequalities.

37. $|x| < 3$

38. $|x| \geqslant 3$

39. $|x-4| < 1$

40. $|x-6| < 0.1$

41. $|x+5| \geqslant 2$

42. $|x+1| \geqslant 3$

43. $|2x-3| \leqslant 0.4$

44. $|5x-2| < 6$

45. $1 \leqslant |x| \leqslant 4$

46. $0 < |x-5| < \frac{1}{2}$

47. $|x| > |x-1|$

48. $|2x-5| \leqslant |x+4|$

49. $\left|\dfrac{x}{2+x}\right| < 1$

50. $\left|\dfrac{2-3x}{1+2x}\right| \leqslant 4$

51. $|x| + |x-2| < 3$

52. $|x+1| + |x-2| < 7$

In Exercises 53 and 54 solve for x assuming a, b, and c are positive constants.

53. $a(bx - c) \geqslant bc$

54. $a \leqslant bx + c < 2a$

In Exercises 55 and 56 solve for x assuming a, b, and c are negative constants.

55. $ax + b < c$

56. $\dfrac{ax+b}{c} \leqslant b$

57. Show that if $a < b$, then $a < \dfrac{a+b}{2} < b$.

58. Use Rule (c) to prove Rule (e) of (1.2).

59. Prove that $|ab| = |a|\,|b|$. (*Hint:* Use Equation 1.4.)

60. Prove that $\left|\dfrac{a}{b}\right| = \dfrac{|a|}{|b|}$.

61. Show that if $0 < a < b$, then $a^2 < b^2$.

62. Show that $-|a| \leqslant a \leqslant |a|$.

63. Prove the Triangle Inequality: $|a+b| \leqslant |a| + |b|$. (*Hint:* Use Exercise 62 applied to both a and b.)

64. Prove that $|x-y| \geqslant |x| - |y|$. (*Hint:* Use the Triangle Inequality with $a = x - y$ and $b = y$.)

65. Suppose that $|x-2| < 0.01$ and $|y-3| < 0.04$. Use the Triangle Inequality to show that $|(x+y) - 5| < 0.05$.

The Cartesian Plane

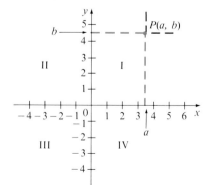

Figure 1.8

Just as the points on a line can be identified with real numbers by assigning them coordinates, as described in Section 1.1, so the points in a plane can be identified with ordered pairs of real numbers. We start by drawing two perpendicular coordinate lines that intersect at the origin O on each line. Usually one line is horizontal with positive direction to the right and is called the x-axis; the other line is vertical with positive direction upward and is called the y-axis.

Any point P in the plane can be located by a unique ordered pair of numbers as follows. Draw lines through P perpendicular to the x- and y-axes. These lines will intersect the axes in points with coordinates a and b as shown in Figure 1.8. Then the point P is assigned the ordered pair (a, b). The first number a is called the **x-coordinate** (or **abscissa**) of P; the second number b is called the **y-coordinate** (or **ordinate**) of P. We say that P is the

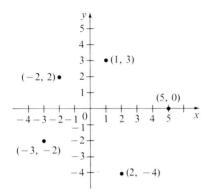

Figure 1.9

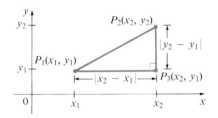

Figure 1.10

point with coordinates (a, b) and we denote the point by the symbol $P(a, b)$. Several points are labeled with their coordinates in Figure 1.9.

By reversing the preceding process we can start with an ordered pair (a, b) and arrive at the corresponding point P. Often we identify the point P with the ordered pair (a, b) and refer to "the point (a, b)." [Although the notation used for an open interval (a, b) is the same as the notation used for a point (a, b), you will be able to tell from the context which meaning is intended.]

This coordinate system is called the **rectangular coordinate system** or the **Cartesian coordinate system** in honor of the French mathematician René Descartes (1596–1650), even though another Frenchman, Pierre Fermat (1601–1665), invented the principles of analytic geometry at about the same time as Descartes. The plane, supplied with this coordinate system, is called the **coordinate plane** or the **Cartesian plane** and is denoted by R^2.

The x- and y-axes are called the **coordinate axes** and divide the Cartesian plane into four quadrants, which are labeled I, II, III, and IV in Figure 1.8. Notice that the first quadrant consists of those points whose x- and y-coordinates are both positive.

Recall from Section 1.1 that the distance between points a and b on a number line is $|a - b| = |b - a|$. Thus the distance between points $P_1(x_1, y_1)$ and $P_3(x_2, y_1)$ on a horizontal line must be $|x_2 - x_1|$ and the distance between $P_2(x_2, y_2)$ and $P_3(x_2, y_1)$ on a vertical line must be $|y_2 - y_1|$ (see Figure 1.10).

To find the distance $|P_1 P_2|$ between any two points $P_1(x_1, y_1)$ and $P_2(x_2, y_2)$, we note that triangle $P_1 P_2 P_3$ in Figure 1.10 is a right triangle, and so by the Pythagorean Theorem we have

$$|P_1 P_2| = \sqrt{|P_1 P_3|^2 + |P_2 P_3|^2} = \sqrt{|x_2 - x_1|^2 + |y_2 - y_1|^2}$$
$$= \sqrt{(x_2 - x_1)^2 + (y_2 - y_1)^2}$$

Distance Formula (1.7)

> The distance between the points $P_1(x_1, y_1)$ and $P_2(x_2, y_2)$ is
>
> $$|P_1 P_2| = \sqrt{(x_2 - x_1)^2 + (y_2 - y_1)^2}$$

● **Example 1**   The distance between $(1, -2)$ and $(5, 3)$ is

$$\sqrt{(5 - 1)^2 + [3 - (-2)]^2} = \sqrt{4^2 + 5^2} = \sqrt{41}$$ ●

Equations and Graphs

Suppose we have an equation involving the variables x and y, such as

$$x^2 + y^2 = 1 \quad \text{or} \quad y = \frac{2}{x} \quad \text{or} \quad x^3 + y^3 = 6xy$$

The **graph** of such an equation in x and y is the set of all points (x, y) that satisfy the equation. The graph gives a visual representation of the equation. For example, we shall see that the first equation represents a circle (later in this section), the second equation represents a hyperbola (Section 1.4), and the third represents a curve called the folium of Descartes (Figure 3.12).

● **Example 2**   Sketch the graph of the equation $x = y^2$.

Solution   There are infinitely many points on the graph and it is impossible to plot all of them. But we obtain some of these points in the following table and plot them in Figure 1.11. Then we connect the points by a smooth curve as shown.

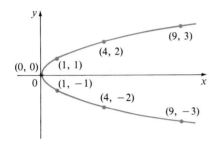

Figure 1.11

THE GRAPH OF $x = y^2$

y	0	1	2	3	-1	-2	-3
$x = y^2$	0	1	4	9	1	4	9
(x, y)	$(0, 0)$	$(1, 1)$	$(4, 2)$	$(9, 3)$	$(1, -1)$	$(4, -2)$	$(9, -3)$

For the time being we cannot be sure that Figure 1.11 is an accurate graph of the equation $x = y^2$ because we have plotted only a finite number of points. In fact, one of the major achievements of calculus is a procedure (given in Chapter 4) that guarantees that a sketch is an accurate depiction of a graph.   ●

So far we have discussed how to find the graph of an equation in x and y. The converse problem is how to find an *equation of a graph,* that is, an equation that represents a given curve in the xy-plane. Such an equation is satisfied by the coordinates of the points on the curve and by no other points. This is the other half of the basic principle of analytic geometry as formulated by Descartes and Fermat. The idea is that if a geometric curve can be represented by an algebraic equation, then the rules of algebra can be used to analyze the geometric problem.

As an example of this type of problem, let us find the equation of a circle with radius r and center (h, k). By definition, the circle is the set of all points $P(x, y)$ whose distance from the center $C(h, k)$ is r (see Figure 1.12). Thus P is on the circle if and only if $|PC| = r$. From the distance formula (1.7) we have

$$\sqrt{(x - h)^2 + (y - k)^2} = r$$

or equivalently, squaring both sides,

$$(x - h)^2 + (y - k)^2 = r^2$$

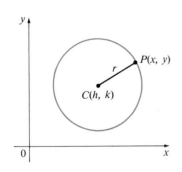

Figure 1.12   This is the desired equation.

Equation of a Circle (1.8)	The equation of a circle with center (h, k) and radius r is

$$(x - h)^2 + (y - k)^2 = r^2$$

In particular, if the center is the origin $(0, 0)$, the equation is

$$x^2 + y^2 = r^2$$

● **Example 3**   Find the equation of the circle with radius 3 and center $(2, -5)$.

Solution   From Equation 1.8 with $r = 3$, $h = 2$, and $k = -5$, we obtain

$$(x - 2)^2 + (y + 5)^2 = 9$$ ●

● **Example 4**   Sketch the graph of the equation $x^2 + y^2 + 2x - 6y + 7 = 0$ by first showing that it represents a circle and then finding its center and radius.

Solution   We first group the x-terms and y-terms as follows:

$$(x^2 + 2x) + (y^2 - 6y) = -7$$

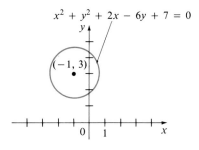

$x^2 + y^2 + 2x - 6y + 7 = 0$

Then we complete the square within each grouping, adding the appropriate constants to both sides of the equation:

$$(x^2 + 2x + 1) + (y^2 - 6y + 9) = -7 + 1 + 9$$

or

$$(x + 1)^2 + (y - 3)^2 = 3$$

(See Appendix A for a review of completing the square.) Comparing this equation with the standard equation of a circle (1.8), we see that $h = -1$, $k = 3$, and $r = \sqrt{3}$, so the given equation represents a circle with center $(-1, 3)$ and radius $\sqrt{3}$. It is sketched in Figure 1.13.   ●

Figure 1.13

EXERCISES 1.2

In Exercises 1–6 find the distance between the given points.

1. $(1, 1), (4, 5)$ **2.** $(1, -3), (5, 7)$

3. $(6, -2), (-1, 3)$ **4.** $(1, -6), (-1, -3)$

5. $(2, 5), (4, -7)$ **6.** $(a, b), (b, a)$

7. Show that the triangle with vertices $A(0, 2)$, $B(-3, -1)$, and $C(-4, 3)$ is isosceles.

8. Show that the triangle with vertices $A(6, -7)$, $B(11, -3)$, and $C(2, -2)$ is a right triangle using the converse of the Pythagorean Theorem. Find the area of the triangle.

9. Show that the points $(-2, 9)$, $(4, 6)$, $(1, 0)$, and $(-5, 3)$ are the vertices of a square.

10. Show that the points $A(-1, 3)$, $B(3, 11)$, and $C(5, 15)$ are collinear by showing that $|AB| + |BC| = |AC|$.

11. Find a point on the y-axis that is equidistant from $(5, -5)$ and $(1, 1)$.

12. Show that the midpoint of the line segment from $P_1(x_1, y_1)$ to $P_2(x_2, y_2)$ is

$$\left(\frac{x_1 + x_2}{2}, \frac{y_1 + y_2}{2} \right)$$

13. Find the midpoint of the line segment joining the points
(a) $(1, 3)$ and $(7, 15)$　　(b) $(-1, 6)$ and $(8, -12)$

14. Find the lengths of the medians of the triangle with vertices $A(1, 0)$, $B(3, 6)$, and $C(8, 2)$. (A median is a line segment from a vertex to the midpoint of the opposite side.)

In Exercises 15–24 sketch the graph of the given equation.

15. $x = 3$　　　**16.** $y = -2$　　　**17.** $xy = 0$

18. $|y| = 1$　　　**19.** $y = x$　　　**20.** $y = 2x + 5$

21. $xy = 2$　　　**22.** $y = x^3$　　　**23.** $x + y^2 = 4$

24. $x + \sqrt{y} = 4$

In Exercises 25–32 sketch the given region in the xy-plane.

25. $\{(x, y) | x < 0\}$

26. $\{(x, y) | y > 0\}$

27. $\{(x, y) | xy < 0\}$

28. $\{(x, y) | x \geqslant 1 \text{ and } y < 3\}$

29. $\{(x, y) | |x| \leqslant 2\}$

30. $\{(x, y) | |x| < 3 \text{ and } |y| < 2\}$

31. $\{(x, y) | x^2 + y^2 \leqslant 1\}$

32. $\{(x, y) | x^2 + y^2 > 4\}$

In Exercises 33–36 find an equation of a circle that satisfies the given conditions.

33. Center $(3, -1)$; radius 5

34. Center $(-2, -8)$; radius 10

35. Center at the origin; passes through $(4, 7)$

36. Center $(-1, 5)$; passes through $(-4, -6)$

In Exercises 37–41 show that the given equation represents a circle and find the center and radius.

37. $x^2 + y^2 - 4x + 10y + 13 = 0$

38. $x^2 + y^2 + 6y + 2 = 0$

39. $x^2 + y^2 + x = 0$

40. $16x^2 + 16y^2 + 8x + 32y + 1 = 0$

41. $2x^2 + 2y^2 - x + y = 1$

42. Under what condition on the coefficients a, b, and c does the equation $x^2 + y^2 + ax + by + c = 0$ represent a circle? When that condition is satisfied, find the center and radius of the circle.

SECTION 1.3

Lines

In mathematics the word *line* always means "straight line." We can find the equation of a line in terms of its slope, which is a measure of the steepness of the line.

Definition (1.9)

> The **slope** of a nonvertical line that passes through the points $P_1(x_1, y_1)$ and $P_2(x_2, y_2)$ is
>
> $$m = \frac{y_2 - y_1}{x_2 - x_1}$$
>
> The slope of a vertical line is not defined.

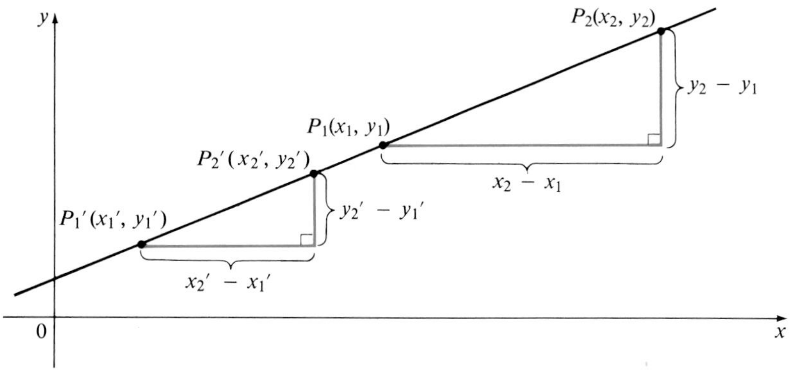

Figure 1.14

Thus the slope of a line is the ratio of the change in y to the change in x. From the similar triangles in Figure 1.14 we see that the slope is independent of which two points are chosen on the line:

$$\frac{y_2 - y_1}{x_2 - x_1} = \frac{y_2' - y_1'}{x_2' - x_1'}$$

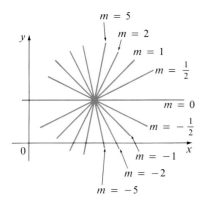

Figure 1.15

Figure 1.15 shows several lines labeled with their slopes. Notice that lines with positive slope slant upward to the right, whereas lines with negative slope slant downward to the right. Notice also that the steepest lines are the ones where the absolute value of the slope is largest, and a horizontal line has slope 0.

Now let us find the equation of the line that passes through a given point $P_1(x_1, y_1)$ and has slope m. A point $P(x, y)$ with $x \neq x_1$ lies on this line if and only if the slope of the line through P_1 and P is equal to m; that is,

$$\frac{y - y_1}{x - x_1} = m$$

This equation can be rewritten in the form

$$y - y_1 = m(x - x_1)$$

and we observe that this equation is also satisfied when $x = x_1$ and $y = y_1$. Therefore it is the equation of the given line.

Point-Slope Form of the Equation of a Line (1.10)

> An equation of the line passing through the point $P_1(x_1, y_1)$ and having slope m is
>
> $$y - y_1 = m(x - x_1)$$

● **Example 1**   Find an equation of the line through $(1, -7)$ with slope $-\frac{1}{2}$.

Solution   Using (1.10) with $m = -\frac{1}{2}$, $x_1 = 1$, and $y_1 = -7$, we obtain an equation of the line as

$$y + 7 = -\tfrac{1}{2}(x - 1)$$

which we can rewrite as

$$2y + 14 = -x + 1 \qquad \text{or} \qquad x + 2y + 13 = 0 \qquad\qquad ●$$

● **Example 2**   Find an equation of the line through the points $(-1, 2)$ and $(3, -4)$.

Solution   By Definition 1.9 the slope of the line is

$$m = \frac{-4 - 2}{3 - (-1)} = -\frac{3}{2}$$

Using the point-slope form with $x_1 = -1$ and $y_1 = 2$, we obtain

$$y - 2 = -\tfrac{3}{2}(x + 1)$$

which simplifies to

$$3x + 2y = 1 \qquad\qquad ●$$

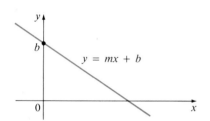

Figure 1.16

Suppose a nonvertical line has slope m and y-intercept b (see Figure 1.16). This means it intersects the y-axis in the point $(0, b)$, so the point-slope form of the equation of the line, with $x_1 = 0$ and $y_1 = b$, becomes

$$y - b = m(x - 0)$$

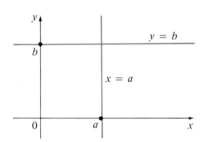

Figure 1.17

This simplifies to

(1.11)

$$\boxed{\, y = mx + b \,}$$

which is called the **slope-intercept form** of the equation of a line.

     In particular, if a line is horizontal, its slope is $m = 0$, so its equation is $y = b$, where b is the y-intercept (see Figure 1.17). A vertical line does not have a slope, but we can write its equation as $x = a$, where a is the x-intercept, because the x-coordinate of every point on the line is a.

     Observe that the equation of every line can be written in the form

(1.12)

$$\boxed{\, Ax + By + C = 0 \,}$$

because a vertical line has the equation $x = a$ or $x - a = 0$ ($A = 1$, $B = 0$, $C = -a$) and a nonvertical line has the equation $y = mx + b$ or

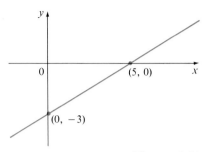

Figure 1.18

THE LINE $3x - 5y = 15$

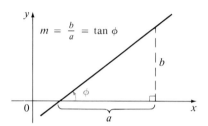

Figure 1.19

ANGLE OF INCLINATION

$-mx + y - b = 0$ $(A = -m, B = 1, C = -b)$. Conversely, if we start with a general first-degree equation, that is, an equation of the form (1.12) where A, B, and C are constants and A and B are not both 0, then we can show that it is the equation of a line. If $B = 0$, it becomes $Ax + C = 0$ or $x = -C/A$, which represents a vertical line with x-intercept $-C/A$. If $B \neq 0$, it can be rewritten by solving for y:

$$ y = -\frac{A}{B}x - \frac{C}{B} $$

and we recognize this as being the slope-intercept form of the equation of a line ($m = -A/B, b = -C/B$). Therefore an equation of the form (1.12) is called a **linear equation** or the **general equation of a line.** For brevity, we often refer to "the line $Ax + By + C = 0$" instead of saying "the line whose equation is $Ax + By + C = 0$."

● **Example 3**   Sketch the graph of the equation $3x - 5y = 15$.

Solution   Since the equation is linear, its graph is a line. To draw the graph it suffices to find two points on the line. It is easiest to find the intercepts. Substituting $y = 0$ (the equation of the x-axis) in the given equation, we get $3x = 15$, so $x = 5$ is the x-intercept. Substituting $x = 0$ in the equation, we see that the y-intercept is -3. This allows us to sketch the graph as in Figure 1.18.   ●

Definition (1.13)

> The **angle of inclination** of a line is the angle ϕ ($0 \leqslant \phi < 180°$) that it makes with the positive direction of the x-axis.

You can see from Figure 1.19 that the slope m and the angle of inclination ϕ are related by the equation

(1.14)

> $$ m = \tan \phi $$

(Although Figure 1.19 shows this for $0 \leqslant \phi < 90°$, you can check that it is also true when $90° < \phi < 180°$. See Appendix B for a review of trigonometry.) Note that $\phi = 90°$ ($\pi/2$ radians) for a vertical line and $\tan 90°$ is undefined. This corresponds to the fact that the slope of a vertical line is undefined.

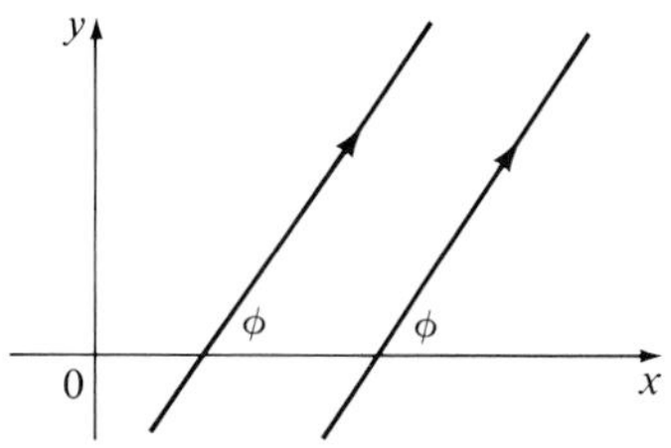

It follows from elementary geometry that parallel lines have the same angle of inclination (see Figure 1.20). Because of Equation 1.14 we have the following:

Figure 1.20

(1.15)

> Two lines are parallel if and only if they have the same slope.

● **Example 4**   Find an equation of the line through the point $(5, 2)$ that is parallel to the line $4x + 6y + 5 = 0$.

Solution   The given line can be written in the form

$$y = -\tfrac{2}{3}x - \tfrac{5}{6}$$

which is in slope-intercept form with $m = -\tfrac{2}{3}$. Parallel lines have the same slope, so the required line has slope $-\tfrac{2}{3}$ and its equation in point-slope form is

$$y - 2 = -\tfrac{2}{3}(x - 5)$$

This simplifies to $2x + 3y = 16$.   ●

Suppose that two nonvertical lines L_1 and L_2, which are not parallel, intersect at an angle θ as in Figure 1.21. If these lines have slopes m_1 and m_2 and angles of inclination ϕ_1 and ϕ_2, then $m_1 = \tan\phi_1$ and $m_2 = \tan\phi_2$. Figure 1.21 shows that $\phi_2 = \phi_1 + \theta$ and so $\theta = \phi_2 - \phi_1$. Therefore, using the identity for $\tan(x - y)$ (Equation 14b in Appendix B), we have

$$\tan\theta = \tan(\phi_2 - \phi_1) = \frac{\tan\phi_2 - \tan\phi_1}{1 + \tan\phi_2 \tan\phi_1}$$

and so

(1.16)

$$\boxed{\tan\theta = \frac{m_2 - m_1}{1 + m_1 m_2}}$$

In particular, L_1 and L_2 are perpendicular when $\theta = 90°$ and so $\tan\theta$ is undefined. This is the case when the denominator in Equation 1.16 is 0; that is, $1 + m_1 m_2 = 0$, or $m_1 m_2 = -1$.

(1.17)

> Two lines with slopes m_1 and m_2 are perpendicular if and only if $m_1 m_2 = -1$, that is, their slopes are negative reciprocals:
>
> $$m_2 = -\frac{1}{m_1}$$

● **Example 5**   Show that the points $P(3, 3)$, $Q(8, 17)$, and $R(11, 5)$ are the vertices of a right triangle.

Solution   The slopes of the lines PR and QR are

$$m_1 = \frac{5 - 3}{11 - 3} = \frac{1}{4} \quad \text{and} \quad m_2 = \frac{5 - 17}{11 - 8} = -4$$

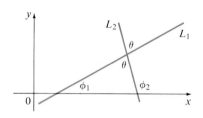

Figure 1.21

Since $m_1 m_2 = -1$, these lines are perpendicular and so PQR is a right triangle. ●

● **Example 6**   (a)   Find the point of intersection of the lines $3x - y = 3$ and $2x + 3y = 13$. (b)   Find, correct to the nearest degree, the angle at which they intersect.

Solution   (a)   By definition, a point of intersection of two curves lies on both curves, so its coordinates satisfy both equations. Therefore a point of intersection is found by solving the two equations. We solve the linear equations $3x - y = 3$ and $2x + 3y = 13$ by substituting $y = 3x - 3$ from the first equation into the second. This gives

$$2x + 3(3x - 3) = 11x - 9 = 13$$

So $11x = 22$, $x = 2$, and $y = 3$. Thus $(2, 3)$ is the point of intersection.
    (b)   The second line can be written as

$$y = -\tfrac{2}{3}x + \tfrac{13}{3}$$

so its slope is $m_2 = -\tfrac{2}{3}$. The first line is $y = 3x - 3$, so its slope is $m_1 = 3$. If θ is the angle of intersection, then Equation 1.16 gives

$$\tan\theta = \frac{m_2 - m_1}{1 + m_1 m_2} = \frac{(-\tfrac{2}{3}) - 3}{1 + 3(-\tfrac{2}{3})}$$

$$= \frac{-\tfrac{11}{3}}{-1} = \frac{11}{3}$$

and so θ is given by

$$\theta = \tan^{-1}(\tfrac{11}{3}) \approx 75°$$

(If we had taken the lines in the other order, we would have obtained $\tan\theta = -\tfrac{11}{3}$, which gives $\theta \approx 105°$, the supplementary angle of 75°. Either answer is acceptable.) ●

EXERCISES 1.3

In Exercises 1–4 find the slope of the line through P and Q.

1. $P(1, 5)$, $Q(4, 11)$ 2. $P(-1, 6)$, $Q(4, -3)$

3. $P(-3, 3)$, $Q(-1, -6)$ 4. $P(-1, -4)$, $Q(6, 0)$

In Exercises 5–22 find an equation of the line that satisfies the given conditions.

5. Through $(2, -3)$, slope 6

6. Through $(-1, 4)$, slope -3

7. Through $(1, 7)$, slope $\tfrac{2}{3}$

8. Through $(-3, -5)$, slope $-\tfrac{7}{2}$

9. Through $(2, 1)$ and $(1, 6)$

10. Through $(-1, -2)$ and $(4, 3)$

11. Slope 3, y-intercept -2

12. Slope $\frac{2}{5}$, y-intercept 4

13. x-intercept 1, y-intercept -3

14. x-intercept -8, y-intercept 6

15. Through $(2, -4)$, angle of inclination $30°$

16. Through $(-5, 3)$, angle of inclination $135°$

17. Through $(4, 5)$, parallel to the x-axis

18. Through $(4, 5)$, parallel to the y-axis

19. Through $(1, -6)$, parallel to the line $x + 2y = 6$

20. y-intercept 6, parallel to the line $2x + 3y + 4 = 0$

21. Through $(-1, -2)$, perpendicular to the line $2x + 5y + 8 = 0$

22. Through $(\frac{1}{2}, -\frac{2}{3})$, perpendicular to the line $4x - 8y = 1$

23. Show that $A(1, 1)$, $B(7, 4)$, $C(5, 10)$, and $D(-1, 7)$ are vertices of a parallelogram.

24. Show that $A(-3, -1)$, $B(3, 3)$, and $C(-9, 8)$ are vertices of a right triangle.

25. Show that $A(1, 1)$, $B(11, 3)$, $C(10, 8)$, and $D(0, 6)$ are vertices of a rectangle.

26. Use slopes to determine whether the given points are collinear (lie on a line).
(a) $(1, 1)$, $(3, 9)$, and $(6, 21)$
(b) $(-1, 3)$, $(1, 7)$, and $(4, 15)$

27. Find the equation of the perpendicular bisector of the line segment joining the points $A(1, 4)$ and $B(7, -2)$.

28. (a) Find equations for the sides of the triangle with vertices $P(1, 0)$, $Q(3, 4)$, and $R(-1, 6)$.
(b) Find equations for the medians of this triangle. Where do they intersect?

In Exercises 29–34 find the slope and y-intercept of the line and draw its graph.

29. $x + 3y = 0$ **30.** $2x - 5y = 0$

31. $y = -2$ **32.** $2x - 3y + 6 = 0$

33. $3x - 4y = 12$ **34.** $4x + 5y = 10$

In Exercises 35 and 36 find the angle of inclination of the line.

35. $y = x + 1$ **36.** $\sqrt{3}x - y = 1$

In Exercises 37 and 38 find the angle of inclination of the line correct to the nearest degree.

37. $2x + 3y = 4$ **38.** $3x - 4y + 5 = 0$

In Exercises 39–42 find the point of intersection of the given lines and the angle between them (correct to the nearest degree).

39. $y = x$, $y = 3 - 2x$

40. $y = 2x + 3$, $y = 3x + 2$

41. $x - 2y = 5$, $x - 3y = 6$

42. $2x - y + 1 = 0$, $x - 2y + 5 = 0$

43. Show that if the x- and y-intercepts of a line are nonzero numbers a and b, then the equation of the line can be put in the form

$$\frac{x}{a} + \frac{y}{b} = 1$$

In Exercises 44–48 sketch the graph of the given set.

44. $\{(x, y) | y > 2x - 1\}$

45. $\{(x, y) | 1 + x \leqslant y \leqslant 1 - 2x\}$

46. $\left\{(x, y) \Big| -x \leqslant y < \dfrac{x + 3}{2}\right\}$

47. $\{(x, y) | |x| + |y| \leqslant 1\}$

48. $\{(x, y) | |x - y| + |x| - |y| \leqslant 2\}$

SECTION 1.4

Second-Degree Equations

Second-degree equations of the form

$$y = ax^2 + bx + c \qquad \frac{x^2}{a^2} + \frac{y^2}{b^2} = 1 \qquad \frac{x^2}{a^2} - \frac{y^2}{b^2} = 1$$

occur frequently and their graphs are called parabolas, ellipses, and hyper-bolas, respectively. A study of the geometric properties of these curves, which are called conics, is deferred to Chapter 9. For now it is enough to be able to sketch quickly the graphs of such equations.

Parabolas

The graph of the equation

(1.18)

$$y = ax^2$$

for any $a \neq 0$, is called a **parabola** with vertex the origin. It is illustrated, for several values of the constant a, in the following example.

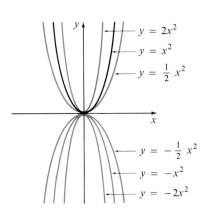

● **Example 1**   Sketch the graphs of the equations $y = x^2$, $y = 2x^2$, $y = \frac{1}{2}x^2$, $y = -x^2$, $y = -2x^2$, and $y = -\frac{1}{2}x^2$.

Solution   By plotting points we sketch the graphs in Figure 1.22.   ●

From Example 1 we see that the parabola $y = ax^2$ opens upward if $a > 0$ and downward if $a < 0$ (see Figure 1.23). To get the graph of $y = ax^2$ from the graph of $y = x^2$ we stretch (in the y-direction) if $|a| > 1$ and we shrink if $|a| < 1$. Notice that if (x, y) satisfies $y = ax^2$, then so does $(x, -y)$. This corresponds to the geometric fact that if the right half of the graph is reflected in the y-axis, then the left half of the graph is obtained. We say that the graph is **symmetric with respect to the y-axis.**

Figure 1.22

The graph of an equation is symmetric with respect to the y-axis if the equation is unchanged when x is replaced by $-x$.

Figure 1.23

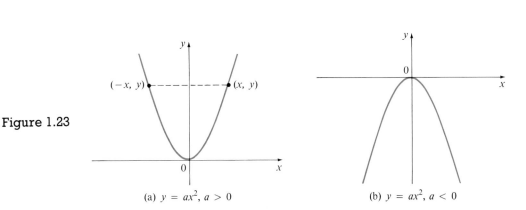

(a) $y = ax^2$, $a > 0$

(b) $y = ax^2$, $a < 0$

Figure 1.24

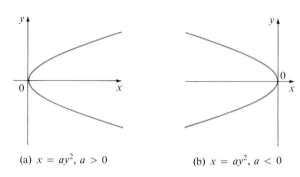

(a) $x = ay^2$, $a > 0$ (b) $x = ay^2$, $a < 0$

If x and y are interchanged in Equation 1.18, the resulting equation is $x = ay^2$, which also represents a parabola. (Interchanging x and y amounts to reflecting in the diagonal line $y = x$. The special case $x = y^2$ was sketched in Example 2 in Section 1.2.) The parabola $x = ay^2$ opens to the right if $a > 0$ and to the left if $a < 0$ (see Figure 1.24). This time the parabola is symmetric with respect to the x-axis because if (x, y) satisfies $x = ay^2$, then so does $(x, -y)$.

> The graph of an equation is symmetric with respect to the x-axis if the equation is unchanged when y is replaced by $-y$.

● **Example 2**   Sketch the region bounded by the parabola $x = y^2$ and the line $y = x - 2$.

Solution   First we find the points of intersection by solving the two equations. Substituting $x = y + 2$ into the equation $x = y^2$, we get $y + 2 = y^2$, which gives

$$0 = y^2 - y - 2 = (y - 2)(y + 1)$$

so $y = 2$ or -1. Thus the points of intersection are $(4, 2)$ and $(1, -1)$ and we draw the line $y = x - 2$ passing through these points. We then sketch the parabola $x = y^2$ by referring to Figure 1.24(a) and having the parabola pass through $(4, 2)$ and $(1, -1)$. The region bounded by $x = y^2$ and $y = x - 2$ means the finite region whose boundaries are these curves. It is sketched in Figure 1.25.   ●

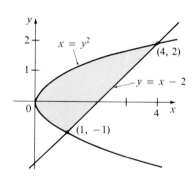

Figure 1.25

Ellipses

The curve with equation

(1.19)

$$\frac{x^2}{a^2} + \frac{y^2}{b^2} = 1$$

where a and b are positive numbers, is called an **ellipse** in standard position. Observe that Equation 1.19 is unchanged if x is replaced by $-x$ or y is replaced by $-y$, so the ellipse is symmetric with respect to both axes. As a further aid to sketching the ellipse we find its intercepts.

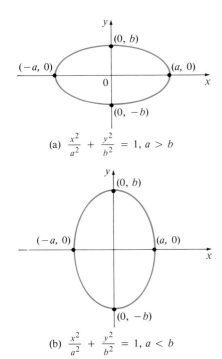

(a) $\dfrac{x^2}{a^2} + \dfrac{y^2}{b^2} = 1, \ a > b$

(b) $\dfrac{x^2}{a^2} + \dfrac{y^2}{b^2} = 1, \ a < b$

Figure 1.26

> The **x-intercepts** of a graph are the x-coordinates of the points where the graph intersects the x-axis. They are found by setting $y = 0$ in the equation of the graph. The **y-intercepts** are the y-coordinates of the points where the graph intersects the y-axis. They are found by setting $x = 0$ in its equation.

If we set $y = 0$ in Equation 1.19 we get $x^2 = a^2$ and so the x-intercepts are $\pm a$. Setting $x = 0$, we get $y^2 = b^2$, so the y-intercepts are $\pm b$. Using this information, together with symmetry, we sketch the ellipse in Figure 1.26. Notice that if $a > b$, the ellipse is wider than it is high; if $a < b$, it is higher than it is wide. If $a = b$, the ellipse is a circle with radius a.

● **Example 3**   Sketch the graph of $9x^2 + 16y^2 = 144$.

Solution   Divide both sides of the equation by 144:

$$\frac{x^2}{16} + \frac{y^2}{9} = 1$$

The equation is now in the standard form for an ellipse (1.19), so we have $a^2 = 16, b^2 = 9, a = 4,$ and $b = 3$. The x-intercepts are ± 4; the y-intercepts are ± 3. The graph is sketched in Figure 1.27.   ●

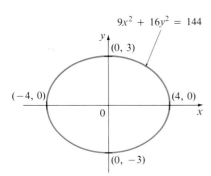

$9x^2 + 16y^2 = 144$

Figure 1.27

Hyperbolas

The curve with equation

(1.20)

$$\frac{x^2}{a^2} - \frac{y^2}{b^2} = 1$$

is called a **hyperbola** in standard position. Again, Equation 1.20 is unchanged when x is replaced by $-x$ or y is replaced by $-y$, so the hyperbola is symmetric with respect to both axes. To find the x-intercepts we set $y = 0$ and obtain $x^2 = a^2$ and $x = \pm a$. However, if we put $x = 0$ in Equation 1.20, we get $y^2 = -b^2$, which is impossible, so there is no y-intercept. In fact, from Equation 1.20 we obtain

$$\frac{x^2}{a^2} = 1 + \frac{y^2}{b^2} \geqslant 1$$

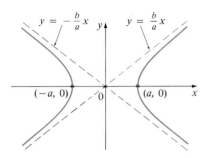

Figure 1.28

THE HYPERBOLA $\dfrac{x^2}{a^2} - \dfrac{y^2}{b^2} = 1$

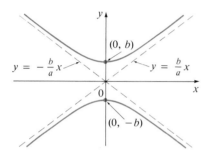

Figure 1.29

THE HYPERBOLA $\dfrac{y^2}{b^2} - \dfrac{x^2}{a^2} = 1$

which shows that $x^2 \geqslant a^2$ and so $|x| = \sqrt{x^2} \geqslant a$. Therefore, we have $x \geqslant a$ or $x \leqslant -a$. This means that the hyperbola consists of two parts, called its *branches*. It is sketched in Figure 1.28.

In drawing a hyperbola it is useful to draw first its *asymptotes,* which are the lines $y = (b/a)x$ and $y = -(b/a)x$ shown in Figure 1.28. Both branches of the hyperbola approach the asymptotes; that is, they come arbitrarily close to the asymptotes. (This involves the idea of a limit, which is discussed in Chapter 2. See Exercise 44 in Section 9.6.)

By interchanging the roles of x and y we get an equation of the form

$$\frac{y^2}{b^2} - \frac{x^2}{a^2} = 1$$

which also represents a hyperbola and is sketched in Figure 1.29. It has the same asymptotes as the hyperbola given by Equation 1.20.

● **Example 4**   Sketch the curve $9x^2 - 4y^2 = 36$.

Solution   Dividing both sides by 36, we obtain

$$\frac{x^2}{4} - \frac{y^2}{9} = 1$$

which is in the standard form of the equation of a hyperbola (Equation 1.20). Since $a^2 = 4$, the x-intercepts are ± 2. Since $b^2 = 9$, we have $b = 3$ and the asymptotes are $y = \pm(\frac{3}{2})x$. The hyperbola is sketched in Figure 1.30.   ●

If $b = a$, a hyperbola has the equation $x^2 - y^2 = a^2$ (or $y^2 - x^2 = a^2$) and is called an *equilateral hyperbola* (see Figure 1.31). Its asymptotes are $y = \pm x$, which are perpendicular. If an equilateral hyperbola is rotated by $45°$, the asymptotes become the x- and y-axes, and it can be shown (see Section 9.8) that the new equation of the hyperbola is $xy = k$, where k is a constant (see Figure 1.31).

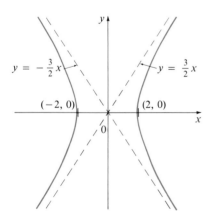

Figure 1.30

THE HYPERBOLA $9x^2 - 4y^2 = 36$

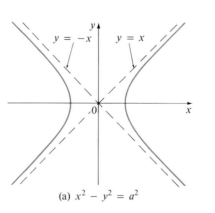

(a) $x^2 - y^2 = a^2$

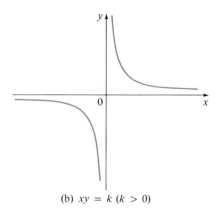

(b) $xy = k$ $(k > 0)$

Figure 1.31

EQUILATERAL HYPERBOLAS

Shifted Conics

Recall that the equation of a circle with center the origin and radius r is $x^2 + y^2 = r^2$, but if the center is the point (h, k), then the equation of the circle becomes

$$(x - h)^2 + (y - k)^2 = r^2$$

Similarly, if we take the ellipse with equation

(1.21)
$$\frac{x^2}{a^2} + \frac{y^2}{b^2} = 1$$

and translate it (shift it) so that its center is the point (h, k), then its equation becomes

(1.22)
$$\frac{(x - h)^2}{a^2} + \frac{(y - k)^2}{b^2} = 1$$

(See Figure 1.32.)

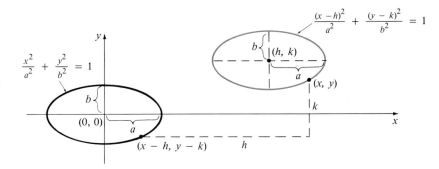

Figure 1.32

Notice that in shifting the ellipse, we replaced x by $x - h$ and y by $y - k$ in Equation 1.21 to obtain Equation 1.22. We use the same procedure to shift the parabola $y = ax^2$ so that its vertex (the origin) becomes the point (h, k) as in Figure 1.33. Replacing x by $x - h$ and y by $y - k$, we see that the new equation is

$$y - k = a(x - h)^2 \qquad \text{or} \qquad y = a(x - h)^2 + k$$

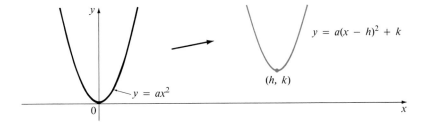

Figure 1.33

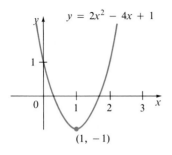

$y = 2x^2 - 4x + 1$

(1, −1)

Figure 1.34

● **Example 5**   Sketch the graph of the equation $y = 2x^2 - 4x + 1$.

Solution   First we complete the square:

$$y = 2(x^2 - 2x) + 1 = 2(x - 1)^2 - 1$$

In this form we see that the equation represents the parabola obtained by shifting $y = 2x^2$ so that its vertex is at the point $(1, -1)$. The graph is sketched in Figure 1.34.   ●

● **Example 6**   Sketch the curve $x = 1 - y^2$.

Solution   This time we start with the parabola $x = -y^2$ (as in Figure 1.24 with $a = -1$) and shift one unit to the right to get the graph of $x = 1 - y^2$ (see Figure 1.35).   ●

Figure 1.35

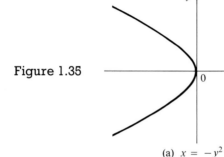

(a) $x = -y^2$

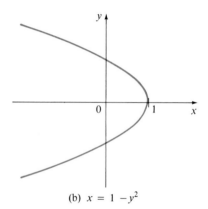

(b) $x = 1 - y^2$

EXERCISES 1.4

In Exercises 1–22 identify the type of curve and sketch the graph. Do not plot points. Just use the standard graphs given in Figures 1.23, 1.24, 1.26, 1.28, and 1.29 and shift if necessary.

1. $y = -x^2$

2. $y^2 - x^2 = 1$

3. $x^2 + 4y^2 = 16$

4. $x = -2y^2$

5. $16x^2 - 25y^2 = 400$

6. $25x^2 + 4y^2 = 100$

7. $4x^2 + y^2 = 1$

8. $y = x^2 + 2$

9. $x = y^2 - 1$

10. $9x^2 - 25y^2 = 225$

11. $9y^2 - x^2 = 9$

12. $2x^2 + 5y^2 = 10$

13. $xy = 4$

14. $y = x^2 + 2x$

15. $9(x - 1)^2 + 4(y - 2)^2 = 36$

16. $16x^2 + 9y^2 - 36y = 108$

17. $y = x^2 - 6x + 13$

18. $x^2 - y^2 - 4x + 3 = 0$

19. $x = 4 - y^2$

20. $y^2 - 2x + 6y + 5 = 0$

21. $x^2 + 4y^2 - 6x + 5 = 0$

22. $4x^2 + 9y^2 - 16x + 54y + 61 = 0$

In Exercises 23 and 24 sketch the region bounded by the given curves.

23. $y = 3x, y = x^2$

24. $y = 4 - x^2, x - 2y = 2$

25. Find the equation of the parabola with vertex $(1, -1)$ that passes through the points $(-1, 3)$ and $(3, 3)$.

26. Find the equation of the ellipse with center at the origin that passes through the points $(1, -10\sqrt{2}/3)$ and $(-2, 5\sqrt{5}/3)$.

In Exercises 27 and 28 sketch the graph of the given set.

27. $\{(x, y) | y \geqslant x^2 - 1\}$ **28.** $\{(x, y) | x^2 + 4y^2 \leqslant 4\}$

SECTION 1.5

Functions and Their Graphs

The area A of a circle depends on the radius r of the circle. The rule that connects r and A is given by the equation $A = \pi r^2$. With each positive number r there is associated one value of A, and we say that A is a function of r.

The number N of bacteria in a culture depends on the time t. If the culture starts with 5000 bacteria and the population doubles every hour, then after t hours the number of bacteria will be $N = (5000)2^t$. This is the rule that connects t and N. For each value of t there is a corresponding value of N, and we say that N is a function of t.

The cost C of mailing a first-class letter depends on the mass m of the letter. Although there is no single neat formula connecting m and C, the post office has a rule for determining C when m is known.

In each of these examples there is a rule whereby, given a number $(r, t,$ or $m)$, another number $(A, N,$ or $C)$ is assigned. In each case we say that the second number is a function of the first number.

Definition (1.23)

> A **function** f is a rule that assigns to each element x in a set A exactly one element, called $f(x)$, in a set B.

We usually consider functions for which the sets A and B are sets of real numbers. The set A in Definition 1.23 is called the **domain** of the function. The symbol $f(x)$ is read "f of x" and is called the **value of f at x,** or the **image of x under f.** The **range** of f is the set of all possible values of $f(x)$ as x varies throughout the domain, that is, $\{f(x) | x \in A\}$.

It is helpful to think of a function as a **machine** (see Figure 1.36). If x is in the domain of the function f, then when x enters the machine, it is accepted as an input and the machine produces an output $f(x)$ according to the rule of the function. Thus, we can think of the domain as the set of all possible inputs and the range as the set of all possible outputs.

The preprogrammed functions in a calculator are good examples of a function as a machine. For example, the $\sqrt{x}$ key on your calculator is such a function. First you input x into the display. Then you press the key labeled $\sqrt{x}$. If $x < 0$, then x is not in the domain of this function; that is, x is not an

Figure 1.36

MACHINE DIAGRAM FOR A
FUNCTION f

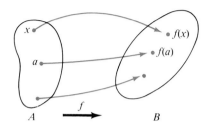

Figure 1.37

ARROW DIAGRAM FOR f

acceptable input, and the calculator will indicate an error. If $x \geqslant 0$, then an approximation to $\sqrt{x}$ will appear in the display. Thus the $\sqrt{x}$ key on your calculator is not quite the same as the exact mathematical function f defined by $f(x) = \sqrt{x}$. (See Appendix D.)

Another way to picture a function is by an **arrow diagram** as in Figure 1.37. Each arrow connects an element of A to an element of B. The arrow indicates that $f(x)$ is associated with x, $f(a)$ is associated with a, and so on.

● **Example 1**   The squaring function assigns to each real number x its square x^2. It is defined by the equation

$$f(x) = x^2$$

The values of f are found by substituting for x in this equation. For example,

$$f(3) = 3^2 = 9 \qquad\qquad f(-2) = (-2)^2 = 4$$

The domain of f is the set R of all real numbers. The range of f consists of all values of $f(x)$, that is, all numbers of the form x^2. But $x^2 \geqslant 0$ for all numbers x, and any nonnegative number c is a square since $c = (\sqrt{c})^2 = f(\sqrt{c})$. Therefore the range of f is $\{y \mid y \geqslant 0\} = [0, \infty)$. The machine diagram and arrow diagram for this function are shown in Figure 1.38.

Figure 1.38

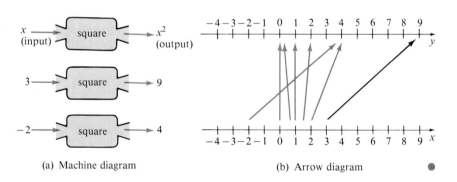

(a) Machine diagram   (b) Arrow diagram

● **Example 2**   If we define a function g by

$$g(x) = x^2 \qquad 0 \leqslant x \leqslant 3$$

then the domain of g is given as the closed interval $[0, 3]$. This is different from the function f given in Example 1 because in considering g we are restricting our attention to those values of x between 0 and 3. The range of g is

$$\{x^2 \mid 0 \leqslant x \leqslant 3\} = \{y \mid 0 \leqslant y \leqslant 9\} = [0, 9]$$

In Examples 1 and 2 the domain of the function was given explicitly. But if a function is given by a formula and the domain is not stated explicitly, the convention is that the domain is the set of all numbers for which the formula makes sense and defines a real number.

We should distinguish between a function f and the number $f(x)$, which is the value of f at x. Nonetheless, it is common to abbreviate an expression such as

"the function f defined by $f(x) = x^2 + x$"

to   "the function $f(x) = x^2 + x$"

● **Example 3**   Find the domain of the function $f(x) = \dfrac{1}{x^2 - x}$.

Solution   Since

$$f(x) = \frac{1}{x^2 - x} = \frac{1}{x(x - 1)}$$

and division by 0 is not allowed, we see that $f(x)$ is not defined when $x = 0$ or $x = 1$. Thus the domain of f is

$$\{x \mid x \neq 0, x \neq 1\}$$

which could also be written in interval notation as

$$(-\infty, 0) \cup (0, 1) \cup (1, \infty)$$ ●

● **Example 4**   Find the domain of $h(x) = \sqrt{2 - x - x^2}$.

Solution   Since the square root of a negative number is not defined (as a real number), the domain of h consists of all values of x such that

$$2 - x - x^2 \geqslant 0$$

We solve this inequality using the methods of Section 1.1. Since $2 - x - x^2 = (2 + x)(1 - x)$, the product will change sign when $x = -2$ or 1 as indicated in the following chart:

Interval	$2 + x$	$1 - x$	$(2 + x)(1 - x)$
$x \leqslant -2$	$-$	$+$	$-$
$-2 \leqslant x \leqslant 1$	$+$	$+$	$+$
$x \geqslant 1$	$+$	$-$	$-$

Therefore the domain of h is

$$\{x \mid -2 \leqslant x \leqslant 1\} = [-2, 1]$$ ●

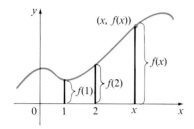

Figure 1.39

THE GRAPH OF f

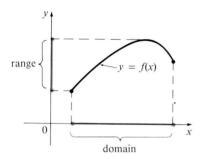

Figure 1.40

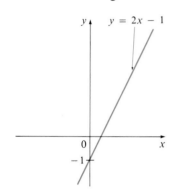

Figure 1.41

GRAPH OF $f(x) = 2x - 1$

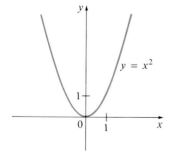

Figure 1.42

GRAPH OF $f(x) = x^2$

The symbol that represents an arbitrary number in the *domain* of a function f is called an **independent variable.** The symbol that represents a number in the *range* of f is called a **dependent variable.** For example, the squaring function of Example 1 could be defined by saying that each number x is assigned the number y by the rule $y = x^2$. Then x is the independent variable and y is the dependent variable. In the example on bacteria at the beginning of this section, t is the independent variable, N is the dependent variable, and they are connected by the equation $N = (5000)2^t$.

We have seen how to picture functions using machine diagrams and arrow diagrams. A third method for visualizing a function is its graph. If f is a function with domain A, its **graph** is the set of ordered pairs

$$\{(x, f(x)) \mid x \in A\}$$

In other words, the graph of f consists of all points (x, y) in the coordinate plane such that $y = f(x)$ and x is in the domain of f. Thus the graph of a function f is the same as the graph of the equation $y = f(x)$ as discussed in Section 1.2.

The graph of a function f gives us a useful picture of the behavior or "life history" of a function. Since the y-coordinate of any point (x, y) on the graph is $y = f(x)$, we can read the value of $f(x)$ from the graph as being the height of the graph above the point x (see Figure 1.39). The graph of f also allows us to picture the domain and range of f on the x-axis and y-axis as in Figure 1.40.

● **Example 5**   Sketch the graph of the function $f(x) = 2x - 1$.

Solution   The equation of the graph is $y = 2x - 1$, and we recognize this as being the equation of a line with slope 2 and y-intercept -1. This enables us to sketch the graph of f in Figure 1.41.   ●

● **Example 6**   Sketch the graph of $f(x) = x^2$.

Solution   The equation of the graph is $y = x^2$, which we recognize from Section 1.4 as a parabola. The graph is shown in Figure 1.42.   ●

● **Example 7**   Sketch the graph of $f(x) = x^3$.

Solution   We list some functional values and the corresponding points on the graph in the following table:

x	0	$\frac{1}{2}$	1	2	$-\frac{1}{2}$	-1	-2
$f(x) = x^3$	0	$\frac{1}{8}$	1	8	$-\frac{1}{8}$	-1	-8
(x, x^3)	$(0, 0)$	$(\frac{1}{2}, \frac{1}{8})$	$(1, 1)$	$(2, 8)$	$(-\frac{1}{2}, -\frac{1}{8})$	$(-1, -1)$	$(-2, -8)$

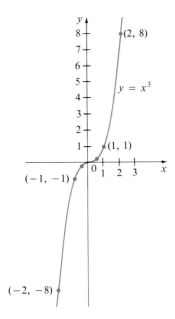

Figure 1.43

GRAPH OF $f(x) = x^3$

Then we plot these points and join them by a smooth curve to obtain the graph shown in Figure 1.43. At the present stage we cannot be absolutely certain that the graph is exactly as shown, but we shall later develop calculus techniques that will confirm the picture.   ●

● **Example 8**   Sketch the graph of $f(x) = \sqrt{x + 2}$.

Solution 1   We first observe that $\sqrt{x + 2}$ is defined when $x + 2 \geqslant 0$, so the domain of f is $\{x \mid x \geqslant -2\} = [-2, \infty)$. Then we plot the points given by the following table and use them to produce the sketch in Figure 1.44.

x	-2	-1	0	1	2	3
$f(x) = \sqrt{x + 2}$	0	1	$\sqrt{2}$	$\sqrt{3}$	2	$\sqrt{5}$

Solution 2   The equation of the graph of f is $y = \sqrt{x + 2}$ and we note that this implies that $y \geqslant 0$. Squaring, we obtain $y^2 = x + 2$, or $x = y^2 - 2$, which represents a shifted parabola as discussed in Section 1.4. [See Figure 1.45(a).] But since $y \geqslant 0$, the graph of $y = \sqrt{x + 2}$ is just the top half of this parabola [Figure 1.45(b)]. [The bottom half of the parabola would be the graph of the function $y = -\sqrt{x + 2}$ shown in Figure 1.45(c).]

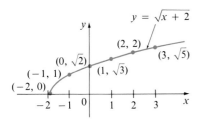

Figure 1.44

GRAPH OF $f(x) = \sqrt{x + 2}$

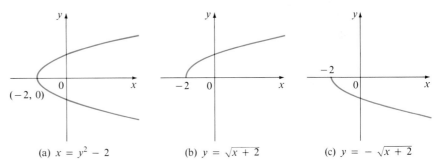

(a) $x = y^2 - 2$ (b) $y = \sqrt{x + 2}$ (c) $y = -\sqrt{x + 2}$

Figure 1.45   ●

The graph of a function is a curve in the xy-plane. But the question arises: Which curves in the xy-plane are graphs of functions? This is answered by the following test.

The Vertical Line Test (1.24)

> A curve in the plane is the graph of a function if and only if no vertical line intersects the curve more than once.

The reason for the truth of the Vertical Line Test can be seen in Figure 1.46. If each vertical line $x = a$ intersects a curve only once at (a, b), then

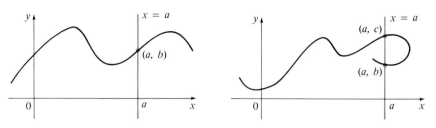

Figure 1.46

exactly one functional value is defined by $f(a) = b$. But if a line $x = a$ intersects the curve twice at (a, b) and (a, c), then the curve cannot represent a function because a function cannot assign two different values to a.

For example, the parabola $x = y^2 - 2$ shown in Figure 1.45(a) is not the graph of a function of x because, as you can see, there are vertical lines that intersect the parabola twice. The parabola, however, does represent *two* functions of x; the upper and lower halves of the parabola are the graphs of the functions $f(x) = \sqrt{x + 2}$ and $g(x) = -\sqrt{x + 2}$ [Figure 1.45(b) and (c)]. We observe that if we reverse the roles of x and y, then the equation $x = h(y) = y^2 - 2$ does define x as a function of y (with y as the independent variable and x as the dependent variable) and the parabola now appears as the graph of the function h.

The functions we have looked at so far have been defined by means of simple formulas. But there are many functions that are not given by such formulas. Here are some examples: the cost of mailing a first-class letter as a function of its mass, the population of New York City as a function of time, and the cost of a taxi ride as a function of distance. The following examples give further illustrations.

● **Example 9**   A function f is defined by

$$f(x) = \begin{cases} 1 - x & \text{if } x \leqslant 1 \\ x^2 & \text{if } x > 1 \end{cases}$$

Evaluate $f(0)$, $f(1)$, and $f(2)$ and sketch the graph.

Solution   Remember that a function is a rule. For this particular function the rule is the following: First look at the value of the input x. If it happens that $x \leqslant 1$, then the value of $f(x)$ is $1 - x$. On the other hand, if $x > 1$, then the value of $f(x)$ is x^2.

Since $0 \leqslant 1$, we have $f(0) = 1 - 0 = 1$.

Since $1 \leqslant 1$, we have $f(1) = 1 - 1 = 0$.

Since $2 > 1$, we have $f(2) = 2^2 = 4$.

How do we draw the graph of f? We observe that if $x \leqslant 1$, then $f(x) = 1 - x$, so the part of the graph of f that lies to the left of the vertical line $x = 1$ must coincide with the line $y = 1 - x$, which has slope -1 and y-intercept 1. If $x > 1$, then $f(x) = x^2$, so the part of the graph of f that lies to the right of the line $x = 1$ must coincide with the graph of $y = x^2$, which is a parabola. This enables us to sketch the graph in Figure 1.47. The solid

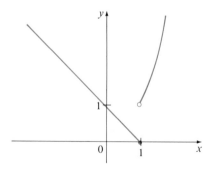

Figure 1.47

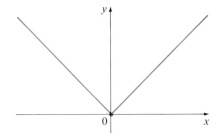

Figure 1.48

GRAPH OF $f(x) = |x|$

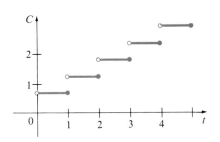

Figure 1.49

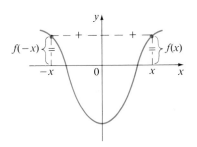

Figure 1.50

AN EVEN FUNCTION

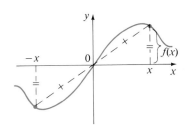

Figure 1.51

AN ODD FUNCTION

dot indicates that the point is included on the graph; the open dot indicates that the point is excluded from the graph.   ●

● **Example 10**   Sketch the graph of the absolute value function $f(x) = |x|$.

Solution   Recall from Equation 1.3 that

$$|x| = \begin{cases} x & \text{if } x \geq 0 \\ -x & \text{if } x < 0 \end{cases}$$

Using the same method as in Example 9, we see that the graph of f coincides with the line $y = x$ to the right of the y-axis and coincides with the line $y = -x$ to the left of the y-axis (see Figure 1.48).   ●

● **Example 11**   The cost of a long-distance daytime phone call from Toronto to New York City is 69 cents for the first minute and 58 cents for each additional minute (or part of a minute). Draw the graph of the cost C (in dollars) of the phone call as a function of the time t (in minutes).

Solution   Let $C(t)$ be the cost for t minutes. Since $t > 0$, the domain of the function is $(0, \infty)$. From the given information we have

$$C(t) = 0.69 \qquad\qquad\qquad\quad \text{if } 0 < t \leq 1$$
$$C(t) = 0.69 + 0.58 = 1.27 \qquad \text{if } 1 < t \leq 2$$
$$C(t) = 0.69 + 2(0.58) = 1.85 \quad \text{if } 2 < t \leq 3$$
$$C(t) = 0.69 + 3(0.58) = 2.43 \quad \text{if } 3 < t \leq 4$$

and so on. The graph is shown in Figure 1.49.   ●

Symmetry

If a function f satisfies $f(-x) = f(x)$ for every number x in its domain, then f is called an **even function.** For instance, the function $f(x) = x^2$ is even because

$$f(-x) = (-x)^2 = x^2 = f(x)$$

The geometric significance of an even function is that its graph is symmetric with respect to the y-axis (see Figure 1.50). This means that if we have plotted the graph of f for $x \geq 0$, we obtain the entire graph simply by reflecting in the y-axis.

If f satisfies $f(-x) = -f(x)$ for every number x in its domain, then f is called an **odd function.** For example, the function $f(x) = x^3$ is odd because

$$f(-x) = (-x)^3 = -x^3 = -f(x)$$

The graph of an odd function is symmetric about the origin (see Figure 1.51). If we already have the graph of f for $x \geq 0$, we can obtain the entire graph by rotating through $180°$ about the origin. For instance, in Example 7 we need only have plotted the graph of $y = x^3$ for $x \geq 0$ and then rotated that part about the origin.

EXERCISES 1.5

1. If $f(x) = x^2 - 3x + 2$, find $f(1)$, $f(-2)$, $f(\frac{1}{2})$, $f(\sqrt{5})$, $f(a)$, and $f(-a)$.

2. If $f(x) = x^3 + 2x^2 - 3$, find $f(0)$, $f(3)$, $f(-3)$, $f(-x)$, and $f(1/a)$.

3. If $g(x) = \dfrac{1 - x}{1 + x}$, find $g(2)$, $g(-2)$, $g(\pi)$, $g(a)$, $g(a - 1)$, and $g(-a)$.

4. If $h(t) = t + \dfrac{1}{t}$, find $h(1)$, $h(\pi)$, $h(t + 1)$, $h(t) + h(1)$, and $h(x)$.

In Exercises 5 and 6 draw a machine diagram, an arrow diagram, and a graph for the given function.

5. $f(x) = \sqrt{x}$, $0 \leqslant x \leqslant 4$

6. $f(x) = \dfrac{2}{x}$, $1 \leqslant x \leqslant 4$

7. The domain of f is $A = \{1, 2, 3, 4, 5, 6\}$ and $f(1) = 2$, $f(2) = 1$, $f(3) = 0$, $f(4) = 1$, $f(5) = 2$, and $f(6) = 4$. What is the range of f? Draw an arrow diagram and a graph for f.

In Exercises 8–16 find the domain and range of the given function.

8. $f(x) = 2x + 7$, $-1 \leqslant x \leqslant 6$

9. $f(x) = 6 - 4x$, $-2 \leqslant x \leqslant 3$

10. $g(x) = \dfrac{1}{x + 4}$

11. $g(x) = \dfrac{2}{3x - 5}$

12. $h(x) = \sqrt[4]{7 - 3x}$

13. $h(x) = \sqrt{2x - 5}$

14. $F(x) = 1 - \sqrt{x}$

15. $F(x) = \sqrt{1 - x^2}$

16. $G(x) = \sqrt{x^2 - 9}$

In Exercises 17–24 find the domain of the given function.

17. $f(x) = \dfrac{x + 2}{x^2 - 1}$

18. $f(x) = \dfrac{x^4}{x^2 + x - 6}$

19. $g(x) = \sqrt[4]{x^2 - 6x}$

20. $g(x) = \sqrt{x^2 - 2x - 8}$

21. $\phi(x) = \sqrt{\dfrac{x}{\pi - x}}$

22. $\phi(x) = \sqrt{\dfrac{x^2 - 2x}{x - 1}}$

23. $f(t) = \sqrt[3]{t - 1}$

24. $f(t) = \sqrt{t^2 + 1}$

In Exercises 25–60 find the domain and sketch the graph of the function.

25. $f(x) = 2$

26. $f(x) = -3$

27. $f(x) = 3 - 2x$

28. $f(x) = \dfrac{x + 3}{2}$, $-2 \leqslant x \leqslant 2$

29. $f(x) = -x^2$

30. $f(x) = x^2 - 4$

31. $f(x) = x^2 + 2x - 1$

32. $f(x) = -x^2 + 6x - 7$

33. $g(x) = x^4$

34. $g(x) = x^5$

35. $g(x) = \sqrt{-x}$

36. $g(x) = \sqrt{6 - 2x}$

37. $h(x) = \sqrt{4 - x^2}$

38. $h(x) = \sqrt{x^2 - 4}$

39. $F(x) = \dfrac{1}{x}$

40. $F(x) = \dfrac{2}{x + 4}$

41. $G(x) = |x| + x$

42. $G(x) = |x| - x$

43. $H(x) = |2x|$

44. $H(x) = |2x - 3|$

45. $f(x) = \dfrac{x}{|x|}$

46. $f(x) = |x^2 - 1|$

47. $f(x) = \dfrac{x^2 - 1}{x - 1}$

48. $f(x) = \dfrac{x^2 + 5x + 6}{x + 2}$

49. $f(x) = \begin{cases} x + 1 & \text{if } x \neq 1 \\ 1 & \text{if } x = 1 \end{cases}$

50. $f(x) = \begin{cases} x + 3 & \text{if } x \neq -2 \\ 4 & \text{if } x = -2 \end{cases}$

51. $f(x) = \begin{cases} 0 & \text{if } x < 2 \\ 1 & \text{if } x \geqslant 2 \end{cases}$

52. $f(x) = \begin{cases} -1 & \text{if } x < -1 \\ 1 & \text{if } -1 \leqslant x \leqslant 1 \\ -1 & \text{if } x > 1 \end{cases}$

53. $f(x) = \begin{cases} x & \text{if } x \leqslant 0 \\ x + 1 & \text{if } x > 0 \end{cases}$

54. $f(x) = \begin{cases} 2x + 3 & \text{if } x < -1 \\ 3 - x & \text{if } x \geqslant -1 \end{cases}$

55. $f(x) = \begin{cases} -1 & \text{if } x < -1 \\ x & \text{if } -1 \leqslant x \leqslant 1 \\ 1 & \text{if } x > 1 \end{cases}$

56. $f(x) = \begin{cases} |x| & \text{if } |x| \leqslant 1 \\ 1 & \text{if } |x| > 1 \end{cases}$

57. $f(x) = \begin{cases} x + 2 & \text{if } x \leqslant -1 \\ x^2 & \text{if } x > -1 \end{cases}$

58. $f(x) = \begin{cases} 1 - x^2 & \text{if } x \leqslant 2 \\ 2x - 7 & \text{if } x > 2 \end{cases}$

59. $f(x) = \begin{cases} -1 & \text{if } x \leqslant -1 \\ 3x + 2 & \text{if } |x| < 1 \\ 7 - 2x & \text{if } x \geqslant 1 \end{cases}$

60. $f(x) = \begin{cases} \sqrt{-x} & \text{if } x < 0 \\ x & \text{if } 0 \leqslant x \leqslant 2 \\ \sqrt{x - 2} & \text{if } x > 2 \end{cases}$

In Exercises 61–64 state whether the given curve is the graph of a function of x. If it is, state the domain and range of the function.

61.

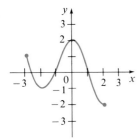

62.

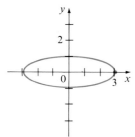

63.

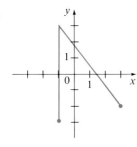

64.

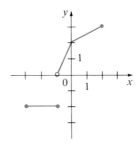

In Exercises 65–68 find a function whose graph is the given curve.

65. The line segment joining the points $(-2, 1)$ and $(4, -6)$

66. The line segment joining the points $(-3, -2)$ and $(6, 3)$

67. The bottom half of the parabola $x + (y - 1)^2 = 0$

68. The top half of the circle $(x - 1)^2 + y^2 = 1$

In Exercises 69–73 find a formula for the described function and state its domain.

69. A rectangle has a perimeter of 20 m. Express the area of the rectangle as a function of the length of one of its sides.

70. A rectangle has an area of 16 m². Express the perimeter of the rectangle as a function of the length of one of its sides.

71. Express the area of an equilateral triangle as a function of the length of a side.

72. Express the surface area of a cube as a function of its volume.

73. An open rectangular box with a volume of 2 m³ has a square base. Express the surface area of the box as a function of the length of a side of the base.

74. A taxi company charges two dollars for the first mile (or part of a mile) and 20 cents for each succeeding tenth of a mile (or part). Express the cost C (in dollars) of a ride as a function of the distance x traveled (in miles) for $0 < x < 2$ and sketch the graph of this function.

In Exercises 75–80 determine whether f is even, odd, or neither. If f is even or odd, use symmetry to sketch its graph.

75. $f(x) = x^{-2}$

76. $f(x) = x^{-3}$

77. $f(x) = x^2 + x$

78. $f(x) = x^4 - 4x^2$

79. $f(x) = x^3 - x$

80. $f(x) = 3x^3 + 2x^2 + 1$

Combinations of Functions

Two functions f and g can be combined to form new functions $f + g$, $f - g$, fg, and f/g in a manner similar to the way we add, subtract, multiply, and divide real numbers.

If we define the sum $f + g$ by the equation

(1.25)
$$(f + g)(x) = f(x) + g(x)$$

then the right side of Equation 1.25 makes sense if both $f(x)$ and $g(x)$ are defined, that is, if x belongs to the domain of f and also to the domain of g. If the domain of f is A and the domain of g is B, then the domain of $f + g$ is the intersection of these domains, that is, $A \cap B$.

Notice that the $+$ sign on the left side of Equation 1.25 stands for the operation of addition of *functions*, but the $+$ sign on the right side of the equation stands for addition of the *numbers* $f(x)$ and $g(x)$.

Similarly, we can define the difference $f - g$ and the product fg, and their domains will also be $A \cap B$. But in defining the quotient f/g we must remember not to divide by 0.

Algebra of Functions (1.26)

> Let f and g be functions with domains A and B. Then the functions $f + g$, $f - g$, fg, and f/g are defined as follows:
>
> $$(f + g)(x) = f(x) + g(x) \quad \text{domain} = A \cap B$$
>
> $$(f - g)(x) = f(x) - g(x) \quad \text{domain} = A \cap B$$
>
> $$(fg)(x) = f(x)g(x) \qquad \text{domain} = A \cap B$$
>
> $$\left(\frac{f}{g}\right)(x) = \frac{f(x)}{g(x)} \qquad\qquad \text{domain} = \{x \in A \cap B \,|\, g(x) \neq 0\}$$

● **Example 1**   If $f(x) = \sqrt{x}$ and $g(x) = \sqrt{4 - x^2}$, find the functions $f + g$, $f - g$, fg, and f/g.

Solution   The domain of $f(x) = \sqrt{x}$ is $[0, \infty)$. The domain of $g(x) = \sqrt{4 - x^2}$ consists of all numbers x such that $4 - x^2 \geq 0$, that is, $x^2 \leq 4$. Taking square roots of both sides, we get $|x| \leq 2$, or $-2 \leq x \leq 2$, so the domain of g is the interval $[-2, 2]$. The intersection of the domains of f and g is

$$[0, \infty) \cap [-2, 2] = [0, 2]$$

Using the definitions in (1.26), we have

$$(f + g)(x) = \sqrt{x} + \sqrt{4 - x^2} \qquad\qquad 0 \leq x \leq 2$$

$$(f - g)(x) = \sqrt{x} - \sqrt{4 - x^2} \qquad 0 \leqslant x \leqslant 2$$

$$(fg)(x) = \sqrt{x}\sqrt{4 - x^2} = \sqrt{4x - x^3} \qquad 0 \leqslant x \leqslant 2$$

$$\left(\frac{f}{g}\right)(x) = \frac{\sqrt{x}}{\sqrt{4 - x^2}} = \sqrt{\frac{x}{4 - x^2}} \qquad 0 \leqslant x < 2$$

Notice that the domain of f/g is the interval $[0, 2)$ because we must exclude the points where $g(x) = 0$, that is, $x = \pm 2$.   ●

The graph of the function $f + g$ is obtained from the graphs of f and g by **graphical addition.** This means that we add corresponding y-coordinates as in Figure 1.52. Figure 1.53 shows the result of using this procedure to graph the function $f + g$ from Example 1.

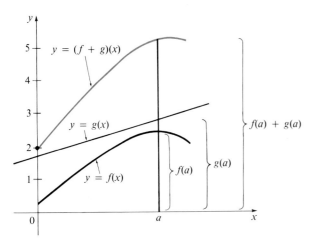

Figure 1.52

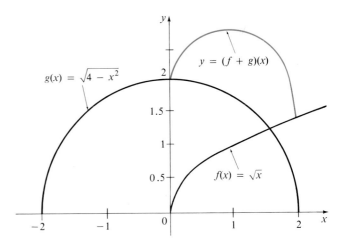

Figure 1.53

Composition of Functions

There is another way of combining two functions to get a new function. For example, suppose that $y = f(u) = \sqrt{u}$ and $u = g(x) = x^2 + 1$. Since y is a function of u and u is, in turn, a function of x, it follows that y is ultimately a function of x. We compute this by substitution:

$$y = f(u) = f(g(x)) = f(x^2 + 1) = \sqrt{x^2 + 1}$$

The procedure is called **composition** because the new function is composed of the two given functions f and g.

In general, given any two functions f and g, we start with a number x in the domain of g and find its image $g(x)$. If this number $g(x)$ is in the domain of f, then we can calculate the value of $f(g(x))$. The result is a new function $h(x) = f(g(x))$ obtained by substituting g into f. It is called the **composition** (or **composite**) of f and g and is denoted by $f \circ g$ ("f circle g").

Definition (1.27)

Given two functions f and g, the **composite function** $f \circ g$ (also called the **composition** of f and g) is defined by

$$(f \circ g)(x) = f(g(x))$$

The domain of $f \circ g$ is the set of all x in the domain of g such that $g(x)$ is in the domain of f. In other words, $(f \circ g)(x)$ is defined whenever both $g(x)$ and $f(g(x))$ are defined. The best way to picture $f \circ g$ is by a machine diagram (Figure 1.54) or an arrow diagram (Figure 1.55).

Figure 1.54

THE $f \circ g$ MACHINE IS COMPOSED OF THE g MACHINE (FIRST) AND THEN THE f MACHINE

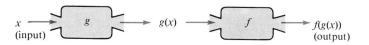

Figure 1.55

ARROW DIAGRAM FOR $f \circ g$

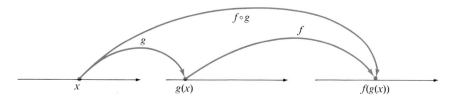

● **Example 2**   If $f(x) = x^2$ and $g(x) = x - 3$, find the composite functions $f \circ g$ and $g \circ f$ and their domains.

Solution   We have

$$(f \circ g)(x) = f(g(x)) = f(x - 3) = (x - 3)^2$$

$$(g \circ f)(x) = g(f(x)) = g(x^2) = x^2 - 3$$

The domains of both $f \circ g$ and $g \circ f$ are R (the set of all real numbers).   ●

Note: You can see from Example 2 that, in general, $f \circ g \ne g \circ f$. Remember, the notation $f \circ g$ means that the function g is applied first and then f is applied second. In Example 2, $f \circ g$ is the function that *first* subtracts 3 and *then* squares; $g \circ f$ is the function that *first* squares and *then* subtracts 3.

● **Example 3**   If $f(x) = \sqrt{x}$ and $g(x) = \sqrt{2 - x}$, find the functions $f \circ g$, $g \circ f$, $f \circ f$, and $g \circ g$ and their domains.

Solution

$$(f \circ g)(x) = f(g(x)) = f(\sqrt{2 - x}) = \sqrt{\sqrt{2 - x}} = \sqrt[4]{2 - x}$$

The domain of $f \circ g$ is $\{x \,|\, 2 - x \geqslant 0\} = \{x \,|\, x \leqslant 2\} = (-\infty, 2]$.

$$(g \circ f)(x) = g(f(x)) = g(\sqrt{x}) = \sqrt{2 - \sqrt{x}}$$

For $\sqrt{x}$ to be defined we must have $x \geqslant 0$. For $\sqrt{2 - \sqrt{x}}$ to be defined we must have $2 - \sqrt{x} \geqslant 0$, that is, $\sqrt{x} \leqslant 2$, or $x \leqslant 4$. Thus we have $0 \leqslant x \leqslant 4$, so the domain of $g \circ f$ is the closed interval $[0, 4]$.

$$(f \circ f)(x) = f(f(x)) = f(\sqrt{x}) = \sqrt{\sqrt{x}} = \sqrt[4]{x}$$

The domain of $f \circ f$ is $[0, \infty)$.

$$(g \circ g)(x) = g(g(x)) = g(\sqrt{2 - x}) = \sqrt{2 - \sqrt{2 - x}}$$

This expression is defined when $2 - x \geqslant 0$, that is, $x \leqslant 2$ and $2 - \sqrt{2 - x} \geqslant 0$. This latter inequality is equivalent to $\sqrt{2 - x} \leqslant 2$, or $2 - x \leqslant 4$, that is, $x \geqslant -2$. Thus $-2 \leqslant x \leqslant 2$, so the domain of $g \circ g$ is the closed interval $[-2, 2]$. ●

It is possible to take the composition of three or more functions. For instance, the composite function $f \circ g \circ h$ is found by first applying h, then g, and then f as follows:

$$(f \circ g \circ h)(x) = f(g(h(x)))$$

● **Example 4**   Find $f \circ g \circ h$ if $f(x) = x/(x + 1)$, $g(x) = x^{10}$, and $h(x) = x + 3$.

Solution

$$(f \circ g \circ h)(x) = f(g(h(x))) = f(g(x + 3)) = f((x + 3)^{10}) = \frac{(x + 3)^{10}}{(x + 3)^{10} + 1}$$

●

So far we have used composition to build up complicated functions from simpler ones. But in calculus it is sometimes useful to be able to decompose a complicated function into simpler ones, as in the following example.

● **Example 5**   Given $F(x) = \sqrt[4]{x + 9}$, find functions f and g such that $F = f \circ g$.

Solution   Since the formula for F says to first add 9 and then take the fourth root, we let

$$g(x) = x + 9 \quad \text{and} \quad f(x) = \sqrt[4]{x}$$

Then $(f \circ g)(x) = f(g(x)) = f(x + 9) = \sqrt[4]{x + 9} = F(x)$ ●

EXERCISES 1.6

In Exercises 1–6 find $f + g$, $f - g$, fg, and f/g and their domains.

1. $f(x) = x^2 - x$, $g(x) = x + 5$

2. $f(x) = x^3 + 2x^2$, $g(x) = 3x^2 - 1$

3. $f(x) = \sqrt{1 + x}$, $g(x) = \sqrt{1 - x}$

4. $f(x) = \sqrt{9 - x^2}$, $g(x) = \sqrt{x^2 - 1}$

5. $f(x) = \sqrt{x}$, $g(x) = \sqrt[3]{x}$

6. $f(x) = \sqrt[4]{x + 1}$, $g(x) = \sqrt{x + 2}$

In Exercises 7 and 8 find the domain of the given function.

7. $F(x) = \dfrac{\sqrt{4-x} + \sqrt{3+x}}{x^2 - 2}$

8. $F(x) = \sqrt{1-x} + \sqrt{x-2}$

In Exercises 9–12 use the graphs of f and g and the method of graphical addition to sketch the graph of $f + g$.

9. $f(x) = x^3, g(x) = 1$ **10.** $f(x) = \sqrt{x}, g(x) = 3$

11. $f(x) = x, g(x) = \dfrac{1}{x}$ **12.** $f(x) = x^3, g(x) = -x^2$

In Exercises 13–22 find the functions $f \circ g$, $g \circ f$, $f \circ f$, and $g \circ g$ and their domains.

13. $f(x) = 2x + 3, g(x) = 4x - 1$

14. $f(x) = 6x - 5, g(x) = \dfrac{x}{2}$

15. $f(x) = 2x^2 - x, g(x) = 3x + 2$

16. $f(x) = \sqrt{x-1}, g(x) = x^2$

17. $f(x) = \dfrac{1}{x}, g(x) = x^3 + 2x$

18. $f(x) = \dfrac{1}{x-1}, g(x) = \dfrac{x-1}{x+1}$

19. $f(x) = \sqrt[3]{x}, g(x) = 1 - \sqrt{x}$

20. $f(x) = \sqrt{x^2 - 1}, g(x) = \sqrt{1 - x}$

21. $f(x) = \dfrac{x+2}{2x+1}, g(x) = \dfrac{x}{x-2}$

22. $f(x) = \dfrac{1}{\sqrt{x}}, g(x) = x^2 - 4x$

In Exercises 23–26 find $f \circ g \circ h$.

23. $f(x) = x - 1, g(x) = \sqrt{x}, h(x) = x - 1$

24. $f(x) = \dfrac{1}{x}, g(x) = x^3, h(x) = x^2 + 2$

25. $f(x) = x^4 + 1, g(x) = x - 5, h(x) = \sqrt{x}$

26. $f(x) = \sqrt{x}, g(x) = \dfrac{x}{x-1}, h(x) = \sqrt[3]{x}$

In Exercises 27–30 express the given function in the form $f \circ g$.

27. $F(x) = (x - 9)^5$ **28.** $F(x) = \sqrt{x} + 1$

29. $G(x) = \dfrac{x^2}{x^2 + 4}$ **30.** $G(x) = \dfrac{1}{x+3}$

In Exercises 31 and 32 express the given function in the form $f \circ g \circ h$.

31. $H(x) = \dfrac{1}{x^2 + 1}$ **32.** $H(x) = \sqrt[3]{\sqrt{x} - 1}$

33. If $f(x) = 3x + 5$ and $h(x) = 3x^2 + 3x + 2$, find a function g such that $f \circ g = h$.

34. If $f(x) = x + 4$ and $h(x) = 4x - 1$, find a function g such that $g \circ f = h$.

SECTION 1.7

Types of Functions and Transformed Functions

In solving calculus problems you will find that it is helpful to be familiar with the graphs of some commonly occurring functions. We classify various types of functions as follows.

1. Constant Functions. The constant function $f(x) = c$ has domain R and its range consists of the single number c. Its graph is a horizontal line and is illustrated in Figure 1.56 for $c = 2$.

2. Power Functions. A function of the form $f(x) = x^a$, where a is a constant, is called a **power function.** A general study of such functions will have to wait until Chapter 6, but we can graph the following special cases now.

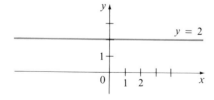

Figure 1.56

GRAPH OF $f(x) = 2$

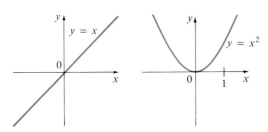
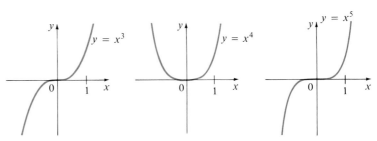

Figure 1.57

GRAPHS OF $f(x) = x^n$, $n = 1, 2, 3, 4, 5$

(a) $a = n$, *a positive integer.* The graphs of $f(x) = x^n$ for $n = 1, 2, 3, 4,$ and 5 are shown in Figure 1.57. We have already seen the graphs of $y = x$ (a line through the origin with slope 1), $y = x^2$ (a parabola), and $y = x^3$ (Example 7 in Section 1.5).

The general shape of the graph of $f(x) = x^n$ depends on whether n is even or odd. If n is even, then $f(x) = x^n$ is an even function and its graph is similar to the parabola $y = x^2$. If n is odd, then $f(x) = x^n$ is an odd function and its graph is similar to that of $y = x^3$. Notice from Figure 1.58, however, that as n increases, the graph of $y = x^n$ becomes flatter near 0 and steeper when $|x| \geq 1$. (If x is small, then x^2 is smaller, x^3 is even smaller, x^4 is smaller still, and so on.)

(b) $a = -1$. The graph of the reciprocal function $f(x) = x^{-1} = 1/x$ is shown in Figure 1.59. Its graph has the equation $y = 1/x$ or $xy = 1$. This is an equilateral hyperbola with the coordinate axes as its asymptotes. (See Section 1.4.)

(c) $a = 1/n$, *n a positive integer.* The function $f(x) = x^{1/n} = \sqrt[n]{x}$ is a **root function.** For $n = 2$ it is the square root function $f(x) = \sqrt{x}$ whose domain is $[0, \infty)$ and whose graph is the upper half of the parabola $x = y^2$ [see Figure 1.60(a)]. For other even values of n,

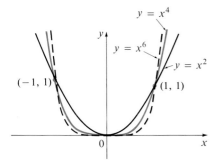

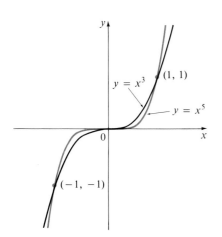

Figure 1.58

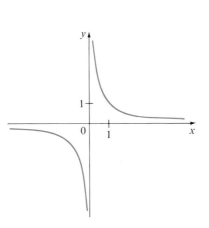

Figure 1.59

GRAPH OF $f(x) = 1/x$

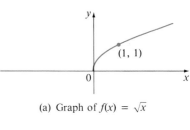

(a) Graph of $f(x) = \sqrt{x}$

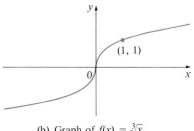

(b) Graph of $f(x) = \sqrt[3]{x}$

Figure 1.60

ROOT FUNCTIONS

the graph of $y = \sqrt[n]{x}$ is similar to that of $y = \sqrt{x}$. For $n = 3$ we have the cube root function $f(x) = \sqrt[3]{x}$ whose domain is R (recall that every real number has a cube root) and whose graph is shown in Figure 1.60(b). The graph of $y = \sqrt[n]{x}$ for n odd ($n > 3$) is similar to that of $y = \sqrt[3]{x}$.

3. *Polynomials.*   A function P is called a **polynomial** if

$$P(x) = a_n x^n + a_{n-1} x^{n-1} + \cdots + a_2 x^2 + a_1 x + a_0$$

where n is a nonnegative integer and the numbers $a_0, a_1, a_2, \ldots, a_n$ are constants called the **coefficients** of the polynomial. The domain of any polynomial is $R = (-\infty, \infty)$. If the leading coefficient $a_n \neq 0$, then the **degree** of the polynomial is n. For example, the function

$$P(x) = 2x^6 - x^4 + \tfrac{2}{5}x^3 + \sqrt{2}$$

is a polynomial of degree 6.

A polynomial of degree 1 is of the form $P(x) = ax + b$ and is called a **linear function** because its graph is the line $y = ax + b$ (slope a, y-intercept b).

A polynomial of degree 2 is of the form $P(x) = ax^2 + bx + c$ and is called a **quadratic function.** The graph of a quadratic function is always a parabola obtained by shifting the parabola $y = ax^2$. (See Example 2.)

A polynomial of degree 3 is of the form

$$P(x) = ax^3 + bx^2 + cx + d$$

and is called a **cubic function.** The graphs of cubic functions and higher-degree polynomials will be discussed in Chapter 4.

4. *Rational Functions.*   A **rational function** f is a ratio of two polynomials:

$$f(x) = \frac{P(x)}{Q(x)}$$

where P and Q are polynomials. The domain consists of all values of x such that $Q(x) \neq 0$. For example, the function

$$f(x) = \frac{2x^4 - x^2 + 1}{x^2 - 4}$$

is a rational function with domain $\{x \mid x \neq \pm 2\}$. We shall learn how to graph rational functions in Chapter 4.

5. *Algebraic Functions.*   A function f is called **algebraic** if it can be constructed using algebraic operations (addition, subtraction, multiplication, division, and taking roots) starting with polynomials. Any rational function is automatically an algebraic function. Here are two more examples:

$$f(x) = \sqrt{x^2 + 1} \qquad g(x) = \frac{x^4 - 16x^2}{x + \sqrt{x}} + (x - 2)\sqrt[3]{x + 1}$$

6. *Transcendental Functions.* Functions that are not algebraic are called **transcendental.** Some examples of transcendental functions are the following.

(a) *Exponential functions.* These are functions of the form $f(x) = a^x$, where a is a positive constant. They will be studied in Chapter 6.

(b) *Logarithmic functions.* These are functions $f(x) = \log_a x$, where the base a is a positive constant. We shall study logarithmic functions in Chapter 6.

(c) *Trigonometric functions.* A review of trigonometry and the trigonometric functions is given in Appendix B. In calculus the convention is that radian measure is always used (except when otherwise indicated). For example, when we use the function $f(x) = \sin x$, it is understood that $\sin x$ means the sine of the angle whose radian measure is x. Thus the graphs of the sine and cosine functions are as shown in Figures 1.61 and 1.62.

Figure 1.61

GRAPH OF $f(x) = \sin x$

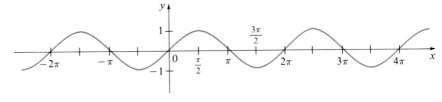

Figure 1.62

GRAPH OF $f(x) = \cos x$

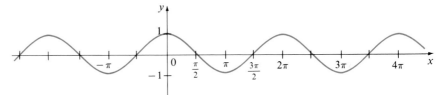

Transformed Functions

By applying certain transformations to the graph of a given function we can obtain the graphs of certain related functions and thereby reduce the amount of work in graphing. Let us first consider **translations.** By adding the constant function $g(x) = c > 0$ to a given function f by graphical addition, we see that the graph of $y = f(x) + c$ is just the graph of $y = f(x)$ shifted upward a distance of c units. Likewise, if $g(x) = f(x - c)$, where $c > 0$, then the value of g at x is the same as the value of f at $x - c$ (c units to the left of x). Therefore the graph of $y = f(x - c)$ is just the graph of $y = f(x)$ shifted c units to the right (see Figure 1.63).

Vertical and Horizontal Shifts (1.28)

Suppose $c > 0$. To obtain the graph of
$y = f(x) + c$, shift the graph of $y = f(x)$ c units upward
$y = f(x) - c$, shift the graph of $y = f(x)$ c units downward
$y = f(x - c)$, shift the graph of $y = f(x)$ c units to the right
$y = f(x + c)$, shift the graph of $y = f(x)$ c units to the left

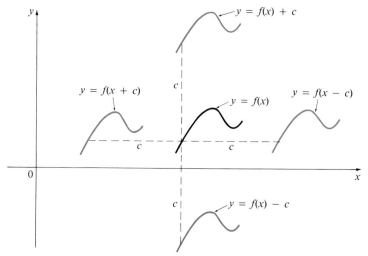

Figure 1.63

Now let us consider the **stretching** and **reflecting** transformations. By multiplying the given function f by the constant function $g(x) = c$, where $c > 1$, we see that the graph of $y = cf(x)$ is the graph of $y = f(x)$ stretched by a factor of c in the vertical direction. The graph of $y = -f(x)$ is the graph of $y = f(x)$ reflected in the x-axis because the point (x, y) is replaced by the point $(x, -y)$. [See Figure 1.64 and Table (1.29), where the results of other stretching, compressing, and reflecting transformations are also given.]

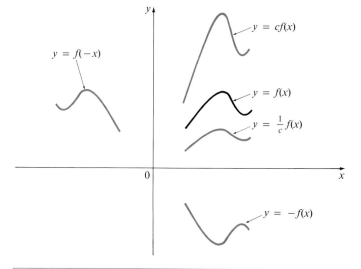

Figure 1.64

Vertical and Horizontal Stretching and Reflecting (1.29)	Suppose $c > 1$. To obtain the graph of $y = cf(x)$, stretch the graph of $y = f(x)$ vertically by a factor of c $y = (1/c)f(x)$, compress the graph of $y = f(x)$ vertically by a factor of c $y = f(cx)$, compress the graph of $y = f(x)$ horizontally by a factor of c $y = f(x/c)$, stretch the graph of $y = f(x)$ horizontally by a factor of c $y = -f(x)$, reflect the graph of $y = f(x)$ in the x-axis $y = f(-x)$, reflect the graph of $y = f(x)$ in the y-axis

● **Example 1**   Given the graph of $y = \sqrt{x}$, use transformations to graph $y = \sqrt{x} - 2$, $y = \sqrt{x - 2}$, $y = -\sqrt{x}$, $y = 2\sqrt{x}$, and $y = \sqrt{-x}$.

Solution   The graph of the square root function $y = \sqrt{x}$, obtained from Figure 1.60, is shown in Figure 1.65(a). Then in the other parts of the figure we sketch $y = \sqrt{x} - 2$ by shifting 2 units down, $y = \sqrt{x - 2}$ by shifting 2 units to the right, $y = -\sqrt{x}$ by reflecting in the x-axis, $y = 2\sqrt{x}$ by stretching vertically by a factor of 2, and $y = \sqrt{-x}$ by reflecting in the y-axis.

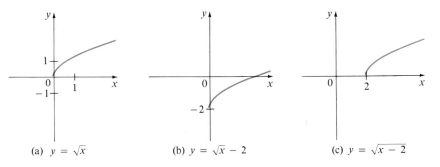

(a) $y = \sqrt{x}$ (b) $y = \sqrt{x} - 2$ (c) $y = \sqrt{x - 2}$

Figure 1.65

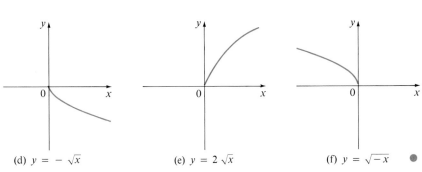

(d) $y = -\sqrt{x}$ (e) $y = 2\sqrt{x}$ (f) $y = \sqrt{-x}$ ●

● **Example 2**   Sketch the graph of the function $f(x) = x^2 + 6x + 10$.

Solution   Completing the square, we write the equation of the graph as

$$y = x^2 + 6x + 10 = (x + 3)^2 + 1$$

According to (1.28), this means we obtain the desired graph by starting with the parabola $y = x^2$ and shifting 3 units to the left and then 1 unit upward (see Figure 1.66).   ●

● **Example 3**   Sketch the graph of the functions (a) $y = \sin 2x$ and (b) $y = 1 - \sin x$.

Solution   (a) According to (1.29) we obtain the graph of $y = \sin 2x$ from that of $y = \sin x$ by compressing horizontally by a factor of 2 (see Figure 1.67).
        (b) To obtain the graph of $y = 1 - \sin x$, we again start with $y = \sin x$. We reflect in the x-axis to get $y = -\sin x$ and then we shift 1 unit upward to get $y = 1 - \sin x$ (see Figure 1.68).

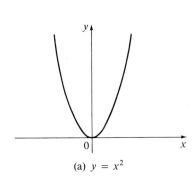

(a) $y = x^2$

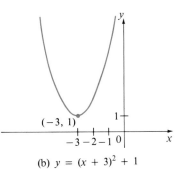

$(-3, 1)$

(b) $y = (x + 3)^2 + 1$

Figure 1.66

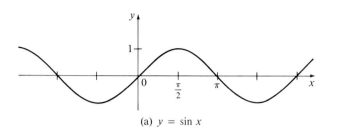

(a) $y = \sin x$ (b) $y = \sin 2x$

Figure 1.67

Figure 1.68

$y = 1 - \sin x$

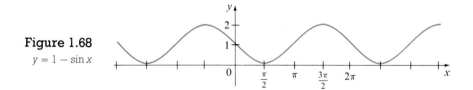

EXERCISES 1.7

Graph the following functions, not by plotting points, but by starting with the graphs of the standard functions given in this section and then applying the appropriate transformations.

1. $y = x^8$

2. $y = \sqrt[4]{x}$

3. $y = -\dfrac{1}{x}$

4. $y = -x^3$

5. $y = 2 \sin x$

6. $y = 1 + \sqrt{x}$

7. $y = (x - 1)^3 + 2$

8. $y = -\cos x$

9. $y = 1 + \sqrt[6]{-x}$

10. $y = \sqrt[3]{x + 2}$

11. $y = \cos\left(\dfrac{x}{2}\right)$

12. $y = x^2 + x + 1$

13. $y = \dfrac{1}{x - 3}$

14. $y = -2 \sin \pi x$

15. $y = \dfrac{1}{3}\sin\left(x - \dfrac{\pi}{6}\right)$

16. $y = 2 + \dfrac{1}{x + 1}$

17. $y = 1 + 2x - x^2$

18. $y = \dfrac{1}{2}\sqrt{x + 4} - 3$

19. $y = 2 - \sqrt{x + 1}$

20. $y = 1 - (x - 8)^6$

REVIEW OF CHAPTER 1

Define, state, or discuss the following.

1. Rational and irrational numbers

2. Real number line

3. Open interval, closed interval

4. Rules for inequalities

5. Absolute value of a number

6. Properties of absolute values

7. Triangle inequality

8. Rectangular coordinate system

9. Distance formula

10. Equation of a circle

11. Slope of a line

12. Point-slope form of the equation of a line

13. Slope-intercept form of the equation of a line

14. Angle of inclination of a line

15. Slope relationships for parallel and perpendicular lines

16. Angle between two lines

17. Equations of parabolas, ellipses, and hyperbolas

18. Function

19. Domain and range of a function

20. Independent and dependent variables

21. Graph of a function

22. Vertical Line Test

23. Even function, odd function

24. Sum, difference, product, and quotient of functions

25. Composition of functions

26. Constant function

27. Power function

28. Polynomial

29. Linear function

30. Quadratic function

31. Rational function

32. Algebraic and transcendental functions

33. Rules for vertical and horizontal shifts of functions

34. Rules for stretching and reflecting functions

REVIEW EXERCISES FOR CHAPTER 1

Solve the inequalities in Exercises 1–6 in terms of intervals.

1. $2 + 5x \leqslant 9 - 2x$

2. $-5 \leqslant 1 - 2x \leqslant 3$

3. $|x + 3| < 7$

4. $|2x - 7| > 1$

5. $x^2 - 5x + 6 > 0$

6. $\dfrac{2x^2 + x}{x^2 + 2} \leqslant 1$

7. Find the distance between the points $(2, 4)$ and $(-4, -4)$.

8. Sketch the region $\{(x, y)\,|\,|x| \leqslant 2, |y| > 1\}$.

9. Find an equation of the circle with center $(2, 1)$ and radius 3.

10. Find an equation of the circle that passes through the origin and has center $(-6, 4)$.

11. Find the center and radius of the circle $x^2 + y^2 + 2x - 8y + 8 = 0$.

In Exercises 12–16 find an equation of the line that satisfies the given conditions.

12. Through $(2, 1)$, with slope -3

13. Through $(-1, -6)$ and $(2, -1)$

14. With slope $-\frac{1}{3}$, y-intercept 2

15. Through $(2, 3)$, parallel to $x + 2y = 1$

16. Through $(-1, 1)$, perpendicular to $3x - 4y = 6$

17. Find the slope and y-intercept of the line $3x + 5y = 10$.

■ 18. (a) Find the point of intersection of the lines $y = 3x + 1$ and $2x + y = 9$.
(b) Find the angle between these lines, correct to the nearest degree.

Identify the curves in Exercises 19–24 and sketch their graphs.

19. $y = 8 - 2x^2$

20. $x^2 + 4y^2 = 16$

21. $x^2 - 4y^2 = 4$

22. $x + 4y^2 = 0$

23. $2x^2 + y^2 - 16x + 30 = 0$

24. $y^2 - x^2 = 4$

25. If $f(x) = 1 + \sqrt{x - 1}$, find $f(5)$, $f(9)$, $f(-x)$, $f(x^2)$, and $[f(x)]^2$.

In Exercises 26–28 find the domain of the function.

26. $f(x) = \dfrac{\sqrt[3]{2x + 1}}{\sqrt[3]{2x + 2}}$

27. $g(x) = \dfrac{x^2 + x + 1}{x^2 + x - 1}$

28. $h(x) = \sqrt{5 - 4x - x^2}$

In Exercises 29–36 sketch the graph of the function.

29. $f(x) = -1$

30. $f(x) = 2 + 3x$

31. $g(x) = x^2 + 2$

32. $g(x) = 4x - x^2$

33. $h(x) = \sqrt{x - 5}$

34. $h(x) = 1 - x^5$

35. $y = -\sin 2x$

36. $F(x) = \begin{cases} 1 - \dfrac{x}{2} & \text{if } x < 2 \\ x - 3 & \text{if } x \geqslant 2 \end{cases}$

In Exercises 37 and 38 find the following functions and their domains: (a) $f + g$, (b) f/g, (c) $f \circ g$, (d) $g \circ f$.

37. $f(x) = x^2$, $g(x) = x + 2$

38. $f(x) = x^2 + 3x$, $g(x) = \sqrt{x + 2}$

39. If $h(x) = \sqrt{x^2 + x + 9}$, find functions f and g such that $h = f \circ g$.

40. If $F(x) = 1/\sqrt[3]{x^2 + 3}$, find functions f, g, and h such that $F = f \circ g \circ h$.

2 Limits and Rates of Change

The calculus was the first achievement of modern mathematics and it is difficult to overestimate its importance. I think it defines more unequivocally than anything else the inception of modern mathematics; and the system of mathematical analysis, which is its logical development, still constitutes the greatest technical advance in exact thinking.

John von Neumann

The idea of a limit underlies all of calculus. In this chapter we begin our study of limits after first seeing how limits arise as slopes of tangents and rates of change.

Tangents and Velocities

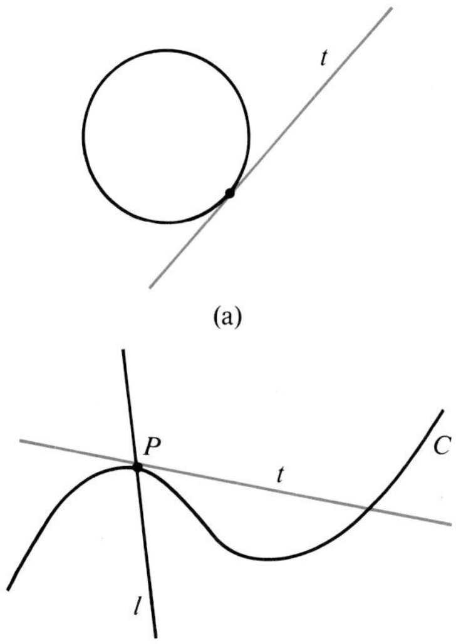

The word *tangent* is derived from the Latin word *tangens,* which means "touching." Thus a tangent to a curve is a line that touches the curve. How can this idea be made precise?

For a circle we could simply follow Euclid and say that a tangent is a line that intersects the circle once and only once as in Figure 2.1(a).

For more complicated curves this definition is inadequate. Figure 2.1(b) shows two lines l and t passing through a point P on a curve C. The line l intersects C only once, but it certainly does not look like what we think of as a tangent. The line t, on the other hand, looks like a tangent but it intersects C twice.

To be specific, let us look at the problem of trying to find the tangent line t to the parabola $y = x^2$ at the point $P(1, 1)$. This problem amounts to finding the slope m of t. The difficulty is that we know only one point, P, on t, whereas we need two points to compute the slope. But observe that we can compute an approximation to m by choosing a nearby point $Q(x, x^2)$ on the parabola (as in Figure 2.2) and computing the slope m_{PQ} of the secant line PQ.

We choose $x \neq 1$ so that $Q \neq P$. Then

(2.1)
$$m_{PQ} = \frac{x^2 - 1}{x - 1}$$

For instance, for the point $Q(1.5, 2.25)$ we have

$$m_{PQ} = \frac{2.25 - 1}{1.5 - 1} = \frac{1.25}{0.5} = 2.5$$

The tables in (2.2) show the values of m_{PQ} for several values of x close to 1. The closer Q is to P, the closer x is to 1, and, it appears from (2.2), the closer m_{PQ} is to 2.

(a)

(b)

Figure 2.1

(2.2)

x	m_{PQ}	x	m_{PQ}
2	3	0	1
1.5	2.5	0.5	1.5
1.1	2.1	0.9	1.9
1.01	2.01	0.99	1.99
1.001	2.001	0.999	1.999

In order to confirm this impression, we simplify the expression in Equation 2.1 as follows, noting that $x \neq 1$ so that $x - 1 \neq 0$:

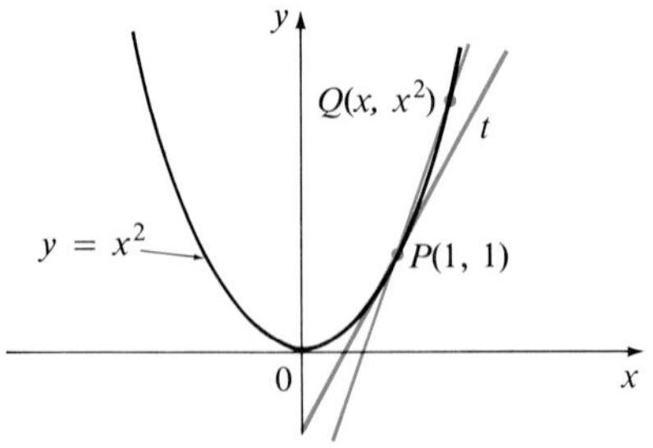

$$m_{PQ} = \frac{x^2 - 1}{x - 1} = \frac{(x - 1)(x + 1)}{x - 1} = x + 1$$

Figure 2.2

Thus if x is very close to 1, $m_{PQ} = x + 1$ is very close to 2. In fact, we can make $x + 1$ as close as we like to 2 by taking x close enough to 1.

This suggests that the slope of the tangent line t should be $m = 2$. We say that the slope of the tangent line is the *limit* of the slopes of the secant lines, and we express this symbolically by writing

$$\lim_{Q \to P} m_{PQ} = m$$

and

$$\lim_{x \to 1} \frac{x^2 - 1}{x - 1} = 2$$

Now that we know that the slope of the tangent line is 2 and that it passes through $P(1, 1)$, we can use the point-slope form of the equation of a line (1.10) to write the equation of the tangent as

$$y - 1 = 2(x - 1) \qquad \text{or} \qquad y = 2x - 1$$

In general, if a curve C has equation $y = f(x)$ and we want to find the tangent to C at the point $P(a, f(a))$, then we consider a nearby point $Q(x, f(x))$, where $x \neq a$, and compute the slope of the secant line PQ:

$$m_{PQ} = \frac{f(x) - f(a)}{x - a}$$

Then we let Q approach P along the curve C by letting x approach a. If m_{PQ} approaches a number m, then we define the **tangent** t to be the line through P with slope m. (This amounts to saying that the tangent line is the limiting position of the secant line PQ as Q approaches P. See Figure 2.3.) In the notation of limits we write

(2.3)

$$m = \lim_{x \to a} m_{PQ} = \lim_{x \to a} \frac{f(x) - f(a)}{x - a}$$

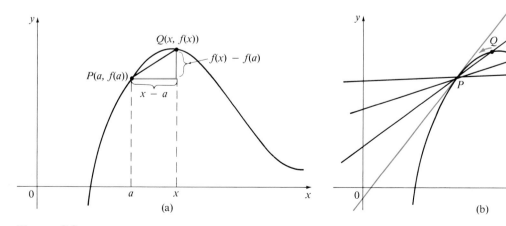

Figure 2.3

For some purposes it is convenient to let

$$h = x - a$$

Then

$$x = a + h$$

so the slope of the secant line PQ is

$$m_{PQ} = \frac{f(a + h) - f(a)}{h}$$

(See Figure 2.4 where the case $h > 0$ is illustrated and Q is to the right of P. If $h < 0$, however, Q would be to the left of P.) Notice that as x approaches a, h approaches 0 and so the expression (2.3) for the slope of the tangent line becomes

(2.4)

$$m = \lim_{h \to 0} m_{PQ} = \lim_{h \to 0} \frac{f(a + h) - f(a)}{h}$$

Figure 2.4

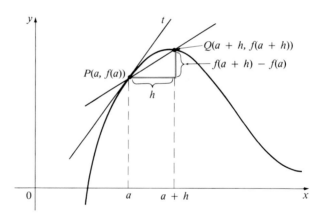

The CN Tower is currently the highest freestanding building in the world.
Courtesy of CN Tower.

Velocity

Suppose that a ball is dropped from the upper observation deck of the CN Tower in Toronto, 450 meters above the ground, and that we want to know the answers to two problems:

(1)   What is the velocity of the ball after 5 seconds?
(2)   How fast is it traveling when it hits the ground?

In trying to solve these problems we use the fact, discovered by Galileo almost four centuries ago, that the distance fallen by any freely falling body is proportional to the square of the time it has been falling. (This neglects air resistance.) If the distance fallen after t seconds is denoted by $s(t)$ and measured in meters, then Galileo's law is expressed by the equation

$$s(t) = 4.9t^2$$

The difficulty in finding the velocity after 5 s is that we are dealing with a single instant of time ($t = 5$) so there is no time interval involved. However, we can approximate the desired quantity by computing the average velocity over the brief time interval of a tenth of a second from $t = 5$ to $t = 5.1$:

$$\text{average velocity} = \frac{\text{distance traveled}}{\text{time elapsed}}$$

$$= \frac{s(5.1) - s(5)}{0.1}$$

$$= \frac{4.9(5.1)^2 - 4.9(5)^2}{0.1} = 49.49 \text{ m/s}$$

The following table shows the results of similar calculations of the average velocity over successively smaller time periods.

time interval	average velocity (m/s)
$5 \leqslant t \leqslant 6$	53.9
$5 \leqslant t \leqslant 5.1$	49.49
$5 \leqslant t \leqslant 5.05$	49.245
$5 \leqslant t \leqslant 5.01$	49.049
$5 \leqslant t \leqslant 5.001$	49.0049

It appears that as we shorten the time period, the average velocity is becoming closer to 49 m/s. Let us compute the average velocity over the general time interval $5 \leqslant t \leqslant 5 + h$:

$$\text{average velocity} = \frac{\text{distance traveled}}{\text{time elapsed}}$$

$$= \frac{s(5 + h) - s(5)}{h}$$

$$= \frac{4.9(5 + h)^2 - 4.9(5)^2}{h}$$

$$= \frac{4.9(25 + 10h + h^2 - 25)}{h}$$

$$= \frac{4.9(10h + h^2)}{h} = 49 + 4.9h \qquad \text{if } h \neq 0$$

If the time interval is very short, then h is small, so $4.9h$ is close to 0 and the average velocity is close to 49 m/s. The **instantaneous velocity** when $t = 5$ is defined to be the limiting value of these average velocities as h approaches 0. Thus the (instantaneous) velocity after 5 s is

$$v = 49 \text{ m/s}$$

Notice that we did not put $h = 0$ in the expression for the average velocity because that would have resulted in the expression $\frac{0}{0}$, which has no meaning. What we have done is compute the instantaneous velocity as the *limit* of the average velocities as h approaches 0.

We use a similar argument to find the velocity $v(t)$ at any time t before the ball hits the ground. We consider a brief period from time t to time $t + h$. The average velocity during this interval is

$$\frac{s(t + h) - s(t)}{h} = \frac{4.9(t + h)^2 - 4.9t^2}{h}$$

$$= \frac{4.9(2th + h^2)}{h}$$

$$= 4.9(2t + h) = 9.8t + 4.9h$$

The velocity at time t is the limit of this average velocity as h approaches 0, and so

$$v(t) = 9.8t$$

We are now in a position to answer the second question: How fast is the ball traveling when it hits the ground? Since the observation deck is 450 m above the ground, the ball will hit the ground at the time t_1 when $s(t_1) = 450$, that is,

$$4.9t_1^2 = 450$$

This gives

$$t_1^2 = \frac{450}{4.9} \quad \text{and} \quad t_1 = \sqrt{\frac{450}{4.9}} \approx 9.6 \text{ s}$$

The velocity of the ball as it hits the ground is therefore

$$v(t_1) = 9.8t_1 = 9.8\sqrt{\frac{450}{4.9}} \approx 94 \text{ m/s}$$

You may have the feeling that the calculations used in solving this problem are very similar to those used earlier in this section to find tangents. In fact, there is a close connection between the tangent problem and the problem of finding velocities. If we draw the graph of the distance function of the ball (as in Figure 2.5) and we consider the points $P(a, 4.9a^2)$ and $Q(a + h, 4.9(a + h)^2)$ on the graph, then the slope of the secant line PQ is

$$m_{PQ} = \frac{4.9(a + h)^2 - 4.9a^2}{(a + h) - a}$$

which is the same as the average velocity over the time interval $[a, a + h]$. Therefore the velocity at time t (the limit of these average velocities as h approaches 0) must be equal to the slope of the tangent line at P (the limit of the slopes of the secant lines).

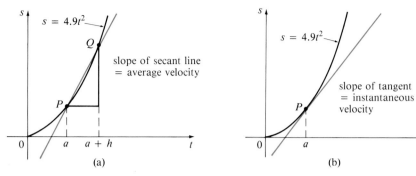

Figure 2.5

In general, suppose an object moves along a straight line according to an equation of motion $s = f(t)$, where s is the displacement (directed distance) of the object from the origin at time t. The function f that describes the motion is called the **position function** of the object. In the time interval from $t = a$ to $t = a + h$ the change in position is $f(a + h) - f(a)$ (see Figure 2.6). The average velocity over this time interval is

$$\text{average velocity} = \frac{\text{displacement}}{\text{time}} = \frac{f(a + h) - f(a)}{h}$$

which is the same as the slope of the secant line PQ in Figure 2.7.

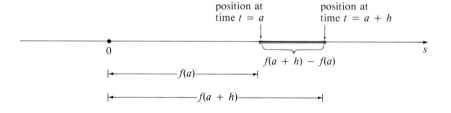

Figure 2.6

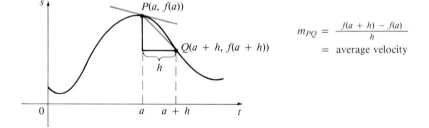

Figure 2.7

Now suppose we compute the average velocities over shorter and shorter time intervals $[a, a + h]$. In other words, we let h approach 0. As in the example of the falling ball, we define the **velocity** (or **instantaneous velocity**) $v(a)$ at time $t = a$ to be the limit of these average velocities:

(2.5)

$$v(a) = \lim_{h \to 0} \frac{f(a + h) - f(a)}{h}$$

This means that the velocity at time $t = a$ is equal to the slope of the tangent line at P. (Compare Equations 2.4 and 2.5.)

Other Rates of Change

Suppose y is a quantity that depends on another quantity x. Thus y is a function of x and we write $y = f(x)$. If x changes from x_1 to x_2, then the change in x (also called the **increment** of x) is

$$\Delta x = x_2 - x_1$$

and the corresponding change in y is

$$\Delta y = f(x_2) - f(x_1)$$

The difference quotient

$$\frac{\Delta y}{\Delta x} = \frac{f(x_2) - f(x_1)}{x_2 - x_1}$$

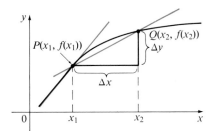

average rate of change $= m_{PQ}$
instantaneous rate of change $=$ slope of tangent at P

Figure 2.8

is called the **average rate of change of y with respect to x** over the interval $[x_1, x_2]$ and can be interpreted as the slope of the secant line PQ in Figure 2.8. By analogy with velocity, we consider the average rate of change over smaller and smaller intervals by letting x_2 approach x_1 and therefore letting Δx approach 0. The limit of these average rates of change is the **(instantaneous) rate of change of y with respect to x** at $x = x_1$, which is interpreted as the slope of the tangent to the curve $y = f(x)$ at $P(x_1, f(x_1))$:

(2.6)
$$\text{instantaneous rate of change} = \lim_{\Delta x \to 0} \frac{\Delta y}{\Delta x}$$

$$= \lim_{x_2 \to x_1} \frac{f(x_2) - f(x_1)}{x_2 - x_1}$$

● **Example 1**   Temperature readings T (in degrees Celsius) were recorded every hour starting at midnight on a day in April in Whitefish, Montana, as in the following table. The time x is measured in hours from midnight.

x (h)	0	1	2	3	4	5	6	7	8	9	10	11	12
T (°C)	6.5	6.1	5.6	4.9	4.2	4.0	4.0	4.8	6.1	8.3	10.0	12.1	14.3

x (h)	13	14	15	16	17	18	19	20	21	22	23	24
T (°C)	16.0	17.3	18.2	18.8	17.6	16.0	14.1	11.5	10.2	9.0	7.9	7.0

(a) Find the average rate of change of temperature with respect to time
   (i)   from noon to 3 P.M.,   (ii)   from noon to 2 P.M.,
   (iii) from noon to 1 P.M.
(b) Estimate the instantaneous rate of change at noon.

Solution   (a)(i)   From noon to 3 P.M. the temperature changes from 14.3°
to 18.2°, so

$$\Delta T = T(15) - T(12) = 18.2 - 14.3 = 3.9°$$

while the change in time is $\Delta x = 3$ h. Therefore the average rate of change
of temperature with respect to time is

$$\frac{\Delta T}{\Delta x} = \frac{3.9}{3} = 1.3°/h$$

(ii)   From noon to 2 P.M. the average rate of change is

$$\frac{\Delta T}{\Delta x} = \frac{T(14) - T(12)}{14 - 12} = \frac{17.3 - 14.3}{2} = 1.5°/h$$

(iii)   From noon to 1 P.M. the average rate of change is

$$\frac{\Delta T}{\Delta x} = \frac{T(13) - T(12)}{13 - 12} = \frac{16.0 - 14.3}{1} = 1.7°/h$$

   (b)   We plot the given data in Figure 2.9 and use them to sketch a
smooth curve that approximates the graph of the temperature function.
Then we draw the tangent at the point P where $x = 12$ and, after measuring
the sides of triangle ABC, we estimate that the slope of the tangent line is

$$\frac{|BC|}{|AC|} = \frac{10.3}{5.5} \approx 1.9$$

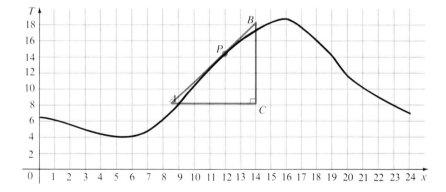

Figure 2.9

Therefore the instantaneous rate of change of temperature with respect to
time at noon is about 1.9°/h.   ●

    The velocity of a particle is the rate of change of displacement with
respect to time. Physicists are interested in other rates of change as well—
for instance, the rate of change of work with respect to time (which is called

power). Chemists who study a chemical reaction are interested in the rate of change in the concentration of a reactant with respect to time (called the rate of reaction). A steel manufacturer is interested in the rate of change of the cost of producing x tons of steel per day with respect to x (called the marginal cost). A biologist is interested in the rate of change of the population of a colony of bacteria with respect to time. In fact, the computation of rates of change is important in all of the natural sciences, engineering, and even the social sciences. (See Section 3.7.)

All these rates of change can be interpreted as slopes of tangents. This gives added significance to the solution of the tangent problem. Whenever we solve a problem involving tangent lines, we are not just solving a problem in geometry. We are also implicitly solving a great variety of problems involving rates of change in science and engineering.

EXERCISES 2.1

For each of the curves C and points $P(a, f(a))$ in Exercises 1–10, (a) find the slope m_{PQ} of the secant line in terms of x; (b) find the values of m_{PQ} when $x - a = \pm 0.5, \pm 0.1, \pm 0.01$, and ± 0.001; (c) guess the value of m, the slope of the tangent at a; and (d) using the value of m from part (c), find the equation of the tangent to C at P.

1. $y = 1 + x + x^2, (1, 3)$ 2. $y = x^2 - 2x, (4, 8)$

3. $y = x^3, (2, 8)$ 4. $y = x^3 + x^2, (1, 2)$

5. $y = 4x^2 - x + 1, (0, 1)$ 6. $y = 1 - 2x^3, (-1, 3)$

7. $y = \dfrac{1}{x}, (1, 1)$ 8. $y = \dfrac{x}{1 - x}, (0, 0)$

9. $y = \sqrt{x}, (4, 2)$ 10. $y = \cos x, \left(\dfrac{\pi}{6}, \dfrac{\sqrt{3}}{2}\right)$

11. If a ball is thrown into the air with a velocity of 40 ft/s, its height in feet after t seconds is given by $y = 40t - 16t^2$.
   (a) Find the average velocity for the time period beginning when $t = 2$ and lasting
      (i)   0.5 s   (ii)   0.1 s
      (iii) 0.05 s (iv)   0.01 s
   (b) Find the instantaneous velocity when $t = 2$.

12. If an arrow is shot upward on the moon with a velocity of 58 m/s, its height in meters after t seconds is given by $h = 58t - 0.83t^2$.
   (a) Find the average velocity over the given time intervals: $[1, 2]$, $[1, 1.5]$, $[1, 1.1]$, $[1, 1.01]$, and $[1, 1.001]$.

(b) Find the instantaneous velocity after 1 s.
(c) Find the velocity of the arrow after t seconds.
(d) When will the arrow hit the moon?
(e) With what velocity will the arrow hit the moon?

13. The displacement in meters of a particle moving in a straight line is given by $s = t^2 + t$, where t is measured in seconds.
   (a) Find the average velocity over the following time periods.
      (i) $[0, 2]$ (ii) $[0, 1]$
      (iii) $[0, 0.5]$ (iv) $[0, 0.1]$
   (b) Find the instantaneous velocity when $t = 0$.
   (c) Draw the graph of s as a function of t and draw the secant lines whose slopes are the average velocities in part (a).
   (d) Draw the tangent line whose slope is the instantaneous velocity in part (b).

14. The displacement in feet of a particle moving in a straight line is given by $s = t^3/6$, where t is measured in seconds.
   (a) Find the average velocity over the following time periods.
      (i) $[1, 3]$ (ii) $[1, 2]$
      (iii) $[1, 1.5]$ (iv) $[1, 1.1]$
   (b) Find the instantaneous velocity when $t = 1$.
   (c) Draw the graph of s as a function of t and draw the secant lines whose slopes are the average velocities in part (a).
   (d) Draw the tangent line whose slope is the instantaneous velocity in part (b).

15. (a)   Use the data in Example 1 to find the average
rate of change of temperature with respect to time
   (i)   from 8 P.M. to 11 P.M.
   (ii)   from 8 P.M. to 10 P.M.
   (iii)   from 8 P.M. to 9 P.M.
(b)   Estimate the instantaneous rate of change of T
with respect to x by measuring the slope of a
tangent.

16. The population P (in thousands) of a city from 1980
to 1986 is given in the following table.
(a)   Find the average rate of growth
   (i)   from 1982 to 1986
   (ii)   from 1982 to 1985
   (iii)   from 1982 to 1984
   (iv)   from 1982 to 1983
(b)   Estimate the instantaneous rate of growth in
1982 by measuring the slope of a tangent.

Year	1980	1981	1982	1983	1984	1985	1986
P in thousands	105	110	117	126	137	150	164

17. The cost (in dollars) of producing x units of a certain
commodity is $C(x) = 5000 + 10x + 0.05x^2$.
(a)   Find the average rate of change of C with respect
to x when the production level is changed (i) from
$x = 100$ to $x = 105$ and (ii) from $x = 100$ to
$x = 101$.
(b)   Find the instantaneous rate of change of C with
respect to x when $x = 100$. (This is called the
marginal cost.)

18. If a cylindrical tank holds 100,000 gallons of water,
which takes an hour to drain from the bottom of the
tank, then Torricelli's Law gives the volume V of
water remaining in the tank after t minutes as

$$V(t) = 100{,}000\left(1 - \frac{t}{60}\right)^2 \qquad 0 \le t \le 60$$

Find the rate at which the water is flowing out of the
tank (the instantaneous rate of change of V with
respect to t) after 20 minutes.

SECTION 2.2

The Limit of a Function

Having seen in the preceding section how limits arise when we want to find
the tangent to a curve, or the velocity of a particle, or the rate of change of
one quantity with respect to another, we now turn our attention to limits in
general and methods for computing them.

   Let us investigate the behavior of the function f defined by
$f(x) = x^2 - x + 2$ for values of x near 2. The following table gives values
of $f(x)$ for values of x close to 2 but not equal to 2.

x	$f(x)$	x	$f(x)$
1.0	2.000000	3.0	8.000000
1.5	2.750000	2.5	5.750000
1.8	3.440000	2.2	4.640000
1.9	3.710000	2.1	4.310000
1.95	3.852500	2.05	4.152500
1.99	3.970100	2.01	4.030100
1.995	3.985025	2.005	4.015025
1.999	3.997001	2.001	4.003001

From the table and the graph of f (a parabola) shown in Figure 2.10 you can see that when x is close to 2 (on either side of 2), $f(x)$ is close to 4. In fact, it appears that we can make the values of $f(x)$ as close as we like to 4 by taking x sufficiently close to 2. We express this by saying "the limit of the function $f(x) = x^2 - x + 2$ as x approaches 2 is equal to 4." The notation for this is

$$\lim_{x \to 2}(x^2 - x + 2) = 4$$

In general, we use the following notation.

Definition (2.7)

We write

$$\lim_{x \to a} f(x) = L$$

and say

"the limit of $f(x)$, as x approaches a, equals L"

if we can make the values of $f(x)$ arbitrarily close to L (as close to L as we like) by taking x to be sufficiently close to a but not equal to a.

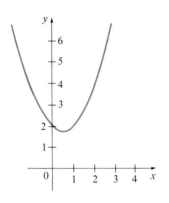

Figure 2.10

Roughly speaking, this says that the values of $f(x)$ get closer and closer to the number L as x gets closer and closer to the number a (from either side of a) but $x \neq a$. A more precise definition will be given in the next section.

An alternative notation for

$$\lim_{x \to a} f(x) = L$$

is $$f(x) \to L \quad \text{as} \quad x \to a$$

which is usually read "$f(x)$ approaches L as x approaches a."

Notice the phrase "but $x \neq a$" in the definition of limit. This means that in finding the limit of $f(x)$ as x approaches a, we never consider $x = a$. In fact, $f(x)$ need not even be defined when $x = a$. The only thing that matters is how f is defined *near* a.

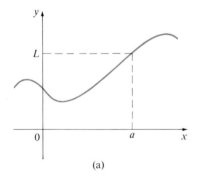

(a)

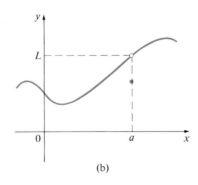

(b)

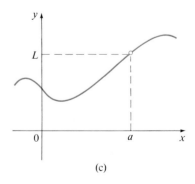

(c)

Figure 2.11

Figure 2.11 shows the graphs of three functions. Note that in part (c) $f(a)$ is not defined and in part (b) $f(a) \neq L$. But in each case, regardless of what happens at a, $\lim_{x \to a} f(x) = L$.

● **Example 1**   Find $\lim_{x \to 3}(x^3 - 2x + 1)$.

Solution   It seems clear that when x is close to 3, x^3 is close to $3^3 = 27$ and $2x$ is close to $2 \cdot 3 = 6$. Thus it appears that as x approaches 3, the function $f(x) = x^3 - 2x + 1$ approaches $27 - 6 + 1 = 22$. Therefore

$$\lim_{x \to 3}(x^3 - 2x + 1) = 22$$

(Here we have arrived at the answer in an intuitive fashion, but in the next sections, when we have formulated a precise definition of limit, we will be able to prove conclusively that the correct answer is indeed 22.)   ●

Notice that in Example 1 the value of the limit is the same as the value obtained by substituting the value $x = 3$ in the expression for $f(x)$ since $f(3) = 3^3 - 2 \cdot 3 + 1 = 22$. However, as the following examples show, it is not always possible to evaluate $\lim_{x \to a} f(x)$ simply by finding $f(a)$.

● **Example 2**   Find $\lim_{x \to 1} \dfrac{x^2 - 1}{x - 1}$.

Solution   Let

$$f(x) = \frac{x^2 - 1}{x - 1}$$

We cannot find the limit by substituting $x = 1$ because $f(1)$ is not defined ($\frac{0}{0}$ is meaningless). Remember that the definition of $\lim_{x \to a} f(x)$ says that we consider values of x that are close to a but not equal to a. Therefore, in this example we have $x \neq 1$ and so we can write

$$\lim_{x \to 1} \frac{x^2 - 1}{x - 1} = \lim_{x \to 1} \frac{(x - 1)(x + 1)}{x - 1}$$

$$= \lim_{x \to 1}(x + 1)$$

$$= 1 + 1 = 2$$

The limit in this example arose in Section 2.1 when we were trying to find the tangent to the parabola $y = x^2$ at the point $(1, 1)$.   ●

● **Example 3**   Find $\lim_{x \to 1} g(x)$ where

$$g(x) = \begin{cases} x + 1 & \text{if } x \neq 1 \\ \pi & \text{if } x = 1 \end{cases}$$

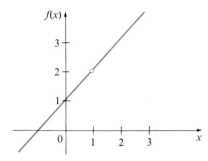

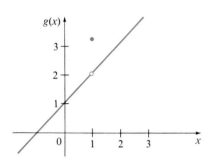

Figure 2.12

Solution   Here g is defined at $x = 1$ and $g(1) = \pi$, but the value of a limit as x approaches 1 does not depend on the value of the function at 1. Since $g(x) = x + 1$ for $x \neq 1$, we have

$$\lim_{x \to 1} g(x) = \lim_{x \to 1}(x + 1) = 2$$ ●

Note that the values of the functions in Examples 2 and 3 are identical except when $x = 1$ (see Figure 2.12) and so they have the same limit as x approaches 1.

● **Example 4**   Evaluate $\lim\limits_{h \to 0} \dfrac{(3 + h)^2 - 9}{h}$.

Solution   If we define

$$F(h) = \frac{(3 + h)^2 - 9}{h}$$

then, as in Example 2, we cannot compute $\lim_{h \to 0} F(h)$ by letting $h = 0$ since $F(0)$ is undefined. But if we simplify $F(h)$ algebraically, we find

$$F(h) = \frac{(9 + 6h + h^2) - 9}{h} = \frac{6h + h^2}{h} = 6 + h$$

(Recall that we consider only $h \neq 0$ when letting h approach 0.) Thus

$$\lim_{h \to 0} \frac{(3 + h)^2 - 9}{h} = \lim_{h \to 0}(6 + h) = 6$$ ●

● **Example 5**   Find $\lim\limits_{t \to 4} \dfrac{\sqrt{t} - 2}{t - 4}$.

Solution   The function

$$f(t) = \frac{\sqrt{t} - 2}{t - 4}$$

has domain $\{t \in R \mid t \geqslant 0, t \neq 4\}$. In calculating its limit as $t \to 4$ we consider $t \neq 4$. It helps to factor the denominator:

$$t - 4 = (\sqrt{t} - 2)(\sqrt{t} + 2)$$

Therefore

$$\lim_{t \to 4} \frac{\sqrt{t} - 2}{t - 4} = \lim_{t \to 4} \frac{\sqrt{t} - 2}{(\sqrt{t} - 2)(\sqrt{t} + 2)} = \lim_{t \to 4} \frac{1}{\sqrt{t} + 2}$$

As t gets closer to 4, $\sqrt{t}$ gets closer to $\sqrt{4} = 2$, so $f(t)$ gets closer to $\frac{1}{4}$. Thus

$$\lim_{t \to 4} \frac{\sqrt{t} - 2}{t - 4} = \frac{1}{4}$$

Notice that instead of factoring the denominator we could have achieved the same result by rationalizing the numerator, that is, by multiplying numerator and denominator by $\sqrt{t} + 2$:

$$\frac{\sqrt{t} - 2}{t - 4} = \frac{(\sqrt{t} - 2)(\sqrt{t} + 2)}{(t - 4)(\sqrt{t} + 2)} = \frac{t - 4}{(t - 4)(\sqrt{t} + 2)} = \frac{1}{\sqrt{t} + 2}$$ ●

● **Example 6**   Find $\lim\limits_{x \to 0} \dfrac{\sin x}{x}$.

Solution   Again the function $f(x) = \sin x / x$ is not defined when $x = 0$. Using a calculator (and remembering that, if $x \in R$, $\sin x$ means the sine of the angle whose *radian* measure is x), we construct the following table of values correct to eight decimal places.

x	$f(x) = \dfrac{\sin x}{x}$
± 1.0	0.84147098
± 0.5	0.95885108
± 0.4	0.97354586
± 0.3	0.98506736
± 0.2	0.99334665
± 0.1	0.99833417
± 0.05	0.99958339
± 0.01	0.99998333
± 0.005	0.99999583
± 0.001	0.99999983

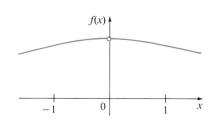

Figure 2.13

From the table of values and Figure 2.13 (drawn with the aid of the table) we make the guess

$$\lim_{x \to 0} \frac{\sin x}{x} = 1$$

This guess is in fact correct, as will be proved in Chapter 3 using a geometric argument.   ●

The method used in Example 6—namely, the computation of specific values of the function—allows us only to *guess* the value of a limit. Sometimes the guess may even be wrong. The next example illustrates the pitfalls in such an approach.

● **Example 7**   Find $\lim\limits_{x \to 0} \sin \dfrac{\pi}{x}$.

Solution   Once again the function $f(x) = \sin(\pi/x)$ is undefined at 0. Let us try the same approach as in Example 6. Evaluating the function for some small values of x, we get

$$f(1) = \sin \pi = 0 \qquad\qquad f(\tfrac{1}{2}) = \sin 2\pi = 0$$

$$f(\tfrac{1}{3}) = \sin 3\pi = 0 \qquad\qquad f(\tfrac{1}{4}) = \sin 4\pi = 0$$

$$f(0.1) = \sin 10\pi = 0 \qquad f(0.01) = \sin 100\pi = 0$$

Similarly, $f(0.001) = f(0.0001) = 0$. On the basis of this information we might be tempted to make the guess

$$\lim_{x \to 0} \sin \frac{\pi}{x} = 0$$

but this time our guess is wrong. Note that although $f(1/n) = \sin n\pi = 0$ for any integer n, it is also true that $f(x) = 1$ for infinitely many values of x that approach 0. [In fact, $\sin(\pi/x) = 1$ when

$$\frac{\pi}{x} = \frac{\pi}{2} + 2n\pi$$

and, solving for x, we get $x = 2/(4n + 1)$.] The graph of f is given in Figure 2.14. The values of $\sin(\pi/x)$ oscillate between 1 and -1 infinitely often as x approaches 0. Since the values of $f(x)$ do not approach a fixed number as x approaches 0,

$$\lim_{x \to 0} \sin \frac{\pi}{x}$$

does not exist.

Figure 2.14

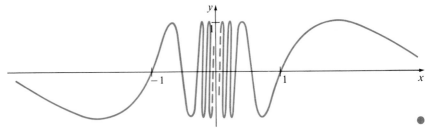

Example 8   Find $\lim_{x \to 0} \dfrac{1}{x^2}$ if it exists.

Solution   As x becomes close to 0, x^2 also becomes close to 0, so $1/x^2$ becomes very large. In fact, you can see from the graph of the function $f(x) = 1/x^2$ shown in Figure 2.15 that the values of $f(x)$ can be made arbitrarily large by taking x close enough to 0. Thus the values of $f(x)$ do not approach a number, so $\lim_{x \to 0}(1/x^2)$ does not exist.

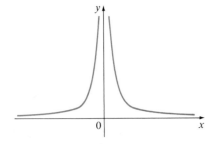

Figure 2.15

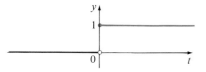

Figure 2.16

● Example 9   The Heaviside function H is defined by

$$H(t) = \begin{cases} 0 & \text{if } t < 0 \\ 1 & \text{if } t \geq 0 \end{cases}$$

[This function is named after the electrical engineer Oliver Heaviside (1850–1925) and can be used to describe an electric current that is switched on at time $t = 0$.] Its graph is shown in Figure 2.16.

    As t approaches 0 from the left, $H(t)$ approaches 0. As t approaches 0 from the right, $H(t)$ approaches 1. There is no single number that $H(t)$ approaches as t approaches 0. Therefore $\lim_{t \to 0} H(t)$ does not exist.   ●

One-Sided Limits

We noticed in Example 9 that $H(t)$ approaches 0 as t approaches 0 from the left and $H(t)$ approaches 1 as t approaches 0 from the right. We indicate this situation symbolically by writing

$$\lim_{t \to 0^-} H(t) = 0 \quad \text{and} \quad \lim_{t \to 0^+} H(t) = 1$$

The symbol "$t \to 0^-$" indicates that we consider only values of t that are less than 0. Likewise "$t \to 0^+$" indicates that we consider only values of t that are greater than 0.

Definition (2.8)

We write

$$\lim_{x \to a^-} f(x) = L$$

and say the **left-hand limit of $f(x)$ as x approaches a** (or the **limit of $f(x)$ as x approaches a from the left**) is equal to L if we can make the values of $f(x)$ arbitrarily close to L by taking x to be sufficiently close to a and $x < a$.

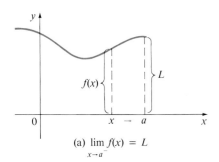

(a) $\lim_{x \to a^-} f(x) = L$

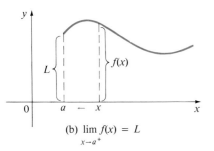

(b) $\lim_{x \to a^+} f(x) = L$

Figure 2.17

    Notice that Definition 2.8 differs from Definition 2.7 only in that we require x to be less than a. Similarly, if we require that x be greater than a, we get "the **right-hand limit of $f(x)$ as x approaches a** is equal to L" and we write

$$\lim_{x \to a^+} f(x) = L$$

Thus the symbol "$x \to a^+$" means that we consider only $x > a$. These definitions are illustrated in Figure 2.17.

● Example 10   Find $\lim_{x \to 0^+} \sqrt{x}$.

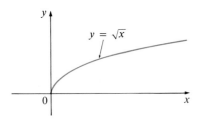

Figure 2.18

Solution   The function $f(x) = \sqrt{x}$ is defined only for $x \geqslant 0$, so the two-sided limit $\lim_{x \to 0} \sqrt{x}$ does not make sense. If we let x approach 0 while restricting x to be a positive number, we see that $\sqrt{x}$ approaches 0, and so

$$\lim_{x \to 0^+} \sqrt{x} = 0$$

(See Figure 2.18.)   ●

● **Example 11**   State the values (if they exist) of

(a) $\displaystyle\lim_{x \to 2^-} g(x)$ (b) $\displaystyle\lim_{x \to 2^+} g(x)$ (c) $\displaystyle\lim_{x \to 2} g(x)$

for the function g whose graph is shown in Figure 2.19.

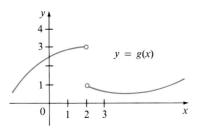

Figure 2.19

Solution   From the graph we see that

(a) $\displaystyle\lim_{x \to 2^-} g(x) = 3$ (b) $\displaystyle\lim_{x \to 2^+} g(x) = 1$

Since the left and right limits are different, there is no single number that $g(x)$ approaches and we conclude that

(c) $\displaystyle\lim_{x \to 2} g(x)$ does not exist.   ●

EXERCISES 2.2

In Exercises 1–12 state the value of the limit, if it exists, from the given graphs.

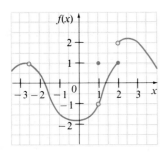

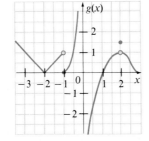

1. $\displaystyle\lim_{x \to 3} f(x)$ **2.** $\displaystyle\lim_{x \to 1} f(x)$ **3.** $\displaystyle\lim_{x \to -3} f(x)$

4. $\displaystyle\lim_{x \to 2^-} f(x)$ **5.** $\displaystyle\lim_{x \to 2^+} f(x)$ **6.** $\displaystyle\lim_{x \to 2} f(x)$

7. $\displaystyle\lim_{x \to 1} g(x)$ **8.** $\displaystyle\lim_{x \to 0} g(x)$ **9.** $\displaystyle\lim_{x \to 2} g(x)$

10. $\displaystyle\lim_{x \to -2} g(x)$ **11.** $\displaystyle\lim_{x \to -1^-} g(x)$ **12.** $\displaystyle\lim_{x \to -1} g(x)$

In Exercises 13–44 find the limit if it exists. Explain why there is no limit if it does not exist.

13. $\displaystyle\lim_{x \to 2}(x^2 - 2x + 7)$ **14.** $\displaystyle\lim_{x \to -1}(5 - x^2)$

15. $\displaystyle\lim_{x \to 12} \sqrt{x - 3}$ **16.** $\displaystyle\lim_{x \to 3} \frac{6 + x}{9 - x}$

17. $\displaystyle\lim_{x \to 1} \frac{x^2 + 1}{x^2 - x + 1}$ **18.** $\displaystyle\lim_{x \to 4} \frac{\sqrt{x}}{1 + 2x - x^2}$

19. $\displaystyle\lim_{x \to 0} \frac{x + x^2}{x}$ **20.** $\displaystyle\lim_{x \to 3} \frac{x^2 - 9}{x - 3}$

21. $\displaystyle\lim_{x \to 1} \frac{x^2 - x - 2}{x + 1}$ **22.** $\displaystyle\lim_{x \to -1} \frac{x^2 - x - 2}{x + 1}$

23. $\displaystyle\lim_{x \to -1} \frac{x^2 - x - 3}{x + 1}$ **24.** $\displaystyle\lim_{x \to 1} \frac{x^3 - 1}{x^2 - 1}$

25. $\displaystyle\lim_{t \to 1} \frac{t^3 - t}{t^2 - 1}$ **26.** $\displaystyle\lim_{h \to 0} \frac{(h - 5)^2 - 25}{h}$

27. $\displaystyle\lim_{h\to 0} \frac{(2 + h)^3 - 8}{h}$

28. $\displaystyle\lim_{h\to 0} \frac{(a + h)^3 - a^3}{h}$

29. $\displaystyle\lim_{h\to 0} \frac{(1 + h)^4 - 1}{h}$

30. $\displaystyle\lim_{x\to 1} \frac{x^2 + x - 2}{x^2 - 3x + 2}$

31. $\displaystyle\lim_{x\to -2} \frac{x + 2}{x^2 - x - 6}$

32. $\displaystyle\lim_{t\to 2} \frac{t^2 + t - 6}{t^2 - 4}$

33. $\displaystyle\lim_{x\to -2} \frac{1}{x^4}$

34. $\displaystyle\lim_{x\to 0} \frac{1}{x^4}$

35. $\displaystyle\lim_{x\to 3} \frac{1}{(x - 3)^2}$

36. $\displaystyle\lim_{x\to -3} \frac{1}{x + 3}$

37. $\displaystyle\lim_{x\to 3^+} \sqrt{x - 3}$

38. $\displaystyle\lim_{x\to 5^-} \sqrt[4]{5 - x}$

39. (a) $\displaystyle\lim_{x\to 2^-} f(x)$ (b) $\displaystyle\lim_{x\to 2^+} f(x)$ (c) $\displaystyle\lim_{x\to 2} f(x)$

where $f(x) = \begin{cases} -1 & \text{if } 0 \leqslant x < 2 \\ 1 & \text{if } x \geqslant 2 \end{cases}$

40. (a) $\displaystyle\lim_{x\to 1^-} g(x)$ (b) $\displaystyle\lim_{x\to 1^+} g(x)$ (c) $\displaystyle\lim_{x\to 1} g(x)$

where $g(x) = \begin{cases} x + 2 & \text{if } x \neq 1 \\ 1 & \text{if } x = 1 \end{cases}$

41. (a) $\displaystyle\lim_{x\to 1^-} h(x)$ (b) $\displaystyle\lim_{x\to 1^+} h(x)$ (c) $\displaystyle\lim_{x\to 1} h(x)$

where $h(x) = \begin{cases} x + 2 & \text{if } x < 1 \\ 1 - x & \text{if } x \geqslant 1 \end{cases}$

42. (a) $\displaystyle\lim_{x\to 1^-} p(x)$ (b) $\displaystyle\lim_{x\to 1^+} p(x)$ (c) $\displaystyle\lim_{x\to 1} p(x)$

where $p(x) = \begin{cases} x + 2 & \text{if } x < 1 \\ 4 - x & \text{if } x \geqslant 1 \end{cases}$

43. $\displaystyle\lim_{x\to 2} \frac{\frac{1}{x} - \frac{1}{2}}{x - 2}$

44. $\displaystyle\lim_{x\to 0} \frac{|x|}{x}$

45. (a) Evaluate $f(x) = \dfrac{1 - \cos x}{x^2}$ correct to six decimal places for $x = 1, 0.5, 0.4, 0.3, 0.2, 0.1, 0.05,$ and 0.01.

(b) Guess the value of

$$\lim_{x\to 0} \frac{1 - \cos x}{x^2}$$

46. (a) Evaluate $g(x) = \dfrac{\cos x - 1}{\sin x}$ for the same values of x as in Exercise 45.

(b) Guess the value of

$$\lim_{x\to 0} \frac{\cos x - 1}{\sin x}$$

47. The existence of the limit

$$\lim_{x\to 0} (1 + x)^{1/x}$$

will be established in Chapter 10. Estimate the value of the limit to five decimal places by first evaluating the function for $x = 1, 0.1, 0.01, 0.001, 0.0001, 0.00001, 0.000001, 0.0000001, 0.00000001,$ and 0.000000001.

48. (a) Evaluate $h(x) = \dfrac{\tan x - x}{x^3}$ for $x = 1, 0.5, 0.1, 0.05, 0.01,$ and 0.005.

(b) Guess the value of $\displaystyle\lim_{x\to 0} \frac{\tan x - x}{x^3}$.

(c) Evaluate h for $x = 0.001, 0.0005, 0.0004, 0.0003, 0.0002, 0.0001, 0.00005, 0.00001, 0.000001,$ and 0.0000001. Are you still confident that your guess in part (b) is correct? In Section 6.10 a method for evaluating the limit will be explained. See Appendix D for an explanation of what went wrong with the calculator computations.

SECTION 2.3

The Precise Definition of a Limit

The intuitive definition of a limit given in the preceding section is inadequate for some purposes because such phrases as "x is close to 2" and "$f(x)$ gets closer and closer to L" are vague. If we want to be able to prove conclusively that

$$\lim_{x\to 3} (x^3 - 2x + 1) = 22 \quad \text{or} \quad \lim_{x\to 0} \frac{\sin x}{x} = 1$$

then we must make the definition of a limit more precise.

In order to motivate the precise definition of a limit, let us consider the function

$$f(x) = \begin{cases} 2x - 1 & \text{if } x \neq 3 \\ 6 & \text{if } x = 3 \end{cases}$$

Intuitively it is clear that when x is close to 3 but $x \neq 3$, $f(x)$ is close to 5, and so $\lim_{x \to 3} f(x) = 5$.

To obtain more detailed information as to how $f(x)$ varies when x is close to 3, let us ask the following question.

How close to 3 does x have to be so that $f(x)$ differs from 5 by less than 0.1?

The distance from x to 3 is $|x - 3|$ and the distance from $f(x)$ to 5 is $|f(x) - 5|$, so our problem is to find a number δ such that

$$|f(x) - 5| < 0.1 \quad \text{if} \quad |x - 3| < \delta \text{ but } x \neq 3$$

[It is traditional to use the Greek letter δ (delta) in this situation.] If $|x - 3| > 0$, then $x \neq 3$, so an equivalent formulation of our problem is to find a number δ such that

$$|f(x) - 5| < 0.1 \quad \text{if} \quad 0 < |x - 3| < \delta$$

Notice that if $0 < |x - 3| < 0.1/2 = 0.05$, then

$$|f(x) - 5| = |(2x - 1) - 5| = |2x - 6| = 2|x - 3| < 0.1$$

that is,

$$|f(x) - 5| < 0.1 \quad \text{if} \quad 0 < |x - 3| < 0.05$$

Thus the answer to the problem is given by $\delta = 0.05$; that is, if x is within a distance of 0.05 from 3, then $f(x)$ will be within a distance of 0.1 from 5.

If we change the number 0.1 in our problem to the smaller number 0.01, then by using the same method we find that $f(x)$ will differ from 5 by less than 0.01 provided that x differs from 3 by less than $0.01/2 = 0.005$:

$$|f(x) - 5| < 0.01 \quad \text{if} \quad 0 < |x - 3| < 0.005$$

Similarly,

$$|f(x) - 5| < 0.001 \quad \text{if} \quad 0 < |x - 3| < 0.0005$$

If, instead of tolerating an error of 0.1 or 0.01 or 0.001, we want accuracy to within a tolerance of an arbitrary positive number ε (the Greek letter epsilon), then we find as before that

(2.9)
$$|f(x) - 5| < \varepsilon \quad \text{if} \quad 0 < |x - 3| < \delta = \frac{\varepsilon}{2}$$

This is a precise way of saying that $f(x)$ is close to 5 when x is close to 3 because (2.9) says that we can make the values of $f(x)$ within an arbitrary distance ε from 5 by taking the values of x within a distance $\varepsilon/2$ from 3 (but $x \neq 3$).

Note that (2.9) can be rewritten as

(2.10) $5 - \varepsilon < f(x) < 5 + \varepsilon$ whenever $3 - \delta < x < 3 + \delta$ $(x \neq 3)$

and this is illustrated in Figure 2.20. By taking the values of x ($\neq 3$) to lie in the interval $(3 - \delta, 3 + \delta)$ we can make the values of $f(x)$ lie in the interval $(5 - \varepsilon, 5 + \varepsilon)$.

Using (2.9) as a model, we give a precise definition of a limit.

Definition (2.11)

Let f be a function defined on some open interval that contains the number a, except possibly at a itself. Then we say that the **limit of $f(x)$ as x approaches a is L,** and we write

$$\lim_{x \to a} f(x) = L$$

if for every number $\varepsilon > 0$ there is a corresponding number $\delta > 0$ such that

$$|f(x) - L| < \varepsilon \qquad \text{whenever} \qquad 0 < |x - a| < \delta$$

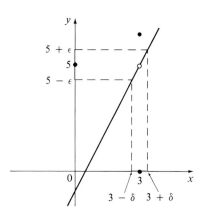

Figure 2.20

Another way of writing the last line of this definition is

"if $0 < |x - a| < \delta$, then $|f(x) - L| < \varepsilon$"

Another notation for $\lim_{x \to a} f(x) = L$ is

$$f(x) \to L \quad \text{as} \quad x \to a$$

Since $|x - a|$ is the distance from x to a and $|f(x) - L|$ is the distance from $f(x)$ to L, and since ε can be arbitrarily small, the definition of a limit can be expressed in words as follows:

$\lim_{x \to a} f(x) = L$ means that the distance between $f(x)$ and L can be made arbitrarily small by taking the distance from x to a sufficiently small (but not 0).

Alternatively,

$\lim_{x \to a} f(x) = L$ means that the values of $f(x)$ can be made as close as we please to L by taking x close enough to a (but not equal to a).

We can also reformulate Definition 2.11 in terms of intervals by observing that the inequality $|x - a| < \delta$ is equivalent to $-\delta < x - a < \delta$, which in turn can be written as $a - \delta < x < a + \delta$. Also $0 < |x - a|$ is true if and only if $x - a \neq 0$, that is, $x \neq a$. Similarly, the inequality $|f(x) - L| < \varepsilon$ is equivalent to $L - \varepsilon < f(x) < L + \varepsilon$. Therefore, in terms of intervals Definition 2.11 can be stated as follows:

$\lim_{x \to a} f(x) = L$ means that for every $\varepsilon > 0$ (no matter how small ε is) we can find $\delta > 0$ such that if x lies in the open interval $(a - \delta, a + \delta)$ and $x \neq a$, then $f(x)$ lies in the open interval $(L - \varepsilon, L + \varepsilon)$.

Figure 2.21

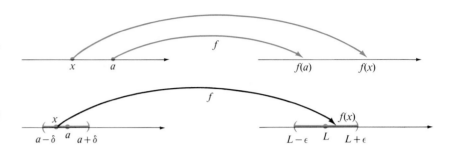

Figure 2.22

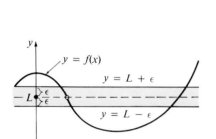

We interpret this statement geometrically by representing a function by an arrow diagram as in Figure 2.21, where f maps a subset of R onto another subset of R.

The definition of limit says that if any small interval $(L - \varepsilon, L + \varepsilon)$ is given around L, then we can find an interval $(a - \delta, a + \delta)$ around a such that f maps all the points in $(a - \delta, a + \delta)$ (except possibly a) into the interval $(L - \varepsilon, L + \varepsilon)$ (see Figure 2.22).

Another geometric interpretation of limits can be given in terms of the graph of a function. If $\varepsilon > 0$ is given, then we draw the horizontal lines $y = L + \varepsilon$ and $y = L - \varepsilon$ and the graph of f (see Figure 2.23). If $\lim_{x \to a} f(x) = L$, then we can find a number $\delta > 0$ such that if we restrict x to lie in the interval $(a - \delta, a + \delta)$ and take $x \neq a$, then the curve $y = f(x)$ lies between the lines $y = L - \varepsilon$ and $y = L + \varepsilon$ (see Figure 2.24). You can see that if such a δ has been found, then any smaller δ will also work.

It is important to realize that the process illustrated in Figures 2.23 and 2.24 must work for *every* positive number ε no matter how small it is chosen. Figure 2.25 shows that if a smaller ε is chosen, then a smaller δ may be required.

In proving limit statements it may be helpful to think of the definition of limit as a challenge. First it challenges you with a number ε. Then you must be able to produce a suitable δ. You have to be able to do this for *every* $\varepsilon > 0$, not just a particular ε.

Imagine a contest between two people, A and B, and imagine yourself to be B. A stipulates that the fixed number L should be approximated by the values of $f(x)$ to within a degree of accuracy ε (say, 0.01). B then responds by finding a number δ such that $|f(x) - L| < \varepsilon$ whenever $0 < |x - a| < \delta$. Then A may become more exacting and challenge B with a smaller value of ε (say, 0.0001). Again B has to respond by finding a corresponding δ. Usually the smaller the value of ε, the smaller the corresponding value of δ must be. If B always wins, no matter how small A makes ε, then $\lim_{x \to a} f(x) = L$.

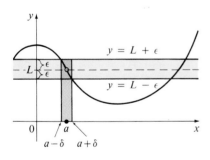

Figure 2.23

● **Example 1**   Prove that $\lim_{x \to 3} (4x - 5) = 7$.

Solution   *1. Preliminary analysis of the problem (guessing a value for δ).* Let ε be a given positive number. We want to find a number δ such that

$$|(4x - 5) - 7| < \varepsilon \qquad \text{whenever} \qquad 0 < |x - 3| < \delta$$

But $|(4x - 5) - 7| = |4x - 12| = |4(x - 3)| = 4|x - 3|$. Therefore we want

$$4|x - 3| < \varepsilon \qquad \text{whenever} \qquad 0 < |x - 3| < \delta$$

Figure 2.24

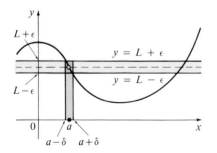

Figure 2.25

that is,

$$|x - 3| < \frac{\varepsilon}{4} \qquad \text{whenever} \qquad 0 < |x - 3| < \delta$$

This suggests that we should choose $\delta = \varepsilon/4$.

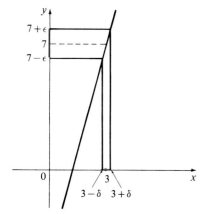

Figure 2.26

2. Proof (showing that the δ works).   Given $\varepsilon > 0$, choose $\delta = \varepsilon/4$. If $0 < |x - 3| < \delta$, then

$$|(4x - 5) - 7| = |4x - 12| = 4|x - 3| < 4\delta = 4\left(\frac{\varepsilon}{4}\right) = \varepsilon$$

Thus

$$|(4x - 5) - 7| < \varepsilon \qquad \text{whenever} \qquad 0 < |x - 3| < \delta$$

Therefore, by definition of a limit,

$$\lim_{x \to 3}(4x - 5) = 7$$

This example is illustrated by Figure 2.26.   ●

Note that in the solution of Example 1 there were two stages—guessing and proving. We made a preliminary analysis that enabled us to guess a value for δ. But then in the second stage we had to go back and prove in a careful logical fashion that we had made a correct guess. This procedure is typical of much of mathematics. Sometimes it is necessary to first make an intelligent guess about the answer to a problem and then prove that the guess is correct.

For a more complicated function than $f(x) = 4x - 5$, for instance, $f(x) = (6x^2 - 8x + 9)/(x^2 + 1)$, it might require much ingenuity to prove a limit statement using the ε, δ definition directly. Fortunately this is not necessary because in the next section you will learn some properties of limits that will enable you to find limits rigorously without resorting to the definition.

● **Example 2**   Show that $\displaystyle\lim_{x \to 0} \frac{1}{x^2}$ does not exist.

Solution   The function $f(x) = 1/x^2$ does not have a value at $x = 0$, but that does not matter. [Recall that $\lim_{x \to a} f(x)$ can exist even if $f(a)$ is not defined.]

To show that a limit does not exist we use an indirect proof. Suppose that

$$\lim_{x \to 0} \frac{1}{x^2} = L$$

for some number L. Then, if we take $\varepsilon = 1$, for example, the definition of

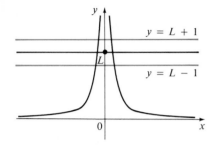

Figure 2.27

limit says that there exists a number δ such that

$$\left| \frac{1}{x^2} - L \right| < 1 \qquad \text{whenever} \qquad -\delta < x < \delta \qquad (x \neq 0)$$

that is,

$$L - 1 < \frac{1}{x^2} < L + 1 \qquad \text{whenever} \qquad -\delta < x < \delta \qquad (x \neq 0)$$

This is impossible, since $1/x^2$ takes on arbitrarily large values in any interval $(-\delta, \delta)$ (see Figure 2.27). In fact,

$$\frac{1}{x^2} > L + 1 \qquad \text{if} \qquad 0 < x < \frac{1}{\sqrt{L+1}}$$

Therefore $\lim_{x \to 0} 1/x^2$ does not exist.   ●

One-Sided Limits

The intuitive definitions of one-sided limits that were given in Section 2.2 can be precisely reformulated as follows:

Definition of Left-Hand Limit (2.12)

$$\lim_{x \to a^-} f(x) = L$$

if for every number $\varepsilon > 0$ there is a corresponding number $\delta > 0$ such that

$$|f(x) - L| < \varepsilon \qquad \text{whenever} \qquad a - \delta < x < a$$

Definition of Right-Hand Limit (2.13)

$$\lim_{x \to a^+} f(x) = L$$

if for every number $\varepsilon > 0$ there is a corresponding number $\delta > 0$ such that

$$|f(x) - L| < \varepsilon \qquad \text{whenever} \qquad a < x < a + \delta$$

Notice that Definition 2.12 is the same as Definition 2.11 except that x is restricted to lie in the *left* half $(a - \delta, a)$ of the interval $(a - \delta, a + \delta)$. In Definition 2.13 x is restricted to lie in the *right* half $(a, a + \delta)$ of the interval $(a - \delta, a + \delta)$.

● **Example 3**   Use Definition 2.13 to prove that $\lim_{x \to 0^+} \sqrt{x} = 0$.

Solution   1. *(Guessing a value for δ)*.   Let ε be a given positive number. Here $a = 0$ and $L = 0$, so we want to find a number δ such that

$$\left|\sqrt{x} - 0\right| < \varepsilon \qquad \text{whenever} \qquad 0 < x < \delta$$

that is,

$$\sqrt{x} < \varepsilon \qquad \text{whenever} \qquad 0 < x < \delta$$

or, squaring both sides of the inequality $\sqrt{x} < \varepsilon$,

$$x < \varepsilon^2 \qquad \text{whenever} \qquad 0 < x < \delta$$

This suggests that we should choose $\delta = \varepsilon^2$.

2. *(Showing that δ works)*.   Given $\varepsilon > 0$, let $\delta = \varepsilon^2$. If $0 < x < \delta$, then

$$\sqrt{x} < \sqrt{\delta} = \sqrt{\varepsilon^2} = \varepsilon$$

so

$$\left|\sqrt{x} - 0\right| < \varepsilon$$

According to Definition 2.13, this shows that $\lim_{x \to 0^+} \sqrt{x} = 0$.   ●

EXERCISES 2.3

1. How close to 3 do we have to take x so that $6x + 1$ is within a distance of
(a) 0.1   (b) 0.01 from 19?

2. How close to 2 do we have to take x so that $8x - 5$ is within a distance of
(a) 0.01   (b) 0.001   (c) 0.0001 from 11?

In Exercises 3–6 prove the statements using the ε, δ definition of limit and illustrate with a diagram like Figure 2.26.

3. $\lim\limits_{x \to 2}(3x - 2) = 4$

4. $\lim\limits_{x \to 4}(5 - 2x) = -3$

5. $\lim\limits_{x \to -1}(5x + 8) = 3$

6. $\lim\limits_{x \to -1}(3 - 4x) = 7$

In Exercises 7–18 prove the statements using the ε, δ definition of limit.

7. $\lim\limits_{x \to 2}\dfrac{x}{7} = \dfrac{2}{7}$

8. $\lim\limits_{x \to 4}\left(\dfrac{x}{3} + 1\right) = \dfrac{7}{3}$

9. $\lim\limits_{x \to -5}\left(4 - \dfrac{3x}{5}\right) = 7$

10. $\lim\limits_{x \to 3} x = 3$

11. $\lim\limits_{x \to a} x = a$

12. $\lim\limits_{x \to 1} \pi = \pi$

13. $\lim\limits_{x \to 2} c = c$

14. $\lim\limits_{x \to a} c = c$

15. $\lim\limits_{x \to 0} x^2 = 0$

16. $\lim\limits_{x \to 0} x^3 = 0$

17. $\lim\limits_{x \to 0}|x| = 0$

18. $\lim\limits_{x \to 9^-} \sqrt[4]{9 - x} = 0$

19. Prove that the following limits do not exist.

(a) $\lim\limits_{x \to 0} \dfrac{1}{x}$ (b) $\lim\limits_{x \to -4} \dfrac{1}{(x + 4)^2}$

20. Draw a diagram like Figure 2.24 to show graphically that

$$\lim\limits_{x \to 3} x^2 = 9$$

21. Prove that $\lim_{x \to 3} x^2 = 9$ by using Definition 2.11.

22. If H is the Heaviside function defined in Example 9 in Section 2.2, prove, using Definition 2.11, that $\lim_{t \to 0} H(t)$ does not exist.

23. If the function f is defined by

$$f(x) = \begin{cases} 0 & \text{if } x \text{ is rational} \\ 1 & \text{if } x \text{ is irrational} \end{cases}$$

prove that $\lim_{x \to 0} f(x)$ does not exist.

24. By comparing Definitions 2.11, 2.12, and 2.13, prove that

$$\lim_{x \to a} f(x) = L \quad \text{if and only if} \quad \lim_{x \to a^-} f(x) = L = \lim_{x \to a^+} f(x)$$

SECTION 2.4

Properties of Limits

In this section we present several basic properties of limits. These properties will enable us to calculate limits without having to use the ε, δ definition of a limit.

Properties of Limits (2.14)

Suppose that c is a constant and the limits

$$\lim_{x \to a} f(x) \quad \text{and} \quad \lim_{x \to a} g(x)$$

exist. Then

1. $\lim_{x \to a} [f(x) + g(x)] = \lim_{x \to a} f(x) + \lim_{x \to a} g(x)$

2. $\lim_{x \to a} [f(x) - g(x)] = \lim_{x \to a} f(x) - \lim_{x \to a} g(x)$

3. $\lim_{x \to a} [cf(x)] = c \lim_{x \to a} f(x)$

4. $\lim_{x \to a} [f(x)g(x)] = \lim_{x \to a} f(x) \cdot \lim_{x \to a} g(x)$

5. $\lim_{x \to a} \dfrac{f(x)}{g(x)} = \dfrac{\lim_{x \to a} f(x)}{\lim_{x \to a} g(x)} \quad \text{if } \lim_{x \to a} g(x) \neq 0$

These five properties can be stated verbally as follows:

1. The limit of a sum is the sum of the limits.
2. The limit of a difference is the difference of the limits.
3. The limit of a constant times a function is the constant times the limit of the function.
4. The limit of a product is the product of the limits.
5. The limit of a quotient is the quotient of the limits (provided that the limit of the denominator is not 0).

It is easy to believe that these properties are true. For instance, if $f(x)$ is close to L and $g(x)$ is close to M, it is reasonable to conclude that

$f(x) + g(x)$ is close to $L + M$. This gives us an intuitive basis for believing that Property 1 is true. Proofs based on the precise definition of a limit are given in Appendix C.

In order to compute limits, we shall also need to use the following properties.

Further Properties of Limits (2.15)

6. $\lim\limits_{x \to a}[f(x)]^n = [\lim\limits_{x \to a} f(x)]^n$ if n is a positive integer

7. $\lim\limits_{x \to a} c = c$ 8. $\lim\limits_{x \to a} x = a$

9. $\lim\limits_{x \to a} x^n = a^n$ where n is a positive integer

10. $\lim\limits_{x \to a} \sqrt[n]{x} = \sqrt[n]{a}$ where $a > 0$ and n is a positive integer (or if $a \leqslant 0$ and n is an odd positive integer)

11. $\lim\limits_{x \to a} \sqrt[n]{f(x)} = \sqrt[n]{\lim\limits_{x \to a} f(x)}$ where $L = \lim\limits_{x \to a} f(x) > 0$ and n is a positive integer (or $L \leqslant 0$ and n is an odd positive integer)

Properties 7 and 8 were given as Exercises 14 and 11 in Section 2.3. By using Property 4 repeatedly with $g(x) = f(x)$, we obtain Property 6. Property 9 then follows from Properties 6 and 8. Property 10 will follow from our work in Chapter 5 and Property 11 will be proved in Section 2.5.

● **Example 1**   Find $\lim\limits_{x \to -2} \dfrac{x^3 + 2x^2 - 1}{5 - 3x}$.

Solution

$$
\lim_{x \to -2} \frac{x^3 + 2x^2 - 1}{5 - 3x} = \frac{\lim\limits_{x \to -2} (x^3 + 2x^2 - 1)}{\lim\limits_{x \to -2} (5 - 3x)} \qquad \textit{(by Property 5)}
$$

$$
= \frac{\lim\limits_{x \to -2} x^3 + 2 \lim\limits_{x \to -2} x^2 - \lim\limits_{x \to -2} 1}{\lim\limits_{x \to -2} 5 - 3 \lim\limits_{x \to -2} x} \qquad \textit{(by 1, 2, and 3)}
$$

$$
= \frac{(-2)^3 + 2(-2)^2 - 1}{5 - 3(-2)} \qquad \textit{(by 9, 8, and 7)}
$$

$$
= -\frac{1}{11}
$$

● **Example 2**   Calculate $\lim\limits_{x \to 1}[\sqrt[5]{x^2 - x} + (x^3 + x)^9]$.

Solution

$$\lim_{x \to 1}\left[\sqrt[5]{x^2 - x} + (x^3 + x)^9\right]$$

$$= \lim_{x \to 1}\sqrt[5]{x^2 - x} + \lim_{x \to 1}(x^3 + x)^9 \qquad \textit{(by Property 1)}$$

$$= \sqrt[5]{\lim_{x \to 1}(x^2 - x)} + \left[\lim_{x \to 1}(x^3 + x)\right]^9 \qquad \textit{(by 11 and 6)}$$

$$= \sqrt[5]{\lim_{x \to 1}x^2 - \lim_{x \to 1}x} + \left[\lim_{x \to 1}x^3 + \lim_{x \to 1}x\right]^9 \qquad \textit{(by 2 and 1)}$$

$$= \sqrt[5]{1^2 - 1} + \left[1^3 + 1\right]^9 \qquad \textit{(by 9)}$$

$$= 2^9 = 512 \qquad\qquad\qquad\qquad\qquad\qquad \bullet$$

Note 1:   It would be extremely difficult to solve Examples 1 and 2 using the ε, δ definition of limit directly. The properties listed in (2.14) and (2.15) provide a simple method for computing limits rigorously.

Note 2:   If we let $f(x) = (x^3 + 2x^2 - 1)/(5 - 3x)$ we find that $f(-2) = -\frac{1}{11}$. In other words, we would have gotten the correct answer in Example 1 by substituting -2 for x. Similarly, direct substitution provides the correct answer in Example 2. Functions with this property, that is, $\lim_{x \to a} f(x) = f(a)$, are called *continuous at a* and will be studied in Section 2.5. However, not all limits can be evaluated by direct substitution, as we saw in Examples 2–6 in Section 2.2. The following example will also illustrate this fact.

● **Example 3**   Evaluate $\displaystyle\lim_{x \to 0} \frac{\sqrt[3]{1 + x} - 1}{x}$.

Solution   Using   the   identity $a^3 - 1 = (a - 1)(a^2 + a + 1)$ with $a = \sqrt[3]{1 + x}$ (see Appendix A, Equation 4), we have

$$\lim_{x \to 0} \frac{\sqrt[3]{1 + x} - 1}{x}$$

$$= \lim_{x \to 0} \frac{\sqrt[3]{1 + x} - 1}{(1 + x) - 1}$$

$$= \lim_{x \to 0} \frac{\sqrt[3]{1 + x} - 1}{(\sqrt[3]{1 + x} - 1)[(1 + x)^{2/3} + (1 + x)^{1/3} + 1]}$$

$$= \lim_{x \to 0} \frac{1}{(1 + x)^{2/3} + (1 + x)^{1/3} + 1}$$

$$= \frac{1}{\lim_{x \to 0}(1 + x)^{2/3} + \lim_{x \to 0}(1 + x)^{1/3} + 1} \qquad \textit{(by Properties 5, 1, and 7)}$$

$$= \frac{1}{1 + 1 + 1} \qquad \textit{(by 11 and 6)}$$

$$= \frac{1}{3}$$

The next two theorems give two additional properties of limits. Their proofs can be found in Appendix C.

Theorem (2.16)

If $f(x) \leqslant g(x)$ for all x in an open interval that contains a (except possibly at a) and the limits of f and g both exist as x approaches a, then

$$\lim_{x \to a} f(x) \leqslant \lim_{x \to a} g(x)$$

The Squeeze Theorem (2.17)

If $f(x) \leqslant g(x) \leqslant h(x)$ for all x in an open interval that contains a (except possibly at a) and

$$\lim_{x \to a} f(x) = \lim_{x \to a} h(x) = L$$

then $\lim_{x \to a} g(x) = L$

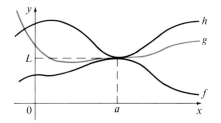

Figure 2.28

The Squeeze Theorem is illustrated by Figure 2.28. It says that if $g(x)$ is squeezed between $f(x)$ and $h(x)$ near a, and if f and h have the same limit L at a, then g is forced to have the same limit L at a.

● **Example 4**   Show that $\lim_{x \to 0} x \sin \dfrac{1}{x} = 0$.

Solution   First note that we *cannot* use

$$\lim_{x \to 0} x \sin \frac{1}{x} = \lim_{x \to 0} x \cdot \lim_{x \to 0} \sin \frac{1}{x}$$

because $\lim_{x \to 0} \sin(1/x)$ does not exist. (See Example 7 in Section 2.2.) Since

$$-1 \leqslant \sin \frac{1}{x} \leqslant 1$$

we have, as illustrated by Figure 2.29,

$$-|x| \leqslant x \sin \frac{1}{x} \leqslant |x|$$

We know that

$$\lim_{x \to 0} |x| = 0 \qquad \text{and} \qquad \lim_{x \to 0} -|x| = 0$$

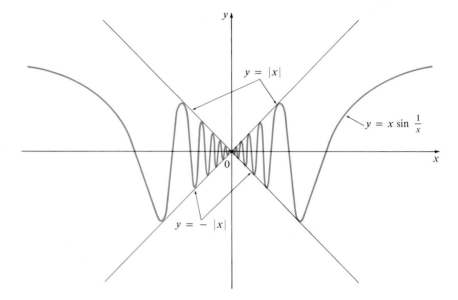

Figure 2.29

(see Exercise 17 in Section 2.3 or Example 5). Taking $f(x) = -|x|$, $g(x) = x \sin(1/x)$ and $h(x) = |x|$ in the Squeeze Theorem, we obtain

$$\lim_{x \to 0} x \sin \frac{1}{x} = 0 \qquad \bullet$$

Some limits are best calculated by first finding the left- and right-hand limits. The following theorem makes precise what we discovered in Section 2.2. It says that a two-sided limit exists if and only if both of the one-sided limits exist and are equal. (For the proof, see Exercise 24 in Section 2.3.)

Theorem (2.18)

$$\lim_{x \to a} f(x) = L \qquad \text{if and only if} \qquad \lim_{x \to a^-} f(x) = L = \lim_{x \to a^+} f(x)$$

When computing one-sided limits we use the fact that Properties 1–11 of limits in (2.14) and (2.15) also hold for one-sided limits.

● **Example 5**   Show that $\lim_{x \to 0} |x| = 0$.

Solution   Since

$$|x| = \begin{cases} x & \text{if } x \geqslant 0 \\ -x & \text{if } x < 0 \end{cases}$$

we have

$$\lim_{x \to 0^+} |x| = \lim_{x \to 0^+} x = 0$$

and

$$\lim_{x \to 0^-} |x| = \lim_{x \to 0^-} (-x) = 0$$

Therefore, by Theorem 2.18,

$$\lim_{x \to 0} |x| = 0$$

● **Example 6**   Prove that $\lim\limits_{x \to 0} \dfrac{|x|}{x}$ does not exist.

Solution

$$\lim_{x \to 0^+} \frac{|x|}{x} = \lim_{x \to 0^+} \frac{x}{x} = \lim_{x \to 0^+} 1 = 1$$

$$\lim_{x \to 0^-} \frac{|x|}{x} = \lim_{x \to 0^-} \frac{-x}{x} = \lim_{x \to 0^-} (-1) = -1$$

Since the right- and left-hand limits are different, it follows from Theorem 2.18 that $\lim_{x \to 0} |x|/x$ does not exist. The graph of the function $f(x) = |x|/x$ is shown in Figure 2.30.

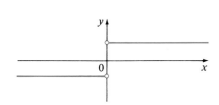

Figure 2.30

$y = |x|/x$

● **Example 7**   If $f(x) = \begin{cases} \sqrt{x - 4} & \text{if } x > 4 \\ 8 - 2x & \text{if } x < 4 \end{cases}$

determine whether $\lim_{x \to 4} f(x)$ exists.

Solution

$$\lim_{x \to 4^+} f(x) = \lim_{x \to 4^+} \sqrt{x - 4} = \sqrt{4 - 4} = 0$$

$$\lim_{x \to 4^-} f(x) = \lim_{x \to 4^-} (8 - 2x) = 8 - 2 \cdot 4 = 0$$

The right- and left-hand limits are equal. Thus the limit exists and

$$\lim_{x \to 4} f(x) = 0$$

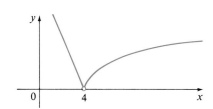

Figure 2.31   The graph of f is shown in Figure 2.31.

● **Example 8**   The *greatest integer function* is defined by $[\![x]\!]$ = the largest integer that is less than or equal to x. (For instance, $[\![4]\!] = 4$, $[\![4.8]\!] = 4$, $[\![\pi]\!] = 3$, $[\![\sqrt{2}]\!] = 1$, $[\![-\frac{1}{2}]\!] = -1$.) Show that $\lim_{x \to 3} [\![x]\!]$ does not exist.

Solution   The graph of the greatest integer function is shown in Figure 2.32. Since $[\![x]\!] = 3$ for $3 \leqslant x < 4$, we have

$$\lim_{x \to 3^+} [\![x]\!] = \lim_{x \to 3^+} 3 = 3$$

Since $[\![x]\!] = 2$ for $2 \leqslant x < 3$, we have

$$\lim_{x \to 3^-} [\![x]\!] = \lim_{x \to 3^-} 2 = 2$$

Because these one-sided limits are not equal, $\lim_{x \to 3} [\![x]\!]$ does not exist by Theorem 2.18.

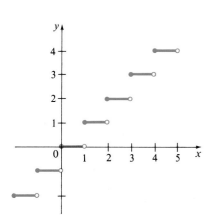

Figure 2.32

Note: If you have a calculator with an "integer part" key INT, note that $\text{INT}(x) = [\![x]\!]$ for $x \geqslant 0$, but be warned that, in some calculators, $\text{INT}(x) = -[\![-x]\!]$ for $x < 0$. For instance, these calculators would compute $\text{INT}(-3.6) = -3$, whereas $[\![-3.6]\!] = -4$.

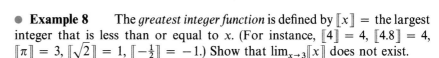

EXERCISES 2.4

In Exercises 1–14 evaluate the given limit and justify each step by indicating the appropriate properties of limits.

1. $\lim\limits_{x \to 4}(5x^2 - 2x + 3)$

2. $\lim\limits_{x \to -3}(x^3 + 2x^2 + 6)$

3. $\lim\limits_{x \to 2}(x^2 + 1)(x^2 + 4x)$

4. $\lim\limits_{x \to -2}(x^2 + x + 1)^5$

5. $\lim\limits_{x \to -1}\dfrac{x - 2}{x^2 + 4x - 3}$

6. $\lim\limits_{t \to -2}\dfrac{t^3 - t^2 - t + 10}{t^2 + 3t + 2}$

7. $\lim\limits_{x \to -1}\sqrt{x^3 + 2x + 7}$

8. $\lim\limits_{x \to 64}(\sqrt[3]{x} + 3\sqrt{x})$

9. $\lim\limits_{t \to -2}(t + 1)^9(t^2 - 1)$

10. $\lim\limits_{r \to 3}(r^4 - 7r + 4)^{2/3}$

11. $\lim\limits_{w \to -2}\sqrt[3]{\dfrac{4w + 3w^3}{3w + 10}}$

12. $\lim\limits_{y \to 3}\dfrac{3(8y^2 - 1)}{2y^2(y - 1)^4}$

13. $\lim\limits_{h \to 1/2}\dfrac{2h}{h + \dfrac{1}{h}}$

14. $\lim\limits_{t \to 16}t^{-1/2}(t^2 - 14t)^{3/5}$

15. Given that

$$\lim\limits_{x \to a}f(x) = -3 \qquad \lim\limits_{x \to a}g(x) = 0 \qquad \lim\limits_{x \to a}h(x) = 8$$

find the limits that exist.

(a) $\lim\limits_{x \to a}[f(x) + h(x)]$

(b) $\lim\limits_{x \to a}[f(x)]^2$

(c) $\lim\limits_{x \to a}\sqrt[3]{h(x)}$

(d) $\lim\limits_{x \to a}\dfrac{1}{f(x)}$

(e) $\lim\limits_{x \to a}\dfrac{f(x)}{h(x)}$

(f) $\lim\limits_{x \to a}\dfrac{g(x)}{f(x)}$

(g) $\lim\limits_{x \to a}\dfrac{f(x)}{g(x)}$

(h) $\lim\limits_{x \to a}\dfrac{2f(x)}{h(x) - f(x)}$

In Exercises 16–31 evaluate the limits, if they exist.

16. $\lim\limits_{x \to 3}\dfrac{x^2 - x + 12}{x + 3}$

17. $\lim\limits_{x \to -3}\dfrac{x^2 - x + 12}{x + 3}$

18. $\lim\limits_{x \to -3}\dfrac{x^2 - x - 12}{x + 3}$

19. $\lim\limits_{t \to 9}\dfrac{9 - t}{3 - \sqrt{t}}$

20. $\lim\limits_{x \to 2}\dfrac{x^4 - 16}{x - 2}$

21. $\lim\limits_{t \to 1}\dfrac{\sqrt[3]{t}}{t - 1}$

22. $\lim\limits_{t \to 1}\dfrac{\sqrt[3]{t} - 1}{t - 1}$

23. $\lim\limits_{x \to -2}\left[\dfrac{x^2}{x + 2} + \dfrac{2x}{x + 2}\right]$

24. $\lim\limits_{h \to 0}\dfrac{\sqrt[4]{1 + h} - 1}{h}$

25. $\lim\limits_{x \to -8}\dfrac{x + 8}{\sqrt[3]{x} + 2}$

26. $\lim\limits_{x \to 1}\left[\dfrac{1}{x - 1} - \dfrac{2}{x^2 - 1}\right]$

27. $\lim\limits_{x \to 9}\dfrac{x^2 - 81}{\sqrt{x} - 3}$

28. $\lim\limits_{h \to 0}\dfrac{(3 + h)^{-1} - 3^{-1}}{h}$

29. $\lim\limits_{t \to 0}\left[\dfrac{1}{t\sqrt{1 + t}} - \dfrac{1}{t}\right]$

30. $\lim\limits_{x \to 2}\dfrac{\sqrt{2x} - x}{x - 2}$

31. $\lim\limits_{x \to 0}\dfrac{x}{\sqrt{1 + 3x} - 1}$

32. If $1 \leqslant f(x) \leqslant x^2 + 2x + 2$ for all x, find $\lim\limits_{x \to -1}f(x)$.

33. If $3x \leqslant f(x) \leqslant x^3 + 2$ for $0 \leqslant x \leqslant 2$, evaluate $\lim\limits_{x \to 1}f(x)$.

34. Prove that $\lim\limits_{x \to 0}x^2 \sin\dfrac{1}{x} = 0$.

35. Prove that $\lim\limits_{x \to 0}\sqrt[3]{x}\sin\dfrac{1}{\sqrt[3]{x}} = 0$.

In Exercises 36–55, find the limit if it exists. If the limit does not exist, explain why. The symbol $[\![\ ]\!]$ denotes the greatest integer function, defined in Example 8.

36. $\lim\limits_{x \to 0^+}(\sqrt[4]{x} - 1)$

37. $\lim\limits_{x \to 5^+}(\sqrt{x - 5} + \sqrt{5x})$

38. $\lim\limits_{x \to 4^-}\sqrt{16 - x^2}$

39. $\lim\limits_{x \to -1.5^+}(\sqrt{3 + 2x} + x)$

40. $\lim\limits_{x \to -4}|x + 4|$

41. $\lim\limits_{x \to -4^-}\dfrac{|x + 4|}{x + 4}$

42. $\lim\limits_{x \to 2}\dfrac{|x - 2|}{x - 2}$

43. $\lim\limits_{x \to 2^+}\dfrac{1}{x - 2}$

44. $\lim\limits_{x \to 9^-}[\![x]\!]$

45. $\lim\limits_{x \to 9^+}[\![x]\!]$

46. $\lim\limits_{x \to -2^+}[\![x]\!]$

47. $\lim\limits_{x \to -2}[\![x]\!]$

48. $\lim\limits_{x \to -2.4}[\![x]\!]$

49. $\lim\limits_{x \to 8^+}(\sqrt{x - 8} + [\![x + 1]\!])$

50. $\lim\limits_{x \to 1^+}\sqrt{x^2 + x - 2}$

51. $\lim\limits_{x \to -2^-}\sqrt{x^2 + x - 2}$

52. $\lim\limits_{x \to 0^+}\dfrac{2 + x^{3/2}}{\sqrt{x + 2}}$

53. $\lim\limits_{x \to 5^+}x\sqrt[6]{x^2 - 25}$

54. $\lim\limits_{x \to 0^-} \left(\dfrac{1}{x} - \dfrac{1}{|x|} \right)$

55. $\lim\limits_{x \to 0^+} \left(\dfrac{1}{x} - \dfrac{1}{|x|} \right)$

56. The *signum* (or *sign*) *function*, denoted by sgn, is defined by

$$\operatorname{sgn} x = \begin{cases} -1 & \text{if } x < 0 \\ 0 & \text{if } x = 0 \\ 1 & \text{if } x > 0 \end{cases}$$

(a) Sketch the graph of this function.

(b) Find the following limits or explain why they do not exist.

   (i) $\lim\limits_{x \to 0^+} \operatorname{sgn} x$ (ii) $\lim\limits_{x \to 0^-} \operatorname{sgn} x$

   (iii) $\lim\limits_{x \to 0} \operatorname{sgn} x$ (iv) $\lim\limits_{x \to 0} |\operatorname{sgn} x|$

57. The Heaviside function H was defined in Example 9 in Section 2.2. Use Theorem 2.18 to show that $\lim_{t \to 0} H(t)$ does not exist.

58. Let

$$f(x) = \begin{cases} x^2 - 2x + 2 & \text{if } x < 1 \\ 3 - x & \text{if } x \geqslant 1 \end{cases}$$

(a) Find $\lim\limits_{x \to 1^-} f(x)$ and $\lim\limits_{x \to 1^+} f(x)$.

(b) Does $\lim\limits_{x \to 1} f(x)$ exist?

(c) Sketch the graph of f.

59. Let

$$g(x) = \begin{cases} -x^3 & \text{if } x < -1 \\ (x + 2)^2 & \text{if } x > -1 \end{cases}$$

(a) Find $\lim\limits_{x \to -1^-} g(x)$ and $\lim\limits_{x \to -1^+} g(x)$.

(b) Does $\lim\limits_{x \to -1} g(x)$ exist?

(c) Sketch the graph of g.

60. Let

$$h(x) = \begin{cases} x & \text{if } x < 0 \\ x^2 & \text{if } 0 < x \leqslant 2 \\ 8 - x & \text{if } x > 2 \end{cases}$$

(a) Evaluate the following limits if they exist.

   (i) $\lim\limits_{x \to 0^+} h(x)$ (ii) $\lim\limits_{x \to 0} h(x)$ (iii) $\lim\limits_{x \to 1} h(x)$

   (iv) $\lim\limits_{x \to 2^-} h(x)$ (v) $\lim\limits_{x \to 2^+} h(x)$ (vi) $\lim\limits_{x \to 2} h(x)$

(b) Sketch the graph of h.

61. (a) If n is an integer, evaluate

   (i) $\lim\limits_{x \to n^-} [\![x]\!]$ (ii) $\lim\limits_{x \to n^+} [\![x]\!]$

(b) For what values of a does $\lim\limits_{x \to a} [\![x]\!]$ exist?

62. Let $f(x) = x - [\![x]\!]$.

(a) Sketch the graph of f.

(b) If n is an integer, evaluate

   (i) $\lim\limits_{x \to n^-} f(x)$ (ii) $\lim\limits_{x \to n^+} f(x)$

(c) For what values of a does $\lim\limits_{x \to a} f(x)$ exist?

(d) If you have a calculator with a "fractional part" key FRAC, how is FRAC related to f? (See the note after Example 8.) Sketch the graphs of INT and FRAC.

63. Let $F(x) = \dfrac{x^2 - 1}{|x - 1|}$.

(a) Find

   (i) $\lim\limits_{x \to 1^+} F(x)$ (ii) $\lim\limits_{x \to 1^-} F(x)$

(b) Does $\lim\limits_{x \to 1} F(x)$ exist?

(c) Sketch the graph of F.

64. Let $g(x) = [\![x/2]\!]$.

(a) Sketch the graph of g.

(b) Evaluate the following limits if they exist.

   (i) $\lim\limits_{x \to 1^+} g(x)$ (ii) $\lim\limits_{x \to 1^-} g(x)$ (iii) $\lim\limits_{x \to 1} g(x)$

   (iv) $\lim\limits_{x \to 2^+} g(x)$ (v) $\lim\limits_{x \to 2^-} g(x)$ (vi) $\lim\limits_{x \to 2} g(x)$

(c) For what values of a does $\lim\limits_{x \to a} g(x)$ exist?

65. If

$$f(x) = \begin{cases} x^2 & \text{if } x \text{ is rational} \\ 0 & \text{if } x \text{ is irrational} \end{cases}$$

prove that $\lim\limits_{x \to 0} f(x) = 0$.

66. Show by means of an example that $\lim_{x \to a} [f(x) + g(x)]$ may exist even though neither $\lim_{x \to a} f(x)$ nor $\lim_{x \to a} g(x)$ exists.

67. Show by means of an example that $\lim_{x \to a} [f(x)g(x)]$ may exist even though neither $\lim_{x \to a} f(x)$ nor $\lim_{x \to a} g(x)$ exists.

68. Evaluate $\lim\limits_{x \to 0} \dfrac{\sqrt[3]{1 + cx} - 1}{x}$.

69. Evaluate $\lim\limits_{x \to 1} \dfrac{\sqrt[3]{x} - 1}{\sqrt{x} - 1}$.

Continuity

We have noticed that the limit of a function as x approaches a can often be found simply by calculating the value of the function at a (see Note 2 in Section 2.4). Functions with this property are called *continuous at a*. We shall see that the mathematical definition of continuity corresponds closely with the ordinary meaning of the word *continuity* in everyday language. (A continuous process is one that takes place gradually, without interruption or abrupt change.)

Definition (2.19)

> A function f is **continuous at a number a** if
> $$\lim_{x \to a} f(x) = f(a)$$

If f is not continuous at a, we say f is **discontinuous at a,** or f has a **discontinuity** at a.

Notice that Definition 2.19 implicitly requires three things if f is continuous at a:

1. $f(a)$ is defined (that is, a is in the domain of f).
2. $\lim_{x \to a} f(x)$ exists (so f must be defined on an open interval that contains a).
3. $\lim_{x \to a} f(x) = f(a)$

Intuitively speaking, f is continuous at a if $f(x)$ gets closer and closer to $f(a)$ as x gets closer and closer to a. Thus a continuous function f has the property that a small change in x produces only a small change in $f(x)$. In fact, the change in $f(x)$ can be kept as small as we please by keeping the change in x sufficiently small.

Physical phenomena are usually continuous. For instance, the displacement or velocity of a vehicle varies continuously with time, as does a person's height. But discontinuities do occur in such situations as electric currents. (See Example 9 in Section 2.2, where the Heaviside function is discontinuous at 0 because $\lim_{t \to 0} H(t)$ does not exist.)

Geometrically, you can think of a continuous function as a function whose graph has no breaks in it. The graph can be drawn without removing your pen from the paper.

●   **Example 1** $f(x) = \dfrac{x^2 - x - 2}{x - 2}$

This function is discontinuous at $a = 2$ since $f(2)$ is not defined.   ●

● **Example 2** $f(x) = \begin{cases} \dfrac{1}{x^2} & \text{if } x \neq 0 \\ 1 & \text{if } x = 0 \end{cases}$

This function is discontinuous at $a = 0$. Here $f(0) = 1$ is defined but

$$\lim_{x \to 0} f(x) = \lim_{x \to 0} \frac{1}{x^2}$$

does not exist. (See Example 8 in Section 2.2 or Example 2 in Section 2.3.)   ●

● **Example 3** $f(x) = \begin{cases} \dfrac{x^2 - x - 2}{x - 2} & \text{if } x \neq 2 \\ 1 & \text{if } x = 2 \end{cases}$

This function has a discontinuity at $a = 2$. Here $f(2) = 1$ is defined and

$$\lim_{x \to 2} f(x) = \lim_{x \to 2} \frac{x^2 - x - 2}{x - 2} = \lim_{x \to 2}(x + 1) = 3$$

exists. But

$$\lim_{x \to 2} f(x) \neq f(2)$$

so f is not continuous at 2.   ●

● **Example 4**   The greatest integer function $f(x) = [\![x]\!]$ has discontinuities at all of the integers because $\lim_{x \to n}[\![x]\!]$ does not exist if n is an integer. (See Example 8 in Section 2.4 and Exercise 61 in Section 2.4.)   ●

Figure 2.33(a)–(d) shows the graphs of the functions in Examples 1–4. In each case the graph cannot be drawn without lifting the pen from the paper because there is a hole or break or jump in the graph. The kind of discontinuity illustrated in parts (a) and (c) is called **removable** because we could remove the discontinuity by redefining f at 2. [The function $g(x) = x + 1$ is continuous.] The discontinuity in part (b) is called an **infinite discontinuity.** The discontinuities in part (d) are called **jump discontinuities** because the function "jumps" from one value to another.

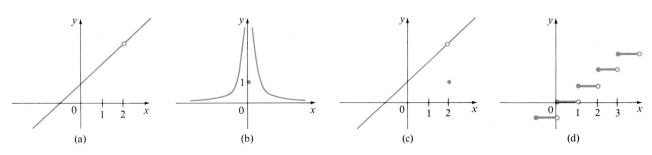

(a)   (b)   (c)   (d)

Figure 2.33

Definition (2.20)

A function f is **continuous from the right at a number** a if

$$\lim_{x \to a^+} f(x) = f(a)$$

and f is **continuous from the left at** a if

$$\lim_{x \to a^-} f(x) = f(a)$$

● **Example 5**   At each integer n, the function $f(x) = [\![x]\!]$ [see Figure 2.33(d)] is continuous from the right but discontinuous from the left because

$$\lim_{x \to n^+} f(x) = \lim_{x \to n^+} [\![x]\!] = n = f(n)$$

but

$$\lim_{x \to n^-} f(x) = \lim_{x \to n^-} [\![x]\!] = n - 1 \neq f(n)$$ ●

Definition (2.21)

A function f is **continuous on the open interval** (a, b) [or (a, ∞) or $(-\infty, b)$ or $(-\infty, \infty)$] if it is continuous at every number in the interval. It is **continuous on the closed interval** $[a, b]$ if it is continuous on (a, b) and is also continuous from the right at a and continuous from the left at b.

Similarly, one can define continuity of functions on other types of intervals such as $[a, b)$, $(a, b]$, $[a, \infty)$, and $(-\infty, b]$. For instance, f is continuous on $[a, b)$ if it is continuous on (a, b) and continuous from the right at a.

● **Example 6**   Show that the function $f(x) = 1 - \sqrt{1 - x^2}$ is continuous on the interval $[-1, 1]$.

Solution   If $-1 < a < 1$, then using the properties of limits, we have

$$\lim_{x \to a} f(x) = \lim_{x \to a}(1 - \sqrt{1 - x^2})$$

$$= 1 - \lim_{x \to a}\sqrt{1 - x^2} \qquad \text{(by Properties 2 and 7)}$$

$$= 1 - \sqrt{\lim_{x \to a}(1 - x^2)} \qquad \text{(by 11)}$$

$$= 1 - \sqrt{1 - a^2} \qquad \text{(by 2, 7, and 9)}$$

$$= f(a)$$

Thus, by Definition 2.19, f is continuous at a if $-1 < a < 1$. We must also calculate the right-hand limit at -1 and the left-hand limit at 1.

$$\lim_{x \to -1^+} f(x) = \lim_{x \to -1^+} (1 - \sqrt{1 - x^2})$$

$$= 1 - \sqrt{\lim_{x \to -1^+} (1 - x^2)} \quad (as\ above)$$

$$= 1 - \sqrt{1 - 1^2}$$

$$= 1 = f(-1)$$

So f is continuous from the right at -1. Similarly,

$$\lim_{x \to 1^-} f(x) = \lim_{x \to 1^-} (1 - \sqrt{1 - x^2})$$

$$= 1 - \sqrt{\lim_{x \to 1^-} (1 - x^2)} = 1 - 0 = 1 = f(1)$$

So f is continuous from the left at 1. Therefore, according to Definition 2.21, f is continuous on $[-1, 1]$.

The graph of f is sketched in Figure 2.34. It is the lower half of the circle $x^2 + (y - 1)^2 = 1$. ●

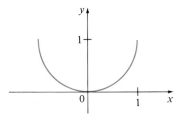

Figure 2.34

Instead of always using Definitions 2.19, 2.20, and 2.21 to verify the continuity of a function as we did in Example 6, it is often convenient to use the next theorem, which shows how to build up complicated continuous functions from simple ones.

Theorem (2.22)

If f and g are continuous at a and c is a constant, then the following functions are also continuous at a:
(a) $f + g$
(b) $f - g$
(c) cf
(d) fg
(e) $\dfrac{f}{g}$ if $g(a) \neq 0$

Proof   Each of the five parts of this theorem follows from the corresponding part of (2.14). For instance we give the proof of (a). Since f and g are continuous at a, we have

$$\lim_{x \to a} f(x) = f(a) \quad \text{and} \quad \lim_{x \to a} g(x) = g(a)$$

Therefore

$$\lim_{x \to a} (f + g)(x) = \lim_{x \to a} [f(x) + g(x)]$$

$$= \lim_{x \to a} f(x) + \lim_{x \to a} g(x) \quad (by\ Property\ 1)$$

$$= f(a) + g(a)$$

$$= (f + g)(a)$$

This shows that $f + g$ is continuous at a. ●

It follows from Theorem 2.22 and Definition 2.21 that if f and g are continuous on an interval, then so are the functions $f + g, f - g, cf, fg,$ and (if g is never 0) f/g.

Theorem (2.23)

(a)   Any polynomial is continuous everywhere; that is, it is continuous on $R = (-\infty, \infty)$.

(b)   Any rational function is continuous wherever it is defined; that is, it is continuous on its domain.

Proof   (a)   A polynomial is a function of the form

$$P(x) = c_n x^n + c_{n-1} x^{n-1} + \cdots + c_1 x + c_0$$

where $c_0, c_1, \ldots, c_n$ are constants. We know that

$$\lim_{x \to a} c_0 = c_0 \qquad \textit{(by Property 7)}$$

(2.24)   and

$$\lim_{x \to a} x^m = a^m \qquad m = 1, 2, \ldots, n \qquad \textit{(by 9)}$$

Equation 2.24 is precisely the statement that the function $f(x) = x^m$ is a continuous function. Thus, by Theorem 2.22(c), the function $g(x) = cx^m$ is continuous. Since P is a sum of functions of this form and a constant function, it follows from Theorem 2.22(a) that P is continuous.

(b)   A rational function is a function of the form

$$f(x) = \frac{P(x)}{Q(x)}$$

where P and Q are polynomials. The domain of f is $D = \{x \in R \mid Q(x) \neq 0\}$. We know from part (a) that P and Q are continuous everywhere. Thus, by Theorem 2.22(e), f is continuous at every number in D.   ●

As an illustration of Theorem 2.23, observe that the volume of a sphere varies continuously with its radius because the formula

$$V(r) = \tfrac{4}{3}\pi r^3$$

shows that V is a polynomial function of r. Likewise, if a ball is thrown vertically into the air with a velocity of 50 ft/s, then the height of the ball in feet after t seconds is given by the formula $h = 50t - 16t^2$. Again this is a polynomial function, so the height is a continuous function of the elapsed time.

Knowledge of which functions are continuous enables us to evaluate some limits very quickly, as the following example shows. Compare it with Example 1 in Section 2.4.

● **Example 7**   Find $\displaystyle\lim_{x \to -2} \frac{x^3 + 2x^2 - 1}{5 - 3x}$.

Solution   The function

$$f(x) = \frac{x^3 + 2x^2 - 1}{5 - 3x}$$

is rational, so by Theorem 2.23 it is continuous on its domain. Therefore

$$\lim_{x \to -2} \frac{x^3 + 2x^2 - 1}{5 - 3x} = \lim_{x \to -2} f(x) = f(-2)$$

$$= \frac{(-2)^3 + 2(-2)^2 - 1}{5 - 3(-2)} = -\frac{1}{11} \qquad \bullet$$

The following theorem is a consequence of Property 10 of limits.

Theorem (2.25)

> If n is a positive even integer, then $f(x) = \sqrt[n]{x}$ is continuous on $[0, \infty)$.
> If n is a positive odd integer, then f is continuous on $(-\infty, \infty)$.

● **Example 8**   On what intervals are the following functions continuous?

(a) $f(x) = x^{100} - 2x^{37} + 75$ (b) $g(x) = \dfrac{x^2 + 2x + 17}{x^2 - 1}$

(c) $h(x) = \sqrt{x} + \dfrac{x + 1}{x - 1} - \dfrac{x + 1}{x^2 + 1}$

Solution   (a) f is a polynomial, so it is continuous on $(-\infty, \infty)$ by Theorem 2.23(a).

(b) g is a rational function, so by Theorem 2.23(b) it is continuous on its domain, which is $D = \{x \mid x^2 - 1 \neq 0\} = \{x \mid x \neq \pm 1\}$. Thus g is continuous on the intervals $(-\infty, -1), (-1, 1)$, and $(1, \infty)$.

(c)   We can write $h(x) = F(x) + G(x) - H(x)$, where

$$F(x) = \sqrt{x} \qquad G(x) = \frac{x + 1}{x - 1} \qquad H(x) = \frac{x + 1}{x^2 + 1}$$

F is continuous on $[0, \infty)$ by Theorem 2.25. G is a rational function, so it is continuous everywhere except when $x - 1 = 0$, that is, $x = 1$. H is also a rational function, but its denominator is never 0, so H is continuous everywhere. Thus, by Theorem 2.22(a) and (b), h is continuous on the intervals $[0, 1)$ and $(1, \infty)$. $\qquad \bullet$

Another way of combining continuous functions f and g to get a new continuous function is to form the composite function $f \circ g$. This fact is a consequence of the following theorem.

Theorem (2.26)

> If f is continuous at b and $\lim_{x \to a} g(x) = b$, then
>
> $$\lim_{x \to a} f(g(x)) = f(b) = f\left(\lim_{x \to a} g(x)\right)$$

Intuitively this theorem is reasonable because if x is close to a, then $g(x)$ is close to b, and since f is continuous at b, if $g(x)$ is close to b, then $f(g(x))$ is close to $f(b)$. A proof of Theorem 2.26 is given in Appendix C.

Let us now apply Theorem 2.26 in the special case where $f(x) = \sqrt[n]{x}$, with n being a positive integer. Then

$$f(g(x)) = \sqrt[n]{g(x)}$$

and
$$f\left(\lim_{x \to a} g(x)\right) = \sqrt[n]{\lim_{x \to a} g(x)}$$

If we put these expressions into Theorem 2.26, we get

$$\lim_{x \to a} \sqrt[n]{g(x)} = \sqrt[n]{\lim_{x \to a} g(x)}$$

and so Property 11 of limits has now been proved. (We assume that the roots exist.)

Theorem (2.27)

> If g is continuous at a and f is continuous at $g(a)$, then $(f \circ g)(x) = f(g(x))$ is continuous at a.

This theorem is often expressed informally by saying "a continuous function of a continuous function is a continuous function."

Proof   Since g is continuous at a, we have

$$\lim_{x \to a} g(x) = g(a)$$

Since f is continuous at $b = g(a)$, we can apply Theorem 2.26 to obtain

$$\lim_{x \to a} f(g(x)) = f(g(a))$$

which is precisely the statement that the function $h(x) = f(g(x))$ is continuous at a; that is, $f \circ g$ is continuous at a.   ●

● **Example 9**   Where are the following functions continuous?

(a) $h(x) = |x|$ (b) $F(x) = 1/(\sqrt{x^2 + 7} - 4)$

Solution   (a)   Since $|x| = \sqrt{x^2}$ for all x, we have $h(x) = f(g(x))$, where

$$g(x) = x^2 \quad \text{and} \quad f(x) = \sqrt{x}$$

Now g is continuous on R since it is a polynomial and f is continuous on $[0, \infty)$ by Theorem 2.25. Thus $h = f \circ g$ is continuous on R by Theorem 2.27.

(b)   Notice that F can be broken up as the composition of four continuous functions:

$$F = f \circ g \circ h \circ k \quad \text{or} \quad F(x) = f(g(h(k(x))))$$

where $f(x) = \dfrac{1}{x}$ $g(x) = x - 4$ $h(x) = \sqrt{x}$ $k(x) = x^2 + 7$

We know that each of these functions is continuous on its domain (by Theorems 2.23 and 2.25), so by Theorem 2.27 F is continuous on its domain, which is

$$\{x \in R \mid \sqrt{x^2 + 7} \neq 4\} = \{x \mid x \neq \pm 3\} = (-\infty, -3) \cup (-3, 3) \cup (3, \infty) \quad \bullet$$

An important property of continuous functions is expressed by the following theorem. It is geometrically plausible but its proof requires sophisticated concepts and so is omitted.

The Intermediate Value Theorem (2.28)

> Suppose that f is continuous on the closed interval $[a, b]$ and let N be any number strictly between $f(a)$ and $f(b)$. Then there exists a number c in (a, b) such that $f(c) = N$.

The Intermediate Value Theorem states that a continuous function takes on every intermediate value between the function values $f(a)$ and $f(b)$. It is illustrated by Figure 2.35. Note that the value N can be taken on once [as in part (a)] or more than once [as in part (b)].

Figure 2.35

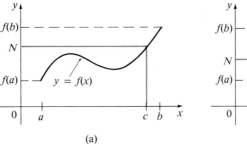

(a)   (b)

If we think of a continuous function as a function whose graph has no holes or breaks, then it is easy to believe that the Intermediate Value Theorem is true. In geometric terms it says that if any horizontal line $y = N$ is given between $y = f(a)$ and $y = f(b)$ as in Figure 2.36, then the graph of f cannot jump over the line. It must intersect $y = N$ somewhere.

It is important that the function f in the theorem be continuous. The Intermediate Value Theorem is not true in general for discontinuous functions (see Exercise 43).

One use of the Intermediate Value Theorem is in locating roots of equations as in the following example.

Figure 2.36

● **Example 10**   Show that there is a root of the equation

$$4x^3 - 6x^2 + 3x - 2 = 0$$

between 1 and 2.

Solution   Let $f(x) = 4x^3 - 6x^2 + 3x - 2$ and let $a = 1$ and $b = 2$. Then

$$f(1) = 4 - 6 + 3 - 2 = -1 < 0 \quad \text{and} \quad f(2) = 32 - 24 + 6 - 2 = 12 > 0$$

Thus $f(1) < 0 < f(2)$; that is, $N = 0$ is a number between $f(1)$ and $f(2)$. Now f is continuous since it is a polynomial, so the Intermediate Value Theorem says there is a number c between 1 and 2 such that $f(c) = 0$. In other words, the equation $4x^3 - 6x^2 + 3x - 2 = 0$ has a root c in the interval $(1, 2)$.

In fact, we can locate the root more precisely by using the Intermediate Value Theorem again. Since

$$f(1.2) = -0.128 < 0 \quad \text{and} \quad f(1.3) = 0.548 > 0$$

the root must lie between 1.2 and 1.3. A calculator gives, by trial and error,

$$f(1.22) = -0.007008 < 0 \quad \text{and} \quad f(1.23) = 0.056068 > 0$$

so the root lies in the interval $(1.22, 1.23)$.   ●

EXERCISES 2.5

1. (a) From the graph of f, state the numbers at which f is discontinuous.
    (b) For each of the numbers stated in part (a), state whether f is continuous from the right or from the left or neither.

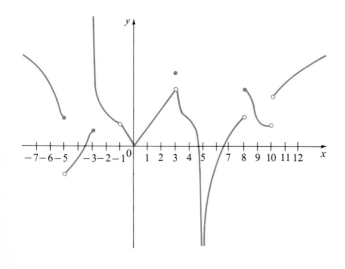

In Exercises 2–5 use the definition of continuity and the properties of limits to show that the given function is continuous at the given point.

2. $f(x) = x^2 + (x - 1)^9$, $a = 2$

3. $f(x) = 1 + \sqrt{x^2 - 9}$, $a = 5$

4. $g(x) = \dfrac{x + 1}{2x^2 - 1}$, $a = 4$

5. $g(t) = \dfrac{\sqrt[3]{t}}{(t + 1)^4}$, $a = -8$

In Exercises 6–9 use the definition of continuity and the properties of limits to show that the given function is continuous on the given interval.

6. $f(x) = x + \sqrt{x - 1}$, $[1, \infty)$

7. $f(x) = x\sqrt{16 - x^2}$, $[-4, 4]$

8. $F(x) = \dfrac{x + 1}{x - 3}$, $(-\infty, 3)$

9. $f(x) = (x^2 - 1)^8$, $(-\infty, \infty)$

In Exercises 10–18 explain why the given function is discontinuous at the given point. In each case sketch the graph of the function.

10. $f(x) = \dfrac{x^2 - 1}{x + 1}, a = -1$

11. $f(x) = \dfrac{3x^2 - 5x - 2}{x - 2}, a = 2$

12. $f(x) = \dfrac{1}{x - 1}, a = 1$

13. $f(x) = -\dfrac{1}{(x - 1)^2}, a = 1$

14. $f(x) = \begin{cases} \dfrac{x^2 - 1}{x + 1} & \text{if } x \neq -1 \\ 6 & \text{if } x = -1 \end{cases}$

$a = -1$

15. $f(x) = \begin{cases} -\dfrac{1}{(x - 1)^2} & \text{if } x \neq 1 \\ 0 & \text{if } x = 1 \end{cases}$

$a = 1$

16. $f(x) = \begin{cases} \dfrac{x^2 - 2x - 8}{x - 4} & \text{if } x \neq 4 \\ 3 & \text{if } x = 4 \end{cases}$

$a = 4$

17. $f(x) = \begin{cases} x^2 - 2 & \text{if } x \neq -3 \\ 5 & \text{if } x = -3 \end{cases}$

$a = -3$

18. $f(x) = \operatorname{sgn} x, a = 0$ (see Exercise 56 in Section 2.4)

In Exercises 19–30 use Theorems 2.22, 2.23, 2.25, and 2.27 to show that the given functions are continuous on their domains. State these domains.

19. $f(x) = (x + 1)(x^3 + 8x + 9)$

20. $g(x) = (x^6 - x^4 + 8x^2 - 7)^{10}$

21. $h(x) = \dfrac{x^2 + 2x - 1}{x + 1}$ **22.** $G(x) = \dfrac{x^4 + 17}{6x^2 + x - 1}$

23. $H(x) = \dfrac{1}{\sqrt{x + 1}}$ **24.** $f(t) = 2t + \sqrt{25 - t^2}$

25. $h(x) = \sqrt[5]{x - 1}(x^2 - 2)$ **26.** $g(t) = \dfrac{1}{t + \sqrt{t^2 - 4}}$

27. $F(t) = (t^2 + t + 1)^{3/2}$ **28.** $G(x) = (x^2 - 1)^{-5/2}$

29. $H(x) = \sqrt{\dfrac{x - 2}{5 + x}}$

30. $f(x) = \sqrt{x} + \sqrt[4]{x^2 - 3x + 2}$

31. Let
$$f(x) = \begin{cases} x - 1 & \text{for } x < 3 \\ 5 - x & \text{for } x \geqslant 3 \end{cases}$$
Show that f is continuous on $(-\infty, \infty)$.

32. Let
$$g(x) = \begin{cases} x & \text{if } x < 0 \\ x^2 & \text{if } 0 \leqslant x \leqslant 1 \\ x^3 & \text{if } x > 1 \end{cases}$$
Show that g is continuous on $(-\infty, \infty)$.

In Exercises 33–40 find the points at which f is discontinuous. At which of these points is f continuous from the right, continuous from the left, or neither? Sketch the graph of f.

33. $f(x) = \begin{cases} (x - 1)^3 & \text{if } x < 0 \\ (x + 1)^3 & \text{if } x \geqslant 0 \end{cases}$

34. $f(x) = \begin{cases} 2x + 1 & \text{if } x \leqslant -1 \\ 3x & \text{if } -1 < x < 1 \\ 2x - 1 & \text{if } x \geqslant 1 \end{cases}$

35. $f(x) = \begin{cases} \dfrac{1}{x} & \text{if } x < -1 \\ x & \text{if } -1 \leqslant x \leqslant 1 \\ \dfrac{1}{x^2} & \text{if } x > 1 \end{cases}$

36. $f(x) = \begin{cases} \sqrt{-x} & \text{if } x < 0 \\ 1 & \text{if } 0 < x \leqslant 1 \\ \sqrt{x} & \text{if } x > 1 \end{cases}$

37. $f(x) = H(x) - H(x - 1)$ (see Example 9 in Section 2.2)

38. $f(x) = [\![x]\!] - x$ (see Example 8 in Section 2.4)

39. $f(x) = [\![2x]\!]$ **40.** $f(x) = [\![1/x]\!]$

41. For what value of the constant c is the function
$$f(x) = \begin{cases} cx + 1 & \text{if } x \leqslant 3 \\ cx^2 - 1 & \text{if } x > 3 \end{cases}$$
continuous on $(-\infty, \infty)$?

42. Which of the following functions f has a removable discontinuity at a? If the discontinuity is removable, find a function g that agrees with f for $x \neq a$ and is continuous on R.

(a) $f(x) = \dfrac{x^2 - 2x - 8}{x + 2}$, $a = -2$

(b) $f(x) = \dfrac{x - 7}{|x - 7|}$, $a = 7$

(c) $f(x) = \dfrac{x^3 + 64}{x + 4}$, $a = -4$

(d) $f(x) = \dfrac{3 - \sqrt{x}}{9 - x}$, $a = 9$

43. Let

$$f(x) = \begin{cases} 1 - x^2 & \text{if } 0 \leqslant x \leqslant 1 \\ 1 + \dfrac{x}{2} & \text{if } 1 < x \leqslant 2 \end{cases}$$

(a) Show that f is not continuous on $[0, 2]$.
(b) Show that f does not take on all values between $f(0)$ and $f(2)$.

44. Use the Intermediate Value Theorem to prove that there is a positive number c such that $c^2 = 2$. (This proves the existence of the number $\sqrt{2}$.)

45. If $f(x) = x^3 - x^2 + x$, show that there is a number c such that $f(c) = 10$.

46. If $g(x) = x^5 - 2x^3 + x^2 + 2$, show that there is a number c such that $g(c) = -1$.

In Exercises 47–52 use the Intermediate Value Theorem to show that there is a root of the given equation in the given interval.

47. $x^3 - 3x + 1 = 0$, $(0, 1)$

48. $x^3 - 3x + 1 = 0$, $(1, 2)$

49. $x^4 - 3x^3 - 2x^2 - 1 = 0$, $(3, 4)$

50. $x^5 - 2x^4 - x - 3 = 0$, $(2, 3)$

51. $x^3 + 2x = x^2 + 1$, $(0, 1)$

52. $x^2 = \sqrt{x + 1}$, $(1, 2)$

In Exercises 53 and 54, (a) prove that the given equation has at least one real root, and (b) use your calculator to find an interval of length 0.01 containing a root.

53. $x^3 - x + 1 = 0$ **54.** $x^5 - x^2 + 2x + 3 = 0$

55. For what values of x is the function

$$f(x) = \begin{cases} 0 & \text{if } x \text{ is rational} \\ 1 & \text{if } x \text{ is irrational} \end{cases}$$

continuous?

56. For what values of x is the function

$$g(x) = \begin{cases} 0 & \text{if } x \text{ is rational} \\ x & \text{if } x \text{ is irrational} \end{cases}$$

continuous?

57. Prove that f is continuous at a if and only if

$$\lim_{h \to 0} f(a + h) = f(a)$$

58. (a) Prove Theorem 2.22(c).
   (b) Prove Theorem 2.22(e).

REVIEW OF CHAPTER 2

Define, state, or discuss the following.

1. Tangent
2. Velocity
3. Rate of change
4. Limit of $f(x)$ as x approaches a
5. One-sided limits
6. Properties of limits
7. The Squeeze Theorem
8. Continuous at a number a
9. Discontinuities of a function
10. Continuous on an interval
11. Intermediate Value Theorem

REVIEW EXERCISES FOR CHAPTER 2

1. (a) Find the slope of the tangent line to the curve $y = 9 - 2x^2$ at the point $(2, 1)$.
   (b) Find the equation of this tangent line.

2. The displacement (in meters) of an object moving in a straight line is given by $s = 1 + 2t + t^2/4$, where t is measured in seconds.

(a) Find the average velocity over the following time periods.

(i) $[1, 3]$ (ii) $[1, 2]$
(iii) $[1, 1.5]$ (iv) $[1, 1.1]$

(b) Find the instantaneous velocity when $t = 1$.

In Exercises 3–14 find the given limits.

3. $\lim\limits_{x \to 4} \sqrt{x + \sqrt{x}}$

4. $\lim\limits_{x \to 0^-} \sqrt{-x}$

5. $\lim\limits_{t \to -1} \dfrac{t + 1}{t^3 - t}$

6. $\lim\limits_{t \to 4} \dfrac{t - 4}{t^2 - 3t - 4}$

7. $\lim\limits_{h \to 0} \dfrac{(1 + h)^2 - 1}{h}$

8. $\lim\limits_{h \to 0} \dfrac{(1 + h)^{-2} - 1}{h}$

9. $\lim\limits_{x \to -1} \dfrac{x^2 - x - 2}{x^2 + 3x - 2}$

10. $\lim\limits_{x \to -1} \dfrac{x^2 - x - 2}{x^2 + 3x + 2}$

11. $\lim\limits_{s \to 16} \dfrac{4 - \sqrt{s}}{s - 16}$

12. $\lim\limits_{v \to 2} \dfrac{v^2 + 2v - 8}{v^4 - 16}$

13. $\lim\limits_{x \to 8^-} \dfrac{|x - 8|}{x - 8}$

14. $\lim\limits_{x \to 9^+} (\sqrt{x - 9} + [\![x + 1]\!])$

15. Prove that $\lim_{x \to 5}(7x - 27) = 8$ from the definition of a limit.

16. If $2x - 1 \leqslant f(x) \leqslant x^2$ for $0 < x < 3$, find $\lim_{x \to 1} f(x)$.

17. Let

$$f(x) = \begin{cases} \sqrt{-x} & \text{if } x < 0 \\ 3 - x & \text{if } 0 \leqslant x < 3 \\ (x - 3)^2 & \text{if } x > 3 \end{cases}$$

(a) Evaluate the following limits if they exist.

(i) $\lim\limits_{x \to 0^+} f(x)$ (ii) $\lim\limits_{x \to 0^-} f(x)$ (iii) $\lim\limits_{x \to 0} f(x)$

(iv) $\lim\limits_{x \to 3^-} f(x)$ (v) $\lim\limits_{x \to 3^+} f(x)$ (vi) $\lim\limits_{x \to 3} f(x)$

(b) Where is f discontinuous?

(c) Sketch the graph of f.

18. Let

$$g(x) = \begin{cases} 2x - x^2 & \text{if } 0 \leqslant x \leqslant 2 \\ 2 - x & \text{if } 2 < x \leqslant 3 \\ x - 4 & \text{if } 3 < x < 4 \\ \pi & \text{if } x \geqslant 4 \end{cases}$$

(a) For each of the numbers 2, 3, and 4, discover whether g is continuous from the left, continuous from the right, or continuous, at the number.

(b) Sketch the graph of g.

Show that the functions in Exercises 19 and 20 are continuous on their domains. State these domains.

19. $f(x) = \dfrac{x + 1}{x^2 + x + 1}$

20. $g(x) = \dfrac{\sqrt{x^2 - 9}}{x^2 - 2}$

In Exercises 21 and 22 use the Intermediate Value Theorem to show that there is a root of the given equation in the given interval.

21. $2x^3 + x^2 + 2 = 0$, $(-2, -1)$

22. $x^4 + 1 = \dfrac{1}{x}$, $(0.5, 1)$

3 Derivatives

Mathematics compares the most
diverse phenomena and discovers the
secret analogies that unite them.
Joseph Fourier

In this chapter we begin our study of differential calculus, which is concerned with how one quantity changes in relation to another quantity. The central concept of differential calculus is the *derivative*. After learning how to calculate derivatives, we use them to solve problems involving rates of change.

SECTION 3.1

Derivatives

In Section 2.1 we saw that the slope of the tangent to a curve with equation $y = f(x)$ at the point where $x = a$ is

(3.1)
$$m = \lim_{h \to 0} \frac{f(a + h) - f(a)}{h}$$

We also saw that the velocity of an object with position function $s = f(t)$ at time $t = a$ is

$$v(a) = \lim_{h \to 0} \frac{f(a + h) - f(a)}{h}$$

In fact, limits of the form

$$\lim_{h \to 0} \frac{f(a + h) - f(a)}{h}$$

arise whenever we calculate a rate of change in any of the sciences or engineering, such as a rate of reaction in chemistry or a marginal cost in economics. Since this type of limit occurs so widely, it is given a special name and notation.

Definition (3.2)

> The **derivative of a function f at a number a,** denoted by $f'(a)$, is
>
> $$f'(a) = \lim_{h \to 0} \frac{f(a + h) - f(a)}{h}$$
>
> if this limit exists.

If we write $x = a + h$, then $h = x - a$ and h approaches 0 if and only if x approaches a. Therefore an equivalent way of stating the definition of derivative is

(3.3)
$$f'(a) = \lim_{x \to a} \frac{f(x) - f(a)}{x - a}$$

● **Example 1**   Find the derivative of the function $f(x) = x^2 - 8x + 9$ at the number a.

Solution   From Definition 3.2 we have

$$f'(a) = \lim_{h \to 0} \frac{f(a + h) - f(a)}{h}$$

$$= \lim_{h \to 0} \frac{[(a + h)^2 - 8(a + h) + 9] - [a^2 - 8a + 9]}{h}$$

$$= \lim_{h \to 0} \frac{a^2 + 2ah + h^2 - 8a - 8h + 9 - a^2 + 8a - 9}{h}$$

$$= \lim_{h \to 0} \frac{2ah + h^2 - 8h}{h} = \lim_{h \to 0}(2a + h - 8)$$

$$= 2a - 8 \qquad\qquad\qquad\qquad\bullet$$

Interpretations of the Derivative

I. The Interpretation of the Derivative As the Slope of a Tangent.   In Section 2.1 we defined the tangent line to the curve $y = f(x)$ at the point $P(a, f(a))$ to be the line that passes through P and has slope m given by Equation 3.1. Since, by Definition 3.2, this is the same as the derivative $f'(a)$, we can now say that *the tangent line to $y = f(x)$ at $(a, f(a))$ is the line through $(a, f(a))$ whose slope is equal to $f'(a)$, the derivative of f at a.* Thus the geometric interpretation of a derivative [as defined by either (3.2) or (3.3)] is as shown in Figure 3.1.

Figure 3.1

GEOMETRIC INTERPRETATION OF
THE DERIVATIVE

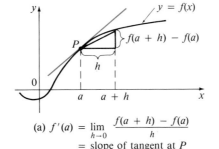

(a) $f'(a) = \lim\limits_{h \to 0} \dfrac{f(a + h) - f(a)}{h}$
    $= $ slope of tangent at P

(b) $f'(a) = \lim\limits_{x \to a} \dfrac{f(x) - f(a)}{x - a}$
    $= $ slope of tangent at P

Using the point-slope form of the equation of a line (Equation 1.10), we have the following:

(3.4)

If $f'(a)$ exists, then the equation of the tangent line to the curve $y = f(x)$ at the point $(a, f(a))$ is

$$y - f(a) = f'(a)(x - a)$$

● **Example 2**   Find the equation of the tangent line to the hyperbola $y = 3/x$ at the point $(3, 1)$.

Solution   Let $f(x) = 3/x$. Then the slope of the tangent at $(3, 1)$ is

$$f'(3) = \lim_{h \to 0} \frac{f(3 + h) - f(3)}{h}$$

$$= \lim_{h \to 0} \frac{\dfrac{3}{3 + h} - 1}{h} = \lim_{h \to 0} \frac{\dfrac{3 - (3 + h)}{3 + h}}{h}$$

$$= \lim_{h \to 0} \frac{-h}{h(3 + h)} = \lim_{h \to 0} -\frac{1}{3 + h}$$

$$= -\frac{1}{3}$$

Therefore the equation of the tangent at the point $(3, 1)$ is

$$y - 1 = -\tfrac{1}{3}(x - 3)$$

which simplifies to $x + 3y - 6 = 0$

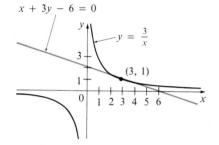

$x + 3y - 6 = 0$

$y = \dfrac{3}{x}$

$(3, 1)$

Figure 3.2   The hyperbola and its tangent are shown in Figure 3.2.   ●

II. The Interpretation of the Derivative As a Rate of Change. In Section 2.1 we defined the instantaneous rate of change of $y = f(x)$ with respect to x at $x = x_1$ as the limit of the average rate of change over smaller and smaller intervals. If the interval is $[x_1, x_2]$, then the change in x is $\Delta x = x_2 - x_1$, the corresponding change in y is

$$\Delta y = f(x_2) - f(x_1)$$

and

(3.5)   instantaneous rate of change $= \lim\limits_{\Delta x \to 0} \dfrac{\Delta y}{\Delta x} = \lim\limits_{x_2 \to x_1} \dfrac{f(x_2) - f(x_1)}{x_2 - x_1}$

From Equation 3.3 we recognize this limit as being the derivative of f at x_1, that is, $f'(x_1)$.

    This gives a second interpretation of the *derivative $f'(a)$ as the instantaneous rate of change of $y = f(x)$ with respect to x when $x = a$*. [The connection with the first interpretation is that if we sketch the curve $y = f(x)$, then the instantaneous rate of change is the slope of the tangent to this curve at the point where $x = a$.]

    In particular, if $s = f(t)$ is the position function of a particle that moves along a straight line, then $f'(a)$ is the rate of change of the displacement s with respect to the time t. In other words, $f'(a)$ is the *velocity of the particle at time $t = a$*. (See Section 2.1.) The speed of the particle is the absolute value of the velocity, that is, $|f'(a)|$.

● **Example 3**   The position of a particle is given by the equation

$$s = f(t) = t^3 - 6t^2 + 9t$$

where t is measured in seconds and s in meters. (a) Find the velocity at time $t = a$. (b) What is the velocity after 2 seconds? after 4 seconds? (c) When is the particle at rest? (d) When is the particle moving in the positive direction?

Solution   (a)   The velocity at time $t = a$ is

$$v(a) = f'(a) = \lim_{h \to 0} \frac{f(a + h) - f(a)}{h}$$

$$= \lim_{h \to 0} \frac{[(a + h)^3 - 6(a + h)^2 + 9(a + h)] - [a^3 - 6a^2 + 9a]}{h}$$

(See Equation 8 in Appendix A for the expansion of $(a + h)^3$.)

$$= \lim_{h \to 0} \frac{a^3 + 3a^2h + 3ah^2 + h^3 - 6a^2 - 12ah - 6h^2 + 9a + 9h - a^3 + 6a^2 - 9a}{h}$$

$$= \lim_{h \to 0} \frac{(3a^2 - 12a + 9)h + (3a - 6)h^2 + h^3}{h}$$

$$= \lim_{h \to 0} [3a^2 - 12a + 9 + (3a - 6)h + h^2]$$

$$= 3a^2 - 12a + 9$$

(b)   The velocity after 2 s means the instantaneous velocity when $t = 2$, that is,

$$v(2) = 3(2)^2 - 12(2) + 9 = -3 \text{ m/s}$$

The velocity after 4 s is

$$v(4) = 3(4)^2 - 12(4) + 9 = 9 \text{ m/s}$$

(c)   The particle is at rest when $v(a) = 0$, that is,

$$3a^2 - 12a + 9 = 3(a^2 - 4a + 3) = 3(a - 1)(a - 3) = 0$$

and this is true when $a = 1$ or $a = 3$. Thus the particle is at rest after 1 s and after 3 s.

(d)   The particle moves in the positive direction when $v(a) > 0$, that is,

$$3a^2 - 12a + 9 = 3(a - 1)(a - 3) > 0$$

This inequality is true when both factors are positive ($a > 3$) or when both factors are negative ($a < 1$). Thus the particle moves in the positive direction in the time intervals $t < 1$ and $t > 3$. It moves in the negative direction when $1 < t < 3$. The motion of the particle is illustrated schematically in Figure 3.3.

Figure 3.3

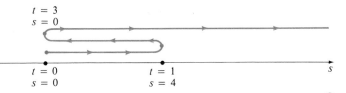

The Derivative As a Function

If we replace a by x in Definition 3.2, we obtain

(3.6)

$$f'(x) = \lim_{h \to 0} \frac{f(x + h) - f(x)}{h}$$

Given a function f, we associate with it a new function f', called the **derivative of f**, defined by Equation 3.6. We know that the value of f' at x, $f'(x)$, can be interpreted geometrically as the slope of the tangent line to the graph of f at the point $(x, f(x))$.

The function f' is called the derivative of f because it has been "derived" from f by the limiting operation in Equation 3.6. The domain of f' is the set $\{x \mid f'(x) \text{ exists}\}$ and may be smaller than the domain of f.

● **Example 4**   If $f(x) = \sqrt{x - 5}$ for $x \geq 5$, find the derivative of f. What is the domain of f'?

Solution   When using Equation 3.6 to compute a derivative we must remember that the variable is h and that x is temporarily regarded as a constant during the calculation of the limit.

$$f'(x) = \lim_{h \to 0} \frac{f(x + h) - f(x)}{h}$$

$$= \lim_{h \to 0} \frac{\sqrt{x + h - 5} - \sqrt{x - 5}}{h}$$

$$= \lim_{h \to 0} \frac{\sqrt{x + h - 5} - \sqrt{x - 5}}{h} \cdot \frac{\sqrt{x + h - 5} + \sqrt{x - 5}}{\sqrt{x + h - 5} + \sqrt{x - 5}}$$

$$= \lim_{h \to 0} \frac{(x + h - 5) - (x - 5)}{h(\sqrt{x + h - 5} + \sqrt{x - 5})}$$

$$= \lim_{h \to 0} \frac{1}{\sqrt{x + h - 5} + \sqrt{x - 5}}$$

$$= \frac{1}{\sqrt{x - 5} + \sqrt{x - 5}} = \frac{1}{2\sqrt{x - 5}}$$

We see that $f'(x)$ exists if $x > 5$. Thus the domain of f' is $(5, \infty)$, which is smaller than the domain of f.   ●

● **Example 5**   Find f' if $f(x) = \dfrac{1 - x}{2 + x}$

Solution

$$f'(x) = \lim_{h \to 0} \frac{f(x + h) - f(x)}{h}$$

$$= \lim_{h \to 0} \frac{\dfrac{1 - (x + h)}{2 + (x + h)} - \dfrac{1 - x}{2 + x}}{h}$$

$$= \lim_{h \to 0} \frac{(1 - x - h)(2 + x) - (1 - x)(2 + x + h)}{h(2 + x + h)(2 + x)}$$

$$= \lim_{h \to 0} \frac{(2 - x - 2h - x^2 - xh) - (2 - x + h - x^2 - xh)}{h(2 + x + h)(2 + x)}$$

$$= \lim_{h \to 0} \frac{-3h}{h(2 + x + h)(2 + x)}$$

$$= \lim_{h \to 0} \frac{-3}{(2 + x + h)(2 + x)} = -\frac{3}{(2 + x)^2}$$

Other Notations

If we use the traditional notation $y = f(x)$ to indicate that the independent variable is x and the dependent variable is y, then some common alternative notations for the derivative are as follows:

$$f'(x) = y' = \frac{dy}{dx} = \frac{df}{dx} = \frac{d}{dx} f(x) = Df(x) = D_x f(x)$$

The symbols D and d/dx are called **differentiation operators** because they indicate the operation of **differentiation,** which is the process of calculating a derivative.

The symbol dy/dx, which was introduced by Leibniz, should not be regarded as a ratio (for the time being); it is simply a synonym for $f'(x)$. Nonetheless, it is a very useful and suggestive notation, especially when used in conjunction with increment notation. Referring to Equation 3.5, we can rewrite the definition of derivative in Leibniz notation in the form

$$\frac{dy}{dx} = \lim_{\Delta x \to 0} \frac{\Delta y}{\Delta x}$$

If we want to indicate the value of a derivative dy/dx in Leibniz notation at a specific number a, we use the notation

$$\frac{dy}{dx}\Bigg]_{x=a}$$

which is a synonym for $f'(a)$.

Definition (3.7)

> A function f is **differentiable at a** if $f'(a)$ exists. It is **differentiable on an open interval** (a, b) [or (a, ∞) or $(-\infty, a)$ or $(-\infty, \infty)$] if it is differentiable at every number in the interval.

● **Example 6**   Where is the function $f(x) = |x|$ differentiable?

Solution   If $x > 0$, then $|x| = x$ and we can choose h small enough that $x + h > 0$ and hence $|x + h| = x + h$. Therefore for $x > 0$ we have

$$f'(x) = \lim_{h \to 0} \frac{|x + h| - |x|}{h}$$

$$= \lim_{h \to 0} \frac{(x + h) - x}{h} = \lim_{h \to 0} \frac{h}{h} = \lim_{h \to 0} 1 = 1$$

Similarly, for $x < 0$ we have $|x| = -x$ and h can be chosen small enough that $|x + h| = -(x + h)$. Therefore

$$f'(x) = \lim_{h \to 0} \frac{|x + h| - |x|}{h}$$

$$= \lim_{h \to 0} \frac{-(x + h) - (-x)}{h} = \lim_{h \to 0} \frac{-h}{h} = \lim_{h \to 0}(-1) = -1$$

For $x = 0$ we have to investigate

$$f'(0) = \lim_{h \to 0} \frac{f(0 + h) - f(0)}{h} = \lim_{h \to 0} \frac{|0 + h| - |0|}{h} \qquad \text{(if it exists)}$$

Let us compute the left and right limits separately:

$$\lim_{h \to 0^+} \frac{|0 + h| - |0|}{h} = \lim_{h \to 0^+} \frac{|h|}{h} = \lim_{h \to 0^+} \frac{h}{h} = \lim_{h \to 0^+} 1 = 1$$

and $$\lim_{h \to 0^-} \frac{|0 + h| - |0|}{h} = \lim_{h \to 0^-} \frac{|h|}{h} = \lim_{h \to 0^-} \frac{-h}{h} = -1$$

Since these limits are different, $f'(0)$ does not exist. Thus f is not differentiable at 0.

A formula for f' is given by

$$f'(x) = \begin{cases} 1 & \text{if } x > 0 \\ -1 & \text{if } x < 0 \end{cases}$$

and its graph is shown in Figure 3.4(b). The fact that $f'(0)$ does not exist is reflected geometrically in the fact that the curve $y = |x|$ does not have a tangent line at $(0, 0)$.   ●

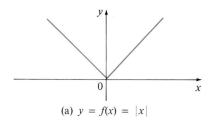

(a) $y = f(x) = |x|$

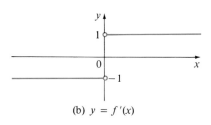

(b) $y = f'(x)$

Figure 3.4

In general if the graph of a function f has corners or kinks in it, then the graph has no tangent at those points and f is not differentiable there. If, on the other hand, f is differentiable on an interval, we expect its graph to be "smooth."

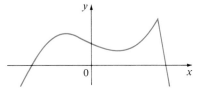

Figure 3.5

● **Example 7**   Sketch the graph of f' from the graph of f given in Figure 3.5.

Solution   Using the interpretation of $f'(x)$ as the slope of the tangent line at $(x, f(x))$, and remembering that a horizontal line has slope 0 and there is no tangent at a corner point, we can sketch the graph of f' directly beneath the graph of f (see Figure 3.6).

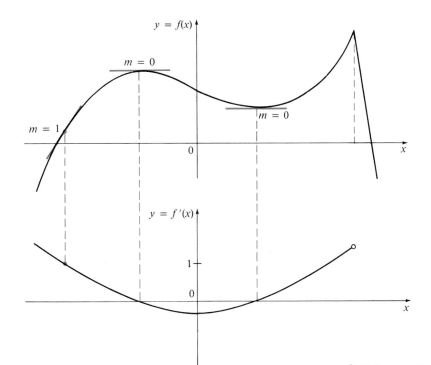

Figure 3.6

Theorem (3.8)   If f is differentiable at a, then f is continuous at a.

Proof   If $x \neq a$ and x is in the domain of f, then we can write

$$f(x) = f(a) + \frac{f(x) - f(a)}{x - a}(x - a)$$

Thus, using the properties of limits, we have

$$\lim_{x \to a} f(x) = \lim_{x \to a}\left[f(a) + \frac{f(x) - f(a)}{x - a}(x - a)\right]$$

$$= \lim_{x \to a} f(a) + \lim_{x \to a}\frac{f(x) - f(a)}{x - a} \cdot \lim_{x \to a}(x - a)$$

$$= f(a) + f'(a) \cdot 0 \qquad\qquad (by\ Equation\ 3.3)$$

$$= f(a)$$

So, by Definition 2.19, f is continuous at a.   ●

Note: The converse of Theorem 3.8 is false; that is, there are functions that are continuous but not differentiable. For instance, the function $f(x) = |x|$ is continuous at 0 because

$$\lim_{x \to 0} f(x) = \lim_{x \to 0}|x| = 0 = f(0)$$

(See Example 5 in Section 2.4.) But in Example 6 we showed that f is not differentiable at 0.

EXERCISES 3.1

In Exercises 1–4 find the slope and the equation of the tangent line to the given curve at the given point. Sketch the curve and the tangent line.

1. $y = x^2 + 2x,\ (-3, 3)$

2. $y = \dfrac{1}{x},\ \left(2, \dfrac{1}{2}\right)$

3. $y = \sqrt{x + 7},\ (2, 3)$

4. $y = 1 - x^3,\ (0, 1)$

In Exercises 5–8 find the equation of the tangent line to the graph of the given function at the given point.

5. $f(x) = 1 - 2x - 3x^2,\ (-2, -7)$

6. $f(x) = x - x^3,\ (2, -6)$

7. $g(x) = \dfrac{2}{x + 3},\ \left(1, \dfrac{1}{2}\right)$

8. $h(x) = \dfrac{1}{\sqrt{x}},\ (1, 1)$

In Exercises 9–12 a particle moves along a straight line with equation of motion $s = f(t)$, where s is measured in meters and t in seconds. Find the average velocities of the particle over the given time intervals and the instantaneous velocity when $t = a$.

9. $f(t) = t^2 - 4t;\ [5, 6],\ [5, 5.1],\ [5, 5.01];\ a = 5$

10. $f(t) = 2 - t + 3t^2;\ [2, 2.5],\ [2, 2.1],\ [2, 2.01];\ a = 2$

11. $f(t) = t^3;\ [1, 1.1],\ [1, 1.01],\ [1, 1.001],\ [1, 1 + h];\ a = 1$

12. $f(t) = 2t^3 - t + 1;\ [1, 1.1],\ [1, 1.01],\ [1, 1.001],\ [1, 1 + h];\ a = 1$

13. If a ball is dropped from the top of the CN Tower in Toronto, 550 meters above the ground, then its height in meters after t seconds is $h = 550 - 4.9t^2$. Find its speed (a) after 1 s, (b) after 2 s, (c) after 5 s, and (d) when it hits the ground.

14. If a ball is thrown directly upward from the ground with an initial velocity of 80 ft/s, then its height after t seconds, in feet, is $y = 80t - 16t^2$.
(a)   Find its velocity after 1 s, 2 s, 3 s, and 4 s.
(b)   Find its speed after 1 s, 2 s, 3 s, and 4 s.

TM

(c)   When does the ball reach its maximum height?
(d)   What is its maximum height?
(e)   When does it hit the ground?
(f)   With what velocity does it hit the ground?

In Exercises 15–18 a particle moves according to a law of motion $s = f(t)$ where t is measured in seconds and s in feet. (a) Find the velocity at times $t = a$, $t = 1$, and $t = 2$. (b) When is the particle at rest? (c) When is the particle moving in the positive direction? (d) Find the total distance traveled during the first four seconds. (e) Draw a diagram like Figure 3.3 to illustrate the motion of the particle.

15. $f(t) = t^2 - 6t + 9$

16. $f(t) = 4t^3 - 9t^2 + 6t + 2$

17. $f(t) = 2t^3 - 9t^2 + 12t + 1$

18. $f(t) = t^4 - 4t + 1$

In Exercises 19–22 find $f'(a)$.

19. $f(x) = \dfrac{x}{2x - 1}$

20. $f(x) = \dfrac{x}{x^2 - 1}$

21. $f(x) = \dfrac{2}{\sqrt{3 - x}}$

22. $f(x) = \sqrt{x - 1}$

23. If $f(x) = \sqrt[3]{x}$, find $f'(a)$ using (a) Equation 3.3 and (b) Definition 3.2.

24. If $f(x) = x^{2/3}$, show that $f'(0)$ does not exist. Sketch the curve $y = x^{2/3}$.

In Exercises 25–30 each limit represents the derivative of some function f at some number a. State f and a in each case.

25. $\lim\limits_{h \to 0} \dfrac{\sqrt{1 + h} - 1}{h}$

26. $\lim\limits_{h \to 0} \dfrac{(2 + h)^3 - 8}{h}$

27. $\lim\limits_{x \to 1} \dfrac{x^9 - 1}{x - 1}$

28. $\lim\limits_{x \to 3\pi} \dfrac{\cos x + 1}{x - 3\pi}$

29. $\lim\limits_{t \to 0} \dfrac{\sin\left(\dfrac{\pi}{2} + t\right) - 1}{t}$

30. $\lim\limits_{t \to 0} \dfrac{3(5 + t)^2 + (5 + t) - 80}{t}$

In Exercises 31–44 find the derivative of the given function using the definition of derivative. State the domain of the function and the domain of its derivative.

31. $f(x) = 5x + 3$

32. $f(x) = 18$

33. $f(x) = x^3 - x^2 + 2x$

34. $f(x) = x^4$

35. $f(x) = x - \dfrac{2}{x}$

36. $f(x) = \dfrac{x + 1}{x - 1}$

37. $g(x) = \sqrt{1 + 2x}$

38. $g(x) = \dfrac{1}{x^2}$

39. $G(x) = \dfrac{4 - 3x}{2 + x}$

40. $F(x) = \dfrac{1}{\sqrt{x - 1}}$

41. $f(t) = (t + 1)^3$

42. $f(t) = \dfrac{6}{1 + t^2}$

43. $f(x) = ax^2 + bx + c$

44. $f(x) = \sqrt{c - x}$

45. Compute the derivatives of the functions $f(x) = x$, $f(x) = x^2$, and $f(x) = x^3$ and observe the result of Exercise 34. On the basis of these results, guess the derivative of $f(x) = x^n$ for n a positive integer. Test your guess by computing the derivative of $f(x) = x^5$.

In Exercises 46–51 trace or copy the graph of the given function f. Then sketch the graph of f' below it.

46.

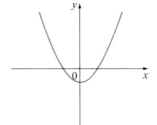

47.

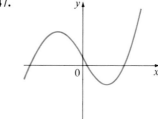

48.

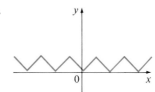

49.

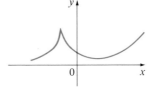

50.

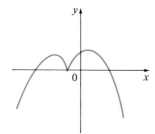

51.

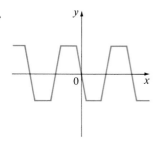

52. Make a careful sketch of the graph of the sine function and below it sketch the graph of its derivative in the same manner as in Exercises 46–51. Can you guess what the derivative of the sine function is from its graph?

53. Show that the function $f(x) = |x - 6|$ is not differentiable at 6. Find a formula for f' and sketch its graph.

54. Where is the greatest integer function $f(x) = [\![x]\!]$ not differentiable? Find a formula for f' and sketch its graph.

55. (a) Sketch the graph of the function $f(x) = x|x|$.
(b) For what values of x is f differentiable?
(c) Find a formula for f'.

56. Where (and why) are the following functions discontinuous? Where are they not differentiable? Sketch the graph in each case.

(a) $f(x) = \begin{cases} 0 & \text{if } x \leqslant 0 \\ 5 - x & \text{if } 0 < x < 4 \\ \dfrac{1}{5 - x} & \text{if } x \geqslant 4 \end{cases}$

(b) $g(x) = \begin{cases} \dfrac{x^3 - x}{x^2 + x} & \text{if } x < 1 \quad (x \neq 0) \\ 0 & \text{if } x = 0 \\ 1 - x & \text{if } x \geqslant 1 \end{cases}$

In Exercises 57 and 58 determine whether or not $f'(0)$ exists.

57. $f(x) = \begin{cases} x \sin \dfrac{1}{x} & \text{if } x \neq 0 \\ 0 & \text{if } x = 0 \end{cases}$

58. $f(x) = \begin{cases} x^2 \sin \dfrac{1}{x} & \text{if } x \neq 0 \\ 0 & \text{if } x = 0 \end{cases}$

59. A function f is called *even* if $f(-x) = f(x)$ for all x in its domain and *odd* if $f(-x) = -f(x)$ for all such x. Prove that (a) the derivative of an even function is an odd function, and (b) the derivative of an odd function is an even function.

SECTION 3.2

Differentiation Formulas

If it were always necessary to compute derivatives directly from the definition, as we did in the preceding section, it would be tedious and require ingenuity in the evaluation of some limits. Fortunately there are several rules for finding derivatives without having to use the definition directly. These formulas greatly simplify the task of differentiation.

Theorem (3.9)

> If f is a constant function, $f(x) = c$, then $f'(x) = 0$.

This result is geometrically evident because the graph of a constant function is a horizontal line that has slope 0, but a formal proof is also easy.

Proof
$$f'(x) = \lim_{h \to 0} \frac{f(x + h) - f(x)}{h} = \lim_{h \to 0} \frac{c - c}{h}$$
$$= \lim_{h \to 0} 0 = 0$$

In Leibniz notation, Theorem 3.9 can be written as

$$\boxed{\frac{d}{dx} c = 0}$$

The next theorem gives a formula for differentiating the power function $f(x) = x^n$. In the preceding section we found that $D(x) = 1$, $D(x^2) = 2x$, $D(x^3) = 3x^2$, $D(x^4) = 4x^3$, and so it seems reasonable to make the guess that $D(x^n) = nx^{n-1}$. We give two proofs of this fact, the second of which uses the Binomial Theorem.

The Power Rule (3.10)

> If n is a positive integer and $f(x) = x^n$, then
> $$f'(x) = nx^{n-1}$$

First Proof   The formula

(3.11)
$$x^n - a^n = (x - a)(x^{n-1} + x^{n-2}a + \cdots + xa^{n-2} + a^{n-1})$$

can be verified simply by multiplying out the right-hand side (or by summing the second factor as a geometric series). Thus, by using Equation 3.3 for $f'(a)$ and then using Equation 3.11, we get

$$f'(a) = \lim_{x \to a} \frac{f(x) - f(a)}{x - a} = \lim_{x \to a} \frac{x^n - a^n}{x - a}$$
$$= \lim_{x \to a}(x^{n-1} + x^{n-2}a + \cdots + xa^{n-2} + a^{n-1})$$
$$= a^{n-1} + a^{n-2}a + \cdots + aa^{n-2} + a^{n-1}$$
$$= na^{n-1}$$

Second Proof

$$f'(x) = \lim_{h \to 0} \frac{f(x + h) - f(x)}{h} = \lim_{h \to 0} \frac{(x + h)^n - x^n}{h}$$

Expanding $(x + h)^n$ by the Binomial Theorem (see Appendix A), we get

$$f'(x) = \lim_{h \to 0} \frac{\left[x^n + nx^{n-1}h + \dfrac{n(n-1)}{2} x^{n-2}h^2 + \cdots + nxh^{n-1} + h^n \right] - x^n}{h}$$

$$= \lim_{h \to 0} \frac{nx^{n-1}h + \dfrac{n(n-1)}{2} x^{n-2}h^2 + \cdots + nxh^{n-1} + h^n}{h}$$

$$= \lim_{h \to 0} \left[nx^{n-1} + \dfrac{n(n-1)}{2} x^{n-2}h + \cdots + nxh^{n-2} + h^{n-1} \right]$$

$$= nx^{n-1}$$

because every term, except the first, has h as a factor and therefore approaches 0. ●

The Power Rule can be written in Leibniz notation as

$$\frac{d}{dx}(x^n) = nx^{n-1}$$

We illustrate the Power Rule using various notations:

If $f(x) = x^6$, then $f'(x) = 6x^5$.

If $y = x^{1000}$, then $y' = 1000x^{999}$.

If $y = t^4$, then $\dfrac{dy}{dt} = 4t^3$.

$$\frac{d}{dr}(r^3) = 3r^2 \qquad D_u(u^m) = mu^{m-1}$$

The next differentiation formulas tell us that the derivative of a constant times a function is the constant times the derivative of the function, and the derivative of a sum of functions is the sum of the derivatives (assuming these derivatives exist).

Theorem (3.12)

Suppose c is a constant and $f'(x)$ and $g'(x)$ exist.
(a)   If $F(x) = cf(x)$, then $F'(x) = cf'(x)$.
(b)   If $G(x) = f(x) + g(x)$, then $G'(x) = f'(x) + g'(x)$.
(c)   If $H(x) = f(x) - g(x)$, then $H'(x) = f'(x) - g'(x)$.
In short:
(a) $(cf)' = cf'$ (b) $(f + g)' = f' + g'$ (c) $(f - g)' = f' - g'$

Proof   For part (a):

$$F'(x) = \lim_{h \to 0} \frac{F(x + h) - F(x)}{h} = \lim_{h \to 0} \frac{cf(x + h) - cf(x)}{h}$$

$$= \lim_{h \to 0} c \left[\frac{f(x + h) - f(x)}{h} \right]$$

$$= c \lim_{h \to 0} \frac{f(x + h) - f(x)}{h} \qquad \text{(by Property 3 of limits)}$$

$$= cf'(x)$$

For part (b):

$$G'(x) = \lim_{h \to 0} \frac{G(x + h) - G(x)}{h}$$

$$= \lim_{h \to 0} \frac{[f(x + h) + g(x + h)] - [f(x) + g(x)]}{h}$$

$$= \lim_{h \to 0} \left[\frac{f(x + h) - f(x)}{h} + \frac{g(x + h) - g(x)}{h} \right]$$

$$= \lim_{h \to 0} \frac{f(x + h) - f(x)}{h} + \lim_{h \to 0} \frac{g(x + h) - g(x)}{h} \qquad \begin{array}{l} \textit{(by Property 1} \\ \textit{of limits)} \end{array}$$

$$= f'(x) + g'(x)$$

Part (c) can be proved similarly using Property 2 of limits.   ●

Using Leibniz notation, the results of Theorem 3.12 can be summarized as follows.

(a) $\dfrac{d}{dx}(cf) = c\dfrac{df}{dx}$		(b) $\dfrac{d}{dx}(f + g) = \dfrac{df}{dx} + \dfrac{dg}{dx}$
(c) $\dfrac{d}{dx}(f - g) = \dfrac{df}{dx} - \dfrac{dg}{dx}$		

The result of Theorem 3.12(b) can be extended to the sum of any number of functions. For instance, using this theorem twice, we get

$$(f + g + h)' = [(f + g) + h]' = (f + g)' + h' = f' + g' + h'$$

Theorem 3.12 can be combined with the Power Rule to differentiate any polynomial, as the following examples demonstrate.

● **Example 1**

$$\frac{d}{dx}(x^8 + 12x^5 - 4x^4 + 10x^3 - 6x + 5)$$

$$= \frac{d}{dx}(x^8) + 12\frac{d}{dx}(x^5) - 4\frac{d}{dx}(x^4) + 10\frac{d}{dx}(x^3) - 6\frac{d}{dx}(x) + \frac{d}{dx}(5)$$

$$= 8x^7 + 12(5x^4) - 4(4x^3) + 10(3x^2) - 6(1) + 0$$

$$= 8x^7 + 60x^4 - 16x^3 + 30x^2 - 6$$ ●

● **Example 2**   If $f(x) = x^4 - x^3 + x^2 - x + 1$, find the equation of the tangent to the graph of f at the point $(1, 1)$.

Solution

$$f'(x) = 4x^3 - 3x^2 + 2x - 1$$

$$f'(1) = 4 - 3 + 2 - 1 = 2$$

Therefore the equation of the tangent at $(1, 1)$ is

$$y - 1 = 2(x - 1)$$

or $$2x - y - 1 = 0$$ ●

Next we need a formula for the derivative of a product of two functions. By analogy with Theorem 3.12(b) and (c), one might be tempted to guess, as Leibniz did three centuries ago, that the derivative of a product is the product of the derivatives. We can see, however, that this guess is wrong by looking at a particular example. Let $f(x) = x$ and $g(x) = x^2$. Then the Power Rule gives $f'(x) = 1$ and $g'(x) = 2x$. But $(fg)(x) = x^3$, so $(fg)'(x) = 3x^2$. Thus $(fg)' \neq f'g'$. The correct formula was discovered by Leibniz (soon after his false start) and is called the Product Rule.

The Product Rule (3.13)

> If $F(x) = f(x)g(x)$, and $f'(x)$ and $g'(x)$ both exist, then
>
> $$F'(x) = f'(x)g(x) + f(x)g'(x)$$
>
> In short:
>
> $$(fg)' = fg' + f'g$$

Proof $$F'(x) = \lim_{h \to 0} \frac{F(x + h) - F(x)}{h}$$

$$= \lim_{h \to 0} \frac{f(x + h)g(x + h) - f(x)g(x)}{h}$$

In order to evaluate this limit, we would like to separate the functions f and g as in the proof of Theorem 3.12(b). We can achieve this separation by adding and subtracting the term $f(x + h)g(x)$ in the numerator:

$$F'(x) = \lim_{h \to 0} \frac{f(x + h)g(x + h) - f(x + h)g(x) + f(x + h)g(x) - f(x)g(x)}{h}$$

$$= \lim_{h \to 0} \left[f(x + h) \frac{g(x + h) - g(x)}{h} + g(x) \frac{f(x + h) - f(x)}{h} \right]$$

$$= \lim_{h \to 0} f(x + h) \cdot \lim_{h \to 0} \frac{g(x + h) - g(x)}{h} + \lim_{h \to 0} g(x) \lim_{h \to 0} \frac{f(x + h) - f(x)}{h}$$

$$= f(x)g'(x) + g(x)f'(x)$$

Note that $\lim_{h \to 0} g(x) = g(x)$ because $g(x)$ is a constant with respect to the variable h. Also, since f is differentiable at x, it is continuous at x by Theorem 3.8, and so $\lim_{h \to 0} f(x + h) = f(x)$. (See Exercise 57 in Section 2.5.)

When expressed in Leibniz notation, the Product Rule becomes

$$\frac{d}{dx}(fg) = f \frac{dg}{dx} + g \frac{df}{dx}$$

In words, this says that the derivative of a product of two functions is the first function times the derivative of the second function plus the second function times the derivative of the first function.

● **Example 3**   If $y = x^2 \sqrt{x - 5}$, then

$$\frac{dy}{dx} = x^2 \frac{d}{dx} \sqrt{x - 5} + \sqrt{x - 5} \frac{d}{dx} (x^2)$$

$$= \frac{x^2}{2\sqrt{x - 5}} + 2x\sqrt{x - 5} \qquad \textit{(by Example 4 in Section 3.1)}$$

Taking the terms to a common denominator, the answer can be simplified to read

$$\frac{dy}{dx} = \frac{5x(x - 4)}{2\sqrt{x - 5}}$$

The Quotient Rule (3.14)

If $F(x) = f(x)/g(x)$ and both $f'(x)$ and $g'(x)$ exist, then

$$F'(x) = \frac{f'(x)g(x) - f(x)g'(x)}{[g(x)]^2}$$

In short:

$$\left(\frac{f}{g} \right)' = \frac{gf' - fg'}{g^2}$$

Proof $\displaystyle F'(x) = \lim_{h \to 0} \frac{F(x + h) - F(x)}{h} = \lim_{h \to 0} \frac{\dfrac{f(x + h)}{g(x + h)} - \dfrac{f(x)}{g(x)}}{h}$

$\displaystyle = \lim_{h \to 0} \frac{f(x + h)g(x) - f(x)g(x + h)}{hg(x + h)g(x)}$

We can separate f and g in this expression by adding and subtracting the term $f(x)g(x)$ in the numerator:

$\displaystyle F'(x) = \lim_{h \to 0} \frac{f(x + h)g(x) - f(x)g(x) + f(x)g(x) - f(x)g(x + h)}{hg(x + h)g(x)}$

$\displaystyle = \lim_{h \to 0} \frac{g(x)\dfrac{f(x + h) - f(x)}{h} - f(x)\dfrac{g(x + h) - g(x)}{h}}{g(x + h)g(x)}$

$\displaystyle = \frac{\displaystyle \lim_{h \to 0} g(x) \cdot \lim_{h \to 0} \frac{f(x + h) - f(x)}{h} - \lim_{h \to 0} f(x) \cdot \lim_{h \to 0} \frac{g(x + h) - g(x)}{h}}{\displaystyle \lim_{h \to 0} g(x + h) \cdot \lim_{h \to 0} g(x)}$

$\displaystyle = \frac{g(x)f'(x) - f(x)g'(x)}{[g(x)]^2}$

Again g is continuous by Theorem 3.8, so $\lim_{h \to 0} g(x + h) = g(x)$.   ●

The Quotient Rule, in Leibniz notation, becomes

$$\frac{d}{dx}\left(\frac{f(x)}{g(x)}\right) = \frac{g(x)\dfrac{d}{dx}f(x) - f(x)\dfrac{d}{dx}g(x)}{[g(x)]^2}$$

In words this says that the derivative of a quotient is the denominator times the derivative of the numerator minus the numerator times the derivative of the denominator, all divided by the square of the denominator.

The theorems of this section show that any polynomial is differentiable on R and any rational function is differentiable on its domain. Furthermore, the Quotient Rule and the other differentiation formulas enable us to compute the derivative of any rational function, as the next example illustrates.

● **Example 4**   Let $\displaystyle y = \frac{x^2 + x - 2}{x^3 + 6}$

Then $y' = \dfrac{(x^3 + 6)D(x^2 + x - 2) - (x^2 + x - 2)D(x^3 + 6)}{(x^3 + 6)^2}$

$= \dfrac{(x^3 + 6)(2x + 1) - (x^2 + x - 2)(3x^2)}{(x^3 + 6)^2}$

$= \dfrac{(2x^4 + x^3 + 12x + 6) - (3x^4 + 3x^3 - 6x^2)}{(x^3 + 6)^2}$

$= \dfrac{-x^4 - 2x^3 + 6x^2 + 12x + 6}{(x^3 + 6)^2}$ ●

The Quotient Rule can also be used to extend the Power Rule to the case where the exponent is a negative integer.

Theorem (3.15)

If $f(x) = x^{-n}$, where n is a positive integer, then

$$f'(x) = -nx^{-n-1}$$

Proof $f'(x) = \dfrac{d}{dx}(x^{-n}) = \dfrac{d}{dx}\left(\dfrac{1}{x^n}\right)$

$= \dfrac{x^n D(1) - 1 \cdot D(x^n)}{(x^n)^2}$

$= \dfrac{x^n \cdot 0 - 1 \cdot nx^{n-1}}{x^{2n}}$

$= \dfrac{-nx^{n-1}}{x^{2n}} = -nx^{n-1-2n} = -nx^{-n-1}$ ●

● **Example 5**

(a)   If $y = \dfrac{1}{x}$, then $\dfrac{dy}{dx} = \dfrac{d}{dx}(x^{-1}) = -x^{-2} = -\dfrac{1}{x^2}$.

(b) $\dfrac{d}{dt}\left(\dfrac{6}{t^3}\right) = 6\dfrac{d}{dt}(t^{-3}) = 6(-3)t^{-4} = -\dfrac{18}{t^4}$ ●

By (3.10) and (3.15) the Power Rule holds if the exponent n is a positive or negative integer. If $n = 0$, then $x^0 = 1$, which we know has a derivative of 0. Thus the Power Rule holds for any integer n. In fact, it also holds for *any real number n*, as we shall prove in Chapter 6. (A proof for rational values of n is indicated in Exercise 36 in Section 3.5.) In the meantime we shall state the general version and use it in the examples and exercises.

The Power Rule
(General Version)
(3.16)

Let n be any real number.

$$\text{If } f(x) = x^n, \text{ then } f'(x) = nx^{n-1}.$$

Or, in Leibniz notation,

$$\frac{d}{dx}(x^n) = nx^{n-1}$$

● **Example 6**

(a)   If $f(x) = x^\pi$, then $f'(x) = \pi x^{\pi-1}$.

(b) $\dfrac{d}{dx}\sqrt{x} = \dfrac{d}{dx}(x^{1/2}) = \dfrac{1}{2}x^{(1/2)-1} = \dfrac{1}{2}x^{-1/2} = \dfrac{1}{2\sqrt{x}}$

(c)   Let

$$y = \frac{1}{\sqrt[3]{x^2}}$$

Then
$$\frac{dy}{dx} = \frac{d}{dx}(x^{-2/3}) = -\frac{2}{3}x^{-(2/3)-1}$$

$$= -\frac{2}{3}x^{-5/3}$$

● **Example 7**   Differentiate the function $f(t) = \sqrt{t}(1 - t)$.

Solution 1   Using the Product Rule, we have

$$f'(t) = \sqrt{t}\,\frac{d}{dt}(1 - t) + (1 - t)\frac{d}{dt}\sqrt{t}$$

$$= \sqrt{t}(-1) + (1 - t)\cdot\frac{1}{2}t^{-1/2}$$

$$= -\sqrt{t} + \frac{1 - t}{2\sqrt{t}} = \frac{1 - 3t}{2\sqrt{t}}$$

Solution 2   If we first use the laws of exponents, then we can proceed directly, without using the Product Rule.

$$f(t) = \sqrt{t} - t\sqrt{t} = t^{1/2} - t^{3/2}$$
$$f'(t) = \tfrac{1}{2}t^{-1/2} - \tfrac{3}{2}t^{1/2}$$

which is equivalent to the answer given in Solution 1.

We summarize the differentiation formulas we have learned so far in the following table.

Table of Differentiation
Formulas (3.17)

$$(cf)' = cf' \qquad\qquad (f + g)' = f' + g'$$

$$(f - g)' = f' - g' \qquad\qquad (fg)' = f'g + fg'$$

$$\left(\frac{f}{g}\right)' = \frac{f'g - fg'}{g^2} \qquad\qquad \frac{d}{dx}c = 0$$

$$\frac{d}{dx}(x^n) = nx^{n-1}$$

EXERCISES 3.2

Differentiate the functions given in Exercises 1–36.

1. $f(x) = x^2 - 10x + 100$

2. $g(x) = x^{100} + 50x + 1$

3. $V(r) = \frac{4}{3}\pi r^3$

4. $s(t) = t^8 + 6t^7 - 18t^2 + 2t$

5. $F(x) = (16x)^3$

6. $G(y) = (y^2 + 1)(2y - 7)$

7. $Y(t) = 6t^{-9}$

8. $R(x) = \frac{\sqrt{10}}{x^7}$

9. $g(x) = x^2 + \frac{1}{x^2}$

10. $f(t) = \sqrt{t} - \frac{1}{\sqrt{t}}$

11. $h(x) = \frac{x + 2}{x - 1}$

12. $f(u) = \frac{1 - u^2}{1 + u^2}$

13. $G(s) = (s^2 + s + 1)(s^2 + 2)$

14. $H(t) = \sqrt[3]{t}(t + 2)$

15. $y = \frac{x^2 + 4x + 3}{\sqrt{x}}$

16. $y = \frac{\sqrt{x} - 1}{\sqrt{x} + 1}$

17. $y = \sqrt{5x}$

18. $y = x^{4/3} - x^{2/3}$

19. $y = \frac{1}{x^4 + x^2 + 1}$

20. $y = x^2 + x + x^{-1} + x^{-2}$

21. $y = ax^2 + bx + c$

22. $y = A + \frac{B}{x} + \frac{C}{x^2}$

23. $y = \frac{3t - 7}{t^2 + 5t - 4}$

24. $y = \frac{4t + 5}{2 - 3t}$

25. $y = x + \sqrt[5]{x^2}$

26. $y = x^4 - \sqrt[4]{x}$

27. $u = x^{\sqrt{2}}$

28. $u = \sqrt[3]{t^2} + 2\sqrt{t^3}$

29. $v = x\sqrt{x} + \frac{1}{x^2\sqrt{x}}$

30. $v = \frac{6}{\sqrt[3]{t^5}}$

31. $f(x) = \frac{x}{x + \dfrac{c}{x}}$

32. $f(x) = \frac{ax + b}{cx + d}$

33. $f(x) = \frac{x^5}{x^3 - 2}$

34. $g(x) = \sqrt[5]{x} - \frac{6}{x^{1.8}} + 0.1(x^{1.8})$

35. $s = \frac{2 - \dfrac{1}{t}}{t + 1}$

36. $s = \sqrt{t}(t^3 - \sqrt{t} + 1)$

37. The general polynomial of degree n has the form

$$P(x) = a_n x^n + a_{n-1}x^{n-1} + \cdots + a_2 x^2 + a_1 x + a_0$$

where $a_n \neq 0$. Find the derivative of P.

In Exercises 38–41, find the equation of the tangent line to the given curve at the given point.

38. $y = \frac{x}{x - 3}$, $(6, 2)$

39. $y = x + \frac{4}{x}$, $(2, 4)$

40. $y = x^{5/2}$, $(4, 32)$

41. $y = \frac{1}{x^2 + 1}$, $\left(-1, \frac{1}{2}\right)$

42. Find the equations of both lines through the point $(2, -3)$ that are tangent to the parabola $y = x^2 + x$.

43. How many tangent lines to the curve $y = x/(x + 1)$ pass through the point $(1, 2)$? At which points do these tangent lines touch the curve?

44. For what values of x does the graph of $f(x) = 2x^3 - 3x^2 - 6x + 87$ have a horizontal tangent?

45. Find the points on the curve $y = x^3 - x^2 - x + 1$ where the tangent is horizontal.

46. Find the equations of the tangent lines to the curve $y = (x - 1)/(x + 1)$ that are parallel to the line $x - 2y = 1$.

47. At what point on the curve $y = x\sqrt{x}$ is the tangent line parallel to the line $3x - y + 6 = 0$?

48. A manufacturer of cartridges for stereo systems has designed a stylus with parabolic cross-section as shown in Figure 3.7. The equation of the parabola is $y = 16x^2$ where x and y are measured in millimeters. If the stylus sits in a record groove whose sides make an angle of θ with the horizontal direction, where $\tan \theta = 1.75$, find the points of contact P and Q of the stylus with the groove.

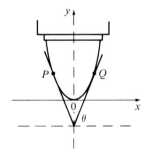

Figure 3.7

In Exercises 49–52 find the equation of the normal line to the given curve at the given point. (The *normal line* to a curve C at a point P is, by definition, the line that passes through P and is perpendicular to the tangent line to C at P.) Also sketch the curve and the normal line in Exercises 49–51.

49. $y = 1 - x^2, (2, -3)$ **50.** $y = \dfrac{1}{x - 1}, (2, 1)$

51. $y = \sqrt[3]{x}, (-8, -2)$ **52.** $y = f(x), (a, f(a))$

53. At what point on the curve $y = x^4$ does the normal line have slope 16?

In Exercises 54–57 a particle moves in a straight line with position function $s = f(t)$, $t \geqslant 0$. (a) Find the velocity v as a function of time. (b) When is the particle at rest? (c) When is it moving in the positive direction?

54. $f(t) = 3 - 6t + t^3$

55. $f(t) = 3t^4 - 16t^3 + 18t^2$

56. $s = \sqrt{t}(5 - 5t + 2t^2)$ **57.** $s = \dfrac{t}{t^2 + 1}$

58. (a) For what values of x is the function $f(x) = |x^2 - 9|$ differentiable? Find a formula for f'.
   (b) Sketch the graphs of both f and f'.

59. Let

$$f(x) = \begin{cases} 2 - x & \text{if } x \leqslant 1 \\ x^2 - 2x + 2 & \text{if } x > 1 \end{cases}$$

Is f differentiable at 1? Sketch the graphs of f and f'.

60. At what numbers is the function g given by

$$g(x) = \begin{cases} -1 - 2x & \text{if } x < -1 \\ x^2 & \text{if } -1 \leqslant x \leqslant 1 \\ x & \text{if } x > 1 \end{cases}$$

differentiable? Give a formula for g' and sketch the graphs of g and g'.

61. Suppose that $f(5) = 1$, $f'(5) = 6$, $g(5) = -3$, and $g'(5) = 2$. Find the values of (a) $(fg)'(5)$, (b) $(f/g)'(5)$, and (c) $(g/f)'(5)$.

62. Where is the function $h(x) = |x - 1| + |x + 2|$ differentiable? Give a formula for h' and sketch the graphs of h and h'.

63. (a) Use the Product Rule twice to prove that if f, g, and h are differentiable, then

$$(fgh)' = f'gh + fg'h + fgh'$$

   (b) Taking $f = g = h$ in part (a), show that

$$\frac{d}{dx}[f(x)]^3 = 3[f(x)]^2 f'(x)$$

Use Exercise 63 to differentiate the functions in Exercises 64–66.

64. $y = (x + 5)(x^2 + 7)(x - 3)$

65. $y = \sqrt{x}(x^4 + x + 1)(2x - 3)$

66. $y = (x^4 + 3x^3 + 17x + 82)^3$

67. An easy proof of the Quotient Rule (3.14) can be given if we make the prior assumption that $F'(x)$

exists, where $F = f/g$. Write $f = Fg$; then differentiate using the Product Rule and solve the resulting equation for F'.

68. Let P be a point on the curve $y = x^3$ and suppose the tangent line at P intersects the curve again at Q. Prove that the slope at Q is four times the slope at P.

Derivatives of Trigonometric Functions

Before starting this section, the reader might need to review the trigonometric functions in Appendix B. In particular it is important to remember that when we talk about the function f defined for all real numbers x by

$$f(x) = \sin x$$

it is understood that $\sin x$ means the sine of the angle whose *radian* measure is x. A similar convention holds for the other trigonometric functions cos, tan, csc, sec, and cot.

In order to compute the derivatives of these functions, we first have to evaluate some trigonometric limits.

Theorem (3.18)

$$\lim_{\theta \to 0} \sin \theta = 0$$

If we let $f(\theta) = \sin \theta$, then $f(0) = \sin 0 = 0$, so Theorem 3.18 tells us that $\lim_{\theta \to 0} f(\theta) = f(0)$. This is precisely the statement that the sine function is continuous at 0.

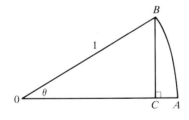

Figure 3.8

Proof   In calculating $\lim_{\theta \to 0^+} \sin \theta$, we may assume that $0 < \theta < \pi/2$. Figure 3.8 shows a sector of a circle with center O, central angle θ, and radius 1. BC is drawn perpendicular to OA. By definition of radian measure, arc $AB = \theta$. Also $|BC| = |OB| \sin \theta = \sin \theta$. From the diagram we see that

$$|BC| < \text{arc } AB$$

(3.19)   Therefore $$0 < \sin \theta < \theta$$

Since we know that $\lim_{\theta \to 0^+} 0 = 0$ and $\lim_{\theta \to 0^+} \theta = 0$, it follows from the Squeeze Theorem that

$$\lim_{\theta \to 0^+} \sin \theta = 0$$

If $-\pi/2 < \theta < 0$, then $0 < -\theta < \pi/2$, so by (3.19) we have

$$0 < \sin(-\theta) < -\theta$$

or $$0 < -\sin \theta < -\theta$$

which implies $$\theta < \sin \theta < 0$$

This inequality, together with the fact that $\lim_{\theta \to 0^-} \theta = 0$ and the Squeeze Theorem, shows that $\lim_{\theta \to 0^-} \sin \theta = 0$. Thus

$$\lim_{\theta \to 0} \sin \theta = 0$$

●

Corollary (3.20)

$$\boxed{\lim_{\theta \to 0} \cos \theta = 1}$$

Again, by the definition of continuity and the fact that $\cos 0 = 1$, Corollary 3.20 says that the cosine function is continuous at 0.

Proof   Using $\sin^2 \theta + \cos^2 \theta = 1$ together with $\cos \theta \geqslant 0$ for $-\pi/2 \leqslant \theta \leqslant \pi/2$, we have $\cos \theta = \sqrt{1 - \sin^2 \theta}$ for those values of θ. Thus, using the properties of limits, we have

$$\lim_{\theta \to 0} \cos \theta = \lim_{\theta \to 0} \sqrt{1 - \sin^2 \theta}$$

$$= \sqrt{\lim_{\theta \to 0}(1 - \sin^2 \theta)}$$

$$= \sqrt{1 - 0} \quad \text{(by Theorem 3.18)}$$

$$= 1$$

●

In Example 6 in Section 2.2 we made the guess that $\lim_{x \to 0}(\sin x)/x = 1$. We are now in a position to prove that the guess is correct.

Theorem (3.21)

$$\boxed{\lim_{\theta \to 0} \frac{\sin \theta}{\theta} = 1}$$

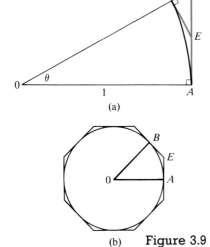

(a)

(b)   **Figure 3.9**

Proof   First suppose that $0 < \theta < \pi/2$. Again Figure 3.9 shows a sector of a circle with center O, central angle θ, and radius 1. Let the tangents at A and B intersect at E. You can see from Figure 3.9(b) that the circumference of a circle is smaller than the length of a circumscribed polygon, so arc $AB < |AE| + |EB|$. Thus

$$\theta = \text{arc } AB < |AE| + |EB|$$

$$< |AE| + |ED|$$

$$= |AD| = |OA| \tan \theta$$

$$= \tan \theta$$

(In Appendix C it is proved that $\theta \leqslant \tan \theta$ directly from the definition of the length of an arc without resorting to geometric intuition as we did above.)

Therefore we have

$$\theta < \frac{\sin \theta}{\cos \theta}$$

or
$$\cos \theta < \frac{\sin \theta}{\theta}$$

From the proof of Theorem 3.18 we have $\sin \theta < \theta$, so

(3.22)
$$\cos \theta < \frac{\sin \theta}{\theta} < 1$$

If $-\pi/2 < \theta < 0$, then $0 < -\theta < \pi/2$, so (3.22) gives

(3.23)
$$\cos(-\theta) < \frac{\sin(-\theta)}{-\theta} < 1$$

But $\cos(-\theta) = \cos \theta$ and $\sin(-\theta) = -\sin \theta$, so (3.23), when simplified, becomes (3.22). Thus (3.22) holds whenever $\theta \in (-\pi/2, \pi/2)$ and $\theta \neq 0$. We know that $\lim_{\theta \to 0} 1 = 1$ and $\lim_{\theta \to 0} \cos \theta = 1$ by Corollary 3.20. Hence, by the Squeeze Theorem,

$$\lim_{\theta \to 0} \frac{\sin \theta}{\theta} = 1 \qquad \bullet$$

Corollary (3.24)

$$\boxed{\lim_{\theta \to 0} \frac{\cos \theta - 1}{\theta} = 0}$$

Proof   Using the identity $\sin^2 x = \frac{1}{2}(1 - \cos 2x)$, or equivalently,

$$1 - \cos \theta = 2 \sin^2 \left(\frac{\theta}{2} \right)$$

we have

$$\lim_{\theta \to 0} \frac{\cos \theta - 1}{\theta} = \lim_{\theta \to 0} \left[-\frac{2 \sin^2(\theta/2)}{\theta} \right]$$
$$= \lim_{\theta \to 0} \left(-\sin \frac{\theta}{2} \right) \frac{\sin(\theta/2)}{\theta/2}$$
$$= \lim_{\theta \to 0} \left[-\sin \frac{\theta}{2} \right] \lim_{\theta \to 0} \frac{\sin(\theta/2)}{\theta/2}$$
$$= 0 \cdot 1 = 0$$

Here we have used Theorems 3.18 and 3.21. (Note that $\theta/2 \to 0$ as $\theta \to 0$.)

●

Note: We could also have proved Corollary 3.24 by multiplying numerator and denominator by $\cos \theta + 1$.

● **Example 1**   Find $\lim\limits_{x \to 0} \dfrac{\sin 7x}{4x}$.

Solution   In order to apply Theorem 3.21, we first rewrite the function as follows:

$$\frac{\sin 7x}{4x} = \frac{7}{4}\left(\frac{\sin 7x}{7x}\right)$$

Notice that as $x \to 0$, we have $7x \to 0$, and so, by Theorem 3.21,

$$\lim_{x \to 0} \frac{\sin 7x}{7x} = \lim_{7x \to 0} \frac{\sin(7x)}{7x} = 1$$

Thus
$$\lim_{x \to 0} \frac{\sin 7x}{4x} = \lim_{x \to 0} \frac{7}{4}\left(\frac{\sin 7x}{7x}\right)$$

$$= \frac{7}{4} \lim_{x \to 0} \frac{\sin 7x}{7x} = \frac{7}{4} \cdot 1 = \frac{7}{4}$$

●

● **Example 2**   Calculate $\lim\limits_{x \to 0} x \cot x$.

Solution $\lim\limits_{x \to 0} x \cot x = \lim\limits_{x \to 0} x \dfrac{\cos x}{\sin x}$

$$= \lim_{x \to 0} \frac{\cos x}{\dfrac{\sin x}{x}} = \frac{\lim\limits_{x \to 0} \cos x}{\lim\limits_{x \to 0} \dfrac{\sin x}{x}}$$

$$= \frac{1}{1} \qquad \textit{(by Corollary 3.20 and Theorem 3.21)}$$

$$= 1$$

●

Derivatives

If you sketch the graph of the function $f(x) = \sin x$ and use the interpretation of $f'(x)$ as the slope of the tangent to the sine curve in order to sketch the graph of f' (see Exercise 52 in Section 3.1), then it looks as if the graph of f' may be the same as the cosine curve (see Figure 3.10). This is confirmed in the next theorem by using the definition of derivative to calculate $f'(x)$.

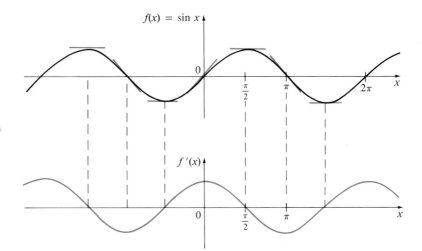

Figure 3.10

Theorem (3.25)

$$\frac{d}{dx}\sin x = \cos x$$

Proof   If $f(x) = \sin x$, then

$$f'(x) = \lim_{h\to 0} \frac{f(x+h) - f(x)}{h}$$

$$= \lim_{h\to 0} \frac{\sin(x+h) - \sin x}{h}$$

$$= \lim_{h\to 0} \frac{\sin x \cos h + \cos x \sin h - \sin x}{h}$$

$$= \lim_{h\to 0}\left[\sin x\left(\frac{\cos h - 1}{h}\right) + \cos x\left(\frac{\sin h}{h}\right)\right]$$

$$= \lim_{h\to 0}\sin x \cdot \lim_{h\to 0}\frac{\cos h - 1}{h} + \lim_{h\to 0}\cos x \cdot \lim_{h\to 0}\frac{\sin h}{h}$$

$$= \sin x \cdot 0 + \cos x \cdot 1 \qquad \textit{(by Corollary 3.24 and Theorem 3.21)}$$

$$= \cos x$$

● **Example 3**   Differentiate $y = x^2 \sin x$.

Solution   Using the Product Rule and Theorem 3.25, we have

$$\frac{dy}{dx} = x^2 \frac{d}{dx}\sin x + \sin x \frac{d}{dx}(x^2)$$

$$= x^2 \cos x + 2x \sin x$$

Using the same method as in the proof of Theorem 3.25, it can be proved (see Exercise 38) that

(3.26)
$$\frac{d}{dx} \cos x = -\sin x$$

The tangent function can also be differentiated from first principles but it is easier to use the Quotient Rule together with Theorem 3.25 and Equation 3.26:

$$\frac{d}{dx} \tan x = \frac{d}{dx} \left(\frac{\sin x}{\cos x} \right)$$

$$= \frac{\cos x \, D \sin x - \sin x \, D \cos x}{\cos^2 x}$$

$$= \frac{\cos x \cdot \cos x - \sin x(-\sin x)}{\cos^2 x}$$

$$= \frac{\cos^2 x + \sin^2 x}{\cos^2 x}$$

$$= \frac{1}{\cos^2 x} = \sec^2 x$$

(3.27)
$$\frac{d}{dx} \tan x = \sec^2 x$$

The derivatives of the remaining trigonometric functions csc, sec, and cot can also be easily found using the Quotient Rule (see Exercises 19–21). We collect all of the differentiation formulas for trigonometric functions in Table 3.28.

Table of Derivatives of Trigonometric Functions (3.28)

$\dfrac{d}{dx}(\sin x) = \cos x$	$\dfrac{d}{dx}(\csc x) = -\csc x \cot x$
$\dfrac{d}{dx}(\cos x) = -\sin x$	$\dfrac{d}{dx}(\sec x) = \sec x \tan x$
$\dfrac{d}{dx}(\tan x) = \sec^2 x$	$\dfrac{d}{dx}(\cot x) = -\csc^2 x$

In memorizing Table 3.28 it is helpful to notice that the minus signs go with the derivatives of the "cofunctions," that is, cosine, cosecant, and cotangent.

These differentiation formulas show that each trigonometric function is differentiable at every number in its domain. It follows from Theorem 3.8 that *all of the trigonometric functions are continuous on their domains.*

● **Example 4**   Differentiate $f(x) = \dfrac{\sec x}{1 + \tan x}$.

Solution   The Quotient Rule gives

$$f'(x) = \frac{(1 + \tan x)D \sec x - \sec x\, D(1 + \tan x)}{(1 + \tan x)^2}$$

$$= \frac{(1 + \tan x)\sec x \tan x - \sec x \cdot \sec^2 x}{(1 + \tan x)^2}$$

$$= \frac{\sec x[\tan x + \tan^2 x - \sec^2 x]}{(1 + \tan x)^2}$$

$$= \frac{\sec x(\tan x - 1)}{(1 + \tan x)^2}$$

In simplifying the answer we have used the identity $\tan^2 x + 1 = \sec^2 x$.

●

EXERCISES 3.3

Find the limits in Exercises 1–18.

1. $\lim\limits_{x \to 0}(x^2 + \cos x)$

2. $\lim\limits_{x \to 0}\cos(\sin x)$

3. $\lim\limits_{x \to \pi/3}(\sin x - \cos x)$

4. $\lim\limits_{x \to \pi}x^2 \sec x$

5. $\lim\limits_{x \to \pi/4}\dfrac{\sin x}{3x}$

6. $\lim\limits_{x \to 0}\dfrac{\sin x}{3x}$

7. $\lim\limits_{t \to -3\pi}t^3 \sin^4 t$

8. $\lim\limits_{t \to \pi/6}\csc t \cot^2 t$

9. $\lim\limits_{t \to 0}\dfrac{\sin 5t}{t}$

10. $\lim\limits_{t \to 0}\dfrac{\sin 8t}{\sin 9t}$

11. $\lim\limits_{\theta \to 0}\dfrac{\sin(\cos \theta)}{\sec \theta}$

12. $\lim\limits_{\theta \to 0}\dfrac{\cos \theta - 1}{\sin \theta}$

13. $\lim\limits_{x \to \pi/4}\dfrac{\tan x}{4x}$

14. $\lim\limits_{x \to 0}\dfrac{\tan x}{4x}$

15. $\lim\limits_{x \to 0}\dfrac{\tan 3x}{3 \tan 2x}$

16. $\lim\limits_{x \to 0}\dfrac{\sec x}{1 - \sin x}$

17. $\lim\limits_{t \to 0}\dfrac{\sin^2 3t}{t^2}$

18. $\lim\limits_{t \to 0}\dfrac{t^3}{\tan^3 2t}$

19. Prove that $\dfrac{d}{dx}(\csc x) = -\csc x \cot x$.

20. Prove that $\dfrac{d}{dx}(\sec x) = \sec x \tan x$.

21. Prove that $\dfrac{d}{dx}(\cot x) = -\csc^2 x$.

In Exercises 22–33 find $\dfrac{dy}{dx}$.

22. $y = \cos x - 2 \tan x$

23. $y = \sin x + \cos x$

24. $y = x \csc x$

25. $y = \csc x \cot x$

26. $y = \dfrac{\sin x}{1 + \cos x}$

27. $y = \dfrac{\tan x}{x}$

28. $y = \dfrac{\tan x - 1}{\sec x}$

29. $y = \dfrac{x}{\sin x + \cos x}$

30. $y = 2x(\sqrt{x} - \cot x)$

31. $y = x^{-3} \sin x \tan x$

32. $y = x \sin x \cos x$

33. $y = \dfrac{x^2 \tan x}{\sec x}$

In Exercises 34–36 find the equation of the tangent line to the given curve at the given point.

34. $y = 2 \sin x, \left(\dfrac{\pi}{6}, 1\right)$

35. $y = \tan x, \left(\dfrac{\pi}{4}, 1\right)$

36. $y = \sec x - 2 \cos x, \left(\dfrac{\pi}{3}, 1\right)$

37. For what values of x does the graph of $f(x) = x + 2 \sin x$ have a horizontal tangent?

38. Prove, using the definition of derivative, that if $f(x) = \cos x$, then $f'(x) = -\sin x$.

Find the limits in Exercises 39–46.

39. $\displaystyle\lim_{x \to 0} \dfrac{\cot 2x}{\csc x}$

40. $\displaystyle\lim_{x \to 0} \dfrac{1 - \cos x}{2x^2}$

41. $\displaystyle\lim_{x \to \pi} \dfrac{\tan x}{\sin 2x}$

42. $\displaystyle\lim_{x \to \pi/4} \dfrac{\sin x - \cos x}{\cos 2x}$

43. $\displaystyle\lim_{\theta \to 0} \dfrac{\sin \theta}{\theta + \tan \theta}$

44. $\displaystyle\lim_{x \to 0} \dfrac{x}{\sin\left(\dfrac{x}{2}\right)}$

45. $\displaystyle\lim_{x \to 0^+} \sqrt{x} \csc \sqrt{x}$

46. $\displaystyle\lim_{x \to 1} \dfrac{\sin(x - 1)}{x^2 + x - 2}$

47. Differentiate the following trigonometric identities to obtain new (or familiar) identities.

(a) $\tan x = \dfrac{\sin x}{\cos x}$

(b) $\sec x = \dfrac{1}{\cos x}$

(c) $\sin x + \cos x = \dfrac{1 + \cot x}{\csc x}$

The Chain Rule

Suppose that you were asked to differentiate the function

$$F(x) = \sqrt{x^2 + 1}$$

The differentiation formulas you learned in the preceding section will not enable you to calculate $F'(x)$.

Observe that F is a composite function. If we let $y = f(u) = \sqrt{u}$ and $u = g(x) = x^2 + 1$, then we can write $y = F(x) = f(g(x))$, that is, $F = f \circ g$. (See Section 1.6 for a review of composite functions.) We know how to differentiate both f and g, so it would be useful to have a rule that tells us how to find the derivative of $F = f \circ g$ in terms of the derivatives of f and g.

It turns out that the derivative of the composite function $f \circ g$ is the product of the derivatives of f and g. This fact is one of the most important of the differentiation rules and is called the Chain Rule. It seems plausible if we interpret derivatives as rates of change. Regard du/dx as the rate of change of u with respect to x, dy/du as the rate of change of y with respect to u, and dy/dx as the rate of change of y with respect to x. If u changes twice as fast as x and y changes three times as fast as u, then it seems reasonable that y changes six times as fast as x, and so we expect that

$$\frac{dy}{dx} = \frac{dy}{du}\frac{du}{dx}$$

**The Chain Rule
(3.29)**

> If the derivatives $g'(x)$ and $f'(g(x))$ both exist, and $F = f \circ g$ is the composite function defined by $F(x) = f(g(x))$, then $F'(x)$ exists and is given by the product
>
> $$F'(x) = f'(g(x))g'(x)$$
>
> In Leibniz notation, if $y = f(u)$ and $u = g(x)$ are both differentiable functions, then
>
> $$\frac{dy}{dx} = \frac{dy}{du}\frac{du}{dx}$$

In the proof of the Chain Rule, the increment notation will be useful. Let Δu be the change in u corresponding to a change of Δx in x, that is,

$$\Delta u = g(x + \Delta x) - g(x)$$

Then the corresponding change in y is

$$\Delta y = f(u + \Delta u) - f(u)$$

It is tempting to write

$$\frac{dy}{dx} = \lim_{\Delta x \to 0} \frac{\Delta y}{\Delta x}$$

(3.30)

$$= \lim_{\Delta x \to 0} \frac{\Delta y}{\Delta u} \cdot \frac{\Delta u}{\Delta x}$$

$$= \lim_{\Delta x \to 0} \frac{\Delta y}{\Delta u} \cdot \lim_{\Delta x \to 0} \frac{\Delta u}{\Delta x}$$

$$= \lim_{\Delta u \to 0} \frac{\Delta y}{\Delta u} \cdot \lim_{\Delta x \to 0} \frac{\Delta u}{\Delta x} \qquad \textit{(Note that } \Delta u \to 0 \textit{ as } \Delta x \to 0 \\ \textit{since } g \textit{ is continuous.)}$$

$$= \frac{dy}{du}\frac{du}{dx}$$

The only thing wrong with this reasoning is that in (3.30) it might happen that $\Delta u = 0$ for infinitely many values of Δx. Most functions g have the property that there is an open interval I containing x such that $g(t) \neq g(x)$ for all t in I ($t \neq x$). For such functions we have $\Delta u \neq 0$, so division by Δu causes no problem in (3.30). The above is a perfectly good proof of the Chain Rule for such functions. However, there are some differentiable functions that do not have this property (for example, constant functions and functions like the one in Exercise 58 in Section 3.1), and so the above reasoning does not work. The following proof works for all functions.

Proof of the Chain Rule   If f is differentiable at a given number u and we define a function h by

(3.31)
$$h(t) = \frac{f(u + t) - f(u)}{t} - f'(u)$$

then by the definition of derivative we have

$$\lim_{t \to 0} h(t) = \lim_{t \to 0} \frac{f(u + t) - f(u)}{t} - \lim_{t \to 0} f'(u)$$

$$= f'(u) - f'(u) = 0$$

In Equation 3.31 if we multiply both sides by t we get

(3.32)
$$f(u + t) - f(u) = [f'(u) + h(t)]t$$

If we define $h(0) = 0$, then Equation 3.32 remains true even for $t = 0$. Replacing t by Δu in Equation 3.32, we get

$$\Delta y = f(u + \Delta u) - f(u)$$

$$= [f'(u) + h(\Delta u)]\Delta u$$

Divide both sides of this equation by Δx:

(3.33)
$$\frac{\Delta y}{\Delta x} = [f'(u) + h(\Delta u)]\frac{\Delta u}{\Delta x}$$

Since g is differentiable at x, it must be continuous there (by Theorem 3.8) and so

$$\lim_{\Delta x \to 0} \Delta u = \lim_{\Delta x \to 0} [g(x + \Delta x) - g(x)] = 0$$

Therefore
$$\lim_{\Delta x \to 0} h(\Delta u) = \lim_{\Delta u \to 0} h(\Delta u) = 0$$

Thus Equation 3.33 gives

$$\frac{dy}{dx} = \lim_{\Delta x \to 0} \frac{\Delta y}{\Delta x} = \lim_{\Delta x \to 0} [f'(u) + h(\Delta u)]\frac{\Delta u}{\Delta x}$$

$$= [f'(u) + \lim_{\Delta x \to 0} h(\Delta u)] \lim_{\Delta x \to 0} \frac{\Delta u}{\Delta x}$$

$$= [f'(u) + 0]\frac{du}{dx} = \frac{dy}{du}\frac{du}{dx}$$

The Chain Rule can be written either in the prime notation

(3.34)
$$(f \circ g)'(x) = f'(g(x))g'(x)$$

or, if $y = f(u)$ and $u = g(x)$, in Leibniz notation:

(3.35)
$$\frac{dy}{dx} = \frac{dy}{du}\frac{du}{dx}$$

Equation 3.35 is easy to remember because if dy/du and du/dx were quotients, then we could cancel du. Remember, however, that du has not been defined and du/dx should not be thought of as an actual quotient.

● **Example 1**   Find $F'(x)$ if $F(x) = \sqrt{x^2 + 1}$.

Solution 1   (using Equation 3.34): At the beginning of this section we expressed F as $F(x) = (f \circ g)(x) = f(g(x))$ where $f(u) = \sqrt{u}$ and $g(x) = x^2 + 1$. Since

$$f'(u) = \frac{1}{2} u^{-1/2} = \frac{1}{2\sqrt{u}} \quad \text{and} \quad g'(x) = 2x$$

we have

$$F'(x) = f'(g(x))g'(x)$$

$$= \frac{1}{2\sqrt{x^2 + 1}} \cdot 2x = \frac{x}{\sqrt{x^2 + 1}}$$

Solution 2   (using Equation 3.35): If we let $u = x^2 + 1$ and $y = \sqrt{u}$, then

$$F'(x) = \frac{dy}{du}\frac{du}{dx} = \frac{1}{2\sqrt{u}}(2x)$$

$$= \frac{1}{2\sqrt{x^2 + 1}}(2x) = \frac{x}{\sqrt{x^2 + 1}}$$ ●

Notice that in using the Chain Rule we work from the outside to the inside. Formula 3.34 says that we differentiate the outer function f [at the inner function $g(x)$] and then we multiply by the derivative of the inner function.

When using Formula 3.35 it should be realized that dy/dx refers to the derivative of y when y is considered as a function of x (called the *derivative of y with respect to x*), whereas dy/du refers to the derivative of y when considered as a function of u (the derivative of y with respect to u). For instance, in Example 1, y can be considered as a function of x ($y = \sqrt{x^2 + 1}$) and also as a function of u ($y = \sqrt{u}$). Note that

$$\frac{dy}{dx} = F'(x) = \frac{x}{\sqrt{x^2 + 1}} \quad \text{whereas} \quad \frac{dy}{du} = f'(u) = \frac{1}{2\sqrt{u}}$$

● **Example 2**   If $y = u^3 + u^2 + 1$, where $u = 2x^2 - 1$, find

$$\frac{dy}{dx}\bigg]_{x=2}$$

Solution 1   Using the Chain Rule, we have

$$\frac{dy}{dx} = \frac{dy}{du}\frac{du}{dx} = (3u^2 + 2u)(4x)$$

When $x = 2$, $u = 2(2)^2 - 1 = 7$, so

$$\frac{dy}{dx}\bigg]_{x=2} = (3 \cdot 7^2 + 2 \cdot 7)(4 \cdot 2) = 161 \cdot 8 = 1288$$

Solution 2   We could solve this problem without using the Chain Rule by expressing y explicitly as a function of x:

$$y = u^3 + u^2 + 1 = (2x^2 - 1)^3 + (2x^2 - 1)^2 + 1$$
$$= (8x^6 - 12x^4 + 6x^2 - 1) + (4x^4 - 4x^2 + 1) + 1$$
$$= 8x^6 - 8x^4 + 2x^2 + 1$$

Therefore $\dfrac{dy}{dx} = 48x^5 - 32x^3 + 4x$

$$\frac{dy}{dx}\bigg]_{x=2} = 48 \cdot 2^5 - 32 \cdot 2^3 + 4 \cdot 2 = 1288$$

Solution 1 is clearly preferable.   ●

Let us make explicit the special case of the Chain Rule where the outer function f is a power function. If $y = [g(x)]^n$, then we can write $y = f(u) = u^n$ where $u = g(x)$. By using the Chain Rule and then the Power Rule, we get

$$\frac{dy}{dx} = \frac{dy}{du}\frac{du}{dx} = nu^{n-1}\frac{du}{dx} = n[g(x)]^{n-1}g'(x)$$

The Power Rule Combined with the Chain Rule (3.36)

If n is any real number and $u = g(x)$ is differentiable, then

$$\frac{d}{dx}(u^n) = nu^{n-1}\frac{du}{dx}$$

Alternatively,

$$\frac{d}{dx}[g(x)]^n = n[g(x)]^{n-1} \cdot g'(x)$$

Notice that the derivative in Example 1 could be calculated by taking $n = \frac{1}{2}$ in (3.36).

● **Example 3**   Differentiate $y = (x^3 - 1)^{100}$.

Solution   Taking $u = g(x) = x^3 - 1$ and $n = 100$ in (3.36), we have

$$\frac{dy}{dx} = \frac{d}{dx}(x^3 - 1)^{100} = 100(x^3 - 1)^{99}\frac{d}{dx}(x^3 - 1)$$
$$= 100(x^3 - 1)^{99} \cdot 3x^2$$
$$= 300x^2(x^3 - 1)^{99}$$ ●

● **Example 4**   Find $f'(x)$ if

$$f(x) = \frac{1}{\sqrt[3]{x^2 + x + 1}}$$

Solution   First rewrite f: $f(x) = (x^2 + x + 1)^{-1/3}$. Thus

$$f'(x) = -\frac{1}{3}(x^2 + x + 1)^{-4/3}\frac{d}{dx}(x^2 + x + 1)$$

$$= -\frac{1}{3}(x^2 + x + 1)^{-4/3}(2x + 1)$$

● **Example 5**   Find the derivative of the function

$$g(t) = \left(\frac{t - 2}{2t + 1}\right)^9$$

Solution   Combining the Power Rule, Chain Rule, and Quotient Rule, we get

$$g'(t) = 9\left(\frac{t - 2}{2t + 1}\right)^8 \frac{d}{dt}\left(\frac{t - 2}{2t + 1}\right)$$

$$= 9\left(\frac{t - 2}{2t + 1}\right)^8 \frac{(2t + 1)\cdot 1 - 2(t - 2)}{(2t + 1)^2}$$

$$= \frac{45(t - 2)^8}{(2t + 1)^{10}}$$

● **Example 6**   Differentiate $y = (2x + 1)^5(x^3 - x + 1)^4$.

Solution   In this example we must use the Product Rule before using the Chain Rule:

$$\frac{dy}{dx} = (2x + 1)^5 \frac{d}{dx}(x^3 - x + 1)^4 + (x^3 - x + 1)^4 \frac{d}{dx}(2x + 1)^5$$

$$= (2x + 1)^5 \cdot 4(x^3 - x + 1)^3 \frac{d}{dx}(x^3 - x + 1)$$

$$+ (x^3 - x + 1)^4 \cdot 5(2x + 1)^4 \frac{d}{dx}(2x + 1)$$

$$= 4(2x + 1)^5(x^3 - x + 1)^3(3x^2 - 1) + 5(x^3 - x + 1)^4(2x + 1)^4 \cdot 2$$

By using common factors, we could write the answer as

$$\frac{dy}{dx} = 2(2x + 1)^4(x^3 - x + 1)^3(17x^3 + 6x^2 - 9x + 3)$$

● **Example 7**   Differentiate (a) $y = \sin(x^2)$ and (b) $y = \sin^2 x$.

Solution   (a)   If $u = x^2$, then $y = \sin u$, and the Chain Rule gives

$$\frac{dy}{dx} = \frac{dy}{du}\frac{du}{dx} = \cos u \cdot 2x = 2x\cos(x^2)$$

(b)   Note that $y = \sin^2 x = (\sin x)^2$. Here $u = \sin x$ and $y = u^2$, so the Chain Rule gives

$$\frac{dy}{dx} = \frac{dy}{du}\frac{du}{dx} = 2u \cdot \cos x = 2\sin x \cos x$$

The answer can be left as above or written as

$$\frac{dy}{dx} = \sin 2x$$

●

In Example 7(a) we combined the Chain Rule with Theorem 3.25. In general, if $y = \sin u$, where u is a differentiable function of x, then, by the Chain Rule,

$$\frac{dy}{dx} = \frac{dy}{du}\frac{du}{dx} = \cos u\,\frac{du}{dx}$$

Thus

(3.37)

$$\boxed{\frac{d}{dx}(\sin u) = \cos u\,\frac{du}{dx}}$$

In a similar fashion, all of the formulas for differentiating trigonometric functions can be combined with the Chain Rule.

The reason for the name "Chain Rule" becomes clear when we make a longer chain by adding another link. Suppose that $y = f(u)$, $u = g(x)$, and $x = h(t)$, where f, g, and h are differentiable functions. Then to compute the derivative of y with respect to t, we use the Chain Rule twice:

$$\frac{dy}{dt} = \frac{dy}{dx}\frac{dx}{dt} = \frac{dy}{du}\frac{du}{dx}\frac{dx}{dt}$$

● **Example 8**   If $f(x) = \sin(\cos(\tan x))$, then

$$f'(x) = \cos(\cos(\tan x))\frac{d}{dx}\cos(\tan x)$$

$$= \cos(\cos(\tan x))[-\sin(\tan x)]\frac{d}{dx}(\tan x)$$

$$= -\cos(\cos(\tan x))\sin(\tan x)\sec^2 x$$

Notice that the Chain Rule has been used twice.   ●

● **Example 9**   Let

$$y = (x^4 + \sqrt{x + \sqrt[3]{x - 1}})^{\sqrt{2}}$$

Then

$$\frac{dy}{dx} = \sqrt{2}(x^4 + \sqrt{x + \sqrt[3]{x - 1}})^{\sqrt{2} - 1} \frac{d}{dx}(x^4 + \sqrt{x + \sqrt[3]{x - 1}})$$

$$= \sqrt{2}(x^4 + \sqrt{x + \sqrt[3]{x - 1}})^{\sqrt{2} - 1} \left[4x^3 + \frac{1}{2}(x + \sqrt[3]{x - 1})^{-1/2} \frac{d}{dx}(x + \sqrt[3]{x - 1})\right]$$

$$= \sqrt{2}(x^4 + \sqrt{x + \sqrt[3]{x - 1}})^{\sqrt{2} - 1} \left\{4x^3 + \frac{1}{2\sqrt{x + \sqrt[3]{x - 1}}} \left[1 + \frac{1}{3}(x - 1)^{-2/3} \frac{d}{dx}(x - 1)\right]\right\}$$

$$= \sqrt{2}(x^4 + \sqrt{x + \sqrt[3]{x - 1}})^{\sqrt{2} - 1} \left[4x^3 + \frac{1}{2\sqrt{x + \sqrt[3]{x - 1}}} \left(1 + \frac{1}{3(x - 1)^{2/3}}\right)\right]$$

Notice that (3.36) has been used three times, with $n = \sqrt{2}$, $\frac{1}{2}$, and $\frac{1}{3}$, respectively.   ●

EXERCISES 3.4

In Exercises 1–4 find dy/dx and $dy/dx]_{x=1}$ in two ways: (a) using the Chain Rule and (b) without using the Chain Rule, as in Example 2.

1. $y = u^2, u = x^2 + 2x + 3$

2. $y = u^2 - 2u + 3, u = 5 - 6x$

3. $y = u^3, u = x + \dfrac{1}{x}$

4. $y = u - u^2, u = \sqrt{x} + \sqrt[3]{x}$

In Exercises 5–52 find the derivatives of the given functions.

5. $F(x) = (x^2 + 4x + 6)^5$

6. $F(x) = (x^3 - 5x)^4$

7. $G(x) = (3x - 2)^{10}(5x^2 - x + 1)^{12}$

8. $g(t) = (6t^2 + 5)^3(t^3 - 7)^4$

9. $f(t) = (2t^2 - 6t + 1)^{-8}$

10. $f(t) = \dfrac{1}{(t^2 - 2t - 5)^4}$

11. $g(x) = \sqrt{x^2 - 7x}$

12. $k(x) = \sqrt[3]{1 + \sqrt{x}}$

13. $h(t) = \left(t - \dfrac{1}{t}\right)^{3/2}$

14. $F(s) = \sqrt{s^3 + 1}(s^2 + 1)^4$

15. $F(y) = \left(\dfrac{y - 6}{y + 7}\right)^3$

16. $s(t) = \sqrt[4]{\dfrac{t^3 + 1}{t^3 - 1}}$

17. $f(z) = \dfrac{1}{\sqrt[5]{2z - 1}}$

18. $f(x) = \dfrac{x}{\sqrt{7 - 3x}}$

19. $y = (2x - 5)^4(8x^2 - 5)^{-3}$

20. $y = (x^2 + 1)\sqrt[3]{x^2 + 2}$

21. $y = \tan 3x$

22. $y = 4\sec 5x$

23. $y = \cos(x^3)$

24. $y = \cos^3 x$

25. $y = (1 + \cos^2 x)^6$

26. $y = \tan(x^2) + \tan^2 x$

27. $y = \cos(\tan x)$

28. $y = \sin(\sin x)$

29. $y = \sec^2 2x - \tan^2 2x$

30. $y = \sqrt{1 + 2\tan x}$

31. $y = \csc \dfrac{x}{3}$

32. $y = \cot \sqrt[3]{1 + x^2}$

33. $y = \sin^3 x + \cos^3 x$

34. $y = \sin^2(\cos 4x)$

35. $y = \sin \dfrac{1}{x}$

36. $y = \dfrac{\sin^2 x}{\cos x}$

37. $y = \dfrac{1 + \sin 2x}{1 - \sin 2x}$

38. $y = x \sin \dfrac{1}{x}$

39. $y = \tan^2(x^3)$

40. $y = (\sin \sqrt{x^2 + 1})^{\sqrt{2}}$

41. $y = \cos^2\left(\dfrac{1 - \sqrt{x}}{1 + \sqrt{x}}\right)$

42. $y = \sqrt{1 + \tan\left(x + \dfrac{1}{x}\right)}$

43. $y = \cos^2(\cos x) + \sin^2(\cos x)$

44. $y = \sin(\sin(\sin x))$

45. $y = \sqrt{x + \sqrt{x}}$

46. $y = \sqrt{x + \sqrt{x + \sqrt{x}}}$

47. $f(x) = [x^3 + (2x - 1)^3]^3$

48. $g(t) = \sqrt[4]{(1 - 3t)^4 + t^4}$

49. $p(t) = \left[\left(1 + \dfrac{2}{t}\right)^{-1} + 3t\right]^{-2}$

50. $N(y) = (y + \sqrt[3]{y + \sqrt{2y - 9}})^8$

51. $y = \sin(\tan \sqrt{\sin x})$

52. $y = \sqrt{\cos(\sin^2 x)}$

In Exercises 53–58 find the equations of the tangent lines to the given curves at the given points.

53. $y = (x^3 - x^2 + x - 1)^{10}$, $(1, 0)$

54. $y = \sqrt{x + \dfrac{1}{x}}$, $(1, \sqrt{2})$

55. $y = \dfrac{8}{\sqrt{4 + 3x}}$, $(4, 2)$

56. $y = \dfrac{x}{(3 - x^2)^5}$, $(2, -2)$

57. $y = \cot^2 x$, $\left(\dfrac{\pi}{4}, 1\right)$

58. $y = \sin x + \cos 2x$, $\left(\dfrac{\pi}{6}, 1\right)$

59. Suppose that $F(x) = f(g(x))$ and $g(3) = 6$, $g'(3) = 4$, $f'(3) = 2$, and $f'(6) = 7$. Find $F'(3)$.

60. Suppose that $w = u \circ v$ and $u(0) = 1$, $v(0) = 2$, $u'(0) = 3$, $u'(2) = 4$, $v'(0) = 5$, and $v'(2) = 6$. Find $w'(0)$.

In Exercises 61–64 find f' and state the domains of f and f'.

61. $f(x) = x^2 \sec^2 3x$

62. $f(x) = \sin \sqrt{2x + 1}$

63. $f(x) = \sqrt{\cos \sqrt{x}}$

64. $f(x) = \cos \sqrt{x} + \sqrt{\cos x}$

65. The displacement of a particle on a vibrating string is given by the equation

$$s(t) = 10 + \frac{1}{4}\sin(10\pi t)$$

where s is measured in centimeters and t in seconds. Find the velocity of the particle after t seconds.

66. If the equation of motion of a particle is given by $s = A\cos(\omega t + \delta)$, it is said to undergo *simple harmonic motion*.
(a) Find the velocity of the particle at time t.
(b) When is the velocity 0?

67. Let h be differentiable on $[0, \infty)$ and define G by $G(x) = h(\sqrt{x})$.
(a) Where is G differentiable?
(b) Find an expression for $G'(x)$.

68. Suppose f is differentiable on R and α is a real number. Let $F(x) = f(x^\alpha)$ and $G(x) = [f(x)]^\alpha$. Find expressions for (a) $F'(x)$ and (b) $G'(x)$.

69. Suppose f is differentiable on R. Let $F(x) = f(\cos x)$ and $G(x) = \cos(f(x))$. Find expressions for (a) $F'(x)$ and (b) $G'(x)$.

70. If $g(t) = [f(\sin t)]^2$, where f is a differentiable function, find $g'(t)$.

71. Use the Chain Rule to give easier proofs of Exercise 59 in Section 3.1.

72. (a) If n is a positive integer, prove that

$$\frac{d}{dx}(\sin^n x \cos nx) = n\sin^{n-1} x \cos(n + 1)x$$

(b) Find a similar formula for $D(\cos^n x \cos nx)$.

Find the derivatives of the functions in Exercises 73–75.
(*Hint:* $|x| = \sqrt{x^2}$.)

73. $f(x) = |x|$

74. $g(x) = |x|/x$

75. $h(x) = x|2x - 1|$

76. (a) Sketch the graph of the function $f(x) = |\sin x|$.
  (b) At what points is f not differentiable?
  (c) Give a formula for f' and sketch its graph.
  (d) Do the same for $g(x) = \sin|x|$.

77. Use the Chain Rule to show that

$$\frac{d}{d\theta}(\sin \theta°) = \frac{\pi}{180} \cos \theta°$$

(This gives one reason for the convention that radian measure is always used when dealing with trigonometric functions in calculus: the differentiation formulas would not be as simple if we were to use degrees.)

SECTION 3.5

Implicit Differentiation; Angles between Curves

The functions that we have met so far can be described by expressing one variable explicitly in terms of another variable—for example,

$$y = \sqrt{x^3 + 1} \quad \text{or} \quad y = x \sin x$$

or in general $y = f(x)$. Some functions, however, are defined implicitly by a relation between x and y such as

(3.38)
$$x^2 + y^2 = 25$$

(3.39)   or
$$x^3 + y^3 = 6xy$$

Sometimes it may be possible to solve such an equation for y as an explicit function (or several functions) of x. For instance, if we solve Equation 3.38 for y we get $y = \pm\sqrt{25 - x^2}$, so two functions determined by the implicit Equation 3.38 are $f(x) = \sqrt{25 - x^2}$ and $g(x) = -\sqrt{25 - x^2}$. The graphs of f and g are the upper and lower semicircles of the circle $x^2 + y^2 = 25$ (see Figure 3.11).

It would be very difficult (though possible) to solve Equation 3.39 for y explicitly as a function of x. Nonetheless, it is the equation of a curve called the *folium of Descartes* shown in Figure 3.12 and it implicitly defines y as

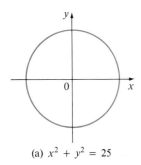

(a) $x^2 + y^2 = 25$

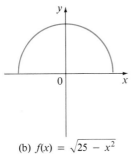

(b) $f(x) = \sqrt{25 - x^2}$

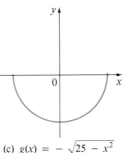

(c) $g(x) = -\sqrt{25 - x^2}$

Figure 3.11

Figure 3.12

THE FOLIUM OF DESCARTES

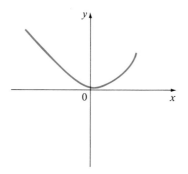

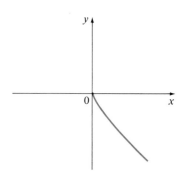

Figure 3.13

several functions of x. The graphs of three such functions are shown in Figure 3.13. When we say that f is a function defined implicitly by Equation 3.39, what we mean is that the equation

$$x^3 + [f(x)]^3 = 6xf(x)$$

is true for all values of x in the domain of f.

Fortunately it is not necessary to solve an equation for y in terms of x in order to find the derivative of y. Instead we can use the method of **implicit differentiation.** This consists of differentiating both sides of the relation with respect to x and then solving the resulting equation for y'. In the examples and exercises of this section it will always be assumed that the given equation determines y implicitly as a differentiable function of x so that the method of implicit differentiation can be applied.

● **Example 1**   (a)   If $x^2 + y^2 = 25$, find $\dfrac{dy}{dx}$.

(b)   Find the equation of the tangent to the circle $x^2 + y^2 = 25$ at the point $(3, 4)$.

Solution 1   (a)   Differentiate both sides of the equation $x^2 + y^2 = 25$:

$$\frac{d}{dx}(x^2 + y^2) = \frac{d}{dx}(25)$$

$$\frac{d}{dx}(x^2) + \frac{d}{dx}(y^2) = 0$$

Remembering that y is a function of x and using the Chain Rule, we have

$$\frac{d}{dx}(y^2) = 2y\frac{dy}{dx}$$

Thus $$2x + 2y\frac{dy}{dx} = 0$$

Now solve this equation for dy/dx:

$$\frac{dy}{dx} = -\frac{x}{y}$$

(b)   At the point $(3, 4)$ we have $x = 3$ and $y = 4$, so

$$\frac{dy}{dx} = -\frac{3}{4}$$

The equation of the tangent to the circle $(3, 4)$ is therefore

$$y - 4 = -\tfrac{3}{4}(x - 3)$$

or $$3x + 4y = 25$$

Solution 2   (b) Solving the equation $x^2 + y^2 = 25$ we get $y = \pm\sqrt{25 - x^2}$. The point $(3, 4)$ lies on the upper semicircle $y = \sqrt{25 - x^2}$, so we consider the function $f(x) = \sqrt{25 - x^2}$. Differentiating f using the Chain Rule, we have

$$f'(x) = \frac{1}{2}(25 - x^2)^{-1/2}\frac{d}{dx}(25 - x^2)$$

$$= \frac{1}{2}(25 - x^2)^{-1/2}(-2x)$$

$$= -\frac{x}{\sqrt{25 - x^2}}$$

$$f'(3) = -\frac{3}{\sqrt{25 - 3^2}} = -\frac{3}{4}$$

and, as in Solution 1, the equation of the tangent is $3x + 4y = 5$.   ●

   Note 1:   Example 1 illustrates the fact that even when it is possible to solve an equation explicitly for y in terms of x, it may be easier to use implicit differentiation.
   Note 2:   The expression $dy/dx = -x/y$ gives the derivative in terms of both x and y. It is correct no matter which function y is determined by the given equation. For instance, for $y = f(x) = \sqrt{25 - x^2}$ we have

$$\frac{dy}{dx} = -\frac{x}{y} = -\frac{x}{\sqrt{25 - x^2}}$$

whereas for $y = g(x) = -\sqrt{25 - x^2}$ we have

$$\frac{dy}{dx} = -\frac{x}{y} = -\frac{x}{-\sqrt{25 - x^2}} = \frac{x}{\sqrt{25 - x^2}}$$

● **Example 2**   (a)   Find y' if $x^3 + y^3 = 6xy$.
   (b)   Find the tangent to the folium of Descartes $x^3 + y^3 = 6xy$ at the point $(3, 3)$.

Solution   (a)   Differentiating both sides of $x^3 + y^3 = 6xy$ with respect to x, regarding y as a function of x, and using the Chain Rule on the y^3 term and the Product Rule on the $6xy$ term, we get

$$3x^2 + 3y^2y' = 6y + 6xy'$$

or

$$x^2 + y^2y' = 2y + 2xy'$$

We now solve for y':

$$(y^2 - 2x)y' = 2y - x^2$$

or

$$y' = \frac{2y - x^2}{y^2 - 2x}$$

(b)   When $x = y = 3$,

$$y' = \frac{2 \cdot 3 - 3^2}{3^2 - 2 \cdot 3} = -1$$

so the equation of the tangent to the folium at $(3, 3)$ is

$$y - 3 = -1(x - 3) \quad \text{or} \quad x + y = 6 \qquad \bullet$$

Note: There is a formula for the three roots of a cubic equation that is like the quadratic formula but much more complicated. If this formula is used to solve the equation $x^3 + y^3 = 6xy$ for y in terms of x, then three functions determined by the equation are

$$y = f(x) = \sqrt[3]{-\frac{x^3}{2} + \sqrt{\frac{x^6}{4} - 8x^3}} + \sqrt[3]{-\frac{x^3}{2} - \sqrt{\frac{x^6}{4} - 8x^3}}$$

and

$$y = \frac{1}{2}\left[-f(x) \pm \sqrt{-3}\left(\sqrt[3]{-\frac{x^3}{2} + \sqrt{\frac{x^6}{4} - 8x^3}} - \sqrt[3]{-\frac{x^3}{2} - \sqrt{\frac{x^6}{4} - 8x^3}}\right)\right]$$

You can see that the method of implicit differentiation saves a lot of work in cases such as this. Moreover, implicit differentiation works just as easily for equations such as

$$y^5 + 3x^2y^2 + 5x^4 = 12$$

which are *impossible* to solve for y in terms of x. [The Norwegian mathematician Niels Abel proved in 1824 that there is no general formula for the roots of a fifth-degree equation. Later the French mathematician Evariste Galois proved that it is impossible to find a general formula for the roots of an nth-degree equation (in terms of algebraic operations on the coefficients) if n is any integer larger than 4.]

● **Example 3**   Find y' if $\sin(x + y) = y^2 \cos x$.

Solution   Differentiating implicitly with respect to x and remembering that y is a function of x, we get

$$\cos(x + y) \cdot (1 + y') = 2yy' \cos x + y^2(-\sin x)$$

(Note that we have used the Chain Rule on the left side and the Product Rule and Chain Rule on the right side.) Solving for y' gives

$$y' = \frac{y^2 \sin x + \cos(x + y)}{2y \cos x - \cos(x + y)}$$ ●

The Angle between Two Curves

Suppose that two smooth curves C_1 and C_2 intersect at a point P as in Figure 3.14(a). Notice that as Q_1 approaches P along C_1 and Q_2 approaches P along C_2, the angle Q_1PQ_2 seems to approach a definite angle that we call the angle between the curves C_1 and C_2. But the limiting direction of the line PQ_1 is the direction of the tangent to C_1 at P and the limiting direction of the line PQ_2 is that of the tangent to C_2 at P. So we define **the angle between the curves C_1 and C_2 at P** as the angle between the tangent lines t_1 and t_2 to these curves at P (if these tangent lines exist) [see Figure 3.14(b)].

Figure 3.14

THE ANGLE α BETWEEN
TWO CURVES

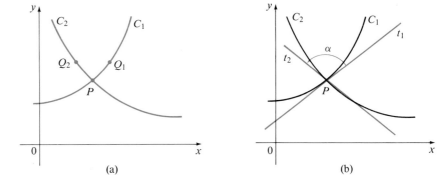

(a)   (b)

If α is the angle between C_1 and C_2 and if t_1 and t_2 have slopes m_1 and m_2, then from Equation 1.16 we have

(3.40)

$$\tan \alpha = \frac{m_2 - m_1}{1 + m_1 m_2}$$

There are two special cases that can occur. If $m_1 = m_2$ (corresponding to $\alpha = 0$), then C_1 and C_2 have the same tangent line at P and we say the curves **C_1 and C_2 are tangent to each other at P.** If $1 + m_1 m_2 = 0$ (corresponding to $\alpha = \pi/2$), then $m_1 = -1/m_2$ and the tangent lines t_1 and t_2 are perpendicular. In this case we say the curves C_1 and C_2 are **orthogonal.**

● **Example 4**   Find the angle α between the parabolas $y = x^2$ and $y = (x - 2)^2$ correct to the nearest degree.

Solution   First we must find the point where the parabolas intersect. This happens when $x^2 = (x - 2)^2$, which gives $x = 1$. If $y = x^2$, then $y' = 2x$, so the slope of the tangent to $y = x^2$ at $(1, 1)$ is $m_1 = 2(1) = 2$. If $y = (x - 2)^2$, then $y' = 2(x - 2)$, so the slope of the tangent to $y = (x - 2)^2$ at $(1, 1)$ is $m_2 = 2(1 - 2) = -2$. Using Formula 3.40, we have

$$\tan \alpha = \frac{m_2 - m_1}{1 + m_1 m_2} = \frac{-2 - 2}{1 + 2(-2)} = \frac{4}{3}$$

and so

$$\alpha = \tan^{-1}\left(\frac{4}{3}\right) \approx 53°$$

This example is illustrated in Figure 3.15.   ●

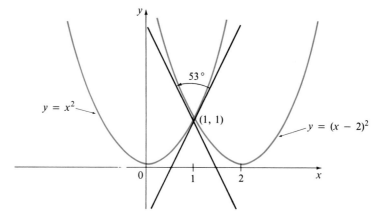

Figure 3.15

● **Example 5**   The equation

(3.41) $xy = c \qquad c \neq 0$

represents a family of hyperbolas. (Different values of the constant c give different hyperbolas. See Figure 3.16.) The equation

(3.42) $x^2 - y^2 = k \qquad k \neq 0$

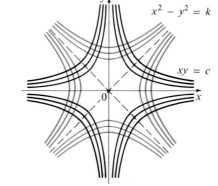

Figure 3.16

represents another family of hyperbolas with asymptotes $y = \pm x$. Show that every curve in the family (3.41) is orthogonal to every curve in the family (3.42); that is, the curves intersect at right angles.

Solution   Implicit differentiation of Equation 3.41 gives

(3.43)
$$y + x\frac{dy}{dx} = 0 \quad \text{so} \quad \frac{dy}{dx} = -\frac{y}{x}$$

Implicit differentiation of Equation 3.42 gives

(3.44)
$$2x - 2y\frac{dy}{dx} = 0 \quad \text{so} \quad \frac{dy}{dx} = \frac{x}{y}$$

From (3.43) and (3.44) we see that at any point of intersection of curves from each family, the slopes of the tangents will be negative reciprocals of each other. Therefore the curves will intersect at right angles.   ●

Two families of curves that are mutually orthogonal (as in Example 5) are called **orthogonal trajectories** of each other and arise in several areas of physics. For example, the lines of force in an electrostatic field are orthogonal to the lines of constant potential.

EXERCISES 3.5

In Exercises 1–6, (a) find y' by implicit differentiation, (b) solve the given equation explicitly for y and differentiate to get y' in terms of x, and (c) check that your solutions to parts (a) and (b) are consistent by substituting the expressions for y into your solution for part (a).

1. $x^2 + 3x + xy = 5$

2. $\dfrac{x^2}{2} + \dfrac{y^2}{4} = 1$

3. $\dfrac{1}{x} + \dfrac{1}{y} = 3$

4. $x^2 + xy - y^2 = 3$

5. $2y^2 + xy = x^2 + 3$

6. $\sqrt{x} + \sqrt{y} = 4$

In Exercises 7–20 find dy/dx by implicit differentiation.

7. $x^2 - xy + y^3 = 8$

8. $\sqrt{xy} - 2x = \sqrt{y}$

9. $2y^2 + \sqrt[3]{xy} = 3x^2 + 17$

10. $y^5 + 3x^2y^2 + 5x^4 = 12$

11. $x^4 + y^4 = 16$

12. $\sqrt{x + y} + \sqrt{xy} = 6$

13. $2xy = (x^2 + y^2)^{3/2}$

14. $x^2 = \dfrac{y^2}{y^2 - 1}$

15. $\dfrac{y}{x - y} = x^2 + 1$

16. $x\sqrt{1 + y} + y\sqrt{1 + 2x} = 2x$

17. $\cos(x - y) = y \sin x$

18. $x \sin y + \cos 2y = \cos y$

19. $xy = \cot(xy)$

20. $x \cos y + y \cos x = 1$

In Exercises 21 and 22 regard y as the independent variable and x as the dependent variable, and use implicit differentiation to find dx/dy.

21. $y^4 + x^2y^2 + yx^4 = y + 1$

22. $(x^2 + y^2)^2 = ax^2y$

23. If $x[f(x)]^3 + xf(x) = 6$ and $f(3) = 1$, find $f'(3)$.

24. If $[g(x)]^2 + 12x = x^2g(x)$ and $g(4) = 12$, find $g'(4)$.

In Exercises 25–30 find an equation of the tangent line to the given curve at the given point. In each case sketch the curve and the tangent. The curves in Exercises 27–30 are sketched in Figures 3.17–3.20.

25. $\dfrac{x^2}{16} - \dfrac{y^2}{9} = 1, \left(-5, \dfrac{9}{4}\right)$ **26.** $\dfrac{x^2}{9} + \dfrac{y^2}{36} = 1, (-1, 4\sqrt{2})$
(hyperbola)   (ellipse)

27. $y^2 = x^3(2 - x), (1, 1)$
(piriform)

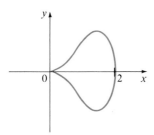

Figure 3.17   A PIRIFORM

28. $x^{2/3} + y^{2/3} = 4, (-3\sqrt{3}, 1)$
(astroid)

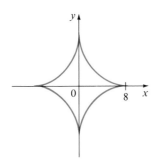

Figure 3.18   AN ASTROID

29. $2(x^2 + y^2)^2 = 25(x^2 - y^2), (3, 1)$
(lemniscate)

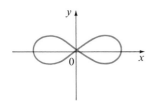

Figure 3.19   A LEMNISCATE

30. $x^2y^2 = (y + 1)^2(4 - y^2), (0, -2)$
(conchoid of Nicomedes)

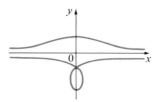

Figure 3.20   A CONCHOID OF NICOMEDES

31. Find the points on the lemniscate in Exercise 29 where the tangent is horizontal.

32. Show by implicit differentiation that the tangent to the ellipse

$$\frac{x^2}{a^2} + \frac{y^2}{b^2} = 1$$

at the point (x_0, y_0) is

$$\frac{x_0 x}{a^2} + \frac{y_0 y}{b^2} = 1$$

33. Find the equation of the tangent line to the hyperbola

$$\frac{x^2}{a^2} - \frac{y^2}{b^2} = 1$$

at the point (x_0, y_0).

34. Show that the sum of the x- and y-intercepts of any tangent line to the curve $\sqrt{x} + \sqrt{y} = \sqrt{c}$ is equal to c.

35. Show, using implicit differentiation, that any tangent line at a point P to a circle with center O is perpendicular to the radius OP.

36. The Power Rule (3.16) can be proved using implicit differentiation for the case where n is a rational number, $n = p/q$, and $y = f(x) = x^n$ is assumed beforehand to be a differentiable function. If $y = x^{p/q}$, then $y^q = x^p$. Use implicit differentiation to show that

$$y' = \frac{p}{q} x^{(p/q) - 1}$$

In Exercises 37–50 find, correct to the nearest degree, the angle between the given curves at each point of intersection.

37. $y = 2x, 3x + y = 5$ **38.** $y = x, y = x^2$

39. $y = x^2, y = x^3$ **40.** $y = x^2, y = 8 - x^2$

41. $y = x^2, xy = 1$ **42.** $y = x^3, y = 2x - x^2$

43. $x - 2y + 5 = 0, x^2 + y^2 = 25$

44. $x^2 + y^2 = 16, x^2 = 6y$

45. $x^2 - y^2 = 3, x^2 - 4x + y^2 + 3 = 0$

46. $x^2 + y^2 = 1, x^2 + y^2 - 2y = 0$

47. $xy = 1, x^2y = 1$

48. $y = x^2 + x, y = x^3 + 1$

49. $y = \sin x, y = \cos x$ **50.** $y = \sin x, y = \sin 2x$

In Exercises 51 and 52 find the point where the given curves are tangent to each other.

51. $y = x^3 + x, y = 2x^2 + x$

52. $y = x - 2x^2, y = x^3 + 2x$

In Exercises 53 and 54 show that the given curves are orthogonal.

53. $2x^2 + y^2 = 3, x = y^2$

54. $x^2 - y^2 = 5, 4x^2 + 9y^2 = 72$

In Exercises 55–58 show that the given families of curves are orthogonal trajectories of each other. Sketch both families of curves.

55. $x^2 + y^2 = r^2, ax + by = 0$

56. $x^2 + y^2 = ax, x^2 + y^2 = by$

57. $y = cx^2, x^2 + 2y^2 = k$

58. $y = ax^3, x^2 + 3y^2 = b$

59. For what value of c is the parabola $y = cx^2$ orthogonal to the line $y = 1 - x$ at one of their points of intersection?

60. For what values of a and b will the parabola $y = x^2 + ax + b$ be tangent to the curve $y = x^3$ at the point $(1, 1)$?

SECTION 3.6

Higher Derivatives

If f is a differentiable function, then its derivative f' is also a function, so f' may have a derivative of its own, denoted by $(f')' = f''$. This new function f'' is called the **second derivative** of f because it is the derivative of the derivative of f. Thus

$$f''(x) = \frac{d}{dx}(f'(x)) = \frac{d}{dx}\left(\frac{d}{dx}f(x)\right)$$

For instance, if $f(x) = x^8$, then $f'(x) = 8x^7$, so

$$f''(x) = \frac{d}{dx}(f'(x)) = \frac{d}{dx}(8x^7) = 56x^6$$

Notation: If $y = f(x)$, then

$$y'' = f''(x) = \frac{d}{dx}\left(\frac{dy}{dx}\right) = \frac{d^2y}{dx^2} = D^2f(x) = D_x^2 f(x)$$

The symbol D^2 indicates that the operation of differentiation is performed twice.

Similarly the *third derivative* f''' is the derivative of the second derivative: $f''' = (f'')'$. If $y = f(x)$, then alternative notations are

$$y''' = f'''(x) = \frac{d}{dx}\left(\frac{d^2y}{dx^2}\right) = \frac{d^3y}{dx^3} = D^3f(x) = D_x^3 f(x)$$

The process can be continued. The fourth derivative f'''' is usually denoted by $f^{(4)}$. In general the nth derivative of f is denoted by $f^{(n)}$ and is obtained from f by differentiating n times. If $y = f(x)$, we write

$$y^{(n)} = f^{(n)}(x) = \frac{d^n y}{dx^n} = D^n f(x) = D_x^n f(x)$$

● **Example 1**   If

$$y = x^3 - 6x^2 - 5x + 3$$

then

$$y' = 3x^2 - 12x - 5$$

$$y'' = 6x - 12$$

$$y''' = 6$$

$$y^{(4)} = 0$$

and in fact $y^{(n)} = 0$ for all $n \geqslant 4$.   ●

● **Example 2**   If $f(x) = \dfrac{1}{x}$, find $f^{(n)}(x)$.

Solution $f(x) = \dfrac{1}{x} = x^{-1}$

$$f'(x) = -x^{-2} = \frac{-1}{x^2}$$

$$f''(x) = (-2)(-1)x^{-3} = \frac{2}{x^3}$$

$$f'''(x) = -3 \cdot 2 \cdot 1 \cdot x^{-4}$$

$$f^{(4)}(x) = 4 \cdot 3 \cdot 2 \cdot 1 \cdot x^{-5}$$

$$f^{(5)}(x) = -5 \cdot 4 \cdot 3 \cdot 2 \cdot 1 \cdot x^{-6} = -5! x^{-6}$$

$$\vdots$$

$$f^{(n)}(x) = (-1)^n n(n-1)(n-2) \cdots 2 \cdot 1 \cdot x^{-(n+1)}$$

or $f^{(n)}(x) = \dfrac{(-1)^n n!}{x^{n+1}}$

Here we have used the factorial symbol $n!$ for the product of the first n positive integers.

$$\boxed{n! = 1 \cdot 2 \cdot 3 \cdots (n-1)n}$$ ●

If $s = f(t)$ is the position function of an object that moves in a straight line, we know that its first derivative represents the velocity $v(t)$ of the object as a function of time:

$$v(t) = f'(t) = \frac{ds}{dt}$$

The instantaneous rate of change of velocity with respect to time is called the **acceleration** $a(t)$ of the object. Thus the acceleration function is the derivative of the velocity function and is therefore the second derivative of the position function:

$$a(t) = v'(t) = f''(t)$$

or, in Leibniz notation,

$$a = \frac{dv}{dt} = \frac{d^2s}{dt^2}$$

● **Example 3**   The equation of motion of a particle is $s = 2t^3 - 5t^2 + 3t + 4$, where s is measured in centimeters and t in seconds. Find the acceleration as a function of time. What is the acceleration after 2 s?

Solution   The velocity and acceleration are

$$v(t) = \frac{ds}{dt} = 6t^2 - 10t + 3$$

$$a(t) = \frac{dv}{dt} = 12t - 10$$

The acceleration after 2 s is $a(2) = 14 \text{ cm/s}^2$.   ●

● **Example 4**   Find y'' if $x^4 + y^4 = 16$.

Solution   Differentiating the equation implicitly with respect to x, we get

$$4x^3 + 4y^3 y' = 0$$

Solving for y' gives

(3.45)
$$y' = -\frac{x^3}{y^3}$$

To find y'' we differentiate this expression for y' using the Quotient Rule and remembering that y is a function of x:

$$y'' = \frac{d}{dx}\left(-\frac{x^3}{y^3}\right) = -\frac{y^3 D(x^3) - x^3 D(y^3)}{(y^3)^2}$$

$$= -\frac{y^3 \cdot 3x^2 - x^3(3y^2 y')}{y^6}$$

If we now substitute Equation 3.45 into this expression for y'' we get

$$y'' = -\frac{3x^2 y^3 - 3x^3 y^2\left(\dfrac{-x^3}{y^3}\right)}{y^6}$$

$$= -\frac{3(x^2 y^4 + x^6)}{y^7} = -\frac{3x^2(y^4 + x^4)}{y^7}$$

But the values of x and y must satisfy the original equation $x^4 + y^4 = 16$. So the answer simplifies to

$$y'' = -\frac{3x^2(16)}{y^7} = -48\frac{x^2}{y^7}$$
●

● **Example 5**   Find $D^{27}\cos x$.

Solution   The first few derivatives of $\cos x$ are as follows:

$$D\cos x = -\sin x$$
$$D^2\cos x = -\cos x$$
$$D^3\cos x = \sin x$$
$$D^4\cos x = \cos x$$
$$D^5\cos x = -\sin x$$

We see that the successive derivatives occur in a cycle of length 4 and, in particular, $D^n\cos x = \cos x$ whenever n is a multiple of 4. Therefore

$$D^{24}\cos x = \cos x$$

and, differentiating three more times, we have

$$D^{27}\cos x = \sin x$$
●

EXERCISES 3.6

In Exercises 1–14 find the first and second derivatives of the given function.

1. $f(x) = x^4 - 3x^3 + 16x$

2. $f(t) = t^{10} - 2t^7 + t^4 - 6t + 8$

3. $h(x) = \sqrt{x^2 + 1}$

4. $G(r) = \sqrt{r} + \sqrt[3]{r}$

5. $F(s) = (3s + 5)^8$

6. $g(u) = \dfrac{1}{\sqrt{1 - u}}$

7. $y = \dfrac{x}{1 - x}$

8. $y = x^\pi$

9. $y = (1 - x^2)^{3/4}$

10. $y = \dfrac{x^2}{x + 1}$

11. $H(t) = \tan^3(2t - 1)$

12. $f(x) = \csc^2(5x)$

13. $F(r) = \sec\sqrt{r}$

14. $g(s) = s^2\cos s$

In Exercises 15–18 find y'''.

15. $y = ax^2 + bx + c$

16. $y = \dfrac{1 - x}{1 + x}$

17. $y = \sqrt{5t - 1}$

18. $y = \dfrac{1}{1 + x^2}$

19. If $f(x) = (2 - 3x)^{-1/2}$, find $f(0)$, $f'(0)$, $f''(0)$, and $f'''(0)$.

20. If $g(t) = (2 - t^2)^6$, find $g(0)$, $g'(0)$, $g''(0)$, and $g'''(0)$.

21. If $f(\theta) = \cot\theta$, find $f'''\left(\dfrac{\pi}{6}\right)$.

22. If $g(x) = \sec x$, find $g^{(4)}\left(\dfrac{\pi}{4}\right)$.

In Exercises 23–26 find y'' by implicit differentiation.

23. $x^3 + y^3 = 1$ **24.** $\sqrt{x} + \sqrt{y} = 1$

25. $x^2 + 6xy + y^2 = 8$ **26.** $\dfrac{x^2}{a^2} - \dfrac{y^2}{b^2} = 1$

27. Find the first 73 derivatives of
$$f(x) = x - x^2 + x^3 - x^4 + x^5 - x^6$$

In Exercises 28–31 find a formula for $f^{(n)}(x)$.

28. $f(x) = \sqrt{x}$ **29.** $f(x) = x^n$

30. $f(x) = \dfrac{1}{(1 - x)^2}$ **31.** $f(x) = \dfrac{1}{3x^3}$

32. Find $D^{99} \sin x$.

33. Find $D^{50} \cos 2x$.

34. Find $D^{35} x \sin x$.

In Exercises 35–38 the equation of motion is given for a particle, where s is in meters and t is in seconds. Find (a) the velocity and acceleration as functions of t, (b) the acceleration after 1 s, and (c) the acceleration at the instants when the velocity is 0.

35. $s = t^3 - 3t$

36. $s = t^2 - t + 1$

37. $s = At^2 + Bt + C$

38. $s = 2t^3 - 7t^2 + 4t + 1$

In Exercises 39 and 40 an equation of motion is given, where s is in meters and t in seconds. (a) At what time is the acceleration 0? (b) Find the displacement and velocity at these times.

39. $s = t^4 - 4t^3 + 2$ **40.** $s = 2t^3 - 9t^2$

41. A mass attached to a vertical spring has position function given by $y(t) = A \sin \omega t$, where A is the amplitude of its oscillations and ω is a constant.
 (a) Find the velocity and acceleration as functions of time.
 (b) Show that the acceleration is proportional to the displacement y.
 (c) Show that the speed is a maximum when the acceleration is 0.

42. If $f(x) = |x^2 - x|$, find f' and f''. What are their domains? Sketch the graphs of all three functions.

43. Find a second-degree polynomial P such that $P(2) = 5$, $P'(2) = 3$, and $P''(2) = 2$.

44. Find a third-degree polynomial Q such that $Q(1) = 1$, $Q'(1) = 3$, $Q''(1) = 6$, and $Q'''(1) = 12$.

45. If P is a polynomial of degree n, show that $P^{(m)}(x) = 0$ for $m > n$.

46. (a) If $F(x) = f(x)g(x)$, where f and g have derivatives of all orders, show that
$$F'' = f''g + 2f'g' + fg''$$
 (b) Find similar formulas for F''' and $F^{(4)}$.
 (c) Guess a formula for $F^{(n)}$.

47. If $y = f(u)$ and $u = g(x)$, where f and g are twice differentiable functions, show that
$$\frac{d^2y}{dx^2} = \frac{d^2y}{du^2}\left(\frac{du}{dx}\right)^2 + \frac{dy}{du}\frac{d^2u}{dx^2}$$

SECTION 3.7

Rates of Change in the Natural and Social Sciences

Recall from Section 3.1 that if $y = f(x)$, then the derivative dy/dx can be interpreted as the rate of change of y with respect to x. In this section we examine some of the applications of this idea to physics, chemistry, biology, economics, and other sciences.

Let us recall the basic idea behind rates of change. If x changes from x_1 to x_2, then the change in x is
$$\Delta x = x_2 - x_1$$
and the corresponding change in y is
$$\Delta y = f(x_2) - f(x_1)$$

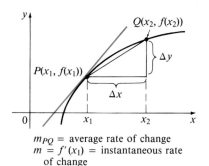

m_{PQ} = average rate of change
$m = f'(x_1)$ = instantaneous rate
of change

Figure 3.21

The difference quotient

$$\frac{\Delta y}{\Delta x} = \frac{f(x_2) - f(x_1)}{x_2 - x_1}$$

is the **average rate of change of y with respect to x** over the interval $[x_1, x_2]$ and can be interpreted as the slope of the secant line PQ in Figure 3.21. Its limit as $\Delta x \to 0$ is the derivative $f'(x_1)$, which can therefore be interpreted as the instantaneous rate of change of y with respect to x or the slope of the tangent line at $P(x_1, f(x_1))$. Using the Leibniz notation, we write the process in the form

$$\frac{dy}{dx} = \lim_{\Delta x \to 0} \frac{\Delta y}{\Delta x}$$

Whenever the function $y = f(x)$ has a specific interpretation in one of the sciences, its derivative will have a specific interpretation as a rate of change. We now look at some of these interpretations in the natural and social sciences.

Physics

If $s = f(t)$ is the position function of a particle that is moving in a straight line, then $\Delta s / \Delta t$ represents the average velocity over a time period Δt, and $v = ds/dt$ represents the instantaneous velocity (the rate of change of displacement with respect to time). This has already been discussed in Sections 2.1 and 3.1. The rate of change of velocity with respect to time is acceleration $(a = dv/dt = d^2s/dt^2)$, which was discussed in Section 3.6.

● **Example 1**   If a rod or piece of wire is homogeneous, then its linear density is uniform and is defined as the mass per unit length $(\rho = m/l)$ and measured in kilograms per meter. Suppose, however, that the rod is not homogeneous but that its mass measured from its left end to a point x is $m = f(x)$ as shown in Figure 3.22.

Figure 3.22

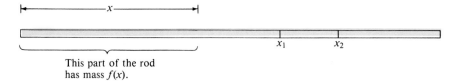

This part of the rod
has mass $f(x)$.

The mass of the part of the rod that lies between $x = x_1$ and $x = x_2$ is $\Delta m = f(x_2) - f(x_1)$ so the average density of that part of the rod is given by

$$\text{average density} = \frac{\Delta m}{\Delta x} = \frac{f(x_2) - f(x_1)}{x_2 - x_1}$$

If we now let $\Delta x \to 0$ (that is, $x_2 \to x_1$), we are computing the average density over a smaller and smaller interval. The **linear density** ρ at x_1 is the limit of these average densities as $\Delta x \to 0$; that is, the linear density is the rate of change of mass with respect to length. Symbolically,

$$\rho = \lim_{\Delta x \to 0} \frac{\Delta m}{\Delta x} = \frac{dm}{dx}$$

Thus the linear density of the rod is the derivative of mass with respect to length.

For instance, if $m = f(x) = \sqrt{x}$, where x is measured in meters and m in kilograms, then the average density of the part of the rod given by $1 \leqslant x \leqslant 1.2$ is

$$\frac{\Delta m}{\Delta x} = \frac{f(1.2) - f(1)}{1.2 - 1} = \frac{\sqrt{1.2} - 1}{0.2} \approx 0.48 \text{ kg/m}$$

while the density right at $x = 1$ is

$$\rho = \frac{dm}{dx}\bigg]_{x=1} = \frac{1}{2\sqrt{x}}\bigg]_{x=1} = 0.50 \text{ kg/m}$$

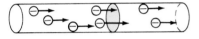

Figure 3.23

● **Example 2**   A current exists whenever electric charges move. Figure 3.23 shows part of a wire and electrons moving through a shaded plane surface. If ΔQ is the net charge that passes through this surface during a time period Δt, then the average current during this time interval is defined as

$$\text{average current} = \frac{\Delta Q}{\Delta t} = \frac{Q_2 - Q_1}{t_2 - t_1}$$

If we take the limit of this average current over smaller and smaller time intervals, we get what is called the **current** I at a given time t_1:

$$I = \lim_{\Delta t \to 0} \frac{\Delta Q}{\Delta t} = \frac{dQ}{dt}$$

Thus the current is the rate at which charge flows through a surface.   ●

Velocity, acceleration, density, and current are not the only rates of change that are important in physics. Others include power (the rate at which work is done), the rate of heat flow, temperature gradient (the rate of change of temperature with respect to position), and the rate of decay of a radioactive substance in nuclear physics.

Chemistry

● **Example 3**   A chemical reaction results in the formation of one or more substances (called products) from one or more starting materials (called reactants). For instance, the "equation"

$$2H_2 + O_2 \to 2H_2O$$

indicates that two molecules of hydrogen and one molecule of oxygen form two molecules of water. Let us consider the reaction

$$A + B \to C$$

where A and B are the reactants and C is the product. The **concentration** of a reactant A is the number of moles (6.022×10^{23} molecules) per liter and is denoted by $[A]$. The concentration will vary during a reaction, so $[A]$, $[B]$, and $[C]$ are all functions of time (t). The average rate of reaction of the product C over a time interval $t_1 \leqslant t \leqslant t_2$ is

$$\frac{\Delta[C]}{\Delta t} = \frac{[C](t_2) - [C](t_1)}{t_2 - t_1}$$

But chemists are more interested in the **instantaneous rate of reaction,** which is obtained by taking the limit of the average rate of reaction as the time interval Δt approaches 0:

$$\text{rate of reaction} = \lim_{\Delta t \to 0} \frac{\Delta[C]}{\Delta t} = \frac{d[C]}{dt}$$

Since the concentration of the product will be increasing as the reaction proceeds, the derivative $d[C]/dt$ will be positive. (You can see intuitively that the slope of the tangent to the graph of an increasing function will be positive.) Thus the rate of reaction of C is positive. The concentrations of the reactants, however, will decrease during the reaction, so, to make the rates of reaction of A and B positive numbers, we put minus signs in front of the derivatives $d[A]/dt$ and $d[B]/dt$. Since $[A]$ and $[B]$ each decrease at the same rate that $[C]$ increases, we have

$$\text{rate of reaction} = \frac{d[C]}{dt} = -\frac{d[A]}{dt} = -\frac{d[B]}{dt}$$

More generally, it turns out that for a reaction of the form

$$aA + bB \to cC + dD$$

we have

$$-\frac{1}{a}\frac{d[A]}{dt} = -\frac{1}{b}\frac{d[B]}{dt} = \frac{1}{c}\frac{d[C]}{dt} + \frac{1}{d}\frac{d[D]}{dt}$$

The rate of reaction can be determined by graphical methods (see Exercise 12). In some cases we can use the rate of reaction to find explicit formulas for the concentrations as functions of time (see Exercise 7 in Section 6.7 and Exercise 25 in Section 8.8). ●

● **Example 4**   One of the quantities of interest in thermodynamics is compressibility. If a given substance is kept at a constant temperature, then its volume V will depend on its pressure P. We can consider the rate of change of volume with respect to pressure—namely, the derivative dV/dP.

As P increases, V will decrease, so $dV/dP < 0$. The compressibility is defined by introducing a minus sign and dividing this derivative by the volume V:

$$\text{isothermal compressibility} = \beta = -\frac{1}{V}\frac{dV}{dP}$$

Thus β measures how fast, per unit volume, the volume of a substance decreases as the pressure on it increases at constant temperature.

For instance, the volume V (in cubic meters) of a sample of air at $25°$ C was found to be related to the pressure P (in kilopascals) by the equation

$$V = \frac{5.3}{P}$$

The rate of change of V with respect to P when $P = 50$ kPa is

$$\frac{dV}{dP}\bigg]_{P=50} = -\frac{5.3}{P^2}\bigg]_{P=50}$$

$$= -\frac{5.3}{2500} = -0.00212 \text{ m}^3/\text{kPa}$$

The compressibility at that pressure is

$$\beta = -\frac{1}{V}\frac{dV}{dP}\bigg]_{P=50}$$

$$= \frac{0.00212}{\dfrac{5.3}{50}} = 0.02 \ (\text{m}^3/\text{kPa})/\text{m}^3$$

Biology

● **Example 5**   Let $n = f(t)$ be the number of individuals in an animal or plant population at time t. The change in the population size between $t = t_1$ and $t = t_2$ is $\Delta n = f(t_2) - f(t_1)$, and so the average rate of growth during the time period $t_1 \leqslant t \leqslant t_2$ is

$$\text{average rate of growth} = \frac{\Delta n}{\Delta t} = \frac{f(t_2) - f(t_1)}{t_2 - t_1}$$

The **instantaneous rate of growth** is obtained from this average rate of growth by letting the time period Δt approach 0:

$$\text{growth rate} = \lim_{\Delta t \to 0} \frac{\Delta n}{\Delta t} = \frac{dn}{dt}$$

Strictly speaking, this is not quite accurate because the actual graph of a population function $n = f(t)$ would be a step function that is discontinuous whenever a birth or death occurs and therefore not differentiable. However

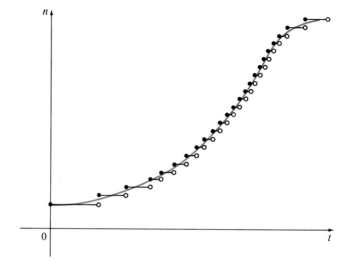

Figure 3.24

A SMOOTH CURVE
APPROXIMATING
A GROWTH FUNCTION

for a large animal or plant population we can replace the graph by a smooth approximating curve as in Figure 3.24.

To be more specific, consider a population of the bacteria cells in a homogeneous nutrient medium. Suppose that by sampling the population at certain intervals it is determined that the population doubles every hour. If the initial population is n_0 and the time t is measured in hours, then

$$f(1) = 2n_0$$
$$f(2) = 2f(1) = 2^2 n_0$$
$$f(3) = 2f(2) = 2^3 n_0$$

and, in general,

$$f(t) = 2^t n_0$$

The population function is $n = n_0 2^t$.

This is an example of an exponential function. In Chapter 6 we discuss exponential functions in general and at that time we will be able to compute their derivatives and thereby determine the rate of growth of the bacteria population.   ●

● **Example 6**   When we consider the flow of blood through a blood vessel, such as a vein or artery, we can take the shape of the blood vessel to be a cylindrical tube with radius R and length l as illustrated in Figure 3.25.

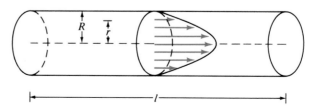

Figure 3.25

BLOOD FLOW IN AN ARTERY

Because of friction at the walls of the tube, the velocity v of the blood is greatest along the central axis of the tube and decreases as the distance r from the axis increases until v becomes 0 at the wall. The relationship between v and r is given by the **law of laminar flow** discovered by the French physician Poiseuille in 1840. This states that

(3.46)
$$v = \frac{P}{4\eta l}(R^2 - r^2)$$

where η is the viscosity of the blood and P is the pressure difference between the ends of the tube. If P and l are constant, then v is a function of r with domain $[0, R]$. [For more detailed information see D. A. McDonald, *Blood Flow in Arteries* (London: Arnold, 1960).]

The average rate of change of the velocity as we move from $r = r_1$ outward to $r = r_2$ is

$$\frac{\Delta v}{\Delta r} = \frac{v(r_2) - v(r_1)}{r_2 - r_1}$$

and if we let $\Delta r \to 0$ we obtain the instantaneous rate of change of velocity with respect to r:

$$\text{velocity gradient} = \lim_{\Delta r \to 0} \frac{\Delta v}{\Delta r} = \frac{dv}{dr}$$

Using Equation 3.46, we obtain

$$\frac{dv}{dr} = \frac{P}{4\eta l}(0 - 2r) = -\frac{Pr}{2\eta l}$$

In a typical human artery we can take $\eta = 0.027$, $R = 0.008$ cm, $l = 2$ cm, and $P = 4000$ dynes/cm^2, which gives

$$v = \frac{4000}{4(0.027)2}(0.000064 - r^2)$$

$$\approx 1.85 \times 10^4(6.4 \times 10^{-5} - r^2)$$

At $r = 0.002$ cm the blood is flowing at a speed of

$$v(0.002) \approx 1.85 \times 10^4(64 \times 10^{-6} - 4 \times 10^{-6})$$

$$= 1.11 \text{ cm/s}$$

and the velocity gradient at that point is

$$\frac{dv}{dr}\bigg]_{r=0.002} = -\frac{4000(0.002)}{2(0.027)2}$$

$$\approx -74 \text{ (cm/s)/cm}$$

Economics

● **Example 7**   Suppose that $C(x)$ is the total cost that a company incurs in producing x units of a certain commodity. The function C is called a **cost function.** If the number of items produced is increased from x_1 to x_2, the additional cost is $\Delta C = C(x_2) - C(x_1)$, and the average rate of change of the cost is

$$\frac{\Delta C}{\Delta x} = \frac{C(x_2) - C(x_1)}{x_2 - x_1} = \frac{C(x_1 + \Delta x) - C(x_1)}{\Delta x}$$

The limit of this quantity as $\Delta x \to 0$, that is, the instantaneous rate of change of cost with respect to the number of items produced, is called the **marginal cost** by economists:

$$\text{marginal cost} = \lim_{\Delta x \to 0} \frac{\Delta C}{\Delta x} = \frac{dC}{dx}$$

[Since x can usually take on only integer values, it may not make literal sense to let Δx approach 0, but we can always replace $C(x)$ by a smooth approximating function as in Example 5.]
    Taking $\Delta x = 1$ and n large, we have

$$C'(n) \approx C(n + 1) - C(n)$$

Thus the marginal cost of producing n units is approximately equal to the cost of producing one more unit [the $(n + 1)$st unit].
    It is often appropriate to represent a total cost function by a polynomial

$$C(x) = a + bx + cx^2 + dx^3$$

where a represents the overhead cost (rent, heat, maintenance) and the other terms represent the cost of raw materials, labor, and so on. (The cost of raw materials may be proportional to x, but labor costs might depend partly on higher powers of x because of overtime costs and inefficiencies involved in large-scale operations.)
    For instance, suppose a company has estimated that the cost (in dollars) of producing x items is

$$C(x) = 10,000 + 5x + 0.01x^2$$

Then the marginal cost function is

$$C'(x) = 5 + 0.02x$$

The marginal cost at the production level of 500 items is

$$C'(500) = 5 + (0.02)500 = \$15/\text{item}$$

This gives the rate at which costs are increasing with respect to the production level when $x = 500$.

The cost of producing the 501st item is

$$C(501) - C(500) = [10{,}000 + 5(501) + 0.01(501)^2]$$
$$- [10{,}000 + 5(500) + 0.01(500)^2]$$
$$= \$15.01$$

Notice that $C'(500) \approx C(501) - C(500)$.   ●

Economists also study marginal demand, marginal revenue, and marginal profit, which are the derivatives of the demand, revenue, and profit functions. These will be considered in Chapter 4 after we have developed techniques for finding the maximum and minimum values of functions.

Other Sciences

Rates of change occur in all the sciences. A geologist is interested in knowing the rate at which an intruded body of molten rock cools by conduction of heat into surrounding rocks. An engineer wants to know the rate at which water flows into or out of a reservoir. An urban geographer is interested in the rate of change of the population density in a city as the distance from the city center increases. A meteorologist is concerned with the rate of change of atmospheric pressure with respect to height. (See Exercise 13 in Section 6.7.)

In psychology, those interested in learning theory study the so-called learning curve, which graphs the performance $P(t)$ of someone learning a skill as a function of the training time t. Of particular interest is the rate at which performance improves as time passes, that is, dP/dt.

In sociology, differential calculus is used in analyzing the spread of rumors (or innovations or fads or fashions). If $p(t)$ denotes the proportion of a population that knows a rumor by time t, then the derivative dp/dt represents the rate of spread of the rumor. (See Exercise 60 in Section 6.5.)

Summary

Velocity, density, current, power, and temperature gradient in physics, rate of reaction and compressibility in chemistry, rate of growth and blood velocity gradient in biology, marginal cost and marginal profit in economics, rate of heat flow in geology, rate of improvement of performance in psychology, rate of spread of a rumor in sociology—these are all special cases of a single mathematical concept, the derivative.

This is an illustration of the fact that part of the power of mathematics lies in its abstractness. A single abstract mathematical concept (such as the derivative) can have different interpretations in each of the sciences. When we develop the properties of the mathematical concept once and for all, we can then turn around and apply these results to all of the sciences. This is much more efficient than developing properties of special concepts in each separate science. The French mathematician Joseph Fourier (1768–1830) put it succinctly: "Mathematics compares the most diverse phenomena and discovers the secret analogies which unite them."

EXERCISES 3.7

1. (a) Find the average rate of change of the volume of a cube with respect to its edge length x as x changes from
   (i) 5 to 6   (ii) 5 to 5.1   (iii) 5 to 5.01
   (b) Find the instantaneous rate of change when $x = 5$.
   (c) Show that the rate of change of the volume of a cube with respect to its edge length (at any x) is equal to half the surface area of the cube.

2. (a) Find the average rate of change of the area of a circle with respect to its radius r as r changes from
   (i) 2 to 3   (ii) 2 to 2.5   (iii) 2 to 2.1
   (b) Find the instantaneous rate of change when $r = 2$.
   (c) Show that the rate of change of the area of a circle with respect to its radius (at any r) is equal to the circumference of the circle.

3. A stone is dropped into a lake, creating a circular ripple that travels outward at a speed of 60 cm/s. Find the rate at which the area within the circle is increasing after (a) 1 s, (b) 3 s, and (c) 5 s.

4. (a) The volume of a growing spherical cell is $V = \frac{4}{3}\pi r^3$, where the radius r is measured in micrometers ($1\ \mu\text{m} = 10^{-6}$ m). Find the average rate of change of V with respect to r when r changes from
   (i)   5 to 8 μm   (ii) 5 to 6 μm
   (iii) 5 to 5.1 μm
   (b) Find the instantaneous rate of change of V with respect to r when $r = 5\ \mu$m.

5. A spherical balloon is being inflated. Find the rate of increase of the surface area ($S = 4\pi r^2$) with respect to the radius r when r is (a) 1 ft, (b) 2 ft, and (c) 3 ft.

6. Show that the rate of change of the volume of a sphere with respect to its radius is equal to its surface area.

7. The mass of the part of a metal rod that lies between its left end and a point x meters to the right is $3x^2$ kg. Find the linear density (see Example 1) when x is (a) 1 m, (b) 2 m, and (c) 3 m.

8. If a tank holds 5000 gallons of water that drain from the bottom of the tank in 40 min, then Torricelli's Law gives the volume V of water remaining in the tank after t minutes as

$$V = 5000\left(1 - \frac{t}{40}\right)^2 \qquad 0 \leqslant t \leqslant 40$$

Find the rate at which water is draining from the tank after (a) 5 min, (b) 10 min, and (c) 20 min.

9. The quantity of charge Q in coulombs (C) that passes through a surface at time t (measured in seconds) is given by $Q(t) = t^3 - 2t^2 + 6t + 2$. Find the current when (a) $t = 0.5$ s and (b) $t = 1$ s. [See Example 2. The unit of current is an ampere (1 A = 1 C/s).]

10. The frequency of vibrations of a vibrating violin string is given by

$$f = \frac{1}{2L}\sqrt{\frac{T}{\rho}}$$

where L is the length of the string, T is its tension, and ρ is its linear density. [See Chapter 11 in D. E. Hall, *Musical Acoustics* (Belmont, Calif.: Wadsworth, 1980).] Find the rate of change of the frequency with respect to (a) the length (when T and ρ are constant), (b) the tension (when L and ρ are constant), and (c) the linear density (when L and T are constant).

11. Boyle's Law states that when a sample of gas is compressed at a constant temperature, the product of the pressure and the volume remains constant: $PV = C$. (a) Find the rate of change of volume with respect to pressure. (b) Prove that the isothermal compressibility (see Example 4) is given by $\beta = 1/P$.

12. The data in the following table concern the lactonization of hydroxyvaleric acid at 25° C. They give the concentration $C(t)$ of this acid in moles per liter after t minutes.

t	0	2	4	6	8
$C(t)$	0.0800	0.0570	0.0408	0.0295	0.0210

(a) Find the average rate of reaction for the following
     time intervals:
     (i) $2 \leqslant t \leqslant 6$ (ii) $2 \leqslant t \leqslant 4$
     (iii) $0 \leqslant t \leqslant 2$
(b) Plot the points from the table and draw a smooth
     curve through them as an approximation to the
     graph of the concentration function. Then draw
     the tangent at $t = 2$ and use it to estimate the
     instantaneous rate of reaction when $t = 2$.

13. The mass of glucose in a metabolic experiment
     decreased according to the equation
     $m = 5 - (0.02)t^2$, where t is measured in hours. Find
     the rate of change of the amount of glucose after 1 h.

14. The population of a slowly growing bacteria colony
     after t hours is given by $n = 100 + 24t + 2t^2$. Find
     the growth rate after 2 h.

15. Use the law of laminar flow (see Example 6) to find
     the velocity gradient at a point 0.005 cm from the
     central axis of a blood vessel with radius 0.01 cm,

length 3 cm, pressure difference 3000 dynes/cm^2, and
viscosity $\eta = 0.027$.

16. The population of a bacteria culture is 5000 initially
     and doubles every 20 min. Find an expression for the
     population function $n = f(t)$ where t is measured in
     hours.

In Exercises 17–20 a cost function is given for a certain
commodity. Find the marginal cost function. Then compare
the marginal cost at the production level of 100 units with
the cost of producing the 101st item.

17. $C(x) = 420 + 1.5x + 0.002x^2$

18. $C(x) = 1200 + \dfrac{x}{10} + \dfrac{x^2}{10,000}$

19. $C(x) = 2000 + 3x + 0.01x^2 + 0.0002x^3$

20. $C(x) = 2500 + 2\sqrt{x}$

SECTION 3.8

Related Rates

In a related rates problem the idea is to compute the rate of change of one
quantity in terms of the rate of change of another quantity (which may be
more easily measured). The procedure is to find an equation that relates the
two quantities and then use the Chain Rule to differentiate both sides with
respect to time.

● **Example 1**   Air is being pumped into a spherical balloon so that its
volume increases at a rate of 100 cm^3/s. How fast is the radius of the balloon
increasing when the diameter is 50 cm?

Solution   If we let V be the volume of the balloon and r the radius, then
V and r are related by the equation

(3.47)
$$V = \tfrac{4}{3}\pi r^3$$

We are given that

$$\frac{dV}{dt} = 100 \text{ cm}^3/\text{s}$$

and we are asked to find dr/dt when $r = 25$ cm. We use the Chain Rule to
differentiate both sides of Equation 3.47:

$$\frac{dV}{dt} = \frac{dV}{dr}\frac{dr}{dt} = 4\pi r^2 \frac{dr}{dt}$$

and so
$$\frac{dr}{dt} = \frac{1}{4\pi r^2}\frac{dV}{dt}$$

If we put $r = 25$ and $dV/dt = 100$ in this equation, we obtain

$$\frac{dr}{dt} = \frac{1}{4\pi(25)^2}\, 100 = \frac{1}{25\pi}$$

The radius of the balloon is increasing at the rate of $1/25\pi$ cm/s. ●

● **Example 2**   A ladder 10 ft long rests against a vertical wall. If the bottom of the ladder slides away from the wall at a rate of 1 ft/s, how fast is the top of the ladder sliding down the wall when the bottom of the ladder is 6 ft from the wall?

Solution   We first draw a diagram and label it as in Figure 3.26. Let x meters be the distance from the bottom of the ladder to the wall and y meters the distance from the top of the ladder to the ground. Note that x and y are both functions of t (time).

We are given that $dx/dt = 1$ ft/s and we are asked to find dy/dt when $x = 6$ ft. In this question, the relationship between x and y is given by the Pythagorean Theorem:

$$x^2 + y^2 = 100$$

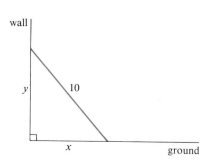

Figure 3.26

Differentiating both sides with respect to t using the Chain Rule, we have

$$2x\frac{dx}{dt} + 2y\frac{dy}{dt} = 0$$

and solving this equation for the desired rate, we obtain

$$\frac{dy}{dt} = -\frac{x}{y}\frac{dx}{dt}$$

When $x = 6$, the Pythagorean Theorem gives $y = 8$ and so, substituting these values and $dx/dt = 1$, we have

$$\frac{dy}{dt} = -\frac{6}{8}(1) = -\frac{3}{4}\ \text{ft/s}$$ ●

● **Example 3**   A water tank has the shape of an inverted circular cone with base radius 2 m and height 4 m. If water is being pumped into the tank at a rate of 2 m³/min, find the rate at which the water level is rising when the water is 3 m deep.

Solution   We first sketch the cone and label it as in Figure 3.27. Let V, r, and h be the volume of the water, the radius of the surface, and the height at time t, where t is measured in minutes.

We are given that $dV/dt = 2$ m³/min and we are asked to find dh/dt when $h = 3$ m. The quantities V and h are related by the equation

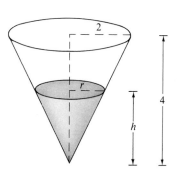

Figure 3.27

$$V = \tfrac{1}{3}\pi r^2 h$$

but we must express V as a function of h alone. In order to eliminate r we use the similar triangles in Figure 3.27 to write

$$\frac{r}{h} = \frac{2}{4} \qquad r = \frac{h}{2}$$

and the expression for V becomes

$$V = \frac{1}{3}\pi\left(\frac{h}{2}\right)^2 h = \frac{\pi}{12}h^3$$

Now we can differentiate both sides with respect to t:

$$\frac{dV}{dt} = \frac{\pi}{4}h^2\frac{dh}{dt}$$

so

$$\frac{dh}{dt} = \frac{4}{\pi h^2}\frac{dV}{dt}$$

Substituting $h = 3$ m and $dV/dt = 2$ m³/min, we have

$$\frac{dh}{dt} = \frac{4}{\pi(3)^2} \cdot 2 = \frac{8}{9\pi} \approx 0.28 \text{ m/min} \qquad \bullet$$

Strategy: Examples 1–3 suggest the following steps in solving problems in related rates:

1. Read the problem carefully.
2. Draw a diagram if possible.
3. Introduce notation. Assign symbols to all quantities that are functions of time.
4. Express the given information and the required rate in terms of derivatives.
5. Write an equation that relates the various quantities of the problem. If necessary, use the geometry of the situation to eliminate one of the variables by substitution (as in Example 3).
6. Use the Chain Rule to differentiate both sides of the equation with respect to t.
7. Substitute the given information into the resulting equation and solve for the unknown rate.

 Warning: A common error is to substitute the given numerical information (for quantities that vary with time) at too early a stage. This should only be done *after* the differentiation. (Step 7 follows Step 6.) For instance, in Example 3 we dealt with general values of h until we finally substituted $h = 3$ at the last stage. (If we had put $h = 3$ earlier, we would have gotten $dV/dt = 0$, which is clearly wrong.)

The following example gives a further illustration of the strategy.

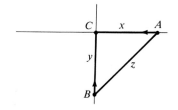

Figure 3.28

● **Example 4**   Car A is going west at 50 mi/h and car B is going north at 60 mi/h. Both are headed for the intersection of the two roads. At what rate are the cars approaching each other when car A is 0.3 mi and car B is 0.4 mi from the intersection?

Solution   We draw Figure 3.28 where C is the intersection of the roads. At a given time t, let x be the distance from car A to C, let y be the distance from car B to C, and let z be the distance between the cars, where x, y, and z are measured in miles.

We are given that $dx/dt = -50$ mi/h and $dy/dt = -60$ mi/h. (We take the derivatives to be negative since x and y are decreasing.) We are asked to find dz/dt. The equation that relates x, y, and z is given by the Pythagorean Theorem:

$$z^2 = x^2 + y^2$$

Differentiating both sides with respect to t, we have

$$2z\frac{dz}{dt} = 2x\frac{dx}{dt} + 2y\frac{dy}{dt}$$

$$\frac{dz}{dt} = \frac{1}{z}\left(x\frac{dx}{dt} + y\frac{dy}{dt}\right)$$

When $x = 0.3$ mi and $y = 0.4$ mi, the Pythagorean Theorem gives $z = 0.5$ mi, so

$$\frac{dz}{dt} = \frac{1}{0.5}[0.3(-50) + 0.4(-60)]$$

$$= -78 \text{ mi/h}$$

The cars are approaching each other at a rate of 78 mi/h.   ●

EXERCISES 3.8

1. If V is the volume of a cube with edge length x, find dV/dt in terms of dx/dt.

2. If A is the area of a circle with radius r, find dA/dt in terms of dr/dt.

3. If $xy = 1$ and $dx/dt = 4$, find dy/dt when $x = 2$.

4. If $x^2 + 3xy + y^2 = 1$ and $dy/dt = 2$, find dx/dt when $y = 1$.

5. A spherical snowball is melting in such a way that its volume is decreasing at a rate of 1 cm³/min. At what

rate is the diameter decreasing when the diameter is 10 cm?

6. If a snowball melts so that its surface area decreases at a rate of 1 cm²/min, find the rate at which the diameter decreases when the diameter is 10 cm.

7. A street light is at the top of a 15-ft pole. A man 6 ft tall walks away from the pole with a speed of 5 ft/s along a straight path. (a) How fast is the tip of his shadow moving when he is 40 ft from the pole? (b) How fast is his shadow lengthening at that point?

8. A spotlight on the ground shines on a wall 12 m away. If a man 2 m tall walks from the spotlight toward the building at a speed of 1.6 m/s, how fast is his shadow on the building decreasing when he is 4 m from the building?

9. A plane flying horizontally at an altitude of 1 mi and a speed of 500 mi/h passes directly over a radar station. Find the rate at which the distance from the plane to the station is increasing when it is 2 mi away from the station.

10. A baseball diamond is a square with side 90 ft. A batter hits the ball and runs toward first base with a speed of 24 ft/s. (a) At what rate is his distance from second base decreasing when he is halfway to first base? (b) At what rate is his distance from third base increasing at the same moment?

11. Two cars start from the same point. One travels south at 60 mi/h and the other travels west at 25 mi/h. At what rate is the distance between them increasing two hours later?

12. At noon, ship A is 150 km west of ship B. Ship A is sailing east at 35 km/h and ship B is sailing north at 25 km/h. How fast is the distance between them changing at 4:00 P.M.?

13. At noon, ship A is 100 km west of ship B. Ship A is sailing south at 35 km/h and ship B is sailing north at 25 km/h. How fast is the distance between them changing at 4:00 P.M.?

14. A man starts walking north at 4 ft/s from a point P. Five minutes later a woman starts walking south at 5 ft/s from a point 500 ft due east of P. At what rate are they separating 15 min after the woman starts?

15. The altitude of a triangle is increasing at a rate of 1 cm/min while the area of the triangle is increasing at a rate of 2 cm^2/min. At what rate is the base of the triangle changing when the altitude is 10 cm and the area is 100 cm^2?

16. A boat is pulled into a dock by a rope attached to the bow of the boat and passing through a pulley on the dock that is 1 m higher than the bow of the boat. If the rope is pulled in at a rate of 1 m/s, how fast is the boat approaching the dock when it is 8 m from the dock?

17. Water is leaking out of an inverted conical tank at a rate of 10,000 cm^3/min at the same time that water is being pumped into the tank at a constant rate. The tank has height 6 m and the diameter at the top is 4 m. If the water level is rising at a rate of 20 cm/min when the height of the water is 2 m, find the rate at which water is being pumped into the tank.

18. A trough is 10 ft long and its ends have the shape of isosceles triangles that are 3 ft across at the top and have a height of 1 ft. If the trough is filled with water at a rate of 12 ft^3/min, how fast does the water level rise when the water is 6 inches deep?

19. A water trough is 10 m long and a cross-section has the shape of an isosceles trapezoid that is 30 cm wide at the bottom, 80 cm wide at the top, and has height 50 cm. If it is being filled with water at the rate of 0.2 m^3/min, how fast is the water level rising when the water is 30 cm deep?

20. A swimming pool is 20 ft wide, 40 ft long, 3 ft deep at the shallow end, and 9 ft deep at its deepest point. A cross-section is shown in Figure 3.29. If the pool is being filled at a rate of 0.8 ft^3/min, how fast is the water level rising when the depth at the deepest point is 5 ft?

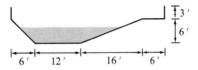

Figure 3.29

21. Boyle's Law states that when a sample of gas is compressed at a constant temperature, the pressure P and volume V satisfy the equation $PV = C$, where C is a constant. Suppose that at a certain instant the volume is 600 cm^3, the pressure is 150 kPa, and the pressure is increasing at a rate of 20 kPa/min. At what rate is the volume decreasing at this instant?

22. When air expands adiabatically (without gaining or losing heat), its pressure P and volume V are related by the equation $PV^{1.4} = C$, where C is a constant. Suppose that at a certain instant the volume is 400 cm^3 and the pressure is 80 kPa and is decreasing at a rate of 10 kPa/min. At what rate is the volume increasing at this instant?

23. A plane flying with a constant speed of 300 km/h passes over a ground radar station at an altitude of 1 km and climbs at an angle of 30°. At what rate is the distance from the plane to the radar station increasing 1 min later?

24. Two people start from the same point. One walks east at 3 mi/h and the other walks northeast at 2 mi/h. How fast is the distance between them changing after 15 min?

25. A ladder 10 ft long rests against a vertical wall. If the bottom of the ladder slides away from the wall at a speed of 2 ft/s, how fast is the angle between the top

of the ladder and the wall changing when the angle is
$\pi/4$ radians?

makes four revolutions per minute. How fast is the
beam of light moving along the shoreline when it is
1 km from P?

26. A lighthouse is on a small island 3 km away from the
nearest point P on a straight shoreline and its light

Differentials

We have used the Leibniz notation dy/dx to denote the derivative of y with
respect to x, but we have regarded it as a single entity and not a ratio. In this
section we shall give the quantities dy and dx separate meanings in such a
way that their ratio is equal to the derivative. We shall also see that these
quantities, called differentials, are useful in finding approximate values of
functions.

Definition (3.48)

> Let $y = f(x)$, where f is a differentiable function. Then the **differential**
> dx is an independent variable; that is, dx can be given the value of
> any real number. The **differential** dy is then defined in terms of dx
> by the equation
>
> $$dy = f'(x)\,dx$$

Note 1:   The differentials dx and dy are both variables, but dx is an
independent variable whereas dy is a dependent variable—it depends on the
values of x and dx. If dx is given a specific value and x is taken to be some
specific number in the domain of f, then the numerical value of dy is
determined.

Note 2:   If $dx \neq 0$, we can divide both sides of the equation in Definition 3.48 by dx to obtain

$$\frac{dy}{dx} = f'(x)$$

We have seen similar equations before, but now the left side can genuinely
be interpreted as a ratio of differentials.

● **Example 1**   (a) Find dy if $y = x^3 + 2x^2$. (b) Find the value of dy when
$x = 2$ and $dx = 0.1$.

Solution   (a)   If $f(x) = x^3 + 2x^2$, then $f'(x) = 3x^2 + 4x$, so

$$dy = (3x^2 + 4x)\,dx$$

(b)   Substituting $x = 2$ and $dx = 0.1$ in the expression for dy, we have

$$dy = (3 \cdot 2^2 + 4 \cdot 2)0.1 = 2$$ ●

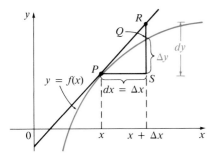

Figure 3.30

The geometric meaning of differentials is shown in Figure 3.30. Let $P(x, f(x))$ and $Q(x + \Delta x, f(x + \Delta x))$ be points on the graph of f and set $dx = \Delta x$. The corresponding change in y is

$$|QS| = \Delta y = f(x + \Delta x) - f(x)$$

The slope of the tangent line PR is the derivative $f'(x)$. But from triangle PRS you can see that the slope of the tangent line can also be written as $|RS|/|PS|$. Thus

$$|RS| = f'(x)|PS| = f'(x)\,dx = dy$$

Therefore dy represents the amount that the tangent line rises or falls, whereas Δy represents the amount that the curve $y = f(x)$ rises or falls when x changes by an amount dx.

Since

$$\frac{dy}{dx} = \lim_{\Delta x \to 0} \frac{\Delta y}{\Delta x}$$

we have

(3.49)
$$\frac{\Delta y}{\Delta x} \approx \frac{dy}{dx}$$

when Δx is small. (Geometrically, this says that the slope of the secant line PQ is very close to the slope of the tangent line at P when Δx is small.) If we take $dx = \Delta x$, then (3.49) becomes

(3.50)
$$\Delta y \approx dy$$

which says that if Δx is small, then the actual change in y is approximately equal to the differential dy. (Again this is geometrically evident in the case illustrated by Figure 3.30.)

The approximation given by (3.50) can be used in computing approximate values of functions. Suppose that $f(x_1)$ is a known number and it is desired to calculate an approximate value for $f(x_1 + \Delta x)$ where Δx is small. Since

$$f(x_1 + \Delta x) = f(x_1) + \Delta y$$

(3.50) gives

(3.51)
$$f(x_1 + \Delta x) \approx f(x_1) + dy$$

When using the approximation in (3.51) we are, in effect, using the tangent line at $P(x_1, f(x_1))$ as an approximation to the curve $y = f(x)$ when x is near x_1.

● **Example 2**   Compare the values of Δy and dy if $y = f(x) = x^3 + x^2 - 2x + 1$ and x changes (a) from 2 to 2.05 and (b) from 2 to 2.01.

Solution   (a)   We have

$$f(2) = 2^3 + 2^2 - 2(2) + 1 = 9$$

$$f(2.05) = (2.05)^3 + (2.05)^2 - 2(2.05) + 1 = 9.717625$$

$$\Delta y = f(2.05) - f(2) = 0.717625$$

In general,

$$dy = f'(x)\,dx = (3x^2 + 2x - 2)\,dx$$

When $x = 2$ and $dx = \Delta x = 0.05$, this becomes

$$dy = [3(2)^2 + 2(2) - 2]0.05 = 0.7$$

(b) $$f(2.01) = (2.01)^3 + (2.01)^2 - 2(2.01) + 1 = 9.140701$$

$$\Delta y = f(2.01) - f(2) = 0.140701$$

When $dx = \Delta x = 0.01$,

$$dy = [3(2)^2 + 2(2) - 2]0.01 = 0.14 \qquad \bullet$$

Notice that the approximation $\Delta y \approx dy$ becomes better as Δx becomes smaller in Example 2. Notice also that dy was easier to compute than Δy. For more complicated functions it may be impossible to compute Δy exactly. In such cases the approximation by differentials is especially useful.

● **Example 3**   Use differentials to find an approximate value for $\sqrt[3]{65}$.

Solution   Let $y = f(x) = \sqrt[3]{x} = x^{1/3}$. Then

$$dy = \tfrac{1}{3}x^{-2/3}\,dx$$

Since $f(64) = 4$, we take $x_1 = 64$ and $dx = \Delta x = 1$. This gives

$$dy = \frac{1}{3}(64)^{-2/3}(1) = \frac{1}{3 \cdot 16} = \frac{1}{48}$$

Therefore (3.51) gives

$$\sqrt[3]{65} = f(64 + 1) \approx f(64) + dy$$

$$= 4 + \tfrac{1}{48} \approx 4.021 \qquad \bullet$$

Note: The actual value of $\sqrt[3]{65}$ is 4.0207257. . . . Thus the approximation by differentials in Example 3 is accurate to three decimal places even when $\Delta x = 1$.

● **Example 4**   The radius of a sphere was measured and found to be 21 cm with a possible error in measurement of at most 0.05 cm. What is the maximum possible error in using this value of the radius to compute the volume?

Solution   If the radius of the sphere is r, then its volume is $V = \frac{4}{3}\pi r^3$. If the error in the measured value of r is denoted by $dr = \Delta r$, then the corresponding error in the calculated value of V is ΔV, which can be approximated by the differential

$$dV = 4\pi r^2 \, dr$$

When $x = 21$ and $dr = 0.05$, this becomes

$$dV = 4\pi(21)^2 0.05 \approx 277$$

The maximum possible error in the calculated volume is about 277 cm^3.

Note: Although the possible error in Example 4 may appear to be rather large, a better picture of the error is given by the **relative error,** which is computed by dividing the error by the total volume:

$$\frac{\Delta V}{V} \approx \frac{dV}{V} \approx \frac{277}{38{,}792} \approx 0.00714$$

Thus a relative error of $dr/r = 0.05/21 \approx 0.0024$ in the radius produces a relative error of about 0.007 in the volume. The errors could also be expressed as **percentage errors** of 0.24% in the radius and 0.7% in the volume.

EXERCISES 3.9

In Exercises 1–8 find the differential of the given function.

1. $y = x^5$

2. $y = \sqrt[4]{x}$

3. $y = \sqrt{x^4 + x^2 + 1}$

4. $y = (x^2 - 2x - 3)^{10}$

5. $y = \dfrac{x - 2}{2x + 3}$

6. $y = \sqrt{x + \sqrt{2x - 1}}$

7. $y = \sin 2x$

8. $y = x \tan x$

In Exercises 9–16, (a) find the differential dy and (b) evaluate dy for the given values of x and dx.

9. $y = 1 - x^2$, $x = 5$, $dx = \frac{1}{2}$

10. $y = x^4 - 3x^3 + x - 1$, $x = 2$, $dx = 0.1$

11. $y = (x^2 + 5)^3$, $x = 1$, $dx = 0.05$

12. $y = \sqrt{1 - x}$, $x = 0$, $dx = 0.02$

13. $y = \dfrac{1}{\sqrt[3]{3x + 2}}$, $x = 2$, $dx = -0.04$

14. $y = \dfrac{x - 1}{x^2 + 1}$, $x = 1$, $dx = -0.3$

15. $y = \cos x$, $x = \dfrac{\pi}{6}$, $dx = 0.05$

16. $y = \sin x$, $x = \dfrac{\pi}{6}$, $dx = -0.1$

In Exercises 17–20 compute Δy and dy for the given values of x and $dx = \Delta x$. Then sketch a diagram like Figure 3.30 showing the line segments with lengths dx, dy, and Δy.

17. $y = x^2$, $x = 1$, $\Delta x = 0.5$

18. $y = \sqrt{x}$, $x = 1$, $\Delta x = 1$

19. $y = 6 - x^2$, $x = -2$, $\Delta x = 0.4$

20. $y = \dfrac{16}{x}$, $x = 4$, $\Delta x = -1$

In Exercises 21 and 22 compute Δy, dy, and $\Delta y - dy$ for the given values of x and for each of the following values of $dx = \Delta x$: 1, 0.5, 0.1, and 0.01.

21. $y = 2x^3 + 3x - 4$, $x = 3$

22. $y = x^4 + x^2 + 1$, $x = 1$

In Exercises 23–32 use differentials to find an approximate value for the given number.

23. $\sqrt{36.1}$ **24.** $\sqrt{99}$

25. $\sqrt[3]{218}$ **26.** $\sqrt[3]{1.02} + \sqrt[4]{1.02}$

27. $\dfrac{1}{10.1}$ **28.** $(1.97)^6$

29. $\sin 59°$ **30.** $\cos 31.5°$

31. $\tan 47°$ **32.** $\sec 62°$

33. The edge of a cube was found to be 30 cm with a possible error in measurement of 0.1 cm. Use differentials to estimate the maximum possible error in computing (a) the volume of the cube and (b) the surface area of the cube.

34. The radius of a circular disk is given as 24 cm with a maximum error in measurement of 0.2 cm. (a) Use differentials to estimate the maximum error in the calculated area of the disk. (b) What is the relative error?

35. The circumference of a sphere was measured to be 84 cm with a possible error of 0.5 cm. (a) Use differentials to estimate the maximum error in the calculated surface area. (b) What is the relative error?

36. Do Exercise 35 with "surface area" replaced by "volume."

37. (a) Use differentials to find a formula for the approximate volume of a thin cylindrical shell with height h, inner radius r, and thickness Δr.
(b) What is the error involved in using this formula?

38. Use differentials to estimate the amount of paint needed to apply a coat of paint 0.05 cm thick to a hemispherical dome with diameter 50 m.

39. Establish the following rules for working with differentials (where c denotes a constant and u and v are functions of x).
(a) $dc = 0$ (b) $d(cu) = c\,du$
(c) $d(u + v) = du + dv$ (d) $d(uv) = u\,dv + v\,du$
(e) $d\left(\dfrac{u}{v}\right) = \dfrac{v\,du - u\,dv}{v^2}$ (f) $d(x^n) = nx^{n-1}\,dx$

SECTION 3.10

Newton's Method

Many problems in science, engineering, and mathematics lead to the problem of finding the roots of an equation of the form $f(x) = 0$ where f is a differentiable function. For a quadratic equation $ax^2 + bx + c = 0$ there is a well-known formula for the roots. For third- and fourth-degree equations there are also formulas for the roots but they are extremely complicated. If f is a polynomial of degree 5 or higher, there is no such formula (see the note after Example 2 in Section 3.5). Likewise there is no formula that will enable us to find the exact roots of a transcendental equation such as $\cos x = x$. However there are methods that give *approximations* to the roots of such equations.

One such method is called **Newton's method** or the **Newton-Raphson method.** The idea behind it is shown in Figure 3.31 where the root that we are trying to find is labeled r. We start with a first approximation x_1, which is obtained by guessing or from a rough sketch of the graph of f. Consider the tangent line L to the curve $y = f(x)$ at the point $(x_1, f(x_1))$ and look at the x-intercept of L, labeled x_2. If x_1 is close to r, it appears that x_2 is even closer to r and we use it as the second approximation to r. To find a formula

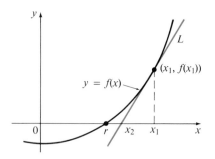

Figure 3.31

for x_2 in terms of x_1 we use the fact that the slope of L is $f'(x_1)$, so its equation is

$$y - f(x_1) = f'(x_1)(x - x_1)$$

Since the x-intercept of L is x_2, we set $y = 0$ and obtain

$$0 - f(x_1) = f'(x_1)(x_2 - x_1)$$

If $f'(x_1) \neq 0$, we can solve this equation for x_2:

$$x_2 = x_1 - \frac{f(x_1)}{f'(x_1)}$$

Next we repeat this procedure with x_1 replaced by x_2, using the tangent line at $(x_2, f(x_2))$. This gives a third approximation:

$$x_3 = x_2 - \frac{f(x_2)}{f'(x_2)}$$

If we keep repeating this process we obtain a sequence of approximations $x_1, x_2, x_3, x_4, \ldots$ as shown in Figure 3.32. In general, if the nth approximation is x_n and $f'(x_n) \neq 0$, then the next approximation is given by

(3.52)

$$x_{n+1} = x_n - \frac{f(x_n)}{f'(x_n)}$$

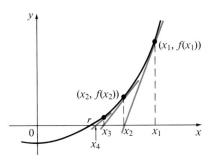

Figure 3.32

If the numbers x_n become closer and closer to r as n becomes large, then we say that the sequence converges to r and we write

$$\lim_{n \to \infty} x_n = r$$

(See Section 10.1 for a discussion of general sequences.) Although the sequence of successive approximations will converge to the desired root for functions of the type illustrated in Figure 3.32, it may happen in certain circumstances that the sequence does not converge. For example, consider the situation shown in Figure 3.33. You can see that x_2 is a worse approximation than x_1. This is likely to be the case when $f'(x_1)$ is close to 0. It might

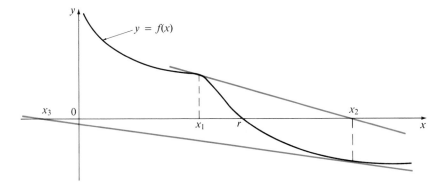

Figure 3.33

even happen that an approximation (such as x_3 in Figure 3.33) falls outside the domain of f. Then Newton's method breaks down and a better initial approximation x_1 should be chosen.

● **Example 1**   Starting with $x_1 = 2$, find the third approximation x_3 to the root of the equation $x^3 - 2x - 5 = 0$.

Solution   We apply Newton's method with

$$f(x) = x^3 - 2x - 5 \qquad f'(x) = 3x^2 - 2$$

Newton himself used this equation to illustrate his method and he chose $x_1 = 2$ after some experimentation because $f(1) = -6$, $f(2) = -1$, and $f(3) = 16$. Equation 3.52 becomes

$$x_{n+1} = x_n - \frac{x_n^3 - 2x_n - 5}{3x_n^2 - 2}$$

With $n = 1$ we have

$$x_2 = x_1 - \frac{x_1^3 - 2x_1 - 5}{3x_1^2 - 2}$$

$$= 2 - \frac{2^3 - 2(2) - 5}{3(2)^2 - 2} = 2.1$$

Then with $n = 2$ we obtain

$$x_3 = x_2 - \frac{x_2^3 - 2x_2 - 5}{3x_2^2 - 2}$$

$$= 2.1 - \frac{(2.1)^3 - 2(2.1) - 5}{3(2.1)^2 - 2} \approx 2.0946$$

It turns out that this third approximation $x_3 \approx 2.0946$ is accurate to four decimal places.   ●

Suppose that we want to achieve a given accuracy, say to eight decimal places, using Newton's method. How do we know when to stop? The rule of thumb that is generally used is that we can stop when successive approximations x_n and x_{n+1} agree to eight decimal places.

Notice that the procedure in going from n to $n + 1$ is the same for all values of n. (It is called an *iterative* process.) This means that Newton's method is particularly convenient for use with a programmable calculator or a computer.

● **Example 2**   Use Newton's method to find $\sqrt[6]{2}$ correct to eight decimal places.

Solution   First we observe that finding $\sqrt[6]{2}$ is equivalent to finding the positive root of the equation

$$x^6 - 2 = 0$$

so we take $f(x) = x^6 - 2$. Then $f'(x) = 6x^5$ and Formula 3.52 (Newton's method) becomes

$$x_{n+1} = x_n - \frac{x_n^6 - 2}{6x_n^5}$$

If we choose $x_1 = 1$ as the initial approximation, then we obtain

$$x_2 \approx 1.166666667$$
$$x_3 \approx 1.126443678$$
$$x_4 \approx 1.122497067$$
$$x_5 \approx 1.122462051$$
$$x_6 \approx 1.122462048$$

Since x_5 and x_6 agree to eight decimal places, we conclude that

$$\sqrt[6]{2} \approx 1.12246205$$

to eight decimal places.

● **Example 3**   Find, correct to six decimal places, the root of the equation $\cos x = x$.

Solution   We first rewrite the equation in standard form:

$$\cos x - x = 0$$

Therefore we let $f(x) = \cos x - x$. Then $f'(x) = -\sin x - 1$, so Formula 3.52 becomes

$$x_{n+1} = x_n - \frac{\cos x_n - x_n}{-\sin x_n - 1} = x_n + \frac{\cos x_n - x_n}{\sin x_n + 1}$$

In order to guess a suitable value for x_1 we sketch the graphs of $y = \cos x$ and $y = x$ in Figure 3.34. It appears that they intersect at a point whose x-coordinate is somewhat less than 1, so let us take $x_1 = 1$ as a convenient first approximation. Then

$$x_2 \approx 0.75036387$$
$$x_3 \approx 0.73911289$$
$$x_4 \approx 0.73908513$$
$$x_5 \approx 0.73908513$$

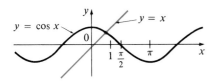

Figure 3.34

Since x_4 and x_5 agree to six decimal places (eight, in fact), we conclude that the root of the equation, correct to six decimal places, is 0.739085.   ●

▦ EXERCISES 3.10

In Exercises 1–4 use Newton's method with the given initial approximation x_1 to find x_3, the third approximation to the root of the given equation. (Give your answer to four decimal places.)

1. $x^3 + x + 1 = 0$, $x_1 = -1$

2. $x^3 + x^2 + 2 = 0$, $x_1 = -2$

3. $x^5 - 10 = 0$, $x_1 = 1.5$

4. $x^7 - 100 = 0$, $x_1 = 2$

In Exercises 5 and 6 use Newton's method to approximate the given number correct to eight decimal places.

5. $\sqrt[4]{22}$ 6. $\sqrt[10]{100}$

In Exercises 7–12 use Newton's method to approximate the indicated root of the equation correct to six decimal places.

7. The root of $x^3 - 2x - 1 = 0$ in the interval $[1, 2]$

8. The root of $x^3 + x^2 + x - 2 = 0$ in the interval $[0, 1]$

9. The root of $x^4 + x^3 - 22x^2 - 2x + 41 = 0$ in the interval $[3, 4]$

10. The root of $x^4 + x^3 - 22x^2 - 2x + 41 = 0$ in the interval $[1, 2]$

11. The positive root of $2 \sin x = x$

12. The root of $\tan x = x$ in the interval $(\pi/2, 3\pi/2)$

In Exercises 13–18 use Newton's method to find all roots of the given equation correct to six decimal places.

13. $x^3 = 4x - 1$

14. $x^5 = 5x - 2$

15. $x^4 = 1 + x - x^2$

16. $(x - 2)^4 = \dfrac{x}{2}$

17. $2 \cos x = 2 - x$

18. $\sin \pi x = x$

19. (a) Apply Newton's method to the equation $x^2 - a = 0$ to derive the following square-root algorithm (used by the ancient Babylonians to compute $\sqrt{a}$):

$$x_{n+1} = \frac{1}{2}\left(x_n + \frac{a}{x_n}\right)$$

(b) Use part (a) to compute $\sqrt{1000}$ correct to six decimal places.

20. (a) Apply Newton's method to the equation $1/x - a = 0$ to derive the following reciprocal algorithm:

$$x_{n+1} = 2x_n - ax_n^2$$

(This algorithm enables a computer to find reciprocals without actually dividing.)

(b) Use part (a) to compute $1/1.6984$ correct to six decimal places.

REVIEW OF CHAPTER 3

Define, state, or discuss the following.

1. Derivative of a function
2. Interpretations of the derivative
3. Equation of the tangent line to $y = f(x)$ at $(a, f(a))$
4. Differentiable function
5. Relation between differentiability and continuity
6. Power Rule
7. Product Rule
8. Quotient Rule
9. Derivatives of the trigonometric functions
10. Chain Rule
11. Implicit differentiation
12. The angle between two curves
13. Orthogonal curves
14. Second derivative; higher derivatives
15. Position function, velocity, acceleration
16. Average and instantaneous rate of change
17. Related rates
18. Differentials
19. Newton's method

REVIEW EXERCISES FOR CHAPTER 3

In Exercises 1–4 find $f'(x)$ from first principles—that is, directly from the definition of a derivative.

1. $f(x) = x^3 + 5x + 4$

2. $f(x) = \dfrac{4 - x}{3 + x}$

3. $f(x) = \sqrt{3 - 5x}$

4. $f(x) = x \sin x$

In Exercises 5–32 calculate y'.

5. $y = (x + 2)^8 (x + 3)^6$

6. $y = \sqrt[3]{x} + \dfrac{1}{\sqrt[3]{x}}$

7. $y = \dfrac{x}{\sqrt{9 - 4x}}$

8. $y = \left(x + \dfrac{1}{x^2}\right)^{\sqrt{7}}$

9. $x^2y^3 + 3y^2 = x - 4y$

10. $y = (1 - x^{-1})^{-1}$

11. $y = \sqrt{x\sqrt{x\sqrt{x}}}$

12. $y = \dfrac{-2}{\sqrt[4]{x^3}}$

13. $y = \dfrac{x}{8 - 3x}$

14. $y\sqrt{x - 1} + x\sqrt{y - 1} = xy$

15. $y = \sqrt[5]{x \tan x}$

16. $y = \sin(\cos x)$

17. $x^2 = y(y + 1)$

18. $y = \dfrac{1}{\sqrt[3]{x + \sqrt{x}}}$

19. $y = \dfrac{(x - 1)(x - 4)}{(x - 2)(x - 3)}$

20. $y = \sqrt{\sin\sqrt{x}}$

21. $y = \tan\sqrt{1 - x}$

22. $y = \dfrac{1}{\sin(x - \sin x)}$

23. $y = \sin(\tan\sqrt{1 + x^3})$

24. $y = \dfrac{(x + \lambda)^4}{x^4 + \lambda^4}$

25. $y = \cot(3x^2 + 5)$

26. $y = \dfrac{\cos^2 x}{1 + \tan x} + \dfrac{\sin^2 x}{1 + \cot x}$

27. $y = \dfrac{\sin mx}{x}$

28. $y = \sqrt[10]{4 + 7\sqrt[8]{x^5}}$

29. $y = \cos^2(\tan x)$

30. $x \tan y = y - 1$

31. $y = \sqrt{7 - x^2}\,(x^3 + 7)^5$

32. $y = \sqrt{x}\sec\sqrt{x}$

33. If $f(x) = \dfrac{1}{(2x - 1)^5}$, find $f''(0)$.

34. If $g(t) = \csc 2t$, find $g'''\left(\dfrac{-\pi}{8}\right)$.

35. Find y'' if $x^6 + y^6 = 1$.

36. Find $f^{(n)}(x)$ if $f(x) = \dfrac{1}{2 - x}$.

In Exercises 37–40 find the equation of the tangent to the given curve at the given point.

37. $y = \dfrac{x}{x^2 - 2}$, $(2, 1)$

38. $\sqrt{x} + \sqrt{y} = 3$, $(4, 1)$

39. $y = \tan x$, $\left(\dfrac{\pi}{3}, \sqrt{3}\right)$

40. $y = x\sqrt{1 + x^2}$, $(1, \sqrt{2})$

41. At what points on the curve $y = \sin x + \cos x$, $0 \leqslant x \leqslant 2\pi$, is the tangent line horizontal?

42. Find the points on the ellipse $x^2 + 2y^2 = 1$ where the tangent line has slope 1.

43. A particle moves on a vertical line so that its coordinate at time t is $y = t^3 - 12t + 3$, $t \geqslant 0$.
  (a) Find the velocity and acceleration functions.
  (b) When is the particle moving upward and when is it moving downward?
  (c) Find the distance that it travels in the time interval $0 \leqslant t \leqslant 3$.

44. A particle moves along a horizontal line so that its coordinate at time t is $x = \sqrt{b^2 + c^2t^2}$, $t \geqslant 0$, where b and c are positive constants.
  (a) Find the velocity and acceleration functions.
  (b) Show that the particle always moves in the positive direction.

In Exercises 45–49 find f' in terms of g'.

45. $f(x) = x^2g(x)$

46. $f(x) = g(x^2)$

47. $f(x) = [g(x)]^2$

48. $f(x) = x^ag(x^b)$

49. $f(x) = g(g(x))$

50. The volume of a right circular cone is $V = \pi r^2 h/3$, where r is the radius of the base and h is the height.
  (a) Find the rate of change of the volume with respect to the height if the radius is constant.
  (b) Find the rate of change of the volume with respect to the radius if the height is constant.

51. The mass of part of a wire is $x(1 + \sqrt{x})$ kilograms, where x is measured in meters from one end of the wire. Find the linear density of the wire when $x = 4$ m.

52. The cost, in dollars, of manufacturing x units of a given item is $950 + 12x + 0.01x^2$. Find the marginal cost at a production level of 200 units. Compare this with the cost of producing the 201st unit.

53. The volume of a cube is increasing at a rate of $10 \text{ cm}^3/\text{min}$. How fast is the surface area increasing when the length of an edge is 30 cm?

54. A paper cup has the shape of a cone with height 10 cm and radius 3 cm (at the top). If water is poured into the cup at a rate of $2 \text{ cm}^3/\text{s}$, how fast is the water level rising when the water is 5 cm deep?

55. A balloon is rising at a constant speed of 5 ft/s. A boy is cycling along a straight road at a speed of 15 ft/s. When he passes under the balloon it is 45 ft above him. How fast is the distance between the boy and the balloon increasing 3 s later?

56. A man walks along a straight path with a speed of 4 ft/s. A searchlight is located on the ground 20 ft from the path and is kept focused on the man. At what rate (in radians per second) is the searchlight rotating when the man is 15 ft from the point on the path closest to the searchlight?

57. Find dy if $y = (4 - x^2)^{3/2}$.

58. Evaluate dy if $y = x^3 - 2x^2 + 1$, $x = 2$, and $dx = 0.2$.

59. Use differentials to approximate $8 + \sqrt{143.6}$.

60. A window has the shape of a square surmounted by a semicircle. The base of the window is measured as having width 60 cm with a possible error in measurement of 0.1 cm. Use differentials to estimate the maximum possible error in computing the area of the window.

61. Use Newton's method to find the root of the equation $x^4 + x - 1 = 0$ in the interval $[0, 1]$ correct to six decimal places.

62. Use Newton's method to find all roots of the equation $6 \cos x = x$ correct to six decimal places.

63. Find the coordinates (correct to four decimal places) of the point on the curve $y = x^6 + 2x^2 - 8x + 3$ where the tangent is horizontal.

64. Sketch the graph of $f(x) = |x| + |x + 1| + |x - 1|$. Where is f not differentiable?

4 The Mean Value Theorem and Its Applications

The greatest mathematicians, as
Archimedes, Newton, and Gauss,
always united theory and applications
in equal measure.
Felix Klein

For since the fabric of the universe is
most perfect and the work of a most
wise Creator, nothing at all takes
place in the universe in which some
rule of maximum or minimum does
not appear.
Leonhard Euler

The Mean Value Theorem, which will be stated and proved in Section 4.2, is one of the most important theorems in calculus. It will enable us to develop the tools that are needed to graph functions using derivatives. In particular we shall learn how to find maximum and minimum values of functions. This is a useful skill because many practical problems require us to minimize a cost or maximize an area or somehow find the best possible outcome of a situation.

SECTION 4.1

Maximum and Minimum Values

Some of the most important applications of differential calculus are *optimization problems* in which we are required to find the optimal (best) way of doing something. In many cases these problems can be reduced to finding the maximum or minimum values of a function. Let us first explain exactly what we mean by maximum and minimum values.

Definition (4.1)

A function f has an **absolute maximum** at c if $f(c) \geqslant f(x)$ for all x in D, where D is the domain of f, and the number $f(c)$ is called the **maximum value** of f on D. Similarly, f has an **absolute minimum** at c if $f(c) \leqslant f(x)$ for all x in D and the number $f(c)$ is called the **minimum value** of f on D. The maximum and minimum values of f are called the **extreme values** of f.

Figure 4.1 shows the graph of a function f with absolute maximum at d and absolute minimum at a. Note that $(d, f(d))$ is the highest point on the graph and $(a, f(a))$ is the lowest point.

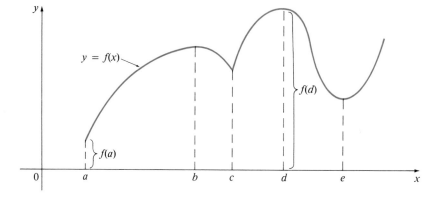

Figure 4.1

MINIMUM VALUE $f(a)$,
MAXIMUM VALUE $f(d)$

In Figure 4.1 if we consider only values of x near b [for instance, if we restrict our attention to the interval (a, c)], then $f(b)$ is the largest of those values of $f(x)$ and is called a *local maximum value* of f. Likewise $f(c)$ is called a *local minimum value* of f because $f(c) \leqslant f(x)$ for x near c [in the interval (b, d), for instance]. The function f also has a local minimum at e. In general we have the following definition:

Definition (4.2)

A function f has a **local maximum** (or **relative maximum**) at c if there is an open interval I containing c such that $f(c) \geqslant f(x)$ for all x in I. Similarly, f has a **local minimum** at c if there is an open interval I containing c such that $f(c) \leqslant f(x)$ for all x in I.

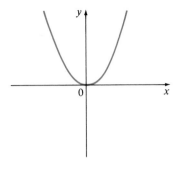

Figure 4.2

$f(x) = x^2$, MINIMUM VALUE 0, NO MAXIMUM

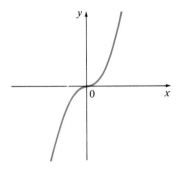

Figure 4.3

$f(x) = x^3$, NO MINIMUM, NO MAXIMUM

● **Example 1**   The function $f(x) = \cos x$ takes on its (local and absolute) maximum value of 1 infinitely many times since $\cos 2n\pi = 1$ for any integer n and $-1 \leqslant \cos x \leqslant 1$ for all x. Likewise $\cos(2n + 1)\pi = -1$ is its minimum value, where n is any integer.   ●

● **Example 2**   If $f(x) = x^2$, then $f(x) \geqslant f(0)$ because $x^2 \geqslant 0$ for all x. Therefore $f(0) = 0$ is the absolute (and local) minimum value of f. This corresponds to the fact that the origin is the lowest point on the parabola $y = x^2$ (see Figure 4.2). However there is no highest point on the parabola and so this function has no maximum value.   ●

● **Example 3**   From the graph of the function $f(x) = x^3$, shown in Figure 4.3, we see that this function has neither an absolute maximum value nor an absolute minimum value. In fact, it has no local extreme values either.   ●

● **Example 4**   The graph of the function

$$f(x) = 3x^4 - 16x^3 + 18x^2 \qquad -1 \leqslant x \leqslant 4$$

is shown in Figure 4.4. You can see that $f(1) = 5$ is a local maximum, whereas the absolute maximum is $f(-1) = 37$. Also $f(0) = 0$ is a local minimum and $f(3) = -27$ is both a local and an absolute minimum.   ●

We have seen that some functions have extreme values, while some do not. The following theorem gives conditions under which a function is guaranteed to possess extreme values.

The Extreme Value Theorem (4.3)

> If f is continuous on a closed interval $[a, b]$, then f attains an absolute maximum value $f(c)$ and an absolute minimum value $f(d)$ at some numbers c and d in $[a, b]$.

The Extreme Value Theorem is illustrated in Figure 4.5. Note that an extreme value can be taken on more than once. Although the Extreme Value Theorem is intuitively very plausible, it is difficult to prove and so we omit the proof.

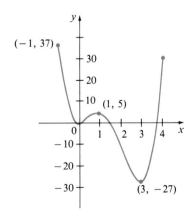

Figure 4.4

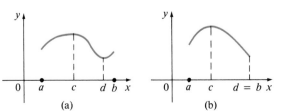

Figure 4.5

The next two examples show that a function need not possess extreme values if either hypothesis (continuity or closed interval) is omitted from the Extreme Value Theorem.

● **Example 5**   The function

$$f(x) = \begin{cases} x^2 & \text{if } 0 \leqslant x < 1 \\ 0 & \text{if } 1 \leqslant x \leqslant 2 \end{cases}$$

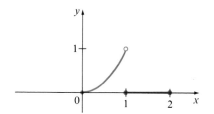

is defined on the closed interval $[0, 2]$ but has no maximum value. Notice that the range of f is the interval $[0, 1)$. The function takes on values arbitrarily close to 1 but never actually attains the value 1. This does not contradict the Extreme Value Theorem because f is not continuous on $[0, 2]$. In fact, it has a discontinuity at $x = 1$ (see Figure 4.6).   ●

Figure 4.6

● **Example 6**   The function $f(x) = x^2$, $0 < x < 2$, is continuous on the finite interval $(0, 2)$ but has neither a maximum nor a minimum value. The range of f is the interval $(0, 4)$; the values 0 and 4 are never taken on by f. This does not contradict the Extreme Value Theorem since the interval $(0, 2)$ is not closed.

If we alter the function by including either endpoint of the interval $(0, 2)$, then we get the situations shown in Figure 4.7. In particular the function $k(x) = x^2$, $0 \leqslant x \leqslant 2$, is continuous on the closed interval $[0, 2]$, so the Extreme Value Theorem says that there will be an absolute maximum and an absolute minimum.

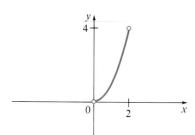

(a) $f(x) = x^2$, $0 < x < 2$
no maximum, no minimum

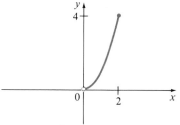

(b) $g(x) = x^2$, $0 < x \leqslant 2$
maximum $g(2) = 4$, no minimum

Figure 4.7

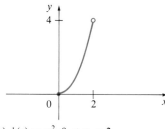

(c) $h(x) = x^2$, $0 \leqslant x < 2$
no maximum, minimum $h(0) = 0$

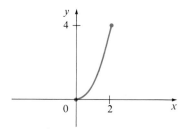

(d) $k(x) = x^2$, $0 \leqslant x \leqslant 2$
maximum $k(2) = 4$, minimum $k(0) = 0$ ●

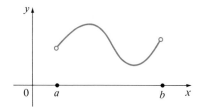

Figure 4.8

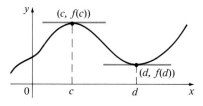

Figure 4.9

In spite of Example 6 we point out that a continuous function, like the one whose graph is shown in Figure 4.8, *could* have a maximum or minimum value even when defined on an open interval. Likewise a discontinuous function *might* have maximum and minimum values (see Exercise 63), but there is no guarantee.

The Extreme Value Theorem says that a continuous function on a closed interval has a maximum value and a minimum value, but it does not tell us how to find these extreme values. We start by looking for local extreme values.

Figure 4.9 shows the graph of a function f with a local maximum at c and a local minimum at d. It appears that at the maximum and minimum points the tangent line is horizontal and therefore has slope 0. We know that the derivative is the slope of the tangent line, so it appears that $f'(c) = 0$ and $f'(d) = 0$. The following theorem shows that this is always true for differentiable functions. It is named after the French mathematician Pierre Fermat (1601–1665).

Fermat's Theorem (4.4)

> If f has a local extremum (that is, maximum or minimum) at c, and if $f'(c)$ exists, then $f'(c) = 0$.

Proof   Suppose, for the sake of definiteness, that f has a local maximum at c. Then, according to Definition 4.2, $f(c) \geqslant f(x)$ if x is sufficiently close to c. This implies that if h is sufficiently close to 0, with h being positive or negative, then

$$f(c) \geqslant f(c + h)$$

and therefore

(4.5)
$$f(c + h) - f(c) \leqslant 0$$

We can divide both sides of an inequality by a positive number. Thus, if $h > 0$ and h is sufficiently small, we have

$$\frac{f(c + h) - f(c)}{h} \leqslant 0$$

Taking the right-hand limit of both sides of this inequality (using Theorem 2.16), we get

$$\lim_{h \to 0^+} \frac{f(c + h) - f(c)}{h} \leqslant \lim_{h \to 0^+} 0 = 0$$

But since $f'(c)$ exists, we have

$$f'(c) = \lim_{h \to 0} \frac{f(c + h) - f(c)}{h} = \lim_{h \to 0^+} \frac{f(c + h) - f(c)}{h}$$

and so we have shown that $f'(c) \leqslant 0$.

If $h < 0$, then the direction of the inequality (4.5) is reversed when we divide by h:

$$\frac{f(c + h) - f(c)}{h} \geqslant 0 \qquad h < 0$$

So, taking the left-hand limit, we have

$$f'(c) = \lim_{h \to 0} \frac{f(c + h) - f(c)}{h} = \lim_{h \to 0^-} \frac{f(c + h) - f(c)}{h} \geqslant 0$$

We have shown that $f'(c) \geqslant 0$ and also that $f'(c) \leqslant 0$. Since both of these inequalities must be true, the only possibility is that $f'(c) = 0$.

We have proved Fermat's Theorem for the case of a local maximum. The case of a local minimum can be proved in a similar manner, or it can be deduced from the case that we have proved by using Exercise 64 (see Exercise 65). ●

The following examples caution us against reading too much into Fermat's Theorem. We cannot expect to locate extreme values simply by setting $f'(x) = 0$ and solving for x.

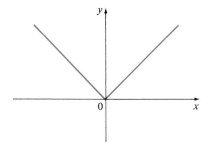

Figure 4.10

$f(x) = |x|$

● **Example 7**   The function $f(x) = |x|$ has its minimum value (local and absolute) at 0 but it cannot be found by setting $f'(x) = 0$ because, as was shown in Example 6 in Section 3.1, $f'(0)$ does not exist. (See Figure 4.10.) ●

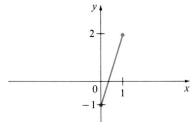

Figure 4.11

$f(x) = 3x - 1, \, 0 \leqslant x \leqslant 1$

● **Example 8**   The function $f(x) = 3x - 1$, $0 \leqslant x \leqslant 1$, whose graph is shown in Figure 4.11, has its maximum value when $x = 1$, but $f'(1) = 3 \neq 0$. This does not contradict Fermat's Theorem since $f(1) = 2$ is not a *local* maximum. (Note that the number 1 is not contained in an *open* interval in the domain of f.) ●

● **Example 9**   If $f(x) = x^3$, then $f'(x) = 3x^2$, so $f'(0) = 0$. But f has no maximum or minimum at 0 as you can see from its graph in Figure 4.3. (Or observe that $x^3 > 0$ for $x > 0$ but $x^3 < 0$ for $x < 0$.) The fact that $f'(0) = 0$ simply means that the curve $y = x^3$ has a horizontal tangent at $(0, 0)$. Instead of having a maximum or minimum at $(0, 0)$, the curve crosses its horizontal tangent there. ●

Warning: Examples 7–9 show that we must be careful when using Fermat's Theorem. Example 9 demonstrates that even when $f'(c) = 0$ there need not be a maximum or minimum at c. (In other words, the converse of Fermat's Theorem is false in general.) Furthermore, there may be an extreme value even when $f'(c) \neq 0$ (as in Example 8) or when $f'(c)$ does not exist (as in Example 7).

Fermat's Theorem does suggest that we should at least start looking for extreme values of f at the numbers c where $f'(c) = 0$ or where $f'(c)$ does not exist. Such numbers are given a special name.

Definition (4.6)

> A **critical number** of a function f is a number c in the domain of f such that either $f'(c) = 0$ or $f'(c)$ does not exist.

● **Example 10**   Find the critical numbers of $f(x) = x^{3/5}(4 - x)$.

Solution   The Product Rule gives

$$f'(x) = \frac{3}{5} x^{-2/5}(4 - x) + x^{3/5}(-1)$$

$$= \frac{3(4 - x) - 5x}{5x^{2/5}} = \frac{12 - 8x}{5x^{2/5}}$$

[The same result could be obtained by first writing $f(x) = 4x^{3/5} - x^{8/5}$.] Therefore $f'(x) = 0$ if $12 - 8x = 0$, that is, $x = \frac{3}{2}$, and $f'(x)$ does not exist when $x = 0$. Thus the critical numbers are $\frac{3}{2}$ and 0.   ●

In terms of critical numbers, Fermat's Theorem can be rephrased as follows (compare Definition 4.6 with Theorem 4.4):

(4.7)

> If f has a local extremum at c, then c is a critical number of f.

To find an absolute maximum or minimum of a continuous function on a closed interval, we note that either it is a local extremum [in which case it occurs at a critical number by (4.7)] or it occurs at an endpoint of the interval. Thus the following three-step procedure always works.

(4.8)

> To find the *absolute* maximum and minimum values of a continuous function f on a closed interval $[a, b]$:
> 1. Find the values of f at the critical numbers of f in (a, b).
> 2. Find the values $f(a)$ and $f(b)$.
> 3. The largest of the values from steps 1 and 2 is the absolute maximum value; the smallest of these values is the absolute minimum value.

● **Example 11**   Find the absolute maximum and minimum values of the function

$$f(x) = x^3 - 3x^2 + 1 \qquad -\tfrac{1}{2} \leqslant x \leqslant 4$$

Solution   Since f is continuous on $\left[-\frac{1}{2},4\right]$ we can use the procedure outlined in (4.8):

$$f(x) = x^3 - 3x^2 + 1$$

$$f'(x) = 3x^2 - 6x = 3x(x - 2)$$

Since $f'(x)$ exists for all x, the only critical numbers of f occur when $f'(x) = 0$, that is, $x = 0$ or $x = 2$. Notice that each of these critical numbers lies in the interval $\left[-\frac{1}{2},4\right]$. The values of f at these critical numbers are

$$f(0) = 1 \qquad f(2) = -3$$

The values of f at the endpoints of the interval are

$$f(-\tfrac{1}{2}) = \tfrac{1}{8} \qquad f(4) = 17$$

Comparing these four numbers, we see that the absolute maximum value is $f(4) = 17$ and the absolute minimum value is $f(2) = -3$.

Note that in this example the absolute maximum occurs at an endpoint, whereas the absolute minimum occurs at a critical number. The graph of f is sketched in Figure 4.12.   ●

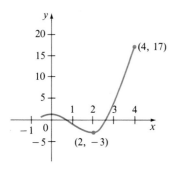

Figure 4.12

EXERCISES 4.1

For the functions whose graphs are shown in Exercises 1 and 2, state whether the function has an absolute or local maximum or minimum at the numbers $a, b, c, d, e, r, s,$ and t.

1.

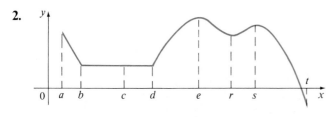

2.

In Exercises 3–22 find the absolute and local maximum and minimum values of the given function. Include a sketch of the graph.

3. $f(x) = 1 + 2x, \; x \geqslant -1$

4. $f(x) = 4x - 1, \; x \leqslant 8$

5. $f(x) = |x|, \; -2 \leqslant x \leqslant 1$

6. $f(x) = |4x - 1|, \; 0 \leqslant x \leqslant 2$

7. $f(x) = 1 - x^2, \; 0 < x < 1$

8. $f(x) = 1 - x^2, \; 0 < x \leqslant 1$

9. $f(x) = 1 - x^2, \; 0 \leqslant x < 1$

10. $f(x) = 1 - x^2, \; 0 \leqslant x \leqslant 1$

11. $f(x) = 1 - x^2, \; -2 \leqslant x \leqslant 1$

12. $f(x) = 1 + (x + 1)^2, \; -2 \leqslant x < 5$

13. $f(t) = \dfrac{1}{t}, \ 0 < t < 1$ **14.** $f(t) = \dfrac{1}{t}, \ 0 < t \leqslant 1$

15. $f(\theta) = \sin\theta, \ -2\pi \leqslant \theta \leqslant 2\pi$

16. $f(\theta) = \tan\theta, \ -\dfrac{\pi}{4} \leqslant \theta < \dfrac{\pi}{2}$

17. $f(\theta) = \cos\left(\dfrac{\theta}{2}\right), \ -\pi < \theta < \pi$

18. $f(\theta) = \sec\theta, \ -\dfrac{\pi}{2} < \theta \leqslant \dfrac{\pi}{3}$

19. $f(x) = x^5$

20. $f(x) = 2 - x^4$

21. $f(x) = \begin{cases} 2x & \text{if } 0 \leqslant x < 1 \\ 2 - x & \text{if } 1 \leqslant x \leqslant 2 \end{cases}$

22. $f(x) = \begin{cases} x^2 & \text{if } -1 \leqslant x < 0 \\ 2 - x^2 & \text{if } \ \ 0 \leqslant x \leqslant 1 \end{cases}$

In Exercises 23–42 find the critical numbers of the given function.

23. $f(x) = 2x - 3x^2$

24. $f(x) = 5 + 8x$

25. $f(x) = x^3 - 3x + 1$

26. $f(x) = 4x^3 - 9x^2 - 12x + 3$

27. $f(t) = 2t^3 + 3t^2 + 6t + 4$

28. $f(t) = t^3 + 6t^2 + 3t - 1$

29. $s(t) = 2t^3 + 3t^2 - 6t + 4$

30. $s(t) = t^4 + 4t^3 + 2t^2$

31. $g(x) = \sqrt[9]{x}$ **32.** $g(x) = |x + 1|$

33. $g(t) = 5t^{2/3} + t^{5/3}$ **34.** $g(t) = \sqrt{t}(1 - t)$

35. $f(r) = \dfrac{r}{r^2 + 1}$ **36.** $f(z) = \dfrac{z + 1}{z^2 + z + 1}$

37. $F(x) = x^{4/5}(x - 4)^2$ **38.** $G(x) = \sqrt[3]{x^2 - x}$

39. $V(x) = x\sqrt{x - 2}$ **40.** $T(x) = x^2(2x - 1)^{2/3}$

41. $f(\theta) = \sin^2(2\theta)$ **42.** $g(\theta) = \theta + \sin\theta$

In Exercises 43–58 find the absolute maximum and absolute minimum values of f on the given interval.

43. $f(x) = x^2 - 2x + 2, \ [0, 3]$

44. $f(x) = 1 - 2x - x^2, \ [-4, 1]$

45. $f(x) = x^3 - 12x + 1, \ [-3, 5]$

46. $f(x) = 4x^3 - 15x^2 + 12x + 7, \ [0, 3]$

47. $f(x) = 2x^3 + 3x^2 + 4, \ [-2, 1]$

48. $f(x) = 18x + 15x^2 - 4x^3, \ [-3, 4]$

49. $f(x) = x^4 - 4x^2 + 2, \ [-3, 2]$

50. $f(x) = 3x^5 - 5x^3 - 1, \ [-2, 2]$

51. $f(x) = x^2 + \dfrac{2}{x}, \ \left[\dfrac{1}{2}, 2\right]$

52. $f(x) = \sqrt{9 - x^2}, \ [-1, 2]$

53. $f(x) = x^{4/5}, \ [-32, 1]$

54. $f(x) = \dfrac{x}{x + 1}, \ [1, 2]$

55. $f(x) = |x - 1| - 1, \ [-1, 2]$

56. $f(x) = |x^2 + x|, \ [-2, 1]$

57. $f(x) = \sin x + \cos x, \ \left[0, \dfrac{\pi}{3}\right]$

58. $f(x) = x - 2\cos x, \ [-\pi, \pi]$

59. Show that 0 is a critical number of the function $f(x) = x^5$, but f does not have a local extremum at 0.

60. Show that 5 is a critical number of the function $g(x) = 2 + (x - 5)^3$, but g does not have a local extremum at 5.

61. Prove that the function $f(x) = x^{101} + x^{51} + x + 1$ has neither a local maximum nor a local minimum.

62. A cubic function is a polynomial of degree 3; that is, it has the form $f(x) = ax^3 + bx^2 + cx + d$, where $a \neq 0$.
(a) Show that a cubic function can have two, one, or no critical numbers. Give examples and sketches to illustrate the three possibilities.
(b) How many local extreme values can a cubic function have?

63. Sketch the graph of a discontinuous function on $[0, 1]$ that has both an absolute maximum and an absolute minimum.

64. If f has a minimum value at c, show that the function $g(x) = -f(x)$ has a maximum value at c.

65. Prove Fermat's Theorem for the case where f has a local minimum at c.

SECTION 4.2

The Mean Value Theorem

We shall see that many of the results of this chapter depend on one central fact, which is called the Mean Value Theorem. But to arrive at the Mean Value Theorem we first need a result that is named after the French mathematician Michel Rolle (1652–1719).

Rolle's Theorem (4.9)

Let f be a function that satisfies the following three hypotheses:

1. f is continuous on the closed interval $[a, b]$
2. f is differentiable on the open interval (a, b)
3. $f(a) = f(b)$

Then there is a number c in (a, b) such that $f'(c) = 0$.

Before giving the proof let us take a look at the graphs of some typical functions that satisfy the three hypotheses. Figure 4.13 shows the graphs of four such functions. In each case it appears that there is at least one point $(c, f(c))$ on the graph where the tangent is horizontal and therefore $f'(c) = 0$. Thus Rolle's Theorem is plausible.

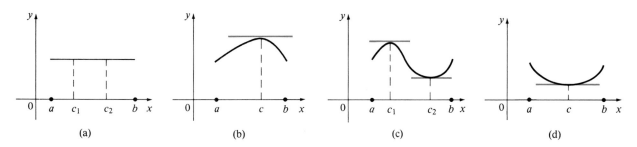

(a)   (b)   (c)   (d)

Figure 4.13

Proof   There are three cases:

CASE I:

$f(x) = k$, a constant: Then $f'(x) = 0$, so the number c can be taken to be *any* number in (a, b).

CASE II:

$f(x) > f(a)$ for some x in (a, b) [as in Figure 4.13(b) or (c)]: By the Extreme Value Theorem (which we can apply by hypothesis 1) f has a maximum value somewhere in $[a, b]$. Since $f(a) = f(b)$, it must attain this maximum value at a number c in the open interval (a, b). Then f has a *local* maximum at c and, by hypothesis 2, f is differentiable at c. Therefore $f'(c) = 0$ by Fermat's Theorem.

CASE III:

$f(x) < f(a)$ for some x in (a, b) [as in Figure 4.13(c) or (d)]: By the Extreme Value Theorem, f has a minimum value in $[a, b]$ and, since $f(a) = f(b)$, it attains this minimum value at a number c in (a, b). Again $f'(c) = 0$ by Fermat's Theorem.   ●

● **Example 1**   Let us apply Rolle's Theorem to the position function $s = f(t)$ of a moving object. If the object is in the same place at two different instants $t = a$ and $t = b$, then $f(a) = f(b)$. Rolle's Theorem says that there is some instant of time $t = c$ between a and b when $f'(c) = 0$; that is, the velocity is 0. (In particular you can see that this will be true in the situation where a ball is thrown directly upward.)   ●

● **Example 2**   Prove that the equation $x^3 + x - 1 = 0$ has exactly one real root.

Solution   First we use the Intermediate Value Theorem (2.28) to show that there is a root. Let $f(x) = x^3 + x - 1$. Then $f(0) = -1 < 0$ and $f(1) = 1 > 0$. Since f is a polynomial, it is continuous, so the Intermediate Value Theorem states that there is a number c between 0 and 1 such that $f(c) = 0$. Thus the given equation has a root.

To show that there is no other real root we use Rolle's Theorem and argue by contradiction. Suppose that the equation had two roots a and b. Then $f(a) = 0 = f(b)$ and, since f is a polynomial, it is differentiable on (a, b) and continuous on $[a, b]$. Thus, by Rolle's Theorem, there is a number c between a and b such that $f'(c) = 0$. But

$$f'(x) = 3x^2 + 1 \geqslant 1 \qquad \text{for all } x$$

(since $x^2 \geqslant 0$) so $f'(x)$ can never be 0. This gives a contradiction. Therefore the equation cannot have two roots.   ●

Our main use of Rolle's Theorem is in proving the following important theorem, which was first stated by another French mathematician, Joseph-Louis Lagrange (1736–1813).

The Mean Value Theorem (4.10)

Let f be a function that satisfies the following hypotheses:

1. f is continuous on the closed interval $[a, b]$
2. f is differentiable on the open interval (a, b)

Then there is a number c in (a, b) such that

(4.11)
$$f'(c) = \frac{f(b) - f(a)}{b - a}$$

or, equivalently,

(4.12)
$$f(b) - f(a) = f'(c)(b - a)$$

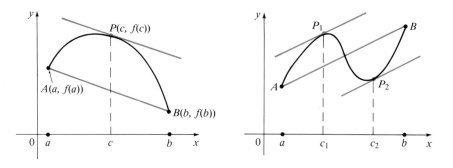

Figure 4.14

Before proving this theorem, we can see that it is reasonable by interpreting it geometrically. Figure 4.14 shows the points $A(a, f(a))$ and $B(b, f(b))$ on the graphs of two differentiable functions. The slope of the secant line AB is

$$(4.13) \qquad m_{AB} = \frac{f(b) - f(a)}{b - a}$$

which is the same expression as on the right side of Equation 4.11. Since $f'(c)$ is the slope of the tangent line at the point $(c, f(c))$, the Mean Value Theorem, in the form given by Equation 4.11, says that there is at least one point $P(c, f(c))$ on the graph where the slope of the tangent line is the same as the slope of the secant line AB. In other words, there is a point P where the tangent line is parallel to the secant line AB.

Proof   We apply Rolle's Theorem to a new function h defined by

$$(4.14) \qquad h(x) = f(x) - f(a) - \frac{f(b) - f(a)}{b - a}(x - a)$$

[See the note after this proof for a geometric interpretation of $h(x)$.] First we must verify that h satisfies the three hypotheses of Rolle's Theorem.

1. The function h is continuous on $[a, b]$ because it is the sum of f and a first-degree polynomial, both of which are continuous.

2. The function h is differentiable on (a, b) because both f and the first-degree polynomial are differentiable. In fact, we can compute h' directly from Equation 4.14:

$$h'(x) = f'(x) - \frac{f(b) - f(a)}{b - a}$$

    {Note that $f(a)$ and $[f(b) - f(a)]/(b - a)$ are constants.}

3. $$h(a) = f(a) - f(a) - \frac{f(b) - f(a)}{b - a}(a - a) = 0$$

$$h(b) = f(b) - f(a) - \frac{f(b) - f(a)}{b - a}(b - a)$$

$$= f(b) - f(a) - [f(b) - f(a)] = 0$$

Therefore $h(a) = h(b)$.

Since h satisfies the hypotheses of Rolle's Theorem, that theorem says there is a number c in (a, b) such that $h'(c) = 0$. Therefore

$$0 = h'(c) = f'(c) - \frac{f(b) - f(a)}{b - a}$$

and so

$$f'(c) = \frac{f(b) - f(a)}{b - a} \qquad \bullet$$

Note: The function h used in the proof of the Mean Value Theorem can be interpreted as follows. Using Equation 4.13 we see that the equation of the line AB can be written as

$$y - f(a) = \frac{f(b) - f(a)}{b - a}(x - a)$$

or as

$$y = f(a) + \frac{f(b) - f(a)}{b - a}(x - a)$$

Thus $|h(x)|$ can be interpreted as the vertical distance between the line AB and the curve $y = f(x)$ (see Figure 4.15).

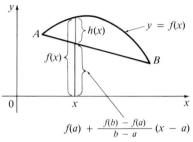

Figure 4.15

● **Example 3**   To illustrate the Mean Value Theorem with a specific function, let us consider $f(x) = x^3 - x$, $a = 0$, $b = 2$. Since f is a polynomial it is continuous and differentiable for all x, so it is certainly continuous on $[0, 2]$ and differentiable on $(0, 2)$. Therefore, by the Mean Value Theorem, there is a number c in $(0, 2)$ such that

$$f(2) - f(0) = f'(c)(2 - 0)$$

Now $f(2) = 6$, $f(0) = 0$, and $f'(x) = 3x^2 - 1$, so this equation becomes

$$6 = (3c^2 - 1)2 = 6c^2 - 2$$

which gives $c^2 = \frac{4}{3}$, that is, $c = \pm 2/\sqrt{3}$. But c must lie in $(0, 2)$, so $c = 2/\sqrt{3}$. ●

● **Example 4**   If an object moves in a straight line with position function $s = f(t)$, then the average velocity between $t = a$ and $t = b$ is

$$\frac{f(b) - f(a)}{b - a}$$

and the velocity at $t = c$ is $f'(c)$. Thus the Mean Value Theorem tells us that at some time $t = c$ between a and b the instantaneous velocity $f'(c)$ is equal to that average velocity. For instance, if a car traveled 180 km in 2 hours then the speedometer must have read 90 km/h at least once. ●

The main significance of the Mean Value Theorem is that it can be used to establish some of the basic facts of differential calculus. One of these basic facts is the following theorem. Others will be found in the following sections.

Theorem (4.15)

> If $f'(x) = 0$ for all x in an interval (a, b), then f is constant on (a, b).

Proof   Let x_1 and x_2 be any two numbers in (a, b) with $x_1 < x_2$. Since f is differentiable on (a, b) it must be differentiable on (x_1, x_2) and continuous on $[x_1, x_2]$. By applying the Mean Value Theorem to f on the interval $[x_1, x_2]$, we get a number c such that $x_1 < c < x_2$ and

(4.16)
$$f(x_2) - f(x_1) = f'(c)(x_2 - x_1)$$

Since $f'(x) = 0$ for all x, we have $f'(c) = 0$, and so Equation 4.16 becomes

$$f(x_2) - f(x_1) = 0 \qquad \text{or} \qquad f(x_2) = f(x_1)$$

Therefore f has the same value at *any* two numbers x_1 and x_2 in (a, b). This means that f is constant on (a, b).   ●

Corollary (4.17)

> If $f'(x) = g'(x)$ for all x in an interval (a, b), then $f - g$ is constant on (a, b); that is, $f(x) = g(x) + c$ where c is a constant.

Proof   Let $F(x) = f(x) - g(x)$. Then

$$F'(x) = f'(x) - g'(x) = 0$$

for all x in (a, b). Thus by Theorem 4.15, F is constant; that is, $f - g$ is constant.   ●

Note: Care must be taken in applying Theorem 4.15. Let

$$f(x) = \frac{x}{|x|} = \begin{cases} 1 & \text{if } x > 0 \\ -1 & \text{if } x < 0 \end{cases}$$

The domain of f is $D = \{x \mid x \neq 0\}$ and $f'(x) = 0$ for all x in D. But f is obviously not a constant function. This does not contradict Theorem 4.15 because D is not an interval. Notice that f is constant on the interval $(0, \infty)$ and also on the interval $(-\infty, 0)$.

EXERCISES 4.2

In Exercises 1–4 verify that the given function satisfies the three hypotheses of Rolle's Theorem on the given interval. Then find all numbers c that satisfy the conclusion of Rolle's Theorem.

1. $f(x) = x^3 - x$, $[-1, 1]$

2. $f(x) = x^3 + x^2 - 2x + 1$, $[-2, 0]$

3. $f(x) = \cos 2x$, $[0, \pi]$

4. $f(x) = \sin x + \cos x$, $[0, 2\pi]$

5. Let $f(x) = 1 - x^{2/3}$. Show that $f(-1) = f(1)$ but there is no number c in $(-1, 1)$ such that $f'(c) = 0$. Why does this not contradict Rolle's Theorem?

6. Let $f(x) = (x - 1)^{-2}$. Show that $f(0) = f(2)$ but there is no number c in $(0, 2)$ such that $f'(c) = 0$. Why does this not contradict Rolle's Theorem?

In Exercises 7–12 verify that the given function satisfies the hypotheses of the Mean Value Theorem on the given interval. Then find all numbers c that satisfy the conclusion of the Mean Value Theorem.

7. $f(x) = 1 - x^2, [0, 3]$

8. $f(x) = x^2 - 4x + 5, [1, 5]$

9. $f(x) = x^3 - 2x + 1, [-2, 3]$

10. $f(x) = 2x^3 + x^2 - x - 1, [0, 2]$

11. $f(x) = \dfrac{1}{x}, [1, 2]$

12. $f(x) = \sqrt{x}, [1, 4]$

13. Let $f(x) = |x - 1|$. Show that there is no value of c such that $f(3) - f(0) = f'(c)(3 - 0)$. Why does this not contradict the Mean Value Theorem?

14. Let $f(x) = (x + 1)/(x - 1)$. Show that there is no value of c such that $f(2) - f(0) = f'(c)(2 - 0)$. Why does this not contradict the Mean Value Theorem?

15. Show that the equation $x^5 + 10x + 3 = 0$ has exactly one real root.

16. Show that the equation $x^7 + 5x^3 + x - 6 = 0$ has exactly one real root.

17. Show that the equation $x^5 - 6x + c = 0$ has at most one root in the interval $[-1, 1]$.

18. Show that the equation $x^4 + 4x + c = 0$ has at most two real roots.

19. (a) Show that a polynomial of degree 3 has at most three real roots.
   (b) Show that a polynomial of degree n has at most n real roots.

20. (a) Suppose that f is differentiable on R and has two roots. Show that f' has at least one root.
   (b) Suppose f is twice differentiable on R and has three roots. Show that f'' has at least one real root.
   (c) Can you generalize parts (a) and (b)?

21. Does there exist a function f such that $f(0) = -1$, $f(2) = 4$, and $f'(x) \leqslant 2$ for all x?

22. Suppose f is continuous on $[2, 5]$ and $1 \leqslant f'(x) \leqslant 4$ for all x in $(2, 5)$. Show that $3 \leqslant f(5) - f(2) \leqslant 12$.

23. Use the Mean Value Theorem to prove the inequality
$$|\sin a - \sin b| \leqslant |a - b| \qquad \text{for all } a \text{ and } b$$

24. If $f'(x) = c$, a constant, for all x, use Corollary 4.17 to show that $f(x) = cx + d$ for some constant d.

25. Let $f(x) = \dfrac{1}{x}$ and
$$g(x) = \begin{cases} \dfrac{1}{x} & \text{if } x > 0 \\[2mm] 1 + \dfrac{1}{x} & \text{if } x < 0 \end{cases}$$

Show that $f'(x) = g'(x)$ for all x in their domains. Can we conclude from Corollary 4.17 that $f - g$ is constant?

26. At 2:00 P.M. a car's speedometer reads 30 mi/h. At 2:10 P.M. it reads 50 mi/h. Show that at some time between 2:00 and 2:10 the acceleration was exactly 120 mi/h^2.

27. Two runners start a race at the same time and finish in a tie. Prove that at some time during the race they had the same velocity. [*Hint:* Consider $f(t) = g(t) - h(t)$ where g and h are the position functions of the two runners.]

SECTION 4.3

Monotonic Functions and the First Derivative Test

In sketching the graph of a function it is very useful to know where it rises and where it falls. The graph shown in Figure 4.16 rises from A to B, falls from B to C, and rises again from C to D. The function f is said to be increasing on the interval $[a, b]$, decreasing on $[b, c]$, and increasing again on $[c, d]$. Notice that if x_1 and x_2 are any two numbers between a and b

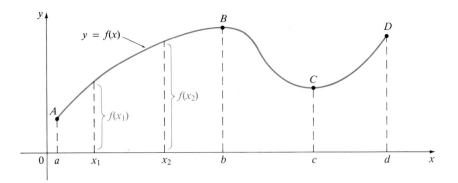

Figure 4.16

with $x_1 < x_2$, then $f(x_1) < f(x_2)$. We use this as the defining property of an increasing function.

Definition (4.18)

A function f is called **increasing** on an interval I if

$$f(x_1) < f(x_2) \qquad \text{whenever } x_1 < x_2 \text{ in } I$$

It is called **decreasing** on I if

$$f(x_1) > f(x_2) \qquad \text{whenever } x_1 < x_2 \text{ in } I$$

It is called **monotonic** on I if it is either increasing or decreasing on I.

In the definition of an increasing function it is important to realize that the inequality $f(x_1) < f(x_2)$ must be satisfied for *every* pair of numbers x_1 and x_2 in I with $x_1 < x_2$.

The function $f(x) = x^2$ is decreasing on $(-\infty, 0]$ and increasing on $[0, \infty)$. (Make a sketch.) Therefore f is monotonic on $(-\infty, 0]$ and on $[0, \infty)$, but it is not monotonic on $(-\infty, \infty)$.

To see how the derivative can help us determine where a function is increasing or decreasing, look at Figure 4.17. It appears that where the slope of the tangent is positive the function is increasing, and where the slope of the tangent is negative the function is decreasing. We know that $f'(x)$ is the

Figure 4.17

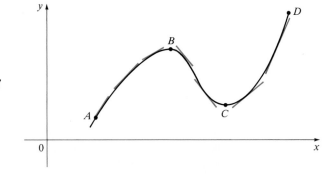

slope of the tangent at $(x, f(x))$. So it appears that the function increases when $f'(x) > 0$ and decreases when $f'(x) < 0$. To prove that this is always the case we use the Mean Value Theorem.

Test for Monotonic Functions (4.19)

> Suppose f is continuous on $[a, b]$ and differentiable on (a, b).
> (a)   If $f'(x) > 0$ for all x in (a, b), then f is increasing on $[a, b]$.
> (b)   If $f'(x) < 0$ for all x in (a, b), then f is decreasing on $[a, b]$.

Proof of (a)   Let x_1 and x_2 be any two numbers in $[a, b]$ with $x_1 < x_2$. Then f is continuous on $[x_1, x_2]$ and differentiable on (x_1, x_2), so by the Mean Value Theorem there is a number c between x_1 and x_2 such that

(4.20)
$$f(x_2) - f(x_1) = f'(c)(x_2 - x_1)$$

Now $f'(c) > 0$ by assumption and $x_2 - x_1 > 0$ because $x_1 < x_2$. Thus the right side of Equation 4.20 is positive, and so

$$f(x_2) - f(x_1) > 0 \qquad \text{or} \qquad f(x_1) < f(x_2)$$

This shows that f is increasing on $[a, b]$.
   Part (b) is proved similarly (see Exercise 43).   ●

● **Example 1**   Find where the function $f(x) = 3x^4 - 4x^3 - 12x^2 + 5$ is increasing and where it is decreasing.

Solution $f'(x) = 12x^3 - 12x^2 - 24x = 12x(x - 2)(x + 1)$

To use Theorem 4.19 we have to know where $f'(x) > 0$ and where $f'(x) < 0$. This depends on the signs of the three factors of $f'(x)$—namely, $12x$, $x - 2$, and $x + 1$. We divide the real line into intervals whose endpoints are the critical numbers -1, 0, and 2 and arrange our work in a chart. A plus sign indicates that the given expression is positive, and a minus sign indicates that it is negative. The last column of the chart gives the conclusion based on Theorem 4.19. From this information and the values of f at the critical numbers, the graph of f is sketched in Figure 4.18.

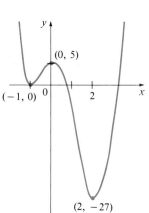

Figure 4.18

Interval	$12x$	$x - 2$	$x + 1$	$f'(x)$	f
$x < -1$	$-$	$-$	$-$	$-$	decreasing on $(-\infty, -1]$
$-1 < x < 0$	$-$	$-$	$+$	$+$	increasing on $[-1, 0]$
$0 < x < 2$	$+$	$-$	$+$	$-$	decreasing on $[0, 2]$
$x > 2$	$+$	$+$	$+$	$+$	increasing on $[2, \infty)$

●

Recall from Section 4.1 that if f has a local maximum or minimum at c, then c must be a critical number of f (by Fermat's Theorem), but not every critical number gives rise to an extremum. We therefore need a test that will tell us whether or not f has a local extremum at a critical number.

You can see from Figure 4.18 that $f(0) = 5$ is a local maximum value of f because f increases on $[-1,0]$ and decreases on $[0,2]$. Or, in terms of derivatives, $f'(x) > 0$ for $-1 < x < 0$ and $f'(x) < 0$ for $0 < x < 2$. In other words, the sign of $f'(x)$ changes from positive to negative at 0. This observation is the basis of the following test.

The First Derivative Test (4.21)

Suppose that c is a critical number of a function f that is continuous on $[a,b]$.

(a) If $f'(x) > 0$ for $a < x < c$ and $f'(x) < 0$ for $c < x < b$ (that is, f' changes from positive to negative at c), then f has a local maximum at c.

(b) If $f'(x) < 0$ for $a < x < c$ and $f'(x) > 0$ for $c < x < b$ (that is, f' changes from negative to positive at c), then f has a local minimum at c.

(c) If f' does not change sign at c, then f has no local extremum at c.

Proof of (a)   Let $x \in (a,b)$. If $a < x < c$, then $f(x) < f(c)$ since $f' > 0$ implies that f is increasing on $[a,c]$. If $c < x < b$, then $f(c) > f(x)$ since $f' < 0$ implies that f is decreasing on $[c,b]$. Therefore $f(c) \geqslant f(x)$ for every x in (a,b). Thus, by Definition 4.2, f has a local maximum at c.

Parts (b) and (c) are proved similarly and are left as Exercise 44.   ●

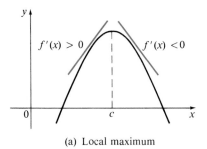

(a) Local maximum

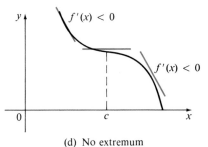

(b) Local minimum

Figure 4.19

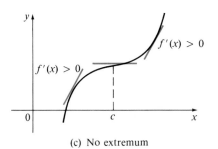

(c) No extremum

(d) No extremum

It is easy to remember the First Derivative Test by visualizing diagrams such as those in Figure 4.19.

● **Example 2**   Find the local extrema of $f(x) = x(1 - x)^{2/5}$ and sketch its graph.

Solution   First we find the critical numbers of f:

$$f'(x) = (1 - x)^{2/5} + x \cdot \frac{2}{5}(1 - x)^{-3/5}(-1)$$

$$= \frac{5(1 - x) - 2x}{5(1 - x)^{3/5}} = \frac{5 - 7x}{5(1 - x)^{3/5}}$$

The derivative $f'(x) = 0$ when $5 - 7x = 0$, that is, $x = \frac{5}{7}$. Also $f'(x)$ does not exist when $x = 1$. So the critical numbers are $\frac{5}{7}$ and 1.

Next we set up a chart as in Example 1, dividing the real line into intervals with the critical numbers as endpoints.

Interval	$5 - 7x$	$(1 - x)^{3/5}$	$f'(x)$	f
$x < \frac{5}{7}$	+	+	+	increasing on $(-\infty, \frac{5}{7}]$
$\frac{5}{7} < x < 1$	−	+	−	decreasing on $[\frac{5}{7}, 1]$
$x > 1$	−	−	+	increasing on $[1, \infty)$

You can see from the chart that $f'(x)$ changes from positive to negative at $\frac{5}{7}$. So, by the First Derivative Test, f has a local maximum at $\frac{5}{7}$ and the local maximum value is

$$f(\tfrac{5}{7}) = \tfrac{5}{7}(\tfrac{2}{7})^{2/5}$$

Since $f'(x)$ changes from negative to positive at 1, $f(1) = 0$ is a local minimum. We use these maximum and minimum values and the information in the chart to sketch the graph in Figure 4.20. Note that the curve is not smooth at $(1, 0)$ but has a "corner" there (called a cusp). This is because $f'(1)$ does not exist.   ●

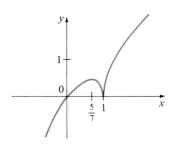

Figure 4.20

● **Example 3**   Find the local and absolute extreme values of the function $f(x) = x^3(x - 2)^2$, $-1 \leqslant x \leqslant 3$. Sketch its graph.

Solution   Using the Product Rule, we have

$$f'(x) = 3x^2(x - 2)^2 + x^3 \cdot 2(x - 2) = x^2(x - 2)(5x - 6)$$

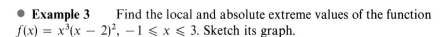

To find the critical numbers we set $f'(x) = 0$ and obtain $x = 0, 2, \frac{6}{5}$. To determine whether these give rise to extreme values we set up the following chart.

Interval	x^2	$x - 2$	$5x - 6$	$f'(x)$	f
$-1 \leqslant x < 0$	+	−	−	+	increasing on $[-1, 0]$
$0 < x < \frac{6}{5}$	+	−	−	+	increasing on $[0, \frac{6}{5}]$
$\frac{6}{5} < x < 2$	+	−	+	−	decreasing on $[\frac{6}{5}, 2]$
$2 < x \leqslant 3$	+	+	+	+	increasing on $[2, 3]$

Notice that $f'(x)$ does not change sign at 0, so by part (c) of the First Derivative Test, f has neither a maximum nor a minimum at 0. [The significance of $f'(0) = 0$ is just that the tangent is horizontal there.] Since f' changes from positive to negative at $\frac{6}{5}$, $f(\frac{6}{5}) = (1.2)^3(-0.8)^2 = 1.10592$ is a local maximum. Since f' changes from negative to positive at 2, $f(2) = 0$ is a local minimum.

To find the absolute extreme values we evaluate f at the endpoints of the interval:

$$f(-1) = -9 \qquad f(3) = 27$$

Using the procedure in (4.8) we compare these values with the values of f at $\frac{6}{5}$ and 2 and find that the absolute maximum value is $f(3) = 27$ and the absolute minimum value is $f(-1) = -9$. The graph of f is sketched in Figure 4.21. ●

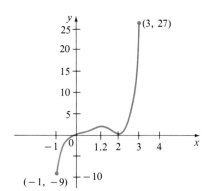

Figure 4.21

● **Example 4**   Prove that the inequality $(1 + x)^n > 1 + nx$ is true whenever $x > 0$ and $n > 1$.

Solution   Consider the difference

$$f(x) = (1 + x)^n - (1 + nx)$$

Then

$$f'(x) = n(1 + x)^{n-1} - n$$

$$= n[(1 + x)^{n-1} - 1]$$

If $x > 0$, then $(1 + x)^{n-1} > 1$, so $f'(x) > 0$. Therefore f is increasing on $[0, \infty)$. In particular, $f(0) < f(x)$ when $0 < x$. But $f(0) = 0$, so

$$0 < (1 + x)^n - (1 + nx)$$

and therefore

$$(1 + x)^n > 1 + nx$$

when $x > 0$. ●

EXERCISES 4.3

In Exercises 1–24 (a) find the intervals on which f is increasing or decreasing, (b) find the local maximum and minimum values of f, and (c) sketch the graph of f.

1. $f(x) = 20 - x - x^2$

2. $f(x) = x^3 - x + 1$

3. $f(x) = x^3 + x + 1$

4. $f(x) = 4x^3 - 3x^2 - 18x + 5$

5. $f(x) = x^3 - 2x^2 + x$

6. $f(x) = 2x^3 - 6x^2 - 18x + 7$

7. $f(x) = 1 - 3x + 5x^2 - x^3$

8. $f(x) = x^4 - 4x^3 - 8x^2 + 3$

9. $f(x) = 2x^2 - x^4$

10. $f(x) = x^2(1 - x)^2$

11. $f(x) = x^4 + 4x + 1$

12. $f(x) = 3x^5 - 25x^3 + 60x$

13. $f(x) = x^3(x - 4)^4$ **14.** $f(x) = x\sqrt{1 - x^2}$

15. $f(x) = x\sqrt{6 - x}$ **16.** $f(x) = x^{2/3}(x - 2)^2$

17. $f(x) = x^{1/5}(x + 1)$ **18.** $f(x) = \sqrt[3]{x} - \sqrt[3]{x^2}$

19. $f(x) = x\sqrt{x - x^2}$ **20.** $f(x) = x^2\sqrt[3]{6x - 7}$

21. $f(x) = x - 2\sin x, 0 \leqslant x \leqslant 2\pi$

22. $f(x) = x + \cos x, 0 \leqslant x \leqslant 2\pi$

23. $f(x) = \sin^4 x + \cos^4 x, 0 \leqslant x \leqslant 2\pi$

24. $f(x) = x\sin x + \cos x, -\pi \leqslant x \leqslant \pi$

In Exercises 25–28 find the intervals on which the given function is increasing or decreasing.

25. $f(x) = x^3 + 2x^2 - x + 1$

26. $f(x) = x^5 + 4x^3 - 6$

27. $f(x) = x^6 + 192x + 17$

28. $f(x) = 2\tan x - \tan^2 x$

In Exercises 29–34 find the local and absolute extreme values of the given function and sketch its graph.

29. $f(x) = x^3 - 3x^2 + 6x - 2, -1 \leqslant x \leqslant 1$

30. $f(x) = x + \dfrac{1}{x}, 0.5 \leqslant x \leqslant 3$

31. $f(x) = x + \sqrt{1 - x}, 0 \leqslant x \leqslant 1$

32. $f(x) = x^3 + 6x^2 + 9x + 2, -4 \leqslant x \leqslant 0$

33. $g(x) = \dfrac{x}{x^2 + 1}, -5 \leqslant x \leqslant 5$

34. $g(x) = \sin x - \cos x, -\dfrac{\pi}{2} \leqslant x \leqslant \dfrac{\pi}{2}$

35. Prove that

$$a + \frac{1}{a} < b + \frac{1}{b} \qquad \text{whenever } 1 < a < b$$

[*Hint:* Show that the function $f(x) = x + 1/x$ is increasing on $[1, \infty)$.]

36. Prove that

$$\frac{\tan b}{\tan a} > \frac{b}{a} \qquad \text{whenever } 0 < a < b < \frac{\pi}{2}$$

In Exercises 37–40 prove the given inequality using the method of Example 4.

37. $2\sqrt{x} > 3 - \dfrac{1}{x}, x > 1$ **38.** $\cos x > 1 - \dfrac{x^2}{2}, x > 0$

39. $\sin x > x - \dfrac{x^3}{6}, x > 0$ **40.** $\tan x > x, 0 < x < \dfrac{\pi}{2}$

41. Find a cubic function $f(x) = ax^3 + bx^2 + cx + d$ that has a local maximum value of 3 at -2 and a local minimum value of 0 at 1.

42. Sketch the graph of a function that satisfies all of the following conditions.
(a) $f'(4) = f'(10) = 0$
(b) $f'(x) < 0$ if $|x| < 2$, $f'(x) \geqslant 0$ if $2 < x < 10$, $f'(x) < 0$ if $x > 10$
(c) $f'(x) = 1$ if $x < -2$, $\displaystyle\lim_{x \to -2^+} f'(x) = -1$

43. Prove Theorem 4.19(b).

44. (a) Prove Theorem 4.21(b).
(b) Prove Theorem 4.21(c).

Concavity and Points of Inflection

We have seen that knowledge of the first derivative of a function is useful in sketching its graph. In this section we shall see, with the help of the Mean Value Theorem, that the second derivative gives us additional information that enables us to sketch a better picture of the graph.

Figure 4.22 shows the graphs of two increasing functions on $[a, b]$. Both graphs join point A to point B but they look different because they bend in different directions. How can we distinguish between these two types of behavior? In Figure 4.23 tangents to these curves have been drawn at several points. In (a) the curve lies above the tangents and f is called *concave upward* on $[a, b]$. In (b) the curve lies below the tangents and g is called *concave downward* on $[a, b]$.

Figure 4.22

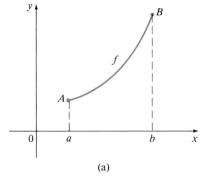

(a)

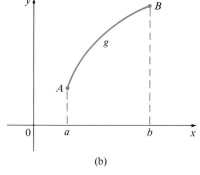

(b)

Figure 4.23

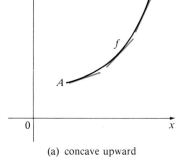

(a) concave upward

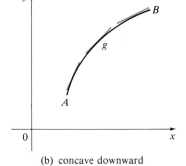

(b) concave downward

Definition (4.22)

> If the graph of f lies above all of its tangents on an interval I, then it is called **concave upward** on I. If the graph of f lies below all of these tangents, it is called **concave downward** on I.

Figure 4.24 shows the graph of a function that is concave upward (abbreviated CU) on the intervals $[b, c]$, $[d, e)$, and $(e, p]$, and concave downward (CD) on the intervals $[a, b]$, $[c, d]$, and $[p, q]$.

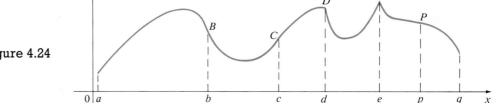

Figure 4.24

Let us see how the second derivative helps determine the intervals of concavity. Looking at Figure 4.23(a) you can see that, going from left to right, the slope of the tangent increases. This means that the derivative $f'(x)$ is an increasing function and therefore its derivative $f''(x)$ is positive. Likewise in Figure 4.23(b) the slope of the tangent decreases from left to right, so $f'(x)$ decreases and therefore $f''(x) < 0$. This reasoning can be reversed and suggests that the following theorem is true.

The Test for Concavity (4.23)

Suppose f is twice differentiable on an interval I.

(a) If $f''(x) > 0$ for all x in I, then the graph of f is concave upward on I.

(b) If $f''(x) < 0$ for all x in I, then the graph of f is concave downward on I.

Proof of (a)   Let a be any number in I. We must show that the curve $y = f(x)$ lies above the tangent line at the point $(a, f(a))$. The equation of this tangent is

$$y = f(a) + f'(a)(x - a)$$

So we must show that

$$f(x) > f(a) + f'(a)(x - a)$$

whenever $x \in I$ ($x \neq a$). (See Figure 4.25.)

Figure 4.25

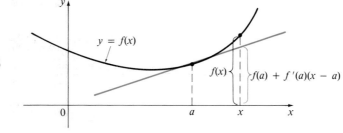

First let us take the case where $x > a$. Applying the Mean Value Theorem to f on the interval $[a, x]$, we get a number c, with $a < c < x$, such that

$$(4.24) \qquad f(x) - f(a) = f'(c)(x - a)$$

Since $f'' > 0$ on I we know from Theorem 4.19 that f' is increasing on I. Thus, since $a < c$, we have

$$f'(a) < f'(c)$$

and so, multiplying this inequality by the positive number $x - a$, we get

$$(4.25) \qquad f'(a)(x - a) < f'(c)(x - a)$$

Now add $f(a)$ to both sides of this inequality:

$$f(a) + f'(a)(x - a) < f(a) + f'(c)(x - a)$$

But from Equation 4.24 we have $f(x) = f(a) + f'(c)(x - a)$. So this inequality becomes

$$(4.26) \qquad f(x) > f(a) + f'(a)(x - a)$$

which is what we wanted to prove.

For the case where $x < a$ we have $f'(c) < f'(a)$, but multiplication by the negative number $x - a$ reverses the inequality, so we get (4.25) and (4.26) as before. ●

Definition (4.27)

A point P on a curve is called a **point of inflection** if the curve changes from concave upward to concave downward or from concave downward to concave upward at P.

For instance, in Figure 4.24, B, C, D, and P are the points of inflection. Notice that if a curve has a tangent at a point of inflection, then the curve crosses its tangent there.

In view of the Test for Concavity, there will be a point of inflection at any point where the second derivative changes sign.

● **Example 1**   Determine where the curve $y = x^3 - 3x + 1$ is concave upward and where it is concave downward. Find the inflection points and sketch the curve.

Solution   If $f(x) = x^3 - 3x + 1$, then

$$f'(x) = 3x^2 - 3 = 3(x^2 - 1)$$

Since $f'(x) = 0$ when $x^2 = 1$, the critical numbers are ± 1. Also

$$f'(x) < 0 \ \Leftrightarrow \ x^2 - 1 < 0 \ \Leftrightarrow \ x^2 < 1 \ \Leftrightarrow \ |x| < 1$$
$$f'(x) > 0 \ \Leftrightarrow \ x^2 > 1 \ \Leftrightarrow \ x > 1 \ \text{ or } \ x < -1$$

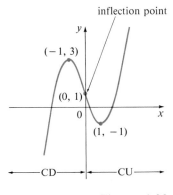

Figure 4.26

$$y = x^3 - 3x + 1$$

Therefore f is increasing on the intervals $(-\infty, -1]$ and $[1, \infty)$ and is decreasing on $[-1, 1]$. By the First Derivative Test, $f(-1) = 3$ is a local maximum value and $f(1) = -1$ is a local minimum value.

To determine the concavity we compute the second derivative:

$$f''(x) = 6x$$

Thus $f''(x) > 0$ when $x > 0$ and $f''(x) < 0$ when $x < 0$. The Test for Concavity then tells us that the curve is concave downward on $(-\infty, 0)$ and concave upward on $(0, \infty)$. Since the curve changes from concave downward to concave upward when $x = 0$, the point $(0, 1)$ is a point of inflection. We use this information to sketch the curve in Figure 4.26. ●

Another application of the second derivative is in finding maximum and minimum values of a function.

The Second Derivative Test (4.28)

> Suppose f'' is continuous on an open interval that contains c.
> (a) If $f'(c) = 0$ and $f''(c) > 0$, then f has a local minimum at c.
> (b) If $f'(c) = 0$ and $f''(c) < 0$, then f has a local maximum at c.

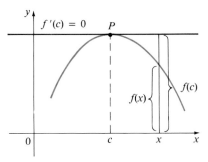

(a) $f''(c) > 0$, concave upward

(b) $f''(c) < 0$, concave downward

Figure 4.27

Proof of (a)   If $f''(c) > 0$, then $f''(x) > 0$ on some open interval I that contains c because f'' is continuous. So, by the Test for Concavity, f is concave upward on I. Therefore the graph of f lies *above* its tangent at $P(c, f(c))$. But since $f'(c) = 0$, the tangent at P is horizontal. This shows that $f(x) \geqslant f(c)$ whenever x is in I [see Figure 4.27(a)] and so f has a local minimum at c.
Part (b) is proved similarly [see Figure 4.27(b)]. ●

● **Example 2**   Discuss the curve $y = x^4 - 4x^3$ with respect to concavity, points of inflection, and local extrema. Use this information to sketch the curve.

Solution   If $f(x) = x^4 - 4x^3$, then

$$f'(x) = 4x^3 - 12x^2 = 4x^2(x - 3)$$

$$f''(x) = 12x^2 - 24x = 12x(x - 2)$$

To find the critical numbers we set $f'(x) = 0$ and obtain $x = 0$ and $x = 3$. To use the Second Derivative Test we evaluate f'' at these critical numbers:

$$f''(0) = 0 \qquad f''(3) = 36 > 0$$

Since $f'(3) = 0$ and $f''(3) > 0$, $f(3) = -27$ is a local minimum. Since $f''(0) = 0$, the Second Derivative Test gives no information about the critical number 0. But since $f'(x) < 0$ for $x < 0$ and also for $0 < x < 3$, the First Derivative Test tells us that f does not have a local extremum at 0.

Since $f''(x) = 0$ when $x = 0$ or 2, we divide the real line into intervals with these numbers as endpoints and complete the following chart.

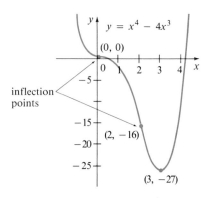

Figure 4.28

Interval	$f''(x) = 12x(x-2)$	Concavity
$(-\infty, 0)$	$+$	upward
$(0, 2)$	$-$	downward
$(2, \infty)$	$+$	upward

The point $(0,0)$ is an inflection point since the curve changes from concave upward to concave downward there. Also $(2, -16)$ is an inflection point since the curve changes from concave downward to concave upward there.

Using the local minimum, the intervals of concavity, and the inflection points, we sketch the curve in Figure 4.28.   ●

Note: Example 2 illustrates the fact that the Second Derivative Test gives no information when $f''(c) = 0$. This test also fails when $f''(c)$ does not exist. In such cases the First Derivative Test must be used.

● **Example 3**   Sketch the graph of the function $f(x) = x^{2/3}(6-x)^{1/3}$.

Solution   Calculation of the first two derivatives gives

$$f'(x) = \frac{4-x}{x^{1/3}(6-x)^{2/3}} \qquad f''(x) = \frac{-8}{x^{4/3}(6-x)^{5/3}}$$

Since $f'(x) = 0$ when $x = 4$ and $f'(x)$ does not exist when $x = 0$ or $x = 6$, the critical numbers are 0, 4, and 6.

Interval	$4 - x$	$x^{1/3}$	$(6-x)^{2/3}$	$f'(x)$	f
$x < 0$	$+$	$-$	$+$	$-$	decreasing on $(-\infty, 0]$
$0 < x < 4$	$+$	$+$	$+$	$+$	increasing on $[0, 4]$
$4 < x < 6$	$-$	$+$	$+$	$-$	decreasing on $[4, 6]$
$x > 6$	$-$	$+$	$+$	$-$	decreasing on $[6, \infty)$

To find the local extreme values we use the First Derivative Test. Since f' changes from negative to positive at 0, $f(0) = 0$ is a local minimum. Since f' changes from positive to negative at 4, $f(4) = 2^{5/3}$ is a local maximum. The sign of f' does not change at 6, so there is no extremum there. (The Second Derivative Test could be used at 4, but not at 0 or 6 since f'' does not exist there.)

Looking at the expression for $f''(x)$ and noting that $x^{4/3} \geq 0$ for all x, we have $f''(x) < 0$ for $x < 0$ and for $0 < x < 6$ and $f''(x) > 0$ for $x > 6$. So f is concave downward on $(-\infty, 0)$ and $(0, 6)$ and concave upward on $(6, \infty)$, and the only inflection point is $(6, 0)$. The graph is sketched in Figure 4.29. Note that the curve has vertical tangents at $(0, 0)$ and $(6, 0)$.   ●

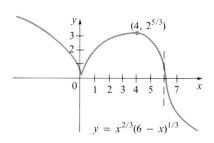

Figure 4.29

EXERCISES 4.4

In Exercises 1–20 find (a) the intervals of increase or decrease, (b) the local maximum and minimum values, (c) the intervals of concavity, and (d) the x-coordinates of the points of inflection. Then use this information to sketch the graph.

1. $f(x) = x^3 - x$

2. $f(x) = 2x^3 + 5x^2 - 4x$

3. $f(x) = x^3 - x^2 - x + 1$

4. $f(x) = x^4 - 6x^2$

5. $g(x) = x^4 - 3x^3 + 3x^2 - x$

6. $g(x) = 4 + 72x - 3x^2 - x^3$

7. $h(x) = 3x^5 - 5x^3 + 3$

8. $h(x) = (x + 2)^3(x - 1)^2$

9. $F(x) = (x^2 - x)^3$

10. $F(x) = x^3(x + 6)^4$

11. $G(x) = 8 - \sqrt[3]{x}$

12. $G(x) = x^{4/3} - 4x^{1/3}$

13. $P(x) = x\sqrt{x^2 + 1}$

14. $P(x) = x\sqrt{x + 1}$

15. $Q(x) = x^{1/3}(x + 3)^{2/3}$

16. $Q(x) = \sqrt[3]{x} - \sqrt[5]{x}$

17. $f(\theta) = \sin^2 \theta$

18. $f(\theta) = \cos^2 \theta$

19. $f(t) = t + \sin t$

20. $f(t) = t + \cos t$

In Exercises 21–24 sketch the graph of a function that satisfies all of the given conditions.

21. $f'(-1) = f'(1) = 0$, $f'(x) < 0$ if $|x| < 1$, $f'(x) > 0$ if $|x| > 1$, $f(-1) = 4$, $f(1) = 0$, $f''(x) < 0$ if $x < 0$, $f''(x) > 0$ if $x > 0$

22. $f'(-1) = 0$, $f'(1)$ does not exist, $f'(x) < 0$ if $|x| < 1$, $f'(x) > 0$ if $|x| > 1$, $f(-1) = 4$, $f(1) = 0$, $f''(x) < 0$ if $x \neq 1$

23. $f'(-1) = 0, f'(2) = 0, f(-1) = f(2) = -1, f(-3) = 4$, $f'(x) = 0$ if $x < -3$, $f'(x) < 0$ on $(-3, -1)$ and $(0, 2)$,

$f'(x) > 0$ on $(-1, 0)$ and $(2, \infty)$, $f''(x) > 0$ on $(-3, 0)$ and $(0, 5)$, $f''(x) < 0$ on $(5, \infty)$

24. $f(0) = 0, f(-1) = 1, f'(-1) = 0, f''(x) > 0$ on $(-\infty, -1), f''(x) < 0$ on $(-1, 0)$ and $(0, \infty), f'(x) > 0$ for $x > 0$

25. Show that a cubic function $f(x) = ax^3 + bx^2 + cx + d$, $a \neq 0$, has exactly one point of inflection.

26. Show that if a function f is concave upward on an interval I, then the function $-f$ is concave downward on I.

27. Prove that if $(c, f(c))$ is a point of inflection of the graph of f and f'' exists in an open interval that contains c, then $f''(c) = 0$. (*Hint:* Apply the First Derivative Test and Fermat's Theorem to the function $g = f'$.)

28. Show that if $f(x) = x^4$, then $f''(0) = 0$, but $(0, 0)$ is not an inflection point of the graph of f.

In Exercises 29–32 assume that all of the functions are twice differentiable.

29. If f and g are concave upward on I, show that $f + g$ is concave upward on I.

30. If f is positive and concave upward on I, show that the function $g(x) = [f(x)]^2$ is concave upward on I.

31. If f and g are positive increasing concave upward functions on I, show that the product function fg is concave upward on I.

32. Suppose f and g are both concave upward on $(-\infty, \infty)$. Under what condition on f will the composite function $h(x) = f(g(x))$ be concave upward?

SECTION 4.5

Limits at Infinity: Horizontal Asymptotes

Let us investigate the behavior of the function f defined by

$$f(x) = \frac{x^2 - 1}{x^2 + 1}$$

as x becomes large. The following table gives values of this function correct to six decimal places and the graph of f has been sketched (with the aid of the table) in Figure 4.30.

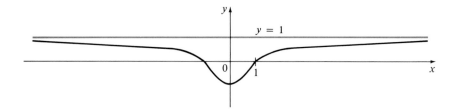

Figure 4.30

x	0	± 1	± 2	± 3	± 4
$f(x)$	-1	0	0.600000	0.800000	0.882353

x	± 5	± 10	± 50	± 100	± 1000
$f(x)$	0.923077	0.980198	0.999200	0.999800	0.999998

As x grows larger and larger you can see that the values of $f(x)$ get closer and closer to 1. In fact, it seems that we can make the values of $f(x)$ as close as we like to 1 by taking x sufficiently large. This situation is expressed symbolically by writing

$$\lim_{x \to \infty} \frac{x^2 - 1}{x^2 + 1} = 1$$

In general we use the symbolism

$$\lim_{x \to \infty} f(x) = L$$

to indicate that the values of $f(x)$ become closer and closer to L as x becomes larger and larger.

Definition (4.29)

Let f be a function defined on some interval (a, ∞). Then

$$\lim_{x \to \infty} f(x) = L$$

means that the values of $f(x)$ can be made arbitrarily close to L by taking x sufficiently large.

Another notation for $\lim_{x \to \infty} f(x) = L$ is

$$f(x) \to L \qquad \text{as } x \to \infty$$

The symbol ∞ does not represent a number. Nonetheless the expression $\lim_{x \to \infty} f(x) = L$ is often read as

"the limit of $f(x)$, as x approaches infinity, is L"

or   "the limit of $f(x)$, as x becomes infinite, is L"

or   "the limit of $f(x)$, as x increases without bound, is L"

The meaning of such phrases is given by Definition 4.29. A more precise definition, similar to the ε, δ definition of Section 2.3, is given at the end of this section.

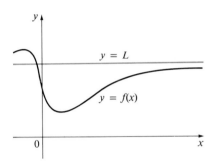
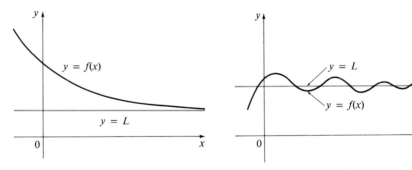

Figure 4.31

EXAMPLES ILLUSTRATING $\lim\limits_{x \to \infty} f(x) = L$

Geometric illustrations of Definition 4.29 are shown in Figure 4.31. Notice that there are many ways for the graph of f to approach the line $y = L$ (which is called a horizontal asymptote).

Referring back to Figure 4.30 we see that for numerically large negative values of x, the values of $f(x)$ are close to 1. By letting x decrease through negative values without bound, we can make $f(x)$ as close as we like to 1. This is expressed by writing

$$\lim_{x \to -\infty} \frac{x^2 - 1}{x^2 + 1} = 1$$

The general definition is as follows:

Definition (4.30)

> Let f be a function defined on some interval $(-\infty, a)$. Then
>
> $$\lim_{x \to -\infty} f(x) = L$$
>
> means that the values of $f(x)$ can be made arbitrarily close to L by taking x sufficiently large negative.

Again, the symbol $-\infty$ does not represent a number, but the expression $\lim_{x \to -\infty} f(x) = L$ is often read as

"the limit of $f(x)$, as x approaches negative infinity, is L"

Definition 4.30 is illustrated in Figure 4.32.

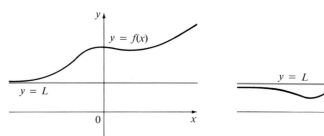

Figure 4.32

EXAMPLES ILLUSTRATING
$\lim\limits_{x \to -\infty} f(x) = L$

Definition (4.31)

> The line $y = L$ is called a **horizontal asymptote** of the curve $y = f(x)$ if either
>
> $$\lim_{x \to \infty} f(x) = L \qquad \text{or} \qquad \lim_{x \to -\infty} f(x) = L$$

For instance, the curve illustrated in Figure 4.30 has the line $y = 1$ as a horizontal asymptote because

$$\lim_{x \to \infty} \frac{x^2 - 1}{x^2 + 1} = 1$$

The curve $y = f(x)$ sketched in Figure 4.33 has both $y = -1$ and $y = 2$ as horizontal asymptotes because

$$\lim_{x \to \infty} f(x) = -1 \quad \text{and} \quad \lim_{x \to -\infty} f(x) = 2$$

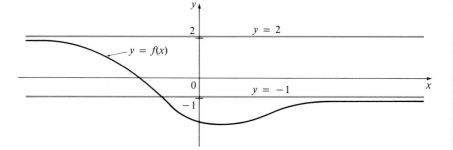

Figure 4.33

● **Example 1**   Find $\lim\limits_{x \to \infty} \dfrac{1}{x}$ and $\lim\limits_{x \to -\infty} \dfrac{1}{x}$.

Solution   Observe that when x is large, $1/x$ is small. For instance,

$$\frac{1}{100} = 0.01 \qquad \frac{1}{10,000} = 0.0001 \qquad \frac{1}{1,000,000} = 0.000001$$

In fact, by taking x large enough, we can make $1/x$ as close to 0 as we please. Therefore, according to Definition 4.29, we have

$$\lim_{x \to \infty} \frac{1}{x} = 0$$

Similar reasoning shows that when x is large negative, $1/x$ is small negative, so we also have

$$\lim_{x \to -\infty} \frac{1}{x} = 0$$

It follows that the line $y = 0$ (the x-axis) is a horizontal asymptote of the curve $y = 1/x$. (This is an equilateral hyperbola; see Figure 4.34.)   ●

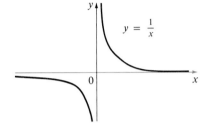

Figure 4.34

$\lim\limits_{x \to \infty}(1/x) = 0, \ \lim\limits_{x \to -\infty}(1/x) = 0$

Most of the properties of limits that were given in Section 2.4 also hold for limits at infinity. It can be proved that the *properties of limits listed in (2.14) and (2.15) (with the exception of Properties 8, 9, and 10) are also valid if "$x \to a$" is replaced by "$x \to \infty$" or "$x \to -\infty$."* In particular, if we combine Properties 6 and 11 with the results of Example 1 we obtain the following important rule for calculating limits.

Theorem (4.32)

> If $r > 0$ is a rational number, then
>
> $$\lim_{x \to \infty} \frac{1}{x^r} = 0$$
>
> If $r > 0$ is a rational number such that x^r is defined for all x, then
>
> $$\lim_{x \to -\infty} \frac{1}{x^r} = 0$$

● **Example 2**   Evaluate

$$\lim_{x \to \infty} \frac{3x^2 - x - 2}{5x^2 + 4x + 1}$$

and indicate which properties of limits are used at each stage.

Solution   To evaluate the limit at infinity of a rational function, we first divide both the numerator and denominator by the highest power of x that occurs. (We may assume that $x \neq 0$ since we are interested only in large values of x.) In this case the highest power of x is x^2, so we have

$$\lim_{x \to \infty} \frac{3x^2 - x - 2}{5x^2 + 4x + 1} = \lim_{x \to \infty} \frac{3 - \dfrac{1}{x} - \dfrac{2}{x^2}}{5 + \dfrac{4}{x} + \dfrac{1}{x^2}}$$

$$= \frac{\lim\limits_{x \to \infty} \left(3 - \dfrac{1}{x} - \dfrac{2}{x^2} \right)}{\lim\limits_{x \to \infty} \left(5 + \dfrac{4}{x} + \dfrac{1}{x^2} \right)} \qquad \textit{(by Property 5)}$$

$$= \frac{\lim\limits_{x \to \infty} 3 - \lim\limits_{x \to \infty} \dfrac{1}{x} - 2 \lim\limits_{x \to \infty} \dfrac{1}{x^2}}{\lim\limits_{x \to \infty} 5 + 4 \lim\limits_{x \to \infty} \dfrac{1}{x} + \lim\limits_{x \to \infty} \dfrac{1}{x^2}} \qquad \textit{(by 1, 2, and 3)}$$

$$= \frac{3 - 0 - 0}{5 + 0 + 0} \qquad \textit{(by 7 and Theorem 4.32)}$$

$$= \frac{3}{5}$$

● **Example 3**   Find the horizontal asymptotes of the graph of the function

$$f(x) = \frac{\sqrt{2x^2 + 1}}{3x - 5}$$

Solution   Dividing numerator and denominator by x and using the properties of limits, we have

$$\lim_{x \to \infty} \frac{\sqrt{2x^2 + 1}}{3x - 5} = \lim_{x \to \infty} \frac{\sqrt{2 + \dfrac{1}{x^2}}}{3 - \dfrac{5}{x}} \qquad \textit{(since } \sqrt{x^2} = x \textit{ for } x > 0)$$

$$= \frac{\lim\limits_{x \to \infty} \sqrt{2 + \dfrac{1}{x^2}}}{\lim\limits_{x \to \infty} \left(3 - \dfrac{5}{x} \right)}$$

$$= \frac{\sqrt{\lim\limits_{x \to \infty} 2 + \lim\limits_{x \to \infty} \dfrac{1}{x^2}}}{\lim\limits_{x \to \infty} 3 - 5 \lim\limits_{x \to \infty} \dfrac{1}{x}}$$

$$= \frac{\sqrt{2 + 0}}{3 - 5 \cdot 0} = \frac{\sqrt{2}}{3}$$

Therefore the line $y = \sqrt{2}/3$ is a horizontal asymptote of the graph of f.

In computing the limit as $x \to -\infty$, we must remember that for $x < 0$, we have $\sqrt{x^2} = |x| = -x$, so when we divide the numerator by x, when $x < 0$, we get

$$\frac{1}{x}\sqrt{2x^2 + 1} = -\frac{1}{\sqrt{x^2}}\sqrt{2x^2 + 1} = -\sqrt{2 + \frac{1}{x^2}}$$

Therefore
$$\lim_{x \to -\infty} \frac{\sqrt{2x^2 + 1}}{3x - 5} = \lim_{x \to -\infty} \frac{-\sqrt{2 + \dfrac{1}{x^2}}}{3 - \dfrac{5}{x}}$$

$$= \frac{-\sqrt{2 + \lim\limits_{x \to -\infty} \dfrac{1}{x^2}}}{3 - 5\lim\limits_{x \to -\infty} \dfrac{1}{x}} = -\frac{\sqrt{2}}{3}$$

Thus the line $y = -\sqrt{2}/3$ is also a horizontal asymptote. The graph of f is sketched in Figure 4.35.   ●

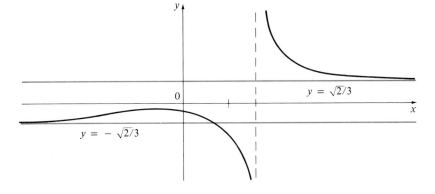

Figure 4.35

● **Example 4**   Compute $\lim\limits_{x \to \infty}(\sqrt{x^2 + 1} - x)$.

Solution   We first multiply numerator and denominator by the conjugate radical:

$$\lim_{x \to \infty}(\sqrt{x^2 + 1} - x) = \lim_{x \to \infty}(\sqrt{x^2 + 1} - x)\frac{\sqrt{x^2 + 1} + x}{\sqrt{x^2 + 1} + x}$$

$$= \lim_{x \to \infty} \frac{(x^2 + 1) - x^2}{\sqrt{x^2 + 1} + x} = \lim_{x \to \infty} \frac{1}{\sqrt{x^2 + 1} + x}$$

The Squeeze Theorem could be used to show that this limit is 0. But an easier method is to divide numerator and denominator by x. Doing this and using the properties of limits, we obtain

$$\lim_{x \to \infty} (\sqrt{x^2 + 1} - x) = \lim_{x \to \infty} \frac{\dfrac{1}{x}}{\sqrt{1 + \dfrac{1}{x^2}} + 1}$$

$$= \frac{0}{\sqrt{1 + 0} + 1} = 0 \qquad \bullet$$

● **Example 5**   Use concavity, inflection points, and asymptotes to sketch the curve $y = 1/(1 + x^2)$.

Solution   If $f(x) = 1/(1 + x^2)$, then

$$f'(x) = -\frac{2x}{(1 + x^2)^2}$$

$$f''(x) = -\frac{2(1 + x^2)^2 - (2x) \cdot 2(1 + x^2) \cdot 2x}{(1 + x^2)^4} = \frac{2(3x^2 - 1)}{(1 + x^2)^3}$$

Since $f'(x) > 0$ when $x < 0$ and $f'(x) < 0$ when $x > 0$, f is increasing on $(-\infty, 0]$ and decreasing on $[0, \infty)$. Also $f(0) = 1$ is a local maximum by the First Derivative Test.

Now observe that

$$f''(x) > 0 \quad \Leftrightarrow \quad 3x^2 - 1 > 0 \quad \Leftrightarrow \quad x^2 > \frac{1}{3} \quad \Leftrightarrow \quad |x| > \frac{1}{\sqrt{3}}$$

This allows us to summarize the information concerning concavity in the following chart.

Interval	$f''(x)$	Concavity
$(-\infty, -1/\sqrt{3})$	$+$	upward
$(-1/\sqrt{3}, 1/\sqrt{3})$	$-$	downward
$(1/\sqrt{3}, \infty)$	$+$	upward

Since the curve changes from concave upward to concave downward when $x = -1/\sqrt{3}$, the point $(-1/\sqrt{3}, \frac{3}{4})$ is a point of inflection. The curve changes from concave downward to concave upward when $x = 1/\sqrt{3}$, so $(1/\sqrt{3}, \frac{3}{4})$ is also a point of inflection.

To complete the picture, note that

$$\lim_{x \to \pm\infty} \frac{1}{1 + x^2} = \lim_{x \to \pm\infty} \frac{\dfrac{1}{x^2}}{\dfrac{1}{x^2} + 1} = \frac{0}{0 + 1} = 0$$

so $y = 0$ (the x-axis) is a horizontal asymptote. The curve is sketched in Figure 4.36. ●

Figure 4.36

$y = 1/(1 + x^2)$

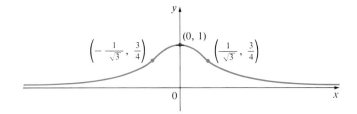

● **Example 6**   Evaluate (a) $\lim_{x \to \infty} \sin x$ and (b) $\lim_{x \to \infty} \sin \dfrac{1}{x}$.

Solution   (a)   As x increases, the values of $\sin x$ oscillate between 1 and -1 infinitely often. Thus $\lim_{x \to \infty} \sin x$ does not exist.

(b)   If we let $t = 1/x$, then $t \to 0^+$ as $x \to \infty$, so we have

$$\lim_{x \to \infty} \sin \frac{1}{x} = \lim_{t \to 0^+} \sin t = 0$$

(See Exercise 45.)   ●

Precise Definitions

Definition 4.29 can be stated precisely as follows:

Definition (4.33)

> Let f be a function defined on some interval (a, ∞). Then
>
> $$\lim_{x \to \infty} f(x) = L$$
>
> means that for every $\varepsilon > 0$ there is a corresponding number N such that
>
> $$|f(x) - L| < \varepsilon \qquad \text{whenever} \qquad x > N$$

In words, this says that the values of $f(x)$ can be made arbitrarily close to L (within a distance ε, where ε is any positive number) by taking x sufficiently large (larger than N, where N depends on ε). Graphically it says that by choosing x large enough (larger than some number N) we can make the graph of f lie between the given horizontal lines $y = L - \varepsilon$ and $y = L + \varepsilon$ as in Figure 4.37. This must be true no matter how small we choose ε. Figure 4.38 shows that if a smaller value of ε is chosen, then a larger value of N may be required.

Similarly, a precise version of Definition 4.30 is given by Definition 4.34, which is illustrated in Figure 4.39.

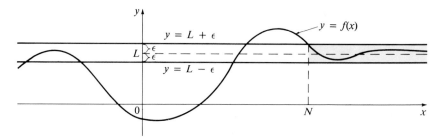

Figure 4.37

$\lim\limits_{x \to \infty} f(x) = L$

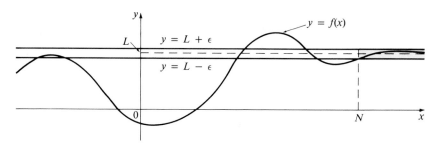

Figure 4.38

$\lim\limits_{x \to \infty} f(x) = L$

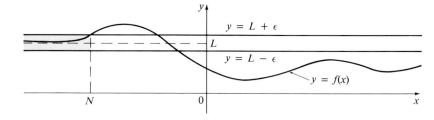

Figure 4.39

$\lim\limits_{x \to -\infty} f(x) = L$

Definition (4.34)

Let f be a function defined on some interval $(-\infty, a)$. Then

$$\lim_{x \to -\infty} f(x) = L$$

means that for every $\varepsilon > 0$ there is a corresponding number N such that

$$|f(x) - L| < \varepsilon \qquad \text{whenever} \qquad x < N$$

● **Example 7**   Use Definition 4.33 to prove that $\lim\limits_{x \to \infty} \dfrac{1}{x} = 0$.

Solution   *(a) (Preliminary analysis of the problem: guessing a value for N)*
Given $\varepsilon > 0$, we want to find N such that

$$\left|\frac{1}{x} - 0\right| < \varepsilon \qquad \text{whenever} \qquad x > N$$

In computing the limit we may assume $x > 0$, in which case

$$\left|\frac{1}{x} - 0\right| = \left|\frac{1}{x}\right| = \frac{1}{x}$$

Therefore we want

$$\frac{1}{x} < \varepsilon \qquad \text{whenever} \qquad x > N$$

that is,

$$x > \frac{1}{\varepsilon} \qquad \text{whenever} \qquad x > N$$

This suggests that we should take $N = 1/\varepsilon$.

(b) Proof (showing that N works)   Given $\varepsilon > 0$, we choose $N = 1/\varepsilon$. Let $x > N$. Then

$$\left| \frac{1}{x} - 0 \right| = \frac{1}{|x|} = \frac{1}{x} < \frac{1}{N} = \varepsilon$$

Thus

$$\left| \frac{1}{x} - 0 \right| < \varepsilon \qquad \text{whenever} \qquad x > N$$

Therefore, by Definition 4.33,

$$\lim_{x \to \infty} \frac{1}{x} = 0$$

Figure 4.40 illustrates the proof by showing some values of ε and the corresponding values of N.

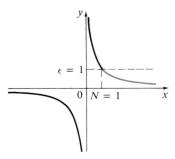
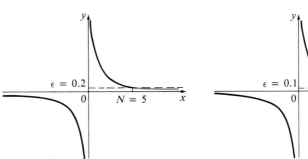
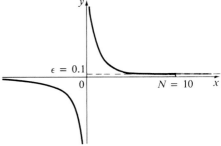

Figure 4.40

EXERCISES 4.5

In Exercises 1–10 evaluate the given limit and justify each step by indicating the appropriate properties of limits.

1. $\displaystyle \lim_{x \to \infty} \frac{1}{x\sqrt{x}}$

2. $\displaystyle \lim_{x \to -\infty} 5x^{-2/3}$

3. $\displaystyle \lim_{x \to \infty} \frac{5 + 2x}{3 - x}$

4. $\displaystyle \lim_{x \to \infty} \frac{x + 4}{x^2 - 2x + 5}$

5. $\displaystyle \lim_{x \to -\infty} \frac{2x^2 - x - 1}{4x^2 + 7}$

6. $\displaystyle \lim_{t \to \infty} \frac{7t^3 + 4t}{2t^3 - t^2 + 3}$

7. $\displaystyle \lim_{x \to -\infty} \frac{(1 - x)(2 + x)}{(1 + 2x)(2 - 3x)}$

8. $\displaystyle \lim_{x \to \infty} \sqrt{\frac{2x^2 - 1}{x + 8x^2}}$

9. $\displaystyle \lim_{x \to \infty} \frac{1}{3 + \sqrt{x}}$

10. $\displaystyle \lim_{x \to \infty} \frac{\sin^2 x}{x^2}$

In Exercises 11–26 find the given limits.

11. $\displaystyle\lim_{r \to \infty} \frac{r^4 - r^2 + 1}{r^5 + r^3 - r}$

12. $\displaystyle\lim_{t \to -\infty} \frac{6t^2 + 5t}{(1 - t)(2t - 3)}$

13. $\displaystyle\lim_{x \to \infty} \frac{\sqrt{1 + 4x^2}}{4 + x}$

14. $\displaystyle\lim_{x \to -\infty} \frac{\sqrt{x^2 + 4x}}{4x + 1}$

15. $\displaystyle\lim_{x \to \infty} \frac{\sqrt[3]{x^2 + 8}}{x + 2}$

16. $\displaystyle\lim_{x \to \infty} \frac{\sqrt[3]{x^3 + 8}}{x + 2}$

17. $\displaystyle\lim_{x \to \infty} \frac{1 - \sqrt{x}}{1 + \sqrt{x}}$

18. $\displaystyle\lim_{x \to \infty} (\sqrt{x^2 + 3x + 1} - x)$

19. $\displaystyle\lim_{x \to \infty} (\sqrt{x^2 + 1} - \sqrt{x^2 - 1})$

20. $\displaystyle\lim_{x \to -\infty} (x + \sqrt{x^2 + 2x})$

21. $\displaystyle\lim_{x \to \infty} (\sqrt{1 + x} - \sqrt{x})$

22. $\displaystyle\lim_{x \to \infty} (\sqrt[3]{1 + x} - \sqrt[3]{x})$

23. $\displaystyle\lim_{x \to -\infty} (\sqrt{x^2 + x + 1} + x)$

24. $\displaystyle\lim_{x \to \infty} \cos \frac{1}{x}$

25. $\displaystyle\lim_{x \to \infty} \cos x$

26. $\displaystyle\lim_{x \to \infty} \left(x - x \cos \frac{1}{x} \right)$

In Exercises 27–30 find the horizontal asymptotes of the graphs of the given functions.

27. $f(x) = \dfrac{1 + 2x}{2 + x}$

28. $g(x) = \dfrac{3x^4 + 6x^2 - x}{2x^4 - 5x^3 + 9}$

29. $h(x) = \dfrac{x}{\sqrt[4]{x^4 + 1}}$

30. $F(x) = \dfrac{x - 9}{\sqrt{4x^2 + 3x + 2}}$

In Exercises 31–34 find the horizontal asymptotes of the curve and use them, together with concavity and intervals of increase and decrease, to sketch the curve.

31. $y = \dfrac{x}{x^2 + 1}$

32. $y = \dfrac{2x^2 - x + 2}{x^2 + 1}$

33. $y = 1 - \dfrac{1}{\sqrt{x^2 + 1}}$

34. $y = \dfrac{1}{x^2 + x + 1}$

35. (a) Find $\lim_{x \to \infty} x \sin(1/x)$.
   (b) Use part (a) and Example 4 in Section 2.4 to sketch the graph of the function $f(x) = x \sin(1/x)$.

36. (a) Use the Squeeze Theorem to evaluate

$$\lim_{x \to \infty} \frac{\sin x}{x}$$

   (b) Sketch the graph of the function $f(x) = (\sin x)/x$.

In Exercises 37 and 38 sketch the graph of a function that satisfies all of the given conditions.

37. $f'(2) = 0$, $f(2) = -1$, $f(0) = 0$, $f'(x) < 0$ if $0 < x < 2$, $f'(x) > 0$ if $x > 2$, $f''(x) < 0$ if $0 \leqslant x < 1$ or if $x > 4$, $f''(x) > 0$ if $1 < x < 4$, $\lim_{x \to \infty} f(x) = 1$, $f(-x) = f(x)$ for all x

38. $f'(2) = 0$, $f'(0) = 1$, $f'(x) > 0$ if $0 < x < 2$, $f'(x) < 0$ if $x > 2$, $f''(x) < 0$ if $0 < x < 4$, $f''(x) > 0$ if $x > 4$, $\lim_{x \to \infty} f(x) = 0$, $f(-x) = -f(x)$ for all x

39. Guess the value of the limit

$$\lim_{x \to \infty} \frac{x^2}{2^x}$$

by evaluating the function $f(x) = x^2/2^x$ for $x = 0, 1, 2, 3, 4, 5, 6, 7, 8, 9, 10, 20, 50,$ and 100.

40. How large do we have to take x so that $1/x^2 < 0.0001$?

41. Taking $r = 2$ in Theorem 4.32 we have the statement

$$\lim_{x \to \infty} \frac{1}{x^2} = 0$$

Prove this directly using Definition 4.33.

42. How large do we have to take x so that $1/\sqrt{x} < 0.0001$?

43. Taking $r = \frac{1}{2}$ in Theorem 4.32, we have the statement

$$\lim_{x \to \infty} \frac{1}{\sqrt{x}} = 0$$

Prove this directly using Definition 4.33.

44. Use Definition 4.34 to prove that

$$\lim_{x \to -\infty} \frac{1}{x} = 0$$

45. Prove that

$$\lim_{x \to \infty} f(x) = \lim_{t \to 0^+} f\left(\frac{1}{t}\right) \quad \text{and} \quad \lim_{x \to -\infty} f(x) = \lim_{t \to 0^-} f\left(\frac{1}{t}\right)$$

if these limits exist.

SECTION 4.6

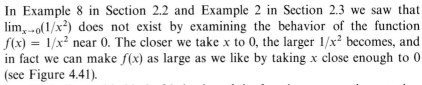

Infinite Limits: Vertical Asymptotes

In Example 8 in Section 2.2 and Example 2 in Section 2.3 we saw that $\lim_{x\to 0}(1/x^2)$ does not exist by examining the behavior of the function $f(x) = 1/x^2$ near 0. The closer we take x to 0, the larger $1/x^2$ becomes, and in fact we can make $f(x)$ as large as we like by taking x close enough to 0 (see Figure 4.41).

To indicate this kind of behavior of the function we use the notation

$$\lim_{x\to 0}\frac{1}{x^2} = \infty$$

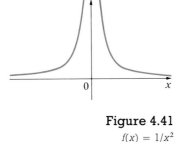

Figure 4.41

$f(x) = 1/x^2$

In general, we write symbolically

$$\lim_{x\to a} f(x) = \infty$$

to indicate that the values of $f(x)$ become larger and larger (or "increase without bound") as x becomes closer and closer to a.

Definition (4.35)

Let f be a function defined on both sides of a, except possibly at a itself. Then

$$\lim_{x\to a} f(x) = \infty$$

means that the values of $f(x)$ can be made arbitrarily large (as large as we please) by taking x sufficiently close to a ($x \neq a$).

Another notation for $\lim_{x\to a} f(x) = \infty$ is

$$f(x) \to \infty \qquad \text{as} \qquad x \to a$$

Again the symbol ∞ is not a number, but the expression $\lim_{x\to a} f(x) = \infty$ is often read as

"the limit of $f(x)$, as x approaches a, is infinity"

or   "$f(x)$ becomes infinite as x approaches a"

or   "$f(x)$ increases without bound as x approaches a"

This definition is illustrated graphically in Figure 4.42. (A more precise definition is given at the end of this section.)

A similar sort of limit, for functions that become large negative as x gets close to a, is defined in Definition 4.36 and is illustrated in Figure 4.43.

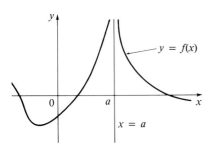

Figure 4.42

$\lim_{x\to a} f(x) = \infty$

Definition (4.36)

Let f be defined on both sides of a, except possibly at a itself. Then

$$\lim_{x \to a} f(x) = -\infty$$

means that the values of $f(x)$ can be made arbitrarily large negative by taking x sufficiently close to a ($x \neq a$).

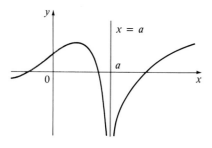

Figure 4.43

$\lim\limits_{x \to a} f(x) = -\infty$

The symbol $\lim_{x \to a} f(x) = -\infty$ can be read as "the limit of $f(x)$, as x approaches a, is negative infinity" or "$f(x)$ decreases without bound as x approaches a." As an example we have

$$\lim_{x \to 0}\left(-\frac{1}{x^2}\right) = -\infty$$

Similar definitions can be given for the one-sided infinite limits

$$\lim_{x \to a^-} f(x) = \infty \qquad \lim_{x \to a^+} f(x) = \infty$$

$$\lim_{x \to a^-} f(x) = -\infty \qquad \lim_{x \to a^+} f(x) = -\infty$$

remembering that "$x \to a^-$" means that we consider only values of x that are less than a, and similarly "$x \to a^+$" means that we consider only $x > a$. Illustrations of these four cases are given in Figure 4.44.

Definition (4.37)

The line $x = a$ is called a **vertical asymptote** of the curve $y = f(x)$ if at least one of the following statements is true:

$$\lim_{x \to a} f(x) = \infty \qquad \lim_{x \to a^-} f(x) = \infty \qquad \lim_{x \to a^+} f(x) = \infty$$

$$\lim_{x \to a} f(x) = -\infty \qquad \lim_{x \to a^-} f(x) = -\infty \qquad \lim_{x \to a^+} f(x) = -\infty$$

For instance, the y-axis is a vertical asymptote of the curve $y = 1/x^2$ because $\lim_{x \to 0}(1/x^2) = \infty$. In Figure 4.44 the line $x = a$ is a vertical asymptote in each of the four cases.

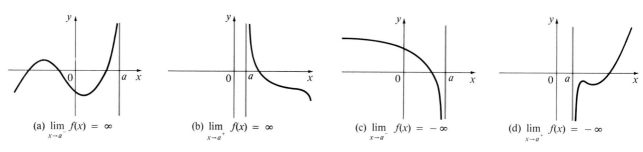

(a) $\lim\limits_{x \to a^-} f(x) = \infty$ (b) $\lim\limits_{x \to a^+} f(x) = \infty$ (c) $\lim\limits_{x \to a^-} f(x) = -\infty$ (d) $\lim\limits_{x \to a^+} f(x) = -\infty$

Figure 4.44

● **Example 1**   Find $\lim\limits_{x\to 3^+} \dfrac{2}{x-3}$ and $\lim\limits_{x\to 3^-} \dfrac{2}{x-3}$.

Solution   If x is close to 3 but larger than 3, then $x-3$ is a small positive number and so $2/(x-3)$ is a large positive number. Thus intuitively we see that

$$\lim\limits_{x\to 3^+} \dfrac{2}{x-3} = \infty$$

Likewise if x is close to 3 but smaller than 3, then $x-3$ is a small negative number and so $2/(x-3)$ is a numerically large negative number. Thus

$$\lim\limits_{x\to 3^-} \dfrac{2}{x-3} = -\infty$$

The graph of the curve $y = 2/(x-3)$ is given in Figure 4.45. The line $x = 3$ is a vertical asymptote.   ●

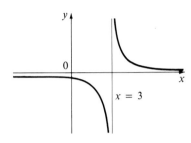

Figure 4.45

The following theorem can be established intuitively by arguing as in Example 1 or proved from Definition 4.39.

Theorem (4.38)

(a)   If n is a positive even integer, then

$$\lim\limits_{x\to a} \dfrac{1}{(x-a)^n} = \infty$$

(b)   If n is a positive odd integer, then

$$\lim\limits_{x\to a^+} \dfrac{1}{(x-a)^n} = \infty \quad \text{and} \quad \lim\limits_{x\to a^-} \dfrac{1}{(x-a)^n} = -\infty$$

● **Example 2**   (a)   Find the vertical and horizontal asymptotes of the function

$$f(x) = \dfrac{x}{x^2+x-2}$$

(b)   Use this information to sketch the graph of f.

Solution   (a)   Write

$$f(x) = \dfrac{x}{(x-1)(x+2)}$$

The vertical asymptotes are likely to occur when the denominator is 0, that is, when $x = 1$ or -2. If x is close to 1 but $x > 1$, then the denominator is close to 0, but $x > 0$, $x - 1 > 0$, and $x + 2 > 0$, so $f(x) > 0$. Thus

$$\lim_{x \to 1^+} \frac{x}{(x - 1)(x + 2)} = \infty$$

Similarly we see that

$$\lim_{x \to 1^-} \frac{x}{(x - 1)(x + 2)} = -\infty$$

If x is close to -2 but $x < -2$, then the denominator is close to 0, but $x < 0$, $x - 1 < 0$, and $x + 2 < 0$, so $f(x) < 0$. Thus

$$\lim_{x \to -2^-} \frac{x}{(x - 1)(x + 2)} = -\infty$$

Similarly we see that

$$\lim_{x \to -2^+} \frac{x}{(x - 1)(x + 2)} = \infty$$

Thus the vertical asymptotes are $x = 1$ and $x = -2$. Since

$$\lim_{x \to \infty} \frac{x}{x^2 + x - 2} = \lim_{x \to \infty} \frac{\dfrac{1}{x}}{1 + \dfrac{1}{x} - \dfrac{2}{x^2}}$$

$$= \frac{\displaystyle\lim_{x \to \infty} \frac{1}{x}}{1 + \displaystyle\lim_{x \to \infty} \frac{1}{x} - 2 \lim_{x \to \infty} \frac{1}{x^2}}$$

$$= \frac{0}{1 + 0 - 0} = 0$$

the horizontal asymptote is $y = 0$. A similar computation shows that

$$\lim_{x \to -\infty} \frac{x}{x^2 + x - 2} = 0$$

(b)   Now we compute the derivative of f:

$$f'(x) = \frac{(x^2 + x - 2)1 - x(2x + 1)}{(x^2 + x - 2)^2} = -\frac{x^2 + 2}{(x^2 + x - 2)^2}$$

Since $x^2 + 2 > 0$ for all x, there is no critical number and $f'(x) < 0$ for all x in the domain of f. Therefore f is decreasing on each of the intervals $(-\infty, -2)$, $(-2, 1)$, and $(1, \infty)$. Also

$$f''(x) = \frac{-(x^2 + x - 2)^2 2x + (x^2 + 2)2(x^2 + x - 2)(2x + 1)}{(x^2 + x - 2)^4}$$

$$= \frac{2(x^3 + 6x + 2)}{(x - 1)^3(x + 2)^3}$$

To analyze the sign of f'' we let $g(x) = x^3 + 6x + 2$. Since $g(0) = 2 > 0$ and $g(-1) = -5 < 0$, the Intermediate Value Theorem gives a number α between -1 and 0 such that $g(\alpha) = 0$. Also $g'(x) = 3x^2 + 6 > 0$ for all x, so g is always increasing. This means that $g(x) < 0$ for $x < \alpha$ and $g(x) > 0$ for $x > \alpha$. We can now analyze the concavity of f in the accompanying chart. The only inflection point occurs when $x = \alpha$. Using this information we sketch the graph of f in Figure 4.46.

Interval	$g(x)$	$x + 2$	$x - 1$	$f''(x)$	f
$x < -2$	$-$	$-$	$-$	$-$	CD on $(-\infty, -2)$
$-2 < x < \alpha$	$-$	$+$	$-$	$+$	CU on $(-2, \alpha)$
$\alpha < x < 1$	$+$	$+$	$-$	$-$	CD on $(\alpha, 1)$
$x > 1$	$+$	$+$	$+$	$+$	CU on $(1, \infty)$

Figure 4.46

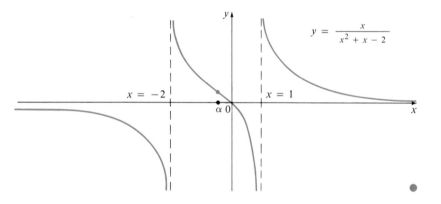

● **Example 3**   Find $\displaystyle\lim_{x \to (\pi/2)^-} \tan x$.

Solution   Write

$$\tan x = \frac{\sin x}{\cos x} \cdot$$

Since sine and cosine are continuous functions, we have $\sin x \to \sin(\pi/2) = 1$ as $x \to (\pi/2)^-$ and $\cos x \to \cos(\pi/2) = 0$ through positive values as $x \to (\pi/2)^-$. Thus

$$\lim_{x \to (\pi/2)^-} \tan x = \infty$$

This shows that the line $x = \pi/2$ is a vertical asymptote of the graph of the tangent function. Similarly any line of the form $x = (2n + 1)\pi/2$, with n an integer, is also a vertical asymptote (see Figure 4.47).   ●

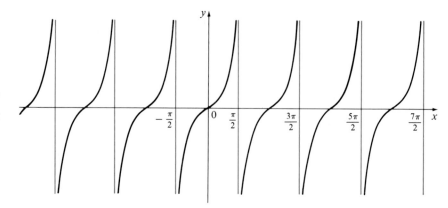

Figure 4.47

$y = \tan x$

Infinite Limits at Infinity

The notation

$$\lim_{x \to \infty} f(x) = \infty$$

is used to indicate that the values of $f(x)$ become large as x becomes large. Similar meanings are attached to the following symbols:

$$\lim_{x \to -\infty} f(x) = \infty \qquad \lim_{x \to \infty} f(x) = -\infty \qquad \lim_{x \to -\infty} f(x) = -\infty$$

● **Example 4**   Find $\lim\limits_{x \to \infty} x^3$ and $\lim\limits_{x \to -\infty} x^3$.

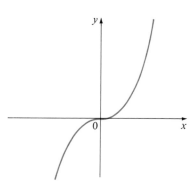

Solution   When x becomes large, x^3 also becomes large. For instance,

$$10^3 = 1000 \qquad 100^3 = 1,000,000 \qquad 1000^3 = 1,000,000,000$$

In fact we can make x^3 as big as we like by taking x large enough. Therefore we can write

$$\lim_{x \to \infty} x^3 = \infty$$

Similarly, when x is large negative, so is x^3. Thus

$$\lim_{x \to -\infty} x^3 = -\infty$$

Figure 4.48

$\lim\limits_{x \to \infty} x^3 = \infty, \ \lim\limits_{x \to -\infty} x^3 = -\infty$

These limit statements can also be seen from the graph of $y = x^3$ in Figure 4.48.   ●

● **Example 5**   Find $\lim\limits_{x \to \infty} (x^2 - x)$.

Solution   Note that we *cannot* write

$$\lim_{x \to \infty} (x^2 - x) = \lim_{x \to \infty} x^2 - \lim_{x \to \infty} x$$

$$= \infty - \infty$$

The properties of limits cannot always be applied to infinite limits because ∞ is not a number ($\infty - \infty$ cannot be defined). However we can write

$$\lim_{x \to \infty} (x^2 - x) = \lim_{x \to \infty} x(x - 1) = \infty$$

because both x and $x - 1$ become arbitrarily large.   •

● **Example 6**   Find $\displaystyle\lim_{x \to \infty} \frac{x^2 + x}{3 - x}$.

Solution 1   As in Example 2 in Section 4.5 we can evaluate the limit at infinity of a rational function by dividing numerator and denominator by the highest power of x that occurs:

$$\lim_{x \to \infty} \frac{x^2 + x}{3 - x} = \lim_{x \to \infty} \frac{1 + \dfrac{1}{x}}{\dfrac{3}{x^2} - \dfrac{1}{x}} = -\infty$$

because $1 + 1/x \to 1$ and $3/x^2 - 1/x \to 0$ through negative values as $x \to \infty$. (Note that $3/x^2 - 1/x < 0$ for $x > 3$.)

Solution 2   It is also possible to find the given limit by dividing numerator and denominator by x instead of x^2:

$$\lim_{x \to \infty} \frac{x^2 + x}{3 - x} = \lim_{x \to \infty} \frac{x + 1}{\dfrac{3}{x} - 1} = -\infty$$

because $x + 1 \to \infty$ and $3/x - 1 \to -1$ as $x \to \infty$.   •

Precise Definitions

Definition 4.35 can be stated more precisely as follows:

Definition (4.39)

> Let f be a function defined on some open interval that contains the number a, except possibly at a itself. Then
>
> $$\lim_{x \to a} f(x) = \infty$$
>
> means that for every positive number M there is a corresponding number $\delta > 0$ such that
>
> $$f(x) > M \qquad \text{whenever} \qquad 0 < |x - a| < \delta$$

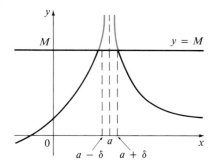

Figure 4.49

This says that the values of $f(x)$ can be made arbitrarily large (larger than any given number M) by taking x close enough to a (within a distance δ, where δ depends on M). A geometric illustration is shown in Figure 4.49.

Given any horizontal line $y = M$, we can find a number $\delta > 0$ such that if we restrict x to lie in the interval $(a - \delta, a + \delta)$, then the curve $y = f(x)$ lies above the line $y = M$. You can see that if a larger M is chosen, then a smaller δ may be required.

● **Example 7**   Use Definition 4.39 to prove that

$$\lim_{x \to 0} \frac{1}{x^2} = \infty$$

Solution   *(a) (Preliminary analysis of the problem: guessing a value for δ).* Given $M > 0$, we want to find $\delta > 0$ such that

$$\frac{1}{x^2} > M \qquad \text{whenever} \qquad 0 < |x - 0| < \delta$$

that is,

$$x^2 < \frac{1}{M} \qquad \text{whenever} \qquad 0 < |x| < \delta$$

or

$$|x| < \frac{1}{\sqrt{M}} \qquad \text{whenever} \qquad 0 < |x| < \delta$$

This suggests that we should take $\delta = 1/\sqrt{M}$.

(b) Proof (showing that this δ works).   If $M > 0$ is given, let $\delta = 1/\sqrt{M}$. If $0 < |x - 0| < \delta$, then

$$|x| < \delta \Rightarrow x^2 < \delta^2$$

$$\Rightarrow \frac{1}{x^2} > \frac{1}{\delta^2} = M$$

Thus $\dfrac{1}{x^2} > M \qquad \text{whenever} \qquad 0 < |x - 0| < \delta$

Therefore, by Definition 4.39,

$$\lim_{x \to 0} \frac{1}{x^2} = \infty \qquad ●$$

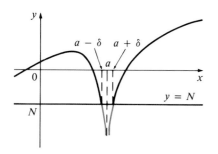

Figure 4.50

Similarly, a precise version of Definition 4.36 is given by Definition 4.40, which is illustrated by Figure 4.50.

Definition (4.40)

Let f be a function defined on some open interval that contains the number a, except possibly at a itself. Then

$$\lim_{x \to a} f(x) = -\infty$$

means that for every negative number N there is a corresponding number $\delta > 0$ such that

$$f(x) < N \qquad \text{whenever} \qquad 0 < |x - a| < \delta$$

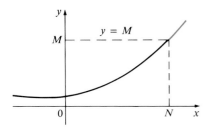

Figure 4.51

$$\lim_{x \to \infty} f(x) = \infty$$

Finally we note that an infinite limit at infinity can be defined as follows. The geometric illustration is given in Figure 4.51.

Definition (4.41)

Let f be a function defined on some interval (a, ∞). Then

$$\lim_{x \to \infty} f(x) = \infty$$

means that for every positive number M there is a corresponding number $N > 0$ such that

$$f(x) > M \qquad \text{whenever} \qquad x > N$$

EXERCISES 4.6

1. State the equations of the vertical and horizontal asymptotes of the curve shown.

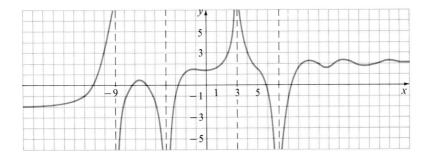

In Exercises 2–35 find each limit.

2. $\lim\limits_{x \to 0} \dfrac{2}{x^4}$

3. $\lim\limits_{x \to -1} \dfrac{-2}{(x + 1)^6}$

4. $\lim\limits_{x \to 3} \dfrac{1}{(x - 3)^8}$

5. $\lim\limits_{x \to 5^+} \dfrac{6}{x - 5}$

6. $\lim\limits_{x \to 5^-} \dfrac{6}{x - 5}$

7. $\lim\limits_{x \to -1^-} \dfrac{x^2}{x + 1}$

8. $\lim\limits_{x \to -6^+} \dfrac{x}{x + 6}$

9. $\lim\limits_{t \to 3^-} \dfrac{t + 3}{t^2 - 9}$

10. $\lim\limits_{x \to 0^+} \dfrac{3}{x^2 + 2x}$

11. $\lim\limits_{x \to -3^-} \dfrac{x^2}{x^2 - 9}$

12. $\lim\limits_{x \to 0} \dfrac{x - 1}{x^2(x + 2)}$

13. $\lim\limits_{x \to -2^+} \dfrac{x - 1}{x^2(x + 2)}$

14. $\lim\limits_{x \to 0^+} \left(\dfrac{1}{x} - \dfrac{1}{x^2} \right)$

15. $\lim\limits_{x \to -3^+} \dfrac{x^3}{x^2 + 5x + 6}$

16. $\lim\limits_{x \to 5^-} \dfrac{2}{\sqrt{5 - x}}$

17. $\lim\limits_{x \to 2^+} \dfrac{-3}{\sqrt[3]{x - 2}}$

18. $\lim\limits_{x \to 0^+} \cot x$

19. $\lim\limits_{x \to 0^-} \cot x$

20. $\lim\limits_{t \to (\pi/4)^+} \sec 2t$

21. $\lim\limits_{x \to \pi^-} \csc x$

22. $\lim\limits_{x \to \pi} \csc x$

23. $\lim\limits_{t \to (-3\pi/2)^-} \sec t$

24. $\lim\limits_{x \to \infty} \sqrt{x}$

25. $\lim\limits_{x \to -\infty} \sqrt[3]{x}$

26. $\lim\limits_{x \to \infty} (x - \sqrt{x})$

27. $\lim\limits_{x \to \infty} (x + \sqrt{x})$

28. $\lim\limits_{x \to -\infty} (x^3 - 5x^2)$

29. $\lim\limits_{x \to \infty} (x^2 - x^4)$

30. $\lim\limits_{x \to \infty} \dfrac{x^7 - 1}{x^6 + 1}$

31. $\lim\limits_{x \to \infty} \dfrac{x^3 - 1}{x^4 + 1}$

32. $\lim\limits_{x \to \infty} \dfrac{\sqrt{x} + 3}{x + 3}$

33. $\lim\limits_{x \to \infty} \dfrac{x}{\sqrt{x - 1}}$

34. $\lim\limits_{x \to \infty} \dfrac{\sqrt{x}}{\sqrt[3]{x}}$

35. $\lim\limits_{x \to -\infty} \dfrac{x^8 + 3x^4 + 2}{x^5 + x^3}$

In Exercises 36–39 find the vertical and horizontal asymptotes of the given curves.

36. $y = \dfrac{x}{x + 4}$

37. $y = \dfrac{x^2 + 4}{x^2 - 1}$

38. $y = \dfrac{x^3}{x^2 + 3x - 10}$

39. $y = \dfrac{x^3 + 1}{x^3 + x}$

In Exercises 40–43 find the vertical and horizontal asymptotes of the given curve. Use this information, together with information about concavity and intervals of increase and decrease, to sketch the curve.

40. $y = \dfrac{x - 2}{x + 2}$

41. $y = \dfrac{4}{x - 4}$

42. $y = \dfrac{x}{(x + 1)^2}$

43. $y = \dfrac{1}{1 - x^2}$

44. Make a rough sketch of the curve $y = x^n$ (n an integer) for the following five cases: (i) $n = 0$, (ii) $n > 0$, n odd, (iii) $n > 0$, n even, (iv) $n < 0$, n odd, and (v) $n < 0$, n even. Then use these sketches to find the following limits.

(a) $\lim\limits_{x \to 0^+} x^n$ (b) $\lim\limits_{x \to 0^-} x^n$

(c) $\lim\limits_{x \to \infty} x^n$ (d) $\lim\limits_{x \to -\infty} x^n$

45. Let P and Q be polynomials. What is

$$\lim\limits_{x \to \infty} \dfrac{P(x)}{Q(x)}$$

if (a) the degree of P is less than the degree of Q? (b) the degree of P is greater than the degree of Q?

46. The graph of a function f has a **vertical tangent line** at the point $(a, f(a))$ if f is continuous at a and

$$\lim\limits_{x \to a} |f'(x)| = \infty$$

(a) Show that the curve $y = \sqrt[3]{x}$ has a vertical tangent line at $(0, 0)$ and sketch the curve.
(b) Show that the curve $y = x^{2/3}$ has a vertical tangent line at $(0, 0)$ and sketch the curve.

47. How close to -3 do we have to take x so that

$$\dfrac{1}{(x + 3)^4} > 10{,}000$$

48. Prove, using Definition 4.39, that

$$\lim\limits_{x \to -3} \dfrac{1}{(x + 3)^4} = \infty$$

49. Prove that

$$\lim\limits_{x \to -1^-} \dfrac{5}{(x + 1)^3} = -\infty$$

50. Prove, using Definition 4.41, that $\lim\limits_{x \to \infty} x^3 = \infty$.

51. If $\lim\limits_{x \to a} f(x) = \infty$ and $\lim\limits_{x \to a} g(x) = c$, where c is a real number, prove that

$$\lim\limits_{x \to a} [f(x) + g(x)] = \infty$$

52. Suppose $\lim_{x \to a} f(x) = \infty$ and $\lim_{x \to a} g(x) = c$, where c is a real number.

(a)   If $c > 0$, prove that

$$\lim_{x \to a} [f(x)g(x)] = \infty$$

(b)   If $c < 0$, prove that

$$\lim_{x \to a} [f(x)g(x)] = -\infty$$

SECTION 4.7

Curve Sketching

The graph of a function tells us a great deal about the behavior of the function. Therefore one of the aims of calculus is to provide a procedure for sketching curves with reasonable accuracy.

So far we have been concerned with some particular aspects of curve sketching: domain, range, and symmetry in Chapter 1; limits and continuity in Chapter 2; derivatives and tangents in Chapter 3; and extreme values, monotonicity, concavity, points of inflection, and asymptotes in this chapter. It is now time to assemble all of this information and put it together to produce sketches of graphs that reveal the important features of functions.

You may ask: What is wrong with just using a calculator to plot points and then joining these points with a smooth curve? To see the pitfalls of this approach, suppose you have used a calculator to produce the table of values and corresponding points in Figure 4.52.

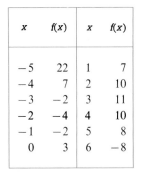
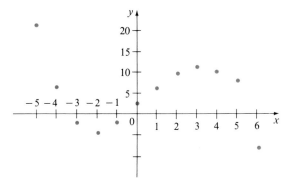

x	$f(x)$	x	$f(x)$
-5	22	1	7
-4	7	2	10
-3	-2	3	11
-2	-4	4	10
-1	-2	5	8
0	3	6	-8

Figure 4.52

You might then join these points to produce the curve shown in Figure 4.53, but the correct graph might be the one shown in Figure 4.54. You can see the drawbacks of the method of plotting points. Certain essential features of the graph may be missed, such as the maximum and minimum values between -2 and -1 or between 2 and 5. If you just plot points, you don't know when to stop. (How far should you go to the left or right?) But the use of calculus ensures that all the important aspects of the curve are illustrated.

Before sketching a curve it is wise to consider first the following items.

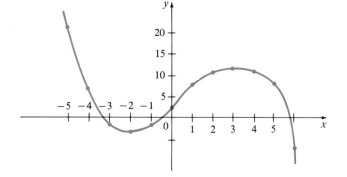

Figure 4.53

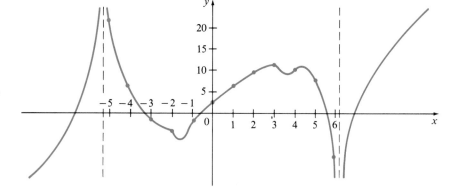

Figure 4.54

Checklist of Information for Sketching a Curve $y = f(x)$

A. **Domain.** The first step is to determine the domain D of f, that is, the set of values of x for which $f(x)$ is defined.

B. **Intercepts.** The y-intercept is $f(0)$ and tells us where the curve intersects the y-axis. To find the x-intercepts, we set $y = 0$ and solve for x, if this can be done easily. Sometimes it is difficult, and not worthwhile, to find the x-intercepts. For instance, to find the x-intercept of $f(x) = x^3 - x^2 + 2x - 1$ we would have to solve the equation $x^3 - x^2 + 2x - 1 = 0$. This is not easy, so we do not bother with it.

C. **Symmetry.**
  (i) If $f(-x) = f(x)$ for all x in D, that is, the equation of the curve is unchanged when x is replaced by $-x$, then f is an **even function** and the curve is symmetric about the y-axis. This means that our work is cut in half. If we know what the curve looks like for $x \geqslant 0$, then we need only reflect in the y-axis to obtain the complete curve [see Figure 4.55(a)]. Here are some examples: $y = x^2$, $y = x^4$, $y = |x|$, and $y = \cos x$.
  (ii) If $f(-x) = -f(x)$ for all x in D, then f is an **odd function** and the curve is symmetric about the origin. Again we can obtain the complete curve if we know what it looks like for

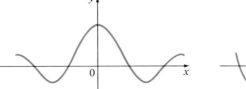

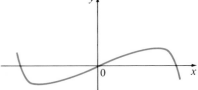

Figure 4.55

(a) Even function: reflectional symmetry   (b) Odd function: rotational symmetry

$x > 0$ [see Figure 4.55(b)]. Some simple examples of odd functions are: $y = x$, $y = x^3$, $y = x^5$, and $y = \sin x$.

(iii) If $f(x + p) = f(x)$ for all x in D, where p is a positive constant, then f is called a **periodic function** and the smallest such number p is called the **period.** For instance, $y = \sin x$ has period 2π and $y = \tan x$ has period π. If we know what the graph looks like in an interval of length p, then we can use translation to sketch the entire graph (see Figure 4.56).

Figure 4.56

PERIODIC FUNCTION:
TRANSLATIONAL SYMMETRY

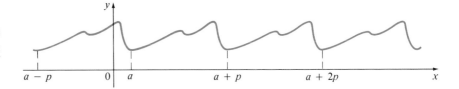

D. **Asymptotes.**

(i)   *Horizontal Asymptotes.* Recall from Section 4.5 that if either

$$\lim_{x \to \infty} f(x) = L \quad \text{or} \quad \lim_{x \to -\infty} f(x) = L$$

then the line $y = L$ is a horizontal asymptote of the curve $y = f(x)$. If it turns out that $\lim_{x \to \infty} f(x) = \infty$ (or $-\infty$), then we do not have an asymptote to the right but that is still useful information for sketching the curve.

(ii)   *Vertical Asymptotes.* Recall from Section 4.6 that the line $x = a$ is a vertical asymptote if at least one of the following statements is true:

(4.42)

$$\lim_{x \to a^+} f(x) = \infty \quad \lim_{x \to a^-} f(x) = \infty \quad \lim_{x \to a^+} f(x) = -\infty \quad \lim_{x \to a^-} f(x) = -\infty$$

(For rational functions you can locate the vertical asymptotes by equating the denominator to 0 after canceling any common factors. But for other functions this method does not apply.) Furthermore, in sketching the curve it is very useful to know exactly which of the statements in (4.42) is true. If $f(a)$ is not defined but a is an endpoint of the domain of f, then you should compute $\lim_{x \to a^-} f(x)$ or $\lim_{x \to a^+} f(x)$, whether or not this limit is infinite.

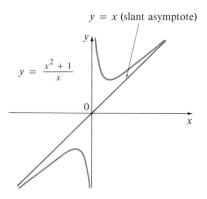

$y = x$ (slant asymptote)

$y = \dfrac{x^2 + 1}{x}$

Figure 4.57

(iii)   *Slant Asymptotes.* These are asymptotes that are oblique, that is, neither horizontal nor vertical. If

$$\lim_{x \to \infty} [f(x) - (mx + b)] = 0$$

then the line $y = mx + b$ is a **slant asymptote** because the vertical distance between the curve $y = f(x)$ and the line $y = mx + b$ approaches 0. (A similar situation exists if we let $x \to -\infty$.) For rational functions, slant asymptotes occur when the degree of the numerator is one more than the degree of the denominator. For instance, if

$$f(x) = \frac{x^2 + 1}{x} = x + \frac{1}{x}$$

then $f(x) - x = \dfrac{1}{x} \to 0$ as $x \to \pm \infty$

so the line $y = x$ is a slant asymptote (see Figure 4.57).

E.   *Intervals of Increase or Decrease.* Use the Test for Monotonic Functions (Theorem 4.19). Compute $f'(x)$ and find the intervals on which $f'(x)$ is positive (f is increasing) and the intervals on which $f'(x)$ is negative (f is decreasing).

F.   *Local Maximum and Minimum Values.* Find the critical numbers of f [the numbers c where $f'(c) = 0$ or $f'(c)$ does not exist]. Then use the First Derivative Test (Theorem 4.21). If f' changes from positive to negative at a critical number c, then $f(c)$ is a local maximum. If f' changes from negative to positive at c, then $f(c)$ is a local minimum. Although it is usually preferable to use the First Derivative Test, you can use the Second Derivative Test (Theorem 4.28) if c is a critical number such that $f''(c) \neq 0$. Then $f''(c) > 0$ implies that $f(c)$ is a local minimum, whereas $f''(c) < 0$ implies that $f(c)$ is a local maximum.

G.   *Concavity and Points of Inflection.* Compute $f''(x)$ and use the Test for Concavity (Theorem 4.23). The curve is concave upward where $f''(x) > 0$ and concave downward where $f''(x) < 0$. Inflection points occur where the direction of concavity changes.

H.   *Sketch the Curve.* Using the information in items A–G, draw the graph. Draw in the asymptotes as broken lines. Plot the intercepts, maximum and minimum points, and inflection points. Then make the curve pass through these points, rising and falling according to E, with concavity according to G, and approaching the asymptotes. If additional accuracy is desired near any point, you could compute the value of the derivative there. The tangent will indicate the direction in which the curve proceeds.

● **Example 1**   Discuss the curve $y = \dfrac{2x^2}{x^2 - 1}$ under the headings A–H.

Solution

A.   The domain is

$$\{x|x^2 - 1 \neq 0\} = \{x|x \neq \pm 1\} = (-\infty, -1) \cup (-1, 1) \cup (1, \infty)$$

B.   The x- and y-intercepts are both 0.

C.   Since $f(-x) = f(x)$, f is even. The curve is symmetric about the y-axis.

D.
$$\lim_{x \to \pm \infty} \frac{2x^2}{x^2 - 1} = \lim_{x \to \pm \infty} \frac{2}{1 - 1/x^2} = 2$$

Therefore the line $y = 2$ is a horizontal asymptote. Since the denominator is 0 when $x = \pm 1$, we compute the following limits:

$$\lim_{x \to 1^+} \frac{2x^2}{x^2 - 1} = \infty \qquad \lim_{x \to 1^-} \frac{2x^2}{x^2 - 1} = -\infty$$

$$\lim_{x \to -1^+} \frac{2x^2}{x^2 - 1} = -\infty \qquad \lim_{x \to -1^-} \frac{2x^2}{x^2 - 1} = \infty$$

Therefore the lines $x = 1$ and $x = -1$ are vertical asymptotes.

E.
$$f'(x) = \frac{4x(x^2 - 1) - 2x^2 \cdot 2x}{(x^2 - 1)^2} = \frac{-4x}{(x^2 - 1)^2}$$

Since $f'(x) > 0$ when $x < 0$ $(x \neq -1)$ and $f'(x) < 0$ when $x > 0$ $(x \neq 1)$, f is increasing on $(-\infty, -1)$ and $(-1, 0]$ and decreasing on $[0, 1)$ and $(1, \infty)$.

F.   The only critical number is $x = 0$. Since f' changes from positive to negative at 0, $f(0) = 0$ is a local maximum by the First Derivative Test.

G.
$$f''(x) = \frac{-4(x^2 - 1)^2 + 4x \cdot 2(x^2 - 1)2x}{(x^2 - 1)^4} = \frac{12x^2 + 4}{(x^2 - 1)^3}$$

Since $12x^2 + 4 > 0$ for all x, we have

$$f''(x) > 0 \quad \Leftrightarrow \quad x^2 - 1 > 0 \quad \Leftrightarrow \quad |x| > 1$$

and $f''(x) < 0 \Leftrightarrow |x| < 1$. Thus the curve is concave upward on the intervals $(-\infty, -1)$ and $(1, \infty)$ and concave downward on $(-1, 1)$. There is no point of inflection since 1 and -1 are not in the domain of f.

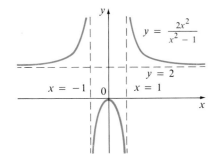

Figure 4.58   H.   Using the information in A–G, we sketch the curve in Figure 4.58.

● **Example 2**   Sketch the graph of $y = \dfrac{x^2 + 2x - 4}{x^2}$.

A.   The domain is $\{x|x \neq 0\} = (-\infty, 0) \cup (0, \infty)$.

B.   There is no y-intercept because of A. To find the x-intercepts we set $y = 0$. Using the quadratic formula, we find that the roots of the equation $x^2 + 2x - 4 = 0$ are $x = -1 \pm \sqrt{5}$. These are the x-intercepts.

C.   Symmetry: none.

D.
$$\lim_{x \to \pm \infty} \frac{x^2 + 2x - 4}{x^2} = \lim_{x \to \pm \infty} \left(1 + \frac{2}{x} - \frac{4}{x^2}\right) = 1$$

Therefore the line $y = 1$ is a horizontal asymptote. Since $x^2 \to 0$ as $x \to 0$ and $x^2 + 2x - 4 < 0$ when x is near 0, we have

$$\lim_{x \to 0} \frac{x^2 + 2x - 4}{x^2} = -\infty$$

This shows that $x = 0$ (the y-axis) is a vertical asymptote.

E.
$$f'(x) = -\frac{2}{x^2} + \frac{8}{x^3} = 2\left(\frac{4 - x}{x^3}\right)$$

Interval	$4 - x$	x^3	$f'(x)$	f
$x < 0$	$+$	$-$	$-$	decreasing on $(-\infty, 0)$
$0 < x < 4$	$+$	$+$	$+$	increasing on $(0, 4]$
$x > 4$	$-$	$+$	$-$	decreasing on $[4, \infty)$

F.   Since $f'(4) = 0$ and f' changes from positive to negative at 4, $f(4) = \frac{5}{4}$ is a local maximum by the First Derivative Test.

G.
$$f''(x) = \frac{4}{x^3} - \frac{24}{x^4} = 4\left(\frac{x - 6}{x^4}\right)$$

Thus $f''(x) > 0$ when $x > 6$ and $f''(x) < 0$ when $x < 6$ ($x \neq 0$). The curve is concave upward on $(6, \infty)$ and concave downward on $(-\infty, 0)$ and $(0, 6)$. Therefore $(6, \frac{11}{9})$ is a point of inflection.

H.   The curve is sketched in Figure 4.59.

Figure 4.59

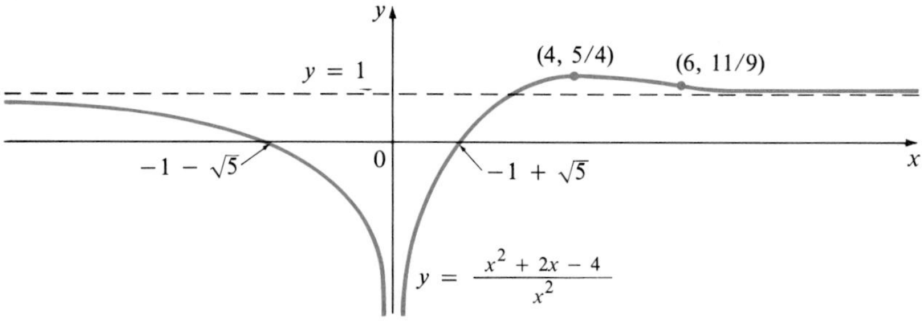

● **Example 3**   Sketch the graph of $f(x) = \dfrac{x^3}{x^2 + 1}$.

A.   The domain is $R = (-\infty, \infty)$.

B.   The x- and y-intercepts are both 0.

C. Since $f(-x) = -f(x)$, f is odd and its graph is symmetric about the origin.

D. Since $x^2 + 1$ is never 0, there is no vertical asymptote. Since $f(x) \to \infty$ as $x \to \infty$ and $f(x) \to -\infty$ as $x \to -\infty$, there is no horizontal asymptote. But long division gives

$$f(x) = \frac{x^3}{x^2 + 1} = x - \frac{x}{x^2 + 1}$$

$$f(x) - x = -\frac{x}{x^2 + 1} = -\frac{\dfrac{1}{x}}{1 + \dfrac{1}{x^2}} \to 0 \qquad \text{as } x \to \pm\infty$$

So the line $y = x$ is a slant asymptote.

E. $$f'(x) = \frac{3x^2(x^2 + 1) - x^3 \cdot 2x}{(x^2 + 1)^2} = \frac{x^2(x^2 + 3)}{(x^2 + 1)^2}$$

Since $f'(x) > 0$ for all x (except 0), f is increasing on $(-\infty, \infty)$.

F. Although $f'(0) = 0$, f' does not change sign at 0, so there is no local maximum or minimum.

G. $$f''(x) = \frac{(4x^3 + 6x)(x^2 + 1)^2 - (x^4 + 3x^2) \cdot 2(x^2 + 1)2x}{(x^2 + 1)^4}$$

$$= \frac{2x(3 - x^2)}{(x^2 + 1)^3}$$

Since $f''(x) = 0$ when $x = 0$ or $x = \pm\sqrt{3}$, we set up the following chart:

Interval	x	$3 - x^2$	$(x^2 + 1)^3$	$f''(x)$	f
$x < -\sqrt{3}$	$-$	$-$	$+$	$+$	concave upward on $(-\infty, -\sqrt{3})$
$-\sqrt{3} < x < 0$	$-$	$+$	$+$	$-$	concave downward on $(-\sqrt{3}, 0)$
$0 < x < \sqrt{3}$	$+$	$+$	$+$	$+$	concave upward on $(0, \sqrt{3})$
$x > \sqrt{3}$	$+$	$-$	$+$	$-$	concave downward on $(\sqrt{3}, \infty)$

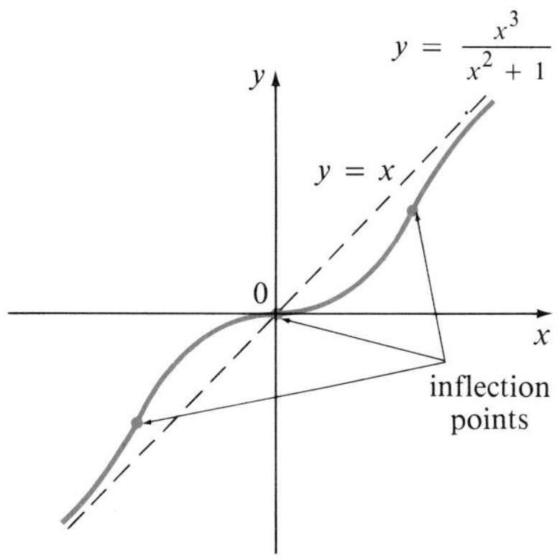

Figure 4.60

The points of inflection are $(-\sqrt{3}, -3\sqrt{3}/4)$, $(0, 0)$, and $(\sqrt{3}, 3\sqrt{3}/4)$.

H. The graph of f is sketched in Figure 4.60.

● **Example 4**   Sketch the graph of $f(x) = \dfrac{x^2}{\sqrt{x+1}}$.

A.   Domain $= \{x \mid x + 1 > 0\} = \{x \mid x > -1\} = (-1, \infty)$

B.   The x- and y-intercepts are both 0.

C.   Symmetry: none.

D.   Since

$$\lim_{x \to \infty} \frac{x^2}{\sqrt{x+1}} = \infty$$

there is no horizontal asymptote. Since $\sqrt{x+1} \to 0$ as $x \to -1^+$ and $f(x)$ is always positive, we have

$$\lim_{x \to -1^+} \frac{x^2}{\sqrt{x+1}} = \infty$$

and so the line $x = -1$ is a vertical asymptote.

E. $$f'(x) = \frac{2x\sqrt{x+1} - x^2 \cdot 1/2\sqrt{x+1}}{x+1} = \frac{x(3x+4)}{2(x+1)^{3/2}}$$

Note that $f'(x) = 0$ when $x = 0$ or $-\frac{4}{3}$. But $-\frac{4}{3}$ is not in the domain of f, so the only critical number is 0. Since $f'(x) < 0$ when $-1 < x < 0$ and $f'(x) > 0$ when $x > 0$, f is decreasing on $(-1, 0]$ and increasing on $[0, \infty)$.

F.   Since $f'(0) = 0$ and f' changes from negative to positive at 0, $f(0) = 0$ is a local (and absolute) minimum by the First Derivative Test.

G. $$f''(x) = \frac{2(x+1)^{3/2}(6x+4) - (3x^2+4x)3(x+1)^{1/2}}{4(x+1)^3}$$

$$= \frac{3x^2 + 8x + 8}{4(x+1)^{5/2}}$$

Note that the denominator is always positive. The numerator is the quadratic $3x^2 + 8x + 8$, which is always positive because its discriminant is $b^2 - 4ac = -32 < 0$ and the coefficient of x^2 is positive. Thus $f''(x) > 0$ for all x, which means that f is concave upward on $(-1, \infty)$ and there is no point of inflection.

H.   The curve is sketched in Figure 4.61.

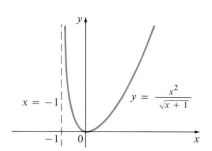

Figure 4.61

● **Example 5**   Sketch the graph of $f(x) = 2\cos x + \sin 2x$.

A.   The domain is R.

B.   The y-intercept is $f(0) = 2$. The x-intercepts occur when

$$2\cos x + \sin 2x = 2\cos x + 2\sin x \cos x = 2\cos x(1 + \sin x) = 0$$

that is, when $\cos x = 0$ or $\sin x = -1$. Thus, in the interval $[0, 2\pi]$, the x-intercepts are $\pi/2$ and $3\pi/2$.

C. f is neither even nor odd, but $f(x + 2\pi) = f(x)$ for all x and so f is periodic and has period 2π. Thus in what follows we need to consider only $0 \leqslant x \leqslant 2\pi$ and then extend the curve by translation in H.

D.   Asymptotes: none.

E. $f'(x) = -2\sin x + 2\cos 2x = -2\sin x + 2(1 - 2\sin^2 x)$

$$= -2(2\sin^2 x + \sin x - 1) = -2(2\sin x - 1)(\sin x + 1)$$

Thus $f'(x) = 0$ when $\sin x = \frac{1}{2}$ or $\sin x = -1$, so in $[0, 2\pi]$, $x = \pi/6$, $5\pi/6$, $3\pi/2$.

Interval	$f'(x)$	f
$0 < x < \pi/6$	$+$	increasing on $[0, \pi/6]$
$\pi/6 < x < 5\pi/6$	$-$	decreasing on $[\pi/6, 5\pi/6]$
$5\pi/6 < x < 3\pi/2$	$+$	increasing on $[5\pi/6, 3\pi/2]$
$3\pi/2 < x < 2\pi$	$+$	increasing on $[3\pi/2, 2\pi]$

F.   From the chart in E the First Derivative Test says that $f(\pi/6) = 3\sqrt{3}/2$ is a local maximum and $f(5\pi/6) = -3\sqrt{3}/2$ is a local minimum but f has no extremum at $3\pi/2$, only a horizontal tangent.

G. $f''(x) = -2\cos x - 4\sin 2x = -2\cos x(1 + 4\sin x)$

Thus $f''(x) = 0$ when $\cos x = 0$ (so $x = \pi/2$ or $3\pi/2$) and when $\sin x = -\frac{1}{4}$. From Figure 4.62 we see that there are two values of x between 0 and 2π for which $\sin x = -\frac{1}{4}$. Let us call them α_1 and α_2. Then $f''(x) > 0$ on $(\pi/2, \alpha_1)$ and $(3\pi/2, \alpha_2)$, so f is concave upward there. Also $f''(x) < 0$ on $(0, \pi/2)$, $(\alpha_1, 3\pi/2)$, and $(\alpha_2, 2\pi)$, so f is concave downward there. Inflection points occur when $x = \pi/2$, α_1, $3\pi/2$, and α_2.

Figure 4.62

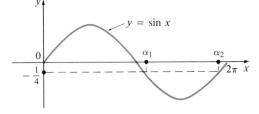

H.   The graph of the function restricted to $0 \leqslant x \leqslant 2\pi$ is shown in Figure 4.63. Then it is extended, using periodicity, to the complete graph in Figure 4.64.

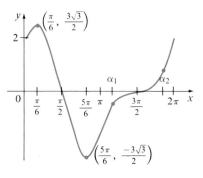

Figure 4.63

Figure 4.64

$$y = 2 \cos x + \sin 2x$$

EXERCISES 4.7

Discuss the following curves under the headings A–H given in this section.

1. $y = \dfrac{1}{x - 1}$

2. $y = \dfrac{x}{x - 1}$

3. $y = \dfrac{1}{x^2 - 9}$

4. $y = \dfrac{x}{x^2 - 9}$

5. $y = \dfrac{x^3}{x^2 - 1}$

6. $y = \dfrac{4}{(x - 5)^2}$

7. $y = \dfrac{x}{(2x - 3)^2}$

8. $y = x - \dfrac{1}{x}$

9. $xy = x^2 + 4$

10. $xy = x^2 + x + 1$

11. $y = \dfrac{x - 3}{x + 3}$

12. $y = \dfrac{x^2}{2x + 5}$

13. $y = \dfrac{1}{(x - 1)(x + 2)}$

14. $y = \dfrac{1}{x^2(x + 3)}$

15. $y = \dfrac{1 + x^2}{1 - x^2}$

16. $y = \dfrac{1}{4x^3 - 9x}$

17. $y = \dfrac{x^3 - 1}{x^3 + 1}$

18. $y = \dfrac{x^3 - 1}{x}$

19. $y = \dfrac{1}{x - 1} - x$

20. $y = \dfrac{1}{(x - 1)^2(x + 5)}$

21. $y = \dfrac{1}{x^3 - x}$

22. $y = \dfrac{1 - x^2}{x^3}$

23. $y = \sqrt{2 - x}$

24. $y = x\sqrt{x + 3}$

25. $y = x + \sqrt{x}$

26. $y = \sqrt{x} - \sqrt{x - 1}$

27. $y = \sqrt{x^2 + 1} - x$

28. $y = \sqrt{\dfrac{x}{x - 5}}$

29. $y = \dfrac{1}{x} - \sqrt{x}$

30. $y = \sqrt{x} - \dfrac{1}{\sqrt{x}}$

31. $y = \sqrt[4]{x^2 - 25}$

32. $y = \sqrt{x} - \sqrt{x}$

33. $y = x\sqrt{x^2 - 9}$

34. $y = \dfrac{x^2}{\sqrt{1 - x^2}}$

35. $y = \dfrac{\sqrt{1 - x^2}}{x}$

36. $y = \dfrac{2x + 1}{\sqrt{3x^2 + 4}}$

37. $y = x + 3x^{2/3}$

38. $y = x^{5/3} - 5x^{2/3}$

39. $y = x + \sqrt{|x|}$

40. $y = \sqrt[3]{(x^2 - 1)^2}$

41. $y = \cos x - \sin x$

42. $y = \sin x + \cos x$

43. $y = \sin x - \tan x$

44. $y = \sin x + \sqrt{3} \cos x$

45. $y = x \tan x,\ -\dfrac{\pi}{2} < x < \dfrac{\pi}{2}$

46. $y = 2x + \cot x,\ 0 < x < \pi$

47. $y = \dfrac{x}{2} - \sin x, 0 < x < 3\pi$

48. $y = 2\sin x + \sin^2 x$

49. $y = 2\cos x + \sin^2 x$

50. $y = \sin x - x$

51. $y = \sin 2x - 2\sin x$

52. $y = \dfrac{\cos x}{2 + \sin x}$

SECTION 4.8

Applied Maximum and Minimum Problems

The methods we have learned in this chapter for finding extreme values have practical applications in many areas of life. A businessperson wants to minimize costs and maximize profits. A chemist wants to maximize a rate of reaction. Fermat's Principle in optics states that light follows the path that takes the least time.

In solving such practical problems the greatest challenge is often to convert the word problem into a maximum–minimum problem by setting up the function that is to be maximized or minimized. The following steps may be useful.

Steps in Solving Applied Maximum and Minimum Problems

1. *Understand the problem.* The first step is to read the problem carefully until it is clearly understood. Ask yourself: What is the unknown? What are the given quantities? What are the given conditions?

2. *Draw a diagram.* In most problems it is useful to draw a diagram and identify the given and required quantities on the diagram.

3. *Introduce notation.* Assign a symbol to the quantity that is to be maximized or minimized (let us call it Q for now). Also select symbols $(a, b, c, \ldots, x, y)$ for other unknown quantities and label the diagram with these symbols. It may help to use initials as suggestive symbols— for example, A for area, h for height, t for time.

4. Express Q in terms of some of the other symbols from Step 3.

5. If Q has been expressed as a function of more than one variable in Step 4, use the given information to find relationships (in the form of equations) among these variables. Then use these equations to eliminate all but one of the variables in the expression for Q. Thus Q will be given as a function of *one* variable x—say, $Q = f(x)$. Write the domain of this function.

6. Use the methods of Sections 4.1, 4.3, and 4.4 to find the *absolute* maximum or minimum value of f. In particular, if the domain of f is a closed interval, then the procedure outlined in (4.8) can be used.

● **Example 1**   A farmer has 2400 ft of fencing and wants to fence off a rectangular field that borders a straight river. He needs no fence along the river. What are the dimensions of the field that has the largest area?

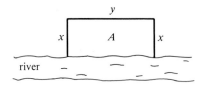

Figure 4.65

Solution   First we draw a diagram (Figure 4.65). We wish to maximize the area A of the rectangle. Let x and y be the width and length of the rectangle (in feet).

Then we express A in terms of x and y:

$$A = xy$$

We want to express A as a function of just one variable, so we eliminate y by expressing it in terms of x. To do this we use the given information that the total length of the fencing is 2400 ft. Thus

$$2x + y = 2400$$

From this equation we have $y = 2400 - 2x$, which gives

$$A = x(2400 - 2x) = 2400x - 2x^2$$

Note that $x \geqslant 0$ and $x \leqslant 1200$ (otherwise $A < 0$). So the function that we wish to maximize is

$$A(x) = 2400x - 2x^2 \qquad 0 \leqslant x \leqslant 1200$$

To find the critical numbers we solve the equation

$$A'(x) = 2400 - 4x = 0$$

which gives $x = 600$. The maximum value of A must occur either at this critical number or at an endpoint of the interval. Since $A(0) = 0$, $A(600) = 720,000$, and $A(1200) = 0$, the procedure of (4.8) gives the maximum value as $A(600) = 720,000$.

[Alternatively, we could have observed that $A''(x) = -4 < 0$ for all x, so A is always concave downward and the local maximum at $x = 600$ must be an absolute maximum.]

Thus the rectangular field should be 600 ft wide and 1200 ft long.   ●

● **Example 2**   A can is to be made to hold one liter of oil. Find the dimensions that will minimize the cost of the metal to make the can.

Solution   Draw the diagram as in Figure 4.66 where r is the radius and h the height (in centimeters). In order to minimize the cost of the metal, we minimize the total surface area of the cylinder (top, bottom, and sides), which is

$$A = 2\pi r^2 + 2\pi r h$$

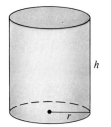

Figure 4.66

To eliminate h we use the fact that the volume is given as 1 L, which we take to be 1000 cm^3. Thus

$$\pi r^2 h = 1000$$

which gives $h = 1000/\pi r^2$. Substitution of this into the expression for A gives

$$A = 2\pi r^2 + 2\pi r \left(\frac{1000}{\pi r^2} \right) = 2\pi r^2 + \frac{2000}{r}$$

Therefore the function that we want to minimize is

$$A(r) = 2\pi r^2 + \frac{2000}{r} \qquad r > 0$$

To find the critical numbers, we differentiate:

$$A'(r) = 4\pi r - \frac{2000}{r^2} = \frac{4(\pi r^3 - 500)}{r^2}$$

Then $A'(r) = 0$ when $\pi r^3 = 500$, so the only critical number is $r = \sqrt[3]{500/\pi}$.

Since the domain of A is $(0, \infty)$, we cannot use the argument of Example 1 concerning endpoint extrema. But we can observe that $A'(r) < 0$ for $r < \sqrt[3]{500/\pi}$ and $A'(r) > 0$ for $r > \sqrt[3]{500/\pi}$, so A is decreasing for *all* r to the left of the critical number and increasing for *all* r to the right. Thus $r = \sqrt[3]{500/\pi}$ must give rise to an *absolute* minimum.

[Alternatively, we could argue that $A(r) \to \infty$ as $r \to 0^+$ and $A(r) \to \infty$ as $r \to \infty$, so there must be a minimum value of $A(r)$ that must occur at the critical number.]

The value of h corresponding to $r = \sqrt[3]{500/\pi}$ is

$$h = \frac{1000}{\pi r^2} = \frac{1000}{\pi \left(\dfrac{500}{\pi}\right)^{2/3}} = 2\sqrt[3]{\frac{500}{\pi}} = 2r$$

Thus to minimize the cost of the can, the radius should be $\sqrt[3]{500/\pi}$ cm and the height should be equal to twice the radius—namely, the diameter.   ●

Note: The argument used in Example 2 to justify the absolute minimum is a variant of the First Derivative Test (which applies only to local extrema) and is stated here for future reference.

First Derivative Test for Absolute Extrema
(4.43)

Suppose that c is a critical number of a continuous function f defined on an interval.

(a)   If $f'(x) > 0$ for all $x < c$ and $f'(x) < 0$ for all $x > c$, then $f(c)$ is the absolute maximum value of f.

(b)   If $f'(x) < 0$ for all $x < c$ and $f'(x) > 0$ for all $x > c$, then $f(c)$ is the absolute minimum value of f.

● **Example 3**   Find the point on the parabola $y^2 = 2x$ that is closest to the point $(1, 4)$.

Solution   The distance between the point $(1, 4)$ and the point (x, y) is

$$d = \sqrt{(x - 1)^2 + (y - 4)^2}$$

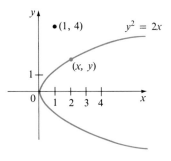

Figure 4.67

(see Figure 4.67). But if (x, y) lies on the parabola, then $x = y^2/2$, so the expression for d becomes

$$d = \sqrt{\left(\frac{y^2}{2} - 1\right)^2 + (y - 4)^2}$$

(Alternatively, we could have substituted $y = \sqrt{2x}$ to get d in terms of x alone.) Instead of minimizing d, we shall minimize its square:

$$d^2 = f(y) = \left(\frac{y^2}{2} - 1\right)^2 + (y - 4)^2$$

(You should convince yourself that the minimum of d occurs at the same point as the minimum of d^2, but d^2 is easier to work with.) Differentiating, we obtain

$$f'(y) = 2\left(\frac{y^2}{2} - 1\right)y + 2(y - 4) = y^3 - 8$$

so $f'(y) = 0$ when $y = 2$. Observe that $f'(y) < 0$ when $y < 2$ and $f'(y) > 0$ when $y > 2$, so by the First Derivative Test for Absolute Extrema (4.43), the absolute minimum occurs when $y = 2$. (Or we could simply say that because of the geometric nature of the problem, it is obvious that there is a closest point but not a farthest point.) The corresponding value of x is $x = y^2/2 = 2$. Thus the point on $y^2 = 2x$ closest to $(1, 4)$ is $(2, 2)$.   ●

● **Example 4**   A man is at point A on a bank of a straight river, 3 km wide, and wants to reach point B, 8 km downstream on the opposite bank, as quickly as possible (see Figure 4.68). He could row his boat directly across the river to point C and then run to B, or he could row directly to B, or he could row to some point D between C and B and then run to B. If he can row at 6 km/h and run at 8 km/h, where should he land to reach B as soon as possible?

Solution   Let x be the distance from C to D. Then the walking distance is $|DB| = 8 - x$ and the Pythagorean Theorem gives the rowing distance as $|AD| = \sqrt{x^2 + 9}$. We assume the speed of the water is 0 km/h and use the equation

$$\text{time} = \frac{\text{distance}}{\text{rate}}$$

Then the rowing time is $\sqrt{x^2 + 9}/6$ and the walking time is $(8 - x)/8$, so the total time T as a function of x is

$$T(x) = \frac{\sqrt{x^2 + 9}}{6} + \frac{8 - x}{8}$$

The domain of this function T is $[0, 8]$. Notice that if $x = 0$ he rows to C and if $x = 8$ he rows directly to B. The derivative of T is

$$T'(x) = \frac{x}{6\sqrt{x^2 + 9}} - \frac{1}{8}$$

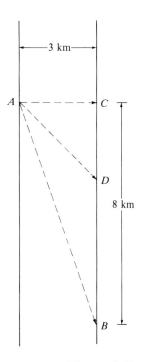

Figure 4.68

Thus, using the fact that $x \geqslant 0$, we have

$$T'(x) = 0 \quad \Leftrightarrow \quad \frac{x}{6\sqrt{x^2 + 9}} = \frac{1}{8} \quad \Leftrightarrow \quad 4x = 3\sqrt{x^2 + 9}$$

$$\Leftrightarrow \quad 16x^2 = 9(x^2 + 9) \quad \Leftrightarrow \quad 7x^2 = 81$$

$$\Leftrightarrow \quad x = \frac{9}{\sqrt{7}}$$

The only critical number is $x = 9/\sqrt{7}$. To see whether the minimum occurs at this critical number or at an endpoint of the domain $[0, 8]$, we evaluate T at all three points:

$$T(0) = 1.5 \qquad T\left(\frac{9}{\sqrt{7}}\right) = 1 + \frac{\sqrt{7}}{8} \approx 1.33 \qquad T(8) = \frac{\sqrt{73}}{6} \approx 1.42$$

Since the smallest of these values of T occurs when $x = 9/\sqrt{7}$, the absolute minimum value of T must occur there by (4.8).

Thus the man should land the boat at a point $9/\sqrt{7}$ km (≈ 3.4 km) downstream from his starting point.   ●

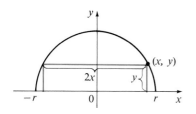

Figure 4.69

● **Example 5**   Find the area of the largest rectangle that can be inscribed in a semicircle of radius r.

Solution   Let us take the semicircle to be the upper half of the circle $x^2 + y^2 = r^2$ with center the origin. Then the word *inscribed* means that the rectangle has two vertices on the semicircle and two vertices on the x-axis as shown in Figure 4.69.

Let (x, y) be the vertex that lies in the first quadrant. Then the rectangle has sides of lengths $2x$ and y, so its area is

$$A = 2xy$$

To eliminate y we use the fact that (x, y) lies on the circle $x^2 + y^2 = r^2$, so $y = \sqrt{r^2 - x^2}$. Thus

$$A = 2x\sqrt{r^2 - x^2}$$

The domain of this function is $0 \leqslant x \leqslant r$. Its derivative is

$$A' = 2\sqrt{r^2 - x^2} - \frac{2x^2}{\sqrt{r^2 - x^2}} = \frac{2(r^2 - 2x^2)}{\sqrt{r^2 - x^2}}$$

which is 0 when $2x^2 = r^2$, that is, $x = r/\sqrt{2}$ (since $x \geqslant 0$). This value of x gives a maximum value of A since $A(0) = 0$ and $A(r) = 0$. Therefore the area of the largest inscribed rectangle is

$$A\left(\frac{r}{\sqrt{2}}\right) = 2\frac{r}{\sqrt{2}}\sqrt{r^2 - \frac{r^2}{2}} = r^2$$ ●

EXERCISES 4.8

1. Find two numbers whose sum is 100 and whose product is a maximum.

2. Find two numbers whose difference is 100 and whose product is a minimum.

3. Find two positive numbers whose product is 100 and whose sum is a minimum.

4. Find two positive numbers whose sum is 100 and the sum of whose squares is a minimum.

5. Show that of all the rectangles with a given perimeter, the one with greatest area is a square.

6. Show that of all the rectangles with a given area, the one with smallest perimeter is a square.

7. Show that of all the isosceles triangles with a given perimeter, the one with greatest area is equilateral.

8. A farmer wants to fence an area of 1.5 million square feet in a rectangular field and then divide it in half with a fence parallel to one of the sides of the rectangle. How can he do this so as to minimize the cost of the fence?

9. A farmer with 750 ft of fencing wants to enclose a rectangular area and then divide it into four pens with fencing parallel to one side of the rectangle. What is the largest possible total area of the four pens?

10. A box with a square base and open top must have a volume of 32,000 cm^3. Find the dimensions of the box that minimize the amount of material used.

11. If 1200 cm^2 of material is available to make a box with a square base and open top, find the largest possible volume of the box.

12. A rectangular storage container with open top is to have a volume of 10 m^3. The length of its base is twice the width. Material for the base costs $10 per square meter. Material for the sides costs $6 per square meter. Find the cost of materials for the cheapest such container.

13. Do Exercise 12 assuming the container has a lid that is made from the same material as the sides.

14. A box with an open top is to be constructed from a square piece of cardboard, 3 ft wide, by cutting out a square from each of the four corners and bending up the sides. Find the largest volume that such a box can have.

15. Find the point on the line $y = 2x - 3$ that is closest to the origin.

16. Find the point on the line $2x + 3y + 5 = 0$ that is closest to the point $(-1, -2)$.

17. Find the points on the hyperbola $y^2 - x^2 = 4$ that are closest to the point $(2, 0)$.

18. Find the point on the parabola $x + y^2 = 0$ that is closest to the point $(0, -3)$.

19. Find the dimensions of the rectangle of largest area that can be inscribed in a circle of radius r.

20. Find the area of the largest rectangle that can be inscribed in the ellipse $x^2/a^2 + y^2/b^2 = 1$.

21. Find the dimensions of the rectangle of largest area that can be inscribed in an equilateral triangle of side L if one side of the rectangle lies on the base of the triangle.

22. Find the dimensions of the rectangle of largest area that has its base on the x-axis and its other two vertices above the x-axis and lying on the parabola $y = 8 - x^2$.

23. Find the dimensions of the isosceles triangle of largest area that can be inscribed in a circle of radius r.

24. Find the area of the largest rectangle that can be inscribed in a right triangle with legs of lengths 3 cm and 4 cm if two sides of the rectangle lie along the legs.

25. A right circular cylinder is inscribed in a sphere of radius r. Find the largest possible volume of such a cylinder.

26. A right circular cylinder is inscribed in a cone with height h and base radius r. Find the largest possible volume of such a cylinder.

27. A right circular cylinder is inscribed in a sphere of radius r. Find the largest possible surface area of such a cylinder.

28. A Norman window has the shape of a rectangle surmounted by a semicircle. (Thus the diameter of the semicircle is equal to the width of the rectangle.) If the perimeter of the window is 30 ft, find the dimensions of the window so that the greatest possible amount of light is admitted.

29. The top and bottom margins of a poster are each 6 cm and the side margins are each 4 cm. If the area of printed material on the poster is fixed at 384 cm^2, find the dimensions of the poster with the smallest area.

30. A poster is to have an area of 180 in.^{2} with 1-inch margins at the bottom and sides and a 2-inch margin at the top. What dimensions will give the largest printed area?

31. A piece of wire 10 m long is cut into two pieces. One piece is bent into a square and the other is bent into an equilateral triangle. How should the wire be cut so that the total area enclosed is (a) a maximum and (b) a minimum?

32. Answer Exercise 31 if one piece is bent into a square and the other into a circle.

33. A cylindrical can without a top is made to contain V cm^3 of liquid. Find the dimensions that will minimize the cost of the metal to make the can.

34. A fence 8 ft tall runs parallel to a tall building at a distance of 4 ft from the building. What is the length of the shortest ladder that will reach from the ground over the fence to the wall of the building?

35. What is the maximum possible volume of a right circular cone with slant height l?

36. The velocity of a wave of length L in deep water is

$$v = K\sqrt{\frac{L}{C} + \frac{C}{L}}$$

where K and C are known positive constants. What is the length of the wave that gives the minimum velocity?

37. The effectiveness of a television commercial depends on how many times a viewer watches it. After some experiments, an advertising agency found that if the effectiveness E is put on a scale of 0 to 10, then

$$E(n) = \tfrac{2}{3}n - \tfrac{1}{90}n^2$$

where n is the number of times a viewer watches a given commercial. For maximum effectiveness of a commercial, how many times should a viewer watch it?

38. A boat leaves a dock at 2:00 P.M. and travels due south at a speed of 20 km/h. Another boat has been heading due east at 15 km/h and reaches the same dock at 3:00 P.M. At what time were the two boats closest together?

39. The illumination of an object by a light source is directly proportional to the strength of the source and inversely proportional to the square of the distance from the source. If two light sources, one three times as strong as the other, are placed 10 ft apart, where should an object be placed on the line between the sources so as to receive the least illumination?

40. Solve the problem in Example 4 if the river is 5 km wide and point B is only 5 km downstream from A.

41. Let v_1 be the velocity of light in air and v_2 the velocity of light in water. According to Fermat's Principle, a ray of light will travel from a point A in the air to a point B in the water by a path ACB that minimizes the time taken. Show that

$$\frac{\sin\theta_1}{\sin\theta_2} = \frac{v_1}{v_2}$$

where θ_1 (the angle of incidence) and θ_2 (the angle of refraction) are shown in Figure 4.70. This equation is known as Snell's Law.

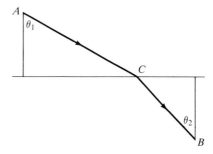

Figure 4.70

42. Two vertical poles PQ and ST are secured by a rope PRS going from the top of the first pole to a point R on the ground between the poles and then to the top of the second pole as in Figure 4.71. Show that the shortest length of such a rope occurs when $\theta_1 = \theta_2$.

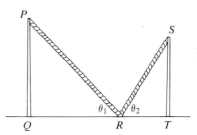

Figure 4.71

43. The upper left-hand corner of a piece of paper 8 in. wide by 12 in. long is folded over to the right-hand edge as in Figure 4.72. How would you fold it so as

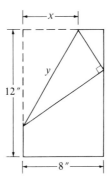

Figure 4.72

to minimize the length of the fold? In other words, how would you choose x to minimize y?

44. A steel pipe is being carried down a hallway 9 ft wide. At the end of the hall there is a right-angled turn into a narrower hallway 6 ft wide. What is the length of the longest pipe that can be carried horizontally around the corner? (See Figure 4.73.)

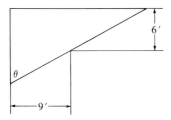

Figure 4.73

45. A hockey team plays in an arena with a seating capacity of 15,000 spectators. With ticket prices at $12, average attendance at a game has been 11,000. A market survey indicates that for each dollar that ticket prices are lowered, the average attendance will increase by 1000. How should the owners of the team set ticket prices so as to maximize their revenue from ticket sales?

46. A rain gutter is to be constructed from a metal sheet of width 30 cm by bending up one-third of the sheet on each side through an angle θ. (See Figure 4.74.) How should θ be chosen so that the gutter will carry the maximum amount of water?

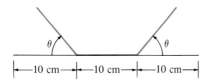

Figure 4.74

47. Find the maximum area of a rectangle that can be circumscribed about a given rectangle with length L and width W. (See Figure 4.75.)

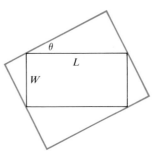

Figure 4.75

48. The blood vascular system consists of blood vessels (arteries, arterioles, capillaries, and veins) that convey blood from the heart to the organs and back to the heart. This system should work so as to minimize the energy expended by the heart in pumping the blood. In particular, this energy is reduced when the resistance of the blood is lowered. One of Poiseuille's Laws gives the resistance R of the blood as

$$R = C \frac{L}{r^4}$$

where L is the length of the blood vessel, r is the radius, and C is a positive constant determined by the viscosity of the blood. (Poiseuille established this law experimentally but it also follows from Example 6 in Section 5.8.) Figure 4.76 shows a main blood vessel with radius r_1 branching at an angle θ into a smaller vessel with radius r_2.

(a) Use Poiseuille's Law to show that the total resistance of the blood along the path ABC is

$$R = C \left(\frac{a - b \cot \theta}{r_1^4} + \frac{b \csc \theta}{r_2^4} \right)$$

Figure 4.76
VASCULAR BRANCHING

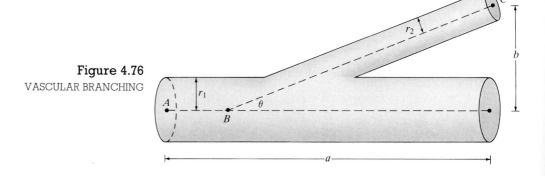

© Manfred Kage/Peter Arnold

where a and b are the distances shown in the figure.

(b) Prove that this resistance is minimized when

$$\cos \theta = \frac{r_2^4}{r_1^4}$$

(c) Find the optimal branching angle (correct to the nearest degree) when the radius of the smaller blood vessel is two-thirds of the radius of the larger vessel.

SECTION 4.9

Applications to Economics

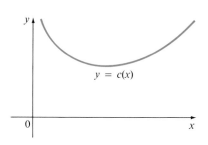

Figure 4.77

COST FUNCTION

In Section 3.7 we introduced the idea of marginal cost. Recall that if $C(x)$, the **cost function,** is the cost of producing x units of a certain product, then the **marginal cost** is the rate of change of C with respect to x. In other words, the marginal cost function is the derivative, $C'(x)$, of the cost function.

The graph of a typical cost function is shown in Figure 4.77. The marginal cost $C'(x)$ is the slope of the tangent to the cost curve at $(x, C(x))$. Notice that the cost curve is initially concave downward (the marginal cost is decreasing) because of economies of scale (more efficient use of the fixed costs of production). But eventually there is an inflection point and the cost curve becomes concave upward (the marginal cost is increasing) perhaps because of overtime costs or the inefficiencies of a large-scale operation.

The **average cost function**

(4.44)

$$c(x) = \frac{C(x)}{x}$$

represents the cost per unit when x units are produced. We sketch a typical average cost function in Figure 4.78 by noting that $C(x)/x$ is the slope of the line that joins the origin to the point $(x, C(x))$ in Figure 4.77. It appears that there will be an absolute minimum. To find it we locate the critical point of c by using the Quotient Rule to differentiate Equation 4.44:

$$c'(x) = \frac{xC'(x) - C(x)}{x^2}$$

Now $c'(x) = 0$ when $xC'(x) - C(x) = 0$ and this gives

$$C'(x) = \frac{C(x)}{x} = c(x)$$

Figure 4.78

AVERAGE COST FUNCTION

Therefore:

> If the average cost is a minimum, then
>
> $$\text{marginal cost} = \text{average cost}$$

● **Example 1**   A company estimates that the cost (in dollars) of producing x items is $C(x) = 2600 + 2x + 0.001x^2$. (See Example 7 in Section 3.7 for an explanation of why it is reasonable to model a cost function by a polynomial.) (a) Find the cost, average cost, and marginal cost of producing 1000 items, 2000 items, and 3000 items. (b) At what production level will the average cost be smallest, and what is this minimum average cost?

Solution   (a)   The average cost function is

$$c(x) = \frac{C(x)}{x} = \frac{2600}{x} + 2 + 0.001x$$

The marginal cost function is

$$C'(x) = 2 + 0.002x$$

We use these expressions to fill in the following table giving the cost, average cost, and marginal cost (in dollars, or dollars per item, rounded to the nearest cent).

x	$C(x)$	$c(x)$	$C'(x)$
1000	5,600.00	5.60	4.00
2000	10,600.00	5.30	6.00
3000	17,600.00	5.87	8.00

(b)   To minimize the average cost we must have

$$\text{marginal cost} = \text{average cost}$$

$$C'(x) = c(x)$$

$$2 + 0.002x = \frac{2600}{x} + 2 + 0.001x$$

This equation simplifies to

$$0.001x = \frac{2600}{x}$$

so $$x^2 = \frac{2600}{0.001} = 2,600,000$$

and $$x = \sqrt{2,600,000} \approx 1612$$

To see that this production level actually gives a minimum we note that $c''(x) = 5200/x^3 > 0$, so c is concave upward on its entire domain. The minimum average cost is

$$c(1612) = \tfrac{2600}{1612} + 2 + 0.001(1612) = \$5.22/\text{item} \qquad \bullet$$

Now let us consider marketing. Let $p(x)$ be the price per unit that the company can charge if it sells x units. Then p is called the **demand function** (or **price function**) and we would expect it to be a decreasing function of x. If x units are sold and the price per unit is $p(x)$, then the total revenue is

$$R(x) = xp(x)$$

and R is called the **revenue function** (or **sales function**). The derivative R' of the revenue function is called the **marginal revenue function** and is the rate of change of revenue with respect to the number of units sold.

If x units are sold, then the total profit is

$$P(x) = R(x) - C(x)$$

and P is called the **profit function.** The **marginal profit function** is P', the derivative of the profit function. In order to maximize profit we look for the critical numbers of P—that is, the numbers where the marginal profit is 0. But if

$$P'(x) = R'(x) - C'(x) = 0$$

then $$R'(x) = C'(x)$$

Therefore:

If the profit is a maximum, then

$$\text{marginal revenue} = \text{marginal cost}$$

To ensure that this condition gives a maximum we could use the Second Derivative Test. Note that

$$P''(x) = R''(x) - C''(x) < 0$$

when $$R''(x) < C''(x)$$

and this condition says that the rate of increase of marginal revenue is less than the rate of increase of marginal cost. Thus the profit will be a maximum when

$$R'(x) = C'(x) \quad \text{and} \quad R''(x) < C''(x)$$

● **Example 2**   What production level will maximize profits for a company with cost and demand functions

$$C(x) = 3800 + 5x - \frac{x^2}{1000} \qquad p(x) = 50 - \frac{x}{100}$$

Solution   The revenue function is

$$R(x) = xp(x) = 50x - \frac{x^2}{100}$$

so the marginal revenue function is

$$R'(x) = 50 - \frac{x}{50}$$

and the marginal cost function is

$$C'(x) = 5 - \frac{x}{500}$$

Thus marginal revenue is equal to marginal cost when

$$50 - \frac{x}{50} = 5 - \frac{x}{500}$$

Solving, we get

$$x = 2500$$

To check that this gives a maximum we compute the second derivatives:

$$R''(x) = -\tfrac{1}{50} \qquad C''(x) = -\tfrac{1}{500}$$

Thus $R''(x) < C''(x)$ for all values of x. Therefore a production level of 2500 units will maximize profits.   ●

EXERCISES 4.9

For each of the cost functions in Exercises 1–6 find (a) the cost, average cost, and marginal cost of producing 1000 units; (b) the production level that will minimize the average cost; and (c) the minimum average cost.

1. $C(x) = 10{,}000 + 25x + x^2$

2. $C(x) = 1600 + 8x + 0.01x^2$

3. $C(x) = 45 + \dfrac{x}{2} + \dfrac{x^2}{560}$

4. $C(x) = 2000 + 10x + 0.001x^3$

5. $C(x) = 2\sqrt{x} + \dfrac{x^2}{8000}$

6. $C(x) = 1000 + 96x + 2x^{3/2}$

For each of the cost and demand functions in Exercises 7–12, find the production level that will maximize profits.

7. $C(x) = 680 + 4x + 0.01x^2, \, p(x) = 12$

8. $C(x) = 680 + 4x + 0.01x^2$, $p(x) = 12 - \dfrac{x}{500}$

9. $C(x) = 1200 + 25x - 0.0001x^2$, $p(x) = 55 - \dfrac{x}{1000}$

10. $C(x) = 900 + 110x - 0.1x^2 + 0.02x^3$, $p(x) = 260 - 0.1x$

11. $C(x) = 1450 + 36x - x^2 + 0.001x^3$, $p(x) = 60 - 0.01x$

12. $C(x) = 10{,}000 + 28x - 0.01x^2 + 0.002x^3$, $p(x) = 90 - 0.02x$

In Exercises 13 and 14 find the production level at which the marginal cost function starts to increase.

13. $C(x) = 0.001x^3 - 0.3x^2 + 6x + 900$

14. $C(x) = 0.0002x^3 - 0.25x^2 + 4x + 1500$

15. A baseball team plays in a stadium that holds 55,000 spectators. With ticket prices at $10, the average attendance has been 27,000. When ticket prices were lowered to $8, the average attendance rose to 33,000. (a) Find the demand function, assuming that it is linear. (b) How should ticket prices be set to maximize revenue?

16. During the summer months Terry makes and sells necklaces on the beach. Last summer he sold the necklaces for $10 each and his sales averaged 20 per day. When he increased the price by $1, he found that he lost two sales per day. (a) Find the demand function, assuming that it is linear. (b) If the material for each necklace costs Terry $6, what should the selling price be to maximize profits?

17. A manufacturer has been selling 1000 television sets a week at $450 each. A market survey indicates that for each $10 rebate offered to the buyer, the number of sets sold will increase by 100 per week. (a) Find the demand function. (b) How large a rebate should the company offer the buyer in order to maximize its revenue? (c) If its weekly cost function is $C(x) = 68{,}000 + 150x$, how should it set the size of the rebate in order to maximize its profits?

SECTION 4.10

Antiderivatives

We know how to solve the derivative problem: given a function, find its derivative. But many problems in mathematics and its applications require us to solve the inverse of the derivative problem: given a function f, find a function F whose derivative is f. If such a function F exists, it is called an *antiderivative* of f.

Definition (4.45)

> A function F is called an **antiderivative** of f on an interval I if $F'(x) = f(x)$ for all x in I.

For instance, let $f(x) = x^2$. It is not difficult to discover an antiderivative of f if we keep the Power Rule in mind. In fact, if $F(x) = \frac{1}{3}x^3$, then $F'(x) = x^2 = f(x)$. But the function $G(x) = \frac{1}{3}x^3 + 100$ also satisfies $G'(x) = x^2$. Therefore both F and G are antiderivatives of f. Indeed any function of the form $H(x) = \frac{1}{3}x^3 + C$, where C is a constant, is an antiderivative of f. The question arises: Are there any others?

To answer this question, recall that in Section 4.2 we used the Mean Value Theorem to prove that if two functions have identical derivatives on an interval, then they must differ by a constant (Corollary 4.17). Thus if F

and G are any two antiderivatives of f, then

$$F'(x) = f(x) = G'(x)$$

so $G(x) - F(x) = C$, where C is a constant. We can write this as $G(x) = F(x) + C$, so we have the following result:

Theorem (4.46)

> If F is an antiderivative of f on an interval I, then the most general antiderivative of f on I is
>
> $$F(x) + C$$
>
> where C is an arbitrary constant.

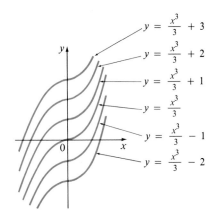

$y = \dfrac{x^3}{3} + 3$

$y = \dfrac{x^3}{3} + 2$

$y = \dfrac{x^3}{3} + 1$

$y = \dfrac{x^3}{3}$

$y = \dfrac{x^3}{3} - 1$

$y = \dfrac{x^3}{3} - 2$

Figure 4.79

THE ANTIDERIVATIVES OF $f(x) = x^2$

Going back to the function $f(x) = x^2$, we see that the general antiderivative of f is $x^3/3 + C$. By assigning specific values to the constant C we obtain a family of functions whose graphs are vertical translates of each other (see Figure 4.79).

● **Example 1**   Find the most general antiderivatives of the following functions: (a) $f(x) = \sin x$, (b) $f(x) = x^n$, $n \geqslant 0$, and (c) $f(x) = x^{-3}$.

Solution   (a)   If $F(x) = -\cos x$, then $F'(x) = \sin x$, so an antiderivative of sine is $-\cosine$. By Theorem 4.46, the most general antiderivative is $G(x) = -\cos x + C$.

(b)
$$\frac{d}{dx}\left(\frac{x^{n+1}}{n+1}\right) = \frac{(n+1)x^n}{n+1} = x^n$$

Thus the general antiderivative of $f(x) = x^n$ is

$$F(x) = \frac{x^{n+1}}{n+1} + C$$

This is valid for $n \geqslant 0$ since then $f(x) = x^n$ is defined on the interval $(-\infty, \infty)$.

(c)   If we put $n = -3$ in part (b) we get the particular antiderivative $F(x) = x^{-2}/(-2)$ by the same calculation. But notice that $f(x) = x^{-3}$ is not defined at $x = 0$. Thus Theorem 4.46 tells us only that the general antiderivative of f is $x^{-2}/(-2) + C$ on any interval that does not contain 0. So the general antiderivative of $f(x) = 1/x^3$ is

$$F(x) = \begin{cases} -\dfrac{1}{2x^2} + C_1 & \text{if } x > 0 \\[2ex] -\dfrac{1}{2x^2} + C_2 & \text{if } x < 0 \end{cases}$$

●

As in Example 1, every differentiation formula, when read from right to left, gives rise to an antidifferentiation formula. In Table 4.47 we list some particular antiderivatives. Each formula in the table is true because the derivative of the function in the right column appears in the left column. In particular, the first formula says that the antiderivative of a constant times a function is the constant times the antiderivative of the function. The second formula says that the antiderivative of a sum is the sum of the antiderivatives. (We use the notation $F' = f$, $G' = g$.)

Table of Antidifferentiation Formulas (4.47)

Function	Particular antiderivative
$cf(x)$	$cF(x)$
$f(x) + g(x)$	$F(x) + G(x)$
$x^n \ (n \neq -1)$	$\dfrac{x^{n+1}}{n+1}$
$\cos x$	$\sin x$
$\sin x$	$-\cos x$
$\sec^2 x$	$\tan x$
$\sec x \tan x$	$\sec x$

To obtain the most general antiderivative from the particular ones in Table 4.47 we have to add a constant (or constants) as in Example 1.

● **Example 2**   Find all functions g such that

$$g'(x) = 4 \sin x - 3x^5 + 6\sqrt[4]{x^3}$$

Solution   We want to find the antiderivative of

$$f(x) = g'(x) = 4 \sin x - 3x^5 + 6x^{3/4}$$

Using the formulas in Table 4.47 together with Theorem 4.46, we obtain

$$g(x) = 4(-\cos x) - 3\frac{x^6}{6} + 6\frac{x^{7/4}}{\frac{7}{4}} + C$$

$$= -4 \cos x - \frac{x^6}{2} + \frac{24}{7}x^{7/4} + C \qquad ●$$

In applications of calculus it is very common to have a situation as in Example 2, where it is required to find a function, given knowledge about its derivatives. An equation that involves the derivatives of a function is called a **differential equation.** These will be studied in some detail in Section 8.8 and Chapter 15 but for the present we can solve some elementary differential equations. The general solution of a differential equation will involve an arbitrary constant (or constants) as in Example 2. However there may be some extra conditions given that will determine the constants and therefore uniquely specify the solution.

● **Example 3**   Find f if $f'(x) = x\sqrt{x}$ and $f(1) = 2$.

Solution   The general antiderivative of

$$f'(x) = x^{3/2}$$

is $$f(x) = \frac{x^{5/2}}{\frac{5}{2}} + C = \frac{2}{5} x^{5/2} + C$$

To determine C we use the fact that $f(1) = 2$:

$$f(1) = \tfrac{2}{5} + C = 2$$

Solving for C, we get $C = 2 - \tfrac{2}{5} = \tfrac{8}{5}$, so the particular solution is

$$f(x) = \frac{2x^{5/2} + 8}{5}$$ ●

● **Example 4**   Find f if $f''(x) = 12x^2 + 6x - 4$, $f(0) = 4$, and $f(1) = 1$.

Solution   The general antiderivative of $f''(x) = 12x^2 + 6x - 4$ is

$$f'(x) = 12\frac{x^3}{3} + 6\frac{x^2}{2} - 4x + C = 4x^3 + 3x^2 - 4x + C$$

Using the antidifferentiation rules once more, we find that

$$f(x) = 4\frac{x^4}{4} + 3\frac{x^3}{3} - 4\frac{x^2}{2} + Cx + D = x^4 + x^3 - 2x^2 + Cx + D$$

To determine C and D we use the given conditions that $f(0) = 4$ and $f(1) = 1$. Since $f(0) = 0 + D = 4$, we have $D = 4$. Since

$$f(1) = 1 + 1 - 2 + C + 4 = 1$$

we have $C = -3$. Therefore the required function is

$$f(x) = x^4 + x^3 - 2x^2 - 3x + 4$$ ●

Antidifferentiation is particularly useful in analyzing the motion of an object moving in a straight line. Recall that if the object has position function $s = f(t)$, then the velocity function is $v(t) = s'(t)$. This means that the position function is an antiderivative of the velocity function. Likewise the acceleration function is $a(t) = v'(t)$, so the velocity function is an antiderivative of the acceleration. If the acceleration and the initial values $s(0)$ and $v(0)$ are known, then the position function can be found by antidifferentiating twice.

● **Example 5**   A particle moves in a straight line and has acceleration given by $a(t) = 6t + 4$. Its initial velocity is $v(0) = -6$ cm/s and its initial displacement is $s(0) = 9$ cm. Find its position function $s(t)$.

Solution   Since $v'(t) = a(t) = 6t + 4$, antidifferentiation gives

$$v(t) = 6\frac{t^2}{2} + 4t + C = 3t^2 + 4t + C$$

Note that $v(0) = C$. But we are given that $v(0) = -6$, so $C = -6$ and

$$v(t) = 3t^2 + 4t - 6$$

Since $v(t) = s'(t)$, s is the antiderivative of v:

$$s(t) = 3\frac{t^3}{3} + 4\frac{t^2}{2} - 6t + D = t^3 + 2t^2 - 6t + D$$

This gives $s(0) = D$. We are given that $s(0) = 9$, so $D = 9$ and the required position function is

$$s(t) = t^3 + 2t^2 - 6t + 9$$ ●

An object near the surface of the earth is subject to a gravitational force that produces a downward acceleration denoted by g. For motion close to the earth we may assume that g is constant, its value being about 9.8 m/s² (or 32 ft/s²).

● **Example 6**   A ball is thrown upward with a speed of 48 ft/s from the edge of a cliff 432 ft above the ground. Find its height above the ground t seconds later. When does it reach its maximum height? When does it hit the ground?

Solution   The motion is vertical and we choose the positive direction to be upward. At time t the distance above the ground is $s(t)$ and the velocity $v(t)$ is decreasing. Therefore the acceleration must be negative and we have

$$a(t) = \frac{dv}{dt} = -32$$

Taking antiderivatives, we have

$$v(t) = -32t + C$$

To determine C we use the given information that $v(0) = 48$. This gives $48 = 0 + C$, so

$$v(t) = -32t + 48$$

The maximum height is reached when $v(t) = 0$, that is, after 1.5 s. Since $s'(t) = v(t)$, we antidifferentiate again and obtain

$$s(t) = -16t^2 + 48t + D$$

Using the fact that $s(0) = 432$, we have $432 = 0 + D$, and so

$$s(t) = -16t^2 + 48t + 432$$

The expression for $s(t)$ is valid until the ball hits the ground. This happens when $s(t) = 0$, that is, when

$$-16t^2 + 48t + 432 = 0$$

or, equivalently, $t^2 - 3t - 27 = 0$

Using the quadratic formula to solve this equation, we get

$$t = \frac{3 \pm 3\sqrt{13}}{2}$$

We reject the solution with the minus sign since it gives a negative value for t. Therefore the ball hits the ground after $3(1 + \sqrt{13})/2$ s.   ●

EXERCISES 4.10

In Exercises 1–16 find the most general antiderivative of the given function. (Check your answer by differentiation.)

1. $f(x) = 12x^2 + 6x - 5$

2. $f(x) = x^3 - 4x^2 + 17$

3. $f(x) = 6x^9 - 4x^7 + 3x^2 + 1$

4. $f(x) = x^{99} - 2x^{49} - 1$

5. $f(x) = \sqrt{x} + \sqrt[3]{x}$

6. $f(x) = \sqrt[3]{x^2} - \sqrt{x^3}$

7. $f(x) = \dfrac{6}{x^5}$

8. $f(x) = \dfrac{3}{x^2} - \dfrac{5}{x^4}$

9. $f(x) = \sqrt{x} + \dfrac{1}{\sqrt{x}}$

10. $f(x) = x^{2/3} + 2x^{-1/3}$

11. $g(t) = \dfrac{t^3 + 2t^2}{\sqrt{t}}$

12. $g(t) = \sqrt[5]{t^4} + t^{-6}$

13. $h(x) = \sin x - 2\cos x$

14. $f(t) = \sin t - 2\sqrt{t}$

15. $f(t) = \sec^2 t + t^2$

16. $f(\theta) = \theta + \sec\theta\tan\theta$

In Exercises 17–42 find $f(x)$.

17. $f'(x) = x^4 - 2x^2 + x - 1$

18. $f'(x) = \sin x - \sqrt[5]{x^2}$

19. $f''(x) = x^2 + x^3$

20. $f''(x) = 60x^4 - 45x^2$

21. $f''(x) = 1$

22. $f''(x) = \sin x$

23. $f'''(x) = 24x$

24. $f'''(x) = \sqrt{x}$

25. $f'(x) = 4x + 3, \ f(0) = -9$

26. $f'(x) = 12x^2 - 24x + 1, \ f(1) = -2$

27. $f'(x) = 3\sqrt{x} - \dfrac{1}{\sqrt{x}}, f(1) = 2$

28. $f'(x) = 1 + \dfrac{1}{x^2}, x > 0, f(1) = 1$

29. $f'(x) = 3\cos x + 5\sin x, f(0) = 4$

30. $f'(x) = \sin x - 2\sqrt{x}, f(0) = 0$

31. $f'(x) = 2 + \sqrt[5]{x^3}, f(1) = 3$

32. $f'(x) = 3x^{-2}, f(1) = f(-1) = 0$

33. $f''(x) = -8, f(0) = 6, f'(0) = 5$

34. $f''(x) = x, f(0) = -3, f'(0) = 2$

35. $f''(x) = 20x^3 - 10, f(1) = 1, f'(1) = -5$

36. $f''(x) = 12x^2 - 6x + 8, f(-1) = 5, f'(-1) = -21$

37. $f''(x) = x^2 + 3\cos x, f(0) = 2, f'(0) = 3$

38. $f''(x) = x + \sqrt{x}, f(1) = 1, f'(1) = 2$

39. $f''(x) = 6x + 6, f(0) = 4, f(1) = 3$

40. $f''(x) = 12x^2 - 6x + 2, f(0) = 1, f(2) = 11$

41. $f''(x) = \dfrac{1}{x^3}, x > 0, f(1) = 0, f(2) = 0$

42. $f'''(x) = \sin x, f(0) = 1, f'(0) = 1, f''(0) = 1$

In Exercises 43–48 a particle is moving with the given data. Find the position function of the particle.

43. $v(t) = 3 - 2t, s(0) = 4$

44. $v(t) = 3\sqrt{t}, s(1) = 5$

45. $a(t) = 3t + 8, s(0) = 1, v(0) = -2$

46. $a(t) = \cos t + \sin t, s(0) = 0, v(0) = 5$

47. $a(t) = t^2 - t, s(0) = 0, s(6) = 12$

48. $a(t) = 10 + 3t - 3t^2, s(0) = 0, s(2) = 10$

49. A stone is dropped from the upper observation deck (the Space Deck) of the CN Tower, 450 m above the ground.
   (a) Find the distance of the stone above ground level at time t.
   (b) How long does it take to reach the ground?
   (c) With what velocity does it strike the ground?

50. Answer Exercise 49 if the stone is thrown downward with a speed of 5 m/s.

51. Answer Exercise 49 if the stone is thrown upward with a speed of 5 m/s.

52. Show that for motion in a straight line with constant acceleration a, initial velocity v_0, and initial displacement s_0, the displacement after time t is

$$s = \tfrac{1}{2}at^2 + v_0 t + s_0$$

53. An object is projected upward with initial velocity v_0 meters per second from a point s_0 meters above the ground. Show that

$$[v(t)]^2 = v_0^2 - 19.6[s(t) - s_0]$$

54. A car is traveling at 50 mi/h and the brakes are fully applied, producing a constant deceleration of 40 ft/s². What is the distance covered before the car comes to a stop?

55. What constant acceleration is required to increase the speed of a car from 30 mi/h to 50 mi/h in 5 s?

56. A car braked with a constant deceleration of 40 ft/s², producing skid marks measuring 160 ft before coming to a stop. How fast was the car traveling when the brakes were first applied?

57. A stone was dropped off a cliff and hit the ground with a speed of 120 ft/s. What is the height of the cliff?

REVIEW OF CHAPTER 4

Define, state, or discuss the following.

1. Absolute maximum and minimum values
2. Local maximum and minimum values
3. Extreme Value Theorem
4. Fermat's Theorem
5. Critical number
6. Procedure for absolute extreme values of f on $[a, b]$
7. Rolle's Theorem
8. Mean Value Theorem
9. Increasing function; decreasing function; monotonic function
10. Test for Monotonic Functions
11. First Derivative Test
12. Concave upward; concave downward
13. Test for Concavity
14. Point of inflection

15. Second Derivative Test

16. Limit at infinity

17. Horizontal asymptote

18. Infinite limit

19. Vertical asymptote

20. Infinite limit at infinity

21. Procedure for curve sketching

22. First Derivative Test for Absolute Extrema

23. Antiderivative

REVIEW EXERCISES FOR CHAPTER 4

In Exercises 1–6 find the local and absolute extreme values of the given function on the given interval.

1. $f(x) = x^3 - 12x + 5, [-5, 3]$

2. $f(x) = 3x^5 - 25x^3 + 60x, [-1, 3]$

3. $f(x) = \dfrac{x-2}{x+2}, [0, 4]$

4. $f(x) = \sqrt{x^2 + 4x + 8}, [-3, 0]$

5. $f(x) = x - \sqrt{2}\sin x, [0, \pi]$

6. $f(x) = 2x + 2\cos x - 4\sin x - \cos 2x, [0, \pi]$

In Exercises 7–20 find each limit.

7. $\lim\limits_{t \to 6} \dfrac{17}{(t-6)^2}$

8. $\lim\limits_{x \to 9^-} \left(\dfrac{-2}{x+9} \right)$

9. $\lim\limits_{x \to 2^-} \dfrac{3}{\sqrt{2-x}}$

10. $\lim\limits_{x \to \infty} \dfrac{1 + 2x - x^2}{1 - x + 2x^2}$

11. $\lim\limits_{x \to \infty} \dfrac{1 + x}{1 - x^2}$

12. $\lim\limits_{x \to \infty} x \tan \dfrac{1}{x}$

13. $\lim\limits_{x \to \infty} \dfrac{\sqrt{x^2 - 9}}{2x - 6}$

14. $\lim\limits_{x \to 3^+} \dfrac{\sqrt{x^2 - 9}}{2x - 6}$

15. $\lim\limits_{x \to -\infty} \dfrac{4x^5 + x^2 + 3}{2x^5 + x^3 + 1}$

16. $\lim\limits_{x \to -\infty} \dfrac{\cos^2 x}{x^2}$

17. $\lim\limits_{x \to \infty} \left(\sqrt[3]{x} - \dfrac{x}{3} \right)$

18. $\lim\limits_{x \to 2\pi^-} \dfrac{\cot x}{1 + x^2}$

19. $\lim\limits_{x \to -2^+} \dfrac{x}{x^2 - 4}$

20. $\lim\limits_{x \to \infty} (\sqrt{x^2 + x + 1} - \sqrt{x^2 - x})$

In Exercises 21–34 discuss the given curve under the headings A–H of Section 4.7.

21. $y = 1 + x + x^3$

22. $y = 3x^4 - 4x^3 - 12x^2 + 2$

23. $y = \dfrac{1}{x(x-3)^2}$

24. $y = \dfrac{1}{x^2 - x - 6}$

25. $y = x\sqrt{5 - x}$

26. $y = \sqrt{1 - x} + \sqrt{x - 2}$

27. $y = \dfrac{x^2}{x + 8}$

28. $y = \dfrac{1}{x} + \dfrac{1}{x + 1}$

29. $y = (x + 3)^4(x - 2)^5$

30. $y = x + \sqrt{1 - x}$

31. $y = x\sqrt[3]{x^2 + 5}$

32. $y = \sqrt{x} - \sqrt[3]{x}$

33. $y = 4x - \tan x, -\dfrac{\pi}{2} < x < \dfrac{\pi}{2}$

34. $y = |\cos x - \sin x|$

35. Show that the equation $x^{101} + x^{51} + x - 1 = 0$ has exactly one real root.

36. Suppose that f is continuous on $[0, 4]$, $f(0) = 1$, and $2 \leqslant f'(x) \leqslant 5$ for all x in $(0, 4)$. Show that $9 \leqslant f(4) \leqslant 21$.

37. Sketch the graph of a function that satisfies the following conditions: $f(0) = 0$, $f'(-2) = f'(1) = f'(9) = 0$, $\lim_{x \to \infty} f(x) = 0$, $\lim_{x \to 6} f(x) = -\infty$, $f'(x) < 0$ on $(-\infty, -2)$, $(1, 6)$, and $(9, \infty)$, $f'(x) > 0$ on $(-2, 1)$ and $(6, 9)$, $f''(x) > 0$ on $(-\infty, 0)$ and $(12, \infty)$, $f''(x) < 0$ on $(0, 6)$ and $(6, 12)$.

38. Sketch the graph of a function that satisfies the following conditions: $f(0) = 0$, f is continuous and even, $f'(x) = 2x$ if $0 < x < 1$, $f'(x) = -1$ if $1 < x < 3$, and $f'(x) = 1$ if $x > 3$.

39. Show that the shortest distance from the point (x_1, y_1) to the straight line $Ax + By + C = 0$ is

$$\frac{|Ax_1 + By_1 + C|}{\sqrt{A^2 + B^2}}$$

40. Find the point on the hyperbola $xy = 8$ that is closest to the point $(3, 0)$.

41. Find the smallest possible area of an isosceles triangle that is circumscribed about a circle of radius r.

42. Find the volume of the largest circular cone that can be inscribed in a sphere of radius r.

43. In $\triangle ABC$, D lies on AB, $CD \perp AB$, $|AD| = |BD| = 4$ cm, and $|CD| = 5$ cm. Where should a point P be chosen on CD so that the sum $|PA| + |PB| + |PC|$ is a minimum?

44. Do Exercise 43 when $|CD| = 2$ cm.

45. A rectangular beam is cut from a cylindrical log of radius R. The strength of the beam is proportional to the width and the square of the depth of a cross-section. Find the dimensions of the strongest such beam.

46. A metal storage tank with volume V is to be constructed in the shape of a right circular cylinder surmounted by a hemisphere. What dimensions will require the least amount of metal?

In Exercises 47–52 find $f(x)$.

47. $f'(x) = x - \sqrt[4]{x}$

48. $f'(x) = 2/\sqrt{x^5}$

49. $f'(x) = \dfrac{1 + x}{\sqrt{x}}$, $f(1) = 0$

50. $f'(x) = 1 + 2\sin x - \cos x$, $f(0) = 3$

51. $f''(x) = x^3 + x$, $f(0) = -1$, $f'(0) = 1$

52. $f''(x) = x^4 - 4x^2 + 3x - 2$, $f(0) = 0$, $f(1) = 1$

53. If the acceleration function of a particle is given by $a(t) = \sqrt{t}$ and its initial position is $s(0) = 0$ and its initial velocity is $v(0) = 2$, find its position function.

54. In an automobile race along a straight road, car A passed car B twice. Prove that at some time during the race their accelerations were equal.

5 Integrals

Calculus is the most powerful weapon of thought yet devised by the wit of man.
W. B. Smith

The experimental verification of a theory concerning any natural phenomenon generally rests on the result of an integration.
J. W. Mellor

Chapters 2–4 were concerned with the branch of calculus called differential calculus. We now turn our attention to another branch of calculus, called integral calculus, whose central concept is the definite integral.

In differential calculus, the tangent problem led us to formulate, in terms of limits, the idea of a derivative, which later turned out to be applicable, through velocities and other rates of change, to a variety of applied problems.

In integral calculus, the area problem will lead us to formulate, again in terms of limits, the idea of an integral, which will later be used to find volumes, lengths of curves, work, and forces, and to solve problems in chemistry, physics, biology, and economics.

There is a connection between the two branches of calculus. It is called the Fundamental Theorem of Calculus and we shall see in this chapter that it greatly simplifies the solution of many problems.

Sigma Notation

In finding areas and evaluating integrals we often encounter sums with many terms. There is a convenient way of writing such sums that uses the Greek letter $\sum$ (capital sigma, corresponding to our S) and is called **sigma notation.**

Definition (5.1)

If $a_m, a_{m+1}, \ldots, a_n$ are real numbers and m and n are integers such that $m \leqslant n$, then

$$\sum_{i=m}^{n} a_i = a_m + a_{m+1} + a_{m+2} + \cdots + a_{n-1} + a_n$$

With function notation, Definition 5.1 can be written as

$$\sum_{i=m}^{n} f(i) = f(m) + f(m+1) + f(m+2) + \cdots + f(n-1) + f(n)$$

Thus the symbol $\sum_{i=m}^{n}$ indicates a summation in which the letter i (called the **index of summation**) takes on the values $m, m+1, \ldots, n$. Other letters can also be used as the index of summation.

● **Example 1**

(a) $\displaystyle\sum_{i=1}^{4} i^2 = 1^2 + 2^2 + 3^2 + 4^2 = 30$

(b) $\displaystyle\sum_{i=3}^{n} i = 3 + 4 + 5 + \cdots + (n-1) + n$

(c) $\displaystyle\sum_{j=0}^{5} 2^j = 2^0 + 2^1 + 2^2 + 2^3 + 2^4 + 2^5 = 63$

(d) $\displaystyle\sum_{k=1}^{n} \frac{1}{k} = 1 + \frac{1}{2} + \frac{1}{3} + \cdots + \frac{1}{n}$

(e) $\displaystyle\sum_{i=1}^{3} \frac{i-1}{i^2 + 3} = \frac{1-1}{1^2 + 3} + \frac{2-1}{2^2 + 3} + \frac{3-1}{3^2 + 3} = 0 + \frac{1}{7} + \frac{1}{6}$

(f) $\displaystyle\sum_{i=1}^{4} 2 = 2 + 2 + 2 + 2 = 8$ ●

● **Example 2**   Write the sum $2^3 + 3^3 + \cdots + n^3$ in sigma notation.

Solution   There is no unique way of writing a sum in sigma notation. We could write

$$2^3 + 3^3 + \cdots + n^3 = \sum_{i=2}^{n} i^3$$

or
$$2^3 + 3^3 + \cdots + n^3 = \sum_{j=1}^{n-1} (j + 1)^3$$

or
$$2^3 + 3^3 + \cdots + n^3 = \sum_{k=0}^{n-2} (k + 2)^3 \qquad \bullet$$

The following theorem gives three simple rules for working with sigma notation.

Theorem (5.2)

If c is any constant (that is, it does not depend on i), then

(a) $\displaystyle\sum_{i=m}^{n} ca_i = c \sum_{i=m}^{n} a_i$

(b) $\displaystyle\sum_{i=m}^{n} (a_i + b_i) = \sum_{i=m}^{n} a_i + \sum_{i=m}^{n} b_i$

(c) $\displaystyle\sum_{i=m}^{n} (a_i - b_i) = \sum_{i=m}^{n} a_i - \sum_{i=m}^{n} b_i$

Proof   To see why these rules are true, all we have to do is write both sides in expanded form. Rule (a) is just the distributive property of real numbers:

$$ca_m + ca_{m+1} + \cdots + ca_n = c(a_m + a_{m+1} + \cdots + a_n)$$

Rule (b) follows from the associative property:

$$(a_m + b_m) + (a_{m+1} + b_{m+1}) + \cdots + (a_n + b_n)$$
$$= (a_m + a_{m+1} + \cdots + a_n) + (b_m + b_{m+1} + \cdots + b_n)$$

Rule (c) is proved similarly. $\bullet$

● **Example 3**   Find $\displaystyle\sum_{i=1}^{n} 1$.

Solution
$$\sum_{i=1}^{n} 1 = \underbrace{1 + 1 + \cdots + 1}_{n \text{ terms}} = n \qquad \bullet$$

● **Example 4**   Prove the formula for the sum of the first n positive integers:

$$\sum_{i=1}^{n} i = 1 + 2 + 3 + \cdots + n = \frac{n(n + 1)}{2}$$

Solution   This formula can be proved by mathematical induction (see Appendix A) or by the following method used by the German mathematician Karl Friedrich Gauss (1777–1855) when he was ten years old.

Write the sum S twice, once in the usual order and once in reverse order:

$$S = 1 + \quad 2 \quad + \quad 3 \quad + \cdots + (n - 1) + n$$
$$S = n + (n - 1) + (n - 2) + \cdots + \quad 2 \quad + 1$$

Adding all columns vertically, we get

$$2S = (n + 1) + (n + 1) + (n + 1) + \cdots + (n + 1) + (n + 1)$$

On the right side there are n terms, each of which is $n + 1$, so

$$2S = n(n + 1) \qquad \text{or} \qquad S = \frac{n(n + 1)}{2}$$

Example 5   Prove the formula for the sum of the squares of the first n positive integers:

$$\sum_{i=1}^{n} i^2 = 1^2 + 2^2 + 3^2 + \cdots + n^2 = \frac{n(n + 1)(2n + 1)}{6}$$

Solution   This formula can be proved by mathematical induction (see Exercise 40) or as follows: Let S be the desired sum. We start with the telescoping sum

$$\sum_{i=1}^{n} [(1 + i)^3 - i^3] = (2^3 - 1^3) + (3^3 - 2^3) + (4^3 - 3^3) + \cdots$$

$$+ [(n + 1)^3 - n^3]$$

$$= (n + 1)^3 - 1^3 = n^3 + 3n^2 + 3n$$

On the other hand, using Theorem 5.2 and Examples 3 and 4, we have

$$\sum_{i=1}^{n} [(1 + i)^3 - i^3] = \sum_{i=1}^{n} [3i^2 + 3i + 1]$$

$$= 3 \sum_{i=1}^{n} i^2 + 3 \sum_{i=1}^{n} i + \sum_{i=1}^{n} 1$$

$$= 3S + 3 \frac{n(n + 1)}{2} + n$$

$$= 3S + \frac{3}{2} n^2 + \frac{5}{2} n$$

Thus we have

$$n^3 + 3n^2 + 3n = 3S + \tfrac{3}{2}n^2 + \tfrac{5}{2}n$$

Solving this equation for S, we obtain

$$3S = n^3 + \tfrac{3}{2}n^2 + \tfrac{1}{2}n$$

$$\text{or} \qquad S = \frac{2n^3 + 3n^2 + n}{6} = \frac{n(n + 1)(2n + 1)}{6}$$

We list the results of Examples 3, 4, and 5 together with similar results for cubes and fourth powers (see Exercises 41 and 42) as Theorem 5.3. These formulas will be needed for finding areas in the next section.

Theorem (5.3)

Let c be a constant and n a positive integer. Then

(a) $\displaystyle\sum_{i=1}^{n} 1 = n$ \qquad\qquad\qquad (b) $\displaystyle\sum_{i=1}^{n} c = nc$

(c) $\displaystyle\sum_{i=1}^{n} i = \frac{n(n+1)}{2}$ \qquad\qquad (d) $\displaystyle\sum_{i=1}^{n} i^2 = \frac{n(n+1)(2n+1)}{6}$

(e) $\displaystyle\sum_{i=1}^{n} i^3 = \left[\frac{n(n+1)}{2}\right]^2$

(f) $\displaystyle\sum_{i=1}^{n} i^4 = \frac{n(n+1)(2n+1)(3n^2+3n-1)}{30}$

● **Example 6**   Evaluate $\displaystyle\sum_{i=1}^{n} i(4i^2 - 3)$.

Solution   Using Theorems 5.2 and 5.3, we have

$$\sum_{i=1}^{n} i(4i^2 - 3) = \sum_{i=1}^{n} (4i^3 - 3i) = 4\sum_{i=1}^{n} i^3 - 3\sum_{i=1}^{n} i$$

$$= 4\left[\frac{n(n+1)}{2}\right]^2 - 3\frac{n(n+1)}{2}$$

$$= \frac{n(n+1)[2n(n+1) - 3]}{2}$$

$$= \frac{n(n+1)(2n^2 + 2n - 3)}{2}$$

● **Example 7**   Find $\displaystyle\lim_{n\to\infty} \sum_{i=1}^{n} \frac{3}{n}\left[\left(\frac{i}{n}\right)^2 + 1\right]$.

Solution

$$\lim_{n\to\infty} \sum_{i=1}^{n} \frac{3}{n}\left[\left(\frac{i}{n}\right)^2 + 1\right] = \lim_{n\to\infty} \sum_{i=1}^{n}\left[\frac{3}{n^3} i^2 + \frac{3}{n}\right]$$

$$= \lim_{n\to\infty}\left[\frac{3}{n^3}\sum_{i=1}^{n} i^2 + \frac{3}{n}\sum_{i=1}^{n} 1\right]$$

$$= \lim_{n\to\infty}\left[\frac{3}{n^3}\frac{n(n+1)(2n+1)}{6} + \frac{3}{n}\cdot n\right]$$

$$= \lim_{n\to\infty}\left[\frac{1}{2}\cdot\frac{n}{n}\cdot\left(\frac{n+1}{n}\right)\left(\frac{2n+1}{n}\right) + 3\right]$$

$$= \lim_{n\to\infty}\left[\frac{1}{2}\cdot 1\left(1 + \frac{1}{n}\right)\left(2 + \frac{1}{n}\right) + 3\right]$$

$$= \frac{1}{2}\cdot 1\cdot 1\cdot 2 + 3 = 4$$

EXERCISES 5.1

In Exercises 1–10 write the sum in expanded form.

1. $\displaystyle\sum_{i=1}^{5} \sqrt{i}$

2. $\displaystyle\sum_{i=1}^{6} \frac{1}{i+1}$

3. $\displaystyle\sum_{i=4}^{6} 3^i$

4. $\displaystyle\sum_{i=4}^{6} i^3$

5. $\displaystyle\sum_{k=0}^{4} \frac{2k-1}{2k+1}$

6. $\displaystyle\sum_{k=5}^{8} x^k$

7. $\displaystyle\sum_{i=1}^{n} i^{10}$

8. $\displaystyle\sum_{j=n}^{n+3} j^2$

9. $\displaystyle\sum_{j=0}^{n-1} (-1)^j$

10. $\displaystyle\sum_{i=1}^{n} f(x_i)\,\Delta x_i$

In Exercises 11–20 write the sum in sigma notation.

11. $1 + 2 + 3 + 4 + \cdots + 10$

12. $\sqrt{3} + \sqrt{4} + \sqrt{5} + \sqrt{6} + \sqrt{7}$

13. $\frac{1}{2} + \frac{2}{3} + \frac{3}{4} + \frac{4}{5} + \cdots + \frac{19}{20}$

14. $\frac{3}{7} + \frac{4}{8} + \frac{5}{9} + \frac{6}{10} + \cdots + \frac{23}{27}$

15. $2 + 4 + 6 + 8 + \cdots + 2n$

16. $1 + 3 + 5 + 7 + \cdots + (2n-1)$

17. $1 + 2 + 4 + 8 + 16 + 32$

18. $\frac{1}{1} + \frac{1}{4} + \frac{1}{9} + \frac{1}{16} + \frac{1}{25} + \frac{1}{36}$

19. $x + x^2 + x^3 + \cdots + x^n$

20. $1 - x + x^2 - x^3 + \cdots + (-1)^n x^n$

In Exercises 21–38 find the value of the sum.

21. $\displaystyle\sum_{i=4}^{8} (3i-2)$

22. $\displaystyle\sum_{i=3}^{6} i(i+2)$

23. $\displaystyle\sum_{j=1}^{6} 3^{j+1}$

24. $\displaystyle\sum_{k=0}^{8} \cos k\pi$

25. $\displaystyle\sum_{n=1}^{20} (-1)^n$

26. $\displaystyle\sum_{i=1}^{100} 4$

27. $\displaystyle\sum_{i=0}^{4} (2^i + i^2)$

28. $\displaystyle\sum_{i=-2}^{4} 2^{3-i}$

29. $\displaystyle\sum_{i=1}^{n} \left(\frac{1}{i} - \frac{1}{i+1} \right)$

30. $\displaystyle\sum_{k=1}^{n} (\sqrt{k} - \sqrt{k-1})$

31. $\displaystyle\sum_{i=1}^{n} 2i$

32. $\displaystyle\sum_{i=1}^{n} (2-5i)$

33. $\displaystyle\sum_{i=1}^{n} (i^2 + 3i + 4)$

34. $\displaystyle\sum_{i=1}^{n} (3 + 2i)^2$

35. $\displaystyle\sum_{i=1}^{n} (i+1)(i+2)$

36. $\displaystyle\sum_{i=1}^{n} i(i+1)(i+2)$

37. $\displaystyle\sum_{i=1}^{n} (i^3 - i - 2)$

38. $\displaystyle\sum_{k=1}^{n} k^2(k^2 - k + 1)$

39. Prove Formula (b) of Theorem 5.3.

40. Prove Formula (d) of Theorem 5.3 using mathematical induction. (See Appendix A for the Principle of Mathematical Induction.)

41. Prove Formula (e) of Theorem 5.3
(a) using mathematical induction,
(b) using a method similar to that of Example 5 [start with $(1 + i)^4 - i^4$],
(c) using the following method published by Abu Bekr Mohammed ibn Alhusain Alkarchi in about A.D. 1010. Figure 5.1 shows a square $ABCD$ in which sides AB and AD have been divided into segments of lengths $1, 2, 3, \ldots, n$. Thus the side of the square has length $n(n+1)/2$ so the area is $[n(n+1)/2]^2$. But the area is also the sum of the areas of the n "gnomons" $G_1, G_2, \ldots, G_n$ shown in the figure. Show that the area of G_i is i^3 and conclude that Formula (e) is true.

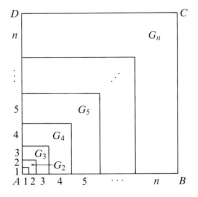

Figure 5.1

42. Prove Formula (f) of Theorem 5.3
(a) using mathematical induction,
(b) using the method of Example 5.

43. Prove that

$$\sum_{i=1}^{n} (a_i - a_{i-1}) = a_n - a_0$$

(This is called a telescoping sum.)

44. Prove the generalized triangle inequality

$$\left| \sum_{i=1}^{n} a_i \right| \le \sum_{i=1}^{n} |a_i|$$

Find the limits in Exercises 45–48.

45. $\displaystyle\lim_{n\to\infty}\sum_{i=1}^{n}\frac{1}{n}\left(\frac{i}{n}\right)^2$

46. $\displaystyle\lim_{n\to\infty}\sum_{i=1}^{n}\frac{1}{n}\left[\left(\frac{i}{n}\right)^3+1\right]$

47. $\displaystyle\lim_{n\to\infty}\sum_{i=1}^{n}\frac{2}{n}\left[\left(\frac{2i}{n}\right)^3+5\left(\frac{2i}{n}\right)\right]$

48. $\displaystyle\lim_{n\to\infty}\sum_{i=1}^{n}\frac{3}{n}\left[\left(1+\frac{3i}{n}\right)^3-2\left(1+\frac{3i}{n}\right)\right]$

49. Prove the formula for the sum of a finite geometric series with first term a and common ratio r:

$$\sum_{i=1}^{n}ar^{i-1}=a+ar+ar^2+\cdots+ar^{n-1}=\frac{a(r^n-1)}{r-1}$$

50. (a) Use the product formula (18a in Appendix B) to show that

$$2\sin\tfrac{1}{2}x\cos ix=\sin(i+\tfrac{1}{2})x-\sin(i-\tfrac{1}{2})x$$

(b) Use the identity in part (a) together with Exercise 43 to prove the formula

$$\sum_{i=1}^{n}\cos ix=\frac{\sin(n+\tfrac{1}{2})x-\sin\tfrac{1}{2}x}{2\sin\tfrac{1}{2}x}$$

where x is not an integer multiple of 2π. Deduce that

$$\sum_{i=1}^{n}\cos ix=\frac{\sin\tfrac{1}{2}nx\cos\tfrac{1}{2}(n+1)x}{\sin\tfrac{1}{2}x}$$

51. Use the method of Exercise 50 to prove the formula

$$\sum_{i=1}^{n}\sin ix=\frac{\sin\tfrac{1}{2}nx\sin\tfrac{1}{2}(n+1)x}{\sin\tfrac{1}{2}x}$$

where x is not an integer multiple of 2π.

SECTION 5.2

Area

We begin by attempting to solve the *area problem:* Find the area of the region S that lies under the curve $y=f(x)$ from a to b. This means that S, illustrated in Figure 5.2, is bounded by the graph of a function f [where $f(x)\geq 0$], the vertical lines $x=a$ and $x=b$, and the x-axis.

In trying to solve the area problem we have to ask ourselves: What is the meaning of the word *area?* This question is easy to answer for regions with straight sides. For a rectangle, the area is defined as the product of the length and the width. The area of a triangle is half the base times the height. The area of a polygon is found by dividing it into triangles (as in Figure 5.3) and adding the areas of the triangles.

However it is not so easy to find the area of a region with curved sides. We all have an intuitive idea of what the area of a region is. But part of the area problem is to make this intuitive idea precise by giving an exact definition of area.

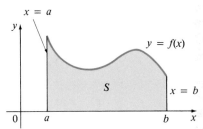

Figure 5.2

$S=\{(x,y)\,|\,a\leq x\leq b,0\leq y\leq f(x)\}$

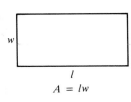
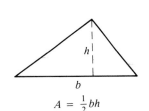
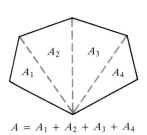

Figure 5.3

$A=lw$ $A=\tfrac{1}{2}bh$ $A=A_1+A_2+A_3+A_4$

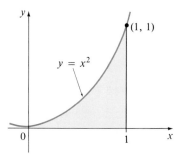

Figure 5.4

Recall that in defining a tangent we first approximated the slope of the tangent line by slopes of secant lines and then we took the limit of these approximations. We shall pursue a similar idea for areas. We first approximate the region S by polygons and then we take the limit of the areas of these polygons. The following example will illustrate the procedure.

● **Example 1**   Let us try to find the area under the parabola $y = x^2$ from 0 to 1 (the parabolic segment illustrated in Figure 5.4). One method of approximating the desired area is to divide the interval $[0, 1]$ into subintervals of equal length and consider the rectangles whose bases are these subintervals and whose heights are the values of the function at the right-hand endpoints of these subintervals. Figure 5.5 shows the approximation of the parabolic segment by four, eight, and n rectangles.

Let S_n be the sum of the areas of the n rectangles in Figure 5.5(c). Each rectangle has width $1/n$ and the heights are the values of the function $f(x) = x^2$ at the points $1/n, 2/n, 3/n, \ldots, n/n$; that is, the heights are $(1/n)^2$, $(2/n)^2, (3/n)^2, \ldots, (n/n)^2$. Thus

$$S_n = \frac{1}{n}\left(\frac{1}{n}\right)^2 + \frac{1}{n}\left(\frac{2}{n}\right)^2 + \frac{1}{n}\left(\frac{3}{n}\right)^2 + \cdots + \frac{1}{n}\left(\frac{n}{n}\right)^2$$

$$= \frac{1}{n}\frac{1}{n^2}(1^2 + 2^2 + 3^2 + \cdots + n^2)$$

$$= \frac{1}{n^3}\sum_{i=1}^{n} i^2$$

Using the formula for the sum of the squares of the first n integers [Equation (d) in Theorem 5.3] we can write

$$S_n = \frac{1}{n^3}\frac{n(n + 1)(2n + 1)}{6} = \frac{(n + 1)(2n + 1)}{6n^2}$$

For instance, the sum of the areas of the four shaded rectangles in Figure 5.5(a) is

$$S_4 = \frac{5(9)}{6(16)} = 0.46875$$

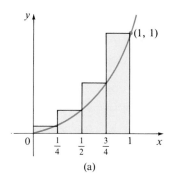

(a)

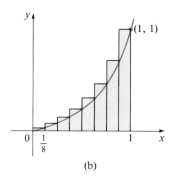

(b)

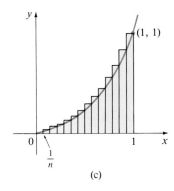

(c)

Figure 5.5

and the sum of the areas of the eight rectangles in Figure 5.5(b) is

$$S_8 = \frac{9(17)}{6(64)} = 0.3984375$$

Similarly we find that

$$S_{100} \approx 0.33835 \qquad S_{1000} \approx 0.33383$$

It looks as if S_n is becoming closer to $\frac{1}{3}$ as n increases. In fact,

$$\lim_{n \to \infty} S_n = \lim_{n \to \infty} \frac{(n + 1)(2n + 1)}{6n^2}$$

$$= \lim_{n \to \infty} \frac{1}{6}\left(\frac{n + 1}{n}\right)\left(\frac{2n + 1}{n}\right)$$

$$= \lim_{n \to \infty} \frac{1}{6}\left(1 + \frac{1}{n}\right)\left(2 + \frac{1}{n}\right)$$

$$= \frac{1}{6} \cdot 1 \cdot 2 = \frac{1}{3}$$

From Figure 5.5 it appears that, as n increases, S_n becomes a better and better approximation to the area of the parabolic segment. Therefore we *define* the area A to be the limit of the sums of the areas of the approximating rectangles, that is,

$$A = \lim_{n \to \infty} S_n = \tfrac{1}{3}$$ ●

In applying the idea of Example 1 to the more general region S of Figure 5.2, there is no need to use rectangles of equal width. We start by subdividing the interval $[a, b]$ into n smaller subintervals by choosing partition points $x_0, x_1, x_2, \ldots, x_n$ so that

$$a = x_0 < x_1 < x_2 < \cdots < x_{n-1} < x_n = b$$

Then the n subintervals are

$$[x_0, x_1], \ [x_1, x_2], \ [x_2, x_3], \ldots, [x_{n-1}, x_n]$$

This subdivision is called a **partition** of $[a, b]$ and we denote it by P. We use the notation Δx_i for the length of the ith subinterval $[x_{i-1}, x_i]$. Thus

$$\Delta x_i = x_i - x_{i-1}$$

The length of the longest subinterval is denoted by $\|P\|$ and is called the **norm** of P. Thus

$$\|P\| = \max\{\Delta x_1, \Delta x_2, \ldots, \Delta x_n\}$$

Figure 5.6 illustrates one possible partition of $[a, b]$.

Figure 5.6

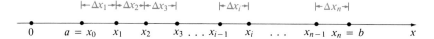

By drawing the lines $x = a, x = x_1, x = x_2, \ldots, x = b$, we use the partition P to divide the region S into strips $S_1, S_2, \ldots, S_n$ as in Figure 5.7. Next we approximate these strips S_i by rectangles R_i. To do this we choose a number x_i^* in each subinterval $[x_{i-1}, x_i]$ and construct a rectangle R_i with base Δx_i and height $f(x_i^*)$ as in Figure 5.8. Each point x_i^* can be anywhere in its subinterval—at the right endpoint (as in Example 1) or at the left endpoint or somewhere in the middle. The area of the ith rectangle R_i is

$$A_i = f(x_i^*)\,\Delta x_i$$

Figure 5.7

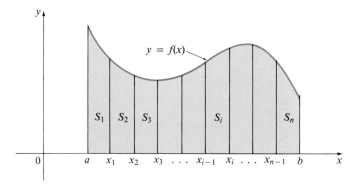

Figure 5.8

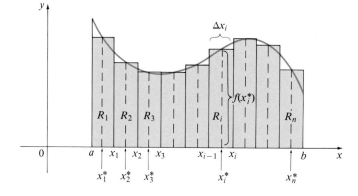

The n rectangles $R_1, \ldots, R_n$ form a polygonal approximation to the region S. What we think of intuitively as the area of S is approximated by the sum of the areas of these rectangles, which is

(5.4)
$$\sum_{i=1}^{n} A_i = \sum_{i=1}^{n} f(x_i^*)\,\Delta x_i = f(x_1^*)\,\Delta x_1 + \cdots + f(x_n^*)\,\Delta x_n$$

Figure 5.9 shows this approximation for partitions with $n = 2, 4, 8$, and 12. Notice that this approximation appears to become better and better as the strips become thinner and thinner, that is, as $\|P\| \to 0$. Therefore we define

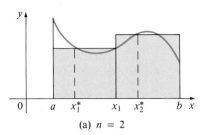

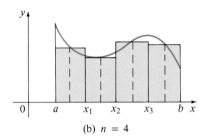

(a) $n = 2$ (b) $n = 4$

Figure 5.9

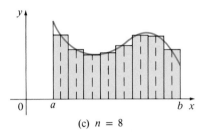

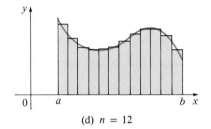

(c) $n = 8$ (d) $n = 12$

the **area** A of the region S as the limiting value (if it exists) of the areas of the approximating polygons, that is, the limit of the sum (5.4) of the areas of the approximating rectangles. In symbols:

(5.5)

$$A = \lim_{\|P\| \to 0} \sum_{i=1}^{n} f(x_i^*)\, \Delta x_i$$

The preceding discussion and the diagrams in Figures 5.8 and 5.9 show that the definition of area in (5.5) corresponds to our intuitive feeling of what area ought to be.

The limit in (5.5) may or may not exist. It can be shown that if f is continuous, then this limit will exist; that is, the region will have an area. [The precise meaning of the limit in Definition 5.5 is that for every $\varepsilon > 0$ there is a corresponding number $\delta > 0$ such that

$$\left| A - \sum_{i=1}^{n} f(x_i^*)\, \Delta x_i \right| < \varepsilon \qquad \text{whenever} \qquad \|P\| < \delta$$

In other words, the area can be approximated by a sum of areas of rectangles to within an arbitrary degree of accuracy (ε) by taking the norm of the partition sufficiently small.]

● **Example 2**   (a) If the interval $[0, 3]$ is divided into subintervals by the partition P and the set of partition points is $\{0, 0.6, 1.2, 1.6, 2, 2.5, 3\}$, find $\|P\|$. (b) If $f(x) = x^2 - 4x + 5$ and x_i^* is chosen to be the left endpoint of the ith subinterval, find the sum of the areas of the approximating rectangles. (c) Sketch the approximating rectangles.

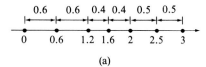

(a)

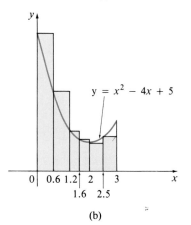

$y = x^2 - 4x + 5$

(b)

Figure 5.10

Solution   (a)   We are given $x_0 = 0, x_1 = 0.6, x_2 = 1.2, x_3 = 1.6, x_4 = 2,$ $x_5 = 2.5$, and $x_6 = 3$, so

$$\Delta x_1 = 0.6 - 0 = 0.6 \qquad \Delta x_2 = 1.2 - 0.6 = 0.6$$

$$\Delta x_3 = 1.6 - 1.2 = 0.4 \qquad \Delta x_4 = 2 - 1.6 = 0.4$$

$$\Delta x_5 = 2.5 - 2 = 0.5 \qquad \Delta x_6 = 3 - 2.5 = 0.5$$

[See Figure 5.10(a).] Therefore

$$\|P\| = \max\{0.6, 0.6, 0.4, 0.4, 0.5, 0.5\} = 0.6$$

(b)   Since $x_i^* = x_{i-1}$, the sum of the areas of the approximating rectangles is, by (5.4),

$$\sum_{i=1}^{6} f(x_i^*)\,\Delta x_i = \sum_{i=1}^{6} f(x_{i-1})\,\Delta x_i$$

$$= f(0)\,\Delta x_1 + f(0.6)\,\Delta x_2 + f(1.2)\,\Delta x_3 + f(1.6)\,\Delta x_4 + f(2)\,\Delta x_5$$

$$+ f(2.5)\,\Delta x_6$$

$$= 5(0.6) + 2.96(0.6) + 1.64(0.4) + 1.16(0.4) + 1(0.5)$$

$$+ 1.25(0.5)$$

$$= 7.021$$

(c)   The graph of f and the approximating rectangles are sketched in Figure 5.10(b).   ●

● **Example 3**   Find the area under the parabola $y = x^2 + 1$ from 0 to 2.

Solution   Since $f(x) = x^2 + 1$ is continuous, the limit (5.5) that defines the area must exist for all possible partitions P of the interval $[0,2]$ as long as $\|P\| \to 0$. To simplify things let us take the partition P that divides $[0,2]$ into n subintervals of equal length. (This is called a regular partition.) Then the partition points are

$$x_0 = 0, \; x_1 = \frac{2}{n}, \; x_2 = \frac{4}{n}, \ldots, \; x_i = \frac{2i}{n}, \ldots, \; x_n = \frac{2n}{n} = 2$$

and $$\Delta x_1 = \Delta x_2 = \ldots = \Delta x_i = \ldots = \Delta x_n = \frac{2}{n}$$

so the norm of P is

$$\|P\| = \max\{\Delta x_i\} = \frac{2}{n}$$

The point x_i^* can be chosen to be anywhere in the ith subinterval. For the sake of definiteness, let us choose it to be the right-hand endpoint:

$$x_i^* = x_i = \frac{2i}{n}$$

Since $\|P\| = 2/n$, the condition $\|P\| \to 0$ is equivalent to $n \to \infty$. So the definition of area (5.5) becomes

$$A = \lim_{\|P\| \to 0} \sum_{i=1}^{n} f(x_i^*) \Delta x_i$$

$$= \lim_{n \to \infty} \sum_{i=1}^{n} f\left(\frac{2i}{n}\right) \frac{2}{n}$$

$$= \lim_{n \to \infty} \sum_{i=1}^{n} \left[\left(\frac{2i}{n}\right)^2 + 1\right] \frac{2}{n}$$

$$= \lim_{n \to \infty} \sum_{i=1}^{n} \left[\frac{8i^2}{n^3} + \frac{2}{n}\right]$$

$$= \lim_{n \to \infty} \left[\frac{8}{n^3} \sum_{i=1}^{n} i^2 + \frac{2}{n} \sum_{i=1}^{n} 1\right] \quad \textit{(by Theorem 5.2)}$$

$$= \lim_{n \to \infty} \left[\frac{8}{n^3} \cdot \frac{n(n+1)(2n+1)}{6} + \frac{2}{n} \cdot n\right] \quad \textit{(by Theorem 5.3)}$$

$$= \lim_{n \to \infty} \left[\frac{4}{3} \cdot 1 \cdot \left(1 + \frac{1}{n}\right)\left(2 + \frac{1}{n}\right) + 2\right]$$

$$= \frac{4}{3} \cdot 1 \cdot 1 \cdot 2 + 2 = \frac{14}{3}$$

The sum in this calculation is represented by the areas of the shaded rectangles in Figure 5.11(a). Notice that in this case, with our choice of x_i^* as the right-hand endpoint and since f is increasing, $f(x_i^*)$ is the maximum value of f on $[x_{i-1}, x_i]$ and the required area is approximated at the nth stage by the area of a circumscribed polygon.

We could just as well have chosen x_i^* to be the left-hand endpoint, $x_i^* = x_{i-1} = 2(i-1)/n$. Then $f(x_i^*)$ is the minimum value of f on $[x_{i-1}, x_i]$ and A is approximated by the area of an inscribed polygon [see Figure

Figure 5.11

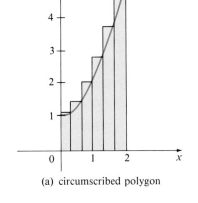

(a) circumscribed polygon

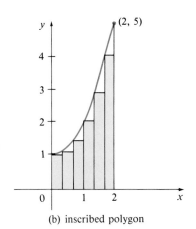

(b) inscribed polygon

5.11(b)]. The calculation with this choice is as follows:

$$A = \lim_{\|P\| \to 0} \sum_{i=1}^{n} f(x_i^*) \, \Delta x_i$$

$$= \lim_{n \to \infty} \sum_{i=1}^{n} f\left[\frac{2(i-1)}{n}\right] \frac{2}{n}$$

$$= \lim_{n \to \infty} \sum_{i=1}^{n} \left\{ \left[\frac{2(i-1)}{n}\right]^2 + 1 \right\} \frac{2}{n}$$

$$= \lim_{n \to \infty} \sum_{i=1}^{n} \left[\frac{8}{n^3}(i^2 - 2i + 1) + \frac{2}{n} \right]$$

$$= \lim_{n \to \infty} \left[\frac{8}{n^3} \sum_{i=1}^{n} i^2 - \frac{16}{n^3} \sum_{i=1}^{n} i + \frac{8}{n^3} \sum_{i=1}^{n} 1 + \frac{2}{n} \sum_{i=1}^{n} 1 \right]$$

$$= \lim_{n \to \infty} \left[\frac{8}{n^3} \frac{n(n+1)(2n+1)}{6} - \frac{16}{n^3} \frac{n(n+1)}{2} + \frac{8}{n^3} n + \frac{2}{n} n \right]$$

$$= \lim_{n \to \infty} \left[\frac{4}{3} \cdot 1 \cdot \left(1 + \frac{1}{n}\right)\left(2 + \frac{1}{n}\right) - \frac{8}{n}\left(1 + \frac{1}{n}\right) + \frac{8}{n^2} + 2 \right]$$

$$= \frac{4}{3} \cdot 1 \cdot 1 \cdot 2 - 0 \cdot 1 + 0 + 2 = \frac{14}{3}$$

Notice that we have obtained the same answer with the different choice of x_i^*. In fact, we would obtain the same answer if x_i^* were chosen to be the midpoint of $[x_{i-1}, x_i]$ (see Exercise 11) or indeed any other point of this interval.   ●

● **Example 4**   Find the area under the cosine curve from 0 to b, where $0 \leqslant b \leqslant \pi/2$.

Solution   As in the first part of Example 3, we choose a regular partition P so that

$$\|P\| = \Delta x_1 = \Delta x_2 = \ldots = \Delta x_n = \frac{b}{n}$$

and we choose x_i^* to be the right-hand endpoint of the ith subinterval:

$$x_i^* = x_i = \frac{ib}{n}$$

Since $\|P\| = b/n \to 0$ as $n \to \infty$, the area under the cosine curve from 0 to b is

(5.6)

$$A = \lim_{\|P\| \to 0} \sum_{i=1}^{n} f(x_i^*) \, \Delta x_i = \lim_{n \to \infty} \sum_{i=1}^{n} \cos\left(i \frac{b}{n}\right) \frac{b}{n}$$

$$= \lim_{n \to \infty} \frac{b}{n} \sum_{i=1}^{n} \cos\left(i \frac{b}{n}\right)$$

To evaluate this limit we use the formula of Exercise 50 in Section 5.1:

$$\sum_{i=1}^{n} \cos ix = \frac{\sin \frac{1}{2}nx \cos \frac{1}{2}(n+1)x}{\sin \frac{1}{2}x}$$

with $x = b/n$. Then Equation 5.6 becomes

(5.7)
$$A = \lim_{n \to \infty} \frac{b}{n} \frac{\sin \frac{1}{2} b \cos\left[\dfrac{(n+1)b}{2n}\right]}{\sin \dfrac{b}{2n}}$$

Now
$$\cos\left[\frac{(n+1)b}{2n}\right] = \cos\left(1 + \frac{1}{n}\right)\frac{b}{2} \to \cos \frac{b}{2} \quad \text{as } n \to \infty$$

since cosine is continuous. Letting $t = b/n$ and using Theorem 3.21, we have

$$\lim_{n \to \infty} \frac{b}{n} \cdot \frac{1}{\sin \dfrac{b}{2n}} = \lim_{t \to 0^+} \frac{t}{\sin \dfrac{t}{2}} = \lim_{t \to 0^+} 2 \cdot \frac{\dfrac{t}{2}}{\sin \dfrac{t}{2}} = 2$$

Putting these limits in Equation 5.7, we obtain

$$A = 2 \sin \frac{b}{2} \cos \frac{b}{2} = \sin b$$

In particular, taking $b = \pi/2$, we have proved that the area under the cosine curve from 0 to $\pi/2$ is $\sin(\pi/2) = 1$ (see Figure 5.12).   ●

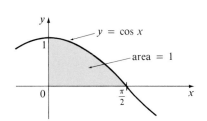

$y = \cos x$

area = 1

Figure 5.12

EXERCISES 5.2

In Exercises 1–10 you are given a function f, an interval, partition points, and a description of the point x_i^* within the ith subinterval. (a) Find $\|P\|$. (b) Find the sum of the areas of the approximating rectangles (5.4). (c) Sketch the graph of f and the approximating rectangles.

1. $f(x) = 16 - x^2$, $[0,4]$, $\{0,1,2,3,4\}$, $x_i^* = $ left endpoint

2. $f(x) = 16 - x^2$, $[0,4]$, $\{0,1,2,3,4,\}$, $x_i^* = $ right endpoint

3. $f(x) = 16 - x^2$, $[0,4]$, $\{0,1,2,3,4\}$, $x_i^* = $ midpoint

4. $f(x) = 16 - x^2$, $[-4,4]$, $\{-4,-3,-2,-1,0,1,2,3,4\}$, $x_i^* = $ midpoint

5. $f(x) = 2x + 1$, $[0,5]$, $\{0,1,2,3,4,5\}$, $x_i^* = $ right endpoint

6. $f(x) = 2x + 1$, $[0,4]$, $\{0,0.5,1,2,4\}$, $x_i^* = $ left endpoint

7. $f(x) = x^3 + 2$, $[-1,2]$, $\{-1,-0.5,0,0.5,1.0,1.5,2\}$, $x_i^* = $ right endpoint

8. $f(x) = \dfrac{1}{x+1}$, $[0,2]$, $\{0,0.5,1.0,1.5,2\}$, $x_1^* = 0.25$, $x_2^* = 1$, $x_3^* = 1.25$, $x_4^* = 2$

9. $f(x) = 2 \sin x$, $[0,\pi]$, $\left\{0, \dfrac{\pi}{4}, \dfrac{\pi}{2}, \dfrac{3\pi}{4}, \pi\right\}$, $x_1^* = \dfrac{\pi}{6}$, $x_2^* = \dfrac{\pi}{3}$, $x_3^* = \dfrac{2\pi}{3}$, $x_4^* = \dfrac{5\pi}{6}$

10. $f(x) = 4\cos x$, $\left[0, \dfrac{\pi}{2}\right]$, $\left\{0, \dfrac{\pi}{6}, \dfrac{\pi}{4}, \dfrac{\pi}{3}, \dfrac{\pi}{2}\right\}$, $x_i^* = $ left endpoint

11. Find the area of Example 3 taking x_i^* to be the midpoint of $[x_{i-1}, x_i]$. Illustrate the approximating rectangles with a sketch like Figure 5.11.

In Exercises 12–14 find the area under the given curve from a to b using subintervals of equal length and taking x_i^* in (5.5) to be the (a) left endpoint, (b) right endpoint, and (c) midpoint of the ith subinterval. In each case sketch the approximating rectangles.

12. $y = 3 - \dfrac{x}{2}$, $a = -2$, $b = 2$

13. $y = 16 - x^2$, $a = -4$, $b = 4$

14. $y = x^3$, $a = 0$, $b = 1$

In Exercises 15–23 use Definition 5.5 to find the area under the given curve from a to b. Use equal subintervals and take

x_i^* to be the right endpoint of the ith subinterval. Sketch the region.

15. $y = 5$, $a = -2$, $b = 2$

16. $y = 2x + 1$, $a = 0$, $b = 5$

17. $y = x^2 + 3x - 2$, $a = 1$, $b = 4$

18. $y = 2x^2 - 4x + 5$, $a = -3$, $b = 2$

19. $y = x^3 + 2x$, $a = 0$, $b = 2$

20. $y = x^3 + 2x^2 + x$, $a = 0$, $b = 1$

21. $y = 1 - 2x^2 + x^4$, $a = -1$, $b = 1$

22. $y = x^4 + 3x + 2$, $a = 0$, $b = 3$

23. $y = \sin x$, $a = 0$, $b = \pi$ (*Hint:* Use Exercise 51 in Section 5.1.)

SECTION 5.3

The Definite Integral

We saw in the preceding section that a limit of the form

(5.8)
$$\lim_{\|P\| \to 0} \sum_{i=1}^{n} f(x_i^*) \, \Delta x_i$$

arises when we compute an area. It turns out that this same type of limit occurs in a wide variety of situations even when f is not necessarily a positive function. In Chapter 8 we shall see that limits of the form (5.8) also arise in finding lengths of curves, volumes of solids, areas of surfaces, centers of mass, fluid pressure, and work, as well as other quantities. We therefore give this type of limit a special name and notation.

Definition of a Definite Integral (5.9)

If f is a function defined on a closed interval $[a, b]$, let P be a partition of $[a, b]$ with partition points $x_0, x_1, \ldots, x_n$, where

$$a = x_0 < x_1 < x_2 < \cdots < x_n = b$$

Choose points x_i^* in $[x_{i-1}, x_i]$ and let $\Delta x_i = x_i - x_{i-1}$ and $\|P\| = \max\{\Delta x_i\}$. Then the **definite integral of f from a to b** is

$$\int_a^b f(x) \, dx = \lim_{\|P\| \to 0} \sum_{i=1}^{n} f(x_i^*) \, \Delta x_i$$

if this limit exists. If the limit does exist, then f is called **integrable** on the interval $[a, b]$.

Note 1:   The symbol $\int$ was introduced by Leibniz and is called an **integral sign.** It is an elongated *S* and was chosen because an integral is a limit of sums. In the notation $\int_a^b f(x)\, dx$, $f(x)$ is called the **integrand** and a and b are called the **limits of integration;** a is the **lower limit** and b is the **upper limit.** The symbol dx has no meaning by itself; $\int_a^b f(x)\, dx$ is all one symbol. The procedure of calculating an integral is called **integration.**

Note 2:   The definite integral $\int_a^b f(x)\, dx$ is a number; it does not depend on x. In fact, we could use any letter in place of x without changing the value of the integral:

$$\int_a^b f(x)\, dx = \int_a^b f(t)\, dt = \int_a^b f(r)\, dr$$

For this reason, x is called a **dummy variable** and the notation $\int_a^b f$ is sometimes used for the definite integral:

$$\int_a^b f = \int_a^b f(x)\, dx$$

Note 3:   The sum

(5.10)
$$\sum_{i=1}^n f(x_i^*)\, \Delta x_i$$

that occurs in Definition 5.9 is called a **Riemann sum** after the German mathematician Bernhard Riemann (1826–1866). The definite integral is sometimes called the **Riemann integral.** If f happens to be positive, then the Riemann sum can be interpreted as a sum of areas of approximating rectangles. [Compare (5.10) and (5.4).] But, in general, f is not positive and the Riemann sum cannot be interpreted in that way.

● **Example 1**   Let $f(x) = 1 + 5x$ and consider the partition P of the interval $[-2, 1]$ by means of the set of partition points $\{-2, -1.5, -1, -0.3, 0.2, 1\}$. In this example, $a = -2$, $b = 1$, $n = 5$, and $x_0 = -2$, $x_1 = -1.5$, $x_2 = -1$, $x_3 = -0.3$, $x_4 = 0.2$, and $x_5 = 1$. The lengths of the subintervals are

$$\Delta x_1 = -1.5 - (-2) = 0.5 \qquad \Delta x_2 = -1 - (-1.5) = 0.5$$

$$\Delta x_3 = -0.3 - (-1) = 0.7 \qquad \Delta x_4 = 0.2 - (-0.3) = 0.5$$

$$\Delta x_5 = 1 - 0.2 = 0.8$$

Thus the norm of the partition P is

$$\|P\| = \max\{0.5, 0.5, 0.7, 0.5, 0.8\} = 0.8$$

Suppose we choose $x_1^* = -1.8$, $x_2^* = -1.2$, $x_3^* = -0.3$, $x_4^* = 0$, and $x_5^* = 0.7$. Then the corresponding Riemann sum is

$$\sum_{i=1}^5 f(x_i^*)\, \Delta x_i = f(-1.8)\, \Delta x_1 + f(-1.2)\, \Delta x_2 + f(-0.3)\, \Delta x_3$$

$$+ f(0)\, \Delta x_4 + f(0.7)\, \Delta x_5$$

$$= (-8)(0.5) + (-5)(0.5) + (-0.5)(0.7) + 1(0.5) + (4.5)(0.8)$$

$$= -2.75$$

Notice that, in this example, f is not a positive function and so the Riemann sum does not represent a sum of areas of rectangles.   ●

Note 4:   In general an integral need not represent an area. But for *positive* functions, an integral can be interpreted as an area. In fact, comparing Equation 5.5 and Definition 5.9, we see the following:

For the special case where $f(x) \geqslant 0$,

$$\int_a^b f(x)\,dx = \text{the area under the graph of } f \text{ from } a \text{ to } b$$

Note 5:   The precise meaning of the limit that defines the integral in Definition 5.9 is as follows.

$\int_a^b f(x)\,dx = I$ means that for every $\varepsilon > 0$ there is a corresponding number $\delta > 0$ such that

$$\left| I - \sum_{i=1}^{n} f(x_i^*)\,\Delta x_i \right| < \varepsilon$$

for all partitions P of $[a, b]$ with $\|P\| < \delta$ and for all possible choices of x_i^* in $[x_{i-1}, x_i]$.

This means that a definite integral can be approximated to within any desired degree of accuracy by a Riemann sum.

Note 6:   In Definition 5.9 we are dealing with a function f defined on an interval $[a, b]$, so we are implicitly assuming that $a < b$. But for some purposes it is useful to extend the definition of $\int_a^b f(x)\,dx$ to the case where $a > b$ or $a = b$ as follows:

If $a > b$, then $\int_a^b f(x)\,dx = -\int_b^a f(x)\,dx.$

If $a = b$, then $\int_a^a f(x)\,dx = 0.$

● **Example 2**   Express

$$\lim_{\|P\|\to 0} \sum_{i=1}^{n} \left[(x_i^*)^3 + x_i^* \sin x_i^* \right] \Delta x_i$$

as an integral on the interval $[0, \pi]$.

Solution   Comparing the given limit with the limit in Definition 5.9, we see that they will be identical if we choose

$$f(x) = x^3 + x\sin x$$

We are given that $a = 0$ and $b = \pi$. Therefore, by Definition 5.9, we have

$$\lim_{\|P\| \to 0} \sum_{i=1}^{n} [(x_i^*)^3 + x_i^* \sin x_i^*] \Delta x_i = \int_0^\pi (x^3 + x\sin x)\, dx \qquad \bullet$$

The question arises: Which functions are integrable? A partial answer is given by the following theorem, which is proved in courses on advanced calculus.

Theorem (5.11)

> If f is either continuous or monotonic on $[a, b]$, then f is integrable on $[a, b]$; that is, the definite integral $\int_a^b f(x)\, dx$ exists.

If f is discontinuous at some points in $[a, b]$, then $\int_a^b f(x)\, dx$ might exist or it might not exist (see Exercises 30 and 31). If f has only a finite number of discontinuities and these are all jump discontinuities, then f is called **piecewise continuous** and it turns out that f is integrable.

It can be shown that if f is integrable on $[a, b]$, then f must be a **bounded function** on $[a, b]$; that is, there exists a number M such that $|f(x)| \leqslant M$ for all x in $[a, b]$. Geometrically, this means that the graph of f lies between the horizontal lines $y = M$ and $y = -M$. In particular, if f has an infinite discontinuity at some point in $[a, b]$, then f is not bounded and is therefore not integrable. (See Exercise 30.)

If f is integrable on $[a, b]$, then the Riemann sums (5.10) must approach $\int_a^b f$ as $\|P\| \to 0$ no matter how the partitions P are chosen and no matter how the points x_i^* are chosen in $[x_{i-1}, x_i]$. Therefore if it is known beforehand that f is integrable on $[a, b]$ (for instance, if it is known that f is continuous or monotonic), then in calculating the value of an integral we are free to choose partitions P and points x_i^* in any way we like as long as $\|P\| \to 0$. For purposes of calculation, it is convenient to take P to be a **regular partition;** that is, all the subintervals have the same length Δx. Then

$$\Delta x = \Delta x_1 = \Delta x_2 = \ldots = \Delta x_n = \frac{b-a}{n}$$

and $x_0 = a, \quad x_1 = a + \Delta x, \quad x_2 = a + 2\Delta x, \quad \ldots, \quad x_i = a + i\Delta x$

If we choose x_i^* to be the right endpoint of the ith subinterval, then

$$x_i^* = x_i = a + i\Delta x = a + i\frac{b-a}{n}$$

Since $\|P\| = \Delta x = (b - a)/n$, we have $\|P\| \to 0$ as $n \to \infty$, so Definition 5.9 gives

$$\int_a^b f(x)\,dx = \lim_{\|P\| \to 0} \sum_{i=1}^n f(x_i^*)\,\Delta x$$

$$= \lim_{n \to \infty} \sum_{i=1}^n f\left(a + i\,\frac{b - a}{n}\right)\frac{b - a}{n}$$

Since $(b - a)/n$ does not depend on i, Theorem 5.2 allows us to take it in front of the sigma sign, and we have the following formula for calculating integrals.

Theorem (5.12)

> If f is integrable on $[a, b]$, then
>
> $$\int_a^b f(x)\,dx = \lim_{n \to \infty} \frac{b - a}{n} \sum_{i=1}^n f\left(a + i\,\frac{b - a}{n}\right)$$

● **Example 3**   Evaluate $\int_1^4 (2x^3 - 5x)\,dx$.

Solution   Here we have $f(x) = 2x^3 - 5x$, $a = 1$, and $b = 4$. Since f is continuous, we know it is integrable and so Theorem 5.12 gives

$$\int_1^4 (2x^3 - 5x)\,dx = \lim_{n \to \infty} \frac{3}{n} \sum_{i=1}^n f\left(1 + \frac{3i}{n}\right)$$

$$= \lim_{n \to \infty} \frac{3}{n} \sum_{i=1}^n \left[2\left(1 + \frac{3i}{n}\right)^3 - 5\left(1 + \frac{3i}{n}\right)\right]$$

$$= \lim_{n \to \infty} \frac{3}{n} \sum_{i=1}^n \left[-3 + 3\frac{i}{n} + 54\frac{i^2}{n^2} + 54\frac{i^3}{n^3}\right]$$

$$= \lim_{n \to \infty} \left[\frac{3}{n} \sum_{i=1}^n (-3) + \frac{9}{n^2} \sum_{i=1}^n i + \frac{162}{n^3} \sum_{i=1}^n i^2 + \frac{162}{n^4} \sum_{i=1}^n i^3\right]$$

$$= \lim_{n \to \infty} \left\{\frac{3}{n}(-3)n + \frac{9}{n^2}\frac{n(n + 1)}{2} + \frac{162}{n^3}\frac{n(n + 1)(2n + 1)}{6}\right.$$

$$\left. + \frac{162}{n^4}\left[\frac{n(n + 1)}{2}\right]^2\right\}$$

$$= \lim_{n \to \infty} \left\{-9 + \frac{9}{2} \cdot 1 \cdot \left(1 + \frac{1}{n}\right) + 27 \cdot 1 \cdot \left(1 + \frac{1}{n}\right)\left(2 + \frac{1}{n}\right)\right.$$

$$\left. + \frac{81}{2}\left[1\left(1 + \frac{1}{n}\right)^2\right]\right\}$$

$$= -9 + \frac{9}{2} + 27 \cdot 2 + \frac{81}{2} = 90$$

Note that this integral cannot be interpreted as an area since f takes on both positive and negative values.   ●

EXERCISES 5.3

In Exercises 1–6 you are given a function f, an interval, partition points that define a partition P, and points x_i^* in the ith subinterval. (a) Find $\|P\|$. (b) Find the Riemann sum (5.10).

1. $f(x) = 7 - 2x$, $[1, 5]$, $\{1, 1.6, 2.2, 3.0, 4.2, 5\}$, $x_i^* = $ midpoint

2. $f(x) = 3x - 1$, $[-2, 2]$, $\{-2, -1.2, -0.6, 0, 0.8, 1.6, 2\}$, $x_i^* = $ midpoint

3. $f(x) = 2 - x^2$, $[-2, 2]$, $\{-2, -1.4, -1, 0, 0.8, 1.4, 2\}$, $x_i^* = $ right endpoint

4. $f(x) = x + x^2$, $[-2, 0]$, $\{-2, -1.5, -1, -0.7, -0.4, 0\}$, $x_i^* = $ left endpoint

5. $f(x) = x^3$, $[-1, 1]$, $\{-1, -0.5, 0, 0.5, 1\}$, $x_1^* = -1$, $x_2^* = -0.4$, $x_3^* = 0.2$, $x_4^* = 1$

6. $f(x) = \sin x$, $\left[-\dfrac{\pi}{2}, \pi \right]$, $\left\{ -\dfrac{\pi}{2}, -1, 0, 1, 2, \pi \right\}$,

    $x_1^* = -1.5$, $x_2^* = -0.5$, $x_3^* = 0.5$, $x_4^* = 1.5$, $x_5^* = 3$

In Exercises 7–18 use Theorem 5.12 to evaluate the integral.

7. $\displaystyle\int_a^b c \, dx$

8. $\displaystyle\int_{-2}^{7}(6 - 2x) \, dx$

9. $\displaystyle\int_1^4 (x^2 - 2) \, dx$

10. $\displaystyle\int_0^1 (ax + b) \, dx$

11. $\displaystyle\int_{-3}^{0}(2x^2 - 3x - 4) \, dx$

12. $\displaystyle\int_1^5 (2 + 3x - x^2) \, dx$

13. $\displaystyle\int_{-1}^{1}(t^3 - t^2 + 1) \, dt$

14. $\displaystyle\int_a^b (Px^2 + Qx + R) \, dx$

15. $\displaystyle\int_0^b (x^3 + 4x) \, dx$

16. $\displaystyle\int_2^5 (t^3 - 2t + 3) \, dt$

17. $\displaystyle\int_0^2 (x^4 - x + 1) \, dx$

18. $\displaystyle\int_0^1 (x^3 - 5x^4) \, dx$

19. Prove that $\displaystyle\int_a^b x \, dx = \dfrac{b^2 - a^2}{2}$

20. Prove that $\displaystyle\int_a^b x^2 \, dx = \dfrac{b^3 - a^3}{3}$

In Exercises 21–24 express the given limit as a definite integral on the given interval.

21. $\displaystyle\lim_{\|P\| \to 0} \sum_{i=1}^{n} [2(x_i^*)^2 - 5x_i^*] \Delta x_i$, $[0, 1]$

22. $\displaystyle\lim_{\|P\| \to 0} \sum_{i=1}^{n} \sqrt{x_i^*} \, \Delta x_i$, $[1, 4]$

23. $\displaystyle\lim_{\|P\| \to 0} \sum_{i=1}^{n} \cos x_i \, \Delta x_i$, $[0, \pi]$

24. $\displaystyle\lim_{\|P\| \to 0} \sum_{i=1}^{n} \dfrac{\tan x_i}{x_i} \Delta x_i$, $[2, 4]$

In Exercises 25–27 express the given limit as a definite integral.

25. $\displaystyle\lim_{n \to \infty} \sum_{i=1}^{n} \dfrac{i^4}{n^5}$ [*Hint:* Consider $f(x) = x^4$.]

26. $\displaystyle\lim_{n \to \infty} \dfrac{1}{n} \sum_{i=1}^{n} \dfrac{1}{1 + \left(\dfrac{i}{n}\right)^2}$

27. $\displaystyle\lim_{n \to \infty} \sum_{i=1}^{n} \left[3\left(1 + \dfrac{2i}{n}\right)^5 - 6 \right] \dfrac{2}{n}$

28. Evaluate $\displaystyle\int_1^1 x^2 \cos x \, dx$.

29. Given that $\displaystyle\int_4^9 \sqrt{x} \, dx = \dfrac{38}{3}$, what is $\displaystyle\int_9^4 \sqrt{t} \, dt$?

30. Let

$$f(x) = \begin{cases} \dfrac{1}{x} & \text{if } 0 < x \leqslant 1 \\ 0 & \text{if } x = 0 \end{cases}$$

(a) Show that f is not continuous on $[0, 1]$.
(b) Show that f is unbounded on $[0, 1]$.
(c) Show that $\int_0^1 f(x) \, dx$ does not exist—that is, f is not integrable on $[0, 1]$. [*Hint:* Show that the first term in the Riemann sum, $f(x_1^*) \Delta x_1$, can be made arbitrarily large.]

31. Let

$$f(x) = \begin{cases} 0 & \text{if } x \text{ is rational} \\ 1 & \text{if } x \text{ is irrational} \end{cases}$$

Show that f is bounded but not integrable on $[a, b]$. (*Hint:* Show that, no matter how small $\|P\|$ is, some Riemann sums are 0 whereas others are equal to $b - a$.)

32. Evaluate $\int_1^2 x^3 \, dx$ using a partition of $[1, 2]$ by points of a geometric progression: $x_0 = 1$, $x_1 = 2^{1/n}$, $x_2 = 2^{2/n}, \ldots, x_i = 2^{i/n}, \ldots, x_n = 2^{n/n} = 2$. Take $x_i^* = x_i$ and use the formula in Exercise 49 in Section 5.1 for the sum of a geometric series.

33. Find $\int_1^2 x^{-2} \, dx$. *Hint:* Use a regular partition but choose x_i^* to be the geometric mean of x_{i-1} and x_i ($x_i^* = \sqrt{x_{i-1} x_i}$) and use the identity

$$\dfrac{1}{m(m + 1)} = \dfrac{1}{m} - \dfrac{1}{m + 1}$$

Properties of the Definite Integral

In this section we develop some basic properties of integrals that will later help us to evaluate integrals in a simple manner.

Properties of the
Integral (5.13)

> Suppose that all of the following integrals exist. Then
>
> 1. $\int_a^b c\,dx = c(b - a)$, where c is any constant
>
> 2. $\int_a^b (f(x) + g(x))\,dx = \int_a^b f(x)\,dx + \int_a^b g(x)\,dx$
>
> 3. $\int_a^b cf(x)\,dx = c\int_a^b f(x)\,dx$, where c is any constant
>
> 4. $\int_a^b (f(x) - g(x))\,dx = \int_a^b f(x)\,dx - \int_a^b g(x)\,dx$
>
> 5. $\int_a^b f(x)\,dx = \int_a^c f(x)\,dx + \int_c^b f(x)\,dx$

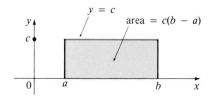

Figure 5.13
$\int_a^b c\,dx = c(b - a)$

Property 1 was given as Exercise 7 in Section 5.3. It says that the integral of a constant function $f(x) = c$ is the constant times the length of the interval. If $c > 0$ and $a < b$, this is to be expected because $c(b - a)$ is the area of the shaded rectangle in Figure 5.13.

Proof of Property 2   Since $\int_a^b (f + g)$ exists, we can compute it using a regular partition and choosing x_i^* to be the right endpoint of the ith subinterval, that is, $x_i^* = x_i$. Using the fact that the limit of a sum is the sum of the limits, we have

$$\int_a^b (f(x) + g(x))\,dx = \lim_{n \to \infty} \sum_{i=1}^n (f(x_i) + g(x_i))\,\Delta x$$

$$= \lim_{n \to \infty} \left[\sum_{i=1}^n f(x_i)\,\Delta x + \sum_{i=1}^n g(x_i)\,\Delta x \right] \quad \text{(by Theorem 5.2)}$$

$$= \lim_{n \to \infty} \sum_{i=1}^n f(x_i)\,\Delta x + \lim_{n \to \infty} \sum_{i=1}^n g(x_i)\,\Delta x$$

$$= \int_a^b f(x)\,dx + \int_a^b g(x)\,dx \qquad \bullet$$

Property 2 says that the integral of a sum is the sum of the integrals. Property 3 can be proved in a similar manner (see Exercise 54) and says that the integral of a constant times a function is the constant times the integral of the function. In other words, a constant (but *only* a constant) can be taken in front of an integral sign. Property 4 is proved by writing $f - g = f + (-g)$ and using Properties 2 and 3 with $c = -1$.

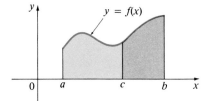

Figure 5.14

Property 5 is somewhat more complicated and is proved in Appendix C, but for the case where $f(x) \geqslant 0$ and $a < c < b$, it can be seen from the geometric interpretation in Figure 5.14. For positive functions f, $\int_a^b f$ is the total area under $y = f(x)$ from a to b, which is the sum of $\int_a^c f$ (the area from a to c) and $\int_c^b f$ (the area from c to b).

● **Example 1**   Use Theorem 5.13 together with the results of Sections 5.2 and 5.3 to evaluate the following integrals:

(a) $\displaystyle\int_0^{\pi/2}(x + 3\cos x)\,dx$ (b) $\displaystyle\int_a^b \cos x\,dx$ (c) $\displaystyle\int_{-4}^5 |x|\,dx$

Solution   (a) We know that $\int_0^{\pi/2} x\,dx = \pi^2/8$ (see Exercise 19 in Section 5.3) and $\int_0^{\pi/2} \cos x\,dx = 1$ (see Example 4 in Section 5.2). Therefore, using Properties 2 and 3 of integrals, we get

$$\int_0^{\pi/2}(x + 3\cos x)\,dx = \int_0^{\pi/2} x\,dx + 3\int_0^{\pi/2}\cos x\,dx = \frac{\pi^2}{8} + 3$$

(b)   The result of Example 4 in Section 5.2 can be extended to show that $\int_0^b \cos x\,dx = \sin b$. By Property 5 of integrals we have

$$\int_0^b \cos x\,dx = \int_0^a \cos x\,dx + \int_a^b \cos x\,dx$$

and so $\displaystyle\int_a^b \cos x\,dx = \int_0^b \cos x\,dx - \int_0^a \cos x\,dx = \sin b - \sin a$

(c)   Since $|x| = \begin{cases} x & \text{if } x \geqslant 0 \\ -x & \text{if } x < 0 \end{cases}$

we use Property 5 to split the integral at 0:

$$\int_{-4}^5 |x|\,dx = \int_{-4}^0 |x|\,dx + \int_0^5 |x|\,dx$$

$$= \int_{-4}^0 (-x)\,dx + \int_0^5 x\,dx = -\int_{-4}^0 x\,dx + \int_0^5 x\,dx$$

We know from Exercise 19 in Section 5.3 that

$$\int_a^b x\,dx = \frac{b^2 - a^2}{2}$$

so $\displaystyle\int_{-4}^5 |x|\,dx = -\int_{-4}^0 x\,dx + \int_0^5 x\,dx$

$$= -\tfrac{1}{2}[0^2 - (-4)^2] + \tfrac{1}{2}[5^2 - 0^2]$$

$$= 20.5 \qquad\qquad ●$$

Notice that Properties 1–5 are true whether $a < b$, $a = b$, or $a > b$. The properties in the next theorem, however, are true only if $a \leqslant b$.

**Order Properties of the
Integral (5.14)**

Suppose the following integrals exist and $a \leqslant b$.

6. If $f(x) \geqslant 0$ for $a \leqslant x \leqslant b$, then $\int_a^b f(x)\,dx \geqslant 0$.

7. If $f(x) \geqslant g(x)$ for $a \leqslant x \leqslant b$, then $\int_a^b f(x)\,dx \geqslant \int_a^b g(x)\,dx$.

8. If $m \leqslant f(x) \leqslant M$ for $a \leqslant x \leqslant b$, then
$$m(b - a) \leqslant \int_a^b f(x)\,dx \leqslant M(b - a)$$

9. $\left| \int_a^b f(x)\,dx \right| \leqslant \int_a^b |f(x)|\,dx$

If $f(x) \geqslant 0$, then $\int_a^b f$ represents the area under the graph of f, so the geometric interpretation of Property 6 is simply that areas are positive. Property 7 says that a bigger function has a bigger integral.

Proof of Property 6   As in the proof of Property 2, we write
$$\int_a^b f(x)\,dx = \lim_{n \to \infty} \sum_{i=1}^n f(x_i)\,\Delta x$$

Now $f(x_i) \geqslant 0$ and $\Delta x \geqslant 0$ so $f(x_i)\,\Delta x \geqslant 0$ and therefore
$$\sum_{i=1}^n f(x_i)\,\Delta x \geqslant 0$$

But the limit of nonnegative quantities is nonnegative (from Theorem 2.16) and so $\int_a^b f(x)\,dx \geqslant 0$.   ●

Proof of Property 7   If $f(x) \geqslant g(x)$, then $f(x) - g(x) \geqslant 0$ and so $\int_a^b (f - g) \geqslant 0$ by Property 6. Then Property 4 gives
$$\int_a^b f - \int_a^b g = \int_a^b (f - g) \geqslant 0$$

and so $\int_a^b f \geqslant \int_a^b g$.   ●

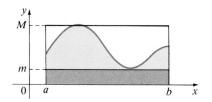

Figure 5.15

Property 8 is illustrated by Figure 5.15 for the case where $f(x) \geqslant 0$. If f is continuous we could take m and M to be the absolute minimum and maximum values of f on the interval $[a, b]$. In this case Property 8 says that the area under the graph of f is greater than the area of the rectangle with height m and less than the area of the rectangle with height M.

Proof of Property 8   Since $m \leqslant f(x) \leqslant M$, Property 7 gives
$$\int_a^b m\,dx \leqslant \int_a^b f(x)\,dx \leqslant \int_a^b M\,dx$$

Using Property 1 to evaluate the integrals on the left and right sides, we obtain

$$m(b - a) \leqslant \int_a^b f(x)\,dx \leqslant M(b - a)$$

The proof of Property 9 is left as Exercise 55.

● **Example 2**   Show that $\int_1^3 (2x^4 + 1)\,dx \geqslant \int_1^3 (x^4 + 2)\,dx$.

Solution   In order to use Property 7 we must show that

(5.15) $x^4 + 2 \leqslant 2x^4 + 1$ for $1 \leqslant x \leqslant 3$

To do this, observe that

$$(2x^4 + 1) - (x^4 + 2) = x^4 - 1 \geqslant 0 \qquad \text{when } x \geqslant 1$$

This shows that (5.15) is true, so Property 7 gives

$$\int_1^3 (x^4 + 2)\,dx \leqslant \int_1^3 (2x^4 + 1)\,dx$$

● **Example 3**   Use Property 8 to estimate the value of $\int_1^4 \sqrt{x}\,dx$.

Solution   Since $f(x) = \sqrt{x}$ is an increasing function, its absolute minimum on $[1, 4]$ is $m = f(1) = 1$ and its absolute maximum on $[1, 4]$ is $M = f(4) = \sqrt{4} = 2$. Thus Property 8 gives

$$1(4 - 1) \leqslant \int_1^4 \sqrt{x}\,dx \leqslant 2(4 - 1)$$

or $3 \leqslant \int_1^4 \sqrt{x}\,dx \leqslant 6$

● **Example 4**   Show that $\int_1^4 \sqrt{1 + x^2}\,dx \geqslant 7.5$.

Solution   The minimum value of $f(x) = \sqrt{1 + x^2}$ on $[0, 4]$ is $m = f(0) = 1$ since f is increasing. Thus Property 8 gives

$$\int_1^4 \sqrt{1 + x^2}\,dx \geqslant 1(4 - 1) = 3$$

This is not good enough, so instead we use Property 7. Notice that

$$1 + x^2 > x^2 \quad \Rightarrow \quad \sqrt{1 + x^2} > \sqrt{x^2} = |x|$$

Since $|x| = x$ for $x > 0$, we have $\sqrt{1 + x^2} > x$ for $1 \leqslant x \leqslant 4$. Thus, by Property 7,

$$\int_1^4 \sqrt{1 + x^2}\,dx \geqslant \int_1^4 x\,dx = \tfrac{1}{2}(4^2 - 1^2) = 7.5$$

[Here we have used the fact that $\int_a^b x\,dx = (b^2 - a^2)/2$ from Exercise 19 in Section 5.3.]

EXERCISES 5.4

In Exercises 1–22 use the properties of integrals to evaluate the given integrals. You may assume from Section 5.3 that

$$\int_a^b x \, dx = \tfrac{1}{2}(b^2 - a^2) \qquad \int_a^b x^2 \, dx = \tfrac{1}{3}(b^3 - a^3)$$

$$\int_0^b \cos x \, dx = \sin b$$

1. $\int_2^6 3 \, dx$

2. $\int_{-1}^4 \pi \, dx$

3. $\int_{-4}^{-1} \sqrt{3} \, dx$

4. $\int_{-1}^{-\sqrt{2}} (\sqrt{2} - 1) \, dx$

5. $\int_0^2 (5x + 3) \, dx$

6. $\int_3^6 (4 - 7x) \, dx$

7. $\int_1^4 (2x^2 - 3x + 1) \, dx$

8. $\int_{-2}^0 (3x^2 + 2x - 4) \, dx$

9. $\int_{-1}^1 (x - 1)^2 \, dx$

10. $\int_1^3 (x - 2)(x + 3) \, dx$

11. $\int_0^{\pi/3} (1 - 2\cos x) \, dx$

12. $\int_0^1 (5\cos x + 4x) \, dx$

13. $\int_{-2}^2 |x + 1| \, dx$

14. $\int_0^2 |2x - 3| \, dx$

15. $\int_{-2}^5 [\![x]\!] \, dx$

16. $\int_0^2 [\![2x]\!] \, dx$

17. $\int_{-1}^1 f(x) \, dx$ where $f(x) = \begin{cases} -2x & \text{if } -1 \leqslant x < 0 \\ 3x^2 & \text{if } 0 \leqslant x \leqslant 1 \end{cases}$

18. $\int_0^3 g(x) \, dx$ where $g(x) = \begin{cases} 1 - x & \text{if } 0 \leqslant x \leqslant 1 \\ x - 1 & \text{if } 1 < x \leqslant 3 \end{cases}$

19. $\int_{-2}^1 F(x) \, dx$ where $F(x) = \begin{cases} 1 + x^2 & \text{if } -2 \leqslant x < 0 \\ 1 + 2x & \text{if } 0 \leqslant x \leqslant 1 \end{cases}$

20. $\int_{-\pi/2}^{\pi/2} G(x) \, dx$ where $G(x) = \begin{cases} 1 + x & \text{if } -\frac{\pi}{2} \leqslant x \leqslant 0 \\ \cos x & \text{if } 0 < x \leqslant \frac{\pi}{4} \\ 1 & \text{if } \frac{\pi}{4} < x \leqslant \frac{\pi}{2} \end{cases}$

21. $\int_0^4 x^2 \, dx + \int_4^{10} x^2 \, dx$

22. $\int_3^4 f(x) \, dx + \int_1^3 f(x) \, dx + \int_4^1 f(x) \, dx$

In Exercises 23–26 write the given sum or difference as a single integral in the form $\int_a^b f(x) \, dx$.

23. $\int_1^3 f(x) \, dx + \int_3^6 f(x) \, dx + \int_6^{12} f(x) \, dx$

24. $\int_5^8 f(x) \, dx + \int_0^5 f(x) \, dx$

25. $\int_2^{10} f(x) \, dx - \int_2^7 f(x) \, dx$

26. $\int_{-3}^5 f(x) \, dx - \int_{-3}^0 f(x) \, dx + \int_5^6 f(x) \, dx$

In Exercises 27–38 use the properties of integrals to verify the given inequality without evaluating the integrals.

27. $\int_0^1 x \, dx \geqslant \int_0^1 x^2 \, dx$

28. $\int_1^2 x \, dx \leqslant \int_1^2 x^2 \, dx$

29. $\int_2^6 (x^2 - 1) \, dx \geqslant 0$

30. $\int_{-2}^8 (x^2 - 3x + 4) \, dx \geqslant 0$

31. $\int_0^{\pi/4} \sin^3 x \, dx \leqslant \int_0^{\pi/4} \sin^2 x \, dx$

32. $\int_0^5 (4x^4 - 3) \, dx \geqslant \int_0^5 (3x^4 - 4) \, dx$

33. $\int_1^2 \sqrt{5 - x} \, dx \geqslant \int_1^2 \sqrt{x + 1} \, dx$

34. $\int_4^6 \frac{1}{x} \, dx \leqslant \int_4^6 \frac{1}{8 - x} \, dx$

35. $8 \leqslant \int_2^4 x^2 \, dx \leqslant 32$

36. $\frac{\pi}{6} \leqslant \int_{\pi/6}^{\pi/2} \sin x \, dx \leqslant \frac{\pi}{3}$

37. $2 \leqslant \int_{-1}^1 \sqrt{1 + x^2} \, dx \leqslant 2\sqrt{2}$

38. $3 \leqslant \int_1^4 (x^2 - 4x + 5) \, dx \leqslant 15$

In Exercises 39–46 use Property 8 to estimate the value of the integral.

39. $\int_1^3 x^3 \, dx$

40. $\int_0^2 \sqrt{x^3 + 1} \, dx$

41. $\int_1^2 \frac{1}{x} \, dx$

42. $\int_{\pi/4}^{\pi/3} \cos x \, dx$

43. $\int_{-3}^0 (x^2 + 2x) \, dx$

44. $\int_{-1}^3 (x^2 - 3x) \, dx$

45. $\int_{-1}^1 \sqrt{1 + x^4} \, dx$

46. $\int_{\pi/4}^{3\pi/4} \sin^2 x \, dx$

In Exercises 47–50 use Theorem 5.14, together with the list of integrals in the instructions for Exercises 1–22, to prove the given inequality.

47. $\int_1^3 \sqrt{x^4 + 1} \, dx \geqslant \frac{26}{3}$

48. $\int_2^5 \sqrt{x^2 - 1} \, dx \leqslant 10.5$

49. $\int_0^{\pi/2} x \sin x \, dx \leqslant \frac{\pi^2}{8}$

50. $\left| \int_0^\pi x^2 \cos x \, dx \right| \leqslant \frac{\pi^3}{3}$

51. Use Properties 2 and 3 of integrals to show that if c and d are constants, then

$$\int_a^b (cf(x) + dg(x)) \, dx = c \int_a^b f(x) \, dx + d \int_a^b g(x) \, dx$$

52. Extend Property 2 to the sum of any number of functions; that is, given integrable functions $f_1, \ldots, f_n$, prove that

$$\int_a^b \left[\sum_{i=1}^n f_i(x) \right] dx = \sum_{i=1}^n \int_a^b f_i(x) \, dx$$

53. If $f(x) \leqslant 0$ and f is integrable on $[a, b]$, show that $\int_a^b f(x) \, dx \leqslant 0$.

54. Prove Property 3 of Theorem 5.13.

55. Prove Property 9 of Theorem 5.14.
[*Hint:* $-|f(x)| \leqslant f(x) \leqslant |f(x)|$.]

56. Suppose that f is continuous on $[a, b]$ and $f(x) > 0$ for all x in $[a, b]$. Prove that $\int_a^b f(x) \, dx > 0$. (*Hint:* Use the Extreme Value Theorem and Property 8.)

SECTION 5.5

The Fundamental Theorem of Calculus

The Fundamental Theorem of Calculus is appropriately named because it establishes a connection between the two branches of calculus: differential calculus and integral calculus. Differential calculus arose from the tangent problem, whereas integral calculus arose from a seemingly unrelated problem, the area problem. Newton's teacher at Cambridge, Isaac Barrow (1630–1677), discovered that these two problems are actually closely related. In fact, he realized that differentiation and integration are inverse processes. The Fundamental Theorem of Calculus gives the precise inverse relationship between the derivative and the integral. It was Newton and Leibniz who exploited this relationship and used it to develop calculus into a systematic mathematical method. In particular, they saw that the Fundamental Theorem enabled them to compute areas and integrals very easily without having to compute them as limits of sums as we did in Sections 5.2 and 5.3.

In order to motivate the Fundamental Theorem, let f be a continuous function on $[a, b]$ and define a new function g by

(5.16)
$$g(x) = \int_a^x f(t) \, dt$$

Observe that g depends only on x, which appears as the variable upper limit in the integral. To make this clear we could use the other notation for an integral and write $g(x) = \int_a^x f$. If x is a fixed number, then the integral $\int_a^x f$ is a definite number. If we then let x vary, the number $\int_a^x f$ also varies and defines a function of x denoted by $g(x)$. For instance, if we take $f(t) = t$ and $a = 0$, then, using Exercise 19 in Section 5.3, we have

$$g(x) = \int_0^x t \, dt = \frac{x^2}{2}$$

Notice that $g'(x) = x$, that is, $g' = f$. In other words, if g is defined as the integral of f by Equation 5.16, then g turns out to be an antiderivative of f, at least in this case.

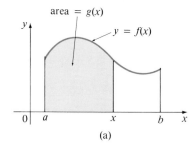

area = g(x)

y = f(x)

(a)

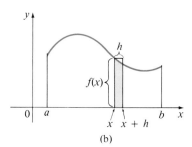

(b)

Figure 5.16

To see why this might be generally true we consider any continuous function f with $f(x) \geq 0$. Then $g(x) = \int_a^x f$ can be interpreted as the area under the graph of f from a to x, where x can vary from a to b. [Think of g as the "area so far" function; see Figure 5.16(a).]

In order to compute $g'(x)$ from the definition of derivative we first observe that, for $h > 0$, $g(x + h) - g(x)$ is obtained by subtracting areas, so it is the area under the graph of f from x to $x + h$ [the shaded area in Figure 5.16(b)]. For small h you can see from the figure that this area is approximately equal to the area of the rectangle with height $f(x)$ and width h:

$$g(x + h) - g(x) \approx hf(x) \qquad \frac{g(x + h) - g(x)}{h} \approx f(x)$$

Intuitively, we therefore expect that

$$g'(x) = \lim_{h \to 0} \frac{g(x + h) - g(x)}{h} = f(x)$$

The fact that this is true, even when f is not necessarily positive, is the first part of the Fundamental Theorem of Calculus.

The Fundamental Theorem of Calculus, Part 1 (5.17)

If f is continuous on $[a, b]$, then the function g defined by

$$g(x) = \int_a^x f(t)\, dt \qquad a \leq x \leq b$$

is continuous on $[a, b]$ and differentiable on (a, b), and $g'(x) = f(x)$.

Proof   If x and $x + h$ are in (a, b), then

$$g(x + h) - g(x) = \int_a^{x+h} f - \int_a^x f$$

$$= \left(\int_a^x f + \int_x^{x+h} f \right) - \int_a^x f \qquad (by\ Property\ 5)$$

$$= \int_x^{x+h} f$$

and so, for $h \neq 0$,

(5.18)
$$\frac{g(x + h) - g(x)}{h} = \frac{1}{h} \int_x^{x+h} f$$

For now let us assume that $h > 0$. Since f is continuous on $[x, x + h]$, the Extreme Value Theorem says that there are numbers u and v in $[x, x + h]$ such that $f(u) = m$ and $f(v) = M$, where m and M are the absolute minimum and maximum values of f on $[x, x + h]$ (see Figure 5.17).

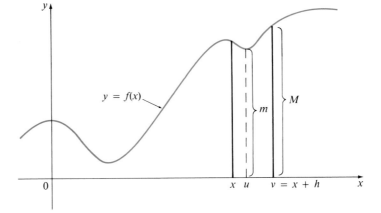

Figure 5.17

By Property 8 of integrals we have

$$mh \leqslant \int_x^{x+h} f \leqslant Mh$$

that is,

$$f(u)h \leqslant \int_x^{x+h} f \leqslant f(v)h$$

Since $h > 0$, we can divide this inequality by h:

$$f(u) \leqslant \frac{1}{h} \int_x^{x+h} f \leqslant f(v)$$

Now we use Equation 5.18 to replace the middle part of this inequality:

(5.19)
$$f(u) \leqslant \frac{g(x+h) - g(x)}{h} \leqslant f(v)$$

Inequality 5.19 can be proved in a similar manner for the case where $h < 0$ (see Exercise 69).

Now we let $h \to 0$. Then $u \to x$ and $v \to x$ since u and v lie between x and $x + h$. Therefore

$$\lim_{h \to 0} f(u) = \lim_{u \to x} f(u) = f(x) \qquad \lim_{h \to 0} f(v) = \lim_{v \to x} f(v) = f(x)$$

because f is continuous at x. We conclude, from (5.19) and the Squeeze Theorem, that

(5.20)
$$g'(x) = \lim_{h \to 0} \frac{g(x+h) - g(x)}{h} = f(x)$$

If $x = a$ or b, Equation 5.20 can be interpreted as a one-sided limit. Then Theorem 3.8 (modified for one-sided limits) shows that f is continuous on $[a, b]$. ●

Using the Leibniz notation for derivatives, we can write Theorem 5.17 as

(5.21)
$$\frac{d}{dx} \int_a^x f(t)\, dt = f(x)$$

when f is continuous. Roughly speaking, Equation 5.21 says that if we first integrate f and then differentiate the result, we get back to the original function f.

● **Example 1**   Find the derivative of the function $g(x) = \int_0^x \sqrt{1 + t^2}\, dt$.

Solution   Since $f(t) = \sqrt{1 + t^2}$ is continuous, Part 1 of the Fundamental Theorem of Calculus gives

$$g'(x) = \sqrt{1 + x^2} \qquad\qquad ●$$

● **Example 2**   Differentiate $y = \int_1^x \sin \sqrt{u}\, du$.

Solution   Again, Part 1 of the Fundamental Theorem gives

$$\frac{dy}{dx} = \sin \sqrt{x} \qquad\qquad ●$$

● **Example 3**   Find $\dfrac{d}{dx} \displaystyle\int_1^{x^4} \sec t\, dt$.

Solution   Here we have to be careful to use the Chain Rule in conjunction with Part 1 of the Fundamental Theorem. Let $u = x^4$. Then

$$\frac{d}{dx} \int_1^{x^4} \sec t\, dt = \frac{d}{dx} \int_1^u \sec t\, dt$$

$$= \frac{d}{du} \left[\int_1^u \sec t\, dt \right] \frac{du}{dx} \qquad \textit{(by the Chain Rule)}$$

$$= \sec u \, \frac{du}{dx} \qquad \textit{(by Theorem 5.17)}$$

$$= \sec(x^4) \cdot 4x^3 \qquad\qquad ●$$

● **Example 4**   Find the derivative of $h(x) = \int_{x^2}^{\pi} (r^3 + 1)^{10}\, dr$.

Solution   Here we reverse the order of integration and use the Chain Rule and the Fundamental Theorem:

$$h'(x) = \frac{d}{dx} \int_{x^2}^{\pi} (r^3 + 1)^{10}\, dr = \frac{d}{dx} \left[-\int_{\pi}^{x^2} (r^3 + 1)^{10}\, dr \right]$$

$$= -\frac{d}{dx} \int_{\pi}^{x^2} (r^3 + 1)^{10}\, dr$$

$$= -[(x^2)^3 + 1]^{10}\, \frac{d}{dx}(x^2) = -2x(x^6 + 1)^{10} \qquad ●$$

● **Example 5**   If $G(x) = \int_x^{\sec x} \dfrac{\sin t}{t^3}\, dt$, $0 < x < \dfrac{\pi}{2}$, find $G'(x)$.

Solution   First we split the integral into two parts using Property 5. This enables us to apply the methods of Examples 3 and 4 to each integral:

$$G'(x) = \frac{d}{dx} \int_x^{\sec x} \frac{\sin t}{t^3}\, dt = \frac{d}{dx}\left(\int_x^1 \frac{\sin t}{t^3}\, dt + \int_1^{\sec x} \frac{\sin t}{t^3}\, dt \right)$$

$$= -\frac{d}{dx} \int_1^x \frac{\sin t}{t^3}\, dt + \frac{d}{dx} \int_1^{\sec x} \frac{\sin t}{t^3}\, dt$$

$$= -\frac{\sin x}{x^3} + \frac{\sin(\sec x)}{\sec^3 x}\, \sec x \tan x \qquad \textit{(by the Chain Rule)}$$

$$= -\frac{\sin x}{x^3} + \sin x \cos x \sin(\sec x)$$

Although we chose to split the integral at the number 1, any other number in $(0, \pi/2)$ would also have worked.   ●

In Section 5.3 we computed integrals from the definition as a limit of Riemann sums and we saw that this procedure is sometimes long and difficult. The second part of the Fundamental Theorem of Calculus, which follows easily from the first part, provides us with a much simpler method for the evaluation of integrals.

The Fundamental Theorem of Calculus, Part 2 (5.22)

If f is continuous on $[a, b]$, then

$$\int_a^b f(x)\, dx = F(b) - F(a)$$

where F is any antiderivative of f, that is, $F' = f$.

Proof   Let $g(x) = \int_a^x f$. We know from Part 1 that $g'(x) = f(x)$; that is, g is an antiderivative of f. If F is any other antiderivative of f on $[a, b]$, then we know from Corollary 4.17 that F and g differ by a constant:

(5.23)
$$F(x) = g(x) + C$$

for $a < x < b$. But both F and g are continuous on $[a, b]$ and so, by taking limits of both sides of Equation 5.23 (as $x \to a^+$ and $x \to b^-$), we see that it also holds when $x = a$ and $x = b$.

If we put $x = a$ in the formula for $g(x)$, we get

$$g(a) = \int_a^a f = 0$$

So, using Equation 5.23 with $x = b$ and $x = a$, we have

$$F(b) - F(a) = [g(b) + C] - [g(a) + C]$$

$$= g(b) - g(a) = g(b) = \int_a^b f \qquad ●$$

Part 2 of the Fundamental Theorem states that if we know an anti-derivative F of f, then we can evaluate $\int_a^b f$ simply by subtracting the values of F at the endpoints of the interval $[a, b]$. It is very surprising that $\int_a^b f$, which was defined by a complicated procedure involving all of the values of $f(x)$ for $a \leqslant x \leqslant b$, can be found by knowing the values of $F(x)$ at only two points, a and b.

We use the notation

$$F(x)\Big]_a^b = F(b) - F(a)$$

and so the equation of Theorem 5.22 can be written as

$$\int_a^b f(x)\,dx = F(x)\Big]_a^b \qquad \text{where} \qquad F' = f$$

● **Example 6**   Find the area under the parabola $y = x^2 + 1$ from 0 to 2.

Solution   Since we know from Section 4.10 that an antiderivative of $f(x) = x^2 + 1$ is $F(x) = \frac{1}{3}x^3 + x$, the required area A is found using Part 2 of the Fundamental Theorem:

$$A = \int_0^2 (x^2 + 1)\,dx = \frac{x^3}{3} + x\Big]_0^2 = \left(\frac{2^3}{3} + 2\right) - \left(\frac{0^3}{3} + 0\right) = \frac{14}{3}$$

Notice that in applying the Fundamental Theorem we used a particular anti-derivative F of f. It is not necessary to use the most general antiderivative.

●

If you compare the calculation in Example 6 with the one in Example 3 in Section 5.2, you will see that the Fundamental Theorem gives us a much shorter method.

● **Example 7**   Find the area under the cosine curve from 0 to b, where $0 \leqslant b \leqslant \pi/2$.

Solution   Since an antiderivative of $f(x) = \cos x$ is $F(x) = \sin x$, we have

$$A = \int_0^b \cos x\,dx = \sin x\Big]_0^b = \sin b - \sin 0 = \sin b \qquad ●$$

A comparison of Example 7 with Example 4 in Section 5.2 will reveal the power of the Fundamental Theorem of Calculus. When the French mathematician Gilles de Roberval first found the area under the sine and cosine curves in 1635, this was a very challenging problem that required a great deal of ingenuity. (Recall that the calculation in Example 4 in Section 5.2 depended on some little-known trigonometric identities.) But in the 1660s and 1670s when the Fundamental Theorem was discovered by Barrow and exploited by Newton and Leibniz, such problems became very easy, as you can see from Example 7.

Notation: Because of the relation given by the Fundamental Theorem between antiderivatives and integrals, the notation $\int f(x)\,dx$ is traditionally used for an antiderivative of f and is called an **indefinite integral.** Thus

(5.24)

$$\int f(x)\,dx = F(x) \quad \text{means} \quad F'(x) = f(x)$$

⊘   You should distinguish carefully between definite and indefinite integrals. A definite integral $\int_a^b f(x)\,dx$ is a number, whereas an indefinite integral $\int f(x)\,dx$ is a function. The connection between them is given by Part 2 of the Fundamental Theorem. If f is continuous on $[a, b]$, then

(5.25)

$$\int_a^b f(x)\,dx = \int f(x)\,dx\Big]_a^b$$

The effectiveness of the Fundamental Theorem depends on having a supply of antiderivatives of functions. We therefore restate the Table of Antidifferentiation Formulas (4.47), together with a few others, in the notation of indefinite integrals. Any formula can be verified by differentiating the function on the right side and obtaining the integrand. For instance,

$$\int \sec^2 x\,dx = \tan x \quad \text{since} \quad \frac{d}{dx}(\tan x) = \sec^2 x$$

The list in Table 5.26 contains *particular* antiderivatives.

Table of Indefinite Integrals (5.26)

Constants of integration have been omitted from the following integrals.

$$\int c\,f(x)\,dx = c\int f(x)\,dx$$

$$\int (f(x) + g(x))\,dx = \int f(x)\,dx + \int g(x)\,dx$$

$$\int x^n\,dx = \frac{x^{n+1}}{n+1} \quad (n \neq -1)$$

$$\int \sin x\,dx = -\cos x \qquad\qquad \int \cos x\,dx = \sin x$$

$$\int \sec^2 x\,dx = \tan x \qquad\qquad \int \csc^2 x\,dx = -\cot x$$

$$\int \sec x \tan x\,dx = \sec x \qquad \int \csc x \cot x\,dx = -\csc x$$

Recall from Theorem 4.46 that the most general antiderivative *on a given interval* is obtained by adding a constant to a particular antiderivative. **We adopt the convention that when a formula for a general indefinite integral**

is given, it is valid only on an interval. Thus we write

$$\int \frac{1}{x^2}\, dx = -\frac{1}{x} + C$$

with the understanding that it is valid on the interval $(0, \infty)$ or on the interval $(-\infty, 0)$. This is true despite the fact that the general antiderivative of the function $f(x) = 1/x^2$, $x \neq 0$, is

$$F(x) = \begin{cases} -\dfrac{1}{x} + C_1 & \text{if } x < 0 \\[2ex] -\dfrac{1}{x} + C_2 & \text{if } x > 0 \end{cases}$$

● **Example 8**   Find the general indefinite integral

$$\int (10x^4 + 3\sec^2 x)\, dx$$

Solution   Using our convention and Table 5.26, we have

$$\int (10x^4 + 3\sec^2 x)\, dx = 10\frac{x^5}{5} + 3\tan x + C$$

$$= 2x^5 + 3\tan x + C \qquad ●$$

● **Example 9**   Evaluate $\int_1^4 (2x^3 - 5x)\, dx$.

Solution   Using the Fundamental Theorem, Part 2, and Table 5.26, we have

$$\int_1^4 (2x^3 - 5x)\, dx = 2\frac{x^4}{4} - 5\frac{x^2}{2}\Big]_1^4$$

$$= \left(\frac{1}{2}\cdot 4^4 - \frac{5}{2}\cdot 4^2\right) - \left(\frac{1}{2}\cdot 1^4 - \frac{5}{2}\cdot 1^2\right)$$

$$= 128 - 40 - \frac{1}{2} + \frac{5}{2} = 90$$

This calculation should be compared with Example 3 in Section 5.3.   ●

● **Example 10**

$$\int_1^9 \frac{2t^2 + t^2\sqrt{t} - 1}{t^2}\, dt = \int_1^9 (2 + t^{1/2} - t^{-2})\, dt$$

$$= 2t + \frac{t^{3/2}}{\frac{3}{2}} - \frac{t^{-1}}{-1}\Big]_1^9$$

$$= 2t + \tfrac{2}{3}t^{3/2} + (1/t)\big]_1^9$$

$$= [2\cdot 9 + \tfrac{2}{3}(9)^{3/2} + \tfrac{1}{9}] - (2\cdot 1 + \tfrac{2}{3}\cdot 1^{3/2} + \tfrac{1}{1})$$

$$= 18 + 18 + \tfrac{1}{9} - 2 - \tfrac{2}{3} - 1 = 32\tfrac{4}{9} \qquad ●$$

● **Example 11**   What is wrong with the following calculation?

$$\int_{-1}^{3} \frac{1}{x^2} \, dx = \left. \frac{x^{-1}}{-1} \right]_{-1}^{3} = -\frac{1}{3} - 1 = -\frac{4}{3}$$

Solution   To start, notice that this calculation must be wrong because $f(x) = 1/x^2 \geq 0$ and Property 6 of integrals says that $\int_a^b f \geq 0$ when $f \geq 0$. The Fundamental Theorem of Calculus applies to continuous functions. It cannot be applied here because $f(x) = 1/x^2$ is not continuous on $[-1, 3]$. In fact, $f(0)$ is not defined, but even if it were defined [say, by $f(0) = 0$], the integral would not exist because f is unbounded on $[-1, 3]$. In fact, we know that $\lim_{x \to 0}(1/x^2) = \infty$. Thus

$$\int_{-1}^{3} \frac{1}{x^2} \, dx \quad \text{does not exist} \qquad ●$$

We end this section by bringing the two parts of the Fundamental Theorem together.

The Fundamental Theorem of Calculus (5.27)

Suppose f is continuous on $[a, b]$.

(a)   If $g(x) = \int_a^x f(t) \, dt$, then $g'(x) = f(x)$.

(b) $\int_a^b f(x) \, dx = F(b) - F(a)$, where F is any antiderivative of f, that is, $F' = f$.

We have previously noted that Part 1 can be rewritten as

$$\frac{d}{dx} \int_a^x f(t) \, dt = f(x)$$

which says that if f is integrated and then the result is differentiated, we arrive back at the original function f. Since $F'(x) = f(x)$, Part 2 can be rewritten as

$$\int_a^b F'(x) \, dx = F(b) - F(a)$$

This version says that if we take a function F, first differentiate it, and then integrate the result, we arrive back at the original function F, but in the form $F(b) - F(a)$. Taken together, the two parts of the Fundamental Theorem of Calculus say that differentiation and integration are inverse processes. Each undoes what the other does.

The Fundamental Theorem of Calculus is unquestionably the most important theorem in calculus and, indeed, it ranks as one of the great accomplishments of the human mind. Before it was discovered, from the time of Eudoxus and Archimedes to the time of Galileo and Fermat, problems of finding areas, volumes, and lengths of curves were so difficult that only a genius could meet the challenge. But now, armed with the systematic method that Newton and Leibniz fashioned out of the Fundamental Theorem, we shall see in the chapters to come that these challenging problems are accessible to all of us.

EXERCISES 5.5

In Exercises 1–16 use Part 1 of the Fundamental Theorem of Calculus to find the derivative of the given function.

1. $g(x) = \int_1^x (t^2 - 1)^{20} dt$

2. $g(x) = \int_{-1}^x \sqrt{t^3 + 1} \, dt$

3. $g(u) = \int_\pi^u \frac{1}{1 + t^4} dt$

4. $g(t) = \int_0^t \sin(x^2) dx$

5. $F(x) = \int_x^2 \cos(t^2) dt$

6. $F(x) = \int_x^4 (2 + \sqrt{u})^8 du$

7. $h(x) = \int_2^{1/x} \sin^4 t \, dt$

8. $h(x) = \int_1^{\sqrt{x}} \frac{s^2}{s^2 + 1} ds$

9. $y = \int_{\tan x}^{17} \sin(t^4) dt$

10. $y = \int_{x^2}^\pi \frac{\sin t}{t} dt$

11. $y = \int_0^{5x+1} \frac{1}{u^2 - 5} du$

12. $y = \int_{-5}^{\sin x} t \cos(t^3) dt$

13. $g(x) = \int_{2x}^{3x} \frac{u - 1}{u + 1} du$

14. $g(x) = \int_{\tan x}^{x^2} \frac{1}{\sqrt{2 + t^4}} dt$

15. $y = \int_{\sqrt{x}}^{x^3} \sqrt{t} \sin t \, dt$

16. $y = \int_{\cos x}^{5x} \cos(u^2) du$

In Exercises 17–50 use Part 2 of the Fundamental Theorem of Calculus to evaluate the given integral if it exists.

17. $\int_{-1}^8 6 \, dx$

18. $\int_{-2}^4 (3x - 5) dx$

19. $\int_0^1 (1 - 2x - 3x^2) dx$

20. $\int_1^2 (5x^2 - 4x + 3) dx$

21. $\int_{-3}^0 (5y^4 - 6y^2 + 14) dy$

22. $\int_0^1 (y^9 - 2y^5 + 3y) dy$

23. $\int_1^3 \left(\frac{1}{t^2} - \frac{1}{t^4} \right) dt$

24. $\int_1^2 \frac{t^6 - t^2}{t^4} dt$

25. $\int_1^2 \frac{x^2 + 1}{\sqrt{x}} dx$

26. $\int_0^2 (x^3 - 1)^2 dx$

27. $\int_0^1 u(\sqrt{u} + \sqrt[3]{u}) du$

28. $\int_{-1}^1 \frac{3}{t^4} dt$

29. $\int_{-2}^3 |x^2 - 1| dx$

30. $\int_1^2 \left(x + \frac{1}{x} \right)^2 dx$

31. $\int_3^3 \sqrt{x^5 + 2} \, dx$

32. $\int_{-1}^2 |x - x^2| dx$

33. $\int_{-4}^2 \frac{2}{x^6} dx$

34. $\int_1^{-1} (x - 1)(3x + 2) dx$

35. $\int_1^4 \left(\sqrt{t} - \frac{2}{\sqrt{t}} \right) dt$

36. $\int_1^8 \left(\sqrt[3]{r} + \frac{1}{\sqrt[3]{r}} \right) dr$

37. $\int_{-1}^0 (x + 1)^3 dx$

38. $\int_{-5}^{-2} \frac{x^4 - 1}{x^2 + 1} dx$

39. $\int_{\pi/4}^{\pi/3} \sin t \, dt$

40. $\int_0^{\pi/2} (\cos \theta + 2 \sin \theta) d\theta$

41. $\int_{\pi/2}^\pi \sec x \tan x \, dx$

42. $\int_{\pi/3}^{\pi/2} \csc x \cot x \, dx$

43. $\int_{\pi/6}^{\pi/3} \csc^2 \theta \, d\theta$

44. $\int_{\pi/4}^\pi \sec^2 \theta \, d\theta$

45. $\int_0^1 (\sqrt[4]{x^5} + \sqrt[5]{x^4}) dx$

46. $\int_1^8 \frac{x - 1}{\sqrt[3]{x^2}} dx$

47. $\int_{-1}^2 (x - 2|x|) dx$

48. $\int_0^2 (x^2 - |x - 1|) dx$

49. $\int_0^2 f(x) dx$ where $f(x) = \begin{cases} x^4 & \text{if } 0 \leqslant x < 1 \\ x^5 & \text{if } 1 \leqslant x \leqslant 2 \end{cases}$

50. $\int_{-\pi}^\pi f(x) dx$ where $f(x) = \begin{cases} x & \text{if } -\pi \leqslant x \leqslant 0 \\ \sin x & \text{if } 0 < x \leqslant \pi \end{cases}$

In Exercises 51–58 find the area of the region that lies beneath the given curve.

51. $y = 4x^2 - 4x + 3, 0 \leqslant x \leqslant 2$

52. $y = 1 + x^3, 0 \leqslant x \leqslant 4$

53. $y = \sqrt[3]{x}, 0 \leqslant x \leqslant 27$ **54.** $y = x^{-4}, 1 \leqslant x \leqslant 6$

55. $y = \sin x, 0 \leqslant x \leqslant \pi$ **56.** $y = \sec^2 x, 0 \leqslant x \leqslant \frac{\pi}{3}$

57. $y = x^{0.8}, a = -1, b = 2$ **58.** $y = \dfrac{1}{x^2}, a = 1, b = 2$

Verify by differentiation that the formulas in Exercises 59–62 are correct.

59. $\displaystyle\int \frac{1}{\sqrt{(a^2 - x^2)^3}}\, dx = \frac{x}{a^2\sqrt{a^2 - x^2}} + C$

60. $\displaystyle\int \frac{1}{x^2\sqrt{x^2 + a^2}}\, dx = -\frac{\sqrt{x^2 + a^2}}{a^2 x} + C$

61. $\displaystyle\int \sin^2 x\, dx = \frac{x}{2} - \frac{\sin 2x}{4} + C$

62. $\displaystyle\int x^2 \sin x\, dx = -x^2 \cos x + 2\int x \cos x\, dx$

In Exercises 63–68 find the general indefinite integral.

63. $\displaystyle\int x\sqrt{x}\, dx$

64. $\displaystyle\int \sqrt{x}\left(x^2 - \frac{1}{x}\right) dx$

65. $\displaystyle\int (2 - \sqrt{x})^2\, dx$

66. $\displaystyle\int (\cos x - 2\sin x)\, dx$

67. $\displaystyle\int (2x + \sec x \tan x)\, dx$

68. $\displaystyle\int \left(x^2 + 1 + \frac{1}{x^2}\right) dx$

69. Justify (5.19) for the case where $h < 0$.

70. If f is continuous and g and h are differentiable functions, find a formula for

$$\frac{d}{dx}\int_{g(x)}^{h(x)} f(t)\, dt$$

The following exercises are intended only for those who have already covered Chapter 6.

Evaluate the integrals in Exercises 71–79.

71. $\displaystyle\int_4^8 \frac{1}{x}\, dx$

72. $\displaystyle\int_{\ln 3}^{\ln 6} 8 e^x\, dx$

73. $\displaystyle\int_8^9 2^t\, dt$

74. $\displaystyle\int_{-e^2}^{-e} \frac{3}{x}\, dx$

75. $\displaystyle\int_1^{\sqrt{3}} \frac{6}{1 + x^2}\, dx$

76. $\displaystyle\int_0^{0.5} \frac{dx}{\sqrt{1 - x^2}}$

77. $\displaystyle\int_1^e \frac{x^2 + x + 1}{x}\, dx$

78. $\displaystyle\int_4^9 \left(\sqrt{x} + \frac{1}{\sqrt{x}}\right)^2 dx$

79. $\displaystyle\int \left(x^2 + 1 + \frac{1}{x^2 + 1}\right) dx$

80. Find the area of the region bounded by $y = 1/x$, $y = 0$, $x = 1$, and $x = 2$.

SECTION 5.6

The Substitution Rule

The Fundamental Theorem of Calculus reduces the problem of integration to the problem of antidifferentiation. But the antidifferentiation formulas in Table 5.26 do not suffice to evaluate integrals such as

(5.28)
$$\int 2x\sqrt{1 + x^2}\, dx$$

In such cases the task is simplified by changing from the variable x to a new variable. Suppose that we let u be the quantity under the root sign in (5.28),

$u = 1 + x^2$. Then the differential of u is $du = 2x\,dx$. Notice that if the dx in the notation for an integral were to be interpreted as a differential, then the differential $2x\,dx$ would occur in (5.28) so, formally, without justifying our calculation, we could write

(5.29)
$$\int 2x\sqrt{1 + x^2}\,dx = \int \sqrt{1 + x^2}\,2x\,dx = \int \sqrt{u}\,du = \tfrac{2}{3}u^{3/2} + C$$
$$= \tfrac{2}{3}(x^2 + 1)^{3/2} + C$$

But now we could check that we have the correct answer by using the Chain Rule to differentiate the function on the right side of Equation 5.29:

$$\frac{d}{dx}\left[\frac{2}{3}(x^2 + 1)^{3/2} + C\right] = \frac{2}{3}\cdot\frac{3}{2}(x^2 + 1)^{1/2}\cdot 2x = 2x\sqrt{x^2 + 1}$$

In general this method works when we have an integral of the form $\int f(g(x))g'(x)\,dx$. Observe that if $F' = f$, then

(5.30)
$$\int F'(g(x))g'(x)\,dx = F(g(x)) + C$$

because, by the Chain Rule,

$$\frac{d}{dx}[F(g(x))] = F'(g(x))g'(x)$$

If we make the "change of variable" or "substitution" $u = g(x)$, then from Equation 5.30 we have

$$\int F'(g(x))g'(x)\,dx = F(g(x)) + C = F(u) + C = \int F'(u)\,du$$

or, writing $F' = f$,

$$\int f(g(x))g'(x)\,dx = \int f(u)\,du$$

Thus we have proved the following rule:

The Substitution Rule
(5.31)

> If $u = g(x)$ is a differentiable function whose range is an interval I and f is continuous on I, then
>
> $$\int f(g(x))g'(x)\,dx = \int f(u)\,du$$

Notice that the Substitution Rule for integration was proved using the Chain Rule for differentiation. Notice also that if $u = g(x)$, then $du = g'(x)\,dx$, so a way of remembering the Substitution Rule is to think of dx and du in Theorem 5.31 as differentials.

Thus the Substitution Rule says: **It is permissible to operate with dx and du after integral signs as if they were differentials.**

● **Example 1**   Find $\int x^3 \cos(x^4 + 2)\,dx$.

Solution   We make the substitution $u = x^4 + 2$ because its differential is $du = 4x^3\,dx$, which, apart from the constant factor 4, occurs in the integral. Thus, using $x^3\,dx = du/4$ and the Substitution Rule, we have

$$\int x^3 \cos(x^4 + 2)\,dx = \int \cos u \cdot \tfrac{1}{4}\,du$$

$$= \tfrac{1}{4}\int \cos u\,du$$

$$= \tfrac{1}{4}\sin u + C$$

$$= \tfrac{1}{4}\sin(x^4 + 2) + C$$

Notice that at the final stage we had to return to the original variable x.
●

The idea behind the Substitution Rule is to replace a relatively complicated integral by a simpler integral. This is accomplished by changing from the original variable x to a new variable u that is a function of x. Thus in Example 1 the integral $\int x^3 \cos(x^4 + 2)\,dx$ was replaced by the simpler integral $\tfrac{1}{4}\int \cos u\,du$.

The main challenge in using the Substitution Rule is to think of an appropriate substitution. You should try to choose u to be some function in the integrand whose differential also occurs (except for a constant factor). This was the case in Example 1. If that is not possible, try choosing u to be some complicated part of the integrand.

● **Example 2**   Evaluate $\int \sqrt{3x + 4}\,dx$.

Solution 1   Let $u = 3x + 4$. Then $du = 3\,dx$, so $dx = du/3$. Thus the Substitution Rule gives

$$\int \sqrt{3x + 4}\,dx = \int \sqrt{u}\,\frac{du}{3} = \frac{1}{3}\int u^{1/2}\,du$$

$$= \frac{1}{3}\cdot\frac{u^{3/2}}{3/2} + C = \frac{2}{9}u^{3/2} + C$$

$$= \frac{2}{9}(3x + 4)^{3/2} + C$$

Solution 2   Another possible substitution is $u = \sqrt{3x + 4}$. Then

$$du = \frac{3\,dx}{2\sqrt{3x + 4}} \quad \text{and} \quad dx = \frac{2}{3}\sqrt{3x + 4}\,du = \frac{2}{3}u\,du$$

(Or observe that $u^2 = 3x + 4$, so $2u\,du = 3\,dx$.) Therefore

$$\int \sqrt{3x + 4}\,dx = \int u \cdot \frac{2}{3}\,u\,du = \frac{2}{3}\int u^2\,du$$

$$= \frac{2}{3} \cdot \frac{u^3}{3} + C = \frac{2}{9}\,u^3 + C$$

$$= \frac{2}{9}\,(3x + 4)^{3/2} + C \qquad \bullet$$

● **Example 3**   Find $\displaystyle \int \frac{x}{\sqrt{1 - 4x^2}}\,dx$.

Solution   Let $u = 1 - 4x^2$. Then $du = -8x\,dx$, so $x\,dx = -\frac{1}{8}\,du$ and

$$\int \frac{x}{\sqrt{1 - 4x^2}}\,dx = -\frac{1}{8}\int \frac{du}{\sqrt{u}} = -\frac{1}{8}\int u^{-1/2}\,du$$

$$= -\frac{1}{8}\,(2\sqrt{u}) + C = -\frac{1}{4}\sqrt{1 - 4x^2} + C \qquad \bullet$$

● **Example 4**   Calculate $\displaystyle \int \cos 5x\,dx$.

Solution   If we let $u = 5x$, then $du = 5\,dx$, so $dx = \frac{1}{5}\,du$. Therefore

$$\int \cos 5x\,dx = \frac{1}{5}\int \cos u\,du = \frac{1}{5}\sin u + C = \frac{1}{5}\sin 5x + C \qquad \bullet$$

● **Example 5**   Find $\displaystyle \int \sqrt{1 + x^2}\,x^5\,dx$.

Solution   An appropriate substitution becomes more obvious if we factor x^5 as $x^4 \cdot x$. Let $u = 1 + x^2$. Then $du = 2x\,dx$, so $x\,dx = du/2$. Also $x^2 = u - 1$, so $x^4 = (u - 1)^2$:

$$\int \sqrt{1 + x^2}\,x^5\,dx = \int \sqrt{1 + x^2}\,x^4 \cdot x\,dx$$

$$= \int \sqrt{u}(u - 1)^2\,\frac{du}{2} = \frac{1}{2}\int \sqrt{u}(u^2 - 2u + 1)\,du$$

$$= \frac{1}{2}\int (u^{5/2} - 2u^{3/2} + u^{1/2})\,du$$

$$= \frac{1}{2}\left(\frac{2}{7}\,u^{7/2} - 2 \cdot \frac{2}{5}\,u^{5/2} + \frac{2}{3}\,u^{3/2}\right) + C$$

$$= \frac{1}{7}(1 + x^2)^{7/2} - \frac{2}{5}(1 + x^2)^{5/2} + \frac{1}{3}(1 + x^2)^{3/2} + C \qquad \bullet$$

When evaluating a *definite* integral by substitution, there are two possible methods. One method is to evaluate the indefinite integral first and then use the Fundamental Theorem. For instance, using the result of Example 2, we have

$$\int_0^4 \sqrt{3x + 4}\, dx = \left[\int \sqrt{3x + 4}\, dx\right]_0^4 = \frac{2}{9}(3x + 4)^{3/2}\Big]_0^4$$

$$= \frac{2}{9}(16)^{3/2} - \frac{2}{9}(4)^{3/2} = \frac{2}{9}(64 - 8) = \frac{112}{9}$$

Another method, which is usually preferable, is to change the limits of integration when the variable is changed.

The Substitution Rule for Definite Integrals
(5.32)

If g' is continuous on $[a, b]$ and f is continuous on the range of g, then

$$\int_a^b f(g(x))g'(x)\, dx = \int_{g(a)}^{g(b)} f(u)\, du$$

Proof

$$\int_a^b f(g(x))g'(x)\, dx = \left[\int f(g(x))g'(x)\, dx\right]_{x=a}^{x=b} \quad (by\ Equation\ 5.25)$$

$$= \left[\int f(u)\, du\right]_{x=a}^{x=b} \quad (by\ Rule\ 5.31)$$

$$= \left[\int f(u)\, du\right]_{u=g(a)}^{u=g(b)} \quad [since\ u = g(x)]$$

$$= \int_{g(a)}^{g(b)} f(u)\, du \quad (by\ Equation\ 5.25) \qquad \bullet$$

Rule 5.32 says that when using a substitution in a definite integral, we must put everything in terms of the new variable u—not only x and dx but also the limits of integration. The new limits of integration are the values of u that correspond to $x = a$ and $x = b$.

● **Example 6**   Evaluate $\int_0^4 \sqrt{3x + 4}\, dx$ using Theorem 5.32.

Solution   Using the substitution of Solution 1 of Example 2, we have $u = 3x + 4$ and $dx = du/3$. To find the new limits of integration we note that

$$\text{when } x = 0, u = 4 \quad \text{and} \quad \text{when } x = 4, u = 16$$

Therefore $\int_0^4 \sqrt{3x + 4}\, dx = \frac{1}{3}\int_4^{16} \sqrt{u}\, du$

$$= \frac{1}{3} \cdot \frac{2}{3} u^{3/2}\Big]_4^{16}$$

$$= \frac{2}{9}(16^{3/2} - 4^{3/2}) = \frac{112}{9} \qquad \bullet$$

Observe that when using Theorem 5.32 we do not return to the variable x after integrating. We simply evaluate the expression in u between the appropriate values of u.

● **Example 7**   Evaluate $\int_1^2 \dfrac{dx}{(3 - 5x)^2}$.

Solution   Let $u = 3 - 5x$. Then $du = -5dx$, so $dx = -du/5$. When $x = 1$, $u = -2$; when $x = 2$, $u = -7$.

$$\int_1^2 \frac{dx}{(3 - 5x)^2} = -\frac{1}{5}\int_{-2}^{-7} \frac{du}{u^2}$$

$$= -\frac{1}{5}\left[-\frac{1}{u}\right]_{-2}^{-7} = \frac{1}{5u}\bigg]_{-2}^{-7}$$

$$= \frac{1}{5}\left(-\frac{1}{7} + \frac{1}{2}\right) = \frac{1}{14}$$ ●

The next theorem uses the Substitution Rule for Definite Integrals to simplify the calculation of integrals of functions that possess symmetry properties.

Integrals of Symmetric Functions (5.33)

Suppose f is continuous on $[-a, a]$.

(a)   If f is even $[f(-x) = f(x)]$, then $\int_{-a}^a f(x)\,dx = 2\int_0^a f(x)\,dx$.

(b)   If f is odd $[f(-x) = -f(x)]$, then $\int_{-a}^a f(x)\,dx = 0$.

Proof   Using Property 5 of integrals we have

(5.34) $\displaystyle\int_{-a}^a f(x)\,dx = \int_{-a}^0 f(x)\,dx + \int_0^a f(x)\,dx = -\int_0^{-a} f(x)\,dx + \int_0^a f(x)\,dx$

In the first integral on the right side we make the substitution $u = -x$. Then $du = -dx$ and when $x = -a$, $u = a$. Therefore

(5.35) $\displaystyle -\int_0^{-a} f(x)\,dx = -\int_0^a f(-u)(-du) = \int_0^a f(-u)\,du$

(a)   If f is even, then $f(-u) = f(u)$ so Equations 5.34 and 5.35 give

$$\int_{-a}^a f(x)\,dx = \int_0^a f(-u)\,du + \int_0^a f(x)\,dx$$

$$= \int_0^a f(u)\,du + \int_0^a f(x)\,dx$$

$$= 2\int_0^a f$$

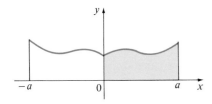

(a) f even, $\int_{-a}^{a} f = 2 \int_{0}^{a} f$

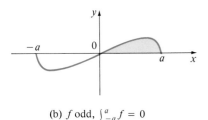

(b) f odd, $\int_{-a}^{a} f = 0$

Figure 5.18

(b)   If f is odd, then $f(-u) = -f(u)$ and so

$$\int_{-a}^{a} f(x)\, dx = \int_{0}^{a} f(-u)\, du + \int_{0}^{a} f(x)\, dx$$

$$= -\int_{0}^{a} f(u)\, du + \int_{0}^{a} f(x)\, dx = 0 \quad\bullet$$

Theorem 5.33 is illustrated by Figure 5.18. For the case where f is positive and even, part (a) says that the area under $y = f(x)$ from $-a$ to a is twice the area from 0 to a because of symmetry. We shall see in the next section that an integral $\int_{a}^{b} f$ can be expressed as the area above the x-axis and below $y = f(x)$ minus the area below the axis and above the curve. Thus part (b) says the integral is 0 because the areas cancel.

● **Example 8**   Since $f(x) = x^6 + 1$ satisfies $f(-x) = f(x)$, it is even and so

$$\int_{-2}^{2} (x^6 + 1)\, dx = 2 \int_{0}^{2} (x^6 + 1)\, dx$$

$$= 2 \left[\frac{x^7}{7} + x \right]_{0}^{2} = 2 \left(\frac{128}{7} + 2 \right) = \frac{284}{7} \quad\bullet$$

● **Example 9**   Since $f(x) = \tan x/(1 + x^2 + x^4)$ satisfies $f(-x) = -f(x)$, it is odd and so

$$\int_{-1}^{1} \frac{\tan x}{1 + x^2 + x^4}\, dx = 0 \quad\bullet$$

EXERCISES 5.6

Evaluate the integrals in Exercises 1–6 by making the given substitution.

1. $\int x(x^2 - 1)^{99}\, dx, \ u = x^2 - 1$

2. $\int \frac{x^2}{\sqrt{2 + x^3}}\, dx, \ u = 2 + x^3$

3. $\int \sin 4x\, dx, \ u = 4x$

4. $\int \frac{dx}{(2x + 1)^2}, \ u = 2x + 1$

5. $\int \frac{x + 3}{(x^2 + 6x)^2}\, dx, \ u = x^2 + 6x$

6. $\int \sec a\theta \tan a\theta\, d\theta, \ u = a\theta$

Evaluate the integrals in Exercises 7–56 if they exist.

7. $\int (2x + 1)(x^2 + x + 1)^3\, dx$

8. $\int x^3(1 - x^4)^5\, dx$

9. $\int \sqrt{x - 1}\, dx$

10. $\int \sqrt[3]{1 - x}\, dx$

11. $\int x^3 \sqrt{2 + x^4}\, dx$

12. $\int x(x^2 + 1)^{3/2}\, dx$

13. $\int \dfrac{3x - 1}{(3x^2 - 2x + 1)^4}\, dx$

14. $\int \dfrac{x}{\sqrt{x^2 + 1}}\, dx$

15. $\int \dfrac{2}{(t + 1)^6}\, dt$
    16. $\int \dfrac{1}{(1 - 3t)^4}\, dt$

17. $\int (1 - 2y)^{1.3}\, dy$
    18. $\int \sqrt[5]{3 - 5y}\, dy$

19. $\int \cos 2\theta\, d\theta$
    20. $\int \sec^2 3\theta\, d\theta$

21. $\int \dfrac{x}{\sqrt[4]{x + 2}}\, dx$
    22. $\int \dfrac{x^2}{\sqrt{1 - x}}\, dx$

23. $\int t \sin(t^2)\, dt$
    24. $\int \dfrac{(1 + \sqrt{x})^9}{\sqrt{x}}\, dx$

25. $\int x^3 (1 - x^2)^{3/2}\, dx$
    26. $\int t^2 \cos(1 - t^3)\, dt$

27. $\int \sin^3 x \cos x\, dx$
    28. $\int \dfrac{\cos \sqrt{x}}{\sqrt{x}}\, dx$

29. $\int \sec x \tan x \sqrt{1 + \sec x}\, dx$

30. $\int \cos^4 x \sin x\, dx$

31. $\int \dfrac{ax + b}{\sqrt{ax^2 + 2bx + c}}\, dx$

32. $\int \tan^2 \theta \sec^2 \theta\, d\theta$

33. $\int \sin(2x + 3)\, dx$

34. $\int \cos(7 - 3x)\, dx$

35. $\int (\sin 3\alpha - \sin 3x)\, dx$

36. $\int \sqrt[3]{x^3 + 1}\, x^5\, dx$

37. $\int x^a \sqrt{b + cx^{a+1}}\, dx$ $(c \neq 0,\, a \neq -1)$

38. $\int \cos x \cos(\sin x)\, dx$

39. $\int_0^1 (2x - 1)^{100}\, dx$

40. $\int_0^{-4} \sqrt{1 - 2x}\, dx$

41. $\int_0^1 (x^4 + x)^5 (4x^3 + 1)\, dx$

42. $\int_2^3 \dfrac{3x^2 - 1}{(x^3 - x)^2}\, dx$

43. $\int_1^2 x\sqrt{x - 1}\, dx$
    44. $\int_0^4 \dfrac{x}{\sqrt{1 + 2x}}\, dx$

45. $\int_0^1 \cos \pi t\, dt$
    46. $\int_0^{\pi/4} \sin 4t\, dt$

47. $\int_1^4 \dfrac{1}{x^2} \sqrt{1 + \dfrac{1}{x}}\, dx$
    48. $\int_0^2 \dfrac{dx}{(2x - 3)^2}$

49. $\int_0^{\pi/3} \dfrac{\sin \theta}{\cos^2 \theta}\, d\theta$
    50. $\int_{-\pi/2}^{\pi/2} \dfrac{x^2 \sin x}{1 + x^6}\, dx$

51. $\int_0^{13} \dfrac{dx}{\sqrt[3]{(1 + 2x)^2}}$
    52. $\int_{-\pi/3}^{\pi/3} \sin^5 \theta\, d\theta$

53. $\int_0^a x\sqrt{a^2 - x^2}\, dx$

54. $\int_0^a x\sqrt{x^2 + a^2}\, dx$ $(a > 0)$

55. $\int_{-a}^a x\sqrt{x^2 + a^2}\, dx$

56. $\int_0^4 \dfrac{dx}{(x - 2)^3}$

In Exercises 57–62 find the area under the given curve.

57. $y = \sqrt{x + 1},\, 0 \leqslant x \leqslant 3$

58. $y = \sqrt{2px},\, 0 \leqslant x \leqslant 1$

59. $y = \sin\left(\dfrac{x}{2}\right),\, 0 \leqslant x \leqslant \dfrac{\pi}{3}$

60. $y = 3\cos\left(\dfrac{x}{3}\right),\, -\pi \leqslant x \leqslant \pi$

61. $y = \dfrac{1}{(x + 1)^2},\, 0 \leqslant x \leqslant 10$

62. $y = x(x^2 + 1)^4,\, 1 \leqslant x \leqslant 2$

63. If f is continuous on R, prove that
$$\int_a^b f(-x)\, dx = \int_{-b}^{-a} f(x)\, dx$$

64. If f is continuous on R, prove that
$$\int_a^b f(x + c)\, dx = \int_{a+c}^{b+c} f(x)\, dx$$

The following exercises are intended only for those who have already covered Chapter 6.

Evaluate the integrals in Exercises 65–86.

65. $\displaystyle\int \frac{dx}{2x-1}$

66. $\displaystyle\int \frac{x}{x^2+1}\,dx$

67. $\displaystyle\int \frac{(\ln x)^2}{x}\,dx$

68. $\displaystyle\int xe^{x^2}\,dx$

69. $\displaystyle\int e^x(1+e^x)^{10}\,dx$

70. $\displaystyle\int \frac{\tan^{-1}x}{1+x^2}\,dx$

71. $\displaystyle\int \frac{dx}{x\ln x}$

72. $\displaystyle\int e^x\sin(e^x)\,dx$

73. $\displaystyle\int \frac{e^x+1}{e^x}\,dx$

74. $\displaystyle\int \frac{e^x}{e^x+1}\,dx$

75. $\displaystyle\int \frac{x+1}{x^2+2x}\,dx$

76. $\displaystyle\int \frac{\sin x}{1+\cos^2 x}\,dx$

77. $\displaystyle\int \frac{e^x}{e^{2x}+1}\,dx$

78. $\displaystyle\int \frac{dx}{1+9x^2}$

79. $\displaystyle\int \frac{1+x}{1+x^2}\,dx$

80. $\displaystyle\int \frac{x^2+1}{x^3+3x+1}\,dx$

81. $\displaystyle\int \frac{x}{1+x^4}\,dx$

82. $\displaystyle\int \frac{x}{x+1}\,dx$

83. $\displaystyle\int_0^3 \frac{dx}{2x+3}$

84. $\displaystyle\int_0^1 t^2 2^{-t^3}\,dt$

85. $\displaystyle\int_e^{e^4} \frac{dx}{x\sqrt{\ln x}}$

86. $\displaystyle\int_0^{1/2} \frac{\sin^{-1}x}{\sqrt{1-x^2}}\,dx$

In Exercises 87 and 88 find the area under the given curve.

87. $y=2e^{-2x}, 0 \leqslant x \leqslant 1$

88. $y=\dfrac{2}{x-2}, 3 \leqslant x \leqslant 5$

SECTION 5.7

Areas between Curves

So far we have defined and calculated areas of regions that lie under the graphs of functions. In this section we shall use integrals to find areas of more general regions.

Consider the region S that lies between two curves $y=f(x)$ and $y=g(x)$ and between the vertical lines $x=a$ and $x=b$, where f and g are continuous functions and $f(x) \geqslant g(x)$ for all x in $[a,b]$ (see Figure 5.19).

Let P be a partition of $[a,b]$ by points x_i and choose points x_i^* in $[x_{i-1}, x_i]$. Let $\Delta x_i = x_i - x_{i-1}$ and $\|P\| = \max\{\Delta x_i\}$. Figure 5.20 shows the approximating rectangles with base Δx_i and height $f(x_i^*) - g(x_i^*)$. The Riemann sum

$$\sum_{i=1}^{n} (f(x_i^*) - g(x_i^*))\,\Delta x_i$$

Figure 5.19

$S = \{(x,y)\,|\,a \leqslant x \leqslant b,\ g(x) \leqslant y \leqslant f(x)\}$

Figure 5.20

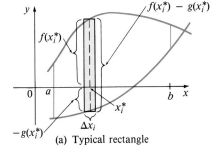

(a) Typical rectangle

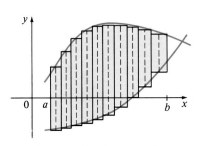

(b) Approximating rectangles

is therefore an approximation to what we intuitively think of as the area of S.

This approximation appears to become better and better as $\|P\| \to 0$. Therefore we define the **area** A of S as the limiting value of the areas of these approximating rectangles.

(5.36)

$$A = \lim_{\|P\| \to 0} \sum_{i=1}^{n} (f(x_i^*) - g(x_i^*)) \Delta x_i$$

We recognize the limit in Equation 5.36 as a Riemann integral that exists because $f - g$ is continuous. Therefore:

(5.37)

> The area of the region bounded by the curves $y = f(x)$, $y = g(x)$, and the lines $x = a$ and $x = b$, where f and g are continuous and $f(x) \geqslant g(x)$ for all x in $[a, b]$, is
>
> $$A = \int_a^b (f(x) - g(x))\, dx$$

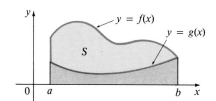

Figure 5.21
$A = \int_a^b f - \int_a^b g$

Notice that in the special case where $g(x) = 0$, S is the region under the graph of f and our general definition of area (5.36) reduces to our previous definition (5.5).

In the case where both f and g are positive, you can see from Figure 5.21 why (5.37) is true:

$$A = [\text{area under } y = f(x)] - [\text{area under } y = g(x)]$$
$$= \int_a^b f(x)\, dx - \int_a^b g(x)\, dx = \int_a^b (f(x) - g(x))\, dx$$

● **Example 1**   Find the area of the region bounded by the parabolas $y = x^2$ and $y = 2x - x^2$.

Solution   We first find the points of intersection of the parabolas by solving their equations simultaneously. This gives $x^2 = 2x - x^2$ or $2x^2 = 2x$. Thus $x(x - 1) = 0$, so $x = 0$ or 1. The points of intersection are $(0, 0)$ and $(1, 1)$ and the region is shown in Figure 5.22.

When using the formula for area (5.37) it is important to ensure that $f(x) \geqslant g(x)$ when $a \leqslant x \leqslant b$. In this case you can see from the diagram that

$$2x - x^2 \geqslant x^2 \qquad \text{for } 0 \leqslant x \leqslant 1$$

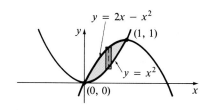

Figure 5.22

and so we choose $f(x) = 2x - x^2$, $g(x) = x^2$, $a = 0$, and $b = 1$. Then (5.37) gives the required area as

$$A = \int_0^1 [(2x - x^2) - x^2]\,dx$$

$$= \int_0^1 2(x - x^2)\,dx = 2\left[\frac{x^2}{2} - \frac{x^3}{3}\right]_0^1$$

$$= 2\left(\frac{1}{2} - \frac{1}{3}\right) = \frac{1}{3}$$

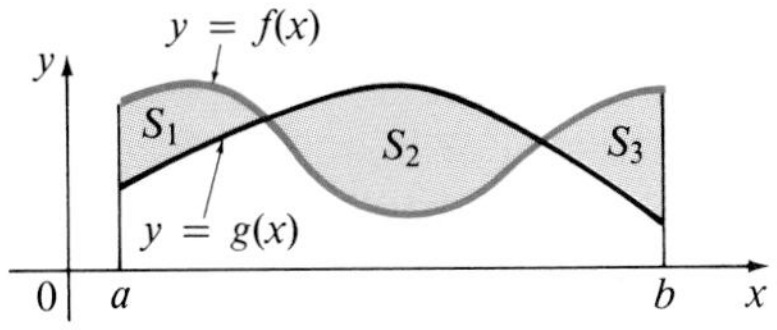

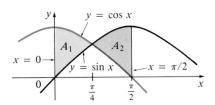

$y = f(x)$

S_1 S_2 S_3

$y = g(x)$

0 a b x

Figure 5.23

If we are asked to find the area between curves $y = f(x)$ and $y = g(x)$ where $f(x) \geqslant g(x)$ for some values of x but $g(x) \geqslant f(x)$ for other values of x, then we split the given region S into several regions $S_1, S_2, \ldots$ with areas $A_1, A_2, \ldots$ as shown in Figure 5.23.

We then define the area of the region S to be the sum of the areas of the smaller regions $S_1, S_2, \ldots$: $A = A_1 + A_2 + \cdots$. Since

$$|f(x) - g(x)| = \begin{cases} f(x) - g(x) & \text{when } f(x) \geqslant g(x) \\ g(x) - f(x) & \text{when } g(x) \geqslant f(x) \end{cases}$$

we have the following expression for A:

(5.38)

> The area between the curves $y = f(x)$ and $y = g(x)$ and between $x = a$ and $x = b$ is
> $$A = \int_a^b |f(x) - g(x)|\,dx$$

When evaluating the integral in (5.38), however, it is still necessary to split it into integrals corresponding to $A_1, A_2, \ldots$.

● **Example 2**   Find the area of the region bounded by the curves $y = \sin x$, $y = \cos x$, $x = 0$, and $x = \pi/2$.

Solution   The points of intersection occur when $\sin x = \cos x$, that is, when $x = \pi/4$ (since $0 \leqslant x \leqslant \pi/2$). The region is sketched in Figure 5.24. Observe that $\cos x \geqslant \sin x$ when $0 \leqslant x \leqslant \pi/4$ but $\sin x \geqslant \cos x$ when $\pi/4 \leqslant x \leqslant \pi/2$. Therefore the required area is

$y = \cos x$

A_1 A_2

$x = 0$

$y = \sin x$ $x = \pi/2$

0 $\dfrac{\pi}{4}$ $\dfrac{\pi}{2}$ x

Figure 5.24

$$A = \int_0^{\pi/2} |\cos x - \sin x|\,dx$$

$$= A_1 + A_2$$

$$= \int_0^{\pi/4} (\cos x - \sin x)\,dx + \int_{\pi/4}^{\pi/2} (\sin x - \cos x)\,dx$$

$$= \left[\sin x + \cos x\right]_0^{\pi/4} + \left[-\cos x - \sin x\right]_{\pi/4}^{\pi/2}$$

$$= \left(\frac{1}{\sqrt{2}} + \frac{1}{\sqrt{2}} - 0 - 1\right) + \left(-0 - 1 + \frac{1}{\sqrt{2}} + \frac{1}{\sqrt{2}}\right)$$

$$= 2\sqrt{2} - 2$$

In this particular example we could have saved some work by noticing that the region is symmetric about $x = \pi/4$ and so

$$A = 2A_1 = 2 \int_0^{\pi/4} (\cos x - \sin x)\, dx$$

● **Example 3**   Find the area of the region bounded by the line $y = x - 1$ and the parabola $y^2 = 2x + 6$.

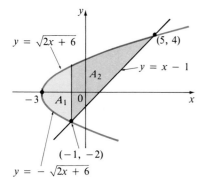

$y = \sqrt{2x + 6}$

(5, 4)

$y = x - 1$

A_2

-3 A_1 0 x

$(-1, -2)$

$y = -\sqrt{2x + 6}$

Figure 5.25

Solution   Solving $y^2 = 2x + 6$ and $y = x - 1$, we get $2x + 6 = (x - 1)^2$, which gives $x^2 - 4x - 5 = (x - 5)(x + 1) = 0$. Thus $x = 5$ or -1 and the points of intersection are $(5, 4)$ and $(-1, -2)$. The region is sketched in Figure 5.25. Notice that just because we found the x-coordinates of the points of intersection to be -1 and 5, we cannot simply integrate a difference of functions between -1 and 5 or we miss the area labeled A_1.

In order to use Formula 5.37 we note that the parabola $y^2 = 2x + 6$ defines two functions given by $y = \sqrt{2x + 6}$ (whose graph is the upper half of the parabola) and $y = -\sqrt{2x + 6}$ (the lower half). Since the lower boundary of the region consists of part of a parabola and part of a line, we break up the area as $A = A_1 + A_2$, where the lower and upper boundaries for A_1 are given by

$$y = -\sqrt{2x + 6} \quad \text{and} \quad y = \sqrt{2x + 6}$$

and the lower and upper boundaries for A_2 are

$$y = x - 1 \quad \text{and} \quad y = \sqrt{2x + 6}$$

Thus the required area is

$$A = A_1 + A_2$$
$$= \int_{-3}^{-1} [\sqrt{2x + 6} - (-\sqrt{2x + 6})]\, dx + \int_{-1}^{5} [\sqrt{2x + 6} - (x - 1)]\, dx$$
$$= 2 \int_{-3}^{-1} \sqrt{2x + 6}\, dx + \int_{-1}^{5} \sqrt{2x + 6}\, dx - \int_{-1}^{5} (x - 1)\, dx$$

In the first two integrals we make the substitution $u = 2x + 6$ (so $du = 2\, dx$) and change the limits of integration accordingly:

$$A = 2 \int_0^4 \sqrt{u}\, \frac{du}{2} + \int_4^{16} \sqrt{u}\, \frac{du}{2} - \int_{-1}^5 (x - 1)\, dx$$
$$= \left[\frac{2}{3} u^{3/2}\right]_0^4 + \left[\frac{1}{3} u^{3/2}\right]_4^{16} - \left[\frac{x^2}{2} - x\right]_{-1}^5$$
$$= \frac{2}{3}(8) + \frac{1}{3}(64 - 8) - \left(\frac{25}{2} - 5 - \frac{1}{2} - 1\right) = 18$$ ●

There is an easier method for solving Example 3. Instead of regarding y as a function of x, let us regard x as a function of y. In general, if a region is bounded by curves with equations $x = f(y)$, $x = g(y)$, $y = c$, and $y = d$, where f and g are continuous and $f(y) \geqslant g(y)$ for $c \leqslant y \leqslant d$ (see **Figure 5.26**), then its area is

(5.39)

$$A = \int_c^d (f(y) - g(y))\, dy$$

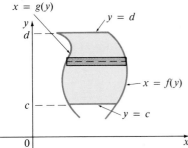

$x = g(y)$

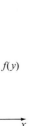

Figure 5.26

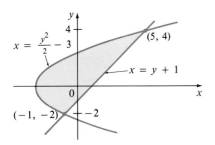

$x = \dfrac{y^2}{2} - 3$

$(5, 4)$

$x = y + 1$

$(-1, -2)$

Figure 5.27

● **Example 4**   Use Equation 5.39 to find the area in Example 3.

Solution   This time we do not have to split the region into two parts. Solving the equation of the parabola for x, we get $x = (y^2/2) - 3$, which represents a single function of y. Also, we write the equation of the line as $x = y + 1$. Notice that

$$y + 1 \geqslant \frac{y^2}{2} - 3 \qquad \text{for } -2 \leqslant y \leqslant 4$$

This means that the left boundary of the region is $x = (y^2/2) - 3$ and the right boundary is $x = y + 1$ (see **Figure 5.27**). We must integrate between the appropriate y-values, $y = -2$ and $y = 4$. Thus Formula 5.39 gives

$$A = \int_{-2}^4 \left[(y + 1) - \left(\frac{y^2}{2} - 3 \right) \right] dy$$

$$= \int_{-2}^4 \left[-\frac{y^2}{2} + y + 4 \right] dy$$

$$= -\frac{1}{2}\left(\frac{y^3}{3} \right) + \frac{y^2}{2} + 4y \,\Big]_{-2}^4$$

$$= -\frac{1}{6}(64) + 8 + 16 - \left(\frac{4}{3} + 2 - 8 \right) = 18 \qquad ●$$

Recall that the definite integral of a function can be interpreted as an area only when the function is positive. However we can use Formula 5.37 to give an interpretation of an integral in terms of certain areas as follows: For the case where $f(x) = 0$ and $g(x) < 0$, Formula 5.37 becomes

$$A = \int_a^b (0 - g(x))\, dx = -\int_a^b g(x)\, dx$$

and so

$$\int_a^b g(x)\, dx = -A$$

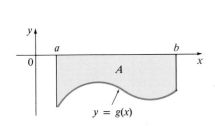

Figure 5.28

$$\int_a^b g = -A$$

This says that the integral of a negative function can be interpreted as the *negative* of the area of the region *above* its graph between a and b (see **Figure 5.28**).

In general, if f takes on both positive and negative values, then we can interpret $\int_a^b f$ as a difference of areas:

$$\int_a^b f(x)\, dx = A_1 - A_2$$

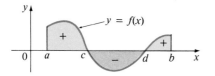

Figure 5.29

where A_1 is the area of the region between a and b that lies above the x-axis and below the graph of f, and A_2 is the area below the x-axis and above the graph of f. For instance, if, as illustrated in Figure 5.29, $f(x) \geqslant 0$ on $[a, c]$ and $[d, b]$ and $f(x) \leqslant 0$ on $[c, d]$, then

$$\int_a^b f = \int_a^c f + \int_c^d f + \int_d^b f$$

$$= \left(\int_a^c f + \int_d^b f \right) - \left(-\int_c^d f \right)$$

$$= A_1 - A_2$$

EXERCISES 5.7

In Exercises 1–32 sketch the region bounded by the given curves and find the area of the region.

1. $y = x^2 + 3, y = x, x = -1, x = 1$

2. $y = x^4, y = -x - 1, x = -2, x = 0$

3. $y^2 = x, y = x + 5, y = -1, y = 2$

4. $x + y^2 = 0, x = y^2 + 1, y = 0, y = 3$

5. $y = x^2 - 4x, y = 2x$

6. $x^2 + 2x + y = 0, x + y + 2 = 0$

7. $y = x, y = x^2$ **8.** $y = x, y = x^3$

9. $y = x^2, y^2 = x$ **10.** $y = x^2, y = x^4$

11. $y = 4x^2, y = x^2 + 3$

12. $y = x^4 - x^2, y = 1 - x^2$

13. $y = x^2 + 2, y = 2x + 5, x = 0, x = 6$

14. $y = 4 - x^2, y = x + 2, x = -3, x = 0$

15. $y = x^2 + 1, y = 3 - x^2, x = -2, x = 2$

16. $y = x^2 + 2x + 2, y = x + 4, x = -3, x = 2$

17. $y^2 = x, x - 2y = 3$

18. $x + y^2 = 2, x + y = 0$

19. $x = 1 - y^2, x = y^2 - 1$

20. $y = x^3 - 4x^2 + 3x, y = x^2 - x$

21. $y = 2x - x^2, y = x^3$

22. $x = 1 - y^4, x = y^3 - y$

23. $y = x, y = \sin x, x = -\dfrac{\pi}{4}, x = \dfrac{\pi}{2}$

24. $y = \cos x, y = \sec^2 x, x = -\dfrac{\pi}{4}, x = \dfrac{\pi}{4}$

25. $y = \cos x, y = \sin 2x, x = 0, x = \dfrac{\pi}{2}$

26. $y = \sin x, y = \sin 2x, x = 0, x = \dfrac{\pi}{2}$

27. $y = \cos x, y = \sin 2x, x = \dfrac{\pi}{2}, x = \pi$

28. $y = \sin x, y = \cos 2x, x = 0, x = \dfrac{\pi}{4}$

29. $y = |x|, y = (x + 1)^2 - 7, x = -4$

30. $y = |x - 1|, y = x^2 - 3, x = 0$

31. $x = 3y, x + y = 0, 7x + 3y = 24$

32. $y = x + 5, y = |x^2 - 1|$

In Exercises 33 and 34 find the area of the region bounded by the given curves by two methods: (a) integrating with respect to x, and (b) integrating with respect to y.

33. $4x + y^2 = 0, y = 2x + 4$

34. $x + 1 = 2(y - 2)^2, x + 6y = 7$

In Exercises 35 and 36 use calculus to find the area of the triangle with the given vertices.

35. $(0, 0), (1, 8), (4, 3)$ **36.** $(-2, 5), (0, -3), (5, 2)$

In Exercises 37 and 38 evaluate the integral and interpret it as the area of a region. Sketch the region.

37. $\int_0^2 |x^2 - x^3| \, dx$ **38.** $\int_0^\pi \left| \sin x - \dfrac{2}{\pi} x \right| dx$

In Exercises 39–42 evaluate the integral and interpret it as a difference of areas. Illustrate with a sketch like Figure 5.29.

39. $\int_{-1}^2 x^3 \, dx$ **40.** $\int_{-2}^2 (x + x^2) \, dx$

41. $\int_0^\pi \cos x \, dx$ **42.** $\int_{\pi/4}^{5\pi/2} \sin x \, dx$

The following exercises are intended only for those who have already covered Chapter 6.

In Exercises 43–48 sketch the region bounded by the given curves and find the area of the region.

43. $y = \dfrac{1}{x}, y = \dfrac{1}{x^2}, x = 1, x = 2$

44. $y = \dfrac{1}{x}, x = 0, y = 1, y = 2$

45. $y = x^2, y = \dfrac{2}{x^2 + 1}$

46. $y = 2^x, y = 5^x, x = -1, x = 1$

47. $y = e^x, y = e^{3x}, x = 1$

48. $y = e^x, y = \dfrac{e^{2x}}{2}, x = 0, x = 1$

SECTION 5.8

Applications to Science

In Section 3.7 we showed how the derivative arises in the natural and social sciences. Now that we are equipped with the Fundamental Theorem of Calculus, we can look at some of the applications of the definite integral to these sciences. Recall that Part 2 of the Fundamental Theorem of Calculus says that if f is continuous on $[a, b]$, then

$$\int_a^b f(x) \, dx = F(b) - F(a) \qquad \text{where } F'(x) = f(x)$$

● **Example 1**   Suppose a particle is moving along a straight line with position function $s(t)$, velocity function $v(t)$, and acceleration function $a(t)$. Since $s'(t) = v(t)$, the Fundamental Theorem of Calculus gives

(5.40)
$$\int_{t_1}^{t_2} v(t) \, dt = s(t_2) - s(t_1)$$

The right side of Equation 5.40 is the change of position, or *displacement*, of the particle during the time interval $[t_1, t_2]$. Thus Equation 5.40 enables us to compute the displacement, by integration, if the velocity function is known. Similarly, since $v'(t) = a(t)$, the Fundamental Theorem of Calculus gives

(5.41)
$$\int_{t_1}^{t_2} a(t) \, dt = v(t_2) - v(t_1)$$

If we want to calculate the distance traveled during the time interval, we have to take into account the intervals when $v(t) \geq 0$ (the particle moves to the right) and also the intervals when $v(t) \leq 0$ (the particle moves to the left). In both cases the distance is computed by integrating $|v(t)|$ (the speed). Therefore

(5.42)
$$\text{total distance traveled} = \int_{t_1}^{t_2} |v(t)|\, dt$$

In view of the discussion in Section 5.7, Figure 5.30 shows how both displacement and distance traveled can be interpreted in terms of areas under a velocity curve.

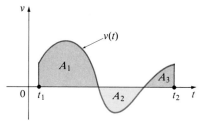

$$\text{displacement} = \int_{t_1}^{t_2} v(t)\, dt$$
$$= A_1 - A_2 + A_3$$
$$\text{distance} = \int_{t_1}^{t_2} |v(t)|\, dt$$
$$= A_1 + A_2 + A_3$$

Figure 5.30

● **Example 2**   A particle moves along a line so that its velocity at time t is $v(t) = t^2 - t - 6$ (measured in meters per second). (a) Find the displacement of the particle during the time period $1 \leq t \leq 4$. (b) Find the distance traveled during this time period.

Solution   (a)   By Equation 5.40, the displacement is

$$s(4) - s(1) = \int_1^4 v(t)\, dt = \int_1^4 (t^2 - t - 6)\, dt$$
$$= \left[\frac{t^3}{3} - \frac{t^2}{2} - 6t \right]_1^4 = -\frac{9}{2}$$

This means that the particle moved 4.5 m to the left.
   (b)   Note that $v(t) = t^2 - t - 6 = (t - 3)(t + 2)$ and so $v(t) \leq 0$ on the interval $[1, 3]$ and $v(t) \geq 0$ on $[3, 4]$. Thus, from Equation 5.42, the distance traveled is

$$\int_1^4 |v(t)|\, dt = \int_1^3 (-v(t))\, dt + \int_3^4 v(t)\, dt$$
$$= \int_1^3 (-t^2 + t + 6)\, dt + \int_3^4 (t^2 - t - 6)\, dt$$
$$= \left[-\frac{t^3}{3} + \frac{t^2}{2} + 6t \right]_1^3 + \left[\frac{t^3}{3} - \frac{t^2}{2} - 6t \right]_3^4$$
$$= \frac{61}{6}\ \text{m}$$

● **Example 3**   If the mass of a rod measured from the left end to a point x is $m(x)$ and the linear density is $\rho(x)$, then $m'(x) = \rho(x)$ (see Example 1 in Section 3.7) and so, by the Fundamental Theorem of Calculus, we have

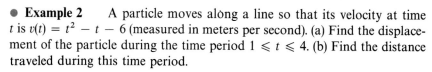

$$m(x_2) - m(x_1) = \int_{x_1}^{x_2} \rho(x)\, dx$$

This enables us to compute the mass of a segment of the rod if the linear density is known.

● **Example 4**   If the rate of growth dn/dt of a population is known (see Example 5 in Section 3.7), then the Fundamental Theorem of Calculus allows us to compute the population at time t, in terms of the initial population n_0, as

$$n(t) = n_0 + \int_0^t \frac{dn}{dt}\, dt$$

● **Example 5**   The marginal cost function is defined as the derivative of the cost function (see Example 7 in Section 3.7). Therefore if the marginal cost function $C'(x)$ is known, then the Fundamental Theorem of Calculus tells us how to compute the increase in cost of production when the production level is increased from $x = x_1$ to $x = x_2$:

$$C(x_2) - C(x_1) = \int_{x_1}^{x_2} C'(x)\, dx$$

For instance, if the marginal cost of manufacturing x units of a product is

$$C'(x) = 0.006x^2 - 1.5x + 8$$

(measured in dollars per unit) and the fixed start-up cost is

$$C(0) = \$1,500,000$$

then the cost of producing 2000 units is

$$
\begin{aligned}
C(2000) &= C(0) + \int_0^{2000} C'(x)\, dx \\
&= 1,500,000 + \int_0^{2000} (0.006x^2 - 1.5x + 8)\, dx \\
&= 1,500,000 + \left[0.002x^3 - 0.75x^2 + 8x\right]_0^{2000} \\
&= \$14,516,000
\end{aligned}
$$

● **Example 6**   In Example 6 in Section 3.7 we discussed the law of laminar flow

$$v(r) = \frac{P}{4\eta l}(R^2 - r^2)$$

which gives the velocity v of blood that flows along a blood vessel with radius R and length l at a distance r from the central axis, where P is the pressure difference between the ends of the vessel and η is the viscosity of the blood. Now in order to compute the flux (volume per unit time) we consider radii r_i where

$$0 = r_0 < r_1 < r_2 < \cdots < r_n = R$$

The approximate area of the annulus with inner radius r_{i-1} and outer radius r_i is

$$2\pi r_i \Delta r_i \qquad \text{where } \Delta r_i = r_i - r_{i-1}$$

If Δr_i is small, then the velocity is almost constant throughout this annulus and can be approximated by $v(r_i)$. Thus the volume of blood per unit time that flows across the annulus is approximately

$$(2\pi r_i \Delta r_i)v(r_i) = 2\pi r_i v(r_i) \Delta r_i$$

and the total volume of blood that flows across a cross-section per unit time is approximately

$$\sum_{i=1}^{n} 2\pi r_i v(r_i) \Delta r_i$$

This approximation is illustrated in Figure 5.31. Notice that the velocity (and hence the volume per unit time) increases toward the center of the blood vessel.

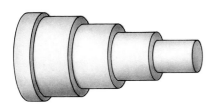

Figure 5.31

The approximation gets better as we take finer subdivisions—that is, as the norm $\|P\| \to 0$. When we take the limit we get the exact value of the *flux* (or *discharge*), which is the volume of blood that passes a cross-section per unit time:

$$
\begin{aligned}
Q &= \lim_{\|P\| \to 0} \sum_{i=1}^{n} 2\pi r_i v(r_i) \Delta r_i \\
&= \int_0^R 2\pi r v(r)\, dr \qquad \textit{(by definition of definite integral)} \\
&= \int_0^R 2\pi r \frac{P}{4\eta l}(R^2 - r^2)\, dr \\
&= \frac{\pi P}{2\eta l}\int_0^R (R^2 r - r^3)\, dr = \frac{\pi P}{2\eta l}\left[R^2 \frac{r^2}{2} - \frac{r^4}{4} \right]_{r=0}^{r=R} \\
&= \frac{\pi P}{2\eta l}\left[\frac{R^4}{2} - \frac{R^4}{4} \right] = \frac{\pi P R^4}{8\eta l}
\end{aligned}
$$

The resulting equation

$$Q = \frac{\pi P R^4}{8\eta l}$$

is called *Poiseuille's Law* and shows that the flux is proportional to the fourth power of the radius of the blood vessel. ●

The examples in this section provide just a glimpse of some of the applications of integrals that are made possible by the Fundamental Theorem of Calculus. After learning more about techniques of integration in the next two chapters, we shall explore the deeper applications of integration in Chapter 8.

EXERCISES 5.8

In Exercises 1–6 the velocity function (in meters per second) is given for a particle moving along a line. Find (a) the displacement and (b) the distance traveled by the particle during the given time interval.

1. $v(t) = 15 - 2t^2, 0 \leqslant t \leqslant 2$

2. $v(t) = 3t - 5, 0 \leqslant t \leqslant 3$

3. $v(t) = t^2 - 2t - 8, 1 \leqslant t \leqslant 6$

4. $v(t) = \sin\left(\dfrac{\pi t}{4}\right), 0 \leqslant t \leqslant 3$

5. $v(t) = 3 - \sqrt{t + 4}, 1 \leqslant t \leqslant 12$

6. $v(t) = 4t - t^3, 1 \leqslant t \leqslant 4$

In Exercises 7–10 the acceleration function (in meters per square second) and the initial velocity are given for a particle moving along a line. Find (a) the velocity at time t and (b) the distance traveled during the given time interval.

7. $a(t) = t + 4, v(0) = 5, 0 \leqslant t \leqslant 10$

8. $a(t) = 2t + 3, v(0) = -4, 0 \leqslant t \leqslant 3$

9. $a(t) = 3t^2 - 2t - 2, v(1) = -2, 1 \leqslant t \leqslant 3$

10. $a(t) = \dfrac{3t + 1}{2\sqrt{t}}, v(1) = 2, 1 \leqslant t \leqslant 4$

11. The linear density of a 5-m long metal rod is given by $\rho(x) = 9 + 2\sqrt{x + 2}$, measured in kilograms per meter, where x is measured in meters from one end of the rod. Find the total mass of the rod.

12. An animal population is increasing at a rate of $200 + 50t$ per year (where t is measured in years). By how much does the animal population increase between the fourth and tenth years?

13. The marginal cost of producing x units of a certain product is $140 - 0.5x + 0.012x^2$ (in dollars per unit). Find the increase in cost if the production level is raised from 3000 units to 5000 units.

14. The marginal revenue from selling x items is $90 - 0.02x$. The revenue from the sale of the first 100 items was \$8800. What is the revenue from the sale of the first 200 items?

15. If the amount of capital that a company has at time t is $f(t)$, then the derivative, $f'(t)$, is called the *net investment flow*. Suppose that the net investment flow is $\sqrt{t}$ million dollars per year (where t is measured in years). Find the increase in capital (the capital formation) from the fourth year to the eighth year.

16. Alabama Instruments Company has set up a production line to manufacture a new calculator. The rate of production of these calculators after t weeks is

$$\frac{dx}{dt} = 5000\left(1 - \frac{100}{(t + 10)^2}\right) \quad \text{calculators/week}$$

(Notice that production approaches 5000 per week as time goes on, but the initial production is less because of the workers' unfamiliarity with the new techniques.) Find the number of calculators produced during the third and fourth weeks.

17. Breathing is cyclic and a full respiratory cycle from the beginning of inhalation to the end of exhalation takes about 5 seconds. The maximum rate of air flow into the lungs is about 0.5 liter per second. This explains, in part, why the function $f(t) = \frac{1}{2}\sin(2\pi t/5)$ has often been used to model the rate of air flow into the lungs. Use this model to find the volume of inhaled air in the lungs at time t.

SECTION 5.9

Average Value of a Function

It is easy to calculate the average value of finitely many numbers $y_1, y_2, \ldots, y_n$:

$$y_{\text{ave}} = \frac{y_1 + y_2 + \cdots + y_n}{n}$$

But how would we compute the average temperature during a day if there are infinitely many possible temperature readings? Figure 5.32 shows the

Figure 5.32

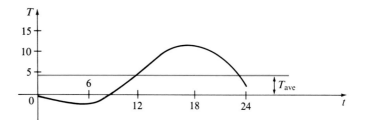

graph of a temperature function $T(t)$ (where t is measured in hours, T in °C) and a guess at the average temperature, T_{ave}.

In general, let us try to compute the average value of a function $y = f(x)$, $a \leqslant x \leqslant b$. We start by dividing the interval $[a, b]$ into n equal subintervals, each with length $\Delta x = (b - a)/n$. Then we choose points $x_1^*, \ldots, x_n^*$ in successive subintervals and calculate the average of the numbers $f(x_1^*), \ldots, f(x_n^*)$:

$$\frac{f(x_1^*) + \cdots + f(x_n^*)}{n}$$

(For example, if f represents a temperature function and $n = 24$, this would mean that we take temperature readings every hour and average them.) Since $\Delta x = (b - a)/n$, we can write $n = (b - a)/\Delta x$ and the average value becomes

$$\frac{f(x_1^*) + \cdots + f(x_n^*)}{\dfrac{b - a}{\Delta x}} = \frac{1}{b - a} \left[f(x_1^*) \Delta x + \cdots + f(x_n^*) \Delta x \right]$$

$$= \frac{1}{b - a} \sum_{i=1}^{n} f(x_i^*) \Delta x$$

If we let n increase, we will be computing the average value of a large number of closely spaced values. (For example, we would be averaging temperature readings taken every minute or even every second.) The limiting value is

$$\lim_{n \to \infty} \frac{1}{b - a} \sum_{i=1}^{n} f(x_i^*) \Delta x = \frac{1}{b - a} \int_a^b f(x)\, dx$$

by the definition of a definite integral.

Therefore we define the **average value of f** on the interval $[a, b]$ as

(5.43)

$$\boxed{f_{ave} = \frac{1}{b - a} \int_a^b f(x)\, dx}$$

● **Example 1**   Find the average value of the function $f(x) = 1 + x^2$ over the interval $[-1, 2]$.

Solution   With $a = -1$ and $b = 2$ we have

$$f_{ave} = \frac{1}{b-a} \int_a^b f(x)\,dx = \frac{1}{2-(-1)} \int_{-1}^2 (1 + x^2)\,dx$$

$$= \frac{1}{3}\left[x + \frac{x^3}{3}\right]_{-1}^2 = 2 \qquad \bullet$$

The question arises: Is there a number c at which the value of f is exactly equal to the average value of the function, that is, $f(c) = f_{ave}$? The following theorem says that this is true for continuous functions.

Mean Value Theorem for Integrals (5.44)

If f is continuous on $[a,b]$, then there exists a number c in $[a,b]$ such that

$$\int_a^b f(x)\,dx = f(c)(b-a)$$

Proof   By the Extreme Value Theorem (4.3) f has an absolute minimum value m and an absolute maximum value M. Then, by Property 8 of integrals, we have

$$m(b-a) \leqslant \int_a^b f(x)\,dx \leqslant M(b-a)$$

Dividing through by the positive number $b - a$, we get

$$m \leqslant \frac{1}{b-a} \int_a^b f(x)\,dx \leqslant M$$

This says that f_{ave} lies between m and M. Since f is continuous and takes on both of the values m and M, the Intermediate Value Theorem says that there is a number c in $[a,b]$ such that

$$f(c) = \frac{1}{b-a} \int_a^b f(x)\,dx$$

or equivalently,

$$\int_a^b f(x)\,dx = f(c)(b-a)$$

(An alternative proof, using the Mean Value Theorem for derivatives, is outlined in Exercise 19.) $\bullet$

The geometric interpretation of the Mean Value Theorem for Integrals is that, for *positive* functions f, there is a number c such that the rectangle with base $[a,b]$ and height $f(c)$ has the same area as the region under the graph of f from a to b (see Figure 5.33).

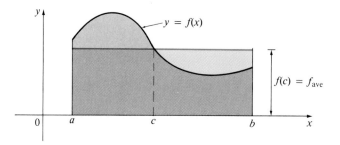

Figure 5.33

● **Example 2**   Since $f(x) = 1 + x^2$ is continuous on the interval $[-1, 2]$, the Mean Value Theorem for Integrals says there is a number c in $[-1, 2]$ such that

$$\int_{-1}^{2} (1 + x^2)\,dx = f(c)[2 - (-1)]$$

In this particular case we can find c explicitly. From Example 1 we know that

$$f(c) = f_{\text{ave}} = 2$$

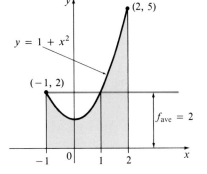

Figure 5.34

Therefore $1 + c^2 = 2$ $c^2 = 1$

Thus in this case there happen to be two numbers $c = \pm 1$ in the interval $[-1, 2]$ that work in the Mean Value Theorem for Integrals.   ●

Examples 1 and 2 are illustrated by Figure 5.34.

EXERCISES 5.9

In Exercises 1–8 find the average value of the given function on the given interval.

1. $f(x) = 1 - 2x$, $[0, 3]$

2. $f(x) = x^2 - 2x$, $[0, 3]$

3. $f(x) = x^2 + 2x - 5$, $[-2, 2]$

4. $f(x) = \sin x$, $[0, \pi]$

5. $f(x) = x^4$, $[-1, 1]$

6. $f(x) = x^3 - x$, $[1, 3]$

7. $f(x) = \sin^2 x \cos x$, $\left[-\dfrac{\pi}{2}, \dfrac{\pi}{4}\right]$

8. $f(x) = \sqrt{x}$, $[4, 9]$

In Exercises 9–12, (a) find the average value of f on the given interval, (b) find c such that $f_{\text{ave}} = f(c)$, and (c) sketch the graph of f and a rectangle whose area is the same as the area under the graph of f.

9. $f(x) = 2x$, $[0, 3]$

10. $f(x) = x^3$, $[-1, 2]$

11. $f(x) = 4 - x^2$, $[0, 2]$

12. $f(x) = 4x - x^2$, $[0, 3]$

13. The temperature (in °F) in a certain city t hours after 9 A.M. was approximated by the function

$$T(t) = 50 + 14 \sin \frac{\pi t}{12}$$

Find the average temperature during the period from 9 A.M. to 9 P.M.

14. The temperature of a metal rod, 5 m long, is $4x$ (in °C) at a distance x meters from one end of the rod. What is the average temperature of the rod?

15. The linear density in a rod 8 m long is $12/\sqrt{x+1}$ kg/m, where x is measured in meters from one end of the rod. Find the average density of the rod.

16. If $s(t)$ is the position function of a particle moving along a line, then its average velocity over the time interval $[t_1, t_2]$ is

$$\frac{\text{displacement}}{\text{time}} = \frac{s(t_2) - s(t_1)}{t_2 - t_1}$$

Prove that this is the same as the average value of the velocity function on the interval $[t_1, t_2]$.

17. Use the model given in Exercise 17 in Section 5.8 to compute the average volume of inhaled air in the lungs in one respiratory cycle.

18. The velocity v of blood that flows in a blood vessel with radius r and length l at a distance r from the central axis is

$$v(r) = \frac{P}{4\eta l}(R^2 - r^2)$$

where P is the pressure difference between the ends of the vessel and η is the viscosity of the blood (see Example 6 in Section 3.7). Find the average velocity (with respect to r) over the interval $0 \leqslant r \leqslant R$. Compare the average velocity with the maximum velocity.

19. Prove the Mean Value Theorem for Integrals (5.44) by applying the Mean Value Theorem for derivatives (4.10) to the function $F(x) = \int_a^x f(t)\,dt$.

REVIEW OF CHAPTER 5

Define, state, or discuss the following.

1. Sigma notation

2. Partition of $[a, b]$

3. Norm of a partition

4. Area under a curve

5. Riemann sum

6. Definite integral of f from a to b

7. Integrand; limits of integration

8. Properties of the definite integral

9. Fundamental Theorem of Calculus

10. Indefinite integral

11. Substitution Rule

12. Integrals of even or odd functions

13. Area between curves

14. Average value of a function

15. Mean Value Theorem for Integrals

REVIEW EXERCISES FOR CHAPTER 5

Find the value of the sum in Exercises 1–4.

1. $\displaystyle\sum_{i=2}^{5} \frac{1}{(-10)^i}$

2. $\displaystyle\sum_{j=1}^{5} \sin\left(\frac{j\pi}{2}\right)$

3. $\displaystyle\sum_{k=1}^{n} (2 + k^3)$

4. $\displaystyle\sum_{i=1}^{n} i(i - 2)$

5. Find the Riemann sum for the function $f(x) = 2 + (x - 2)^2$ on the interval $[0, 2]$ using a regular partition with $n = 4$ and choosing x_i^* to be the left endpoint of the ith subinterval. Sketch the graph of f and the approximating rectangles.

6. Do Exercise 5 if x_i^* is the midpoint of the ith subinterval.

Evaluate the integrals in Exercises 7–10 *without* using the Fundamental Theorem of Calculus.

7. $\displaystyle\int_2^4 (3 - 4x)\,dx$

8. $\displaystyle\int_1^2 (x + 3x^2)\,dx$

9. $\displaystyle\int_0^5 (x^3 - 2x^2)\,dx$

10. $\displaystyle\int_0^b (x^3 + 4x - 1)\,dx$

In Exercises 11–32 evaluate the integral if it exists.

11. $\displaystyle\int_0^5 (x^3 - 2x^2)\,dx$

12. $\displaystyle\int_0^b (x^3 + 4x - 1)\,dx$

13. $\displaystyle\int_0^1 (1 - x^9)\,dx$

14. $\displaystyle\int_0^1 (1 - x)^9\,dx$

15. $\displaystyle\int_1^8 \sqrt[3]{x}(x - 1)\,dx$

16. $\displaystyle\int_1^4 \frac{x^2 - x + 1}{\sqrt{x}}\,dx$

17. $\int_0^2 x^2(1 + 2x^3)^3 \, dx$

18. $\int_0^4 x\sqrt{16 - 3x} \, dx$

19. $\int_3^{11} \frac{dx}{\sqrt{2x + 3}}$

20. $\int_0^2 \frac{x}{(x^2 - 1)^2} \, dx$

21. $\int_{-2}^{-1} \frac{dx}{(2x + 3)^4}$

22. $\int_{-1}^1 \frac{x + x^3 + x^5}{1 + x^2 + x^4} \, dx$

23. $\int \frac{x^4}{(2 + x^5)^6} \, dx$

24. $\int (1 - x)\sqrt{2x - x^2} \, dx$

25. $\int \sin \pi x \, dx$

26. $\int \frac{\cos x}{\sqrt{1 + \sin x}} \, dx$

27. $\int \frac{\cos(1/t)}{t^2} \, dt$

28. $\int \csc^2 3t \, dt$

29. $\int \sin x \sec^2(\cos x) \, dx$

30. $\int \frac{x^3}{\sqrt{x^2 + 1}} \, dx$

31. $\int_0^{2\pi} |\sin x| \, dx$

32. $\int_0^8 |x^2 - 6x + 8| \, dx$

Find the derivatives of the functions in Exercises 33–38.

33. $F(x) = \int_1^x \sqrt{1 + t^2 + t^4} \, dt$

34. $F(x) = \int_\pi^x \tan(s^2) \, ds$

35. $g(x) = \int_0^{x^3} \frac{t}{\sqrt{1 + t^3}} \, dt$

36. $g(x) = \int_1^{\cos x} \sqrt[3]{1 - t^2} \, dt$

37. $y = \int_{\sqrt{x}}^x \frac{\cos \theta}{\theta} \, d\theta$

38. $y = \int_{2x}^{3x + 1} \sin(t^4) \, dt$

In Exercises 39–46 find the area of the region bounded by the given curves.

39. $y = x^2 - 4x + 3, \, y = 0$

40. $y = 4 + 3x - x^2, \, y = 0$

41. $y = x^2 - 6x, \, y = 12x - 2x^2$

42. $y = x^2 - 6, \, y = 12 - x^2, \, x = -5, \, x = 5$

43. $y = x^3, \, x = y^3$

44. $x - 2y + 7 = 0, \, y^2 - 6y - x = 0$

45. $y = \sin x, \, y = -\cos x, \, x = 0, \, x = \pi$

46. $y = x^3, \, y = x^2 - 4x + 4, \, x = 0, \, x = 2$

In Exercises 47 and 48 use Property 8 of integrals to estimate the value of the given integral.

47. $\int_1^3 \sqrt{x^2 + 3} \, dx$

48. $\int_3^5 \frac{1}{x + 1} \, dx$

In Exercises 49–52 use the properties of integrals to verify the given inequalities.

49. $\int_0^\pi \cos^8 x \, dx \leqslant \int_0^\pi \cos^6 x \, dx$

50. $\int_0^{\sqrt{\pi/2}} \sin(x^2) \, dx \leqslant \int_0^{\sqrt{\pi/2}} \cos(x^2) \, dx$

51. $\int_0^1 x^2 \cos x \, dx \leqslant \frac{1}{3}$

52. $\int_{\pi/4}^{\pi/2} \frac{\sin x}{x} \, dx \leqslant \frac{\sqrt{2}}{2}$

53. Find the average value of the function $f(x) = x^3$ on the interval $[2, 4]$.

54. A particle moves along a line with velocity function $v(t) = t^2 - t$. Find (a) the displacement and (b) the distance traveled by the particle during the time interval $[0, 5]$.

55. If f is continuous on $[0, 1]$, prove that
$$\int_0^1 f(x) \, dx = \int_0^1 f(1 - x) \, dx$$

56. Evaluate
$$\lim_{n \to \infty} \frac{1}{n} \left[\left(\frac{1}{n}\right)^9 + \left(\frac{2}{n}\right)^9 + \left(\frac{3}{n}\right)^9 + \cdots + \left(\frac{n}{n}\right)^9 \right]$$

The following exercises are intended only for those who have already covered Chapter 6.

Evaluate the integrals in Exercises 57–62.

57. $\int \frac{e^{\sqrt{x}}}{\sqrt{x}} \, dx$

58. $\int \frac{\cos(\ln x)}{x} \, dx$

59. $\int \tan x \ln(\cos x) \, dx$

60. $\int \frac{x}{\sqrt{1 - x^4}} \, dx$

61. $\int \frac{x^3}{1 + x^4} \, dx$

62. $\int \frac{e^x}{(e^x + 1)\ln(e^x + 1)} \, dx$

6

Inverse Functions: Exponential, Logarithmic, and Inverse Trigonometric Functions

Each problem that I solved became a rule which served afterwards to solve other problems.

René Descartes

Two of the most important functions in mathematics and its applications are the exponential function $f(x) = a^x$ and its inverse function, the logarithmic function $g(x) = \log_a x$. In this chapter we investigate their properties, compute their derivatives, and use them to describe exponential growth and decay in such sciences as chemistry, physics, biology, and economics. We also study the inverses of trigonometric and hyperbolic functions. Finally, we look at a method (l'Hospital's Rule) for computing difficult limits and apply it to sketching curves.

Exponential Functions

An **exponential function** is a function of the form

$$f(x) = a^x$$

where a is a positive constant. It is defined in five stages:

1. If $x = n$, a positive integer, then

$$a^n = \underbrace{a \cdot a \cdots\cdots a}_{n \text{ factors}}$$

2. If $x = 0$, $a^0 = 1$.

3. If $x = -n$, n a positive integer, then

$$a^{-n} = \frac{1}{a^n}$$

4. If x is a rational number, $x = p/q$, where p and q are integers and $q > 0$, then

$$a^x = a^{p/q} = \sqrt[q]{a^p}$$

5. If x is an irrational number, we wish to define a^x so as to fill in the holes of the graph of the function $y = a^x$, where x is rational. In other words, we want to make $f(x) = a^x$, $x \in R$, a continuous function. Since any irrational number can be approximated as closely as we like by a rational number (see Section 1.1), we define

(6.1)

$$a^x = \lim_{r \to x} a^r \qquad r \text{ rational}$$

It can be shown that this definition uniquely specifies a^x and makes the function $f(x) = a^x$ continuous. [See Chapter 9 in M. P. Dolciani et al., *Modern Introductory Analysis* (Boston: Houghton Mifflin, 1964).]

To illustrate Equation 6.1 let us take $x = \sqrt{2}$. The number $\sqrt{2}$ is an irrational number that has a nonrepeating decimal representation:

$$\sqrt{2} = 1.414213562373095\ldots$$

This means that $\sqrt{2}$ is the limit of the sequence of rational numbers

$$1, \quad 1.4, \quad 1.41, \quad 1.414, \quad 1.4142, \quad 1.41421, \quad 1.414213, \quad \ldots$$

According to Equation 6.1, $a^{\sqrt{2}}$ is the limit of the sequence

$$a^1, \quad a^{1.4}, \quad a^{1.41}, \quad a^{1.414}, \quad a^{1.4142}, \quad a^{1.41421}, \quad a^{1.414213}, \quad \ldots$$

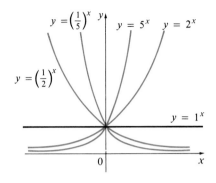

Figure 6.1

and each term of this sequence has been defined in stage 4. In fact, if $a > 1$, $a^{\sqrt{2}}$ is the unique number that satisfies all of the following inequalities:

$$a^1 < a^{\sqrt{2}} < a^2$$

$$a^{1.4} < a^{\sqrt{2}} < a^{1.5}$$

$$a^{1.41} < a^{\sqrt{2}} < a^{1.42}$$

$$a^{1.414} < a^{\sqrt{2}} < a^{1.415}$$

$$a^{1.4142} < a^{\sqrt{2}} < a^{1.4143}$$

$$\vdots$$

The graphs of the function $y = a^x$ are shown in Figure 6.1 for various values of the base a. Notice that all of these graphs pass through the same point $(0, 1)$ because $a^0 = 1$ for $a \neq 0$.

You can see from Figure 6.1 that there are basically three kinds of exponential functions $y = a^x$. If $0 < a < 1$, the exponential function decreases rapidly; if $a = 1$, it is constant; and if $a > 1$, it increases rapidly. These three cases are illustrated in Figure 6.2. Since $(1/a)^x = 1/a^x = a^{-x}$, the graph of $y = (1/a)^x$ is just the reflection of the graph of $y = a^x$ in the y-axis.

The properties of the exponential function are summarized in the following theorem.

Theorem (6.2)

> If $a > 0$ and $a \neq 1$, then $f(x) = a^x$ is a continuous function with domain R and range $(0, \infty)$. In particular, $a^x > 0$ for all x. If $0 < a < 1$, $f(x) = a^x$ is a decreasing function; if $a > 1$, f is an increasing function. If $a, b > 0$ and $x, y \in R$, then
>
> (a) $a^{x+y} = a^x a^y$ (b) $a^{x-y} = \dfrac{a^x}{a^y}$
>
> (c) $(a^x)^y = a^{xy}$ (d) $(ab)^x = a^x b^x$

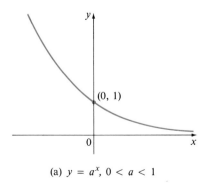

(a) $y = a^x$, $0 < a < 1$

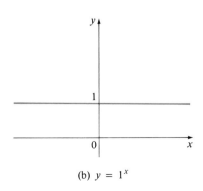

(b) $y = 1^x$

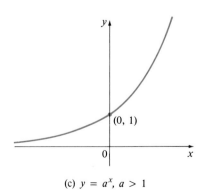

(c) $y = a^x$, $a > 1$

Figure 6.2

The reason for the importance of the exponential function lies in properties (a)–(d), which are called the Laws of Exponents. If x and y are rational numbers, then these laws are well known from elementary algebra (see Appendix A). For arbitrary real numbers x and y these laws can be deduced from the special case where the exponents are rational by using Definition 6.1.

The following limits can be read from the graphs shown in Figure 6.2 or proved from the definition of a limit at infinity. (See Exercise 85 in Section 6.3.)

(6.3)

> If $a > 1$, then
>
> $$\lim_{x \to \infty} a^x = \infty \qquad \text{and} \qquad \lim_{x \to -\infty} a^x = 0$$
>
> If $0 < a < 1$, then
>
> $$\lim_{x \to \infty} a^x = 0 \qquad \text{and} \qquad \lim_{x \to -\infty} a^x = \infty$$

In particular, if $a \neq 1$, then the x-axis is a horizontal asymptote of the graph of the exponential function $y = a^x$.

● **Example 1**   (a)   Find $\lim_{x \to \infty} (2^{-x} - 1)$.

(b)   Sketch the graph of the function $y = 2^{-x} - 1$.

Solution   (a)

$$\lim_{x \to \infty} (2^{-x} - 1) = \lim_{x \to \infty} \left[(\tfrac{1}{2})^x - 1 \right]$$

$$= 0 - 1 \qquad [by\ (6.3)\ with\ a = \tfrac{1}{2} < 1]$$

$$= -1$$

(b)   We write $y = (\tfrac{1}{2})^x - 1$ as above. The graph of $y = (\tfrac{1}{2})^x$ is shown in Figure 6.1, so we shift it down one unit to obtain the graph of $y = (\tfrac{1}{2})^x - 1$ shown in Figure 6.3. (For a review of shifting graphs, see Section 1.7 and, in particular, Table 1.28.)   ●

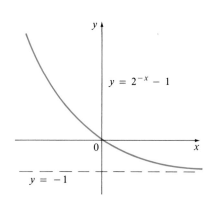

$y = 2^{-x} - 1$

$y = -1$

Figure 6.3

● **Example 2**   Find $\lim_{x \to 0^-} 2^{1/x}$.

Solution   As $x \to 0^-$, $1/x \to -\infty$, so by (6.3),

$$\lim_{x \to 0^-} 2^{1/x} = 0 \qquad\qquad ●$$

EXERCISES 6.1

(*Note:* For a review of the Laws of Exponents, see the exercises in Appendix A.)

1. Sketch, using the same axes, the graphs of the following functions.
   (a) $y = 3^x$ (b) $y = 10^x$
   (c) $y = (\frac{1}{3})^x$ (d) $y = (\frac{1}{10})^x$

2. Sketch, using the same axes, the graphs of the following functions.
   (a) $y = 4^x$ (b) $y = 4^{-x}$ (c) $y = 1^x$

In Exercises 3–16 make a rough sketch of the graph of the given function. Do not use a calculator. Just use the graphs given in Figures 6.1 and 6.2 and, if necessary, the transformations of Section 1.7.

3. $y = 100^x$

4. $y = (1.1)^x$

5. $y = (0.9)^x$

6. $y = (0.1)^x$

7. $y = 2^x + 1$

8. $y = 2^{x+1}$

9. $y = 3^{-x}$

10. $y = -3^x$

11. $y = -3^{-x}$

12. $y = 2^{|x|}$

13. $y = 5^{-3x}$

14. $y = 5^{x-3}$

15. $y = 3 - 2^{x-1}$

16. $y = 2 + 5(1 - 10^{-x})$

Find the limits in Exercises 17–36.

17. $\lim\limits_{x \to \infty} (1.1)^x$

18. $\lim\limits_{x \to -\infty} (1.1)^x$

19. $\lim\limits_{x \to \infty} \pi^{-x}$

20. $\lim\limits_{x \to -\infty} \pi^{-x}$

21. $\lim\limits_{x \to \infty} 5^{2x+5}$

22. $\lim\limits_{x \to -\infty} 2^{3x+1}$

23. $\lim\limits_{x \to \infty} \left(2^{-0.8x} + \dfrac{1}{x} \right)$

24. $\lim\limits_{x \to \infty} (0.8)^{x+1}$

25. $\lim\limits_{x \to -\infty} \left(\dfrac{\pi}{4} \right)^x$

26. $\lim\limits_{x \to \infty} \left(\dfrac{2\pi}{7} \right)^x$

27. $\lim\limits_{x \to (\pi/2)^-} 2^{\tan x}$

28. $\lim\limits_{x \to -(\pi/2)^+} 2^{\tan x}$

29. $\lim\limits_{x \to \infty} 3^{1/x}$

30. $\lim\limits_{x \to -\infty} 3^{1/x}$

31. $\lim\limits_{x \to 0^+} 3^{1/x}$

32. $\lim\limits_{x \to 0^-} 3^{1/x}$

33. $\lim\limits_{x \to \infty} \dfrac{10^x}{10^x + 1}$

34. $\lim\limits_{x \to \infty} \dfrac{2^x - 2^{-x}}{2^x + 3 \cdot 2^{-x}}$

35. $\lim\limits_{t \to 0^+} \pi^{\csc t}$

36. $\lim\limits_{t \to 0^-} (3 - 2^{\csc t})$

SECTION 6.2

Inverse Functions

Let us compare the functions f and g whose arrow diagrams are shown in Figure 6.4. Note that f never takes on the same value twice (any two numbers in A have different images), whereas g does take on the same value twice (both 2 and 3 have the same image, 4). In symbols,

$$g(2) = g(3)$$

but $f(x_1) \neq f(x_2)$ whenever $x_1 \neq x_2$

Figure 6.4

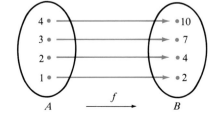
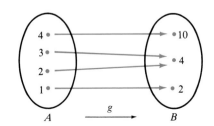

Functions that have this latter property are called *one-to-one*.

Definition (6.4)

> A function f with domain A is called a **one-to-one function** if no two elements of A have the same image; that is,
>
> $$f(x_1) \neq f(x_2) \qquad \text{whenever } x_1 \neq x_2$$

If a horizontal line intersects the graph of f in more than one point, then we see from Figure 6.5 that there are numbers x_1 and x_2 such that $f(x_1) = f(x_2)$. This means that f is not one-to-one. Therefore we have the following geometric method for determining whether or not a function is one-to-one.

Figure 6.5

THIS FUNCTION IS NOT ONE-TO-ONE
BECAUSE $f(x_1) = f(x_2)$

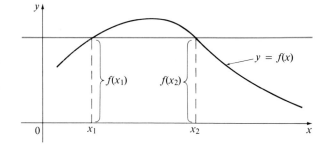

Horizontal Line Test (6.5)

> A function is one-to-one if and only if no horizontal line intersects its graph more than once.

● **Example 1**   Is the function $f(x) = x^3$ one-to-one?

Solution 1   If $x_1 \neq x_2$, then $x_1^3 \neq x_2^3$ (two different numbers cannot have the same cube). Therefore, by Definition 6.4, $f(x) = x^3$ is one-to-one.

Solution 2   From Figure 6.6 we see that no horizontal line intersects the graph of $f(x) = x^3$ more than once. Therefore, by the Horizontal Line Test, f is one-to-one.   ●

● **Example 2**   Is the function $g(x) = x^2$ one-to-one?

Solution 1   This function is not one-to-one because, for instance,

$$g(1) = 1 = g(-1)$$

and so 1 and -1 have the same image.

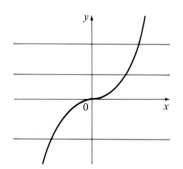

Figure 6.6

$f(x) = x^3$ IS ONE-TO-ONE

Solution 2   From Figure 6.7 we see that there are horizontal lines that intersect the graph of g more than once. Therefore, by the Horizontal Line Test, g is not one-to-one.   ●

Notice that the function f of Example 1 is increasing and is also one-to-one. More generally we have the following theorem (see Exercise 27):

Theorem (6.6)

> Every decreasing function and every increasing function are one-to-one functions.

One-to-one functions are important because they are precisely the functions that possess inverse functions according to the following definition.

Definition (6.7)

> Let f be a one-to-one function with domain A and range B. Then its **inverse function** f^{-1} has domain B and range A and is defined by
>
> $$f^{-1}(y) = x \quad \Leftrightarrow \quad f(x) = y$$
>
> for any y in B.

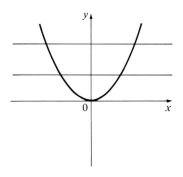

Figure 6.7

$g(x) = x^2$ IS NOT ONE-TO-ONE

This definition says that if f maps x into y, then f^{-1} maps y back into x. (If f were not one-to-one, then f^{-1} would not be uniquely defined.) The arrow diagram in Figure 6.8 indicates that f^{-1} reverses the effect of f. Note that

> domain of f^{-1} = range of f
>
> range of f^{-1} = domain of f

For example, the inverse function of $f(x) = x^3$ is $f^{-1}(x) = x^{1/3}$ because if $y = x^3$, then

$$f^{-1}(y) = f^{-1}(x^3) = (x^3)^{1/3} = x$$

Figure 6.8

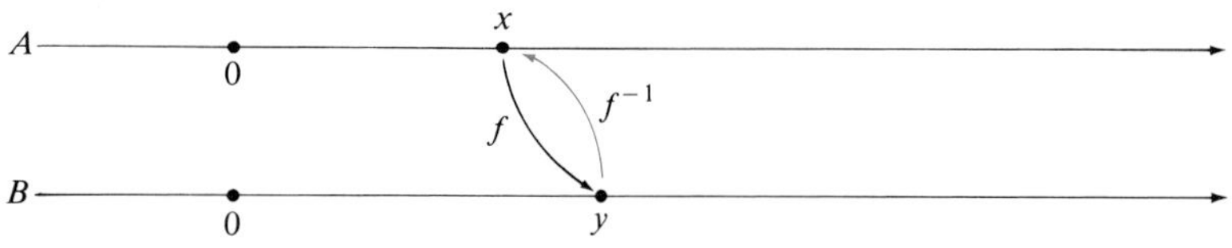

Caution: Do not mistake the -1 in f^{-1} for an exponent. Thus

$$f^{-1}(x) \quad \text{does } not \text{ mean} \quad \frac{1}{f(x)}$$

The reciprocal $1/f(x)$ could, however, be written as $[f(x)]^{-1}$.

The letter x is traditionally used as the independent variable, so when we concentrate on f^{-1} rather than on f we usually reverse the roles of x and y in Definition 6.7 and write

(6.8)
$$f^{-1}(x) = y \quad \Leftrightarrow \quad f(y) = x$$

By substituting for y in Definition 6.7 and substituting for x in (6.8) we get the following **cancellation equations:**

(6.9)
$$f^{-1}(f(x)) = x \quad \text{for every } x \text{ in } A$$
$$f(f^{-1}(x)) = x \quad \text{for every } x \text{ in } B$$

The first cancellation equation says that if we start with x, apply f, and then apply f^{-1}, we arrive back at x, where we started. Thus f^{-1} undoes what f does. The second equation says that f undoes what f^{-1} does.

For example, if $f(x) = x^3$, then $f^{-1}(x) = x^{1/3}$ and the cancellation equations become

$$f^{-1}(f(x)) = (x^3)^{1/3} = x$$
$$f(f^{-1}(x)) = (x^{1/3})^3 = x$$

These equations simply say that the cube function and the cube root function cancel each other out.

Let us now see how to compute inverse functions. If we have a function $y = f(x)$ and are able to solve this equation for x in terms of y, then according to Definition 6.7 we must have $x = f^{-1}(y)$. If we then interchange x and y, we have $y = f^{-1}(x)$, which is the desired equation.

How to Find the Inverse Function of a One-to-One Function f   (6.10)

(a) Write $y = f(x)$.
(b) Solve this equation for x in terms of y (if possible).
(c) Interchange x and y. The resulting equation is $y = f^{-1}(x)$.

● **Example 3**   Find the inverse function of $f(x) = x^3 + 2$.

Solution   According to (6.10) we first write

$$y = x^3 + 2$$

Then we solve this equation for x:

$$x^3 = y - 2$$
$$x = \sqrt[3]{y - 2}$$

Finally we interchange x and y:

$$y = \sqrt[3]{x - 2}$$

Therefore the inverse function is $f^{-1}(x) = \sqrt[3]{x - 2}$.   ●

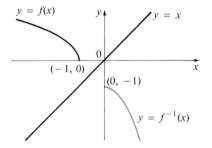

Figure 6.9

The principle of interchanging x and y to find the inverse function also gives us the method for obtaining the graph of f^{-1} from the graph of f. If $f(a) = b$, then $f^{-1}(b) = a$. Thus the point (a, b) is on the graph of f if and only if the point (b, a) is on the graph of f^{-1}. But we get the point (b, a) from (a, b) by reflecting in the line $y = x$ (see Figure 6.9). Therefore, as illustrated by Figure 6.10:

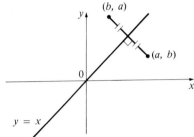

Figure 6.10

> The graph of f^{-1} is obtained by reflecting the graph of f in the line $y = x$.

● **Example 4**   Sketch the graphs of $f(x) = \sqrt{-1 - x}$ and its inverse function using the same coordinate axes.

Solution   First we sketch the curve $y = \sqrt{-1 - x}$ (the top half of the parabola $y^2 = -1 - x$ or $x = -y^2 - 1$) and then we reflect in the line $y = x$ to get the graph of f^{-1} (see Figure 6.11).   ●

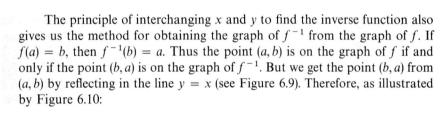
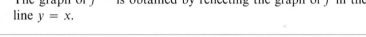

Figure 6.11

Now let us look at inverse functions from the point of view of calculus. Suppose that f is both one-to-one and continuous. We think of a continuous function as one whose graph has no breaks in it. (It consists of just one piece.) Since the graph of f^{-1} is obtained from the graph of f by reflecting in the line $y = x$, the graph of f^{-1} has no breaks in it either (see Figure 6.10). Thus we would expect that f^{-1} is also a continuous function.

This geometrical argument does not prove the following theorem but at least it makes the theorem plausible. A proof can be found in Appendix C.

Theorem (6.11)

> If f is a one-to-one continuous function defined on an interval, then its inverse function f^{-1} is also continuous.

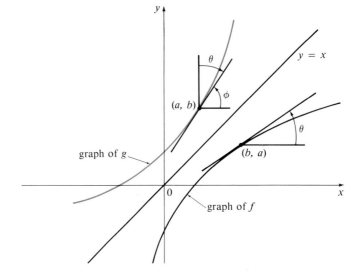

Figure 6.12

Now suppose that f is a one-to-one differentiable function. Geometrically we can think of a differentiable function as one whose graph has no corners or kinks in it. We get the graph of f^{-1} by reflecting the graph of f in the line $y = x$, so the graph of f^{-1} has no corners or kinks in it either. We therefore expect that f^{-1} is also differentiable (except where its tangents are vertical). In fact, we can predict the value of the derivative of f^{-1} at a given point by a geometric argument. In Figure 6.12 the graphs of f and its inverse $g = f^{-1}$ are shown. If $f(b) = a$, then $g(a) = f^{-1}(a) = b$ and $g'(a)$ is the slope of the tangent to the graph of g at (a, b), which is $\tan \phi$. Likewise $f'(b) = \tan \theta$. From Figure 6.12 we see that $\theta + \phi = \pi/2$, so

$$g'(a) = \tan \phi = \tan\left(\frac{\pi}{2} - \theta\right) = \frac{1}{\tan \theta} = \frac{1}{f'(b)}$$

that is, $$g'(a) = \frac{1}{f'(g(a))}$$

Theorem (6.12)

> If f is a one-to-one differentiable function with inverse function $g = f^{-1}$ and $f'(g(a)) \neq 0$, then the inverse function is differentiable at a and
>
> $$g'(a) = \frac{1}{f'(g(a))}$$

Proof   Write the definition of derivative in the form given by (3.3):

$$g'(a) = \lim_{x \to a} \frac{g(x) - g(a)}{x - a}$$

By (6.8) we have

$$g(x) = y \quad \Leftrightarrow \quad f(y) = x$$

and

$$g(a) = b \quad \Leftrightarrow \quad f(b) = a$$

Since f is differentiable, it is continuous, so $g = f^{-1}$ is continuous by Theorem 6.11. Thus if $x \to a$, then $g(x) \to g(a)$, that is, $y \to b$. Therefore

$$
\begin{aligned}
g'(a) &= \lim_{x \to a} \frac{g(x) - g(a)}{x - a} \\
&= \lim_{y \to b} \frac{y - b}{f(y) - f(b)} \\
&= \lim_{y \to b} \frac{1}{\dfrac{f(y) - f(b)}{y - b}} = \frac{1}{\displaystyle\lim_{y \to b} \frac{f(y) - f(b)}{y - b}} \\
&= \frac{1}{f'(b)} = \frac{1}{f'(g(a))}
\end{aligned}
$$

Note 1:   Replacing a by the general number x in the formula of Theorem 6.12, we get

(6.13)

$$g'(x) = \frac{1}{f'(g(x))}$$

If we write $y = g(x)$, then $f(y) = x$, so Equation 6.13, when expressed in Leibniz notation, becomes

$$\frac{dy}{dx} = \frac{1}{\dfrac{dx}{dy}}$$

Note 2:   If it is known in advance that f^{-1} is differentiable, then its derivative can be computed more easily than in the proof of Theorem 6.12 by using implicit differentiation. If $y = f^{-1}(x)$, then $f(y) = x$. Differentiating the equation $f(y) = x$ implicitly with respect to x, remembering that y is a function of x, and using the Chain Rule, we get

$$f'(y) \frac{dy}{dx} = 1$$

Therefore

$$\frac{dy}{dx} = \frac{1}{f'(y)} = \frac{1}{\dfrac{dx}{dy}}$$

● **Example 5**   Although the function $y = x^2$, $x \in R$, is not one-to-one and therefore does not have an inverse function, we can make it one-to-one by restricting its domain. For instance, the function $f(x) = x^2$, $0 \leqslant x \leqslant 2$, is

Figure 6.13

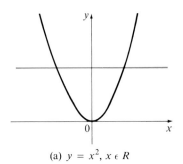

(a) $y = x^2, x \in R$

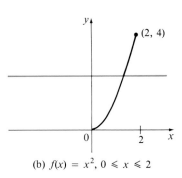

(b) $f(x) = x^2, 0 \leqslant x \leqslant 2$

one-to-one (by the Horizontal Line Test) and has domain $[0, 2]$ and range $[0, 4]$ (see Figure 6.13). Thus f has an inverse function $g = f^{-1}$ with domain $[0, 4]$ and range $[0, 2]$.

Without computing a formula for g' we can still calculate $g'(1)$. Since $f(1) = 1$, we have $g(1) = 1$. Also $f'(x) = 2x$. So by Theorem 6.12 we have

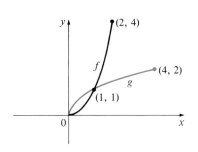

Figure 6.14

$$g'(1) = \frac{1}{f'(g(1))} = \frac{1}{f'(1)} = \frac{1}{2}$$

In this case it is easy to find g explicitly. In fact, $g(x) = \sqrt{x}, 0 \leqslant x \leqslant 4$. [In general we could use the method given by (6.10).] Then $g'(x) = 1/2\sqrt{x}$, so $g'(1) = \frac{1}{2}$, which agrees with the preceding computation. The functions f and g are graphed in Figure 6.14. ●

EXERCISES 6.2

In Exercises 1–6 the graph of a function f is shown. Determine whether or not f is one-to-one.

1.

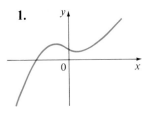

2.

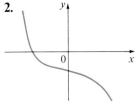

5.

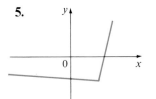

6.

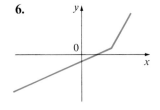

3.

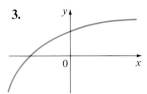

4.

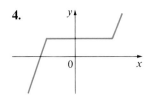

In Exercises 7–12 determine whether or not the given function is one-to-one.

7. $f(x) = 7x - 3$

8. $f(x) = x^2 - 2x + 5$

9. $g(x) = \sqrt{x}$

10. $g(x) = |x|$

11. $h(x) = x^4 + 5$

12. $h(x) = x^4 + 5, 0 \leqslant x \leqslant 2$

In Exercises 13–18 show that f is one-to-one and find its inverse function.

13. $f(x) = 4x + 7$

14. $f(x) = \dfrac{x - 2}{x + 2}$

15. $f(x) = \dfrac{1 + 3x}{5 - 2x}$

16. $f(x) = 5 - 4x^3$

17. $f(x) = \sqrt{2 + 5x}$

18. $f(x) = x^2 + x, \; x \geqslant -\frac{1}{2}$

In Exercises 19–24, (a) show that f is one-to-one, (b) use Theorem 6.12 to find $g'(a)$, where $g = f^{-1}$, (c) calculate $g(x)$ and state the domain and range of g, (d) calculate $g'(a)$ from the formula in part (c) and check that it agrees with the result of part (b), and (e) sketch the graphs of f and g.

19. $f(x) = 2x + 1, \; a = 3$

20. $f(x) = 6 - x, \; a = 2$

21. $f(x) = x^3, \; a = 8$

22. $f(x) = \sqrt{x - 2}, \; a = 2$

23. $f(x) = 9 - x^2, \; 0 \leqslant x \leqslant 3, \; a = 8$

24. $f(x) = \dfrac{1}{x - 1}, \; x > 1, \; a = 2$

25. If n is a positive integer, then $y = \sqrt[n]{x}$ has domain R if n is odd and domain $[0, \infty)$ if n is even. It is the inverse function of the power function. Use the method of Note 2 to find dy/dx.

26. Show that $h(x) = \sin x, \; x \in R$, is not one-to-one, but its restriction $f(x) = \sin x, \; -\pi/2 \leqslant x \leqslant \pi/2$ is one-to-one. Compute the derivative of $f^{-1} = \sin^{-1}$ by the method of Note 2.

27. Prove that every increasing function is one-to-one.

SECTION 6.3

Logarithmic Functions

If $a > 0$ and $a \neq 1$, Theorem 6.2 tells us that the exponential function $f(x) = a^x$ is either increasing or decreasing. Thus, by Theorem 6.6, f is one-to-one. It therefore has an inverse function f^{-1} that is called the **logarithmic function to the base a** and is denoted by $\log_a$. If we use the formulation of an inverse function given by (6.8),

$$f^{-1}(x) = y \quad \Leftrightarrow \quad f(y) = x$$

then we have

(6.14)
$$\log_a x = y \quad \Leftrightarrow \quad a^y = x$$

Thus, if $x > 0$, $\log_a x$ is the exponent to which the base a must be raised to give x.

● **Example 1**   Evaluate (a) $\log_3 81$, (b) $\log_{25} 5$, and (c) $\log_{10} 0.001$.

Solution

(a) $\log_3 81 = 4$ because $3^4 = 81$

(b) $\log_{25} 5 = \frac{1}{2}$ because $25^{1/2} = 5$

(c) $\log_{10} 0.001 = -3$ because $10^{-3} = 0.001$ ●

The cancellation equations (6.9), when applied to $f(x) = a^x$ and $f^{-1}(x) = \log_a x$, become

(6.15)

$$\log_a(a^x) = x \quad \text{for every } x \in R$$

$$a^{\log_a x} = x \quad \text{for every } x > 0$$

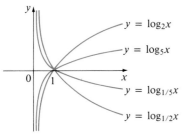

$y = \log_2 x$

$y = \log_5 x$

$y = \log_{1/5} x$

$y = \log_{1/2} x$

Figure 6.15

The logarithmic function, $\log_a$, has domain $(0, \infty)$ and range R and is continuous by Theorem 6.11 since it is the inverse of the exponential function. Its graph is the reflection of the graph of $y = a^x$ in the line $y = x$. Thus Figure 6.15 shows the graphs of $y = \log_a x$ for various values of the base a and is obtained from Figure 6.1.

Since $\log_a 1 = 0$ (because $a^0 = 1$), the graphs of all logarithmic functions pass through the point $(1, 0)$. If $a > 1$, the logarithmic function $y = \log_a x$ increases; if $0 < a < 1$, it decreases. These two cases are illustrated in Figure 6.16.

Figure 6.16

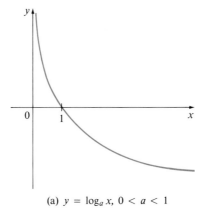

(a) $y = \log_a x, \ 0 < a < 1$

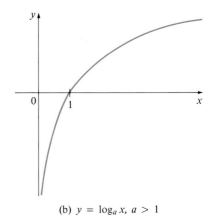

(b) $y = \log_a x, \ a > 1$

The following theorem summarizes the properties of logarithmic functions.

Theorem (6.16)

If $a > 0$, $a \neq 1$, the function $f(x) = \log_a x$ is a one-to-one continuous function with domain $(0, \infty)$ and range R. If $0 < a < 1$, $\log_a$ is decreasing; if $a > 1$, it is increasing. If $x, y > 0$, then

(a) $\log_a(xy) = \log_a x + \log_a y$

(b) $\log_a\left(\dfrac{x}{y}\right) = \log_a x - \log_a y$

(c) $\log_a(x^y) = y \log_a x$

Properties (a), (b), and (c) follow from the corresponding properties in Theorem 6.2 (see Exercise 53).

● **Example 2**   Use the properties of logarithms in Theorem 6.16 to evaluate the following: (a) $\log_4 2 + \log_4 32$ and (b) $\log_2 80 - \log_2 5$.

Solution   (a)   Using (a) in Theorem 6.16 we have

$$\log_4 2 + \log_4 32 = \log_4(2 \cdot 32) = \log_4 64 = 3$$

since $4^3 = 64$.
     (b)   Using (b) we have

$$\log_2 80 - \log_2 5 = \log_2(\tfrac{80}{5}) = \log_2 16 = 4$$

since $2^4 = 16$.   ●

The limits of exponential functions given in (6.3) are reflected in the following limits of logarithmic functions.

(6.17)

> If $a > 1$, then
>
> $$\lim_{x \to \infty} \log_a x = \infty \qquad \text{and} \qquad \lim_{x \to 0^+} \log_a x = -\infty$$
>
> If $0 < a < 1$, then
>
> $$\lim_{x \to \infty} \log_a x = -\infty \qquad \text{and} \qquad \lim_{x \to 0^+} \log_a x = \infty$$

In particular, the y-axis is a vertical asymptote of the curve $y = \log_a x$.

● **Example 3**   Find $\lim_{x \to 0^+} \log_{10}(\tan^2 x)$.

Solution   As $x \to 0$, we know that $\tan x \to \tan 0 = 0$, so by (6.17) with $a = 10 > 1$, we have

$$\lim_{x \to 0^+} \log_{10}(\tan^2 x) = -\infty$$ ●

The Number *e*

Of all possible bases a for logarithms, we shall see in the next section that the most convenient choice of a base is the number e, which is defined as follows:

Definition (6.18)

$$e = \lim_{x \to 0}(1 + x)^{1/x}$$

The graph of the function $y = (1 + x)^{1/x}$ is shown in Figure 6.17. It is not defined when $x = 0$ but its behavior when x is near 0 is indicated by the table of values correct to eight decimal places (compare with Exercise 47 in Section 2.2).

x	$(1 + x)^{1/x}$
0.1	2.59374246
0.01	2.70481383
0.001	2.71692393
0.0001	2.71814593
0.00001	2.71826824
0.000001	2.71828047
0.0000001	2.71828169
0.00000001	2.71828182

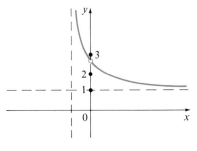

Figure 6.17

These values suggest (but do not prove) that the limit in Definition 6.18 exists and that $e \approx 2.71828$. The existence of the limit will be proved in Chapter 10 (see Exercises 72 and 73 in Section 10.1). The approximate value to 20 decimal places is

$$e \approx 2.71828182845904523536$$

The decimal expansion of e is nonrepeating because e is an irrational number (see Exercise 49 in Section 10.11). In fact, e, like π, is a *transcendental* number, which means that it is not the root of any polynomial equation with integer coefficients. (The notation e for this number was chosen by the Swiss mathematician Leonhard Euler in 1727 probably because it is the first letter of the word *exponential*.)

If we put $n = 1/x$ in Definition 6.18, then $n \to \infty$ as $x \to 0$, so an alternative expression for e is

(6.19)
$$e = \lim_{n \to \infty} \left(1 + \frac{1}{n}\right)^n$$

The logarithm with base e is called the **natural logarithm** and has a special notation:

$$\log_e x = \ln x$$

(Another notation that is sometimes used is $\log_e x = \log x$; that is, the base is omitted because e is the most frequently used base.)

If we put $a = e$ and $\log_e = \ln$ in (6.14) and (6.15), then the defining properties of the natural logarithm function become

(6.20)

$$\ln x = y \quad \Leftrightarrow \quad e^y = x$$

(6.21)   and

$$\ln(e^x) = x \qquad x \in R$$
$$e^{\ln x} = x \qquad x > 0$$

In particular, if we set $x = 1$, we get

$$\ln e = 1$$

● **Example 4**   Find x if $\ln x = 5$.

Solution 1   From (6.20) we see that

$$\ln x = 5 \quad \text{means} \quad e^5 = x$$

Therefore $x = e^5$.

(If you have trouble working with the "ln" notation, just replace it by $\log_e$. Then the equation becomes $\log_e x = 5$; so by the definition of logarithm, $e^5 = x$.)

Solution 2   Start with the equation

$$\ln x = 5$$

and apply the exponential function to both sides of the equation:

$$e^{\ln x} = e^5$$

But the second cancellation equation in (6.21) says that $e^{\ln x} = x$. Therefore $x = e^5$.   ●

● **Example 5**   Express $\ln a + \frac{1}{2}\ln b$ as a single logarithm.

Solution   Using properties (c) and (a) of logarithms, we have

$$\ln a + \tfrac{1}{2}\ln b = \ln a + \ln b^{1/2}$$
$$= \ln a + \ln \sqrt{b}$$
$$= \ln(a\sqrt{b})$$ ●

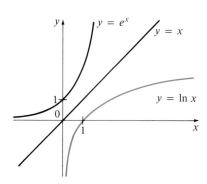

Figure 6.18

The graphs of the exponential function $y = e^x$ and its inverse function, the natural logarithm function, are shown in Figure 6.18.

If we put $a = e$ in (6.3) and (6.17), then we have the following limits:

(6.22)

$$\lim_{x \to \infty} e^x = \infty \qquad \lim_{x \to -\infty} e^x = 0$$

$$\lim_{x \to \infty} \ln x = \infty \qquad \lim_{x \to 0^+} \ln x = -\infty$$

● **Example 6**   Sketch the graph of the function $y = \ln(x - 2) - 1$.

Solution   We start with the graph of $y = \ln x$ as given in Figure 6.18. Using the transformations of Section 1.7, we shift it two units to the right to get the graph of $y = \ln(x - 2)$ and then we shift it one unit downward to get the graph of $y = \ln(x - 2) - 1$ (see Figure 6.19). Notice that the line $x = 2$ is a vertical asymptote since

$$\lim_{x \to 2^+} [\ln(x - 2) - 1] = -\infty$$

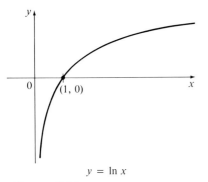

$y = \ln x$

Figure 6.19

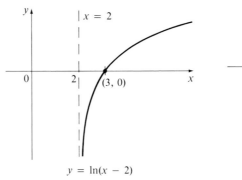

$y = \ln(x - 2)$

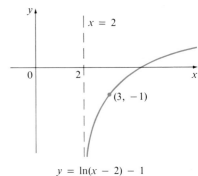

$y = \ln(x - 2) - 1$

●

Exponential functions of the form $f(x) = e^{cx}$, where c is a constant, occur frequently in mathematics and most of the sciences, and so it is wise to become familiar with them. Notice that $f(x) = e^{cx} = (e^c)^x = a^x$ where $a = e^c$. If $c > 0$, then $a > 1$; if $c < 0$, then $0 < a < 1$. Therefore:

(6.23)

If $c > 0$, then

$$\lim_{x \to \infty} e^{cx} = \infty \qquad \text{and} \qquad \lim_{x \to -\infty} e^{cx} = 0$$

If $c < 0$, then

$$\lim_{x \to \infty} e^{cx} = 0 \qquad \text{and} \qquad \lim_{x \to -\infty} e^{cx} = \infty$$

The two cases are illustrated in Figure 6.20.

Figure 6.20

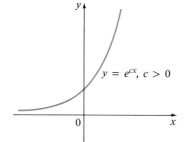

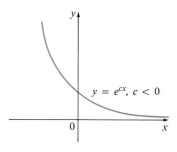

● **Example 7**   Find $\lim\limits_{x \to \infty} \dfrac{e^{2x}}{e^{2x} + 1}$.

Solution   Divide numerator and denominator by e^{2x}:

$$\lim_{x \to \infty} \frac{e^{2x}}{e^{2x} + 1} = \lim_{x \to \infty} \frac{1}{1 + e^{-2x}}$$

$$= \frac{1}{1 + \lim\limits_{x \to \infty} e^{-2x}}$$

$$= \frac{1}{1 + 0} \qquad [by\ (6.23)]$$

$$= 1 \qquad\qquad ●$$

EXERCISES 6.3

In Exercises 1–14 evaluate the given expression.

1. $\log_2 64$

2. $\log_6 \frac{1}{36}$

3. $\log_8 2$

4. $\log_8 4$

5. $\log_3 \frac{1}{27}$

6. $e^{\ln 6}$

7. $\ln e^{\sqrt{2}}$

8. $\log_3 3^{\sqrt{5}}$

9. $\log_{10} 1.25 + \log_{10} 80$

10. $\log_3 108 - \log_3 4$

11. $\log_8 6 - \log_8 3 + \log_8 4$

12. $\log_5 10 + \log_5 20 - 3\log_5 2$

13. $2^{(\log_2 3 + \log_2 5)}$

14. $e^{3 \ln 2}$

In Exercises 15–18 express the given quantity as a single logarithm.

15. $\log_5 a + \log_5 b - \log_5 c$

16. $\log_2 x + 5\log_2(x + 1) + \frac{1}{2}\log_2(x - 1)$

17. $\frac{1}{3}\ln x - 4\ln(2x + 3)$ **18.** $\ln x + a \ln y - b \ln z$

19. Sketch, using the same axes, the graphs of the following functions.
  (a) $y = \log_3 x$ (b) $y = \log_{10} x$
  (c) $y = \log_{1/3} x$ (d) $y = \log_{1/10} x$

20. Sketch, using the same axes, the graphs of the following functions.
  (a) $y = \log_2 x$ (b) $y = \ln x$ (c) $y = \log_{10} x$

In Exercises 21–38 make a rough sketch of the graph of the given function. Do not use a calculator. Just use the graphs given in Figures 6.15, 6.16, 6.18, and 6.20 and, if necessary, the transformations of Section 1.7.

21. $y = \log_{1.1} x$

22. $y = \log_{100} x$

23. $y = \log_{0.1} x$

24. $y = \log_{0.9} x$

25. $y = \log_{10}(x + 5)$ **26.** $y = 1 + \log_5(x - 1)$

27. $y = -\ln x$ **28.** $y = \ln(-x)$

29. $y = -\ln(-x)$ **30.** $y = \ln|x|$

31. $y = \ln(x^2)$ **32.** $y = \ln \dfrac{1}{x}$

33. $y = \ln(x + 3)$ **34.** $y = \ln|x + 3|$

35. $y = 2e^{-3x}$ **36.** $y = e^{|x - 2|}$

37. $y = 1 - e^{2x}$ **38.** $y = e^{-x} \sin x$

In Exercises 39–52 solve the given equations for x.

39. $\log_2 x = 3$ **40.** $x = e^{\ln \pi}$

41. $y = \log_5(x - 1)$ **42.** $\log_{10} x^2 = A$

43. $3^{x+2} = m$ **44.** $y = e^{3x-4}$

45. $\ln x = 2$ **46.** $5^{\log_5(2x)} = 6$

47. $\ln x = \ln 5 + \ln 8$ **48.** $\ln x^2 = 2 \ln 4 - 4 \ln 2$

49. $\ln(e^{2x-1}) = 5$ **50.** $\ln x + \ln(x - 1) = 1$

51. $y = \ln(\ln x)$ **52.** $y = \log_2(\log_3(\log_4 x))$

53. Prove properties (a), (b), and (c) of logarithms in Theorem 6.16 by deducing them from properties (a), (b), and (c) of exponential functions in Theorem 6.2.

54. (a) Prove that $\log_a b \, \log_b c = \log_a c$.

     (b) Deduce that $\log_a b = \dfrac{1}{\log_b a}$.

     (c) Deduce that $\log_a e = \dfrac{1}{\ln a}$.

55. Deduce from Exercise 54(a) that

$$\log_b c = \frac{\ln c}{\ln b}$$

56. Use Exercise 55 to evaluate the following logarithms correct to six decimal places.
(a) $\log_2 5$ (b) $\log_5 26.05$
(c) $\log_3 e$ (d) $\log_{0.7} 14$

57. The geologist C. F. Richter defined the magnitude of an earthquake to be $\log_{10}(I/S)$ where I is the intensity of the earthquake (measured by the amplitude of a seismograph 100 km from the earthquake) and S is the intensity of a "standard" earthquake (where the amplitude is only 1 micron $= 10^{-4}$ cm). The 1906 earthquake in San Francisco had a magnitude of 8.3 on the Richter scale. In the same year the strongest earthquake ever recorded occurred on the Colombia–

Ecuador border and was four times as intense. What was the magnitude of the Colombia–Ecuador earthquake on the Richter scale?

58. A sound so faint that it can just be heard has intensity $I_0 = 10^{-12}$ watt/m^2 at a frequency of 1000 hertz. The loudness, in decibels, of a sound with intensity I is then defined to be $L = 10 \log_{10}(I/I_0)$. Rock music with amplifiers is measured at 120 decibels. The noise from a power mower is measured at 106 decibels. Find the ratio of the intensity of the rock music to that of the power mower.

Find the limits in Exercises 59–74.

59. $\displaystyle \lim_{x \to 5^+} \ln(x - 5)$ **60.** $\displaystyle \lim_{x \to 0^+} \log_{0.5}(4x)$

61. $\displaystyle \lim_{x \to \infty} \log_2(x^2 - x)$ **62.** $\displaystyle \lim_{x \to 0^+} \ln(\sin x)$

63. $\displaystyle \lim_{x \to \infty} \frac{e^{3x} - e^{-3x}}{e^{3x} + e^{-3x}}$ **64.** $\displaystyle \lim_{x \to -\infty} \frac{e^{3x} - e^{-3x}}{e^{3x} + e^{-3x}}$

65. $\displaystyle \lim_{x \to (\pi/2)^-} \log_{10}(\cos x)$ **66.** $\displaystyle \lim_{x \to \infty} \frac{\ln x}{1 + \ln x}$

67. $\displaystyle \lim_{x \to 1^-} e^{2/(x-1)}$ **68.** $\displaystyle \lim_{x \to 1^+} e^{2/(x-1)}$

69. $\displaystyle \lim_{x \to (\pi/2)^-} \frac{2}{1 + e^{\tan x}}$ **70.** $\displaystyle \lim_{x \to 0^-} \frac{2}{1 + e^{\cot x}}$

71. $\displaystyle \lim_{x \to \infty} \left[\ln(2 + x) - \ln(1 + x) \right]$

72. $\displaystyle \lim_{x \to 0} (1 + x)^{2/x}$

73. $\displaystyle \lim_{x \to \infty} \frac{e^x}{e^{2x} + e^{-x}}$ **74.** $\displaystyle \lim_{x \to -\infty} \frac{e^x + 2e^{-x}}{e^x - 3e^{-x}}$

In Exercises 75–78 find the domain and range of the given function.

75. $f(x) = \log_{10}(1 - x)$ **76.** $g(x) = \ln(4 - x^2)$

77. $F(t) = \sqrt{t} \ln(t^2 - 1)$ **78.** $G(t) = \ln(t^3 - t)$

In Exercises 79–84 find the inverse functions of the given functions.

79. $y = \ln(x + 3)$ **80.** $y = 2^{10^x}$

81. $y = e^{\sqrt{x}}$ **82.** $y = (\ln x)^2, \; x \geqslant 1$

83. $y = \dfrac{10^x}{10^x + 1}$ **84.** $y = \dfrac{1 + e^x}{1 - e^x}$

85. Let $a > 1$. Prove, using Definitions 4.34 and 4.41, that

(a) $\displaystyle \lim_{x \to -\infty} a^x = 0$ (b) $\displaystyle \lim_{x \to \infty} a^x = \infty$

86. Any function of the form $f(x) = [g(x)]^{h(x)}$, where $h(x) > 0$, can be analyzed as a power of e by writing $g(x) = e^{\ln g(x)}$ so that $f(x) = e^{h(x) \ln g(x)}$. Using this device, calculate

(a) $\lim\limits_{x \to \infty} x^{\ln x}$ (b) $\lim\limits_{x \to 0^+} x^{-\ln x}$

(c) $\lim\limits_{x \to 0^+} x^{1/x}$ (d) $\lim\limits_{x \to \infty} (\ln 2x)^{-\ln x}$

87. Find $\lim_{x \to 0}(1 + 5x)^{1/x}$. (*Hint:* Make the change of variable $t = 5x$.)

88. Prove that $\log_2 5$ is an irrational number.

SECTION 6.4

Derivatives of Logarithmic Functions

In this section we calculate the derivative of the function $y = \log_a x$. We shall see that the differentiation formula is simplest when $a = e$.

Theorem (6.24)

> The function $f(x) = \log_a x$ is differentiable and
> $$f'(x) = \frac{1}{x} \log_a e$$

Proof

$$f'(x) = \lim_{h \to 0} \frac{f(x + h) - f(x)}{h}$$

$$= \lim_{h \to 0} \frac{\log_a(x + h) - \log_a x}{h}$$

$$= \lim_{h \to 0} \frac{\log_a\left(\dfrac{x + h}{x}\right)}{h}$$

$$= \lim_{h \to 0} \frac{1}{h} \log_a\left(1 + \frac{h}{x}\right)$$

$$= \lim_{h \to 0} \frac{1}{x} \cdot \frac{x}{h} \log_a\left(1 + \frac{h}{x}\right)$$

$$= \frac{1}{x} \lim_{h \to 0} \frac{x}{h} \log_a\left(1 + \frac{h}{x}\right) \qquad \text{(by Property 3 of limits)}$$

$$= \frac{1}{x} \lim_{h \to 0} \log_a\left(1 + \frac{h}{x}\right)^{x/h} \qquad \text{[by (6.16) (c)]}$$

$$= \frac{1}{x} \log_a\left[\lim_{h \to 0}\left(1 + \frac{h}{x}\right)^{x/h}\right] \qquad \text{(since } \log_a \text{ is continuous)}$$

$$= \frac{1}{x} \log_a\left[\lim_{h \to 0}\left(1 + \frac{h}{x}\right)^{1/(h/x)}\right] = \frac{1}{x} \log_a e$$

The final step may be seen more clearly by making the change of variable $t = h/x$. As $h \to 0$, we also have $t \to 0$, so

$$\lim_{h \to 0}\left(1 + \frac{h}{x}\right)^{1/(h/x)} = \lim_{t \to 0}(1 + t)^{1/t} = e$$

by the definition of e. Thus

$$f'(x) = \frac{1}{x}\log_a e$$

● **Example 1**   If $f(x) = \log_{10}(1 + x^2)$, find $f'(x)$.

Solution   Combining Theorem 6.24 with the Chain Rule, we have

$$f'(x) = \frac{1}{1 + x^2}\log_{10} e \frac{d}{dx}(1 + x^2)$$

$$= \frac{2x}{1 + x^2}\log_{10} e$$

Note: We know from Exercise 54 in Section 6.3 that $\log_a e = 1/\ln a$. (More directly, if $y = \log_a e$, then $a^y = e$ gives $y \ln a = \ln e = 1$, so $y = 1/\ln a$.) Thus Theorem 6.24 can be rewritten as

(6.25)

$$\boxed{\frac{d}{dx}\log_a x = \frac{1}{x \ln a}}$$

If we now put $a = e$ in Theorem 6.24 or Equation 6.25, we obtain the important formula

(6.26)

$$\boxed{\frac{d}{dx}\ln x = \frac{1}{x}}$$

You can now see one of the main reasons that natural logarithms, that is, logarithms to the base e, are used in calculus: the constant $\log_a e = 1/\ln a$ disappears from the differentiation formula when $a = e$.

When combined with the Chain Rule, Equation 6.26 becomes

(6.27)

$$\boxed{\frac{d}{dx}\ln u = \frac{1}{u}\frac{du}{dx}} \quad \text{or} \quad \boxed{\frac{d}{dx}\ln g(x) = \frac{g'(x)}{g(x)}}$$

● **Example 2**   Find $\dfrac{d}{dx}\ln(\sin x)$.

Solution   Using (6.27) we have

$$\frac{d}{dx}\ln(\sin x) = \frac{1}{\sin x}\frac{d}{dx}\sin x$$

$$= \frac{1}{\sin x}\cos x = \cot x$$ ●

● **Example 3**   Find $\dfrac{d}{dx}\ln\dfrac{x+1}{\sqrt{x-2}}$.

Solution 1

$$\frac{d}{dx}\ln\frac{x+1}{\sqrt{x-2}} = \frac{1}{\dfrac{x+1}{\sqrt{x-2}}}\frac{d}{dx}\frac{x+1}{\sqrt{x-2}}$$

$$= \frac{\sqrt{x-2}}{x+1}\frac{1\cdot\sqrt{x-2} - (x+1)(\frac{1}{2})(x+2)^{-1/2}}{x-2}$$

$$= \frac{x-2 - (\frac{1}{2})(x+1)}{(x+1)(x-2)} = \frac{x-5}{2(x+1)(x-2)}$$

Solution 2   If we first simplify the given function using the laws of logarithms, then the differentiation becomes easier:

$$\frac{d}{dx}\ln\frac{x+1}{\sqrt{x-2}} = \frac{d}{dx}\left[\ln(x+1) - \frac{1}{2}\ln(x-2)\right]$$

$$= \frac{1}{x+1} - \frac{1}{2}\left(\frac{1}{x-2}\right)$$

(This answer can be left as it is, but if we were to use a common denominator we would see that it gives the same answer as in Solution 1.)   ●

● **Example 4**   Find $f'(x)$ if $f(x) = \ln|x|$.

Solution   Since

$$f(x) = \begin{cases} \ln x & \text{if } x > 0 \\ \ln(-x) & \text{if } x < 0 \end{cases}$$

it follows that

$$f'(x) = \begin{cases} \dfrac{1}{x} & \text{if } x > 0 \\ \dfrac{1}{-x}(-1) = \dfrac{1}{x} & \text{if } x < 0 \end{cases}$$

Thus $f'(x) = 1/x$ for all $x \neq 0$.   ●

The result of this example is worth remembering:

(6.28)

$$\frac{d}{dx} \ln|x| = \frac{1}{x}$$

The corresponding integration formula is

(6.29)

$$\int \frac{1}{x} dx = \ln|x| + C$$

Notice that this fills the gap in the rule for integrating power functions:

$$\int x^n dx = \frac{x^{n+1}}{n+1} + C \qquad \text{if } n \neq -1$$

The missing case ($n = -1$) is supplied by Formula 6.29.

● Example 5   Find, correct to three decimal places, the area of the region under the hyperbola $xy = 1$ from $x = 1$ to $x = 2$.

Solution   The given region is shown in Figure 6.21. Using Formula 6.29 (without the absolute value sign since $x > 0$) we see that the area is

$$A = \int_1^2 \frac{1}{x} dx = \ln x \Big]_1^2$$

$$= \ln 2 - \ln 1 = \ln 2 \approx 0.693 \qquad ●$$

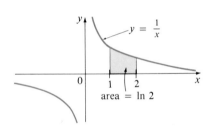

Figure 6.21

● Example 6   Evaluate $\int \frac{x}{x^2 + 1} dx$.

Solution   We make the substitution $u = x^2 + 1$ because the differential $du = 2x\,dx$ occurs (except for the constant factor 2). Thus $x\,dx = \frac{1}{2}du$ and

$$\int \frac{x}{x^2 + 1} dx = \frac{1}{2} \int \frac{du}{u} = \frac{1}{2} \ln|u| + C$$

$$= \frac{1}{2} \ln|x^2 + 1| + C = \frac{1}{2} \ln(x^2 + 1) + C$$

Notice that we removed the absolute value signs because $x^2 + 1 > 0$ for all x. We could use the laws of logarithms to write the answer as

$$\ln \sqrt{x^2 + 1} + C$$

but this is not necessary.   ●

● **Example 7**   Calculate $\int_1^e \dfrac{\ln x}{x}\, dx$.

Solution   We let $u = \ln x$ because its differential $du = dx/x$ occurs in the integral. When $x = 1$, $u = \ln 1 = 0$; when $x = e$, $u = \ln e = 1$. Thus

$$\int_1^e \frac{\ln x}{x}\, dx = \int_0^1 u\, du = \frac{u^2}{2}\bigg]_0^1 = \frac{1}{2}$$

● **Example 8**   Calculate $\int \tan x\, dx$.

Solution   First we write tangent in terms of sine and cosine:

$$\int \tan x\, dx = \int \frac{\sin x}{\cos x}\, dx$$

This suggests that we substitute $u = \cos x$ since then $du = -\sin x\, dx$ and so $\sin x\, dx = -du$:

$$\int \tan x\, dx = \int \frac{\sin x}{\cos x}\, dx = -\int \frac{du}{u}$$
$$= -\ln|u| + C = -\ln|\cos x| + C$$

Thus we have the formula

(6.30)
$$\boxed{\int \tan x\, dx = -\ln|\cos x| + C}$$

Since $-\ln|\cos x| = \ln(1/|\cos x|) = \ln|\sec x|$, Formula 6.30 can also be written as

(6.31)
$$\boxed{\int \tan x\, dx = \ln|\sec x| + C}$$

Logarithmic Differentiation

● **Example 9**   Differentiate

$$y = \frac{x^{3/4}\sqrt{x^2 + 1}}{(3x + 2)^5}$$

Solution   Although this function contains no logarithms, the differentiation can be simplified by taking logarithms of both sides:

$$\ln y = \tfrac{3}{4}\ln x + \tfrac{1}{2}\ln(x^2 + 1) - 5\ln(3x + 2)$$

Differentiating implicitly with respect to x gives

$$\frac{1}{y}\frac{dy}{dx} = \frac{3}{4}\cdot\frac{1}{x} + \frac{1}{2}\cdot\frac{2x}{x^2 + 1} - 5\cdot\frac{3}{3x + 2}$$

Solving for dy/dx, we get

$$\frac{dy}{dx} = y\left(\frac{3}{4x} + \frac{x}{x^2 + 1} - \frac{15}{3x + 2}\right)$$

$$= \frac{x^{3/4}\sqrt{x^2 + 1}}{(3x + 2)^5}\left(\frac{3}{4x} + \frac{x}{x^2 + 1} - \frac{15}{3x + 2}\right) \qquad \bullet$$

The procedure used in Example 9 is called **logarithmic differentiation.** It is useful when differentiating a function that is simplified by the use of logarithms.

Steps in Logarithmic Differentiation

1. Take logarithms of both sides of an equation $y = f(x)$.
2. Differentiate implicitly with respect to x.
3. Solve the resulting equation for y'.

If $f(x) < 0$ for some values of x, then $\ln f(x)$ is not defined, but we can write $|y| = |f(x)|$ and use Equation 6.28 as in the following example.

● **Example 10**   Find y' if $y = \dfrac{1 + 2\cos x}{x^3(2x + 1)^7}$.

Solution   Since the various parts of this function can be negative, we first write

$$|y| = \left|\frac{1 + 2\cos x}{x^3(2x + 1)^7}\right| = \frac{|1 + 2\cos x|}{|x|^3|2x + 1|^7}$$

Taking logarithms of both sides, we have

$$\ln|y| = \ln|1 + 2\cos x| - 3\ln|x| - 7\ln|2x + 1|$$

Then, using Equation 6.28 to differentiate this equation with respect to x, we get

$$\frac{1}{y}y' = \frac{-2\sin x}{1 + 2\cos x} - \frac{3}{x} - \frac{7(2)}{2x + 1}$$

$$y' = -\frac{1 + 2\cos x}{x^3(2x + 1)^7}\left(\frac{2\sin x}{1 + 2\cos x} + \frac{3}{x} + \frac{14}{2x + 1}\right) \qquad \bullet$$

Curve Sketching

● **Example 11**   Discuss the curve $y = \ln(4 - x^2)$ under the headings A–H of Section 4.7.

Solution

A.   The domain is

$$\{x\,|\,4 - x^2 > 0\} = \{x\,|\,x^2 < 4\} = \{x\,|\,|x| < 2\} = (-2, 2)$$

B.   The y-intercept is $f(0) = \ln 4$. To find the x-intercept we set

$$y = \ln(4 - x^2) = 0$$

We know that $\ln 1 = \log_e 1 = 0$ (since $e^0 = 1$), so we have

$$4 - x^2 = 1$$
$$x^2 = 3$$

and therefore the x-intercepts are $\pm\sqrt{3}$.

C.   Since $f(-x) = f(x)$, f is even and the curve is symmetric about the y-axis.

D.   We look for vertical asymptotes at the endpoints of the domain. Since $4 - x^2 \to 0^+$ as $x \to 2^-$ and also as $x \to -2^+$, we have

$$\lim_{x \to 2^-} \ln(4 - x^2) = -\infty \qquad \lim_{x \to -2^+} \ln(4 - x^2) = -\infty$$

by (6.22). Thus the lines $x = 2$ and $x = -2$ are vertical asymptotes.

E. $$f'(x) = \frac{-2x}{4 - x^2}$$

Since $f'(x) > 0$ when $-2 < x < 0$ and $f'(x) < 0$ when $0 < x < 2$, f is increasing on $(-2, 0]$ and decreasing on $[0, 2)$.

F.   The only critical number is $x = 0$. Since f' changes from positive to negative at 0, $f(0) = \ln 4$ is a local maximum by the First Derivative Test.

G. $$f''(x) = \frac{(4 - x^2)(-2) + 2x(-2x)}{(4 - x^2)^2} = \frac{-8 - 2x^2}{(4 - x^2)^2}$$

Since $f''(x) < 0$ for all x, the curve is concave downward on $(-2, 2)$ and there is no inflection point.

H.   Using this information, we sketch the curve in Figure 6.22.

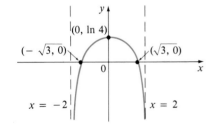

Figure 6.22

EXERCISES 6.4

In Exercises 1–10 find f' and state the domains of f and f'.

1. $f(x) = \ln(x + 1)$

2. $f(x) = \log_{10}(5 - 2x)$

3. $f(x) = \log_3(x^2 - 4)$

4. $f(x) = \ln \ln x$

5. $f(x) = \ln(\cos x)$

6. $f(x) = \cos(\ln x)$

7. $f(x) = \ln(2 - x - x^2)$

8. $f(x) = \ln \ln \ln x$

9. $f(x) = x^2 \ln(1 - x^2)$

10. $f(x) = \ln(\sqrt{x} - \sqrt{x - 1})$

In Exercises 11–14 find y' and y''.

11. $y = \log_{10} x$ **12.** $y = \ln(ax)$

13. $y = x \ln x$ **14.** $y = \ln(\sec x + \tan x)$

Differentiate the functions in Exercises 15–36.

15. $f(x) = \sqrt{x}\, \ln x$ **16.** $f(x) = \log_{10}\left(\dfrac{x}{x-1}\right)$

17. $g(x) = \ln \dfrac{a-x}{a+x}$ **18.** $h(x) = \ln(x + \sqrt{x^2 - 1})$

19. $F(x) = \ln \sqrt{x}$ **20.** $G(x) = \sqrt{\ln x}$

21. $f(t) = \log_2(t^4 - t^2 + 1)$

22. $g(t) = \sin(\ln t)$

23. $h(y) = \ln(y^3 \sin y)$

24. $k(r) = r \sin r \ln r$

25. $g(u) = \dfrac{1 - \ln u}{1 + \ln u}$ **26.** $G(u) = \ln \sqrt{\dfrac{3u + 2}{3u - 2}}$

27. $y = (\ln \sin x)^3$ **28.** $y = \log_3(\log_2 x)$

29. $y = \dfrac{\ln x}{1 + x^2}$ **30.** $y = \ln(x \sqrt{1 - x^2} \sin x)$

31. $y = \ln\left(\dfrac{x+1}{x-1}\right)^{3/5}$ **32.** $y = \ln|\tan 2x|$

33. $y = \ln|x^3 - x^2|$ **34.** $y = \tan[\ln(ax + b)]$

35. $y = \ln(x + \ln x)$ **36.** $y = \log_x e$

Evaluate the integrals in Exercises 37–48.

37. $\displaystyle \int_4^8 \frac{1}{x}\, dx$ **38.** $\displaystyle \int_{-e^2}^{-e} \frac{3}{x}\, dx$

39. $\displaystyle \int_1^e \frac{x^2 + x + 1}{x}\, dx$ **40.** $\displaystyle \int_4^9 \left(\sqrt{x} + \frac{1}{\sqrt{x}}\right)^2 dx$

41. $\displaystyle \int \frac{dx}{2x - 1}$ **42.** $\displaystyle \int \frac{x^2 + 1}{x^3 + 3x + 1}\, dx$

43. $\displaystyle \int \frac{x + 1}{x^2 + 2x}\, dx$ **44.** $\displaystyle \int \frac{dx}{x \ln x}$

45. $\displaystyle \int \frac{(\ln x)^2}{x}\, dx$ **46.** $\displaystyle \int \frac{\sec^2 x}{2 - \tan x}\, dx$

47. $\displaystyle \int \frac{\sin x}{1 + \cos x}\, dx$ **48.** $\displaystyle \int \frac{(1 + \ln x)^4}{x}\, dx$

49.   Show that $\int \cot x\, dx = \ln|\sin x| + C$ (a) by differentiating the right side of the equation and (b) by using the method of Example 8.

50.   Find, correct to three decimal places, the area of the region above the hyperbola $y = 2/(x - 2)$, below the x-axis, and between the lines $x = -4$ and $x = -1$.

In Exercises 51–56 use logarithmic differentiation to find the derivatives of the given functions.

51. $y = (3x - 7)^4(8x^2 - 1)^3$

52. $y = x^{2/5}(x^2 + 8)^4 e^{x^2 + x}$

53. $y = \dfrac{(x + 1)^4(x - 5)^3}{(x - 3)^8}$ **54.** $y = \sqrt{\dfrac{x^2 + 1}{x + 1}}$

55. $y = \dfrac{e^x \sqrt{x^5 + 2}}{(x + 1)^4(x^2 + 3)^2}$ **56.** $y = \dfrac{(x^3 + 1)^4 \sin^2 x}{\sqrt[3]{x}}$

57.   Find y' if $y = \ln(x^2 + y^2)$.

58.   Find y' if $\ln(x + y) = x^2 + y^2$.

59.   If $f(x) = \dfrac{x}{\ln x}$, find $f'(e)$.

60.   If $f(x) = x^2 \ln x$, find $f'(1)$.

In Exercises 61 and 62 find the equation of the tangent line to the given curve at the given point.

61. $y = \ln \ln x$, $(e, 0)$ **62.** $y = \ln x$, $(a, \ln a)$

63.   Find a formula for $f^{(n)}(x)$ if $f(x) = \ln(x - 1)$.

64.   Find $\dfrac{d^9}{dx^9}(x^8 \ln x)$.

65.   If g is the inverse function of $f(x) = 2x + \ln x$, find $g'(2)$.

66.   If $h = f^{-1}$, where $f(x) = \cos x + 5\ln(x + 1)$, find $h'(1)$.

In Exercises 67 and 68 find the intervals of concavity and the points of inflection of the given function.

67. $f(x) = \dfrac{\ln x}{\sqrt{x}}$ **68.** $f(x) = \sqrt{x}\, \ln x$

Discuss the curves in Exercises 69–76 under the headings A–H of Section 4.7.

69. $y = \ln(\cos x)$ **70.** $y = \ln(\sin x)$

71. $y = x + \ln x$ **72.** $y = x^2 + \ln x$

73. $y = \ln(x + \sqrt{1 + x^2})$ **74.** $y = \ln(\tan^2 x)$

75. $y = \ln(1 + x^2)$ **76.** $y = \ln(x^2 - x)$

77. Find the most general antiderivative of the function $f(x) = 1/x$.

78. Find f if $f''(x) = x^{-2}$, $x > 0$, $f(1) = 0$, and $f(2) = 0$.

79. Prove that $\ln(1 + x) > x - x^2/2$ for $x > 0$.

80. Use the definition of derivative to prove that

$$\lim_{x \to 0} \frac{\ln(1 + x)}{x} = 1$$

SECTION 6.5

Derivatives of Exponential Functions

In order to compute the derivative of the exponential function $y = a^x$ we can either use logarithmic differentiation or use the fact that exponential and logarithmic functions are inverse functions.

Theorem (6.32)

The exponential function $f(x) = a^x$, $a > 0$, is differentiable and

$$f'(x) = a^x \ln a$$

Proof   We know that the logarithmic function $y = \log_a x$ is differentiable by Theorem 6.24 so its inverse function $y = a^x$ is differentiable by Theorem 6.12. There are two possible methods for finding y'.

   Method (a) (inverse function method): If $y = a^x$, then $\log_a y = x$. Differentiating this equation implicitly with respect to x, we get

$$\frac{1}{y \ln a} \frac{dy}{dx} = 1$$

Thus $\dfrac{dy}{dx} = y \ln a = a^x \ln a$

   Method (b) (logarithmic differentiation): If $y = a^x$, then $\ln y = x \ln a$. We differentiate this equation with respect to x:

$$\frac{1}{y} \frac{dy}{dx} = \ln a$$

Therefore $\dfrac{dy}{dx} = y \ln a = a^x \ln a$ ●

● **Example 1** $\dfrac{d}{dx} 5^x = 5^x \ln 5$ ●

If we put $a = e$ in Theorem 6.32, the differentiation formula for exponential functions takes on a particularly simple form:

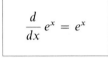

(6.33)

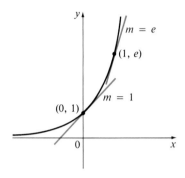

Figure 6.23

$y = e^x$

This equation says that the exponential function $f(x) = e^x$ is its own derivative. This is another reason for the importance of the number e. In fact, it turns out that the most general function that is equal to its own derivative is $f(x) = ce^x$ where c is a constant. (See Theorem 6.59.)

The geometric significance of Equation 6.33 is that the slope of a tangent to the curve $y = e^x$ is equal to the y-coordinate of the point. In particular, if $f(x) = e^x$, then $f'(0) = e^0 = 1$. This means that of all the possible exponential functions $y = a^x$, $y = e^x$ is the one that crosses the y-axis with a slope of 1 (see Figure 6.23).

Combining Theorem 6.32 and Equation 6.33 with the Chain Rule gives

(6.34)

$$\frac{d}{dx} a^u = a^u \ln a \frac{du}{dx} \quad \text{and} \quad \frac{d}{dx} e^u = e^u \frac{du}{dx}$$

● **Example 2**

$$\frac{d}{dx}(\pi^{x^3}) = \pi^{x^3} \ln \pi \frac{d}{dx}(x^3) = (3\ln \pi)x^2 \pi^{x^3}$$

● **Example 3**   If $y = xe^{x^3}$, then

$$y' = 1 \cdot e^{x^3} + x \cdot e^{x^3} \frac{d}{dx}(x^3)$$

$$= e^{x^3} + xe^{x^3}(3x^2) = e^{x^3}(1 + 3x^3)$$

● **Example 4**   In Example 5 in Section 3.7 we considered a population of bacteria cells in a homogeneous nutrient medium. We showed that if the population doubles every hour then the population after t hours is

$$n = n_0 2^t$$

where n_0 is the initial population. Now we can use Theorem 6.32 to compute the growth rate:

$$\frac{dn}{dt} = n_0 2^t \ln 2$$

For instance, if the initial population is $n_0 = 1000$ cells, then the growth rate after 2 h is

$$\frac{dn}{dt}\Bigg]_{t=2} = 1000 \cdot 2^t \ln 2]_{t=2}$$

$$= 4000 \ln 2 \approx 2773 \text{ cells/h} \qquad \bullet$$

Having defined arbitrary powers of positive numbers in this chapter, we are now in a position to prove the general version of the Power Rule.

Theorem (6.35)

> If n is any real number and $f(x) = x^n$, $x > 0$, then
>
> $$f'(x) = nx^{n-1}$$

Proof 1   Let $y = x^n$ and use logarithmic differentiation:

$$\ln y = \ln(x^n) = n \ln x$$

Therefore
$$\frac{y'}{y} = \frac{n}{x}$$

Hence
$$y' = n\frac{y}{x} = n\frac{x^n}{x} = nx^{n-1} \qquad \bullet$$

Proof 2   Since $x = e^{\ln x}$, we have $x^n = (e^{\ln x})^n = e^{n \ln x}$:

$$\frac{d}{dx}(x^n) = \frac{d}{dx}(e^{n \ln x})$$

$$= e^{n \ln x}\frac{d}{dx}(n \ln x)$$

$$= x^n \cdot \frac{n}{x} = nx^{n-1} \qquad \bullet$$

Note 1:   If n is irrational, then the power function $f(x) = x^n$ is defined only for $x > 0$. In those cases where f is defined for $x \in R$, $x \neq 0$, the proof of Theorem 6.35 can be modified by writing $|y| = |x|^n$ and using Equation 6.28.

 Note 2:   You should distinguish carefully between the Power Rule $(Dx^n = nx^{n-1})$, where the base is variable and the exponent is constant, and the rule for differentiating exponential functions $(Da^x = a^x \ln a)$, where the base is constant and the exponent is variable. In general there are four cases for exponents and bases:

1. $\dfrac{d}{dx}(a^b) = 0$ (*a* and *b* are constants)

2. $\dfrac{d}{dx}[f(x)]^b = b[f(x)]^{b-1}f'(x)$

3. $\dfrac{d}{dx}[a^{g(x)}] = a^{g(x)}(\ln a)g'(x)$

4. To find $(d/dx)[f(x)]^{g(x)}$, logarithmic differentiation can be used as in the next example.

● **Example 5**   Differentiate $y = x^{\sqrt{x}}$.

Solution 1   Using logarithmic differentiation, we have

$$\ln y = \ln x^{\sqrt{x}} = \sqrt{x}\,\ln x$$

$$\frac{y'}{y} = \frac{1}{2\sqrt{x}}\ln x + \sqrt{x}\cdot\frac{1}{x}$$

$$y' = y\left(\frac{\ln x}{2\sqrt{x}} + \frac{1}{\sqrt{x}}\right)$$

$$= x^{\sqrt{x}}\left(\frac{\ln x + 2}{2\sqrt{x}}\right)$$

Solution 2   Another method is to write $x^{\sqrt{x}} = (e^{\ln x})^{\sqrt{x}}$:

$$\frac{d}{dx}(x^{\sqrt{x}}) = \frac{d}{dx}(e^{\sqrt{x}\ln x})$$

$$= e^{\sqrt{x}\ln x}\frac{d}{dx}(\sqrt{x}\ln x)$$

$$= x^{\sqrt{x}}\left(\frac{\ln x + 2}{2\sqrt{x}}\right) \qquad (as\ above) \qquad ●$$

Integration

Because exponential functions have simple derivatives, their integrals are also simple. The integration formulas corresponding to Theorem 6.32 and Equation 6.33 are

(6.36)

$$\int a^x\,dx = \frac{a^x}{\ln a} + C$$

(6.37)

$$\int e^x\,dx = e^x + C$$

● **Example 6**   Find, correct to three decimal places, the area under the curve $y = 2^x$ from $x = 0$ to $x = 1$.

Solution   Using Formula 6.36 with $a = 2$, we find that the area (shown in Figure 6.24) is

$$A = \int_0^1 2^x \, dx = \frac{2^x}{\ln 2}\Big]_0^1 = \frac{2 - 1}{\ln 2}$$

$$= \frac{1}{\ln 2} \approx 1.443$$

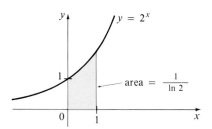

$y = 2^x$

area $= \dfrac{1}{\ln 2}$

Figure 6.24

● **Example 7**   Evaluate $\int x^2 e^{x^3} \, dx$.

Solution   We substitute $u = x^3$. Then $du = 3x^2 \, dx$, so $x^2 \, dx = \tfrac{1}{3} du$ and

$$\int x^2 e^{x^3} \, dx = \tfrac{1}{3}\int e^u \, du = \tfrac{1}{3} e^u + C = \tfrac{1}{3} e^{x^3} + C$$

Curve Sketching

● **Example 8**   Discuss the curve $y = e^{1/x}$ under the headings A–H of Section 4.7.

A.   The domain is $\{x \mid x \neq 0\} = (-\infty, 0) \cup (0, \infty)$.

B.   There is no y-intercept since $f(0)$ is undefined. There is no x-intercept since $e^{1/x} > 0$ for all x.

C.   Symmetry: none.

D.   Since $1/x \to 0$ as $x \to \pm\infty$, we have

$$\lim_{x \to \pm\infty} e^{1/x} = 1$$

so $y = 1$ is a horizontal asymptote. Since $1/x \to \infty$ as $x \to 0^+$, we have

$$\lim_{x \to 0^+} e^{1/x} = \infty$$

so $x = 0$ is a vertical asymptote. Also

$$\lim_{x \to 0^-} e^{1/x} = 0$$

since $1/x \to -\infty$ as $x \to 0^-$.

E. $$f'(x) = -\frac{e^{1/x}}{x^2}$$

Since $e^{1/x} > 0$ and $x^2 > 0$ for all $x \neq 0$, we have $f'(x) < 0$ for all $x \neq 0$. Thus f is decreasing on $(-\infty, 0)$ and on $(0, \infty)$.

F.   There is no critical number, so there is no extremum.

G.
$$f''(x) = \frac{e^{1/x}\dfrac{1}{x^2}x^2 + e^{1/x}(2x)}{x^4} = \frac{e^{1/x}(2x+1)}{x^4}$$

Since $e^{1/x} > 0$ and $x^4 > 0$ we have $f''(x) > 0$ when $x > -\frac{1}{2}$ ($x \neq 0$) and $f''(x) < 0$ when $x < -\frac{1}{2}$. So the curve is concave downward on $(-\infty, -\frac{1}{2})$ and concave upward on $(-\frac{1}{2}, 0)$ and on $(0, \infty)$. The inflection point is $(-\frac{1}{2}, e^{-2})$.

H.   The curve is sketched in Figure 6.25.   ●

Figure 6.25

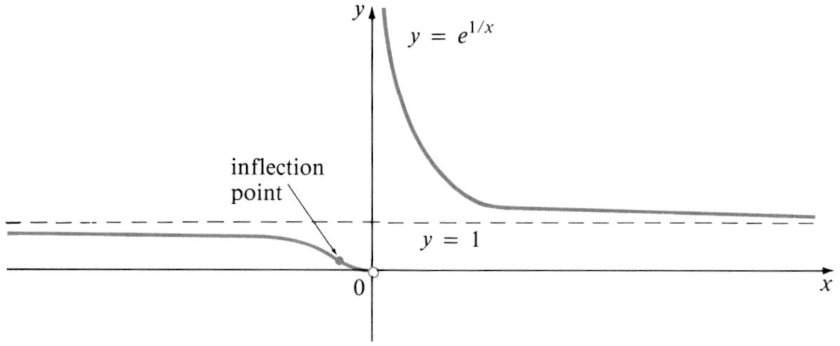

● **Example 9**   Discuss the curve $y = x^{-\ln x}$ under the headings A–H of Section 4.7.

A.   Since $\ln x$ is defined only when $x > 0$, the domain is $(0, \infty)$.

B.   Intercepts: none.

C.   Symmetry: none.

D.
$$\lim_{x \to \infty} x^{-\ln x} = \lim_{x \to \infty} (e^{\ln x})^{-\ln x} = \lim_{x \to \infty} e^{-(\ln x)^2} = 0$$

since $-(\ln x)^2 \to -\infty$ as $x \to \infty$. Thus $y = 0$ (the x-axis) is a horizontal asymptote. We also compute the limit at the left endpoint of the domain:
$$\lim_{x \to 0^+} x^{-\ln x} = \lim_{x \to 0^+} e^{-(\ln x)^2} = 0$$

since $-(\ln x)^2 \to -\infty$ as $x \to 0^+$.

E.   By using logarithmic differentiation or by using $f(x) = e^{-(\ln x)^2}$, we get
$$f'(x) = \frac{x^{-\ln x}(-2\ln x)}{x}$$

Since $x > 0$ and $x^{-\ln x} > 0$, we have
$$f'(x) > 0 \quad \Leftrightarrow \quad \ln x < 0 \quad \Leftrightarrow \quad 0 < x < 1$$
$$f'(x) < 0 \quad \Leftrightarrow \quad \ln x > 0 \quad \Leftrightarrow \quad x > 1$$

Thus f is increasing on $(0, 1]$ and decreasing on $[1, \infty)$.

F.   Since $f'(1) = 0$ and f' changes from positive to negative at 1, $f(1) = 1$ is a local maximum by the First Derivative Test.

G.
$$f''(x) = \frac{2x^{-\ln x}}{x^2}\left[2(\ln x)^2 + \ln x - 1\right]$$

$$= \frac{2x^{-\ln x}}{x^2}(2\ln x - 1)(\ln x + 1)$$

$$2\ln x - 1 = 0 \quad \Leftrightarrow \quad \ln x = \tfrac{1}{2} \quad \Leftrightarrow \quad x = e^{1/2} = \sqrt{e}$$

$$\ln x + 1 = 0 \quad \Leftrightarrow \quad \ln x = -1 \quad \Leftrightarrow \quad x = e^{-1} = \frac{1}{e}$$

Thus $f''(x) > 0$ if $0 < x < 1/e$ or if $x > \sqrt{e}$ and $f''(x) < 0$ if $1/e < x < \sqrt{e}$. Therefore f is concave upward on $(0, 1/e)$ and $(\sqrt{e}, \infty)$ and concave downward on $(1/e, \sqrt{e})$. The inflection points are $(1/e, 1/e)$ and $(\sqrt{e}, 1/\sqrt[4]{e})$.

H.   The graph is sketched in Figure 6.26.

Figure 6.26

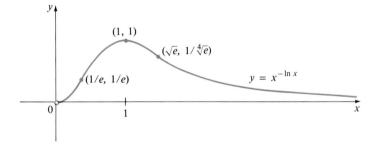

EXERCISES 6.5

In Exercises 1–8 find f' and state the domains of f and f'.

1. $f(x) = e^{\sqrt{x}}$

2. $f(x) = 10^{\sqrt{1-x}}$

3. $f(x) = 7^{x^4}$

4. $f(x) = \sqrt{3 - 2^x}$

5. $f(x) = \sqrt{2}^{\sin x}$

6. $f(x) = (\sin x)^{\sqrt{2}}$

7. $f(x) = (\ln x)^{\sqrt{5}}$

8. $f(x) = 3^{\ln(x-1)}$

In Exercises 9–12 find y' and y''.

9. $y = e^{-x}$

10. $y = 2^{3x}$

11. $y = e^{\tan x}$

12. $y = x^2 e^x$

Differentiate the functions in Exercises 13–46.

13. $f(x) = 2^{x^2}$

14. $f(x) = xe^{-x^2}$

15. $F(x) = e^x \ln x$

16. $G(x) = 5^{\tan x}$

17. $f(t) = \pi^{-t}$

18. $g(x) = 1.6^x + x^{1.6}$

19. $h(t) = t^3 - 3^t$

20. $h(\theta) = 10^{\sec \theta}$

21. $y = xe^{2x}$

22. $y = e^{x \cos x}$

23. $y = 2^{3^x}$

24. $y = \dfrac{e^{-x^2}}{x}$

25. $y = \sin(3^x)$ **26.** $y = e^{\sin \lambda x}$

27. $y = \ln|e^{2x} - 2|$

28. $y = \ln(e^{2x} + \sqrt{e^{4x} + 1})$

29. $y = e^{-1/x}$ **30.** $y = a^x x^a$

31. $y = e^{x + e^x}$ **32.** $y = e^x \cos(1 - \ln x)$

33. $y = \ln(e^{-x} + xe^{-x})$ **34.** $y = x^x$

35. $y = x^{\sin x}$ **36.** $y = (\sin x)^x$

37. $y = x^{e^x}$ **38.** $y = (x^e)^x$

39. $y = (\ln x)^x$ **40.** $y = x^{\ln x}$

41. $y = x^{1/\ln x}$ **42.** $y = (\sin x)^{\cos x}$

43. $y = x^{1/x}$ **44.** $y = x^{x^x}$

45. $y = (x^x)^x$ **46.** $y = \cos(x^{\sqrt{x}})$

47. Find y' if $\cos(x - y) = xe^x$.

48. Find y' if $x^y = y^x$.

49. If $g(t) = \sin(1 - e^{2t})$, find $g'(0)$.

In Exercises 50–53 find the equation of the tangent line to the given curve at the given point.

50. $y = 10^x$, $(1, 10)$

51. $y = e^{-x} \sin x$, $(\pi, 0)$

52. $y = (x + 1)^x$, $(1, 2)$

53. $y = \ln(1 + e^x)$, $(0, \ln 2)$

54. Find the equation of the tangent to the curve $y = e^{-x}$ that is perpendicular to the line $2x - y = 8$.

55. Show that the function $y = e^{2x} + e^{-3x}$ satisfies the differential equation $y'' + y' - 6y = 0$.

56. Show that the function $y = Ae^{-x} + Bxe^{-x}$ satisfies the differential equation $y'' + 2y' + y = 0$.

57. If $f(x) = e^{-2x}$, find $f^{(8)}(x)$.

58. Find the thousandth derivative of $f(x) = xe^{-x}$.

59. The velocity of a particle that moves in a straight line under the influence of viscous forces is $v(t) = ce^{-kt}$ where c and k are positive constants.
(a) Show that the acceleration is proportional to the velocity.
(b) What is the significance of the number c?
(c) At what time is the velocity equal to half the initial velocity?

60. Under certain circumstances a rumor spreads according to the equation

$$p(t) = \frac{1}{1 + ae^{-kt}}$$

where $p(t)$ is the proportion of the population that knows the rumor at time t and a and k are positive constants.
(a) Find $\lim_{t \to \infty} p(t)$.
(b) Find the rate of spread of the rumor.

61. A biologist makes a sample count of bacteria in a culture and finds that it doubles every 3 hours. The estimated count after 6 hours was 10,000.
(a) Find an expression for the population function $n = f(t)$.
(b) Find an expression for the growth rate.
(c) After how many hours will the population be 15,000?

62. A bacteria culture starts with 2000 bacteria. After an hour the count is 5000.
(a) What is the doubling period?
(b) Find the growth rate after 1 h and after 2 h.

63. The function $C(t) = K(e^{-at} - e^{-bt})$, where a, b, and K are positive constants and $b > a$, is used to model the concentration at time t of a drug injected into the bloodstream.
(a) Show that $\lim_{t \to \infty} C(t) = 0$.
(b) Find $C'(t)$, the rate at which the drug is cleared from circulation.
(c) When is this rate equal to 0?

64. On what intervals is the function $f(x) = x - e^x$ increasing or decreasing?

65. Use the Intermediate Value Theorem to show that there is a root of the equation $e^x + x = 0$.

66. Use Newton's Method to find the root of the equation in Exercise 65 correct to six decimal places.

Evaluate the integrals in Exercises 67–78.

67. $\displaystyle\int_3^4 5^t \, dt$ **68.** $\displaystyle\int_{\ln 3}^{\ln 6} 8e^x \, dx$

69. $\displaystyle\int e^{-6x} \, dx$ **70.** $\displaystyle\int xe^{x^2} \, dx$

71. $\displaystyle\int e^x (1 + e^x)^{10} \, dx$ **72.** $\displaystyle\int \frac{10^{\sqrt{x}}}{\sqrt{x}} \, dx$

73. $\displaystyle\int \frac{e^x + 1}{e^x} \, dx$ **74.** $\displaystyle\int \frac{e^x}{e^x + 1} \, dx$

75. $\displaystyle\int \frac{x}{e^{x^2}} \, dx$ **76.** $\displaystyle\int \frac{e^{1/x}}{x^2} \, dx$

77. $\displaystyle\int (x - 2)e^{x^2 - 4x - 3} \, dx$ **78.** $\displaystyle\int e^x \sin(e^x) \, dx$

In Exercises 79–82 find, correct to three decimal places, the area of the region bounded by the given curves.

79. $y = 2e^{-2x}$, $y = 0$, $x = 0$, $x = 1$

80. $y = 2^x$, $y = 5^x$, $x = -1$, $x = 1$

81. $y = e^x$, $y = e^{3x}$, $x = 1$

82. $y = e^x$, $y = e^{2x}/2$, $x = 0$, $x = 1$

In Exercises 83 and 84 find (a) the intervals of increase or decrease, (b) the intervals of concavity, and (c) the points of inflection.

83. $f(x) = xe^x$

84. $f(x) = x^2 e^x$

In Exercises 85–96 discuss the given curve under the headings A–H of Section 4.7.

85. $y = e^x - 2e^{-x}$

86. $y = e^x + 2e^{-x}$

87. $y = e^{-1/(x+1)}$

88. $y = e^{x/(x-2)}$

89. $y = e^{-x^2}$

90. $y = \ln(e^x - 2e^{-x})$

91. $y = xe^{x^2}$

92. $y = e^x + e^{-2x}$

93. $y = x^{\ln x}$

94. $y = (2x)^{-\ln 3x}$

95. $y = e^{1/x^2}$

96. $y = e^{1/(1-x^2)}$

97. Find the most general antiderivative of $f(x) = 2e^x + 3\sqrt{x}$.

98. Find $f(x)$ if $f''(x) = 3e^x + 5\sin x$, $f(0) = 1$, and $f'(0) = 2$.

99. If $f(x) = x + x^2 + e^x$ and $g(x) = f^{-1}(x)$, find $g'(1)$.

100. If $f(x) = e^x - \ln x$ and $h(x) = f^{-1}(x)$, find $h'(e)$.

101. Prove that $e^x > 1 + x$ if $x > 0$.

102. If f and g are differentiable, find a formula for the derivative of $h(x) = [f(x)]^{g(x)}$.

SECTION 6.6

An Alternative Approach to Logarithmic and Exponential Functions

In this section we give an alternative treatment of the functions $\ln x$, e^x, a^x, $\log_a x$, and their derivatives. Instead of starting with a^x and defining $\log_a x$ as its inverse, this time we start by defining $\ln x$ as an integral and then define the exponential function as its inverse. In this section the reader should bear in mind that we do not use the definitions and results of Sections 6.1, 6.3, 6.4, and 6.5.

The Natural Logarithm

Definition (6.38)

> The **natural logarithmic function** is the function defined by
>
> $$\ln x = \int_1^x \frac{1}{t}\, dt \qquad x > 0$$

The existence of this function depends on Theorem 6.11, which guarantees that a continuous function is integrable. If $x > 1$, $\ln x$ can be interpreted geometrically as the area under the hyperbola $y = 1/t$ from $t = 1$ to $t = x$ [see Figure 6.27(a)]. For $x = 1$, we have

$$\ln 1 = \int_1^1 \frac{1}{t}\, dt = 0$$

Figure 6.27

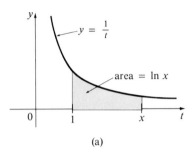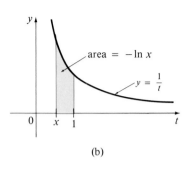

(a)   (b)

For $0 < x < 1$,

$$\ln x = \int_1^x \frac{1}{t}\, dt = -\int_x^1 \frac{1}{t}\, dt < 0$$

and so $\ln x$ is the negative of the area shown in Figure 6.27(b).

Notice that the integral that defines $\ln x$ is exactly the type of integral discussed in Part 1 of the Fundamental Theorem of Calculus (5.17). In fact, using that theorem, we have

$$\frac{d}{dx} \int_1^x \frac{1}{t}\, dt = \frac{1}{x}$$

and so

(6.39)

$$\boxed{\frac{d}{dx} \ln x = \frac{1}{x}}$$

We now use this differentiation rule to prove the following properties of the logarithm function.

Laws of Logarithms (6.40)

If x and y are positive numbers and r is a rational number, then

(a) $\ln(xy) = \ln x + \ln y$

(b) $\ln\!\left(\dfrac{x}{y}\right) = \ln x - \ln y$

(c) $\ln(x^r) = r \ln x$

Proof of (a)   Let $f(x) = \ln(ax)$, where a is a positive constant. Then, using Equation 6.39 and the Chain Rule, we have

$$f'(x) = \frac{1}{ax} \frac{d}{dx}(ax) = \frac{1}{ax} \cdot a = \frac{1}{x}$$

Therefore $f(x)$ and $\ln x$ have the same derivative and so they must differ by a constant:

$$\ln(ax) = \ln x + C$$

Putting $x = 1$ in this equation, we get $\ln a = \ln 1 + C = 0 + C = C$. Thus

$$\ln(ax) = \ln x + \ln a$$

If we now replace the constant a by any number y, we have

$$\ln(xy) = \ln x + \ln y$$ ●

Proof of (b)   Using law (a) with $x = 1/y$, we have

$$\ln\frac{1}{y} + \ln y = \ln\!\left(\frac{1}{y} \cdot y\right) = \ln 1 = 0$$

and so

$$\ln\frac{1}{y} = -\ln y$$

Using (a) again, we have

$$\ln\!\left(\frac{x}{y}\right) = \ln\!\left(x \cdot \frac{1}{y}\right) = \ln x + \ln\frac{1}{y} = \ln x - \ln y$$

The proof of (c) is left to the reader. ●

Theorem (6.41)

(a) $\displaystyle\lim_{x \to \infty} \ln x = \infty$ (b) $\displaystyle\lim_{x \to 0^+} \ln x = -\infty$

Proof of (a)   Using (6.40) law (c) with $x = 2$ and $r = n$ (any positive integer), we have $\ln(2^n) = n\ln 2$. Now $\ln 2 > 0$, so this shows that $\ln(2^n) \to \infty$ as $n \to \infty$. But $\ln x$ is an increasing function since its derivative $1/x > 0$. Therefore $\ln x \to \infty$ as $x \to \infty$. ●

Proof of (b)   If we let $t = 1/x$, then $t \to \infty$ as $x \to 0^+$. Thus, using (a), we have

$$\lim_{x \to 0^+} \ln x = \lim_{t \to \infty} \ln\!\left(\frac{1}{t}\right) = \lim_{t \to \infty}(-\ln t) = -\infty$$ ●

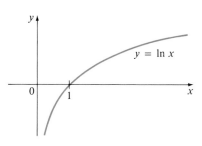

If $y = \ln x$, $x > 0$, then

$$\frac{dy}{dx} = \frac{1}{x} > 0 \quad\text{and}\quad \frac{d^2 y}{dx^2} = -\frac{1}{x^2} < 0$$

which shows that $\ln x$ is increasing and concave downward on $(0, \infty)$. Putting this information together with Theorem 6.41, we draw the graph of $y = \ln x$ in Figure 6.28.

Figure 6.28

Since $\ln 1 = 0$ and $\ln x$ is an increasing continuous function that takes on arbitrarily large values, the Intermediate Value Theorem (2.28) shows that there is a number where $\ln x$ takes on the value 1 (see Figure 6.29). This important number is denoted by e.

Definition (6.42)

$$e \text{ is the number such that } \ln e = 1.$$

We shall show (in Theorem 6.57) that this definition is consistent with our previous definition, that is, $e = \lim_{x \to 0}(1 + x)^{1/x}$.

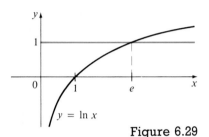

Figure 6.29

The Natural Exponential Function

Since $\ln$ is an increasing function, it is one-to-one and therefore has an inverse function, which we denote by exp. Thus, according to the definition of an inverse function,

(6.43)

$$\exp(x) = y \quad \Leftrightarrow \quad \ln y = x$$

and the cancellation equations are

(6.44)

$$\exp(\ln x) = x \quad \text{and} \quad \ln(\exp x) = x$$

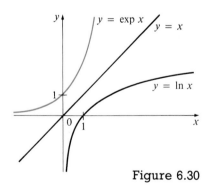

Figure 6.30

In particular we have

$$\exp(0) = 1 \quad \text{since } \ln 1 = 0$$

$$\exp(1) = e \quad \text{since } \ln e = 1$$

We obtain the graph of $y = \exp x$ by reflecting the graph of $y = \ln x$ in the line $y = x$ (see Figure 6.30). The domain of exp is the range of ln, that is, $(-\infty, \infty)$; the range of exp is the domain of ln, that is, $(0, \infty)$.

If r is any rational number, then (6.40) law (c) gives

$$\ln(e^r) = r \ln e = r$$

Therefore, by (6.43), $\exp(r) = e^r$

Thus $\exp(x) = e^x$ whenever x is a rational number. This leads us to define e^x, even for irrational values of x, by the equation

(6.45)

$$e^x = \exp(x)$$

In other words, for the reasons given, we define e^x to be the inverse of the function $\ln x$. In this notation (6.43) becomes

(6.46)
$$e^x = y \iff \ln y = x$$

and the cancellation equations (6.44) become

(6.47)
$$e^{\ln x} = x \qquad x > 0$$

(6.48)
$$\ln(e^x) = x \qquad \text{all } x$$

The function $f(x) = e^x$ has domain $(-\infty, \infty)$ and range $(0, \infty)$, that is, $e^x > 0$ for all x. It follows from Theorem 6.41 that

(6.49)
$$\lim_{x \to \infty} e^x = \infty \quad \text{and} \quad \lim_{x \to -\infty} e^x = 0$$

We now verify that f has the properties expected of an exponential function.

Laws of Exponents (6.50)

If x and y are real numbers and r is rational, then

(a) $e^{x+y} = e^x e^y$ 　　　　　　　(b) $e^{x-y} = \dfrac{e^x}{e^y}$

(c) $(e^x)^r = e^{rx}$

Proof of (a)　　From (6.40) and (6.48) we have

$$\ln(e^x e^y) = \ln(e^x) + \ln(e^y) = x + y = \ln(e^{x+y})$$

Since $\ln$ is a one-to-one function, it follows that $e^x e^y = e^{x+y}$.

Parts (b) and (c) are proved similarly. As we shall see in (6.53), part (c) actually holds when r is any real number. ●

Theorem (6.51)
$$\frac{d}{dx} e^x = e^x$$

Proof   The function $y = e^x$ is differentiable because it is the inverse function of $y = \ln x$, which we know is differentiable. To find its derivative, we use the inverse function method. Let $y = e^x$. Then $\ln y = x$ and, differentiating this latter equation implicitly with respect to x, we get

$$\frac{1}{y}\frac{dy}{dx} = 1$$

$$\frac{dy}{dx} = y = e^x$$ ●

General Exponential Functions

If $a > 0$ and r is any rational number, then by Equation 6.47 and (6.50) law (c),

$$a^r = (e^{\ln a})^r = e^{r \ln a}$$

Therefore, even for irrational numbers x, we *define*

(6.52)

$$\boxed{a^x = e^{x \ln a}}$$

The function $f(x) = a^x$ is called the **exponential function with base a.** Notice that a^x is positive for all x because e^x is positive for all x. The general laws of exponents follow from Definition 6.52 together with the laws of exponents for e^x.

Laws of Exponents (6.53)

If x and y are real numbers and $a, b > 0$, then

(a) $a^{x+y} = a^x a^y$ (b) $a^{x-y} = a^x/a^y$

(c) $(a^x)^y = a^{xy}$ (d) $(ab)^x = a^x b^x$

Proof of (a)   Using Definition 6.52 and (6.50), we have

$$a^{x+y} = e^{(x+y)\ln a} = e^{x \ln a + y \ln a}$$

$$= e^{x \ln a} e^{y \ln a} = a^x a^y$$ ●

Proof of (c) $(a^x)^y = e^{y \ln(a^x)} = e^{y \ln(e^{x \ln a})}$

$$= e^{yx \ln a} = e^{xy \ln a} = a^{xy}$$ ●

The remaining proofs are left as exercises.

Theorem (6.54)

$$\frac{d}{dx} a^x = a^x \ln a$$

Proof

$$\frac{d}{dx} a^x = \frac{d}{dx}(e^{x \ln a}) = e^{x \ln a}\left(\frac{d}{dx} x \ln a\right)$$

$$= a^x \ln a \qquad \bullet$$

If $a > 1$, then $\ln a > 0$, so $D_x a^x = a^x \ln a > 0$, which shows that $y = a^x$ is increasing (see Figure 6.31). If $0 < a < 1$, then $\ln a < 0$ and so $y = a^x$ is decreasing (see Figure 6.32).

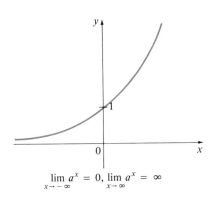

$$\lim_{x \to -\infty} a^x = 0, \lim_{x \to \infty} a^x = \infty$$

Figure 6.31

$y = a^x, a > 1$

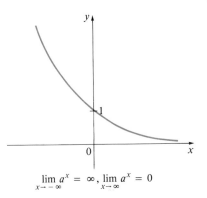

$$\lim_{x \to -\infty} a^x = \infty, \lim_{x \to \infty} a^x = 0$$

Figure 6.32

$y = a^x, 0 < a < 1$

General Logarithmic Functions

If $a > 0$ and $a \neq 1$, then $f(x) = a^x$ is a one-to-one function. Its inverse function is called the **logarithmic function with base** a and is denoted by $\log_a$. Thus

(6.55)

$$\log_a x = y \quad \Leftrightarrow \quad a^y = x$$

In particular, we see that

$$\log_e x = \ln x$$

The laws of logarithms are similar to those for the natural logarithm and can be deduced from the laws of exponents (see Exercise 9).

Theorem (6.56)

$$\frac{d}{dx}\log_a x = \frac{1}{x\ln a}$$

Proof   First note that $\log_a x$ is differentiable because it is the inverse function of the differentiable function a^x. Let $y = \log_a x$. Then $a^y = x$, and using Theorem 6.54 to differentiate both sides of the latter equation, we have

$$a^y \ln a \frac{dy}{dx} = 1$$

$$\frac{dy}{dx} = \frac{1}{a^y \ln a} = \frac{1}{x\ln a}$$

An alternative proof is indicated in Exercise 8.   ●

The Number e Expressed As a Limit

In this section we defined e as the number such that $\ln e = 1$. The next theorem shows that this is the same as the number e defined in Section 6.3.

Theorem (6.57)

$$\lim_{x \to 0}(1 + x)^{1/x} = e$$

Proof   Let $f(x) = \ln x$. Then $f'(x) = 1/x$, so $f'(1) = 1$. But, by the definition of derivative,

$$f'(1) = \lim_{h \to 0}\frac{f(1 + h) - f(1)}{h} = \lim_{x \to 0}\frac{f(1 + x) - f(1)}{x}$$

$$= \lim_{x \to 0}\frac{\ln(1 + x) - \ln 1}{x}$$

$$= \lim_{x \to 0}\frac{1}{x}\ln(1 + x)$$

$$= \lim_{x \to 0}\ln(1 + x)^{1/x}$$

$$= \ln\left[\lim_{x \to 0}(1 + x)^{1/x}\right] \qquad \textit{(since ln is continuous)}$$

Therefore $$\ln\left[\lim_{x \to 0}(1 + x)^{1/x}\right] = 1$$

It follows that

$$\lim_{x \to 0}(1 + x)^{1/x} = e \qquad\qquad ●$$

EXERCISES 6.6

In these exercises use only the definitions and results of *this* section.

1. By comparing areas, show that

$$\frac{1}{2} + \frac{1}{3} + \cdots + \frac{1}{n} < \ln n < 1 + \frac{1}{2} + \frac{1}{3} + \cdots + \frac{1}{n-1}$$

2. (a) By comparing areas, show that $\ln 2 < 1 < \ln 3$.
   (b) Deduce that $2 < e < 3$.

3. Prove (6.40) law (c). (*Hint:* Start by showing that both sides of the equation have the same derivative.)

4. Prove (6.50) law (b).   5. Prove (6.50) law (c).

6. Prove (6.53) law (b).   7. Prove (6.53) law (d).

8. (a) Show that

$$\log_a x = \frac{\ln x}{\ln a}$$

   [*Hint:* Let $y = \log_a x$ and use (6.55).]

   (b) Deduce that $D_x \log_a x = 1/x \ln a$.

9. Deduce the following Laws of Logarithms from (6.53):
   (a) $\log_a(xy) = \log_a x + \log_a y$
   (b) $\log_a(\frac{x}{y}) = \log_a x - \log_a y$
   (c) $\log_a(x^y) = y \log_a x$

SECTION 6.7

Exponential Growth and Decay

In many natural phenomena, quantities occur that grow or decay at a rate proportional to their size. For instance, if $y = f(t)$ is the number of individuals in a population of animals or bacteria cells at time t, then it seems reasonable to expect that the rate of growth $f'(t)$ is proportional to the population $f(t)$; that is, $f'(t) = kf(t)$ for some constant k. Indeed, under ideal conditions (unlimited environment, adequate nutrition, freedom from disease) that is what happens. Another example occurs in nuclear physics where the mass of a radioactive substance decays at a rate proportional to the mass. In chemistry, the rate of a unimolecular first-order reaction is proportional to the concentration of the substance. In finance, the value of a savings account with continuously compounded interest increases at a rate proportional to that value.

In general, if $y = y(t)$ is the value of a quantity y at time t and if the rate of change of y with respect to t is proportional to its size $y(t)$ at any time, then

(6.58)

$$\frac{dy}{dt} = ky$$

where k is a constant. Equation 6.58 is sometimes called the **law of natural growth** (if $k > 0$) or the **law of natural decay** (if $k < 0$). It is called a **differential equation** because it involves an unknown function y and its derivative dy/dt.

It is not hard to think of a solution of Equation 6.58. This equation asks us to find a function whose derivative is a constant multiple of itself.

We have met such functions in this chapter. Any exponential function of the form $y(t) = Ce^{kt}$, where C is a constant, satisfies

$$y'(t) = Cke^{kt} = k(Ce^{kt}) = ky(t)$$

Conversely we can show that *any* function that satisfies $dy/dt = ky$ must be of the form $y = Ce^{kt}$.

To see that this is true we let f be any function that satisfies $f'(t) = kf(t)$ and we consider the quotient function

$$g(t) = \frac{f(t)}{e^{kt}} = f(t)e^{-kt}$$

Then
$$g'(t) = f'(t)e^{-kt} + f(t)(-k)e^{-kt}$$
$$= [kf(t)]e^{-kt} - kf(t)e^{-kt} = 0$$

Since $g'(t) = 0$, it follows from Theorem 4.15 that g is a constant function: $g(t) = C$. Then $f(t) = Ce^{kt}$ and this is what we wanted to show. To see the significance of the constant C, we observe that

$$y(0) = Ce^{k0} = C$$

Therefore C is the initial value of the function and we have proved the following:

Theorem (6.59)

> The only solutions of the differential equation $dy/dt = ky$ are the exponential functions
>
> $$y(t) = y(0)e^{kt}$$

● **Example 1**   A bacteria culture starts with 1000 bacteria and after 2 hours there are 2500 bacteria. Assuming that the culture grows at a rate proportional to its size, find the population after 6 hours.

Solution   Let $y(t)$ be the number of bacteria after t hours. Then $y(0) = 1000$ and $y(2) = 2500$. Since we are assuming $dy/dt = ky$, Theorem 6.59 gives

$$y(t) = y(0)e^{kt} = 1000e^{kt}$$
$$y(2) = 1000e^{2k} = 2500$$

Therefore $e^{2k} = 2.5 \quad \text{and} \quad 2k = \ln 2.5$

Substituting the value $k = \frac{1}{2}\ln 2.5$ back into the expression for $y(t)$, we have

(6.60) $$y(t) = 1000e^{\ln 2.5(t/2)}$$

Since $e^{\ln 2.5} = 2.5$, an alternative expression for Equation 6.60 is

$$y(t) = 1000(2.5)^{t/2}$$

and so $\qquad\qquad\qquad y(6) = 1000(2.5)^3 = 15,625 \qquad\qquad\qquad\bullet$

● **Example 2**   The *half-life* of radium-226 ($^{226}_{88}\text{Ra}$) is 1590 years. This means that the rate of decay is proportional to the amount present and half of any given quantity will disintegrate in 1590 years. (a) A sample of radium-226 has a mass of 100 mg. Find a formula for the mass that remains after t years. (b) Find the mass after 1000 years correct to the nearest milligram. (c) When will the mass be reduced to 30 mg?

Solution   (a)   Let $y(t)$ be the mass of radium (in milligrams) that remains after t years. Then $dy/dt = ky$ and $y(0) = 100$, so Theorem 6.59 gives

$$y(t) = y(0)e^{kt} = 100e^{kt}$$

In order to determine the value of k, we use the fact that $y(1590) = 50$. Thus

$$100e^{1590k} = 50 \quad \text{so} \quad e^{1590k} = \tfrac{1}{2}$$

and $\qquad\qquad\qquad 1590k = \ln\!\left(\frac{1}{2}\right) = -\ln 2$

$$k = -\frac{\ln 2}{1590}$$

Therefore $\qquad\qquad y(t) = 100e^{-(\ln 2/1590)t}$

As in Example 1 we could use the fact that $e^{\ln 2} = 2$ to write the expression for $y(t)$ in the alternative form

$$y(t) = 100 \times 2^{-t/1590}$$

(b)   The mass after 1000 years is

$$y(1000) = 100e^{-(\ln 2/1590)1000} \approx 65 \text{ mg}$$

(c)   We want to find the value of t such that $y(t) = 30$, that is,

$$100e^{-(\ln 2/1590)t} = 30$$

or $\qquad\qquad\qquad e^{-(\ln 2/1590)t} = 0.3$

We solve this equation for t by taking the natural logarithm of both sides:

$$-\frac{\ln 2}{1590}t = \ln 0.3$$

Thus $\qquad\qquad\qquad t = -1590\frac{\ln 0.3}{\ln 2} \approx 2762 \text{ years} \qquad\qquad\bullet$

● **Example 3**   Newton's Law of Cooling states that the rate of cooling of an object is proportional to the temperature difference between the object and its surroundings, provided that this difference is not too large. Suppose the object takes 40 minutes to cool from 30° C to 24° C in a room that is kept at 20° C. (a) What was the temperature of the object 15 minutes after it was 30° C? (b) How long would it take the object to cool down to 21° C?

Solution   (a) Let $y(t)$ be the temperature t minutes after it was 30° C. Then Newton's Law of Cooling states that

$$\frac{dy}{dt} = k(y - 20)$$

This differential equation is not quite the same as Equation 5.33, so we introduce a new function $u(t) = y(t) - 20$. Then $du/dt = dy/dt$, so

$$\frac{du}{dt} = ku$$

Therefore, by Theorem 6.59, we have

$$u(t) = u(0)e^{kt} = 10e^{kt}$$

We are given that $y(40) = 24$, so $u(40) = 4$ and

$$10e^{40k} = 4 \qquad e^{40k} = 0.4$$

Taking logarithms, we have

$$k = \frac{\ln 0.4}{40}$$

Thus

$$u(t) = 10e^{(\ln 0.4)t/40}$$

(6.61)

$$y(t) = 20 + 10e^{(\ln 0.4)t/40}$$

$$y(15) = 20 + 10e^{(\ln 0.4)15/40} \approx 27.1° \text{ C}$$

(b)   We have $y(t) = 21$ when

$$20 + 10e^{(\ln 0.4)t/40} = 21$$

$$e^{(\ln 0.4)t/40} = \frac{1}{10}$$

$$\frac{(\ln 0.4)t}{40} = \ln\left(\frac{1}{10}\right) = -\ln 10$$

$$t = -40\frac{\ln 10}{\ln 0.4} \approx 100.5$$

The object will cool down to 21° after about 1 hour and 41 minutes.   ●

Notice that, in Example 3, $\ln(0.4) < 0$ since $0.4 < 1$. Thus, from Equation 6.61 we have

$$\lim_{t \to \infty} y(t) = 20 + \lim_{t \to \infty} 10e^{(\ln 0.4)t/40}$$

$$= 20 + 0 = 20$$

which is to be expected.

● **Example 4**   If $1000 is invested at 12% interest, compounded annually, then after 1 year the investment is worth $1000(1.12) = 1120; after 2 years it is worth $[1000(1.12)]1.12 = 1254.40; and after t years it is worth $1000(1.12)^t$. In general, if an amount A_0 is invested at an interest rate i ($i = 0.12$ in the example just considered), then after t years it is worth $A_0(1 + i)^t$. Usually, however, interest is compounded more frequently—say, n times a year. Then in each compounding period the interest rate is i/n and there are nt compounding periods in t years, so the investment is worth

$$A_0\left(1 + \frac{i}{n}\right)^{nt}$$

For instance, after 3 years at 12% interest a $1000 investment will be worth

$$\$1000(1.12)^3 = \$1404.93 \quad \text{with annual compounding}$$

$$\$1000(1.06)^6 = \$1418.52 \quad \text{with semiannual compounding}$$

$$\$1000(1.03)^{12} = \$1425.76 \quad \text{with quarterly compounding}$$

$$\$1000(1.01)^{36} = \$1430.77 \quad \text{with monthly compounding}$$

$$\$1000\left(1 + \frac{0.12}{365}\right)^{365 \cdot 3} = \$1433.24 \quad \text{with daily compounding}$$

You can see that the interest paid increases as the number of compounding periods (n) increases. If we let $n \to \infty$, then we will be compounding the interest continuously and the value of the investment will be

$$A(t) = \lim_{n \to \infty} A_0\left(1 + \frac{i}{n}\right)^{nt} = \lim_{n \to \infty} A_0\left[\left(1 + \frac{i}{n}\right)^{n/i}\right]^{it}$$

$$= A_0\left[\lim_{n \to \infty}\left(1 + \frac{i}{n}\right)^{n/i}\right]^{it}$$

$$= A_0\left[\lim_{m \to \infty}\left(1 + \frac{1}{m}\right)^{m}\right]^{it} \quad (\text{where } m = n/i)$$

$$= A_0 e^{it} \quad [\text{by the definition of } e \ (6.19)]$$

This formula for continuous compounding $[A(t) = A_0 e^{it}]$ should be compared with the expression for exponential growth in Theorem 6.59:

$y(t) = y(0)e^{kt}$. When differentiated, it becomes

$$\frac{dA}{dt} = iA_0e^{it} = iA(t)$$

which says that, with continuous compounding of interest, the rate of increase of an investment is proportional to its size.

Returning to the example of $1000 invested for 3 years at 12% interest, we see that with continuous compounding of interest the value of the investment will be

$$A(3) = \$1000e^{(0.12)3}$$
$$= \$1000e^{0.36} = \$1433.33$$

EXERCISES 6.7

In Exercises 1–6 assume that the population grows at a rate proportional to its size.

1. A cell of the bacterium *Escherichia coli* in a nutrient broth medium divides into two cells every 20 minutes. If there are initially 100 cells, find (a) an expression for the number of cells after t hours, (b) the number of cells after 10 hours, and (c) the time required for the population to reach 10,000 cells.

2. A bacteria culture starts with 4000 bacteria and the population triples every half-hour. (a) Find an expression for the number of bacteria after t hours. (b) Find the number of bacteria after 20 minutes. (c) When will the population reach 20,000?

3. A bacteria culture starts with 500 bacteria and after 3 hours there are 8000 bacteria. (a) Find an expression for the number of bacteria after t hours. (b) Find the number of bacteria after 4 hours. (c) When will the population reach 30,000?

4. The count in a bacteria culture was 400 after 2 hours and 25,600 after 6 hours. (a) What was the initial size of the culture? (b) Find an expression for the population after t hours. (c) In what period of time does the population double? (d) When will the population be 100,000?

5. The population of a certain city grows at a rate of 5% per year. The population in 1985 was 307,000. What will the population be (a) in 1995 and (b) in 2001?

6. Another city had a population of 450,000 in 1980 and 500,000 in 1985. What will its population be (a) in 1995 and (b) in 2001?

7. Experiments show that if the chemical reaction

$$N_2O_5 \rightarrow 2NO_2 + \tfrac{1}{2}O_2$$

takes place at 45° C, the rate of reaction of dinitrogen pentoxide is proportional to its concentration as follows:

$$-\frac{d[N_2O_5]}{dt} = 0.0005[N_2O_5]$$

(See Example 3 in Section 3.7.) (a) Find an expression for the concentration $[N_2O_5]$ after t seconds if the initial concentration is C. (b) How long will the reaction take to reduce the concentration of N_2O_5 to 90% of its original value?

8. Polonium-210 has a half-life of 140 days (see Example 2). (a) If a sample has a mass of 200 mg, find a formula for the mass that remains after t days. (b) Find the mass after 100 days. (c) When will the mass be reduced to 10 mg?

9. Polonium-214 has a very short half-life of 1.4×10^{-4} s. (a) If a sample has a mass of 50 mg, find a formula for the mass that remains after t seconds. (b) Find the mass that remains after a hundredth of a second. (c) How long would it take for the mass to decay to 40 mg?

10. After three days a sample of Radon-222 decayed to 58% of its original amount. (a) What is the half-life

of Radon-222? (b) How long would it take the sample to decay to 10% of its original amount?

11. A certain object cools at a rate (in degrees Celsius per minute) equal to one-tenth of the difference between its temperature and that of the surrounding air. If a room is kept at $21°$ C and the temperature of the object is $33°$ C, find an expression for the temperature of the object t minutes later.

12. A thermometer is taken from a room where the temperature is $20°$ C to the outdoors where the temperature is $5°$ C. After one minute the thermometer reads $12°$ C. Use Newton's Law of Cooling to answer the following questions: (a) What will the reading on the thermometer be after one more minute? (b) When will the thermometer read $6°$ C?

13. The rate of change of atmospheric pressure P with respect to altitude h is proportional to P, provided that the temperature is constant. At $15°$ C the pressure is $101.3\,\mathrm{kPa}$ at sea level and $87.14\,\mathrm{kPa}$ at $h = 1000$ m. (a) What is the pressure at an altitude of 3000 m? (b) What is the pressure at the top of Mount McKinley (6187 m)?

14. If \$500 is borrowed at 14% interest, find the amount due at the end of 2 years if the interest is compounded (a) annually, (b) quarterly, (c) monthly, (d) daily, (e) hourly, and (f) continuously.

15. If \$3000 is invested at 9% interest, find the value of the investment at the end of 5 years if the interest is compounded (a) annually, (b) semiannually, (c) monthly, (d) weekly, (e) daily, and (f) continuously.

16. How long will it take an investment to double in value if the interest rate is 10% compounded continuously?

17. A tank contains 1500 L of brine with a concentration of 0.3 kg of salt per liter. In order to dilute the solution, pure water is run into the tank at a rate of 20 L/min and the resulting solution, which is continuously stirred, runs out at the same rate. (a) How many kilograms of salt will remain after half an hour? (b) When will the concentration be reduced to 0.2 kg of salt per liter?

18. Do Exercise 17 if, instead of pure water, brine with a concentration of 0.1 kg of salt per liter is used.

SECTION 6.8

Inverse Trigonometric Functions

In this section we apply the ideas of Section 6.2 to find the derivatives of the so-called inverse trigonometric functions. There is a slight difficulty here because the trigonometric functions are not one-to-one so they do not have inverse functions. The difficulty is overcome by restricting the domains of these functions so that they are one-to-one.

Figure 6.33

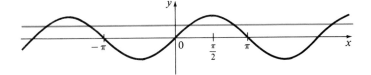

You can see from Figure 6.33 that the sine function $y = \sin x$ is not one-to-one (use the Horizontal Line Test). But the function $f(x) = \sin x$, $-\pi/2 \leqslant x \leqslant \pi/2$ (see Figure 6.34), *is* one-to-one. The inverse function of this restricted sine function f exists and is denoted by $\sin^{-1}$ or arcsin. It is called the **inverse sine function** or the **arcsine function.** Using the definition of inverse function given by (6.8),

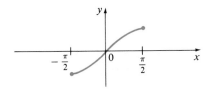

Figure 6.34

$$f^{-1}(x) = y \quad \Leftrightarrow \quad f(y) = x$$

we have

$$\sin^{-1} x = y \quad \Leftrightarrow \quad \sin y = x \quad \text{and} \quad -\frac{\pi}{2} \leqslant y \leqslant \frac{\pi}{2}$$

(6.62)

Thus if $-1 \leqslant x \leqslant 1$, $\sin^{-1} x$ is the number between $-\pi/2$ and $\pi/2$ whose sine is x.

$$\sin^{-1}x \neq \frac{1}{\sin x}$$

● **Example 1**   Evaluate (a) $\sin^{-1}\frac{1}{2}$ and (b) $\tan(\arcsin\frac{1}{3})$.

Solution   (a)   We have

$$\sin^{-1}\frac{1}{2} = \frac{\pi}{6}$$

because $\sin(\pi/6) = \frac{1}{2}$ and $\pi/6$ lies between $-\pi/2$ and $\pi/2$.

(b)   Let $\theta = \arcsin\frac{1}{3}$. Then we can draw a right triangle with angle θ as in Figure 6.35 and deduce from the Pythagorean Theorem that the third side has length $\sqrt{9-1} = 2\sqrt{2}$. This enables us to read from the triangle that

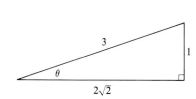

Figure 6.35

$$\tan\left(\arcsin\frac{1}{3}\right) = \tan\theta = \frac{1}{2\sqrt{2}}$$

●

The cancellation equations (6.9) for inverse functions become, in this case,

(6.63)

$$\sin^{-1}(\sin x) = x \quad \text{for} \quad -\frac{\pi}{2} \leqslant x \leqslant \frac{\pi}{2}$$

$$\sin(\sin^{-1} x) = x \quad \text{for} \quad -1 \leqslant x \leqslant 1$$

We must be careful when using the first cancellation equation because it is valid only when x lies in the interval $[-\pi/2, \pi/2]$. The following example shows how to proceed when x lies outside this interval.

● **Example 2**   Evaluate: (a) $\sin(\sin^{-1} 0.6)$, (b) $\sin^{-1}(\sin(\pi/12))$, and (c) $\sin^{-1}(\sin(2\pi/3))$.

Solution   (a)   Since 0.6 lies between -1 and 1, the second cancellation equation in (6.63) gives

$$\sin(\sin^{-1} 0.6) = 0.6$$

(b)   Since $\pi/12$ lies between $-\pi/2$ and $\pi/2$, the first cancellation equation gives

$$\sin^{-1}\left(\sin\frac{\pi}{12}\right) = \frac{\pi}{12}$$

(c)   Since $2\pi/3$ does not lie in the interval $[-\pi/2, \pi/2]$, we cannot use the cancellation equation. Instead we note that $\sin(2\pi/3) = \sqrt{3}/2$ and $\sin^{-1}(\sqrt{3}/2) = \pi/3$ because $\pi/3$ lies between $-\pi/2$ and $\pi/2$. Therefore

$$\sin^{-1}\left(\sin\frac{2\pi}{3}\right) = \sin^{-1}\left(\frac{\sqrt{3}}{2}\right) = \frac{\pi}{3}$$ ●

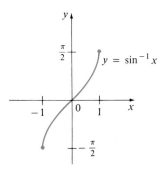

Figure 6.36

The inverse sine function, $\sin^{-1}$, has domain $[-1, 1]$ and range $[-\pi/2, \pi/2]$, and its graph, shown in Figure 6.36, is obtained from that of the restricted sine function (Figure 6.34) by reflection in the line $y = x$.

We know that the sine function f is continuous, so by Theorem 6.11 the inverse sine function is also continuous. We also know from Section 3.3 that the sine function is differentiable, so by Theorem 6.12 the inverse sine function is differentiable. We could calculate the derivative of $\sin^{-1}$ by the formula in Theorem 6.12, but since we know that $\sin^{-1}$ is differentiable we can just as easily calculate it by implicit differentiation as follows.

Let $y = \sin^{-1} x$. Then $\sin y = x$ and $-\pi/2 \leq y \leq \pi/2$. Differentiating the latter equation implicitly with respect to x, we obtain

$$\cos y \frac{dy}{dx} = 1$$

and

$$\frac{dy}{dx} = \frac{1}{\cos y}$$

Now $\cos y \geq 0$ since $-\pi/2 \leq y \leq \pi/2$, so

$$\cos y = \sqrt{1 - \sin^2 y} = \sqrt{1 - x^2}$$

Therefore

$$\frac{dy}{dx} = \frac{1}{\cos y} = \frac{1}{\sqrt{1 - x^2}}$$

(6.64)

$$\boxed{\frac{d}{dx}(\sin^{-1} x) = \frac{1}{\sqrt{1 - x^2}} \qquad -1 < x < 1}$$

● **Example 3**   If $f(x) = \sin^{-1}(x^2 - 1)$, find (a) the domain of f, (b) $f'(x)$, and (c) the domain of f'.

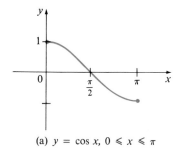

(a) $y = \cos x, \ 0 \leqslant x \leqslant \pi$

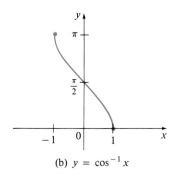

(b) $y = \cos^{-1} x$

Figure 6.37

Solution   (a)   Since the domain of the inverse sine function is $[-1, 1]$, the domain of f is

$$\{x \mid -1 \leqslant x^2 - 1 \leqslant 1\} = \{x \mid 0 \leqslant x^2 \leqslant 2\}$$
$$= \{x \mid |x| \leqslant \sqrt{2}\} = [-\sqrt{2}, \sqrt{2}]$$

(b)   Combining Formula 6.64 with the Chain Rule, we have

$$f'(x) = \frac{1}{\sqrt{1 - (x^2 - 1)^2}} \frac{d}{dx}(x^2 - 1)$$

$$= \frac{1}{\sqrt{1 - (x^4 - 2x^2 + 1)}} 2x = \frac{2x}{\sqrt{2x^2 - x^4}}$$

(c)   The domain of f' is

$$\{x \mid -1 < x^2 - 1 < 1\} = \{x \mid 0 < x^2 < 2\}$$
$$= \{x \mid 0 < |x| < \sqrt{2}\} = (-\sqrt{2}, 0) \cup (0, \sqrt{2}) \quad \bullet$$

The inverse cosine function is handled similarly. The restricted cosine function $f(x) = \cos x, \ 0 \leqslant x \leqslant \pi$, is one-to-one [see Figure 6.37(a)] and so it has an inverse function $\cos^{-1} = \arccos$.

(6.65)

$$\cos^{-1} x = y \quad \Leftrightarrow \quad \cos y = x \quad \text{and} \quad 0 \leqslant y \leqslant \pi$$

The cancellation equations are

(6.66)

$$\cos^{-1}(\cos x) = x \quad \text{for } 0 \leqslant x \leqslant \pi$$
$$\cos(\cos^{-1} x) = x \quad \text{for } -1 \leqslant x \leqslant 1$$

The inverse cosine function, $\cos^{-1}$, has domain $[-1, 1]$ and range $[0, \pi]$ and is a continuous function whose graph is shown in Figure 6.37(b). Since cos is differentiable, $\cos^{-1}$ is differentiable by Theorem 6.12 and its derivative is given by

(6.67)

$$\frac{d}{dx}(\cos^{-1} x) = -\frac{1}{\sqrt{1 - x^2}} \qquad -1 < x < 1$$

Formula 6.67 can be proved by the same method as for Formula 6.64 and is left as Exercise 31.

The tangent function can be made one-to-one by restricting it to the interval $(-\pi/2, \pi/2)$. Thus the **inverse tangent function** is defined as the inverse of the function $f(x) = \tan x$, $-\pi/2 < x < \pi/2$ [see Figure 6.39(a)].

(6.68)

$$\tan^{-1} x = y \quad \Leftrightarrow \quad \tan y = x \quad \text{and} \quad -\frac{\pi}{2} < y < \frac{\pi}{2}$$

● **Example 4**   Simplify the expression $\cos(\tan^{-1} x)$.

Solution 1   Let $y = \tan^{-1} x$. Then $\tan y = x$ and $-\pi/2 < y < \pi/2$. We want to find $\cos y$ but, since $\tan y$ is known, it is easier to find $\sec y$ first:

$$\sec^2 y = 1 + \tan^2 y = 1 + x^2$$

$$\sec y = \sqrt{1 + x^2} \quad \textit{(since } \sec y > 0 \textit{ for } -\pi/2 < y < \pi/2\text{)}$$

Thus $\qquad \cos(\tan^{-1} x) = \cos y = \dfrac{1}{\sec y} = \dfrac{1}{\sqrt{1 + x^2}}$

Solution 2   Instead of using trigonometric identities as in Solution 1, it is perhaps easier to use a diagram. If $y = \tan^{-1} x$, then $\tan y = x$, and we can read from Figure 6.38 (which illustrates the case where $y > 0$) that

$$\cos(\tan^{-1} x) = \cos y = \frac{1}{\sqrt{1 + x^2}} \qquad ●$$

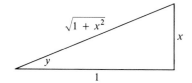

Figure 6.38

The inverse tangent function, $\tan^{-1} = \arctan$, has domain R and range $(-\pi/2, \pi/2)$. Its graph is shown in Figure 6.39(b).

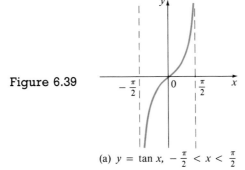

(a) $y = \tan x$, $-\dfrac{\pi}{2} < x < \dfrac{\pi}{2}$

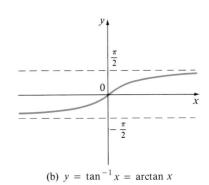

(b) $y = \tan^{-1} x = \arctan x$

Figure 6.39

We showed in Example 3 in Section 4.6 that

$$\lim_{x \to (\pi/2)^-} \tan x = \infty \quad \text{and} \quad \lim_{x \to -(\pi/2)^+} \tan x = -\infty$$

and so the lines $x = \pm\pi/2$ are vertical asymptotes of the graph of tan. Since the graph of $\tan^{-1}$ is obtained by reflecting the graph of the restricted tangent function in the line $y = x$, it follows that the lines $y = \pi/2$ and $y = -\pi/2$ are horizontal asymptotes of the graph of $\tan^{-1}$. This fact is expressed by the following limits:

(6.69)

$$\lim_{x \to \infty} \tan^{-1} x = \frac{\pi}{2} \qquad \lim_{x \to -\infty} \tan^{-1} x = -\frac{\pi}{2}$$

● **Example 5**   Evaluate $\displaystyle\lim_{x \to 2^+} \arctan\left(\frac{1}{x-2}\right)$.

Solution   Since

$$\frac{1}{x-2} \to \infty \qquad \text{as } x \to 2^+$$

(6.69) gives

$$\lim_{x \to 2^+} \arctan\left(\frac{1}{x-2}\right) = \frac{\pi}{2} \qquad\qquad\qquad ●$$

Since tan is differentiable, $\tan^{-1}$ is also differentiable. To find its derivative, let $y = \tan^{-1} x$. Then $\tan y = x$. Differentiating this latter equation implicitly with respect to x, we have

$$\sec^2 y \frac{dy}{dx} = 1$$

and so

$$\frac{dy}{dx} = \frac{1}{\sec^2 y} = \frac{1}{1 + \tan^2 y} = \frac{1}{1 + x^2}$$

(6.70)

$$\frac{d}{dx}(\tan^{-1} x) = \frac{1}{1 + x^2}$$

The remaining inverse trigonometric functions are not used as frequently and are summarized here.

(6.71)

$$y = \csc^{-1} x \, (|x| \geq 1) \quad \Leftrightarrow \quad \csc y = x \quad \text{and} \quad y \in \left(0, \frac{\pi}{2}\right] \cup \left(\pi, \frac{3\pi}{2}\right]$$

$$y = \sec^{-1} x \, (|x| \geq 1) \quad \Leftrightarrow \quad \sec y = x \quad \text{and} \quad y \in \left[0, \frac{\pi}{2}\right) \cup \left[\pi, \frac{3\pi}{2}\right)$$

$$y = \cot^{-1} x \, (x \in R) \quad \Leftrightarrow \quad \cot y = x \quad \text{and} \quad y \in (0, \pi)$$

The graphs in part (a) of Figures 6.40–6.42 can be compared with the complete graphs of cosecant, secant, and cotangent in Figure 13 of Appendix B to see how the functions have been restricted to make them one-to-one. It should be mentioned that the choice of intervals for y in the definitions of $\csc^{-1}$ and $\sec^{-1}$ is not universally accepted. For instance, some authors use $y \in [0, \pi/2) \cup (\pi/2, \pi]$ in the definition of $\sec^{-1}$. The reason for the choice in

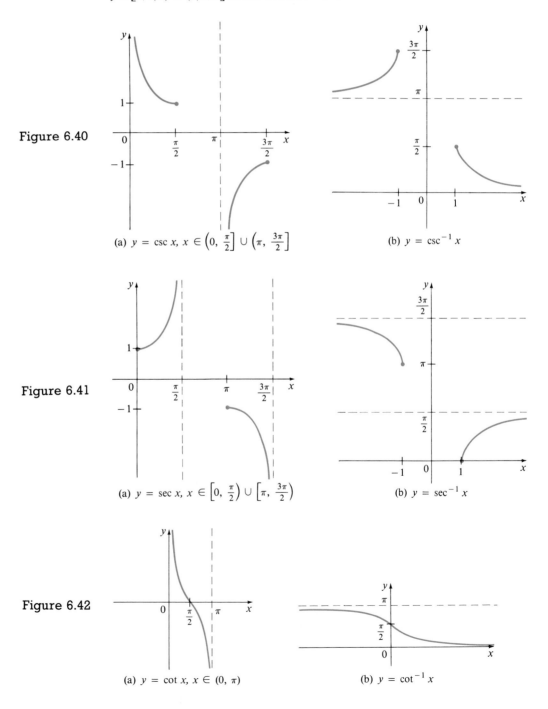

Figure 6.40

(a) $y = \csc x$, $x \in \left(0, \frac{\pi}{2}\right] \cup \left(\pi, \frac{3\pi}{2}\right]$

(b) $y = \csc^{-1} x$

Figure 6.41

(a) $y = \sec x$, $x \in \left[0, \frac{\pi}{2}\right) \cup \left[\pi, \frac{3\pi}{2}\right)$

(b) $y = \sec^{-1} x$

Figure 6.42

(a) $y = \cot x$, $x \in (0, \pi)$

(b) $y = \cot^{-1} x$

(6.71) is that the differentiation formulas turn out to be simpler (see Exercise 108).

We collect in Table 6.72 the differentiation formulas for all of the inverse trigonometric functions. The proofs of the formulas for the derivatives of $\csc^{-1}$, $\sec^{-1}$, and $\cot^{-1}$ are left as Exercises 33–35.

Table of Derivatives of Inverse Trigonometric Functions (6.72)

$$\frac{d}{dx}(\sin^{-1} x) = \frac{1}{\sqrt{1 - x^2}} \qquad \frac{d}{dx}(\csc^{-1} x) = -\frac{1}{x\sqrt{x^2 - 1}}$$

$$\frac{d}{dx}(\cos^{-1} x) = -\frac{1}{\sqrt{1 - x^2}} \qquad \frac{d}{dx}(\sec^{-1} x) = \frac{1}{x\sqrt{x^2 - 1}}$$

$$\frac{d}{dx}(\tan^{-1} x) = \frac{1}{1 + x^2} \qquad \frac{d}{dx}(\cot^{-1} x) = -\frac{1}{1 + x^2}$$

All of these formulas can be combined with the Chain Rule. For instance, if u is a differentiable function of x, then

$$\frac{d}{dx}(\sin^{-1} u) = \frac{1}{\sqrt{1 - u^2}}\frac{du}{dx} \quad \text{and} \quad \frac{d}{dx}(\tan^{-1} u) = \frac{1}{1 + u^2}\frac{du}{dx}$$

● **Example 6**   Differentiate (a) $y = \dfrac{1}{\sin^{-1} x}$ and (b) $f(x) = x\tan^{-1}\sqrt{x}$.

Solution   (a) $\dfrac{dy}{dx} = \dfrac{d}{dx}(\sin^{-1} x)^{-1} = -(\sin^{-1} x)^{-2}\dfrac{d}{dx}(\sin^{-1} x)$

$$= -\frac{1}{(\sin^{-1} x)^2\sqrt{1 - x^2}}$$

(b) $f'(x) = \tan^{-1}\sqrt{x} + x\dfrac{1}{1 + (\sqrt{x})^2}\dfrac{1}{2}x^{-1/2}$

$$= \tan^{-1}\sqrt{x} + \frac{\sqrt{x}}{2(1 + x)} \qquad ●$$

Each of the formulas in Table 6.72 gives rise to an integration formula. The two most useful of these are the following:

(6.73)

$$\int \frac{1}{\sqrt{1 - x^2}}\,dx = \sin^{-1} x + C$$

(6.74)

$$\int \frac{1}{x^2 + 1}\,dx = \tan^{-1} x + C$$

● **Example 7**   Find $\int \dfrac{1}{\sqrt{1-4x^2}}\,dx$.

Solution   If we write

$$\int \frac{1}{\sqrt{1-4x^2}}\,dx = \int \frac{1}{\sqrt{1-(2x)^2}}\,dx$$

then the integral resembles Equation 6.73 and the substitution $u = 2x$ is suggested. This gives $du = 2\,dx$, so $dx = du/2$ and

$$\int \frac{1}{\sqrt{1-4x^2}}\,dx = \frac{1}{2}\int \frac{du}{\sqrt{1-u^2}}$$

$$= \frac{1}{2}\sin^{-1}u + C = \frac{1}{2}\sin^{-1}(2x) + C \qquad ●$$

● **Example 8**   Evaluate $\int \dfrac{1}{x^2+a^2}\,dx$.

Solution   To make the given integral more like Equation 6.74 we write

$$\int \frac{dx}{x^2+a^2} = \int \frac{dx}{a^2\left(\dfrac{x^2}{a^2}+1\right)} = \frac{1}{a^2}\int \frac{dx}{\left(\dfrac{x}{a}\right)^2+1}$$

This suggests that we substitute $u = x/a$. Then $du = dx/a$, $dx = a\,du$, and

$$\int \frac{dx}{x^2+a^2} = \frac{1}{a^2}\int \frac{a\,du}{u^2+1} = \frac{1}{a}\int \frac{du}{u^2+1} = \frac{1}{a}\tan^{-1}u + C$$

Thus we have the formula

(6.75)

$$\boxed{\int \frac{1}{x^2+a^2}\,dx = \frac{1}{a}\tan^{-1}\left(\frac{x}{a}\right) + C} \qquad ●$$

● **Example 9**   Find $\int \dfrac{x}{x^4+9}\,dx$.

Solution   We substitute $u = x^2$ because then $du = 2x\,dx$ and we can use Equation 6.75 with $a = 3$:

$$\int \frac{x}{x^4+9}\,dx = \frac{1}{2}\int \frac{du}{u^2+9} = \frac{1}{2}\cdot\frac{1}{3}\tan^{-1}\left(\frac{u}{3}\right) + C$$

$$= \frac{1}{6}\tan^{-1}\left(\frac{x^2}{3}\right) + C \qquad ●$$

EXERCISES 6.8

Find the exact value of the expressions in Exercises 1–26.

1. $\cos^{-1}(-1)$

2. $\sin^{-1}(0.5)$

3. $\tan^{-1}\sqrt{3}$

4. $\arctan(-1)$

5. $\csc^{-1}\sqrt{2}$

6. $\arcsin 1$

7. $\cot^{-1}(-\sqrt{3})$

8. $\cos^{-1}\left(\dfrac{\sqrt{3}}{2}\right)$

9. $\sin^{-1}\left(-\dfrac{1}{\sqrt{2}}\right)$

10. $\sec^{-1} 2$

11. $\arctan\left(-\dfrac{\sqrt{3}}{3}\right)$

12. $\arccos(-0.5)$

13. $\sin(\sin^{-1} 0.7)$

14. $\sin^{-1}(\sin 1)$

15. $\tan(\tan^{-1} 10)$

16. $\tan^{-1}\left(\tan\dfrac{4\pi}{3}\right)$

17. $\cos\left(\sin^{-1}\dfrac{\sqrt{3}}{2}\right)$

18. $\tan(\cos^{-1} 0.5)$

19. $\sin(\cos^{-1}\frac{4}{5})$

20. $\sec(\arctan 2)$

21. $\arcsin\left(\sin\dfrac{5\pi}{4}\right)$

22. $\sin(2\sin^{-1}\frac{3}{5})$

23. $\cos(2\sin^{-1}\frac{5}{13})$

24. $\cos(2\sin^{-1} x),\ |x| \leqslant 1$

25. $\sin[\sin^{-1}\frac{1}{3} + \sin^{-1}\frac{2}{3}]$

26. $\cos[\sin^{-1}\frac{3}{4} + \cos^{-1}\frac{1}{4}]$

27. Prove that $\cos(\sin^{-1} x) = \sqrt{1 - x^2}$ for $-1 \leqslant x \leqslant 1$.

In Exercises 28–30 simplify each expression as in Exercise 27.

28. $\tan(\sin^{-1} x)$

29. $\sin(\tan^{-1} x)$

30. $\sin(2\cos^{-1} x)$

31. Prove Formula 6.67 for the derivative of $\cos^{-1}$ by the same method as for Formula 6.64.

32. (a) Prove that $\sin^{-1} x + \cos^{-1} x = \pi/2$ for all x.
(b) Use part (a) to prove Formula 6.67.

33. Prove that $D\cot^{-1} x = -\dfrac{1}{1 + x^2}$.

34. Prove that $D\sec^{-1} x = \dfrac{1}{x\sqrt{x^2 - 1}},\ |x| > 1$.

35. Prove that $D\csc^{-1} x = -\dfrac{1}{x\sqrt{x^2 - 1}},\ |x| > 1$.

In Exercises 36–59 find the derivatives of the given functions. Simplify where possible.

36. $f(x) = \sin^{-1}(2x - 1)$

37. $g(x) = \tan^{-1}(x^3)$

38. $y = (\sin^{-1} x)^2$

39. $y = \sin^{-1}(x^2)$

40. $G(x) = \sin^{-1}\left(\dfrac{x}{a}\right),\ a > 0$

41. $F(x) = \tan^{-1}\left(\dfrac{x}{a}\right)$

42. $h(x) = \arcsin x \ln x$

43. $H(x) = (1 + x^2)\arctan x$

44. $f(t) = \dfrac{\cos^{-1} t}{t}$

45. $g(t) = \sin^{-1}\dfrac{4}{t}$

46. $F(t) = \sqrt{1 - t^2} + \sin^{-1} t$

47. $G(t) = \cos^{-1}\sqrt{2t - 1}$

48. $y = \tan^{-1}\left(\dfrac{x}{a}\right) + \ln\sqrt{\dfrac{x - a}{x + a}}$

49. $y = \sec^{-1}\sqrt{1 + x^2}$

50. $y = x\cos^{-1} x - \sqrt{1 - x^2}$

51. $y = \tan^{-1}(\sin x)$

52. $y = \sin^{-1}\left(\dfrac{\cos x}{1 + \sin x}\right)$

53. $y = \dfrac{\sin^{-1} x}{\cos^{-1} x}$

54. $y = \sqrt[4]{\arcsin\sqrt{x^2 + 2x}}$

55. $y = (\tan^{-1} x)^{-1}$

56. $y = \tan^{-1}(x - \sqrt{1 + x^2})$

57. $y = x^2\cot^{-1}(3x)$

58. $y = x\sin x\csc^{-1} x$

59. $y = \arccos\left(\dfrac{b + a\cos x}{a + b\cos x}\right)$

In Exercises 60–67 find the derivative of the given function. State the domain of the function and the domain of its derivative.

60. $f(x) = \cos^{-1}(\sin^{-1} x)$

61. $g(x) = \sin^{-1}(3x + 1)$

62. $h(x) = x\sec^{-1}(3x^2)$

63. $S(x) = \sin^{-1}(\tan^{-1} x)$

64. $F(x) = \sqrt{\sin^{-1}\left(\dfrac{2}{x}\right)}$

65. $G(x) = \sqrt{\csc^{-1} x}$

66. $R(t) = \arcsin(2^t)$

67. $U(t) = 2^{\arctan t}$

68. If $f(x) = x \tan^{-1} x$, find $f'(1)$.

69. If $g(x) = x \sin^{-1}\left(\dfrac{x}{4}\right) + \sqrt{16 - x^2}$, find $g'(2)$.

70. If $h(x) = (3 \tan^{-1} x)^4$, find $h'(3)$.

In Exercises 71–80 find the given limits.

71. $\displaystyle\lim_{x \to -1^+} \sin^{-1} x$

72. $\displaystyle\lim_{x \to \infty} \sin^{-1}\left(\dfrac{x + 1}{2x + 1}\right)$

73. $\displaystyle\lim_{x \to -\infty} \cot^{-1} x$

74. $\displaystyle\lim_{x \to \infty} \sec^{-1} x$

75. $\displaystyle\lim_{x \to -\infty} \csc^{-1} x$

76. $\displaystyle\lim_{x \to \infty} \tan^{-1}(x^2)$

77. $\displaystyle\lim_{x \to \infty} (\tan^{-1} x)^2$

78. $\displaystyle\lim_{x \to \infty} \tan^{-1}(x - x^2)$

79. $\displaystyle\lim_{x \to 1^-} \dfrac{\arcsin x}{\tan\left(\dfrac{\pi x}{2}\right)}$

80. $\displaystyle\lim_{x \to \infty} \dfrac{\tan^{-1} x}{x}$

81. Where should the point P be chosen on the line segment AB in Figure 6.43 so as to maximize the angle θ?

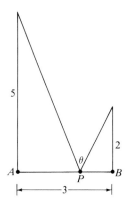

Figure 6.43

82. A painting in an art gallery has height h and is hung so that its lower edge is a distance d above the eye of an observer (as in Figure 6.44). How far from the wall should the observer stand to get the best view? (In other

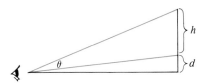

Figure 6.44

words, where should he stand so as to maximize the angle θ subtended at his eye by the painting?)

In Exercises 83–90 discuss the given curve under the headings A–H of Section 4.7.

83. $y = \sin^{-1}\left(\dfrac{x}{x + 1}\right)$

84. $y = \tan^{-1}\left(\dfrac{x - 1}{x + 1}\right)$

85. $y = \cos^{-1}(1 + x^2)$

86. $y = \arctan(\tan x)$

87. $y = \arctan x + \arctan\left(\dfrac{1}{x}\right)$

88. $y = \tan^{-1}(e^x)$

89. $y = x - \tan^{-1} x$

90. $y = \tan^{-1}(\ln x)$

91. Find the most general antiderivative of $f(x) = 2x + 5(1 - x^2)^{-1/2}$.

92. Find $f(x)$ if $f'(x) = 4 - 3(1 + x^2)^{-1}$ and $f(\pi/4) = 0$.

Evaluate the integrals in Exercises 93–104.

93. $\displaystyle\int_1^{\sqrt{3}} \dfrac{6}{1 + x^2}\, dx$

94. $\displaystyle\int_0^{0.5} \dfrac{dx}{\sqrt{1 - x^2}}$

95. $\displaystyle\int \dfrac{x^2}{\sqrt{1 - x^6}}\, dx$

96. $\displaystyle\int \dfrac{\tan^{-1} x}{1 + x^2}\, dx$

97. $\displaystyle\int \dfrac{x + 9}{x^2 + 9}\, dx$

98. $\displaystyle\int \dfrac{\sin x}{1 + \cos^2 x}\, dx$

99. $\displaystyle\int \dfrac{dx}{1 + 9x^2}$

100. $\displaystyle\int \dfrac{1}{x\sqrt{x^2 - 4}}\, dx$

101. $\displaystyle\int \dfrac{e^x}{e^{2x} + 1}\, dx$

102. $\displaystyle\int \dfrac{e^{2x}}{\sqrt{1 - e^{4x}}}\, dx$

103. $\displaystyle\int_0^{1/2} \dfrac{\sin^{-1} x}{\sqrt{1 - x^2}}\, dx$

104. $\displaystyle\int \dfrac{dx}{x[4 + (\ln x)^2]}$

105. Use the method of Example 8 to show that

$$\int \dfrac{1}{\sqrt{a^2 - x^2}}\, dx = \sin^{-1}\left(\dfrac{x}{a}\right) + C$$

106. (a) Sketch the graph of the function $f(x) = \sin(\sin^{-1} x)$.
  (b) Sketch the graph of the function $g(x) = \sin^{-1}(\sin x)$, $x \in R$.
  (c) Show that

$$g'(x) = \dfrac{\cos x}{|\cos x|}$$

(d)   Sketch the graph of $h(x) = \cos^{-1}(\sin x), x \in R,$ and find its derivative.

107.   Apply Theorem 4.15 to the function $f(x) = \tan^{-1} x + \cot^{-1} x$ to prove the identity $\tan^{-1} x + \cot^{-1} x = \pi/2.$

108.   Some authors define $y = \sec^{-1} x \Leftrightarrow \sec y = x$ and $y \in [0, \pi/2) \cup (\pi/2, \pi].$ Show that if this definition

is adopted, then, instead of the formula given in Exercise 34, we have

$$\frac{d}{dx}(\sec^{-1} x) = \frac{1}{|x|\sqrt{x^2 - 1}} \qquad |x| > 1$$

Hyperbolic Functions

There are certain combinations of the exponential functions e^x and e^{-x} that arise so frequently in mathematics and its applications that they deserve to be given special names. In many ways they are analogous to the trigonometric functions, and they have the same relationship to the hyperbola that the trigonometric functions have to the circle. For this reason they are collectively called **hyperbolic functions** and individually called **hyperbolic sine, hyperbolic cosine,** and so on.

Definition of the Hyperbolic Functions (6.76)

$$\sinh x = \frac{e^x - e^{-x}}{2} \qquad \operatorname{csch} x = \frac{1}{\sinh x}$$

$$\cosh x = \frac{e^x + e^{-x}}{2} \qquad \operatorname{sech} x = \frac{1}{\cosh x}$$

$$\tanh x = \frac{\sinh x}{\cosh x} \qquad \coth x = \frac{\cosh x}{\sinh x}$$

The graphs of hyperbolic sine and cosine can be sketched using graphical addition as in Figure 6.45.

Figure 6.45

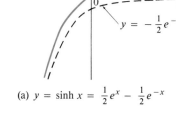

(a) $y = \sinh x = \frac{1}{2}e^x - \frac{1}{2}e^{-x}$

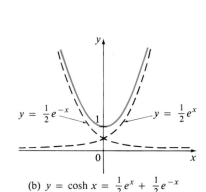

(b) $y = \cosh x = \frac{1}{2}e^x + \frac{1}{2}e^{-x}$

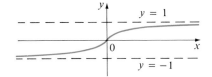

Figure 6.46

$y = \tanh x$

Note that sinh has domain R and range R, while cosh has domain R and range $[1, \infty)$. The graph of tanh is shown in Figure 6.46. It has the horizontal asymptotes $y = \pm 1$. (See Exercise 19.)

Some of the mathematical uses of hyperbolic functions will be seen in Chapter 7. Applications to science and engineering occur whenever an entity such as light, velocity, electricity, or radioactivity is gradually absorbed or extinguished, for the decay can be represented by hyperbolic functions. The most famous application is the use of hyperbolic cosine to describe the shape of a hanging wire. It can be proved that if a heavy flexible cable (such as a telephone or power line) is suspended between two points at the same height, then it takes the shape of a curve with equation $y = a \cosh(x/a)$ called a *catenary* (see Figure 6.47). (The Latin word *catena* means "chain.")

The hyperbolic functions satisfy a number of identities that are analogues of well-known trigonometric identities. We list some of them here and leave most of the proofs to the exercises.

Hyperbolic Identities (6.77)

$$\sinh(-x) = -\sinh x \qquad \cosh(-x) = \cosh x$$
$$\cosh^2 x - \sinh^2 x = 1 \qquad 1 - \tanh^2 x = \operatorname{sech}^2 x$$
$$\sinh(x + y) = \sinh x \cosh y + \cosh x \sinh y$$
$$\cosh(x + y) = \cosh x \cosh y + \sinh x \sinh y$$

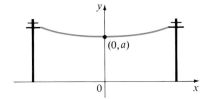

Figure 6.47

A CATENARY

● **Example 1**   Prove (a) $\cosh^2 x - \sinh^2 x = 1$ and
(b) $1 - \tanh^2 x = \operatorname{sech}^2 x$.

Solution   (a)

$$\cosh^2 x - \sinh^2 x = \left(\frac{e^x + e^{-x}}{2}\right)^2 - \left(\frac{e^x - e^{-x}}{2}\right)^2$$

$$= \frac{e^{2x} + 2 + e^{-2x}}{4} - \frac{e^{2x} - 2 + e^{-2x}}{4}$$

$$= \frac{4}{4} = 1$$

(b)   If we start with the identity proved in part (a):

$$\cosh^2 x - \sinh^2 x = 1$$

and we divide both sides by $\cosh^2 x$, we get

$$1 - \frac{\sinh^2 x}{\cosh^2 x} = \frac{1}{\cosh^2 x}$$

or

$$1 - \tanh^2 x = \operatorname{sech}^2 x \qquad \qquad ●$$

The identity proved in Example 1(a) gives a clue to the reason for the name "hyperbolic" functions.

Figure 6.48

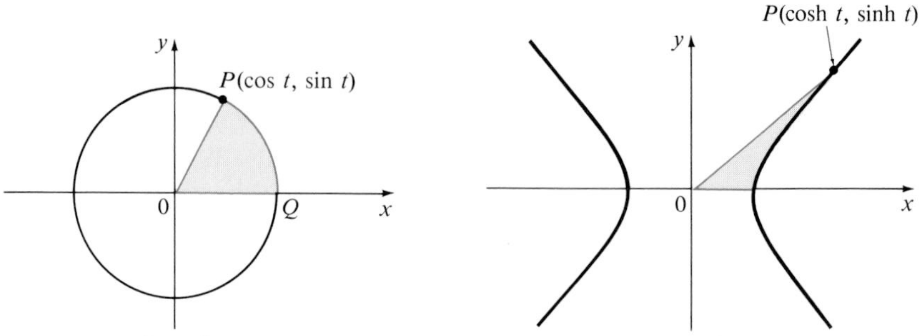

(a) $x^2 + y^2 = 1$ (b) $x^2 - y^2 = 1$

If t is any real number, then the point $P(\cos t, \sin t)$ lies on the unit circle $x^2 + y^2 = 1$ because $\cos^2 t + \sin^2 t = 1$. In fact, t can be interpreted as the radian measure of $\angle POQ$ in Figure 6.48(a). For this reason the trigonometric functions are sometimes called circular functions.

Likewise if t is any real number, then the point $P(\cosh t, \sinh t)$ lies on the right branch of the hyperbola $x^2 - y^2 = 1$ because $\cosh^2 t - \sinh^2 t = 1$ and $\cosh t \geqslant 1$. This time, t does not represent the measure of an angle. However it turns out that t represents twice the area of the shaded hyperbolic sector in Figure 6.48(b), just as in the trigonometric case t represents twice the area of the shaded circular sector in Figure 6.48(a).

The derivatives of the hyperbolic functions are easily computed. For example,

$$\frac{d}{dx}(\sinh x) = \frac{d}{dx}\left(\frac{e^x - e^{-x}}{2}\right) = \frac{e^x + e^{-x}}{2} = \cosh x$$

We list the differentiation formulas for the hyperbolic functions as Table 6.78. The remaining proofs are left as exercises. Note the analogy with the differentiation formulas for trigonometric functions, but beware that the signs are sometimes different.

Table of Derivatives of Hyperbolic Functions (6.78)

$$\frac{d}{dx}\sinh x = \cosh x \qquad\qquad \frac{d}{dx}\operatorname{csch} x = -\operatorname{csch} x \coth x$$

$$\frac{d}{dx}\cosh x = \sinh x \qquad\qquad \frac{d}{dx}\operatorname{sech} x = -\operatorname{sech} x \tanh x$$

$$\frac{d}{dx}\tanh x = \operatorname{sech}^2 x \qquad\qquad \frac{d}{dx}\coth x = -\operatorname{csch}^2 x$$

● **Example 2**   Any of these differentiation rules can be combined with the Chain Rule. For instance,

$$\frac{d}{dx}(\cosh \sqrt{x}) = \sinh \sqrt{x} \cdot \frac{d}{dx}\sqrt{x} = \frac{\sinh \sqrt{x}}{2\sqrt{x}}$$

   ●

Inverse Hyperbolic Functions

You can see from Figures 6.45(a) and 6.46 that sinh and tanh are one-to-one functions and so they have inverse functions denoted by $\sinh^{-1}$ and $\tanh^{-1}$. Figure 6.45(b) shows that cosh is not one-to-one, but when restricted to the domain $[0, \infty)$ it becomes one-to-one. The inverse hyperbolic cosine function is defined as the inverse of this restricted function.

(6.79)

$$
\begin{aligned}
y &= \sinh^{-1} x \quad \Leftrightarrow \quad \sinh y = x \\
y &= \cosh^{-1} x \quad \Leftrightarrow \quad \cosh y = x \quad \text{and} \quad y \geq 0 \\
y &= \tanh^{-1} x \quad \Leftrightarrow \quad \tanh y = x
\end{aligned}
$$

The remaining inverse hyperbolic functions are defined similarly (see Exercise 24).

We can sketch the graphs of $\sinh^{-1}$, $\cosh^{-1}$, and $\tanh^{-1}$ in Figure 6.49 by using Figures 6.45 and 6.46.

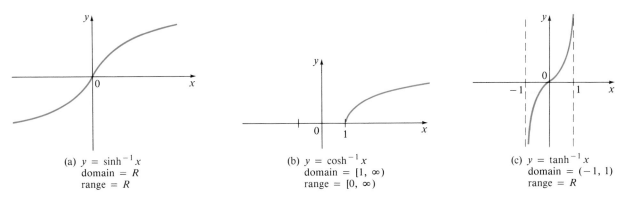

(a) $y = \sinh^{-1} x$
domain $= R$
range $= R$

(b) $y = \cosh^{-1} x$
domain $= [1, \infty)$
range $= [0, \infty)$

(c) $y = \tanh^{-1} x$
domain $= (-1, 1)$
range $= R$

Figure 6.49

Since the hyperbolic functions are defined in terms of exponential functions, it is not surprising to learn that the inverse hyperbolic functions can be expressed in terms of logarithms. In particular we have:

(6.80)

(6.81)

(6.82)

$$
\begin{aligned}
\sinh^{-1} x &= \ln(x + \sqrt{x^2 + 1}) \qquad x \in R \\
\cosh^{-1} x &= \ln(x + \sqrt{x^2 - 1}) \qquad x \geq 1 \\
\tanh^{-1} x &= \frac{1}{2} \ln\left(\frac{1 + x}{1 - x}\right) \qquad -1 < x < 1
\end{aligned}
$$

● **Example 3**   Show that $\sinh^{-1} x = \ln(x + \sqrt{x^2 + 1})$.

Solution   Let $y = \sinh^{-1} x$. Then

$$x = \sinh y = \frac{e^y - e^{-y}}{2}$$

so $$e^y - 2x - e^{-y} = 0$$

or, multiplying by e^y,

$$e^{2y} - 2xe^y - 1 = 0$$

This is really a quadratic equation in e^y:

$$(e^y)^2 - 2x(e^y) - 1 = 0$$

Solving by the quadratic formula, we get

$$e^y = \frac{2x \pm \sqrt{4x^2 + 4}}{2} = x \pm \sqrt{x^2 + 1}$$

Note that $e^y > 0$, but $x - \sqrt{x^2 + 1} < 0$ (because $x < \sqrt{x^2 + 1}$). Thus the minus sign is inadmissible and we have

$$e^y = x + \sqrt{x^2 + 1}$$

Therefore $$y = \ln(e^y) = \ln(x + \sqrt{x^2 + 1})$$

(See Exercise 21 for another method.)   ●

Table of Derivatives of Inverse Hyperbolic Functions (6.83)

$$\frac{d}{dx} \sinh^{-1} x = \frac{1}{\sqrt{1 + x^2}} \qquad \frac{d}{dx} \operatorname{csch}^{-1} x = -\frac{1}{|x|\sqrt{x^2 + 1}}$$

$$\frac{d}{dx} \cosh^{-1} x = \frac{1}{\sqrt{x^2 - 1}} \qquad \frac{d}{dx} \operatorname{sech}^{-1} x = -\frac{1}{x\sqrt{1 - x^2}}$$

$$\frac{d}{dx} \tanh^{-1} x = \frac{1}{1 - x^2} \qquad \frac{d}{dx} \coth^{-1} x = \frac{1}{1 - x^2}$$

The inverse hyperbolic functions are all differentiable by Theorem 6.12 because the hyperbolic functions are differentiable. The formulas in Table 6.83 can be proved either by the method for inverse functions or by differentiating Formulas 6.80, 6.81, and 6.82.

● **Example 4**   Prove that $\dfrac{d}{dx} \sinh^{-1} x = \dfrac{1}{\sqrt{1 + x^2}}$.

Solution 1   Let $y = \sinh^{-1} x$. Then $\sinh y = x$. If we differentiate this equation implicitly with respect to x, we get

$$\cosh y \frac{dy}{dx} = 1$$

Since $\cosh^2 y - \sinh^2 y = 1$ and $\cosh y \geq 0$, we have $\cosh y = \sqrt{1 + \sinh^2 y}$,

so $$\frac{dy}{dx} = \frac{1}{\cosh y} = \frac{1}{\sqrt{1 + \sinh^2 y}} = \frac{1}{\sqrt{1 + x^2}}$$

Solution 2   From (6.80) (proved in Example 3), we have

$$\frac{d}{dx} \sinh^{-1} x = \frac{d}{dx} \ln(x + \sqrt{x^2 + 1})$$

$$= \frac{1}{x + \sqrt{x^2 + 1}} \frac{d}{dx} (x + \sqrt{x^2 + 1})$$

$$= \frac{1}{x + \sqrt{x^2 + 1}} \left(1 + \frac{x}{\sqrt{x^2 + 1}}\right)$$

$$= \frac{\sqrt{x^2 + 1} + x}{(x + \sqrt{x^2 + 1})\sqrt{x^2 + 1}}$$

$$= \frac{1}{\sqrt{x^2 + 1}}$$

● **Example 5**   Find $\dfrac{d}{dx} \tanh^{-1}(\sin x)$.

Solution   Using Table 6.83 and the Chain Rule, we have

$$\frac{d}{dx} \tanh^{-1}(\sin x) = \frac{1}{1 - (\sin x)^2} \frac{d}{dx} (\sin x)$$

$$= \frac{1}{1 - \sin^2 x} \cos x = \frac{\cos x}{\cos^2 x} = \sec x$$

● **Example 6**   Evaluate $\displaystyle\int_0^1 \frac{dx}{\sqrt{1 + x^2}}$.

Solution   From Table 6.83 (or Example 4) we know that an antiderivative of $1/\sqrt{1 + x^2}$ is $\sinh^{-1} x$. Therefore

$$\int_0^1 \frac{dx}{\sqrt{1 + x^2}} = \sinh^{-1} x \Big]_0^1$$

$$= \sinh^{-1} 1$$

$$= \ln(1 + \sqrt{2}) \qquad [\text{ from Equation (6.80)}]$$

EXERCISES 6.9

Prove the identities in Exercises 1–15.

1. $\sinh(-x) = -\sinh x$ (This shows that sinh is an odd function.)

2. $\cosh(-x) = \cosh x$ (This shows that cosh is an even function.)

3. $\cosh x + \sinh x = e^x$

4. $\cosh x - \sinh x = e^{-x}$

5. $\sinh(x + y) = \sinh x \cosh y + \cosh x \sinh y$

6. $\cosh(x + y) = \cosh x \cosh y + \sinh x \sinh y$

7. $\coth^2 x - 1 = \operatorname{csch}^2 x$

8. $\tanh(x + y) = \dfrac{\tanh x + \tanh y}{1 + \tanh x \tanh y}$

9. $\sinh 2x = 2 \sinh x \cosh x$

10. $\cosh 2x = \cosh^2 x + \sinh^2 x$

11. $\sinh \dfrac{x}{2} = \pm \sqrt{\dfrac{\cosh x - 1}{2}}$

12. $\cosh \dfrac{x}{2} = \sqrt{\dfrac{\cosh x + 1}{2}}$

13. $\tanh(\ln x) = \dfrac{x^2 - 1}{x^2 + 1}$

14. $\dfrac{1 + \tanh x}{1 - \tanh x} = e^{2x}$

15. $(\cosh x + \sinh x)^n = \cosh nx + \sinh nx$ (*n* any real number)

16. If $\sinh x = \frac{3}{4}$, find the values of the other hyperbolic functions at x.

17. If $\tanh x = \frac{4}{5}$, find the values of the other hyperbolic functions at x.

18. Use the graphs of sinh, cosh, and tanh in Figures 6.45 and 6.46 to draw the graphs of csch, sech, and coth.

19. Use the definitions of the hyperbolic functions to find the following limits:

(a) $\lim\limits_{x \to \infty} \tanh x$ (b) $\lim\limits_{x \to -\infty} \tanh x$

(c) $\lim\limits_{x \to \infty} \sinh x$ (d) $\lim\limits_{x \to -\infty} \sinh x$

(e) $\lim\limits_{x \to \infty} \operatorname{sech} x$ (f) $\lim\limits_{x \to \infty} \coth x$

(g) $\lim\limits_{x \to 0^+} \coth x$ (h) $\lim\limits_{x \to 0^-} \coth x$

(i) $\lim\limits_{x \to -\infty} \operatorname{csch} x$

20. Prove the formulas given in Table 6.78 for the derivatives of the following functions: (a) cosh, (b) tanh, (c) csch, (d) sech, and (e) coth.

21. Give an alternative solution to Example 3 by letting $y = \sinh^{-1} x$ and then using Exercise 3 and Example 1(a) (with x replaced by y).

22. Prove Equation 6.81.

23. Prove Equation 6.82 (a) using the method of Example 3, and (b) using Exercise 14 with x replaced by y.

24. For each of the following functions (i) give a definition like those in (6.79), (ii) sketch the graph, and (iii) find a formula similar to Equation 6.80.
(a) csch^{-1} (b) sech^{-1} (c) $\coth^{-1}$

25. Prove the formulas given in Table 6.83 for the derivatives of the following functions.
(a) $\cosh^{-1}$ (b) $\tanh^{-1}$ (c) csch^{-1}
(d) sech^{-1} (e) $\coth^{-1}$

In Exercises 26–45 find the derivatives of the given functions.

26. $f(x) = e^x \sinh x$ **27.** $f(x) = \tanh 3x$

28. $g(x) = \cosh^4 x$ **29.** $h(x) = \cosh(x^4)$

30. $F(x) = e^{\coth 2x}$ **31.** $G(x) = x^2 \operatorname{sech} x$

32. $f(t) = \ln(\sinh t)$ **33.** $H(t) = \tanh(e^t)$

34. $y = \cos(\sinh x)$ **35.** $y = x^{\cosh x}$

36. $y = e^{\tanh x} \cosh(\cosh x)$ **37.** $y = \cosh^{-1}(x^2)$

38. $y = \sqrt{x}\,\sinh^{-1}\sqrt{x}$ **39.** $y = x \ln(\operatorname{sech} 4x)$

40. $y = x \tanh^{-1} x + \ln\sqrt{1 - x^2}$

41. $y = \tanh^{-1}\left(\dfrac{x}{a}\right)$

42. $y = x \sinh^{-1}\left(\dfrac{x}{3}\right) - \sqrt{9 + x^2}$

43. $y = \operatorname{csch}^{-1}(x^4)$

44. $y = \operatorname{sech}^{-1}\sqrt{1 - x^2}$

45. $y = \coth^{-1}\sqrt{x^2 + 1}$

Evaluate the integrals in Exercises 46–53.

46. $\displaystyle\int \operatorname{sech}^2 x \, dx$

47. $\displaystyle\int \sinh 2x \, dx$

48. $\displaystyle\int \tanh x \, dx$

49. $\displaystyle\int \coth x \, dx$

50. $\displaystyle\int \frac{\sinh x}{1 + \cosh x} \, dx$

51. $\displaystyle\int \frac{1}{\sqrt{4 + x^2}} \, dx$

52. $\displaystyle\int_2^3 \frac{1}{\sqrt{x^2 - 1}} \, dx$

53. $\displaystyle\int_0^{1/2} \frac{1}{1 - x^2} \, dx$

54. Evaluate:

$$\lim_{x \to \infty} \frac{\sinh x}{e^x}$$

55. At what point of the curve $y = \cosh x$ does the tangent have slope 1?

56. (a) Show that any function of the form $y = A \sinh mx + B \cosh mx$ satisfies the differential equation $y'' = m^2 y$.

(b) Find $y = y(x)$ such that $y'' = 9y$, $y(0) = -4$, and $y'(0) = 6$.

SECTION 6.10

Indeterminate Forms and L'Hospital's Rule

Suppose we are trying to sketch the graph of the function

$$F(x) = \frac{2^x - 1}{x}$$

Although F is not defined when $x = 0$, we need to know how F behaves *near* 0. In particular, we would like to know the value of the limit

(6.84)

$$\lim_{x \to 0} \frac{2^x - 1}{x}$$

But we cannot apply Property 5 of limits (the limit of a quotient is the quotient of the limits) to (6.84) because the limit of the denominator is 0. In fact, although the limit in (6.84) exists, its value is not obvious because both numerator and denominator approach 0 and $\frac{0}{0}$ is not defined.

In general if we have a limit of the form

$$\lim_{x \to a} \frac{f(x)}{g(x)}$$

where both $f(x) \to 0$ and $g(x) \to 0$ as $x \to a$, then this limit may or may not exist and is called an **indeterminate form of type $\frac{0}{0}$**. We have already met some limits of this type in Chapter 2. For rational functions, cancellation of common factors can be used:

$$\lim_{x \to 1} \frac{x^2 - x}{x^2 - 1} = \lim_{x \to 1} \frac{x(x - 1)}{(x + 1)(x - 1)} = \lim_{x \to 1} \frac{x}{x + 1} = \frac{1}{2}$$

We used a geometric argument to show that

$$\lim_{x \to 0} \frac{\sin x}{x} = 1$$

But these methods do not work for limits such as (6.84), so in this section we introduce a systematic method, known as l'Hospital's Rule, for the evaluation of indeterminate forms.

Another situation in which a limit is not obvious occurs when we try to sketch the curve $y = \ln x/x$. In searching for horizontal asymptotes or other aspects of the curve for large values of x, we need to evaluate the limit

(6.85)
$$\lim_{x \to \infty} \frac{\ln x}{x}$$

It is not obvious how to evaluate this limit because both numerator and denominator become large as $x \to \infty$. There is a struggle between numerator and denominator. If the numerator wins, the limit will be ∞; if the denominator wins, the answer will be 0. Or there may be some compromise, in which case the answer may be some finite positive number.

In general if we have a limit of the form

$$\lim_{x \to a} \frac{f(x)}{g(x)}$$

where both $f(x) \to \infty$ (or $-\infty$) and $g(x) \to \infty$ (or $-\infty$), then the limit may or may not exist and is called an **indeterminate form of type ∞/∞.** We have already seen in Section 4.5 that this type of limit can be evaluated for certain functions, including rational functions, by dividing numerator and denominator by the highest power of x that occurs. For instance,

$$\lim_{x \to \infty} \frac{x^2 - 1}{2x^2 + 1} = \lim_{x \to \infty} \frac{1 - \dfrac{1}{x^2}}{2 + \dfrac{1}{x^2}} = \frac{1 - 0}{2 + 0} = \frac{1}{2}$$

This method will not work for limits such as (6.85) but l'Hospital's Rule also applies to this type of indeterminate form.

L'Hospital's Rule is named after a French nobleman, the Marquis de l'Hospital (1661–1704), but was discovered by a Swiss mathematician, John Bernoulli (1667–1748).

L'Hospital's Rule (6.86)

Suppose f and g are differentiable and $g'(x) \neq 0$ on an open interval I that contains a (except possibly at a). Suppose that

$$\lim_{x \to a} f(x) = 0 \qquad \text{and} \quad \lim_{x \to a} g(x) = 0$$

or that $\qquad \lim_{x \to a} f(x) = \pm\infty \quad \text{and} \quad \lim_{x \to a} g(x) = \pm\infty$

(In other words, we have an indeterminate form of type $\frac{0}{0}$ or ∞/∞.) Then

$$\lim_{x \to a} \frac{f(x)}{g(x)} = \lim_{x \to a} \frac{f'(x)}{g'(x)}$$

if the limit on the right side exists (or is ∞ or $-\infty$).

Note 1:   L'Hospital's Rule says that the limit of a quotient of functions is equal to the limit of the quotient of their derivatives, provided that the conditions stated in Theorem 6.86 are satisfied. It is especially important to verify the conditions regarding the limits of f and g before using l'Hospital's Rule.

Note 2:   L'Hospital's Rule is also valid for one-sided limits and for limits at infinity or negative infinity; that is, in Theorem 6.86, "$x \to a$" can be replaced by any of the following symbols: $x \to a^+$, $x \to a^-$, $x \to \infty$, $x \to -\infty$.

Note 3:   For the special case where $f(a) = g(a) = 0$, f' and g' are continuous, and $g'(a) \neq 0$, it is easy to see why l'Hospital's Rule is true. In fact, using (3.3) we have

$$\lim_{x \to a} \frac{f(x)}{g(x)} = \lim_{x \to a} \frac{f(x) - f(a)}{g(x) - g(a)}$$

$$= \lim_{x \to a} \frac{\dfrac{f(x) - f(a)}{x - a}}{\dfrac{g(x) - g(a)}{x - a}} = \frac{\lim\limits_{x \to a} \dfrac{f(x) - f(a)}{x - a}}{\lim\limits_{x \to a} \dfrac{g(x) - g(a)}{x - a}}$$

$$= \frac{f'(a)}{g'(a)} = \lim_{x \to a} \frac{f'(x)}{g'(x)}$$

The general version of l'Hospital's Rule for the indeterminate form $\frac{0}{0}$ is somewhat more difficult and its proof is deferred to the end of this section. The proof for the indeterminate form ∞/∞ can be found in more advanced books.

● **Example 1**   Find $\lim\limits_{x \to 0} \dfrac{2^x - 1}{x}$.

Solution   Since $\lim_{x \to 0}(2^x - 1) = 0$ and $\lim_{x \to 0} x = 0$, we can apply l'Hospital's Rule:

$$\lim_{x \to 0} \frac{2^x - 1}{x} = \lim_{x \to 0} \frac{\dfrac{d}{dx}(2^x - 1)}{\dfrac{d}{dx}(x)} = \lim_{x \to 0} \frac{2^x \ln 2}{1} = \ln 2 \qquad ●$$

● **Example 2**   Calculate $\lim\limits_{x \to \infty} \dfrac{e^x}{x^2}$.

Solution   We have $\lim_{x \to \infty} e^x = \infty$ and $\lim_{x \to \infty} x^2 = \infty$, so l'Hospital's Rule gives

$$\lim_{x \to \infty} \frac{e^x}{x^2} = \lim_{x \to \infty} \frac{e^x}{2x}$$

Since $e^x \to \infty$ and $2x \to \infty$ as $x \to \infty$, the limit on the right side is also indeterminate, but a second application of l'Hospital's Rule gives

$$\lim_{x \to \infty} \frac{e^x}{x^2} = \lim_{x \to \infty} \frac{e^x}{2x} = \lim_{x \to \infty} \frac{e^x}{2} = \infty$$

● **Example 3**   Calculate $\displaystyle\lim_{x \to \infty} \frac{\ln x}{\sqrt[3]{x}}$.

Solution   Since $\ln x \to \infty$ and $\sqrt[3]{x} \to \infty$ as $x \to \infty$, l'Hospital's Rule applies:

$$\lim_{x \to \infty} \frac{\ln x}{\sqrt[3]{x}} = \lim_{x \to \infty} \frac{\dfrac{1}{x}}{\dfrac{1}{3} x^{-2/3}}$$

Notice that the limit on the right side is now indeterminate of type $\frac{0}{0}$. But instead of applying l'Hospital's Rule a second time as we did in Example 2, we simplify the expression and see that a second application is unnecessary:

$$\lim_{x \to \infty} \frac{\ln x}{\sqrt[3]{x}} = \lim_{x \to \infty} \frac{\dfrac{1}{x}}{\dfrac{1}{3} x^{-2/3}} = \lim_{x \to \infty} \frac{3}{\sqrt[3]{x}} = 0$$

● **Example 4**   Find $\displaystyle\lim_{x \to 0} \frac{\tan x - x}{x^3}$. (See Exercise 48 in Section 2.2.)

Solution   Noting that both $\tan x - x \to 0$ and $x^3 \to 0$ as $x \to 0$, we use l'Hospital's Rule:

$$\lim_{x \to 0} \frac{\tan x - x}{x^3} = \lim_{x \to 0} \frac{\sec^2 x - 1}{3x^2}$$

Since the limit on the right side is still indeterminate of type $\frac{0}{0}$, we apply l'Hospital's Rule again:

$$\lim_{x \to 0} \frac{\sec^2 x - 1}{3x^2} = \lim_{x \to 0} \frac{2 \sec^2 x \tan x}{6x}$$

Again both numerator and denominator approach 0, so a third application of l'Hospital's Rule is necessary. Putting together all three steps, we get

$$\lim_{x \to 0} \frac{\tan x - x}{x^3} = \lim_{x \to 0} \frac{\sec^2 x - 1}{3x^2} = \lim_{x \to 0} \frac{2 \sec^2 x \tan x}{6x}$$

$$= \lim_{x \to 0} \frac{4 \sec^2 x \tan^2 x + 2 \sec^4 x}{6} = \frac{2}{6} = \frac{1}{3}$$

● **Example 5**   Find $\displaystyle\lim_{x \to \pi^-} \frac{\sin x}{1 - \cos x}$.

Solution   If we were to blindly attempt to use l'Hospital's Rule, we would get

$$\lim_{x \to \pi^-} \frac{\sin x}{1 - \cos x} = \lim_{x \to \pi^-} \frac{\cos x}{\sin x} = -\infty$$

This is *wrong!* Although the numerator $\sin x \to 0$ as $x \to \pi^-$, notice that the denominator $(1 - \cos x)$ does not approach 0, so l'Hospital's Rule cannot be applied here.

The required limit is, in fact, easy to find because the function is continuous and the denominator is nonzero at π:

$$\lim_{x \to \pi^-} \frac{\sin x}{1 - \cos x} = \frac{\sin \pi}{1 - \cos \pi} = \frac{0}{1 - (-1)} = 0 \qquad \bullet$$

Example 5 shows what can go wrong if you use l'Hospital's Rule without thinking. There are other limits that *can* be found using l'Hospital's Rule but are more easily found by other methods. (See Examples 2 and 4 in Section 2.2, Examples 2 and 3 in Section 4.5, Examples 1 and 2 in Section 3.3, and the discussion at the beginning of this section.) So when evaluating any limit, you should try other methods before using l'Hospital's Rule.

Indeterminate Products

If $\lim_{x \to a} f(x) = 0$ and $\lim_{x \to a} g(x) = \infty$ (or $-\infty$), then it is not clear what the value of $\lim_{x \to a} f(x)g(x)$, if any, will be. There is a struggle between f and g. If f wins, the answer will be 0; if g wins, the answer will be ∞ (or $-\infty$). Or there may be a compromise where the answer is a finite nonzero number. This kind of limit is called an **indeterminate form of type $0 \cdot \infty$.** We can deal with it by writing the product fg as a quotient:

$$fg = \frac{f}{1/g} \quad \text{or} \quad fg = \frac{g}{1/f}$$

This converts the given limit into an indeterminate form of type $\frac{0}{0}$ or ∞/∞ so that we can use l'Hospital's Rule.

● **Example 6**   Evaluate $\lim_{x \to 0^+} x \ln x$.

Solution   The given limit is indeterminate because $x \to 0^+$ while $\ln x \to -\infty$. Writing $x = 1/(1/x)$, we have $1/x \to \infty$ as $x \to 0^+$, so l'Hospital's Rule gives

$$\lim_{x \to 0^+} x \ln x = \lim_{x \to 0^+} \frac{\ln x}{\dfrac{1}{x}} = \lim_{x \to 0^+} \frac{\dfrac{1}{x}}{\dfrac{-1}{x^2}}$$

$$= \lim_{x \to 0^+} (-x) = 0 \qquad \bullet$$

Indeterminate Differences

If $\lim_{x \to a} f(x) = \infty$ and $\lim_{x \to a} g(x) = \infty$, then the limit

$$\lim_{x \to a} (f(x) - g(x))$$

is called an **indeterminate form of type $\infty - \infty$.** Again there is a contest between f and g. Will the answer be ∞ (f wins) or will it be $-\infty$ (g wins) or will they compromise on a finite number? To find out, we convert the difference into a quotient (for instance, by using a common denominator or rationalization, or factoring out a common factor) so that we have an indeterminate form of type $\frac{0}{0}$ or ∞/∞.

● **Example 7**   Compute $\displaystyle\lim_{x \to (\pi/2)^-} (\sec x - \tan x)$.

Solution   First notice that $\sec x \to \infty$ and $\tan x \to \infty$ as $x \to (\pi/2)^-$, so the limit is indeterminate. Here we use a common denominator:

$$\lim_{x \to \pi/2^-} (\sec x - \tan x) = \lim_{x \to \pi/2^-} \left(\frac{1}{\cos x} - \frac{\sin x}{\cos x} \right)$$

$$= \lim_{x \to \pi/2^-} \frac{1 - \sin x}{\cos x} = \lim_{x \to \pi/2^-} \frac{-\cos x}{-\sin x} = 0$$

Note that the use of l'Hospital's Rule was justified because $1 - \sin x \to 0$ and $\cos x \to 0$ as $x \to (\pi/2)^-$.   ●

Indeterminate Powers

There are several indeterminate forms that arise from the limit

$$\lim_{x \to a} [f(x)]^{g(x)}$$

1. $\displaystyle\lim_{x \to a} f(x) = 0$ and $\displaystyle\lim_{x \to a} g(x) = 0$ type 0^0
2. $\displaystyle\lim_{x \to a} f(x) = \infty$ and $\displaystyle\lim_{x \to a} g(x) = 0$ type ∞^0
3. $\displaystyle\lim_{x \to a} f(x) = 1$ and $\displaystyle\lim_{x \to a} g(x) = \pm\infty$ type 1^∞

Each of these three cases can be treated either by taking the natural logarithm:

$$\text{let } y = [f(x)]^{g(x)}, \text{ then } \ln y = g(x) \ln f(x)$$

or by writing the function as an exponential:

$$[f(x)]^{g(x)} = e^{g(x) \ln f(x)}$$

(Recall that both of these methods were used in differentiating such functions.) In either method we are led to the indeterminate product $g(x) \ln f(x)$, which will be of type $0 \cdot \infty$.

● **Example 8**   Calculate $\lim\limits_{x \to 0^+} (1 + \sin 4x)^{\cot x}$.

Solution   First notice that as $x \to 0^+$, we have $1 + \sin 4x \to 1$ and $\cot x \to \infty$, so the given limit is indeterminate. Let

$$y = (1 + \sin 4x)^{\cot x}$$

Then $\ln y = \ln[(1 + \sin 4x)^{\cot x}] = \cot x \ln(1 + \sin 4x)$

so l'Hospital's Rule gives

$$\lim_{x \to 0^+} \ln y = \lim_{x \to 0^+} \frac{\ln(1 + \sin 4x)}{\tan x}$$

$$= \lim_{x \to 0^+} \frac{\dfrac{4 \cos 4x}{1 + \sin 4x}}{\sec^2 x} = 4$$

So far we have computed the limit of $\ln y$, but what we want is the limit of y. To find this we use the fact that $y = e^{\ln y}$:

$$\lim_{x \to 0^+} (1 + \sin 4x)^{\cot x} = \lim_{x \to 0^+} y$$

$$= \lim_{x \to 0^+} e^{\ln y}$$

$$= e^4$$ ●

● **Example 9**   Use l'Hospital's Rule to help sketch the curve $y = x^x$, $x > 0$.

Solution

A.   The domain of $f(x) = x^x$ is given as $(0, \infty)$.

B.   Intercepts: none.

C.   Symmetry: none.

D.   There is no horizontal asymptote since

$$\lim_{x \to \infty} x^x = \infty$$

To determine the behavior of the function near 0 we need to compute $\lim_{x \to 0^+} x^x$. (Notice that this limit is indeterminate since $0^x = 0$ for any $x > 0$ but $x^0 = 1$ for any $x \ne 0$.) We could proceed as in Example 8 or by writing the function as an exponential:

$$x^x = (e^{\ln x})^x = e^{x \ln x}$$

Now l'Hospital's Rule gives

$$\lim_{x \to 0^+} x \ln x = \lim_{x \to 0^+} \frac{\ln x}{\dfrac{1}{x}} = \lim_{x \to 0^+} \frac{\dfrac{1}{x}}{\dfrac{-1}{x^2}}$$

$$= \lim_{x \to 0^+} (-x) = 0$$

Therefore
$$\lim_{x \to 0^+} x^x = \lim_{x \to 0^+} e^{x \ln x} = e^0 = 1$$

E.
$$f'(x) = e^{x \ln x}\left(\ln x + x \cdot \frac{1}{x}\right) = x^x(\ln x + 1)$$

$$f'(x) > 0 \quad \Leftrightarrow \quad \ln x + 1 > 0 \quad \Leftrightarrow \quad \ln x > -1 \quad \Leftrightarrow \quad x > e^{-1}$$
$$f'(x) < 0 \quad \Leftrightarrow \quad 0 < x < e^{-1}$$

Thus f is increasing on $[1/e, \infty)$ and decreasing on $(0, 1/e]$.

F. $f(1/e) = e^{-1/e}$ is a local minimum by the First Derivative Test since $f'(1/e) = 0$ and f' changes from negative to positive at $1/e$.

G.
$$f''(x) = x^x(\ln x + 1)^2 + x^x \cdot \frac{1}{x}$$

Thus $f''(x) > 0$ for all $x > 0$ and so the curve is concave upward on $(0, \infty)$.

H. As additional information we note that

$$\lim_{x \to 0^+} f'(x) = \lim_{x \to 0^+} x^x(\ln x + 1) = -\infty$$

because $x^x \to 1$ and $\ln x \to -\infty$ as $x \to 0^+$.

The curve $y = x^x$ is sketched in Figure 6.50.   ●

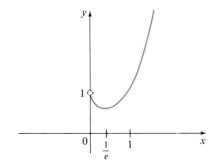

Figure 6.50
$y = x^x, \, x > 0$

In order to give the promised proof of l'Hospital's Rule we first need a generalization of the Mean Value Theorem, which is due to yet another French mathematician, Augustin Cauchy (1789–1867).

Cauchy's Mean Value Theorem (6.87)

Suppose that the functions f and g are continuous on $[a, b]$ and differentiable on (a, b) and $g'(x) \neq 0$ for all x in (a, b). Then there is a number c in (a, b) such that

$$\frac{f'(c)}{g'(c)} = \frac{f(b) - f(a)}{g(b) - g(a)}$$

Notice that if we take the special case where $g(x) = x$, then $g'(c) = 1$ and Theorem 6.87 is just the ordinary Mean Value Theorem (Theorem 4.10). Furthermore, Theorem 6.87 can be proved in a manner very similar to that of Theorem 4.10. You can verify that all we have to do is change the function h given by (4.14) to the function

$$h(x) = f(x) - f(a) - \frac{f(b) - f(a)}{g(b) - g(a)} [g(x) - g(a)]$$

and apply Rolle's Theorem as before.

Proof of L'Hospital's Rule   We are assuming that $\lim_{x \to a} f(x) = 0$ and $\lim_{x \to a} g(x) = 0$. Let

$$L = \lim_{x \to a} \frac{f'(x)}{g'(x)}$$

We must show that $\lim_{x \to a} f(x)/g(x) = L$. Define

$$F(x) = \begin{cases} f(x) & \text{if } x \neq a \\ 0 & \text{if } x = a \end{cases} \qquad G(x) = \begin{cases} g(x) & \text{if } x \neq a \\ 0 & \text{if } x = a \end{cases}$$

Then F is continuous on I since f is continuous on $\{x \in I \,|\, x \neq a\}$ and

$$\lim_{x \to a} F(x) = \lim_{x \to a} f(x) = 0 = F(a)$$

Likewise G is continuous on I. Let $x \in I$ and $x > a$. Then F and G are continuous on $[a, x]$ and differentiable on (a, x) and $G' \neq 0$ there (since $F' = f'$ and $G' = g'$). Therefore by Cauchy's Mean Value Theorem there is a number y such that $a < y < x$ and

$$\frac{F'(y)}{G'(y)} = \frac{F(x) - F(a)}{G(x) - G(a)} = \frac{F(x)}{G(x)}$$

Here we have used the fact that, by definition, $F(a) = 0$ and $G(a) = 0$. Now let $x \to a^+$. Then $y \to a^+$ (since $a < y < x$) so

$$\lim_{x \to a^+} \frac{f(x)}{g(x)} = \lim_{x \to a^+} \frac{F(x)}{G(x)} = \lim_{y \to a^+} \frac{F'(y)}{G'(y)} = \lim_{y \to a^+} \frac{f'(y)}{g'(y)} = L$$

A similar argument shows that the left-hand limit is also L. Therefore

$$\lim_{x \to a} \frac{f(x)}{g(x)} = L$$

This proves l'Hospital's Rule for the case where a is finite.

If a is infinite, we let $t = 1/x$. Then $t \to 0^+$ as $x \to \infty$, so we have

$$\lim_{x \to \infty} \frac{f(x)}{g(x)} = \lim_{t \to 0^+} \frac{f\left(\dfrac{1}{t}\right)}{g\left(\dfrac{1}{t}\right)}$$

$$= \lim_{t \to 0^+} \frac{f'\left(\dfrac{1}{t}\right)\dfrac{-1}{t^2}}{g'\left(\dfrac{1}{t}\right)\dfrac{-1}{t^2}} \quad \textit{(by l'Hospital's Rule for finite a)}$$

$$= \lim_{t \to 0^+} \frac{f'\left(\dfrac{1}{t}\right)}{g'\left(\dfrac{1}{t}\right)}$$

$$= \lim_{x \to \infty} \frac{f'(x)}{g'(x)} \qquad \bullet$$

EXERCISES 6.10

Find the limits in Exercises 1–78.

1. $\displaystyle\lim_{x \to 2} \frac{x-2}{x^2-4}$

2. $\displaystyle\lim_{x \to 1} \frac{x^2+3x-4}{x-1}$

3. $\displaystyle\lim_{x \to -1} \frac{x^6-1}{x^4-1}$

4. $\displaystyle\lim_{x \to 1} \frac{x^a-1}{x^b-1}$

5. $\displaystyle\lim_{x \to 0} \frac{e^x-1}{\sin x}$

6. $\displaystyle\lim_{x \to 1} \frac{\ln x}{x-1}$

7. $\displaystyle\lim_{x \to 0} \frac{\sin x}{x^3}$

8. $\displaystyle\lim_{x \to \pi} \frac{\tan x}{x}$

9. $\displaystyle\lim_{x \to 0} \frac{\tan x}{x+\sin x}$

10. $\displaystyle\lim_{x \to 3\pi/2} \frac{\cos x}{x-\dfrac{3\pi}{2}}$

11. $\displaystyle\lim_{x \to \infty} \frac{\ln x}{x}$

12. $\displaystyle\lim_{x \to 0^+} \frac{\ln x}{\sqrt{x}}$

13. $\displaystyle\lim_{x \to \infty} \frac{e^x}{x^3}$

14. $\displaystyle\lim_{x \to \infty} \frac{(\ln x)^3}{x^2}$

15. $\displaystyle\lim_{x \to a} \frac{\sqrt[3]{x}-\sqrt[3]{a}}{x-a}, a \neq 0$

16. $\displaystyle\lim_{x \to 0} \frac{6^x-2^x}{x}$

17. $\displaystyle\lim_{x \to 0} \frac{e^x-1-x}{x^2}$

18. $\displaystyle\lim_{x \to 0} \frac{e^x-1-x-\dfrac{x^2}{2}}{x^3}$

19. $\displaystyle\lim_{x \to 0} \frac{\sin x}{e^x}$

20. $\displaystyle\lim_{x \to 0} \frac{\sin^2 x}{\tan(x^2)}$

21. $\displaystyle\lim_{x \to 0} \frac{1-\cos x}{x^2}$

22. $\displaystyle\lim_{x \to 0} \frac{\sin x - x}{x^3}$

23. $\displaystyle\lim_{x \to 2^-} \frac{\ln x}{\sqrt{2-x}}$

24. $\displaystyle\lim_{x \to 0} \frac{\sin x}{\sinh x}$

25. $\displaystyle\lim_{x \to \infty} \frac{\ln \ln x}{\sqrt{x}}$

26. $\displaystyle\lim_{x \to \infty} \frac{\ln(1+e^x)}{5x}$

27. $\displaystyle\lim_{x \to 0} \frac{\tan^{-1}(2x)}{3x}$

28. $\displaystyle\lim_{x \to 0} \frac{x}{\sin^{-1}(3x)}$

29. $\lim\limits_{x\to 0} \dfrac{\tan\alpha x}{x}$

30. $\lim\limits_{x\to 0} \dfrac{\sin mx}{\sin nx}$

31. $\lim\limits_{x\to 0} \dfrac{\tan 2x}{\tanh 3x}$

32. $\lim\limits_{x\to 0} \dfrac{\sin^{10} x}{\sin(x^{10})}$

33. $\lim\limits_{x\to 0} \dfrac{x + \sin 3x}{x - \sin 3x}$

34. $\lim\limits_{x\to 0} \dfrac{2x - \sin^{-1} x}{2x + \cos^{-1} x}$

35. $\lim\limits_{x\to 0} \dfrac{e^{4x} - 1}{\cos x}$

36. $\lim\limits_{x\to 0} \dfrac{2x - \sin^{-1} x}{2x + \tan^{-1} x}$

37. $\lim\limits_{x\to 0} \dfrac{\tan x - \sin x}{x^3}$

38. $\lim\limits_{x\to 0} \dfrac{\cos mx - \cos nx}{x^2}$

39. $\lim\limits_{x\to 0^+} \sqrt{x}\, \ln x$

40. $\lim\limits_{x\to -\infty} xe^x$

41. $\lim\limits_{x\to \infty} e^{-x} \ln x$

42. $\lim\limits_{x\to \pi/2^-} \sec 7x \cos 3x$

43. $\lim\limits_{x\to \infty} x^3 e^{-x^2}$

44. $\lim\limits_{x\to 0^+} \sqrt{x}\, \sec x$

45. $\lim\limits_{x\to \pi}(x - \pi)\cot x$

46. $\lim\limits_{x\to 1^+}(x - 1)\tan\left(\dfrac{\pi x}{2}\right)$

47. $\lim\limits_{x\to 0}\left(\dfrac{1}{x^4} - \dfrac{1}{x^2}\right)$

48. $\lim\limits_{x\to 0}(\csc x - \cot x)$

49. $\lim\limits_{x\to 0}\left(\dfrac{1}{x} - \csc x\right)$

50. $\lim\limits_{x\to 1}\left(\dfrac{1}{\ln x} - \dfrac{1}{x - 1}\right)$

51. $\lim\limits_{x\to \infty}(x - \sqrt{x^2 - 1})$

52. $\lim\limits_{x\to \infty}(\sqrt{x^2 + x + 1} - \sqrt{x^2 - x})$

53. $\lim\limits_{x\to \infty}\left(\dfrac{x^3}{x^2 - 1} - \dfrac{x^3}{x^2 + 1}\right)$

54. $\lim\limits_{x\to \infty}(xe^{1/x} - x)$

55. $\lim\limits_{x\to 0^+} x^{\sin x}$

56. $\lim\limits_{x\to 0^+}(\sin x)^{\tan x}$

57. $\lim\limits_{x\to 0^+} x^{1/\ln x}$

58. $\lim\limits_{x\to 0^+} x^{-1/\sqrt{-\ln x}}$

59. $\lim\limits_{x\to 0}(1 - 2x)^{1/x}$

60. $\lim\limits_{x\to \infty}\left(1 + \dfrac{a}{x}\right)^{bx}$

61. $\lim\limits_{x\to \infty}\left(1 + \dfrac{3}{x} + \dfrac{5}{x^2}\right)^x$

62. $\lim\limits_{x\to \infty}\left(1 + \dfrac{1}{x^2}\right)^x$

63. $\lim\limits_{x\to \infty} x^{1/x}$

64. $\lim\limits_{x\to \infty}(e^x + x)^{1/x}$

65. $\lim\limits_{x\to 0^+}(\cot x)^{\sin x}$

66. $\lim\limits_{x\to \infty}\left(1 + \dfrac{1}{x}\right)^{x^2}$

67. $\lim\limits_{x\to \infty}\left(\dfrac{x}{x + 1}\right)^x$

68. $\lim\limits_{x\to 0}(\cos 3x)^{5/x}$

69. $\lim\limits_{x\to 0^+} x^{x^x}$

70. $\lim\limits_{x\to 0^+}(-\ln x)^x$

71. $\lim\limits_{x\to \infty} \dfrac{x}{2x + 3\sin x}$

72. $\lim\limits_{x\to \infty}\left(\dfrac{2x - 3}{2x + 5}\right)^{2x + 1}$

73. $\lim\limits_{x\to 0^+} \dfrac{x + 1 - e^x}{x^3}$

74. $\lim\limits_{x\to \infty} x[\ln(x + 5) - \ln x]$

75. $\lim\limits_{x\to 0} \dfrac{2x \sin x}{\sec x - 1}$

76. $\lim\limits_{x\to 0} \dfrac{1 - \cos 2x + \tan^2 x}{x \sin x}$

77. $\lim\limits_{x\to 0} \dfrac{\cos x - 1 + \dfrac{x^2}{2}}{x^4}$

78. $\lim\limits_{x\to -\infty}(\sqrt{x^2 + 1} - \sqrt{x^2 - 4x})$

In Exercises 79–98 sketch the curve under the headings A–H of Section 4.7 using l'Hospital's Rule where appropriate.

79. $y = xe^{-x}$

80. $y = x^2 e^{-x}$

81. $y = x\ln x$

82. $y = \dfrac{\ln x}{x}$

83. $y = x^2 \ln x$

84. $y = x(\ln x)^2$

85. $y = xe^{-x^2}$

86. $y = x^2 e^{-x^2}$

87. $y = \dfrac{e^x}{x}$

88. $y = \dfrac{e^x}{x^2}$

89. $y = \dfrac{e^x}{x^3}$

90. $y = \dfrac{e^{1/x}}{x}$

91. $y = xe^{1/x}$

92. $y = x^2 e^{-1/x}$

93. $y = e^x + x$

94. $y = e^x - x$

95. $y = x - \ln(1 + x)$

96. $y = e^x - 3e^{-x} - 4x$

97. $y = x^{1/\ln x}$

98. $y = (2x)^{3x}$

In Exercises 99–104 discuss the curve under the headings A–F and H of Section 4.7; that is, omit the discussion of concavity.

99. $y = x^{\sqrt{x}}$

100. $y = x^{1/x}$

101. $y = x^{x^2}$

102. $y = (\sin x)^{\sin x}$

103. $y = (1 + x)^{1/x}$

104. $y = \dfrac{2^x - 1}{x}$

105. Prove that

$$\lim_{x \to \infty} \frac{e^x}{x^n} = \infty$$

for any integer n. This shows that the exponential function approaches infinity faster than any power of x.

106. Prove that

$$\lim_{x \to \infty} \frac{\ln x}{x^p} = 0$$

for any number $p > 0$. This shows that the logarithmic function approaches ∞ more slowly than any power of x.

107. Prove that $\lim_{x \to 0^+} x^\alpha \ln x = 0$ for any $\alpha > 0$.

108. Evaluate

$$\lim_{x \to 0} \frac{1}{x^3} \int_0^x \sin(t^2)\, dt$$

109. Let

$$f(x) = \begin{cases} e^{-1/x^2} & \text{if } x \neq 0 \\ 0 & \text{if } x = 0 \end{cases}$$

(a) Use the definition of derivative to compute $f'(0)$.

(b) Show that f has derivatives of all orders that are defined on R.

[*Hint for part (b):* First show, by induction, that there is a polynomial $p_n(x)$ and a nonnegative integer k_n such that $f^{(n)}(x) = p_n(x)f(x)/x^{k_n}$ for $x \neq 0$.]

REVIEW OF CHAPTER 6

Define, state, or discuss the following.

1. Exponential functions
2. Graphs of the exponential functions
3. Properties of the exponential functions
4. Limits of exponential functions
5. One-to-one function
6. Horizontal Line Test
7. Inverse function
8. Cancellation equations
9. Procedure for finding an inverse function
10. Graph of an inverse function
11. Continuity and differentiability of an inverse function
12. Formula for the derivative of an inverse function
13. Logarithmic function to the base a
14. Graphs of logarithmic functions
15. Properties of logarithms
16. Limits of logarithmic functions
17. The number e
18. Natural logarithm
19. Derivatives of logarithmic and exponential functions
20. Logarithmic differentiation
21. Law of natural growth or decay
22. Inverse trigonometric functions
23. Derivatives of inverse trigonometric functions
24. Hyperbolic functions
25. Hyperbolic identities
26. Derivatives of hyperbolic functions
27. Inverse hyperbolic functions
28. L'Hospital's Rule
29. Indeterminate quotients, products, differences, and powers

REVIEW EXERCISES FOR CHAPTER 6

In Exercises 1–8 sketch the graph of the given function.

1. $y = 7^x$

2. $y = (0.7)^x$

3. $y = -e^{-x}$

4. $y = \pi^{-x}$

5. $y = \log_5 x$

6. $y = \ln(x - 1)$

7. $y = \log_{1/5} x$

8. $y = e^x \cos x$

In Exercises 9–16 solve the given equation for x.

9. $e^x = 5$

10. $\ln x = 2$

11. $\log_{10}(e^x) = 1$

12. $e^{e^x} = 2$

13. $\ln(x^\pi) = 2$

14. $\ln(x + 1) - \ln x = 1$

15. $\tan x = 4$

16. $\sin^{-1} x = 1$

In Exercises 17–45 calculate y'.

17. $y = \log_{10}(x^2 - x)$

18. $y = \sqrt{2^x}$

19. $y = \dfrac{\sqrt{x + 1}\,(2 - x)^5}{(x + 3)^7}$

20. $y = \ln(\csc 5x)$

21. $y = e^{cx}(c \sin x - \cos x)$

22. $y = \sin^{-1}(e^x)$

23. $y = \ln(\sec^2 x)$

24. $y = \ln(x^2 e^x)$

25. $y = xe^{-1/x}$

26. $y = \ln|\csc 3x + \cot 3x|$

27. $y = (\cos^{-1} x)^{\sin^{-1} x}$

28. $y = x^r e^{sx}$

29. $y = e^{e^x}$

30. $y = 5^{x \tan x}$

31. $y = \ln\left(\dfrac{1}{x}\right) + \dfrac{1}{\ln x}$

32. $xe^y = y - 1$

33. $y = 7^{\sqrt{2x}}$

34. $y = e^{\cos x} + \cos(e^x)$

35. $y = xe^{\sec^{-1} x}$

36. $y = \ln\left|\dfrac{x^2 - 4}{2x + 5}\right|$

37. $y = \ln(\cosh 3x)$

38. $y = \log_3 \sqrt{(1 + cx^a)^s}$

39. $y = \cosh^{-1}(\sinh x)$

40. $y = x \tanh^{-1} \sqrt{x}$

41. $y = \ln \sin x - \tfrac{1}{2}\sin^2 x$

42. $y = \left(\dfrac{c}{x}\right)^x$

43. $y = \sin^{-1}\left(\dfrac{x - 1}{x + 1}\right)$

44. $y = \arctan(\arcsin \sqrt{x})$

45. $y = \ln \sqrt[4]{\dfrac{x^2 + x + 1}{x^2 - x + 1}}$

$\quad + \dfrac{1}{2\sqrt{3}}\left[\tan^{-1}\left(\dfrac{2x + 1}{\sqrt{3}}\right) + \tan^{-1}\left(\dfrac{2x - 1}{\sqrt{3}}\right)\right]$

46. If

$$y = \dfrac{x}{\sqrt{a^2 - 1}} - \dfrac{2}{\sqrt{a^2 - 1}}\arctan\dfrac{\sin x}{a + \sqrt{a^2 - 1} + \cos x}$$

show that $y' = \dfrac{1}{a + \cos x}$.

Find $f^{(n)}(x)$ in Exercises 47 and 48.

47. $f(x) = 2^x$

48. $f(x) = \ln(2x)$

In Exercises 49 and 50 find the equation of the tangent to the given curve at the given point.

49. $y = \ln(e^x + e^{2x})$, $(0, \ln 2)$

50. $y = x \ln x$, (e, e)

51. At what point on the curve $y = [\ln(x + 4)]^2$ is the tangent horizontal?

52. Find the equation of the tangent to the curve $y = e^x$ that is parallel to the line $x - 4y = 1$.

Evaluate the limits in Exercises 53–72.

53. $\displaystyle\lim_{x \to -\infty} 10^{-x}$

54. $\displaystyle\lim_{x \to \infty} \dfrac{4^x}{4^x - 1}$

55. $\displaystyle\lim_{x \to 0^+} \ln(\tan x)$

56. $\displaystyle\lim_{x \to -\infty} e^{x/2} \cos x$

57. $\displaystyle\lim_{x \to -4^+} e^{1/(x + 4)}$

58. $\displaystyle\lim_{x \to -1^+} e^{\tanh^{-1} x}$

59. $\displaystyle\lim_{x \to 1} \cos^{-1}\left(\dfrac{x}{x + 1}\right)$

60. $\displaystyle\lim_{x \to -\infty} \tan^{-1}(x^4)$

61. $\displaystyle\lim_{x \to \pi} \dfrac{\sin x}{x^2 - \pi^2}$

62. $\displaystyle\lim_{x \to 0} \dfrac{e^{ax} - e^{bx}}{x}$

63. $\displaystyle\lim_{x \to \infty} \dfrac{\ln(\ln x)}{\ln x}$

64. $\displaystyle\lim_{x \to 0} \dfrac{1 + \sin x - \cos x}{1 - \sin x - \cos x}$

65. $\displaystyle\lim_{x \to 0} \dfrac{\ln(1 - x) + x + \dfrac{x^2}{2}}{x^3}$

66. $\displaystyle\lim_{x \to \pi/2} \left(\dfrac{\pi}{2} - x\right)\tan x$

67. $\displaystyle\lim_{x \to 0^+} \sin x(\ln x)^2$

68. $\displaystyle\lim_{x \to 0} (\csc^2 x - x^{-2})$

69. $\displaystyle\lim_{x \to 1}(\ln x)^{\sin x}$

70. $\displaystyle\lim_{x \to 1} x^{1/(1 - x)}$

71. $\displaystyle\lim_{x \to 0^+} \dfrac{\sqrt[3]{x} - 1}{\sqrt[4]{x} - 1}$

72. $\displaystyle\lim_{x \to \infty} \tan^{-1}\left(\dfrac{\sqrt{x}}{\ln x}\right)$

In Exercises 73–80 discuss the given curve under the headings A–H of Section 4.7.

73. $y = \tan^{-1}\left(\dfrac{1}{x}\right)$

74. $y = \sin^{-1}\left(\dfrac{1}{x}\right)$

75. $y = 2^{1/(x - 1)}$

76. $y = e^{2x - x^2}$

77. $y = e^x + e^{-3x}$

78. $y = \ln(x^2 - 1)$

79. $y = 2x^2 - \ln x$

80. $y = e^{-1/x^2}$

81. A bacteria culture starts with 1000 bacteria and the growth rate is proportional to the number of bacteria. After 2 hours the population is 9000 (a) Find an expression for the number of bacteria after t hours. (b) Find the number of bacteria after 3 hours. (c) In what period of time does the number of bacteria double?

82. An isotope of strontium, Sr^{90}, has a half-life of 25 years. (a) Find the mass of Sr^{90} that remains from a

396

Chapter 6   Inverse Functions

sample of 18 mg after t years. (b) How long would it take for the mass to decay to 2 mg?

83. An equation of motion of the form $s = Ae^{-ct}\cos(\omega t + \delta)$ represents damped oscillation of an object. Find the velocity and acceleration of the object.

In Exercises 84–87 find f' in terms of g'.

84. $f(x) = g(e^x)$

85. $f(x) = e^{g(x)}$

86. $f(x) = g(\ln x)$

87. $f(x) = \ln|g(x)|$

In Exercises 88–101 evaluate the integral.

88. $\int_1^2 \dfrac{1}{2 - 3x}\,dx$

89. $\int_0^{2\sqrt{3}} \dfrac{1}{x^2 + 4}\,dx$

90. $\int_0^1 e^{\pi t}\,dt$

91. $\int_2^4 \dfrac{1 + x - x^2}{x^2}\,dx$

92. $\int \dfrac{\cos(\ln x)}{x}\,dx$

93. $\int \dfrac{e^{\sqrt{x}}}{\sqrt{x}}\,dx$

94. $\int \dfrac{x}{\sqrt{1 - x^4}}\,dx$

95. $\int \tan x \ln(\cos x)\,dx$

96. $\int \dfrac{e^x}{(e^x + 1)\ln(e^x + 1)}\,dx$

97. $\int \dfrac{x^3}{1 + x^4}\,dx$

98. $\int \dfrac{1}{\sqrt{x}\,(1 + x)}\,dx$

99. $\int \dfrac{\sec\theta\tan\theta}{1 + \sec\theta}\,d\theta$

100. $\int \dfrac{x^2}{2^{x^3}}\,dx$

101. $\int \cosh 3t\,dt$

In Exercises 102–104 use the properties of integrals to prove the inequalities.

102. $\int_0^1 \sqrt{1 + e^{2x}}\,dx \geqslant e - 1$

103. $\int_0^1 e^x \cos x\,dx \leqslant e - 1$

104. $\int_0^1 x \sin^{-1} x\,dx \leqslant \dfrac{\pi}{4}$

In Exercises 105 and 106 find $f'(x)$.

105. $f(x) = \int_1^{\sqrt{x}} \dfrac{e^s}{s}\,ds$

106. $f(x) = \int_{\ln x}^{2x} e^{-t^2}\,dt$

107. Find the average value of the function $f(x) = 1/x$ on the interval $[1, 4]$.

108. Find the area of the region bounded by the curves $y = e^x$, $y = e^{-x}$, $x = -2$, and $x = 1$.

109. If g is the inverse function of $f(x) = \ln x + \tan^{-1} x$, find $g'(\pi/4)$.

110. Evaluate $\int_0^1 e^x\,dx$ without using the Fundamental Theorem of Calculus. (*Hint:* Use Theorem 5.12, sum a geometric series, and then use l'Hospital's Rule.)

111. If $F(x) = \int_a^b t^x\,dt$, where $a, b > 0$, then, by the Fundamental Theorem,

$$F(x) = \dfrac{b^{x+1} - a^{x+1}}{x + 1} \quad (x \neq -1)$$

$$F(-1) = \ln b - \ln a$$

Use l'Hospital's Rule to show that F is continuous at -1.

7 Techniques of Integration

Common integration is only the memory of differentiation. The different devices by which integration is accomplished are changes, not from the known to the unknown, but from forms in which memory will not serve us to those in which it will.

Augustus de Morgan

Because of the Fundamental Theorem of Calculus, we can integrate a function if we know an antiderivative, that is, an indefinite integral. We summarize here the most important integrals that we have learned so far.

$$\int x^n \, dx = \frac{x^{n+1}}{n+1} + C \quad (n \neq -1) \qquad \int \frac{1}{x} \, dx = \ln|x| + C$$

$$\int e^x \, dx = e^x + C \qquad\qquad \int a^x \, dx = \frac{a^x}{\ln a} + C$$

$$\int \sin x \, dx = -\cos x + C \qquad\qquad \int \cos x \, dx = \sin x + C$$

$$\int \sec^2 x \, dx = \tan x + C \qquad\qquad \int \csc^2 x \, dx = -\cot x + C$$

$$\int \sec x \tan x \, dx = \sec x + C \qquad\qquad \int \csc x \cot x \, dx = -\csc x + C$$

$$\int \sinh x \, dx = \cosh x + C \qquad\qquad \int \cosh x \, dx = \sinh x + C$$

$$\int \tan x \, dx = \ln|\sec x| + C \quad \text{(see Equation 6.31)}$$

$$\int \cot x \, dx = \ln|\sin x| + C \quad \text{(see Exercise 49 in Section 6.4)}$$

$$\int \frac{1}{x^2 + a^2}\, dx = \frac{1}{a} \tan^{-1}\left(\frac{x}{a}\right) + C \quad \text{(see Equation 6.75)}$$

$$\int \frac{1}{\sqrt{a^2 - x^2}}\, dx = \sin^{-1}\left(\frac{x}{a}\right) + C \quad \text{(see Exercise 105 in Section 6.8)}$$

In this chapter we shall develop techniques for using these basic integration formulas to obtain indefinite integrals of more complicated functions. We have already learned the most important method of integration, the Substitution Rule, in Section 5.6. The other general technique, integration by parts, will be presented in Section 7.1. Then we shall learn methods that are special to particular classes of functions such as trigonometric functions and rational functions.

Integration is not as straightforward as differentiation; there are no rules that absolutely guarantee obtaining an indefinite integral of a function. Therefore in Section 7.6 we discuss a strategy for integration.

SECTION 7.1

Integration by Parts

For every differentiation rule there is a corresponding integration rule. For instance, the Substitution Rule for integration corresponds to the Chain Rule for differentiation. The rule that corresponds to the Product Rule for differentiation is called the rule for integration by parts.

The Product Rule states that if f and g are differentiable functions, then

$$\frac{d}{dx}(f(x)g(x)) = f'(x)g(x) + f(x)g'(x)$$

In the notation for indefinite integrals this equation becomes

$$\int (f'(x)g(x) + f(x)g'(x))\, dx = f(x)g(x)$$

or $\qquad \int f'(x)g(x)\, dx + \int f(x)g'(x)\, dx = f(x)g(x)$

We can rearrange this latter equation as

(7.1)
$$\boxed{\int f(x)g'(x)\, dx = f(x)g(x) - \int f'(x)g(x)\, dx}$$

Formula 7.1 is called the formula for **integration by parts.** It is perhaps easier to remember in the following notation. Let $u = f(x)$ and $v = g(x)$. Then $du = f'(x)\, dx$ and $dv = g'(x)\, dx$, so, by the Substitution Rule, the formula for integration by parts becomes

(7.2)

$$\int u\,dv = uv - \int v\,du$$

● **Example 1**   Find $\int x \sin x\,dx$.

Solution using Formula 7.1   Suppose we choose $f(x) = x$ and $g'(x) = \sin x$. Then $f'(x) = 1$ and $g(x) = -\cos x$. (For g we can choose *any* antiderivative of g'.) Thus, using Formula 7.1, we have

$$\int x \sin x\,dx = f(x)g(x) - \int f'(x)g(x)\,dx$$

$$= x(-\cos x) - \int(-\cos x)\,dx$$

$$= -x\cos x + \int \cos x\,dx$$

$$= -x\cos x + \sin x + C$$

It is helpful to use the pattern:

$u = \square \qquad dv = \square$

$du = \square \qquad v = \square$

Solution using Formula 7.2   Let

$$u = x \qquad dv = \sin x\,dx$$

Then

$$du = dx \qquad v = -\cos x$$

and so

$$\int x \sin x\,dx = \int u\,dv = uv - \int v\,du$$

$$= -x\cos x + \int \cos x\,dx$$

$$= -x\cos x + \sin x + C \qquad ●$$

Note: The object in using integration by parts is to obtain a simpler integral than the one we started with. Thus in Example 1 we started with $\int x \sin x\,dx$ and expressed it in terms of the simpler integral $\int \cos x\,dx$. If we had chosen $u = \sin x$ and $dv = x\,dx$, then $du = \cos x\,dx$ and $v = x^2/2$, so integration by parts gives

$$\int x \sin x\,dx = (\sin x)\frac{x^2}{2} - \frac{1}{2}\int x^2 \cos x\,dx$$

But $\int x^2 \cos x\,dx$ is a more difficult integral than the one we started with. In general, when deciding on a choice for u and dv, we usually try to choose $u = f(x)$ to be a function that becomes simpler when differentiated (or at least not more complicated) as long as $dv = g'(x)\,dx$ can be readily integrated to give v.

● **Example 2**   Evaluate $\int \ln x \, dx$.

Solution   Here there is not much choice. Let

$$u = \ln x \qquad dv = dx$$

Then

$$du = \frac{1}{x} \, dx \qquad v = x$$

Integrating by parts, we get

$$\int \ln x \, dx = x \ln x - \int x \, \frac{dx}{x}$$

$$= x \ln x - \int dx$$

$$= x \ln x - x + C$$

Integration by parts was effective in this example because the derivative of the function $f(x) = \ln x$ is simpler than f.   ●

● **Example 3**   Find $\int x^2 e^x \, dx$.

Solution   Let

$$u = x^2 \qquad dv = e^x \, dx$$

Then

$$du = 2x \, dx \qquad v = e^x$$

Integration by parts gives

(7.3)
$$\int x^2 e^x \, dx = x^2 e^x - 2 \int x e^x \, dx$$

The integral that we have obtained, $\int x e^x \, dx$, is simpler than the original integral but is still not obvious. Therefore we use integration by parts a second time, this time with $u = x$ and $dv = e^x \, dx$. Then $du = dx$, $v = e^x$, and

$$\int x e^x \, dx = x e^x - \int e^x \, dx$$

$$= x e^x - e^x + C$$

Putting this in Equation 7.3, we get

$$\int x^2 e^x \, dx = x^2 e^x - 2 \int x e^x \, dx$$

$$= x^2 e^x - 2(x e^x - e^x + C)$$

$$= x^2 e^x - 2x e^x + 2e^x + C_1$$

where $C_1 = -2C$.   ●

● **Example 4**   Evaluate $\int e^x \sin x \, dx$.

Solution Let $u = e^x$ and $dv = \sin x \, dx$. Then $du = e^x \, dx$ and $v = -\cos x$, so integration by parts gives

(7.4)
$$\int e^x \sin x \, dx = -e^x \cos x + \int e^x \cos x \, dx$$

The integral that we have obtained, $\int e^x \cos x \, dx$, is no simpler than the original one, but at least it is no more difficult. Having had success in the preceding example integrating by parts twice, we persevere and integrate by parts again. This time we use $u = e^x$ and $dv = \cos x \, dx$. Then $du = e^x \, dx$, $v = \sin x$, and

(7.5)
$$\int e^x \cos x \, dx = e^x \sin x - \int e^x \sin x \, dx$$

At first glance it appears as if we have accomplished nothing because we have arrived at $\int e^x \sin x \, dx$, which is where we started. However if we put Equation 7.5 into Equation 7.4 we get

$$\int e^x \sin x \, dx = -e^x \cos x + e^x \sin x - \int e^x \sin x \, dx$$

This can be regarded as an equation to be solved for the unknown integral. Solving, we obtain

$$2 \int e^x \sin x \, dx = -e^x \cos x + e^x \sin x$$

and, dividing by 2 and adding the constant of integration,

$$\int e^x \sin x \, dx = \tfrac{1}{2} e^x (\sin x - \cos x) + C \qquad \bullet$$

If we combine the formula for integration by parts with Part 2 of the Fundamental Theorem of Calculus, we can evaluate definite integrals by parts. Evaluating both sides of Formula 7.1 between a and b, assuming f' and g' are continuous, and using the Fundamental Theorem in the form of Equation 5.25, we obtain

(7.6)
$$\int_a^b f(x)g'(x) \, dx = f(x)g(x) \Big]_a^b - \int_a^b f'(x)g(x) \, dx$$

● **Example 5** Calculate $\int_0^1 \tan^{-1} x \, dx$.

Solution Let

$$u = \tan^{-1} x \qquad dv = dx$$

Then
$$du = \frac{dx}{1 + x^2} \qquad v = x$$

So Formula 7.6 gives

$$\int_0^1 \tan^{-1} x \, dx = x \tan^{-1} x \Big]_0^1 - \int_0^1 \frac{x}{1+x^2} \, dx$$

$$= 1 \cdot \tan^{-1} 1 - 0 \cdot \tan^{-1} 0 - \int_0^1 \frac{x}{1+x^2} \, dx$$

$$= \frac{\pi}{4} - \int_0^1 \frac{x}{1+x^2} \, dx$$

To evaluate this integral we use the substitution $t = 1 + x^2$ (since u has another meaning in this example). Then $dt = 2x \, dx$ so $x \, dx = dt/2$. When $x = 0$, $t = 1$; when $x = 1$, $t = 2$; so

$$\int_0^1 \frac{x}{1+x^2} \, dx = \frac{1}{2} \int_1^2 \frac{dt}{t} = \frac{1}{2} \ln|t| \Big]_1^2$$

$$= \frac{1}{2} (\ln 2 - \ln 1) = \frac{1}{2} \ln 2$$

Therefore $\int_0^1 \tan^{-1} x \, dx = \frac{\pi}{4} - \int_0^1 \frac{x}{1+x^2} \, dx = \frac{\pi}{4} - \frac{\ln 2}{2}$ ●

● **Example 6**   Prove the reduction formula

(7.7)
$$\int \sin^n x \, dx = -\frac{1}{n} \cos x \sin^{n-1} x + \frac{n-1}{n} \int \sin^{n-2} x \, dx$$

where $n \geqslant 2$ is an integer.

Solution   Let

$$u = \sin^{n-1} x \qquad\qquad dv = \sin x \, dx$$

Then $du = (n-1)\sin^{n-2} x \cos x \, dx \qquad v = -\cos x$

so integration by parts gives

$$\int \sin^n x \, dx = -\cos x \sin^{n-1} x + (n-1) \int \sin^{n-2} x \cos^2 x \, dx$$

Since $\cos^2 x = 1 - \sin^2 x$, we have

$$\int \sin^n x \, dx = -\cos x \sin^{n-1} x + (n-1) \int \sin^{n-2} x \, dx - (n-1) \int \sin^n x \, dx$$

As in Example 4, we solve this equation for the desired integral by taking the last term on the right side to the left side. Thus we have

$$n \int \sin^n x \, dx = -\cos x \sin^{n-1} x + (n-1) \int \sin^{n-2} x \, dx$$

or $\int \sin^n x \, dx = -\frac{1}{n} \cos x \sin^{n-1} x + \frac{n-1}{n} \int \sin^{n-2} x \, dx$ ●

The reduction formula (7.7) is useful because by using it repeatedly we could eventually express $\int \sin^n x \, dx$ in terms of $\int \sin x \, dx$ (if n is odd) or $\int (\sin x)^0 \, dx = \int dx$ (if n is even).

EXERCISES 7.1

Evaluate the integrals in Exercises 1–36.

1. $\int x e^{2x} \, dx$

2. $\int x \cos x \, dx$

3. $\int x \sin 4x \, dx$

4. $\int x \ln x \, dx$

5. $\int x^2 \cos 3x \, dx$

6. $\int x^2 \sin 2x \, dx$

7. $\int (\ln x)^2 \, dx$

8. $\int \sin^{-1} x \, dx$

9. $\int \theta \sin \theta \cos \theta \, d\theta$

10. $\int \theta \sec^2 \theta \, d\theta$

11. $\int t^2 \ln t \, dt$

12. $\int t^3 e^t \, dt$

13. $\int e^{2\theta} \sin 3\theta \, d\theta$

14. $\int e^{-\theta} \cos 3\theta \, d\theta$

15. $\int y \sinh y \, dy$

16. $\int y \cosh ay \, dy$

17. $\int_0^1 t e^{-t} \, dt$

18. $\int_1^4 \sqrt{t} \ln t \, dt$

19. $\int_0^{\pi/2} x \cos 2x \, dx$

20. $\int_0^1 x^2 e^{-x} \, dx$

21. $\int_0^1 \cos^{-1} x \, dx$

22. $\int_{\pi/4}^{\pi/2} x \csc^2 x \, dx$

23. $\int \sin 3x \cos 5x \, dx$

24. $\int \sin 2x \sin 4x \, dx$

25. $\int \cos x \ln(\sin x) \, dx$

26. $\int x^3 e^{x^2} \, dx$

27. $\int (2x + 3) e^x \, dx$

28. $\int x 5^x \, dx$

29. $\int \cos(\ln x) \, dx$

30. $\int_1^4 e^{\sqrt{x}} \, dx$

31. $\int_1^4 \ln \sqrt{x} \, dx$

32. $\int \sin(\ln x) \, dx$

33. $\int \sin \sqrt{x} \, dx$

34. $\int x^5 \cos(x^3) \, dx$

35. $\int x^5 e^{x^2} \, dx$

36. $\int x \tan^{-1} x \, dx$

37. (a) Use the reduction formula in Example 6 to show that

$$\int \sin^2 x \, dx = \frac{x}{2} - \frac{\sin 2x}{4} + C$$

(b) Use part (a) and the reduction formula to evaluate $\int \sin^4 x \, dx$.

38. (a) Prove the reduction formula

$$\int \cos^n x \, dx = \frac{1}{n} \cos^{n-1} x \sin x + \frac{n-1}{n} \int \cos^{n-2} x \, dx$$

(b) Use part (a) to evaluate $\int \cos^2 x \, dx$.
(c) Use parts (a) and (b) to evaluate $\int \cos^4 x \, dx$.

39. (a) Use the reduction formula in Example 6 to show that

$$\int_0^{\pi/2} \sin^n x \, dx = \frac{n-1}{n} \int_0^{\pi/2} \sin^{n-2} x \, dx$$

where $n \geq 2$ is an integer.

(b) Use part (a) to evaluate $\int_0^{\pi/2} \sin^3 x \, dx$ and $\int_0^{\pi/2} \sin^5 x \, dx$.

(c) Use part (a) to show that, if n is odd,

$$\int_0^{\pi/2} \sin^n x \, dx = \frac{2 \cdot 4 \cdot 6 \cdot \cdots \cdot (n-1)}{3 \cdot 5 \cdot 7 \cdot \cdots \cdot n}$$

This formula is called a *Wallis product*.

40. Prove that, if n is even,

$$\int_0^{\pi/2} \sin^n x \, dx = \frac{1 \cdot 3 \cdot 5 \cdot \cdots \cdot (n-1)}{2 \cdot 4 \cdot 6 \cdot \cdots \cdot n} \frac{\pi}{2}$$

Use integration by parts to prove the reduction formulas in Exercises 41–44.

41. $\int (\ln x)^n \, dx = x(\ln x)^n - n \int (\ln x)^{n-1} \, dx$

42. $\int x^n e^x \, dx = x^n e^x - n \int x^{n-1} e^x \, dx$

43. $\int (x^2 + a^2)^n \, dx$

$$= \frac{x(x^2 + a^2)^n}{2n + 1} + \frac{2na^2}{2n + 1} \int (x^2 + a^2)^{n-1} \, dx \quad \left(n \neq -\frac{1}{2} \right)$$

44. $\int \sec^n x \, dx$

$$= \frac{\tan x \sec^{n-2} x}{n - 1} + \frac{n - 2}{n - 1} \int \sec^{n-2} x \, dx \quad (n \neq 1)$$

45. Use Exercise 41 to find $\int (\ln x)^3 \, dx$.

46. Use Exercise 42 to find $\int x^4 e^x \, dx$.

In Exercises 47–50 find the area of the region bounded by the given curves.

47. $y = \sin^{-1} x, \ y = 0, \ x = 1$

48. $y = xe^{-x}, \ y = 0, \ x = 5$

49. $y = \ln x, \ y = 0, \ x = e$

50. $y = 5 \ln x, \ y = x \ln x$

51. A particle that moves along a straight line has velocity $v(t) = t^2 e^{-t}$ meters per second after t seconds. How far will it travel during the first t seconds?

52. If $f(0) = g(0) = 0$, show that

$$\int_0^a f(x) g''(x) \, dx = f(a) g'(a) - f'(a) g(a) + \int_0^a f''(x) g(x) \, dx$$

53. Use integration by parts to show that

$$\int f(x) \, dx = x f(x) - \int x f'(x) \, dx$$

54. If f and g are inverse functions and f' is continuous, prove that

$$\int_a^b f(x) \, dx = bf(b) - af(a) - \int_{f(a)}^{f(b)} g(y) \, dy$$

[*Hint:* Use Exercise 53 and make the substitution $y = f(x)$.]

55. Use Exercise 54 to evaluate $\int_1^e \ln x \, dx$.

56. In the case where f and g are positive functions and $b > a > 0$, draw a diagram to give a geometric interpretation of Exercise 54.

SECTION 7.2

Trigonometric Integrals

In this section we use trigonometric identities to integrate certain combinations of trigonometric functions. We start with powers of sine and cosine.

(7.8)

To evaluate $\int \sin^m x \cos^n x \, dx$ ($m \geq 0$ and $n \geq 0$ are integers):

(a) If m and n are both even, use the half-angle formulas (Appendix B, Equation 17):

$$\sin^2 x = \tfrac{1}{2}(1 - \cos 2x) \qquad \cos^2 x = \tfrac{1}{2}(1 + \cos 2x)$$

(b) If m is odd, use $\sin^2 x = 1 - \cos^2 x$ and substitute $u = \cos x$.

(c) If n is odd, use $\cos^2 x = 1 - \sin^2 x$ and substitute $u = \sin x$.

● **Example 1**   Find $\int \sin^2 x \, dx$.

Solution   Since $m = 2$ and $n = 0$, this integral falls into the category of (7.8)(a) so we use the half-angle formula for $\sin^2 x$:

$$\int \sin^2 x \, dx = \frac{1}{2} \int (1 - \cos 2x) \, dx$$

$$= \frac{1}{2} \left(x - \frac{1}{2} \sin 2x \right) + C$$

$$= \frac{x}{2} - \frac{\sin 2x}{4} + C$$

Notice that we have mentally made the substitution $u = 2x$ when integrating $\cos 2x$. Another method for evaluating this integral was given in Exercise 37 in Section 7.1. ●

● **Example 2**   Find $\int \sin^4 x \, dx$.

Solution   It would be possible to evaluate this integral using the reduction formula for $\int \sin^n x \, dx$ (Equation 7.7) together with Example 1 (as in Exercise 37 in Section 7.1), but an easier method is to write $\sin^4 x = (\sin^2 x)^2$ and use (7.8)(a):

$$\int \sin^4 x \, dx = \int (\sin^2 x)^2 \, dx$$

$$= \int \left(\frac{1 - \cos 2x}{2} \right)^2 dx$$

$$= \frac{1}{4} \int (1 - 2\cos 2x + \cos^2 2x) \, dx$$

Since $\cos^2 2x$ occurs we must use another half-angle formula

$$\cos^2 2x = \tfrac{1}{2}(1 + \cos 4x)$$

This gives

$$\int \sin^4 x \, dx = \tfrac{1}{4} \int [1 - 2\cos 2x + \tfrac{1}{2}(1 + \cos 4x)] \, dx$$

$$= \tfrac{1}{4} \int (\tfrac{3}{2} - 2\cos 2x + \tfrac{1}{2}\cos 4x) \, dx$$

$$= \tfrac{1}{4}(\tfrac{3}{2}x - \sin 2x + \tfrac{1}{8}\sin 4x) + C$$ ●

● **Example 3**   Evaluate $\int_0^{\pi/2} \cos^3 x \, dx$.

Solution   Here $m = 0$ and $n = 3$. We write $\cos^3 x = \cos^2 x \cdot \cos x$, enabling us to use $\cos^2 x = 1 - \sin^2 x$. It is appropriate to have the extra factor of $\cos x$ because when we make the substitution $u = \sin x$ we have $du = \cos x \, dx$:

$$\int_0^{\pi/2} \cos^3 x \, dx = \int_0^{\pi/2} \cos^2 x \cos x \, dx$$

$$= \int_0^{\pi/2} (1 - \sin^2 x) \cos x \, dx$$

$$= \int_0^1 (1 - u^2) \, du$$

$$= u - \frac{u^3}{3} \Big]_0^1 = 1 - \frac{1}{3} = \frac{2}{3}$$ ●

● **Example 4**　Find $\int \sin^5 x \cos^2 x \, dx$.

Solution　Here m is odd and n is even, so according to the advice in (7.8)(b) we substitute $u = \cos x$. Since $du = -\sin x \, dx$, we factor $\sin^5 x$ as $\sin^4 x \cdot \sin x$:

$$\int \sin^5 x \cos^2 x \, dx = \int \sin^4 x \cos^2 x \sin x \, dx$$

$$= \int (1 - \cos^2 x)^2 \cos^2 x \sin x \, dx$$

$$= \int (1 - u^2)^2 u^2 (-du)$$

$$= -\int (u^2 - 2u^4 + u^6) \, du$$

$$= -\left(\frac{u^3}{3} - 2\frac{u^5}{5} + \frac{u^7}{7} \right) + C$$

$$= -\frac{1}{3} \cos^3 x + \frac{2}{5} \cos^5 x - \frac{1}{7} \cos^7 x + C \qquad ●$$

(7.9)

To evaluate $\int \tan^m x \sec^n x \, dx$:

(a)　If n is even and $n \geqslant 2$, use $\sec^2 x = 1 + \tan^2 x$ and substitute $u = \tan x$.

(b)　If m is odd and $n \geqslant 1$, use $\tan^2 x = \sec^2 x - 1$ and substitute $u = \sec x$.

(c)　If n is odd and m is even, express the integral totally in terms of $\sec x$. Powers of $\sec x$ may require integration by parts.

(d)　If $n = 0$, use $\tan^2 x = \sec^2 x - 1$ and, if necessary, the formula

$$\int \tan x \, dx = \ln|\sec x| + C$$

● **Example 5**　Find $\int \tan^3 x \, dx$.

Solution　$\int \tan^3 x \, dx = \int \tan x \tan^2 x \, dx$

$$= \int \tan x (\sec^2 x - 1) \, dx$$

$$= \int \tan x \sec^2 x \, dx - \int \tan x \, dx$$

$$= \frac{\tan^2 x}{2} - \ln|\sec x| + C$$

In the first integral we mentally substituted $u = \tan x$ so that $du = \sec^2 x \, dx$.

●

● **Example 6**   Evaluate $\int \tan^6 x \sec^4 x \, dx$.

Solution   Since there is an even power of the secant, we factor off $\sec^2 x$ and substitute $u = \tan x$ so that $du = \sec^2 x \, dx$. The rest of the integrand is then expressed totally in terms of $\tan x$ by means of the identity $\sec^2 x = 1 + \tan^2 x$:

$$\int \tan^6 x \sec^4 x \, dx = \int \tan^6 x \sec^2 x \sec^2 x \, dx$$

$$= \int \tan^6 x (1 + \tan^2 x) \sec^2 x \, dx$$

$$= \int u^6 (1 + u^2) \, du = \int (u^6 + u^8) \, du$$

$$= \frac{u^7}{7} + \frac{u^9}{9} + C$$

$$= \frac{1}{7} \tan^7 x + \frac{1}{9} \tan^9 x + C \qquad ●$$

● **Example 7**   Find $\int \tan^5 x \sec^7 x \, dx$.

Solution   Since there is an odd power of $\tan x$, we factor $\tan x \sec x$ from the integrand and substitute $u = \sec x$ so that $du = \sec x \tan x \, dx$. Since an even power of $\tan x$ remains, we use the identity $\tan^2 x = \sec^2 x - 1$ to express the remainder of the integral totally in terms of $\sec x$:

$$\int \tan^5 x \sec^7 x \, dx = \int \tan^4 x \sec^6 x \sec x \tan x \, dx$$

$$= \int (\sec^2 x - 1)^2 \sec^6 x \sec x \tan x \, dx$$

$$= \int (u^2 - 1)^2 u^6 \, du = \int (u^{10} - 2u^8 + u^6) \, du$$

$$= \frac{u^{11}}{11} - 2\frac{u^9}{9} + \frac{u^7}{7} + C$$

$$= \frac{1}{11} \sec^{11} x - \frac{2}{9} \sec^9 x + \frac{1}{7} \sec^7 x + C \qquad ●$$

● **Example 8**   Find $\int \sec x \, dx$.

Solution   We multiply numerator and denominator by $\sec x + \tan x$:

$$\int \sec x \, dx = \int \sec x \, \frac{\sec x + \tan x}{\sec x + \tan x} \, dx$$

$$= \int \frac{\sec^2 x + \sec x \tan x}{\sec x + \tan x} \, dx$$

If we substitute $u = \sec x + \tan x$, then $du = (\sec x \tan x + \sec^2 x)\,dx$, so the integral becomes $\int (1/u)\,du = \ln|u| + C$. Thus we have

(7.10)

$$\int \sec x\,dx = \ln|\sec x + \tan x| + C$$

The method of Example 8 was admittedly very tricky, but we shall need Formula 7.10 for our future work.

● **Example 9**   Find $\int \sec^3 x\,dx$.

Solution   Here we integrate by parts with

$$u = \sec x \qquad\qquad dv = \sec^2 x\,dx$$
$$du = \sec x \tan x\,dx \qquad v = \tan x$$

$$\int \sec^3 x\,dx = \sec x \tan x - \int \sec x \tan^2 x\,dx$$
$$= \sec x \tan x - \int \sec x(\sec^2 x - 1)\,dx$$
$$= \sec x \tan x - \int \sec^3 x\,dx + \int \sec x\,dx$$

Using Formula 7.10 and solving for the required integral, we get

$$\int \sec^3 x\,dx = \tfrac{1}{2}(\sec x \tan x + \ln|\sec x + \tan x|) + C$$

Integrals such as the one in Example 9 and others of the form (7.9) may seem very special but they occur frequently in applications of integration as we shall see in Chapter 8. Integrals of the form $\int \cot^m x \csc^n x\,dx$ can be found by similar methods because of the identity $1 + \cot^2 x = \csc^2 x$.

(7.11)

To evaluate the integrals (a) $\int \sin mx \cos nx\,dx$, (b) $\int \sin mx \sin nx\,dx$, and (c) $\int \cos mx \cos nx\,dx$ use the identities (Appendix B, Equation 18):

(a) $\sin A \cos B = \tfrac{1}{2}[\sin(A - B) + \sin(A + B)]$

(b) $\sin A \sin B = \tfrac{1}{2}[\cos(A - B) - \cos(A + B)]$

(c) $\cos A \cos B = \tfrac{1}{2}[\cos(A - B) + \cos(A + B)]$

● **Example 10**   Evaluate $\int \sin 4x \cos 5x\,dx$.

Solution   This integral could be evaluated using integration by parts but it is easier to use the identity in (7.11)(a) as follows:

$$\int \sin 4x \cos 5x \, dx = \int \tfrac{1}{2}[\sin(-x) + \sin 9x] \, dx$$
$$= \tfrac{1}{2} \int (-\sin x + \sin 9x) \, dx$$
$$= \tfrac{1}{2}(\cos x - \tfrac{1}{9}\cos 9x) + C \qquad \bullet$$

EXERCISES 7.2

In Exercises 1–56 evaluate the integral.

1. $\int_0^{\pi/2} \sin^2 3x \, dx$

2. $\int_0^{\pi/2} \cos^2 x \, dx$

3. $\int \cos^4 x \, dx$

4. $\int \sin^3 x \, dx$

5. $\int \sin^3 x \cos^4 x \, dx$

6. $\int \sin^4 x \cos^3 x \, dx$

7. $\int_0^{\pi/4} \sin^4 x \cos^2 x \, dx$

8. $\int_0^{\pi/2} \sin^2 x \cos^2 x \, dx$

9. $\int (1 - \sin 2x)^2 \, dx$

10. $\int \sin\left(x + \dfrac{\pi}{6}\right) \cos x \, dx$

11. $\int \cos^5 x \sin^5 x \, dx$

12. $\int \sin^6 x \, dx$

13. $\int \cos^6 x \, dx$

14. $\int \sin^5 2x \cos^4 2x \, dx$

15. $\int \sin^5 x \, dx$

16. $\int \sin^4 x \cos^4 x \, dx$

17. $\int \sin^3 x \sqrt{\cos x} \, dx$

18. $\int \dfrac{\cos^3 x}{\sqrt{\sin x}} \, dx$

19. $\int \dfrac{\cos^2(\sqrt{x})}{\sqrt{x}} \, dx$

20. $\int x \sin^3(x^2) \, dx$

21. $\int \cos^2 x \tan^3 x \, dx$

22. $\int \cot^5 x \sin^2 x \, dx$

23. $\int \dfrac{1 - \sin x}{\cos x} \, dx$

24. $\int \dfrac{dx}{1 - \sin x}$

25. $\int \tan^2 x \, dx$

26. $\int \tan^4 x \, dx$

27. $\int \sec^4 x \, dx$

28. $\int \sec^6 x \, dx$

29. $\int_0^{\pi/4} \tan^4 x \sec^2 x \, dx$

30. $\int_0^{\pi/4} \tan^2 x \sec^4 x \, dx$

31. $\int \tan x \sec^3 x \, dx$

32. $\int \tan^3 x \sec^3 x \, dx$

33. $\int \tan^5 x \, dx$

34. $\int \tan^6 x \, dx$

35. $\int_0^{\pi/3} \tan^5 x \sec x \, dx$

36. $\int_0^{\pi/3} \tan^5 x \sec^3 x \, dx$

37. $\int \tan x \sec^6 x \, dx$

38. $\int \tan^3 x \sec^6 x \, dx$

39. $\int \dfrac{\sec^2 x}{\cot x} \, dx$

40. $\int \tan^2 x \sec x \, dx$

41. $\int_{\pi/6}^{\pi/2} \cot^2 x \, dx$

42. $\int_{\pi/4}^{\pi/2} \cot^3 x \, dx$

43. $\int \cot^4 x \csc^4 x \, dx$

44. $\int \cot^3 x \csc^4 x \, dx$

45. $\int \csc x \, dx$

46. $\int \csc^3 x \, dx$

47. $\int \dfrac{\cos^2 x}{\sin x} \, dx$

48. $\int \dfrac{dx}{\sin^4 x}$

49. $\int \sin 5x \sin 2x \, dx$

50. $\int \sin 3x \cos x \, dx$

51. $\int \cos 3x \cos 4x \, dx$

52. $\int \sin 3x \sin 6x \, dx$

53. $\int \sin x \cos 5x \, dx$

54. $\int \cos x \cos 2x \cos 3x \, dx$

55. $\int \dfrac{1 - \tan^2 x}{\sec^2 x} \, dx$

56. $\int \dfrac{\cos x + \sin x}{\sin 2x} \, dx$

57. Find the average value of the function
$f(x) = \sin^2 x \cos^3 x$ on the interval $[-\pi, \pi]$.

58. Evaluate $\int \sin x \cos x \, dx$ by four methods: (a) the substitution $u = \cos x$, (b) the substitution $u = \sin x$, (c) the identity $\sin 2x = 2 \sin x \cos x$, and (d) integration by parts. How do you explain the different appearances of the answers?

In Exercises 59 and 60 find the area of the region bounded by the given curves.

59. $y = \sin x$, $y = \sin^3 x$, $x = 0$, $x = \dfrac{\pi}{2}$

60. $y = \sin x$, $y = 2 \sin^2 x$, $x = 0$, $x = \dfrac{\pi}{2}$

61. A particle moves on a straight line with velocity function $v(t) = \sin \omega t \cos^2 \omega t$. Find its position function $s = f(t)$ if $f(0) = 0$.

Prove the formulas in Exercises 62–64, where m and n are positive integers.

62. $\displaystyle\int_{-\pi}^{\pi} \sin mx \cos nx \, dx = 0$

63. $\displaystyle\int_{-\pi}^{\pi} \sin mx \sin nx \, dx = \begin{cases} 0 & \text{if } m \neq n \\ \pi & \text{if } m = n \end{cases}$

64. $\displaystyle\int_{-\pi}^{\pi} \cos mx \cos nx \, dx = \begin{cases} 0 & \text{if } m \neq n \\ \pi & \text{if } m = n \end{cases}$

65. If

$$f(x) = \sum_{n=1}^{N} a_n \sin nx$$

$$= a_1 \sin x + a_2 \sin 2x + \cdots + a_N \sin Nx$$

show that the mth coefficient a_m is given by the formula

$$a_m = \frac{1}{\pi} \int_{-\pi}^{\pi} f(x) \sin mx \, dx$$

(The sum above is called a *finite Fourier series.*)

SECTION 7.3

Trigonometric Substitution

In finding the area of a circle or an ellipse, an integral of the form $\int \sqrt{a^2 - x^2} \, dx$ arises, where $a > 0$. If it were $\int x\sqrt{a^2 - x^2} \, dx$, the substitution $u = a^2 - x^2$ would be effective but, as it stands, $\int \sqrt{a^2 - x^2} \, dx$ is more difficult. If we change the variable from x to θ by the substitution $x = a \sin \theta$, then the identity $1 - \sin^2 \theta = \cos^2 \theta$ allows us to get rid of the root sign because

$$\sqrt{a^2 - x^2} = \sqrt{a^2 - a^2 \sin^2 \theta} = \sqrt{a^2(1 - \sin^2 \theta)} = \sqrt{a^2 \cos^2 \theta} = a|\cos \theta|$$

Notice that there is a difference between the substitution $u = a^2 - x^2$ (where the new variable is a function of the old one) and the substitution $x = a \sin \theta$ (where the old variable is a function of the new one).

In general we can make a substitution of the form $x = g(t)$ by using the Substitution Rule in reverse, but only if g has an inverse function, that is, if g is one-to-one. In this case if we replace u by x and x by t in Theorem 5.31, we obtain

$$\int f(x) \, dx = \int f(g(t)) g'(t) \, dt$$

This kind of substitution is called *inverse substitution*.

We can make the inverse substitution $x = a \sin \theta$ provided that it defines a one-to-one function. This can be accomplished by restricting θ to lie in the interval $[-\pi/2, \pi/2]$.

In the following table we list trigonometric substitutions that are effective for the given radical expressions because of the given trigonometric identities. In each case the restriction on θ is imposed to ensure that the function that defines the substitution is one-to-one. (These are the same intervals used in Section 6.8 in defining the inverse functions.)

Table of Trigonometric Substitutions (7.12)

Expression	Substitution	Identity
$\sqrt{a^2 - x^2}$	$x = a \sin \theta, \ -\frac{\pi}{2} \leqslant \theta \leqslant \frac{\pi}{2}$	$1 - \sin^2 \theta = \cos^2 \theta$
$\sqrt{a^2 + x^2}$	$x = a \tan \theta, \ -\frac{\pi}{2} < \theta < \frac{\pi}{2}$	$1 + \tan^2 \theta = \sec^2 \theta$
$\sqrt{x^2 - a^2}$	$x = a \sec \theta, \ 0 \leqslant \theta < \frac{\pi}{2}$	$\sec^2 \theta - 1 = \tan^2 \theta$
	or $\pi \leqslant \theta < \frac{3\pi}{2}$	

● **Example 1**   Evaluate $\displaystyle\int \frac{\sqrt{9 - x^2}}{x^2} \, dx$.

Solution   Let $x = 3 \sin \theta$, where $-\pi/2 \leqslant \theta \leqslant \pi/2$. Then $dx = 3 \cos \theta \, d\theta$ and

$$\sqrt{9 - x^2} = \sqrt{9 - 9 \sin^2 \theta} = \sqrt{9 \cos^2 \theta} = 3|\cos \theta| = 3 \cos \theta$$

(Note that $\cos \theta \geqslant 0$ because $-\pi/2 \leqslant \theta \leqslant \pi/2$.) Thus the Inverse Substitution Rule gives

$$\int \frac{\sqrt{9 - x^2}}{x^2} \, dx = \int \frac{3 \cos \theta}{9 \sin^2 \theta} \, 3 \cos \theta \, d\theta$$

$$= \int \frac{\cos^2 \theta}{\sin^2 \theta} \, d\theta = \int \cot^2 \theta \, d\theta$$

$$= \int (\csc^2 \theta - 1) \, d\theta$$

$$= -\cot \theta - \theta + C$$

Since this is an indefinite integral we must return to the original variable x. This can be done either by using trigonometric identities to express $\cot \theta$ in terms of $\sin \theta = x/3$ or by drawing a diagram, as in Figure 7.1, where θ is interpreted as an angle of a right triangle. Since $\sin \theta = x/3$, we label the opposite side and the hypotenuse as having lengths x and 3. Then the Pythagorean Theorem gives the length of the adjacent side as $\sqrt{9 - x^2}$ so we can simply read the value of $\cot \theta$ from the figure:

$$\cot \theta = \frac{\sqrt{9 - x^2}}{x}$$

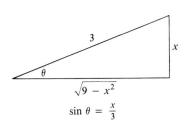

$$\sin \theta = \frac{x}{3}$$

Figure 7.1

(Although $\theta > 0$ in the diagram, this expression for $\cot \theta$ is valid even when $\theta < 0$.) Since $\sin \theta = x/3$, we have $\theta = \sin^{-1}(x/3)$ and so

$$\int \frac{\sqrt{9 - x^2}}{x^2} \, dx = -\frac{\sqrt{9 - x^2}}{x} - \sin^{-1}\left(\frac{x}{3}\right) + C$$ ●

● **Example 2**   Find the area enclosed by the ellipse

$$\frac{x^2}{a^2} + \frac{y^2}{b^2} = 1$$

Solution   Solving the equation of the ellipse for y, we get

$$\frac{y^2}{b^2} = 1 - \frac{x^2}{a^2} = \frac{a^2 - x^2}{a^2} \quad \text{or} \quad y = \pm\frac{b}{a}\sqrt{a^2 - x^2}$$

Because the ellipse is symmetric with respect to both axes, the total area A is four times the area in the first quadrant (see Figure 7.2). The part of the ellipse in the first quadrant is given by the function

$$y = \frac{b}{a}\sqrt{a^2 - x^2} \qquad 0 \leqslant x \leqslant a$$

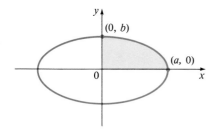

and so

$$\frac{1}{4} A = \int_0^a \frac{b}{a}\sqrt{a^2 - x^2} \, dx$$

Figure 7.2
$$\frac{x^2}{a^2} + \frac{y^2}{b^2} = 1$$

To evaluate this integral we substitute $x = a \sin \theta$. Then $dx = a \cos \theta \, d\theta$. To change the limits of integration we note that when $x = 0$, $\sin \theta = 0$, so $\theta = 0$; when $x = a$, $\sin \theta = 1$, so $\theta = \pi/2$. Also

$$\sqrt{a^2 - x^2} = \sqrt{a^2 - a^2 \sin^2 \theta} = \sqrt{a^2 \cos^2 \theta} = a|\cos \theta| = a \cos \theta$$

since $0 \leqslant \theta \leqslant \pi/2$. Therefore

$$A = 4\frac{b}{a}\int_0^a \sqrt{a^2 - x^2} \, dx = 4\frac{b}{a}\int_0^{\pi/2} a \cos \theta \cdot a \cos \theta \, d\theta$$

$$= 4ab \int_0^{\pi/2} \cos^2 \theta \, d\theta = 4ab \int_0^{\pi/2} \frac{1}{2}(1 + \cos 2\theta) \, d\theta$$

$$= 2ab\left[\theta + \frac{1}{2}\sin 2\theta\right]_0^{\pi/2} = 2ab\left[\frac{\pi}{2} + 0 - 0\right]$$

$$= \pi ab$$

We have shown that the area of an ellipse with semiaxes a and b is πab. In particular, taking $a = b = r$, we have proved the famous formula that the area of a circle with radius r is πr^2. ●

Note: Since the integral in Example 2 was a definite integral, we changed the limits of integration and did not have to convert back to the original variable x.

● **Example 3**   Find $\int \dfrac{1}{x^2 \sqrt{x^2 + 4}}\, dx.$

Solution   Let $x = 2 \tan \theta,\ -\pi/2 < \theta < \pi/2$. Then $dx = 2 \sec^2 \theta\, d\theta$ and

$$\sqrt{x^2 + 4} = \sqrt{4(\tan^2 \theta + 1)} = \sqrt{4 \sec^2 \theta} = 2|\sec \theta| = 2 \sec \theta$$

Thus we have

$$\int \frac{dx}{x^2 \sqrt{x^2 + 4}} = \int \frac{2 \sec^2 \theta\, d\theta}{4 \tan^2 \theta \cdot 2 \sec \theta} = \frac{1}{4} \int \frac{\sec \theta}{\tan^2 \theta}\, d\theta$$

To evaluate this trigonometric integral we put everything in terms of $\sin \theta$ and $\cos \theta$:

$$\frac{\sec \theta}{\tan^2 \theta} = \frac{1}{\cos \theta} \cdot \frac{\cos^2 \theta}{\sin^2 \theta} = \frac{\cos \theta}{\sin^2 \theta}$$

Therefore, making the substitution $u = \sin \theta$, we have

$$\int \frac{dx}{x^2 \sqrt{x^2 + 4}} = \frac{1}{4} \int \frac{\cos \theta}{\sin^2 \theta}\, d\theta = \frac{1}{4} \int \frac{du}{u^2}$$

$$= \frac{1}{4} \left(-\frac{1}{u} \right) + C = -\frac{1}{4 \sin \theta} + C$$

$$= -\frac{\csc \theta}{4} + C$$

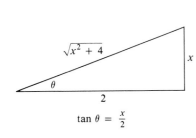

$\sqrt{x^2 + 4}$

x

θ

2

$\tan \theta = \dfrac{x}{2}$

Figure 7.3

We use Figure 7.3 to determine that $\csc \theta = \sqrt{x^2 + 4}/x$ and so

$$\int \frac{dx}{x^2 \sqrt{x^2 + 4}} = -\frac{\sqrt{x^2 + 4}}{4x} + C$$ ●

● **Example 4**   Find $\int \dfrac{x}{\sqrt{x^2 + 4}}\, dx.$

Solution   It would be possible to use the trigonometric substitution $x = 2 \tan \theta$ here (as in Example 3). But the direct substitution $u = x^2 + 4$ is simpler because then $du = 2x\, dx$ and

$$\int \frac{x}{\sqrt{x^2 + 4}}\, dx = \frac{1}{2} \int \frac{du}{\sqrt{u}} = \sqrt{u} + C = \sqrt{x^2 + 4} + C$$ ●

Note: Example 4 illustrates the fact that even when trigonometric substitutions are possible, they may not give the easiest solution. You should look for a simpler method first.

● **Example 5**   Evaluate $\int \dfrac{dx}{\sqrt{x^2 - a^2}}$, where $a > 0$.

Solution 1   Let $x = a \sec \theta$ where $0 < \theta < \pi/2$ or $\pi < \theta < 3\pi/2$. Then $dx = a \sec \theta \tan \theta \, d\theta$ and

$$\sqrt{x^2 - a^2} = \sqrt{a^2(\sec^2 \theta - 1)} = \sqrt{a^2 \tan^2 \theta} = a|\tan \theta| = a \tan \theta$$

Therefore

$$\int \frac{dx}{\sqrt{x^2 - a^2}} = \int \frac{a \sec \theta \tan \theta}{a \tan \theta} \, d\theta$$

$$= \int \sec \theta \, d\theta = \ln|\sec \theta + \tan \theta| + C$$

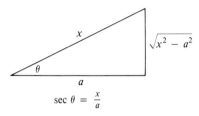

$\sec \theta = \dfrac{x}{a}$

Figure 7.4   The triangle in Figure 7.4 gives $\tan \theta = \sqrt{x^2 - a^2}/a$, so we have

$$\int \frac{dx}{\sqrt{x^2 - a^2}} = \ln\left| \frac{x}{a} + \frac{\sqrt{x^2 - a^2}}{a} \right| + C$$

$$= \ln|x + \sqrt{x^2 - a^2}| - \ln a + C$$

Writing $C_1 = C - \ln a$, we have

(7.13)
$$\int \frac{dx}{\sqrt{x^2 - a^2}} = \ln|x + \sqrt{x^2 - a^2}| + C_1 \qquad ●$$

Solution 2   For $x > 0$ the hyperbolic substitution $x = a \cosh t$ can also be used. Using the identity $\cosh^2 y - \sinh^2 y = 1$, we have

$$\sqrt{x^2 - a^2} = \sqrt{a^2(\cosh^2 t - 1)} = \sqrt{a^2 \sinh^2 t} = a \sinh t$$

Since $dx = a \sinh t \, dt$, we obtain

$$\int \frac{dx}{\sqrt{x^2 - a^2}} = \int \frac{a \sinh t \, dt}{a \sinh t} = \int dt = t + C$$

Since $\cosh t = x/a$, we have $t = \cosh^{-1}(x/a)$ and

(7.14)
$$\int \frac{dx}{\sqrt{x^2 - a^2}} = \cosh^{-1}\left(\frac{x}{a}\right) + C$$

Although Formulas 7.13 and 7.14 look quite different, they are actually equivalent by Formula 6.81.   ●

   Note: As Example 5 illustrates, hyperbolic substitutions can be used in place of trigonometric substitutions and sometimes they lead to simpler answers. But we usually use trigonometric substitutions because trigonometric identities are more familiar than hyperbolic identities.

● **Example 6**   Find $\int_0^{3\sqrt{3}/2} \dfrac{x^3}{(4x^2 + 9)^{3/2}}\, dx$.

Solution   First we note that $(4x^2 + 9)^{3/2} = (\sqrt{4x^2 + 9})^3$ so trigonometric substitution is appropriate. Although $\sqrt{4x^2 + 9}$ is not quite one of the expressions in (7.12), it becomes one of them if we make the preliminary substitution $u = 2x$. When we combine this with the tangent substitution we have $x = \frac{3}{2}\tan\theta$, which gives $dx = \frac{3}{2}\sec^2\theta\, d\theta$ and

$$\sqrt{4x^2 + 9} = \sqrt{9\tan^2\theta + 9} = 3\sec\theta$$

When $x = 0$, $\tan\theta = 0$, so $\theta = 0$; when $x = 3\sqrt{3}/2$, $\tan\theta = \sqrt{3}$, so $\theta = \pi/3$.

$$\int_0^{3\sqrt{3}/2} \frac{x^3}{(4x^2 + 9)^{3/2}}\, dx = \int_0^{\pi/3} \frac{\frac{27}{8}\tan^3\theta}{27\sec^3\theta} \frac{3}{2}\sec^2\theta\, d\theta$$

$$= \frac{3}{16}\int_0^{\pi/3} \frac{\tan^3\theta}{\sec\theta}\, d\theta = \frac{3}{16}\int_0^{\pi/3} \frac{\sin^3\theta}{\cos^2\theta}\, d\theta$$

$$= \frac{3}{16}\int_0^{\pi/3} \frac{(1 - \cos^2\theta)}{\cos^2\theta}\sin\theta\, d\theta$$

Now we substitute $u = \cos\theta$ so that $du = -\sin\theta\, d\theta$. When $\theta = 0$, $u = 1$; when $\theta = \pi/3$, $u = 1/2$. Therefore

$$\int_0^{3\sqrt{3}/2} \frac{x^3}{(4x^2 + 9)^{3/2}}\, dx = -\frac{3}{16}\int_1^{1/2} \frac{1 - u^2}{u^2}\, du = \frac{3}{16}\int_1^{1/2}(1 - u^{-2})\, du$$

$$= \frac{3}{16}\left[u + \frac{1}{u}\right]_1^{1/2} = \frac{3}{16}\left[\left(\frac{1}{2} + 2\right) - (1 + 1)\right]$$

$$= \frac{3}{32} \qquad\qquad ●$$

● **Example 7**   Evaluate $\int \dfrac{x}{\sqrt{3 - 2x - x^2}}\, dx$.

Solution   We can transform the integrand into a function where trigonometric substitution is appropriate by first completing the square under the root sign:

$$3 - 2x - x^2 = 3 - (x^2 + 2x) = 3 + 1 - (x^2 + 2x + 1)$$
$$= 4 - (x + 1)^2$$

This suggests that we make the substitution $u = x + 1$. Then $du = dx$ and $x = u - 1$, so

$$\int \frac{x}{\sqrt{3 - 2x - x^2}}\, dx = \int \frac{u - 1}{\sqrt{4 - u^2}}\, du$$

We now substitute $u = 2\sin\theta$, giving $du = 2\cos\theta\,d\theta$ and $\sqrt{4 - u^2} = 2\cos\theta$, so

$$\int \frac{x}{\sqrt{3 - 2x - x^2}}\,dx = \int \frac{2\sin\theta - 1}{2\cos\theta}\,2\cos\theta\,d\theta$$

$$= \int (2\sin\theta - 1)\,d\theta$$

$$= -2\cos\theta - \theta + C$$

$$= -\sqrt{4 - u^2} - \sin^{-1}\left(\frac{u}{2}\right) + C$$

$$= -\sqrt{3 - 2x - x^2} - \sin^{-1}\left(\frac{x + 1}{2}\right) + C \quad \bullet$$

EXERCISES 7.3

In Exercises 1–30 evaluate the integral.

1. $\displaystyle\int_{1/2}^{\sqrt{3}/2} \frac{1}{x^2\sqrt{1 - x^2}}\,dx$

2. $\displaystyle\int_0^2 x^3\sqrt{4 - x^2}\,dx$

3. $\displaystyle\int \frac{x}{\sqrt{1 - x^2}}\,dx$

4. $\displaystyle\int x\sqrt{4 - x^2}\,dx$

5. $\displaystyle\int \sqrt{1 - 4x^2}\,dx$

6. $\displaystyle\int_0^2 \frac{x^3}{\sqrt{x^2 + 4}}\,dx$

7. $\displaystyle\int_0^3 \frac{dx}{\sqrt{9 + x^2}}$

8. $\displaystyle\int \sqrt{x^2 + 1}\,dx$

9. $\displaystyle\int \frac{dx}{x^3\sqrt{x^2 - 16}}$

10. $\displaystyle\int \frac{\sqrt{x^2 - a^2}\,dx}{x^4}$

11. $\displaystyle\int \frac{\sqrt{9x^2 - 4}}{x}\,dx$

12. $\displaystyle\int \frac{dx}{x^2\sqrt{16x^2 - 9}}$

13. $\displaystyle\int \frac{x^2}{(a^2 - x^2)^{3/2}}\,dx$

14. $\displaystyle\int \frac{x^2}{\sqrt{5 - x^2}}\,dx$

15. $\displaystyle\int \frac{dx}{x\sqrt{x^2 + 3}}$

16. $\displaystyle\int \frac{x}{(x^2 + 4)^{5/2}}\,dx$

17. $\displaystyle\int_0^{2/3} x^3\sqrt{4 - 9x^2}\,dx$

18. $\displaystyle\int_0^3 x^2\sqrt{9 - x^2}\,dx$

19. $\displaystyle\int 5x\sqrt{1 + x^2}\,dx$

20. $\displaystyle\int \frac{dx}{(4x^2 - 25)^{3/2}}$

21. $\displaystyle\int \frac{dx}{x^4\sqrt{x^2 - 2}}$

22. $\displaystyle\int \frac{dx}{(1 + x^2)^2}$

23. $\displaystyle\int \sqrt{2x - x^2}\,dx$

24. $\displaystyle\int \frac{dx}{\sqrt{x^2 + 4x + 8}}$

25. $\displaystyle\int \frac{1}{\sqrt{9x^2 + 6x - 8}}\,dx$

26. $\displaystyle\int \frac{x^2}{\sqrt{4x - x^2}}\,dx$

27. $\displaystyle\int \frac{dx}{(x^2 + 2x + 2)^2}$

28. $\displaystyle\int \frac{dx}{(5 - 4x - x^2)^{5/2}}$

29. $\displaystyle\int e^t\sqrt{9 - e^{2t}}\,dt$

30. $\displaystyle\int \sqrt{e^{2t} - 9}\,dt$

31. (a) Use trigonometric substitution to show that

$$\int \frac{dx}{\sqrt{x^2 + a^2}} = \ln(x + \sqrt{x^2 + a^2}) + C$$

(b) Use the hyperbolic substitution $x = a\sinh t$ to show that

$$\int \frac{dx}{\sqrt{x^2 + a^2}} = \sinh^{-1}\left(\frac{x}{a}\right) + C$$

These formulas are connected by Formula 6.80.

32. Evaluate $\int \dfrac{x^2}{(x^2 + a^2)^{3/2}}\, dx$

(a) by trigonometric substitution and (b) by the hyperbolic substitution $x = a \sinh t$.

33. Prove the formula $A = \frac{1}{2}r^2\theta$ for the area of a sector of a circle with radius r and central angle θ. [*Hint:* Assume $0 < \theta < \pi/2$ and place the center of the circle at the origin so it has the equation $x^2 + y^2 = r^2$. Then A is the sum of the area of the triangle POQ and the area of the region PQR (see Figure 7.5).]

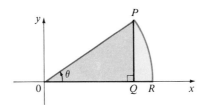

Figure 7.5

34. Find the area of the region bounded by the hyperbola $9x^2 - 4y^2 = 36$ and the line $x = 3$.

Integration of Rational Functions by Partial Fractions

In this section we show how to integrate any rational function (a ratio of polynomials) by expressing it as a sum of simpler fractions, called partial fractions, that we already know how to integrate.

To illustrate the method, observe that by taking the fractions $2/(x - 1)$ and $1/(x + 2)$ to a common denominator we obtain

$$\frac{2}{x - 1} - \frac{1}{x + 2} = \frac{2(x + 2) - (x - 1)}{(x - 1)(x + 2)} = \frac{x + 5}{x^2 + x - 2}$$

If we now reverse the procedure we see how to integrate the function on the right side of this equation:

$$\int \frac{x + 5}{x^2 + x - 2}\, dx = \int \left(\frac{2}{x - 1} - \frac{1}{x + 2} \right) dx$$

$$= 2 \ln|x - 1| - \ln|x + 2| + C$$

To see how the method of partial fractions works in general, let us consider a rational function

$$f(x) = \frac{P(x)}{Q(x)}$$

where P and Q are polynomials. It is possible to express f as a sum of simpler fractions provided that the degree of P is less than the degree of Q. Such a rational function is called *proper*. Recall that if

$$P(x) = a_n x^n + a_{n-1} x^{n-1} + \cdots + a_1 x + a_0$$

where $a_n \neq 0$, then the degree of P is n and we write $\deg(P) = n$.

If f is improper, that is, $\deg(P) \geq \deg(Q)$, then we must take the preliminary step of dividing Q into P (by long division) until a remainder $R(x)$

is obtained with $\deg(R) < \deg(Q)$. The division statement is

(7.15)
$$f(x) = \frac{P(x)}{Q(x)} = S(x) + \frac{R(x)}{Q(x)}$$

where S and R are also polynomials.

As the following example illustrates, sometimes this preliminary step is all that is required.

● **Example 1**   Find $\int \dfrac{x^3 + x}{x - 1}\, dx$.

Solution   Since the degree of the numerator is greater than the degree of the denominator, we first perform the long division:

$$
\begin{array}{r}
x^2 + x + 2 \\
x - 1 \overline{\smash{\big)}\, x^3 \phantom{{}- x^2 {}} + x} \\
\underline{x^3 - x^2} \phantom{{}+ x} \\
x^2 + x \\
\underline{x^2 - x} \\
2x \\
2x - 2 \\
\underline{\phantom{2x -{}} 2}
\end{array}
$$

Thus
$$\int \frac{x^3 + x}{x - 1}\, dx = \int \left(x^2 + x + 2 + \frac{2}{x - 1} \right) dx$$

$$= \frac{x^3}{3} + \frac{x^2}{2} + 2x + 2\ln|x - 1| + C \qquad ●$$

The next step is to factor the denominator $Q(x)$ as far as possible. It can be shown that any polynomial Q can be factored as a product of linear factors (of the form $ax + b$) and irreducible quadratic factors (of the form $ax^2 + bx + c$ where $b^2 - 4ac < 0$). For instance, if $Q(x) = x^4 - 16$, we could factor it as

$$Q(x) = (x^2 - 4)(x^2 + 4) = (x - 2)(x + 2)(x^2 + 4)$$

The third step is to express the proper rational function $R(x)/Q(x)$ (from Equation 7.15) as a sum of **partial fractions** of the form

$$\frac{A}{(ax + b)^i} \quad \text{and} \quad \frac{Ax + B}{(ax^2 + bx + c)^j}$$

A theorem in algebra guarantees that it is always possible to do this. We explain the details for the four cases that occur.

Case 1: The denominator $Q(x)$ is a product of distinct linear factors.   This means that we can write

$$Q(x) = (a_1 x + b_1)(a_2 x + b_2) \cdots (a_k x + b_k)$$

where no factor is repeated. In this case the partial fraction theorem states that there exist constants $A_1, A_2, \ldots, A_k$ such that

(7.16)
$$\frac{R(x)}{Q(x)} = \frac{A_1}{a_1 x + b_1} + \frac{A_2}{a_2 x + b_2} + \cdots + \frac{A_k}{a_k x + b_k}$$

These constants can be determined as in the following example.

● **Example 2**   Evaluate $\displaystyle\int \frac{x^2 + 2x - 1}{2x^3 + 3x^2 - 2x}\, dx$.

Solution   Since the degree of the numerator is less than the degree of the denominator, there is no need to divide. We factor the denominator as

$$2x^3 + 3x^2 - 2x = x(2x^2 + 3x - 2) = x(2x - 1)(x + 2)$$

Since there are three distinct linear factors, the partial fraction decomposition of the integrand (7.16) has the form

(7.17)
$$\frac{x^2 + 2x - 1}{x(2x - 1)(x + 2)} = \frac{A}{x} + \frac{B}{2x - 1} + \frac{C}{x + 2}$$

To determine the values of A, B, and C we multiply both sides of this equation by $x(2x - 1)(x + 2)$, obtaining

(7.18)
$$x^2 + 2x - 1 = A(2x - 1)(x + 2) + Bx(x + 2) + Cx(2x - 1)$$

Expanding the right side of Equation 7.18 and writing it in the standard form for polynomials, we get

(7.19)
$$x^2 + 2x - 1 = (2A + B + 2C)x^2 + (3A + 2B - C)x - 2A$$

The polynomials in Equation 7.19 are equal, so their coefficients must be equal. The coefficient of x^2 on the right side, $2A + B + 2C$, must equal the coefficient of x^2 on the left side—namely, 1. Likewise the coefficients of x are equal and the constant terms are equal. This gives the following system of equations for A, B, and C:

$$\begin{aligned}
2A + B + 2C &= 1 \\
3A + 2B - C &= 2 \\
-2A &= -1
\end{aligned}$$

Solving, we get $A = \frac{1}{2}$, $B = \frac{1}{5}$, and $C = -\frac{1}{10}$, and so

$$\int \frac{x^2 + 2x - 1}{2x^3 + 3x^2 - 2x}\, dx = \int \left[\frac{1}{2}\frac{1}{x} + \frac{1}{5}\frac{1}{2x - 1} - \frac{1}{10}\frac{1}{x + 2} \right] dx$$

$$= \frac{1}{2}\ln|x| + \frac{1}{10}\ln|2x - 1| - \frac{1}{10}\ln|x + 2| + C$$

In integrating the middle term we have made the mental substitution $u = 2x - 1$, which gives $du = 2\, dx$ and $dx = du/2$.   ●

Note: There is another way to find the coefficients A, B, and C in Example 2. Equation 7.18 is an identity; it is true for every value of x. Let us choose values of x that simplify the equation. If we put $x = 0$ in Equation 7.18 then the second and third terms on the right side vanish and the equation becomes $-2A = -1$ or $A = \frac{1}{2}$. Likewise $x = \frac{1}{2}$ gives $5B/4 = \frac{1}{4}$ and $x = -2$ gives $10C = -1$, so $B = \frac{1}{5}$ and $C = -\frac{1}{10}$. (You may object that Equation 7.17 is not valid for $x = 0, \frac{1}{2}$, or -2, so why should Equation 7.18 be valid for those values? In fact, Equation 7.18 is true for all values of x, even $x = 0, \frac{1}{2}$, and -2. See Exercise 83 for the reason.) Although the method of this note is effective for the functions of Case 1, it is less effective for the other cases, so we usually prefer to use the method of comparing coefficients as in Example 2.

● **Example 3**   Find $\int \dfrac{dx}{x^2 - a^2}$, where $a \neq 0$.

Solution   The method of partial fractions gives

$$\frac{1}{x^2 - a^2} = \frac{1}{(x - a)(x + a)} = \frac{A}{x - a} + \frac{B}{x + a}$$

and therefore

$$A(x + a) + B(x - a) = 1$$

Using the method of the note above, we put $x = a$ in this equation and get $A(2a) = 1$, so $A = 1/2a$. If we put $x = -a$, we get $B(-2a) = 1$, so $B = -1/2a$. Thus

$$\int \frac{dx}{x^2 - a^2} = \frac{1}{2a} \int \left[\frac{1}{x - a} - \frac{1}{x + a} \right] dx$$

$$= \frac{1}{2a} \left[\ln|x - a| - \ln|x + a| \right] + C$$

Since $\ln x - \ln y = \ln(x/y)$, we can write the integral as

(7.20)

$$\boxed{\int \frac{dx}{x^2 - a^2} = \frac{1}{2a} \ln\left| \frac{x - a}{x + a} \right| + C}$$

See Exercises 76–79 for ways of using Formula 7.20.   ●

Case 2: Q(x) is a product of linear factors, some of which are repeated.
Suppose the first linear factor $(a_1 x + b_1)$ is repeated r times; that is, $(a_1 x + b_1)^r$ occurs in the factorization of $Q(x)$. Then instead of the single term $A_i/(a_1 x + b_1)$ in Equation 7.16 we would use

(7.21)

$$\frac{A_1}{a_1 x + b_1} + \frac{A_2}{(a_1 x + b_1)^2} + \cdots + \frac{A_r}{(a_1 x + b_1)^r}$$

By way of illustration we could write

$$\frac{x^3 - x + 1}{x^2(x - 1)^3} = \frac{A}{x} + \frac{B}{x^2} + \frac{C}{x - 1} + \frac{D}{(x - 1)^2} + \frac{E}{(x - 1)^3}$$

but we prefer to work out in detail a simpler example.

● **Example 4**   Find $\displaystyle\int \frac{x^4 - 2x^2 + 4x + 1}{x^3 - x^2 - x + 1}\,dx.$

Solution   The first step is to divide. The result of long division is

$$\frac{x^4 - 2x^2 + 4x + 1}{x^3 - x^2 - x + 1} = x + 1 + \frac{4x}{x^3 - x^2 - x + 1}$$

The second step is to factor the denominator $Q(x) = x^3 - x^2 - x + 1$. Since $Q(1) = 0$, we know that $x - 1$ is a factor and we obtain

$$x^3 - x^2 - x + 1 = (x - 1)(x^2 - 1) = (x - 1)(x - 1)(x + 1)$$
$$= (x - 1)^2(x + 1)$$

Since the linear factor $x - 1$ occurs twice, the partial fraction decomposition is

$$\frac{4x}{(x - 1)^2(x + 1)} = \frac{A}{x - 1} + \frac{B}{(x - 1)^2} + \frac{C}{x + 1}$$

Multiplying by $(x - 1)^2(x + 1)$, we get

$$4x = A(x - 1)(x + 1) + B(x + 1) + C(x - 1)^2$$
$$= (A + C)x^2 + (B - 2C)x + (-A + B + C)$$

Now we equate coefficients:

$$A + C = 0$$
$$B - 2C = 4$$
$$-A + B + C = 0$$

Solving, we obtain $A = 1$, $B = 2$, and $C = -1$, so

$$\int \frac{x^4 - 2x^2 + 4x + 1}{x^3 - x^2 - x + 1}\,dx = \int \left[x + 1 + \frac{1}{x - 1} + \frac{2}{(x - 1)^2} - \frac{1}{x + 1} \right] dx$$

$$= \frac{x^2}{2} + x + \ln|x - 1| - \frac{2}{x - 1} - \ln|x + 1| + K$$

$$= \frac{x^2}{2} + x - \frac{2}{x - 1} + \ln\left|\frac{x - 1}{x + 1}\right| + K \qquad ●$$

Case 3: Q(x) contains irreducible quadratic factors, none of which is repeated.
If $Q(x)$ has the factor $ax^2 + bx + c$, where $b^2 - 4ac < 0$, then, in addition to the partial fractions in Equation 7.16 and (7.21), the expression for $R(x)/Q(x)$ will have a term of the form

(7.22)
$$\frac{Ax + B}{ax^2 + bx + c}$$

where A and B are constants to be determined. For instance, the function $f(x) = x/(x - 2)(x^2 + 1)(x^2 + 4)$ has a partial fraction decomposition of the form

$$\frac{x}{(x - 2)(x^2 + 1)(x^2 + 4)} = \frac{A}{x - 2} + \frac{Bx + C}{x^2 + 1} + \frac{Dx + E}{x^2 + 4}$$

The term given in (7.22) can be integrated by completing the square and using the formula

(7.23)
$$\int \frac{dx}{x^2 + a^2} = \frac{1}{a} \tan^{-1}\left(\frac{x}{a}\right) + C$$

● **Example 5**   Evaluate $\int \dfrac{2x^2 - x + 4}{x^3 + 4x} \, dx$.

Solution   Since $x^3 + 4x = x(x^2 + 4)$ cannot be factored further, we write

$$\frac{2x^2 - x + 4}{x(x^2 + 4)} = \frac{A}{x} + \frac{Bx + C}{x^2 + 4}$$

Multiplying by $x(x^2 + 4)$, we have

$$2x^2 - x + 4 = A(x^2 + 4) + (Bx + C)x$$
$$= (A + B)x^2 + Cx + 4A$$

Equating coefficients, we obtain

$$A + B = 2 \qquad C = -1 \qquad 4A = 4$$

Thus $A = 1$, $B = 1$, and $C = -1$ and so

$$\int \frac{2x^2 - x + 4}{x^3 + 4x} \, dx = \int \left[\frac{1}{x} + \frac{x - 1}{x^2 + 4}\right] dx$$

In order to integrate the second term we split it into two parts:

$$\int \frac{x - 1}{x^2 + 4} \, dx = \int \frac{x}{x^2 + 4} \, dx - \int \frac{1}{x^2 + 4} \, dx$$

We make the substitution $u = x^2 + 4$ in the first of these integrals so that $du = 2x\,dx$. We evaluate the second integral by means of Formula 7.23 with $a = 2$:

$$\int \frac{2x^2 - x + 4}{x(x^2 + 4)}\,dx = \int \frac{1}{x}\,dx + \int \frac{x}{x^2 + 4}\,dx - \int \frac{1}{x^2 + 4}\,dx$$

$$= \ln|x| + \frac{1}{2}\ln(x^2 + 4) - \frac{1}{2}\tan^{-1}\left(\frac{x}{2}\right) + C \quad \bullet$$

● **Example 6**   Evaluate $\displaystyle\int \frac{4x^2 - 3x + 2}{4x^2 - 4x + 3}\,dx$.

Solution   Since the degree of the numerator is not less than the degree of the denominator, we first divide and obtain

$$\frac{4x^2 - 3x + 2}{4x^2 - 4x + 3} = 1 + \frac{x - 1}{4x^2 - 4x + 3}$$

Note that the quadratic $4x^2 - 4x + 3$ is irreducible because its discriminant is $b^2 - 4ac = -32 < 0$. This means it cannot be factored, so we do not need to use the partial fraction technique.

To integrate the given function we complete the square in the denominator:

$$4x^2 - 4x + 3 = (2x - 1)^2 + 2$$

This suggests that we make the substitution $u = 2x - 1$. Then $du = 2\,dx$ and $x = (u + 1)/2$, so

$$\int \frac{4x^2 - 3x + 2}{4x^2 - 4x + 3}\,dx = \int \left(1 + \frac{x - 1}{4x^2 - 4x + 3}\right)dx$$

$$= x + \frac{1}{2}\int \frac{\frac{1}{2}(u + 1) - 1}{u^2 + 2}\,du$$

$$= x + \frac{1}{4}\int \frac{u - 1}{u^2 + 2}\,du$$

$$= x + \frac{1}{4}\int \frac{u}{u^2 + 2}\,du - \frac{1}{4}\int \frac{1}{u^2 + 2}\,du$$

$$= x + \frac{1}{8}\ln(u^2 + 2) - \frac{1}{4}\cdot\frac{1}{\sqrt{2}}\tan^{-1}\left(\frac{u}{\sqrt{2}}\right) + C$$

$$= x + \frac{1}{8}\ln(4x^2 - 4x + 3) - \frac{1}{4\sqrt{2}}\tan^{-1}\left(\frac{2x - 1}{\sqrt{2}}\right) + C$$

$\quad\bullet$

Note: Example 6 illustrates the general procedure for integrating a partial fraction of the form

$$\frac{Ax + B}{ax^2 + bx + c} \qquad b^2 - 4ac < 0$$

Complete the square in the denominator and then make a substitution that brings the integral into the form

$$\int \frac{Cu + D}{u^2 + a^2}\, du = C \int \frac{u}{u^2 + a^2}\, du + D \int \frac{1}{u^2 + a^2}\, du$$

Then the first integral is a logarithm and the second is expressed in terms of $\tan^{-1}$.

Case 4: $Q(x)$ contains a repeated irreducible quadratic factor.   If $Q(x)$ has the factor $(ax^2 + bx + c)^r$, where $b^2 - 4ac < 0$, then instead of the single partial fraction (7.22), the sum

(7.24)
$$\frac{A_1 x + B_1}{ax^2 + bx + c} + \frac{A_2 x + B_2}{(ax^2 + bx + c)^2} + \cdots + \frac{A_r x + B_r}{(ax^2 + bx + c)^r}$$

occurs in the partial fraction decomposition of $R(x)/Q(x)$. Each of the terms in (7.24) can be integrated by completing the square and making a tangent substitution.

● **Example 7**   Write out the form of the partial fraction decomposition of the function

$$\frac{x^3 + x^2}{x(x - 1)(x^2 + x + 1)(x^2 + 1)^3}$$

Solution

$$\frac{x^3 + x^2}{x(x - 1)(x^2 + x + 1)(x^2 + 1)^3}$$

$$= \frac{A}{x} + \frac{B}{x - 1} + \frac{Cx + D}{x^2 + x + 1} + \frac{Ex + F}{x^2 + 1} + \frac{Gx + H}{(x^2 + 1)^2} + \frac{Ix + J}{(x^2 + 1)^3} \quad ●$$

● **Example 8**   Evaluate $\displaystyle \int \frac{1 - 3x + 2x^2 - x^3}{x(x^2 + 1)^2}\, dx.$

Solution   The form of the partial fraction decomposition is

$$\frac{1 - 3x + 2x^2 - x^3}{x(x^2 + 1)^2} = \frac{A}{x} + \frac{Bx + C}{x^2 + 1} + \frac{Dx + E}{(x^2 + 1)^2}$$

Multiplying by $x(x^2 + 1)^2$, we have

$$-x^3 + 2x^2 - 3x + 1 = A(x^2 + 1)^2 + (Bx + C)x(x^2 + 1) + (Dx + E)x$$

$$= A(x^4 + 2x^2 + 1) + B(x^4 + x^2) + C(x^3 + x) + Dx^2 + Ex$$

$$= (A + B)x^4 + Cx^3 + (2A + B + D)x^2 + (C + E)x + A$$

If we equate coefficients we get the system

$$A + B = 0 \qquad C = -1 \qquad 2A + B + D = 2 \qquad C + E = -3 \qquad A = 1$$

which has the solution $A = 1, B = -1, C = -1, D = 1,$ and $E = -2.$ Thus

$$\int \frac{1 - 3x + 2x^2 - x^3}{x(x^2 + 1)^2}\, dx$$

$$= \int \left(\frac{1}{x} - \frac{x + 1}{x^2 + 1} + \frac{x - 2}{(x^2 + 1)^2} \right) dx$$

$$= \int \frac{dx}{x} - \int \frac{x}{x^2 + 1}\, dx - \int \frac{dx}{x^2 + 1} + \int \frac{x\, dx}{(x^2 + 1)^2} - 2 \int \frac{dx}{(x^2 + 1)^2}$$

$$= \ln|x| - \frac{1}{2} \ln(x^2 + 1) - \tan^{-1} x - \frac{1}{2(x^2 + 1)} - 2 \int \frac{dx}{(x^2 + 1)^2}$$

To evaluate the final integral we substitute $x = \tan\theta.$ Then $dx = \sec^2\theta\, d\theta$ and $x^2 + 1 = \sec^2\theta,$ so

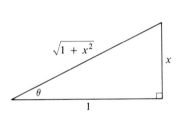

Figure 7.6

$$\int \frac{dx}{(x^2 + 1)^2} = \int \frac{\sec^2\theta}{\sec^4\theta}\, d\theta = \int \cos^2\theta\, d\theta$$

$$= \frac{1}{2} \int (1 + \cos 2\theta)\, d\theta = \frac{\theta}{2} + \frac{\sin 2\theta}{4} + C$$

$$= \frac{\theta}{2} + \frac{\sin\theta\cos\theta}{2} + C$$

$$= \frac{1}{2} \left(\tan^{-1} x + \frac{x}{\sqrt{x^2 + 1}} \cdot \frac{1}{\sqrt{x^2 + 1}} \right) + C \quad \text{(from Figure 7.6)}$$

$$= \frac{1}{2} \left(\tan^{-1} x + \frac{x}{x^2 + 1} \right) + C$$

Thus

$$\int \frac{1 - 3x + 2x^2 - x^3}{x(x^2 + 1)^2}\, dx$$

$$= \ln|x| - \frac{1}{2} \ln(x^2 + 1) - \tan^{-1} x - \frac{1}{2(x^2 + 1)} - \tan^{-1} x - \frac{x}{x^2 + 1} + C$$

$$= \ln \frac{|x|}{\sqrt{x^2 + 1}} - 2\tan^{-1} x - \frac{2x + 1}{2(x^2 + 1)} + C \qquad \bullet$$

Finally we note that sometimes partial fractions can be avoided when integrating a rational function. For instance, although the integral

$$\int \frac{x^2 + 1}{x(x^2 + 3)}\, dx$$

could be evaluated by the method of Case 3, it is easier to observe that if $u = x(x^2 + 3) = x^3 + 3x,$ then $du = (3x^2 + 3)\, dx$ and so

$$\int \frac{x^2 + 1}{x(x^2 + 3)}\, dx = \frac{1}{3} \ln|x^3 + 3x| + C$$

EXERCISES 7.4

In Exercises 1–18 write out the form of the partial fraction decomposition of the given function (as in Example 7). Do not determine the numerical values of the coefficients.

1. $\dfrac{1}{(x - 1)(x + 2)}$

2. $\dfrac{x}{(x + 3)(2x - 5)}$

3. $\dfrac{x + 1}{x^2 + 2x}$

4. $\dfrac{7}{2x^2 + 5x - 12}$

5. $\dfrac{x^2 + 3x - 4}{(2x - 1)^2(2x + 3)}$

6. $\dfrac{x^3 - x^2}{(x - 6)(5x + 3)^3}$

7. $\dfrac{1}{x^4 - x^3}$

8. $\dfrac{1 + x + x^2}{(x + 1)(x + 2)^2(x + 3)^3}$

9. $\dfrac{x^2 + 1}{x^2 - 1}$

10. $\dfrac{x^4 + x^3 - x^2 - x + 1}{x^3 - x}$

11. $\dfrac{x^2 - 2}{x(x^2 + 2)}$

12. $\dfrac{x^3 - 4x^2 + 2}{(x^2 + 1)(x^2 + 2)}$

13. $\dfrac{x^4 + x^2 + 1}{(x^2 + 1)(x^2 + 4)^2}$

14. $\dfrac{1 + 16x}{(2x - 3)(x + 5)^2(x^2 + x + 1)}$

15. $\dfrac{x^4}{(x^2 + 9)^3}$

16. $\dfrac{19x}{(x - 1)^3(4x^2 + 5x + 3)^2}$

17. $\dfrac{x^3 + x^2 + 1}{x^4 + x^3 + 2x^2}$

18. $\dfrac{1}{x^6 - x^3}$

Evaluate the integrals in Exercises 19–74.

19. $\displaystyle\int \frac{x^2}{x + 1}\,dx$

20. $\displaystyle\int \frac{x}{x - 5}\,dx$

21. $\displaystyle\int_2^4 \frac{4x - 1}{(x - 1)(x + 2)}\,dx$

22. $\displaystyle\int_3^7 \frac{1}{(x + 1)(x - 2)}\,dx$

23. $\displaystyle\int \frac{6x - 5}{2x + 3}\,dx$

24. $\displaystyle\int \frac{1}{(x + a)(x + b)}\,dx$

25. $\displaystyle\int \frac{x^2 + 1}{x^2 - x}\,dx$

26. $\displaystyle\int_0^2 \frac{x^3 + x^2 - 12x + 1}{x^2 + x - 12}\,dx$

27. $\displaystyle\int_0^1 \frac{2x + 3}{(x + 1)^2}\,dx$

28. $\displaystyle\int \frac{1}{(x - 1)(x - 2)(x - 3)}\,dx$

29. $\displaystyle\int \frac{1}{x(x + 1)(2x + 3)}\,dx$

30. $\displaystyle\int \frac{3x^2 - 6x + 2}{2x^3 - 3x^2 + x}\,dx$

31. $\displaystyle\int_2^3 \frac{6x^2 + 5x - 3}{x^3 + 2x^2 - 3x}\,dx$

32. $\displaystyle\int_0^1 \frac{x}{x^2 + 4x + 4}\,dx$

33. $\displaystyle\int \frac{1}{(x - 1)^2(x + 4)}\,dx$

34. $\displaystyle\int \frac{x^2}{(x - 3)(x + 2)^2}\,dx$

35. $\displaystyle\int \frac{5x^2 + 3x - 2}{x^3 + 2x^2}\,dx$

36. $\displaystyle\int \frac{18 - 2x - 4x^2}{x^3 + 4x^2 + x - 6}\,dx$

37. $\displaystyle\int \frac{2x^3 + 4x^2 - 12x + 3}{x^3 - 3x + 2}\,dx$

38. $\displaystyle\int \frac{2x^4 - 3x^3 - 10x^2 + 2x + 11}{x^3 - x^2 - 5x - 3}\,dx$

39. $\displaystyle\int \frac{x^2 + 2x}{x^3 + 3x^2 + 4}\,dx$

40. $\displaystyle\int \frac{dx}{x^2(x - 1)^2}$

41. $\displaystyle\int \frac{dx}{x(x - 1)^3}$

42. $\displaystyle\int \frac{dx}{2x^4 + x^3}$

43. $\displaystyle\int \frac{x^2}{(x + 1)^3}\,dx$

44. $\displaystyle\int \frac{x^3}{(x + 1)^3}\,dx$

45. $\displaystyle\int \frac{1}{(x+1)(x^2-1)^2}\,dx$

46. $\displaystyle\int \frac{1}{x^5+2x^4+x^3}\,dx$

47. $\displaystyle\int \frac{dx}{x^4-x^2}$

48. $\displaystyle\int \frac{2x^3-x}{x^4-x^2+1}\,dx$

49. $\displaystyle\int_0^1 \frac{x^3}{x^2+1}\,dx$

50. $\displaystyle\int_0^1 \frac{x-1}{x^2+2x+2}\,dx$

51. $\displaystyle\int_0^1 \frac{x}{x^2+x+1}\,dx$

52. $\displaystyle\int_{-1/2}^{1/2} \frac{4x^2+5x+7}{4x^2+4x+5}\,dx$

53. $\displaystyle\int \frac{3x^2-4x+5}{(x-1)(x^2+1)}\,dx$

54. $\displaystyle\int \frac{2x+3}{x^3+3x}\,dx$

55. $\displaystyle\int \frac{x^2+7x-6}{(x+1)(x^2-4x+7)}\,dx$

56. $\displaystyle\int \frac{4x+1}{(x-3)(x^2+6x+12)}\,dx$

57. $\displaystyle\int \frac{1}{x^3-1}\,dx$

58. $\displaystyle\int \frac{x^3}{x^3+1}\,dx$

59. $\displaystyle\int \frac{x^2-2x-1}{(x-1)^2(x^2+1)}\,dx$

60. $\displaystyle\int \frac{x^4}{x^4-1}\,dx$

61. $\displaystyle\int \frac{3x^3-x^2+6x-4}{(x^2+1)(x^2+2)}\,dx$

62. $\displaystyle\int \frac{x^3-2x^2+x+1}{x^4+5x^2+4}\,dx$

63. $\displaystyle\int \frac{dx}{(x^2+3x+2)(x^2+2x+2)}$

64. $\displaystyle\int \frac{dx}{(x^2+4x+4)(x^2+4x+8)}$

65. $\displaystyle\int \frac{x-3}{(x^2+2x+4)^2}\,dx$

66. $\displaystyle\int \frac{x+1}{(x^2+x+2)^2}\,dx$

67. $\displaystyle\int \frac{x^4+1}{x(x^2+1)^2}\,dx$

68. $\displaystyle\int \frac{3x^4-2x^3+20x^2-5x+34}{(x-1)(x^2+4)^2}\,dx$

69. $\displaystyle\int \frac{x^3+x^2+2x-3}{(x^2+2x+2)^2}\,dx$

70. $\displaystyle\int \frac{x^3-6x^2+13x+2}{(x^2-6x+10)^2}\,dx$

71. $\displaystyle\int \frac{8x}{(x^2+4)^3}\,dx$

72. $\displaystyle\int \frac{x^2+1}{(x^3+3x)^2}\,dx$

73. $\displaystyle\int \frac{(2\sin x-3)\cos x}{\sin^2 x-3\sin x+2}\,dx$

74. $\displaystyle\int \frac{\sin x\cos^2 x}{5+\cos^2 x}\,dx$

75. Formula 7.20 can be rewritten as

$$\int \frac{dx}{a^2-x^2}=\frac{1}{2a}\ln\left|\frac{x+a}{x-a}\right|+K$$

Show by a hyperbolic substitution, or by differentiation, that an alternative formula is

$$\int \frac{dx}{a^2-x^2}=\begin{cases}\dfrac{1}{a}\tanh^{-1}\!\left(\dfrac{x}{a}\right)+C & \text{if } |x|<a \\[2ex] \dfrac{1}{a}\coth^{-1}\!\left(\dfrac{x}{a}\right)+C & \text{if } |x|>a\end{cases}$$

In Exercises 76–79 evaluate the integral by completing the square and using Formula 7.20 or the formula in Exercise 75.

76. $\displaystyle\int \frac{dx}{x^2+2x-3}$

77. $\displaystyle\int \frac{dx}{x^2-2x}$

78. $\displaystyle\int \frac{2x+1}{4x^2+12x-7}\,dx$

79. $\displaystyle\int \frac{x}{x^2+x-1}\,dx$

In Exercises 80–82 find the area of the region under the given curve from a to b.

80. $y=\dfrac{1}{x^2-6x+8}$, $a=5, b=10$

81. $y=\dfrac{x+1}{x-1}$, $a=2, b=3$

82. $y=\dfrac{x}{x^2-2x+10}$, $a=-1, b=2$

83. Suppose that F, G, and Q are polynomials and

$$\frac{F(x)}{Q(x)}=\frac{G(x)}{Q(x)}$$

for all x except when $Q(x)=0$. Prove that $F(x)=G(x)$ for all x. (*Hint:* Use continuity.)

Rationalizing Substitutions

By means of an appropriate substitution, some functions can be changed into rational functions and therefore integrated by the methods of the preceding section. In particular, when an integrand contains an expression of the form $\sqrt[n]{g(x)}$, then the substitution $u = \sqrt[n]{g(x)}$ [or $u = g(x)$] may be effective.

● **Example 1**   Evaluate $\displaystyle\int \frac{\sqrt{x+4}}{x}\,dx$.

Solution   Let $u = \sqrt{x+4}$. Then $u^2 = x + 4$, so $x = u^2 - 4$ and $dx = 2u\,du$. Therefore

$$\int \frac{\sqrt{x+4}}{x}\,dx = \int \frac{u}{u^2 - 4}\,2u\,du$$

$$= 2\int \frac{u^2}{u^2 - 4}\,du$$

$$= 2\int\left(1 + \frac{4}{u^2 - 4}\right)du$$

We can evaluate the integral $\int du/(u^2 - 4)$ either by factoring $u^2 - 4$ as $(u - 2)(u + 2)$ and using partial fractions or by using Formula 7.20 with $a = 2$:

$$\int \frac{\sqrt{x+4}}{x}\,dx = 2\int du + 8\int \frac{du}{u^2 - 4}$$

$$= 2u + 8 \cdot \frac{1}{2 \cdot 2}\ln\left|\frac{u - 2}{u + 2}\right| + C$$

$$= 2\sqrt{x+4} + 2\ln\left|\frac{\sqrt{x+4} - 2}{\sqrt{x+4} + 2}\right| + C \qquad ●$$

● **Example 2**   Find $\displaystyle\int \frac{dx}{\sqrt{x} - \sqrt[3]{x}}$.

Solution   If we were to substitute $u = \sqrt{x}$, then the square root would disappear but a cube root would remain. On the other hand, the substitution $u = \sqrt[3]{x}$ would eliminate the cube root but leave a square root. We can eliminate both roots by means of the substitution $u = \sqrt[6]{x}$. (Note that 6 is the least common multiple of 2 and 3.)

Let $u = \sqrt[6]{x}$. Then $x = u^6$, so $dx = 6u^5\,du$ and $\sqrt{x} = u^3$, $\sqrt[3]{x} = u^2$. Thus

$$\int \frac{dx}{\sqrt{x} - \sqrt[3]{x}} = \int \frac{6u^5\,du}{u^3 - u^2} = 6\int \frac{u^3}{u - 1}\,du$$

$$= 6\int \left(u^2 + u + 1 + \frac{1}{u - 1}\right)du \quad \text{(by long division)}$$

$$= 6\left(\frac{u^3}{3} + \frac{u^2}{2} + u + \ln|u - 1|\right) + C$$

$$= 2\sqrt{x} + 3\sqrt[3]{x} + 6\sqrt[6]{x} + 6\ln|\sqrt[6]{x} - 1| + C \qquad \bullet$$

The German mathematician Karl Weierstrass (1815–1897) noticed that the substitution $t = \tan(x/2)$ will convert any rational function of $\sin x$ and $\cos x$ into an ordinary rational function. Let

$$t = \tan\left(\frac{x}{2}\right) \qquad -\pi < x < \pi$$

Then
$$\cos\left(\frac{x}{2}\right) = \frac{1}{\sec\left(\dfrac{x}{2}\right)} = \frac{1}{\sqrt{1 + \tan^2\left(\dfrac{x}{2}\right)}} = \frac{1}{\sqrt{1 + t^2}}$$

$$\sin\left(\frac{x}{2}\right) = \cos\left(\frac{x}{2}\right)\tan\left(\frac{x}{2}\right) = \frac{t}{\sqrt{1 + t^2}}$$

These expressions can also be seen from Figure 7.7. Therefore

$$\sin x = 2\sin\left(\frac{x}{2}\right)\cos\left(\frac{x}{2}\right) = 2\frac{t}{\sqrt{1 + t^2}}\cdot\frac{1}{\sqrt{1 + t^2}} = \frac{2t}{1 + t^2}$$

$$\cos x = \cos^2\left(\frac{x}{2}\right) - \sin^2\left(\frac{x}{2}\right) = \frac{1 - t^2}{1 + t^2}$$

Since $t = \tan(x/2)$ we have $x = 2\tan^{-1} t$, so

$$dx = \frac{2}{1 + t^2}\,dt$$

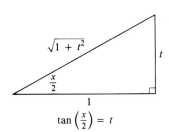

$\tan\left(\dfrac{x}{2}\right) = t$

Figure 7.7

Thus if we make the Weierstrass substitution $t = \tan(x/2)$ then we have

(7.25)
$$\sin x = \frac{2t}{1 + t^2} \qquad \cos x = \frac{1 - t^2}{1 + t^2} \qquad dx = \frac{2}{1 + t^2}\,dt$$

You can see from (7.25) that the Weierstrass substitution will transform any rational function of $\sin x$ and $\cos x$ into a rational function of t.

● **Example 3**   Find $\displaystyle\int \frac{1}{3\sin x - 4\cos x}\,dx$.

Solution   Let $t = \tan(x/2)$. Then, using the expressions in (7.25), we have

$$\int \frac{1}{3 \sin x - 4 \cos x} \, dx = \int \frac{1}{3\left(\dfrac{2t}{1 + t^2}\right) - 4\left(\dfrac{1 - t^2}{1 + t^2}\right)} \frac{2dt}{1 + t^2}$$

$$= 2 \int \frac{dt}{3(2t) - 4(1 - t^2)}$$

$$= \int \frac{dt}{2t^2 + 3t - 2}$$

$$= \int \frac{dt}{(2t - 1)(t + 2)}$$

$$= \int \left[\frac{2}{5} \frac{1}{2t - 1} - \frac{1}{5} \frac{1}{t + 2}\right] dt \qquad \text{(using partial fractions)}$$

$$= \frac{1}{5} \left[\ln|2t - 1| - \ln|t + 2|\right] + C$$

$$= \frac{1}{5} \ln\left|\frac{2t - 1}{t + 2}\right| + C$$

$$= \frac{1}{5} \ln\left|\frac{2 \tan\left(\dfrac{x}{2}\right) - 1}{\tan\left(\dfrac{x}{2}\right) + 2}\right| + C \qquad \bullet$$

EXERCISES 7.5

Evaluate the integrals in Exercises 1–40.

1. $\displaystyle\int_0^1 \frac{1}{1 + \sqrt{x}} \, dx$

2. $\displaystyle\int_0^1 \frac{1}{1 + \sqrt[3]{x}} \, dx$

3. $\displaystyle\int \frac{\sqrt{x}}{x + 1} \, dx$

4. $\displaystyle\int \frac{1}{x\sqrt{x} + 1} \, dx$

5. $\displaystyle\int \frac{1}{x - \sqrt[3]{x}} \, dx$

6. $\displaystyle\int \frac{1}{x - \sqrt{x + 2}} \, dx$

7. $\displaystyle\int_5^{10} \frac{x^2}{\sqrt{x - 1}} \, dx$

8. $\displaystyle\int_1^3 \frac{\sqrt{x - 1}}{x + 1} \, dx$

9. $\displaystyle\int \frac{1}{\sqrt{1 + \sqrt{x}}} \, dx$

10. $\displaystyle\int_{1/3}^3 \frac{\sqrt{x}}{x^2 + x} \, dx$

11. $\displaystyle\int \frac{\sqrt{x} + 1}{\sqrt{x} - 1} \, dx$

12. $\displaystyle\int \frac{\sqrt[3]{x} + 1}{\sqrt[3]{x} - 1} \, dx$

13. $\displaystyle\int \frac{x^3}{\sqrt[3]{x^2 + 1}} \, dx$

14. $\displaystyle\int \frac{x}{x^2 - \sqrt[3]{x^2}} \, dx$

15. $\displaystyle\int \frac{dx}{\sqrt{x} + \sqrt[3]{x}}$

16. $\displaystyle\int \frac{\sqrt{x}}{\sqrt{x} - \sqrt[3]{x}} \, dx$

17. $\displaystyle\int \frac{1}{\sqrt{x} + \sqrt[4]{x}} \, dx$

18. $\displaystyle\int \frac{1}{\sqrt[3]{x} + \sqrt[4]{x}} \, dx$

19. $\displaystyle\int \frac{1}{(bx + c)\sqrt{x}}\, dx \quad (bc > 0)$

20. $\displaystyle\int \frac{1}{(1 + \sqrt{x})^3}\, dx$

21. $\displaystyle\int \frac{\sqrt[3]{x - 1}}{x}\, dx$

22. $\displaystyle\int \frac{\sqrt[3]{1 + \sqrt{x}}}{x}\, dx$

23. $\displaystyle\int \sqrt{\frac{1 - x}{x}}\, dx$

24. $\displaystyle\int \sqrt{\frac{x - 1}{x}}\, dx$

25. $\displaystyle\int \frac{\cos x}{\sin^2 x + \sin x}\, dx$

26. $\displaystyle\int \frac{\sin x}{\cos^2 x + \cos x - 6}\, dx$

27. $\displaystyle\int \frac{e^{2x}}{e^{2x} + 3e^x + 2}\, dx$

28. $\displaystyle\int \frac{1}{\sqrt{1 + e^x}}\, dx$

29. $\displaystyle\int \sqrt{1 - e^x}\, dx$

30. $\displaystyle\int \frac{e^{3x}}{e^{2x} - 1}\, dx$

31. $\displaystyle\int \frac{dx}{1 - \cos x}$

32. $\displaystyle\int \frac{dx}{3 - 5 \sin x}$

33. $\displaystyle\int_0^{\pi/2} \frac{1}{\sin x + \cos x}\, dx$

34. $\displaystyle\int_{\pi/3}^{\pi/2} \frac{1}{1 + \sin x - \cos x}\, dx$

35. $\displaystyle\int \frac{1}{3 \sin x + 4 \cos x}\, dx$

36. $\displaystyle\int \frac{1}{\sin x + \tan x}\, dx$

37. $\displaystyle\int \frac{1}{2 \sin x + \sin 2x}\, dx$

38. $\displaystyle\int \frac{\sec x}{1 + \sin x}\, dx$

39. $\displaystyle\int \frac{dx}{a \sin x + b \cos x}, \quad b > 0$

40. $\displaystyle\int \frac{dx}{a^2 \sin^2 x + b^2 \cos^2 x}, \quad a, b \neq 0$

41. (a) Use the Weierstrass substitution $t = \tan(x/2)$ to prove the formula

$$\int \sec x\, dx = \ln\left| \frac{1 + \tan\left(\dfrac{x}{2}\right)}{1 - \tan\left(\dfrac{x}{2}\right)} \right| + C$$

(b) Show that

$$\int \sec x\, dx = \ln\left| \tan\left(\frac{\pi}{4} + \frac{x}{2}\right) \right| + C$$

by using the formula for $\tan(x + y)$ and part (a).

42. Find a formula for $\int \csc x\, dx$ similar to the one in Exercise 41(a).

SECTION 7.6

Strategy for Integration

Integration is more challenging than differentiation. In finding the derivative of a function it is obvious which differentiation formula should be applied. But it may not be obvious which technique we should use to integrate a given function.

Until now individual techniques have been applied in each section. For instance, we usually used substitution in Exercises 5.6, integration by parts in Exercises 7.1, and partial fractions in Exercises 7.4. But in this section we present a collection of miscellaneous integrals in random order and the main challenge will be to recognize which technique or formula to use. There are no hard and fast rules as to which method applies in a given situation, but we give some advice on strategy that you may find useful.

A prerequisite is a knowledge of the basic integration formulas. In Table 7.26 we have collected the integrals from our previous list together with several additional formulas that we have learned in this chapter. Most of them should be memorized. It is useful to know them all, but the ones marked with an asterisk need not be memorized since they are easily derived. Formulas 19 and 20 can be avoided by using partial fractions, and trigonometric substitutions can be used in place of Formula 21.

Table of Integration Formulas (7.26)

Constants of integration have been omitted.

1. $\int x^n \, dx = \dfrac{x^{n+1}}{n+1} \ (n \neq -1)$ 2. $\int \dfrac{1}{x} \, dx = \ln|x|$

3. $\int e^x \, dx = e^x$ 4. $\int a^x \, dx = \dfrac{a^x}{\ln a}$

5. $\int \sin x \, dx = -\cos x$ 6. $\int \cos x \, dx = \sin x$

7. $\int \sec^2 x \, dx = \tan x$ 8. $\int \csc^2 x \, dx = -\cot x$

9. $\int \sec x \tan x \, dx = \sec x$ 10. $\int \csc x \cot x \, dx = -\csc x$

11. $\int \sec x \, dx = \ln|\sec x + \tan x|$ 12. $\int \csc x \, dx = \ln|\csc x - \cot x|$

13. $\int \tan x \, dx = \ln|\sec x|$ 14. $\int \cot x \, dx = \ln|\sin x|$

15. $\int \sinh x \, dx = \cosh x$ 16. $\int \cosh x \, dx = \sinh x$

17. $\int \dfrac{dx}{x^2 + a^2} = \dfrac{1}{a} \tan^{-1}\left(\dfrac{x}{a}\right)$ 18. $\int \dfrac{dx}{\sqrt{a^2 - x^2}} = \sin^{-1}\left(\dfrac{x}{a}\right)$

*19. $\int \dfrac{dx}{x^2 - a^2} = \dfrac{1}{2a} \ln\left|\dfrac{x-a}{x+a}\right|$

*20. $\int \dfrac{dx}{a^2 - x^2} = \dfrac{1}{2a} \ln\left|\dfrac{x+a}{x-a}\right| = \begin{cases} \dfrac{1}{a} \tanh^{-1}\left(\dfrac{x}{a}\right) & \text{if } |x| < a \\[2mm] \dfrac{1}{a} \coth^{-1}\left(\dfrac{x}{a}\right) & \text{if } |x| > a \end{cases}$

*21. $\int \dfrac{dx}{\sqrt{x^2 \pm a^2}} = \ln\left|x + \sqrt{x^2 \pm a^2}\right| = \begin{cases} \sinh^{-1}\left(\dfrac{x}{a}\right) \\[2mm] \cosh^{-1}\left(\dfrac{x}{a}\right) \end{cases}$

Once you are armed with these basic integration formulas, if you do not immediately see how to attack a given integral, you might try the following four-step strategy.

1. **Simplify the integrand if possible.** Sometimes the use of algebraic manipulation or trigonometric identities will simplify the integrand and make the method of integration obvious. Here are some examples:

$$\int \sqrt{x}(1 + \sqrt{x}) \, dx = \int (\sqrt{x} + x) \, dx$$

$$\int \frac{\tan \theta}{\sec^2 \theta} \, d\theta = \int \frac{\sin \theta}{\cos \theta} \cos^2 \theta \, d\theta = \int \sin \theta \cos \theta \, d\theta$$

$$\int (\sin x + \cos x)^2 \, dx = \int (\sin^2 x + 2 \sin x \cos x + \cos^2 x) \, dx$$

$$= \int (1 + 2 \sin x \cos x) \, dx$$

2. **Look for an obvious substitution.**   Try to find some function $u = g(x)$ in the integrand whose differential $du = g'(x) \, dx$ also occurs, apart from a constant factor. For instance, in the integral

$$\int \frac{x}{x^2 - 1} \, dx$$

we notice that if $u = x^2 - 1$, then $du = 2x \, dx$. Therefore we use the substitution $u = x^2 - 1$ instead of the method of partial fractions.

3. **Classify the integrand according to its form.**   If steps 1 and 2 have not led to the solution, then we take a look at the form of the integrand $f(x)$.

    (a) *Trigonometric functions.* If $f(x)$ is a product of powers of $\sin x$ and $\cos x$, or $\tan x$ and $\sec x$, or $\cot x$ and $\csc x$, then we use the substitutions recommended in Section 7.2. If f is a trigonometric function not of those types but still a rational function of $\sin x$ and $\cos x$, then we use the Weierstrass substitution $t = \tan (x/2)$.

    (b) *Rational functions.* If f is a rational function, we use the procedure of Section 7.4 involving partial fractions.

    (c) *Integration by parts.* If $f(x)$ is a product of a power of x (or a polynomial) and a transcendental function (such as a trigonometric, exponential, or logarithmic function), then we try integration by parts, choosing u and dv according to the advice given in Section 7.1. If you look at the functions in Exercises 7.1 you will see that most of them are of the type just described.

    (d) *Radicals.* Particular kinds of substitutions are recommended when certain radicals appear.

    i.   If $\sqrt{\pm x^2 \pm a^2}$ occurs, we use a trigonometric substitution according to Table 7.12.

    ii. If $\sqrt[n]{ax + b}$ occurs, we use the rationalizing substitution $u = \sqrt[n]{ax + b}$. More generally, this sometimes works for $\sqrt[n]{g(x)}$.

4. **Try again.**   If the first three steps have not produced the answer, remember that there are basically only two methods of integration: substitution and parts.

    (a) *Try substitution.* Even if there is no obvious substitution (step 2), some inspiration or ingenuity (or even desperation) may suggest an appropriate substitution.

(b) *Try parts.* Although integration by parts is used most of the time on products of the form described in step 3(c), it is sometimes effective on single functions. Looking at Section 7.1, we see that it worked on $\tan^{-1} x$, $\sin^{-1} x$, and $\ln x$, and these are all inverse functions.

(c) *Manipulate the integrand.* Algebraic manipulations (perhaps rationalizing the denominator or using trigonometric identities) may be useful in transforming the integral into an easier form. These manipulations may be more substantial than in step 1 and may involve some ingenuity. Here is an example:

$$\int \frac{dx}{1 - \cos x} = \int \frac{1}{1 - \cos x} \cdot \frac{1 + \cos x}{1 + \cos x} \, dx = \int \frac{1 + \cos x}{1 - \cos^2 x} \, dx$$

$$= \int \frac{1 + \cos x}{\sin^2 x} \, dx = \int \left(\csc^2 x + \frac{\cos x}{\sin^2 x} \right) dx$$

(d) *Relate the problem to previous problems.* When you have built up some experience in integration you may be able to use a method on a given integral that is similar to a method you have already used on a previous integral. Or you may even be able to express the given integral in terms of a previous one. For instance, $\int \tan^2 x \sec x \, dx$ is a challenging integral, but using the identity $\tan^2 x = \sec^2 x - 1$ we can write

$$\int \tan^2 x \sec x \, dx = \int \sec^3 x \, dx - \int \sec x \, dx$$

and if $\int \sec^3 x \, dx$ has previously been evaluated (see Example 9 in Section 7.2), then that calculation can be used in the present problem.

(e) *Use several methods.* Sometimes two or three methods are required to evaluate an integral. The evaluation could involve several successive substitutions of different types or it might combine integration by parts with one or more substitutions.

In the following examples we shall indicate a method of attack but not fully work out the integral.

● **Example 1** $\int \dfrac{\tan^3 x}{\cos^3 x} \, dx$

In step 1 we rewrite the integral:

$$\int \frac{\tan^3 x}{\cos^3 x} \, dx = \int \tan^3 x \sec^3 x \, dx$$

The integral is now of the form $\int \tan^m x \sec^n x \, dx$ with m odd, so we can use the advice in (7.9) (b).

Alternatively, if in step 1 we had written

$$\int \frac{\tan^3 x}{\cos^3 x} \, dx = \int \frac{\sin^3 x}{\cos^3 x} \frac{1}{\cos^3 x} \, dx = \int \frac{\sin^3 x}{\cos^6 x} \, dx$$

then we could have continued as follows with the substitution $u = \cos x$:

$$\int \frac{\sin^3 x}{\cos^6 x}\, dx = \int \frac{(1 - \cos^2 x)}{\cos^6 x} \sin x\, dx = \int \frac{1 - u^2}{u^6} (-du)$$

$$= \int \frac{u^2 - 1}{u^6}\, du = \int (u^{-4} - u^{-6})\, du \qquad \bullet$$

● **Example 2** $\int e^{\sqrt{x}}\, dx$

According to step 3(d)(ii) we substitute $u = \sqrt{x}$. Then $x = u^2$, so $dx = 2u\, du$ and

$$\int e^{\sqrt{x}}\, dx = 2 \int u e^u\, du$$

The integrand is now a product of u and the transcendental function e^u so it can be integrated by parts. $\qquad \bullet$

● **Example 3** $\int \frac{x^5 + 1}{x^3 - 3x^2 - 10x}\, dx$

There is no algebraic simplification or obvious substitution, so steps 1 and 2 do not apply here. The integrand is a rational function so we apply the procedure of Section 7.4, remembering that the first step is to divide. $\qquad \bullet$

● **Example 4** $\int \frac{dx}{x\sqrt{\ln x}}$

Here step 2 is all that is needed. We substitute $u = \ln x$ because its differential is $du = dx/x$, which occurs in the integral. $\qquad \bullet$

● **Example 5** $\int \sqrt{\frac{1 - x}{1 + x}}\, dx$

Although the rationalizing substitution

$$u = \sqrt{\frac{1 - x}{1 + x}}$$

works here [step 3(d)(ii)], it leads to a very complicated rational function. An easier method is to do some algebraic manipulation [either as step 1 or as step 4(c)]. Multiplying numerator and denominator by $\sqrt{1 - x}$, we have

$$\int \sqrt{\frac{1 - x}{1 + x}}\, dx = \int \frac{1 - x}{\sqrt{1 - x^2}}\, dx$$

$$= \int \frac{1}{\sqrt{1 - x^2}}\, dx - \int \frac{x}{\sqrt{1 - x^2}}\, dx$$

$$= \sin^{-1} x + \sqrt{1 - x^2} + C \qquad \bullet$$

The question arises: Will our strategy for integration enable us to find the integral of every continuous function? In particular, can we use it to evaluate $\int e^{x^2}\,dx$? The answer is no, at least not in terms of the functions that we are familiar with.

The functions that we have been dealing with in this book are called **elementary functions.** These are the polynomials, rational functions, power functions (x^a), exponential functions (a^x), logarithmic functions, trigonometric and inverse trigonometric functions, hyperbolic and inverse hyperbolic functions, and all functions that can be obtained from these by the five operations of addition, subtraction, multiplication, division, and composition. For instance, the function

$$f(x) = \sqrt{\frac{x^2 - 1}{x^3 + 2x - 1}} + \ln(\cosh x) - xe^{\sin 2x}$$

is an elementary function.

If f is an elementary function, then f' is an elementary function but $\int f(x)\,dx$ need not be an elementary function. Consider $f(x) = e^{x^2}$. Since f is continuous its integral exists, and if we define the function F by

$$F(x) = \int_0^x e^{t^2}\,dt$$

then we know from Part 1 of the Fundamental Theorem of Calculus that

$$F'(x) = e^{x^2}$$

Thus $f(x) = e^{x^2}$ has an antiderivative F, but it has been proved that F is not an elementary function. This means that no matter how hard we try, we will never succeed in evaluating $\int e^{x^2}\,dx$ in terms of the functions we know. (In Chapter 10, however, we shall see how to express $\int e^{x^2}\,dx$ as an infinite series.) The same can be said of the following integrals:

$$\int \frac{e^x}{x}\,dx \qquad \int \sin(x^2)\,dx \qquad \int \cos(e^x)\,dx$$

$$\int \sqrt{x^3 + 1}\,dx \qquad \int \frac{1}{\ln x}\,dx \qquad \int \frac{\sin x}{x}\,dx$$

You may rest assured, though, that the integrals in the following exercises are all elementary functions.

EXERCISES 7.6

Evaluate the following integrals.

1. $\int \sin^2 x \cos^3 x\,dx$

2. $\int \dfrac{\sin x - \cos x}{\sin x + \cos x}\,dx$

5. $\int \dfrac{\sqrt{x - 2}}{x + 2}\,dx$

6. $\int \dfrac{x}{(x + 2)^2}\,dx$

3. $\int_0^{1/2} \dfrac{x}{\sqrt{1 - x^2}}\,dx$

4. $\int_1^2 x^3 \ln x\,dx$

7. $\int \ln(1 + x^2)\,dx$

8. $\int \dfrac{\sqrt{1 + \ln x}}{x \ln x}\,dx$

9. $\int_0^1 (1 + \sqrt{x})^8 \, dx$

10. $\int_0^{\pi/4} \tan^3 x \sec^4 x \, dx$

11. $\int \frac{x}{x^2 - 2x + 2} \, dx$

12. $\int x \sin^{-1} x \, dx$

13. $\int \frac{\sqrt{9 - x^2}}{x} \, dx$

14. $\int \frac{x}{x^2 + 3x + 2} \, dx$

15. $\int x^2 \cosh x \, dx$

16. $\int \frac{x^3 + x + 1}{x^4 + 2x^2 + 4x} \, dx$

17. $\int \frac{\cos x}{1 + \sin^2 x} \, dx$

18. $\int \cos \sqrt{x} \, dx$

19. $\int_0^1 \cos \pi x \tan \pi x \, dx$

20. $\int \frac{e^{2x}}{1 + e^x} \, dx$

21. $\int e^{3x} \cos 5x \, dx$

22. $\int \cos 3x \cos 5x \, dx$

23. $\int \frac{dx}{x^3 + x^2 + x + 1}$

24. $\int x^2 \ln(1 + x) \, dx$

25. $\int x^5 e^{-x^3} \, dx$

26. $\int \tan^2 4x \, dx$

27. $\int \frac{1}{\sqrt{9x^2 + 12x - 5}} \, dx$

28. $\int x^2 \tan^{-1} x \, dx$

29. $\int \sqrt[3]{x}(1 - \sqrt{x}) \, dx$

30. $\int \frac{dx}{e^x - e^{-x}}$

31. $\int \frac{2x + 5}{x - 3} \, dx$

32. $\int \frac{1}{x + \sqrt[3]{x}} \, dx$

33. $\int \sin^2 x \cos^4 x \, dx$

34. $\int \frac{1}{\sqrt{5 - 4x - x^2}} \, dx$

35. $\int \frac{x}{1 - x^2 + \sqrt{1 - x^2}} \, dx$

36. $\int \frac{1 + \cos x}{\sin x} \, dx$

37. $\int \frac{e^x}{e^{2x} - 1} \, dx$

38. $\int \frac{1}{x^3 - 8} \, dx$

39. $\int_{-1}^1 x^5 \cosh x \, dx$

40. $\int_{\pi/4}^{\pi/3} \frac{\ln(\tan x)}{\sin x \cos x} \, dx$

41. $\int_{-3}^3 |x^3 + x^2 - 2x| \, dx$

42. $\int_0^{\pi/4} \cos^5 \theta \, d\theta$

43. $\int \cot x \ln(\sin x) \, dx$

44. $\int \frac{1 + e^x}{1 - e^x} \, dx$

45. $\int \frac{x}{(x^2 + 1)(x^2 + 4)} \, dx$

46. $\int \frac{dx}{4 - 5 \sin x}$

47. $\int x \sqrt[3]{x + c} \, dx$

48. $\int e^{\sqrt[3]{x}} \, dx$

49. $\int \frac{1}{x + 4 + 4\sqrt{x + 1}} \, dx$

50. $\int \frac{x^3 + 1}{x^3 - x^2} \, dx$

51. $\int (x^2 + 4x - 3) \sin 2x \, dx$

52. $\int \sin x \cos(\cos x) \, dx$

53. $\int \frac{x}{\sqrt{16 - x^4}} \, dx$

54. $\int \frac{x^3}{(x + 1)^{10}} \, dx$

55. $\int \cot^3 2x \csc^3 2x \, dx$

56. $\int (x + \sin x)^2 \, dx$

57. $\int \frac{e^{\arctan x}}{1 + x^2} \, dx$

58. $\int \frac{dx}{x(x^4 + 1)}$

59. $\int t^3 e^{-2t} \, dt$

60. $\int \frac{\sqrt{t}}{1 + \sqrt[3]{t}} \, dt$

61. $\int \sin x \sin 2x \sin 3x \, dx$

62. $\int_1^3 \left| \ln\left(\frac{x}{2}\right) \right| \, dx$

63. $\int \sqrt{\frac{1 + x}{1 - x}} \, dx$

64. $\int \frac{x \ln x}{\sqrt{x^2 - 1}} \, dx$

65. $\int \frac{x + a}{x^2 + a^2} \, dx$

66. $\int \csc^4 4x \, dx$

67. $\int \frac{x^4}{x^{10} + 16} \, dx$

68. $\int \sqrt{1 + x - x^2} \, dx$

69. $\int \frac{\sin x}{1 + \sin x} \, dx$

70. $\int \frac{x + 2}{x^2 + x + 2} \, dx$

71. $\int x \sec x \tan x \, dx$

72. $\int \frac{x}{x^4 - a^4} \, dx$

73. $\int \frac{1}{\sqrt{x + 1} + \sqrt{x}} \, dx$

74. $\int \frac{1}{1 + 2e^x - e^{-x}} \, dx$

75. $\int \frac{\arctan \sqrt{x}}{\sqrt{x}} \, dx$

76. $\int \frac{\ln(x + 1)}{x^2} \, dx$

77. $\int \frac{1}{\sqrt{4x^2 + 4x + 5}} \, dx$

78. $\int e^{-x} \sinh x \, dx$

79. $\int \frac{1}{e^{3x} - e^x} \, dx$

80. $\int \frac{1 + \cos^2 x}{1 - \cos^2 x} \, dx$

Approximations: The Trapezoidal Rule and Simpson's Rule

In order to evaluate $\int_a^b f(x)\,dx$ using the Fundamental Theorem of Calculus we need to know an antiderivative of f. Sometimes, however, it is difficult, or even impossible, to find an antiderivative (see Section 7.6). For example, it is impossible to evaluate the following integrals exactly:

$$\int_0^1 e^{x^2}\,dx \qquad \int_{-1}^1 \sqrt{1+x^3}\,dx$$

Therefore we need to be able to find approximate values of definite integrals.

We already know one such method. Recall that the definite integral is defined as a limit of Riemann sums, so any Riemann sum could be used as an approximation to the integral. In particular, let us take a partition of $[a,b]$ into n subintervals of equal length $\Delta x = (b-a)/n$. Then we have

$$\int_a^b f(x)\,dx \approx \sum_{i=1}^n f(x_i^*)\,\Delta x$$

where x_i^* is any point in the ith subinterval $[x_{i-1}, x_i]$ of the partition. If x_i^* is chosen to be the left endpoint of the interval, then $x_i^* = x_{i-1}$ and we have

(7.27)
$$\int_a^b f(x)\,dx \approx \sum_{i=1}^n f(x_{i-1})\,\Delta x$$

If $f(x) \geq 0$, then the integral represents an area and (7.27) represents an approximation of this area by the rectangles shown in Figure 7.8(a). If we choose x_i^* to be the right endpoint, then $x_i^* = x_i$ and we have

(7.28)
$$\int_a^b f(x)\,dx \approx \sum_{i=1}^n f(x_i)\,\Delta x$$

[See Figure 7.8(b).]

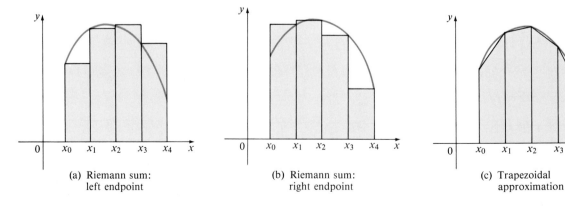

(a) Riemann sum: left endpoint

(b) Riemann sum: right endpoint

(c) Trapezoidal approximation

Figure 7.8

Usually, a more accurate approximation results from averaging the approximations in (7.27) and (7.28):

$$\int_a^b f(x)\,dx \approx \frac{1}{2}\left[\sum_{i=1}^n f(x_{i-1})\,\Delta x + \sum_{i=1}^n f(x_i)\,\Delta x\right]$$

$$= \frac{\Delta x}{2}\left[(f(x_0) + f(x_1)) + (f(x_1) + f(x_2)) + \cdots + (f(x_{n-1}) + f(x_n))\right]$$

$$= \frac{\Delta x}{2}\left[f(x_0) + 2f(x_1) + 2f(x_2) + \cdots + 2f(x_{n-1}) + f(x_n)\right]$$

The Trapezoidal Rule (7.29)

$$\int_a^b f(x)\,dx \approx \frac{\Delta x}{2}\left[f(x_0) + 2f(x_1) + 2f(x_2) + \cdots + 2f(x_{n-1}) + f(x_n)\right]$$

where $\Delta x = (b - a)/n$

The reason for the name Trapezoidal Rule can be seen from Figure 7.8(c), which illustrates the case where $f(x) \geqslant 0$. The area of the trapezoid that lies above the ith subinterval is

$$\Delta x\left(\frac{f(x_{i-1}) + f(x_i)}{2}\right) = \frac{\Delta x}{2}(f(x_{i-1}) + f(x_i))$$

and if we add the areas of all these trapezoids, we get the right side of (7.29).

● **Example 1**   Use the Trapezoidal Rule with $n = 5$ and $n = 10$ to approximate the integral $\int_1^2 (1/x)\,dx$.

Solution   With $n = 5$, $a = 1$, and $b = 2$, we have $\Delta x = (2 - 1)/5 = 0.2$, and so the Trapezoidal Rule gives

$$\int_1^2 \frac{1}{x}\,dx \approx \frac{0.2}{2}\left[f(1) + 2f(1.2) + 2f(1.4) + 2f(1.6) + 2f(1.8) + f(2)\right]$$

$$= 0.1\left[\frac{1}{1} + \frac{2}{1.2} + \frac{2}{1.4} + \frac{2}{1.6} + \frac{2}{1.8} + \frac{1}{2}\right]$$

$$\approx 0.695635$$

With $n = 10$ we have $\Delta x = 0.1$ and the Trapezoidal Rule gives

$$\int_1^2 \frac{1}{x}\,dx \approx \frac{0.1}{2}\left[\frac{1}{1} + \frac{2}{1.1} + \frac{2}{1.2} + \frac{2}{1.3} + \frac{2}{1.4} + \frac{2}{1.5} + \frac{2}{1.6} + \frac{2}{1.7} + \frac{2}{1.8} + \frac{2}{1.9} + \frac{1}{2}\right]$$

$$\approx 0.693771$$

In Example 1 we have deliberately chosen an integral whose value can be computed explicitly so that we can see how accurate the Trapezoidal Rule is. By the Fundamental Theorem of Calculus,

$$\int_1^2 \frac{1}{x}\,dx = \ln x\Big]_1^2 = \ln 2 = 0.693147\ldots$$

so with $n = 5$ the error in using the Trapezoidal Rule is about 0.0025, and with $n = 10$ the error is about 0.0006. In general we get more accurate approximations when we increase the value of n. (But for very large values of n there are so many arithmetic operations that we have to beware of accumulated round-off error.)

For other integrals, however, we may not be able to check the accuracy by evaluating the integral exactly, so we need an upper bound on the error. The following estimate is proved in books on numerical analysis.

Error Estimate for the Trapezoidal Rule (7.30)

> If $|f''(x)| \leqslant M$ for $a \leqslant x \leqslant b$, then the error involved in using the Trapezoidal Rule (7.29) is at most
>
> $$\frac{M(b-a)^3}{12n^2}$$

Let us apply this error estimate to the approximation in Example 1. If $f(x) = 1/x$, then $f'(x) = -1/x^2$ and $f''(x) = 2/x^3$. Since $1 \leqslant x \leqslant 2$, we have $1/x \leqslant 1$, so

$$|f''(x)| = \left|\frac{2}{x^3}\right| \leqslant \frac{2}{1^3} = 2$$

Therefore, taking $M = 2$, $a = 1$, $b = 2$, and $n = 10$ in the error estimate (7.30), we see that the error is at most

$$\frac{2(2-1)^3}{12(10)^2} \approx 0.0017$$

Comparing this error estimate of 0.0017 with the actual error of about 0.0006, we see that it can happen that the actual error is substantially less than the upper bound for the error given by (7.30).

● **Example 2**   How large should we take n in order to guarantee that the Trapezoidal Rule approximation for $\int_1^2 (1/x)\,dx$ is accurate to within 0.0001?

Solution   We have seen in the preceding calculation that $|f''(x)| \leqslant 2$ for $1 \leqslant x \leqslant 2$, so we can take $M = 2$, $a = 1$, and $b = 2$ in (7.30). For an error less than 0.0001 we should choose n so that

$$\frac{2(1)^3}{12n^2} < 0.0001$$

Solving the inequality for n we get

$$n^2 > \frac{2}{12(0.0001)}$$

or

$$n > \frac{1}{\sqrt{0.0006}} \approx 40.8$$

Thus $n = 41$ will ensure the desired accuracy.   ●

● **Example 3**   Use the Trapezoidal Rule with $n = 10$ to approximate the integral

$$\int_0^1 e^{x^2}\, dx$$

and give an upper bound for the error involved in this approximation.

Solution   Since $a = 0$, $b = 1$, and $n = 10$, the Trapezoidal Rule gives

$$\int_0^1 e^{x^2}\, dx \approx \frac{\Delta x}{2}\left[f(0) + 2f(0.1) + 2f(0.2) + \cdots + 2f(0.9) + f(1) \right]$$

$$= \frac{0.1}{2}\left[e^0 + 2e^{0.01} + 2e^{0.04} + 2e^{0.09} + 2e^{0.16} + 2e^{0.25} \right.$$

$$\left. + 2e^{0.36} + 2e^{0.49} + 2e^{0.64} + 2e^{0.81} + e^1 \right]$$

$$\approx 1.467175$$

Since $f(x) = e^{x^2}$, we have $f'(x) = 2xe^{x^2}$ and $f''(x) = (2 + 4x^2)e^{x^2}$. Also, since $0 \leqslant x \leqslant 1$, we have $x^2 \leqslant 1$, and so

$$0 \leqslant f''(x) = (2 + 4x^2)e^{x^2} \leqslant 6e$$

Taking $M = 6e$, $a = 0$, $b = 1$, and $n = 10$ in the error estimate (7.30), we see that an upper bound for the error is

$$\frac{6e(1)^3}{12(10)^2} = \frac{e}{200} \approx 0.014$$ ●

Another rule for approximate integration results from using parabolas instead of trapezoids to approximate the area under a curve. As before we take a partition of $[a, b]$ into n subintervals of equal length $h = \Delta x = (b - a)/n$ but this time we assume that n is an *even* number. Then on each consecutive pair of intervals we approximate the curve $y = f(x) \geqslant 0$ by a parabola as shown in Figure 7.9. If $y_i = f(x_i)$, then $P_i(x_i, y_i)$ is the point on the curve lying above x_i. A typical parabola passes through three consecutive points P_i, P_{i+1}, and P_{i+2}.

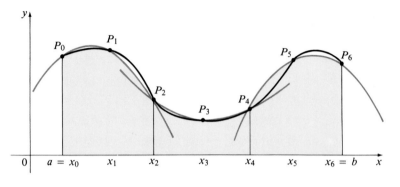

Figure 7.9

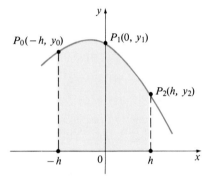

Figure 7.10

In order to simplify our calculations, we first consider the case where $x_0 = -h$, $x_1 = 0$, and $x_2 = h$ (see Figure 7.10). We know that the equation of the parabola through P_0, P_1, and P_2 is of the form $y = Ax^2 + Bx + C$ and so the area under the parabola from $x = -h$ to $x = h$ is

$$\int_{-h}^{h} (Ax^2 + Bx + C)\,dx = A\frac{x^3}{3} + B\frac{x^2}{2} + Cx\Big]_{-h}^{h}$$

$$= A\frac{h^3}{3} + B\frac{h^2}{2} + Ch + A\frac{h^3}{3} - B\frac{h^2}{2} + Ch$$

$$= \frac{h}{3}(2Ah^2 + 6C)$$

But, since the parabola passes through $P_0(-h, y_0)$, $P_1(0, y_1)$, and $P_2(h, y_2)$, we have

$$y_0 = A(-h)^2 + B(-h) + C = Ah^2 - Bh + C$$

$$y_1 = C$$

$$y_2 = Ah^2 + Bh + C$$

and therefore

$$y_0 + 4y_1 + y_2 = 2Ah^2 + 6C$$

Thus we can rewrite the area under the parabola as

$$\frac{h}{3}(y_0 + 4y_1 + y_2)$$

Now by shifting this parabola horizontally we do not change the area under it. This means that the area under the parabola through P_0, P_1, and P_2 from $x = x_0$ to $x = x_2$ in Figure 7.9 is still

$$\frac{h}{3}(y_0 + 4y_1 + y_2)$$

Similarly, the area under the parabola through P_2, P_3, and P_4 from $x = x_2$ to $x = x_4$ is

$$\frac{h}{3}(y_2 + 4y_3 + y_4)$$

If we compute the areas under all the parabolas in this manner and add the results, we get

$$\int_a^b f(x)\,dx \approx \frac{h}{3}(y_0 + 4y_1 + y_2) + \frac{h}{3}(y_2 + 4y_3 + y_4)$$

$$+ \cdots + \frac{h}{3}(y_{n-2} + 4y_{n-1} + y_n)$$

$$= \frac{h}{3}(y_0 + 4y_1 + 2y_2 + 4y_3 + 2y_4$$

$$+ \cdots + 2y_{n-2} + 4y_{n-1} + y_n)$$

Although we have derived this approximation for the case where $f(x) \geqslant 0$, it is a reasonable approximation for any continuous function f and is called Simpson's Rule after the English mathematician Thomas Simpson (1710–1761). Note the pattern of coefficients: $1, 4, 2, 4, 2, 4, 2, \ldots, 4, 2, 4, 1$.

Simpson's Rule (7.31)

$$\int_a^b f(x)\,dx \approx \frac{\Delta x}{3}\left[f(x_0) + 4f(x_1) + 2f(x_2) + 4f(x_3) + \cdots + 2f(x_{n-2})\right.$$

$$\left. + 4f(x_{n-1}) + f(x_n)\right]$$

where n is even and $\Delta x = (b - a)/n$

● **Example 4**   Use Simpson's Rule with $n = 10$ to approximate $\int_1^2 (1/x)\,dx$.

Solution   Putting $f(x) = 1/x$, $n = 10$, and $\Delta x = 0.1$ in (7.31), we obtain

$$\int_1^2 \frac{1}{x}\,dx \approx \frac{\Delta x}{3}\left[f(1) + 4f(1.1) + 2f(1.2) + 4f(1.3) + \cdots + 2f(1.8) + 4f(1.9) + f(2)\right]$$

$$= \frac{0.1}{3}\left[\frac{1}{1} + \frac{4}{1.1} + \frac{2}{1.2} + \frac{4}{1.3} + \frac{2}{1.4} + \frac{4}{1.5} + \frac{2}{1.6} + \frac{4}{1.7} + \frac{2}{1.8} + \frac{4}{1.9} + \frac{1}{2}\right]$$

$$\approx 0.693150$$

Notice that we have obtained a much better approximation to the true value of the integral ($\ln 2 = 0.693147\ldots$) than we did using the Trapezoidal Rule in Example 1 with the same value of n.   ●

The following error estimate provides an upper bound for the error involved when using Simpson's Rule. It is analogous to the estimate given in (7.30) for the Trapezoidal Rule but it uses the fourth derivative of f.

Error Estimate for Simpson's Rule (7.32)

If $\left|f^{(4)}(x)\right| \leqslant M$ for $a \leqslant x \leqslant b$, then the error involved in using Simpson's Rule (7.31) is at most

$$\frac{M(b - a)^5}{180n^4}$$

● **Example 5**   How large should we take n in order to guarantee that the Simpson's Rule approximation for $\int_1^2 (1/x)\,dx$ is accurate to within 0.0001?

Solution   If $f(x) = 1/x$, then $f^{(4)}(x) = 24/x^5$. Since $x \geqslant 1$, we have $1/x \leqslant 1$ and so

$$\left| f^{(4)}(x) \right| = \left| \frac{24}{x^5} \right| \leqslant 24$$

Therefore we can take $M = 24$ in (7.32). Thus for an error less than 0.0001 we should choose n so that

$$\frac{24(1)^5}{180n^4} < 0.0001$$

This gives

$$n^4 > \frac{24}{180(0.0001)}$$

or

$$n > \frac{1}{\sqrt[4]{0.00075}} \approx 6.04$$

Therefore $n = 8$ (n must be even) will give the desired accuracy. (Compare with Example 2 where we obtained $n = 41$ for the Trapezoidal Rule.)   ●

● **Example 6**   Use Simpson's Rule with $n = 10$ to approximate the integral

$$\int_0^1 e^{x^2}\,dx$$

and estimate the error.

Solution   If $n = 10$, then $\Delta x = 0.1$ and Simpson's Rule gives

$$\int_0^1 e^{x^2}\,dx \approx \frac{\Delta x}{3}\left[f(0) + 4f(0.1) + 2f(0.2) + \cdots + 2f(0.8) + 4f(0.9) + f(1) \right]$$

$$= \frac{0.1}{3}\left[e^0 + 4e^{0.01} + 2e^{0.04} + 4e^{0.09} + 2e^{0.16} + 4e^{0.25} + 2e^{0.36} \right.$$

$$\left. + 4e^{0.49} + 2e^{0.64} + 4e^{0.81} + e^1 \right]$$

$$\approx 1.462681$$

The fourth derivative of $f(x) = e^{x^2}$ is

$$f^{(4)}(x) = (12 + 48x^2 + 16x^4)e^{x^2}$$

and so, since $0 \leqslant x \leqslant 1$, we have

$$0 \leqslant f^{(4)}(x) \leqslant (12 + 48 + 16)e^1 = 76e$$

Therefore, putting $M = 76e$, $a = 0$, $b = 1$, and $n = 10$ in (7.32) we see that the error is at most

$$\frac{76e(1)^5}{180(10)^4} \approx 0.000115$$

(Compare with Example 3.) Thus, correct to three decimal places, we have

$$\int_0^1 e^{x^2}\, dx \approx 1.463$$

●

Recall that it is quite possible for y to be a function of x even if no explicit formula is known for y in terms of x. If a scientific experiment establishes values for y corresponding to certain equally spaced values of x, then the Trapezoidal Rule or Simpson's Rule can still be used to find an approximate value for $\int_a^b y\, dx$, the integral of y with respect to x.

● **Example 7**   Suppose the following data were obtained from an experiment:

x	3.0	3.25	3.5	3.75	4.0	4.25	4.5	4.75	5.0
y	6.7	7.4	8.2	9.2	10.4	11.6	12.5	13.3	14.0

Use Simpson's Rule to approximate $\int_3^5 y\, dx$.

Solution   There are $n = 8$ intervals and the interval length is $\Delta x = 0.25$, so Simpson's Rule gives

$$\int_3^5 y\, dx \approx \frac{0.25}{3}[6.7 + 4(7.4) + 2(8.2) + 4(9.2) + 2(10.4)$$

$$+ 4(11.6) + 2(12.5) + 4(13.3) + 14.0]$$

$$\approx 20.7$$

●

EXERCISES 7.7

In Exercises 1–6 use (a) the Trapezoidal Rule and (b) Simpson's Rule to approximate the given integral with the given value of n. (Round your answers to six decimal places.) Then compare these approximations with the exact value of the integral computed using the Fundamental Theorem of Calculus.

1. $\int_0^1 x^3\, dx, \, n = 4$

2. $\int_0^2 e^x\, dx, \, n = 8$

3. $\int_1^4 \sqrt{x}\, dx, \, n = 6$

4. $\int_0^1 \frac{1}{1 + x^2}\, dx, \, n = 10$

5. $\int_2^4 \frac{x}{x^2 + 1}\, dx, \, n = 10$

6. $\int_{-1}^2 xe^x\, dx, \, n = 12$

In Exercises 7–16 use (a) the Trapezoidal Rule and (b) Simpson's Rule to approximate the given integral with the given value of n. (Round your answers to six decimal places.)

7. $\int_{-1}^1 \sqrt{1 + x^3}\, dx, \, n = 8$

8. $\int_0^1 \cos(x^2)\, dx, \, n = 4$

9. $\int_0^1 e^{-x^2}\, dx, \, n = 10$

10. $\int_0^{\pi/4} x \tan x\, dx, \, n = 6$

11. $\int_{\pi/2}^\pi \frac{\sin x}{x}\, dx, \, n = 6$

12. $\int_0^2 \frac{1}{\sqrt{1 + x^3}}\, dx, \, n = 10$

13. $\int_0^{1/2} \cos(e^x)\,dx$, $n = 8$

14. $\int_2^3 \dfrac{1}{\ln x}\,dx$, $n = 10$

15. $\int_0^1 x^5 e^x\,dx$, $n = 10$

16. $\int_0^1 \ln(1 + e^x)\,dx$, $n = 8$

17. Estimate the error involved in parts (a) and (b) of Exercise 9.

18. Estimate the error involved in parts (a) and (b) of Exercise 8.

19. Estimate the error involved in parts (a) and (b) of Exercise 15.

20. Estimate the error·involved in parts (a) and (b) of Exercise 16.

21. How large should n be so that the Trapezoidal Rule approximation for $\int_0^1 e^{-x^2}\,dx$ is accurate to within 0.00001?

22. How large should n be so that the Simpson's Rule approximation for $\int_0^1 e^{-x^2}\,dx$ is accurate to within 0.00001?

23. Estimate $\int_0^1 \sin(x^2)\,dx$ with an error less than 0.001.

24. Estimate $\int_0^1 x e^{x^3}\,dx$ with an error less than 0.01.

25. Use the Trapezoidal Rule and the following data to estimate the value of the integral $\int_1^{3.2} y\,dx$.

x	1.0	1.2	1.4	1.6	1.8	2.0	2.2	2.4	2.6	2.8	3.0	3.2
y	4.9	5.4	5.8	6.2	6.7	7.0	7.3	7.5	8.0	8.2	8.3	8.3

26. Use Simpson's Rule and the following data to estimate the value of the integral $\int_2^6 y\,dx$.

x	2.0	2.5	3.0	3.5	4.0	4.5	5.0	5.5	6.0
y	9.22	9.01	8.76	8.30	7.52	6.83	7.32	7.69	7.91

27. The speedometer reading (v) on a car was observed at 1-min intervals and recorded in the following chart. Use Simpson's Rule to estimate the distance traveled by the car.

t (min)	0	1	2	3	4	5	6	7	8	9	10
v (mi/h)	40	42	45	49	52	54	56	57	57	55	56

28. The widths (in meters) of a kidney-shaped swimming pool were measured at 2-m intervals as indicated in the diagram. Use Simpson's Rule to estimate the area of the pool.

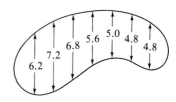

29. Show that if f is a polynomial of degree 3 or lower, then Simpson's Rule gives the exact value of $\int_a^b f(x)\,dx$.

SECTION 7.8

Improper Integrals

In defining a definite integral $\int_a^b f(x)\,dx$ we dealt with a function f defined on a finite interval $[a, b]$ and we observed that if the integral exists then f is a bounded function (see Section 5.3). In this section we extend the concept of a definite integral to the case where the interval is infinite and also to the case where f is unbounded (for instance, where f has an infinite discontinuity in $[a, b]$). In either case the integral is called an improper integral.

Type 1: Infinite Intervals

Consider the infinite region S that lies under the curve $y = 1/x^2$, above the x-axis, and to the right of the line $x = 1$. You might think that, since S is in-

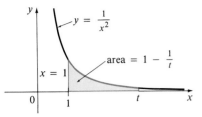

Figure 7.11

finite in extent, its area must be infinite, but let us take a closer look. The area of the part of S that lies to the left of the line $x = t$ (shaded in Figure 7.11) is

$$A(t) = \int_1^t \frac{1}{x^2}\, dx = -\frac{1}{x}\Big]_1^t = 1 - \frac{1}{t}$$

Notice that $A(t) < 1$ no matter how large t is chosen and

$$\lim_{t \to \infty} A(t) = \lim_{t \to \infty}\left(1 - \frac{1}{t}\right) = 1$$

The area of the shaded region approaches 1 as $t \to \infty$ (see Figure 7.12) so we say that the area of the infinite region S is equal to 1 and we write

$$\int_1^\infty \frac{1}{x^2}\, dx = \lim_{t \to \infty}\int_1^t \frac{1}{x^2}\, dx = 1$$

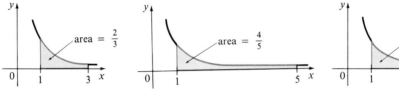

Figure 7.12

Using this example as a guide, we define the integral of f (not necessarily a positive function) as the limit of integrals over finite intervals.

Definition of an Improper Integral of Type 1 (7.33)

(a) If $\int_a^t f(x)\, dx$ exists for every number $t \geqslant a$, then

$$\int_a^\infty f(x)\, dx = \lim_{t \to \infty}\int_a^t f(x)\, dx$$

provided this limit exists (as a finite number).

(b) If $\int_t^b f(x)\, dx$ exists for every number $t \leqslant b$, then

$$\int_{-\infty}^b f(x)\, dx = \lim_{t \to -\infty}\int_t^b f(x)\, dx$$

provided this limit exists (as a finite number).

The improper integrals in (a) and (b) are called **convergent** if the limit exists and **divergent** if the limit does not exist.

(c) If both $\int_a^\infty f(x)\, dx$ and $\int_{-\infty}^a f(x)\, dx$ are convergent, then we define

$$\int_{-\infty}^\infty f(x)\, dx = \int_{-\infty}^a f(x)\, dx + \int_a^\infty f(x)\, dx$$

In part (c) any real number a can be used (see Exercise 60).

Any of the improper integrals in Definition 7.33 can be interpreted as an area provided that f is a positive function. For instance, in case (a) if $f(x) \geqslant 0$ and the integral $\int_a^\infty f(x)\,dx$ is convergent, then we define the area of the region $S = \{(x, y) \mid x \geqslant a, 0 \leqslant y \leqslant f(x)\}$ in Figure 7.13 to be

$$A(S) = \int_a^\infty f(x)\,dx$$

This is appropriate because $\int_a^\infty f(x)\,dx$ is the limit as $t \to \infty$ of the area under the graph of f from a to t.

Figure 7.13

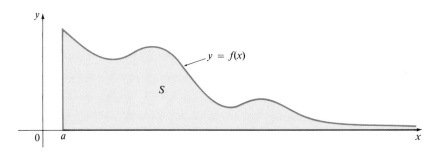

● Example 1   Determine whether the integral $\int_1^\infty (1/x)\,dx$ is convergent or divergent.

Solution   According to part (a) of Definition 7.33, we have

$$\int_1^\infty \frac{1}{x}\,dx = \lim_{t \to \infty} \int_1^t \frac{1}{x}\,dx = \lim_{t \to \infty} \ln|x| \Big]_1^t$$

$$= \lim_{t \to \infty} (\ln t - \ln 1) = \lim_{t \to \infty} \ln t = \infty$$

The limit does not exist as a finite number and so the improper integral $\int_1^\infty (1/x)\,dx$ is divergent.   ●

Let us compare the result of Example 1 with the example at the beginning of this section:

$$\int_1^\infty \frac{1}{x^2}\,dx \text{ converges} \qquad \int_1^\infty \frac{1}{x}\,dx \text{ diverges}$$

Geometrically, this says that although the curves $y = 1/x^2$ and $y = 1/x$ look very similar for $x > 0$, the region under $y = 1/x^2$ to the right of $x = 1$ [the shaded region in Figure 7.14(a)] has finite area whereas the corresponding region under $y = 1/x$ has infinite area. Note that both $1/x^2$ and $1/x$ approach 0 as $x \to \infty$ but $1/x^2$ approaches 0 faster than $1/x$.

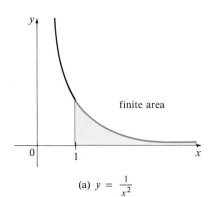

(a) $y = \dfrac{1}{x^2}$

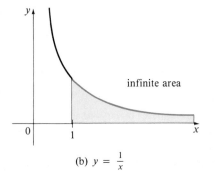

(b) $y = \dfrac{1}{x}$

Figure 7.14

● Example 2   Evaluate $\displaystyle\int_{-\infty}^0 xe^x\,dx$.

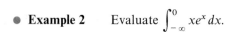

Solution   Using part (b) of Definition 7.33, we have

$$\int_{-\infty}^{0} xe^x \, dx = \lim_{t \to -\infty} \int_{t}^{0} xe^x \, dx$$

We integrate by parts with $u = x$, $dv = e^x \, dx$ so that $du = dx$, $v = e^x$:

$$\int_{t}^{0} xe^x \, dx = xe^x \Big]_{t}^{0} - \int_{t}^{0} e^x \, dx$$

$$= -te^t - 1 + e^t$$

We know that $e^t \to 0$ as $t \to -\infty$, and by l'Hospital's Rule we have

$$\lim_{t \to -\infty} te^t = \lim_{t \to -\infty} \frac{t}{e^{-t}} = \lim_{t \to -\infty} \frac{1}{-e^{-t}}$$

$$= \lim_{t \to -\infty} (-e^t) = 0$$

Therefore $$\int_{-\infty}^{0} xe^x \, dx = \lim_{t \to -\infty} (-te^t - 1 + e^t)$$

$$= -0 - 1 + 0 = -1$$

● **Example 3**   Evaluate $\int_{-\infty}^{\infty} \dfrac{1}{1 + x^2} \, dx$.

Solution   It is convenient to choose $a = 0$ in Definition 7.33(c):

$$\int_{-\infty}^{\infty} \frac{1}{1 + x^2} \, dx = \int_{-\infty}^{0} \frac{1}{1 + x^2} \, dx + \int_{0}^{\infty} \frac{1}{1 + x^2} \, dx$$

We must now evaluate the integrals on the right side separately:

$$\int_{0}^{\infty} \frac{1}{1 + x^2} \, dx = \lim_{t \to \infty} \int_{0}^{t} \frac{dx}{1 + x^2} = \lim_{t \to \infty} \tan^{-1} x \Big]_{0}^{t}$$

$$= \lim_{t \to \infty} (\tan^{-1} t - \tan^{-1} 0) = \lim_{t \to \infty} \tan^{-1} t$$

$$= \frac{\pi}{2}$$

$$\int_{-\infty}^{0} \frac{1}{1 + x^2} \, dx = \lim_{t \to -\infty} \int_{t}^{0} \frac{dx}{1 + x^2} = \lim_{t \to -\infty} \tan^{-1} x \Big]_{t}^{0}$$

$$= \lim_{t \to -\infty} (\tan^{-1} 0 - \tan^{-1} t) = 0 - \left(-\frac{\pi}{2}\right) = \frac{\pi}{2}$$

Since both of these integrals are convergent, the given integral is convergent and

$$\int_{-\infty}^{\infty} \frac{1}{1 + x^2} \, dx = \frac{\pi}{2} + \frac{\pi}{2} = \pi$$

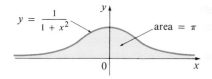

$y = \dfrac{1}{1 + x^2}$ area $= \pi$

Figure 7.15

Since $1/(1 + x^2) > 0$, the given improper integral can be interpreted as the area of the infinite region that lies under the curve $y = 1/(1 + x^2)$ and above the x-axis (see Figure 7.15).

● **Example 4**   For what values of p is the integral

$$\int_1^\infty \frac{1}{x^p}\, dx$$

convergent?

Solution   We know from Example 1 that if $p = 1$ then the integral is divergent, so let us assume that $p \neq 1$. Then

$$\int_1^\infty \frac{1}{x^p}\, dx = \lim_{t \to \infty} \int_1^t \frac{1}{x^p}\, dx$$

$$= \lim_{t \to \infty} \frac{x^{-p+1}}{-p+1}\Bigg]_{x=1}^{x=t}$$

$$= \lim_{t \to \infty} \frac{1}{1-p}\left[\frac{1}{t^{p-1}} - 1\right]$$

If $p > 1$, then $p - 1 > 0$ and so $1/t^{p-1} \to 0$ as $t \to \infty$. Therefore

$$\int_1^\infty \frac{1}{x^p}\, dx = \frac{1}{p-1} \qquad \text{if } p > 1$$

But if $p < 1$, then $p - 1 < 0$ and so

$$\frac{1}{t^{p-1}} = t^{1-p} \to \infty \qquad \text{as } t \to \infty$$

and the integral diverges.   ●

We summarize the result of Example 4 for future reference:

(7.34)

$$\int_1^\infty \frac{1}{x^p}\, dx \text{ is convergent if } p > 1 \text{ and divergent if } p \leqslant 1.$$

Type 2: Unbounded Integrands

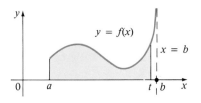

Figure 7.16

Now we suppose that f is a positive function defined on a finite interval $[a, b]$ but has a vertical asymptote at b. Let S be the unbounded region under the graph of f and above the x-axis between a and b. (For Type 1 integrals, the regions extended indefinitely in a horizontal direction. Here they are infinite in a vertical direction.) The area of the part of S between a and t (the shaded region in Figure 7.16) is

$$A(t) = \int_a^t f(x)\, dx$$

If it happens that $A(t)$ approaches a definite number A as $t \to b^-$, then we say that the area of the region S is A and we write

$$\int_a^b f(x)\,dx = \lim_{t \to b^-} \int_a^t f(x)\,dx$$

We use this equation to define an improper integral of Type 2 even when f is not a positive function.

Definition of an Improper Integral of Type 2 (7.35)

(a) If $\int_a^t f(x)\,dx$ exists whenever $a \leqslant t < b$ and f has a vertical asymptote at b, then

$$\int_a^b f(x)\,dx = \lim_{t \to b^-} \int_a^t f(x)\,dx$$

if this limit exists (as a finite number).

(b) If $\int_t^b f(x)\,dx$ exists whenever $a < t \leqslant b$ and f has a vertical asymptote at a, then

$$\int_a^b f(x)\,dx = \lim_{t \to a^+} \int_t^b f(x)\,dx$$

if this limit exists (as a finite number).

The improper integrals in (a) and (b) are called **convergent** if the limit exists and **divergent** if the limit does not exist.

(c) If f has a vertical asymptote at c, where $a < c < b$, and both $\int_a^c f(x)\,dx$ and $\int_c^b f(x)\,dx$ are convergent, then we define

$$\int_a^b f(x)\,dx = \int_a^c f(x)\,dx + \int_c^b f(x)\,dx$$

Parts (b) and (c) of Definition 7.35 are illustrated in Figures 7.17 and 7.18 for the case where $f(x) \geqslant 0$.

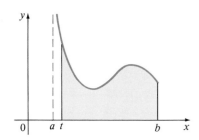

Figure 7.17

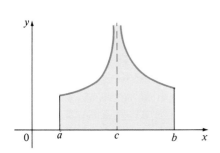

Figure 7.18

● **Example 5**   Find $\displaystyle\int_2^5 \frac{1}{\sqrt{x-2}}\,dx$.

Solution   We note first of all that the given integral is improper because $f(x) = 1/\sqrt{x-2}$ has the vertical asymptote $x = 2$. Since the infinite discontinuity occurs at the left endpoint of $[2, 5]$, we use part (b) of Definition 7.35:

$$\int_2^5 \frac{dx}{\sqrt{x-2}} = \lim_{t \to 2^+} \int_t^5 \frac{dx}{\sqrt{x-2}}$$

$$= \lim_{t \to 2^+} 2\sqrt{x-2}\,\Big]_t^5$$

$$= \lim_{t \to 2^+} 2(\sqrt{3} - \sqrt{t-2})$$

$$= 2\sqrt{3}$$

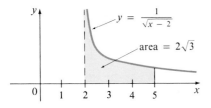

area $= 2\sqrt{3}$

$y = \dfrac{1}{\sqrt{x-2}}$

Figure 7.19

Thus the given improper integral is convergent and since the integrand is positive we can interpret the value of the integral as the area of the shaded region in Figure 7.19.   ●

● **Example 6**   Determine whether $\int_0^{\pi/2} \sec x\, dx$ converges or diverges.

Solution   Note that the given integral is improper because $\lim_{x \to (\pi/2)^-} \sec x = \infty$. Using part (a) of Definition 7.35 we have

$$\int_0^{\pi/2} \sec x\, dx = \lim_{t \to \pi/2^-} \int_0^t \sec x\, dx$$

$$= \lim_{t \to \pi/2^-} \ln|\sec x + \tan x|\Big]_0^t$$

$$= \lim_{t \to \pi/2^-} \big[\ln(\sec t + \tan t) - \ln 1\big]$$

$$= \infty$$

because $\sec t \to \infty$ and $\tan t \to \infty$ as $t \to \pi/2^-$. Thus the given improper integral is divergent.   ●

● **Example 7**   Evaluate $\int_0^3 \dfrac{dx}{x-1}$ if possible.

Solution   Observe that the line $x = 1$ is a vertical asymptote of the integrand. Since it occurs in the middle of the interval $[0,3]$ we must use part (c) of Definition 7.35 with $c = 1$:

$$\int_0^3 \frac{dx}{x-1} = \int_0^1 \frac{dx}{x-1} + \int_1^3 \frac{dx}{x-1}$$

where

$$\int_0^1 \frac{dx}{x-1} = \lim_{t \to 1^-} \int_0^t \frac{dx}{x-1} = \lim_{t \to 1^-} \ln|x-1|\Big]_0^t$$

$$= \lim_{t \to 1^-} \big(\ln|t-1| - \ln|-1|\big)$$

$$= \lim_{t \to 1^-} \ln(1-t) = -\infty$$

because $1 - t \to 0^+$ as $t \to 1^-$. Thus $\int_0^1 dx/(x-1)$ is divergent. This implies that $\int_0^3 dx/(x-1)$ is divergent. [We do not need to evaluate $\int_1^3 dx/(x-1)$.]   ●

⊘   *Warning:*   If we had not noticed the asymptote $x = 1$ in Example 7 and had instead confused the integral with an ordinary integral, then we might have made the following erroneous calculation:

$$\int_0^3 \frac{dx}{x-1} = \ln|x-1|\Big]_0^3 = \ln 2 - \ln 1 = \ln 2$$

This is wrong because the integral is improper and must be calculated in terms of limits. (Compare with Example 11 in Section 5.5.)

From now on, whenever you meet the symbol $\int_a^b f(x)\,dx$ you must decide, by looking at the function f on $[a, b]$, whether it is an ordinary definite integral or an improper integral.

● **Example 8**   Evaluate $\int_0^1 \ln x\,dx$.

Solution   We know that the function $f(x) = \ln x$ has a vertical asymptote at 0 since $\lim_{x \to 0^+} \ln x = -\infty$. Thus the given integral is improper and we have

$$\int_0^1 \ln x\,dx = \lim_{t \to 0^+} \int_t^1 \ln x\,dx$$

Now we integrate by parts with $u = \ln x$, $dv = dx$, $du = dx/x$, and $v = x$:

$$\int_t^1 \ln x\,dx = x \ln x\Big]_t^1 - \int_t^1 dx$$
$$= 1 \ln 1 - t \ln t - (1 - t)$$
$$= -t \ln t - 1 + t$$

To find the limit of the first term we use l'Hospital's Rule:

$$\lim_{t \to 0^+} t \ln t = \lim_{t \to 0^+} \frac{\ln t}{1/t}$$
$$= \lim_{t \to 0^+} \frac{1/t}{-1/t^2}$$
$$= \lim_{t \to 0^+} (-t) = 0$$

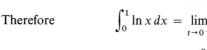

Therefore

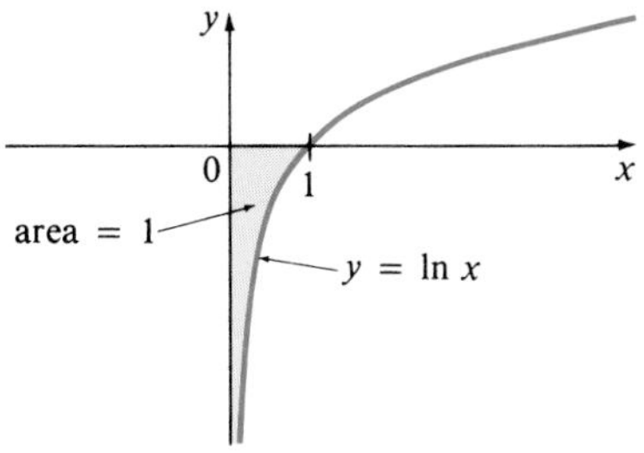

$$\int_0^1 \ln x\,dx = \lim_{t \to 0^+} (-t \ln t - 1 + t)$$
$$= -0 - 1 + 0 = -1$$

Figure 7.20 shows the geometric interpretation of this result. The area of the shaded region above $y = \ln x$ and below the x-axis is 1.   ●

Figure 7.20

area = 1

$y = \ln x$

EXERCISES 7.8

Determine whether the integrals in Exercises 1–46 are convergent or divergent. Evaluate those that are convergent.

1. $\int_2^\infty \dfrac{1}{\sqrt{x+3}}\,dx$

2. $\int_2^\infty \dfrac{1}{(x+3)^{3/2}}\,dx$

3. $\int_{-\infty}^1 \dfrac{1}{(2x-3)^2}\,dx$

4. $\int_{-\infty}^{-1} \dfrac{1}{\sqrt[3]{x-1}}\,dx$

5. $\int_{-\infty}^\infty x\,dx$

6. $\int_{-\infty}^\infty (2x^2 - x + 3)\,dx$

7. $\int_0^\infty e^{-x}\,dx$

8. $\int_{-\infty}^0 e^{3x}\,dx$

9. $\int_{-\infty}^\infty xe^{-x^2}\,dx$

10. $\int_{-\infty}^\infty x^2 e^{-x^3}\,dx$

11. $\int_0^\infty \dfrac{1}{(x+2)(x+3)}\,dx$

12. $\int_0^\infty \dfrac{x}{(x+2)(x+3)}\,dx$

13. $\int_0^\infty \cos x\,dx$

14. $\int_1^\infty \sin \pi x\,dx$

15. $\int_0^\infty \dfrac{5}{2x+3}\,dx$

16. $\int_{-\infty}^3 \dfrac{1}{x^2+9}\,dx$

17. $\int_{-\infty}^1 xe^{2x}\,dx$

18. $\int_0^\infty xe^{-x}\,dx$

19. $\int_1^\infty \dfrac{\ln x}{x}\,dx$

20. $\int_e^\infty \dfrac{1}{x(\ln x)^2}\,dx$

21. $\int_{-\infty}^\infty \dfrac{x}{1+x^2}\,dx$

22. $\int_{-\infty}^\infty e^{-|x|}\,dx$

23. $\int_1^\infty \dfrac{\ln x}{x^2}\,dx$

24. $\int_{-\infty}^\infty \dfrac{dx}{x^2+4x+6}$

25. $\int_0^\infty \dfrac{1}{2^x}\,dx$

26. $\int_1^\infty \dfrac{\ln x}{x^3}\,dx$

27. $\int_0^3 \dfrac{1}{\sqrt{x}}\,dx$

28. $\int_0^3 \dfrac{1}{x\sqrt{x}}\,dx$

29. $\int_{-1}^0 \dfrac{1}{x^2}\,dx$

30. $\int_1^9 \dfrac{1}{\sqrt[3]{x-9}}\,dx$

31. $\int_{-2}^3 \dfrac{1}{x^4}\,dx$

32. $\int_0^2 \dfrac{1}{4x-5}\,dx$

33. $\int_4^5 \dfrac{1}{(5-x)^{2/5}}\,dx$

34. $\int_{\pi/4}^{\pi/2} \sec^2 x\,dx$

35. $\int_{\pi/4}^{\pi/2} \tan^2 x\,dx$

36. $\int_0^{\pi/4} \dfrac{\cos x}{\sqrt{\sin x}}\,dx$

37. $\int_0^2 \dfrac{x}{\sqrt{4-x^2}}\,dx$

38. $\int_0^9 \dfrac{dx}{(x+9)\sqrt{x}}$

39. $\int_0^\pi \sec x\,dx$

40. $\int_0^4 \dfrac{1}{x^2+x-6}\,dx$

41. $\int_{-2}^2 \dfrac{1}{x^2-1}\,dx$

42. $\int_{\pi/4}^{3\pi/4} \tan x\,dx$

43. $\int_1^e \dfrac{1}{x\sqrt[4]{\ln x}}\,dx$

44. $\int_0^2 \dfrac{x-3}{2x-3}\,dx$

45. $\int_0^1 x\ln x\,dx$

46. $\int_0^1 \dfrac{\ln x}{\sqrt{x}}\,dx$

In Exercises 47–52 sketch the given region and find its area (if the area is finite).

47. $S = \{(x,y) \mid x \leqslant 1, 0 \leqslant y \leqslant e^x\}$

48. $S = \{(x,y) \mid x \geqslant -2, 0 \leqslant y \leqslant e^{-x/2}\}$

49. $S = \left\{(x,y) \Big| 0 \leqslant y \leqslant \dfrac{1}{x^2-2x+5}\right\}$

50. $S = \left\{(x,y) \Big| x \geqslant 0, 0 \leqslant y \leqslant \dfrac{1}{\sqrt{x+1}}\right\}$

51. $S = \left\{(x,y) \Big| 0 \leqslant x < \dfrac{\pi}{2}, 0 \leqslant y \leqslant \tan x\right\}$

52. $S = \left\{(x,y) \Big| 3 < x \leqslant 7, 0 \leqslant y \leqslant \dfrac{1}{\sqrt{x-3}}\right\}$

53. The integral

$$\int_0^\infty \dfrac{1}{\sqrt{x}(1+x)}\,dx$$

is improper for two reasons: the interval $[0, \infty)$ is infinite and the integrand is unbounded near 0. Evaluate it by expressing it as a sum of improper integrals of Types 2 and 1 as follows:

$$\int_0^\infty \dfrac{1}{\sqrt{x}(1+x)}\,dx = \int_0^1 \dfrac{1}{\sqrt{x}(1+x)}\,dx + \int_1^\infty \dfrac{1}{\sqrt{x}(1+x)}\,dx$$

54. Evaluate

$$\int_2^\infty \dfrac{1}{x\sqrt{x^2-4}}\,dx$$

by the same method as in Exercise 53.

In Exercises 55–57 find the values of p for which the given integral converges and evaluate the integral for those values of p.

55. $\int_0^1 \dfrac{1}{x^p}\,dx$

56. $\int_e^\infty \dfrac{1}{x(\ln x)^p}\,dx$

57. $\int_0^1 x^p \ln x\,dx$

58. (a) Evaluate the integral $\int_0^\infty x^n e^{-x}\,dx$ for $n = 0, 1, 2,$ and 3.
   (b) Guess the value of $\int_0^\infty x^n e^{-x}\,dx$ when n is an arbitrary positive integer.
   (c) Prove your guess using mathematical induction.

59. (a) Show that $\int_{-\infty}^\infty x\,dx$ is divergent.
   (b) Show that

$$\lim_{t \to \infty} \int_{-t}^t x\,dx = 0$$

This shows that we cannot define

$$\int_{-\infty}^{\infty} f(x)\, dx = \lim_{t \to \infty} \int_{-t}^{t} f(x)\, dx$$

60. If $\int_{-\infty}^{\infty} f(x)\, dx$ is convergent and a and b are real numbers, show that

$$\int_{-\infty}^{a} f(x)\, dx + \int_{a}^{\infty} f(x)\, dx = \int_{-\infty}^{b} f(x)\, dx + \int_{b}^{\infty} f(x)\, dx$$

In Exercises 61–64 state whether the given integral is convergent or divergent by using the following Comparison

Theorem: Suppose f and g are continuous functions with $f(x) \geq g(x) \geq 0$ for $x \geq a$. (i) If $\int_{a}^{\infty} f(x)\, dx$ is convergent, then $\int_{a}^{\infty} g(x)\, dx$ is convergent. (ii) If $\int_{a}^{\infty} g(x)\, dx$ is divergent, then $\int_{a}^{\infty} f(x)\, dx$ is divergent.

61. $\displaystyle \int_{1}^{\infty} \frac{\sin^2 x}{x^2}\, dx$

62. $\displaystyle \int_{1}^{\infty} \frac{\sqrt{1 + \sqrt{x}}}{\sqrt{x}}\, dx$

63. $\displaystyle \int_{-\infty}^{\infty} e^{-x^2}\, dx$

64. $\displaystyle \int_{1}^{\infty} \frac{1}{\sqrt{x^3 + 1}}\, dx$

SECTION 7.9

Using Tables of Integrals

A scientist or engineer often uses tables of integrals when integrating complicated functions. A relatively brief table of 112 integrals is printed on the inside covers of this book. More extensive tables are available in the *CRC Mathematical Tables* (463 entries) or in Gradshteyn and Ryzhik's *Table of Integrals, Series and Products* (New York: Academic Press, 1979), which contains hundreds of pages of integrals. It should be remembered, however, that integrals rarely occur in exactly the form listed in a table. Usually one of the methods of this chapter, such as substitution or integration by parts, is required to transform a given integral into one of the forms in the table.

● **Example 1**   Use the Table of Integrals to find $\displaystyle \int \frac{x^2}{\sqrt{5 - 4x^2}}\, dx$.

Solution   If we look at the section of the table entitled *forms involving* $\sqrt{a^2 - x^2}$, we see that the closest entry is number 34:

$$\int \frac{x^2}{\sqrt{a^2 - x^2}}\, dx = -\frac{x}{2}\sqrt{a^2 - x^2} + \frac{a^2}{2} \sin^{-1}\left(\frac{x}{a}\right) + C$$

This is not exactly what we have, so we make the substitution $u = 2x$:

$$\int \frac{x^2}{\sqrt{5 - 4x^2}} = \int \frac{(u/2)^2}{\sqrt{5 - u^2}} \frac{du}{2} = \frac{1}{8} \int \frac{u^2}{\sqrt{5 - u^2}}\, du$$

Then we use Formula 34 with $a^2 = 5$:

$$\int \frac{x^2}{\sqrt{5 - 4x^2}}\, dx = \frac{1}{8} \int \frac{u^2}{\sqrt{5 - u^2}}\, du = \frac{1}{8}\left[-\frac{u}{2}\sqrt{5 - u^2} + \frac{5}{2} \sin^{-1} \frac{u}{\sqrt{5}} \right] + C$$

$$= -\frac{x}{8}\sqrt{5 - 4x^2} + \frac{5}{16} \sin^{-1}\left(\frac{2x}{\sqrt{5}}\right) + C \qquad ●$$

● **Example 2**   Use the Table of Integrals to find $\int x^3 \sin x \, dx$.

Solution   We look in the section called *trigonometric forms* and use the reduction formula in entry 84 with $n = 3$:

$$\int x^3 \sin x \, dx = -x^3 \cos x + 3 \int x^2 \cos x \, dx$$

Then we use entries 85 and 82:

$$\int x^2 \cos x \, dx = x^2 \sin x - 2 \int x \sin x \, dx$$

$$= x^2 \sin x - 2(\sin x - x \cos x) + C$$

Combining these calculations we get

$$\int x^3 \sin x \, dx = -x^3 \cos x + 3x^2 \sin x + 6x \cos x - 6 \sin x + C \quad ●$$

● **Example 3**   Use the Table of Integrals to find $\int x\sqrt{x^2 + 2x + 4} \, dx$.

Solution   Since there are forms involving $\sqrt{a^2 + x^2}$, $\sqrt{a^2 - x^2}$, and $\sqrt{x^2 - a^2}$, but not $\sqrt{ax^2 + bx + c}$, we first complete the square:

$$x^2 + 2x + 4 = (x + 1)^2 + 3$$

Therefore we make the substitution $u = x + 1$:

$$\int x\sqrt{x^2 + 2x + 4} \, dx = \int (u - 1)\sqrt{u^2 + 3} \, du$$

$$= \int u\sqrt{u^2 + 3} \, du - \int \sqrt{u^2 + 3} \, du$$

The first integral is evaluated using the substitution $t = u^2 + 3$:

$$\int u\sqrt{u^2 + 3} \, du = \frac{1}{2} \int \sqrt{t} \, dt = \frac{1}{2} \cdot \frac{2}{3} t^{3/2} = \frac{1}{3}(u^2 + 3)^{3/2}$$

For the second integral we use Formula 21 with $a = \sqrt{3}$:

$$\int \sqrt{u^2 + 3} \, du = \frac{u}{2}\sqrt{u^2 + 3} + \frac{3}{2} \ln|u + \sqrt{u^2 + 3}|$$

Thus

$$\int x\sqrt{x^2 + 2x + 4} \, dx = \frac{1}{3}(x^2 + 2x + 4)^{3/2} - \frac{x + 1}{2}\sqrt{x^2 + 2x + 4}$$

$$- \frac{3}{2} \ln|x + 1 + \sqrt{x^2 + 2x + 4}| + C \quad ●$$

EXERCISES 7.9

Use the Table of Integrals in the inside covers to evaluate the following integrals.

1. $\int e^{-3x} \cos 4x\, dx$

2. $\int \csc^3\left(\frac{x}{2}\right) dx$

3. $\int \frac{\sqrt{9x^2-1}}{x^2}\, dx$

4. $\int \frac{\sqrt{4-3x^2}}{x}\, dx$

5. $\int x^2 e^{3x}\, dx$

6. $\int \frac{\sin x \cos x}{\sqrt{1+\sin x}}\, dx$

7. $\int x \sin^{-1}(x^2)\, dx$

8. $\int x^3 \sin^{-1}(x^2)\, dx$

9. $\int e^x \operatorname{sech}(e^x)\, dx$

10. $\int x^2 \cos 3x\, dx$

11. $\int \sqrt{5-4x-x^2}\, dx$

12. $\int \frac{x^5}{x^2+\sqrt{2}}\, dx$

13. $\int \sec^5 x\, dx$

14. $\int \sin^6 2x\, dx$

15. $\int \sin^2 x \cos x \ln(\sin x)\, dx$

16. $\int \frac{dx}{e^x(1+2e^x)}$

17. $\int \sqrt{2+3\cos x} \tan x\, dx$

18. $\int \frac{x}{\sqrt{x^2-4x}}\, dx$

19. $\int_0^{\pi/2} \cos^5 x\, dx$

20. $\int_0^{\infty} x^4 e^{-x}\, dx$

REVIEW OF CHAPTER 7

Define, state, or discuss the following.

1. Integration by parts

2. Trigonometric integrals

3. Trigonometric substitution

4. Partial fractions

5. Rationalizing substitutions

6. Strategy for integration

7. The Trapezoidal Rule

8. Simpson's Rule

9. Improper integrals

REVIEW EXERCISES FOR CHAPTER 7

(*Note:* Additional practice in techniques of integration is provided in Exercises 7.6.)

Evaluate the integrals in Exercises 1–32.

1. $\int \frac{x-1}{x+1}\, dx$

2. $\int \frac{\sin^3 x}{\cos x}\, dx$

3. $\int \frac{(\arctan x)^5}{1+x^2}\, dx$

4. $\int x^2 e^{-3x}\, dx$

5. $\int \frac{\cos x}{e^{\sin x}}\, dx$

6. $\int \frac{x^2+1}{x-1}\, dx$

7. $\int x^4 \ln x\, dx$

8. $\int \frac{\sec^2 \theta}{1-\tan \theta}\, d\theta$

9. $\int x \sin(x^2)\, dx$

10. $\int x \sin^2 x\, dx$

11. $\int \frac{dx}{2x^2-5x+2}$

12. $\int \frac{dt}{\sin^2 t + \cos 2t}$

13. $\int \tan^7 x \sec^3 x\, dx$

14. $\int \frac{dx}{\sqrt{8+2x-x^2}}$

15. $\int \frac{dx}{\sqrt{1+2x}+3}$

16. $\int x(\tan^{-1} x)^2\, dx$

17. $\int \frac{e^{\sqrt{x}}}{\sqrt{x}}\, dx$

18. $\int \frac{dx}{x^3-2x^2+x}$

19. $\int \ln(x^2+2x+2)\, dx$

20. $\int \frac{dx}{x^2\sqrt{1+x^2}}$

21. $\int \frac{dx}{x^3+x}$

22. $\int \frac{dx}{1+e^x}$

23. $\int \cot^2 x\, dx$

24. $\int \frac{dx}{5-3\cos x}$

25. $\int \frac{dx}{(x^2-1)^{3/2}}$

26. $\int \sqrt{1+\cos x}\, dx$

27. $\int \frac{2x^2+3x+11}{x^3+x^2+3x-5}\, dx$

28. $\int \frac{x^3}{(x+1)^{10}}\, dx$

29. $\int e^{x+e^x} dx$

30. $\int (\arcsin x)^2 dx$

31. $\int \dfrac{\ln(\ln x)}{x} dx$

32. $\int \dfrac{\sin x}{1 + \sin x} dx$

In Exercises 33–50 evaluate the integral or show that it is divergent.

33. $\int_0^{\pi/2} \cos^3 x \sin 2x \, dx$

34. $\int_{-1}^1 \dfrac{1}{2x + 1} dx$

35. $\int_0^3 \dfrac{dx}{x^2 - x - 2}$

36. $\int_0^{\pi/4} \cos^5(2\theta) \, d\theta$

37. $\int_0^1 \dfrac{t^2 - 1}{t^2 + 1} dt$

38. $\int_2^6 \dfrac{y}{\sqrt{y - 2}} dy$

39. $\int_0^\infty \dfrac{1}{(x + 2)^4} dx$

40. $\int_1^4 \dfrac{e^{1/x}}{x^2} dx$

41. $\int_1^e \dfrac{dx}{x\sqrt{\ln x}}$

42. $\int_0^\infty \dfrac{dx}{(x + 1)^2(x + 2)}$

43. $\int_1^4 \dfrac{\sqrt{x}}{\sqrt{x} + 2} dx$

44. $\int_{-3}^3 x\sqrt{1 + x^4} \, dx$

45. $\int_{-\infty}^\infty \dfrac{dx}{4x^2 + 4x + 5}$

46. $\int_1^e \dfrac{dx}{x[1 + (\ln x)^2]}$

47. $\int_1^2 \dfrac{\sqrt{x^2 - 1}}{x} dx$

48. $\int_{-1}^1 \dfrac{x + 1}{\sqrt[3]{x^4}} dx$

49. $\int_0^\infty e^{ax} \cos bx \, dx$

50. $\int_1^\infty \dfrac{\tan^{-1} x}{x^2} dx$

In Exercises 51–54 use the Table of Integrals in the inside covers to evaluate the integral.

51. $\int e^x \sqrt{1 - e^{2x}} \, dx$

52. $\int \tan^5 x \, dx$

53. $\int \sqrt{x^2 + x + 1} \, dx$

54. $\int \dfrac{\cot x}{\sqrt{1 + 2\sin x}} dx$

In Exercises 55 and 56 use (a) the Trapezoidal Rule and (b) Simpson's Rule with $n = 10$ to approximate the given integral. Round your answers to six decimal places.

55. $\int_0^1 \sqrt{1 + x^4} \, dx$

56. $\int_0^{\pi/2} \sqrt{\sin x} \, dx$

57. Estimate the error involved in Exercise 55(a).

58. Estimate $\int_2^4 dx/(\ln x)$ correct to two decimal places.

59. Use Simpson's Rule with $n = 6$ to estimate the area under the curve $y = e^x/x$ from $x = 1$ to $x = 4$.

60. Find the area of the region bounded by the hyperbola $y^2 - x^2 = 1$ and the line $y = 3$.

61. Find the area bounded by the curves $y = \cos x$ and $y = \cos^2 x$ between $x = 0$ and $x = \pi$.

62. Find the area of the region bounded by the curves $y = 1/(2 + \sqrt{x})$, $y = 1/(2 - \sqrt{x})$, and $x = 1$.

63. Is it possible to find a number n such that $\int_0^\infty x^n \, dx$ is convergent?

64. If n is a positive integer, prove that
$$\int_0^1 (\ln x)^n \, dx = (-1)^n n!$$

65. If f' is continuous on $[0, \infty)$ and $\lim_{x \to \infty} f(x) = 0$, show that
$$\int_0^\infty f'(x) \, dx = -f(0)$$

8 Further Applications of Integration

The calculus is the greatest aid we have to the appreciation of physical truth in the broadest sense of the word.

W. F. Osgood

Mathematics takes us still further from what is human, into the region of absolute necessity, to which not only the actual world, but every possible world, must conform.

Bertrand Russell

In applying the definite integral to compute the various quantities Q of this chapter (volume, arc length, surface area, work, moments) we use the following general method that is similar to the one we used to find areas. We break up the quantity Q into a large number of small parts. Each small part is then approximated by a quantity of the form $f(x_i^*)\,\Delta x_i$ and so Q is approximated by a Riemann sum. The final step is to take the limit and express Q as an integral that we can evaluate by the methods of Chapter 7.

Volume

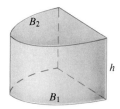

(a) Cylinder
$V = Ah$

(b) Circular cylinder
$V = \pi r^2 h$

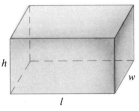

(c) Rectangular box
$V = lwh$

Figure 8.1

In trying to find the volume of a solid we face the same type of problem as in finding areas. We have an intuitive idea of what volume means but we must make this idea precise by using calculus to give an exact definition of volume.

We start with a simple type of solid called a **cylinder** (or, more precisely, a *right cylinder*). As illustrated in Figure 8.1(a), a cylinder is bounded by a plane region B_1, called the **base,** and a congruent region B_2 in a parallel plane. The cylinder consists of all points on line segments perpendicular to the base that join B_1 to B_2. If the area of the base is A and the height of the cylinder (the distance from B_1 to B_2) is h, then the volume V of the cylinder is defined as

$$V = Ah$$

In particular, if the base is a circle with radius r, then the cylinder is a circular cylinder with volume $V = \pi r^2 h$ [see Figure 8.1(b)], and if the base is a rectangle with length l and width w, then the cylinder is a rectangular box (also called a rectangular parallelepiped) with volume $V = lwh$ [see Figure 8.1(c)].

Now let S be any solid. The intersection of S with a plane is a plane region that is called a **cross-section** of S. Suppose that the area of the cross-section of S in a plane P_x perpendicular to the x-axis and passing through the point x is $A(x)$, where $a \leqslant x \leqslant b$. (See Figure 8.2. Think of slicing S with a knife through x and computing the area of this slice.) The cross-sectional area $A(x)$ will vary as x increases from a to b.

Take a partition P of the interval $[a, b]$ by points x_i with $a = x_0 < x_1 < \cdots < x_n = b$. The planes P_{x_i} will slice S into smaller "slabs." (Think of slicing a loaf of bread.) If we choose numbers x_i^* in $[x_{i-1}, x_i]$, we can approximate the ith slab S_i (the part of S that lies between the

Figure 8.2

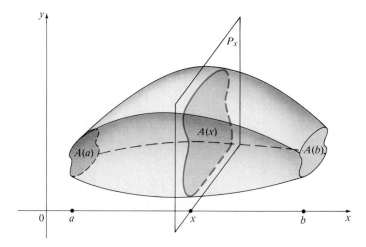

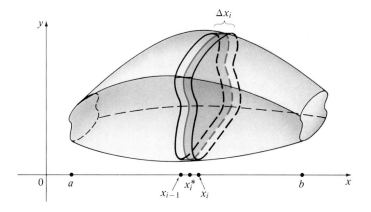

Figure 8.3

planes $P_{x_{i-1}}$ and P_{x_i}) by a cylinder with base area $A(x_i^*)$ and height $\Delta x_i = x_i - x_{i-1}$ (see Figure 8.3).

The volume of this cylinder is $A(x_i^*)\Delta x_i$ so an approximation to our intuitive conception of the volume of the ith slab S_i is

$$V(S_i) \approx A(x_i^*)\Delta x_i$$

Adding the volumes of these slabs, we get an approximation to the total volume (that is, what we think of intuitively as the volume):

$$V \approx \sum_{i=1}^{n} A(x_i^*)\Delta x_i$$

This approximation appears to become better and better as $\|P\| \to 0$. (Think of the slices as becoming thinner and thinner.) Therefore we *define* the volume as the limit of this sum as $\|P\| \to 0$. But we recognize the limit of a Riemann sum as a definite integral and so we have the following definition:

Definition of Volume (8.1)

Let S be a solid that lies between the planes P_a and P_b. If the cross-sectional area of S in the plane P_x is $A(x)$, where A is an integrable function, then the **volume** of S is

$$V = \lim_{\|P\| \to 0} \sum_{i=1}^{n} A(x_i^*)\Delta x_i = \int_a^b A(x)\, dx$$

In using the volume formula $V = \int_a^b A(x)\, dx$ it is important to remember that $A(x)$ is the area of a moving cross-section obtained by slicing through x perpendicular to the x-axis.

● **Example 1**   Find the volume of a pyramid whose base is a square with side L and whose height is h.

Solution   In order to use the formula in Definition 8.1 we place the origin O at the vertex of the pyramid and the x-axis along its central axis as in

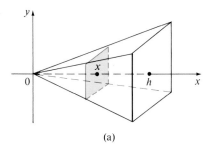

(a)

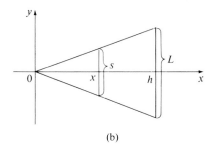

(b)

Figure 8.4

Figure 8.4(a). Any plane P_x that passes through x and is perpendicular to the x-axis will intersect the pyramid in a square with side of length s, say. We can express s in terms of x by observing from the similar triangles in Figure 8.4(b) that

$$\frac{s}{L} = \frac{x}{h}$$

and so $s = Lx/h$. Thus the cross-sectional area is

$$A(x) = s^2 = \frac{L^2}{h^2}x^2$$

Taking $a = 0$ and $b = h$ in Definition 8.1, we see that the volume of the pyramid is

$$V = \int_0^h A(x)\,dx = \int_0^h \frac{L^2}{h^2}x^2\,dx$$

$$= \frac{L^2}{h^2}\frac{x^3}{3}\bigg]_0^h = \frac{L^2 h}{3} \qquad \bullet$$

Note: Usually we do not specify the units of measurement unless some are given in the problem. For instance, in Example 1 if we were given that $h = 150\,\text{m}$ and $L = 250\,\text{m}$, then we would give the volume as $V = 3{,}125{,}000\,\text{m}^3$.

● **Example 2**   Show that the volume of a sphere of radius r is

$$V = \tfrac{4}{3}\pi r^3$$

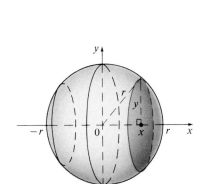

Figure 8.5

Solution   If we place the sphere so that its center is at the origin (see Figure 8.5), then the plane P_x intersects the sphere in a circle whose radius (from the Pythagorean Theorem) is $y = \sqrt{r^2 - x^2}$. So the cross-sectional area is

$$A(x) = \pi y^2 = \pi(r^2 - x^2)$$

Using the volume formula with $a = -r$ and $b = r$, we have

$$V = \int_{-r}^{r} A(x)\,dx = \int_{-r}^{r} \pi(r^2 - x^2)\,dx$$

$$= 2\pi \int_0^r (r^2 - x^2)\,dx \qquad (\textit{the integrand is even})$$

$$= 2\pi\left[r^2 x - \frac{x^3}{3}\right]_0^r = 2\pi\left(r^3 - \frac{r^3}{3}\right)$$

$$= \frac{4}{3}\pi r^3 \qquad \bullet$$

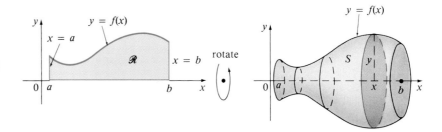

Figure 8.6

The sphere in Example 2 is an example of a **solid of revolution** since it can be obtained by revolving a circle about a diameter. In general let S be the solid obtained by revolving the plane region $\mathcal{R}$ bounded by $y = f(x)$, $y = 0$, $x = a$, and $x = b$ about the x-axis (see Figure 8.6).

Because S is obtained by rotation, a cross-section through x perpendicular to the x-axis is a circular disk with radius $|y| = |f(x)|$ and so the cross-sectional area is

$$A(x) = \pi y^2 = \pi [f(x)]^2$$

Thus, using the basic volume formula $V = \int_a^b A(x)\,dx$, we have the following **formula for a volume of revolution:**

(8.2)
$$V = \int_a^b \pi [f(x)]^2\,dx$$

In particular, since the sphere can be obtained by rotating the region under the semicircle

$$y = \sqrt{r^2 - x^2} \qquad -r \leqslant x \leqslant r$$

about the x-axis, Example 2 could be done using Formula 8.2 with $f(x) = \sqrt{r^2 - x^2}$, $a = -r$, and $b = r$.

Formula 8.2 applies only when the axis of rotation is the x-axis. If, as illustrated in Figure 8.7, the region bounded by the curves $x = g(y)$, $x = 0$, $y = c$, and $y = d$ is rotated about the y-axis, then the corresponding volume

Figure 8.7

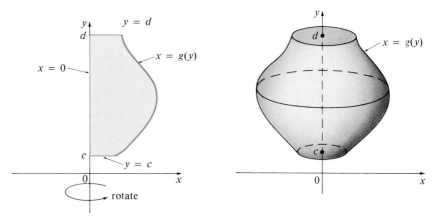

Figure 8.8

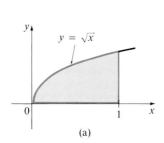

(a)

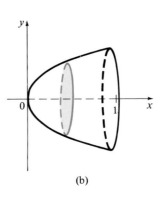

(b)

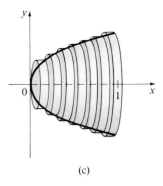

(c)

of revolution would be

(8.3)
$$V = \int_c^d \pi [g(y)]^2 \, dy$$

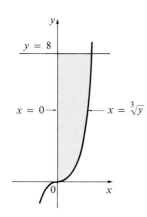

● **Example 3**   (a) Find the volume of the solid obtained by rotating about the x-axis the region under the curve $y = \sqrt{x}$ from 0 to 1.

(b) Illustrate the definition of volume by drawing the approximating cylinders.

Solution   (a) Taking $a = 0$, $b = 1$, and $f(x) = \sqrt{x}$ in Formula 8.2, we find that the volume of the solid [illustrated in Figure 8.8(b)] is

$$V = \int_0^1 \pi(\sqrt{x})^2 \, dx = \pi \int_0^1 x \, dx = \pi \frac{x^2}{2}\Big]_0^1 = \frac{\pi}{2}$$

(b) According to the definition of volume in terms of Riemann sums, we have

$$V = \lim_{||P|| \to 0} \sum_{i=1}^n A(x_i^*)\,\Delta x_i = \lim_{||P|| \to 0} \sum_{i=1}^n \pi(\sqrt{x_i^*})^2\,\Delta x_i$$

Taking x_i^* to be the right endpoint of $[x_{i-1}, x_i]$, the Riemann sum can be interpreted as the approximating volume of the flat circular cylinders (or disks) shown in Figure 8.8(c). You can see why the use of Formula 8.2 is called the **disk method**.   ●

● **Example 4**   Find the volume of the solid obtained by rotating the region bounded by $y = x^3$, $y = 8$, and $x = 0$ around the y-axis.

Solution   Since the axis of rotation is the y-axis, we write the curve $y = x^3$ in the form $x = \sqrt[3]{y}$. The region and the solid with a typical disk are sketched in Figure 8.9. Using Formula 8.3 with $c = 0$, $d = 8$ and $g(y) = \sqrt[3]{y}$, we have

$$V = \int_0^8 \pi(\sqrt[3]{y})^2 \, dy = \pi \int_0^8 y^{2/3} \, dy$$

$$= \pi \left[\frac{3}{5} y^{5/3} \right]_0^8 = \frac{96\pi}{5}$$

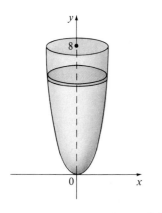

Figure 8.9

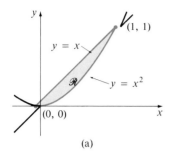

(a)

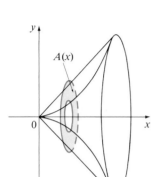

(b)

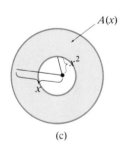

(c)

Figure 8.10

● **Example 5**   The region $\mathscr{R}$ bounded by the curves $y = x$ and $y = x^2$ is rotated about the x-axis. Find the volume of the resulting solid.

Solution 1   The curves $y = x$ and $y = x^2$ intersect at the points $(0, 0)$ and $(1, 1)$. The region between them, the solid of rotation, and a cross-section perpendicular to the x-axis are shown in Figure 8.10. A cross-section in the plane P_x has the shape of an annulus (a ring) with inner radius x^2 and outer radius x, so the cross-sectional area is

$$A(x) = \pi x^2 - \pi(x^2)^2 = \pi(x^2 - x^4)$$

Using Definition 8.1 we have

$$V = \int_0^1 A(x)\,dx = \int_0^1 \pi(x^2 - x^4)\,dx$$

$$= \pi\left[\frac{x^3}{3} - \frac{x^5}{5}\right]_0^1 = \frac{2\pi}{15}$$ ●

Solution 2   Another method is to subtract volumes. If V_1 is the volume of the solid obtained by rotating the region under $y = x$ from 0 to 1 around the x-axis and V_2 is the corresponding volume for the curve $y = x^2$, then

$$V = V_1 - V_2$$

Applying Formula 8.2 to calculate each of V_1 and V_2, we have

$$V = \int_0^1 \pi x^2\,dx - \int_0^1 \pi(x^2)^2\,dx$$

$$= \pi \int_0^1 (x^2 - x^4)\,dx = \frac{2\pi}{15}$$ ●

In general let S be the solid generated when the region bounded by the curves $y = f(x)$, $y = g(x)$, $x = a$, and $x = b$ [where $f(x) \geqslant g(x)$] is rotated about the x-axis (see Figure 8.11). Then, either by the method of Solution 1 of Example 5 (called the **washer method** because the approximating cylinders have the shape of washers) or by the method of Solution 2 (subtracting volumes), we see that the volume of S is

(8.4)

$$V = \pi \int_a^b \left\{[f(x)]^2 - [g(x)]^2\right\}\,dx$$

Figure 8.11

● **Example 6**   Find the volume of the solid obtained by rotating the region in Example 5 about the line $y = 2$.

Solution   The solid and a cross-section are shown in Figure 8.12. Again a cross-section is an annulus, but this time the inner radius is $2 - x$ and the outer radius is $2 - x^2$. The cross-sectional area is

$$A(x) = \pi(2 - x^2)^2 - \pi(2 - x)^2$$

and so the volume of S is

$$V = \int_0^1 A(x)\,dx = \pi \int_0^1 [(2 - x^2)^2 - (2 - x)^2]\,dx$$

$$= \pi \int_0^1 (x^4 - 5x^2 + 4x)\,dx$$

$$= \pi \left[\frac{x^5}{5} - 5\frac{x^3}{3} + 4\frac{x^2}{2} \right]_0^1 = \frac{8\pi}{15}$$

Figure 8.12

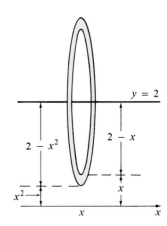

● **Example 7**   Find the volume of the solid obtained by rotating the region in Example 5 about the y-axis.

Solution   Figure 8.13 shows a cross-section perpendicular to the y-axis. It is an annulus with inner radius y and outer radius $\sqrt{y}$, so the cross-sectional area is

$$A(y) = \pi(\sqrt{y})^2 - \pi y^2$$

and the volume is

$$V = \int_0^1 A(y)\,dy = \pi \int_0^1 [(\sqrt{y})^2 - y^2]\,dy$$

$$= \pi \int_0^1 (y - y^2)\,dy$$

$$= \pi \left[\frac{y^2}{2} - \frac{y^3}{3} \right]_0^1 = \frac{\pi}{6}$$

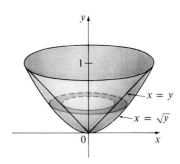

Figure 8.13

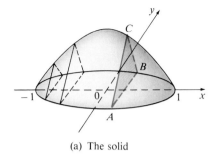

(a) The solid

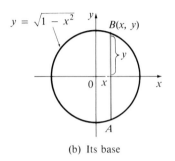

(b) Its base

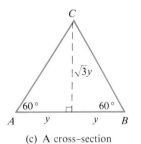

(c) A cross–section

Figure 8.14

Another method would be to subtract volumes as in Solution 2 of Example 5. Notice that the volume in this example is larger than the volume of revolution about the x-axis in Example 5.   ●

Although we have several volume formulas that are applicable in different situations, it is probably best not to memorize Formulas 8.2, 8.3, or 8.4 but rather to remember the basic volume formula $V = \int_a^b A(x)\,dx$, which applies to any situation. We conclude this section by finding the volumes of two solids that, like the one in Example 1, are not solids of revolution.

● **Example 8**   A solid has a circular base of radius 1. Parallel cross-sections perpendicular to the base are equilateral triangles. Find the volume of the solid.

Solution   Let us take the circle to be $x^2 + y^2 = 1$. The solid, its base, and a typical cross-section at a distance x from the origin are shown in Figure 8.14. Since B lies on the circle we have $y = \sqrt{1 - x^2}$ and so the base of the triangle ABC is $|AB| = 2\sqrt{1 - x^2}$. Since the triangle is equilateral, its height is $\sqrt{3}\,y = \sqrt{3}\sqrt{1 - x^2}$. The cross-sectional area is therefore

$$A(x) = \tfrac{1}{2} \cdot 2\sqrt{1 - x^2} \cdot \sqrt{3}\sqrt{1 - x^2} = \sqrt{3}(1 - x^2)$$

and the volume of the solid is

$$V = \int_{-1}^{1} A(x)\,dx = \int_{-1}^{1} \sqrt{3}(1 - x^2)\,dx$$

$$= \sqrt{3}\left[x - \frac{x^3}{3} \right]_{-1}^{1} = \frac{4\sqrt{3}}{3}$$ ●

● **Example 9**   A wedge is cut out of a circular cylinder of radius 4 by two planes. One plane is perpendicular to the axis of the cylinder. The other intersects the first at an angle of $30°$ along a diameter of the cylinder. Find the volume of the wedge.

Solution   If we place the x-axis along the diameter where the planes meet, then the base of the solid is a semicircle with equation $y = \sqrt{16 - x^2}$, $-4 \leqslant x \leqslant 4$. A cross-section perpendicular to the x-axis at a distance x from the origin is a triangle ABC as shown in Figure 8.15 with base

Figure 8.15

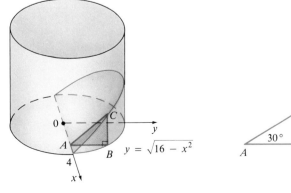

$y = \sqrt{16 - x^2}$ and height $|BC| = y \tan 30° = \sqrt{16 - x^2}/\sqrt{3}$. Thus the cross-sectional area is

$$A(x) = \frac{1}{2}\sqrt{16 - x^2} \cdot \frac{1}{\sqrt{3}}\sqrt{16 - x^2} = \frac{16 - x^2}{2\sqrt{3}}$$

and the volume is

$$V = \int_{-4}^{4} A(x)\,dx = \int_{-4}^{4} \frac{16 - x^2}{2\sqrt{3}}\,dx$$

$$= \frac{1}{\sqrt{3}}\int_{0}^{4}(16 - x^2)\,dx = \frac{1}{\sqrt{3}}\left[16x - \frac{x^3}{3}\right]_{0}^{4}$$

$$= \frac{128}{3\sqrt{3}}$$

For another method see Exercise 59.   ●

EXERCISES 8.1

In Exercises 1–12 find the volume of the solid obtained by rotating the region bounded by the given curves about the given axis. Sketch the region, the solid, and a typical disk or "washer."

1. $y = x^2$, $x = 1$, $y = 0$; about the x-axis

2. $y^2 = x^3$, $x = 4$, $y = 0$; about the x-axis

3. $x + y = 1$, $x = 0$, $y = 0$; about the x-axis

4. $y = \sin x$, $x = \pi/2$, $x = \pi$, $y = 0$; about the x-axis

5. $y = x^2$, $y = 4$, $x = 0$; about the y-axis

6. $x = y - y^2$, $x = 0$; about the y-axis

7. $y = x^2$, $y^2 = x$; about the x-axis

8. $y = x^2 + 1$, $y = 3 - x^2$; about the x-axis

9. $y^2 = x$, $x = 2y$; about the y-axis

10. $y = 2x - x^2$, $y = 0$, $x = 1$; about the y-axis

11. $y = x^4$, $y = 1$; about $y = 2$

12. $y = x$, $y = 0$, $x = 2$, $x = 4$; about $x = 1$

In Exercises 13–24 refer to Figure 8.16 and find the volume generated by rotating the given region about the given line.

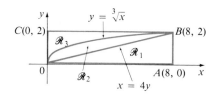

Figure 8.16

13. $\mathcal{R}_1$ about OA

14. $\mathcal{R}_1$ about OC

15. $\mathcal{R}_1$ about AB

16. $\mathcal{R}_1$ about BC

17. $\mathcal{R}_2$ about OA

18. $\mathcal{R}_2$ about OC

19. $\mathcal{R}_2$ about BC

20. $\mathcal{R}_2$ about AB

21. $\mathcal{R}_3$ about OA

22. $\mathcal{R}_3$ about OC

23. $\mathcal{R}_3$ about BC

24. $\mathcal{R}_3$ about AB

In Exercises 25–40 find the volume of the solid obtained by rotating the region bounded by the given curves about the given axis.

25. $y = x^2 - 1$, $y = 0$, $x = 0$, $x = 2$; about the x-axis

26. $y = -1/x$, $y = 0$, $x = 1$, $x = 3$; about the x-axis

27. $y = e^x$, $y = 0$, $x = 0$, $x = 1$; about the x-axis

28. $y = e^x$, $y = 0$, $x = 0$, $x = 1$; about the y-axis

29. $y = \sec x$, $y = 1$, $x = -1$, $x = 1$; about the x-axis

30. $y = \cos x$, $y = \sin x$, $x = 0$, $x = \pi/4$; about the x-axis

31. $y = \sin^{-1} x$, $y = \pi/2$, $x = 0$; about the y-axis

32. $y = \sin^{-1} x$, $y = \pi/2$, $x = 0$; about the x-axis

33. $y = \tan x$, $y = 1$, $x = 0$; about the x-axis

34. $y = e^{-x^2}$, $x = 0$, $x = 1$, $y = 0$; about the y-axis

35. $x - y = 1$, $y = (x - 4)^2 + 1$; about $y = 7$

36. $y = \cos x$, $y = 0$, $x = 0$, $x = \pi/2$; about $y = 1$

37. $y = \cos x$, $y = 0$, $x = 0$, $x = \pi/2$; about $y = -1$

38. $y = \ln x$, $y = 0$, $x = e$; about $x = -1$

39. $y = |x + 2|$, $y = 0$, $x = -3$, $x = 0$; about the x-axis

40. $y = [\![x]\!]$, $x = 1$, $x = 6$, $y = 0$; about the x-axis

In Exercises 41 and 42 sketch and find the volume of the solid obtained by rotating the region under the graph of f about the x-axis.

41. $f(x) = \begin{cases} 3 & \text{if } 0 \leqslant x \leqslant 1 \\ 1 & \text{if } 1 < x < 4 \\ 3 & \text{if } 4 \leqslant x \leqslant 5 \end{cases}$

42. $f(x) = \begin{cases} \frac{1}{2} & \text{if } 0 \leqslant x < 1 \\ x^2 - 2x + 2 & \text{if } 1 \leqslant x \leqslant 2 \end{cases}$

In Exercises 43–57 find the volume of the described solid S.

43. A right circular cone with height h and base radius r (Figure 8.17)

Figure 8.17

44. A frustum of a right circular cone with height h, lower base radius R, and top radius r (Figure 8.18)

Figure 8.18

45. A cap of a sphere with radius r and height h (Figure 8.19)

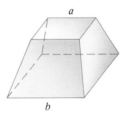

Figure 8.19

46. A frustum of a pyramid with square base of side b, square top of side a, and height h (Figure 8.20)

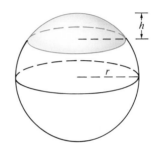

Figure 8.20

47. A pyramid with height h and rectangular base with dimensions b and $2b$

48. A pyramid with height h and base an equilateral triangle with side a (a tetrahedron) (Figure 8.21)

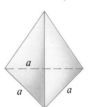

Figure 8.21

49. A solid torus (the donut-shaped solid in Figure 8.22) with radii r and R

Figure 8.22

50. A tetrahedron with three mutually perpendicular faces and three mutually perpendicular edges with lengths 3 cm, 4 cm, and 5 cm

51. The base of S is a circular disk with radius r. Parallel cross-sections perpendicular to the base are squares.

52. The base of S is a circular disk with radius r. Parallel cross-sections perpendicular to the base are isosceles triangles with height h and unequal side in the base.

53. The base of S is an elliptical region with boundary $9x^2 + 4y^2 = 36$. Cross-sections perpendicular to the x-axis are isosceles right triangles with hypotenuse in the base.

54. The base of S is the parabolic region $\{(x, y) \mid x^2 \leqslant y \leqslant 1\}$. Cross-sections perpendicular to the y-axis are equilateral triangles.

55. S has the same base as in Exercise 54 but cross-sections perpendicular to the y-axis are squares.

56. The base of S is the triangular region with vertices $(0, 0)$, $(2, 0)$, and $(0, 1)$. Cross-sections perpendicular to the x-axis are semicircles.

57. S has the same base as in Exercise 56 but cross-sections perpendicular to the x-axis are isosceles triangles with height equal to the base.

58. Find the volume common to two circular cylinders, each with radius r, if the axes of the cylinders intersect at right angles.

59. Solve Example 9 taking cross-sections to be parallel to the line of intersection of the two planes.

60. Solve Example 9 if the planes intersect at an angle of $45°$.

61. Find the volume enclosed by the ellipsoid
$$\frac{x^2}{a^2} + \frac{y^2}{b^2} + \frac{z^2}{c^2} = 1$$
[See the section on quadric surfaces (Section 11.6) for a description and diagram of the ellipsoid.]

62. A hole of radius r is bored through the center of a sphere of radius $R > r$. Find the volume of the remaining portion of the sphere.

63. A hole of radius r is bored through a cylinder of radius $R > r$ at right angles to the axis of the cylinder. Set up, but do not evaluate, an integral for the volume cut out.

64. (a) Cavalieri's Principle states that if a family of parallel planes gives equal cross-sectional areas for two solids S_1 and S_2, then the volumes of S_1 and S_2 are equal. Prove Cavalieri's Principle.
   (b) Use this principle to find the volume of the oblique cylinder in Figure 8.23.

Figure 8.23

65. We know from Section 7.8 that the region $\mathcal{R} = \{(x, y) \mid x \geqslant 1, 0 \leqslant y \leqslant 1/x\}$ has infinite area (see Figure 7.14). Show that by rotating $\mathcal{R}$ about the x-axis we obtain a solid with finite volume.

66. A log 10 m long is cut at 1-m intervals and its cross-sectional areas A (at a distance x from the end of the log) are listed in the following table. Use Simpson's Rule to estimate the volume of the log.

x(m)	0	1	2	3	4	5	6	7	8	9	10
A(m²)	0.68	0.65	0.64	0.61	0.58	0.59	0.53	0.55	0.52	0.50	0.48

SECTION 8.2

Volumes by Cylindrical Shells

There are some volume problems that are very difficult to handle by the methods of the preceding section. For instance, we consider the problem of finding the volume of the solid obtained by rotating about the y-axis the

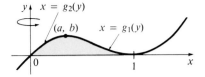

Figure 8.24

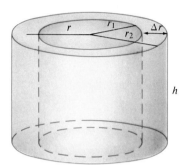

Figure 8.25

region bounded by $y = x(x - 1)^2$ and $y = 0$. If we were to use the "washer method" we would first have to locate the local maximum point (a, b) of $y = x(x - 1)^2$ using the methods of Chapter 4. Then we would have to solve the equation $y = x(x - 1)^2$ for x in terms of y to obtain the functions $x = g_1(y)$ and $x = g_2(y)$ shown in Figure 8.24. This step would be difficult because it involves the cubic formula. Finally we would find the volume using

$$V = \pi \int_0^b \{[g_1(y)]^2 - [g_2(y)]^2\} \, dy$$

Fortunately there is a method, called the method of cylindrical shells, that is easier to use in such a case. Figure 8.25 shows a cylindrical shell with inner radius r_1, outer radius r_2, and height h. Its volume V is calculated by subtracting the volume V_1 of the inner cylinder from the volume V_2 of the outer cylinder:

$$\begin{aligned} V &= V_2 - V_1 \\ &= \pi r_2^2 h - \pi r_1^2 h = \pi (r_2^2 - r_1^2)h \\ &= \pi (r_2 + r_1)(r_2 - r_1)h \\ &= 2\pi \frac{r_2 + r_1}{2} h(r_2 - r_1) \end{aligned}$$

If we let $\Delta r = r_2 - r_1$ (the thickness of the shell) and $r = \frac{1}{2}(r_2 + r_1)$ (the average radius of the shell), then this formula for the volume of a cylindrical shell becomes

(8.5)

$$V = 2\pi rh \, \Delta r$$

and it can be remembered as

$$V = [\text{circumference}][\text{height}][\text{thickness}]$$

Now let S be the solid obtained by rotating about the y-axis the region bounded by $y = f(x)$ [where $f(x) \geqslant 0$], $y = 0$, $x = a$, and $x = b$, where $b > a \geqslant 0$ (see Figure 8.26). Let P be a partition of $[a, b]$ by points x_i with $a = x_0 < x_1 < \cdots < x_n = b$ and let x_i^* be the midpoint of $[x_{i-1}, x_i]$—that is, $x_i^* = \frac{1}{2}(x_{i-1} + x_i)$. If the rectangle with base $[x_{i-1}, x_i]$ and height $f(x_i^*)$ is rotated about the y-axis, then the result is a cylindrical shell with average

Figure 8.26

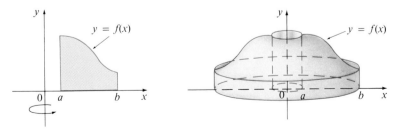

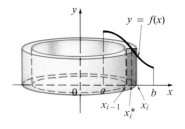

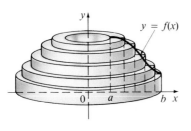

Figure 8.27

radius x_i^*, height $f(x_i^*)$, and thickness $\Delta x_i = x_i - x_{i-1}$ (see Figure 8.27), so by Formula 8.5 its volume is

$$V_i = 2\pi x_i^* f(x_i^*) \Delta x_i$$

Therefore an approximation to the volume V of S is given by the sum of the volumes of these shells:

$$V \approx \sum_{i=1}^{n} V_i = \sum_{i=1}^{n} 2\pi x_i^* f(x_i^*) \Delta x_i$$

This approximation appears to become better as $\|P\| \to 0$. But, from the definition of an integral, we know that

$$\lim_{\|P\| \to 0} \sum_{i=1}^{n} 2\pi x_i^* f(x_i^*) \Delta x_i = \int_a^b 2\pi x f(x)\, dx$$

Thus it appears plausible that

(8.6)
$$\boxed{V = \int_a^b 2\pi x f(x)\, dx \qquad \text{where } 0 \leqslant a < b}$$

The argument using cylindrical shells makes Formula 8.6 seem reasonable but we can now prove it, at least for the case where f is one-to-one and therefore has an inverse function g. Using Formula 8.3 for volumes of rotation, looking at Figure 8.28, and realizing that the curve $y = f(x)$ can also be described as $x = g(y)$, we see that

(8.7)
$$V = \int_0^d \pi b^2\, dy - \int_0^c \pi a^2\, dy - \int_c^d \pi [g(y)]^2\, dy$$
$$= \pi b^2 d - \pi a^2 c - \pi \int_a^b x^2 f'(x)\, dx$$

Here we have used the substitution $y = f(x)$, which gives $dy = f'(x)\, dx$ and $g(y) = x$. Now we integrate by parts with $u = x^2$ and $dv = f'(x)\, dx$. Then $du = 2x\, dx$, $v = f(x)$, and

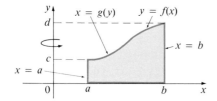

Figure 8.28

$$\int_a^b x^2 f'(x)\, dx = x^2 f(x) \Big]_a^b - \int_a^b 2x f(x)\, dx$$
$$= b^2 f(b) - a^2 f(a) - \int_a^b 2x f(x)\, dx$$

Putting this expression in Equation 8.7 and using the fact that $f(a) = c$ and $f(b) = d$, we have

$$V = \pi b^2 d - \pi a^2 c - \pi \left[b^2 d - a^2 c - \int_a^b 2x f(x)\, dx \right]$$
$$= \int_a^b 2\pi x f(x)\, dx$$

● **Example 1**   Find the volume of the solid obtained by rotating about the y-axis the region bounded by $y = x(x - 1)^2$ and $y = 0$.

Solution   The region was sketched in Figure 8.24. Formula 8.6 gives the volume of rotation as

$$V = \int_0^1 2\pi x[x(x - 1)^2]\,dx$$

$$= 2\pi \int_0^1 (x^4 - 2x^3 + x^2)\,dx$$

$$= 2\pi \left[\frac{x^5}{5} - 2\frac{x^4}{4} + \frac{x^3}{3}\right]_0^1$$

$$= \frac{\pi}{15}$$ ●

Note: Comparing the solution of Example 1 with the remarks at the beginning of this section, we see that the method of cylindrical shells is much easier than the washer method for this problem. We did not have to find the coordinates of the local maximum and we did not have to solve the equation of the curve for x in terms of y. However, in other examples the methods of the preceding section may be easier.

● **Example 2**   Find the volume of the solid obtained by rotating about the y-axis the region between $y = x$ and $y = x^2$.

Solution   We use Formula 8.6 to find the volume obtained by rotating about the y-axis the region under $y = x$ from 0 to 1 and then we subtract the corresponding volume for the region under $y = x^2$. Thus

$$V = \int_0^1 2\pi x \cdot x\,dx - \int_0^1 2\pi x \cdot x^2\,dx$$

$$= 2\pi \int_0^1 (x^2 - x^3)\,dx = 2\pi \left[\frac{x^3}{3} - \frac{x^4}{4}\right]_0^1$$

$$= \frac{\pi}{6}$$

Comparison with Example 7 in Section 8.1 shows that the washer and shell methods take about the same amount of work in this problem.   ●

In general, by subtracting volumes as in Example 2, or by considering cylindrical shells directly, we see that the volume of the solid generated by rotating about the y-axis the region between the curves $y = f(x)$ and $y = g(x)$ from a to b [where $f(x) \geq g(x)$ and $0 \leq a < b$] is

$$V = \int_a^b 2\pi x[f(x) - g(x)]\,dx$$

The method of cylindrical shells also allows us to compute volumes of revolution about the x-axis. If we interchange the roles of x and y in Formula

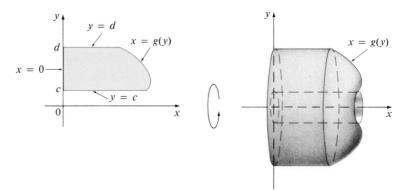

Figure 8.29

8.6, we see that if we rotate the region in Figure 8.29 about the x-axis, then the volume of the resulting solid is

(8.8)

$$V = \int_c^d 2\pi y g(y)\, dy$$

● **Example 3**   Use cylindrical shells to find the volume of the solid in Example 3 in Section 8.1.

Solution   Looking at Figure 8.8(a), we relabel $y = \sqrt{x}$ as $x = y^2$. Then, by using Formula 8.8 and subtracting volumes, we get

$$V = \int_0^1 2\pi y \cdot 1\, dy - \int_0^1 2\pi y \cdot y^2\, dy$$

$$= 2\pi \int_0^1 (y - y^3)\, dy = 2\pi \left[\frac{y^2}{2} - \frac{y^4}{4} \right]_0^1 = \frac{\pi}{2}$$

In this problem the method of Example 3 in Section 8.1 was simpler.   ●

● **Example 4**   Find the volume of the solid obtained by rotating the region bounded by $y = x - x^2$ and $y = 0$ about the line $x = 2$.

Solution   Since the axis of rotation is $x = 2$ instead of the y-axis, we cannot use Formula 8.6 so we go back to the basic method of deriving it using cylindrical shells. Figure 8.30 shows a rectangle with base $[x_{i-1}, x_i]$ rotated about $x = 2$ to form a cylindrical shell with average radius $2 - x_i^*$, height $x_i^* - (x_i^*)^2$, and thickness Δx_i, where x_i^* is the midpoint of $[x_{i-1}, x_i]$. By Formula 8.5 the volume of this shell is

$$\underbrace{2\pi(2 - x_i^*)}_{\text{circumference}}\underbrace{\left[x_i^* - (x_i^*)^2\right]}_{\text{height}}\underbrace{\Delta x_i}_{\text{thickness}}$$

Figure 8.30

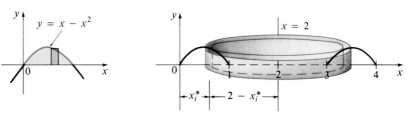

and so the volume of the given solid is

$$V = \lim_{||P|| \to 0} \sum_{i=1}^{n} 2\pi(2 - x_i^*)[x_i^* - (x_i^*)^2]\,\Delta x_i$$

$$= \int_0^1 2\pi(2 - x)(x - x^2)\,dx$$

$$= 2\pi \int_0^1 (x^3 - 3x^2 + 2x)\,dx$$

$$= 2\pi \left[\frac{x^4}{4} - x^3 + x^2\right]_0^1$$

$$= \frac{\pi}{2}$$

EXERCISES 8.2

In Exercises 1–14 use the method of cylindrical shells to find the volume generated by rotating the region bounded by the given curves about the y-axis.

1. $y = x^2$, $y = 0$, $x = 1$, $x = 2$

2. $y = 1/x$, $y = 0$, $x = 1$, $x = 10$

3. $y = \sqrt{4 + x^2}$, $y = 0$, $x = 0$, $x = 4$

4. $y = \sin(x^2)$, $y = 0$, $x = 0$, $x = \sqrt{\pi}$

5. $y = x^2$, $y = 4$, $x = 0$

6. $y^2 = x$, $x = 2y$

7. $y = x^2 - x^3$, $y = 0$

8. $y = -x^2 + 4x - 3$, $y = 0$

9. $y = \sin x$, $y = 0$, $x = 2\pi$, $x = 3\pi$

10. $y = \dfrac{1}{1 + x^2}$, $y = 0$, $x = 0$, $x = 3$

11. $y = x^2 - 6x + 10$, $y = -x^2 + 6x - 6$

12. $y = x - 2$, $y = \sqrt{x - 2}$

13. $y = \ln x$, $y = 0$, $x = e$

14. $y = e^x$, $y = e^{-x}$, $x = 1$

In Exercises 15–22 use the method of cylindrical shells to find the volume of the solid obtained by rotating the region bounded by the given curves about the x-axis.

15. $x = \sqrt[4]{y}$, $x = 0$, $y = 16$

16. $x = y^2$, $x = 0$, $y = 2$, $y = 5$

17. $y = x^2$, $y = 9$

18. $y = e^x$, $x = 0$, $y = \pi$

19. $y^2 - 6y + x = 0$, $x = 0$

20. $x = \cos y$, $x = 0$, $y = 0$, $y = \dfrac{\pi}{4}$

21. $x = y^2$, $y = 0$, $x + y = 2$

22. $y = x$, $x = 0$, $x + y = 2$

In Exercises 23–30 use the method of cylindrical shells as in Example 4 to find the volume generated by rotating the region bounded by the given curves about the given axis. Sketch the region and a typical shell.

23. $y = \sqrt{x}$, $y = 0$, $x = 1$, $x = 4$; about the y-axis

24. $y = x^2$, $y = 0$, $x = -2$, $x = -1$; about the y-axis

25. $y = x^2$, $y = 0$, $x = 1$, $x = 2$; about $x = 1$

26. $y = x^2$, $y = 0$, $x = 1$, $x = 2$; about $x = 4$

27. $y = e^{-x}$, $y = 0$, $x = -1$, $x = 0$; about $x = 1$

28. $y = 4x - x^2$, $y = 8x - 2x^2$; about $x = -2$

29. $y = e^x$, $x = 0$, $y = 2$; about $y = 1$

30. $y = \ln x$, $y = 0$, $x = e$; about $y = 3$

In Exercises 31–38 the region bounded by the given curves is rotated about the given axis. Find the volume of the resulting solid by any method.

31. $y = x^2 + x - 2$, $y = 0$; about the x-axis

32. $y = x^2 - 3x + 2$, $y = 0$; about the y-axis

33. $y = \sec^2 x$, $y = 0$, $x = \dfrac{3\pi}{4}$, $x = \dfrac{5\pi}{4}$; about the y-axis

34. $y = -x^2 + 7x - 10$, $y = x - 2$; about the x-axis

35. $x = 1 - y^2$, $x = 0$; about the y-axis

36. $y = x\sqrt{1 + x^3}$, $y = 0$, $x = 0$, $x = 2$; about the y-axis

37. $x^2 + (y - 1)^2 = 1$; about the y-axis

38. $x^2 + (y - 1)^2 = 1$; about the x-axis

Use cylindrical shells to find the volumes of the solids in Exercises 39–42.

39. A sphere of radius r

40. The solid torus of Exercise 49 in Section 8.1

41. A right circular cone with height h and base radius r

42. The solid of Exercise 62 in Section 8.1

43. Formula 8.6 is valid only if $a \geqslant 0$. Show that if $a < b \leqslant 0$, the volume formula becomes

$$V = -\int_a^b 2\pi x f(x)\,dx$$

44. Find a formula for the volume of the solid obtained when the region under the graph of f from a to b is rotated about the vertical line $x = c$ in the following cases: (a) $c \leqslant a < b$ and (b) $a < b \leqslant c$.

Arc Length

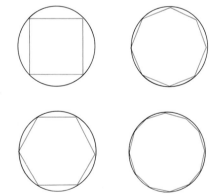

Figure 8.31

We all have an intuitive idea as to what the length of a curve is. But, like the concepts of area and volume, the concept of the length of an arc of a curve is one that requires a careful definition.

For the simple case where the curve is a finite line segment joining the point $P_1(x_1, y_1)$ to the point $P_2(x_2, y_2)$ we know that its length is given by the distance formula

$$|P_1 P_2| = \sqrt{(x_2 - x_1)^2 + (y_2 - y_1)^2}$$

We can also compute the length of a polygon by adding the lengths of the line segments that form the polygon. We shall define the length of a curve by first approximating it by a polygon and then taking a limit. This process is familiar for the case of a circle where the circumference is the limit of inscribed polygons (see Figure 8.31).

Now suppose that a curve C is defined by the equation $y = f(x)$, where $a \leqslant x \leqslant b$. We obtain a polygonal approximation to C by taking a partition P of $[a, b]$ determined by points x_i with $a = x_0 < x_1 < \cdots < x_n = b$. If $y_i = f(x_i)$, then the point $P_i(x_i, y_i)$ lies on C and the polygon with vertices $P_0, P_1, \ldots, P_n$, illustrated in Figure 8.32, is an approximation to C. The length of this polygonal approximation is

$$\sum_{i=1}^n |P_{i-1} P_i|$$

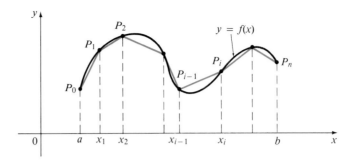

Figure 8.32

and this approximation appears to become better as $\|P\| \to 0$. (See Figure 8.33 where the arc of the curve between P_{i-1} and P_i has been magnified and approximations with successively smaller $\|P\|$ are shown.)

Therefore we define the **length** L of the curve C with equation $y = f(x)$, $a \leqslant x \leqslant b$, as the limit of the lengths of these inscribed polygons (if the limit exists):

(8.9)

$$L = \lim_{\|P\| \to 0} \sum_{i=1}^{n} |P_{i-1}P_i|$$

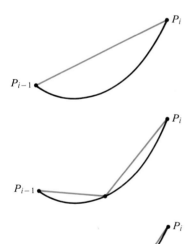

Notice that the procedure for defining arc length is very similar to the procedure we used for defining area and volume. We divided the curve into a large number of small parts. We then found the approximate lengths of the small parts and added them. Finally we took the limit as $\|P\| \to 0$.

The definition of arc length given by Equation 8.9 is not very convenient for computational purposes, but we can derive an integral formula for L in the case where f has a continuous derivative. [Such a function f is called **smooth** because a small change in x produces a small change in $f'(x)$.]

If we let $\Delta y_i = y_i - y_{i-1}$, then

$$|P_{i-1}P_i| = \sqrt{(\Delta x_i)^2 + (\Delta y_i)^2}$$

By applying the Mean Value Theorem (4.10) to f on the interval $[x_{i-1}, x_i]$, we find that there is a number x_i^* between x_{i-1} and x_i such that

$$f(x_i) - f(x_{i-1}) = f'(x_i^*)(x_i - x_{i-1})$$

that is, $$\Delta y_i = f'(x_i^*)\, \Delta x_i$$

Thus we have

$$\begin{aligned}
|P_{i-1}P_i| &= \sqrt{(\Delta x_i)^2 + (\Delta y_i)^2} \\
&= \sqrt{(\Delta x_i)^2 + [f'(x_i^*)\,\Delta x_i]^2} \\
&= \sqrt{1 + [f'(x_i^*)]^2}\; \Delta x_i
\end{aligned}$$

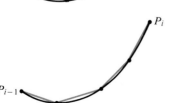

Figure 8.33

Therefore, by Definition 8.9,

$$L = \lim_{||P|| \to 0} \sum_{i=1}^{n} |P_{i-1}P_i|$$

$$= \lim_{||P|| \to 0} \sum_{i=1}^{n} \sqrt{1 + [f'(x_i^*)]^2}\, \Delta x_i$$

We recognize this expression as being equal to

$$\int_a^b \sqrt{1 + [f'(x)]^2}\, dx$$

by the definition of a definite integral. This integral exists because the function $g(x) = \sqrt{1 + [f'(x)]^2}$ is continuous. Thus we have proved the following theorem:

The Arc Length Formula (8.10)

If f' is continuous on $[a, b]$, then the length of the curve $y = f(x)$, $a \leqslant x \leqslant b$, is

$$L = \int_a^b \sqrt{1 + [f'(x)]^2}\, dx$$

If we use the Leibniz notation for derivatives, we can write the arc length formula as follows:

(8.11)

$$L = \int_a^b \sqrt{1 + \left(\frac{dy}{dx}\right)^2}\, dx$$

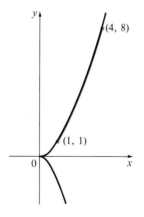

Figure 8.34

● **Example 1**   Find the length of the arc of the semicubical parabola $y^2 = x^3$ between the points $(1, 1)$ and $(4, 8)$ (see Figure 8.34).

Solution   For the top half of the curve we have

$$y = x^{3/2} \qquad \frac{dy}{dx} = \frac{3}{2} x^{1/2}$$

and so the arc length formula gives

$$L = \int_1^4 \sqrt{1 + \left(\frac{dy}{dx}\right)^2}\, dx = \int_1^4 \sqrt{1 + \frac{9}{4}x}\, dx$$

If we substitute $u = 1 + 9x/4$, then $du = 9\, dx/4$. When $x = 1$, $u = \frac{13}{4}$; when $x = 4$, $u = 10$. Therefore

$$L = \frac{4}{9} \int_{13/4}^{10} \sqrt{u}\, du = \frac{4}{9} \cdot \frac{2}{3} u^{3/2} \Big]_{13/4}^{10}$$

$$= \frac{8}{27} \left[10^{3/2} - \left(\frac{13}{4}\right)^{3/2} \right]$$

$$= \frac{80\sqrt{10} - 13\sqrt{13}}{27}$$

●

If a curve has the equation $x = g(y)$, $c \leqslant y \leqslant d$, then by interchanging the roles of x and y in Theorem 8.10 or Equation 8.11, we obtain the following formula for its length:

(8.12)

$$L = \int_c^d \sqrt{1 + [g'(y)]^2} \, dy = \int_c^d \sqrt{1 + \left(\frac{dx}{dy}\right)^2} \, dy$$

● **Example 2**   Find the length of the arc of the parabola $y^2 = x$ from $(0, 0)$ to $(1, 1)$.

Solution   Since $x = y^2$, $dx/dy = 2y$, and Formula 8.12 gives

$$L = \int_0^1 \sqrt{1 + \left(\frac{dx}{dy}\right)^2} \, dy = \int_0^1 \sqrt{1 + 4y^2} \, dy$$

We make the trigonometric substitution $y = \frac{1}{2}\tan\theta$, which gives $dy = \frac{1}{2}\sec^2\theta \, d\theta$ and $\sqrt{1 + 4y^2} = \sqrt{1 + \tan^2\theta} = \sec\theta$. When $y = 0$, $\tan\theta = 0$, so $\theta = 0$; when $y = 1$, $\tan\theta = 2$, so $\theta = \tan^{-1} 2 = \alpha$, say. Thus

$$L = \int_0^\alpha \sec\theta \cdot \tfrac{1}{2} \sec^2\theta \, d\theta = \tfrac{1}{2} \int_0^\alpha \sec^3\theta \, d\theta$$

$$= \tfrac{1}{2} \cdot \tfrac{1}{2} \Big[\sec\theta\tan\theta + \ln|\sec\theta + \tan\theta| \Big]_0^\alpha \qquad \textit{(from Example 9 in Section 7.2)}$$

$$= \tfrac{1}{4} (\sec\alpha\tan\alpha + \ln|\sec\alpha + \tan\alpha|)$$

Since $\tan\alpha = 2$, we have $\sec^2\alpha = 1 + \tan^2\alpha = 5$, so $\sec\alpha = \sqrt{5}$ and

$$L = \frac{\sqrt{5}}{2} + \frac{\ln(\sqrt{5} + 2)}{4} \qquad\qquad ●$$

Because of the presence of the square root sign in Formulas 8.10 and 8.12, the calculation of an arc length often leads to an integral that is very difficult or even impossible to evaluate explicitly. Thus we sometimes have to be content with finding an approximation to the length of a curve as in the following example.

● **Example 3**   (a)   Set up an integral for the length of the arc of the hyperbola $xy = 1$ from the point $(1, 1)$ to the point $(2, \frac{1}{2})$.
        (b)   Use Simpson's Rule with $n = 10$ to estimate the arc length.

Solution   (a)   We have

$$y = \frac{1}{x} \qquad\qquad \frac{dy}{dx} = -\frac{1}{x^2}$$

and so the arc length is

$$L = \int_1^2 \sqrt{1 + \left(\frac{dy}{dx}\right)^2} \, dx = \int_1^2 \sqrt{1 + \frac{1}{x^4}} \, dx = \int_1^2 \frac{\sqrt{x^4 + 1}}{x^2} \, dx$$

(b)   Using Simpson's Rule (7.31) with $a = 1$, $b = 2$, $n = 10$, $\Delta x = 0.1$, and $f(x) = \sqrt{1 + 1/x^4}$, we have

$$L = \int_1^2 \sqrt{1 + \frac{1}{x^4}}\, dx$$

$$\approx \frac{\Delta x}{3} \left[f(1) + 4f(1.1) + 2f(1.2) + 4f(1.3) + \cdots \right.$$

$$\left. + 2f(1.8) + 4f(1.9) + f(2) \right]$$

$$= \frac{0.1}{3} \left[\sqrt{1 + \frac{1}{1^4}} + 4\sqrt{1 + \frac{1}{(1.1)^4}} + 2\sqrt{1 + \frac{1}{(1.2)^4}} + 4\sqrt{1 + \frac{1}{(1.3)^4}} \right.$$

$$\left. + \cdots + 2\sqrt{1 + \frac{1}{(1.8)^4}} + 4\sqrt{1 + \frac{1}{(1.9)^4}} + \sqrt{1 + \frac{1}{2^4}} \right]$$

$$\approx 0.9674$$

The Arc Length Function

If a smooth curve C has the equation $y = f(x)$, $a \leqslant x \leqslant b$, let $s(x)$ be the distance along C from the initial point $P_0(a, f(a))$ to the point $Q(x, f(x))$. Then s is a function, called the **arc length function**, and, by Theorem 8.10,

(8.13)
$$s(x) = \int_a^x \sqrt{1 + [f'(t)]^2}\, dt$$

(We have replaced the dummy variable of integration by t so that x will not have two meanings.) We can use Part 1 of the Fundamental Theorem of Calculus to differentiate Equation 8.13 (since the integrand is continuous):

(8.14)
$$\frac{ds}{dx} = \sqrt{1 + [f'(x)]^2} = \sqrt{1 + \left(\frac{dy}{dx}\right)^2}$$

Equation 8.14 shows that the rate of change of s with respect to x is always at least 1 and is equal to 1 when $f'(x)$, the slope of the curve, is 0. The differential of arc length is

(8.15)
$$ds = \sqrt{1 + \left(\frac{dy}{dx}\right)^2}\, dx$$

and this equation is sometimes written in the symmetric form

(8.16)
$$(ds)^2 = (dx)^2 + (dy)^2$$

The geometric interpretation of Equation 8.16 is shown in Figure 8.35. It can be used as a mnemonic device for remembering both of the formulas 8.11 and 8.12. If we write $L = \int ds$, then from Equation 8.16 either we can solve to get (8.15), which gives (8.11), or we can solve to get

$$ds = \sqrt{1 + \left(\frac{dx}{dy}\right)^2}\, dy$$

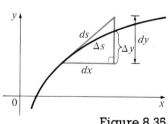

Figure 8.35   which gives (8.12).

● **Example 4**   Find the arc length function for the curve
$y = x^2 - (\ln x)/8$ taking $P_0(1, 1)$ as the starting point.

Solution   If $f(x) = x^2 - (\ln x)/8$, then

$$f'(x) = 2x - \frac{1}{8x}$$

$$1 + [f'(x)]^2 = 1 + \left(2x - \frac{1}{8x}\right)^2 = 1 + 4x^2 - \frac{1}{2} + \frac{1}{64x^2}$$

$$= 4x^2 + \frac{1}{2} + \frac{1}{64x^2} = \left(2x + \frac{1}{8x}\right)^2$$

$$\sqrt{1 + [f'(x)]^2} = 2x + \frac{1}{8x}$$

Thus the arc length function is given by

$$s(x) = \int_1^x \sqrt{1 + [f'(t)]^2}\, dt$$

$$= \int_1^x \left(2t + \frac{1}{8t}\right) dt = t^2 + \frac{1}{8}\ln t\Big]_1^x$$

$$= x^2 + \frac{1}{8}\ln x - 1 \qquad\qquad ●$$

EXERCISES 8.3

1. Use Formula 8.11 to find the length of the arc of the curve $x^2 = 64y^3$ from $(8, 1)$ to $(64, 4)$.

2. Use Formula 8.12 to find the arc length in Exercise 1.

In Exercises 3–6 find the length of the arc of the given curve from point A to point B.

3. $y^2 = (x - 1)^3$, $A(1, 0)$, $B(2, 1)$

4. $y = 1 - x^{2/3}$, $A(-8, -3)$, $B(-1, 0)$

5. $12xy = 4y^4 + 3$, $A(\frac{7}{12}, 1)$, $B(\frac{67}{24}, 2)$

6. $9y^2 = x(x - 3)^2$, $A(0, 0)$, $B(4, \frac{2}{3})$

In Exercises 7–18 find the length of the given curve.

7. $y = \frac{1}{3}(x^2 + 2)^{3/2}$, $0 \leqslant x \leqslant 1$

8. $y = \frac{x^3}{6} + \frac{1}{2x}$, $1 \leqslant x \leqslant 2$

9. $y = \frac{x^4}{4} + \frac{1}{8x^2}$, $1 \leqslant x \leqslant 3$

10. $y = \frac{x^2}{2} - \frac{\ln x}{4}$, $2 \leqslant x \leqslant 4$

11. $y = \ln(\cos x)$, $0 \leqslant x \leqslant \frac{\pi}{4}$

12. $y = \ln(\sin x)$, $\pi/6 \leqslant x \leqslant \pi/3$

13. $y = \ln(1 - x^2)$, $0 \leqslant x \leqslant \frac{1}{2}$

14. $y = \ln\left(\frac{e^x + 1}{e^x - 1}\right)$, $a \leqslant x \leqslant b$, $a > 0$

15. $y = e^x, 0 \leqslant x \leqslant 1$ **16.** $y = \ln x, 1 \leqslant x \leqslant \sqrt{3}$

17. $y = \cosh x, 0 \leqslant x \leqslant 1$ **18.** $y^2 = 4x, 0 \leqslant y \leqslant 2$

In Exercises 19–24 set up, but do not evaluate, an integral for the length of the given curve.

19. $y = x^3, 0 \leqslant x \leqslant 1$

20. $y = x^4 - x^2, -1 \leqslant x \leqslant 2$

21. $y = \sin x, 0 \leqslant x \leqslant \pi$

22. $y = \tan x, 0 \leqslant x \leqslant \pi/4$

23. $y = e^x \cos x, 0 \leqslant x \leqslant \pi/2$

24. $\dfrac{x^2}{a^2} + \dfrac{y^2}{b^2} = 1$

25. Find the arc length function for the curve $y = 2x^{3/2}$ with starting point $P_0(1, 2)$.

26. (a) Sketch the curve $y = x^3/3 + 1/4x$.
(b) Find the arc length function for this curve with starting point $P_0(1, \frac{7}{12})$.

27. If a bomb is dropped from an aircraft flying at 200 m/s at an altitude of 4500 m, then the parabolic trajectory of the bomb is described by the equation

$$y = 4500 - \frac{x^2}{8000}$$

until it hits the ground, where y is its height above the ground and x is the horizontal distance traveled in meters. Calculate the distance traveled by the bomb from the time it is dropped to the time it hits the

ground. Express your answer correct to the nearest meter.

28. Figure 8.36 shows a telephone wire hanging between two poles at $x = -b$ and $x = b$. It takes the shape of a catenary with equation $y = a \cosh(x/a)$. Find the length of the wire.

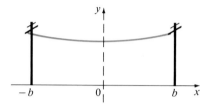

Figure 8.36

29. Sketch the curve with equation $x^{2/3} + y^{2/3} = 1$ and use symmetry to find its length.

30. (a) Sketch the curve $y^3 = x^2$.
(b) Use Formulas 8.11 and 8.12 to set up two integrals for the arc length from $(0, 0)$ to $(1, 1)$. Observe that one of these is an improper integral and evaluate both of them.
(c) Find the length of the arc of this curve from $(-1, 1)$ to $(8, 4)$.

In Exercises 31–34 use Simpson's Rule with $n = 10$ to estimate the arc length of the given curve.

31. $y = x^3, 0 \leqslant x \leqslant 1$ **32.** $y = x^4, 0 \leqslant x \leqslant 2$

33. $y = \sin x, 0 \leqslant x \leqslant \pi$ **34.** $y = \tan x, 0 \leqslant x \leqslant \pi/4$

SECTION 8.4

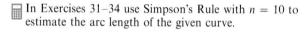

Area of a Surface of Revolution

A surface of revolution is formed when a curve is rotated about a line. Such a surface is the lateral boundary of a solid of a revolution of the type discussed in Sections 8.1 and 8.2.

We want to define the area of a surface of revolution in such a way that it corresponds to our intuition. We can think of peeling away a very thin outer layer of the solid of revolution and laying it out flat so that we can measure its area. Or, if the surface area is A, we can imagine that painting the surface would require the same amount of paint as a flat region with area A.

Let us start with some simple surfaces. The lateral surface area of a circular cylinder with radius r and height h is taken to be $A = 2\pi rh$ because we can imagine cutting the cylinder and unrolling it (as in Figure 8.37) to obtain a rectangle with dimensions $2\pi r$ and h.

Figure 8.37

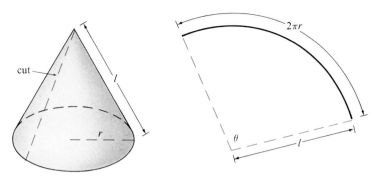

Figure 8.38

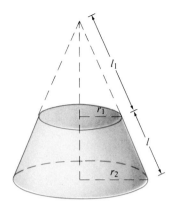

Figure 8.39

Likewise we can take a circular cone with base radius r and slant height l, cut it along the broken line in Figure 8.38, and flatten it to form a sector of a circle with radius l and central angle $\theta = 2\pi r/l$. We know that, in general, the area of a sector of a circle with radius l and angle θ is $\frac{1}{2}l^2\theta$ (see Exercise 33 in Section 7.3) and so in this case it is

$$A = \tfrac{1}{2}l^2\theta = \tfrac{1}{2}l^2\left(\frac{2\pi r}{l}\right) = \pi r l$$

Therefore we define the lateral surface area of a cone to be $A = \pi r l$.

The area of the band (or frustum of a cone) shown in Figure 8.39 with slant height l and upper and lower radii r_1 and r_2 is found by subtracting the areas of two cones:

(8.17) $$A = \pi r_2(l_1 + l) - \pi r_1 l_1 = \pi[(r_2 - r_1)l_1 + r_2 l]$$

From similar triangles we have

$$\frac{l_1}{r_1} = \frac{l_1 + l}{r_2}$$

which gives

$$r_2 l_1 = r_1 l_1 + r_1 l \qquad \text{or} \qquad (r_2 - r_1)l_1 = r_1 l$$

Putting this in Equation 8.17, we get

$$A = \pi(r_1 l + r_2 l)$$

(8.18)   or
$$\boxed{A = 2\pi r l}$$

where $r = \frac{1}{2}(r_1 + r_2)$ is the average radius of the band.

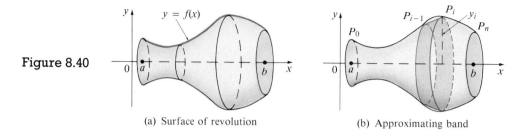

Figure 8.40

(a) Surface of revolution   (b) Approximating band

Now we consider the surface shown in Figure 8.40, which is obtained by rotating the curve $y = f(x)$, $a \leqslant x \leqslant b$, about the x-axis, where f is positive and has a continuous derivative. In order to define its surface area we use a method similar to the one for arc length. We take a partition P of $[a, b]$ by points $a = x_0, x_1, \ldots, x_n = b$, and let $y_i = f(x_i)$ so that the point $P_i(x_i, y_i)$ lies on the curve. The part of the surface between x_{i-1} and x_i is approximated by taking the line segment $P_{i-1}P_i$ and rotating it about the x-axis. The result is a band (a frustum of a cone) with slant height $l = |P_{i-1}P_i|$ and average radius $r = \frac{1}{2}(y_{i-1} + y_i)$ so, by Formula 8.18, its surface area is

$$2\pi \frac{y_{i-1} + y_i}{2} |P_{i-1}P_i|$$

As in the proof of Theorem 8.10, we have

$$|P_{i-1}P_i| = \sqrt{1 + [f'(x_i^*)]^2}\, \Delta x_i$$

where $x_i^* \in [x_{i-1}, x_i]$. When Δx_i is small we have $y_i = f(x_i) \approx f(x_i^*)$ and $y_{i-1} = f(x_{i-1}) \approx f(x_i^*)$ since f is continuous. Therefore

$$2\pi \frac{y_{i-1} + y_i}{2} |P_{i-1}P_i| \approx 2\pi f(x_i^*)\sqrt{1 + [f'(x_i^*)]^2}\, \Delta x_i$$

and so an approximation to what we think of as the area of the complete surface of revolution is

(8.19)
$$\sum_{i=1}^{n} 2\pi f(x_i^*)\sqrt{1 + [f'(x_i^*)]^2}\, \Delta x_i$$

This approximation appears to become better as $\|P\| \to 0$ and, recognizing (8.19) as a Riemann sum for the function $g(x) = 2\pi f(x)\sqrt{1 + [f'(x)]^2}$, we have

$$\lim_{\|P\| \to 0} \sum_{i=1}^{n} 2\pi f(x_i^*)\sqrt{1 + [f'(x_i^*)]^2}\, \Delta x_i = \int_a^b 2\pi f(x)\sqrt{1 + [f'(x)]^2}\, dx$$

Therefore, in the case where f is positive and has a continuous derivative, we define the **surface area** of the surface obtained by rotating the curve $y = f(x)$, $a \leqslant x \leqslant b$, about the x-axis as

(8.20)
$$S = \int_a^b 2\pi f(x)\sqrt{1 + [f'(x)]^2}\, dx$$

With the Leibniz notation for derivatives, this formula becomes

(8.21)

$$S = \int_a^b 2\pi y \sqrt{1 + \left(\frac{dy}{dx}\right)^2}\, dx$$

If the curve is described as $x = g(y)$, $c \leqslant y \leqslant d$, then the formula for surface area becomes

(8.22)

$$S = \int_c^d 2\pi y \sqrt{1 + \left(\frac{dx}{dy}\right)^2}\, dy$$

and both Formula 8.21 and Formula 8.22 can be summarized symbolically, using the notation for arc length given in Section 8.3, as

(8.23)

$$S = \int 2\pi y\, ds$$

For rotation about the y-axis, the surface area formula becomes

(8.24)

$$S = \int 2\pi x\, ds$$

where, as before, we can use either

$$ds = \sqrt{1 + \left(\frac{dy}{dx}\right)^2}\, dx \qquad \text{or} \qquad ds = \sqrt{1 + \left(\frac{dx}{dy}\right)^2}\, dy$$

These formulas can be remembered by thinking of $2\pi y$ or $2\pi x$ as the circumference of a circle traced out by the point (x, y) on the curve as it is rotated about the x-axis or y-axis, respectively (see Figure 8.41).

Figure 8.41

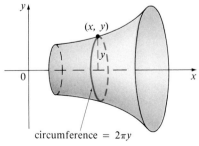

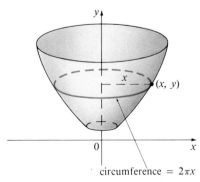

circumference = $2\pi y$ circumference = $2\pi x$

(a) Rotation about x-axis: $S = \int 2\pi y\, ds$ (b) Rotation about y-axis: $S = \int 2\pi x\, ds$

● **Example 1**   The curve $y = \sqrt{4 - x^2}$, $-1 \leqslant x \leqslant 1$, is an arc of the circle $x^2 + y^2 = 4$. Find the area of the surface obtained by rotating this arc about the x-axis. (The surface is a portion of a sphere of radius 2.)

Solution   We have

$$\frac{dy}{dx} = \frac{1}{2}(4 - x^2)^{-1/2}(-2x) = \frac{x}{\sqrt{4 - x^2}}$$

and so, by Formula 8.21, the surface area is

$$S = \int_{-1}^{1} 2\pi y \sqrt{1 + \left(\frac{dy}{dx}\right)^2}\, dx$$

$$= 2\pi \int_{-1}^{1} \sqrt{4 - x^2} \sqrt{1 + \frac{x^2}{4 - x^2}}\, dx$$

$$= 2\pi \int_{-1}^{1} \sqrt{4 - x^2} \frac{2}{\sqrt{4 - x^2}}\, dx$$

$$= 4\pi \int_{-1}^{1} 1\, dx = 4\pi(2) = 8\pi$$

●

● **Example 2**   The arc of the parabola $y = x^2$ from $(1, 1)$ to $(2, 4)$ is rotated about the y-axis. Find the area of the resulting surface.

Solution 1   Using

$$y = x^2 \qquad \text{and} \qquad \frac{dy}{dx} = 2x$$

we have, from Formula 8.24,

$$S = \int 2\pi x\, ds = \int_{1}^{2} 2\pi x \sqrt{1 + \left(\frac{dy}{dx}\right)^2}\, dx$$

$$= 2\pi \int_{1}^{2} x\sqrt{1 + 4x^2}\, dx$$

Substituting $u = 1 + 4x^2$, we have $du = 8x\, dx$. Remembering to change the limits of integration, we have

$$S = \frac{\pi}{4} \int_{5}^{17} \sqrt{u}\, du = \frac{\pi}{4}\left[\frac{2}{3} u^{3/2}\right]_{5}^{17}$$

$$= \frac{\pi}{6}(17\sqrt{17} - 5\sqrt{5})$$

●

Solution 2   Using

$$x = \sqrt{y} \qquad \text{and} \qquad \frac{dx}{dy} = \frac{1}{2\sqrt{y}}$$

we have

$$S = \int 2\pi x \, ds = \int_1^4 2\pi x \sqrt{1 + \left(\frac{dx}{dy}\right)^2} \, dy$$

$$= 2\pi \int_1^4 \sqrt{y} \sqrt{1 + \frac{1}{4y}} \, dy$$

$$= \pi \int_1^4 \sqrt{4y + 1} \, dy$$

$$= \frac{\pi}{4} \int_5^{17} \sqrt{u} \, du \qquad (where \ u = 1 + 4y)$$

$$= \frac{\pi}{6} (17\sqrt{17} - 5\sqrt{5}) \qquad (as \ in \ Solution \ 1)$$

● **Example 3**   Find the area of the surface generated by rotating the curve $y = e^x, 0 \leqslant x \leqslant 1$, about the x-axis.

Solution 1   Using Formula 8.21 with

$$y = e^x \qquad and \qquad \frac{dy}{dx} = e^x$$

we have

$$S = \int_0^1 2\pi y \sqrt{1 + \left(\frac{dy}{dx}\right)^2} \, dx = 2\pi \int_0^1 e^x \sqrt{1 + e^{2x}} \, dx$$

$$= 2\pi \int_1^e \sqrt{1 + u^2} \, du \qquad (where \ u = e^x)$$

$$= 2\pi \int_{\pi/4}^{\alpha} \sec^3 \theta \, d\theta \qquad (where \ u = \tan \theta \ and \ \alpha = \tan^{-1} e)$$

$$= 2\pi \cdot \frac{1}{2} \left[\sec \theta \tan \theta + \ln|\sec \theta + \tan \theta| \right]_{\pi/4}^{\alpha} \qquad (by \ Example \ 9 \ in \ Section \ 7.2)$$

$$= \pi[\sec \alpha \tan \alpha + \ln(\sec \alpha + \tan \alpha) - \sqrt{2} - \ln(\sqrt{2} + 1)]$$

Since $\tan \alpha = e$, we have $\sec^2 \alpha = 1 + \tan^2 \alpha = 1 + e^2$ and

$$S = \pi[e\sqrt{1 + e^2} + \ln(e + \sqrt{1 + e^2}) - \sqrt{2} - \ln(\sqrt{2} + 1)]$$

Solution 2   Using Formula 8.22 with

$$x = \ln y \qquad and \qquad \frac{dx}{dy} = \frac{1}{y}$$

we have

$$S = \int_1^e 2\pi y \sqrt{1 + \left(\frac{dx}{dy}\right)^2} \, dy = 2\pi \int_1^e y \sqrt{1 + \frac{1}{y^2}} \, dy$$

$$= 2\pi \int_1^e \sqrt{y^2 + 1} \, dy$$

which is evaluated as in Solution 1.

EXERCISES 8.4

In Exercises 1–16 find the area of the surface obtained by rotating the given curve about the x-axis.

1. $y = \sqrt{x}, 4 \leqslant x \leqslant 9$

2. $y^2 = 4x + 4, 0 \leqslant x \leqslant 8$

3. $2y = x + 4, 0 \leqslant x \leqslant 2$

4. $y = x^3, 0 \leqslant x \leqslant 2$

5. $y = x^3 + \dfrac{1}{12x}, 1 \leqslant x \leqslant 2$

6. $y = \dfrac{x^2}{4} - \dfrac{\ln x}{2}, 1 \leqslant x \leqslant 4$

7. $y = \sin x, 0 \leqslant x \leqslant \pi$ **8.** $y = \cos x, 0 \leqslant x \leqslant \dfrac{\pi}{3}$

9. $y = \cosh x, 0 \leqslant x \leqslant 1$ **10.** $2y = 3x^{2/3}, 1 \leqslant x \leqslant 8$

11. $x = \dfrac{y^4}{2} + \dfrac{1}{16y^2}, 1 \leqslant y \leqslant 3$

12. $x = \frac{1}{3}(y^2 + 2)^{3/2}, 1 \leqslant y \leqslant 2$

13. $x = 1 + 2y^2, 1 \leqslant y \leqslant 2$

14. $x = 2\ln y, 1 \leqslant y \leqslant \sqrt{3}$

15. $y = \sqrt{x^2 - 1}, 1 \leqslant x \leqslant 2$

16. $y = x^2, 0 \leqslant x \leqslant 1$

In Exercises 17–24 the given curve is rotated about the y-axis. Find the area of the resulting surface.

17. $y = \sqrt[3]{x}, 1 \leqslant y \leqslant 2$

18. $x = \sqrt{2y - y^2}, 0 \leqslant y \leqslant 1$

19. $y^2 = x^3, 1 \leqslant y \leqslant 8$

20. $4x + 3y = 19, 1 \leqslant x \leqslant 4$

21. $x = e^{2y}, 0 \leqslant y \leqslant \frac{1}{2}$

22. $y = 1 - x^2, 0 \leqslant x \leqslant 1$

23. $x = \dfrac{1}{2\sqrt{2}}(y^2 - \ln y), 1 \leqslant y \leqslant 2$

24. $x = a\cosh\left(\dfrac{y}{a}\right), -a \leqslant y \leqslant a$

In Exercises 25 and 26 use Simpson's Rule with $n = 10$ to find the area of the surface obtained by rotating the given curve about the x-axis.

25. $y = x^4, 0 \leqslant x \leqslant 1$

26. $y = \tan x, 0 \leqslant x \leqslant \pi/4$

27. Find the surface area generated by rotating a loop of the curve $8y^2 = x^2(1 - x^2)$ about the x-axis.

28. Find the surface area generated by rotating the loop of the curve $9ay^2 = x(3a - x)^2$ about the y-axis.

29. If the loop in Exercise 28 is rotated about the x-axis, find the area of the resulting surface.

30. If the infinite curve $y = e^{-x}, x \geqslant 0$, is rotated about the x-axis, find the area of the resulting surface.

31. If the region $\mathcal{R} = \{(x, y)\,|\,x \geqslant 1, 0 \leqslant y \leqslant 1/x\}$ is rotated about the x-axis, the volume of the resulting solid is finite (see Exercise 65 in Section 8.1). Show that the surface area is infinite.

32. Find the surface area of the torus in Exercise 49 in Section 8.1.

33. The ellipse

$$\frac{x^2}{a^2} + \frac{y^2}{b^2} = 1 \qquad a > b$$

is rotated about the x-axis to form a surface called an ellipsoid. Find the surface area of this ellipsoid.

34. Show that the surface area of a zone of a sphere that lies between two parallel planes is $S = \pi dh$, where d is the diameter of the sphere and h is the distance between the planes. (Notice that S depends only on the distance between the planes and not on their location, provided that both planes intersect the sphere.)

35. Formula 8.20 is valid only when $f(x) \geqslant 0$. Show that when $f(x)$ is not necessarily positive, the formula for surface area becomes

$$S = \int_a^b 2\pi |f(x)| \sqrt{1 + [f'(x)]^2}\, dx$$

36. If the curve $y = f(x), a \leqslant x \leqslant b$, is rotated about the horizontal line $y = c$, where $f(x) \leqslant c$, find a formula for the area of the resulting surface.

37. Find the area of the surface obtained by rotating the circle $x^2 + y^2 = r^2$ about the line $y = r$.

38. Let L be the length of the curve $y = f(x), a \leqslant x \leqslant b$, where f is positive and has a continuous derivative. Let S_f be the surface area generated by rotating the curve about the x-axis. If c is a positive constant, define $g(x) = f(x) + c$ and let S_g be the corresponding surface area generated by the curve $y = g(x)$, $a \leqslant x \leqslant b$. Express S_g in terms of S_f and L.

SECTION 8.5

Work

The term **work** is used in everyday language to mean the total amount of effort required to perform a task. In physics it has a technical meaning that depends on the idea of a **force.** Intuitively you can think of a force as describing a push or pull on an object—for example, a horizontal push of a book across a table or the downward pull of the earth's gravity on a ball. In general, if an object moves along a straight line with position function $s(t)$, then the force F on the object (in that same direction) is defined by Newton's second law of motion as the product of its mass m and its acceleration:

$$F = m\frac{d^2s}{dt^2}$$

(8.25)

In the SI metric system, the mass is measured in kilograms (kg), the displacement in meters (m), the time in seconds (s), and the force in newtons ($N = \text{kg-m/s}^2$). Thus a force of 1 N acting on a mass of 1 kg produces an acceleration of 1 m/s². In the British engineering system the fundamental unit is chosen to be the unit of force, which is the pound.

In the case of constant acceleration, the force F is also constant and the work done is defined to be the product of the force F and the distance d that the object moves:

$$W = Fd \qquad \text{work} = \text{force} \times \text{distance}$$

(8.26)

If F is measured in newtons and d in meters, then the unit for W is a newton-meter, which is called a joule (J). If F is measured in pounds and d in feet, then the unit for W is a foot-pound (ft-lb), which is about 1.36 J.

● **Example 1**   (a)   How much work is done in lifting a 1.2-kg book off the floor to put it on a desk that is 0.7 m high? Use the fact that the acceleration due to gravity is $g = 9.8$ m/s².   (b)   How much work is done in lifting a 20-lb weight 6 ft off the ground?

Solution   (a)   The force exerted is equal and opposite to that exerted by gravity, so Equation 8.25 gives

$$F = mg = (1.2)(9.8) = 11.76 \text{ N}$$

and then Equation 8.26 gives the work done as

$$W = Fd = (11.76)(0.7) \approx 8.2 \text{ J}$$

(b)   Here the force is given as $F = 20$ lb, so the work done is

$$W = Fd = 20 \cdot 6 = 120 \text{ ft-lb}$$

Notice that in part (b), unlike part (a), we did not have to multiply by g because we were given the *weight* (which is a force) and not the mass of the object.   ●

Equation 8.26 defines work as long as the force is constant, but what happens if the force is variable? Let us suppose that the object moves along the x-axis in the positive direction from $x = a$ to $x = b$ and at each point x between a and b a force $f(x)$ acts on the object, where f is a continuous function. Let P be a partition of $[a, b]$ by points x_i $(i = 1, 2, \ldots, n)$ and let $\Delta x_i = x_i - x_{i-1}$. Choose any x_i^* in the ith subinterval $[x_{i-1}, x_i]$. Then the force at that point is $f(x_i^*)$. If $\|P\|$ is small, then Δx_i is small, and since f is continuous the values of f do not change very much over the interval $[x_{i-1}, x_i]$. In other words, f is almost constant on the interval and so the work W_i that is done in moving the particle from x_{i-1} to x_i is approximately given by Equation 8.26:

$$W_i \approx f(x_i^*)\,\Delta x_i$$

Thus we can approximate the total work by

(8.27)
$$W \approx \sum_{i=1}^{n} f(x_i^*)\,\Delta x_i$$

It seems that this approximation becomes better as we make $\|P\|$ smaller (and therefore n larger). Therefore we define the **work done in moving the object from a to b** as the limit of this quantity as $\|P\| \to 0$. Since the right side of (8.27) is a Riemann sum, we recognize its limit as being a definite integral and so

(8.28)
$$W = \lim_{\|P\| \to 0} \sum_{i=1}^{n} f(x_i^*)\,\Delta x_i = \int_a^b f(x)\,dx$$

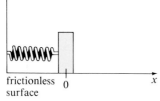

frictionless 0 surface

(a) Natural position of spring

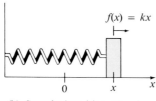

(b) Stretched position of spring

Figure 8.42

HOOKE'S LAW

● **Example 2**   When a particle is at a distance x feet from the origin, a force of $x^2 + 2x$ pounds acts on it. How much work is done in moving it from $x = 1$ to $x = 3$?

Solution $W = \int_1^3 (x^2 + 2x)\,dx = \dfrac{x^3}{3} + x^2 \Big]_1^3 = \dfrac{50}{3}$

The work done is $16\frac{2}{3}$ ft-lb.   ●

In the next example we use **Hooke's Law** from physics, which states that the force required to stretch a spring x units beyond its natural length is proportional to x:

$$f(x) = kx$$

where k is a positive constant (called the **spring constant**). Hooke's Law holds provided that x is not too large (see Figure 8.42).

● **Example 3**   A force of 40 N is required to stretch a spring from its natural length of 10 cm to a length of 15 cm. How much work is done in stretching the spring from 15 cm to 18 cm?

Solution   According to Hooke's Law, the force required to stretch the spring x meters beyond its natural length is $f(x) = kx$. In stretching the spring from 10 cm to 15 cm the amount stretched is 5 cm = 0.05 m. This means that $f(0.05) = 40$, so

$$0.05k = 40 \qquad k = \frac{40}{0.05} = 800$$

Thus $f(x) = 800x$ and the work done in stretching the spring from 15 cm to 18 cm is

$$W = \int_{0.05}^{0.08} 800x \, dx = 800 \frac{x^2}{2}\Bigg]_{0.05}^{0.08}$$

$$= 400[(0.08)^2 - (0.05)^2] = 1.56 \text{ J}$$

● **Example 4**   A tank has the shape of an inverted circular cone with height 10 m and base radius 4 m. It is filled with water to a height of 8 m. Find the work required to empty the tank by pumping all of the water to the top of the tank. (The density of water is 1000 kg/m^3.)

Solution   Let us measure depths from the top of the tank by introducing a vertical coordinate line as in Figure 8.43(a). The water extends from a depth of 2 m to a depth of 10 m so we take a partition P of the interval $[2, 10]$ by points x_i with $2 = x_0 < x_1 < \cdots < x_n = 10$ and choose x_i^* in the ith subinterval. This divides the water into n layers. The ith layer is approximated by a circular cylinder with radius r_i and height Δx_i. We can compute r_i from similar triangles using Figure 8.43(b) as follows:

$$\frac{r_i}{10 - x_i^*} = \frac{4}{10} \qquad r_i = \frac{2}{5}(10 - x_i^*)$$

Thus an approximation to the volume of the ith layer of water is

$$V_i \approx \pi r_i^2 \, \Delta x_i = \frac{4\pi}{25}(10 - x_i^*)^2 \, \Delta x_i$$

and so its mass is

$$m_i = \text{density} \times \text{volume}$$

$$\approx 1000 \cdot \frac{4\pi}{25}(10 - x_i^*)^2 \, \Delta x_i = 160\pi(10 - x_i^*)^2 \, \Delta x_i$$

The force required to raise this layer must overcome the force of gravity and so

$$F_i = m_i g \approx (9.81)160\pi(10 - x_i^*)^2 \, \Delta x_i$$

$$\approx 1570\pi(10 - x_i^*)^2 \, \Delta x_i$$

Each particle in the layer must travel a distance of approximately x_i^*. The work W_i done to raise this layer to the top is approximately the product of

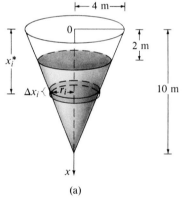

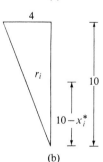

Figure 8.43

the force F_i and the distance x_i^*:

$$W_i \approx F_i x_i^* \approx 1570\pi x_i^*(10 - x_i^*)^2 \, \Delta x_i$$

To find the total work done in emptying the entire tank, we add the contributions of each of the n layers and then take the limit as $\|P\| \to 0$:

$$
\begin{aligned}
W &= \lim_{\|P\| \to 0} \sum_{i=1}^{n} 1570\pi x_i^*(10 - x_i^*)^2 \, \Delta x_i \\
&= \int_2^{10} 1570\pi x(10 - x)^2 \, dx \\
&= 1570\pi \int_2^{10} (100x - 20x^2 + x^3) \, dx \\
&= 1570\pi \left[50x^2 - \frac{20x^3}{3} + \frac{x^4}{4} \right]_2^{10} \\
&= 1570\pi \left(\frac{2048}{3} \right) \approx 3.37 \times 10^6 \text{ J} \qquad \bullet
\end{aligned}
$$

EXERCISES 8.5

1. Find the work done in pushing a car a distance of 8 m while exerting a constant force of 900 N.

2. How much work is done by a weightlifter in raising a 60-kg barbell from the floor to a height of 2 m?

3. A particle is moved along the x-axis by a force that measures $5x^2 + 1$ pounds at a point x feet from the origin. Find the work done in moving the particle from the origin to a distance of 10 ft.

4. When a particle is at a distance x meters from the origin, a force of $\cos(\pi x/3)$ newtons acts on it. How much work is done in moving the particle from $x = 1$ to $x = 2$?

5. A spring is stretched 4 in. beyond its natural length by a force of 10 lb. How much work is done in stretching it from its natural length to 6 in. beyond its natural length?

6. A spring has a natural length of 20 cm. If a 25-N force is required to stretch it to a length of 30 cm, how much work is required to stretch it from 20 cm to 25 cm?

7. Suppose that 2 joules of work are needed to stretch a spring from its natural length of 30 cm to a length of 42 cm. How much work is needed to stretch it from 35 cm to 40 cm?

8. If the work required to stretch a spring 1 ft beyond its natural length is 12 ft-lb, how much work is needed to stretch it 9 in. beyond its natural length?

9. How many centimeters will a force of 30 N stretch the spring in Exercise 7?

10. If 6 joules of work are needed to stretch a spring from 10 cm to 12 cm and another 10 joules are needed to stretch it from 12 cm to 14 cm, what is the natural length of the spring?

11. A heavy rope, 50 ft long, weighs 0.5 lb/ft and hangs over the edge of a building 120 ft high. How much work is done in pulling this rope to the top of the building?

12. A uniform cable hanging over the edge of a tall building is 40 ft long and weighs 60 lb. How much work is required to pull 10 ft of the cable to the top?

13. A cable that weighs 2 lb/ft is used to lift 800 lb of coal up a mineshaft 500 ft deep. Find the work done.

14. A bucket that weighs 4 lb and a rope of negligible weight are used to draw water from a well that is 80 ft deep. The bucket starts with 40 lb of water and is pulled up at a rate of 2 ft/s but water leaks out of a hole at a rate of 0.2 lb/s. Find the work done in pulling the bucket to the top of the well.

15. An aquarium 2 m long, 1 m wide, and 1 m deep is full of water. Find the work needed to pump half of the water out of the aquarium.

16. A circular swimming pool has a diameter of 24 ft, the sides are 5 ft high, and the depth of the water is 4 ft. How much work is required to pump all of the water out over the side?

In Exercises 17–20 a tank is full of water. Find the work required to pump the water out of the outlet. In Exercises 19 and 20 use the fact that water weighs 62.5 lb/ft^3.

17.   **18.**

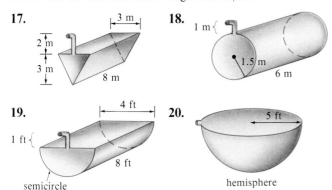

19.   **20.**

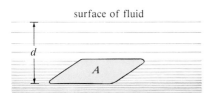

21. Solve Exercise 17 if the tank is filled with gasoline that has a density of 680 kg/m^3.

22. Solve Exercise 18 if the tank is half full of oil that has a density of 920 kg/m^3.

23. When gas expands in a cylinder with radius r, the pressure at any given time is a function of the volume: $P = P(V)$. The force exerted by the gas on the piston (see

the figure) is the product of the pressure and the area: $F = \pi r^2 P$. Show that the work done by the gas when the volume expands from volume V_1 to volume V_2 is

$$W = \int_{V_1}^{V_2} P \, dV$$

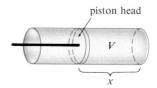
piston head

24. In a steam engine the pressure P and volume V of steam satisfy the equation $PV^{1.4} = k$, where k is a constant. (This is true for adiabatic expansion—that is, expansion in which there is no heat transfer between the cylinder and its surroundings.) Use Exercise 23 to calculate the work done by the engine during a cycle when the steam starts at a pressure of 160 lb/in.^{2} and a volume of 100 in.^{3} and expands to a volume of 800 in.^{3}

25. Newton's Law of Gravitation states that two bodies with masses m_1 and m_2 attract each other with a force

$$F = G \frac{m_1 m_2}{r^2}$$

where r is the distance between the bodies and G is the gravitational constant. If one of the bodies is fixed, find the work needed to move the other from $r = a$ to $r = b$.

26. Use Newton's Law of Gravitation to compute the work required to launch a 1000-kg satellite vertically to an orbit 1000 km high. You may assume that the mass of the earth is 5.98×10^{24} kg and is concentrated at the center of the earth. Take the radius of the earth to be 6.37×10^6 m and $G = 6.67 \times 10^{-11}$ N-m^2/kg^2.

SECTION 8.6

Hydrostatic Pressure

Divers realize that water pressure increases as they dive deeper. This is because the weight of the water above them increases.

In general, suppose that a thin horizontal plate with area A square meters is submerged in a fluid of density ρ kilograms per cubic meter at a depth d meters below the surface of the fluid as in Figure 8.44. The fluid directly above the plate has volume $V = Ad$ so its mass is $m = \rho V = \rho A d$. The force exerted by the fluid on the plate is therefore

$$F = mg = \rho g d A$$

Figure 8.44

where g is the acceleration due to gravity. The pressure P on the plate is defined to be the force per unit area:

$$P = \frac{F}{A} = \rho g d$$

The SI unit for measuring pressure is a newton per square meter, which is called a pascal (abbreviation: $1 \text{ N/m}^2 = 1$ Pa). Since this is a small unit, the kilopascal (kPa) is often used. For instance, since the density of water is $\rho = 1000 \text{ kg/m}^3$, the pressure at the bottom of a swimming pool 2 m deep is

$$P = \rho g d = 1000 \text{ kg/m}^3 \times 9.8 \text{ m/s}^2 \times 2 \text{ m}$$

$$= 19{,}600 \text{ Pa} = 19.6 \text{ kPa}$$

When using British units, we write $P = \rho g d = \delta d$, where $\delta = \rho g$ is the weight density (as opposed to ρ, which is the mass density). For instance, the weight density of water is $\delta = 62.5 \text{ lb/ft}^3$.

An important principle of fluid pressure is the experimentally verified fact that *at any point in a liquid the pressure is the same in all directions.* (A diver feels the same pressure on both ears and his nose.) Thus the pressure in *any* direction at a depth d in a fluid with mass density ρ is given by

(8.29)

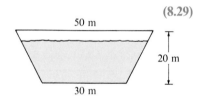

$$P = \rho g d = \delta d$$

This will help us determine the hydrostatic force against a *vertical* plate or wall or dam in a fluid. This is not a straightforward problem since the pressure is not constant but increases as the depth increases.

Figure 8.45

● **Example 1**   A dam has the shape of the trapezoid shown in Figure 8.45. The height is 20 m and the width is 50 m at the top and 30 m at the bottom. Find the force on the dam due to hydrostatic pressure if the water level is 4 m from the top of the dam.

Solution   We choose a vertical x-axis with origin at the surface of the water as in Figure 8.46(a). The depth of the water is 16 m so we consider a partition P of the interval $[0, 16]$ by points x_i and we choose $x_i^* \in [x_{i-1}, x_i]$. The ith horizontal strip of the dam is approximated by a rectangle with height Δx_i and width w_i where, from similar triangles in Figure 8.46(b),

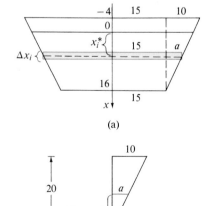

(a)

(b)

Figure 8.46

$$\frac{a}{16 - x_i^*} = \frac{10}{20} \qquad a = \frac{16 - x_i^*}{2} = 8 - \frac{x_i^*}{2}$$

$$w_i = 2(15 + a) = 2\left(15 + 8 - \frac{x_i^*}{2}\right) = 46 - x_i^*$$

If A_i is the area of the ith strip, then

$$A_i \approx w_i \Delta x_i = (46 - x_i^*) \Delta x_i$$

If Δx_i is small, then the pressure P_i on the ith strip is almost constant and we can use Equation 8.29 to write

$$P_i \approx 1000gx_i^*$$

The hydrostatic force F_i acting on the ith strip is the product of the pressure and the area:

$$F_i = P_i A_i \approx 1000gx_i^*(46 - x_i^*)\Delta x_i$$

Adding these forces and taking the limit as $\|P\| \to 0$, we obtain the total hydrostatic force on the dam:

$$
\begin{aligned}
F &= \lim_{\|P\| \to 0} \sum_{i=1}^{n} 1000gx_i^*(46 - x_i^*)\Delta x_i \\
&= \int_0^{16} 1000gx(46 - x)\,dx \\
&= 1000(9.81)\int_0^{16} (46x - x^2)\,dx \\
&= 9810\left[23x^2 - \frac{x^3}{3} \right]_0^{16} \\
&\approx 4.44 \times 10^7 \text{ N}
\end{aligned}
$$

● **Example 2**   Find the hydrostatic force on one end of a cylindrical drum with radius 3 ft if it is submerged in water 10 ft deep.

Solution   In this example it is convenient to choose the axes as in Figure 8.47 so that the origin is placed at the center of the drum. Then the circle has a simple equation, $x^2 + y^2 = 9$. As in Example 1 we divide the circular region into horizontal strips. From the equation of the circle, we see that the length of the ith strip is $2\sqrt{9 - (y_i^*)^2}$ and so its area is

$$A_i = 2\sqrt{9 - (y_i^*)^2}\,\Delta y_i$$

The pressure on this strip is approximately

$$\delta d_i = 62.5(7 - y_i^*)$$

and so the force on the strip is approximately

$$\delta d_i A_i = 62.5(7 - y_i^*)2\sqrt{9 - (y_i^*)^2}\,\Delta y_i$$

The total force is obtained by adding the forces on all the strips and taking the limit:

$$
\begin{aligned}
F &= \lim_{\|P\| \to 0} \sum_{i=1}^{n} 62.5(7 - y_i^*)2\sqrt{9 - (y_i^*)^2}\,\Delta y_i \\
&= 125\int_{-3}^{3}(7 - y)\sqrt{9 - y^2}\,dy \\
&= 125 \cdot 7\int_{-3}^{3}\sqrt{9 - y^2}\,dy - 125\int_{-3}^{3} y\sqrt{9 - y^2}\,dy
\end{aligned}
$$

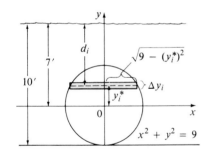

Figure 8.47

The second integral is 0 because the integrand is an odd function (see Theorem 5.33). The first integral can be evaluated using the trigonometric substitution $y = 3\sin\theta$ but it is simpler to observe that it is the area of a semicircular disk with radius 3. Thus

$$F = 875 \int_{-3}^{3} \sqrt{9 - y^2}\, dy = 875 \cdot \frac{1}{2}\pi(3)^2$$

$$= \frac{7875\pi}{2} \approx 12{,}370 \text{ lb}$$

EXERCISES 8.6

1. An aquarium 2 m long, 1 m wide, and 1 m deep is full of water. Find (a) the hydrostatic pressure on the bottom of the aquarium, (b) the hydrostatic force on the bottom, and (c) the hydrostatic force on one end of the aquarium.

2. A swimming pool 5 m wide, 10 m long, and 3 m deep is filled with seawater of density 1030 kg/m³ to a depth of 2.5 m. Find (a) the hydrostatic pressure at the bottom of the pool, (b) the hydrostatic force on the bottom, and (c) the hydrostatic force on one end of the pool.

In Exercises 3–12 a tank contains water. The end of the tank is vertical and has the indicated shape. Find the hydrostatic force against the end of the tank.

3. |←—10 m—→|

4. |←— 10 m—→|
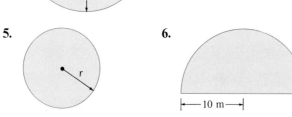
water level
5 m

5.
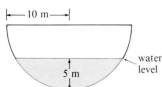

6.
|←— 10 m —→|

7.
6 ft
4 ft

8.
h
b

9.
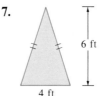
water level
4 ft
4 ft

10.
b
h

11.
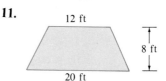
12 ft
20 ft
8 ft

12.
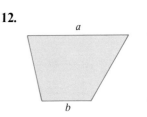
a
b
h

13. A trough is filled with a liquid that has a density of 840 kg/m³. The ends of the trough are equilateral triangles with sides 8 m and vertex at the bottom. Find the hydrostatic force on one end of the trough.

14. Do Exercise 13 if the height of the liquid is 4 m.

15. A cube with side 20 cm is sitting on the bottom of an aquarium in which the water is 1 m deep. Find the hydrostatic force on (a) the top of the cube and (b) one of the sides of the cube.

16. A vertical dam has a semicircular gate as shown in the figure. Find the hydrostatic force against the gate.

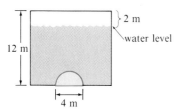

17. A trough 2 m high is filled with water. Its ends are vertical but its sides are 3 m wide and are inclined at an angle of 45° from the vertical. Find the hydrostatic force on one of the sides of the trough.

18. A dam is inclined at an angle of 30° from the vertical and has the shape of an isosceles trapezoid 100 ft wide at the top and 50 ft wide at the bottom and with a slant height of 70 ft. Find the hydrostatic force on the dam when it is full of water.

19. A swimming pool is 20 ft wide and 40 ft long and its bottom is an inclined plane, the shallow end having a depth of 3 ft, the deep end 9 ft. Find the hydrostatic force on (a) the shallow end, (b) the deep end, (c) one of the sides, and (d) the bottom.

20. Suppose that a plate is immersed vertically in a fluid with density ρ and the width of the plate is $w(x)$ at a depth of x meters beneath the surface of the fluid. If the top of the plate is at depth a and the bottom at depth b, find a formula for the hydrostatic force on the plate.

SECTION 8.7

Moments and Centers of Mass

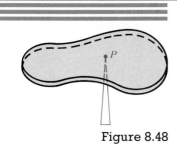

Figure 8.48

The main object of this section is to find the point P on which a thin plate of any given shape will balance horizontally as in Figure 8.48. This point is called the **center of mass** (or center of gravity) of the plate.

We first consider the simpler situation illustrated in Figure 8.49 where two masses m_1 and m_2 are attached to a rod of negligible mass on opposite sides of a fulcrum and at distances d_1 and d_2 from the fulcrum. The rod will balance if

(8.30)
$$m_1 d_1 = m_2 d_2$$

This is an experimental fact discovered by Archimedes and called the Law of the Lever. (Think of a lighter person balancing a heavier one on a seesaw by sitting farther away from the center.)

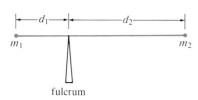

Figure 8.49

Now suppose that the rod lies along the x-axis with m_1 at x_1 and m_2 at x_2 and the center of mass at $\bar{x}$. If we compare Figures 8.49 and 8.50, we see that $d_1 = \bar{x} - x_1$ and $d_2 = x_2 - \bar{x}$ and so Equation 8.30 gives

$$m_1(\bar{x} - x_1) = m_2(x_2 - \bar{x})$$

$$m_1\bar{x} + m_2\bar{x} = m_1 x_1 + m_2 x_2$$

(8.31)
$$\bar{x} = \frac{m_1 x_1 + m_2 x_2}{m_1 + m_2}$$

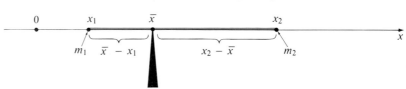

Figure 8.50

The numbers $m_1 x_1$ and $m_2 x_2$ are called the **moments** of the masses m_1 and m_2 (with respect to the origin), and Equation 8.31 says that the center of mass $\bar{x}$ is obtained by adding the moments of the masses and dividing by the total mass $m = m_1 + m_2$.

In general if we have a system of n particles with masses $m_1, m_2, \ldots, m_n$ located at the points $x_1, x_2, \ldots, x_n$ on the x-axis, it can be shown similarly that the center of mass of the system is located at

$$(8.32) \qquad \bar{x} = \frac{\sum\limits_{i=1}^{n} m_i x_i}{\sum\limits_{i=1}^{n} m_i} = \frac{\sum\limits_{i=1}^{n} m_i x_i}{m}$$

where $m = \sum m_i$ is the total mass of the system, and the sum of the individual moments

$$(8.33) \qquad M = \sum_{i=1}^{n} m_i x_i$$

is called the moment of the system with respect to the origin. Then Equation 8.32 could be rewritten as $m\bar{x} = M$, which says that if the total mass were considered as being concentrated at the center of mass $\bar{x}$, then its moment would be the same as the moment of the system.

● **Example 1**   Find the center of mass of a system of four objects with masses 10 g, 45 g, 32 g, and 24 g that are located at the points -4, 1, 3, and 8, respectively, on the x-axis.

Solution   Using Equation 8.32 we have

$$\bar{x} = \frac{10(-4) + 45(1) + 32(3) + 24(8)}{10 + 45 + 32 + 24}$$

$$= \frac{293}{111}$$

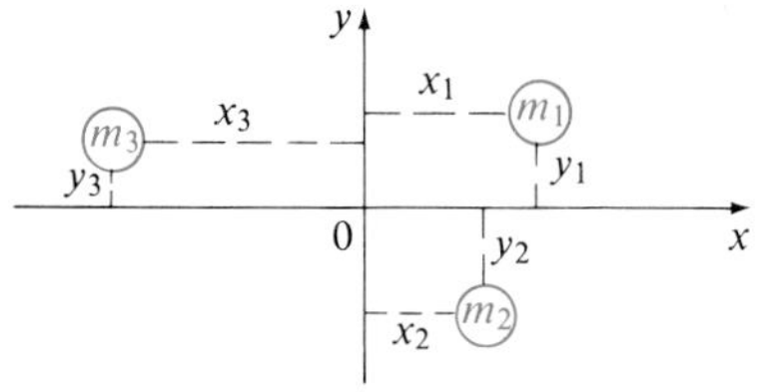

Figure 8.51

Now we consider a system of n particles with masses $m_1, m_2, \ldots, m_n$ located at the points $(x_1, y_1), (x_2, y_2), \ldots, (x_n, y_n)$ in the xy-plane as shown in Figure 8.51. By analogy with the one-dimensional case we define the **moment of the system about the y-axis** to be

$$(8.34) \qquad M_y = \sum_{i=1}^{n} m_i x_i$$

and the **moment of the system about the x-axis** as

$$(8.35) \qquad M_x = \sum_{i=1}^{n} m_i y_i$$

Then M_y measures the tendency to rotate about the y-axis and M_x measures the tendency to rotate about the x-axis.

As in the one-dimensional case, the coordinates $(\bar{x}, \bar{y})$ of the center of mass are given in terms of the moments by the formulas

(8.36)

$$\bar{x} = \frac{M_y}{m} \qquad \bar{y} = \frac{M_x}{m}$$

where $m = \sum m_i$ is the total mass. Since $m\bar{x} = M_y$ and $m\bar{y} = M_x$ the center of mass $(\bar{x}, \bar{y})$ is the point where a single particle of mass m would have the same moments as the system.

● **Example 2**   Find the moments and center of mass of the system of objects that have masses 3, 4, and 8 at the points $(-1, 1)$, $(2, -1)$, and $(3, 2)$.

Solution   We use Equations 8.34 and 8.35 to compute the moments:

$$M_y = 3(-1) + 4(2) + 8(3) = 29$$
$$M_x = 3(1) + 4(-1) + 8(2) = 15$$

Since $m = 3 + 4 + 8 = 15$, we use Equations 8.36 to obtain

$$\bar{x} = \frac{M_y}{m} = \frac{29}{15} \qquad \bar{y} = \frac{M_x}{m} = \frac{15}{15} = 1$$

Thus the center of mass is $(1\frac{14}{15}, 1)$ (see Figure 8.52).   ●

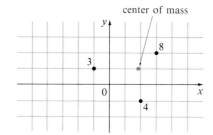

Figure 8.52

Next we consider a flat plate (called a lamina) with uniform density ρ that occupies a region $\mathcal{R}$ of the plane. We wish to locate the center of mass of the plate, which is called the **centroid** of $\mathcal{R}$. In doing so we shall use the following physical principles: The **symmetry principle** says that if $\mathcal{R}$ is symmetric about a line l, then the centroid of $\mathcal{R}$ lies on l. (If $\mathcal{R}$ is reflected about l, then $\mathcal{R}$ remains the same so its centroid remains fixed. But the only fixed points lie on l.) Thus the centroid of a rectangle is its center. Moments should be defined so that if the entire mass of a region is concentrated at the center of mass, then its moments remain unchanged. Also the moment of the union of two nonoverlapping regions should be the sum of the moments of the individual regions.

First we suppose that the region $\mathcal{R}$ is of the type shown in Figure 8.53(a); that is, $\mathcal{R}$ lies between the lines $x = a$ and $x = b$, above the x-axis, and beneath the graph of f, where f is a continuous function. We take a partition P by points x_i with $a = x_0 < x_1 < \cdots < x_n = b$ and choose x_i^* to be the midpoint of the ith subinterval, that is, $x_i^* = (x_{i-1} + x_i)/2$. This determines the polygonal approximation to $\mathcal{R}$ shown in Figure 8.53(b). The centroid of the ith approximating rectangle is its center $C_i(x_i^*, \frac{1}{2}f(x_i^*))$. Its area is $f(x_i^*)\Delta x_i$ so its mass is

$$\rho f(x_i^*)\Delta x_i$$

The moment of R_i about the y-axis is the product of its mass and the distance from C_i to the y-axis, which is x_i^*. Thus

$$M_y(R_i) = (\rho f(x_i^*)\Delta x_i)x_i^* = \rho x_i^* f(x_i^*)\Delta x_i$$

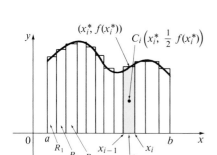

Figure 8.53

Adding these moments we obtain the moment of the polygonal approximation to $\mathcal{R}$, and then by taking the limit as $\|P\| \to 0$ we obtain the moment of $\mathcal{R}$ itself about the y-axis:

(8.37)

$$M_y = \lim_{\|P\| \to 0} \sum_{i=1}^{n} \rho x_i^* f(x_i^*) \Delta x_i = \rho \int_a^b x f(x) \, dx$$

In a similar fashion we compute the moment of R_i about the x-axis as the product of its mass and the distance from C_i to the x-axis:

$$M_x(R_i) = (\rho f(x_i^*) \Delta x_i) \tfrac{1}{2} f(x_i^*) = \rho \cdot \tfrac{1}{2} [f(x_i^*)]^2 \Delta x_i$$

Again we add these moments and take the limit to obtain the moment of $\mathcal{R}$ about the x-axis:

(8.38)

$$M_x = \lim_{\|P\| \to 0} \sum_{i=1}^{n} \rho \cdot \tfrac{1}{2} [f(x_i^*)]^2 \Delta x_i = \rho \int_a^b \tfrac{1}{2} [f(x)]^2 \, dx$$

Just as for systems of particles, the center of mass of the plate is defined so that $m\bar{x} = M_y$ and $m\bar{y} = M_x$. But the mass of the plate is the product of its density and its area:

$$m = \rho A = \rho \int_a^b f(x) \, dx$$

and so
$$\bar{x} = \frac{M_y}{m} = \frac{\rho \int_a^b x f(x) \, dx}{\rho \int_a^b f(x) \, dx} = \frac{\int_a^b x f(x) \, dx}{\int_a^b f(x) \, dx}$$

$$\bar{y} = \frac{M_y}{m} = \frac{\rho \int_a^b \tfrac{1}{2} [f(x)]^2 \, dx}{\rho \int_a^b f(x) \, dx} = \frac{\int_a^b \tfrac{1}{2} [f(x)]^2 \, dx}{\int_a^b f(x) \, dx}$$

Notice the cancellation of the ρ's. The location of the center of mass is independent of the density.

In summary, the center of mass of the plate (or the centroid of $\mathcal{R}$) is located at the point $(\bar{x}, \bar{y})$ where

(8.39)

$$\bar{x} = \frac{1}{A} \int_a^b x f(x) \, dx \qquad \bar{y} = \frac{1}{A} \int_a^b \tfrac{1}{2} [f(x)]^2 \, dx$$

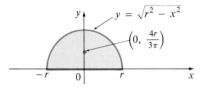

Figure 8.54

● **Example 3**   Find the center of mass of a semicircular plate of radius r.

Solution   In order to use (8.39) we place the semicircle as in Figure 8.54 so that $f(x) = \sqrt{r^2 - x^2}$ and $a = -r$, $b = r$. Here there is no need to use the formula to calculate $\bar{x}$ because, by the symmetry principle, the center of mass must lie on the y-axis, so $\bar{x} = 0$. The area of the semicircle is $A = \pi r^2/2$, so

$$\bar{y} = \frac{1}{A} \int_{-r}^{r} \tfrac{1}{2} [f(x)]^2 \, dx$$

$$= \frac{1}{\pi r^2/2} \cdot \tfrac{1}{2} \int_{-r}^{r} (\sqrt{r^2 - x^2})^2 \, dx$$

$$= \frac{1}{\pi r^2} \int_{-r}^{r} (r^2 - x^2) \, dx = \frac{1}{\pi r^2} \left[r^2 x - \frac{x^3}{3} \right]_{-r}^{r}$$

$$= \frac{1}{\pi r^2} \frac{4r^3}{3} = \frac{4r}{3\pi}$$

The center of mass is located at the point $(0, 4r/3\pi)$.   ●

● **Example 4**   Find the centroid of the region bounded by the curves $y = \cos x$, $y = 0$, $x = 0$, and $x = \pi/2$.

Solution   The area of the region is

$$A = \int_{0}^{\pi/2} \cos x \, dx = \sin x \Big]_{0}^{\pi/2} = 1$$

so Formulas 8.39 give

$$\bar{x} = \frac{1}{A} \int_{0}^{\pi/2} x f(x) \, dx = \int_{0}^{\pi/2} x \cos x \, dx$$

$$= x \sin x \Big]_{0}^{\pi/2} - \int_{0}^{\pi/2} \sin x \, dx \qquad \textit{(by integration by parts)}$$

$$= \frac{\pi}{2} - 1$$

$$\bar{y} = \frac{1}{A} \int_{0}^{\pi/2} \frac{1}{2} [f(x)]^2 \, dx = \frac{1}{2} \int_{0}^{\pi/2} \cos^2 x \, dx$$

$$= \frac{1}{4} \int_{0}^{\pi/2} (1 + \cos 2x) \, dx = \frac{1}{4} \left[x + \frac{1}{2} \sin 2x \right]_{0}^{\pi/2}$$

$$= \frac{\pi}{8}$$

The centroid is $((\pi/2) - 1, \pi/8)$ and is shown in Figure 8.55.   ●

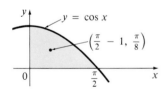

Figure 8.55

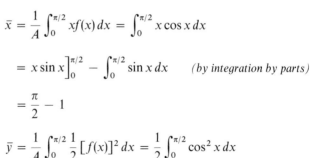

Figure 8.56

If the region $\mathcal{R}$ lies between two curves $y = f(x)$ and $y = g(x)$ where $f(x) \geqslant g(x)$, as illustrated in Figure 8.56, then the same sort of argument that led to Formulas 8.39 can be used to show that the centroid of $\mathscr{R}$ is $(\bar{x}, \bar{y})$, where

(8.40)

$$\bar{x} = \frac{1}{A} \int_a^b x[f(x) - g(x)]\, dx$$

$$\bar{y} = \frac{1}{A} \int_a^b \frac{1}{2}\{[f(x)]^2 - [g(x)]^2\}\, dx$$

(See Exercise 29.)

● **Example 5**   Find the centroid of the region bounded by the line $y = x$ and the parabola $y = x^2$.

Solution   The region is sketched in Figure 8.57. We take $f(x) = x$, $g(x) = x^2$, $a = 0$, and $b = 1$ in Formulas 8.40. First we note that the area of the region is

$$A = \int_0^1 (x - x^2)\, dx = \frac{x^2}{2} - \frac{x^3}{3}\bigg]_0^1 = \frac{1}{6}$$

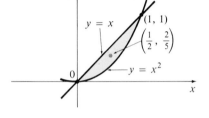

Figure 8.57

Therefore

$$\bar{x} = \frac{1}{A} \int_0^1 x[f(x) - g(x)]\, dx = \frac{1}{\frac{1}{6}} \int_0^1 x(x - x^2)\, dx$$

$$= 6 \int_0^1 (x^2 - x^3)\, dx = 6\left[\frac{x^3}{3} - \frac{x^4}{4}\right]_0^1 = \frac{1}{2}$$

$$\bar{y} = \frac{1}{A} \int_0^1 \frac{1}{2}\{[f(x)]^2 - [g(x)]^2\}\, dx = \frac{1}{\frac{1}{6}} \int_0^1 \frac{1}{2}(x^2 - x^4)\, dx$$

$$= 3\left[\frac{x^3}{3} - \frac{x^5}{5}\right]_0^1 = \frac{2}{5}$$

The centroid is $(\frac{1}{2}, \frac{2}{5})$.   ●

We end this section by showing how centroids can be used in finding volumes of revolution. The following theorem is named after the Greek mathematician Pappus of Alexandria, who lived in the fourth century A.D.

Theorem of Pappus (8.41)

Let $\mathscr{R}$ be a plane region that lies entirely on one side of a line l in the plane. If $\mathscr{R}$ is rotated about l, then the volume of the resulting solid is the product of the area A of $\mathscr{R}$ and the distance d traveled by the centroid of $\mathscr{R}$.

Proof   We give the proof for the special case where the region lies between $y = f(x)$ and $y = g(x)$ as in Figure 8.56 and the line l is the y-axis. Using the method of cylindrical shells (see Section 8.2) we have

$$V = \int_a^b 2\pi x [f(x) - g(x)]\, dx$$

$$= 2\pi \int_a^b x [f(x) - g(x)]\, dx$$

$$= 2\pi(\bar{x}A) \quad \text{(by Formula 8.40)}$$

$$= (2\pi \bar{x})A = dA$$

where $d = 2\pi \bar{x}$ is the distance traveled by the centroid during one rotation about the y-axis.

● **Example 6**   A torus is formed by rotating a circle of radius r about a line in the plane of the circle that is a distance R $(>r)$ from the center of the circle (see Figure 8.22). Find the volume of the torus.

Solution   The circle has area $A = \pi r^2$. By the symmetry principle, its centroid is its center and so the distance traveled by the centroid during a rotation is $d = 2\pi R$. Therefore, by the Theorem of Pappus, the volume of the torus is

$$V = dA = (2\pi R)(\pi r^2) = 2\pi^2 r^2 R$$

The method of Example 6 should be compared with the method of Exercise 49 in Section 8.1.

EXERCISES 8.7

In Exercises 1–4 the masses m_i are located at the points P_i. Find the moments M_x and M_y and the center of mass of the system.

1. $m_1 = 4,\ m_2 = 8;\ P_1(-1, 2),\ P_2(2, 4)$

2. $m_1 = 2,\ m_2 = 3,\ m_3 = 1;\ P_1(5, 1),\ P_2(3, -2),\ P_3(-2, 4)$

3. $m_1 = 4,\ m_2 = 2,\ m_3 = 5;\ P_1(-1, -2),$
    $P_2(-2, 4),\ P_3(5, -3)$

4. $m_1 = 3,\ m_2 = 3,\ m_3 = 8,\ m_4 = 6;\ P_1(0, 0),\ P_2(1, 8),$
    $P_3(3, -4),\ P_4(-6, -5)$

In Exercises 5–20 find the centroid of the region bounded by the given curves.

5. $y = x^2,\ y = 0,\ x = 2$ 6. $y = 1 - x^2,\ y = 0$

7. $y = 3x + 5,\ y = 0,\ x = -1,\ x = 2$

8. $y = \sqrt{x},\ y = 0,\ x = 4$

9. $y = \sqrt{x^2 + 1},\ y = 0,\ x = 0,\ x = 1$

10. $y = \dfrac{1}{x - 1},\ y = 0,\ x = 2,\ x = 4$

11. $y = \cos 2x,\ y = 0,\ x = -\pi/4,\ x = \pi/4$

12. $y = \sin x,\ y = 0,\ x = 0,\ x = \pi/2$

13. $y = e^x,\ y = 0,\ x = 0,\ x = 1$

14. $y = \ln x,\ y = 0,\ x = e$

15. $y = \sqrt{x},\ y = x$

16. $y = x^2,\ y = 8 - x^2$

17. $y = \sin x,\ y = \cos x,\ x = 0,\ x = \pi/4$

18. $y = x,\ y = 0,\ y = 1/x,\ x = 2$

19. $\dfrac{x^2}{a^2} + \dfrac{y^2}{b^2} = 1$ 20. $y^2 = x^3 - x^4$

In Exercises 21–24 calculate the moments M_x and M_y and the center of mass of a lamina with the given density and shape.

21. $\rho = 1$

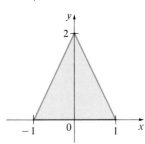

22. $\rho = 2$

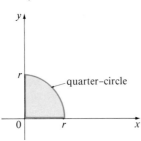

quarter-circle

23. $\rho = 4$

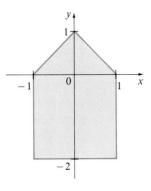

24. $\rho = 5$

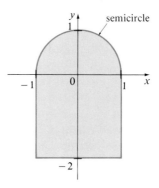

semicircle

25. Prove that the centroid of any triangle is located at the point of intersection of the medians.

In Exercises 26–28 use the Theorem of Pappus to find the volume of the given solid.

26. A sphere of radius r (Use Example 3.)

27. A cone with height h and base radius r

28. The solid obtained by rotating the quadrilateral with vertices $(0,0)$, $(1,4)$, $(7,4)$, and $(6,0)$ about the y-axis

29. Prove Formulas 8.40.

SECTION 8.8

Differential Equations

A **differential equation** is an equation that contains an unknown function and some of its derivatives. Here are some examples:

$$(8.42) \qquad y' = xy$$

$$(8.43) \qquad y'' + 2y' + y = 0$$

$$(8.44) \qquad \frac{d^3y}{dx^3} + x\frac{d^2y}{dx^2} + \frac{dy}{dx} - 2y = e^{-x}$$

In each of these differential equations y is an unknown function of x. The importance of differential equations lies in the fact that when a scientist or engineer formulates a physical law in mathematical terms, it frequently turns out to be a differential equation.

The **order** of a differential equation is the order of the highest derivative that occurs in the equation. Thus Equations 8.42, 8.43, and 8.44 are of order 1, 2, and 3, respectively.

A function f is called a **solution** of a differential equation if the equation is satisfied when $y = f(x)$ and its derivatives are substituted into the equation. Thus f would be a solution of Equation 8.42 if

$$f'(x) = xf(x)$$

for all values of x in some interval.

You can easily verify that both $f(x) = \sin x$ and $g(x) = \cos x$ are solutions of the differential equation

(8.45)
$$y'' + y = 0$$

But when we are asked to *solve* a differential equation we are expected to find all possible solutions of the equation. In Section 15.7 we shall show that any solution of Equation 8.45 is of the form

(8.46)
$$y = A \sin x + B \cos x$$

where A and B are constants. So (8.46) is called the **general solution** of the differential equation and particular solutions are obtained by substituting values for the arbitrary constants A and B.

We have already solved some particularly simple differential equations, namely, those of the form

$$y' = f(x)$$

For instance, we know that the general solution of the differential equation

$$y' = x^3$$

is given by

$$y = \frac{x^4}{4} + C$$

where C is an arbitrary constant.

But, in general, solving a differential equation is not an easy matter. There is no systematic technique that will enable us to solve all differential equations. In this section we shall learn how to solve a certain type of differential equation called a separable equation. Other types of equations are discussed in Chapter 15.

A **separable equation** is a first-order differential equation that can be written in the form

$$\frac{dy}{dx} = g(x)f(y)$$

The name *separable* comes from the fact that the expression on the right side can be "separated" into a function of x and a function of y. Equivalently, we could write

(8.47)
$$\frac{dy}{dx} = \frac{g(x)}{h(y)}$$

To solve this equation we rewrite it in the differential form

$$h(y)\, dy = g(x)\, dx$$

so that all y's are on one side of the equation and all x's are on the other side. Then we integrate both sides of the equation:

$$(8.48) \qquad \int h(y)\,dy = \int g(x)\,dx$$

Equation 8.48 defines y implicitly as a function of x. In some cases we may be able to solve for y in terms of x.

The justification for the step in Equation 8.48 comes from the Substitution Rule (5.31):

$$\int h(y)\,dy = \int h(y(x))\,\frac{dy}{dx}\,dx$$

$$= \int h(y(x))\,\frac{g(x)}{h(y(x))}\,dx \qquad \textit{(from Equation 8.47)}$$

$$= \int g(x)\,dx$$

● **Example 1**   Solve the differential equation $\dfrac{dy}{dx} = \dfrac{6x^2}{2y + \cos y}$.

Solution   Writing the equation in differential form and integrating both sides, we have

$$(2y + \cos y)\,dy = 6x^2\,dx$$

$$\int (2y + \cos y)\,dy = \int 6x^2\,dx$$

$$(8.49) \qquad y^2 + \sin y = 2x^3 + C$$

where C is an arbitrary constant. (We could have used a constant C_1 on the left side and another constant C_2 on the right side. But then we could combine these constants by writing $C = C_2 - C_1$.)

Equation 8.49 gives the general solution implicitly. In this case it is impossible to solve the equation to express y explicitly as a function of x.

●

● **Example 2**   Solve the equation $y' = x^2 y$.

Solution   First we rewrite the equation using Leibniz notation:

$$\frac{dy}{dx} = x^2 y$$

If $y \neq 0$, we can rewrite it in differential notation and integrate:

$$\frac{dy}{y} = x^2\,dx \qquad y \neq 0$$

$$\int \frac{dy}{y} = \int x^2\,dx$$

$$\ln|y| = \frac{x^3}{3} + C$$

This defines y implicitly as a function of x. But in this case we can solve explicitly for y as follows:

$$|y| = e^{\ln|y|} = e^{(x^3/3)+C} = e^C e^{x^3/3}$$

$$y = \pm e^C e^{x^3/3}$$

We note that the function $y = 0$ is also a solution of the given differential equation. So we can write the general solution in the form

$$y = A e^{x^3/3}$$

where A is an arbitrary constant ($A = e^C$ or $A = -e^C$ or $A = 0$).   ●

   In Examples 1 and 2 we found the general solution of the given differential equation. But in many physical problems it is required to find the particular solution that satisfies a condition of the form $y(x_0) = y_0$. This is called an **initial condition** and the problem of finding a solution of the differential equation that satisfies the initial condition is called an **initial-value problem.**

● **Example 3**   Solve the initial-value problem $xy' = -y, x > 0, y(4) = 2$.

Solution   We write the differential equation as

$$x\frac{dy}{dx} = -y \quad\text{or}\quad \frac{dy}{y} = -\frac{dx}{x}$$

Therefore

$$\int \frac{dy}{y} = -\int \frac{dx}{x}$$

$$\ln|y| = -\ln|x| + C$$

$$|y| = \frac{1}{|x|}e^C$$

$$y = \frac{K}{x}$$

where $K = \pm e^C$ is a constant. To determine K we put $x = 4$ and $y = 2$ in this equation:

$$2 = \frac{K}{4} \qquad K = 8$$

The solution of the initial-value problem is

$$y = \frac{8}{x} \qquad x > 0$$

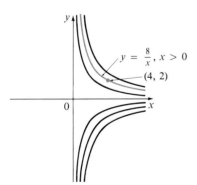

$y = \frac{8}{x}, x > 0$

$(4, 2)$

Figure 8.58

Figure 8.58 shows the family of solutions $xy = K$ for several values of K (equilateral hyperbolas) and, in particular, the solution that satisfies $y(4) = 2$ [the hyperbola that passes through the point $(4, 2)$].   ●

● **Example 4**   Find the solution of $\dfrac{dy}{dx} = \dfrac{6x^2}{2y + \cos y}$ that satisfies $y(1) = \pi$.

Solution   From Example 1 we know that the general solution is

$$y^2 + \sin y = 2x^3 + C$$

We are given that $y(1) = \pi$, so we substitute $x = 1$ and $y = \pi$ in this equation:

$$\pi^2 + \sin \pi = 2(1)^3 + C$$
$$C = \pi^2 - 2$$

Therefore the solution is given implicitly by

$$y^2 + \sin y = 2x^3 + \pi^2 - 2$$ ●

● **Example 5**   Solve $y' = 1 + y^2 - 2x - 2xy^2$, $y(0) = 0$.

Solution   At first glance this does not look like a separable equation, but notice that it is possible to factor the right side as the product of a function of x and a function of y as follows:

$$\frac{dy}{dx} = 1 + y^2 - 2x(1 + y^2) = (1 - 2x)(1 + y^2)$$

$$\int \frac{dy}{1 + y^2} = \int (1 - 2x)\, dx$$

$$\tan^{-1} y = x - x^2 + C$$

Putting $x = 0$ and $y = 0$, we get $C = \tan^{-1} 0 = 0$, so

$$\tan^{-1} y = x - x^2$$

and solving for y we have

$$y = \tan(x - x^2)$$ ●

● **Example 6**   A tank contains 20 kg of salt dissolved in 5000 L of water. Brine that contains 0.03 kg of salt per liter of water enters the tank at a rate of 25 L/min. The solution is kept thoroughly mixed and drains from the tank at the same rate. How much salt remains in the tank after half an hour?

Solution   Let $y(t)$ be the amount of salt (in kilograms) after t minutes. Then $y(0) = 20$ and we want to find $y(30)$. We do this by finding a differential equation satisfied by $y(t)$. Note that dy/dt is the rate of change of the amount of salt, so

(8.50)
$$\frac{dy}{dt} = (\text{rate in}) - (\text{rate out})$$

where (rate in) is the rate at which salt enters the tank and (rate out) is the rate at which salt leaves the tank. We have

$$\text{rate in} = \left(0.03\, \frac{\text{kg}}{\text{L}} \right)\left(25\, \frac{\text{L}}{\text{min}} \right) = 0.75\, \frac{\text{kg}}{\text{min}}$$

The tank always contains 5000 L of liquid, so the concentration at time t is $y(t)/5000$ (measured in kilograms per liter). Since the brine flows out at a rate of 25 L/min, we have

$$\text{rate out} = \left(\frac{y(t)}{5000} \frac{\text{kg}}{\text{L}}\right)\left(25 \frac{\text{L}}{\text{min}}\right) = \frac{y(t)}{200} \frac{\text{kg}}{\text{min}}$$

Thus, from Equation 8.50, we get

$$\frac{dy}{dt} = 0.75 - \frac{y(t)}{200} = \frac{150 - y(t)}{200}$$

Solving this separable differential equation, we obtain

$$\int \frac{dy}{150 - y} = \int \frac{dt}{200}$$

$$-\ln|150 - y| = \frac{t}{200} + C$$

Since $y(0) = 20$, we have $-\ln 130 = C$, so

$$-\ln|150 - y| = \frac{t}{200} - \ln 130$$

Therefore $|150 - y| = 130e^{-t/200}$

Since $y(t)$ is continuous and $y(0) = 20$ and the right side is never 0, we deduce that $150 - y(t)$ is always positive. Thus $|150 - y| = 150 - y$ and

$$y(t) = 150 - 130e^{-t/200}$$

The amount of salt after 30 min is

$$y(30) = 150 - 130e^{-30/200} \approx 38.1 \text{ kg}$$

●

● **Example 7**   Solve the equation $\dfrac{dy}{dt} = ky$.

Solution   This differential equation was studied in Section 6.7, where it was called the law of natural growth (or decay). Since it is a separable equation we can solve it by the methods of this section as follows:

$$\int \frac{dy}{y} = \int k\,dt \qquad y \neq 0$$

$$\ln|y| = kt + C$$

$$|y| = e^{kt+C} = e^C e^{kt}$$

$$y = Ae^{kt}$$

where $A \,(= \pm e^C$ or 0) is an arbitrary constant.   ●

Logistic Growth

The differential equation of Example 7 is appropriate for modeling population growth ($y' = ky$ says that the rate of growth is proportional to the size of the population) under conditions of unlimited environment and food supply. However, in a restricted environment and with limited food supply, the population cannot exceed a maximal size M at which it consumes its entire food supply. If we make the assumption that the rate of growth of population is jointly proportional to the size of the population (y) and the amount by which y falls short of the maximal size ($M - y$), then we have the equation

(8.51)

$$\frac{dy}{dt} = ky(M - y)$$

where k is a constant. Equation 8.51 is called the **logistic differential equation** and was used by the Dutch mathematical biologist Verhulst in the 1840s to model world population growth.

The logistic equation is separable, so we write it in the form

$$\int \frac{dy}{y(M - y)} = \int k\,dt$$

Using partial fractions, we have

$$\frac{1}{y(M - y)} = \frac{1}{M}\left[\frac{1}{y} + \frac{1}{M - y}\right]$$

and so

$$\frac{1}{M}\left[\int \frac{dy}{y} + \int \frac{dy}{M - y}\right] = \int k\,dt = kt + C$$

$$\frac{1}{M}\left[\ln|y| - \ln|M - y|\right] = kt + C$$

But $|y| = y$ and $|M - y| = M - y$ since $0 < y < M$, so we have

$$\ln \frac{y}{M - y} = M(kt + C)$$

$$\frac{y}{M - y} = Ae^{kMt} \qquad (A = e^{MC})$$

If the population at time $t = 0$ is $y(0) = y_0$, then $A = y_0/(M - y_0)$, so

$$\frac{y}{M - y} = \frac{y_0}{M - y_0}e^{kMt}$$

If we solve this equation for y, we get

$$y = \frac{y_0 M e^{kMt}}{M - y_0 + y_0 e^{kMt}} = \frac{y_0 M}{y_0 + (M - y_0)e^{-kMt}}$$

Using the latter expression for y, we see that

$$\lim_{t \to \infty} y(t) = M$$

which is to be expected.

The graph of the logistic growth function is shown in Figure 8.59. At first the graph is concave upward and the growth curve appears to be almost exponential, but then it becomes concave downward and approaches the limiting population M.

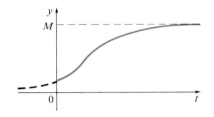

Figure 8.59

LOGISTIC GROWTH FUNCTION

EXERCISES 8.8

Solve the differential equations in Exercises 1–10.

1. $\dfrac{dy}{dx} = y^2$

2. $\dfrac{dy}{dx} = \dfrac{x + \sin x}{3y^2}$

3. $yy' = x$

4. $y' = xy$

5. $x^2 y' + y = 0$

6. $e^{-y} y' + \cos x = 0$

7. $\dfrac{dy}{dx} = \dfrac{x\sqrt{x^2 + 1}}{y e^y}$

8. $y' = \dfrac{\ln x}{xy + xy^3}$

9. $\dfrac{du}{dt} = e^{u + 2t}$

10. $\dfrac{dx}{dt} = 1 + t - x - tx$

In Exercises 11–20 find the solution to the differential equation that satisfies the given initial condition.

11. $\dfrac{dy}{dx} = y^2 + 1,\ y(1) = 0$

12. $xy' = \sqrt{1 - y^2},\ x > 0,\ y(1) = 0$

13. $e^y y' = \dfrac{3x^2}{1 + y},\ y(2) = 0$

14. $\dfrac{dy}{dx} = \dfrac{1 + x}{xy},\ x > 0,\ y(1) = -4$

15. $\dfrac{dy}{dx} = \dfrac{\sin x}{\sin y},\ y(0) = \dfrac{\pi}{2}$

16. $\dfrac{dy}{dx} = e^{x-y},\ y(0) = 1$

17. $xe^{-t} \dfrac{dx}{dt} = t,\ x(0) = 1$

18. $x\,dx + 2y\sqrt{x^2 + 1}\,dy = 0,\ y(0) = 1$

19. $\dfrac{du}{dt} = \dfrac{2t + 1}{2(u - 1)},\ u(0) = -1$

20. $\dfrac{dy}{dt} = \dfrac{ty + 3t}{t^2 + 1},\ y(2) = 2$

21. Find a function f such that $f'(x) = x^3 f(x)$ and $f(0) = 1$.

22. Find a function g such that $g'(x) = g(x)(1 + g(x))$ and $g(0) = 1$.

23. A tank contains 1000 L of brine with 15 kg of dissolved salt. Pure water enters the tank at a rate of 10 L/min. The solution is kept thoroughly mixed and drains from the tank at the same rate. How much salt is in the tank (a) after t minutes and (b) after 20 min?

24. A tank contains 1000 L of pure water. Brine that contains 0.05 kg of salt per liter of water enters the tank at a rate of 5 L/min. Brine that contains 0.04 kg of salt per liter of water enters the tank at a rate of 10 L/min. The solution is kept thoroughly mixed and drains from the tank at a rate of 15 L/min. How much salt is in the tank (a) after t minutes and (b) after an hour?

25. In an elementary chemical reaction, single molecules of two reactants A and B form a molecule of the product C: $A + B \rightarrow C$. The law of mass action states that the rate of reaction is proportional to the product of the concentrations of A and B:

$$\frac{d[C]}{dt} = k[A][B]$$

(See Example 3 in Section 3.7.) Thus, if the initial concentrations are $[A] = a$ moles/L and $[B] = b$ moles/L and we write $x = [C]$, then we have

$$\frac{dx}{dt} = k(a - x)(b - x)$$

Assuming that $a \neq b$, find x as a function of t.

26. (a) Find $x(t)$ in Exercise 25 assuming that $b = a$.
(b) How does this expression for $x(t)$ simplify if it is known that $[C] = a/2$ after 20 seconds?

27. The population of the world was about 5 billion in 1986. Using the exponential model for population growth (see Example 7 or Section 6.7) with the recently observed rate of increase of population of 2% per year, find an expression for the population of the world in the year t. Use this model to predict the population of the world in the following years:
(a) 2000   (b) 2100   (c) 2500
The total land surface area of this planet is about 3.7×10^{11} ft². How many square feet of land per person will there be in the above years under the exponential model?

28. For a more realistic picture of long-term growth than in Exercise 27, we consider the logistic model

for world population growth with $y_0 = 5$ billion in 1986 and an assumed maximum population of $M = 100$ billion. To determine the value of k, use Equation 8.51 and the fact that the population was increasing at a rate of 2% per year when the population was 5 billion. Use this model to predict the population of the world in the following years:
(a) 2000   (b) 2100   (c) 2500
Compare with the results of Exercise 27.

29. One model for the spread of a rumor is that the rate of spread is proportional to the product of the fraction y of the population who have heard the rumor and the fraction who have not heard the rumor.
(a) Write a differential equation that is satisfied by y.
(b) Solve the differential equation.
(c) A small town has 1000 inhabitants. At 8 A.M. 80 people have heard a rumor. By noon half the town has heard it. At what time will 90% of the population have heard the rumor?

30. Another model for a growth function for a limited population is given by the *Gompertz function*, which is a solution of the differential equation

$$\frac{dy}{dt} = c \ln\left(\frac{M}{y}\right) y$$

where c is a constant and M is the maximum size of the population.
(a) Solve this differential equation.
(b) Compute $\lim_{t \to \infty} y(t)$.
(c) Sketch the graph of the Gompertz growth function.

31. Show that the solution to the logistic Equation 8.51 increases most rapidly when $y = M/2$.

32. Snow began to fall during the morning of February 2 and continued steadily into the afternoon. A snowplow began to clear a street at noon at a constant rate (in cubic meters per hour). The plow traveled 6 km from noon to 1 P.M. but only 3 km from 1 P.M. to 2 P.M. When did the snow begin to fall?

REVIEW OF CHAPTER 8

Define, state, or discuss the following:

1. Volume of a general solid
2. Volume of a solid of revolution: disk (or washer) method
3. Volume of a solid of revolution: cylindrical shell method
4. Arc length
5. Area of a surface of revolution

6. Work
7. Force exerted by a liquid
8. Moments and centroid of a plane region
9. Theorem of Pappus
10. Separable differential equation

REVIEW EXERCISES FOR CHAPTER 8

In Exercises 1–8 find the volume of the solid obtained by rotating the region bounded by the given curves about the given axis.

1. $y = \sqrt{x - 1}$, $y = 0$, $x = 3$; about the x-axis

2. $y = x^3$, $y = x^2$; about the x-axis

3. $x + 3 = 4y - y^2$, $x = 0$; about the x-axis

4. $y = x^3$, $y = 8$, $x = 0$; about the y-axis

5. $x^2 - y^2 = a^2$, $x = a + h$ (where $a > 0$, $h > 0$); about the y-axis

6. $y = \cos x$, $y = 0$, $x = \dfrac{3\pi}{2}$, $x = \dfrac{5\pi}{2}$; about the y-axis

7. $y = x^3$, $y = x^2$; about $y = 1$

8. $y = x^3$, $y = 8$, $x = 0$; about $x = 2$

9. The base of a solid is a circular disk with radius 3. Find the volume of the solid if parallel cross-sections perpendicular to the base are isosceles right triangles with hypotenuse lying along the base.

10. The base of a solid is the region bounded by the parabolas $y = x^2$ and $y = 2 - x^2$. Find the volume of the solid if the cross-sections perpendicular to the x-axis are squares with one side lying along the base.

In Exercises 11 and 12 find the length of the curve.

11. $3x = 2(y - 1)^{3/2}$, $2 \leqslant y \leqslant 5$

12. $y = \sqrt{x - x^2} + \sin^{-1} \sqrt{x}$

In Exercises 13 and 14 find the area of the surface obtained by rotating the given curve about the x-axis.

13. $y = \dfrac{x^3}{6} + \dfrac{1}{2x}$, $1 \leqslant x \leqslant 2$

14. $y = e^x$, $0 \leqslant x \leqslant 1$

15. Use Simpson's Rule with $n = 10$ to estimate the length of the arc of the curve $y = 1/x^2$ from $(1, 1)$ to $(2, \frac{1}{4})$.

16. Use Simpson's Rule with $n = 10$ to estimate the area of the surface obtained by rotating the arc of the curve $y = 1/x^2$ from $(1, 1)$ to $(2, \frac{1}{4})$ about the x-axis.

17. A force of 30 N is required to stretch a spring from its natural length of 12 cm to a length of 15 cm. How much work is done in stretching the spring from 12 cm to 20 cm?

18. A 1600-lb elevator is suspended by a 200-ft cable that weighs 10 lb/ft. How much work is required to raise the elevator from the basement to the third floor, a distance of 30 ft?

19. A tank full of water has the shape of a paraboloid of revolution as in Figure 8.60; that is, its shape is obtained by rotating a parabola about a vertical axis. If its height is 4 ft and the radius at the top is 4 ft, find the work required to pump the water out of the tank.

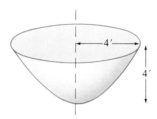

Figure 8.60

20. A trough is filled with water and its vertical ends have the shape of the parabolic region in Figure 8.61. Find the hydrostatic force on one end of the trough.

Figure 8.61

21. Show that the hydrostatic force on a vertical plate submerged in a liquid is equal to the pressure at the centroid of the plate times the area of the plate.

In Exercises 22 and 23 find the centroid of the region bounded by the given curves.

22. $y = 4x - x^2$, $y = 0$

23. $y = 4 - x^2$, $y = x + 2$

24. Let $\mathscr{R}$ be the region in the first quadrant bounded by the curves $y = x^3$ and $y = 2x - x^2$. Calculate the following quantities: (a) the area of $\mathscr{R}$, (b) the centroid of $\mathscr{R}$, (c) the volume obtained by rotating $\mathscr{R}$ about the x-axis, and (d) the volume obtained by rotating $\mathscr{R}$ about the y-axis.

Solve the differential equations in Exercises 25–28.

25. $y^2 \dfrac{dy}{dx} = x + \sin x$ **26.** $\dfrac{dy}{dx} = \dfrac{y^2 + 1}{xy}, \, x > 0$

27. $y' = \dfrac{1}{x^2 y - 2x^2 + y - 2}$

28. $2u \dfrac{du}{dt} = te^t, \, u(0) = 1$

29. Barbara weighs 60 kg and is on a diet of 1600 calories per day of which 850 are used up automatically by basal metabolism. She spends about 15 cal/kg/day times her weight doing exercise. If 1 kg of fat contains 10,000 cal and we assume that the storage of calories in the form of fat is 100% efficient, formulate a differential equation and solve it to find her weight as a function of time. Does her weight ultimately approach an equilibrium weight?

9 Parametric Equations and Polar Coordinates

A mathematician, like a painter or poet,
is a maker of patterns. If his patterns
are more permanent than theirs, it is
because they are made with *ideas*.

G. H. Hardy

So far we have described plane curves by giving y as a function of x $[y = f(x)]$ or x as a function of y $[x = g(y)]$ or by giving a relation between x and y that defines y implicitly as a function of x $[f(x, y) = 0]$. In this chapter we discuss two new methods for describing curves.

Some curves, such as the cycloid, are best handled when both x and y are given in terms of a third variable t called a parameter $[x = f(t), y = g(t)]$. Other curves, such as the cardioid, have their most convenient description when we use a new coordinate system, called the polar coordinate system.

Curves Defined by Parametric Equations

Suppose that x and y are both given as continuous functions of a third variable t (called a **parameter**) by the equations

(9.1)
$$x = f(t) \qquad y = g(t)$$

(called **parametric equations**). Each value of t determines a point (x, y), which we can plot in a coordinate plane. As t varies, the point $(x, y) = (f(t), g(t))$ varies and traces out a curve C. If we interpret t as time and $(x, y) = (f(t), g(t))$ as the position of a particle at time t, then we can imagine the particle moving along the curve C.

● **Example 1**   Sketch and identify the curve defined by the parametric equations $x = t^2 - 2t$ and $y = t + 1$.

Solution   Each value of t gives a point on the curve. For instance, if $t = 0$, then $x = 0$, $y = 1$ and so the corresponding point is $(0, 1)$. In Figure 9.1 we plot the points (x, y) determined by several values of the parameter.

t	x	y
-2	8	-1
-1	3	0
0	0	1
1	-1	2
2	0	3
3	3	4
4	8	5

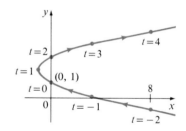

Figure 9.1

A particle whose position is given by the parametric equations will move along the curve in the direction of the arrows as t increases. It appears from Figure 9.1 that the curve traced out by the particle may be a parabola. This can be confirmed by eliminating the parameter t as follows. We obtain $t = y - 1$ from the second equation and substitute into the first equation. This gives

$$x = (y - 1)^2 - 2(y - 1) = y^2 - 4y + 3$$

and so the curve represented by the given parametric equations is the parabola $x = y^2 - 4y + 3$.   ●

● **Example 2**   What curve is represented by the parametric equations $x = \cos t$ and $y = \sin t$, $0 \leqslant t \leqslant 2\pi$?

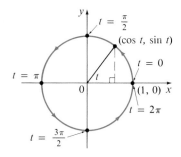

Figure 9.2

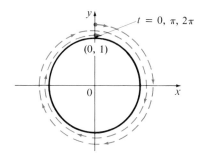

Figure 9.3

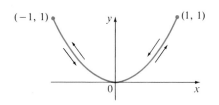

Figure 9.4

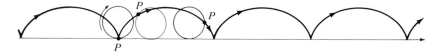

Figure 9.5

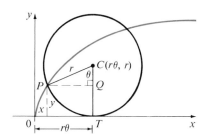

Figure 9.6

Solution   We can eliminate t by noting that

$$x^2 + y^2 = \cos^2 t + \sin^2 t = 1$$

Thus the point (x, y) moves on the unit circle $x^2 + y^2 = 1$. Notice that in this example the parameter t can be interpreted as the angle shown in Figure 9.2. As t increases from 0 to 2π, the point $(x, y) = (\cos t, \sin t)$ moves once around the circle in the counterclockwise direction starting from the point $(1, 0)$.   ●

● **Example 3**   What curve is represented by the parametric equations $x = \sin 2t$ and $y = \cos 2t$, $0 \leqslant t \leqslant 2\pi$?

Solution   Again we have

$$x^2 + y^2 = \sin^2 2t + \cos^2 2t = 1$$

so the parametric equations again represent the unit circle $x^2 + y^2 = 1$. But as t increases from 0 to 2π, the point $(x, y) = (\sin 2t, \cos 2t)$ starts at $(0, 1)$ and moves *twice* around the circle in the clockwise direction as indicated in Figure 9.3.   ●

● **Example 4**   Sketch the curve with parametric equations $x = \sin t$ and $y = \sin^2 t$.

Solution   Observe that $y = x^2$ and so the point (x, y) moves on the parabola $y = x^2$. But note also that, since $-1 \leqslant \sin t \leqslant 1$, we have $-1 \leqslant x \leqslant 1$, so the parametric equations represent only the part of the parabola for which $-1 \leqslant x \leqslant 1$. Since $\sin t$ is periodic, the point $(x, y) = (\sin t, \sin^2 t)$ moves back and forth infinitely often along the parabola from $(-1, 1)$ to $(1, 1)$ (see Figure 9.4).   ●

● **Example 5**   The curve traced out by a point P on the circumference of a circle as the circle rolls along a straight line is called a **cycloid** (see Figure 9.5). If the circle has radius r and rolls along the x-axis and if one position of P is the origin, find parametric equations for the cycloid.

Solution   We choose as parameter the angle of rotation θ of the circle ($\theta = 0$ when P is at the origin). When the circle has rotated through θ radians, the distance it has rolled from the origin is

$$|OT| = \text{arc } PT = r\theta$$

and so the center of the circle is $C(r\theta, r)$. Let the coordinates of P be (x, y). Then from Figure 9.6 we see that

$$x = |OT| - |PQ| = r\theta - r\sin\theta = r(\theta - \sin\theta)$$
$$y = |TC| - |QC| = r - r\cos\theta = r(1 - \cos\theta)$$

Therefore the parametric equations of the cycloid are

(9.2) $$x = r(\theta - \sin\theta) \qquad y = r(1 - \cos\theta) \qquad \theta \in R$$

One arch of the cycloid comes from one rotation of the circle and so is described by $0 \leqslant \theta \leqslant 2\pi$. Although Equations 9.2 were derived from Figure 9.6, which illustrates the case where $0 < \theta < \pi/2$, it can be seen that these equations are still valid for other values of θ (see Exercise 33).

Although it is possible to eliminate the parameter θ from Equations 9.2, the resulting Cartesian equation in x and y is very complicated and not as convenient to work with as the parametric equations.   ●

One of the first people to study the cycloid was Galileo, who proposed that bridges be built in the shape of cycloids and who tried to find the area under one arch of a cycloid. Later this curve arose in connection with the Brachistochrone Problem: Find the curve along which a particle will slide in the shortest time (under the influence of gravity) from a point A to a lower point B not directly beneath A. The Swiss mathematician John Bernoulli, who posed this problem in 1696, showed that among all possible curves that join A to B as in Figure 9.7, the particle will take the least time sliding from A to B if the curve is an inverted arch of a cycloid.

The Dutch physicist Huygens had already shown that the cycloid is also the solution to the Tautochrone Problem; that is, no matter where a particle P is placed on an inverted cycloid, it takes the same time to slide to the bottom (see Figure 9.8). Huygens proposed that pendulum clocks (which he invented) should swing in cycloidal arcs because then the pendulum would take the same time to make a complete oscillation whether it swings through a wide or a small arc.

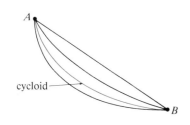

Figure 9.7

Figure 9.8

EXERCISES 9.1

In Exercises 1–20, (a) sketch the curve represented by the given parametric equations, and (b) eliminate the parameter to find the Cartesian equation of the curve.

1. $x = 1 - t, y = 2 + 3t$

2. $x = 2t - 1, y = 2 - t, -3 \leqslant t \leqslant 3$

3. $x = 3t^2, y = 2 + 5t, 0 \leqslant t \leqslant 2$

4. $x = 2t - 1, y = t^2 - 1$

5. $x = \sqrt{t}, y = 1 - t$

6. $x = t^2, y = t^3$

7. $x = \sin\theta, y = \cos\theta, 0 \leqslant \theta \leqslant \pi$

8. $x = 3\cos\theta, y = 2\sin\theta, 0 \leqslant \theta \leqslant 2\pi$

9. $x = \sin^2\theta, y = \cos^2\theta$

10. $x = \sec\theta, y = \tan\theta, -\pi/2 < \theta < \pi/2$

11. $x = e^t, y = e^t$ 12. $x = e^t, y = e^{-t}$

13. $x = \cos^2 t, y = \cos^4 t$ 14. $x = \cos t, y = \cos 2t$

15. $x = \cos^2\theta, y = \sin\theta$

16. $x = \dfrac{1 - t^2}{1 + t^2}, y = \dfrac{2t}{1 + t^2}$

17. $x = e^t, y = \sqrt{t}, 0 \leqslant t \leqslant 1$

18. $x = \dfrac{1 - t}{1 + t}, y = t^2, 0 \leqslant t \leqslant 1$

19. $x = \cosh t, y = \sinh t$

20. $x = 4 \sinh t, y = 3 \cosh t$

In Exercises 21–26 describe the motion of a particle with position (x, y) as t varies in the given time interval.

21. $x = \cos \pi t, y = \sin \pi t, 1 \leqslant t \leqslant 2$

22. $x = 2 + \cos t, y = 3 + \sin t, 0 \leqslant t \leqslant 2\pi$

23. $x = 8t - 3, y = 2 - t, 0 \leqslant t \leqslant 1$

24. $x = \cos^2 t, y = \cos t, 0 \leqslant t \leqslant 4\pi$

25. $x = 2 \sin t, y = 3 \cos t, 0 \leqslant t \leqslant 2\pi$

26. $x = \sin t, y = \csc t, \pi/6 \leqslant t \leqslant 1$

In Exercises 27–30 sketch the curve represented by the given parametric equations.

27. $x = 3(t^2 - 3), y = t^3 - 3t$

28. $x = t \cos t, y = t \sin t$

29. $x = \tan \theta + \sin \theta, y = \cos \theta$

30. $x = \dfrac{3t}{1 + t^3}, y = \dfrac{3t^2}{1 + t^3}$

31. Show that the parametric equations $x = x_1 + (x_2 - x_1)t$ and $y = y_1 + (y_2 - y_1)t, 0 \leqslant t \leqslant 1$, describe the line segment that joins the points $P_1(x_1, y_1)$ and $P_2(x_2, y_2)$.

32. If a projectile is fired with an initial velocity of v_0 meters per second at an angle α above the horizontal, then its position after t seconds is given by the parametric equations

$$x = (v_0 \cos \alpha)t \qquad y = (v_0 \sin \alpha)t - \tfrac{1}{2}gt^2$$

where g is the acceleration due to gravity (9.8 m/s²).
 (a) If a gun is fired with $\alpha = 30°$ and $v_0 = 500$ m/s, when will the bullet hit the ground? How far from the gun will it hit the ground?
 (b) What is the maximum height reached by the bullet?
 (c) Show that the path is parabolic by eliminating the parameter.

33. Derive Equations 9.2 for the case where $\dfrac{\pi}{2} < \theta < \pi$.

34. Let P be a point at a distance d from the center of a circle of radius r. The curve traced out by P as the circle rolls along a straight line is called a **trochoid.** (Think of the motion of a point on a spoke of a bicycle wheel.) The cycloid is the special case of a trochoid with $d = r$. Using the same parameter θ as for the cycloid and assuming the line is the x-axis and $\theta = 0$ when P is at one of its lowest points, show that the parametric equations of the trochoid are

$$x = r\theta - d \sin \theta \qquad y = r - d \cos \theta$$

Sketch the trochoid for the cases $d < r$ and $d > r$.

35. (a) A circle C of radius b rolls on the inside of a larger circle with center 0 and radius a. The curve traced out by a fixed point P on C is called a **hypocycloid.** If the initial position of P is $(a, 0)$ and the parameter θ is chosen as in Figure 9.9, show that the parametric equations of the hypocycloid are

$$x = (a - b) \cos \theta + b \cos\!\left(\frac{a - b}{b}\theta\right)$$

$$y = (a - b) \sin \theta - b \sin\!\left(\frac{a - b}{b}\theta\right)$$

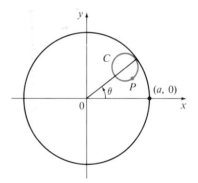

Figure 9.9

 (b) If $b = a/4$, the hypocycloid is called a **hypocycloid of four cusps** or an **astroid.** Show that in this case the parametric equations reduce to

$$x = a \cos^3 \theta \qquad y = a \sin^3 \theta$$

and sketch the curve.

36. If the circle C of Exercise 35 rolls on the outside of the larger circle, the curve traced out by P is called an **epicycloid.** (The special case where $a = b$ is called a **cardioid.**) Find the parametric equations for the epicycloid. Sketch the epicycloid for the case $b = a/4$.

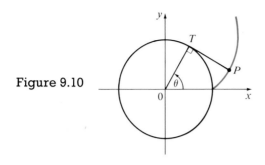

Figure 9.10

37. A string is wound around a circle and then unwound while being held taut. The curve traced by the point P at the end of the string is called the **involute** of the circle. If the circle has radius r and center 0 and the initial position of P is $(r, 0)$, and if the parameter θ is chosen as in Figure 9.10, show that the parametric equations of the involute are

$$x = r(\cos\theta + \theta\sin\theta) \qquad y = r(\sin\theta - \theta\cos\theta)$$

SECTION 9.2

Tangents and Areas

In this section we adapt our previous methods for finding tangents and areas for curves of the form $y = F(x)$ and apply them to curves given by parametric equations.

In the preceding section we saw that some curves defined by parametric equations $x = f(t)$ and $y = g(t)$ can also be expressed, by eliminating the parameter, in the form $y = F(x)$. (See Exercise 38 for general conditions under which this is possible.) If we substitute $x = f(t)$ and $y = g(t)$ in the equation $y = F(x)$, we get

$$g(t) = F(f(t))$$

and so, if g, F, and f are differentiable, the Chain Rule gives

$$g'(t) = F'(f(t))f'(t) = F'(x)f'(t)$$

If $f'(t) \neq 0$, we can solve for $F'(x)$:

(9.3)
$$F'(x) = \frac{g'(t)}{f'(t)}$$

Since the slope of the tangent to the curve $y = F(x)$ at $(x, F(x))$ is $F'(x)$, Equation 9.3 enables us to find tangents to parametric curves without having to eliminate the parameter. Using Leibniz notation we can rewrite Equation 9.3 in the easily remembered form

(9.4)
$$\frac{dy}{dx} = \frac{\dfrac{dy}{dt}}{\dfrac{dx}{dt}} \qquad \text{if} \quad \frac{dx}{dt} \neq 0$$

It can be seen from Equation 9.4 that the curve has a horizontal tangent when $dy/dt = 0$ (provided that $dx/dt \neq 0$) and it has a vertical tangent when

$dx/dt = 0$ (provided that $dy/dt \neq 0$). This information is useful when sketching parametric curves.

As we know from Chapter 4, it is also useful to consider d^2y/dx^2. This can again be found from Equation 9.4 as follows:

$$\frac{d^2y}{dx^2} = \frac{d}{dx}\left(\frac{dy}{dx}\right) = \frac{\dfrac{d}{dt}\left(\dfrac{dy}{dx}\right)}{\dfrac{dx}{dt}}$$

● **Example 1**   (a) Find dy/dx and d^2y/dx^2 for the cycloid $x = r(\theta - \sin\theta)$, $y = r(1 - \cos\theta)$. (See Example 5 in Section 9.1.) (b) Find the tangent to the cycloid at the point where $\theta = \pi/3$.   (c) At what points is the tangent horizontal?   (d) Discuss the concavity.

Solution   (a)

$$\frac{dy}{dx} = \frac{\dfrac{dy}{d\theta}}{\dfrac{dx}{d\theta}} = \frac{r\sin\theta}{r(1 - \cos\theta)} = \frac{\sin\theta}{1 - \cos\theta}$$

$$\frac{d}{d\theta}\left(\frac{dy}{dx}\right) = \frac{d}{d\theta}\left(\frac{\sin\theta}{1 - \cos\theta}\right) = \frac{\cos\theta(1 - \cos\theta) - \sin\theta\sin\theta}{(1 - \cos\theta)^2}$$

$$= \frac{\cos\theta - 1}{(1 - \cos\theta)^2} = -\frac{1}{1 - \cos\theta}$$

$$\frac{d^2y}{dx^2} = \frac{\dfrac{d}{d\theta}\left(\dfrac{dy}{dx}\right)}{\dfrac{dx}{d\theta}} = \frac{-\dfrac{1}{1 - \cos\theta}}{r(1 - \cos\theta)} = -\frac{1}{r(1 - \cos\theta)^2}$$

(b)   When $\theta = \pi/3$, we have

$$x = r\left(\frac{\pi}{3} - \sin\frac{\pi}{3}\right) = r\left(\frac{\pi}{3} - \frac{\sqrt{3}}{2}\right) \qquad y = r\left(1 - \cos\frac{\pi}{3}\right) = \frac{r}{2}$$

and

$$\frac{dy}{dx} = \frac{\sin\left(\dfrac{\pi}{3}\right)}{1 - \cos\left(\dfrac{\pi}{3}\right)} = \frac{\dfrac{\sqrt{3}}{2}}{1 - \dfrac{1}{2}} = \sqrt{3}$$

Therefore the slope of the tangent is $\sqrt{3}$ and its equation is

$$y - \frac{r}{2} = \sqrt{3}\left(x - \frac{r\pi}{3} + \frac{r\sqrt{3}}{2}\right) \quad \text{or} \quad \sqrt{3}x - y = r\left(\frac{\pi}{\sqrt{3}} - 2\right)$$

This tangent is sketched in Figure 9.11.

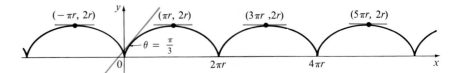

Figure 9.11

(c)   The tangent is horizontal when $dy/dx = 0$, which occurs when $\sin \theta = 0$ and $1 - \cos \theta \neq 0$, that is, $\theta = (2n - 1)\pi$, n an integer. The corresponding point on the cycloid is $((2n - 1)\pi r, 2r)$. (It can be shown that the cycloid has vertical tangents when $\theta = 2n\pi$. See Exercise 28.)

(d)   From part (a) we have $d^2y/dx^2 = -1/r(1 - \cos \theta)^2$, which shows that $d^2y/dx^2 < 0$ except when $\cos \theta = 1$. Thus the cycloid is concave downward on the intervals $(2n\pi, 2(n + 1)\pi)$. ●

● **Example 2**   A curve C is defined by the parametric equations $x = t^2$ and $y = t^3 - 3t$. (a) Show that C has two tangents at the point $(3, 0)$ and find their equations. (b) Find the points on C where the tangent is horizontal or vertical. (c) Sketch the curve.

Solution   (a)   Notice that $y = t^3 - 3t = t(t^2 - 3) = 0$ when $t = 0$ or $t = \pm\sqrt{3}$. Therefore the point $(3, 0)$ on C arises from two values of the parameter, $t = \sqrt{3}$ and $t = -\sqrt{3}$. This indicates that C crosses itself at $(3, 0)$. Since

$$\frac{dy}{dx} = \frac{\dfrac{dy}{dt}}{\dfrac{dx}{dt}} = \frac{3t^2 - 3}{2t}$$

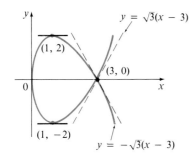

$y = \sqrt{3}(x - 3)$

$(1, 2)$

$(3, 0)$

$(1, -2)$

$y = -\sqrt{3}(x - 3)$

Figure 9.12

the slope of the tangent when $t = \pm\sqrt{3}$ is $dy/dx = \pm 6/2\sqrt{3} = \pm\sqrt{3}$ so the equations of the tangents at $(3, 0)$ are

$$y = \sqrt{3}(x - 3) \qquad \text{and} \qquad y = -\sqrt{3}(x - 3)$$

(b) C has a vertical tangent when $dx/dt = 2t = 0$, that is, $t = 0$. The corresponding point on C is $(0, 0)$. C has a horizontal tangent when $dy/dt = 3t^2 - 3 = 0$, that is, $t = \pm 1$. The corresponding points on C are $(1, -2)$ and $(1, 2)$.

(c)   Using the information in parts (a) and (b) and plotting a few additional points, we sketch C in Figure 9.12. ●

We know that the area under a curve $y = F(x)$ from a to b is given by $A = \int_a^b F(x)\,dx$, where $F(x) \geq 0$. If the curve is given by parametric equations $x = f(t)$ and $y = g(t)$, $\alpha \leq t \leq \beta$, then we can adapt the earlier formula by using the Substitution Rule for definite integrals (5.32) as follows:

$$A = \int_a^b y\,dx = \int_\alpha^\beta g(t)f'(t)\,dt \qquad \left[\text{or} \int_\beta^\alpha g(t)f'(t)\,dt\right]$$

Figure 9.13

● **Example 3**   Find the area under one arch of the cycloid $x = r(\theta - \sin \theta)$, $y = r(1 - \cos \theta)$ (see Figure 9.13).

Solution   One arch of the cycloid is given by $0 \leqslant \theta \leqslant 2\pi$. Using the Substitution Rule with $y = r(1 - \cos\theta)$ and $dx = r(1 - \cos\theta)\,d\theta$, we have

$$A = \int_0^{2\pi r} y\,dx = \int_0^{2\pi} r(1 - \cos\theta)r(1 - \cos\theta)\,d\theta$$

$$= r^2 \int_0^{2\pi}(1 - \cos\theta)^2\,d\theta = r^2 \int_0^{2\pi}(1 - 2\cos\theta + \cos^2\theta)\,d\theta$$

$$= r^2 \int_0^{2\pi}[1 - 2\cos\theta + \tfrac{1}{2}(1 + \cos 2\theta)]\,d\theta$$

$$= r^2 [\tfrac{3}{2}\theta - 2\sin\theta + \tfrac{1}{4}\sin 2\theta]_0^{2\pi}$$

$$= r^2 (\tfrac{3}{2} \cdot 2\pi) = 3\pi r^2$$

The result of Example 3 says that the area under one arch of the cycloid is three times the area of the rolling circle that generates the cycloid (see Example 5 in Section 9.1). Galileo guessed this result but it was first proved by the French mathematician Roberval and the Italian mathematician Torricelli.

EXERCISES 9.2

In Exercises 1–6 find the equation of the tangent to the given curve at the point corresponding to the given value of the parameter.

1. $x = t^2 + t$, $y = t^2 - t$; $t = 0$

2. $x = 1 - t^3$, $y = t^2 - 3t + 1$; $t = 1$

3. $x = t^2 + t$, $y = \sqrt{t}$; $t = 4$

4. $x = \ln t$, $y = te^t$; $t = 1$

5. $x = 2\sin\theta$, $y = 3\cos\theta$; $\theta = \pi/4$

6. $x = t\sin t$, $y = t\cos t$; $t = \pi$

In Exercises 7–10 find the equation of the tangent to the given curve at the given point by two methods: (a) without eliminating the parameter and (b) by first eliminating the parameter.

7. $x = 1 - t$, $y = 1 - t^2$; $(1, 1)$

8. $x = 2t + 3$, $y = t^2 + 2t$; $(5, 3)$

9. $x = 5\cos t$, $y = 5\sin t$; $(3, 4)$

10. $x = t^3$, $y = t^2$; $(1, 1)$

In Exercises 11–18 find dy/dx and d^2y/dx^2.

11. $x = t^2 + t$, $y = t^2 + 1$

12. $x = t^3 + t^2 + 1$, $y = 1 - t^2$

13. $x = \sqrt{t + 1}$, $y = t^2 - 3t$

14. $x = t^4 - t^2 + t$, $y = \sqrt[3]{t}$

15. $x = \sin\pi t$, $y = \cos\pi t$

16. $x = t + 2\cos t$, $y = \sin 2t$

17. $x = e^{-t}$, $y = te^{2t}$

18. $x = 1 + t^2$, $y = t\ln t$

In Exercises 19–24 find the points on the given curve where the tangent is horizontal or vertical. Then sketch the curve.

19. $x = 1 - 2\cos t$, $y = 2 + 3\sin t$

20. $x = t^3 - 3t^2$, $y = t^3 - 3t$

21. $x = t(t^2 - 3)$, $y = 3(t^2 - 3)$

22. $x = \sin 2t$, $y = \sin t$

23. $x = \dfrac{3t}{1 + t^3}$, $y = \dfrac{3t^2}{1 + t^3}$

24. $x = a(\cos\theta - \cos^2\theta)$, $y = a(\sin\theta - \sin\theta\cos\theta)$

25. Show that the curve $x = \cos t$, $y = \sin t \cos t$ has two tangents at $(0,0)$ and find their equations. Sketch the curve.

26. At what point does the curve $x = 1 - 2\cos^2 t$, $y = \tan t(1 - 2\cos^2 t)$ cross itself? Find the equations of both tangents at that point.

27. At what point does the curve of Exercise 21 cross itself? Find the equations of both tangents at that point.

28. Show that the cycloid $x = r(\theta - \sin\theta)$, $y = r(1 - \cos\theta)$ has a vertical tangent at the origin by showing that $dy/dx \to \infty$ as $\theta \to 0^+$.

29. (a) Find the slope of the tangent line to the trochoid $x = r\theta - d\sin\theta$, $y = r - d\cos\theta$ in terms of θ. (See Exercise 34 in Section 9.1.)
   (b) Show that if $d < r$, then the trochoid does not have a vertical tangent.

30. (a) Find the slope of the tangent to the astroid $x = a\cos^3\theta$, $y = a\sin^3\theta$ in terms of θ. (See Exercise 35 in Section 9.1.)
   (b) At what points is the tangent horizontal or vertical?

(c) At what points does the tangent have slope 1 or -1?

31. At what points on the curve $x = t^3 + 4t$, $y = 6t^2$ is the tangent parallel to the line with equations $x = -7t$, $y = 12t - 5$?

32. Find the equations of the tangents to the curve $x = 3t^2 + 1$, $y = 2t^3 + 1$ that pass through the point $(4, 3)$.

33. Use the parametric equations of an ellipse, $x = a\cos\theta$, $y = b\sin\theta$, $0 \leqslant \theta \leqslant 2\pi$, to find the area that it encloses.

34. Find the area bounded by the curve $x = t - 1/t$, $y = t + 1/t$ and the line $y = 2.5$.

35. Find the area bounded by the curve $x = \cos t$, $y = e^t$, $0 \leqslant t \leqslant \pi/2$, and the lines $y = 1$ and $x = 0$.

36. Find the area of the region enclosed by the astroid $x = a\cos^3\theta$, $y = a\sin^3\theta$. (See Exercise 35 in Section 9.1.)

37. Find the area under one arch of the trochoid of Exercise 34 in Section 9.1 for the case where $d < r$.

38. If f' is continuous and $f'(t) \neq 0$ for $a \leqslant t \leqslant b$, show that the parametric curve $x = f(t)$, $y = g(t)$, $a \leqslant t \leqslant b$, can be put in the form $y = F(x)$.

SECTION 9.3

Arc Length and Surface Area

We already know how to find the length L of a curve C given in the form $y = F(x)$, $a \leqslant x \leqslant b$. Formula 8.11 says that if F' is continuous, then

(9.5)
$$L = \int_a^b \sqrt{1 + \left(\frac{dy}{dx}\right)^2}\, dx$$

Suppose that C can also be described by the parametric equations $x = f(t)$ and $y = g(t)$, $\alpha \leqslant t \leqslant \beta$, where $dx/dt = f'(t) > 0$. This means that C is traversed once, from left to right, as t increases from α to β and $f(\alpha) = a$, $f(\beta) = b$. Putting Formula 9.4 into Formula 9.5 and using the Substitution Rule, we obtain

$$L = \int_a^b \sqrt{1 + \left(\frac{dy}{dx}\right)^2}\, dx = \int_\alpha^\beta \sqrt{1 + \left(\frac{dy/dt}{dx/dt}\right)^2}\, \frac{dx}{dt}\, dt$$

Since $dx/dt > 0$, we have

(9.6)
$$L = \int_\alpha^\beta \sqrt{\left(\frac{dx}{dt}\right)^2 + \left(\frac{dy}{dt}\right)^2}\, dt$$

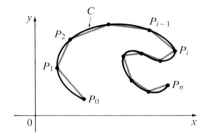

Figure 9.14

Even if C cannot be expressed in the form $y = F(x)$, Formula 9.6 is still valid but we obtain it by polygonal approximations. Let P be a partition of $[\alpha, \beta]$ by points t_i with $\alpha = t_0 < t_1 < \cdots < t_n = \beta$. Let $x_i = f(t_i)$, $y_i = g(t_i)$, $\Delta x_i = x_i - x_{i-1}$, and $\Delta y_i = y_i - y_{i-1}$. Then the point $P_i(x_i, y_i)$ lies on C and the polygon with vertices $P_0, P_1, \ldots, P_n$ approximates C (see Figure 9.14).

As in Section 8.3, we define the length L of C to be the limit of the lengths of these approximating polygons as $\|P\| \to 0$:

$$L = \lim_{\|P\| \to 0} \sum_{i=1}^{n} |P_{i-1}P_i|$$

The Mean Value Theorem, when applied to f on the interval $[t_{i-1}, t_i]$, gives a number t_i^* in (t_{i-1}, t_i) such that

$$f(t_i) - f(t_{i-1}) = f'(t_i^*)(t_i - t_{i-1})$$

that is, $\Delta x_i = f'(t_i^*)\Delta t_i$

Similarly, when applied to g, the Mean Value Theorem gives a number t_i^{**} in (t_{i-1}, t_i) such that

$$\Delta y_i = g'(t_i^{**})\Delta t_i$$

Therefore
$$|P_{i-1}P_i| = \sqrt{(\Delta x_i)^2 + (\Delta y_i)^2}$$
$$= \sqrt{[f'(t_i^*)\Delta t_i]^2 + [g'(t_i^{**})\Delta t_i]^2}$$
$$= \sqrt{[f'(t_i^*)]^2 + [g'(t_i^{**})]^2}\,\Delta t_i$$

and so

(9.7)
$$L = \lim_{\|P\| \to 0} \sum_{i=1}^{n} \sqrt{[f'(t_i^*)]^2 + [g'(t_i^{**})]^2}\,\Delta t_i$$

The sum in (9.7) resembles a Riemann sum for the function $\sqrt{[f'(t)]^2 + [g'(t)]^2}$ but it is not exactly a Riemann sum because $t_i^* \neq t_i^{**}$ in general. Nevertheless if f' and g' are continuous, it can be shown that the limit in (9.7) is the same as if t_i^* and t_i^{**} were equal, namely,

$$L = \int_{\alpha}^{\beta} \sqrt{[f'(t)]^2 + [g'(t)]^2}\,dt$$

Thus, using Leibniz notation, we have the following result, which has the same form as (9.6).

Theorem (9.8)

If a curve C is described by the parametric equations $x = f(t)$, $y = g(t)$, $\alpha \leqslant t \leqslant \beta$, where f' and g' are continuous on $[\alpha, \beta]$ and C is traversed exactly once as t increases from α to β, then the length of C is

$$L = \int_{\alpha}^{\beta} \sqrt{\left(\frac{dx}{dt}\right)^2 + \left(\frac{dy}{dt}\right)^2}\,dt$$

Notice that the formula in Theorem 9.8 is consistent with the general formulas $L = \int ds$ and $(ds)^2 = (dx)^2 + (dy)^2$ of Section 8.3.

● **Example 1**   If we use the representation of the unit circle given in Example 2 in Section 9.1,

$$x = \cos t \qquad y = \sin t \qquad 0 \leqslant t \leqslant 2\pi$$

then $dx/dt = -\sin t$ and $dy/dt = \cos t$, so Theorem 9.8 gives

$$L = \int_0^{2\pi} \sqrt{\left(\frac{dx}{dt}\right)^2 + \left(\frac{dy}{dt}\right)^2}\, dt = \int_0^{2\pi} \sqrt{\sin^2 t + \cos^2 t}\, dt$$

$$= \int_0^{2\pi} dt = 2\pi$$

as expected. If, on the other hand, we use the representation given in Example 3 in Section 9.1,

$$x = \sin 2t \qquad y = \cos 2t \qquad 0 \leqslant t \leqslant 2\pi$$

then $dx/dt = 2\cos 2t$, $dy/dt = -2\sin 2t$, and the integral in Theorem 9.8 gives

$$\int_0^{2\pi} \sqrt{\left(\frac{dx}{dt}\right)^2 + \left(\frac{dy}{dt}\right)^2}\, dt = \int_0^{2\pi} \sqrt{4\cos^2 2t + 4\sin^2 2t}\, dt = \int_0^{2\pi} 2\, dt = 4\pi$$

Notice that the integral gives twice the arc length of the circle because as t increases from 0 to 2π, the point $(\sin 2t, \cos 2t)$ traverses the circle twice. In general, when finding the length of a curve C from a parametric representation, we have to be careful to ensure that C is traversed only once as t increases from α to β.   ●

● **Example 2**   Find the length of one arch of the cycloid $x = r(\theta - \sin\theta)$, $y = r(1 - \cos\theta)$.

Solution   From Example 5 in Section 9.1 we see that one arch is described by the parameter interval $0 \leqslant \theta \leqslant 2\pi$. Since

$$\frac{dx}{d\theta} = r(1 - \cos\theta) \qquad \text{and} \qquad \frac{dy}{d\theta} = r\sin\theta$$

we have

$$L = \int_0^{2\pi} \sqrt{\left(\frac{dx}{d\theta}\right)^2 + \left(\frac{dy}{d\theta}\right)^2}\, d\theta = \int_0^{2\pi} \sqrt{r^2(1 - \cos\theta)^2 + r^2\sin^2\theta}\, d\theta$$

$$= \int_0^{2\pi} \sqrt{r^2(1 - 2\cos\theta + \cos^2\theta + \sin^2\theta)}\, d\theta = r\int_0^{2\pi} \sqrt{2(1 - \cos\theta)}\, d\theta$$

In order to evaluate this integral we use the identity $\sin^2 x = \frac{1}{2}(1 - \cos 2x)$ with $\theta = 2x$, which gives $1 - \cos\theta = 2\sin^2(\theta/2)$. Since $0 \leqslant \theta \leqslant 2\pi$, we have

$0 \leqslant \theta/2 \leqslant \pi$ and so $\sin(\theta/2) \geqslant 0$. Therefore

$$\sqrt{2(1 - \cos\theta)} = \sqrt{4\sin^2\left(\frac{\theta}{2}\right)} = 2\left|\sin\left(\frac{\theta}{2}\right)\right| = 2\sin\left(\frac{\theta}{2}\right)$$

and so

$$L = 2r \int_0^{2\pi} \sin\left(\frac{\theta}{2}\right) d\theta = 2r\left[-2\cos\left(\frac{\theta}{2}\right)\right]_0^{2\pi}$$

$$= 2r[2 + 2] = 8r$$

The result of Example 2 says that the length of one arch of a cycloid is eight times the radius of the generating circle. This was first proved in 1658 by Sir Christopher Wren, who later became the architect of St. Paul's Cathedral in London.

Surface Area

In the same way as for arc length, we can adapt Formula 8.21 to obtain a formula for surface area. If the curve given by the parametric equations $x = f(t)$, $y = g(t)$, $\alpha \leqslant t \leqslant \beta$, is rotated about the x-axis, where f', g' are continuous and $g(t) \geqslant 0$, then the area of the resulting surface is given by

(9.9)
$$S = \int_\alpha^\beta 2\pi y \sqrt{\left(\frac{dx}{dt}\right)^2 + \left(\frac{dy}{dt}\right)^2}\, dt$$

The general symbolic formulas $S = \int 2\pi y\, ds$ and $S = \int 2\pi x\, ds$ (8.23 and 8.24) are still valid, but for parametric curves we use

$$ds = \sqrt{\left(\frac{dx}{dt}\right)^2 + \left(\frac{dy}{dt}\right)^2}\, dt$$

● **Example 3**   Show that the surface area of a sphere of radius r is $4\pi r^2$.

Solution   The sphere is obtained by rotating the semicircle

$$x = r\cos t \qquad y = r\sin t, \qquad 0 \leqslant t \leqslant \pi$$

about the x-axis. Therefore, from Formula 9.9, we get

$$S = \int_0^\pi 2\pi r \sin t \sqrt{(-r\sin t)^2 + (r\cos t)^2}\, dt$$

$$= 2\pi \int_0^\pi r\sin t \sqrt{r^2(\sin^2 t + \cos^2 t)}\, dt$$

$$= 2\pi r^2 \int_0^\pi \sin t\, dt = 2\pi r^2(-\cos t)\Big]_0^\pi = 4\pi r^2$$

● **Example 4**   Find the area of the surface generated by rotating one arch of the cycloid $x = r(\theta - \sin\theta)$, $y = r(1 - \cos\theta)$ about the x-axis.

Solution   Using Formula 9.9 and the same identity as in Example 2, we have

$$S = \int_0^{2\pi} 2\pi y \sqrt{\left(\frac{dx}{d\theta}\right)^2 + \left(\frac{dy}{d\theta}\right)^2} \, d\theta$$

$$= \int_0^{2\pi} 2\pi r(1 - \cos\theta)\sqrt{r^2(1 - \cos\theta)^2 + r^2 \sin^2\theta} \, d\theta$$

$$= 2\pi r^2 \int_0^{2\pi}(1 - \cos\theta)\sqrt{2(1 - \cos\theta)} \, d\theta = 2\pi r^2 \int_0^{2\pi} 2\sin^2\left(\frac{\theta}{2}\right) 2\sin\left(\frac{\theta}{2}\right) d\theta$$

$$= 8\pi r^2 \int_0^{2\pi}\left(1 - \cos^2\left(\frac{\theta}{2}\right)\right)\sin\left(\frac{\theta}{2}\right) d\theta = 16\pi r^2 \int_0^{\pi}(\sin t - \cos^2 t \sin t) \, dt$$

$$= 16\pi r^2\left[-\cos t + \frac{1}{3}\cos^3 t\right]_0^{\pi} = \frac{64\pi r^2}{3} \qquad \bullet$$

The problem in Example 4 was solved by Blaise Pascal (1623–1662) without using the Fundamental Theorem of Calculus.

EXERCISES 9.3

Find the lengths of the curves in Exercises 1–10.

1. $x = 1 + 2\sin \pi t$, $y = 3 - 2\cos \pi t$, $0 \leqslant t \leqslant 1$

2. $x = t^3$, $y = t^2$, $0 \leqslant t \leqslant 4$

3. $x = 5t^2 + 1$, $y = 4 - 3t^2$, $0 \leqslant t \leqslant 2$

4. $x = 3t - t^3$, $y = 3t^2$, $0 \leqslant t \leqslant 2$

5. $x = e^t \cos t$, $y = e^t \sin t$, $0 \leqslant t \leqslant \pi$

6. $x = a(\cos\theta + \theta\sin\theta)$, $y = a(\sin\theta - \theta\cos\theta)$, $0 \leqslant \theta \leqslant \pi$

7. $x = \ln\sin t$, $y = t$, $\pi/4 \leqslant t \leqslant \pi/2$

8. $x = t^2$, $y = 1 + 4t$, $0 \leqslant t \leqslant 2$

9. $x = 2 - 3\sin^2\theta$, $y = \cos 2\theta$, $0 \leqslant \theta \leqslant \pi/2$

10. $x = e^t - t$, $y = 4e^{t/2}$, $0 \leqslant t \leqslant 1$

In Exercises 11 and 12 find the distance traveled by a particle with position (x, y) as t varies in the given time interval. Compare with the length of the curve.

11. $x = \sin^2\theta$, $y = \cos^2\theta$, $0 \leqslant \theta \leqslant 3\pi$

12. $x = \cos^2 t$, $y = \cos t$, $0 \leqslant t \leqslant 4\pi$

13. Show that the total length of the ellipse $x = a\sin\theta$, $y = b\cos\theta$, $a > b > 0$, is

$$L = 4a \int_0^{\pi/2} \sqrt{1 - e^2 \sin^2\theta} \, d\theta$$

where e is the eccentricity of the ellipse ($e = c/a$ where $c = \sqrt{a^2 - b^2}$).

14. Find the total length of the astroid $x = a\cos^3\theta$, $y = a\sin^3\theta$.

In Exercises 15–20 find the area of the surface obtained by rotating the given curve about the x-axis.

15. $x = t^2 + \dfrac{2}{t}$, $y = 8\sqrt{t}$, $1 \leqslant t \leqslant 9$

16. $x = t^3$, $y = t^2$, $0 \leqslant t \leqslant 1$

17. $x = e^t \cos t$, $y = e^t \sin t$, $0 \leqslant t \leqslant \pi/2$

18. $x = 3t - t^3$, $y = 3t^2$, $0 \leqslant t \leqslant 1$

19. $x = 2\cos\theta - \cos 2\theta$, $y = 2\sin\theta - \sin 2\theta$

20. $x = a\cos^3\theta$, $y = a\sin^3\theta$, $0 \leqslant \theta \leqslant \pi/2$

In Exercises 21 and 22 find the surface area generated by rotating the given curve about the *y*-axis.

21. $x = 3t^2, y = 2t^3, 0 \leqslant t \leqslant 5$

22. $x = e^t - t, y = 4e^{t/2}, 0 \leqslant t \leqslant 1$

23.   Find the surface area of the ellipsoid obtained by rotating the ellipse $x = a\cos\theta, y = b\sin\theta \ (a > b)$ about
  (a)   the *x*-axis   (b)   the *y*-axis

24.   Use Formula 9.4 to derive Formula 9.9 from Formula 8.21 for the case where the curve can be represented in the form $y = F(x), a \leqslant x \leqslant b$.

SECTION 9.4

Polar Coordinates

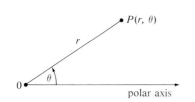

Figure 9.15

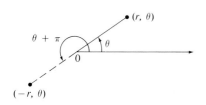

Figure 9.16

A coordinate system represents a point in the plane by an ordered pair of numbers called coordinates. So far we have been using Cartesian coordinates, which are directed distances from two perpendicular axes. In this section we describe a coordinate system introduced by Newton, called the **polar coordinate system,** which is more convenient for many purposes.

We choose a point in the plane that is called the **pole** (or origin) and labeled O. Then we draw a ray (half-line) starting at O called the **polar axis.** This axis is usually drawn horizontally to the right and corresponds to the positive *x*-axis in Cartesian coordinates.

If P is any other point in the plane, let r be the distance from O to P and let θ be the angle (usually measured in radians) between the polar axis and the line OP as in Figure 9.15. Then the point P is represented by the ordered pair (r, θ) and r, θ are called **polar coordinates** of P. We use the convention that an angle is positive if measured in the counterclockwise direction from the polar axis and negative in the clockwise direction. If $P = O$, then $r = 0$ and we agree that $(0, \theta)$ represents the pole for any value of θ.

We extend the meaning of polar coordinates (r, θ) to the case where r is negative by agreeing that, as in Figure 9.16, the points $(-r, \theta)$ and (r, θ) lie on the same line through O and at the same distance $|r|$ from O, but on opposite sides of O. If $r > 0$, the point (r, θ) lies in the same quadrant as θ; if $r < 0$, it lies in the quadrant on the opposite side of the pole. Notice that $(-r, \theta)$ represents the same point as $(r, \theta + \pi)$.

● **Example 1**   Plot the points whose polar coordinates are (a) $(1, 5\pi/4)$, (b) $(2, 3\pi)$, (c) $(2, -2\pi/3)$, and (d) $(-3, 3\pi/4)$.

Solution   The points are plotted in Figure 9.17. In part (d) the point $(-3, 3\pi/4)$ is located three units from the pole in the fourth quadrant because the angle $3\pi/4$ is in the second quadrant and $r = -3$ is negative.   ●

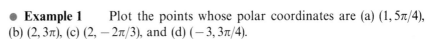

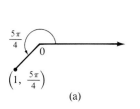

(a)

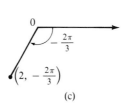

(b)

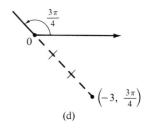

(c)   (d)

Figure 9.17

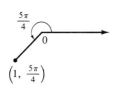
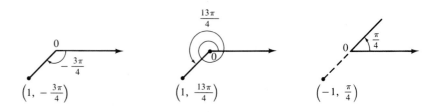

Figure 9.18

In the Cartesian coordinate system every point has only one representation, but in the polar coordinate system each point has many representations. For instance, the point $(1, 5\pi/4)$ in Example 1(a) could be written as $(1, -3\pi/4)$ or $(1, 13\pi/4)$ or $(-1, \pi/4)$ (see Figure 9.18).

In fact, since a complete counterclockwise rotation is given by an angle 2π, the point represented by polar coordinates (r, θ) is also represented by

$$(r, \theta + 2n\pi) \quad \text{and} \quad (-r, \theta + (2n + 1)\pi)$$

where n is any integer.

The connection between polar and Cartesian coordinates can be seen from Figure 9.19 where the pole corresponds to the origin and the polar axis coincides with the positive x-axis.

If the point P has Cartesian coordinates (x, y) and polar coordinates (r, θ), then, from the figure, we have

$$\cos \theta = \frac{x}{r} \qquad \sin \theta = \frac{y}{r}$$

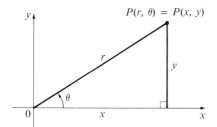

$P(r, \theta) = P(x, y)$

Figure 9.19

(9.10)   and so

$$x = r \cos \theta \qquad y = r \sin \theta$$

Although Equations 9.10 were deduced from Figure 9.19, which illustrates the case where $r > 0$ and $0 < \theta < \pi/2$, these equations are valid for all values of r and θ. (See the general definition of $\sin \theta$ and $\cos \theta$ in Appendix B.)

Equations 9.10 allow us to find the Cartesian coordinates of a point when the polar coordinates are known. To find r and θ when x and y are known we use the equations

(9.11)

$$r^2 = x^2 + y^2 \qquad \tan \theta = \frac{y}{x}$$

which can be deduced from Equations 9.10 or simply read from Figure 9.19.

● **Example 2**   Convert the point $(2, \pi/3)$ from polar to Cartesian coordinates.

Solution   Since $r = 2$ and $\theta = \pi/3$, Equations 9.10 give

$$x = r \cos \theta = 2 \cos \frac{\pi}{3} = 2 \cdot \frac{1}{2} = 1$$

$$y = r \sin \theta = 2 \sin \frac{\pi}{3} = 2 \cdot \frac{\sqrt{3}}{2} = \sqrt{3}$$

Therefore the point is $(1, \sqrt{3})$ in Cartesian coordinates.   ●

● **Example 3**   Represent the point with Cartesian coordinates $(1, -1)$ in terms of polar coordinates.

Solution   If we choose r to be positive, then Equations 9.11 give

$$r = \sqrt{x^2 + y^2} = \sqrt{1^2 + (-1)^2} = \sqrt{2}$$

$$\tan \theta = \frac{y}{x} = -1$$

Since the point $(1, -1)$ lies in the fourth quadrant, we can choose $\theta = -\pi/4$ or $\theta = 7\pi/4$. Thus one possible answer is $(\sqrt{2}, -\pi/4)$. Another is $(\sqrt{2}, 7\pi/4)$.   ●

Note: Equations 9.11 do not uniquely determine θ when x and y are given because, as θ increases through the interval $0 \leqslant \theta < 2\pi$, each value of $\tan \theta$ occurs twice. Therefore, in converting from Cartesian to polar coordinates, it is not good enough just to find r and θ that satisfy Equations 9.11. As in Example 3, we must choose θ so that the point (r, θ) lies in the correct quadrant.

The **graph of a polar equation** $r = f(\theta)$, or more generally $F(r, \theta) = 0$, consists of all points P that have at least one polar representation (r, θ) whose coordinates satisfy the equation.

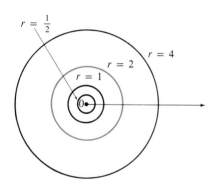

$r = \frac{1}{2}$

$r = 2$

$r = 4$

$r = 1$

Figure 9.20

● **Example 4**   What curve is represented by the polar equation $r = 2$?

Solution   The curve consists of all points (r, θ) with $r = 2$. Since r represents the distance from the point to the pole, the curve $r = 2$ represents the circle with center O and radius 2. In general the equation $r = a$ represents a circle with center O and radius $|a|$ (see Figure 9.20).   ●

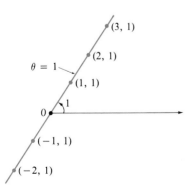

(3, 1)

(2, 1)

$\theta = 1$

(1, 1)

0

(−1, 1)

(−2, 1)

Figure 9.21

● **Example 5**   Sketch the polar curve $\theta = 1$.

Solution   This curve consists of all points (r, θ) such that the polar angle θ is 1 radian. It is the straight line that passes through O and makes an angle of 1 radian with the polar axis (see Figure 9.21). Notice that the points $(r, 1)$ on the line with $r > 0$ are in the first quadrant, whereas those with $r < 0$ are in the third quadrant.   ●

● **Example 6**   (a)   Sketch the curve with polar equation $r = 2 \cos \theta$.
    (b)   Find a Cartesian equation for this curve.

Solution   (a)   In Figure 9.22 we find the values of r for some convenient values of θ and plot the corresponding points (r, θ). Then we join these points to sketch the curve, which appears to be a circle. We have used only values of θ between 0 and π because if we let θ increase beyond π we obtain the same points again.

θ	$r = 2 \cos \theta$
0	2
$\pi/6$	$\sqrt{3}$
$\pi/4$	$\sqrt{2}$
$\pi/3$	1
$\pi/2$	0
$2\pi/3$	-1
$3\pi/4$	$-\sqrt{2}$
$5\pi/6$	$-\sqrt{3}$
π	-2

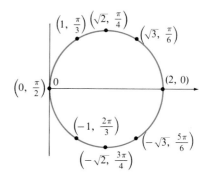

Figure 9.22

(b)   To convert the given equation into a Cartesian equation we use Equations 9.10 and 9.11. From $x = r \cos \theta$ we have $\cos \theta = x/r$, so the equation $r = 2 \cos \theta$ becomes $r = 2x/r$, which gives

$$2x = r^2 = x^2 + y^2 \quad \text{or} \quad x^2 + y^2 - 2x = 0$$

Completing the square, we obtain

$$(x - 1)^2 + y^2 = 1$$

which is the equation of a circle with center $(1, 0)$ and radius 1.   ●

● **Example 7**   Sketch the curve $r = 1 + \sin \theta$.

Solution   Instead of plotting points as in Example 6, we first sketch the graph of $r = 1 + \sin \theta$ in *Cartesian* coordinates in Figure 9.23 by shifting the sine curve up one unit. This will enable us to read at a glance the values of r that correspond to increasing values of θ. For instance, we see that as θ increases from 0 to $\pi/2$, r (the distance from O) increases from 1 to 2, so we sketch the corresponding part of the polar curve in Figure 9.24(a). As θ increases from $\pi/2$ to π, Figure 9.23 shows that r decreases from 2 to 1, so we sketch the next part of the curve in Figure 9.24(b). As θ increases from π to $3\pi/2$, r decreases from 1 to 0 as shown in part (c). Finally, as θ increases from $3\pi/2$ to 2π, r increases from 0 to 1 as shown in part (d). If we were to let θ increase beyond 2π or decrease beyond 0, we would simply retrace our path. Putting together the parts of the curve from Figure 9.24(a)–(d), we sketch the complete curve in Figure 9.24(e). It is called a **cardioid** because it is shaped like a heart.   ●

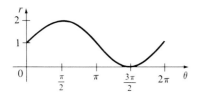

Figure 9.23

$r = 1 + \sin \theta$ IN CARTESIAN COORDINATES, $0 \leqslant \theta \leqslant 2\pi$

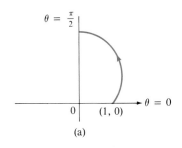

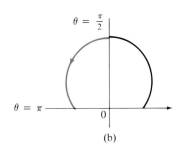

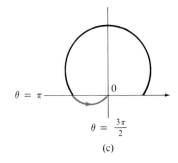

(a)   (b)   (c)

Figure 9.24

STAGES IN SKETCHING THE
CARDIOID
$r = 1 + \sin\theta$

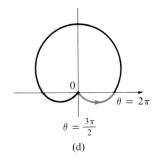

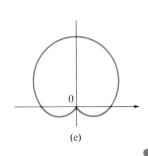

(d)   (e)

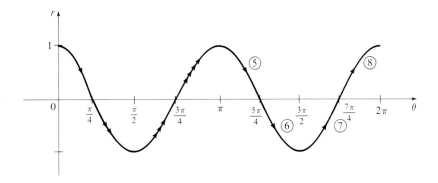

Figure 9.25

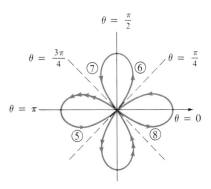

Figure 9.26

FOUR-LEAVED ROSE $r = \cos 2\theta$

● **Example 8**   Sketch the curve $r = \cos 2\theta$.

Solution   As in Example 7, we first sketch $r = \cos 2\theta$, $0 \le \theta \le 2\pi$, in Cartesian coordinates in Figure 9.25. As θ increases from 0 to $\pi/4$, Figure 9.25 shows that r decreases from 1 to 0 and so we draw the corresponding portion of the polar curve in Figure 9.26 (indicated by a single arrow). As θ increases from $\pi/4$ to $\pi/2$, r goes from 0 to -1. This means that the distance from O increases from 0 to 1, but instead of being in the second quadrant this portion of the polar curve (indicated by a double arrow) lies on the opposite side of the pole in the fourth quadrant. The remainder of the curve is drawn in a similar fashion, with the arrows and numbers indicating the order in which the portions are traced out. The resulting curve has four loops and is called a **four-leaved rose.**   ●

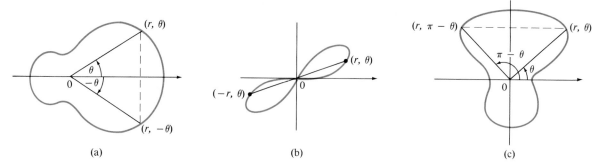

Figure 9.27

(a)   (b)   (c)

In sketching polar curves it is sometimes helpful to take advantage of symmetry. The following three rules are explained by Figure 9.27.

(a) If a polar equation is unchanged when θ is replaced by $-\theta$, the curve is symmetric about the polar axis.

(b) If the equation is unchanged when r is replaced by $-r$, the curve is symmetric about the pole.

(c) If the equation is unchanged when θ is replaced by $\pi - \theta$, the curve is symmetric about the vertical line $\theta = \pi/2$.

The curves in Examples 6 and 8 are symmetric about the polar axis since $\cos(-\theta) = \cos\theta$. The curves in Examples 7 and 8 are symmetric about $\theta = \pi/2$ because $\sin(\pi - \theta) = \sin\theta$ and $\cos 2(\pi - \theta) = \cos 2\theta$. The four-leaved rose is also symmetric about the pole. These symmetry properties could have been used in sketching the curves. For instance, in Example 6 we need only have plotted points for $0 \leqslant \theta \leqslant \pi/2$ and then reflected in the polar axis to obtain the complete circle.

Tangents to Polar Curves

To find a tangent line to a polar curve $r = f(\theta)$ we regard θ as a parameter and write its parametric equations as

$$x = r\cos\theta = f(\theta)\cos\theta \qquad y = r\sin\theta = f(\theta)\sin\theta$$

Then, using the method for finding slopes of parametric curves (Equation 9.4), we have

(9.12)
$$\frac{dy}{dx} = \frac{\dfrac{dy}{d\theta}}{\dfrac{dx}{d\theta}} = \frac{\dfrac{dr}{d\theta}\sin\theta + r\cos\theta}{\dfrac{dr}{d\theta}\cos\theta - r\sin\theta}$$

We locate horizontal tangents by finding the points where $dy/d\theta = 0$ (provided that $dx/d\theta \neq 0$). Likewise we locate vertical tangents at the points where $dx/d\theta = 0$ (provided that $dy/d\theta \neq 0$).

Notice that if we are looking for tangent lines at the pole, then $r = 0$ and Equation 9.12 simplifies to

$$\frac{dy}{dx} = \tan\theta \qquad \text{if} \quad \frac{dr}{d\theta} \neq 0$$

For instance, in Example 8 we found that $r = \cos 2\theta = 0$ when $\theta = \pi/4$ or $3\pi/4$. This means that the lines $\theta = \pi/4$ and $\theta = 3\pi/4$ (or $y = x$ and $y = -x$) are tangent lines to $r = \cos 2\theta$ at the origin.

● **Example 9**   (a) For the cardioid $r = 1 + \sin\theta$ of Example 7, find the slope of the tangent line when $\theta = \pi/3$. (b) Find the points on the cardioid where the tangent line is horizontal or vertical.

Solution   Using Equation 9.12 with $r = 1 + \sin\theta$, we have

$$\frac{dy}{dx} = \frac{\dfrac{dr}{d\theta}\sin\theta + r\cos\theta}{\dfrac{dr}{d\theta}\cos\theta - r\sin\theta} = \frac{\cos\theta\sin\theta + (1 + \sin\theta)\cos\theta}{\cos\theta\cos\theta - (1 + \sin\theta)\sin\theta}$$

$$= \frac{\cos\theta(1 + 2\sin\theta)}{1 - 2\sin^2\theta - \sin\theta} = \frac{\cos\theta(1 + 2\sin\theta)}{(1 + \sin\theta)(1 - 2\sin\theta)}$$

(a)   The slope of the tangent at the point where $\theta = \pi/3$ is

$$\left.\frac{dy}{dx}\right]_{\theta=\pi/3} = \frac{\cos(\pi/3)(1 + 2\sin(\pi/3))}{(1 + \sin(\pi/3))(1 - 2\sin(\pi/3))}$$

$$= \frac{\frac{1}{2}(1 + \sqrt{3})}{\left(1 + \dfrac{\sqrt{3}}{2}\right)(1 - \sqrt{3})} = \frac{1 + \sqrt{3}}{(2 + \sqrt{3})(1 - \sqrt{3})}$$

$$= \frac{1 + \sqrt{3}}{-1 - \sqrt{3}} = -1$$

(b)   Observe that

$$\frac{dy}{d\theta} = \cos\theta(1 + 2\sin\theta) = 0 \qquad \text{when } \theta = \frac{\pi}{2}, \frac{3\pi}{2}, \frac{7\pi}{6}, \frac{11\pi}{6}$$

$$\frac{dx}{d\theta} = (1 + \sin\theta)(1 - 2\sin\theta) = 0 \quad \text{when } \theta = \frac{3\pi}{2}, \frac{\pi}{6}, \frac{5\pi}{6}$$

Therefore there are horizontal tangents at the points $(2, \pi/2)$, $(\frac{1}{2}, 7\pi/6)$, $(\frac{1}{2}, 11\pi/6)$ and vertical tangents at $(\frac{3}{2}, \pi/6)$ and $(\frac{3}{2}, 5\pi/6)$. When $\theta = 3\pi/2$, both $dy/d\theta$ and $dx/d\theta$ are 0, so we must be careful. Using l'Hospital's Rule, we have

$$\lim_{\theta\to 3\pi/2^-}\frac{dy}{dx} = -\frac{1}{3}\lim_{\theta\to 3\pi/2^-}\frac{\cos\theta}{1 + \sin\theta}$$

$$= -\frac{1}{3}\lim_{\theta\to 3\pi/2^-}\frac{-\sin\theta}{\cos\theta} = \infty$$

By symmetry,

$$\lim_{\theta\to 3\pi/2^+}\frac{dy}{dx} = -\infty$$

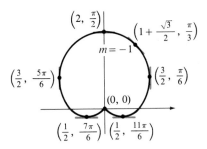

Figure 9.28

TANGENT LINES FOR $r = 1 + \sin\theta$

Thus there is a vertical tangent line at the pole (see Figure 9.28).   ●

EXERCISES 9.4

In Exercises 1–8 plot the points whose polar coordinates are given. Then find two other pairs of polar coordinates of these points, one with $r > 0$ and one with $r < 0$.

1. $\left(1, \dfrac{\pi}{2}\right)$ **2.** $(3, 0)$ **3.** $\left(-1, \dfrac{\pi}{5}\right)$

4. $\left(2, -\dfrac{\pi}{7}\right)$ **5.** $\left(4, -\dfrac{2\pi}{3}\right)$ **6.** $(3, 2)$

7. $(-1, \pi)$ **8.** $\left(-2, \dfrac{3\pi}{2}\right)$

In Exercises 9–16 plot the points whose polar coordinates are given. Then find the Cartesian coordinates of these points.

9. $\left(\sqrt{2}, \dfrac{\pi}{4}\right)$ **10.** $\left(2, \dfrac{2\pi}{3}\right)$ **11.** $\left(1.5, \dfrac{3\pi}{2}\right)$

12. $(4, 3\pi)$ **13.** $\left(4, -\dfrac{7\pi}{6}\right)$ **14.** $\left(-4, \dfrac{5\pi}{4}\right)$

15. $\left(-1, \dfrac{\pi}{3}\right)$ **16.** $\left(-2, -\dfrac{5\pi}{6}\right)$

In Exercises 17–20 the Cartesian coordinates of a point are given. Find the polar coordinates (r, θ) of the given point, where $r > 0$ and $0 \leqslant \theta < 2\pi$.

17. $(-1, 1)$ **18.** $(-1, -\sqrt{3})$ **19.** $(2\sqrt{3}, -2)$

20. $(3, 4)$

In Exercises 21–26 sketch the region in the plane consisting of points whose polar coordinates satisfy the given conditions.

21. $r > 1$ **22.** $0 \leqslant \theta \leqslant \dfrac{\pi}{3}$

23. $0 \leqslant r \leqslant 2, \dfrac{\pi}{2} \leqslant \theta \leqslant \pi$

24. $1 \leqslant r < 3, -\dfrac{\pi}{4} \leqslant \theta \leqslant \dfrac{\pi}{4}$

25. $3 < r < 4, -\dfrac{\pi}{2} \leqslant \theta \leqslant \pi$

26. $-1 \leqslant r \leqslant 1, \dfrac{\pi}{4} \leqslant \theta \leqslant \dfrac{3\pi}{4}$

27. Find the distance between the points with polar coordinates $(1, \pi/6)$ and $(3, 3\pi/4)$.

28. Find a formula for the distance between the points with polar coordinates (r_1, θ_1) and (r_2, θ_2).

In Exercises 29–34 find a Cartesian equation for the curve described by the given polar equation.

29. $r \sin \theta = 2$ **30.** $r = 2 \sin \theta$

31. $r = \dfrac{1}{1 - \cos \theta}$ **32.** $r = \dfrac{5}{3 - 4 \sin \theta}$

33. $r^2 = \sin 2\theta$ **34.** $r^2 = \theta$

In Exercises 35–40 find a polar equation for the curve represented by the given Cartesian equation.

35. $y = 5$ **36.** $y = x + 1$

37. $x^2 + y^2 = 25$ **38.** $x^2 = 4y$

39. $2xy = 1$ **40.** $x^2 - y^2 = 1$

In Exercises 41–74 sketch the curve whose polar equation is given.

41. $r = 5$ **42.** $r = -1$ **43.** $\theta = 3\pi/4$

44. $\theta = -\pi/4$ **45.** $r = 2 \sin \theta$ **46.** $r = -4 \sin \theta$

47. $r = -\cos \theta$ **48.** $r = 2 \sin \theta + 2 \cos \theta$

49. $r = \cos \theta - \sin \theta$ **50.** $r = 2(1 - \sin \theta)$

51. $r = 3(1 - \cos \theta)$ **52.** $r = 1 + \cos \theta$

53. $r = \theta, \theta \geqslant 0$ (spiral)   **54.** $r = \dfrac{\theta}{2}, -4\pi \leqslant \theta \leqslant 4\pi$

55. $r = \dfrac{1}{\theta}$ (reciprocal spiral)

56. $r = e^{\theta}$ (logarithmic spiral)

57. $r = 1 - 2 \cos \theta$ (limaçon)

58. $r = 2 + \cos \theta$ (limaçon)

59. $r = 3 + 2 \sin \theta$ (limaçon)

60. $r = 3 - 4 \sin \theta$ (limaçon)

61. $r = -3 \cos 2\theta$

62. $r = \sin 2\theta$

63. $r = \sin 3\theta$ (three-leaved rose)

64. $r = 2 \cos 3\theta$ (three-leaved rose)

65. $r = 2 \cos 4\theta$ (eight-leaved rose)

66. $r = \sin 4\theta$ (eight-leaved rose)

67. $r = \sin 5\theta$ (five-leaved rose)

68. $r = -\cos 5\theta$ (five-leaved rose)

69. $r^2 = 4\cos 2\theta$ (lemniscate)

70. $r^2 = \sin 2\theta$ (lemniscate)

71. $r = 4 + 2\sec\theta$ (conchoid)

72. $r = 2 - \csc\theta$ (conchoid)

73. $r = \sin\theta\tan\theta$ (cissoid)

74. $r^2\theta = 1$ (lituus)

In Exercises 75–82 find the slope of the tangent line to the given polar curve at the point given by the value of θ.

75. $r = 3\cos\theta,\ \theta = \dfrac{\pi}{3}$ **76.** $r = \cos\theta + \sin\theta,\ \theta = \dfrac{\pi}{4}$

77. $r = 0,\ \theta = \dfrac{\pi}{2}$ **78.** $r = \ln\theta,\ \theta = e$

79. $r = 1 + \cos\theta,\ \theta = \dfrac{\pi}{6}$ **80.** $r = 2 + 4\cos\theta,\ \theta = \dfrac{\pi}{6}$

81. $r = \sin 3\theta,\ \theta = \dfrac{\pi}{3}$ **82.** $r = \sin 3\theta,\ \theta = \dfrac{\pi}{6}$

In Exercises 83–88 find the points on the given curve where the tangent line is horizontal or vertical.

83. $r = 3\cos\theta$ **84.** $r = \cos\theta + \sin\theta$

85. $r = \cos 2\theta$ **86.** $r^2 = \sin 2\theta$

87. $r = 1 + \cos\theta$ **88.** $r = e^\theta$

89. Show that the polar equation $r = a\sin\theta + b\cos\theta$, where $ab \neq 0$, represents a circle and find its center and radius.

90. Show that the curves $r = a\sin\theta$ and $r = a\cos\theta$ intersect at right angles.

91. Let P be any point (except the origin) on the curve $r = f(\theta)$. If ψ is the angle between the tangent line at P and the radial line OP, show that

$$\tan\psi = \frac{r}{dr/d\theta}$$

(*Hint:* Observe that $\psi = \phi - \theta$ in Figure 9.29.)

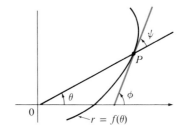

Figure 9.29

Areas and Lengths in Polar Coordinates

In this section we develop the formula for the area of a region whose boundary is given by a polar equation. We shall need to use the formula for the area of a sector of a circle

(9.13)
$$A = \tfrac{1}{2}r^2\theta$$

where, as in Figure 9.30, r is the radius and θ is the radian measure of the central angle. Formula 9.13 can be proved as in Exercise 33 in Section 7.3 or by using the fact that the area of a sector is proportional to its central angle and so $A = (\theta/2\pi)\pi r^2 = \tfrac{1}{2}r^2\theta$.

Let $\mathcal{R}$ be the region, illustrated in Figure 9.31, bounded by the polar curve $r = f(\theta)$ and the rays $\theta = a$ and $\theta = b$, where f is a positive continuous function and $0 < b - a \leqslant 2\pi$. Let P be a partition of $[a, b]$ by numbers θ_i with $a = \theta_0 < \theta_1 < \cdots < \theta_n = b$. The rays $\theta = \theta_i$ then divide $\mathcal{R}$ into n smaller regions with central angles $\Delta\theta_i = \theta_i - \theta_{i-1}$. If we choose θ_i^*

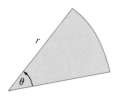

Figure 9.30

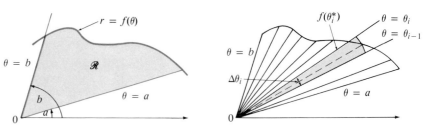

Figure 9.31   Figure 9.32

in the ith subinterval $[\theta_{i-1}, \theta_i]$, then the area ΔA_i of the ith region is approximated by the area of the sector of a circle with central angle $\Delta\theta_i$ and radius $f(\theta_i^*)$ (see Figure 9.32).

Thus from Formula 9.13 we have

$$\Delta A_i \approx \tfrac{1}{2} \left[f(\theta_i^*) \right]^2 \Delta\theta_i$$

and so an approximation to the total area A of $\mathcal{R}$ is

(9.14)
$$A \approx \sum_{i=1}^{n} \tfrac{1}{2} \left[f(\theta_i^*) \right]^2 \Delta\theta_i$$

It appears from Figure 9.32 that the approximation in (9.14) will improve as $\|P\| \to 0$. But the sums in (9.14) are Riemann sums for the function $g(\theta) = \tfrac{1}{2}[f(\theta)]^2$, so

$$\lim_{\|P\| \to 0} \sum_{i=1}^{n} \tfrac{1}{2} \left[f(\theta_i^*) \right]^2 \Delta\theta_i = \int_{a}^{b} \tfrac{1}{2} \left[f(\theta) \right]^2 d\theta$$

It therefore appears plausible (and can in fact be proved) that the formula for the area A of the polar region $\mathcal{R}$ is

(9.15)
$$A = \int_{a}^{b} \tfrac{1}{2} \left[f(\theta) \right]^2 d\theta$$

Formula 9.15 is often written as

(9.16)
$$A = \int_{a}^{b} \tfrac{1}{2} r^2 d\theta$$

with the understanding that $r = f(\theta)$. Note the similarity between Formulas 9.13 and 9.16.

In applying Formula 9.15 or 9.16 it is helpful to think of the area as being swept out by a rotating ray through O that starts with angle a and ends with angle b.

● **Example 1**   Find the area enclosed by one loop of the four-leaved rose $r = \cos 2\theta$.

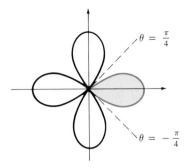

Figure 9.33

Solution   The curve $r = \cos 2\theta$ was sketched in Example 8 in Section 9.4. Notice from Figure 9.33 that the region enclosed by the right loop is swept out by a ray that rotates from $\theta = -\pi/4$ to $\theta = \pi/4$. Therefore Formula 9.16 gives

$$A = \int_{-\pi/4}^{\pi/4} \tfrac{1}{2} r^2 \, d\theta = \tfrac{1}{2} \int_{-\pi/4}^{\pi/4} \cos^2 2\theta \, d\theta$$

$$= \tfrac{1}{2} \int_{-\pi/4}^{\pi/4} \tfrac{1}{2} (1 + \cos 4\theta) \, d\theta$$

$$= \tfrac{1}{4} \left[\theta + \tfrac{1}{4} \sin 4\theta \right]_{-\pi/4}^{\pi/4}$$

$$= \frac{\pi}{8}$$

● **Example 2**   Find the area of the region that lies inside the circle $r = 3 \sin \theta$ and outside the cardioid $r = 1 + \sin \theta$.

Solution   The cardioid (see Example 7 in Section 9.4) and the circle are sketched in Figure 9.34 and the desired region is shaded. The values of a and b in Formula 9.16 are determined by finding the points of intersection of the two curves. They intersect when $3 \sin \theta = 1 + \sin \theta$, which gives $\sin \theta = \tfrac{1}{2}$, so $\theta = \pi/6, 5\pi/6$. The desired area can be found by subtracting the area inside the cardioid between $\theta = \pi/6$ and $\theta = 5\pi/6$ from the area inside the circle from $\pi/6$ to $5\pi/6$. Thus

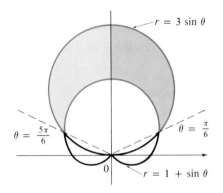

Figure 9.34

$$A = \tfrac{1}{2} \int_{\pi/6}^{5\pi/6} (3 \sin \theta)^2 \, d\theta - \tfrac{1}{2} \int_{\pi/6}^{5\pi/6} (1 + \sin \theta)^2 \, d\theta$$

Since the region is symmetric about the vertical axis $\theta = \pi/2$, we can write

$$A = 2 \left[\tfrac{1}{2} \int_{\pi/6}^{\pi/2} 9 \sin^2 \theta \, d\theta - \tfrac{1}{2} \int_{\pi/6}^{\pi/2} (1 + 2 \sin \theta + \sin^2 \theta) \, d\theta \right]$$

$$= \int_{\pi/6}^{\pi/2} (8 \sin^2 \theta - 1 - 2 \sin \theta) \, d\theta$$

$$= \int_{\pi/6}^{\pi/2} (3 - 4 \cos 2\theta - 2 \sin \theta) \, d\theta$$

$$= 3\theta - 2 \sin 2\theta + 2 \cos \theta \Big]_{\pi/6}^{\pi/2} = \pi$$

Example 2 illustrates the procedure for finding the area of the region bounded by two polar curves. In general let $\mathscr{R}$ be a region, as illustrated in Figure 9.35, bounded by curves with polar equations $r = f(\theta)$, $r = g(\theta)$, $\theta = a$, and $\theta = b$, where $f(\theta) \geq g(\theta) \geq 0$ and $0 < b - a \leq 2\pi$. The area A of $\mathscr{R}$ is found by subtracting the area inside $r = g(\theta)$ from the area inside $r = f(\theta)$, so using Formula 9.15 we have

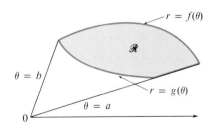

Figure 9.35

$$A = \int_a^b \tfrac{1}{2} [f(\theta)]^2 \, d\theta - \int_a^b \tfrac{1}{2} [g(\theta)]^2 \, d\theta = \tfrac{1}{2} \int_a^b ([f(\theta)]^2 - [g(\theta)]^2) \, d\theta$$

Caution: The fact that a single point has many representations in polar coordinates sometimes makes it difficult to find all of the points of intersection of two polar curves. For instance, it is obvious from Figure 9.34

that the circle and the cardioid have three points of intersection, but in Example 2 we solved the equations $r = 3 \sin \theta$, $r = 1 + \sin \theta$ and found only two such points, $(3/2, \pi/6)$ and $(3/2, 5\pi/6)$. The origin is also a point of intersection but we could not find it by solving the equations of the curves because the origin has no single representation in polar coordinates that satisfies both equations. Notice that, when represented as $(0, 0)$ or $(0, \pi)$, the origin satisfies $r = 3 \sin \theta$ and so it lies on the circle; when represented as $(0, 3\pi/2)$, it satisfies $r = 1 + \sin \theta$ and so it lies on the cardioid. Think of two points moving along the curves as the parameter value θ increases from 0 to 2π. On one curve the origin is reached at $\theta = 0$ and $\theta = \pi$; on the other curve it is reached at $\theta = 3\pi/2$. The points do not collide at the origin because they reach the origin at different times, but the curves intersect there nonetheless.

Thus, to find *all* points of intersection of two polar curves, it is essential to draw the graphs of both curves.

● **Example 3**   Find all points of intersection of the curves $r = \cos 2\theta$ and $r = 1/2$.

Solution   If we solve the equations $r = \cos 2\theta$ and $r = 1/2$, we get $\cos 2\theta = 1/2$ and therefore $2\theta = \pi/3, 5\pi/3, 7\pi/3, 11\pi/3$. Thus the values of θ between 0 and 2π that satisfy both equations are $\theta = \pi/6, 5\pi/6, 7\pi/6$, and $11\pi/6$. We have found four points of intersection $(1/2, \pi/6)$, $(1/2, 5\pi/6)$, $(1/2, 7\pi/6)$, and $(1/2, 11\pi/6)$.

However, you can see from Figure 9.36 that there are four other points of intersection—namely, $(1/2, \pi/3)$, $(1/2, 2\pi/3)$, $(1/2, 4\pi/3)$, and $(1/2, 5\pi/3)$. These can be found using symmetry or by noticing that another equation of the circle is $r = -1/2$ and solving the equations $r = \cos 2\theta$ and $r = -1/2$.
●

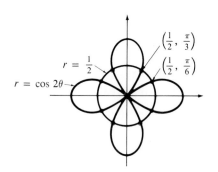

Figure 9.36

Arc Length

To find the length of a polar curve $r = f(\theta)$, $a \le \theta \le b$, we regard θ as a parameter and use Equations 9.10 to write the parametric equations of the curve as

$$x = r \cos \theta = f(\theta) \cos \theta \qquad y = r \sin \theta = f(\theta) \sin \theta$$

Using the Product Rule and differentiating with respect to θ, we obtain

$$\frac{dx}{d\theta} = \frac{dr}{d\theta} \cos \theta - r \sin \theta \qquad \frac{dy}{d\theta} = \frac{dr}{d\theta} \sin \theta + r \cos \theta$$

so, using $\cos^2 \theta + \sin^2 \theta = 1$, we have

$$\left(\frac{dx}{d\theta}\right)^2 + \left(\frac{dy}{d\theta}\right)^2 = \left(\frac{dr}{d\theta}\right)^2 \cos^2 \theta - 2r \frac{dr}{d\theta} \cos \theta \sin \theta + r^2 \sin^2 \theta$$

$$+ \left(\frac{dr}{d\theta}\right)^2 \sin^2 \theta + 2r \frac{dr}{d\theta} \sin \theta \cos \theta + r^2 \cos^2 \theta$$

$$= \left(\frac{dr}{d\theta}\right)^2 + r^2$$

Assuming that f' is continuous, we can use Theorem 9.8 to write the arc length as

$$L = \int_a^b \sqrt{\left(\frac{dx}{d\theta}\right)^2 + \left(\frac{dy}{d\theta}\right)^2}\, d\theta$$

Therefore the length of a curve with polar equation $r = f(\theta)$, $a \leqslant \theta \leqslant b$, is

(9.17)
$$L = \int_a^b \sqrt{r^2 + \left(\frac{dr}{d\theta}\right)^2}\, d\theta$$

● **Example 4**   Find the length of the cardioid $r = 1 + \sin\theta$.

Solution   This cardioid was sketched in Figure 9.24. Notice that it is symmetric about the vertical line $\theta = \pi/2$ so we can compute its total length as being twice the length of its right half:

$$L = 2\int_{-\pi/2}^{\pi/2} \sqrt{r^2 + \left(\frac{dr}{d\theta}\right)^2}\, d\theta = 2\int_{-\pi/2}^{\pi/2} \sqrt{(1 + \sin\theta)^2 + \cos^2\theta}\, d\theta$$

$$= 2\int_{-\pi/2}^{\pi/2} \sqrt{2 + 2\sin\theta}\, d\theta = 2\sqrt{2}\int_{-\pi/2}^{\pi/2} \sqrt{1 + \sin\theta}\,\frac{\sqrt{1 - \sin\theta}}{\sqrt{1 - \sin\theta}}\, d\theta$$

$$= 2\sqrt{2}\int_{-\pi/2}^{\pi/2} \frac{\sqrt{1 - \sin^2\theta}}{\sqrt{1 - \sin\theta}}\, d\theta = 2\sqrt{2}\int_{-\pi/2}^{\pi/2} \frac{\cos\theta}{\sqrt{1 - \sin\theta}}\, d\theta$$

$$= 2\sqrt{2}\left[-2\sqrt{1 - \sin\theta}\,\right]_{-\pi/2}^{\pi/2} = 8 \qquad ●$$

Note: In the solution of Example 4 we used the fact that $\sqrt{\cos^2\theta} = \cos\theta$ when $-\pi/2 \leqslant \theta \leqslant \pi/2$ because $\cos\theta \geqslant 0$ for those values of θ. In general we must write $\sqrt{\cos^2\theta} = |\cos\theta|$. If we had not taken advantage of symmetry, then we would have had to write

$$L = \int_0^{2\pi} \sqrt{r^2 + \left(\frac{dr}{d\theta}\right)^2}\, d\theta \quad \text{or} \quad L = \int_{-\pi}^{\pi} \sqrt{r^2 + \left(\frac{dr}{d\theta}\right)^2}\, d\theta$$

which leads to

(9.18)
$$L = \sqrt{2}\int_0^{2\pi} \frac{\sqrt{\cos^2\theta}}{\sqrt{1 - \sin\theta}}\, d\theta = \sqrt{2}\int_0^{2\pi} \frac{|\cos\theta|}{\sqrt{1 - \sin\theta}}\, d\theta$$

If the absolute value symbol had been forgotten we would have obtained the false equation

$$L = \sqrt{2}\int_0^{2\pi} \frac{\cos\theta}{\sqrt{1 - \sin\theta}}\, d\theta = -2\sqrt{2}\sqrt{1 - \sin\theta}\,\Big]_0^{2\pi} = 0$$

⊘

The integral in Equation 9.18 can be evaluated by observing that $\cos\theta \geqslant 0$ on the intervals $[0, \pi/2]$ and $[3\pi/2, 2\pi]$ but $\cos\theta \leqslant 0$ on the interval $[\pi/2, 3\pi/2]$. Thus

$$L = \sqrt{2}\int_0^{\pi/2} \frac{\cos\theta}{\sqrt{1-\sin\theta}}\,d\theta - \sqrt{2}\int_{\pi/2}^{3\pi/2} \frac{\cos\theta}{\sqrt{1-\sin\theta}}\,d\theta$$

$$+ \sqrt{2}\int_{3\pi/2}^{2\pi} \frac{\cos\theta}{\sqrt{1-\sin\theta}}\,d\theta$$

When these integrals are worked out, it is found that $L = 8$, but the work involved is much greater than in the solution using symmetry.

EXERCISES 9.5

In Exercises 1–8 find the area of the region that is bounded by the given curve and lies in the given sector.

1. $r = \theta, 0 \leqslant \theta \leqslant \pi$

2. $r = e^\theta, -\dfrac{\pi}{2} \leqslant \theta \leqslant \dfrac{\pi}{2}$

3. $r = 2\cos\theta, 0 \leqslant \theta \leqslant \dfrac{\pi}{6}$

4. $r = 3\sin\theta, \dfrac{\pi}{4} \leqslant \theta \leqslant \dfrac{3\pi}{4}$

5. $r = \theta^2, \dfrac{\pi}{2} \leqslant \theta \leqslant \dfrac{3\pi}{2}$

6. $r = \dfrac{1}{\theta}, \dfrac{\pi}{6} \leqslant \theta \leqslant \dfrac{5\pi}{6}$

7. $r = \sin 2\theta, 0 \leqslant \theta \leqslant \dfrac{\pi}{6}$

8. $r = \cos 3\theta, -\dfrac{\pi}{12} \leqslant \theta \leqslant \dfrac{\pi}{12}$

In Exercises 9–18 sketch the given curve and find the area that it encloses.

9. $r = 5\sin\theta$

10. $r = 2\cos\theta$

11. $r = 1 + \sin\theta$

12. $r = 4(1 - \cos\theta)$

13. $r^2 = 4\cos 2\theta$

14. $r^2 = \sin 2\theta$

15. $r = 4 - \sin\theta$

16. $r = 3 - \cos\theta$

17. $r = \sin 4\theta$

18. $r = \sin 3\theta$

In Exercises 19–24 find the area of the region enclosed by one loop of the given curve.

19. $r = \cos 3\theta$

20. $r = 3\sin 2\theta$

21. $r = \sin 5\theta$

22. $r = 2\cos 4\theta$

23. $r = 1 + 2\sin\theta$ (inner loop)

24. $r = 2 + 3\cos\theta$ (inner loop)

In Exercises 25–30 find the area of the region that lies inside the first curve and outside the second curve.

25. $r = 1 - \cos\theta, r = \dfrac{3}{2}$

26. $r = 1 - \sin\theta, r = 1$

27. $r = 4\sin\theta, r = 2$

28. $r = 3\cos\theta, r = 2 - \cos\theta$

29. $r = 3\cos\theta, r = 1 + \cos\theta$

30. $r = 1 + \cos\theta, r = 3\cos\theta$

In Exercises 31–36 find the area of the region that lies inside both curves.

31. $r = \sin\theta, r = \cos\theta$

32. $r = \sin 2\theta, r = \sin\theta$

33. $r = \sin 2\theta, r = \cos 2\theta$

34. $r^2 = 2\sin 2\theta, r = 1$

35. $r = 3 + 2\sin\theta, r = 2$

36. $r = a\sin\theta, r = b\cos\theta, a > 0, b > 0$

37. Find the area inside the larger loop and outside the smaller loop of the limaçon $r = \dfrac{1}{2} + \cos\theta$.

38. Find the area inside the larger loop and outside the smaller loop of the limaçon $r = 3 + 4\sin\theta$.

In Exercises 39–44 find all points of intersection of the given curves.

39. $r = \sin\theta, r = \cos\theta$

40. $r = 2, r = 2\cos 2\theta$

41. $r = \cos\theta, r = 1 - \cos\theta$

42. $r = \cos 3\theta, r = \sin 3\theta$

43. $r = \sin\theta, r = \sin 2\theta$

44. $r^2 = \sin 2\theta, r^2 = \cos 2\theta$

In Exercises 45–52 find the length of the given polar curve.

45. $r = 5\cos\theta, 0 \leqslant \theta \leqslant 3\pi/4$ **46.** $r = e^{-\theta}, 0 \leqslant \theta \leqslant 3\pi$

47. $r = 2^{\theta}, 0 \leqslant \theta \leqslant 2\pi$ **48.** $r = \theta, 0 \leqslant \theta \leqslant 2\pi$

49. $r = \theta^2, 0 \leqslant \theta \leqslant 2\pi$ **50.** $r = 1 + \cos\theta$

51. $r = \cos^4(\theta/4)$ **52.** $r = \cos^2(\theta/2)$

53. (a) Use Formula 9.9 to show that the area of the surface generated by rotating the polar curve

$r = f(\theta), a \leqslant \theta \leqslant b$ (where f' is continuous and $0 \leqslant a < b \leqslant \pi$), about the polar axis is

$$S = \int_a^b 2\pi r \sin\theta \sqrt{r^2 + \left(\frac{dr}{d\theta}\right)^2}\, d\theta$$

(b) Use the formula in part (a) to find the surface area generated by rotating the lemniscate $r^2 = \cos 2\theta$ about the polar axis.

SECTION 9.6

Conic Sections

In Section 1.4 we sketched parabolas, ellipses, and hyperbolas starting with their equations. In this section we give geometric definitions of these curves and derive their standard equations. They are called **conic sections,** or **conics,** because they result from intersecting a cone with a plane as in Figure 9.37.

Figure 9.37

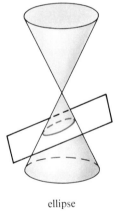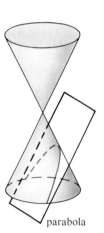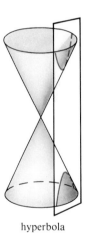

ellipse   parabola   hyperbola

Parabolas

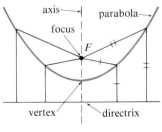

Figure 9.38

A **parabola** is the set of points in a plane that are equidistant from a fixed point F (called the **focus**) and a fixed line (called the **directrix**). This definition is illustrated by Figure 9.38. Notice that the point halfway between the focus and the directrix lies on the parabola; it is called the **vertex.** The line through the focus perpendicular to the directrix is called the **axis** of the parabola.

In the 16th century, Galileo showed that the path of a projectile that is shot into the air at an angle to the ground is a parabola. Since then, parabolic shapes have been used in designing automobile headlights, reflecting telescopes, and suspension bridges.

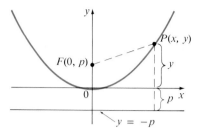

Figure 9.39

We obtain a particularly simple equation for a parabola if we place its vertex at the origin O and its directrix parallel to the x-axis as in Figure 9.39. If the focus is the point $(0, p)$, then the directrix has the equation $y = -p$. If $P(x, y)$ is any point on the parabola, then the distance from P to the focus is

$$|PF| = \sqrt{x^2 + (y - p)^2}$$

and the distance from P to the directrix is $|y + p|$. (Figure 9.39 illustrates the case where $p > 0$.) The defining property of a parabola is that these distances are equal:

$$\sqrt{x^2 + (y - p)^2} = |y + p|$$

We get an equivalent equation by squaring and simplifying:

$$x^2 + (y - p)^2 = |y + p|^2 = (y + p)^2$$
$$x^2 + y^2 - 2py + p^2 = y^2 + 2py + p^2$$
$$x^2 = 4py$$

(9.19)

The equation of a parabola with focus $(0, p)$ and directrix $y = -p$ is

$$x^2 = 4py$$

If we write $a = 1/4p$, then the standard equation of a parabola (9.19) becomes $y = ax^2$ as in Section 1.4. It opens upward if $p > 0$ and downward if $p < 0$ (see Figure 9.40). The graph is symmetric with respect to the y-axis since (9.19) is unchanged when x is replaced by $-x$.

Figure 9.40

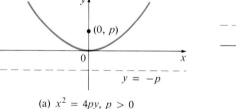

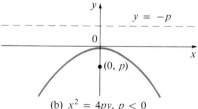

(a) $x^2 = 4py$, $p > 0$ (b) $x^2 = 4py$, $p < 0$

If we interchange x and y in (9.19), we obtain

(9.20)

$$y^2 = 4px$$

which is the equation of a parabola with focus $(p, 0)$ and directrix $x = -p$. (Interchanging x and y amounts to reflecting in the diagonal line $y = x$.) The parabola opens to the right if $p > 0$ and to the left if $p < 0$ (see Figure 9.41).

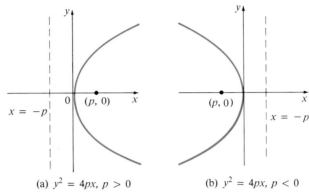

(a) $y^2 = 4px$, $p > 0$ (b) $y^2 = 4px$, $p < 0$

In both cases the graph is symmetric with respect to the x-axis, which is called the axis of the parabola.

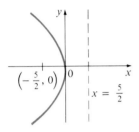

Figure 9.42
$y^2 + 10x = 0$

● **Example 1**   Find the focus and directrix of the parabola $y^2 + 10x = 0$ and sketch the graph.

Solution   If we write the equation as $y^2 = -10x$ and compare with Equation 9.20, we see that $4p = -10$, so $p = -\frac{5}{2}$. Thus the focus is $(p, 0) = (-\frac{5}{2}, 0)$ and the directrix is $x = \frac{5}{2}$. The sketch is shown in Figure 9.42.   ●

Ellipses

An **ellipse** is the set of points in a plane the sum of whose distances from two fixed points F_1 and F_2 is a constant (see Figure 9.43). These two fixed points are called the **foci** (plural of **focus**). One of Kepler's laws is that the orbits of the planets in the solar system are ellipses with the sun at one focus.

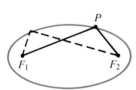

Figure 9.43

       In order to obtain the simplest equation for an ellipse, we place the foci on the x-axis at the points $(-c, 0)$ and $(c, 0)$ as in Figure 9.44 so that the origin is halfway between the foci. Let the sum of the distances from a point on the ellipse to the foci be $2a > 0$. Then $P(x, y)$ will be a point on the ellipse when

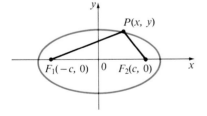

Figure 9.44

$$|PF_1| + |PF_2| = 2a$$

that is, $$\sqrt{(x + c)^2 + y^2} + \sqrt{(x - c)^2 + y^2} = 2a$$

or $$\sqrt{(x - c)^2 + y^2} = 2a - \sqrt{(x + c)^2 + y^2}$$

Squaring both sides, we have

$$x^2 - 2cx + c^2 + y^2 = 4a^2 - 4a\sqrt{(x + c)^2 + y^2} + x^2 + 2cx + c^2 + y^2$$

which simplifies to

$$a\sqrt{(x + c)^2 + y^2} = a^2 + cx$$

Square again:

$$a^2(x^2 + 2cx + c^2 + y^2) = a^4 + 2a^2cx + c^2x^2$$

which becomes

$$(a^2 - c^2)x^2 + a^2y^2 = a^2(a^2 - c^2)$$

From triangle F_1F_2P in Figure 9.44 we see that $2c < 2a$, so $c < a$ and therefore $a^2 - c^2 > 0$. For convenience let $b^2 = a^2 - c^2$. Then the equation of the ellipse becomes $b^2x^2 + a^2y^2 = a^2b^2$ or, if both sides are divided by a^2b^2,

(9.21)
$$\frac{x^2}{a^2} + \frac{y^2}{b^2} = 1$$

Since $b^2 = a^2 - c^2 < a^2$, it follows that $b < a$. The x-intercepts are found by setting $y = 0$. Then $x^2/a^2 = 1$, or $x^2 = a^2$, so $x = \pm a$. The corresponding points $(a, 0)$ and $(-a, 0)$ are called the **vertices** of the ellipse and the line segment joining the vertices is called the **major axis**. To find the y-intercepts we set $x = 0$ and obtain $y^2 = b^2$, so $y = \pm b$. Equation 9.21 is unchanged if x is replaced by $-x$ or y is replaced by $-y$, so the ellipse is symmetric about both axes. Notice that if the foci coincide, then $c = 0$, so $a = b$ and the ellipse becomes a circle with radius $r = a = b$.

We summarize this discussion as follows (see also Figure 9.45):

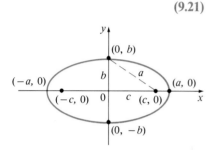

Figure 9.45
$$\frac{x^2}{a^2} + \frac{y^2}{b^2} = 1 \ (a \geqslant b)$$

(9.22)

> The ellipse
> $$\frac{x^2}{a^2} + \frac{y^2}{b^2} = 1 \qquad a \geqslant b > 0$$
> has foci $(\pm c, 0)$, where $c^2 = a^2 - b^2$, and vertices $(\pm a, 0)$.

If the foci of an ellipse are located on the y-axis at $(0, \pm c)$, then we can find its equation by interchanging x and y in (9.22) (see Figure 9.46).

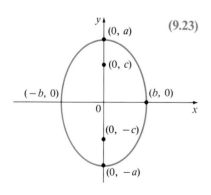

Figure 9.46
$$\frac{x^2}{b^2} + \frac{y^2}{a^2} = 1 \ (a \geqslant b)$$

(9.23)

> The ellipse
> $$\frac{x^2}{b^2} + \frac{y^2}{a^2} = 1 \qquad a \geqslant b > 0$$
> has foci $(0, \pm c)$, where $c^2 = a^2 - b^2$, and vertices $(0, \pm a)$.

● **Example 2**   Sketch the graph of $9x^2 + 16y^2 = 144$ and locate the foci.

Solution   Divide both sides of the equation by 144:

$$\frac{x^2}{16} + \frac{y^2}{9} = 1$$

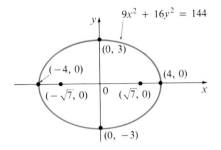

Figure 9.47

The equation is now in the standard form for an ellipse, so we have $a^2 = 16$, $b^2 = 9$, $a = 4$, $b = 3$. The x-intercepts are ± 4; the y-intercepts are ± 3. Also $c^2 = a^2 - b^2 = 7$, so $c = \sqrt{7}$ and the foci are $(\pm\sqrt{7}, 0)$. The graph is sketched in Figure 9.47. ●

● **Example 3**   Find an equation of the ellipse with foci $(0, \pm 2)$ and vertices $(0, \pm 3)$.

Solution   Using the notation of (9.23) we have $c = 2$ and $a = 3$. Then $b^2 = a^2 - c^2 = 9 - 4 = 5$, so the equation of the ellipse is

$$\frac{x^2}{5} + \frac{y^2}{9} = 1$$

Another way of writing the equation is $9x^2 + 5y^2 = 45$. ●

Hyperbolas

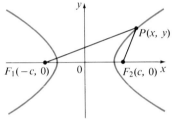

Figure 9.48

P IS ON THE HYPERBOLA WHEN
$|PF_1| - |PF_2| = \pm 2a$

(9.24)

A **hyperbola** is the set of all points in a plane the difference of whose distances from two fixed points F_1 and F_2 (the foci) is a constant. This definition is illustrated in Figure 9.48.

Notice that the definition of a hyperbola is similar to that of an ellipse; the only change is that the sum of distances has become a difference of distances. In fact, the derivation of the equation of a hyperbola is also similar to the one given earlier for an ellipse. It is left as an exercise to show that when the foci are on the x-axis at $(\pm c, 0)$ and the difference of distances is $|PF_1| - |PF_2| = \pm 2a$, then the equation of the hyperbola is

$$\frac{x^2}{a^2} - \frac{y^2}{b^2} = 1$$

where $c^2 = a^2 + b^2$. Notice that the x-intercepts are again $\pm a$ and the points $(a, 0)$ and $(-a, 0)$ are the **vertices** of the hyperbola. But if we put $x = 0$ in Equation 9.24 we get $y^2 = -b^2$, which is impossible, so there is no y-intercept. The hyperbola is symmetric with respect to both axes.

To analyze the hyperbola further, we look at Equation 9.24 and obtain

$$\frac{x^2}{a^2} = 1 + \frac{y^2}{b^2} \geqslant 1$$

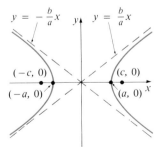

Figure 9.49

$$\frac{x^2}{a^2} - \frac{y^2}{b^2} = 1$$

This shows that $x^2 \geqslant a^2$, so $|x| = \sqrt{x^2} \geqslant a$. Therefore we have $x \geqslant a$ or $x \leqslant -a$. This means that the hyperbola consists of two parts, called its branches.

In drawing a hyperbola it is useful to draw first its **asymptotes**, which are the lines $y = (b/a)x$ and $y = -(b/a)x$ shown in Figure 9.49. Both branches of the hyperbola approach the asymptotes; that is, they come arbitrarily close to the asymptotes. [In Exercise 44 you are asked to verify that $y = (b/a)x$ is a slant asymptote in the sense of Section 4.7.]

(9.25)

> The hyperbola
>
> $$\frac{x^2}{a^2} - \frac{y^2}{b^2} = 1$$
>
> has foci $(\pm c, 0)$, where $c^2 = a^2 + b^2$, vertices $(\pm a, 0)$, and asymptotes $y = \pm(b/a)x$.

If the foci of a hyperbola are on the y-axis, then by reversing the roles of x and y we obtain the following information, which is illustrated in Figure 9.50.

(9.26)

> The hyperbola
>
> $$\frac{y^2}{a^2} - \frac{x^2}{b^2} = 1$$
>
> has foci $(0, \pm c)$, where $c^2 = a^2 + b^2$, vertices $(0, \pm a)$, and asymptotes $y = \pm(a/b)x$.

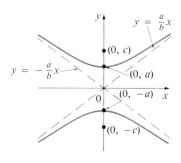

Figure 9.50

$$\frac{y^2}{a^2} - \frac{x^2}{b^2} = 1$$

● **Example 4**   Find the foci and asymptotes of the hyperbola $9x^2 - 16y^2 = 144$ and sketch its graph.

Solution   If we divide both sides of the equation by 144, it becomes

$$\frac{x^2}{16} - \frac{y^2}{9} = 1$$

which is of the form given in (9.25) with $a = 4$ and $b = 3$. Since $c^2 = 16 + 9 = 25$, the foci are $(\pm 5, 0)$. The asymptotes are the lines $y = \frac{3}{4}x$ and $y = -\frac{3}{4}x$. The graph is shown in Figure 9.51.   ●

● **Example 5**   Find the foci and equation of the hyperbola with vertices $(0, \pm 1)$ and asymptote $y = 2x$.

Solution   From (9.26) and the given information we see that $a = 1$ and $a/b = 2$. Thus $b = a/2 = \frac{1}{2}$ and $c^2 = a^2 + b^2 = \frac{5}{4}$. The foci are $(0, \pm\sqrt{5}/2)$ and the equation of the hyperbola is

$$y^2 - 4x^2 = 1$$ ●

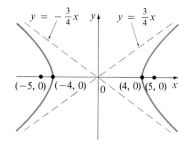

Figure 9.51

$9x^2 - 16y^2 = 144$

Shifted Conics

As in Section 1.4 we shift conics by taking the standard equations 9.19, 9.20, 9.22, 9.23, 9.25, and 9.26 and replacing x and y by $x - h$ and $y - k$.

● **Example 6**   Find an equation of the ellipse with foci $(2, -2)$, $(4, -2)$ and vertices $(1, -2)$, $(5, -2)$.

Solution   The major axis is the line segment that joins the vertices $(1, -2)$ and $(5, -2)$, which has length 4, so $a = 2$. The distance between the foci is 2, so $c = 1$. Thus $b^2 = a^2 - c^2 = 3$. Since the center of the ellipse is $(3, -2)$ we replace x and y in (9.22) by $x - 3$ and $y + 2$ to obtain

$$\frac{(x - 3)^2}{4} + \frac{(y + 2)^2}{3} = 1$$

as the equation of the given ellipse.   ●

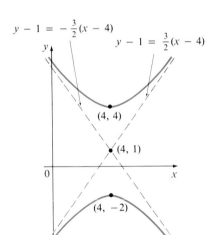

$y - 1 = -\frac{3}{2}(x - 4)$

$y - 1 = \frac{3}{2}(x - 4)$

$(4, 4)$

$(4, 1)$

$(4, -2)$

Figure 9.52

$9x^2 - 4y^2 - 72x + 8y + 176 = 0$

● **Example 7**   Sketch the conic $9x^2 - 4y^2 - 72x + 8y + 176 = 0$ and find its foci.

Solution   We complete the squares as follows:

$$4(y^2 - 2y) - 9(x^2 - 8x) = 176$$

$$4(y^2 - 2y + 1) - 9(x^2 - 8x + 16) = 176 + 4 - 144$$

$$4(y - 1)^2 - 9(x - 4)^2 = 36$$

$$\frac{(y - 1)^2}{9} - \frac{(x - 4)^2}{4} = 1$$

This is in the form (9.26) except that x and y are replaced by $x - 4$ and $y - 1$. Thus $a^2 = 9$, $b^2 = 4$, and $c^2 = 13$. The hyperbola is shifted four units to the right and one unit upward. The foci are $(4, 1 + \sqrt{13})$ and $(4, 1 - \sqrt{13})$ and the vertices are $(4, 4)$ and $(4, -2)$. The asymptotes are $y - 1 = \pm\frac{3}{2}(x - 4)$. The hyperbola is sketched in Figure 9.52.   ●

EXERCISES 9.6

In Exercises 1–8 find the vertex, focus, and directrix of the parabola and sketch the graph.

1. $x^2 = -8y$

2. $x = -5y^2$

3. $y^2 = x$

4. $2x^2 = y$

5. $x + 1 = 2(y - 3)^2$

6. $x^2 - 6x + 8y = 7$

7. $2x + y^2 - 8y + 12 = 0$

8. $x^2 + 12x - y + 39 = 0$

In Exercises 9–20 find the vertices and foci of the conic and sketch the graph. In the case of a hyperbola, find the asymptotes.

9. $\dfrac{x^2}{16} + \dfrac{y^2}{4} = 1$

10. $\dfrac{x^2}{4} + \dfrac{y^2}{25} = 1$

11. $25x^2 + 9y^2 = 225$

12. $x^2 + 4y^2 = 4$

13. $\dfrac{x^2}{144} - \dfrac{y^2}{25} = 1$

14. $\dfrac{y^2}{25} - \dfrac{x^2}{144} = 1$

15. $9y^2 - x^2 = 9$

16. $x^2 - y^2 = 1$

17. $9x^2 - 18x + 4y^2 = 27$

18. $16x^2 - 9y^2 + 64x - 90y = 305$

19. $2y^2 - 3x^2 - 4y + 12x + 8 = 0$

20. $x^2 + 2y^2 - 6x + 4y + 7 = 0$

In Exercises 21–38 find an equation for the conic that satisfies the given conditions.

21. Parabola, focus $(0, 3)$, directrix $y = -3$

22. Parabola, focus $(-2, 0)$, directrix $x = 2$

23. Parabola, focus $(3, 0)$, directrix $x = 1$

24. Parabola, focus $(1, -1)$, directrix $y = 5$

25. Parabola, vertex $(0, 0)$, axis the x-axis, passing through $(1, -4)$

26. Parabola, vertical axis, passing through $(-2, 3)$, $(0, 3)$, and $(1, 9)$

27. Ellipse, foci $(\pm 1, 0)$, vertices $(\pm 2, 0)$

28. Ellipse, foci $(0, \pm 4)$, vertices $(0, \pm 5)$

29. Ellipse, foci $(3, \pm 1)$, vertices $(3, \pm 3)$

30. Ellipse, foci $(\pm 1, 2)$, length of major axis 6

31. Ellipse, center $(2, 2)$, focus $(0, 2)$, vertex $(5, 2)$

32. Ellipse, foci $(\pm 2, 0)$, passing through $(2, 1)$

33. Hyperbola, foci $(0, \pm 3)$, vertices $(0, \pm 1)$

34. Hyperbola, foci $(\pm 6, 0)$, vertices $(\pm 4, 0)$

35. Hyperbola, foci $(1, 3)$ and $(7, 3)$, vertices $(2, 3)$ and $(6, 3)$

36. Hyperbola, foci $(2, -2)$ and $(2, 8)$, vertices $(2, 0)$ and $(2, 6)$

37. Hyperbola, vertices $(\pm 3, 0)$, asymptotes $y = \pm 2x$

38. Hyperbola, foci $(2, 2)$ and $(6, 2)$, asymptotes $y = x - 2$ and $y = 6 - x$

39. The point in a lunar orbit nearest the surface of the moon is called *perilune* and the point farthest from the surface is called *apolune*. The Apollo 11 spacecraft was placed in an elliptical lunar orbit with perilune altitude 110 km and apolune altitude 314 km (above the moon). Find an equation of this ellipse if the radius of the moon is 1728 km and the center of the moon is at one focus.

40. A cross-section of a parabolic reflector is shown in the figure. The bulb is located at the focus and the opening at the focus is 10 cm.

(a) Find an equation of the parabola.
(b) Find the diameter of the opening $|CD|$, 11 cm from the vertex.

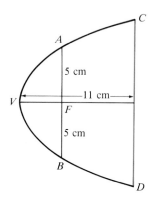

41. In the LORAN (LOng RAnge Navigation) radio navigation system two radio stations located at A and B transmit simultaneous signals to a ship or an aircraft located at P. The onboard computer converts the time difference in receiving these signals into a distance difference $|PA| - |PB|$, and this, according to the definition of a hyperbola, locates the ship or aircraft on one branch of a hyperbola (see the figure). Suppose that station B is located 400 mi due east of station A on a coastline. A ship received the signal from B 1200 microseconds before it received the signal from A.

(a) Assuming that radio signals travel at a speed of 980 ft/microsecond, find the equation of the hyperbola on which the ship lies.

(b) If the ship is due north of B, how far off the coastline is the ship?

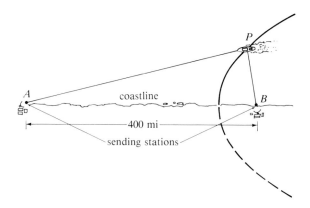

42. Use the definition of a hyperbola to derive Equation 9.24 for a hyperbola with foci $(\pm c, 0)$ and vertices $(\pm a, 0)$.

43. Show that the function defined by the upper branch of the hyperbola $y^2/a^2 - x^2/b^2 = 1$ is concave upward.

44. Show that the line $y = (b/a)x$ is a slant asymptote of the hyperbola $x^2/a^2 - y^2/b^2 = 1$. [*Hint:* According to Section 4.7, part D, we have to show that

$$\lim_{x \to \infty} \left(\frac{b}{a}\sqrt{x^2 - a^2} - \frac{b}{a}x \right) = 0$$

45. What type of curve is represented by the equation

$$\frac{x^2}{k} + \frac{y^2}{k - 16} = 1$$

in the following cases: (a) $k > 16$, (b) $0 < k < 16$, and (c) $k < 0$?

(d) Show that all of the curves in parts (a) and (b) have the same foci, no matter what the value of k is.

46. (a) Show that the equation of the tangent line to the parabola $y^2 = 4px$ at the point (x_0, y_0) can be written as

$$y_0 y = 2p(x + x_0)$$

(b) What is the x-intercept of this tangent line? Use this fact to draw the tangent line.

47. Use Simpson's Rule with $n = 10$ to estimate the length of the ellipse $x^2 + 4y^2 = 4$.

48. Find an equation for the ellipse with foci $(1, 1)$, $(-1, -1)$ and major axis of length 4.

49. Let $P_1(x_1, y_1)$ be a point on the parabola $y^2 = 4px$ with focus $F(p, 0)$. Let α be the angle between the parabola and the line segment FP_1 and let β be the angle between the horizontal line $y = y_1$ and the parabola as in the figure. Prove that $\alpha = \beta$. (Thus, by a principle of geometrical optics, light from a source placed at F will be reflected along a line parallel to the x-axis. This

explains why paraboloids, the surfaces obtained by rotating parabolas about their axes, are used as the shape of some automobile headlights and mirrors for telescopes.)

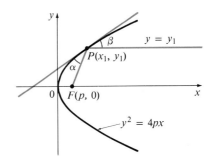

50. Let $P(x_1, y_1)$ be a point on the ellipse $x^2/a^2 + y^2/b^2 = 1$ with foci F_1 and F_2 and let α and β be the angles between the lines PF_1, PF_2 and the ellipse as in the figure. Prove that $\alpha = \beta$. This explains how whispering galleries work. Sound coming from one focus will bounce off the walls of an elliptical room and pass through the other focus. (The hyperbola has a similar reflecting property. See Exercise 47 in the Review Exercises.)

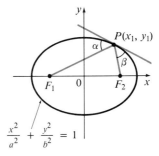

SECTION 9.7

Conic Sections in Polar Coordinates

In the preceding section we defined the parabola in terms of a focus and directrix but we defined the ellipse and hyperbola in terms of two foci. In this section we shall see that it is possible to give a more unified treatment of all three types of conic sections in terms of a focus and directrix. Furthermore, if we place the focus at the origin, then a conic section has a simple polar equation.

Theorem (9.27)

Let F be a fixed point (called the **focus**) and l be a fixed line (called the **directrix**) in a plane. Let e be a fixed positive number (called the **eccentricity**). The set of all points P in the plane such that

$$\frac{|PF|}{|Pl|} = e$$

(that is, the ratio of the distance from F to the distance from l is the constant e) is a conic section. The conic is

(a)   an ellipse if $e < 1$

(b)   a parabola if $e = 1$

(c)   a hyperbola if $e > 1$

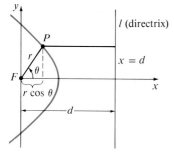

Figure 9.53

Proof   Notice that if the eccentricity is $e = 1$, then $|PF| = |Pl|$ and so the given condition simply becomes the definition of a parabola as in Section 9.6.

Let us place the focus F at the origin and the directrix parallel to the y-axis and d units to the right. Thus the directrix has equation $x = d$ and is perpendicular to the polar axis. If the point P has polar coordinates (r, θ), we see from Figure 9.53 that

$$|PF| = r \qquad |Pl| = d - r\cos\theta$$

Thus the condition $|PF|/|Pl| = e$, or $|PF| = e|Pl|$, becomes

(9.28)
$$r = e(d - r\cos\theta)$$

If we square both sides of this polar equation and convert to rectangular coordinates, we get

$$x^2 + y^2 = e^2(d - x)^2 = e^2(d^2 - 2dx + x^2)$$

or
$$(1 - e^2)x^2 + 2de^2x + y^2 = e^2d^2$$

After completing the square, this becomes

(9.29)
$$\left(x + \frac{e^2d}{1 - e^2}\right)^2 + \frac{y^2}{1 - e^2} = \frac{e^2d^2}{(1 - e^2)^2}$$

If $e < 1$, we recognize Equation 9.29 as being the equation of an ellipse. In fact, it is of the form

$$\frac{(x - h)^2}{a^2} + \frac{y^2}{b^2} = 1$$

(9.30)   where $h = -\dfrac{e^2d}{1 - e^2}$ $a^2 = \dfrac{e^2d^2}{(1 - e^2)^2}$ $b^2 = \dfrac{e^2d^2}{1 - e^2}$

In Section 9.6 we found that the foci of an ellipse are at a distance c from the center, where

(9.31)
$$c^2 = a^2 - b^2 = \frac{e^4 d^2}{(1 - e^2)^2}$$

This shows that
$$c = \frac{e^2 d}{1 - e^2} = -h$$

and confirms that the focus as defined in Theorem 9.27 means the same as the focus defined in Section 9.6. It also follows from the expressions in Equations 9.30 and 9.31 that the eccentricity is given by

$$e = \frac{c}{a}$$

If $e > 1$, then $1 - e^2 < 0$ and we see that Equation 9.29 represents a hyperbola. Just as we did before, we could rewrite Equation 9.29 in the form

$$\frac{x^2}{a^2} - \frac{y^2}{b^2} = 1$$

and see that
$$e = \frac{c}{a} \qquad \text{where } c^2 = a^2 + b^2 \qquad \bullet$$

By solving Equation 9.28 for r, we see that the polar equation of the conic shown in Figure 9.53 can be written as

$$r = \frac{ed}{1 + e \cos \theta}$$

If the directrix is chosen to be to the left of the focus as $x = -d$, or if the directrix is chosen to be parallel to the polar axis as $y = \pm d$, then the polar equation of the conic is given by the following theorem, which is illustrated by Figure 9.54. (See Exercises 21–23.)

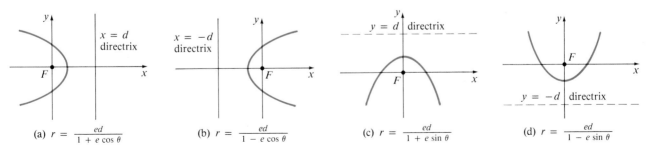

(a) $r = \dfrac{ed}{1 + e \cos \theta}$ (b) $r = \dfrac{ed}{1 - e \cos \theta}$ (c) $r = \dfrac{ed}{1 + e \sin \theta}$ (d) $r = \dfrac{ed}{1 - e \sin \theta}$

Figure 9.54
POLAR EQUATIONS OF CONICS

Theorem (9.32)

A polar equation of the form

$$r = \frac{ed}{1 \pm e\cos\theta} \quad \text{or} \quad r = \frac{ed}{1 \pm e\sin\theta}$$

represents a conic section with eccentricity e. The conic is an ellipse if $e < 1$, a parabola if $e = 1$, or a hyperbola if $e > 1$.

● **Example 1**   Find a polar equation for a parabola that has its focus at the origin and whose directrix is the line $y = -6$.

Solution   Using Theorem 9.32 with $e = 1$ and $d = 6$, and using part (d) of Figure 9.54, we see that the equation of the parabola is

$$r = \frac{6}{1 - \sin\theta}$$

● **Example 2**   A conic is given by the polar equation

$$r = \frac{10}{3 - 2\cos\theta}$$

Find the eccentricity, identify the conic, locate the directrix, and sketch the conic.

Solution   Dividing numerator and denominator by 3, we write the equation as

$$r = \frac{\frac{10}{3}}{1 - \frac{2}{3}\cos\theta}$$

From Theorem 9.32 we see that this represents an ellipse with $e = \frac{2}{3}$. Since $ed = \frac{10}{3}$, we have

$$d = \frac{\frac{10}{3}}{e} = \frac{\frac{10}{3}}{\frac{2}{3}} = 5$$

so the directrix has Cartesian equation $x = -5$. When $\theta = 0, r = 10$; when $\theta = \pi, r = 2$. So the vertices have polar coordinates $(10, 0), (2, \pi)$. The ellipse is sketched in Figure 9.55.   ●

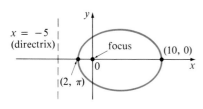

$x = -5$
(directrix)

focus

$(10, 0)$

$(2, \pi)$

Figure 9.55

$r = 10/(3 - 2\cos\theta)$

● **Example 3**   Sketch the conic $r = \dfrac{12}{2 + 4\sin\theta}$.

Solution   Writing the equation in the form

$$r = \frac{6}{1 + 2\sin\theta}$$

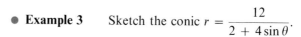

we see that the eccentricity is $e = 2$ and the equation therefore represents a hyperbola. Since $ed = 6, d = 3$ and the directrix has equation $y = 3$. The vertices occur when $\theta = \pi/2$ and $3\pi/2$, so they are $(2, \pi/2)$ and

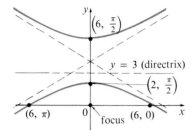

Figure 9.56

$r = 12/(2 + 4\sin\theta)$

$(-6, 3\pi/2) = (6, \pi/2)$. It is also useful to plot the x-intercepts. These occur when $\theta = 0$, π and in both cases $r = 6$. For additional accuracy we could draw the asymptotes. Note that $r \to \pm\infty$ when $2 + 4\sin\theta \to 0^+$ or 0^- and $2 + 4\sin\theta = 0$ when $\sin\theta = -\frac{1}{2}$. Thus the asymptotes are parallel to the rays $\theta = 7\pi/6$ and $\theta = 11\pi/6$. The hyperbola is sketched in Figure 9.56.

     ●

In Figure 9.57 we sketch a number of conics to demonstrate the effect of varying the eccentricity e. Notice that when e is close to 0 the ellipse is nearly circular, whereas it becomes more elongated as $e \to 1^-$. When $e = 1$, of course, the conic is a parabola.

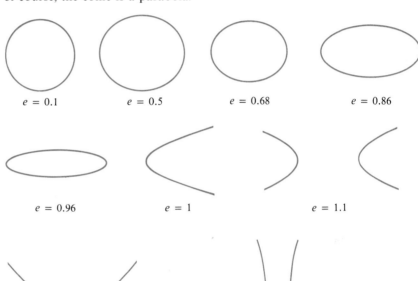

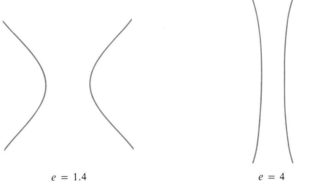

Figure 9.57

EXERCISES 9.7

In Exercises 1–8 write a polar equation of a conic with the focus at the origin and the given data.

1. Ellipse, eccentricity $\frac{2}{3}$, directrix $x = 3$

2. Hyperbola, eccentricity $\frac{4}{3}$, directrix $x = -3$

3. Parabola, directrix $y = 2$

4. Ellipse, eccentricity $\frac{1}{2}$, directrix $y = -4$

5. Hyperbola, eccentricity 4, directrix $r = 5 \sec \theta$

6. Ellipse, eccentricity 0.6, directrix $r = 2 \csc \theta$

7. Parabola, vertex at $\left(5, \dfrac{\pi}{2} \right)$

8. Ellipse, eccentricity 0.4, a vertex at $(2, 0)$

In Exercises 9–20, (a) find the eccentricity, (b) identify the conic, (c) give an equation of the directrix, and (d) sketch the conic.

9. $r = \dfrac{4}{1 + 3 \cos \theta}$

10. $r = \dfrac{8}{3 + 3 \cos \theta}$

11. $r = \dfrac{2}{1 - \cos \theta}$

12. $r = \dfrac{10}{3 - 2 \sin \theta}$

13. $r = \dfrac{6}{2 + \sin \theta}$

14. $r = \dfrac{5}{2 - 3 \sin \theta}$

15. $r = \dfrac{1}{4 - 3 \cos \theta}$

16. $r = \dfrac{6}{1 + 5 \cos \theta}$

17. $r = \dfrac{7}{2 - 5 \sin \theta}$

18. $r = \dfrac{8}{3 + \cos \theta}$

19. $r = \dfrac{5}{2 + 2 \sin \theta}$

20. $r = \dfrac{1}{1 - \sin \theta}$

21. Show that a conic with focus at the origin, eccentricity e, and directrix $x = -d$ has polar equation $r = ed/(1 - e \cos \theta)$.

22. Show that a conic with focus at the origin, eccentricity e, and directrix $y = d$ has polar equation $r = ed/(1 + e \sin \theta)$.

23. Show that a conic with focus at the origin, eccentricity e, and directrix $y = -d$ has polar equation $r = ed/(1 - e \sin \theta)$.

24. Find an approximate polar equation for the elliptical orbit of the earth around the sun (at one focus) given that the eccentricity is about 0.017 and the length of the major axis is about 3×10^8 km.

25. Show that the parabolas $r = c/(1 + \cos \theta)$ and $r = d/(1 - \cos \theta)$ intersect at right angles.

SECTION 9.8

Rotation of Axes

In Section 9.6 we studied conic sections with equations of the form

$$Ax^2 + Cy^2 + Dx + Ey + F = 0$$

In this section we show that the general second-degree equation

(9.33)
$$Ax^2 + Bxy + Cy^2 + Dx + Ey + F = 0$$

can be analyzed by rotating the axes so as to eliminate the term Bxy.

In Figure 9.58 the x and y axes have been rotated about the origin through an acute angle θ to produce the X and Y axes. Thus a given point P has coordinates (x, y) in the first coordinate system and (X, Y) in the new coordinate system. To see how X and Y are related to x and y we observe from Figure 9.59 that

$$X = r \cos \phi \qquad\qquad Y = r \sin \phi$$

$$x = r \cos(\theta + \phi) \qquad y = r \sin(\theta + \phi)$$

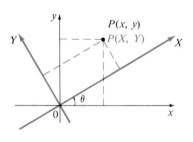

Figure 9.58

The addition formula for the cosine function then gives

$$x = r \cos(\theta + \phi) = r(\cos \theta \cos \phi - \sin \theta \sin \phi)$$

$$= (r \cos \phi) \cos \theta - (r \sin \phi) \sin \theta = X \cos \theta - Y \sin \theta$$

A similar computation gives y in terms of X and Y and so we have the following formulas:

(9.34)

$$x = X \cos \theta - Y \sin \theta \qquad y = X \sin \theta + Y \cos \theta$$

By solving Equations 9.34 for X and Y we obtain

(9.35)

$$X = x \cos \theta + y \sin \theta \qquad Y = -x \sin \theta + y \cos \theta$$

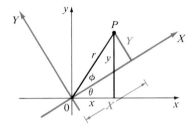

Figure 9.59

● **Example 1**   If the axes are rotated through $60°$, find the XY-coordinates of the point whose xy-coordinates are $(2, 6)$.

Solution   Using Equations 9.35 with $x = 2$, $y = 6$, and $\theta = 60°$, we have

$$X = 2 \cos 60° + 6 \sin 60° = 1 + 3\sqrt{3}$$
$$Y = -2 \sin 60° + 6 \cos 60° = -\sqrt{3} + 3$$

The XY-coordinates are $(1 + 3\sqrt{3}, 3 - \sqrt{3})$.   ●

Now let us try to determine an angle θ such that the term Bxy in Equation 9.33 disappears when the axes are rotated through the angle θ. If we substitute from Equations 9.34 in Equation 9.33, we get

$$A(X \cos \theta - Y \sin \theta)^2 + B(X \cos \theta - Y \sin \theta)(X \sin \theta + Y \cos \theta)$$
$$+ C(X \sin \theta + Y \cos \theta)^2 + D(X \cos \theta - Y \sin \theta)$$
$$+ E(X \sin \theta + Y \cos \theta) + F = 0$$

Expanding and collecting terms, we obtain an equation of the form

(9.36)

$$A'X^2 + B'XY + C'Y^2 + D'X + E'Y + F = 0$$

where the coefficient B' of XY is

$$B' = 2(C - A) \sin \theta \cos \theta + B(\cos^2 \theta - \sin^2 \theta)$$
$$= (C - A) \sin 2\theta + B \cos 2\theta$$

To eliminate the XY term we choose θ so that $B' = 0$, that is,

$$(A - C) \sin 2\theta = B \cos 2\theta$$

(9.37)   or

$$\cot 2\theta = \frac{A - C}{B}$$

● **Example 2**   Show that the graph of the equation $xy = 1$ is a hyperbola.

Solution   Notice that the equation $xy = 1$ is in the form of Equation 9.33 where $A = 0$, $B = 1$, and $C = 0$. According to Equation 9.37, the xy term will be eliminated if we choose θ so that

$$\cot 2\theta = \frac{A - C}{B} = 0$$

This will be true if $2\theta = \pi/2$, that is, $\theta = \pi/4$. Then $\cos\theta = \sin\theta = 1/\sqrt{2}$ and Equations 9.34 become

$$x = \frac{X}{\sqrt{2}} - \frac{Y}{\sqrt{2}} \qquad y = \frac{X}{\sqrt{2}} + \frac{Y}{\sqrt{2}}$$

Substituting these expressions into the original equation gives

$$\left(\frac{X}{\sqrt{2}} - \frac{Y}{\sqrt{2}}\right)\left(\frac{X}{\sqrt{2}} + \frac{Y}{\sqrt{2}}\right) = 1$$

or

$$\frac{X^2}{2} - \frac{Y^2}{2} = 1$$

We recognize this as a hyperbola with vertices $(\pm\sqrt{2}, 0)$ in the XY-coordinate system. The asymptotes are $Y = \pm X$ in the XY-system, which correspond to the coordinate axes in the xy-system (see Figure 9.60).   ●

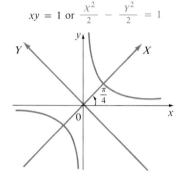

$xy = 1$ or $\dfrac{x^2}{2} - \dfrac{y^2}{2} = 1$

Figure 9.60

● **Example 3**   Identify and sketch the curve

$$73x^2 + 72xy + 52y^2 + 30x - 40y - 75 = 0$$

Solution   This equation is in the form of Equation 9.33 with $A = 73$, $B = 72$, and $C = 52$. Thus

$$\cot 2\theta = \frac{A - C}{B} = \frac{73 - 52}{72} = \frac{7}{24}$$

From the triangle in Figure 9.61 we see that

$$\cos 2\theta = \frac{7}{25}$$

Figure 9.61

The values of $\cos\theta$ and $\sin\theta$ can then be computed from the half-angle formulas:

$$\cos\theta = \sqrt{\frac{1 + \cos 2\theta}{2}} = \sqrt{\frac{1 + \frac{7}{25}}{2}} = \frac{4}{5}$$

$$\sin\theta = \sqrt{\frac{1 - \cos 2\theta}{2}} = \sqrt{\frac{1 - \frac{7}{25}}{2}} = \frac{3}{5}$$

The rotation Equations 9.34 become

$$x = \tfrac{4}{5}X - \tfrac{3}{5}Y \qquad y = \tfrac{3}{5}X + \tfrac{4}{5}Y$$

Substituting into the given equation, we have

$$73(\tfrac{4}{5}X - \tfrac{3}{5}Y)^2 + 72(\tfrac{4}{5}X - \tfrac{3}{5}Y)(\tfrac{3}{5}X + \tfrac{4}{5}Y) + 52(\tfrac{3}{5}X + \tfrac{4}{5}Y)^2$$
$$+ 30(\tfrac{4}{5}X - \tfrac{3}{5}Y) - 40(\tfrac{3}{5}X + \tfrac{4}{5}Y) - 75 = 0$$

which simplifies to

$$4X^2 + Y^2 - 2Y = 3$$

Completing the square gives

$$4X^2 + (Y - 1)^2 = 4 \qquad \text{or} \qquad X^2 + \frac{(Y - 1)^2}{4} = 1$$

and we recognize this as being an ellipse whose center is $(0, 1)$ in XY-coordinates. Since $\theta = \cos^{-1}(\tfrac{4}{5}) \approx 37°$, we can sketch the graph in Figure 9.62. ●

Figure 9.62

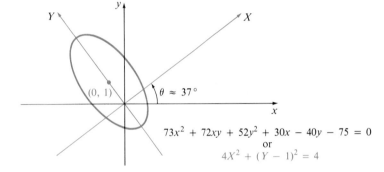

$$73x^2 + 72xy + 52y^2 + 30x - 40y - 75 = 0$$
$$\text{or}$$
$$4X^2 + (Y - 1)^2 = 4$$

EXERCISES 9.8

In Exercises 1–4 find the XY-coordinates of the given point if the axes are rotated through the given angle.

1. $(1, 4)$, $30°$

2. $(4, 3)$, $45°$

3. $(-2, 4)$, $60°$

4. $(1, 1)$, $15°$

In Exercises 5–12 use rotation of axes to identify and sketch the given curve.

5. $x^2 - 2xy + y^2 - x - y = 0$

6. $x^2 - xy + y^2 = 1$

7. $x^2 + xy + y^2 = 1$

8. $\sqrt{3}xy + y^2 = 1$

9. $97x^2 + 192xy + 153y^2 = 225$

10. $3x^2 - 12\sqrt{5}xy + 6y^2 + 9 = 0$

11. $2\sqrt{3}xy - 2y^2 - \sqrt{3}x - y = 0$

12. $16x^2 - 8\sqrt{2}xy + 2y^2 + (8\sqrt{2} - 3)x - (6\sqrt{2} + 4)y = 7$

13. (a) Use rotation of axes to show that the equation

$$36x^2 + 96xy + 64y^2 + 20x - 15y + 25 = 0$$

represents a parabola.

(b) Find the XY-coordinates of the focus. Then find the xy-coordinates of the focus.

(c) Find an equation of the directrix in the xy-coordinate system.

14. (a) Use rotation of axes to show that the equation
$$2x^2 - 72xy + 23y^2 - 80x - 60y = 125$$
represents a hyperbola.
   (b) Find the XY-coordinates of the foci. Then find the xy-coordinates of the foci.
   (c) Find the xy-coordinates of the vertices.
   (d) Find the equations of the asymptotes in the xy-coordinate system.
   (e) Find the eccentricity of the hyperbola.

15. Suppose that a rotation changes Equation 9.33 into Equation 9.36. Show that
$$A' + C' = A + C$$

16. Suppose that a rotation changes Equation 9.33 into Equation 9.36. Show that
$$(B')^2 - 4A'C' = B^2 - 4AC$$

17. Use Exercise 16 to show that Equation 9.33 represents (a) a parabola if $B^2 - 4AC = 0$, (b) an ellipse if $B^2 - 4AC < 0$, and (c) a hyperbola if $B^2 - 4AC > 0$ except in degenerate cases when it reduces to a point, a line, a pair of lines, or no graph at all.

18. Use Exercise 17 to determine the type of curve in Exercises 9–12.

REVIEW OF CHAPTER 9

Define, state, or discuss the following:

1. Parametric equations of a curve
2. Slope of a tangent line to a parametric curve
3. Area under a parametric curve
4. Arc length of a parametric curve
5. Surface area for a rotated parametric curve
6. Polar coordinate system
7. Relation between rectangular and polar coordinates
8. Graph of a polar equation
9. Slope of a tangent line to a polar curve
10. Area formula in polar coordinates
11. Length of a polar curve
12. Definition, focus, directrix, vertex, and axis of a parabola
13. Definition, foci, and vertices of an ellipse
14. Definition, foci, and vertices of a hyperbola
15. Equations of conics in rectangular coordinates
16. Eccentricity
17. Equations of conics in polar coordinates
18. Rotation of axes

REVIEW EXERCISES FOR CHAPTER 9

In Exercises 1–4 sketch the given parametric curve and eliminate the parameter to find the Cartesian equation of the curve.

1. $x = 1 - t^2, y = 1 - t, -1 \leqslant t \leqslant 1$

2. $x = t^2 + 1, y = t^2 - 1$

3. $x = 1 + \sin t, y = 2 + \cos t$

4. $x = 1 + \cos t, y = 1 + \sin^2 t$

In Exercises 5–12 sketch the curve whose polar equation is given.

5. $r = 1 + 3\cos\theta$

6. $r = 3 - \sin\theta$

7. $r^2 = \sec 2\theta$

8. $r = \tan\theta$

9. $r = 2\cos^2\left(\dfrac{\theta}{2}\right)$

10. $r = 2\cos\left(\dfrac{\theta}{2}\right)$

11. $r = \dfrac{1}{1 + \cos\theta}$

12. $r = \dfrac{5}{1 - 3\sin\theta}$

In Exercises 13 and 14 find a polar equation for the curve represented by the given Cartesian equation.

13. $x^2 + y^2 = 4x$

14. $x + y^2 = 0$

In Exercises 15–18 find the slope of the tangent line to the given curve at the point corresponding to the given value of the parameter.

15. $x = t^2 + 2t, y = t^3 - t; t = 1$

16. $x = te^t, y = 1 + \sqrt{1 + t}; t = 0$

17. $r = \theta; \theta = \dfrac{\pi}{4}$

18. $r = 3 - 2\sin\theta; \theta = \dfrac{\pi}{2}$

In Exercises 19 and 20 find dy/dx and d^2y/dx^2.

19. $x = t\cos t, y = t\sin t$

20. $x = t^6 + t^3, y = t^4 + t^2$

21. At what points does the curve $x = 2a\cos t - a\cos 2t$, $y = 2a\sin t - a\sin 2t$ have vertical or horizontal tangents? Use this information to help sketch the curve.

22. Find the area enclosed by the curve in Exercise 21.

23. Find the area enclosed by the curve $r^2 = 9 \cos 5\theta$.

24. Find the area enclosed by the inner loop of the curve $r = 1 - 3 \sin \theta$.

25. Find the points of intersection of the curves $r = 2$ and $r = 4 \cos \theta$.

26. Find the points of intersection of the curves $r = \cot \theta$ and $r = 2 \cos \theta$.

27. Find the area of the region that lies inside both of the circles $r = 2 \sin \theta$ and $r = \sin \theta + \cos \theta$.

28. Find the area of the region that lies inside the curve $r = 2 + \cos 2\theta$ but outside the curve $r = 2 + \sin \theta$.

Find the lengths of the curves in Exercises 29–32.

29. $x = 3t^2$, $y = 2t^3$, $0 \leqslant t \leqslant 2$

30. $x = \cos t + \ln \tan \left(\dfrac{t}{2} \right)$, $y = \sin t$, $\dfrac{\pi}{2} \leqslant t \leqslant \dfrac{3\pi}{4}$

31. $r = \dfrac{1}{\theta}$, $\pi \leqslant \theta \leqslant 2\pi$ **32.** $r = \sin^3 \left(\dfrac{\theta}{3} \right)$, $0 \leqslant \theta \leqslant \pi$

In Exercises 33 and 34 find the area of the surface obtained by rotating the given curve about the x-axis.

33. $x = 4\sqrt{t}$, $y = \dfrac{t^3}{3} + \dfrac{1}{2t^2}$; $1 \leqslant t \leqslant 4$

34. $x = \cos t + \ln \tan \left(\dfrac{t}{2} \right)$, $y = \sin t$; $\dfrac{\pi}{2} \leqslant t \leqslant \dfrac{3\pi}{4}$

In Exercises 35–38 find the foci and vertices and sketch the graph.

35. $\dfrac{x^2}{9} + \dfrac{y^2}{8} = 1$ **36.** $4x^2 - y^2 = 16$

37. $6y^2 + x - 36y + 55 = 0$

38. $25x^2 + 4y^2 + 50x - 16y = 59$

39. Find an equation of the parabola with focus $(0, 6)$ and directrix $y = 2$.

40. Find an equation of the hyperbola with foci $(0, \pm 5)$ and vertices $(0, \pm 2)$.

41. Find an equation of the hyperbola with foci $(\pm 3, 0)$ and asymptotes $2y = \pm x$.

42. Find an equation of the ellipse with foci $(3, \pm 2)$ and major axis with length 8.

43. Use rotation of axes to identify and sketch the curve
$$x^2 + 4xy + y^2 = 1$$

44. (a) Use rotation of axes to show that the equation
$$7x^2 - 6\sqrt{3}xy + 13y^2 - 4\sqrt{3}x - 4y = 0$$
represents an ellipse.
(b) Find the XY-coordinates of the foci. Then find the xy-coordinates of the foci.
(c) Find the xy-coordinates of the vertices.
(d) Find the eccentricity of the ellipse.

45. Find a polar equation for the ellipse with focus at the origin, eccentricity $\frac{1}{3}$, and directrix with equation $r = 4 \sec \theta$.

46. Show that the angles between the polar axis and the asymptotes of the hyperbola $r = ed/(1 - e \cos \theta)$, $e > 1$, are given by $\cos^{-1}(\pm 1/e)$.

47. Let $P(x_1, y_1)$ be a point on the hyperbola $x^2/a^2 - y^2/b^2 = 1$ with foci F_1 and F_2 and let α, β be the angles between the lines PF_1 and PF_2 and the hyperbola as in the figure. Prove that $\alpha = \beta$. (This is the "reflection property" of the hyperbola. It shows that light aimed at a focus F_2 of a hyperbolic mirror will be reflected toward the other focus F_1.)

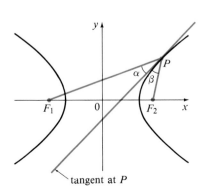

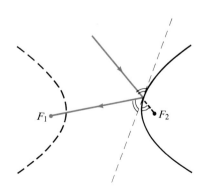

10 Infinite Sequences and Series

Mathematicians are like lovers . . . .
Grant a mathematician the least
principle, and he will draw from it a
consequence which you must also
grant him, and from this consequence
another.

Fontenelle

Our minds are finite, and yet even in
those circumstances of finitude, we
are surrounded by possibilities that
are infinite, and the purpose of human
life is to grasp as much as we can out
of that infinitude.

Alfred North Whitehead

Infinite sequences and series were briefly introduced in the Preview of Calculus (pages xxi–xxix) in connection with Zeno's paradoxes and the decimal representation of numbers. Their importance in calculus stems from Newton's idea of representing functions as sums of infinite series. For instance, in finding areas he often integrated a function by first expressing it as a series and then integrating each term of the series. We shall pursue this idea in Section 10.9 in order to integrate such functions as e^{-x^2}. (Recall that we have previously been unable to do this.) Many of the functions that arise in mathematical physics and chemistry, such as Bessel functions, are defined as sums of series, so it is important to be familiar with the basic concepts of convergence of infinite sequences and series.

Sequences

A **sequence** is a set of numbers written in a definite order:

$$a_1, a_2, a_3, a_4, \ldots, a_n, \ldots$$

The number a_1 is called the *first term,* a_2 is the *second term,* and in general a_n is the *nth term.* We shall be dealing exclusively with infinite sequences and so each term a_n will have a successor a_{n+1}.

Notice that for every positive integer n there is a corresponding number a_n and so a sequence can be regarded as a function whose domain is the set of positive integers. But we usually write a_n instead of the function notation $f(n)$ for the value of the function at the number n.

Notation: The sequence $\{a_1, a_2, a_3, \ldots\}$ is denoted by

$$\{a_n\} \quad \text{or} \quad \{a_n\}_{n=1}^{\infty}$$

● **Example 1**   Some sequences can be defined by giving a formula for the nth term. In the following examples we give three descriptions of the sequence: one by using the above notation, another by using the defining formula, and a third by writing out the terms of the sequence.

(a) $\left\{\dfrac{n}{n+1}\right\}_{n=1}^{\infty}$ $a_n = \dfrac{n}{n+1}$ $\left\{\dfrac{1}{2}, \dfrac{2}{3}, \dfrac{3}{4}, \dfrac{4}{5}, \ldots, \dfrac{n}{n+1}, \ldots\right\}$

(b) $\left\{\dfrac{(-1)^n(n+1)}{3^n}\right\}$ $a_n = \dfrac{(-1)^n(n+1)}{3^n}$ $\left\{-\dfrac{2}{3}, \dfrac{3}{9}, -\dfrac{4}{27}, \dfrac{5}{81}, \ldots, \dfrac{(-1)^n(n+1)}{3^n}, \ldots\right\}$

(c) $\left\{\sqrt{n-3}\right\}_{n=3}^{\infty}$ $a_n = \sqrt{n-3}, n \geq 3$ $\left\{0, 1, \sqrt{2}, \sqrt{3}, \ldots, \sqrt{n-3}, \ldots\right\}$

(d) $\left\{\cos\dfrac{n\pi}{6}\right\}_{n=0}^{\infty}$ $a_n = \cos\dfrac{n\pi}{6}, n \geq 0$ $\left\{1, \dfrac{\sqrt{3}}{2}, \dfrac{1}{2}, 0, \ldots, \cos\dfrac{n\pi}{6}, \ldots\right\}$ ●

● **Example 2**   Here are some sequences that do not have a simple defining equation.

(a)   The sequence $\{p_n\}$, where p_n is the population of the world as of January 1 in the year n.

(b)   If we let a_n be the digit in the nth decimal place of the number e, then $\{a_n\}$ is a well-defined sequence whose first few terms are

$$\{7, 1, 8, 2, 8, 1, 8, 2, 8, 4, 5, \ldots\}$$

(c)   The **Fibonacci sequence** $\{f_n\}$ is defined recursively by the conditions

$$f_1 = 1 \qquad f_2 = 1 \qquad f_n = f_{n-1} + f_{n-2} \quad (n \geq 3)$$

Each term is the sum of the two preceding terms and the first few terms are

$$\{1, 1, 2, 3, 5, 8, 13, 21, \ldots\}$$

This sequence arose when the 13th-century Italian mathematician known as Fibonacci solved a problem concerning the breeding of rabbits (see Exercise 69).   ●

A sequence, such as the one in Example 1(a), $a_n = n/(n + 1)$, can be pictured either by plotting its terms on a number line as in Figure 10.1(a) or by plotting its graph as in Figure 10.1(b). From Figure 10.1(a) or (b) it appears that the terms of the sequence $a_n = n/(n + 1)$ are approaching 1 as n becomes large. In fact, the difference

$$1 - \frac{n}{n + 1} = \frac{1}{n + 1}$$

can be made as small as we like by taking n sufficiently large. We indicate this by writing

$$\lim_{n \to \infty} \frac{n}{n + 1} = 1$$

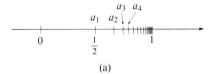

(a)

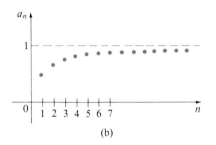

(b)

Figure 10.1

In general, the notation

$$\lim_{n \to \infty} a_n = L$$

means that the terms of the sequence $\{a_n\}$ can be made arbitrarily close to L by taking n sufficiently large. Notice that the following precise definition of the limit of a sequence is very similar to the definition of a limit of a function at infinity given in Section 4.5.

Definition (10.1)

A sequence $\{a_n\}$ has the **limit** L and we write

$$\lim_{n \to \infty} a_n = L \qquad \text{or} \qquad a_n \to L \text{ as } n \to \infty$$

if for every $\varepsilon > 0$ there is a corresponding integer N such that

$$|a_n - L| < \varepsilon \qquad \text{whenever} \qquad n > N$$

If $\lim_{n \to \infty} a_n$ exists, we say the sequence **converges** (or is **convergent**). Otherwise we say the sequence **diverges** (or is **divergent**).

Definition 10.1 is illustrated by Figure 10.2 in which the terms a_1, $a_2, a_3, \ldots$ are plotted on a number line. No matter how small an interval $(L - \varepsilon, L + \varepsilon)$ is chosen, there exists N such that all terms of the sequence from a_{N+1} onward must lie in that interval.

Figure 10.2

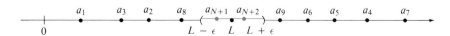

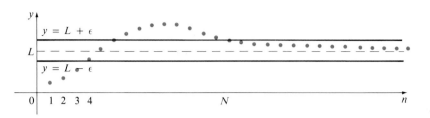

Figure 10.3

Another illustration of Definition 10.1 is given in Figure 10.3. The points on the graph of $\{a_n\}$ must lie between the horizontal lines $y = L + \varepsilon$ and $y = L - \varepsilon$ if $n > N$. This picture must be valid no matter how small ε is chosen, but usually a smaller ε will require a larger N.

Comparison of Definitions 10.1 and 4.33 shows that the only difference between $\lim_{n\to\infty} a_n = L$ and $\lim_{x\to\infty} f(x) = L$ is that n is required to be an integer. Thus we have the following theorem, which is illustrated by Figure 10.4.

Theorem (10.2)

> If $\lim_{x\to\infty} f(x) = L$ and $f(n) = a_n$ when n is an integer, then $\lim_{n\to\infty} a_n = L$.

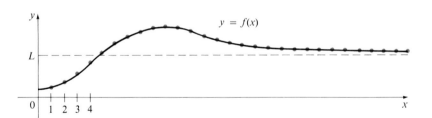

Figure 10.4

In particular, since we know that $\lim_{x\to\infty}(1/x^r) = 0$ when $r > 0$ (Theorem 4.32), we have

(10.3)
$$\lim_{n\to\infty} \frac{1}{n^r} = 0 \qquad \text{if } r > 0$$

The analogue of Definition 4.41 is the following:

Definition (10.4)

> $\lim_{n\to\infty} a_n = \infty$ means that for every positive number M there is an integer N such that
>
> $$a_n > M \qquad \text{whenever} \qquad n > N$$

If $\lim_{n\to\infty} a_n = \infty$, then the sequence $\{a_n\}$ is divergent but in a special way. We say that $\{a_n\}$ diverges to ∞.

The basic properties of limits listed in Tables 2.14 and 2.15 also hold for limits of sequences. (For instance, the limit of a sum is equal to the sum of the limits.) The Squeeze Theorem can also be adapted for sequences as follows. (The proof is similar to the proof of Theorem 2.17 in Appendix C.)

The Squeeze Theorem for Sequences (10.5)

> If $a_n \leqslant b_n \leqslant c_n$ for $n \geqslant n_0$ and
>
> $$\lim_{n \to \infty} a_n = \lim_{n \to \infty} c_n = L$$
>
> then
>
> $$\lim_{n \to \infty} b_n = L$$

A further useful fact about limits of sequences is given by the following theorem, whose proof is requested in Exercise 71.

Theorem (10.6)

> If $\lim_{n \to \infty} |a_n| = 0$, then $\lim_{n \to \infty} a_n = 0$.

● **Example 3**   Find $\lim\limits_{n \to \infty} \dfrac{n}{n + 1}$.

Solution   The method is similar to the one we used in Section 4.5: divide numerator and denominator by the highest power of n:

$$\lim_{n \to \infty} \frac{n}{n + 1} = \lim_{n \to \infty} \frac{1}{1 + \dfrac{1}{n}}$$

$$= \frac{1}{1 + 0} = 1$$

Here we have used Equation 10.3 with $r = 1$.   ●

● **Example 4**   Calculate $\lim\limits_{n \to \infty} \dfrac{\ln n}{n}$.

Solution   Notice that both numerator and denominator approach infinity as $n \to \infty$. We cannot apply l'Hospital's Rule directly since it applies not to sequences but to functions of a real variable. However we can apply l'Hospital's Rule to the related function $f(x) = \ln x / x$ and obtain

$$\lim_{x \to \infty} \frac{\ln x}{x} = \lim_{x \to \infty} \frac{\dfrac{1}{x}}{1} = 0$$

Therefore by Theorem 10.2 we have

$$\lim_{n \to \infty} \frac{\ln n}{n} = 0$$

● **Example 5**   Determine whether the sequence $a_n = (-1)^n$ is convergent or divergent.

Solution   If we write out the terms of the sequence, we obtain

$$\{-1, 1, -1, 1, -1, 1, -1, \dots\}$$

The graph of this sequence is shown in Figure 10.5. Since the terms oscillate between 1 and -1 infinitely often, a_n does not approach any number. Thus $\lim_{n \to \infty}(-1)^n$ does not exist; that is, the sequence $\{(-1)^n\}$ is divergent.   ●

Figure 10.5

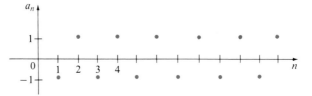

● **Example 6**   Evaluate $\lim\limits_{n \to \infty} \dfrac{(-1)^n}{n}$ if it exists.

Solution
$$\lim_{n \to \infty} \left| \frac{(-1)^n}{n} \right| = \lim_{n \to \infty} \frac{1}{n} = 0$$

Therefore, by Theorem 10.6,

$$\lim_{n \to \infty} \frac{(-1)^n}{n} = 0$$ ●

● **Example 7**   Discuss the convergence of the sequence $a_n = n!/n^n$, where $n! = 1 \cdot 2 \cdot 3 \cdot \dots \cdot n$.

Solution   Both numerator and denominator approach infinity as $n \to \infty$ but here there is no corresponding function for use with l'Hospital's Rule ($x!$ is not defined when x is not an integer). Let us write out a few terms to get a feeling for what happens to a_n as n gets large:

$$a_1 = 1 \qquad a_2 = \frac{1 \cdot 2}{2 \cdot 2} \qquad a_3 = \frac{1 \cdot 2 \cdot 3}{3 \cdot 3 \cdot 3}$$

(10.7)
$$a_n = \frac{1 \cdot 2 \cdot 3 \cdot \dots \cdot n}{n \cdot n \cdot n \cdot \dots \cdot n}$$

It appears that the terms are decreasing and perhaps approach 0. To confirm this, observe from Equation 10.7 that

$$a_n = \frac{1}{n} \left(\frac{2 \cdot 3 \cdot \dots \cdot n}{n \cdot n \cdot \dots \cdot n} \right)$$

so
$$0 < a_n \leqslant \frac{1}{n}$$

We know that $1/n \to 0$ as $n \to \infty$. Therefore, $a_n \to 0$ as $n \to \infty$ by the Squeeze Theorem.   ●

● **Example 8**   For what values of r is the sequence $\{r^n\}$ convergent?

Solution   We know from (6.3) that $\lim_{x \to \infty} a^x = \infty$ for $a > 1$ and $\lim_{x \to \infty} a^x = 0$ for $0 < a < 1$. Therefore, putting $a = r$ and using Theorem 10.2, we have

$$\lim_{n \to \infty} r^n = \begin{cases} \infty & \text{if } r > 1 \\ 0 & \text{if } 0 < r < 1 \end{cases}$$

It is obvious that

$$\lim_{n \to \infty} 1^n = 1 \quad \text{and} \quad \lim_{n \to \infty} 0^n = 0$$

If $-1 < r < 0$, then $0 < |r| < 1$, so

$$\lim_{n \to \infty} |r^n| = \lim_{n \to \infty} |r|^n = 0$$

and therefore $\lim_{n \to \infty} r^n = 0$ by Theorem 10.6. If $r \leqslant -1$, then $\{r^n\}$ diverges as in Example 5.   ●

The results of Example 8 are summarized for future use as follows:

(10.8)

> The sequence $\{r^n\}$ is convergent if $-1 < r \leqslant 1$ and divergent for all other values of r.
>
> $$\lim_{n \to \infty} r^n = \begin{cases} 0 & \text{if } -1 < r < 1 \\ 1 & \text{if } r = 1 \end{cases}$$

Definition (10.9)

> A sequence $\{a_n\}$ is called **increasing** if $a_n \leqslant a_{n+1}$ for all $n \geqslant 1$, that is, $a_1 \leqslant a_2 \leqslant a_3 \leqslant \cdots$. It is called **decreasing** if $a_n \geqslant a_{n+1}$ for all $n \geqslant 1$. It is called **monotonic** if it is either increasing or decreasing.

● **Example 9**   The sequence $\left\{\dfrac{3}{n+5}\right\}$ is decreasing because

$$\frac{3}{n+5} > \frac{3}{n+6}$$

for all $n \geqslant 1$. (The right side is smaller because it has a larger denominator.)   ●

● **Example 10**   Show that the sequence $a_n = \dfrac{n}{n^2 + 1}$ is decreasing.

Solution 1   We must show that $a_{n+1} \leq a_n$, that is,

$$\frac{n + 1}{(n + 1)^2 + 1} \leq \frac{n}{n^2 + 1}$$

This inequality is equivalent to the one we get by cross-multiplication:

$$\frac{n + 1}{(n + 1)^2 + 1} \leq \frac{n}{n^2 + 1} \quad \Leftrightarrow \quad (n + 1)(n^2 + 1) \leq n[(n + 1)^2 + 1]$$

$$\Leftrightarrow \quad n^3 + n^2 + n + 1 \leq n^3 + 2n^2 + 2n$$

$$\Leftrightarrow \quad 1 \leq n^2 + n$$

It is obvious that $n^2 + n \geq 1$ is true for $n \geq 1$. Therefore $a_{n+1} \leq a_n$ and so $\{a_n\}$ is decreasing.

Solution 2   Consider the function $f(x) = x/(x^2 + 1)$:

$$f'(x) = \frac{x^2 + 1 - 2x^2}{(x^2 + 1)^2} = \frac{1 - x^2}{(x^2 + 1)^2} < 0 \qquad \text{when } x^2 > 1$$

Thus f is decreasing on $[1, \infty)$ and so $f(n) > f(n + 1)$. Therefore $\{a_n\}$ is decreasing.   ●

Definition (10.10)

A sequence $\{a_n\}$ is **bounded above** if there is a number M such that

$$a_n \leq M \qquad \text{for all } n \geq 1$$

It is **bounded below** if there is a number m such that

$$m \leq a_n \qquad \text{for all } n \geq 1$$

If it is bounded above and below, $\{a_n\}$ is called **bounded.**

For instance, the sequence $a_n = n$ is bounded below ($a_n > 0$) but not above. The sequence $a_n = n/(n + 1)$ is bounded because $0 < a_n < 1$ for all n.

We know that not all bounded sequences are convergent [$a_n = (-1)^n$ satisfies $-1 \leq a_n \leq 1$ but is divergent from Example 5] and not all monotonic sequences are convergent ($a_n = n \to \infty$). But if a sequence is both bounded *and* monotonic, then it must be convergent. This fact is proved as Theorem 10.11 but intuitively you can understand why it is true by looking at Figure 10.6. If $\{a_n\}$ is increasing and $a_n \leq M$ for all n, then the terms are forced to crowd together and approach some number L.

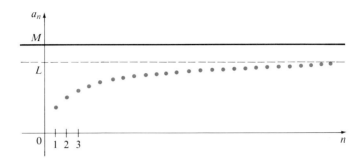

Figure 10.6

The proof of Theorem 10.11 is based on the **Completeness Axiom** for the set R of real numbers, which says that if S is a nonempty set of real numbers that has an upper bound M ($x \leqslant M$ for all x in S), then S has a **least upper bound** b. (This means that b is an upper bound for S, but if M is any other upper bound, then $b \leqslant M$.) The Completeness Axiom is an expression of the fact that there are no gaps or holes in the real number line.

Theorem (10.11)

> Every bounded monotonic sequence is convergent.

Proof   Suppose $\{a_n\}$ is an increasing sequence. Since $\{a_n\}$ is bounded, the set $S = \{a_n \,|\, n \geqslant 1\}$ has an upper bound. By the Completeness Axiom it has a least upper bound L. Given $\varepsilon > 0$, $L - \varepsilon$ is *not* an upper bound for S (since L is the *least* upper bound). Therefore

$$a_N > L - \varepsilon \qquad \text{for some integer } N$$

But the sequence is increasing so $a_n \geqslant a_N$ for every $n > N$. Thus if $n > N$ we have

$$a_n > L - \varepsilon$$

so

$$0 \leqslant L - a_n < \varepsilon$$

since $a_n \leqslant L$. Thus

$$|L - a_n| < \varepsilon \qquad \text{whenever} \qquad n > N$$

so $\lim_{n \to \infty} a_n = L$.

A similar proof (using the greatest lower bound) works if $\{a_n\}$ is decreasing.   ●

The proof of Theorem 10.11 shows that a sequence that is increasing and bounded above is convergent. (Likewise a decreasing sequence that is bounded below is convergent.) This fact will be used many times in dealing with infinite series. It can also be used to prove the existence of the limit that defines the number e (see Exercises 72 and 73).

EXERCISES 10.1

In Exercises 1–10 write the first five terms of the sequence.

1. $a_n = \dfrac{n}{2n + 1}$

2. $a_n = \dfrac{4n - 3}{3n + 4}$

3. $a_n = \dfrac{(-1)^{n-1}n}{2^n}$

4. $a_n = \left(-\dfrac{2}{3}\right)^n$

5. $a_n = \dfrac{1 \cdot 3 \cdot 5 \cdot \cdots \cdot (2n - 1)}{n!}$

6. $\left\{\dfrac{(-7)^{n+1}}{n!}\right\}$

7. $\left\{\sin \dfrac{n\pi}{2}\right\}$

8. $\{\cos n\pi\}$

9. $a_1 = 1, a_{n+1} = \dfrac{1}{1 + a_n}$

10. $a_1 = 0, a_2 = 1, a_n = a_{n-1} - a_{n-2}$

In Exercises 11–18 find a formula for the general term a_n of the given sequence assuming that the pattern of the first few terms continues.

11. $\left\{\frac{1}{2}, \frac{1}{4}, \frac{1}{8}, \frac{1}{16}, \ldots\right\}$

12. $\left\{\frac{1}{2}, \frac{1}{4}, \frac{1}{6}, \frac{1}{8}, \ldots\right\}$

13. $\{1, 4, 7, 10, \ldots\}$

14. $\left\{\frac{3}{16}, \frac{4}{25}, \frac{5}{36}, \frac{6}{49}, \ldots\right\}$

15. $\{-1, 2, -6, 24, \ldots\}$

16. $\left\{\frac{3}{2}, -\frac{9}{4}, \frac{27}{8}, -\frac{81}{16}, \ldots\right\}$

17. $\left\{\frac{2}{3}, -\frac{3}{5}, \frac{4}{7}, -\frac{5}{9}, \ldots\right\}$

18. $\{0, 2, 0, 2, 0, 2, \ldots\}$

In Exercises 19–54 determine whether the given sequence converges or diverges. If it converges, find the limit.

19. $a_n = \dfrac{1}{4n^2}$

20. $a_n = 4\sqrt{n}$

21. $a_n = \dfrac{n^2 - 1}{n^2 + 1}$

22. $a_n = \dfrac{4n - 3}{3n + 4}$

23. $a_n = \dfrac{n^2}{n + 1}$

24. $a_n = \dfrac{\sqrt[3]{n} + \sqrt[4]{n}}{\sqrt{n} + \sqrt[5]{n}}$

25. $a_n = (-1)^n \dfrac{n^2}{1 + n^3}$

26. $a_n = (-1)^{n-1} \dfrac{n^4}{1 + n^2 + n^3}$

27. $a_n = \dfrac{1}{5^n}$

28. $a_n = 2 + \left(-\dfrac{2}{\pi}\right)^n$

29. $a_n = \cos\left(\dfrac{n\pi}{2}\right)$

30. $a_n = \sin\left(\dfrac{n\pi}{2}\right)$

31. $\left\{\dfrac{\pi^n}{3^n}\right\}$

32. $\left\{\arctan\left(\dfrac{2n}{2n + 1}\right)\right\}$

33. $\{\arctan 2n\}$

34. $\left\{\dfrac{\sin n}{\sqrt{n}}\right\}$

35. $\left\{\dfrac{3 + (-1)^n}{n^2}\right\}$

36. $\left\{\dfrac{n!}{(n + 2)!}\right\}$

37. $\left\{\dfrac{\ln(n^2)}{n}\right\}$

38. $\left\{(-1)^n \sin\left(\dfrac{1}{n}\right)\right\}$

39. $\{\sqrt{n + 2} - \sqrt{n}\}$

40. $\left\{\dfrac{\ln(2 + e^n)}{3n}\right\}$

41. $a_n = n2^{-n}$

42. $a_n = \ln(n + 1) - \ln n$

43. $a_n = n^{-1/n}$

44. $a_n = (1 + 3n)^{1/n}$

45. $a_n = \dfrac{\cos^2 n}{2^n}$

46. $a_n = \dfrac{n \cos n}{n^2 + 1}$

47. $a_n = \dfrac{1}{n^2} + \dfrac{2}{n^2} + \cdots + \dfrac{n}{n^2}$

48. $a_n = (\sqrt{n + 1} - \sqrt{n})\sqrt{n + \frac{1}{2}}$

49. $a_n = \dfrac{n!}{2^n}$

50. $a_n = \dfrac{(-3)^n}{n!}$

51. $a_n = \dfrac{n^3}{n!}$

52. $a_n = \sqrt[n]{3^n + 5^n}$

53. $a_n = \dfrac{1 \cdot 3 \cdot 5 \cdot \cdots \cdot (2n - 1)}{(2n)^n}$

54. $a_n = \dfrac{1 \cdot 3 \cdot 5 \cdot \cdots \cdot (2n - 1)}{n!}$

55. For what values of r is the sequence $\{nr^n\}$ convergent?

56. A sequence $\{a_n\}$ is defined by $a_1 = 1$ and $a_{n+1} = 1/(1 + a_n)$ for $n \geqslant 1$. Assuming that $\{a_n\}$ is convergent, find its limit.

In Exercises 57–64 determine whether the given sequence is increasing, decreasing, or not monotonic.

57. $a_n = \dfrac{1}{3n + 5}$

58. $a_n = \dfrac{1}{5^n}$

59. $a_n = \dfrac{n - 2}{n + 2}$

60. $a_n = \dfrac{3n + 4}{2n + 5}$

61. $a_n = \cos\left(\dfrac{n\pi}{2}\right)$ **62.** $a_n = 3 + \dfrac{(-1)^n}{n}$

63. $a_n = \dfrac{n}{n^2 + n - 1}$ **64.** $a_n = \dfrac{\sqrt{n+1}}{5n+3}$

65. Find the limit of the sequence

$$\{\sqrt{2},\ \sqrt{2\sqrt{2}},\ \sqrt{2\sqrt{2\sqrt{2}}},\ \dots\}$$

66. A sequence $\{a_n\}$ is given by $a_1 = \sqrt{2}$, $a_{n+1} = \sqrt{2 + a_n}$.
(a) By induction, or otherwise, show that $\{a_n\}$ is increasing and bounded above by 3. Apply Theorem 10.11 to show that $\lim_{n \to \infty} a_n$ exists.
(b) Find $\lim_{n \to \infty} a_n$.

67. Repeat Exercise 66 for the sequence defined by $a_1 = 1$, $a_{n+1} = 3 - 1/a_n$.

68. (a) Let $a_1 = a$, $a_2 = f(a)$, $a_3 = f(a_2) = f(f(a)), \dots$, $a_{n+1} = f(a_n)$, where f is a continuous function. If $\lim_{n \to \infty} a_n = L$, show that $f(L) = L$.
(b) Illustrate part (a) by taking $f(x) = \cos x$, $a = 1$, and estimating the value of L to five decimal places.

69. (a) Fibonacci posed the following problem: Suppose that rabbits live forever and that every month each pair produces a new pair that becomes productive at age 2 months. If we start with one newborn pair, how many pairs of rabbits will there be in the nth month? Show that the answer is f_n, where $\{f_n\}$ is the Fibonacci sequence defined in Example 2 part (c).

(b) Let $a_n = f_{n+1}/f_n$ and show that $a_{n-1} = 1 + 1/a_{n-2}$. Assuming that $\{a_n\}$ is convergent, find its limit.

70. Use Definition 10.1 directly to prove that $\lim_{n \to \infty} r^n = 0$ when $|r| < 1$.

71. Prove Theorem 10.6. (*Hint:* Use either Definition 10.1 or the Squeeze Theorem.)

72. Let $a_n = \left(1 + \dfrac{1}{n}\right)^n$.

(a) Show that if $0 \leqslant a < b$, then

$$\frac{b^{n+1} - a^{n+1}}{b - a} < (n+1)b^n$$

(b) Deduce that $b^n[(n+1)a - nb] < a^{n+1}$.
(c) Put $a = 1 + 1/(n+1)$ and $b = 1 + 1/n$ in part (b) to show that $\{a_n\}$ is increasing.
(d) Put $a = 1$ and $b = 1 + 1/2n$ in part (b) to show that $a_{2n} < 4$.
(e) Use parts (c) and (d) to show that $a_n < 4$ for all n.
(f) Use Theorem 10.11 to show that $\lim_{n \to \infty}(1 + 1/n)^n$ exists (see Equation 6.19).

73. Show that if $\dfrac{1}{n+1} < x < \dfrac{1}{n}$, then

$$\left(1 + \frac{1}{n+1}\right)^n < (1 + x)^{1/x} < \left(1 + \frac{1}{n}\right)^{n+1}$$

Deduce that $\lim_{x \to 0^+}(1 + x)^{1/x}$ exists.

SECTION 10.2

Series

If we try to add the terms of an infinite sequence $\{a_n\}_{n=1}^{\infty}$ we get an expression of the form

(10.12) $a_1 + a_2 + a_3 + \cdots + a_n + \cdots$

which is called an **infinite series** (or just a **series**) and is denoted, for short, by the symbol

$$\sum_{n=1}^{\infty} a_n \quad \text{or} \quad \sum a_n$$

But does it make sense to talk about the sum of infinitely many terms?
It would be impossible to find a finite sum for the series

$$1 + 2 + 3 + 4 + 5 + \cdots + n + \cdots$$

because if we start adding the terms we get $1, 3, 6, 10, 15, 21, \dots$ and, after the nth term, $n(n+1)/2$, which becomes very large as n increases.

However, if we start to add the terms of the series

$$\frac{1}{2} + \frac{1}{4} + \frac{1}{8} + \frac{1}{16} + \frac{1}{32} + \frac{1}{64} + \cdots + \frac{1}{2^n} + \cdots$$

we get $\frac{1}{2}, \frac{3}{4}, \frac{7}{8}, \frac{15}{16}, \frac{31}{32}, \frac{63}{64}, \ldots, 1 - 1/2^n, \ldots$. As we add more and more terms, these partial sums become closer and closer to 1. (See also Figure 11 on page xxviii.) In fact, by adding sufficiently many terms of the series we can make the partial sums as close as we like to 1. So it seems reasonable to say that the sum of this infinite series is 1 and to write

$$\sum_{n=1}^{\infty} \frac{1}{2^n} = \frac{1}{2} + \frac{1}{4} + \frac{1}{8} + \frac{1}{16} + \cdots + \frac{1}{2^n} + \cdots = 1$$

We use a similar idea to determine whether or not a general series (10.12) has a sum. We consider the **partial sums**

$$s_1 = a_1$$
$$s_2 = a_1 + a_2$$
$$s_3 = a_1 + a_2 + a_3$$
$$s_4 = a_1 + a_2 + a_3 + a_4$$

and, in general,

$$s_n = a_1 + a_2 + a_3 + \cdots + a_n = \sum_{i=1}^{n} a_i$$

These partial sums form a new sequence $\{s_n\}$, which may or may not have a limit. If $\lim_{n \to \infty} s_n = s$ exists (as a finite number) then, as in the preceding example, we call it the sum of the infinite series $\sum a_n$.

Definition (10.13)

> Given a series $\sum_{n=1}^{\infty} a_n = a_1 + a_2 + a_3 + \cdots$, let s_n denote its nth partial sum:
>
> $$s_n = \sum_{i=1}^{n} a_i = a_1 + a_2 + \cdots + a_n$$
>
> If the sequence $\{s_n\}$ is convergent and $\lim_{n \to \infty} s_n = s$ exists as a real number, then the series $\sum a_n$ is called **convergent** and we write
>
> $$a_1 + a_2 + \cdots + a_n + \cdots = s \qquad \text{or} \qquad \sum_{n=1}^{\infty} a_n = s$$
>
> The number s is called the **sum** of the series. Otherwise the series is called **divergent.**

Thus when we write $\sum_{n=1}^{\infty} a_n = s$ we mean that by adding sufficiently many terms of the series we can get as close as we like to the number s.

Notice that

$$\sum_{n=1}^{\infty} a_n = \lim_{n \to \infty} \sum_{i=1}^{n} a_i$$

● **Example 1**   An important example of an infinite series is the **geometric series**

$$a + ar + ar^2 + ar^3 + \cdots + ar^{n-1} + \cdots = \sum_{n=1}^{\infty} ar^{n-1} \qquad a \neq 0$$

Each term is obtained from the preceding one by multiplying by the common ratio r. (We have already considered the special case where $a = \frac{1}{2}$ and $r = \frac{1}{2}$.)

If $r = 1$, then $s_n = a + a + \cdots + a = na \to \pm \infty$. Since $\lim_{n \to \infty} s_n$ does not exist, the geometric series diverges in this case.

If $r \neq 1$, we can write

(10.14) $$s_n = a(1 + r + r^2 + \cdots + r^{n-1}) = \frac{a(1 - r^n)}{1 - r}$$

because, by direct verification, $(1 - r)(1 + r + r^2 + \cdots + r^{n-1}) = 1 - r^n$. If $-1 < r < 1$, we know from (10.8) that $r^n \to 0$ as $n \to \infty$, so

$$\lim_{n \to \infty} s_n = \lim_{n \to \infty} \frac{a(1 - r^n)}{1 - r} = \frac{a}{1 - r} - \frac{a}{1 - r} \lim_{n \to \infty} r^n = \frac{a}{1 - r}$$

Thus when $|r| < 1$ the geometric series is convergent and its sum is $a/(1 - r)$.

If $r \leqslant -1$ or $r > 1$, the sequence $\{r^n\}$ is divergent by (10.8) and so, by Equation 10.14, $\lim_{n \to \infty} s_n$ does not exist. Therefore the geometric series diverges in those cases.   ●

We summarize the results of Example 1 as follows:

(10.15)

> The geometric series
>
> $$\sum_{n=1}^{\infty} ar^{n-1} = a + ar + ar^2 + \cdots$$
>
> is convergent if $|r| < 1$ and its sum is
>
> $$\sum_{n=1}^{\infty} ar^{n-1} = \frac{a}{1 - r} \qquad |r| < 1$$
>
> If $|r| \geqslant 1$, the geometric series is divergent.

● **Example 2**   Find the sum of the geometric series

$$5 - \frac{10}{3} + \frac{20}{9} - \frac{40}{27} + \cdots$$

Solution   The first term is $a = 5$ and the common ratio is $r = -\frac{2}{3}$. Since $|r| = \frac{2}{3} < 1$, the series is convergent by (10.15) and its sum is

$$5 - \frac{10}{3} + \frac{20}{9} - \frac{40}{27} + \cdots = \frac{5}{1 - \left(-\frac{2}{3}\right)} = \frac{5}{\frac{5}{3}} = 3$$

●

● **Example 3**   Is the series $\sum_{n=1}^{\infty} 2^{2n}3^{1-n}$ convergent or divergent?

Solution

$$\sum_{n=1}^{\infty} 2^{2n}3^{1-n} = \sum_{n=1}^{\infty} \frac{4^n}{3^{n-1}} = \sum_{n=1}^{\infty} 4\left(\frac{4}{3}\right)^{n-1}$$

We recognize this series as a geometric series with $a = 4$ and $r = \frac{4}{3}$. Since $r > 1$, the series diverges by (10.15).   ●

● **Example 4**   Write the number $2.3\overline{17} = 2.3171717\ldots$ as a ratio of integers.

Solution

$$2.3171717\ldots = 2.3 + \frac{17}{10^3} + \frac{17}{10^5} + \frac{17}{10^7} + \cdots$$

After the first term we have a geometric series with $a = 17/10^3$ and $r = 1/10^2$. Therefore

$$2.3\overline{17} = 2.3 + \frac{\dfrac{17}{10^3}}{1 - \dfrac{1}{10^2}}$$

$$= 2.3 + \frac{\dfrac{17}{1000}}{\dfrac{99}{100}}$$

$$= \frac{23}{10} + \frac{17}{990} = \frac{1147}{495}$$

●

● **Example 5**   Find the sum of the series $\sum_{n=0}^{\infty} x^n$, where $|x| < 1$.

Solution   Notice that this series starts with $n = 0$ and so the first term is $x^0 = 1$. Thus

$$\sum_{n=0}^{\infty} x^n = 1 + x + x^2 + x^3 + x^4 + \cdots$$

This is a geometric series with $a = 1$ and $r = x$. Since $|r| = |x| < 1$, it converges and (10.15) gives

(10.16)

$$\sum_{n=0}^{\infty} x^n = \frac{1}{1 - x}$$

●

● **Example 6**   Show that the series $\displaystyle\sum_{n=1}^{\infty} \frac{1}{n(n+1)}$ is convergent and find its sum.

Solution   This is not a geometric series so we go back to the definition of a convergent series and compute the partial sums

$$s_n = \sum_{i=1}^{n} \frac{1}{i(i+1)} = \frac{1}{1\cdot 2} + \frac{1}{2\cdot 3} + \frac{1}{3\cdot 4} + \cdots + \frac{1}{n(n+1)}$$

We can simplify this expression if we use the partial fraction decomposition

$$\frac{1}{i(i+1)} = \frac{1}{i} - \frac{1}{i+1}$$

(see Section 7.4). Thus we have

$$s_n = \sum_{i=1}^{n} \frac{1}{i(i+1)} = \sum_{i=1}^{n} \left(\frac{1}{i} - \frac{1}{i+1}\right)$$

$$= \left(1 - \frac{1}{2}\right) + \left(\frac{1}{2} - \frac{1}{3}\right) + \left(\frac{1}{3} - \frac{1}{4}\right) + \cdots + \left(\frac{1}{n} - \frac{1}{n+1}\right)$$

$$= 1 - \frac{1}{n+1}$$

(This is an example of a telescoping sum. See Exercise 43 in Section 5.1.) Therefore the given series is convergent and

$$\sum_{n=1}^{\infty} \frac{1}{n(n+1)} = 1 \qquad\qquad ●$$

● **Example 7**   Show that the **harmonic series**

$$\sum_{n=1}^{\infty} \frac{1}{n} = 1 + \frac{1}{2} + \frac{1}{3} + \frac{1}{4} + \cdots$$

is divergent.

Solution $s_1 = 1 \qquad s_2 = 1 + \frac{1}{2}$

$$s_4 = 1 + \frac{1}{2} + \left(\frac{1}{3} + \frac{1}{4}\right) > 1 + \frac{1}{2} + \left(\frac{1}{4} + \frac{1}{4}\right) = 1 + \frac{2}{2}$$

$$s_8 = 1 + \frac{1}{2} + \left(\frac{1}{3} + \frac{1}{4}\right) + \left(\frac{1}{5} + \frac{1}{6} + \frac{1}{7} + \frac{1}{8}\right)$$

$$> 1 + \frac{1}{2} + \left(\frac{1}{4} + \frac{1}{4}\right) + \left(\frac{1}{8} + \frac{1}{8} + \frac{1}{8} + \frac{1}{8}\right)$$

$$= 1 + \frac{1}{2} + \frac{1}{2} + \frac{1}{2} = 1 + \frac{3}{2}$$

$$s_{16} = 1 + \frac{1}{2} + \left(\frac{1}{3} + \frac{1}{4}\right) + \left(\frac{1}{5} + \cdots + \frac{1}{8}\right) + \left(\frac{1}{9} + \cdots + \frac{1}{16}\right)$$

$$> 1 + \frac{1}{2} + \left(\frac{1}{4} + \frac{1}{4}\right) + \left(\frac{1}{8} + \cdots + \frac{1}{8}\right) + \left(\frac{1}{16} + \cdots + \frac{1}{16}\right)$$

$$= 1 + \frac{1}{2} + \frac{1}{2} + \frac{1}{2} + \frac{1}{2} = 1 + \frac{4}{2}$$

Similarly $s_{32} > 1 + \frac{5}{2}$, $s_{64} > 1 + \frac{6}{2}$, and in general

$$s_{2^n} > 1 + \frac{n}{2}$$

This shows that $s_{2^n} \to \infty$ as $n \to \infty$ and so $\{s_n\}$ is divergent. Therefore the harmonic series diverges.   ●

Theorem (10.17)

> If the series $\sum_{n=1}^{\infty} a_n$ is convergent, then $\lim_{n \to \infty} a_n = 0$.

Proof   Let $s_n = a_1 + a_2 + \cdots + a_n$. Then $a_n = s_n - s_{n-1}$. Since $\sum a_n$ is convergent, the sequence $\{s_n\}$ is convergent. Let $\lim_{n \to \infty} s_n = s$. Since $n - 1 \to \infty$ as $n \to \infty$, we also have $\lim_{n \to \infty} s_{n-1} = s$. Therefore

$$\lim_{n \to \infty} a_n = \lim_{n \to \infty}(s_n - s_{n-1}) = \lim_{n \to \infty} s_n - \lim_{n \to \infty} s_{n-1}$$

$$= s - s = 0 \qquad ●$$

Note 1: With any *series* $\sum a_n$ we associate two *sequences*: the sequence $\{s_n\}$ of its partial sums and the sequence $\{a_n\}$ of its terms. If $\sum a_n$ is convergent, then the limit of the sequence $\{s_n\}$ is s and, as Theorem 10.17 asserts, the limit of the sequence $\{a_n\}$ is 0.

Note 2: The converse of Theorem 10.17 is not true in general. If $\lim_{n \to \infty} a_n = 0$, we cannot conclude that $\sum a_n$ is convergent. Observe that for the harmonic series $\sum 1/n$ we have $a_n = 1/n \to 0$ as $n \to \infty$, but we showed in Example 7 that $\sum 1/n$ is divergent.

The Test for Divergence (10.18)

> If $\lim_{n \to \infty} a_n$ does not exist or if $\lim_{n \to \infty} a_n \neq 0$, then the series $\sum_{n=1}^{\infty} a_n$ is divergent.

Proof   This follows immediately from Theorem 10.17.   ●

● **Example 8**   Show that the series $\sum_{n=1}^{\infty} \dfrac{n^2}{5n^2 + 4}$ diverges.

Solution

$$\lim_{n \to \infty} a_n = \lim_{n \to \infty} \frac{n^2}{5n^2 + 4} = \lim_{n \to \infty} \frac{1}{5 + \dfrac{4}{n^2}} = \frac{1}{5} \neq 0$$

So the series diverges by the Test for Divergence.   ●

Note 3: If we find that $\lim_{n \to \infty} a_n \neq 0$, we know that $\sum a_n$ is divergent. If we find that $\lim_{n \to \infty} a_n = 0$, we know *nothing* about the convergence or divergence of $\sum a_n$. Remember the warning in Note 2: If $\lim_{n\to\infty} a_n = 0$, the series $\sum a_n$ might converge or it might diverge.

Theorem (10.19)

If $\sum a_n$ and $\sum b_n$ are convergent series, then so are the series $\sum ca_n$ (where c is a constant), $\sum(a_n + b_n)$, and $\sum(a_n - b_n)$, and

(a) $\displaystyle\sum_{n=1}^{\infty} ca_n = c \sum_{n=1}^{\infty} a_n$

(b) $\displaystyle\sum_{n=1}^{\infty} (a_n + b_n) = \sum_{n=1}^{\infty} a_n + \sum_{n=1}^{\infty} b_n$

(c) $\displaystyle\sum_{n=1}^{\infty} (a_n - b_n) = \sum_{n=1}^{\infty} a_n - \sum_{n=1}^{\infty} b_n$

Proof   Part (a) is left as Exercise 51. For part (b) let

$$s_n = \sum_{i=1}^{n} a_i \qquad s = \sum_{n=1}^{\infty} a_n \qquad t_n = \sum_{i=1}^{n} b_i \qquad t = \sum_{n=1}^{\infty} b_n$$

The nth partial sum for the series $\sum(a_n + b_n)$ is

$$u_n = \sum_{i=1}^{n} (a_i + b_i)$$

and, using Theorem 5.2, we have

$$\lim_{n\to\infty} u_n = \lim_{n\to\infty} \sum_{i=1}^{n} (a_i + b_i) = \lim_{n\to\infty} \left(\sum_{i=1}^{n} a_i + \sum_{i=1}^{n} b_i \right)$$

$$= \lim_{n\to\infty} \sum_{i=1}^{n} a_i + \lim_{n\to\infty} \sum_{i=1}^{n} b_i$$

$$= \lim_{n\to\infty} s_n + \lim_{n\to\infty} t_n = s + t$$

Therefore $\sum(a_n + b_n)$ is convergent and its sum is

$$\sum_{n=1}^{\infty} (a_n + b_n) = s + t = \sum_{n=1}^{\infty} a_n + \sum_{n=1}^{\infty} b_n$$

Part (c) is proved like part (b) or it can be deduced from parts (a) and (b).
●

● **Example 9**   Find the sum of the series $\displaystyle\sum_{n=1}^{\infty} \left(\frac{3}{n(n+1)} + \frac{1}{2^n} \right)$.

Solution   The series $\sum 1/2^n$ is a geometric series with $a = \frac{1}{2}$ and $r = \frac{1}{2}$, so

$$\sum_{n=1}^{\infty} \frac{1}{2^n} = \frac{\frac{1}{2}}{1 - \frac{1}{2}} = 1$$

In Example 6 we found that

$$\sum_{n=1}^{\infty} \frac{1}{n(n+1)} = 1$$

So, by Theorem 10.19, the given series is convergent and

$$\sum_{n=1}^{\infty} \left(\frac{3}{n(n+1)} + \frac{1}{2^n} \right) = 3 \sum_{n=1}^{\infty} \frac{1}{n(n+1)} + \sum_{n=1}^{\infty} \frac{1}{2^n}$$

$$= 3 \cdot 1 + 1 = 4 \qquad \bullet$$

Note 4: A finite number of terms cannot affect the convergence of a series. For instance, suppose that we were able to show that the series

$$\sum_{n=4}^{\infty} \frac{n}{n^3 + 1}$$

is convergent. Since

$$\sum_{n=1}^{\infty} \frac{n}{n^3 + 1} = \frac{1}{2} + \frac{2}{9} + \frac{3}{28} + \sum_{n=4}^{\infty} \frac{n}{n^3 + 1}$$

it follows that the entire series $\sum_{n=1}^{\infty} n/(n^3 + 1)$ is convergent. Similarly if it is known that the series $\sum_{n=N+1}^{\infty} a_n$ converges, then the full series

$$\sum_{n=1}^{\infty} a_n = \sum_{n=1}^{N} a_n + \sum_{n=N+1}^{\infty} a_n$$

is also convergent.

EXERCISES 10.2

In Exercises 1–34 determine whether the given series is convergent or divergent. If it is convergent, find its sum.

1. $4 + \frac{8}{5} + \frac{16}{25} + \frac{32}{125} + \cdots$

2. $1 - \frac{1}{2} + \frac{1}{4} - \frac{1}{8} + \cdots$

3. $\frac{2}{3} - \frac{2}{9} + \frac{2}{27} - \frac{2}{81} + \cdots$

4. $\frac{1}{2^6} + \frac{1}{2^8} + \frac{1}{2^{10}} + \frac{1}{2^{12}} + \cdots$

5. $\frac{1}{36} + \frac{1}{30} + \frac{1}{25} + \frac{6}{125} + \cdots$

6. $-\frac{81}{100} + \frac{9}{10} - 1 + \frac{10}{9} - \cdots$

7. $\sum_{n=1}^{\infty} 2 \left(\frac{3}{4} \right)^{n-1}$

8. $\sum_{n=1}^{\infty} \left(-\frac{3}{\pi} \right)^{n-1}$

9. $\sum_{n=1}^{\infty} 5 \left(\frac{e}{3} \right)^n$

10. $\sum_{n=1}^{\infty} \frac{1}{e^{2n}}$

11. $\sum_{n=0}^{\infty} \frac{5^n}{8^n}$

12. $\sum_{n=0}^{\infty} \frac{4^{n+1}}{5^n}$

13. $\sum_{n=1}^{\infty} 3^{-n} 8^{n+1}$

14. $\sum_{n=1}^{\infty} 3^{1-n} 5^{n/2}$

15. $\displaystyle\sum_{n=1}^{\infty} \frac{(-2)^{n+3}}{5^{n-2}}$

16. $\displaystyle\sum_{n=1}^{\infty} (-1)^{n-1} \frac{3^{2n}}{2^{3n+1}}$

17. $\displaystyle\sum_{n=1}^{\infty} \frac{n}{n+1}$

18. $\displaystyle\sum_{n=4}^{\infty} \frac{3}{n(n-1)}$

19. $\displaystyle\sum_{n=1}^{\infty} \frac{1}{2n}$

20. $\displaystyle\sum_{n=1}^{\infty} \frac{n^2}{3(n+1)(n+2)}$

21. $\displaystyle\sum_{n=1}^{\infty} \frac{1}{(3n-2)(3n+1)}$

22. $\displaystyle\sum_{n=1}^{\infty} \left(\frac{1}{2^{n-1}} + \frac{2}{3^{n-1}} \right)$

23. $\displaystyle\sum_{n=1}^{\infty} (2(0.1)^n + (0.2)^n)$

24. $\displaystyle\sum_{n=1}^{\infty} \left(\frac{1}{n} + 2^n \right)$

25. $\displaystyle\sum_{n=1}^{\infty} \frac{n}{\sqrt{1+n^2}}$

26. $\displaystyle\sum_{n=1}^{\infty} \frac{1}{4n^2-1}$

27. $\displaystyle\sum_{n=1}^{\infty} \frac{1}{n(n+2)}$

28. $\displaystyle\sum_{n=1}^{\infty} \ln\left(\frac{n}{2n+5} \right)$

29. $\displaystyle\sum_{n=1}^{\infty} \frac{3^n + 2^n}{6^n}$

30. $\displaystyle\sum_{n=1}^{\infty} \frac{2n+1}{n^2(n+1)^2}$

31. $\displaystyle\sum_{n=1}^{\infty} \left(\sin\left(\frac{1}{n}\right) - \sin\left(\frac{1}{n+1}\right) \right)$

32. $\displaystyle\sum_{n=1}^{\infty} \frac{1}{5+2^{-n}}$

33. $\displaystyle\sum_{n=1}^{\infty} \arctan n$

34. $\displaystyle\sum_{n=1}^{\infty} \frac{1}{n(n+1)(n+2)}$

Express the numbers in Exercises 35–40 as ratios of integers.

35. $0.\overline{5} = 0.5555\ldots$

36. $0.\overline{15} = 0.15151515\ldots$

37. $0.\overline{307} = 0.307307307307\ldots$

38. $1.1\overline{23}$

39. $0.123\overline{456}$

40. $4.\overline{1570}$

In Exercises 41–46 find the values of x for which the given series converges. Find the sum of the series for those values of x.

41. $\displaystyle\sum_{n=0}^{\infty} (x-3)^n$

42. $\displaystyle\sum_{n=0}^{\infty} 3^n x^n$

43. $\displaystyle\sum_{n=2}^{\infty} \frac{x^n}{5^n}$

44. $\displaystyle\sum_{n=0}^{\infty} \frac{1}{x^n}$

45. $\displaystyle\sum_{n=0}^{\infty} 2^n \sin^n x$

46. $\displaystyle\sum_{n=0}^{\infty} \tan^n x$

47. A rubber ball is dropped from a height of 4 m and bounces back to half its height after each fall. If it continues to bounce indefinitely, find the total distance it travels.

48. A right triangle ABC is given with $\angle A = \theta$ and $|AC| = b$. CD is drawn perpendicular to AB, DE is drawn perpendicular to BC, $EF \perp AB$, and this process is continued indefinitely as in Figure 10.7. Find the total length of all the perpendiculars

$$|CD| + |DE| + |EF| + |FG| + \cdots$$

in terms of b and θ.

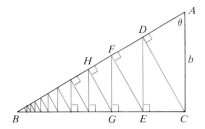

Figure 10.7

49. What is wrong with the following calculation?

$$0 = 0 + 0 + 0 + \cdots$$
$$= (1-1) + (1-1) + (1-1) + \cdots$$
$$= 1 - 1 + 1 - 1 + 1 - 1 + \cdots$$
$$= 1 + (-1+1) + (-1+1) + (-1+1) + \cdots$$
$$= 1 + 0 + 0 + 0 + \cdots = 1$$

(Guido Ubaldus thought that this proved the existence of God because "something has been created out of nothing.")

50. Suppose that $\sum_{n=1}^{\infty} a_n \ (a_n \neq 0)$ is known to be a convergent series. Prove that $\sum_{n=1}^{\infty} 1/a_n$ is a divergent series.

51. Prove Theorem 10.19 part (a).

52. If $\sum a_n$ is divergent and $c \neq 0$, show that $\sum c a_n$ is divergent.

53. If $\sum a_n$ is convergent and $\sum b_n$ is divergent, show that the series $\sum (a_n + b_n)$ is divergent. (*Hint:* Argue by contradiction.)

54. If $\sum a_n$ and $\sum b_n$ are both divergent, is $\sum (a_n + b_n)$ necessarily divergent?

The Integral Test

In general it is difficult to find the exact sum of a series. We were able to accomplish this for geometric series and the series $\sum 1/n(n + 1)$ because in those cases there was a simple formula for the nth partial sum s_n. But usually it is not easy to compute $\lim_{n \to \infty} s_n$. Therefore in the next few sections we shall develop several tests that will enable us to say whether a series is convergent or divergent without explicitly finding its sum. Our first test involves improper integrals.

The Integral Test (10.20)

Suppose f is a continuous, positive, decreasing function on $[1, \infty)$ and let $a_n = f(n)$. Then the series $\sum_{n=1}^{\infty} a_n$ is convergent if and only if the improper integral $\int_1^{\infty} f(x)\,dx$ is convergent. In other words:
(a)   If $\int_1^{\infty} f(x)\,dx$ is convergent, then $\sum_{n=1}^{\infty} a_n$ is convergent.
(b)   If $\int_1^{\infty} f(x)\,dx$ is divergent, then $\sum_{n=1}^{\infty} a_n$ is divergent.

Proof   The basic idea behind the Integral Test can be seen by looking at Figure 10.8. The area of the first shaded rectangle in part (a) is the value of f at the right endpoint of $[1, 2]$, that is, $f(2) = a_2$. So comparing the areas of the shaded rectangles with the area under $y = f(x)$ from 1 to n, we see that

(10.21)
$$a_2 + a_3 + \cdots + a_n \leqslant \int_1^n f(x)\,dx$$

Likewise Figure 10.8(b) shows that

(10.22)
$$\int_1^n f(x)\,dx \leqslant a_1 + a_2 + \cdots + a_{n-1}$$

(a)   If $\int_1^{\infty} f(x)\,dx$ is convergent, then (10.21) gives

$$\sum_{i=2}^{n} a_i \leqslant \int_1^n f(x)\,dx \leqslant \int_1^{\infty} f(x)\,dx$$

since $f(x) \geqslant 0$. Therefore

$$s_n = a_1 + \sum_{i=2}^{n} a_i \leqslant a_1 + \int_1^{\infty} f(x)\,dx = M, \text{ say}$$

Figure 10.8

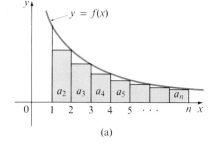

(a)

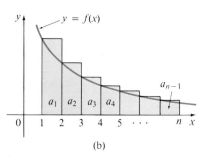

(b)

Since $s_n \leqslant M$ for all n, the sequence $\{s_n\}$ is bounded above. Also

$$s_{n+1} = s_n + a_{n+1} \geqslant s_n$$

since $a_{n+1} = f(n+1) \geqslant 0$. Thus $\{s_n\}$ is an increasing bounded sequence and so it is convergent by Theorem 10.11. This means that $\sum a_n$ is convergent.

(b)   If $\int_1^\infty f(x)\, dx$ is divergent, then $\int_1^n f(x)\, dx \to \infty$ as $n \to \infty$ because $f(x) \geqslant 0$. But (10.22) gives

$$\int_1^n f(x)\, dx \leqslant \sum_{i=1}^{n-1} a_i = s_{n-1}$$

and so $s_{n-1} \to \infty$. This implies that $s_n \to \infty$ and so $\sum a_n$ diverges.   ●

Note: When using the Integral Test it is not necessary to start the series or the integral at $n = 1$. For instance, in testing the series

$$\sum_{n=4}^\infty \frac{1}{(n-3)^2} \quad \text{we use} \quad \int_4^\infty \frac{1}{(x-3)^2}\, dx$$

Also, it is not necessary that f be always decreasing. What is important is that f be *ultimately* decreasing, that is, decreasing for x larger than some number N. Then $\sum_{n=N}^\infty a_n$ will be convergent, so $\sum_{n=1}^\infty a_n$ will be convergent by Note 4 of Section 10.2.

● **Example 1**   Test the series $\displaystyle\sum_{n=1}^\infty \frac{1}{n^2+1}$ for convergence or divergence.

Solution   The function $f(x) = 1/(x^2+1)$ is continuous, positive, and decreasing on $[1, \infty)$ so we use the Integral Test:

$$\int_1^\infty \frac{1}{x^2+1}\, dx = \lim_{t\to\infty} \int_1^t \frac{1}{x^2+1}\, dx = \lim_{t\to\infty} \tan^{-1} x \Big]_1^t$$

$$= \lim_{t\to\infty}\left(\tan^{-1} t - \frac{\pi}{4}\right) = \frac{\pi}{2} - \frac{\pi}{4} = \frac{\pi}{4}$$

Thus $\int_1^\infty 1/(x^2+1)\, dx$ is a convergent integral, so by the Integral Test the series $\sum 1/(n^2+1)$ is convergent.   ●

● **Example 2**   For what values of p is the series $\sum_{n=1}^\infty \dfrac{1}{n^p}$ convergent?

Solution   If $p < 0$, then $\lim_{n\to\infty}(1/n^p) = \infty$. If $p = 0$, then $\lim_{n\to\infty}(1/n^p) = 1$. In either case $\lim_{n\to\infty}(1/n^p) \neq 0$, so the given series diverges by the Test for Divergence (10.18).

If $p > 0$, then the function $f(x) = 1/x^p$ is clearly continuous, positive, and decreasing on $[1, \infty)$. We found in Chapter 7 [see (7.34)] that

$$\int_1^\infty \frac{1}{x^p}\, dx \text{ converges if } p > 1 \text{ and diverges if } p \leqslant 1$$

It follows from the Integral Test that the series $\sum 1/n^p$ converges if $p > 1$ and diverges if $0 < p \leqslant 1$. (For $p = 1$, this series is the harmonic series discussed in Example 7 in Section 10.2.)   ●

The series in Example 2 is called the **p-series.** It will be important in the rest of this chapter so we summarize the results of Example 2 for future reference as follows:

(10.23)

> The *p*-series $\displaystyle\sum_{n=1}^{\infty} \frac{1}{n^p}$ is convergent if $p > 1$ and divergent if $p \leqslant 1$.

● **Example 3**   The series $\sum 1/n^2$ is convergent because it is a *p*-series with $p = 2 > 1$. The exact sum of this series was found by the Swiss mathematician Leonhard Euler (1707–1783) to be

$$\sum_{n=1}^{\infty} \frac{1}{n^2} = \frac{\pi^2}{6}$$

but the proof of this fact is beyond the scope of this book.   ●

Note: We should *not* infer from the Integral Test that the sum of the series is equal to the value of the integral. In fact, we know from Example 3 and from Section 7.8 that

$$\sum_{n=1}^{\infty} \frac{1}{n^2} = \frac{\pi^2}{6} \qquad \text{while} \qquad \int_1^{\infty} \frac{1}{x^2}\,dx = 1$$

Therefore, in general,

$$\sum_{n=1}^{\infty} a_n \neq \int_1^{\infty} f(x)\,dx$$

● **Example 4**   The series $\sum 1/\sqrt{n}$ is divergent because it is a *p*-series with $p = \frac{1}{2} < 1$.   ●

● **Example 5**   Determine whether the series $\displaystyle\sum_{n=1}^{\infty} \frac{\ln n}{n}$ converges or diverges.

Solution   The function $f(x) = \ln x / x$ is positive and continuous for $x > 1$ since the logarithm function is continuous. But it is not obvious whether or not f is decreasing so we compute its derivative:

$$f'(x) = \frac{\dfrac{1}{x}\, x - \ln x}{x^2} = \frac{1 - \ln x}{x^2}$$

Thus $f'(x) < 0$ when $\ln x > 1$, that is, $x > e$. It follows that f is decreasing when $x > e$ and so we can apply the Integral Test:

$$\int_1^\infty \frac{\ln x}{x}\,dx = \lim_{t\to\infty}\int_1^t \frac{\ln x}{x}\,dx = \lim_{t\to\infty}\frac{(\ln x)^2}{2}\Big]_1^t$$

$$= \lim_{t\to\infty}\frac{(\ln t)^2}{2} = \infty$$

Since this improper integral is divergent, the series $\sum \ln n/n$ is also divergent by the Integral Test.   ●

EXERCISES 10.3

Test the series in Exercises 1–22 for convergence or divergence.

1. $\displaystyle\sum_{n=1}^{\infty}\frac{2}{\sqrt[3]{n}}$

2. $\displaystyle\sum_{n=1}^{\infty}\left(\frac{2}{n\sqrt{n}}+\frac{3}{n^3}\right)$

3. $\displaystyle\sum_{n=5}^{\infty}\frac{1}{n^{1.0001}}$

4. $\displaystyle\sum_{n=1}^{\infty}n^{-0.99}$

5. $\displaystyle\sum_{n=5}^{\infty}\frac{1}{(n-4)^2}$

6. $\displaystyle\sum_{n=1}^{\infty}\frac{1}{2n+3}$

7. $\displaystyle\sum_{n=1}^{\infty}\frac{1}{\sqrt{n}+1}$

8. $\displaystyle\sum_{n=2}^{\infty}\frac{1}{n^2-1}$

9. $\displaystyle\sum_{n=1}^{\infty}ne^{-n^2}$

10. $\displaystyle\sum_{n=1}^{\infty}\frac{n}{2^n}$

11. $\displaystyle\sum_{n=1}^{\infty}\frac{n}{n^2+1}$

12. $\displaystyle\sum_{n=2}^{\infty}\frac{1}{2n^2-n-1}$

13. $\displaystyle\sum_{n=2}^{\infty}\frac{1}{n\ln n}$

14. $\displaystyle\sum_{n=1}^{\infty}\frac{1}{4n^2+1}$

15. $\displaystyle\sum_{n=1}^{\infty}\frac{\arctan n}{1+n^2}$

16. $\displaystyle\sum_{n=2}^{\infty}\frac{1}{n(\ln n)^2}$

17. $\displaystyle\sum_{n=1}^{\infty}\frac{\ln n}{n^2}$

18. $\displaystyle\sum_{n=1}^{\infty}\left(\frac{\ln n}{n}\right)^2$

19. $\displaystyle\sum_{n=1}^{\infty}\frac{\sin\left(\frac{1}{n}\right)}{n^2}$

20. $\displaystyle\sum_{n=3}^{\infty}\frac{1}{n\ln n\ln(\ln n)}$

21. $\displaystyle\sum_{n=1}^{\infty}\frac{1}{n^2+2n+2}$

22. $\displaystyle\sum_{n=1}^{\infty}\operatorname{sech}^2 n$

In Exercises 23–26 find the values of p for which the given series is convergent.

23. $\displaystyle\sum_{n=2}^{\infty}\frac{1}{n(\ln n)^p}$

24. $\displaystyle\sum_{n=3}^{\infty}\frac{1}{n\ln n[\ln(\ln n)]^p}$

25. $\displaystyle\sum_{n=1}^{\infty}n(1+n^2)^p$

26. $\displaystyle\sum_{n=1}^{\infty}\frac{\ln n}{n^p}$

27. The Riemann zeta-function ζ is defined by

$$\zeta(x) = \sum_{n=1}^{\infty}\frac{1}{n^x}$$

and is used in number theory to study the distribution of prime numbers. What is the domain of ζ?

28. (a) Use (10.21) to show that if s_n is the nth partial sum of the harmonic series, then

$$s_n \le 1 + \ln n$$

(b) The harmonic series diverges but very slowly. Use part (a) to show that the sum of the first million terms is less than 15 and the sum of the first billion terms is less than 22.

29. Use the following steps to show that the sequence

$$t_n = 1 + \frac{1}{2} + \frac{1}{3} + \cdots + \frac{1}{n} - \ln n$$

has a limit. (The value of the limit is denoted by γ and is called Euler's constant.)

(a)   Draw a picture like Figure 10.8(b) with $f(x) = 1/x$ and interpret t_n as an area [or use (10.22)] to show that $t_n > 0$ for all n.

(b)   Interpret

$$t_n - t_{n+1} = (\ln(n + 1) - \ln n) - \frac{1}{n + 1}$$

as a difference of areas to show that $t_n - t_{n+1} > 0$. Therefore $\{t_n\}$ is a decreasing sequence.

(c)   Use Theorem 10.11 to show that $\{t_n\}$ is convergent.

SECTION 10.4

The Comparison Tests

In the comparison tests the idea is to compare a given series with a series that is known to be convergent or divergent.

The Comparison Test (10.24)

Suppose that $\sum a_n$ and $\sum b_n$ are series with positive terms.

(a)   If $\sum b_n$ is convergent and $a_n \leqslant b_n$ for all n, then $\sum a_n$ is also convergent.

(b)   If $\sum b_n$ is divergent and $a_n \geqslant b_n$ for all n, then $\sum a_n$ is also divergent.

Proof   For part (a) let

$$s_n = \sum_{i=1}^{n} a_i \qquad t_n = \sum_{i=1}^{n} b_i \qquad t = \sum_{n=1}^{\infty} b_n$$

Since both series have positive terms, the sequences $\{s_n\}$ and $\{t_n\}$ are increasing $(s_{n+1} = s_n + a_{n+1} \geqslant s_n)$. Also $t_n \to t$, so $t_n \leqslant t$ for all n. Since $a_i \leqslant b_i$, we have $s_n \leqslant t_n$. Thus $s_n \leqslant t$ for all n. This means that $\{s_n\}$ is increasing and bounded above and therefore converges by Theorem 10.11. Thus $\sum a_n$ converges.

For the proof of part (b), if $\sum b_n$ is divergent, then $t_n \to \infty$ (since $\{t_n\}$ is increasing). But $a_i \geqslant b_i$ so $s_n \geqslant t_n$. Thus $s_n \to \infty$. Therefore $\sum a_n$ diverges.   ●

In using the Comparison Test we must, of course, have some known series $\sum b_n$ for purposes of comparison. Most of the time we use either a p-series $[\sum 1/n^p$ converges if $p > 1$ and diverges if $p \leqslant 1$; see (10.23)] or a geometric series $[\sum ar^{n-1}$ converges if $|r| < 1$ and diverges if $|r| \geqslant 1$; see (10.15)].

● **Example 1**   Determine whether the series $\sum_{n=1}^{\infty} \dfrac{5}{2n^2 + 4n + 3}$ converges or diverges.

Solution   For large n the dominant term in the denominator is $2n^2$ so we compare the given series with the series $\sum 5/2n^2$. Observe that

$$\frac{5}{2n^2 + 4n + 3} < \frac{5}{2n^2}$$

since the left side has a bigger denominator. (In the notation of Theorem 10.24, a_n is the left side and b_n is the right side.) We know that

$$\sum_{n=1}^{\infty} \frac{5}{2n^2} = \frac{5}{2} \sum_{n=1}^{\infty} \frac{1}{n^2}$$

is convergent (p-series with $p = 2 > 1$). Therefore

$$\sum_{n=1}^{\infty} \frac{5}{2n^2 + 4n + 3}$$

is convergent by part (a) of the Comparison Test.   ●

● **Example 2**   Test the series $\displaystyle\sum_{n=1}^{\infty} \frac{\ln n}{n}$ for convergence or divergence.

Solution   This series has already been tested (using the Integral Test) in Example 5 in Section 10.3 but it is also possible to test it by comparing it with the harmonic series. Observe that $\ln n > 1$ for $n \geqslant 3$ and so

$$\frac{\ln n}{n} > \frac{1}{n} \qquad n \geqslant 3$$

We know that $\sum 1/n$ is divergent (p-series with $p = 1$). Thus the given series is divergent by the Comparison Test.   ●

● **Example 3**   Test the series $\displaystyle\sum_{n=1}^{\infty} \frac{1}{2^n + 1}$ for convergence or divergence.

Solution   Notice that

$$\frac{1}{2^n + 1} < \frac{1}{2^n} = \left(\frac{1}{2}\right)^n \qquad n \geqslant 1$$

The series $\sum (\frac{1}{2})^n$ is convergent (geometric series with $r = \frac{1}{2}$) and so the given series converges by the Comparison Test.   ●

  Note: The terms of the series being tested must be smaller than those of a convergent series or bigger than those of a divergent series. If the terms are bigger than the terms of a convergent series or smaller than those of a divergent series, then the Comparison Test does not apply. For instance, suppose that in Example 3 we had been given the similar series

$$\sum_{n=1}^{\infty} \frac{1}{2^n - 1}$$

The inequality

$$\frac{1}{2^n - 1} > \frac{1}{2^n}$$

is useless as far as the Comparison Test is concerned because $\sum b_n = \sum (\frac{1}{2})^n$ is convergent and $a_n > b_n$. Nonetheless we have the feeling that $\sum 1/(2^n - 1)$ ought to be convergent since it is very similar to the convergent geometric series $\sum (\frac{1}{2})^n$. In such cases the following test can be used.

The Limit Comparison Test (10.25)

Suppose that $\sum a_n$ and $\sum b_n$ are series with positive terms.

(a)   If $\lim\limits_{n \to \infty} \dfrac{a_n}{b_n} = c > 0$, then either both series converge or both diverge.

(b)   If $\lim\limits_{n \to \infty} \dfrac{a_n}{b_n} = 0$ and $\sum b_n$ converges, then $\sum a_n$ also converges.

(c)   If $\lim\limits_{n \to \infty} \dfrac{a_n}{b_n} = \infty$ and $\sum b_n$ diverges, then $\sum a_n$ also diverges.

Proof   To prove part (a) we take $\varepsilon = c/2$ in Definition 10.1 and see that, since $\lim_{n \to \infty}(a_n/b_n) = c$, there is an integer N such that

$$\left| \frac{a_n}{b_n} - c \right| < \frac{c}{2} \qquad \text{when} \qquad n > N$$

Thus
$$\frac{c}{2} < \frac{a_n}{b_n} < \frac{3c}{2} \qquad \text{when} \qquad n > N$$

(10.26)   and so
$$\left(\frac{c}{2} \right) b_n < a_n < \left(\frac{3c}{2} \right) b_n \qquad \text{when} \qquad n > N$$

If $\sum b_n$ converges, so does $\sum (3c/2)b_n$. The right half of (10.26) then shows that $\sum_N^\infty a_n$ converges by the Comparison Test. It follows that $\sum_1^\infty a_n$ converges. If $\sum b_n$ diverges, so does $\sum (c/2)b_n$ and the left half of (10.26) together with part (b) of the Comparison Test shows that $\sum a_n$ diverges. The proofs of parts (b) and (c) are similar to that of part (a) and are left as Exercises 42 and 43. ●

● **Example 4**   Test the series $\sum\limits_{n=1}^{\infty} \dfrac{1}{2^n - 1}$ for convergence or divergence.

Solution   We use the Limit Comparison Test with

$$a_n = \frac{1}{2^n - 1} \qquad b_n = \frac{1}{2^n}$$

$$\lim_{n \to \infty} \frac{a_n}{b_n} = \lim_{n \to \infty} \frac{2^n}{2^n - 1} = \lim_{n \to \infty} \frac{1}{1 - 1/2^n} = 1$$

Since this limit exists and $\sum 1/2^n$ is a convergent geometric series, the given series converges by the Limit Comparison Test. ●

● **Example 5**   Solve Example 2 using the Limit Comparison Test.

Solution   Taking $a_n = \ln n / n$ and $b_n = 1/n$, we have

$$\lim_{n \to \infty} \frac{a_n}{b_n} = \lim_{n \to \infty} \frac{\dfrac{\ln n}{n}}{\dfrac{1}{n}} = \lim_{n \to \infty} \ln n = \infty$$

We know the harmonic series $\sum 1/n$ is divergent so by part (c) of the Limit Comparison Test, $\sum \ln n/n$ diverges.   ●

● **Example 6**   Determine whether the series $\displaystyle\sum_{n=1}^{\infty} \frac{2n^2 + 3n}{\sqrt{5 + n^7}}$ converges or diverges.

Solution   The dominant part of the numerator is $2n^2$ and the dominant part of the denominator is $\sqrt{n^7} = n^{7/2}$. This suggests taking

$$a_n = \frac{2n^2 + 3n}{\sqrt{5 + n^7}} \qquad b_n = \frac{2n^2}{n^{7/2}} = \frac{2}{n^{3/2}}$$

$$\lim_{n \to \infty} \frac{a_n}{b_n} = \lim_{n \to \infty} \frac{2n^2 + 3n}{\sqrt{5 + n^7}} \cdot \frac{n^{3/2}}{2} = \lim_{n \to \infty} \frac{2n^{7/2} + 3n^{5/2}}{2\sqrt{5 + n^7}}$$

$$= \lim_{n \to \infty} \frac{2 + \dfrac{3}{n}}{2\sqrt{\dfrac{5}{n^7} + 1}} = \frac{2 + 0}{2\sqrt{0 + 1}} = 1$$

Since $\sum b_n = 2\sum 1/n^{3/2}$ is convergent (*p*-series, $p = \frac{3}{2} > 1$) the given series converges by the Limit Comparison Test.   ●

Notice that in testing many series we find a suitable comparison series $\sum b_n$ by keeping only the highest powers in the numerator and denominator.

EXERCISES 10.4

In Exercises 1–38 determine whether the given series converges or diverges.

1. $\displaystyle\sum_{n=1}^{\infty} \frac{1}{n^3 + n^2}$

2. $\displaystyle\sum_{n=1}^{\infty} \frac{3}{4^n + 5}$

3. $\displaystyle\sum_{n=1}^{\infty} \frac{3}{n2^n}$

4. $\displaystyle\sum_{n=2}^{\infty} \frac{1}{\sqrt{n} - 1}$

5. $\displaystyle\sum_{n=0}^{\infty} \frac{1 + 5^n}{4^n}$

6. $\displaystyle\sum_{n=1}^{\infty} \frac{\sin^2 n}{n\sqrt{n}}$

7. $\displaystyle\sum_{n=1}^{\infty} \frac{3}{n(n + 3)}$

8. $\displaystyle\sum_{n=1}^{\infty} \frac{1}{n^n}$

9. $\displaystyle\sum_{n=1}^{\infty} \frac{1}{\sqrt{n^3 + 1}}$

10. $\displaystyle\sum_{n=1}^{\infty} \frac{1}{\sqrt{n(n + 1)(n + 2)}}$

11. $\displaystyle\sum_{n=2}^{\infty} \frac{\sqrt{n}}{n - 1}$

12. $\displaystyle\sum_{n=1}^{\infty} \frac{1}{\sqrt[3]{n(n + 1)(n + 2)}}$

13. $\displaystyle\sum_{n=1}^{\infty} \frac{n - 1}{n^3 + 1}$

14. $\displaystyle\sum_{n=1}^{\infty} \frac{n}{(n + 1)2^n}$

15. $\displaystyle\sum_{n=1}^{\infty} \frac{3 + \cos n}{3^n}$

16. $\displaystyle\sum_{n=1}^{\infty} \frac{5n}{2n^2 - 5}$

17. $\displaystyle\sum_{n=1}^{\infty} \frac{4}{(n + 1)(2n + 1)}$

18. $\displaystyle\sum_{n=1}^{\infty} \frac{n}{(n+1)(n+2)(n+3)}$

19. $\displaystyle\sum_{n=1}^{\infty} \frac{n}{\sqrt{n^5 + 4}}$

20. $\displaystyle\sum_{n=1}^{\infty} \frac{\arctan n}{n^4}$

21. $\displaystyle\sum_{n=1}^{\infty} \frac{2^n}{1 + 3^n}$

22. $\displaystyle\sum_{n=1}^{\infty} \frac{1 + 2^n}{1 + 3^n}$

23. $\displaystyle\sum_{n=1}^{\infty} \frac{1}{1 + \sqrt{n}}$

24. $\displaystyle\sum_{n=3}^{\infty} \frac{1}{n^2 - 4}$

25. $\displaystyle\sum_{n=1}^{\infty} \frac{n^2 + 1}{n^4 + 1}$

26. $\displaystyle\sum_{n=1}^{\infty} \frac{3n^3 - 2n^2}{n^4 + n^2 + 1}$

27. $\displaystyle\sum_{n=1}^{\infty} \frac{n^2 - n + 2}{\sqrt[4]{n^{10} + n^5 + 3}}$

28. $\displaystyle\sum_{n=1}^{\infty} \frac{n^2 - 3n}{\sqrt[3]{n^{10} - 4n^2}}$

29. $\displaystyle\sum_{n=1}^{\infty} \frac{n + 1}{n2^n}$

30. $\displaystyle\sum_{n=1}^{\infty} \frac{2n^2 + 7n}{3^n(n^2 + 5n - 1)}$

31. $\displaystyle\sum_{n=1}^{\infty} \frac{\ln n}{n^3}$

32. $\displaystyle\sum_{n=2}^{\infty} \frac{1}{\ln n}$

33. $\displaystyle\sum_{n=1}^{\infty} \frac{1}{n!}$

34. $\displaystyle\sum_{n=1}^{\infty} \frac{n!}{(2n)!}$

35. $\displaystyle\sum_{n=1}^{\infty} \frac{n!}{n^2}$

36. $\displaystyle\sum_{n=1}^{\infty} \frac{n!}{n^n}$

37. $\displaystyle\sum_{n=1}^{\infty} \sin\left(\frac{1}{n}\right)$

38. $\displaystyle\sum_{n=1}^{\infty} \frac{1}{n^{1 + 1/n}}$

39. The meaning of the decimal representation of a number $0.d_1 d_2 d_3 \ldots$ (where the digit d_i is one of the numbers $0, 1, 2, \ldots, 9$) is that

$$0.d_1 d_2 d_3 d_4 \ldots = \frac{d_1}{10} + \frac{d_2}{10^2} + \frac{d_3}{10^3} + \frac{d_4}{10^4} + \cdots$$

Show that this series always converges.

40. For what values of p does the series $\sum_{n=2}^{\infty} 1/(n^p \ln n)$ converge?

41. Prove that if $a_n \geqslant 0$ and $\sum a_n$ converges, then $\sum a_n^2$ also converges.

42. Prove part (b) of the Limit Comparison Test.

43. Prove part (c) of the Limit Comparison Test.

44. Give an example of a pair of series $\sum a_n$ and $\sum b_n$ with positive terms where $\lim_{n \to \infty}(a_n/b_n) = 0$, $\sum b_n$ diverges, but $\sum a_n$ converges. [Compare with part (b) of Theorem 10.25.]

Alternating Series

An **alternating series** is a series whose terms are alternately positive and negative. Here are two examples:

$$1 - \frac{1}{2} + \frac{1}{3} - \frac{1}{4} + \frac{1}{5} - \frac{1}{6} + \cdots = \sum_{n=1}^{\infty} \frac{(-1)^{n-1}}{n}$$

$$-\frac{1}{2} + \frac{2}{3} - \frac{3}{4} + \frac{4}{5} - \frac{5}{6} + \frac{6}{7} - \cdots = \sum_{n=1}^{\infty} (-1)^n \frac{n}{n + 1}$$

The following test says that if the terms of an alternating series decrease to 0 in absolute value, then the series converges.

The Alternating Series Test (10.27)

If the alternating series

$$\sum_{n=1}^{\infty} (-1)^{n-1} a_n = a_1 - a_2 + a_3 - a_4 + a_5 - a_6 + \cdots \qquad (a_n > 0)$$

satisfies

(a) $a_{n+1} \leqslant a_n$ for all n

(b) $\lim_{n \to \infty} a_n = 0$

then the series is convergent.

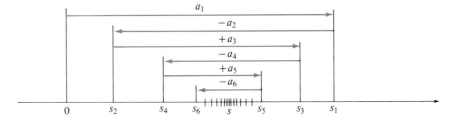

Figure 10.9

Before giving the proof of Theorem 10.27 let us take a look at Figure 10.9, which gives a picture of the idea behind the proof. We first plot $s_1 = a_1$ on a number line. To find s_2 we subtract a_2, so s_2 is to the left of s_1. Then to find s_3 we add a_3, so s_3 is to the right of s_2. But, since $a_3 < a_2$, s_3 is to the left of s_1. Continuing in this manner, we see that the partial sums oscillate back and forth. Since $a_n \to 0$, the successive steps are becoming smaller and smaller. The even partial sums $s_2, s_4, s_6, \ldots$ are increasing and the odd partial sums $s_1, s_3, s_5, \ldots$ are decreasing and it seems plausible that both are converging to some number s. Therefore in the following proof we shall consider the even and odd partial sums separately.

Proof of the Alternating Series Test   We first consider the even partial sums:

$$s_2 = a_1 - a_2 \geqslant 0 \qquad \text{since } a_2 \leqslant a_1$$

$$s_4 = s_2 + (a_3 - a_4) \geqslant s_2 \qquad \text{since } a_4 \leqslant a_3$$

In general

$$s_{2n} = s_{2n-2} + (a_{2n-1} - a_{2n}) \geqslant s_{2n-2} \qquad \text{since } a_{2n} \leqslant a_{2n-1}$$

Thus $$0 \leqslant s_2 \leqslant s_4 \leqslant s_6 \leqslant \cdots \leqslant s_{2n} \leqslant \cdots$$

But we can also write

$$s_{2n} = a_1 - (a_2 - a_3) - (a_4 - a_5) - \cdots - (a_{2n-2} - a_{2n-1}) - a_{2n}$$

Every term in brackets is positive, so $s_{2n} \leqslant a_1$ for all n. Therefore the sequence $\{s_{2n}\}$ of even partial sums is increasing and bounded above. It is therefore convergent by Theorem 10.11. Let us call its limit s, that is,

$$\lim_{n \to \infty} s_{2n} = s$$

Now we compute the limit of the odd partial sums:

$$\lim_{n \to \infty} s_{2n+1} = \lim_{n \to \infty} (s_{2n} + a_{2n+1})$$

$$= \lim_{n \to \infty} s_{2n} + \lim_{n \to \infty} a_{2n+1}$$

$$= s + 0 \qquad [\textit{by condition (b) of (10.27)}]$$

$$= s$$

Since both the even and odd partial sums converge to s, we have $\lim_{n \to \infty} s_n = s$ and so the series is convergent.   ●

● **Example 1**   The alternating harmonic series

$$1 - \frac{1}{2} + \frac{1}{3} - \frac{1}{4} + \cdots = \sum_{n=1}^{\infty} \frac{(-1)^{n-1}}{n}$$

satisfies

(a) $a_{n+1} < a_n$ because $\dfrac{1}{n+1} < \dfrac{1}{n}$

(b) $\displaystyle\lim_{n\to\infty} a_n = \lim_{n\to\infty} \frac{1}{n} = 0$

so the series is convergent by the Alternating Series Test. It can be shown that its sum is ln 2 (see Exercise 35).   ●

● **Example 2**   The series $\displaystyle\sum_{n=1}^{\infty} \frac{(-1)^n 3n}{4n - 1}$ is alternating but

$$\lim_{n\to\infty} a_n = \lim_{n\to\infty} \frac{3n}{4n-1} = \lim_{n\to\infty} \frac{3}{4 - \dfrac{1}{n}} = \frac{3}{4}$$

so condition (b) is not satisfied. In fact, the limit of the nth term of the series does not exist, so the series diverges by the Test for Divergence.   ●

● **Example 3**   Test the series $\displaystyle\sum_{n=1}^{\infty} (-1)^{n+1} \frac{n^2}{n^3 + 1}$ for convergence or divergence.

Solution   The given series is alternating so we try to verify conditions (a) and (b) of Theorem 10.27.

Unlike the situation in Example 1, it is not obvious that the sequence $a_n = n^2/(n^3 + 1)$ is decreasing. But if we consider the related function $f(x) = x^2/(x^3 + 1)$ we find that

$$f'(x) = \frac{x(2 - x^3)}{(x^3 + 1)^2}$$

Since we are considering only positive x, we see that $f'(x) < 0$ if $2 - x^3 < 0$, that is, $x > \sqrt[3]{2}$. Thus f is decreasing on the interval $[\sqrt[3]{2}, \infty)$. This means that $f(n + 1) < f(n)$ and therefore $a_{n+1} < a_n$ when $n \geqslant 2$. (The inequality $a_2 < a_1$ can be verified directly but all that really matters is that the sequence $\{a_n\}$ is eventually decreasing.)

Condition (b) is readily verified:

$$\lim_{n\to\infty} a_n = \lim_{n\to\infty} \frac{n^2}{n^3 + 1} = \lim_{n\to\infty} \frac{\dfrac{1}{n}}{1 + \dfrac{1}{n^3}} = 0$$

Thus the given series is convergent by the Alternating Series Test.   ●

Note: Instead of verifying condition (a) of the Alternating Series Test by computing a derivative as in Example 3, it is possible to verify that

$a_{n+1} < a_n$ directly by using the technique of Solution 1 of Example 10 in Section 10.1.

A partial sum s_n of any convergent series can be used as an approximation to the total sum s but this is not of much use unless we can estimate the accuracy of the approximation. The error involved in using $s \approx s_n$ is $|s - s_n|$. The next theorem says that for series that satisfy the conditions of the Alternating Series Test, the error is smaller than a_{n+1}, which is the magnitude of the first neglected term.

Theorem (10.28)

If $s = \sum(-1)^{n-1}a_n$ is the sum of an alternating series that satisfies
(a) $0 \leqslant a_{n+1} \leqslant a_n$ and (b) $\lim_{n \to \infty} a_n = 0$, then

$$|s - s_n| \leqslant a_{n+1}$$

Proof   The idea is similar to the one for the proof of the Alternating Series Test. (Indeed the result of Theorem 10.28 can be seen geometrically by looking at Figure 10.9.) We have

$$s - s_n = (-1)^n a_{n+1} + (-1)^{n+1}a_{n+2} + (-1)^{n+2}a_{n+3} + \cdots$$

$$= (-1)^n[a_{n+1} - a_{n+2} + a_{n+3} - \cdots]$$

and so

$$|s - s_n| = (a_{n+1} - a_{n+2}) + (a_{n+3} - a_{n+4}) + \cdots$$

$$= a_{n+1} - (a_{n+2} - a_{n+3}) - (a_{n+4} - a_{n+5}) - \cdots$$

Every term in brackets is positive, so $|s - s_n| \leqslant a_{n+1}$.   ●

● **Example 4**   Find the sum of the series $\displaystyle\sum_{n=0}^{\infty} \frac{(-1)^n}{n!}$ correct to three decimal places. (By definition, $0! = 1$.)

Solution   We first observe that the series is convergent by the Alternating Series Test because

(a) $\dfrac{1}{(n+1)!} = \dfrac{1}{n!(n+1)} < \dfrac{1}{n!}$

(b) $0 < \dfrac{1}{n!} < \dfrac{1}{n} \to 0$ so $\dfrac{1}{n!} \to 0$ as $n \to \infty$

To get a feel for how many terms we need to use in our approximation, let us write out the first few terms of the series:

$$s = \frac{1}{0!} - \frac{1}{1!} + \frac{1}{2!} - \frac{1}{3!} + \frac{1}{4!} - \frac{1}{5!} + \frac{1}{6!} - \frac{1}{7!} + \cdots$$

$$= 1 - 1 + \frac{1}{2} - \frac{1}{6} + \frac{1}{24} - \frac{1}{120} + \frac{1}{720} - \frac{1}{5040} + \cdots$$

Notice that $\qquad\qquad a_7 = \frac{1}{5040} < \frac{1}{5000} = 0.0002$

and $\qquad s_6 = 1 - 1 + \frac{1}{2} - \frac{1}{6} + \frac{1}{24} - \frac{1}{120} + \frac{1}{720} \approx 0.368056$

By Theorem 10.28 we know that

$$|s - s_6| \leqslant a_7 < 0.0002$$

This error of less than 0.0002 will not affect the third decimal place, so we have

$$s \approx 0.368$$

correct to three decimal places.

In Section 10.11 we shall prove that $e^x = \sum_{n=0}^{\infty} x^n/n!$ for all x, so what we have obtained in this example is actually an approximation to the number e^{-1}. $\qquad\bullet$

  Note: The rule that the error (in using s_n to approximate s) is smaller than the first neglected term is, in general, valid only for alternating series that satisfy the conditions of Theorem 10.28. The rule does not apply to other types of series. (See the example in Appendix D.)

EXERCISES 10.5

In Exercises 1–22 test the given series for convergence or divergence.

1. $\frac{3}{5} - \frac{3}{6} + \frac{3}{7} - \frac{3}{8} + \frac{3}{9} - \cdots$

2. $-5 - \frac{5}{2} + \frac{5}{5} - \frac{5}{8} + \frac{5}{11} - \frac{5}{14} + \cdots$

3. $-\frac{1}{2} + \frac{2}{3} - \frac{3}{4} + \frac{4}{5} - \frac{5}{6} + \frac{6}{7} - \cdots$

4. $\frac{1}{\ln 2} - \frac{1}{\ln 3} + \frac{1}{\ln 4} - \frac{1}{\ln 5} + \frac{1}{\ln 6} - \cdots$

5. $\sum_{n=1}^{\infty} \frac{(-1)^{n-1}}{n^2}$

6. $\sum_{n=1}^{\infty} \frac{(-1)^n}{\sqrt{n+3}}$

7. $\sum_{n=1}^{\infty} (-1)^{n+1} \cdot \frac{n}{5n+1}$

8. $\sum_{n=2}^{\infty} \frac{(-1)^{n-1}}{n \ln n}$

9. $\sum_{n=1}^{\infty} (-1)^n \frac{n}{n^2 + 1}$

10. $\sum_{n=1}^{\infty} (-1)^n \frac{n^2}{n^2 + 1}$

11. $\sum_{n=1}^{\infty} (-1)^{n-1} \frac{\sqrt{n}}{n + 4}$

12. $\sum_{n=1}^{\infty} (-1)^{n+1} \frac{n}{2^n}$

13. $\sum_{n=2}^{\infty} (-1)^n \frac{n}{\ln n}$

14. $\sum_{n=1}^{\infty} (-1)^{n-1} \frac{\ln n}{n}$

15. $\sum_{n=1}^{\infty} (-1)^{n+1} \frac{n + 10}{n(n + 1)}$

16. $\sum_{n=1}^{\infty} (-1)^{n-1} \frac{(n + 9)(n + 10)}{n(n + 1)}$

17. $\sum_{n=1}^{\infty} \frac{\cos n\pi}{n^{3/4}}$

18. $\sum_{n=1}^{\infty} \frac{\sin\left(\frac{n\pi}{2}\right)}{n!}$

19. $\sum_{n=1}^{\infty} (-1)^n \sin\left(\frac{\pi}{n}\right)$

20. $\sum_{n=1}^{\infty} (-1)^n \cos\left(\frac{\pi}{n}\right)$

21. $\sum_{n=1}^{\infty} \frac{(-1)^n}{\sqrt[n]{n}}$

22. $\sum_{n=1}^{\infty} \frac{(-1)^n}{|n - 10\pi|}$

23. Show that the series $\sum(-1)^{n-1} a_n$, where $a_n = 1/n$ if n is odd and $a_n = 1/n^2$ if n is even, is divergent. Why does the Alternating Series Test not apply?

For what values of p are the series in Exercises 24–26 convergent?

24. $\displaystyle\sum_{n=1}^{\infty} \frac{(-1)^{n-1}}{n^p}$

25. $\displaystyle\sum_{n=1}^{\infty} \frac{(-1)^n}{n+p}$

26. $\displaystyle\sum_{n=1}^{\infty} (-1)^{n-1} \frac{(\ln n)^p}{n}$

In Exercises 27–34 approximate the sum of the given series to the indicated accuracy.

27. $\displaystyle\sum_{n=1}^{\infty} \frac{(-1)^{n-1}}{n^2}$ (error < 0.01)

28. $\displaystyle\sum_{n=1}^{\infty} \frac{(-1)^{n+1}}{n^4}$ (error < 0.001)

29. $\displaystyle\sum_{n=0}^{\infty} \frac{(-2)^n}{n!}$ (error < 0.01)

30. $\displaystyle\sum_{n=0}^{\infty} \frac{(-1)^n n}{4^n}$ (error < 0.002)

31. $\displaystyle\sum_{n=1}^{\infty} \frac{(-1)^{n-1}}{(2n-1)!}$ (four decimal places)

32. $\displaystyle\sum_{n=0}^{\infty} \frac{(-1)^n}{(2n)!}$ (four decimal places)

33. $\displaystyle\sum_{n=0}^{\infty} \frac{(-1)^n}{2^n n!}$ (four decimal places)

34. $\displaystyle\sum_{n=1}^{\infty} \frac{(-1)^{n-1}}{n^6}$ (five decimal places)

35. Use the following steps to show that

$$\sum_{n=1}^{\infty} \frac{(-1)^{n-1}}{n} = \ln 2$$

Let h_n and s_n be the partial sums of the harmonic and alternating harmonic series.
(a) Show that $s_{2n} = h_{2n} - h_n$.
(b) From Exercise 29 in Section 10.3 we have $h_n - \ln n \to \gamma$ as $n \to \infty$ and therefore $h_{2n} - \ln(2n) \to \gamma$ as $n \to \infty$. Use these facts together with part (a) to show that $s_{2n} \to \ln 2$ as $n \to \infty$.

SECTION 10.6

Absolute Convergence and the Ratio and Root Tests

Given any series $\sum a_n$, we can consider the corresponding series

$$\sum_{n=1}^{\infty} |a_n| = |a_1| + |a_2| + |a_3| + \cdots$$

whose terms are the absolute values of the terms of the original series.

Definition (10.29)

> A series $\sum a_n$ is called **absolutely convergent** if the series of absolute values $\sum |a_n|$ is convergent.

Notice that if $\sum a_n$ is a series with positive terms, then $|a_n| = a_n$ and so absolute convergence is the same as convergence.

● **Example 1**   The series

$$\sum_{n=1}^{\infty} \frac{(-1)^{n-1}}{n^2} = 1 - \frac{1}{2^2} + \frac{1}{3^2} - \frac{1}{4^2} + \cdots$$

is absolutely convergent because

$$\sum_{n=1}^{\infty} \left| \frac{(-1)^{n-1}}{n^2} \right| = \sum_{n=1}^{\infty} \frac{1}{n^2} = 1 + \frac{1}{2^2} + \frac{1}{3^2} + \frac{1}{4^2} + \cdots$$

is a convergent p-series ($p = 2$).

● **Example 2**   We know that the alternating harmonic series

$$\sum_{n=1}^{\infty} \frac{(-1)^{n-1}}{n} = 1 - \frac{1}{2} + \frac{1}{3} - \frac{1}{4} + \cdots$$

is convergent (see Example 1 in Section 10.5) but it is not absolutely convergent because the corresponding series of absolute values is

$$\sum_{n=1}^{\infty} \left| \frac{(-1)^{n-1}}{n} \right| = \sum_{n=1}^{\infty} \frac{1}{n} = 1 + \frac{1}{2} + \frac{1}{3} + \frac{1}{4} + \cdots$$

which is the harmonic series (p-series, $p = 1$) and is therefore divergent.

Definition (10.30)

> A series $\sum a_n$ is called **conditionally convergent** if it is convergent but not absolutely convergent.

Example 2 shows that the alternating harmonic series is conditionally convergent. Thus it is possible for a series to be convergent but not absolutely convergent. However the next theorem shows that absolute convergence implies convergence.

Theorem (10.31)

> If a series $\sum a_n$ is absolutely convergent, then it is convergent.

Proof   Observe that the inequality

$$-|a_n| \leq a_n \leq |a_n|$$

is true because a_n is either $-|a_n|$ or $|a_n|$. If we now add $|a_n|$ to each side of this inequality, we get

$$0 \leq a_n + |a_n| \leq 2|a_n|$$

Let $b_n = a_n + |a_n|$. Then $0 \leq b_n \leq 2|a_n|$. If $\sum a_n$ is absolutely convergent, then $\sum |a_n|$ is convergent, so $\sum 2|a_n|$ is convergent by Theorem 10.19 part (a). Therefore $\sum b_n$ is convergent by the Comparison Test. Since $a_n = b_n - |a_n|$,

$$\sum a_n = \sum b_n - \sum |a_n|$$

is convergent by Theorem 10.19 part (c).

● **Example 3**   Determine whether the series

$$\sum_{n=1}^{\infty} \frac{\cos n}{n^2} = \frac{\cos 1}{1^2} + \frac{\cos 2}{2^2} + \frac{\cos 3}{3^2} + \cdots$$

is convergent or divergent.

Solution   This series has both positive and negative terms but it is not alternating. (The first term is positive, the next three are negative, and the following three are positive. The signs change irregularly.) We can apply the Comparison Test to the series of absolute values

$$\sum_{n=1}^{\infty} \left| \frac{\cos n}{n^2} \right| = \sum_{n=1}^{\infty} \frac{|\cos n|}{n^2}$$

Since $|\cos n| \leqslant 1$ for all n we have

$$\frac{|\cos n|}{n^2} \leqslant \frac{1}{n^2}$$

We know that $\sum 1/n^2$ is convergent (p-series, $p = 2$) and therefore $\sum |\cos n|/n^2$ is convergent by the Comparison Test. Thus the given series $\sum (\cos n)/n^2$ is absolutely convergent and therefore convergent by Theorem 10.31.   ●

The following test is very useful in determining whether a given series is absolutely convergent.

The Ratio Test (10.32)

(a)   If $\lim_{n \to \infty} \left| \frac{a_{n+1}}{a_n} \right| = L < 1$, then the series $\sum_{n=1}^{\infty} a_n$ is absolutely convergent (and therefore convergent).

(b)   If $\lim_{n \to \infty} \left| \frac{a_{n+1}}{a_n} \right| = L > 1$ or $\lim_{n \to \infty} \left| \frac{a_{n+1}}{a_n} \right| = \infty$, then the series $\sum_{n=1}^{\infty} a_n$ is divergent.

Proof   For part (a) the idea is to compare the given series with a convergent geometric series. Since $L < 1$ we can choose a number r such that $L < r < 1$. Since

$$\lim_{n \to \infty} \left| \frac{a_{n+1}}{a_n} \right| = L \quad \text{and} \quad L < r$$

the ratio $|a_{n+1}/a_n|$ will eventually be less than r; that is, there exists an integer N such that

$$\left| \frac{a_{n+1}}{a_n} \right| < r \qquad \text{whenever } n \geqslant N$$

or equivalently

(10.33) $|a_{n+1}| < |a_n|r$ whenever $n \geqslant N$

Putting n successively equal to $N, N + 1, N + 2, \ldots$ in (10.33), we obtain

$$|a_{N+1}| < |a_N|r$$
$$|a_{N+2}| < |a_{N+1}|r < |a_N|r^2$$
$$|a_{N+3}| < |a_{N+2}|r < |a_N|r^3$$

and, in general,

(10.34) $|a_{N+k}| < |a_N|r^k$ for all $k \geqslant 1$

Now the series

$$\sum_{k=1}^{\infty} |a_N|r^k = |a_N|r + |a_N|r^2 + |a_N|r^3 + \cdots$$

is convergent since it is a geometric series with $0 < r < 1$. So the inequality (10.34), together with the Comparison Test, shows that the series

$$\sum_{n=N+1}^{\infty} |a_n| = \sum_{k=1}^{\infty} |a_{N+k}| = |a_{N+1}| + |a_{N+2}| + |a_{N+3}| + \cdots$$

is also convergent. It follows that the series $\sum_{n=1}^{\infty} |a_n|$ is convergent. (Recall that a finite number of terms cannot affect convergence.) Therefore $\sum a_n$ is absolutely convergent.

For part (b), if $|a_{n+1}/a_n| \to L > 1$ or $|a_{n+1}/a_n| \to \infty$, then the ratio $|a_{n+1}/a_n|$ will eventually be greater than 1; that is, there exists an integer N such that

$$\left| \frac{a_{n+1}}{a_n} \right| > 1 \qquad \text{whenever } n \geqslant N$$

This means that $|a_{n+1}| > |a_n|$ when $n \geqslant N$ and so

$$\lim_{n \to \infty} a_n \neq 0$$

Therefore $\sum a_n$ diverges by the Test for Divergence.   ●

Note: If $\lim_{n \to \infty} |a_{n+1}/a_n| = 1$, the Ratio Test gives no information. For instance, for the convergent series $\sum 1/n^2$ we have

$$\left| \frac{a_{n+1}}{a_n} \right| = \frac{\dfrac{1}{(n+1)^2}}{\dfrac{1}{n^2}} = \frac{n^2}{(n+1)^2} = \frac{1}{\left(1 + \dfrac{1}{n}\right)^2} \to 1 \qquad \text{as } n \to \infty$$

whereas for the divergent series $\sum 1/n$ we have

$$\left| \frac{a_{n+1}}{a_n} \right| = \frac{\dfrac{1}{n+1}}{\dfrac{1}{n}} = \frac{n}{1+n} = \frac{1}{\dfrac{1}{n}+1} \to 1 \qquad \text{as } n \to \infty$$

Therefore if $\lim_{n \to \infty} |a_{n+1}/a_n| = 1$, the series $\sum a_n$ might converge or it might diverge. In this case the Ratio Test fails and we must use some other test.

● **Example 4**   Test the series $\displaystyle\sum_{n=1}^{\infty} (-1)^n \frac{n^3}{3^n}$ for absolute convergence.

Solution   We use the Ratio Test with $a_n = (-1)^n n^3/3^n$:

$$\left| \frac{a_{n+1}}{a_n} \right| = \left| \frac{\dfrac{(-1)^{n+1}(n+1)^3}{3^{n+1}}}{\dfrac{(-1)^n n^3}{3^n}} \right| = \frac{(n+1)^3}{3^{n+1}} \cdot \frac{3^n}{n^3}$$

$$= \frac{1}{3} \left(\frac{n+1}{n} \right)^3 = \frac{1}{3} \left(1 + \frac{1}{n} \right)^3 \to \frac{1}{3} < 1$$

Thus, by the Ratio Test, the given series is absolutely convergent and therefore convergent.   ●

● **Example 5**   Test the convergence of the series $\displaystyle\sum_{n=1}^{\infty} \frac{n^n}{n!}$.

Solution   Since the terms $a_n = n^n/n!$ are positive we do not need the absolute value signs.

$$\frac{a_{n+1}}{a_n} = \frac{(n+1)^{n+1}}{(n+1)!} \cdot \frac{n!}{n^n} = \frac{(n+1)(n+1)^n}{(n+1)n!} \cdot \frac{n!}{n^n}$$

$$= \left(\frac{n+1}{n} \right)^n = \left(1 + \frac{1}{n} \right)^n \to e \qquad \text{as } n \to \infty$$

by the definition of e (6.19). Since $e > 1$, the given series is divergent by the Ratio Test.   ●

Note: Although the Ratio Test works in Example 5, an easier method is to use the Test for Divergence. Since

$$a_n = \frac{n^n}{n!} = \frac{n \cdot n \cdot n \cdot \cdots \cdot n}{1 \cdot 2 \cdot 3 \cdot \cdots \cdot n} \geq n$$

it follows that a_n does not approach 0 as $n \to \infty$. Therefore the given series is divergent by the Test for Divergence.

The following test is convenient to apply when nth powers occur. Its proof is similar to the proof of the Ratio Test and is left as Exercise 39.

The Root Test (10.35)

(a) If $\lim_{n \to \infty} \sqrt[n]{|a_n|} = L < 1$, then the series $\sum_{n=1}^{\infty} a_n$ is absolutely convergent (and therefore convergent).

(b) If $\lim_{n \to \infty} \sqrt[n]{|a_n|} = L > 1$ or $\lim_{n \to \infty} \sqrt[n]{|a_n|} = \infty$, then the series $\sum_{n=1}^{\infty} a_n$ is divergent.

If $\lim_{n \to \infty} \sqrt[n]{|a_n|} = 1$, then the Root Test gives no information. The series $\sum a_n$ could converge or diverge. (If $L = 1$ in the Ratio Test, do not try the Root Test because L will again be 1.)

● **Example 6**   Test the convergence of the series $\sum_{n=1}^{\infty} \left(\dfrac{2n + 3}{3n + 2} \right)^n$.

Solution
$$a_n = \left(\frac{2n + 3}{3n + 2} \right)^n$$

$$\sqrt[n]{|a_n|} = \frac{2n + 3}{3n + 2} = \frac{2 + \dfrac{3}{n}}{3 + \dfrac{2}{n}} \to \frac{2}{3} < 1$$

Thus the given series converges by the Root Test.   ●

Rearrangements

The question of whether a given convergent series is absolutely convergent or conditionally convergent has a bearing on the question of whether infinite sums behave like finite sums.

If we rearrange the order of the terms in a finite sum, then of course the value of the sum remains unchanged. But this is not always the case for an infinite series. By a **rearrangement** of an infinite series $\sum a_n$ we mean a series obtained by simply changing the order of the terms. For instance, a rearrangement of $\sum a_n$ could start as follows:

$$a_1 + a_2 + a_5 + a_3 + a_4 + a_{15} + a_6 + a_7 + a_{20} + \cdots$$

It turns out that **if $\sum a_n$ is an absolutely convergent series with sum s, then any rearrangement of $\sum a_n$ has the same sum s.** However any conditionally convergent series can be rearranged to give a different sum. To illustrate this fact let us consider the alternating harmonic series

$$1 - \tfrac{1}{2} + \tfrac{1}{3} - \tfrac{1}{4} + \tfrac{1}{5} - \tfrac{1}{6} + \tfrac{1}{7} - \tfrac{1}{8} + \cdots = \ln 2$$

(See Exercise 35 in Section 10.5.) If we rearrange this series by placing two negative terms after each positive term, we obtain the series

$$1 - \tfrac{1}{2} - \tfrac{1}{4} + \tfrac{1}{3} - \tfrac{1}{6} - \tfrac{1}{8} + \tfrac{1}{5} - \tfrac{1}{10} - \tfrac{1}{12} + \cdots$$

$$= (1 - \tfrac{1}{2}) - \tfrac{1}{4} + (\tfrac{1}{3} - \tfrac{1}{6}) - \tfrac{1}{8} + (\tfrac{1}{5} - \tfrac{1}{10}) - \tfrac{1}{12} + \cdots$$

$$= \tfrac{1}{2} - \tfrac{1}{4} + \tfrac{1}{6} - \tfrac{1}{8} + \tfrac{1}{10} - \tfrac{1}{12} + \cdots$$

$$= \tfrac{1}{2}(1 - \tfrac{1}{2} + \tfrac{1}{3} - \tfrac{1}{4} + \tfrac{1}{5} - \tfrac{1}{6} + \cdots)$$

$$= \tfrac{1}{2}\ln 2$$

In fact, Riemann proved that **if $\sum a_n$ is a conditionally convergent series and r is any real number whatsoever, then there is a rearrangement of $\sum a_n$ that has a sum equal to r.** A proof of this fact is outlined in Exercise 41.

EXERCISES 10.6

In Exercises 1–36 determine whether the given series is absolutely convergent, conditionally convergent, or divergent.

1. $\displaystyle\sum_{n=1}^{\infty} \frac{(-1)^{n-1}}{n\sqrt{n}}$

2. $\displaystyle\sum_{n=1}^{\infty} \frac{(-1)^n}{\sqrt{n}}$

3. $\displaystyle\sum_{n=1}^{\infty} \frac{(-3)^n}{n^3}$

4. $\displaystyle\sum_{n=0}^{\infty} \frac{(-3)^n}{n!}$

5. $\displaystyle\sum_{n=1}^{\infty} \frac{(-1)^{n+1}}{2n+1}$

6. $\displaystyle\sum_{n=1}^{\infty} \frac{(-1)^{n-1}}{n^2+1}$

7. $\displaystyle\sum_{n=1}^{\infty} \frac{(-1)^{n-1}}{(2n-1)!}$

8. $\displaystyle\sum_{n=1}^{\infty} e^{-n}n!$

9. $\displaystyle\sum_{n=1}^{\infty} (-1)^n \frac{n}{n^2+4}$

10. $\displaystyle\sum_{n=1}^{\infty} (-1)^{n-1} \frac{\sqrt{n}}{n+1}$

11. $\displaystyle\sum_{n=1}^{\infty} (-1)^n \frac{2n}{3n-4}$

12. $\displaystyle\sum_{n=1}^{\infty} (-1)^n \frac{2^n}{n^2+1}$

13. $\displaystyle\sum_{n=1}^{\infty} \frac{\sin 2n}{n^2}$

14. $\displaystyle\sum_{n=1}^{\infty} \frac{(-1)^n \arctan n}{n^3}$

15. $\displaystyle\sum_{n=1}^{\infty} \frac{(-2)^n}{n3^{n+1}}$

16. $\displaystyle\sum_{n=1}^{\infty} \frac{(-1)^{n+1}5^{n-1}}{(n+1)^2 4^{n+2}}$

17. $\displaystyle\sum_{n=1}^{\infty} \frac{(n+1)5^n}{n3^{2n}}$

18. $\displaystyle\sum_{n=1}^{\infty} \frac{8-n^3}{n!}$

19. $\displaystyle\sum_{n=2}^{\infty} \frac{(-1)^n}{\ln n}$

20. $\displaystyle\sum_{n=1}^{\infty} \frac{\cos\left(\dfrac{n\pi}{6}\right)}{n\sqrt{n}}$

21. $\displaystyle\sum_{n=1}^{\infty} \frac{n!}{(-10)^n}$

22. $\displaystyle\sum_{n=1}^{\infty} \frac{n!}{n^n}$

23. $\displaystyle\sum_{n=1}^{\infty} \frac{\cos\left(\dfrac{n\pi}{3}\right)}{n!}$

24. $\displaystyle\sum_{n=2}^{\infty} \frac{(-1)^n}{(\ln n)^n}$

25. $\displaystyle\sum_{n=1}^{\infty} \frac{(-n)^n}{5^{2n+3}}$

26. $\displaystyle\sum_{n=2}^{\infty} \frac{(-1)^n}{n\ln n}$

27. $\displaystyle\sum_{n=1}^{\infty} \left(\frac{1-3n}{3+4n}\right)^n$

28. $\displaystyle\sum_{n=1}^{\infty} \frac{(-2)^n n^2}{(n+2)!}$

29. $1 - \dfrac{2!}{1\cdot 3} + \dfrac{3!}{1\cdot 3\cdot 5} - \dfrac{4!}{1\cdot 3\cdot 5\cdot 7} + \cdots$

$+ \dfrac{(-1)^{n-1}n!}{1\cdot 3\cdot 5\cdot \cdots \cdot (2n-1)} + \cdots$

30. $\dfrac{1}{3} + \dfrac{1\cdot 4}{3\cdot 5} + \dfrac{1\cdot 4\cdot 7}{3\cdot 5\cdot 7} + \dfrac{1\cdot 4\cdot 7\cdot 11}{3\cdot 5\cdot 7\cdot 9} + \cdots$

$+ \dfrac{1\cdot 4\cdot 7\cdot \cdots \cdot (3n-2)}{3\cdot 5\cdot 7\cdot \cdots \cdot (2n+1)} + \cdots$

31. $\displaystyle\sum_{n=1}^{\infty} \frac{2 \cdot 4 \cdot 6 \cdot \ \cdots \ \cdot (2n)}{n!}$

32. $\displaystyle\sum_{n=1}^{\infty} (-1)^n \frac{2^n n!}{5 \cdot 8 \cdot 11 \cdot \ \cdots \ \cdot (3n+2)}$

33. $\displaystyle\sum_{n=1}^{\infty} \frac{(n+2)!}{n! \, 10^n}$ **34.** $\displaystyle\sum_{n=1}^{\infty} \frac{(n!)^2}{(2n)!}$

35. $\displaystyle\sum_{n=1}^{\infty} \frac{\sin(3n) \cdot n^2}{(1.1)^n}$ **36.** $\displaystyle\sum_{n=1}^{\infty} \frac{(-1)^n}{(\arctan n)^n}$

37. (a) Show that $\sum_{n=0}^{\infty} x^n/n!$ converges for all x.
   (b) Deduce that $\lim_{n \to \infty} x^n/n! = 0$ for all x.

38. Prove that if $\sum a_n$ is absolutely convergent, then

$$\left| \sum_{n=1}^{\infty} a_n \right| \leq \sum_{n=1}^{\infty} |a_n|$$

39. Prove the Root Test (10.35). [*Hint for part (a):* Take any number r such that $L < r < 1$ and use the fact that there is an integer N such that $\sqrt[n]{|a_n|} < r$ when $n \geq N$.]

40. Given any series $\sum a_n$, we define a series $\sum a_n^+$ whose terms are all the positive terms of $\sum a_n$ and a series $\sum a_n^-$ whose terms are all the negative terms of $\sum a_n$. To be specific, we let

$$a_n^+ = \frac{a_n + |a_n|}{2} \qquad a_n^- = \frac{a_n - |a_n|}{2}$$

Notice that if $a_n > 0$, then $a_n^+ = a_n$ and $a_n^- = 0$, whereas if $a_n < 0$, then $a_n^- = a_n$ and $a_n^+ = 0$.
   (a) If $\sum a_n$ is absolutely convergent, show that both of the series $\sum a_n^+$ and $\sum a_n^-$ are convergent.
   (b) If $\sum a_n$ is conditionally convergent, show that both of the series $\sum a_n^+$ and $\sum a_n^-$ are divergent.

41. Prove that if $\sum a_n$ is a conditionally convergent series and r is any real number, then there is a rearrangement of $\sum a_n$ whose sum is r. (*Hints:* Use the notation of Exercise 40. Take just enough positive terms a_n^+ so that their sum is greater than r. Then add just enough negative terms a_n^- so that the cumulative sum is less than r. Continue in this manner and use Theorem 10.17.)

SECTION 10.7

Strategy for Testing Series

We now have several ways of testing a series for convergence or divergence and the problem is to decide which test to use on which series. In this respect testing series is similar to integrating functions. Again there are no hard and fast rules about which test to apply to a given series but you may find the following advice of some use.

It is not wise to have a list of the tests to be applied in a specific order until one finally works. That would be a waste of time and effort. Instead, as with integration, the main strategy is to classify the series according to its *form*.

1. If the series is of the form $\sum 1/n^p$, it is a *p*-series, which we know to be convergent if $p > 1$ and divergent if $p \leq 1$.

2. If the series has the form $\sum ar^{n-1}$ or $\sum ar^n$, it is a geometric series, which converges if $|r| < 1$ and diverges if $|r| \geq 1$. Some preliminary algebraic manipulation may be required to bring the series into this form.

3. If the series has a form that is similar to a *p*-series or a geometric series, then one of the Comparison Tests should be considered. In particular, if a_n is a rational function or algebraic function of n (involving roots of polynomials), then the series should be compared to a *p*-series. Notice that most of the series in Exercises 10.4 have this form. (The value of p should be chosen as in Section 10.4 by keeping only the highest powers of n in the numerator and denominator.) The comparison tests apply only to series with positive terms, but if $\sum a_n$ has

some negative terms then we can apply the Comparison Test to $\sum |a_n|$ and test for absolute convergence.

4. If you can see at a glance that $\lim_{n \to \infty} a_n \neq 0$, then the Test for Divergence should be used.

5. If the series is of the form $\sum (-1)^{n-1} a_n$ or $\sum (-1)^n a_n$, then the Alternating Series Test is an obvious possibility.

6. Series that involve factorials or other products are often conveniently tested using the Ratio Test. Bear in mind that $|a_{n+1}/a_n| \to 1$ as $n \to \infty$ for all p-series and therefore all rational or algebraic functions of n. Thus the Ratio Test should not be used for such series.

7. If a_n is of the form $(b_n)^n$, then the Root Test may be useful.

8. If $a_n = f(n)$, where $\int_1^\infty f(x)\, dx$ is easily evaluated, then the Integral Test is effective (assuming the hypotheses of this test are satisfied).

In the following examples we shall not work out all the details but simply indicate which tests should be used.

● **Example 1** $\displaystyle \sum_{n=1}^{\infty} \frac{n-1}{2n+1}$

Since $a_n \to \frac{1}{2} \neq 0$ as $n \to \infty$, we should use the Test for Divergence.   ●

● **Example 2** $\displaystyle \sum_{n=1}^{\infty} \frac{\sqrt{n^3+1}}{3n^3+4n^2+2}$

Since a_n is an algebraic function of n, we compare the given series with a p-series. The comparison series is $\sum b_n$, where

$$ b_n = \frac{\sqrt{n^3}}{3n^3} = \frac{n^{3/2}}{3n^3} = \frac{1}{3n^{3/2}} $$ ●

● **Example 3** $\displaystyle \sum_{n=1}^{\infty} n e^{-n^2}$

Since the integral $\int_1^\infty x e^{-x^2}\, dx$ is easily evaluated, we use the Integral Test. The Ratio Test also works.   ●

● **Example 4** $\displaystyle \sum_{n=1}^{\infty} (-1)^n \frac{n^3}{n^4+1}$

Since the series is alternating, we use the Alternating Series Test.   ●

● **Example 5** $\displaystyle \sum_{n=1}^{\infty} \frac{2^n}{n!}$

Since the series involves $n!$, we use the Ratio Test.   ●

● **Example 6** $\displaystyle \sum_{n=1}^{\infty} \frac{1}{2+3^n}$

Since the series is closely related to the geometric series $\sum 1/3^n$, we use the Comparison Test.   ●

EXERCISES 10.7

Test the following series for convergence or divergence.

1. $\displaystyle\sum_{n=1}^{\infty} \frac{\sqrt{n}}{n^2 + 1}$

2. $\displaystyle\sum_{n=1}^{\infty} \cos n$

3. $\displaystyle\sum_{n=1}^{\infty} \frac{4^n}{3^{2n-1}}$

4. $\displaystyle\sum_{i=1}^{\infty} \frac{i^4}{4^i}$

5. $\displaystyle\sum_{n=2}^{\infty} \frac{(-1)^n}{(\ln n)^2}$

6. $\displaystyle\sum_{n=1}^{\infty} n^2 e^{-n^3}$

7. $\displaystyle\sum_{k=1}^{\infty} k^{-1.7}$

8. $\displaystyle\sum_{n=0}^{\infty} \frac{10^n}{n!}$

9. $\displaystyle\sum_{n=1}^{\infty} \frac{n}{e^n}$

10. $\displaystyle\sum_{m=1}^{\infty} \frac{2m}{8m - 5}$

11. $\displaystyle\sum_{n=2}^{\infty} \frac{n^3 + 1}{n^4 - 1}$

12. $\displaystyle\sum_{n=1}^{\infty} \left(\frac{n^2 + 1}{2n^2 + 1}\right)^n$

13. $\displaystyle\sum_{n=2}^{\infty} \frac{2}{n(\ln n)^3}$

14. $\displaystyle\sum_{n=1}^{\infty} \frac{\sqrt{n}}{e^{\sqrt{n}}}$

15. $\displaystyle\sum_{n=1}^{\infty} \frac{3^n n^2}{n!}$

16. $\displaystyle\sum_{n=1}^{\infty} \frac{3}{4n - 5}$

17. $\displaystyle\sum_{n=1}^{\infty} \frac{3^n}{5^n + n}$

18. $\displaystyle\sum_{k=1}^{\infty} \frac{k + 5}{5^k}$

19. $\displaystyle\sum_{n=0}^{\infty} \frac{n!}{2 \cdot 5 \cdot 8 \cdot \cdots \cdot (3n + 2)}$

20. $\displaystyle\sum_{n=1}^{\infty} \frac{(-1)^n n}{(n + 1)(n + 2)}$

21. $\displaystyle\sum_{i=1}^{\infty} \frac{1}{\sqrt{i(i + 1)}}$

22. $\displaystyle\sum_{n=1}^{\infty} \frac{n^2}{\sqrt{n^5 + n^2 + 2}}$

23. $\displaystyle\sum_{n=1}^{\infty} (-1)^n 2^{1/n}$

24. $\displaystyle\sum_{n=1}^{\infty} \frac{\cos\left(\dfrac{n}{2}\right)}{n^2 + 4n}$

25. $\displaystyle\sum_{n=1}^{\infty} (-1)^n \frac{\ln n}{\sqrt{n}}$

26. $\displaystyle\sum_{n=1}^{\infty} \frac{\tan\left(\dfrac{1}{n}\right)}{n}$

27. $\displaystyle\sum_{n=0}^{\infty} (-\pi)^n$

28. $\displaystyle\sum_{n=1}^{\infty} \frac{\sqrt[3]{n} + 1}{n(\sqrt{n} + 1)}$

29. $\displaystyle\sum_{n=1}^{\infty} \frac{(-2)^{2n}}{n^n}$

30. $\displaystyle\sum_{n=1}^{\infty} \frac{2^{3n-1}}{n^2 + 1}$

31. $\displaystyle\sum_{k=1}^{\infty} \frac{k \ln k}{(k + 1)^3}$

32. $\displaystyle\sum_{n=1}^{\infty} \frac{e^{1/n}}{n^2}$

33. $\displaystyle\sum_{n=1}^{\infty} \frac{2^n}{(2n + 1)!}$

34. $\displaystyle\sum_{j=1}^{\infty} (-1)^j \frac{\sqrt{j}}{j + 5}$

35. $\displaystyle\sum_{n=1}^{\infty} \frac{\tan^{-1} n}{n\sqrt{n}}$

36. $\displaystyle\sum_{n=1}^{\infty} \frac{(2n)^n}{n^{2n}}$

37. $\displaystyle\sum_{n=1}^{\infty} \frac{1}{1 + \left(\dfrac{3}{\pi}\right)^n}$

38. $\displaystyle\sum_{n=2}^{\infty} \frac{1}{(\ln n)^{\ln n}}$

39. $\displaystyle\sum_{n=1}^{\infty} (\sqrt[n]{2} - 1)^n$

40. $\displaystyle\sum_{n=1}^{\infty} (\sqrt[n]{2} - 1)$

SECTION 10.8

Power Series

A **power series** is a series of the form

(10.36)
$$\sum_{n=0}^{\infty} a_n x^n = a_0 + a_1 x + a_2 x^2 + a_3 x^3 + \cdots$$

where x is a variable and the a_n's are constants called the **coefficients** of the series. For each fixed x, the series (10.36) is a series of constants that we can test for convergence or divergence. A power series may converge for some values of x and diverge for other values of x. The sum of the series is a function

$$f(x) = a_0 + a_1 x + a_2 x^2 + \cdots + a_n x^n + \cdots$$

whose domain is the set of all x for which the series converges. Notice that f resembles a polynomial. The only difference is that f has infinitely many terms.

For instance, if we take $a_n = 1$ for all n, the power series becomes the geometric series

$$\sum_{n=0}^{\infty} x^n = 1 + x + x^2 + \cdots + x^n + \cdots = \frac{1}{1 - x}$$

which converges when $-1 < x < 1$ and diverges when $|x| \geq 1$ (see Equation 10.16).

More generally, a series of the form

(10.37)
$$\sum_{n=0}^{\infty} a_n(x - c)^n = a_0 + a_1(x - c) + a_2(x - c)^2 + \cdots$$

is called a **power series in $(x - c)$** or a **power series centered at c** or a **power series about c.** Notice that in writing out the term corresponding to $n = 0$ in Equations 10.36 and 10.37 we have adopted the convention that $(x - c)^0 = 1$ even when $x = c$. Notice also that when $x = c$ all of the terms are 0 for $n \geq 1$ and so the power series (10.37) always converges when $x = c$.

● **Example 1**   For what values of x is the series $\displaystyle\sum_{n=0}^{\infty} n!\, x^n$ convergent?

Solution   We use the Ratio Test. Since we have been using a_n to mean the coefficient of x^n in this section, we shall use u_n to denote the nth term of the series. Thus $u_n = n!\, x^n$ and, if $x \neq 0$, we have

$$\lim_{n \to \infty} \left| \frac{u_{n+1}}{u_n} \right| = \lim_{n \to \infty} \left| \frac{(n + 1)!\, x^{n+1}}{n!\, x^n} \right|$$

$$= \lim_{n \to \infty} (n + 1)|x| = \infty$$

By the Ratio Test, the series diverges when $x \neq 0$. Thus the given series converges only when $x = 0$.   ●

● **Example 2**   What is the domain of the Bessel function of order 0 defined by

$$J_0(x) = \sum_{n=0}^{\infty} \frac{(-1)^n x^{2n}}{2^{2n}(n!)^2}$$

Solution   Let $u_n = (-1)^n x^{2n}/2^{2n}(n!)^2$. Then

$$\left| \frac{u_{n+1}}{u_n} \right| = \left| \frac{(-1)^{n+1} x^{2(n+1)}}{2^{2(n+1)}[(n + 1)!]^2} \cdot \frac{2^{2n}(n!)^2}{(-1)^n x^{2n}} \right|$$

$$= \frac{x^2}{4(n + 1)^2} \to 0 < 1 \qquad \text{for all } x$$

Thus, by the Ratio Test, the given series converges for all values of x. In other words, the domain of the Bessel function J_0 is $(-\infty, \infty) = R$.   ●

● **Example 3**   For what values of x does the series $\sum\limits_{n=1}^{\infty} \dfrac{(x-3)^n}{n}$ converge?

Solution   Let $u_n = (x-3)^n/n$. Then

$$\left|\frac{u_{n+1}}{u_n}\right| = \left|\frac{(x-3)^{n+1}}{n+1} \cdot \frac{n}{(x-3)^n}\right|$$

$$= \frac{1}{1+\dfrac{1}{n}}|x-3| \to |x-3| \qquad \text{as } n \to \infty$$

By the Ratio Test, the given series is absolutely convergent, and therefore convergent, when $|x-3| < 1$ and divergent when $|x-3| > 1$. Now

$$|x-3| < 1 \quad \Leftrightarrow \quad -1 < x-3 < 1 \quad \Leftrightarrow \quad 2 < x < 4$$

so the series converges when $2 < x < 4$ and diverges when $x < 2$ or $x > 4$.

The Ratio Test gives no information when $|x-3| = 1$ so we must consider $x = 2$ and $x = 4$ separately. If we put $x = 4$ in the series, it becomes $\sum 1/n$, the harmonic series, which is divergent. If $x = 2$, the series is $\sum(-1)^n/n$, which converges by the Alternating Series Test. Thus the given power series converges for $2 \leqslant x < 4$.   ●

For the power series that we have looked at so far, the set of values of x for which the series is convergent has always turned out to be an interval [a finite interval for the geometric series and in Example 3, the infinite interval $(-\infty, \infty)$ in Example 2, and a collapsed interval $[0,0] = \{0\}$ in Example 1]. The following theorem, proved in Appendix C, says that this is true in general.

Theorem (10.38)

> For a given power series $\sum_{n=0}^{\infty} a_n(x-c)^n$ there are only three possibilities:
>
> (a)   The series converges only when $x = c$.
>
> (b)   The series converges for all x.
>
> (c)   There is a positive number R such that the series converges if $|x-c| < R$ and diverges if $|x-c| > R$.

The number R in case (c) is called the **radius of convergence** of the power series. By convention, the radius of convergence is $R = 0$ in case (a) and $R = \infty$ in case (b). The **interval of convergence** of a power series is the interval that consists of all values of x for which the series converges. In case (a) the interval consists of just a single point c. In case (b) the interval is $(-\infty, \infty)$. In case (c) note that the inequality $|x - c| < R$ can be rewritten as $c - R < x < c + R$. When x is an endpoint of the interval, that is, $x = c \pm R$, anything can happen—the series might converge at one or both endpoints or it might diverge at both endpoints. Thus in case (c) there are four possibilities for the interval of convergence:

$$(c - R, c + R) \quad (c - R, c + R] \quad [c - R, c + R) \quad [c - R, c + R]$$

The situation is illustrated in Figure 10.10.

Figure 10.10

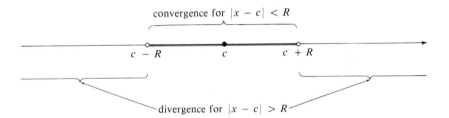

We summarize here the radius and interval of convergence for the examples already considered in this section.

	Series	Radius of convergence	Interval of convergence
Geometric series	$\displaystyle\sum_{n=0}^{\infty} x^n$	$R = 1$	$(-1, 1)$
Example 1	$\displaystyle\sum_{n=0}^{\infty} n!\, x^n$	$R = 0$	$\{0\}$
Example 2	$\displaystyle\sum_{n=0}^{\infty} \frac{(-1)^n x^{2n}}{2^{2n}(n!)^2}$	$R = \infty$	$(-\infty, \infty)$
Example 3	$\displaystyle\sum_{n=1}^{\infty} \frac{(x - 3)^n}{n}$	$R = 1$	$[2, 4)$

In general the Ratio Test (or sometimes the Root Test) should be used to determine the radius of convergence R. The Ratio and Root Tests will always fail when x is an endpoint of the interval of convergence, so the endpoints should be checked using some other test.

● **Example 4**   Find the radius of convergence and interval of convergence of the series $\displaystyle\sum_{n=0}^{\infty} \frac{(-3)^n x^n}{\sqrt{n + 1}}$.

Solution   Let $u_n = (-3)^n x^n / \sqrt{n + 1}$. Then

$$\left| \frac{u_{n+1}}{u_n} \right| = \left| \frac{(-3)^{n+1} x^{n+1}}{\sqrt{n + 2}} \cdot \frac{\sqrt{n + 1}}{(-3)^n x^n} \right|$$

$$= 3 \sqrt{\frac{1 + (1/n)}{1 + (2/n)}} \, |x| \to 3|x| \qquad \text{as } n \to \infty$$

By the Ratio Test, the given series converges if $3|x| < 1$ and diverges if $3|x| > 1$. Thus it converges if $|x| < \frac{1}{3}$ and diverges if $|x| > \frac{1}{3}$. This means that the radius of convergence is $R = \frac{1}{3}$.

We know the series converges in the interval $\left(-\frac{1}{3}, \frac{1}{3}\right)$ but we must now test for convergence at the endpoints of this interval. If $x = -\frac{1}{3}$, the series becomes

$$\sum_{n=0}^{\infty} \frac{(-3)^n \left(-\frac{1}{3}\right)^n}{\sqrt{n + 1}} = \sum_{n=0}^{\infty} \frac{1}{\sqrt{n + 1}} = \frac{1}{\sqrt{1}} + \frac{1}{\sqrt{2}} + \frac{1}{\sqrt{3}} + \frac{1}{\sqrt{4}} + \cdots$$

which diverges. (Use the Integral Test or observe that it is a p-series with $p = \frac{1}{2} < 1$.) If $x = \frac{1}{3}$, the series is

$$\sum_{n=0}^{\infty} \frac{(-3)^n \left(\frac{1}{3}\right)^n}{\sqrt{n + 1}} = \sum_{n=0}^{\infty} \frac{(-1)^n}{\sqrt{n + 1}}$$

which converges by the Alternating Series Test. Therefore the given power series converges when $-\frac{1}{3} < x \leqslant \frac{1}{3}$, so the interval of convergence is $\left(-\frac{1}{3}, \frac{1}{3}\right]$.

● **Example 5**   Find the radius of convergence and interval of convergence of the series $\displaystyle\sum_{n=0}^{\infty} \frac{n(x + 2)^n}{3^{n+1}}$.

Solution   If $u_n = n(x + 2)^n / 3^{n+1}$, then

$$\left| \frac{u_{n+1}}{u_n} \right| = \left| \frac{(n + 1)(x + 2)^{n+1}}{3^{n+2}} \cdot \frac{3^{n+1}}{n(x + 2)^n} \right|$$

$$= \left(1 + \frac{1}{n}\right) \frac{|x + 2|}{3} \to \frac{|x + 2|}{3} \qquad \text{as } n \to \infty$$

By the Ratio Test, the series converges if $|x + 2|/3 < 1$ and diverges if $|x + 2|/3 > 1$. So it converges if $|x + 2| < 3$ and diverges if $|x + 2| > 3$. Thus the radius of convergence is $R = 3$.

The inequality $|x + 2| < 3$ can be written as $-5 < x < 1$, so we test the series at the endpoints -5 and 1. When $x = -5$, the series is

$$\sum_{n=0}^{\infty} \frac{n(-3)^n}{3^{n+1}} = \frac{1}{3} \sum_{n=0}^{\infty} (-1)^n n$$

which diverges by the Test for Divergence $[(-1)^n n$ does not converge to $0]$. When $x = 1$, the series is

$$\sum_{n=0}^{\infty} \frac{n(3)^n}{3^{n+1}} = \frac{1}{3} \sum_{n=0}^{\infty} n$$

which also diverges by the Test for Divergence. Thus the series converges only when $-5 < x < 1$ so the interval of convergence is $(-5, 1)$.   ●

EXERCISES 10.8

Find the radius of convergence and interval of convergence of the series in Exercises 1–28.

1. $\displaystyle\sum_{n=0}^{\infty} \frac{x^n}{n+2}$

2. $\displaystyle\sum_{n=1}^{\infty} \frac{(-1)^n x^n}{\sqrt[3]{n}}$

3. $\displaystyle\sum_{n=0}^{\infty} n x^n$

4. $\displaystyle\sum_{n=1}^{\infty} \frac{x^n}{n^2}$

5. $\displaystyle\sum_{n=0}^{\infty} \frac{x^n}{n!}$

6. $\displaystyle\sum_{n=1}^{\infty} n^n x^n$

7. $\displaystyle\sum_{n=1}^{\infty} \frac{(-1)^n x^n}{n 2^n}$

8. $\displaystyle\sum_{n=1}^{\infty} n 5^n x^n$

9. $\displaystyle\sum_{n=0}^{\infty} \frac{3^n x^n}{(n+1)^2}$

10. $\displaystyle\sum_{n=0}^{\infty} \frac{n^2 x^n}{10^n}$

11. $\displaystyle\sum_{n=2}^{\infty} \frac{x^n}{\ln n}$

12. $\displaystyle\sum_{n=1}^{\infty} \frac{(-1)^n x^{2n-1}}{(2n-1)!}$

13. $\displaystyle\sum_{n=1}^{\infty} (-1)^n \frac{(x-1)^n}{\sqrt{n}}$

14. $\displaystyle\sum_{n=1}^{\infty} \frac{(x-4)^n}{n 5^n}$

15. $\displaystyle\sum_{n=1}^{\infty} \frac{(x-2)^n}{n^n}$

16. $\displaystyle\sum_{n=0}^{\infty} \frac{(-3)^n (x-1)^n}{\sqrt{n+1}}$

17. $\displaystyle\sum_{n=0}^{\infty} \frac{2^n (x-3)^n}{n+3}$

18. $\displaystyle\sum_{n=1}^{\infty} \frac{(x+1)^n}{n(n+1)}$

19. $\displaystyle\sum_{n=0}^{\infty} \frac{n}{(n^2+1)4^n} (x+10)^n$

20. $\displaystyle\sum_{n=0}^{\infty} \frac{n!}{10^n} (x-\pi)^n$

21. $\displaystyle\sum_{n=1}^{\infty} \left(\frac{n}{2}\right)^n (x+6)^n$

22. $\displaystyle\sum_{n=1}^{\infty} \frac{n x^n}{1 \cdot 3 \cdot 5 \cdot \cdots \cdot (2n-1)}$

23. $\displaystyle\sum_{n=1}^{\infty} \frac{(2x-1)^n}{n^3}$

24. $\displaystyle\sum_{n=2}^{\infty} (-1)^n \frac{(2x+3)^n}{n \ln n}$

25. $\displaystyle\sum_{n=0}^{\infty} \frac{n}{\sqrt{n+1}} (x-e)^n$

26. $\displaystyle\sum_{n=2}^{\infty} \frac{x^n}{(\ln n)^n}$

27. $\displaystyle\sum_{n=1}^{\infty} \frac{n! \, x^n}{(2n)!}$

28. $\displaystyle\sum_{n=1}^{\infty} \frac{2 \cdot 4 \cdot 6 \cdot \cdots \cdot (2n)}{1 \cdot 3 \cdot 5 \cdot \cdots \cdot (2n-1)} x^n$

Find the domains of the functions in Exercises 29 and 30.

29. $\displaystyle J_1(x) = \sum_{n=0}^{\infty} \frac{(-1)^n x^{2n+1}}{n!(n+1)! \, 2^{2n+1}}$

30. $\displaystyle A(x) = 1 + \frac{x^3}{2 \cdot 3} + \frac{x^6}{2 \cdot 3 \cdot 5 \cdot 6}$
$\displaystyle \qquad + \frac{x^9}{2 \cdot 3 \cdot 5 \cdot 6 \cdot 8 \cdot 9} + \cdots$

31. Show that if $\lim_{n \to \infty} \sqrt[n]{|a_n|} = a$, then the radius of convergence of the power series $\sum a_n x^n$ is $R = 1/a$.

32. Suppose that the radius of convergence of the power series $\sum a_n x^n$ is R. What is the radius of convergence of the power series $\sum a_n x^{2n}$?

SECTION 10.9

Taylor and Maclaurin Series

The sum of a power series is a function $f(x) = \sum_{n=0}^{\infty} a_n(x-c)^n$ whose domain is the interval of convergence of the series. We would like to be able to differentiate and integrate such functions and the following theorem says that we can do so by differentiating or integrating each individual term in the series, just as we would for a polynomial. This is called **term-by-term differentiation and integration.** The proof is lengthy and is therefore omitted.

Theorem (10.39)

If the power series $\sum a_n(x - c)^n$ has radius of convergence $R > 0$, then the function f defined by

$$f(x) = a_0 + a_1(x - c) + a_2(x - c)^2 + \cdots = \sum_{n=0}^{\infty} a_n(x - c)^n$$

is differentiable (and therefore continuous) on the interval $(c - R, c + R)$ and

(a) $f'(x) = a_1 + 2a_2(x - c) + 3a_3(x - c)^2 + \cdots = \sum_{n=1}^{\infty} na_n(x - c)^{n-1}$

(b) $\int f(x)\,dx = C + a_0(x - c) + a_1 \dfrac{(x - c)^2}{2} + a_2 \dfrac{(x - c)^3}{3} + \cdots$

$$= C + \sum_{n=0}^{\infty} a_n \frac{(x - c)^{n+1}}{n + 1}$$

The radii of convergence of the power series in Equations (a) and (b) are both R.

Note 1: Equations (a) and (b) can be rewritten in the form

(c) $\dfrac{d}{dx}\left[\displaystyle\sum_{n=0}^{\infty} a_n(x - c)^n\right] = \displaystyle\sum_{n=0}^{\infty} \dfrac{d}{dx}\left[a_n(x - c)^n\right]$

(d) $\displaystyle\int\left[\sum_{n=0}^{\infty} a_n(x - c)^n\right]dx = \sum_{n=0}^{\infty} \int a_n(x - c)^n\,dx$

We know that, for finite sums, the derivative of a sum is the sum of the derivatives and the integral of a sum is the sum of the integrals. Equations (c) and (d) assert that the same is true for infinite sums provided that we are dealing with *power series*. (For other types of series of functions the situation is not as simple; see Exercise 60.)

Note 2: Although Theorem 10.39 says that the radius of convergence remains the same when a power series is differentiated or integrated, that does not mean that the *interval* of convergence remains the same. It may happen that the original series converges at an endpoint, whereas the differentiated series diverges there (see Exercise 59).

● **Example 1**   In Example 2 in Section 10.8 we saw that the Bessel function

$$J_0(x) = \sum_{n=0}^{\infty} \frac{(-1)^n x^{2n}}{2^{2n}(n!)^2}$$

is defined for all x. Thus, by Theorem 10.39, J_0 is differentiable for all x and its derivative is found by term-by-term differentiation as follows:

$$J_0'(x) = \sum_{n=1}^{\infty} \frac{(-1)^n 2n x^{2n-1}}{2^{2n}(n!)^2}$$

●

● **Example 2**   Let us apply Theorem 10.39 to the geometric series

$$\frac{1}{1-x} = 1 + x + x^2 + \cdots = \sum_{n=0}^{\infty} x^n \qquad |x| < 1$$

(See Equation 10.16.) Differentiating both sides of this equation, we have

$$\frac{1}{(1-x)^2} = 1 + 2x + 3x^2 + \cdots = \sum_{n=1}^{\infty} nx^{n-1} \qquad |x| < 1$$

Integrating both sides gives

$$-\ln(1-x) = \int \frac{1}{1-x}\, dx = C + x + \frac{x^2}{2} + \frac{x^3}{3} + \cdots$$

$$= C + \sum_{n=0}^{\infty} \frac{x^{n+1}}{n+1} = C + \sum_{n=1}^{\infty} \frac{x^n}{n} \qquad |x| < 1$$

To determine the value of C we put $x = 0$ in this equation and obtain $-\ln(1 - 0) = C$. Thus $C = 0$ and

$$\ln(1-x) = -x - \frac{x^2}{2} - \frac{x^3}{3} - \cdots = -\sum_{n=1}^{\infty} \frac{x^n}{n} \qquad |x| < 1$$

In particular, putting $x = \frac{1}{2}$ and using the fact that $\ln(\frac{1}{2}) = -\ln 2$, we see that

$$\ln 2 = \frac{1}{2} + \frac{1}{8} + \frac{1}{24} + \frac{1}{64} + \cdots = \sum_{n=1}^{\infty} \frac{1}{n2^n} \qquad ●$$

Now let us suppose that f is any function that can be represented by a power series

(10.40) $f(x) = a_0 + a_1(x - c) + a_2(x - c)^2 + a_3(x - c)^3 + a_4(x - c)^4 + \cdots,$

$$|x - c| < R$$

and let us try to determine what the coefficients a_n must be in terms of f. To begin with, notice that if we put $x = c$ in Equation 10.40, then all terms after the first one are 0 and we get

$$f(c) = a_0$$

If we apply Theorem 10.39 to Equation 10.40, we obtain

(10.41) $f'(x) = a_1 + 2a_2(x - c) + 3a_3(x - c)^2 + 4a_4(x - c)^3 + \cdots, \qquad |x - c| < R$

and substitution of $x = c$ in Equation 10.41 gives

$$f'(c) = a_1$$

Now we apply Theorem 10.39 a second time, this time to Equation 10.41, and obtain

(10.42) $f''(x) = 2a_2 + 2 \cdot 3a_3(x - c) + 3 \cdot 4a_4(x - c)^2 + \cdots, \qquad |x - c| < R$

Again we put $x = c$ in Equation 10.42. The result is

$$f''(c) = 2a_2$$

Let us apply the procedure one more time. Differentiation of the series in Equation 10.42 yields

(10.43) $$f'''(x) = 2 \cdot 3a_3 + 2 \cdot 3 \cdot 4a_4(x - c) + 3 \cdot 4 \cdot 5a_5(x - c)^2 + \cdots, \qquad |x - c| < R$$

and putting $x = c$ in Equation 10.43 gives

$$f'''(c) = 2 \cdot 3a_3 = 3!\, a_3$$

By now you can see the pattern. If we continue to differentiate and put $x = c$, we will obtain

$$f^{(n)}(c) = 2 \cdot 3 \cdot 4 \cdot \cdots \cdot na_n = n!\, a_n$$

Solving this equation for the nth coefficient a_n, we get

$$a_n = \frac{f^{(n)}(c)}{n!}$$

This formula will remain valid even for $n = 0$ if we adopt the conventions that $0! = 1$ and $f^{(0)} = f$. Thus we have proved the following theorem:

Theorem (10.44)

> If f has a power series representation (expansion) at c, that is, if
>
> $$f(x) = \sum_{n=0}^{\infty} a_n(x - c)^n \qquad |x - c| < R$$
>
> then its coefficients are given by the formula
>
> $$a_n = \frac{f^{(n)}(c)}{n!}$$

Putting this formula for a_n back into the series, we see that *if f has a power series expansion at c, then it must be of the form*

(10.45)

$$f(x) = \sum_{n=0}^{\infty} \frac{f^{(n)}(c)}{n!}(x - c)^n$$

$$= f(c) + \frac{f'(c)}{1!}(x - c) + \frac{f''(c)}{2!}(x - c)^2 + \frac{f'''(c)}{3!}(x - c)^3 + \cdots$$

The series in Equation 10.45 is called the **Taylor series of the function f at c** after the English mathematician Brook Taylor (1685–1731). For the special case where $c = 0$ the Taylor series becomes

(10.46)

$$f(x) = \sum_{n=0}^{\infty} \frac{f^{(n)}(0)}{n!} x^n = f(0) + \frac{f'(0)}{1!} x + \frac{f''(0)}{2!} x^2 + \cdots$$

and this case arises frequently enough that it is given the special name **Maclaurin series** in honor of the Scottish mathematician Colin Maclaurin (1698–1746).

Note: What we have shown is that *if f can be represented as a power series about c* (such functions are called **analytic at c**), then f is equal to its Taylor series. Theorem 10.39 shows that analytic functions are infinitely differentiable at c; that is, they have derivatives of all orders at c. However not all infinitely differentiable functions are analytic. Exercise 61 gives an example of an infinitely differentiable function that is not analytic at 0. This function is therefore not equal to its Taylor series.

● **Example 3**   Find the Taylor series of the function $f(x) = e^x$ at 0 and at 1 and the associated radii of convergence.

Solution   If $f(x) = e^x$, then $f^{(n)}(x) = e^x$, so $f^{(n)}(0) = e^0 = 1$ for all n. Therefore the Taylor series for f at 0 (that is, the Maclaurin series) is

$$e^x = \sum_{n=0}^{\infty} \frac{f^{(n)}(0)}{n!} x^n = \sum_{n=0}^{\infty} \frac{x^n}{n!} = 1 + \frac{x}{1!} + \frac{x^2}{2!} + \frac{x^3}{3!} + \cdots$$

To find the radius of convergence we let $u_n = x^n/n!$. Then

$$\left| \frac{u_{n+1}}{u_n} \right| = \left| \frac{x^{n+1}}{(n+1)!} \cdot \frac{n!}{x^n} \right| = \frac{|x|}{n+1} \rightarrow 0 < 1$$

so, by the Ratio Test, the series converges for all x and the radius of convergence is $R = \infty$.

What we have proved is that if e^x has a power series expansion at 0, then

(10.47)

$$e^x = \sum_{n=0}^{\infty} \frac{x^n}{n!} \qquad \text{for all } x$$

(In Section 10.11 we shall prove that Equation 10.47 is true without the prior assumption that e^x can be expanded as a power series.)

In particular if we put $x = 1$ in Equation 10.47, we obtain the following expression for the number e as a sum of an infinite series:

(10.48)

$$e = \sum_{n=0}^{\infty} \frac{1}{n!} = 1 + \frac{1}{1!} + \frac{1}{2!} + \frac{1}{3!} + \cdots$$

To find the Taylor series of $f(x) = e^x$ at 1 we observe that $f^{(n)}(1) = e^1 = e$ and so, putting $c = 1$ in Equation 10.45, we have

$$e^x = \sum_{n=0}^{\infty} \frac{f^{(n)}(1)}{n!} (x - 1)^n = \sum_{n=0}^{\infty} \frac{e}{n!} (x - 1)^n$$

Again it can be verified that the radius of convergence is $R = \infty$.   ●

● **Example 4**   Find the Maclaurin series for $\sin x$. For what values of x does it converge?

Solution   We arrange our computation in the following manner:

$$f(x) = \sin x \qquad\qquad f(0) = 0$$
$$f'(x) = \cos x \qquad\qquad f'(0) = 1$$
$$f''(x) = -\sin x \qquad\qquad f''(0) = 0$$
$$f'''(x) = -\cos x \qquad\qquad f'''(0) = -1$$
$$f^{(4)}(x) = \sin x \qquad\qquad f^{(4)}(0) = 0$$

Since the derivatives repeat in a cycle of four, we can write the Maclaurin series as follows:

$$\sin x = f(0) + \frac{f'(0)}{1!} x + \frac{f''(0)}{2!} x^2 + \frac{f'''(0)}{3!} x^3 + \cdots$$

$$= x - \frac{x^3}{3!} + \frac{x^5}{5!} - \frac{x^7}{7!} + \cdots = \sum_{n=0}^{\infty} (-1)^n \frac{x^{2n+1}}{(2n+1)!}$$

If $u_n = (-1)^n x^{2n+1}/(2n+1)!$, then

$$\left| \frac{u_{n+1}}{u_n} \right| = \left| \frac{(-1)^{n+1} x^{2n+3}}{(2n+3)!} \cdot \frac{(2n+1)!}{(-1)^n x^{2n+1}} \right| = \frac{x^2}{(2n+3)(2n+2)} \to 0 < 1$$

so the series converges for all x. Thus, under the assumption that the sine function has a power series expansion, we have shown that

(10.49)

$$\sin x = x - \frac{x^3}{3!} + \frac{x^5}{5!} - \frac{x^7}{7!} + \cdots$$

$$= \sum_{n=0}^{\infty} (-1)^n \frac{x^{2n+1}}{(2n+1)!} \qquad \text{for all } x$$

● **Example 5** Find the Maclaurin series for $\cos x$.

Solution We could proceed directly as in Example 4 but it is easier to use Theorem 10.39 to differentiate the Maclaurin series for $\sin x$ given by Equation 10.49:

$$\cos x = \frac{d}{dx}(\sin x) = \frac{d}{dx}\left(x - \frac{x^3}{3!} + \frac{x^5}{5!} - \frac{x^7}{7!} + \cdots\right)$$

$$= 1 - \frac{3x^2}{3!} + \frac{5x^4}{5!} - \frac{7x^6}{7!} + \cdots$$

$$= 1 - \frac{x^2}{2!} + \frac{x^4}{4!} - \frac{x^6}{6!} + \cdots$$

Since the Maclaurin series for $\sin x$ converges for all x, Theorem 10.39 tells us that the differentiated series for $\cos x$ also converges for all x. Thus

(10.50)

$$\cos x = 1 - \frac{x^2}{2!} + \frac{x^4}{4!} - \frac{x^6}{6!} + \cdots$$

$$= \sum_{n=0}^{\infty} (-1)^n \frac{x^{2n}}{(2n)!} \qquad \text{for all } x$$

● **Example 6** Find the Maclaurin series for the function $f(x) = x \cos x$.

Solution Instead of computing derivatives and substituting in Equation 10.46 it is easier to multiply the known series for $\cos x$ (Equation 10.50) by x:

$$x \cos x = x \sum_{n=0}^{\infty} (-1)^n \frac{x^{2n}}{(2n)!} = \sum_{n=0}^{\infty} (-1)^n \frac{x^{2n+1}}{(2n)!}$$

Notice that we have used Theorem 10.19 part (a).

● **Example 7** Find the Taylor series for $\ln x$ at 1.

Solution Arranging our work in columns as in Example 4, we have

$$f(x) = \ln x \qquad\qquad f(1) = \ln 1 = 0$$

$$f'(x) = x^{-1} \qquad\qquad f'(1) = 1$$

$$f''(x) = -x^{-2} \qquad\qquad f''(1) = -1$$

$$f'''(x) = 2x^{-3} \qquad\qquad f'''(1) = 2$$

$$f^{(4)}(x) = -2 \cdot 3 x^{-4} \qquad\qquad f^{(4)}(1) = -2 \cdot 3$$

$$\vdots \qquad\qquad\qquad \vdots$$

$$f^{(n)}(x) = (-1)^{n-1}(n-1)! \, x^{-n} \qquad f^{(n)}(1) = (-1)^{n-1}(n-1)!$$

So, using Equation 10.45 with $c = 1$, we see that the Taylor expansion of $\ln x$ is

$$\ln x = f(1) + \frac{f'(1)}{1!}(x - 1) + \frac{f''(1)}{2!}(x - 1)^2 + \frac{f'''(1)}{3!}(x - 1)^3 + \cdots$$

$$= (x - 1) - \frac{1}{2!}(x - 1)^2 + \frac{2!}{3!}(x - 1)^3 - \frac{3!}{4!}(x - 1)^4 + \cdots$$

$$= (x - 1) - \frac{(x - 1)^2}{2} + \frac{(x - 1)^3}{3} - \frac{(x - 1)^4}{4} + \cdots$$

$$= \sum_{n=1}^{\infty} (-1)^{n-1} \frac{(x - 1)^n}{n}$$

If $u_n = (-1)^{n-1}(x - 1)^n/n$, then

$$\left| \frac{u_{n+1}}{u_n} \right| = \left| \frac{(-1)^n(x - 1)^{n+1}}{n + 1} \cdot \frac{n}{(-1)^{n-1}(x - 1)^n} \right|$$

$$= \frac{|x - 1|}{1 + \frac{1}{n}} \rightarrow |x - 1| \qquad \text{as } n \rightarrow \infty$$

By the Ratio Test, the series converges if $|x - 1| < 1$ and diverges if $|x - 1| > 1$. So we have convergence if $0 < x < 2$. When $x = 0$ the series is the harmonic series, which diverges. When $x = 2$ the series is the alternating harmonic series, which converges (and represents $\ln 2$ by Exercise 35 in Section 10.5). Thus

$$\ln x = \sum_{n=1}^{\infty} (-1)^{n-1} \frac{(x - 1)^n}{n} \qquad 0 < x \leqslant 2 \qquad \bullet$$

● **Example 8**   Find the Maclaurin series for $f(x) = \dfrac{1}{1 + x^2}$.

Solution   We could proceed as in Examples 4 and 7 but another method is to use a geometric series. Replacing x by $-x^2$ in Equation 10.16 or Example 2, we have

$$\frac{1}{1 + x^2} = \frac{1}{1 - (-x^2)} = \sum_{n=0}^{\infty} (-x^2)^n$$

$$= \sum_{n=0}^{\infty} (-1)^n x^{2n} = 1 - x^2 + x^4 - x^6 + x^8 - \cdots$$

Because this is a geometric series, it converges when $|-x|^2 < 1$, that is, $x^2 < 1$, or $|x| < 1$. ●

● **Example 9**   Find the Maclaurin series for $f(x) = \tan^{-1} x$.

Solution   For this function it is extremely arduous to compute $f^{(n)}(0)$ directly, so we observe that $f'(x) = 1/(1 + x^2)$ and find the required series by integrating the Maclaurin series for $1/(1 + x^2)$ found in Example 8. Theorem 10.39 part (b) gives

$$\tan^{-1} x = \int \frac{1}{1 + x^2}\, dx = \int (1 - x^2 + x^4 - x^6 + \cdots)\, dx$$

$$= C + x - \frac{x^3}{3} + \frac{x^5}{5} - \frac{x^7}{7} + \cdots$$

To find C we put $x = 0$ and obtain $C = \tan^{-1}(0) = 0$. Therefore

$$\tan^{-1} x = x - \frac{x^3}{3} + \frac{x^5}{5} - \frac{x^7}{7} + \cdots = \sum_{n=0}^{\infty} (-1)^n \frac{x^{2n+1}}{2n + 1}$$

Since the radius of convergence of the series for $1/(1 + x^2)$ is 1, the radius of convergence of this Maclaurin series for $\tan^{-1} x$ is also 1.   ●

Note: In Examples 5, 6, 8, and 9 we can be sure that the series we obtained by indirect methods are indeed the Taylor or Maclaurin series of the given functions because Theorem 10.44 asserts that, no matter how a power series representation $f(x) = \sum a_n(x - c)^n$ is obtained, it is always true that $a_n = f^{(n)}(c)/n!$. In other words, the coefficients are uniquely determined.

One of the reasons that Taylor series are important is that they enable us to integrate functions that we could not previously handle. The following example illustrates this for the function $f(x) = e^{-x^2}$. (See the discussion at the end of Section 7.6.)

● **Example 10**   (a) Evaluate $\int e^{-x^2}\, dx$ as an infinite series. (b) Evaluate $\int_0^1 e^{-x^2}\, dx$ correct to within an error of 0.001.

Solution   (a) First we find the Maclaurin series for $f(x) = e^{-x^2}$. Although it is possible to use the direct method, let us find it simply by replacing x by $-x^2$ in Equation 10.47. Thus

$$e^{-x^2} = \sum_{n=0}^{\infty} \frac{(-x^2)^n}{n!} = \sum_{n=0}^{\infty} (-1)^n \frac{x^{2n}}{n!} = 1 - \frac{x^2}{1!} + \frac{x^4}{2!} - \frac{x^6}{3!} + \cdots$$

Now we integrate term by term using Theorem 10.39:

$$\int e^{-x^2}\, dx$$

$$= \int \left(1 - \frac{x^2}{1!} + \frac{x^4}{2!} - \frac{x^6}{3!} + \cdots + (-1)^n \frac{x^{2n}}{n!} + \cdots\right) dx$$

$$= C + x - \frac{x^3}{3 \cdot 1!} + \frac{x^5}{5 \cdot 2!} - \frac{x^7}{7 \cdot 3!} + \cdots + (-1)^n \frac{x^{2n+1}}{(2n + 1)n!} + \cdots$$

This series converges for all x because the original series for e^{-x^2} converges for all x.

(b)   The Fundamental Theorem of Calculus gives

$$\int_0^1 e^{-x^2}\,dx = \left[x - \frac{x^3}{3\cdot 1!} + \frac{x^5}{5\cdot 2!} - \frac{x^7}{7\cdot 3!} + \frac{x^9}{9\cdot 4!} - \cdots \right]_0^1$$

$$= 1 - \frac{1}{3} + \frac{1}{10} - \frac{1}{42} + \frac{1}{216} - \cdots$$

$$\approx 0.7475$$

Since this series is alternating, Theorem 10.28 shows that the error involved in this approximation is less than

$$\frac{1}{11\cdot 5!} = \frac{1}{1320} < 0.001$$

EXERCISES 10.9

In the following exercises, assume that all of the functions possess power series expansions.

In Exercises 1–12 use the direct method of Examples 3, 4, and 7 to find the Taylor series of f at the given value of c. Also find the associated radius of convergence.

1. $f(x) = \cos x$, $c = 0$

2. $f(x) = \sin 2x$, $c = 0$

3. $f(x) = \sin x$, $c = \dfrac{\pi}{4}$

4. $f(x) = \cos x$, $c = -\dfrac{\pi}{4}$

5. $f(x) = \dfrac{1}{(1 + x)^2}$, $c = 0$

6. $f(x) = \dfrac{x}{1 - x}$, $c = 0$

7. $f(x) = \dfrac{1}{x}$, $c = 1$

8. $f(x) = \sqrt{x}$, $c = 4$

9. $f(x) = e^x$, $c = 3$

10. $f(x) = \ln x$, $c = 2$

11. $f(x) = \sinh x$, $c = 0$

12. $f(x) = \cosh x$, $c = 0$

In Exercises 13–24 find the Maclaurin series of f and its radius of convergence by using a geometric series or by differentiating or integrating a geometric series.

13. $f(x) = \dfrac{1}{1 + x}$

14. $f(x) = \ln(1 + x)$

15. $f(x) = \dfrac{1}{(1 + x)^2}$

16. $f(x) = \dfrac{x}{1 - x}$

17. $f(x) = \dfrac{1}{1 + 4x^2}$

18. $f(x) = \tan^{-1}(2x)$

19. $f(x) = \dfrac{1}{4 + x^2}$

20. $f(x) = \dfrac{1}{x^4 + 16}$

21. $f(x) = \dfrac{1}{1 - x^2}$

22. $f(x) = \dfrac{1 + x^2}{1 - x^2}$

23. $f(x) = \ln\!\left(\dfrac{1 + x}{1 - x}\right)$

24. $f(x) = \dfrac{2}{3x + 4}$

In Exercises 25–42 use any method to find the Maclaurin series of f and its radius of convergence. In particular, you may use the known series 10.47, 10.49, and 10.50.

25. $f(x) = e^{3x}$

26. $f(x) = \sin 2x$

27. $f(x) = x^2 \cos x$

28. $f(x) = \cos(x^3)$

29. $f(x) = x \sin\!\left(\dfrac{x}{2}\right)$

30. $f(x) = xe^{-x}$

31. $f(x) = \sin^2 x$ [*Hint:* Use $\sin^2 x = \tfrac{1}{2}(1 - \cos 2x)$.]

32. $f(x) = \cos^2 x$

33. $f(x) = \begin{cases} \dfrac{\sin x}{x} & \text{if } x \neq 0 \\ 1 & \text{if } x = 0 \end{cases}$

34. $f(x) = \begin{cases} \dfrac{1 - \cos x}{x^2} & \text{if } x \neq 0 \\ \dfrac{1}{2} & \text{if } x = 0 \end{cases}$

35. $f(x) = \sqrt{1 + x}$

36. $f(x) = \dfrac{1}{\sqrt{1 + 2x}}$

37. $f(x) = \dfrac{1}{\sqrt[3]{1 - x}}$

38. $f(x) = (1 + x)^{2/3}$

39. $f(x) = (1 + x)^{-3}$

40. $f(x) = 2^x$

41. $f(x) = \ln(5 + x)$

42. $f(x) = \log_{10}(1 + x)$

43. Find the Maclaurin series for $\ln(1 + x)$ and use it to calculate $\ln 1.1$ correct to five decimal places.

44. Use the Maclaurin series for $\sin x$ to compute $\sin 3°$ correct to five decimal places.

In Exercises 45–50 evaluate the given indefinite integral as an infinite series.

45. $\displaystyle\int \sin(x^2)\,dx$

46. $\displaystyle\int \frac{\sin x}{x}\,dx$

47. $\displaystyle\int \frac{1}{1 + x^4}\,dx$

48. $\displaystyle\int e^{x^3}\,dx$

49. $\displaystyle\int \sqrt{x^3 + 1}\,dx$

50. $\displaystyle\int \frac{x}{1 + x^5}\,dx$

In Exercises 51–56 use series to approximate the given definite integral to within the indicated accuracy.

51. $\displaystyle\int_0^1 \sin(x^2)\,dx$ (three decimal places)

52. $\displaystyle\int_0^{0.5} \cos(x^2)\,dx$ (three decimal places)

53. $\displaystyle\int_0^{0.5} \frac{dx}{1 + x^6}$ (four decimal places)

54. $\displaystyle\int_0^{0.1} \frac{dx}{\sqrt{1 + x^3}}$ (error $< 10^{-8}$)

55. $\displaystyle\int_0^{0.5} x^2 e^{-x^2}\,dx$ (error < 0.001)

56. $\displaystyle\int_0^1 \cos(x^3)\,dx$ (four decimal places)

57. Show that J_0 (the Bessel function of order 0 given in Example 1) satisfies the differential equation

$$x^2 J_0''(x) + x J_0'(x) + x^2 J_0(x) = 0$$

58. The Bessel function of order 1 is defined by

$$J_1(x) = \sum_{n=0}^{\infty} \frac{(-1)^n x^{2n+1}}{n!(n + 1)!\, 2^{2n+1}}$$

(a) Show that J_1 satisfies the differential equation

$$x^2 J_1''(x) + x J_1'(x) + (x^2 - 1)J_1(x) = 0$$

(b) Show that $J_0'(x) = -J_1(x)$.

59. Let

$$f(x) = \sum_{n=1}^{\infty} \frac{x^n}{n^2}$$

Find the intervals of convergence for f, f', and f''.

60. Let $f_n(x) = (\sin nx)/n^2$. Show that the series $\sum f_n(x)$ converges for all values of x but the series of derivatives $\sum f_n'(x)$ diverges when $x = 2n\pi$, n an integer. For what values of x does the series $\sum f_n''(x)$ converge?

61. Show that the function defined in Exercise 109 in Section 6.10 is not equal to its Maclaurin series.

SECTION 10.10

The Binomial Series

You may be acquainted with the Binomial Theorem, which states that if a and b are any real numbers and k is a positive integer, then

$$(a + b)^k = a^k + ka^{k-1}b + \frac{k(k - 1)}{2!} a^{k-2}b^2 + \frac{k(k - 1)(k - 2)}{3!} a^{k-3}b^3$$

$$+ \cdots + \frac{k(k - 1)(k - 2) \cdots (k - n + 1)}{n!} a^{k-n}b^n$$

$$+ \cdots + kab^{k-1} + b^k$$

The traditional notation for the binomial coefficients is

$$\binom{k}{0} = 1 \qquad \binom{k}{n} = \frac{k(k-1)(k-2)\cdots(k-n+1)}{n!} \qquad n = 1, 2, \ldots, k$$

which enables us to write the Binomial Theorem in the abbreviated form

$$(a + b)^k = \sum_{n=0}^{k} \binom{k}{n} a^{k-n} b^n$$

In particular, if we put $a = 1$ and $b = x$, we get

(10.51)
$$(1 + x)^k = \sum_{n=0}^{k} \binom{k}{n} x^n$$

One of Newton's accomplishments was to extend the Binomial Theorem (Equation 10.51) to the case where k is no longer a positive integer. In this case the expression for $(1 + x)^k$ is no longer a finite sum; it becomes an infinite series.

Assuming that $(1 + x)^k$ can be expanded as a power series, we compute its Maclaurin series in the usual way:

$$f(x) = (1 + x)^k \qquad\qquad f(0) = 1$$
$$f'(x) = k(1 + x)^{k-1} \qquad\qquad f'(0) = k$$
$$f''(x) = k(k-1)(1 + x)^{k-2} \qquad\qquad f''(0) = k(k-1)$$
$$f'''(x) = k(k-1)(k-2)(1 + x)^{k-3} \qquad\qquad f'''(0) = k(k-1)(k-2)$$
$$\vdots \qquad\qquad\qquad\qquad \vdots$$
$$f^{(n)}(x) = k(k-1)\cdots(k-n+1)(1 + x)^{k-n} \qquad f^{(n)}(0) = k(k-1)\cdots(k-n+1)$$

$$(1 + x)^k = \sum_{n=0}^{\infty} \frac{f^{(n)}(0)}{n!} x^n = \sum_{n=0}^{\infty} \frac{k(k-1)\cdots(k-n+1)}{n!} x^n$$

If u_n is the nth term of this series, then

$$\left| \frac{u_{n+1}}{u_n} \right| = \left| \frac{k(k-1)\cdots(k-n+1)(k-n)x^{n+1}}{(n+1)!} \cdot \frac{n!}{k(k-1)\cdots(k-n+1)x^n} \right|$$

$$= \frac{|k-n|}{n+1}|x| = \frac{\left|1 - \dfrac{k}{n}\right|}{1 + \dfrac{1}{n}}|x| \to |x| \qquad \text{as } n \to \infty$$

Thus, by the Ratio Test, the binomial series converges if $|x| < 1$ and diverges if $|x| > 1$.

The Binomial Series (10.52)

> If k is any real number and $|x| < 1$, then
>
> $$(1 + x)^k = 1 + kx + \frac{k(k-1)}{2!} x^2 + \frac{k(k-1)(k-2)}{3!} x^3 + \cdots$$
>
> $$= \sum_{n=0}^{\infty} \binom{k}{n} x^n$$
>
> where $\displaystyle \binom{k}{n} = \frac{k(k-1)\cdots(k-n+1)}{n!} \; (n \geqslant 1)$ $\displaystyle \binom{k}{0} = 1$

We have proved (10.52) under the assumption that $(1 + x)^k$ has a power series expansion. For a proof without that assumption see Exercise 21.

Although the binomial series always converges when $|x| < 1$, the question of whether or not it converges at the endpoints, ± 1, depends on the value of k. It turns out that the series converges at 1 if $-1 < k \leqslant 0$ and at both endpoints if $k \geqslant 0$. Notice that if k is a positive integer and $n > k$, then the expression for $\binom{k}{n}$ contains a factor $(k - k)$, so $\binom{k}{n} = 0$ for $n > k$. This means that the series terminates and reduces to the ordinary Binomial Theorem (Equation 10.51) when k is a positive integer.

Although, as we have seen, the binomial series is just a special case of the Maclaurin series, it is a special case that occurs frequently and so it is worth remembering.

● **Example 1**   Expand $\dfrac{1}{(1 + x)^2}$ as a power series.

Solution   We use the binomial series with $k = -2$. The binomial coefficient is

$$\binom{-2}{n} = \frac{(-2)(-3)(-4)\cdots(-2-n+1)}{n!}$$

$$= \frac{(-1)^n 2 \cdot 3 \cdot 4 \cdot \cdots \cdot n(n+1)}{n!} = (-1)^n(n+1)$$

and so, when $|x| < 1$,

$$\frac{1}{(1+x)^2} = (1 + x)^{-2} = \sum_{n=0}^{\infty} \binom{-2}{n} x^n$$

$$= \sum_{n=0}^{\infty} (-1)^n(n+1)x^n \qquad ●$$

● **Example 2**   Find the Maclaurin series for the function $f(x) = 1/\sqrt{4 - x}$ and its radius of convergence.

Solution   As given, $f(x)$ is not quite of the form $(1 + x)^k$ so we rewrite it as follows:

$$\frac{1}{\sqrt{4 - x}} = \frac{1}{\sqrt{4\left(1 - \frac{x}{4}\right)}} = \frac{1}{2\sqrt{1 - \frac{x}{4}}} = \frac{1}{2}\left(1 - \frac{x}{4}\right)^{-1/2}$$

Using the binomial series with $k = -\frac{1}{2}$ and with x replaced by $-x/4$, we have

$$\frac{1}{\sqrt{4 - x}} = \frac{1}{2}\left(1 - \frac{x}{4}\right)^{-1/2} = \frac{1}{2}\sum_{n=0}^{\infty}\binom{-\frac{1}{2}}{n}\left(-\frac{x}{4}\right)^n$$

$$= \frac{1}{2}\left[1 + \left(-\frac{1}{2}\right)\left(-\frac{x}{4}\right) + \frac{(-\frac{1}{2})(-\frac{3}{2})}{2!}\left(-\frac{x}{4}\right)^2 + \frac{(-\frac{1}{2})(-\frac{3}{2})(-\frac{5}{2})}{3!}\left(-\frac{x}{4}\right)^3\right.$$

$$\left. + \cdots + \frac{(-\frac{1}{2})(-\frac{3}{2})(-\frac{5}{2})\cdots(-\frac{1}{2} - n + 1)}{n!}\left(-\frac{x}{4}\right)^n + \cdots\right]$$

$$= \frac{1}{2}\left[1 + \frac{1}{8}x + \frac{1\cdot 3}{2!\,8^2}x^2 + \frac{1\cdot 3\cdot 5}{3!\,8^3}x^3 + \cdots + \frac{1\cdot 3\cdot 5\cdot\cdots\cdot(2n - 1)}{n!\,8^n}x^n + \cdots\right]$$

We know from (10.52) that this series converges when $|-x/4| < 1$, that is, $|x| < 4$, so the radius of convergence is $R = 4$.   ●

EXERCISES 10.10

In Exercises 1–12 use the binomial series to expand the given function as a power series. State the radius of convergence.

1. $\sqrt{1 + x}$

2. $\dfrac{1}{(1 + x)^3}$

3. $\dfrac{1}{(1 + 2x)^4}$

4. $\sqrt[3]{1 + x^2}$

5. $\dfrac{x}{\sqrt{1 - x}}$

6. $\dfrac{1}{\sqrt{2 + x}}$

7. $\dfrac{1}{\sqrt[3]{8 + x}}$

8. $(4 + x)^{3/2}$

9. $\sqrt[4]{1 - x^4}$

10. $\dfrac{x^2}{\sqrt{1 - x^3}}$

11. $\left(\dfrac{x}{1 - x}\right)^5$

12. $\sqrt[5]{x - 1}$

13. (a) Use the binomial series to expand $1/\sqrt{1 - x^2}$.
  (b) Use part (a) to find the Maclaurin series for $\sin^{-1} x$.

14. (a) Use the binomial series to expand $1/\sqrt{1 + x^2}$.
  (b) Use part (a) to find the Maclaurin series for $\sinh^{-1} x$.

15. (a) Expand $1/\sqrt{1 + x}$ as a power series.
  (b) Use part (a) to estimate $1/\sqrt{1.1}$ correct to three decimal places.

16. (a) Expand $\sqrt[3]{8 + x}$ as a power series
  (b) Use part (a) to estimate $\sqrt[3]{8.2}$ correct to four decimal places.

17. (a) Expand $f(x) = x/(1 - x)^2$ as a power series.
  (b) Use part (a) to find the sum of the series

$$\sum_{n=1}^{\infty}\frac{n}{2^n}$$

18. (a) Expand $f(x) = (x + x^2)/(1 - x)^3$ as a power series.
    (b) Use part (a) to find the sum of the series

$$\sum_{n=1}^{\infty} \frac{n^2}{2^n}$$

19. (a) Use the binomial series to find the Maclaurin series of $f(x) = \sqrt{1 + x^2}$.
    (b) Use part (a) to evaluate $f^{(10)}(0)$.

20. (a) Use the binomial series to find the Maclaurin series of $f(x) = 1/\sqrt{1 + x^3}$.
    (b) Use part (a) to evaluate $f^{(9)}(0)$.

21. Use the following steps to prove (10.52).
    (a) Let $g(x) = \sum_{n=0}^{\infty} \binom{k}{n} x^n$. Differentiate this series to show that

$$g'(x) = \frac{kg(x)}{1 + x} \qquad -1 < x < 1$$

    (b) Let $h(x) = (1 + x)^{-k} g(x)$ and show that $h'(x) = 0$.
    (c) Deduce that $g(x) = (1 + x)^k$.

SECTION 10.11

Approximation by Taylor Polynomials

Recall that in Section 10.9 we were able to express some functions f as the sum of their Taylor series:

$$f(x) = \sum_{n=0}^{\infty} \frac{f^{(n)}(c)}{n!} (x - c)^n$$

As with any series, the partial sums are approximations to the total sum of the series. In the case of the Taylor series the partial sums are

(10.53)
$$T_n(x) = \sum_{i=0}^{n} \frac{f^{(i)}(c)}{i!} (x - c)^i$$

$$= f(c) + \frac{f'(c)}{1!} (x - c) + \frac{f''(c)}{2!} (x - c)^2 + \cdots + \frac{f^{(n)}(c)}{n!} (x - c)^n$$

Notice that T_n is a polynomial of degree n called the **nth-degree Taylor polynomial of f at c.** If f is the sum of its Taylor series, then $T_n(x) \to f(x)$ as $n \to \infty$ and so T_n can be used as an approximation to $f: f(x) \approx T_n(x)$. It is useful to be able to approximate a function by a polynomial because polynomials are the simplest of functions. We can often gain information about a function by looking at its Taylor polynomials T_n.

We know from Example 3 in Section 10.9 that the Taylor series for $f(x) = e^x$ at $c = 0$ is

$$1 + \frac{x}{1!} + \frac{x^2}{2!} + \frac{x^3}{3!} + \cdots + \frac{x^n}{n!} + \cdots$$

so its first three Taylor polynomials at 0 (or Maclaurin polynomials) are

$$T_1(x) = 1 + x \qquad T_2(x) = 1 + x + \frac{x^2}{2!} \qquad T_3(x) = 1 + x + \frac{x^2}{2!} + \frac{x^3}{3!}$$

The graphs of the exponential function and these three Taylor polynomials are drawn in Figure 10.11.

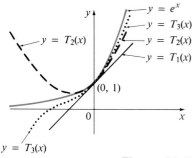

Figure 10.11

The graph of T_1 is the tangent line to $y = e^x$ at $(0, 1)$, which is the best linear approximation to e^x near $(0, 1)$. The graph of T_2 is the parabola $y = 1 + x + x^2/2$ and the graph of T_3 is the cubic curve $y = 1 + x + x^2/2 + x^3/6$, which is a closer fit to the exponential curve $y = e^x$ than T_2. The next Taylor polynomial T_4 would be an even better approximation, and so on.

In general, when using a Taylor polynomial T_n to approximate a function f, we have to ask the question: How good an approximation is it? or How large should we take n to be in order to achieve a desired accuracy? The following theorem can be used to answer these questions.

Taylor's Formula (10.54)

If f has $n + 1$ derivatives in an interval I that contains the number c, then for x in I there is a number z strictly between x and c such that

$$f(x) = f(c) + \frac{f'(c)}{1!}(x - c) + \frac{f''(c)}{2!}(x - c)^2 + \cdots$$

$$+ \frac{f^{(n)}(c)}{n!}(x - c)^n + R_n(x)$$

where $\qquad R_n(x) = \frac{f^{(n+1)}(z)}{(n + 1)!}(x - c)^{n+1}$

Note 1: For the special case $n = 0$, if we put $x = b$, $c = a$, and $z = c$, we get $f(b) = f(a) + f'(c)(b - a)$, which is the Mean Value Theorem. In fact, Theorem 10.54 can be proved by a method similar to the proof of the Mean Value Theorem. The idea is to apply Rolle's Theorem to a specially constructed function. See Exercise 47 for an outline of the steps of the proof.

Note 2: Using the notation (Equation 10.53) for Taylor polynomials, we can write Taylor's Formula as

$$f(x) = T_n(x) + R_n(x)$$

Notice that the remainder term

(10.55) $$R_n(x) = \frac{f^{(n+1)}(z)}{(n + 1)!}(x - c)^{n+1}$$

is very similar to the terms in the Taylor series except that $f^{(n+1)}$ is evaluated at z instead of at c. All we can say about the number z is that it lies somewhere between x and c. The expression for $R_n(x)$ in Equation 10.55 is known as **Lagrange's form of the remainder term.**

The **error** in the approximation $f(x) \approx T_n(x)$ is

$$|f(x) - T_n(x)| = |R_n(x)|$$

and this can often be estimated by using Equation 10.55.

Note 3:   For the special case $n = 1$, this approximation is the same as the approximation by differentials (see Section 3.9) since

$$f(x) \approx T_1(x) = f(c) + f'(c)(x - c) = f(c) + f'(c)\,dx = f(c) + dy$$

but now we have an expression for the error $|R_1(x)|$.

● **Example 1**   (a)   Write Taylor's Formula for the case where $f(x) = \ln(1 + x)$, $c = 0$, $n = 5$.
    (b)   Use part (a) to find an approximate value for $\ln(1.2)$ and to give an estimate for the error involved in this approximation.

Solution   (a)   We arrange our work in columns just as with Taylor series:

$$f(x) = \ln(1 + x) \qquad\qquad f(0) = \ln 1 = 0$$
$$f'(x) = (1 + x)^{-1} \qquad\qquad f'(0) = 1$$
$$f''(x) = -(1 + x)^{-2} \qquad\qquad f''(0) = -1$$
$$f'''(x) = 2(1 + x)^{-3} \qquad\qquad f'''(0) = 2$$
$$f^{(4)}(x) = -6(1 + x)^{-4} \qquad\qquad f^{(4)}(0) = -6$$
$$f^{(5)}(x) = 24(1 + x)^{-5} \qquad\qquad f^{(5)}(0) = 24$$
$$f^{(6)}(x) = -120(1 + x)^{-6}$$

Therefore Taylor's Formula gives

$$\ln(1 + x) = f(0) + \frac{f'(0)}{1!}x + \frac{f''(0)}{2!}x^2 + \frac{f'''(0)}{3!}x^3$$

$$+ \frac{f^{(4)}(0)}{4!}x^4 + \frac{f^{(5)}(0)}{5!}x^5 + R_5(x)$$

$$= x - \frac{1}{2!}x^2 + \frac{2}{3!}x^3 - \frac{6}{4!}x^4 + \frac{24}{5!}x^5 + R_5(x)$$

$$= x - \frac{x^2}{2} + \frac{x^3}{3} - \frac{x^4}{4} + \frac{x^5}{5} + R_5(x)$$

where $$R_5(x) = \frac{f^{(6)}(z)}{6!}x^6 = -\frac{120}{(1 + z)^6}\frac{x^6}{6!} = -\frac{x^6}{6(1 + z)^6}$$

and z is a number between 0 and x.
    (b)   From part (a) we have the approximation

$$\ln(1 + x) \approx T_5(x) = x - \frac{x^2}{2} + \frac{x^3}{3} - \frac{x^4}{4} + \frac{x^5}{5}$$

In particular, with $x = 0.2$, this approximation becomes

$$\ln(1.2) \approx 0.2 - \frac{(0.2)^2}{2} + \frac{(0.2)^3}{3} - \frac{(0.2)^4}{4} + \frac{(0.2)^5}{5}$$

$$\approx 0.18233067$$

The error in this approximation is

$$|R_5(0.2)| = \frac{(0.2)^6}{6(1 + z)^6} \qquad \text{where } 0 < z < 0.2$$

Since $z > 0$ we have

$$\frac{1}{1 + z} < 1 \qquad \text{so} \qquad \frac{1}{(1 + z)^6} < 1$$

and $\qquad |R_5(0.2)| = \dfrac{(0.2)^6}{6(1 + z)^6} < \dfrac{(0.2)^6}{6} = \dfrac{0.000064}{6} < 0.000011$

so the error in this approximation is less than 0.000011.   ●

● **Example 2**   (a)   Approximate the function $f(x) = \sqrt[3]{x}$ by a Taylor polynomial of degree 2 at $c = 8$.

(b)   How accurate is this approximation when $7 \leqslant x \leqslant 9$?

Solution   (a)

$$f(x) = \sqrt[3]{x} = x^{1/3} \qquad f(8) = 2$$
$$f'(x) = \tfrac{1}{3}x^{-2/3} \qquad f'(8) = \tfrac{1}{12}$$
$$f''(x) = -\tfrac{2}{9}x^{-5/3} \qquad f''(8) = -\tfrac{1}{144}$$
$$f'''(x) = \tfrac{10}{27}x^{-8/3}$$

Thus Taylor's Formula becomes

$$\sqrt[3]{x} = f(8) + \frac{f'(8)}{1!}(x - 8) + \frac{f''(8)}{2!}(x - 8)^2 + R_2(x)$$

$$= 2 + \tfrac{1}{12}(x - 8) - \tfrac{1}{288}(x - 8)^2 + R_2(x)$$

The desired approximation is

$$\sqrt[3]{x} \approx T_2(x) = 2 + \tfrac{1}{12}(x - 8) - \tfrac{1}{288}(x - 8)^2$$

(b)   The remainder term is

$$R_2(x) = \frac{f'''(z)}{3!}(x - 8)^3 = \frac{10}{27}z^{-8/3}\frac{(x - 8)^3}{3!} = \frac{5(x - 8)^3}{81z^{8/3}}$$

where z lies between 8 and x. In order to estimate the error $|R_2(x)|$ we note that if $7 \leqslant x \leqslant 9$ then $-1 \leqslant x - 8 \leqslant 1$, so $|x - 8| \leqslant 1$ and therefore

$|x - 8|^3 \leqslant 1$. Also, since $z > 7$, we have

$$z^{8/3} > 7^{8/3} > 179$$

and so

$$|R_2(x)| = \frac{5|x - 8|^3}{81 z^{8/3}} < \frac{5 \cdot 1}{81 \cdot 179} < 0.0004$$

Thus if $7 \leqslant x \leqslant 9$, the approximation in part (a) is accurate to within 0.0004.

Even without a calculator we could have made the cruder estimate

$$z^{8/3} > 7^{8/3} > 7^2 = 49$$

which gives the error estimate

$$|R_2(x)| < \frac{5 \cdot 1}{81 \cdot 49} < 0.002$$

● **Example 3**   Estimate the value of $\sqrt[4]{e}$ to within an accuracy of 0.0001.

Solution   Since $\sqrt[4]{e} = e^{1/4}$, we use Taylor's Formula with $f(x) = e^x$ and $c = 0$. Notice that this example is different from Examples 1 and 2 in that we are given the prescribed accuracy but not the value of n. So we start by writing Taylor's Formula for general n. Since $f^{(n)}(x) = e^x$ for all n, we obtain

$$e^x = 1 + \frac{x}{1!} + \frac{x^2}{2!} + \cdots + \frac{x^n}{n!} + R_n(x)$$

where

$$R_n(x) = \frac{f^{(n+1)}(z)}{(n+1)!} x^{n+1} = \frac{e^z}{(n+1)!} x^{n+1}$$

and z lies between 0 and x. Since we are approximating $e^{1/4}$, we take $x = \frac{1}{4}$. Then $0 < z < \frac{1}{4}$ so, using the fact that $e < 3$, we have

$$e^z < e^{1/4} < 3^{1/4} = \sqrt[4]{3} < 2$$

and

$$\left| R_n\left(\frac{1}{4}\right) \right| = \frac{e^z}{(n+1)!} \left(\frac{1}{4}\right)^{n+1} < \frac{2}{(n+1)! \, 4^{n+1}} = \frac{1}{2 \cdot 4^n (n+1)!}$$

For $n = 3$ we have

$$\left| R_3\left(\frac{1}{4}\right) \right| < \frac{1}{2 \cdot 4^3 \cdot 4!} = \frac{1}{3072} < 0.0004$$

This is not good enough, so we try $n = 4$ instead:

$$\left| R_4\left(\frac{1}{4}\right) \right| < \frac{1}{2 \cdot 4^4 \cdot 5!} = \frac{1}{61,440} < 0.00002$$

Therefore we choose $n = 4$ and the approximation becomes

$$\sqrt[4]{e} \approx T_4\left(\frac{1}{4}\right) = 1 + \frac{1}{4} + \frac{1}{4^2 \cdot 2!} + \frac{1}{4^3 \cdot 3!} + \frac{1}{4^4 \cdot 4!}$$

$$\approx 1.28402$$

● **Example 4**   What is the maximum possible error in using the approximation

$$\sin x \approx x - \frac{x^3}{3!} + \frac{x^5}{5!}$$

when $-0.3 \leqslant x \leqslant 0.3$? Use this approximation to find $\sin 12°$ correct to six decimal places.

Solution   If $f(x) = \sin x$, then $f^{(6)}(0) = -\sin 0 = 0$, so we can use Taylor's Formula with $n = 6$ to write

$$\sin x = x - \frac{x^3}{3!} + \frac{x^5}{5!} + R_6(x)$$

where

$$R_6(x) = \frac{f^{(7)}(z)}{7!} x^7 = -\cos z \frac{x^7}{7!}$$

If $-0.3 \leqslant x \leqslant 0.3$, then $|x| \leqslant 0.3$ so

$$|R_6(x)| = |\cos z| \frac{|x|^7}{7!} \leqslant \frac{|x|^7}{7!} \leqslant \frac{(0.3)^7}{7!} < 0.00000005$$

This means that the maximum possible error is smaller than 0.00000005.
To find $\sin 12°$ we first convert to radian measure.

$$\sin 12° = \sin\left(\frac{12\pi}{180}\right) = \sin\left(\frac{\pi}{15}\right)$$

$$\approx \frac{\pi}{15} - \left(\frac{\pi}{15}\right)^3 \frac{1}{3!} + \left(\frac{\pi}{15}\right)^5 \frac{1}{5!}$$

$$\approx 0.20791169$$

Thus, correct to six decimal places, $\sin 12° \approx 0.207912$.

If we had been asked to approximate $\sin 72°$ instead of $\sin 12°$ in Example 4, it would have been wise to use the Taylor polynomials at $c = \pi/3$ (instead of $c = 0$) because they are better approximations to $\sin x$ for values of x close to $\pi/3$. Notice that $72°$ is close to $60°$ (or $\pi/3$ radians) and the derivatives of $\sin x$ are easy to compute at $\pi/3$.
Figure 10.12 shows the graphs of the Taylor polynomial approximations

$$T_1(x) = x \qquad\qquad T_3(x) = x - \frac{x^3}{3!}$$

$$T_5(x) = x - \frac{x^3}{3!} + \frac{x^5}{5!} \qquad T_7(x) = x - \frac{x^3}{3!} + \frac{x^5}{5!} - \frac{x^7}{7!}$$

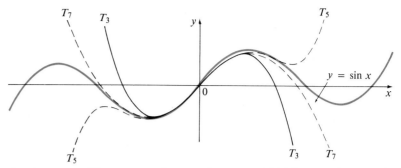

Figure 10.12

to the sine curve. You can see that as n increases, $T_n(x)$ is a good approximation to $\sin x$ on a larger and larger interval.

One use of the type of calculation done in Examples 1–4 occurs in calculators and computers. For instance, when you press the sin or e^x key on your calculator, or when a computer programmer uses a subroutine for a trigonometric or exponential or Bessel function, what happens in many machines is that a polynomial approximation is calculated. The polynomial is often a Taylor polynomial that has been modified so that the error is spread more evenly throughout an interval.

Another use of Taylor's Formula occurs when we try to show that a function is equal to its Taylor series according to the following principle.

Theorem (10.56)

If f has derivatives of all orders and, in the notation of Taylor's Formula,

$$\lim_{n \to \infty} R_n(x) = \lim_{n \to \infty} \frac{f^{(n+1)}(z)}{(n+1)!} (x - c)^{n+1} = 0$$

for $|x - c| < R$, then f is equal to its Taylor series on the interval $|x - c| < R$; that is, f is analytic at c.

Proof   If T_n is the nth Taylor polynomial of f at c, then $f(x) = T_n(x) + R_n(x)$, so

$$\lim_{n \to \infty} T_n(x) = \lim_{n \to \infty}(f(x) - R_n(x)) = f(x) - \lim_{n \to \infty} R_n(x) = f(x)$$

But $T_n(x)$ is just the nth partial sum of the Taylor series. Therefore f is the sum of its Taylor series.   ●

In applying Theorem 10.56 it is often helpful to make use of the fact that

(10.57)

$$\lim_{n \to \infty} \frac{x^n}{n!} = 0 \qquad \text{for every real number } x$$

This is true because we know from Example 3 in Section 10.9 that the series $\sum x^n/n!$ converges for all x and so its nth term approaches 0.

● **Example 5**   Prove that $\cos x = \sum\limits_{n=0}^{\infty} (-1)^n \dfrac{x^{2n}}{(2n)!}$ without assuming $\cos x$ has a power series expansion.

Solution   Using the remainder term with $c = 0$, we have

$$R_n(x) = \frac{f^{(n+1)}(z)}{(n+1)!} x^{n+1}$$

where $f(x) = \cos x$ and z lies between 0 and x. But $f^{(n+1)}(z)$ is $\pm\sin z$ or $\pm\cos z$. In any case $\left|f^{(n+1)}(z)\right| \leqslant 1$ and so

$$\left|R_n(x)\right| = \frac{\left|f^{(n+1)}(z)\right|}{(n+1)!} \left|x^{n+1}\right| \leqslant \frac{|x|^{n+1}}{(n+1)!}$$

By Equation 10.57 the right side of this inequality approaches 0 as $n \to \infty$, so $\left|R_n(x)\right| \to 0$ by the Squeeze Theorem. It follows that $R_n(x) \to 0$ as $n \to \infty$, so $\cos x$ is equal to its Taylor series (which we found in Example 5 in Section 10.9) by Theorem 10.56.   ●

● **Example 6**   Prove that e^x is equal to its Taylor series.

Solution   If $f(x) = e^x$, then $f^{(n+1)}(x) = e^x$, so the remainder term in Taylor's Formula is

$$R_n(x) = \frac{e^z}{(n+1)!} x^{n+1}$$

where z lies between 0 and x. (Note, however, that z depends on n.) If $x > 0$, then $0 < z < x$, so $e^z < e^x$. Therefore

$$0 < R_n(x) = \frac{e^z}{(n+1)!} x^{n+1} < e^x \frac{x^{n+1}}{(n+1)!} \to 0$$

by Equation 10.57, so $R_n(x) \to 0$ as $n \to \infty$ by the Squeeze Theorem. If $x < 0$, then $x < z < 0$, so $e^z < e^0 = 1$ and

$$\left|R_n(x)\right| < \frac{|x|^{n+1}}{(n+1)!} \to 0$$

Again $R_n(x) \to 0$. Thus, by Theorem 10.56, e^x is equal to its Taylor series, that is,

$$e^x = \sum_{n=0}^{\infty} \frac{x^n}{n!} \qquad \text{for all } x \qquad ●$$

EXERCISES 10.11

In Exercises 1–12 find the Taylor polynomial $T_n(x)$ for the given function f at the number c.

1. $f(x) = 1 + 2x + 3x^2 + 4x^3, c = -1, n = 4$

2. $f(x) = x^3 - 1, c = 1, n = 3$

3. $f(x) = \sin x, c = \dfrac{\pi}{6}, n = 3$

4. $f(x) = \cos x, c = \dfrac{2\pi}{3}, n = 4$

5. $f(x) = \tan x, c = 0, n = 4$

6. $f(x) = \tan x, c = \dfrac{\pi}{4}, n = 4$

7. $f(x) = e^x \sin x, c = 0, n = 3$

8. $f(x) = e^x \cos 2x, c = 0, n = 3$

9. $f(x) = \sqrt{x}, c = 9, n = 3$

10. $f(x) = \dfrac{1}{\sqrt[3]{x}}, c = 8, n = 3$

11. $f(x) = \ln \sin x, c = \dfrac{\pi}{2}, n = 3$

12. $f(x) = \sec x, c = \dfrac{\pi}{3}, n = 3$

In Exercises 13 and 14 find the Taylor polynomials $T_n(x)$ at c for the given function. Then sketch the graphs of f and these approximating polynomials.

13. $f(x) = \cos x, c = 0, n = 1, 2, 3, 4$

14. $f(x) = \dfrac{1}{x}, c = 1, n = 1, 2, 3$

In Exercises 15–26, (a) write Taylor's Formula (with remainder term) for the function f at c with the given value of n, and (b) use the remainder term to estimate the accuracy of the approximation $f(x) \approx T_n(x)$ when x lies in the given interval.

15. $f(x) = \sqrt{1 + x}, c = 0, n = 1, 0 \leqslant x \leqslant 0.1$

16. $f(x) = \dfrac{1}{x}, c = 1, n = 3, 0.8 \leqslant x \leqslant 1.2$

17. $f(x) = \sin x, c = \dfrac{\pi}{4}, n = 5, 0 \leqslant x \leqslant \dfrac{\pi}{2}$

18. $f(x) = \cos x, c = \dfrac{\pi}{3}, n = 4, 0 \leqslant x \leqslant \dfrac{2\pi}{3}$

19. $f(x) = \dfrac{1}{(1 + 2x)^4}, c = 0, n = 3, |x| \leqslant 0.1$

20. $f(x) = \sqrt[3]{1 + x^2}, c = 0, n = 2, |x| \leqslant 0.5$

21. $f(x) = \tan x, c = 0, n = 3, 0 \leqslant x \leqslant \dfrac{\pi}{6}$

22. $f(x) = \ln \cos x, c = 0, n = 3, 0 \leqslant x \leqslant \dfrac{\pi}{4}$

23. $f(x) = e^{x^2}, c = 0, n = 3, 0 \leqslant x \leqslant 0.1$

24. $f(x) = \cosh x, c = 0, n = 5, |x| \leqslant 1$

25. $f(x) = x^{3/4}, c = 16, n = 3, 15 \leqslant x \leqslant 17$

26. $f(x) = \ln x, c = 4, n = 3, 3 \leqslant x \leqslant 5$

In Exercises 27–38 use Taylor's Formula to estimate the given number with the indicated accuracy.

27. $e^{0.1}$, error < 0.00001

28. $\sqrt[3]{e}$, to four decimal places

29. $\sqrt[5]{1.1}$, to four decimal places

30. $\sqrt{4.08}$, to four decimal places

31. $\ln 1.4$, error < 0.001

32. $\arctan(0.5)$, error < 0.001

33. $\sin(0.5)$, error < 0.0001

34. $\cos(0.4)$, error < 0.0001

35. $\cos 10°$, to five decimal places

36. $\sin 15°$, to five decimal places

37. $\sin 35°$, to five decimal places

38. $\cos 69°$, to five decimal places

In Exercises 39 and 40 use the remainder term in Taylor's Formula to estimate the range of values of x for which the given approximation is accurate to within the stated error.

39. $\sin x \approx x - \dfrac{x^3}{6}$, error < 0.01

40. $\cos x \approx 1 - \dfrac{x^2}{2} + \dfrac{x^4}{24}$, error < 0.005

In Exercises 41–44 prove that the given function f has the stated Taylor series representation without assuming that f possesses a power series expansion.

41. $\sin x = \displaystyle\sum_{n=0}^{\infty} \dfrac{(-1)^n x^{2n+1}}{(2n + 1)!}$

42. $\displaystyle \sin x = \sum_{n=0}^{\infty} \frac{(-1)^{n(n-1)/2}}{\sqrt{2}} \frac{\left(x - \dfrac{\pi}{4}\right)^n}{n!}$

43. $\displaystyle \sinh x = \sum_{n=0}^{\infty} \frac{x^{2n+1}}{(2n+1)!}$ **44.** $\displaystyle \cosh x = \sum_{n=0}^{\infty} \frac{x^{2n}}{(2n)!}$

45. Show that $y = T_1(x)$ is the equation of the tangent to $y = f(x)$ at $(c, f(c))$.

46. Show that T_n and f have the same derivatives at c up to order n.

47. Use the following steps to prove Theorem 10.54. If x is fixed, let K be the number defined by the equation

$$f(x) = f(c) + \frac{f'(c)}{1!}(x - c) + \frac{f''(c)}{2!}(x - c)^2 + \cdots$$

$$+ \frac{f^{(n)}(c)}{n!}(x - c)^n + \frac{K}{(n+1)!}(x - c)^{n+1}$$

and then let g be the function defined by

$$g(t) = f(x) - f(t) - \frac{f'(t)}{1!}(x - t) - \frac{f''(t)}{2!}(x - t)^2 - \cdots$$

$$- \frac{f^{(n)}(t)}{n!}(x - t)^n - \frac{K}{(n+1)!}(x - t)^{n+1}$$

(a) Verify that $g(x) = 0$ and $g(c) = 0$.
(b) Verify that

$$g'(t) = -\frac{f^{(n+1)}(t)}{n!}(x - t)^n + \frac{K}{n!}(x - t)^n$$

(c) Apply Rolle's Theorem on the interval joining x to c to get a number z such that $K = f^{(n+1)}(z)$.

48. If f has derivatives of all orders on an interval $I = (c - R, c + R)$ and these derivatives have a common bound M ($|f^{(n)}(x)| \leqslant M$ for all x in I and all $n = 1, 2, 3, \ldots$), prove that f is analytic at c.

49. Use the following outline to prove that e is an irrational number.
(a) If e were rational, it would be of the form $e = p/q$, where p and q are positive integers and $q > 2$. Use Taylor's Formula to write

$$\frac{p}{q} = e = 1 + \frac{1}{1!} + \frac{1}{2!} + \cdots + \frac{1}{q!} + \frac{e^z}{(q+1)!}$$

$$= s_q + \frac{e^z}{(q+1)!}$$

where $0 < z < 1$.
(b) Show that $q!(e - s_q)$ is an integer.
(c) Show that $0 < q!\,(e - s_q) < 1$.
(d) Use parts (b) and (c) to deduce that e is irrational.

REVIEW OF CHAPTER 10

Define, state, or discuss the following:

1. Sequence
2. Limit of a sequence
3. Convergent sequence; divergent sequence
4. Increasing, decreasing, and monotonic sequences
5. Bounded sequence
6. Completeness Axiom
7. Convergence of bounded, monotonic sequences
8. Series
9. Convergent series; divergent series
10. Sum of a series
11. Geometric series
12. Harmonic series
13. Test for Divergence
14. Integral Test
15. The convergence of a p-series
16. Comparison Test
17. Limit Comparison Test
18. Alternating Series Test
19. Absolute convergence
20. Conditional convergence
21. Relation between convergence and absolute convergence
22. Ratio Test
23. Root Test
24. Power series
25. Radius of convergence
26. Interval of convergence
27. Differentiation and integration of power series
28. Taylor series
29. Maclaurin series
30. Maclaurin series for e^x, $\sin x$, $\cos x$
31. Binomial series
32. Taylor polynomial
33. Taylor's Formula

REVIEW EXERCISES FOR CHAPTER 10

In Exercises 1–8 determine whether the sequence is convergent or divergent. If it is convergent, find the limit.

1. $a_n = \dfrac{n}{2n + 5}$

2. $a_n = 5 - (0.9)^n$

3. $a_n = 2n + 5$

4. $a_n = \dfrac{n}{\ln n}$

5. $a_n = \sin n$

6. $a_n = \dfrac{\sin n}{n}$

7. $\left\{ \left(1 + \dfrac{3}{n}\right)^{4n} \right\}$

8. $\left\{ \dfrac{(-10)^n}{n!} \right\}$

In Exercises 9–20 determine whether the series is convergent or divergent.

9. $\displaystyle\sum_{n=1}^{\infty} \dfrac{n^2}{n^3 + 1}$

10. $\displaystyle\sum_{n=1}^{\infty} \dfrac{n + n^2}{n + n^4}$

11. $\displaystyle\sum_{n=1}^{\infty} \dfrac{(-1)^n}{\sqrt[4]{n}}$

12. $\displaystyle\sum_{n=1}^{\infty} \dfrac{n^2}{3^n}$

13. $\displaystyle\sum_{n=1}^{\infty} \left(\dfrac{n}{3n + 1}\right)^n$

14. $\displaystyle\sum_{n=1}^{\infty} \sqrt{\dfrac{n - 1}{n}}$

15. $\displaystyle\sum_{n=1}^{\infty} \dfrac{\sin n}{1 + n^2}$

16. $\displaystyle\sum_{n=2}^{\infty} \dfrac{1}{n(\ln n)^2}$

17. $\displaystyle\sum_{n=1}^{\infty} \dfrac{1 \cdot 3 \cdot 5 \cdot \,\cdots\, \cdot (2n - 1)}{5^n n!}$

18. $\displaystyle\sum_{n=1}^{\infty} (-1)^{n+1} \dfrac{\ln n}{\sqrt{n}}$

19. $\displaystyle\sum_{n=1}^{\infty} \dfrac{4^n}{n3^n}$

20. $\displaystyle\sum_{n=1}^{\infty} \dfrac{\sqrt{n + 1} - \sqrt{n - 1}}{n}$

In Exercises 21 and 22 find the sum of the series.

21. $\displaystyle\sum_{n=1}^{\infty} \dfrac{2^{2n+1}}{5^n}$

22. $\displaystyle\sum_{n=1}^{\infty} \dfrac{1}{n(n + 3)}$

23. Express the repeating decimal $1.2345345345\ldots$ as a fraction.

24. For what values of x does the series $\displaystyle\sum_{n=1}^{\infty} (\ln x)^n$ converge?

25. Find the sum of the series $\displaystyle\sum_{n=1}^{\infty} \dfrac{(-1)^{n+1}}{n^5}$ correct to four decimal places.

26. (a) Show that the series $\displaystyle\sum_{n=1}^{\infty} \dfrac{n^n}{(2n)!}$ is convergent.

(b) Deduce that $\displaystyle\lim_{n \to \infty} \dfrac{n^n}{(2n)!} = 0$.

27. Prove that if the series $\displaystyle\sum_{n=1}^{\infty} a_n$ is absolutely convergent, then the series $\displaystyle\sum_{n=1}^{\infty} \left(\dfrac{n + 1}{n}\right) a_n$ is also absolutely convergent.

In Exercises 28–31 find the radius of convergence and interval of convergence of the given series.

28. $\displaystyle\sum_{n=0}^{\infty} \dfrac{(-3)^n x^{2n}}{n + 1}$

29. $\displaystyle\sum_{n=1}^{\infty} \dfrac{x^n}{3^n n^3}$

30. $\displaystyle\sum_{n=1}^{\infty} \dfrac{(x + 1)^n}{n^n}$

31. $\displaystyle\sum_{n=0}^{\infty} \dfrac{2^n (x - 3)^n}{\sqrt{n + 3}}$

32. Find the radius of convergence of the series $\displaystyle\sum_{n=1}^{\infty} \dfrac{(2n)!}{(n!)^2} x^n$.

33. Find the Taylor series of $f(x) = \sin x$ at $c = \pi/6$.

34. Find the Taylor series of $f(x) = \cos x$ at $c = \pi/3$.

In Exercises 35–42 find the Maclaurin series for f and its radius of convergence. You may use either the direct method (definition of a Maclaurin series) or known series such as geometric series, binomial series, or the Maclaurin series for e^x and $\sin x$.

35. $f(x) = \dfrac{x^2}{1 + x}$

36. $f(x) = \sqrt{1 - x^2}$

37. $f(x) = \ln(1 - x)$

38. $f(x) = xe^{2x}$

39. $f(x) = \sin(x^4)$

40. $f(x) = 10^x$

41. $f(x) = \dfrac{1}{\sqrt[4]{16 - x}}$

42. $f(x) = (1 - 3x)^{-5}$

43. Evaluate $\displaystyle\int \dfrac{e^x}{x}\, dx$ as an infinite series.

44. Use series to approximate $\int_0^1 \sqrt{1 - x^4}\, dx$ correct to three decimal places.

In Exercises 45 and 46 use Taylor's Formula to estimate the given number with an error less than 0.0001.

45. $1/\sqrt[4]{e}$

46. $\ln(0.95)$

In Exercises 47 and 48 write out Taylor's Formula (with remainder term) for f at c with the given value of n. Then use the remainder term to estimate the accuracy of the approximation $f(x) \approx T_n(x)$ when x lies in the given interval.

47. $f(x) = \sqrt{x}, c = 1, n = 3, 0.9 \leqslant x \leqslant 1.1$

48. $f(x) = \sec x, c = 0, n = 2, 0 \leqslant x \leqslant \dfrac{\pi}{6}$

49. If $f(x) = e^{x^2}$, show that

$$f^{(2n)}(0) = \frac{(2n)!}{n!}$$

50. If $f(x) = \sum_{m=0}^{\infty} a_m x^m$ has positive radius of convergence and $e^{f(x)} = \sum_{n=0}^{\infty} b_n x^n$, show that

$$nb_n = \sum_{i=1}^{n} i a_i b_{n-i} \qquad n \geqslant 1$$

11

Three-Dimensional Analytic Geometry and Vectors

The advancement and perfection of
mathematics are intimately connected
with the prosperity of the state.
Napoleon

The analytic geometry of three-dimensional space is important not only in its own right but also because it will be needed in the next chapter to study the calculus of functions of several variables. We shall use vectors in our treatment of three-dimensional analytic geometry because vectors give particularly simple descriptions of lines, planes, and curves in space. We shall also see that vector-valued functions can be used to describe the motion of objects through space.

Three-Dimensional Coordinate Systems

To locate a point in a plane, two numbers are necessary. We know that any point in the plane can be represented as an ordered pair (a, b) of real numbers, where a is the x-coordinate and b is the y-coordinate. For this reason, a plane is called two-dimensional. To locate a point in space, three numbers are required. We shall represent any point in space by an ordered triple (a, b, c) of real numbers.

In order to do this we first choose a fixed point O (the origin) and three directed lines through O that are perpendicular to each other, called the **coordinate axes** and labeled the x-axis, y-axis, and z-axis. Usually we think of the x- and y-axes as being horizontal and the z-axis as being vertical and we draw the orientation of the axes as in Figure 11.1(a).

In looking at Figure 11.1 you can think of the y- and z-axes as lying in the plane of the paper and the x-axis as coming out of the paper toward you. The direction of the z-axis is determined by the **right-hand rule:** if you curl the fingers of your right hand around the z-axis in the direction of a 90° counterclockwise rotation from the positive x-axis to the positive y-axis, then your thumb points in the positive direction of the z-axis.

The three coordinate axes determine the three coordinate planes illustrated in Figure 11.2(a). The xy-plane is the plane that contains the x- and y-axes; the yz-plane contains the y- and z-axes; the xz-plane contains the x- and z-axes. These three coordinate planes divide space into eight parts, called **octants.** The **first octant,** in the foreground, is determined by the positive axes.

Since many people have some difficulty visualizing diagrams of three-dimensional figures, you may find it helpful to do the following [see Figure 11.2(b)]. Look at any bottom corner of a room and call the corner the origin. The wall on your left is in the xz-plane, the wall on your right is in the yz-plane, and the floor is in the xy-plane. The x-axis runs along the intersection of the floor and the left wall. The y-axis runs along the intersection of the floor and the right wall. The z-axis runs up from the floor toward the ceiling

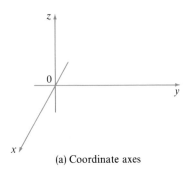

(a) Coordinate axes

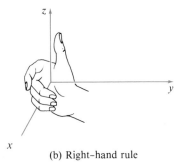

(b) Right–hand rule

Figure 11.1

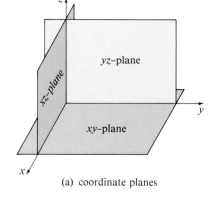

(a) coordinate planes

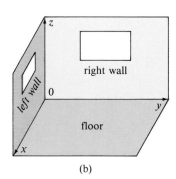

(b)

Figure 11.2

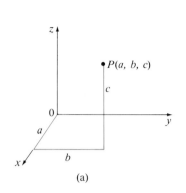

Figure 11.3

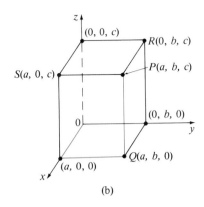

(a)   (b)

along the intersection of the two walls. You are situated in the first octant and you can now imagine seven other rooms situated in the other seven octants (three on the same floor and four on the floor below) all connected by the common corner point O.

Now if P is any point in space, let a be the (directed) distance from the yz-plane to P, let b be the distance from the xz-plane to P, and let c be the distance from the xy-plane to P. We represent the point P by the ordered triple (a, b, c) of real numbers and we call a, b, and c the **coordinates** of P; a is the x-coordinate, b is the y-coordinate, and c is the z-coordinate. Thus to locate the point (a, b, c) we can start at the origin O and move a units along the x-axis, then b units parallel to the y-axis, and then c units parallel to the z-axis as in Figure 11.3(a).

The point $P(a, b, c)$ determines a rectangular box as in Figure 11.3(b). If we drop a perpendicular from P to the xy-plane we get a point Q with coordinates $(a, b, 0)$ called the **projection** of P on the xy-plane. Similarly $R(0, b, c)$ and $S(a, 0, c)$ are the projections of P on the yz- and xz-planes, respectively.

As numerical illustrations, the points $(-4, 3, -5)$ and $(3, -2, -6)$ are plotted in Figure 11.4.

The set of all ordered triples of real numbers is the Cartesian product $R \times R \times R = \{(x, y, z) \,|\, x, y, z \in R\}$, which is denoted by R^3. We have given a one-to-one correspondence between points P in space and ordered triples (a, b, c) in R^3. It is called a **three-dimensional rectangular coordinate system.** Notice that, in terms of coordinates, the first octant can be described as the set $\{(x, y, z) \,|\, x \geqslant 0, y \geqslant 0, z \geqslant 0\}$.

In two-dimensional analytic geometry, the graph of an equation involving x and y is a curve in R^2. In three-dimensional analytic geometry, an equation in x, y, and z represents a **surface** in R^3.

● Example 1   What surfaces in R^3 are represented by the following equations?   (a) $z = 3$ (b) $y = 5$

Solution   (a)   The equation $z = 3$ represents the set $\{(x, y, z) \,|\, z = 3\}$, which is the set of all points in R^3 whose z-coordinate is 3. This is the horizontal plane that is parallel to the xy-plane and three units above it.

Figure 11.4

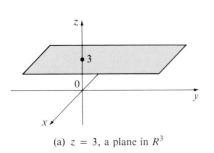

(a) $z = 3$, a plane in R^3

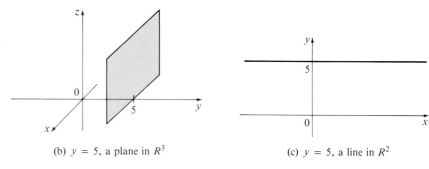

(b) $y = 5$, a plane in R^3

(c) $y = 5$, a line in R^2

Figure 11.5

(b)   The equation $y = 5$ represents the set of all points in R^3 whose y-coordinate is 5. This is the vertical plane that is parallel to the xz-plane and five units to the right of it. (See Figure 11.5.)   ●

Note: When an equation is given, it must be understood from the context whether it represents a curve in R^2 or a surface in R^3. In Example 1, $y = 5$ represents a plane in R^3, but of course $y = 5$ can also represent a line in R^2 if we are dealing with two-dimensional analytic geometry.

In general, if k is a constant, then $x = k$ represents a plane parallel to the yz-plane, $y = k$ is a plane parallel to the xz-plane, and $z = k$ is a plane parallel to the xy-plane. In Figure 11.3(b), the faces of the rectangular box are formed by the three coordinate planes $x = 0$ (the yz-plane), $y = 0$ (the xz-plane), and $z = 0$ (the xy-plane), and the planes $x = a$, $y = b$, and $z = c$.

The familiar formula for the distance between two points in a plane is easily extended to the following three-dimensional formula.

Distance Formula in Three Dimensions (11.1)

> The distance $|P_1P_2|$ between the points $P_1(x_1, y_1, z_1)$ and $P_2(x_2, y_2, z_2)$ is
>
> $$|P_1P_2| = \sqrt{(x_2 - x_1)^2 + (y_2 - y_1)^2 + (z_2 - z_1)^2}$$

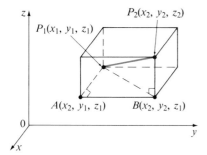

Figure 11.6

Proof   Construct a rectangular box as in Figure 11.6 where P_1 and P_2 are opposite vertices and the faces of the box are parallel to the coordinate planes. If $A(x_2, y_1, z_1)$ and $B(x_2, y_2, z_1)$ are the vertices of the box indicated in the figure, then

$$|P_1A| = |x_2 - x_1| \qquad |AB| = |y_2 - y_1| \qquad |BP_2| = |z_2 - z_1|$$

Since triangles P_1BP_2 and P_1AB are both right-angled, two applications of the Pythagorean Theorem give

$$|P_1P_2|^2 = |P_1B|^2 + |BP_2|^2 \quad \text{and} \quad |P_1B|^2 = |P_1A|^2 + |AB|^2$$

Combining these equations, we get

$$|P_1 P_2|^2 = |P_1 A|^2 + |AB|^2 + |BP_2|^2$$
$$= |x_2 - x_1|^2 + |y_2 - y_1|^2 + |z_2 - z_1|^2$$
$$= (x_2 - x_1)^2 + (y_2 - y_1)^2 + (z_2 - z_1)^2$$

Therefore $|P_1 P_2| = \sqrt{(x_2 - x_1)^2 + (y_2 - y_1)^2 + (z_2 - z_1)^2}$ •

● **Example 2**   The distance from the point $P(2, -1, 7)$ to the point $Q(1, -3, 5)$ is

$$|PQ| = \sqrt{(1 - 2)^2 + (-3 + 1)^2 + (5 - 7)^2}$$
$$= \sqrt{1 + 4 + 4} = 3$$ •

● **Example 3**   Find the equation of a sphere with radius r and center $C(h, k, l)$.

Solution   By definition, the sphere is the set of all points $P(x, y, z)$ whose distance from C is r. Thus P is on the sphere if and only if $|PC| = r$. Squaring both sides, we have $|PC|^2 = r^2$ or

$$(x - h)^2 + (y - k)^2 + (z - l)^2 = r^2$$ •

This result is worth remembering:

Equation of a Sphere (11.2)

> The equation of a sphere with center $C(h, k, l)$ and radius r is
>
> $$(x - h)^2 + (y - k)^2 + (z - l)^2 = r^2$$
>
> In particular, if the center is the origin O, then the equation of the sphere is
>
> $$x^2 + y^2 + z^2 = r^2$$

● **Example 4**   Show that $x^2 + y^2 + z^2 + 4x - 6y + 2z + 6 = 0$ is the equation of a sphere and find its center and radius.

Solution   We can rewrite the given equation in the form given by (11.2) if we complete squares:

$$(x^2 + 4x + 4) + (y^2 - 6y + 9) + (z^2 + 2z + 1) = -6 + 4 + 9 + 1$$
$$(x + 2)^2 + (y - 3)^2 + (z + 1)^2 = 8$$

Comparing this equation with the standard form (11.2), we see that it is the equation of a sphere with center $(-2, 3, -1)$ and radius $\sqrt{8} = 2\sqrt{2}$.   •

EXERCISES 11.1

In Exercises 1–6 draw a rectangular box that has P and Q as opposite vertices and has its faces parallel to the coordinate planes. Then find (a) the coordinates of the other six vertices of the box and (b) the length of the diagonal of the box.

1. $P(0, 0, 0)$, $Q(2, 3, 5)$

2. $P(0, 0, 0)$, $Q(-4, -1, 2)$

3. $P(1, 1, 2)$, $Q(3, 4, 5)$

4. $P(0, -1, -2)$, $Q(5, -3, -6)$

5. $P(4, 3, 0)$, $Q(1, 6, -4)$

6. $P(-1, -1, -1)$, $Q(1, 2, 3)$

In Exercises 7–10 find the lengths of the sides of the triangle ABC and determine whether the triangle is isosceles, a right triangle, both, or neither.

7. $A(2, 1, 0)$, $B(3, 3, 4)$, $C(5, 4, 3)$

8. $A(5, 5, 1)$, $B(3, 3, 2)$, $C(1, 4, 4)$

9. $A(-2, 6, 1)$, $B(5, 4, -3)$, $C(2, -6, 4)$

10. $A(3, -4, 1)$, $B(5, -3, 0)$, $C(6, -7, 4)$

11. Determine whether or not the points $P(1, 2, 3)$, $Q(0, 3, 7)$, $R(3, 5, 11)$ are collinear.

12. Determine whether or not the points $K(0, 3, -4)$, $L(1, 2, -2)$, $M(3, 0, 1)$ are collinear.

In Exercises 13–16, find the equation of the sphere with center C and radius r.

13. $C(0, 1, -1)$, $r = 4$

14. $C(-1, 2, 4)$, $r = \frac{1}{2}$

15. $C(-6, -1, 2)$, $r = 2\sqrt{3}$

16. $C(1, 2, -3)$, $r = 7$

In Exercises 17–22 show that the given equation represents a sphere and find its center and radius.

17. $x^2 + y^2 + z^2 + 2x + 8y - 4z = 28$

18. $x^2 + y^2 + z^2 = 6x + 4y + 10z$

19. $x^2 + y^2 + z^2 + x - 2y + 6z - 2 = 0$

20. $2x^2 + 2y^2 + 2z^2 + 4y - 2z = 1$

21. $x^2 + y^2 + z^2 = x$

22. $x^2 + y^2 + z^2 + ax + by + cz + d = 0$, where $a^2 + b^2 + c^2 > 4d$

23. Find the equation of the locus of all points equidistant from the points $A(-1, 5, 3)$ and $B(6, 2, -2)$. Describe the locus.

24. Find the equation of the locus of all points equidistant from the points $R(0, 6, -5)$ and $S(-3, 4, 2)$. Describe the locus.

25. Prove that the midpoint of the line segment from $P_1(x_1, y_1, z_1)$ to $P_2(x_2, y_2, z_2)$ is

$$\left(\frac{x_1 + x_2}{2}, \frac{y_1 + y_2}{2}, \frac{z_1 + z_2}{2} \right)$$

26. Find the equation of a sphere if one of its diameters has endpoints $(2, 1, 4)$ and $(4, 3, 10)$.

27. Find the lengths of the medians of the triangle with vertices $A(1, 2, 3)$, $B(-2, 0, 5)$, $C(4, 1, 5)$.

In Exercises 28–43 describe in words the region of R^3 represented by the given equation or inequality.

28. $x = 0$

29. $x = 9$

30. $z = -8$

31. $y > 2$

32. $z \leqslant 0$

33. $x = y$

34. $y = z$

35. $x^2 + y^2 = 1$

36. $y^2 + z^2 \leqslant 4$

37. $x^2 + y^2 + z^2 > 1$

38. $1 \leqslant x^2 + y^2 + z^2 \leqslant 25$

39. $x^2 + y^2 + z^2 - 2z < 3$

40. $xy = 0$

41. $xy = 1$

42. $xyz = 0$

43. $|z| \leqslant 2$

SECTION 11.2

Vectors

The term **vector** is used by scientists to indicate a quantity (such as velocity or force) that has both magnitude and direction. A vector is often represented by an arrow or a directed line segment. The length of the arrow represents

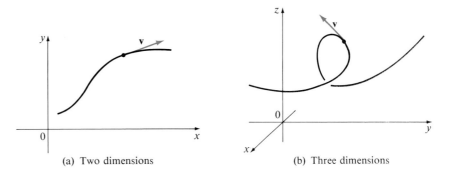

(a) Two dimensions   (b) Three dimensions

Figure 11.7

THE VELOCITY VECTOR
OF A PARTICLE

the magnitude of the vector and the arrow points in the direction of the vector. For instance, Figure 11.7(a) shows a particle moving along a path in the plane and its velocity vector **v** at a specific location of the particle. Here the length of the arrow represents the speed of the particle and it points in the direction that the particle is moving. Figure 11.7(b) shows the path of a particle moving in space. Here the velocity vector **v** is a three-dimensional vector. (This application of vectors will be studied in detail in Section 11.9.)

Notice that all of the arrows in Figure 11.8 are equivalent in the sense that they have the same length and point in the same direction even though they are in different locations. All of the directed line segments have the property that the terminal point is reached from the initial point by a displacement of three units to the right and two upward. We regard each of the directed line segments as equivalent representations of a single entity called a **vector.** In other words, we can regard a vector **v** as a set of equivalent directed line segments. These line segments are characterized by the numbers 3 and 2, and we symbolize this situation by writing $\mathbf{v} = \langle 3, 2 \rangle$. Thus a two-dimensional vector can be thought of as an ordered pair of real numbers. We shall use the notation $\langle a, b \rangle$ for the ordered pair that refers to a vector so as not to confuse it with the ordered pair (a, b) that refers to a point in the plane. A vector can be indicated by printing a letter in boldface (**v**) or by putting an arrow above it ($\vec{v}$).

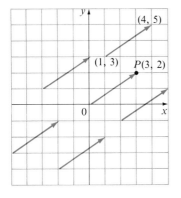

Figure 11.8

REPRESENTATIONS OF
THE VECTOR $\mathbf{v} = \langle 3, 2 \rangle$

Definition (11.3)

> A **two-dimensional vector** is an ordered pair $\mathbf{a} = \langle a_1, a_2 \rangle$ of real numbers. A **three-dimensional vector** is an ordered triple $\mathbf{a} = \langle a_1, a_2, a_3 \rangle$ of real numbers. The numbers a_1, a_2, and a_3 are called the **components** of **a**.

A **representation** of the vector $\mathbf{a} = \langle a_1, a_2 \rangle$ is any directed line segment $\overrightarrow{AB}$ from a point $A(x, y)$ to the point $B(x + a_1, y + a_2)$. A particular representation of **a** is the directed line segment $\overrightarrow{OP}$ from the origin to the point $P(a_1, a_2)$ and $\langle a_1, a_2 \rangle$ is called the **position vector** of the point $P(a_1, a_2)$. Likewise, in three dimensions, the vector $\mathbf{a} = \langle a_1, a_2, a_3 \rangle$ is the position vector of the point $P(a_1, a_2, a_3)$ (see Figure 11.9).

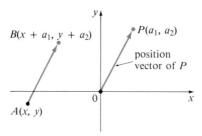

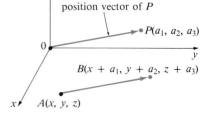

Figure 11.9

(a) Representations of $\mathbf{a} = \langle a_1, a_2 \rangle$ (b) Representations of $\mathbf{a} = \langle a_1, a_2, a_3 \rangle$

Observe that if $\mathbf{a} = \langle a_1, a_2, a_3 \rangle$ is a vector that has the representation $\overrightarrow{AB}$, where the initial point is $A(x_1, y_1, z_1)$ and the terminal point is $B(x_2, y_2, z_2)$, then we must have $x_1 + a_1 = x_2$, $y_1 + a_2 = y_2$, and $z_1 + a_3 = z_2$ and so $a_1 = x_2 - x_1$, $a_2 = y_2 - y_1$, and $a_3 = z_2 - z_1$. Thus we have the following:

(11.4)

> Given the points $A(x_1, y_1, z_1)$ and $B(x_2, y_2, z_2)$, the vector $\mathbf{a}$ with representation $\overrightarrow{AB}$ is
>
> $$\mathbf{a} = \langle x_2 - x_1, y_2 - y_1, z_2 - z_1 \rangle$$

● **Example 1**   Find the vector represented by the directed line segment with initial point $A(2, -3, 4)$ and terminal point $B(-2, 1, 1)$.

Solution   By (11.4), the vector corresponding to $\overrightarrow{AB}$ is

$$\mathbf{a} = \langle -2 - 2, 1 - (-3), 1 - 4 \rangle = \langle -4, 4, -3 \rangle \qquad ●$$

The **magnitude** or **length** of a vector $\mathbf{v}$ is the length of any of its representations and is denoted by the symbol $|\mathbf{v}|$. By using the distance formula (11.1) to compute the length of a segment OP, we obtain the following:

(11.5)

> The length of the two-dimensional vector $\mathbf{a} = \langle a_1, a_2 \rangle$ is
>
> $$|\mathbf{a}| = \sqrt{a_1^2 + a_2^2}$$
>
> The length of the three-dimensional vector $\mathbf{a} = \langle a_1, a_2, a_3 \rangle$ is
>
> $$|\mathbf{a}| = \sqrt{a_1^2 + a_2^2 + a_3^2}$$

The only vector with length 0 is the **zero vector** $\mathbf{0} = \langle 0, 0 \rangle$ (or $\mathbf{0} = \langle 0, 0, 0 \rangle$). This vector is also the only vector with no direction.

According to the following definition, we add vectors by adding the corresponding components of two vectors.

Vector Addition (11.6)

If $\mathbf{a} = \langle a_1, a_2 \rangle$ and $\mathbf{b} = \langle b_1, b_2 \rangle$, then the vector $\mathbf{a} + \mathbf{b}$ is defined by

$$\mathbf{a} + \mathbf{b} = \langle a_1 + b_1, a_2 + b_2 \rangle$$

Similarly, for three-dimensional vectors,

$$\langle a_1, a_2, a_3 \rangle + \langle b_1, b_2, b_3 \rangle = \langle a_1 + b_1, a_2 + b_2, a_3 + b_3 \rangle$$

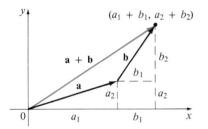

Figure 11.10

TRIANGLE LAW

Definition 11.6 is illustrated geometrically in Figure 11.10 for the two-dimensional case. You can see why the definition of vector addition is sometimes called the **Triangle Law.** Alternatively, another interpretation of vector addition is shown in Figure 11.11 and is called the **Parallelogram Law.**

It is possible to multiply a vector $\mathbf{a}$ by a real number c. (In this context we call the real number c a **scalar** to distinguish it from a vector.) For instance, we want $2\mathbf{a}$ to be the same vector as $\mathbf{a} + \mathbf{a}$, so

$$2\langle a_1, a_2 \rangle = \langle a_1, a_2 \rangle + \langle a_1, a_2 \rangle = \langle 2a_1, 2a_2 \rangle$$

In general, we multiply a vector by a scalar by multiplying each component by that scalar.

Multiplication of a Vector by a Scalar (11.7)

If c is a scalar and $\mathbf{a} = \langle a_1, a_2 \rangle$, then the vector $c\mathbf{a}$ is defined by

$$c\mathbf{a} = \langle ca_1, ca_2 \rangle$$

Similarly, for three-dimensional vectors,

$$c\langle a_1, a_2, a_3 \rangle = \langle ca_1, ca_2, ca_3 \rangle$$

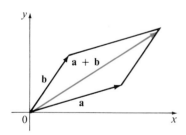

Figure 11.11

PARALLELOGRAM LAW

Definition 11.7 is illustrated by Figure 11.12.

How does the scalar multiple $c\mathbf{a}$ compare with the original vector $\mathbf{a}$? If $\mathbf{a} = \langle a_1, a_2 \rangle$, then

$$|c\mathbf{a}| = \sqrt{(ca_1)^2 + (ca_2)^2} = \sqrt{c^2(a_1^2 + a_2^2)}$$
$$= \sqrt{c^2}\sqrt{a_1^2 + a_2^2} = |c|\,|\mathbf{a}|$$

so the length of $c\mathbf{a}$ is $|c|$ times the length of $\mathbf{a}$.

If $a_1 \neq 0$, we can talk about the slope of $\mathbf{a}$ as being a_2/a_1. But, if $c \neq 0$, then the slope of $c\mathbf{a}$ is $ca_2/ca_1 = a_2/a_1$, the same as the slope of $\mathbf{a}$. Thus $c\mathbf{a}$ is parallel to $\mathbf{a}$. If $c > 0$, then a_1 and ca_1 have the same sign. Also, a_2 and ca_2 have the same sign. This means that $\mathbf{a}$ and $c\mathbf{a}$ have the same direction. On the other hand, if $c < 0$, then a_1 and ca_1 have opposite signs, as do a_2 and ca_2, so $\mathbf{a}$ and $c\mathbf{a}$ have opposite directions. In particular, the vector

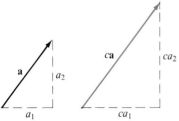

Figure 11.12

Figure 11.13

SCALAR MULTIPLES OF **a**

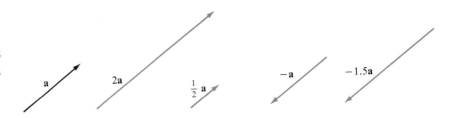

$-\mathbf{a} = (-1)a$ has the same length as **a** but points in the opposite direction. Illustrations of representations are shown in Figure 11.13.

Although we have been considering two-dimensional vectors, it is also true for three-dimensional vectors that $c\mathbf{a}$ is a vector that is $|c|$ times as long as **a** and is parallel to **a**. It has the same direction as **a** if $c > 0$ and the opposite direction if $c < 0$.

By the **difference a** $-$ **b** of two vectors, we mean

$$\mathbf{a} - \mathbf{b} = \mathbf{a} + (-\mathbf{b})$$

so if $\mathbf{a} = \langle a_1, a_2 \rangle$ and $\mathbf{b} = \langle b_1, b_2 \rangle$, then

$$\mathbf{a} - \mathbf{b} = \langle a_1 - b_1, a_2 - b_2 \rangle$$

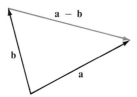

Figure 11.14

Since $(\mathbf{a} - \mathbf{b}) + \mathbf{b} = \mathbf{a}$, $\mathbf{a} - \mathbf{b}$ is the vector that, when added to **b**, gives **a**. This is illustrated in Figure 11.14 by means of the Triangle Law.

● **Example 2**   If $\mathbf{a} = \langle 4, 0, 3 \rangle$ and $\mathbf{b} = \langle -2, 1, 5 \rangle$, find $|\mathbf{a}|$ and the vectors $\mathbf{a} + \mathbf{b}$, $\mathbf{a} - \mathbf{b}$, $3\mathbf{b}$, and $2\mathbf{a} + 5\mathbf{b}$.

Solution $|\mathbf{a}| = \sqrt{4^2 + 0^2 + 3^2} = \sqrt{25} = 5$

$$\mathbf{a} + \mathbf{b} = \langle 4, 0, 3 \rangle + \langle -2, 1, 5 \rangle$$
$$= \langle 4 - 2, 0 + 1, 3 + 5 \rangle = \langle 2, 1, 8 \rangle$$
$$\mathbf{a} - \mathbf{b} = \langle 4, 0, 3 \rangle - \langle -2, 1, 5 \rangle$$
$$= \langle 4 - (-2), 0 - 1, 3 - 5 \rangle = \langle 6, -1, -2 \rangle$$
$$3\mathbf{b} = 3\langle -2, 1, 5 \rangle = \langle 3(-2), 3(1), 3(5) \rangle = \langle -6, 3, 15 \rangle$$
$$2\mathbf{a} + 5\mathbf{b} = 2\langle 4, 0, 3 \rangle + 5\langle -2, 1, 5 \rangle$$
$$= \langle 8, 0, 6 \rangle + \langle -10, 5, 25 \rangle = \langle -2, 5, 31 \rangle \qquad ●$$

We denote by V_2 the set of all two-dimensional vectors and by V_3 the set of all three-dimensional vectors. More generally, we shall later need to consider the set V_n of all n-dimensional vectors. An n-dimensional vector is an ordered n-tuple:

$$\mathbf{a} = \langle a_1, a_2, \ldots, a_n \rangle$$

where $a_1, a_2, \ldots, a_n$ are real numbers that are called the components of **a**. Addition and scalar multiplication are defined in terms of components just as for the cases $n = 2$ and 3.

Properties of Vectors (11.8)

If **a**, **b**, and **c** are vectors in V_n and c and d are scalars, then
1. $\mathbf{a} + \mathbf{b} = \mathbf{b} + \mathbf{a}$ 2. $\mathbf{a} + (\mathbf{b} + \mathbf{c}) = (\mathbf{a} + \mathbf{b}) + \mathbf{c}$
3. $\mathbf{a} + \mathbf{0} = \mathbf{a}$ 4. $\mathbf{a} + (-\mathbf{a}) = \mathbf{0}$
5. $c(\mathbf{a} + \mathbf{b}) = c\mathbf{a} + c\mathbf{b}$ 6. $(c + d)\mathbf{a} = c\mathbf{a} + d\mathbf{a}$
7. $(cd)\mathbf{a} = c(d\mathbf{a})$ 8. $1\mathbf{a} = \mathbf{a}$

The eight properties of vectors in Theorem 11.8 can be readily verified using Definitions 11.6 and 11.7. For instance, here is the verification of Property 1 for the case $n = 2$:

$$\mathbf{a} + \mathbf{b} = \langle a_1, a_2 \rangle + \langle b_1, b_2 \rangle = \langle a_1 + b_1, a_2 + b_2 \rangle$$
$$= \langle b_1 + a_1, b_2 + a_2 \rangle = \langle b_1, b_2 \rangle + \langle a_1, a_2 \rangle$$
$$= \mathbf{b} + \mathbf{a}$$

The remaining proofs are left as exercises.

Whenever we have a set of objects $V = \{\mathbf{a}, \mathbf{b}, \mathbf{c}, \ldots\}$ and a set of scalars $\{c, d, \ldots\}$ that satisfy the eight properties listed in Theorem 11.8, then V is called a **vector space**. In particular, Theorem 11.8 says that V_2 and V_3 are vector spaces.

There are three vectors in V_3 that play a special role. Let

$$\mathbf{i} = \langle 1, 0, 0 \rangle \qquad \mathbf{j} = \langle 0, 1, 0 \rangle \qquad \mathbf{k} = \langle 0, 0, 1 \rangle$$

Then **i**, **j**, and **k** are vectors that have length 1 and point in the directions of the positive x-, y-, and z-axes. Similarly, in two dimensions we define $\mathbf{i} = \langle 1, 0 \rangle$ and $\mathbf{j} = \langle 0, 1 \rangle$ (see Figure 11.15).

If $\mathbf{a} = \langle a_1, a_2, a_3 \rangle$, then we can write

$$\mathbf{a} = \langle a_1, a_2, a_3 \rangle = \langle a_1, 0, 0 \rangle + \langle 0, a_2, 0 \rangle + \langle 0, 0, a_3 \rangle$$
$$= a_1 \langle 1, 0, 0 \rangle + a_2 \langle 0, 1, 0 \rangle + a_3 \langle 0, 0, 1 \rangle$$

(11.9) $$\mathbf{a} = a_1 \mathbf{i} + a_2 \mathbf{j} + a_3 \mathbf{k}$$

Thus any vector in V_3 can be expressed in terms of the **standard basis vectors** **i**, **j**, and **k**. For instance,

$$\langle 1, -2, 6 \rangle = \mathbf{i} - 2\mathbf{j} + 6\mathbf{k}$$

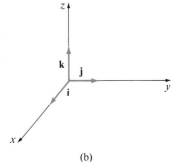

Figure 11.15

STANDARD BASIS VECTORS
IN V_2 AND V_3

(a)   (b)

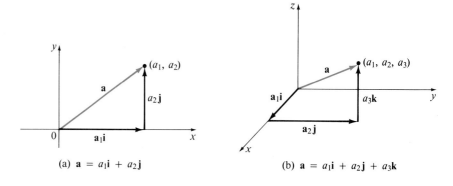

Figure 11.16

(a) $\mathbf{a} = a_1\mathbf{i} + a_2\mathbf{j}$

(b) $\mathbf{a} = a_1\mathbf{i} + a_2\mathbf{j} + a_3\mathbf{k}$

Similarly, in two dimensions, we can write

(11.10)
$$\mathbf{a} = \langle a_1, a_2 \rangle = a_1\mathbf{i} + a_2\mathbf{j}$$

See Figure 11.16 for the geometric interpretation of Equations 11.10 and 11.9 and compare with Figure 11.15.

● **Example 3**   If $\mathbf{a} = \mathbf{i} + 2\mathbf{j} - 3\mathbf{k}$ and $\mathbf{b} = 4\mathbf{i} + 7\mathbf{k}$, express the vector $2\mathbf{a} + 3\mathbf{b}$ in terms of $\mathbf{i}$, $\mathbf{j}$, and $\mathbf{k}$.

Solution   Using Properties 1, 2, 5, 6, and 7 of Theorem 11.8, we have

$$2\mathbf{a} + 3\mathbf{b} = 2(\mathbf{i} + 2\mathbf{j} - 3\mathbf{k}) + 3(4\mathbf{i} + 7\mathbf{k})$$
$$= 2\mathbf{i} + 4\mathbf{j} - 6\mathbf{k} + 12\mathbf{i} + 21\mathbf{k} = 14\mathbf{i} + 4\mathbf{j} + 15\mathbf{k} \qquad ●$$

A **unit vector** is a vector whose length is 1. For instance, $\mathbf{i}$, $\mathbf{j}$, and $\mathbf{k}$ are all unit vectors. In general, if $\mathbf{a} \neq \mathbf{0}$, then the unit vector that has the same direction as $\mathbf{a}$ is

(11.11)
$$\mathbf{u} = \frac{1}{|\mathbf{a}|}\mathbf{a} = \frac{\mathbf{a}}{|\mathbf{a}|}$$

In order to verify this, we let $c = 1/|\mathbf{a}|$. Then $\mathbf{u} = c\mathbf{a}$ and c is a positive scalar, so $\mathbf{u}$ has the same direction as $\mathbf{a}$. Also

$$|\mathbf{u}| = |c\mathbf{a}| = |c|\,|\mathbf{a}| = \frac{1}{|\mathbf{a}|}\,|\mathbf{a}| = 1$$

● **Example 4**   Find the unit vector in the direction of the vector $2\mathbf{i} - \mathbf{j} - 2\mathbf{k}$.

Solution   The given vector has length

$$|2\mathbf{i} - \mathbf{j} - 2\mathbf{k}| = \sqrt{2^2 + (-1)^2 + (-2)^2} = \sqrt{9} = 3$$

so, by Equation 11.11, the unit vector with the same direction is

$$\tfrac{1}{3}(2\mathbf{i} - \mathbf{j} - 2\mathbf{k}) = \tfrac{2}{3}\mathbf{i} - \tfrac{1}{3}\mathbf{j} - \tfrac{2}{3}\mathbf{k} \qquad ●$$

EXERCISES 11.2

In Exercises 1–6 find a vector **a** with representation given by the directed line segment $\overrightarrow{AB}$. Draw $\overrightarrow{AB}$ and the equivalent representation starting at the origin.

1. $A(1, 3)$, $B(4, 4)$ **2.** $A(-3, 4)$, $B(-1, 0)$

3. $A(3, -1)$, $B(3, -3)$ **4.** $A(4, -1)$, $B(1, 2)$

5. $A(0, 3, 1)$, $B(2, 3, -1)$ **6.** $A(1, -2, 0)$, $B(1, -2, 3)$

In Exercises 7–10 find the sum of the given vectors and illustrate geometrically.

7. $\langle 2, 3 \rangle$, $\langle 3, -4 \rangle$ **8.** $\langle -1, 2 \rangle$, $\langle 5, 3 \rangle$

9. $\langle 1, 0, 1 \rangle$, $\langle 0, 0, 1 \rangle$ **10.** $\langle 0, 3, 2 \rangle$, $\langle 1, 0, -3 \rangle$

In Exercises 11–18 find $|\mathbf{a}|$, $\mathbf{a} + \mathbf{b}$, $\mathbf{a} - \mathbf{b}$, $2\mathbf{a}$, and $3\mathbf{a} + 4\mathbf{b}$.

11. $\mathbf{a} = \langle 5, -12 \rangle$, $\mathbf{b} = \langle -2, 8 \rangle$

12. $\mathbf{a} = \langle -1, 2 \rangle$, $\mathbf{b} = \langle 4, 3 \rangle$

13. $\mathbf{a} = \langle 2, -3, 6 \rangle$, $\mathbf{b} = \langle 1, 1, 4 \rangle$

14. $\mathbf{a} = \langle 3, 2, -1 \rangle$, $\mathbf{b} = \langle 0, 6, 7 \rangle$

15. $\mathbf{a} = \mathbf{i} - \mathbf{j}$, $\mathbf{b} = \mathbf{i} + \mathbf{j}$

16. $\mathbf{a} = 2\mathbf{i} + 3\mathbf{j}$, $\mathbf{b} = 3\mathbf{i} - 2\mathbf{j}$

17. $\mathbf{a} = \mathbf{i} + \mathbf{j} + \mathbf{k}$, $\mathbf{b} = 2\mathbf{i} - \mathbf{j} + 3\mathbf{k}$

18. $\mathbf{a} = 6\mathbf{i} + \mathbf{k}$, $\mathbf{b} = \mathbf{i} - 2\mathbf{j} + 7\mathbf{k}$

In Exercises 19–24 find a unit vector that has the same direction as the given vector.

19. $\langle 1, 2 \rangle$ **20.** $\langle 3, -5 \rangle$ **21.** $\langle -2, 4, 3 \rangle$

22. $\langle 1, -4, 8 \rangle$ **23.** $\mathbf{i} + \mathbf{j}$ **24.** $2\mathbf{i} - 4\mathbf{j} + 7\mathbf{k}$

25. Prove Property 2 of Theorem 11.8 for the case $n = 2$.

26. Prove Property 5 of Theorem 11.8 for the case $n = 3$.

27. Prove Property 6 of Theorem 11.8 for the case $n = 3$.

SECTION 11.3

The Dot Product

So far we have added two vectors and multiplied a vector by a scalar. The question arises: Is it possible to multiply two vectors so that their product is a useful quantity? One such product is the dot product whose definition follows. Another is the cross product, which is discussed in the next section.

Definition (11.12)

> If $\mathbf{a} = \langle a_1, a_2, a_3 \rangle$ and $\mathbf{b} = \langle b_1, b_2, b_3 \rangle$, then the **dot product** of $\mathbf{a}$ and $\mathbf{b}$ is the number $\mathbf{a} \cdot \mathbf{b}$ given by
> $$\mathbf{a} \cdot \mathbf{b} = a_1 b_1 + a_2 b_2 + a_3 b_3$$

Thus to find the dot product of **a** and **b** we multiply corresponding components and add. The result is not a vector. It is a real number, that is, a scalar. For this reason, the dot product is sometimes called the **scalar product** (or **inner product**). Although Definition 11.12 was given for three-dimensional vectors, the dot product of two-dimensional vectors is defined in a similar fashion:

$$\langle a_1, a_2 \rangle \cdot \langle b_1, b_2 \rangle = a_1 b_1 + a_2 b_2$$

● **Example 1**

$$\langle 2,4 \rangle \cdot \langle 3, -1 \rangle = 2(3) + 4(-1) = 2$$

$$\langle -1,7,4 \rangle \cdot \langle 6,2,-\tfrac{1}{2} \rangle = (-1)(6) + 7(2) + 4(-\tfrac{1}{2}) = 6$$

$$(\mathbf{i} + 2\mathbf{j} - 3\mathbf{k}) \cdot (2\mathbf{j} - \mathbf{k}) = 1(0) + 2(2) + (-3)(-1) = 7 \qquad ●$$

The dot product obeys many of the laws that hold for ordinary products of real numbers. These are stated in the following theorem.

Properties of the Dot Product
(11.13)

If $\mathbf{a}$, $\mathbf{b}$, and $\mathbf{c}$ are vectors in V_3 and c is a scalar, then

1. $\mathbf{a} \cdot \mathbf{a} = |\mathbf{a}|^2$ 2. $\mathbf{a} \cdot \mathbf{b} = \mathbf{b} \cdot \mathbf{a}$
3. $\mathbf{a} \cdot (\mathbf{b} + \mathbf{c}) = \mathbf{a} \cdot \mathbf{b} + \mathbf{a} \cdot \mathbf{c}$ 4. $(c\mathbf{a}) \cdot \mathbf{b} = c(\mathbf{a} \cdot \mathbf{b}) = \mathbf{a} \cdot (c\mathbf{b})$
5. $\mathbf{0} \cdot \mathbf{a} = 0$

These properties are easily proved using Definition 11.12. For instance, here are the proofs of Properties 1 and 3:

1. $\mathbf{a} \cdot \mathbf{a} = a_1^2 + a_2^2 + a_3^2 = |\mathbf{a}|^2$
3. $\mathbf{a} \cdot (\mathbf{b} + \mathbf{c}) = \langle a_1, a_2, a_3 \rangle \cdot \langle b_1 + c_1, b_2 + c_2, b_3 + c_3 \rangle$
    $$= a_1(b_1 + c_1) + a_2(b_2 + c_2) + a_3(b_3 + c_3)$$
    $$= a_1 b_1 + a_1 c_1 + a_2 b_2 + a_2 c_2 + a_3 b_3 + a_3 c_3$$
    $$= (a_1 b_1 + a_2 b_2 + a_3 b_3) + (a_1 c_1 + a_2 c_2 + a_3 c_3)$$
    $$= \mathbf{a} \cdot \mathbf{b} + \mathbf{a} \cdot \mathbf{c}$$

The proofs of the remaining properties are left as exercises.

The dot product $\mathbf{a} \cdot \mathbf{b}$ can be given a geometric interpretation in terms of the **angle** θ **between a and b**, which is defined to be the angle between the representations of $\mathbf{a}$ and $\mathbf{b}$ that start at the origin, where $0 \leqslant \theta \leqslant \pi$. In other words, θ is the angle between the line segments $\overrightarrow{OA}$ and $\overrightarrow{OB}$ in Figure 11.17. Note that if $\mathbf{a}$ and $\mathbf{b}$ are parallel vectors, then $\theta = 0$ or π.

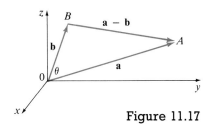

Figure 11.17

Theorem (11.14)

If θ is the angle between the vectors $\mathbf{a}$ and $\mathbf{b}$, then

$$\mathbf{a} \cdot \mathbf{b} = |\mathbf{a}| \, |\mathbf{b}| \cos \theta$$

Proof   If we apply the Law of Cosines to triangle OAB in Figure 11.17, we get

(11.15)
$$|AB|^2 = |OA|^2 + |OB|^2 - 2|OA| \, |OB| \cos \theta$$

(Observe that the Law of Cosines still applies in the limiting cases when $\theta = 0$ or π, or $\mathbf{a} = \mathbf{0}$ or $\mathbf{b} = \mathbf{0}$.) But $|OA| = |\mathbf{a}|$, $|OB| = |\mathbf{b}|$, and $|AB| = |\mathbf{a} - \mathbf{b}|$, so Equation 11.15 becomes

(11.16)
$$|\mathbf{a} - \mathbf{b}|^2 = |\mathbf{a}|^2 + |\mathbf{b}|^2 - 2|\mathbf{a}| \, |\mathbf{b}| \cos \theta$$

Using Properties 1, 2, and 3 of the dot product we can rewrite the left side of this equation as follows:

$$|\mathbf{a} - \mathbf{b}|^2 = (\mathbf{a} - \mathbf{b}) \cdot (\mathbf{a} - \mathbf{b})$$
$$= \mathbf{a} \cdot \mathbf{a} - \mathbf{a} \cdot \mathbf{b} - \mathbf{b} \cdot \mathbf{a} + \mathbf{b} \cdot \mathbf{b}$$
$$= |\mathbf{a}|^2 - 2\mathbf{a} \cdot \mathbf{b} + |\mathbf{b}|^2$$

Therefore Equation 11.16 gives

$$|\mathbf{a}|^2 - 2\mathbf{a} \cdot \mathbf{b} + |\mathbf{b}|^2 = |\mathbf{a}|^2 + |\mathbf{b}|^2 - 2\mathbf{a} \cdot \mathbf{b} \cos \theta$$

Thus
$$-2\mathbf{a} \cdot \mathbf{b} = -2|\mathbf{a}|\,|\mathbf{b}| \cos \theta$$

or
$$\mathbf{a} \cdot \mathbf{b} = |\mathbf{a}|\,|\mathbf{b}| \cos \theta$$

Corollary (11.17)

> If θ is the angle between the nonzero vectors $\mathbf{a}$ and $\mathbf{b}$, then
>
> $$\cos \theta = \frac{\mathbf{a} \cdot \mathbf{b}}{|\mathbf{a}|\,|\mathbf{b}|}$$

● **Example 2**   Find the angle between the vectors $\mathbf{a} = \langle 2, 2, -1 \rangle$ and $\mathbf{b} = \langle 5, -3, 2 \rangle$.

Solution   Since

$$|\mathbf{a}| = \sqrt{2^2 + 2^2 + (-1)^2} = 3 \quad \text{and} \quad |\mathbf{b}| = \sqrt{5^2 + (-3)^2 + 2^2} = \sqrt{38}$$

and since
$$\mathbf{a} \cdot \mathbf{b} = 2(5) + 2(-3) + (-1)(2) = 2$$

we have, from Corollary 11.17,

$$\cos \theta = \frac{\mathbf{a} \cdot \mathbf{b}}{|\mathbf{a}|\,|\mathbf{b}|} = \frac{2}{3\sqrt{38}}$$

So the angle between $\mathbf{a}$ and $\mathbf{b}$ is

$$\theta = \cos^{-1}\left(\frac{2}{3\sqrt{38}}\right) \approx 84°$$

Two nonzero vectors $\mathbf{a}$ and $\mathbf{b}$ are called **perpendicular** or **orthogonal** if the angle between them is $\theta = \pi/2$. Then Theorem 11.14 gives

$$\mathbf{a} \cdot \mathbf{b} = |\mathbf{a}|\,|\mathbf{b}| \cos\left(\frac{\pi}{2}\right) = 0$$

and conversely if $\mathbf{a} \cdot \mathbf{b} = 0$, then $\cos \theta = 0$, so $\theta = \pi/2$. The zero vector $\mathbf{0}$ is considered to be perpendicular to all vectors. Therefore

(11.18)

> $\mathbf{a}$ and $\mathbf{b}$ are orthogonal if and only if $\mathbf{a} \cdot \mathbf{b} = 0$.

● **Example 3**   Show that $2\mathbf{i} + 2\mathbf{j} - \mathbf{k}$ is perpendicular to $5\mathbf{i} - 4\mathbf{j} + 2\mathbf{k}$.

Solution   Since

$$(2\mathbf{i} + 2\mathbf{j} - \mathbf{k}) \cdot (5\mathbf{i} - 4\mathbf{j} + 2\mathbf{k}) = 2(5) + 2(-4) + (-1)(2) = 0$$

these vectors are perpendicular by (11.18).   ●

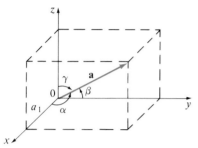

Figure 11.18

Direction Angles and Direction Cosines

The **direction angles** of a nonzero vector $\mathbf{a}$ are the angles α, β, and γ in the interval $[0, \pi]$ that $\mathbf{a}$ makes with the positive x-, y-, and z-axes (see Figure 11.18).

   The cosines of these direction angles, $\cos\alpha$, $\cos\beta$, and $\cos\gamma$, are called the **direction cosines** of the vector $\mathbf{a}$. Using Corollary 11.17 with $\mathbf{b}$ replaced by $\mathbf{i}$, we obtain

$$(11.19) \qquad \cos\alpha = \frac{\mathbf{a} \cdot \mathbf{i}}{|\mathbf{a}|\,|\mathbf{i}|} = \frac{a_1}{|\mathbf{a}|}$$

(This can also be seen directly from Figure 11.18.) Similarly, we also have

$$(11.20) \qquad \cos\beta = \frac{a_2}{|\mathbf{a}|} \qquad \cos\gamma = \frac{a_3}{|\mathbf{a}|}$$

By squaring the expressions in Equations 11.19 and 11.20 and adding, we see that

$$(11.21) \qquad \cos^2\alpha + \cos^2\beta + \cos^2\gamma = 1$$

We can also use Equations 11.19 and 11.20 to write

$$\mathbf{a} = \langle a_1, a_2, a_3 \rangle = \langle |\mathbf{a}|\cos\alpha, |\mathbf{a}|\cos\beta, |\mathbf{a}|\cos\gamma \rangle$$
$$= |\mathbf{a}|\langle \cos\alpha, \cos\beta, \cos\gamma \rangle$$

Therefore

$$(11.22) \qquad \frac{1}{|\mathbf{a}|}\mathbf{a} = \langle \cos\alpha, \cos\beta, \cos\gamma \rangle$$

which says that the direction cosines of $\mathbf{a}$ are the components of the unit vector in the direction of $\mathbf{a}$.

● **Example 4**   Find the direction angles of the vector $\mathbf{a} = \langle 1, 2, 3 \rangle$.

Solution   Since $|\mathbf{a}| = \sqrt{1^2 + 2^2 + 3^2} = \sqrt{14}$, Equations 11.19 and 11.20 give

$$\cos\alpha = \frac{1}{\sqrt{14}} \qquad \cos\beta = \frac{2}{\sqrt{14}} \qquad \cos\gamma = \frac{3}{\sqrt{14}}$$

and so

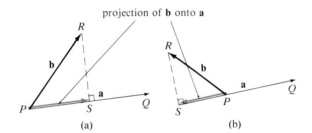

$$\alpha = \cos^{-1}\left(\frac{1}{\sqrt{14}}\right) \approx 74° \qquad \beta = \cos^{-1}\left(\frac{2}{\sqrt{14}}\right) \approx 58°$$

$$\gamma = \cos^{-1}\left(\frac{3}{\sqrt{14}}\right) \approx 37°$$

Projections

Figure 11.19

VECTOR PROJECTIONS

projection of **b** onto **a**

(a)   (b)

Figure 11.19 shows representations $\overrightarrow{PQ}$ and $\overrightarrow{PR}$ of two vectors **a** and **b** with the same initial point P. If S is the foot of the perpendicular from R to the line containing $\overrightarrow{PQ}$, then the vector with representation $\overrightarrow{PS}$ is called the **vector projection** of **b** onto **a**. The **scalar projection** of **b** onto **a** (also called the **component of b along a**) is defined to be the number $|\mathbf{b}|\cos\theta$, where θ is the angle between **a** and **b** (see Figure 11.20). Observe that it is negative if $\pi/2 < \theta \leqslant \pi$. The equation

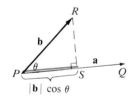

Figure 11.20

SCALAR PROJECTION

$$\mathbf{a} \cdot \mathbf{b} = |\mathbf{a}|\,|\mathbf{b}|\cos\theta = |\mathbf{a}|(|\mathbf{b}|\cos\theta)$$

shows that the dot product of **a** and **b** can be interpreted as the length of **a** times the scalar projection of **b** onto **a**. Since

$$|\mathbf{b}|\cos\theta = \frac{\mathbf{a}\cdot\mathbf{b}}{|\mathbf{a}|} = \frac{\mathbf{a}}{|\mathbf{a}|}\cdot\mathbf{b}$$

the component of **b** along **a** can be computed by taking the dot product of **b** with the unit vector in the direction of **a**.

● **Example 5**   Find the scalar projection and vector projection of $\mathbf{b} = \langle 1, 1, 2 \rangle$ onto $\mathbf{a} = \langle -2, 3, 1 \rangle$.

Solution   Since $|\mathbf{a}| = \sqrt{(-2)^2 + 3^2 + 1^2} = \sqrt{14}$, the scalar projection of **b** onto **a** is

$$|\mathbf{b}|\cos\theta = \frac{\mathbf{a}\cdot\mathbf{b}}{|\mathbf{a}|} = \frac{(-2)(1) + 3(1) + 1(2)}{\sqrt{14}} = \frac{3}{\sqrt{14}}$$

The vector projection is this scalar projection times the unit vector in the direction of **a**:

$$\frac{3}{\sqrt{14}}\frac{\mathbf{a}}{|\mathbf{a}|} = \frac{3}{14}\mathbf{a} = \left\langle -\frac{3}{7}, \frac{9}{14}, \frac{3}{14} \right\rangle$$

One use of projections occurs in physics in calculating work. In Section 8.5 we defined the work done by a constant force F in moving an object through a distance d as $W = Fd$, but this applies only when the force is directed along the line of motion of the object. Suppose, however, that the constant force is a vector $\mathbf{F} = \overrightarrow{PR}$ pointing in some other direction as in Figure 11.21. If the force moves the object from P to Q, then the **displacement vector** is $\mathbf{D} = \overrightarrow{PQ}$. The work done by this force is defined to be the product of the component of the force along $\mathbf{D}$ and the distance moved:

$$W = (|\mathbf{F}|\cos\theta)|\mathbf{D}|$$

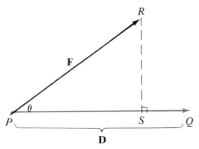

Figure 11.21

But then, from Theorem 11.14, we have

(11.23)
$$W = |\mathbf{F}||\mathbf{D}|\cos\theta = \mathbf{F}\cdot\mathbf{D}$$

Thus the work done by a constant force $\mathbf{F}$ is the dot product $\mathbf{F}\cdot\mathbf{D}$, where $\mathbf{D}$ is the displacement vector.

● **Example 6**   A force is given by a vector $\mathbf{F} = 3\mathbf{i} + 4\mathbf{j} + 5\mathbf{k}$ and moves a particle from the point $P(2,1,0)$ to the point $Q(4,6,2)$. Find the work done.

Solution   By (11.4) the displacement vector is $\mathbf{D} = \langle 2,5,2 \rangle$, so by Equation 11.23, the work done is

$$W = \mathbf{F}\cdot\mathbf{D} = \langle 3,4,5 \rangle \cdot \langle 2,5,2 \rangle$$
$$= 6 + 20 + 10 = 36$$

If the unit of length is meters and the magnitude of the force is measured in newtons, then the work done is 36 joules.   ●

EXERCISES 11.3

In Exercises 1–8 find $\mathbf{a}\cdot\mathbf{b}$.

1. $\mathbf{a} = \langle 2,5 \rangle$, $\mathbf{b} = \langle -3,1 \rangle$

2. $\mathbf{a} = \langle -2,-8 \rangle$, $\mathbf{b} = \langle 6,-4 \rangle$

3. $\mathbf{a} = \langle 4,7,-1 \rangle$, $\mathbf{b} = \langle -2,1,4 \rangle$

4. $\mathbf{a} = \langle -1,-2,-3 \rangle$, $\mathbf{b} = \langle 2,8,-6 \rangle$

5. $\mathbf{a} = \langle 1,-1,1 \rangle$, $\mathbf{b} = \langle \pi, 2\pi, 3\pi \rangle$

6. $\mathbf{a} = \langle 0.7,5,2 \rangle$, $\mathbf{b} = \langle 10, 1.2, 0.5 \rangle$

7. $\mathbf{a} = 2\mathbf{i} + 3\mathbf{j} - 4\mathbf{k}$, $\mathbf{b} = \mathbf{i} - 3\mathbf{j} + \mathbf{k}$

8. $\mathbf{a} = \mathbf{i} - \mathbf{k}, \mathbf{b} = \mathbf{i} + 2\mathbf{j}$

9. If $\mathbf{a} = \langle a_1, a_2, a_3 \rangle$, show that $\mathbf{a} \cdot \mathbf{i} = a_1, \mathbf{a} \cdot \mathbf{j} = a_2$, and $\mathbf{a} \cdot \mathbf{k} = a_3$.

10. (a) Show that $\mathbf{i} \cdot \mathbf{j} = \mathbf{j} \cdot \mathbf{k} = \mathbf{k} \cdot \mathbf{i} = 0$.
(b) Show that $\mathbf{i} \cdot \mathbf{i} = \mathbf{j} \cdot \mathbf{j} = \mathbf{k} \cdot \mathbf{k} = 1$.

In Exercises 11–16 find the angle between the given vectors. (First find an exact expression and then approximate to the nearest degree.)

11. $\mathbf{a} = \langle 1, 2, 2 \rangle, \mathbf{b} = \langle 3, 4, 0 \rangle$

12. $\mathbf{a} = \langle 6, 0, 2 \rangle, \mathbf{b} = \langle 5, 3, -2 \rangle$

13. $\mathbf{a} = \langle 1, 2 \rangle, \mathbf{b} = \langle 12, -5 \rangle$

14. $\mathbf{a} = \langle 3, 1 \rangle, \mathbf{b} = \langle 2, 4 \rangle$

15. $\mathbf{a} = 6\mathbf{i} - 2\mathbf{j} - 3\mathbf{k}, \mathbf{b} = \mathbf{i} + \mathbf{j} + \mathbf{k}$

16. $\mathbf{a} = \mathbf{i} + \mathbf{j} + 2\mathbf{k}, \mathbf{b} = 2\mathbf{j} - 3\mathbf{k}$

In Exercises 17 and 18 find, correct to the nearest degree, the three angles of the triangle with given vertices.

17. $A(1, 2, 3), B(6, 1, 5), C(-1, -2, 0)$

18. $P(0, -1, 6), Q(2, 1, -3), R(5, 4, 2)$

In Exercises 19–24 determine whether the given vectors are orthogonal, parallel, or neither.

19. $\mathbf{a} = \langle 2, -4 \rangle, \mathbf{b} = \langle -1, 2 \rangle$

20. $\mathbf{a} = \langle 2, -4 \rangle, \mathbf{b} = \langle 4, 2 \rangle$

21. $\mathbf{a} = \langle 2, 8, -3 \rangle, \mathbf{b} = \langle -1, 2, 5 \rangle$

22. $\mathbf{a} = \langle -1, 5, 2 \rangle, \mathbf{b} = \langle 4, 2, -3 \rangle$

23. $\mathbf{a} = 3\mathbf{i} + \mathbf{j} - \mathbf{k}, \mathbf{b} = \mathbf{i} - \mathbf{j} + 2\mathbf{k}$

24. $\mathbf{a} = 2\mathbf{i} + 6\mathbf{j} - 4\mathbf{k}, \mathbf{b} = -3\mathbf{i} - 9\mathbf{j} + 6\mathbf{k}$

In Exercises 25 and 26 find the values of x such that the given vectors are orthogonal.

25. $\langle x, 1, 2 \rangle, \langle 3, 4, x \rangle$ **26.** $\langle x, x, -1 \rangle, \langle 1, x, 6 \rangle$

27. Find a unit vector that is orthogonal to both $\mathbf{i} + \mathbf{j}$ and $\mathbf{i} + \mathbf{k}$.

28. For what values of c is the angle between the vectors $\langle 1, 2, 1 \rangle$ and $\langle 1, 0, c \rangle$ equal to $60°$?

In Exercises 29–33 find the direction cosines and direction angles of the given vectors. (Give the direction angles correct to the nearest degree.)

29. $\langle 1, 2, 2 \rangle$ **30.** $\langle -4, -1, 2 \rangle$

31. $-8\mathbf{i} + 3\mathbf{j} + 2\mathbf{k}$ **32.** $3\mathbf{i} + 5\mathbf{j} - 4\mathbf{k}$

33. $\langle 2, 1.2, 0.8 \rangle$

34. If a vector has direction angles $\alpha = \pi/4$ and $\beta = \pi/3$, find the third direction angle γ.

In Exercises 35–38 find the scalar and vector projections of $\mathbf{b}$ onto $\mathbf{a}$.

35. $\mathbf{a} = \langle 4, 2, 0 \rangle, \mathbf{b} = \langle 1, 1, 1 \rangle$

36. $\mathbf{a} = \langle -1, -2, 2 \rangle, \mathbf{b} = \langle 3, 3, 4 \rangle$

37. $\mathbf{a} = \mathbf{i} + \mathbf{k}, \mathbf{b} = \mathbf{i} - \mathbf{j}$

38. $\mathbf{a} = 2\mathbf{i} - 3\mathbf{j} + \mathbf{k}, \mathbf{b} = \mathbf{i} + 6\mathbf{j} - 2\mathbf{k}$

39. A constant force with vector representation $\mathbf{F} = 10\mathbf{i} + 18\mathbf{j} - 6\mathbf{k}$ moves an object along a straight line from the point $(2, 3, 0)$ to the point $(4, 9, 15)$. Find the work done if the distance is measured in meters and the magnitude of the force is measured in newtons.

40. A wagon is pulled a distance of 100 m along a horizontal path by a constant force of 50 N. The handle of the wagon is at an angle of $30°$ above the horizontal. How much work is done?

41. Show that the vector projection of a vector $\mathbf{b}$ onto a vector $\mathbf{a}$ is given by

$$\frac{\mathbf{a} \cdot \mathbf{b}}{|\mathbf{a}|^2} \mathbf{a}$$

42. Show that the vector

$$\mathbf{v} = \mathbf{b} - \frac{\mathbf{a} \cdot \mathbf{b}}{|\mathbf{a}|^2} \mathbf{a}$$

is orthogonal to $\mathbf{a}$.

43. Prove Properties 2 and 5 of the dot product (Theorem 11.13).

44. Prove Property 4 of the dot product (Theorem 11.13).

45. Use Theorem 11.14 to prove the Cauchy-Schwarz Inequality:

$$|\mathbf{a} \cdot \mathbf{b}| \leqslant |\mathbf{a}| \, |\mathbf{b}|$$

46. The Triangle Inequality for vectors is

$$|\mathbf{a} + \mathbf{b}| \leqslant |\mathbf{a}| + |\mathbf{b}|$$

(a) Give a geometric interpretation of the Triangle Inequality.
(b) Use the Cauchy-Schwarz Inequality from Exercise 45 to prove the Triangle Inequality. [*Hint:* Use the fact that $|\mathbf{a} + \mathbf{b}|^2 = (\mathbf{a} + \mathbf{b}) \cdot (\mathbf{a} + \mathbf{b})$ and use Property 3 of the dot product.]

47. The Parallelogram Law states that

$$|\mathbf{a} + \mathbf{b}|^2 + |\mathbf{a} - \mathbf{b}|^2 = 2|\mathbf{a}|^2 + 2|\mathbf{b}|^2$$

(a) Give a geometric interpretation of the Parallelogram Law.

(b) Prove the Parallelogram Law. (See the hint in Exercise 46.)

The Cross Product

The **cross product** $\mathbf{a} \times \mathbf{b}$ of two vectors $\mathbf{a}$ and $\mathbf{b}$, unlike the dot product, is a vector. For this reason it is sometimes called the **vector product.** Note that $\mathbf{a} \times \mathbf{b}$ is defined only when $\mathbf{a}$ and $\mathbf{b}$ are three-dimensional vectors.

Definition (11.24)

> If $\mathbf{a} = \langle a_1, a_2, a_3 \rangle$ and $\mathbf{b} = \langle b_1, b_2, b_3 \rangle$, then the **cross product** of $\mathbf{a}$ and $\mathbf{b}$ is the vector
>
> $$\mathbf{a} \times \mathbf{b} = \langle a_2 b_3 - a_3 b_2, a_3 b_1 - a_1 b_3, a_1 b_2 - a_2 b_1 \rangle$$

This may seem like a strange way of defining a product. The reason for the particular form of Definition 11.24 is that the cross product defined in this way has many useful properties, as we shall soon see. In particular, it turns out that the vector $\mathbf{a} \times \mathbf{b}$ is perpendicular to both $\mathbf{a}$ and $\mathbf{b}$.

In order to make Definition 11.24 easier to remember, we shall use the notation of determinants. A **determinant of order 2** is defined by

$$\begin{vmatrix} a & b \\ c & d \end{vmatrix} = ad - bc$$

For example,

$$\begin{vmatrix} 2 & 1 \\ -6 & 4 \end{vmatrix} = 2(4) - 1(-6) = 14$$

A **determinant of order 3** can be defined in terms of second-order determinants as follows:

(11.25)

$$\begin{vmatrix} a_1 & a_2 & a_3 \\ b_1 & b_2 & b_3 \\ c_1 & c_2 & c_3 \end{vmatrix} = a_1 \begin{vmatrix} b_2 & b_3 \\ c_2 & c_3 \end{vmatrix} - a_2 \begin{vmatrix} b_1 & b_3 \\ c_1 & c_3 \end{vmatrix} + a_3 \begin{vmatrix} b_1 & b_2 \\ c_1 & c_2 \end{vmatrix}$$

Observe that each term on the right side of Equation 11.25 involves a number a_i in the first row of the determinant and a_i is multiplied by the second-order determinant obtained from the left side by deleting the row and column in which a_i appears. Notice also the minus sign in the second term. For example,

$$\begin{vmatrix} 1 & 2 & -1 \\ 3 & 0 & 1 \\ -5 & 4 & 2 \end{vmatrix} = 1 \begin{vmatrix} 0 & 1 \\ 4 & 2 \end{vmatrix} - 2 \begin{vmatrix} 3 & 1 \\ -5 & 2 \end{vmatrix} + (-1) \begin{vmatrix} 3 & 0 \\ -5 & 4 \end{vmatrix}$$

$$= 1(0 - 4) - 2(6 + 5) + (-1)(12 - 0) = -38$$

If we now rewrite Definition 11.24 using second-order determinants and the standard basis vectors $\mathbf{i}$, $\mathbf{j}$, and $\mathbf{k}$, we see that the cross product of $\mathbf{a} = a_1\mathbf{i} + a_2\mathbf{j} + a_3\mathbf{k}$ and $\mathbf{b} = b_1\mathbf{i} + b_2\mathbf{j} + b_3\mathbf{k}$ is

(11.26)
$$\mathbf{a} \times \mathbf{b} = \begin{vmatrix} a_2 & a_3 \\ b_2 & b_3 \end{vmatrix} \mathbf{i} - \begin{vmatrix} a_1 & a_3 \\ b_1 & b_3 \end{vmatrix} \mathbf{j} + \begin{vmatrix} a_1 & a_2 \\ b_1 & b_2 \end{vmatrix} \mathbf{k}$$

In view of the similarity between Equations 11.25 and 11.26, we often write

(11.27)
$$\mathbf{a} \times \mathbf{b} = \begin{vmatrix} \mathbf{i} & \mathbf{j} & \mathbf{k} \\ a_1 & a_2 & a_3 \\ b_1 & b_2 & b_3 \end{vmatrix}$$

Although the first row of the symbolic determinant in Equation 11.27 consists of vectors, if we expand it as if it were an ordinary determinant using the rule in Equation 11.25, we obtain Equation 11.26. The symbolic formula in Equation 11.27 is probably the easiest way of remembering and computing cross products.

● **Example 1**   If $\mathbf{a} = \langle 1, 3, 4 \rangle$ and $\mathbf{b} = \langle 2, 7, -5 \rangle$, then

$$\mathbf{a} \times \mathbf{b} = \begin{vmatrix} \mathbf{i} & \mathbf{j} & \mathbf{k} \\ 1 & 3 & 4 \\ 2 & 7 & -5 \end{vmatrix}$$

$$= \begin{vmatrix} 3 & 4 \\ 7 & -5 \end{vmatrix} \mathbf{i} - \begin{vmatrix} 1 & 4 \\ 2 & -5 \end{vmatrix} \mathbf{j} + \begin{vmatrix} 1 & 3 \\ 2 & 7 \end{vmatrix} \mathbf{k}$$

$$= (-15 - 28)\mathbf{i} - (-5 - 8)\mathbf{j} + (7 - 6)\mathbf{k} = -43\mathbf{i} + 13\mathbf{j} + \mathbf{k} \quad ●$$

● **Example 2**   Show that $\mathbf{a} \times \mathbf{a} = \mathbf{0}$ for any vector $\mathbf{a}$ in V_3.

Solution   If $\mathbf{a} = \langle a_1, a_2, a_3 \rangle$, then

$$\mathbf{a} \times \mathbf{a} = \begin{vmatrix} \mathbf{i} & \mathbf{j} & \mathbf{k} \\ a_1 & a_2 & a_3 \\ a_1 & a_2 & a_3 \end{vmatrix}$$

$$= (a_2 a_3 - a_3 a_2)\mathbf{i} - (a_1 a_3 - a_3 a_1)\mathbf{j} + (a_1 a_2 - a_2 a_1)\mathbf{k}$$

$$= 0\mathbf{i} - 0\mathbf{j} + 0\mathbf{k} = \mathbf{0} \quad ●$$

One of the most important properties of the cross product is given by the following theorem.

Theorem (11.28)

> The vector $\mathbf{a} \times \mathbf{b}$ is orthogonal to both $\mathbf{a}$ and $\mathbf{b}$.

Proof   In order to show that $\mathbf{a} \times \mathbf{b}$ is orthogonal to $\mathbf{a}$, we compute their dot product as follows:

$$(\mathbf{a} \times \mathbf{b}) \cdot \mathbf{a} = \begin{vmatrix} a_2 & a_3 \\ b_2 & b_3 \end{vmatrix} a_1 - \begin{vmatrix} a_1 & a_3 \\ b_1 & b_3 \end{vmatrix} a_2 + \begin{vmatrix} a_1 & a_2 \\ b_1 & b_2 \end{vmatrix} a_3$$

$$= a_1(a_2 b_3 - a_3 b_2) - a_2(a_1 b_3 - a_3 b_1) + a_3(a_1 b_2 - a_2 b_1)$$

$$= a_1 a_2 b_3 - a_1 b_2 a_3 - a_1 a_2 b_3 + b_1 a_2 a_3 + a_1 b_2 a_3 - b_1 a_2 a_3$$

$$= 0$$

A similar computation shows that $(\mathbf{a} \times \mathbf{b}) \cdot \mathbf{b} = 0$. Therefore, by (11.18), $\mathbf{a} \times \mathbf{b}$ is orthogonal to both $\mathbf{a}$ and $\mathbf{b}$.   ●

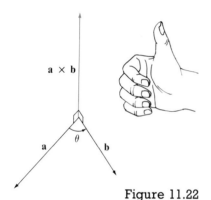

Figure 11.22

If $\mathbf{a}$ and $\mathbf{b}$ are represented by directed line segments with the same initial point (as in Figure 11.22), then Theorem 11.28 says that the cross product $\mathbf{a} \times \mathbf{b}$ points in a direction perpendicular to the plane through $\mathbf{a}$ and $\mathbf{b}$. It turns out that the direction of $\mathbf{a} \times \mathbf{b}$ is given by the right-hand rule: if the fingers of your right hand curl in the direction of a rotation (through an angle less than $180°$) from $\mathbf{a}$ to $\mathbf{b}$, then your thumb points in the direction of $\mathbf{a} \times \mathbf{b}$.

Now that we know the direction of the vector $\mathbf{a} \times \mathbf{b}$, the remaining thing we need to complete its geometric description is its length $|\mathbf{a} \times \mathbf{b}|$. This is given by the following theorem.

Theorem (11.29)

> If θ is the angle between $\mathbf{a}$ and $\mathbf{b}$ (so $0 \leqslant \theta \leqslant \pi$), then
>
> $$|\mathbf{a} \times \mathbf{b}| = |\mathbf{a}|\,|\mathbf{b}|\sin\theta$$

Proof   From the definitions of the cross product and the length of a vector, we have

$$|\mathbf{a} \times \mathbf{b}|^2 = (a_2 b_3 - a_3 b_2)^2 + (a_3 b_1 - a_1 b_3)^2 + (a_1 b_2 - a_2 b_1)^2$$

$$= a_2^2 b_3^2 - 2a_2 a_3 b_2 b_3 + a_3^2 b_2^2 + a_3^2 b_1^2 - 2a_1 a_3 b_1 b_3 + a_1^2 b_3^2$$
$$\quad + a_1^2 b_2^2 - 2a_1 a_2 b_1 b_2 + a_2^2 b_1^2$$

$$= (a_1^2 + a_2^2 + a_3^2)(b_1^2 + b_2^2 + b_3^2) - (a_1 b_1 + a_2 b_2 + a_3 b_3)^2$$

$$= |\mathbf{a}|^2 |\mathbf{b}|^2 - (\mathbf{a} \cdot \mathbf{b})^2$$

$$= |\mathbf{a}|^2 |\mathbf{b}|^2 - |\mathbf{a}|^2 |\mathbf{b}|^2 \cos^2\theta \qquad \textit{(by Theorem 11.14)}$$

$$= |\mathbf{a}|^2 |\mathbf{b}|^2 (1 - \cos^2\theta)$$

$$= |\mathbf{a}|^2 |\mathbf{b}|^2 \sin^2\theta$$

Taking square roots and observing that $\sqrt{\sin^2 \theta} = \sin \theta$ because $\sin \theta \geqslant 0$ when $0 \leqslant \theta \leqslant \pi$, we have

$$|\mathbf{a} \times \mathbf{b}| = |\mathbf{a}| \, |\mathbf{b}| \sin \theta$$ ●

Corollary (11.30)

> Two nonzero vectors $\mathbf{a}$ and $\mathbf{b}$ are parallel if and only if $\mathbf{a} \times \mathbf{b} = \mathbf{0}$.

Proof   Two nonzero vectors $\mathbf{a}$ and $\mathbf{b}$ are parallel if and only if $\theta = 0$ or π. In either case $\sin \theta = 0$.   ●

The geometric interpretation of Theorem 11.29 can be seen by looking at Figure 11.23. If $\mathbf{a}$ and $\mathbf{b}$ are represented by directed line segments with the same initial point, then they determine a parallelogram with base $|\mathbf{a}|$, altitude $|\mathbf{b}| \sin \theta$, and area

$$A = |\mathbf{a}|(|\mathbf{b}| \sin \theta) = |\mathbf{a} \times \mathbf{b}|$$

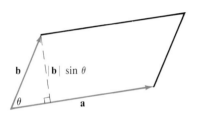

Figure 11.23

Thus the **length of the cross product $\mathbf{a} \times \mathbf{b}$ is equal to the area of the parallelogram determined by $\mathbf{a}$ and $\mathbf{b}$.**

● **Example 3**   Find the area of the triangle with vertices $P(1, 4, 6)$, $Q(-2, 5, -1)$, and $R(1, -1, 1)$.

Solution   By (11.4) the vectors that correspond to the directed line segments $\overrightarrow{PQ}$ and $\overrightarrow{PR}$ are $\mathbf{a} = \langle -3, 1, -7 \rangle$ and $\mathbf{b} = \langle 0, -5, -5 \rangle$. We compute the cross product of these vectors:

$$\mathbf{a} \times \mathbf{b} = \begin{vmatrix} \mathbf{i} & \mathbf{j} & \mathbf{k} \\ -3 & 1 & -7 \\ 0 & -5 & -5 \end{vmatrix}$$

$$= (-5 - 35)\mathbf{i} - (15 - 0)\mathbf{j} + (15 - 0)\mathbf{k} = -40\mathbf{i} - 15\mathbf{j} + 15\mathbf{k}$$

The area A of triangle PQR is half the area of the parallelogram with adjacent sides PQ and PR. Thus

$$A = \frac{1}{2} |\mathbf{a} \times \mathbf{b}| = \frac{1}{2} \sqrt{(-40)^2 + (-15)^2 + 15^2} = \frac{5\sqrt{82}}{2}$$ ●

If we apply Theorems 11.28 and 11.29 to the standard basis vectors $\mathbf{i}$, $\mathbf{j}$, and $\mathbf{k}$ using $\theta = \pi/2$, we obtain

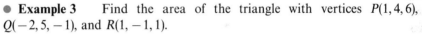

$$\mathbf{i} \times \mathbf{j} = \mathbf{k} \qquad \mathbf{j} \times \mathbf{k} = \mathbf{i} \qquad \mathbf{k} \times \mathbf{i} = \mathbf{j}$$

$$\mathbf{j} \times \mathbf{i} = -\mathbf{k} \qquad \mathbf{k} \times \mathbf{j} = -\mathbf{i} \qquad \mathbf{i} \times \mathbf{k} = -\mathbf{j}$$

Observe that $\mathbf{i} \times \mathbf{j} \neq \mathbf{j} \times \mathbf{i}$

Thus the cross product is not commutative. Also

$$\mathbf{i} \times (\mathbf{i} \times \mathbf{j}) = \mathbf{i} \times \mathbf{k} = -\mathbf{j}$$

whereas $$(\mathbf{i} \times \mathbf{i}) \times \mathbf{j} = \mathbf{0} \times \mathbf{j} = \mathbf{0}$$

So the associative law for multiplication does not usually hold; that is, in general,

$$(\mathbf{a} \times \mathbf{b}) \times \mathbf{c} \neq \mathbf{a} \times (\mathbf{b} \times \mathbf{c})$$

However, some of the usual laws of algebra do hold for cross products. The following theorem summarizes the properties of vector products.

Theorem (11.31)

> If $\mathbf{a}$, $\mathbf{b}$, and $\mathbf{c}$ are vectors and c is a scalar, then
> 1. $\mathbf{a} \times \mathbf{b} = -\mathbf{b} \times \mathbf{a}$
> 2. $(c\mathbf{a}) \times \mathbf{b} = c(\mathbf{a} \times \mathbf{b}) = \mathbf{a} \times (c\mathbf{b})$
> 3. $\mathbf{a} \times (\mathbf{b} + \mathbf{c}) = \mathbf{a} \times \mathbf{b} + \mathbf{a} \times \mathbf{c}$
> 4. $(\mathbf{a} + \mathbf{b}) \times \mathbf{c} = \mathbf{a} \times \mathbf{c} + \mathbf{b} \times \mathbf{c}$
> 5. $\mathbf{a} \cdot (\mathbf{b} \times \mathbf{c}) = (\mathbf{a} \times \mathbf{b}) \cdot \mathbf{c}$
> 6. $\mathbf{a} \times (\mathbf{b} \times \mathbf{c}) = (\mathbf{a} \cdot \mathbf{c})\mathbf{b} - (\mathbf{a} \cdot \mathbf{b})\mathbf{c}$

These properties can be proved by writing the vectors in terms of their components and using the definition of a cross product. We shall give the proof of Property 5 and leave the remaining proofs as exercises.

Proof of Property 5   If $\mathbf{a} = \langle a_1, a_2, a_3 \rangle$, $\mathbf{b} = \langle b_1, b_2, b_3 \rangle$, and $\mathbf{c} = \langle c_1, c_2, c_3 \rangle$, then

(11.32)
$$
\begin{aligned}
\mathbf{a} \cdot (\mathbf{b} \times \mathbf{c}) &= a_1(b_2 c_3 - b_3 c_2) + a_2(b_3 c_1 - b_1 c_3) + a_3(b_1 c_2 - b_2 c_1) \\
&= a_1 b_2 c_3 - a_1 b_3 c_2 + a_2 b_3 c_1 - a_2 b_1 c_3 + a_3 b_1 c_2 - a_3 b_2 c_1 \\
&= (a_2 b_3 - a_3 b_2)c_1 + (a_3 b_1 - a_1 b_3)c_2 + (a_1 b_2 - a_2 b_1)c_3 \\
&= (\mathbf{a} \times \mathbf{b}) \cdot \mathbf{c}
\end{aligned}
$$

The product $\mathbf{a} \cdot (\mathbf{b} \times \mathbf{c})$ that occurs in Property 5 is called the **scalar triple product** of the vectors $\mathbf{a}$, $\mathbf{b}$, and $\mathbf{c}$. Notice from Equation 11.32 that we can write the scalar triple product as a determinant:

(11.33)
$$
\mathbf{a} \cdot (\mathbf{b} \times \mathbf{c}) = \begin{vmatrix} a_1 & a_2 & a_3 \\ b_1 & b_2 & b_3 \\ c_1 & c_2 & c_3 \end{vmatrix}
$$

The geometric significance of the scalar triple product can be seen by considering the parallelepiped determined by the vectors $\mathbf{a}$, $\mathbf{b}$, and $\mathbf{c}$ (Figure 11.24). The area of the base parallelogram is $A = |\mathbf{b} \times \mathbf{c}|$. If θ is the angle

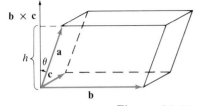

Figure 11.24

between **a** and **b** × **c**, then the height h of the parallelepiped is $h = |\mathbf{a}||\cos\theta|$. (We must use $|\cos\theta|$ instead of $\cos\theta$ in case $\theta > \pi/2$.) Thus the volume of the parallelepiped is

$$V = Ah = |\mathbf{b}\times\mathbf{c}||\mathbf{a}||\cos\theta| = |\mathbf{a}\cdot(\mathbf{b}\times\mathbf{c})|$$

Thus we have proved the following:

(11.34)

> The volume of the parallelepiped determined by the vectors **a**, **b**, and **c** is the magnitude of their scalar triple product:
>
> $$V = |\mathbf{a}\cdot(\mathbf{b}\times\mathbf{c})|$$

● **Example 4**   Use the scalar triple product to show that the vectors $\mathbf{a} = \langle 1, 4, -7 \rangle$, $\mathbf{b} = \langle 2, -1, 4 \rangle$, and $\mathbf{c} = \langle 0, -9, 18 \rangle$ are coplanar; that is, they lie in the same plane.

Solution   We use Equation 11.33 to compute their scalar triple product:

$$\mathbf{a}\cdot(\mathbf{b}\times\mathbf{c}) = \begin{vmatrix} 1 & 4 & -7 \\ 2 & -1 & 4 \\ 0 & -9 & 18 \end{vmatrix}$$

$$= 1\begin{vmatrix} -1 & 4 \\ -9 & 18 \end{vmatrix} - 4\begin{vmatrix} 2 & 4 \\ 0 & 18 \end{vmatrix} - 7\begin{vmatrix} 2 & -1 \\ 0 & -9 \end{vmatrix}$$

$$= 1(18) - 4(36) - 7(-18) = 0$$

Therefore, by (11.34) the volume of the parallelepiped determined by **a**, **b**, and **c** is 0. This means that **a**, **b**, and **c** are coplanar.   ●

The idea of a cross product occurs in physics when we consider a force **F** acting on a rigid body at a point given by a position vector **r** (see Figure 11.25). The **torque** τ (relative to the origin) is defined to be the cross product of the position and force vectors

$$\tau = \mathbf{r}\times\mathbf{F}$$

and measures the tendency of the body to rotate about the origin. The direction of the torque vector indicates the axis of rotation. According to Theorem 11.29, the magnitude of the torque vector is

$$|\tau| = |\mathbf{r}\times\mathbf{F}| = |\mathbf{r}||\mathbf{F}|\sin\theta$$

where θ is the angle between the position and force vectors. Observe that the only component of **F** that can cause a rotation is the one perpendicular to **r**, that is, $|\mathbf{F}|\sin\theta$. The magnitude of the torque is equal to the area of the parallelogram determined by **r** and **F**.

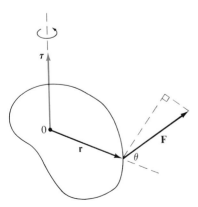

Figure 11.25

EXERCISES 11.4

In Exercises 1–8 find the cross product $\mathbf{a} \times \mathbf{b}$.

1. $\mathbf{a} = \langle 1, 0, 1 \rangle, \mathbf{b} = \langle 0, 1, 0 \rangle$

2. $\mathbf{a} = \langle 2, 4, 0 \rangle, \mathbf{b} = \langle -3, 1, 6 \rangle$

3. $\mathbf{a} = \langle -2, 3, 4 \rangle, \mathbf{b} = \langle 3, 0, 1 \rangle$

4. $\mathbf{a} = \langle 1, 2, -3 \rangle, \mathbf{b} = \langle 5, -1, -2 \rangle$

5. $\mathbf{a} = \mathbf{i} + \mathbf{j} + \mathbf{k}, \mathbf{b} = \mathbf{i} + \mathbf{j} - \mathbf{k}$

6. $\mathbf{a} = \mathbf{i} + 2\mathbf{j} - \mathbf{k}, \mathbf{b} = 3\mathbf{i} - \mathbf{j} + 7\mathbf{k}$

7. $\mathbf{a} = 2\mathbf{i} - \mathbf{k}, \mathbf{b} = \mathbf{i} + 2\mathbf{j}$

8. $\mathbf{a} = \mathbf{j} + 4\mathbf{k}, \mathbf{b} = 6\mathbf{i} - 5\mathbf{k}$

9. If $\mathbf{a} = \langle 0, 1, 2 \rangle$ and $\mathbf{b} = \langle 3, 1, 0 \rangle$, find $\mathbf{a} \times \mathbf{b}$ and $\mathbf{b} \times \mathbf{a}$.

10. If $\mathbf{a} = \langle -4, 0, 3 \rangle$, $\mathbf{b} = \langle 2, -1, 0 \rangle$, and $\mathbf{c} = \langle 0, 2, 5 \rangle$, show that $\mathbf{a} \times (\mathbf{b} \times \mathbf{c}) \neq (\mathbf{a} \times \mathbf{b}) \times \mathbf{c}$.

11. Find two unit vectors orthogonal to both $\langle 1, -1, 1 \rangle$ and $\langle 0, 4, 4 \rangle$.

12. Find two unit vectors orthogonal to both $\mathbf{i} + \mathbf{j}$ and $\mathbf{i} - \mathbf{j} + \mathbf{k}$.

13. Show that $\mathbf{0} \times \mathbf{a} = \mathbf{0} = \mathbf{a} \times \mathbf{0}$ for any vector $\mathbf{a}$ in V_3.

14. Show that $(\mathbf{a} \times \mathbf{b}) \cdot \mathbf{b} = 0$ for all vectors $\mathbf{a}$ and $\mathbf{b}$ in V_3.

15. Prove Property 1 of Theorem 11.31.

16. Prove Property 2 of Theorem 11.31.

17. Prove Property 3 of Theorem 11.31.

18. Prove Property 4 of Theorem 11.31.

In Exercises 19–22, (a) find a vector orthogonal to the plane through the points P, Q, and R, and (b) find the area of triangle PQR.

19. $P(1, 0, 0), Q(0, 2, 0), R(0, 0, 3)$

20. $P(1, 0, -1), Q(2, 4, 5), R(3, 1, 7)$

21. $P(0, 0, 0), Q(1, -1, 1), R(4, 3, 7)$

22. $P(-4, -4, -4), Q(0, 5, -1), R(3, 1, 2)$

In Exercises 23 and 24 find the volume of the parallelepiped determined by the vectors $\mathbf{a}$, $\mathbf{b}$, and $\mathbf{c}$.

23. $\mathbf{a} = \langle 1, 0, 6 \rangle, \mathbf{b} = \langle 2, 3, -8 \rangle, \mathbf{c} = \langle 8, -5, 6 \rangle$

24. $\mathbf{a} = 2\mathbf{i} + 3\mathbf{j} - 2\mathbf{k}, \mathbf{b} = \mathbf{i} - \mathbf{j}, \mathbf{c} = 2\mathbf{i} + 3\mathbf{k}$

In Exercises 25 and 26 find the volume of the parallelepiped with adjacent edges PQ, PR, and PS.

25. $P(1, 1, 1), Q(2, 0, 3), R(4, 1, 7), S(3, -1, -2)$

26. $P(0, 1, 2), Q(2, 4, 5), R(-1, 0, 1), S(6, -1, 4)$

27. Use the scalar triple product to verify that the vectors $\mathbf{a} = 2\mathbf{i} + 3\mathbf{j} + \mathbf{k}$, $\mathbf{b} = \mathbf{i} - \mathbf{j}$, and $\mathbf{c} = 7\mathbf{i} + 3\mathbf{j} + 2\mathbf{k}$ are coplanar.

28. Use the scalar triple product to verify that the points $P(1, 0, 1)$, $Q(2, 4, 6)$, $R(3, -1, 2)$, and $S(6, 2, 8)$ are coplanar.

29. (a) Let P be a point not on the line L that passes through the points Q and R. Show that the distance d from the point P to the line L is

$$d = \frac{|\mathbf{a} \times \mathbf{b}|}{|\mathbf{a}|}$$

where $\mathbf{a} = \overrightarrow{QR}$ and $\mathbf{b} = \overrightarrow{QP}$.

(b) Use the formula in part (a) to find the distance from the point $P(1, 1, 1)$ to the line through $Q(0, 6, 8)$ and $R(-1, 4, 7)$.

30. (a) Let P be a point not on the plane that passes through the points Q, R, and S. Show that the distance d from P to the plane is

$$d = \frac{|\mathbf{a} \cdot (\mathbf{b} \times \mathbf{c})|}{|\mathbf{a} \times \mathbf{b}|}$$

where $\mathbf{a} = \overrightarrow{QR}$, $\mathbf{b} = \overrightarrow{QS}$, and $\mathbf{c} = \overrightarrow{QP}$.

(b) Use the formula in part (a) to find the distance from the point $P(2, 1, 4)$ to the plane through the points $Q(1, 0, 0)$, $R(0, 2, 0)$, and $S(0, 0, 3)$.

31. Prove that

$$(\mathbf{a} - \mathbf{b}) \times (\mathbf{a} + \mathbf{b}) = 2(\mathbf{a} \times \mathbf{b})$$

32. The product $\mathbf{a} \times (\mathbf{b} \times \mathbf{c})$ is called the **vector triple product** of $\mathbf{a}$, $\mathbf{b}$, and $\mathbf{c}$. Prove the following formula for the vector triple product:

$$\mathbf{a} \times (\mathbf{b} \times \mathbf{c}) = (\mathbf{a} \cdot \mathbf{c})\mathbf{b} - (\mathbf{a} \cdot \mathbf{b})\mathbf{c}$$

33. Prove that

$$\mathbf{a} \times (\mathbf{b} \times \mathbf{c}) + \mathbf{b} \times (\mathbf{c} \times \mathbf{a}) + \mathbf{c} \times (\mathbf{a} \times \mathbf{b}) = \mathbf{0}$$

34. Prove that

$$(\mathbf{a} \times \mathbf{b}) \cdot (\mathbf{c} \times \mathbf{d}) = \begin{vmatrix} \mathbf{a} \cdot \mathbf{c} & \mathbf{b} \cdot \mathbf{c} \\ \mathbf{a} \cdot \mathbf{d} & \mathbf{b} \cdot \mathbf{d} \end{vmatrix}$$

SECTION 11.5

Equations of Lines and Planes

A line in the xy-plane is determined when a point on the line and the direction of the line (its slope or angle of inclination) are given. The equation of the line can then be written using the slope-point form.

Likewise, a line L in three-dimensional space is determined when we know a point $P_0(x_0, y_0, z_0)$ on L and the direction of L. In three dimensions the direction of a line is conveniently described by a vector, so we let $\mathbf{v}$ be a vector parallel to L. Let $P(x, y, z)$ be an arbitrary point on L and let $\mathbf{r}_0$ and $\mathbf{r}$ be the position vectors of P_0 and P (that is, they have representations $\overrightarrow{OP_0}$ and $\overrightarrow{OP}$). If $\mathbf{a}$ is the vector with representation $\overrightarrow{P_0P}$ as in Figure 11.26, then the triangle law for vector addition gives $\mathbf{r} = \mathbf{r}_0 + \mathbf{a}$. But, since $\mathbf{a}$ and $\mathbf{v}$ are parallel vectors, there is a scalar t such that $\mathbf{a} = t\mathbf{v}$. Thus

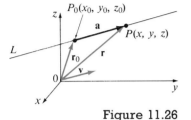

Figure 11.26

(11.35)
$$\boxed{\mathbf{r} = \mathbf{r}_0 + t\mathbf{v}}$$

which is the **vector equation** of L. Each value of the parameter t gives the position vector $\mathbf{r}$ of a point on L.

If the vector $\mathbf{v}$ that gives the direction of L is written in component form as $\mathbf{v} = \langle a, b, c \rangle$, then $t\mathbf{v} = \langle ta, tb, tc \rangle$. We can also write $\mathbf{r} = \langle x, y, z \rangle$ and $\mathbf{r}_0 = \langle x_0, y_0, z_0 \rangle$, so the vector equation (11.35) becomes

$$\langle x, y, z \rangle = \langle x_0 + ta, \ y_0 + tb, \ z_0 + tc \rangle$$

Two vectors are equal if and only if corresponding components are equal. Therefore we have the three scalar equations:

(11.36)
$$\boxed{x = x_0 + at \qquad y = y_0 + bt \qquad z = z_0 + ct}$$

These equations are called the **parametric equations** of the line L through the point $P_0(x_0, y_0, z_0)$ and parallel to the vector $\mathbf{v} = \langle a, b, c \rangle$. Each value of the parameter t gives a point (x, y, z) on L.

● **Example 1**   (a) Find a vector equation and parametric equations for the line that passes through the point $(5, 1, 3)$ and is parallel to the vector $\mathbf{i} + 2\mathbf{j} - \mathbf{k}$.   (b) Find two other points on the line.

Solution   (a)   Here $\mathbf{r}_0 = \langle 5, 1, 3 \rangle = 5\mathbf{i} + \mathbf{j} + 3\mathbf{k}$ and $\mathbf{v} = \mathbf{i} + 2\mathbf{j} - \mathbf{k}$, so the vector equation (11.35) becomes

$$\mathbf{r} = (5\mathbf{i} + \mathbf{j} + 3\mathbf{k}) + t(\mathbf{i} + 2\mathbf{j} - \mathbf{k})$$

or
$$\mathbf{r} = (5 + t)\mathbf{i} + (1 + 2t)\mathbf{j} + (3 - t)\mathbf{k}$$

The parametric equations are

$$x = 5 + t \qquad y = 1 + 2t \qquad z = 3 - t$$

(b)   Choosing the parameter value $t = 1$ gives $x = 6$, $y = 3$, and $z = 2$, so $(6, 3, 2)$ is a point on the line. Similarly, $t = -1$ gives the point $(4, -1, 4)$.

●

The vector equation and parametric equations of a line are not unique. If we change the point or the parameter or choose a different parallel vector, then the equations change. For instance, if, instead of $(5, 1, 3)$, we choose the point $(6, 3, 2)$ in Example 1, then the parametric equations of the line become

$$x = 6 + t \qquad y = 3 + 2t \qquad z = 2 - t$$

Or, if we stay with the point $(5, 1, 3)$ but choose the parallel vector $2\mathbf{i} + 4\mathbf{j} - 2\mathbf{k}$, we arrive at the equations

$$x = 5 + 2t \qquad y = 1 + 4t \qquad z = 3 - 2t$$

In general, if a vector $\mathbf{v} = \langle a, b, c \rangle$ is used to describe the direction of a line L, then the numbers a, b, and c are called **direction numbers** of L. Since any vector parallel to $\mathbf{v}$ could also be used, we see that any three numbers proportional to a, b, and c could also be used as a set of direction numbers for L.

Another way of describing a line L is to eliminate the parameter t from Equations 11.36. If none of a, b, or c is 0, we can solve each of these equations for t, equate the results, and obtain

(11.37)
$$\frac{x - x_0}{a} = \frac{y - y_0}{b} = \frac{z - z_0}{c}$$

These equations are called the **symmetric equations** of L. Notice that the numbers a, b, and c that appear in the denominators of Equations 11.37 are the direction numbers of L, that is, the components of a vector parallel to L. If one of a, b, or c is 0, we can still eliminate t. For instance, if $a = 0$, we could write the equations of L as

$$x = x_0 \qquad \frac{y - y_0}{b} = \frac{z - z_0}{c}$$

This means that L lies in the vertical plane $x = x_0$.

● **Example 2**   (a) Find the parametric equations and symmetric equations of the line that passes through the points $A(2, 4, -3)$ and $B(3, -1, 1)$. (b) At what point does this line intersect the xy-plane?

Solution   (a)   We are not explicitly given a vector parallel to the line, but observe that the vector $\mathbf{v}$ with representation $\overrightarrow{AB}$ is parallel to the line and

$$\mathbf{v} = \langle 3 - 2, -1 - 4, 1 - (-3) \rangle = \langle 1, -5, 4 \rangle$$

Thus the direction numbers are $a = 1$, $b = -5$, and $c = 4$. Taking the point $(2, 4, -3)$ as P_0, we see that the parametric equations (11.36) are

$$x = 2 + t \qquad y = 4 - 5t \qquad z = -3 + 4t$$

and the symmetric equations (11.37) are

$$\frac{x - 2}{1} = \frac{y - 4}{-5} = \frac{z + 3}{4}$$

(b)   The line intersects the xy-plane when $z = 0$, so we put $z = 0$ in the symmetric equations and obtain

$$\frac{x - 2}{1} = \frac{y - 4}{-5} = \frac{3}{4}$$

This gives $x = \frac{11}{4}$ and $y = \frac{1}{4}$, so the line intersects the xy-plane at the point $(\frac{11}{4}, \frac{1}{4}, 0)$. ●

In general, the procedure of Example 2 shows that direction numbers of the line L through the points $P_0(x_0, y_0, z_0)$ and $P_1(x_1, y_1, z_1)$ are $x_1 - x_0$, $y_1 - y_0$, and $z_1 - z_0$ and so the symmetric equations of L are

$$\frac{x - x_0}{x_1 - x_0} = \frac{y - y_0}{y_1 - y_0} = \frac{z - z_0}{z_1 - z_0}$$

● **Example 3**   Show that the lines L_1 and L_2 with parametric equations

$$x = 1 + t \qquad y = -2 + 3t \qquad z = 4 - t$$

$$x = 2s \qquad y = 3 + s \qquad z = -3 + 4s$$

are **skew lines;** that is, they do not intersect and are not parallel (and therefore do not lie in the same plane).

Solution   The lines are not parallel because the corresponding vectors $\langle 1, 3, -1 \rangle$ and $\langle 2, 1, 4 \rangle$ are not parallel. (Their components are not proportional.) If L_1 and L_2 had a point of intersection, there would be values of t and s such that

$$1 + t = 2s$$
$$-2 + 3t = 3 + s$$
$$4 - t = -3 + 4s$$

But if we solve the first two equations, we get $t = \frac{11}{5}$ and $s = \frac{8}{5}$, and these values do not satisfy the third equation. Therefore there are no values of t and s that satisfy the three equations. Thus L_1 and L_2 do not intersect. Hence L_1 and L_2 are skew lines. ●

Planes

A plane in space is determined by a point $P_0(x_0, y_0, z_0)$ in the plane and a vector $\mathbf{n}$ that is orthogonal to the plane. This orthogonal vector $\mathbf{n}$ is called a **normal vector.** Let $P(x, y, z)$ be an arbitrary point in the plane and let $\mathbf{r}_0$ and $\mathbf{r}$ be the position vectors of P_0 and P. Then the vector $\mathbf{r} - \mathbf{r}_0$ is represented by $\overrightarrow{P_0P}$ (see Figure 11.27). The normal vector **n** is orthogonal to every vector in the given plane. In particular, **n** is orthogonal to $\mathbf{r} - \mathbf{r}_0$ and so, by (11.18), we have

(11.38)

$$\mathbf{n} \cdot (\mathbf{r} - \mathbf{r}_0) = 0$$

which can be rewritten as

(11.39)

$$\mathbf{n} \cdot \mathbf{r} = \mathbf{n} \cdot \mathbf{r}_0$$

Either of Equations 11.38 or 11.39 is called the **vector equation of the plane.**
    In order to obtain a scalar equation for the plane, we write $\mathbf{n} = \langle a, b, c \rangle$, $\mathbf{r} = \langle x, y, z \rangle$, and $\mathbf{r}_0 = \langle x_0, y_0, z_0 \rangle$. Then the vector equation (11.38) becomes

$$\langle a, b, c \rangle \cdot \langle x - x_0, y - y_0, z - z_0 \rangle = 0$$

(11.40)   or

$$a(x - x_0) + b(y - y_0) + c(z - z_0) = 0$$

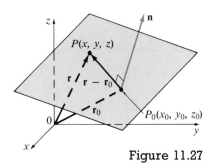

Figure 11.27

Equation 11.40 is the **scalar equation of the plane through** $P_0(x_0, y_0, z_0)$ **with normal vector** $\mathbf{n} = \langle a, b, c \rangle$.

● **Example 4**   Find an equation of the plane through the point $(2, 4, -1)$ with normal vector $\mathbf{n} = \langle 2, 3, 4 \rangle$. Find the intercepts and sketch the plane.

Solution   Putting $a = 2$, $b = 3$, $c = 4$, $x_0 = 2$, $y_0 = 4$, and $z_0 = -1$ in Equation 11.40, we see that the equation of the plane is

$$2(x - 2) + 3(y - 4) + 4(z + 1) = 0$$

or $2x + 3y + 4z = 12$

To find the x-intercept we set $y = z = 0$ in this equation and obtain $x = 6$. Similarly, the y-intercept is 4 and the z-intercept is 3. This enables us to sketch the portion of the plane that lies in the first octant (see Figure 11.28).
                                                                                    ●

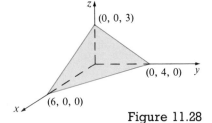

Figure 11.28

By collecting terms in Equation 11.40 as we did in Example 4, we can rewrite the equation of a plane as

(11.41)

$$ax + by + cz = d$$

where $d = ax_0 + by_0 + cz_0$. Equation 11.41 is called a **linear equation** in x, y, and z. Conversely it can be shown that if a, b, and c are not all 0, then the linear equation 11.41 represents a plane with normal vector $\langle a, b, c \rangle$. (See Exercise 46.)

● **Example 5**   Find an equation of the plane that passes through the points $P(1, 0, 2)$, $Q(3, -1, 6)$, and $R(5, 2, 4)$.

Solution   The vectors **a** and **b** corresponding to $\overrightarrow{PQ}$ and $\overrightarrow{PR}$ are

$$\mathbf{a} = \langle 2, -1, 4 \rangle \qquad \mathbf{b} = \langle 4, 2, 2 \rangle$$

Since both **a** and **b** lie in the plane, their cross product $\mathbf{a} \times \mathbf{b}$ is orthogonal to the plane and can be taken as the normal vector. Thus

$$\mathbf{n} = \mathbf{a} \times \mathbf{b} = \begin{vmatrix} \mathbf{i} & \mathbf{j} & \mathbf{k} \\ 2 & -1 & 4 \\ 4 & 2 & 2 \end{vmatrix} = -10\mathbf{i} + 12\mathbf{j} + 8\mathbf{k}$$

and an equation of the plane is

$$-10(x - 1) + 12(y - 0) + 8(z - 2) = 0$$

or

$$5x - 6y - 4z + 3 = 0 \qquad ●$$

Two planes are **parallel** if their normal vectors are parallel. For instance, the planes $x + 2y - 3z = 4$ and $2x + 4y - 6z = 3$ are parallel because their normal vectors are $\mathbf{n}_1 = \langle 1, 2, -3 \rangle$ and $\mathbf{n}_2 = \langle 2, 4, -6 \rangle$ and $\mathbf{n}_2 = 2\mathbf{n}_1$. If two planes are not parallel, then they intersect in a straight line and the angle between the two planes is defined as the angle between their normal vectors (see Figure 11.29).

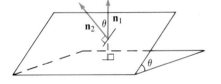

Figure 11.29

● **Example 6**   (a) Find the angle between the planes $x + y + z = 1$ and $x - 2y + 3z = 1$.   (b) Find symmetric equations for the line of intersection L of these two planes.

Solution   (a)   The normal vectors of these planes are

$$\mathbf{n}_1 = \langle 1, 1, 1 \rangle \qquad \mathbf{n}_2 = \langle 1, -2, 3 \rangle$$

and so if θ is the angle between the planes, Corollary 11.17 gives

$$\cos \theta = \frac{\mathbf{n}_1 \cdot \mathbf{n}_2}{|\mathbf{n}_1||\mathbf{n}_2|} = \frac{1(1) + 1(-2) + 1(3)}{\sqrt{1 + 1 + 1}\sqrt{1 + 4 + 9}} = \frac{2}{\sqrt{42}}$$

$$\theta = \cos^{-1}\left(\frac{2}{\sqrt{42}}\right) \approx 72°$$

(b)   We first need to find a point on L. For instance, we can find the point where the line intersects the xy-plane by setting $z = 0$ in the equations of both planes. This gives the equations $x + y = 1$ and $x - 2y = 1$, whose solution is $x = 1$, $y = 0$. So the point $(1, 0, 0)$ lies on L.

Now we observe that since L lies in both planes it is perpendicular to both of the normal vectors. Thus a vector $\mathbf{v}$ parallel to L is given by the cross product

$$\mathbf{v} = \mathbf{n}_1 \times \mathbf{n}_2 = \begin{vmatrix} \mathbf{i} & \mathbf{j} & \mathbf{k} \\ 1 & 1 & 1 \\ 1 & -2 & 3 \end{vmatrix} = 5\mathbf{i} - 2\mathbf{j} - 3\mathbf{k}$$

and so the symmetric equations of L can be written as

$$\frac{x - 1}{5} = \frac{y}{-2} = \frac{z}{-3}$$

●

Note: Since a linear equation in x, y, and z represents a plane and two nonparallel planes intersect in a line, it follows that two linear equations can represent a line. The points (x, y, z) that satisfy both $a_1 x + b_1 y + c_1 z = d_1$ and $a_2 x + b_2 y + c_2 z = d_2$ lie on both of these planes and so the pair of linear equations represents the line of intersection of the planes (if they are not parallel). For instance, in Example 6 the line L was given as the line of intersection of the planes $x + y + z = 1$ and $x - 2y + 3z = 1$. The symmetric equations that we found for L could be written as

$$\frac{x - 1}{5} = \frac{y}{-2} \quad \text{and} \quad \frac{y}{-2} = \frac{z}{-3}$$

which is again a pair of linear equations. They exhibit L as the line of intersection of the planes $(x - 1)/5 = y/(-2)$ and $y/(-2) = z/(-3)$.

In general when we write the equations of a line in the symmetric form

$$\frac{x - x_0}{a} = \frac{y - y_0}{b} = \frac{z - z_0}{c}$$

we can regard the line as the line of intersection of the two planes

$$\frac{x - x_0}{a} = \frac{y - y_0}{b} \quad \text{and} \quad \frac{y - y_0}{b} = \frac{z - z_0}{c}$$

● **Example 7**   Find a formula for the distance D from a point $P_1(x_1, y_1, z_1)$ to the plane $ax + by + cz = d$.

Solution   Let $P_0(x_0, y_0, z_0)$ be any point in the given plane and let $\mathbf{b}$ be the vector corresponding to $\overrightarrow{P_0 P_1}$. Then

$$\mathbf{b} = \langle x_1 - x_0, y_1 - y_0, z_1 - z_0 \rangle$$

From Figure 11.30 you can see that the distance D from P_1 to the plane is equal to the absolute value of the scalar projection of $\mathbf{b}$ onto the normal

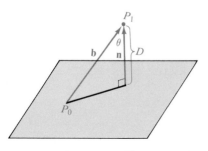

Figure 11.30

vector $\mathbf{n} = \langle a, b, c \rangle$. (See Section 11.3.) Thus

$$D = |\mathbf{b}||\cos\theta| = \frac{|\mathbf{n} \cdot \mathbf{b}|}{|\mathbf{n}|}$$

$$= \frac{|a(x_1 - x_0) + b(y_1 - y_0) + c(z_1 - z_0)|}{\sqrt{a^2 + b^2 + c^2}}$$

$$= \frac{|(ax_1 + by_1 + cz_1) - (ax_0 + by_0 + cz_0)|}{\sqrt{a^2 + b^2 + c^2}}$$

Since P_0 lies in the plane, its coordinates satisfy the equation of the plane and so we have $ax_0 + by_0 + cz_0 = d$. Thus the formula for D can be written as

$$D = \frac{|ax_1 + by_1 + cz_1 - d|}{\sqrt{a^2 + b^2 + c^2}}$$

● **Example 8**   Find the distance from the point $(2, 3, 4)$ to the plane $5x + y - z = 1$.

Solution   The formula derived in Example 7 gives

$$D = \frac{|5(2) + 1(3) - 1(4) - 1|}{\sqrt{5^2 + 1^2 + (-1)^2}} = \frac{8}{3\sqrt{3}}$$

EXERCISES 11.5

In Exercises 1–4 find the vector equation and parametric equations for the line passing through the given point and parallel to the vector **a**.

1. $(3, -1, 8)$, $\mathbf{a} = \langle 2, 3, 5 \rangle$

2. $(-2, 4, 5)$, $\mathbf{a} = \langle 3, -1, 6 \rangle$

3. $(0, 1, 2)$, $\mathbf{a} = 6\mathbf{i} + 3\mathbf{j} + 2\mathbf{k}$

4. $(1, -1, -2)$, $\mathbf{a} = 2\mathbf{i} - 7\mathbf{k}$

In Exercises 5–10 find parametric equations and symmetric equations for the line passing through the given points.

5. $(2, 1, 8)$, $(6, 0, 3)$

6. $(-1, 0, 5)$, $(4, -3, 3)$

7. $(3, 1, -1)$, $(3, 2, -6)$

8. $(3, 1, \frac{1}{2})$, $(-1, 4, 1)$

9. $(-\frac{1}{3}, 1, 1)$, $(0, 5, -8)$

10. $(2, -7, 5)$, $(-4, 2, 5)$

11. Show that the line through the points $(2, -1, -5)$ and $(8, 8, 7)$ is parallel to the line through the points $(4, 2, -6)$ and $(8, 8, 2)$.

12. Show that the line through the points $(0, 1, 1)$ and $(1, -1, 6)$ is perpendicular to the line through the points $(-4, 2, 1)$ and $(-1, 6, 2)$.

13. (a)   Find symmetric equations for the line that passes through the point $(0, 2, -1)$ and is parallel to the line with parametric equations $x = 1 + 2t$, $y = 3t$, and $z = 5 - 7t$.

(b) Find the points in which the required line in part (a) intersects the coordinate planes.

14. (a) Find parametric equations for the line through $(5, 1, 0)$ that is perpendicular to the plane $2x - y + z = 1$.

   (b) In what points does this line intersect the coordinate planes?

In Exercises 15–18 determine whether the lines L_1 and L_2 are parallel, skew, or intersect. If they intersect, find the point of intersection.

15. $L_1: \dfrac{x - 4}{2} = \dfrac{y + 5}{4} = \dfrac{z - 1}{-3}$, $L_2: \dfrac{x - 2}{1} = \dfrac{y + 1}{3} = \dfrac{z}{2}$

16. $L_1: \dfrac{x - 1}{2} = \dfrac{y}{1} = \dfrac{z - 1}{4}$, $L_2: \dfrac{x}{1} = \dfrac{y + 2}{2} = \dfrac{z + 2}{3}$

17. $L_1: x = -6t, y = 1 + 9t, z = -3t,$
    $L_2: x = 1 + 2s, y = 4 - 3s, z = s$

18. $L_1: x = 1 + t, y = 2 - t, z = 3t,$
    $L_2: x = 2 - s, y = 1 + 2s, z = 4 + s$

In Exercises 19–22 find an equation of the plane through the given point and with the given normal vector.

19. $(1, 4, 5)$, $\mathbf{n} = \langle 7, 1, 4 \rangle$

20. $(-5, 1, 2)$, $\mathbf{n} = \langle 3, -5, 2 \rangle$

21. $(1, 2, 3)$, $\mathbf{n} = 15\mathbf{i} + 9\mathbf{j} - 12\mathbf{k}$

22. $(-1, -6, -4)$, $\mathbf{n} = -5\mathbf{i} + 2\mathbf{j} - 2\mathbf{k}$

In Exercises 23–26 find the equation of the plane passing through the given point and parallel to the given plane.

23. $(6, 5, -2)$, $x + y - z + 1 = 0$

24. $(3, 0, 8)$, $2x + 5y + 8z = 17$

25. $(-1, 3, -8)$, $3x - 4y - 6z = 9$

26. $(2, -4, 5)$, $z = 2x + 3y$

In Exercises 27–30 find the equation of the plane passing through the three given points.

27. $(0, 0, 0), (1, 1, 1), (1, 2, 3)$

28. $(-1, 1, -1), (1, -1, 2), (4, 0, 3)$

29. $(1, 0, -3), (0, -2, -4), (4, 1, 6)$

30. $(2, 1, -3), (5, -1, 4), (2, -2, 4)$

In Exercises 31 and 32 find an equation of the plane that passes through the given point and contains the given line.

31. $(1, 6, -4)$; $x = 1 + 2t, y = 2 - 3t, z = 3 - t$

32. $(-1, -3, 2)$; $x = -1 - 2t, y = 4t, z = 2 + t$

In Exercises 33–36 determine whether the given planes are parallel, perpendicular, or neither. If neither, find the angle between them.

33. $x + z = 1, y + z = 1$

34. $-8x - 6y + 2z = 1, z = 4x + 3y$

35. $x + 4y - 3z = 1, -3x + 6y + 7z = 0$

36. $2x + 2y - z = 4, 6x - 3y + 2z = 5$

37. Find symmetric equations for the line of intersection of the planes $x + y - z = 2$ and $3x - 4y + 5z = 6$.

38. Find parametric equations for the line of intersection of the planes $2x + 5z + 3 = 0$ and $x - 3y + z + 2 = 0$.

39. Find an equation of the plane with x-intercept a, y-intercept b, and z-intercept c.

40. Find an equation for the plane consisting of all points that are equidistant from the points $(-4, 2, 1)$ and $(2, -4, 3)$.

In Exercises 41–43 find the distance from the given point to the given plane.

41. $(2, 8, 5)$, $x - 2y - 2z = 1$

42. $(3, -2, 7)$, $4x - 6y + z = 5$

43. $(-1, 4, 6)$, $2x + 3y + 6z = 4$

44. (a) Show that the distance between the parallel planes $ax + by + cz = d_1$ and $ax + by + cz = d_2$ is

$$D = \frac{|d_1 - d_2|}{\sqrt{a^2 + b^2 + c^2}}$$

   (b) Use the formula in part (a) to compute the distance between the planes $3x + 6y - 9z = 4$ and $x + 2y - 3z = 1$.

45. Using Exercise 44, or otherwise, find the distance between the skew lines with parametric equations $x = 1 + t, y = 1 + 6t, z = 2t$ and $x = 1 + 2s, y = 5 + 15s, z = -2 + 6s$.

46. If a, b, and c are not all 0, show that the equation $ax + by + cz = d$ represents a plane and $\langle a, b, c \rangle$ is a normal vector to the plane. [*Hint:* Suppose $a \neq 0$ and rewrite the equation in the form

$$a\left(x - \frac{d}{a}\right) + b(y - 0) + c(z - 0) = 0$$

Quadric Surfaces

A **quadric surface** is the graph of a second-degree equation in three variables x, y, and z. The most general such equation is

$$Ax^2 + By^2 + Cz^2 + Dxy + Eyz + Fxz + Gx + Hy + Iz + J = 0$$

where $A, B, C, \dots, J$ are constants, but by translation and rotation it can be brought into one of the two standard forms

$$Ax^2 + By^2 + Cz^2 + J = 0 \quad \text{or} \quad Ax^2 + By^2 + Iz = 0$$

Quadric surfaces are the analogues in three dimensions of the conic sections in the plane. (See Sections 1.4 and 9.6 for a review of conic sections.)

In order to sketch the graph of a quadric surface (or any surface) it is useful to determine the curves of intersection of the surface with planes parallel to the coordinate planes. These curves are called **traces** (or cross-sections) of the surface.

Ellipsoids.   The quadric surface with equation

(11.42)
$$\frac{x^2}{a^2} + \frac{y^2}{b^2} + \frac{z^2}{c^2} = 1$$

is called an **ellipsoid** because its traces are ellipses. For instance, the horizontal plane $z = k$ (where $-c < k < c$) intersects it in the ellipse

$$\frac{x^2}{a^2} + \frac{y^2}{b^2} = 1 - \frac{k^2}{c^2} \qquad z = k$$

and, in particular, the trace in the xy-plane is just the ellipse $x^2/a^2 + y^2/b^2 = 1$, $z = 0$. Similarly, the traces in the other coordinate planes are the ellipses $y^2/b^2 + z^2/c^2 = 1$, $x = 0$, and $x^2/a^2 + z^2/c^2 = 1$, $y = 0$. The six intercepts of the ellipsoid are $(\pm a, 0, 0)$, $(0, \pm b, 0)$, and $(0, 0, \pm c)$ and it lies in the box

$$|x| \leqslant a \qquad |y| \leqslant b \qquad |z| \leqslant c$$

Since the equation involves only even powers of x, y, and z, the ellipsoid is symmetric with respect to each coordinate plane (see Figure 11.31).

If two of the three semiaxes a, b, and c are equal, then the ellipsoid is a surface of revolution. For instance, if $c = a$, then the ellipsoid could be obtained by revolving the ellipse $x^2/a^2 + y^2/b^2 = 1$, $z = 0$, around the y-axis. If $a = b = c$, the ellipsoid becomes a sphere.

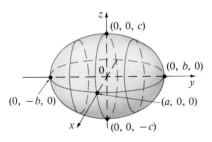

Figure 11.31

THE ELLIPSOID $\dfrac{x^2}{a^2} + \dfrac{y^2}{b^2} + \dfrac{z^2}{c^2} = 1$

Hyperboloids.   The quadric surface

(11.43)
$$\frac{x^2}{a^2} + \frac{y^2}{b^2} - \frac{z^2}{c^2} = 1$$

is also symmetric with respect to the coordinate planes. The trace in any horizontal plane $z = k$ is the ellipse

$$\frac{x^2}{a^2} + \frac{y^2}{b^2} = 1 + \frac{k^2}{c^2} \qquad z = k$$

but the traces in the xz- and yz-planes are the hyperbolas

$$\frac{x^2}{a^2} - \frac{z^2}{c^2} = 1, y = 0 \quad \text{and} \quad \frac{y^2}{b^2} - \frac{z^2}{c^2} = 1, x = 0$$

Furthermore, in contrast to the next example, this surface consists of just one piece, so it is called a **hyperboloid of one sheet.** The z-axis is called the **axis** of this hyperboloid. (If the minus sign in Equation 11.43 occurs in front of the first or second term instead of the third term, then the axis is the x- or y-axis.)

If $a = b$ in Equation 11.43, the surface is a hyperboloid of revolution and is obtained by rotating a hyperbola about the z-axis.

Now consider the surface

(11.44)
$$-\frac{x^2}{a^2} - \frac{y^2}{b^2} + \frac{z^2}{c^2} = 1$$

Traces in the xz- and yz-planes are the hyperbolas

$$-\frac{x^2}{a^2} + \frac{z^2}{c^2} = 1, y = 0 \quad \text{and} \quad -\frac{y^2}{b^2} + \frac{z^2}{c^2} = 1, x = 0$$

If $|k| > c$, the horizontal plane $z = k$ intersects the surface in the ellipse

$$\frac{x^2}{a^2} + \frac{y^2}{b^2} = \frac{k^2}{c^2} - 1, \qquad z = k$$

Figure 11.32

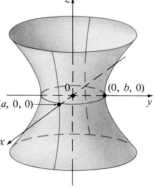

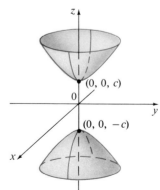

(a) Hyperboloid of one sheet

$$\frac{x^2}{a^2} + \frac{y^2}{b^2} - \frac{z^2}{c^2} = 1$$

(b) Hyperboloid of two sheets

$$-\frac{x^2}{a^2} - \frac{y^2}{b^2} + \frac{z^2}{c^2} = 1$$

whereas if $|k| < c$, the plane $z = k$ does not intersect the surface at all. Thus the surface consists of two parts, one above the plane $z = c$ and one below the plane $z = -c$, and is called a **hyperboloid of two sheets** whose axis is the z-axis. (See Figure 11.32.)

Notice, in comparing Equations 11.43 and 11.44, that the number of minus signs in the equation indicates the number of sheets of the hyperboloid.

Cones. If we replace the right side of Equation 11.43 or 11.44 by 0, we get the surface

(11.45)
$$\frac{z^2}{c^2} = \frac{x^2}{a^2} + \frac{y^2}{b^2}$$

which is a **cone**. This surface has the property that if P is any point on the cone then the line OP lies entirely on the cone.

You can verify that traces in horizontal planes $z = k$ are ellipses and traces in vertical planes $x = k$ or $y = k$ are hyperbolas if $k \neq 0$ but pairs of lines if $k = 0$.

The cone given by Equation 11.45 is asymptotic to both of the hyperboloids given by Equations 11.43 and 11.44. (Compare Figures 11.32 and 11.33.)

Figure 11.33
THE CONE $\dfrac{z^2}{c^2} = \dfrac{x^2}{a^2} + \dfrac{y^2}{b^2}$

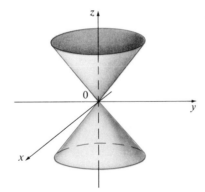

Paraboloids. The surface

(11.46)
$$\frac{z}{c} = \frac{x^2}{a^2} + \frac{y^2}{b^2}$$

is called an **elliptic paraboloid** because its traces in horizontal planes $z = k$ are ellipses, whereas its traces in vertical planes $x = k$ or $y = k$ are parabolas. For instance, its trace in the yz-plane is the parabola

$$z = \frac{c}{b^2} y^2, \qquad x = 0$$

The **axis** of the paraboloid given by Equation 11.46 is the z-axis and its **vertex** is the origin. The case where $c > 0$ is illustrated in Figure 11.34(a).

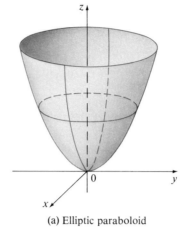

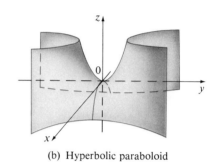

Figure 11.34

(a) Elliptic paraboloid

$$\frac{z}{c} = \frac{x^2}{a^2} + \frac{y^2}{b^2}$$

(b) Hyperbolic paraboloid

$$\frac{z}{c} = \frac{x^2}{a^2} - \frac{y^2}{b^2}$$

If $a = b$, the surface is a **circular paraboloid,** also called a paraboloid of revolution.

The **hyperbolic paraboloid**

(11.47)
$$\frac{z}{c} = \frac{x^2}{a^2} - \frac{y^2}{b^2}$$

also has parabolas as its vertical traces, but it has hyperbolas as its horizontal traces. The case where $c < 0$ is illustrated in Figure 11.34(b).

Quadric Cylinders.   When one of the variables x, y, or z is missing from the equation of a surface, then the surface is a cylinder. For instance, the equation

(11.48)
$$\frac{x^2}{a^2} + \frac{y^2}{b^2} = 1$$

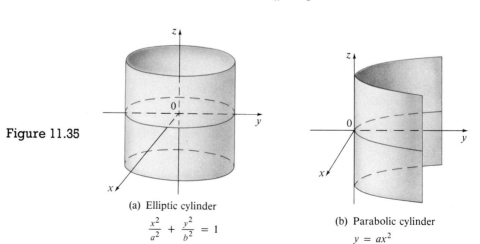

Figure 11.35

(a) Elliptic cylinder

$$\frac{x^2}{a^2} + \frac{y^2}{b^2} = 1$$

(b) Parabolic cylinder

$$y = ax^2$$

represents the **elliptic cylinder** $\{(x, y, z) \mid x^2/a^2 + y^2/b^2 = 1\}$. All horizontal traces are congruent ellipses and the generators of the cylinder are vertical lines. The **parabolic cylinder** $y = ax^2$ is illustrated in Figure 11.35(b).

● **Example 1**   Identify and sketch the surface $4x^2 - y^2 + 2z^2 + 4 = 0$.

Solution   Dividing by -4, we first put the equation in standard form:

$$-x^2 + \frac{y^2}{4} - \frac{z^2}{2} = 1$$

Comparing this equation with Equation 11.44, we see that it represents a hyperboloid of two sheets, the only difference being that in this case the axis of the hyperboloid is the y-axis. The traces in the xy- and yz-planes are the hyperbolas

$$-x^2 + \frac{y^2}{4} = 1, z = 0 \quad \text{and} \quad \frac{y^2}{4} - \frac{z^2}{2} = 1, x = 0$$

There is no trace in the xz-plane, but traces in the vertical planes $y = k$ for $|k| > 2$ are the ellipses

$$x^2 + \frac{z^2}{2} = \frac{k^2}{4} - 1, \qquad y = k$$

which can be written as

$$\frac{x^2}{\dfrac{k^2}{4} - 1} + \frac{z^2}{2\left(\dfrac{k^2}{4} - 1\right)} = 1, \qquad y = k$$

These traces are used to make the sketch in Figure 11.36.   ●

● **Example 2**   Classify the quadric surface $x^2 + 2z^2 - 6x - y + 10 = 0$.

Solution   By completing the square we rewrite the equation as

$$y - 1 = (x - 3)^2 + 2z^2$$

Comparing this equation with Equation 11.46, we see that it represents an elliptic paraboloid. Here, however, the axis of the paraboloid is parallel to the y-axis and its vertex is the point $(3, 1, 0)$. The traces in the plane $y = k$ ($k > 1$) are the ellipses $(x - 3)^2 + 2z^2 = k - 1$, $y = k$. The trace in the xy-plane is the parabola $y = 1 + (x - 3)^2$, $z = 0$. The paraboloid is sketched in Figure 11.37.   ●

● **Example 3**   Identify and sketch the surfaces (a) $x^2 + y^2 = 1$, (b) $x^2 + z^2 = 1$.

Solution   (a) Since z is missing and the equations $x^2 + y^2 = 1$, $z = k$, represent a circle with radius 1 in the plane $z = k$, the surface $x^2 + y^2 = 1$ is a circular cylinder whose axis is the z-axis (see Figure 11.38).

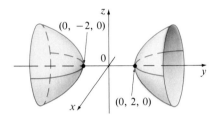

Figure 11.36

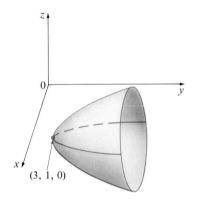

Figure 11.37

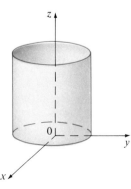

Figure 11.38

$x^2 + y^2 = 1$

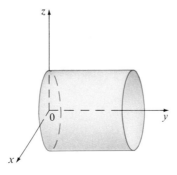

Figure 11.39
$x^2 + z^2 = 1$

(b)   In this case y is missing and the surface is a circular cylinder whose axis is the y-axis (see Figure 11.39). It is obtained by taking the circle $x^2 + z^2 = 1$, $y = 0$, in the xz-plane and moving it parallel to the y-axis. ●

    Note: When dealing with surfaces it is important to recognize that an equation like $x^2 + y^2 = 1$ represents a cylinder and not a circle. The trace of the cylinder $x^2 + y^2 = 1$ in the xy-plane is the circle with equations $x^2 + y^2 = 1$, $z = 0$.

EXERCISES 11.6

In Exercises 1–12 find the traces of the given surface in the planes $x = k$, $y = k$, $z = k$, identify the surface, and sketch it.

1. $x^2 - y^2 + z^2 = 1$ **2.** $x = y^2 + z^2$

3. $4x^2 + 9y^2 + 36z^2 = 36$ **4.** $2x^2 + z^2 = 4$

5. $4z^2 - x^2 - y^2 = 1$ **6.** $z = x^2 - y^2$

7. $z = y^2$ **8.** $25y^2 + z^2 = 100 + 4x^2$

9. $y^2 = x^2 + z^2$ **10.** $9x^2 - y^2 - z^2 = 9$

11. $x^2 + 4z^2 - y = 0$ **12.** $x^2 - y^2 = 1$

In Exercises 13–24 reduce the equation to one of the standard forms, classify the surface, and sketch it.

13. $z^2 = 3x^2 + 4y^2 - 12$

14. $4x^2 - 9y^2 + z^2 + 36 = 0$

15. $z = x^2 + y^2 + 1$

16. $x^2 + 4y^2 + z^2 - 2x = 0$

17. $x^2 + y^2 - 4z^2 + 4x - 6y - 8z = 13$

18. $4x = y^2 - 2z^2$

19. $x^2 + 4y^2 = 100$

20. $9x^2 + y^2 - z^2 - 2y + 2z = 0$

21. $x^2 - y^2 + 4y + z = 4$

22. $yz = 1$

23. $x^2 - y^2 - z^2 - 4x - 4y = 0$

24. $4x^2 - y^2 + z^2 + 8x + 8z + 24 = 0$

25. Find an equation for the surface obtained by rotating the parabola $y = x^2$ about the y-axis.

26. Find an equation for the surface obtained by rotating the line $x = 3y$ about the x-axis.

27. Find an equation for the surface consisting of all points that are equidistant from the point $(-1, 0, 0)$ and the plane $x = 1$. Identify the surface.

28. Find an equation for the surface consisting of all points P for which the distance from P to the x-axis is twice the distance from P to the yz-plane. Identify the surface.

Vector Functions and Space Curves

The functions that we have used so far have been real-valued functions. We now study functions whose values are vectors because such functions are needed to describe curves in space and the motion of particles in space.

In general, a function was defined in Section 1.5 as a rule that assigns to each element in the domain an element in the range. A **vector-valued function,** or **vector function,** is simply a function whose domain is a set of real numbers and whose range is a set of vectors. We are most interested in vector functions $\mathbf{r}$ whose values are three-dimensional vectors. This means that for every number t in the domain of $\mathbf{r}$ there is a unique vector in V_3 denoted by $\mathbf{r}(t)$. If $f(t)$, $g(t)$, and $h(t)$ are the components of the vector $\mathbf{r}(t)$, then f, g, and h are real-valued functions called the **component functions** of $\mathbf{r}$ and we can write

$$\mathbf{r}(t) = \langle f(t), g(t), h(t) \rangle = f(t)\mathbf{i} + g(t)\mathbf{j} + h(t)\mathbf{k}$$

We use the letter t to denote the independent variable because it represents time in most applications of vector functions.

● **Example 1**   If

$$\mathbf{r}(t) = \langle t^3, \ln(3 - t), \sqrt{t} \rangle$$

then the component functions are

$$f(t) = t^3 \qquad g(t) = \ln(3 - t) \qquad h(t) = \sqrt{t}$$

By our usual convention, the domain of $\mathbf{r}$ consists of all values of t for which the expression for $\mathbf{r}(t)$ is defined. The expressions t^3, $\ln(3 - t)$, and $\sqrt{t}$ are all defined when $3 - t > 0$ and $t \geqslant 0$. Therefore the domain of $\mathbf{r}$ is the interval $[0, 3)$.   ●

The **limit** of a vector function $\mathbf{r}$ is defined by taking the limits of its component functions as follows:

(11.49)

> If $\mathbf{r}(t) = \langle f(t), g(t), h(t) \rangle$, then
>
> $$\lim_{t \to a} \mathbf{r}(t) = \left\langle \lim_{t \to a} f(t), \lim_{t \to a} g(t), \lim_{t \to a} h(t) \right\rangle$$
>
> provided the limits of the component functions exist.

Limits of vector functions obey the same rules as limits of real-valued functions. (See Exercise 41.)

● **Example 2**   Find $\lim\limits_{t\to 0} \mathbf{r}(t)$ where $\mathbf{r}(t) = (1 + t^3)\mathbf{i} + te^{-t}\mathbf{j} + \dfrac{\sin t}{t}\mathbf{k}$.

Solution

$$\lim_{t\to 0} \mathbf{r}(t) = \left[\lim_{t\to 0}(1 + t^3)\right]\mathbf{i} + \left[\lim_{t\to 0} te^{-t}\right]\mathbf{j} + \left[\lim_{t\to 0}\frac{\sin t}{t}\right]\mathbf{k}$$

$$= \mathbf{i} + \mathbf{k} \quad \textit{[by (3.21)]}$$

A vector function $\mathbf{r}$ is **continuous at** *a* if

$$\lim_{t\to a} \mathbf{r}(t) = \mathbf{r}(a)$$

In view of Definition 11.49, we see that $\mathbf{r}$ is continuous at *a* if and only if its component functions f, g, and h are continuous at *a*.

There is a close connection between continuous vector functions and space curves. Suppose that f, g, and h are continuous real-valued functions on an interval I. Then the set C of all points (x, y, z) in space, where

(11.50)

$$x = f(t) \qquad y = g(t) \qquad z = h(t)$$

and t varies throughout the interval I, is called a **space curve**. The equations in (11.50) are called the **parametric equations of C** and t is called a **parameter**. We can think of C as being traced out by a moving particle whose position at time t is $(f(t), g(t), h(t))$. If we now consider the vector function $\mathbf{r}(t) = \langle f(t), g(t), h(t)\rangle$, then $\mathbf{r}(t)$ is the position vector of the point $P(f(t), g(t), h(t))$ on C. Thus any continuous vector function $\mathbf{r}$ defines a space curve C that is traced out by the tip of the moving vector $\mathbf{r}(t)$ as in Figure 11.40. The continuity of $\mathbf{r}$ means that there are no breaks or jumps in C.

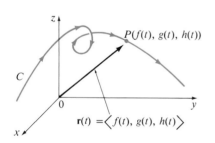

$\mathbf{r}(t) = \langle f(t), g(t), h(t)\rangle$

Figure 11.40

● **Example 3**   Describe the curve defined by the vector function $\mathbf{r}(t) = \langle 1 + t, 2 + 5t, -1 + 6t\rangle$.

Solution   The corresponding parametric equations are

$$x = 1 + t \qquad y = 2 + 5t \qquad z = -1 + 6t$$

which we recognize from Equations 11.36 as the parametric equations of a line passing through the point $(1, 2, -1)$ and parallel to the vector $\langle 1, 5, 6\rangle$. Alternatively, we could observe that the function can be written as $\mathbf{r} = \mathbf{r}_0 + t\mathbf{v}$, where $\mathbf{r}_0 = \langle 1, 2, -1\rangle$ and $\mathbf{v} = \langle 1, 5, 6\rangle$, and this is the vector equation of a line as given by Equation 11.35.   ●

● **Example 4**   Sketch the curve whose vector equation is $\mathbf{r}(t) = 2\cos t\,\mathbf{i} + \sin t\,\mathbf{j} + t\mathbf{k}$.

Solution   The parametric equations for this curve are

$$x = 2\cos t \qquad y = \sin t \qquad z = t$$

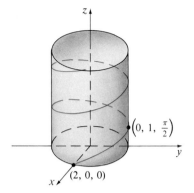

Since $(x/2)^2 + y^2 = \cos^2 t + \sin^2 t = 1$, the curve must lie on the elliptical cylinder $x^2/4 + y^2 = 1$. Since $z = t$, the curve spirals upward around the cylinder as t increases. The curve, shown in Figure 11.41, is called a **helix**.

$\left(0, 1, \frac{\pi}{2}\right)$

$(2, 0, 0)$

Figure 11.41

Plane curves can also be represented in vector notation. For instance, the curve given by the parametric equations $x = \sin t$ and $y = \sin^2 t$ in Example 4 in Section 9.1 could also be described by the vector equation

$$\mathbf{r}(t) = \langle \sin t, \sin^2 t \rangle = \sin t\,\mathbf{i} + \sin^2 t\,\mathbf{j}$$

where $\mathbf{i} = \langle 1, 0 \rangle$ and $\mathbf{j} = \langle 0, 1 \rangle$.

Derivatives and Integrals

The **derivative** $\mathbf{r}'$ of a vector function $\mathbf{r}$ is defined just as for real-valued functions:

(11.51)

$$\frac{d\mathbf{r}}{dt} = \mathbf{r}'(t) = \lim_{h \to 0} \frac{\mathbf{r}(t + h) - \mathbf{r}(t)}{h}$$

if this limit exists. The geometric significance of this definition is shown in Figure 11.42. If P and Q have position vectors $\mathbf{r}(t)$ and $\mathbf{r}(t + h)$, then $\overrightarrow{PQ}$ represents the vector $\mathbf{r}(t + h) - \mathbf{r}(t)$, which can therefore be regarded as a secant vector. If $h > 0$, the scalar multiple $(1/h)(\mathbf{r}(t + h) - \mathbf{r}(t))$ has the same direction as $\mathbf{r}(t + h) - \mathbf{r}(t)$. As $h \to 0$, it appears that this vector approaches a vector that lies on the tangent line. For this reason, the vector $\mathbf{r}'(t)$ is called the **tangent vector** to the curve defined by $\mathbf{r}$ at the point P. The **tangent line** to C at P is defined to be the line through P parallel to the tangent vector $\mathbf{r}'(t)$. If $\mathbf{r}'(t) \neq \mathbf{0}$, we shall also have occasion to consider the **unit tangent**

Figure 11.42

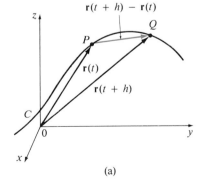

(a)

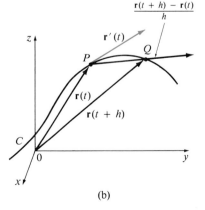

(b)

vector, which is

$$\mathbf{T}(t) = \frac{\mathbf{r}'(t)}{|\mathbf{r}'(t)|}$$

The following theorem gives us a convenient method for computing the derivative of a vector function $\mathbf{r}$: just differentiate each component of $\mathbf{r}$.

Theorem (11.52)

> If $\mathbf{r}(t) = \langle f(t), g(t), h(t) \rangle = f(t)\mathbf{i} + g(t)\mathbf{j} + h(t)\mathbf{k}$, where f, g, and h are differentiable functions, then
>
> $$\mathbf{r}'(t) = \langle f'(t), g'(t), h'(t) \rangle = f'(t)\mathbf{i} + g'(t)\mathbf{j} + h'(t)\mathbf{k}$$

Proof

$$\mathbf{r}'(t) = \lim_{\Delta t \to 0} \frac{1}{\Delta t} \left[\mathbf{r}(t + \Delta t) - \mathbf{r}(t) \right]$$

$$= \lim_{\Delta t \to 0} \frac{1}{\Delta t} \left[\langle f(t + \Delta t), g(t + \Delta t), h(t + \Delta t) \rangle - \langle f(t), g(t), h(t) \rangle \right]$$

$$= \lim_{\Delta t \to 0} \left\langle \frac{f(t + \Delta t) - f(t)}{\Delta t}, \frac{g(t + \Delta t) - g(t)}{\Delta t}, \frac{h(t + \Delta t) - h(t)}{\Delta t} \right\rangle$$

$$= \left\langle \lim_{\Delta t \to 0} \frac{f(t + \Delta t) - f(t)}{\Delta t}, \lim_{\Delta t \to 0} \frac{g(t + \Delta t) - g(t)}{\Delta t}, \lim_{\Delta t \to 0} \frac{h(t + \Delta t) - h(t)}{\Delta t} \right\rangle$$

$$= \langle f'(t), g'(t), h'(t) \rangle \qquad \bullet$$

● **Example 5**   (a) Find the derivative of $\mathbf{r}(t) = (1 + t^3)\mathbf{i} + te^{-t}\mathbf{j} + \sin 2t\mathbf{k}$.
(b) Find the unit tangent vector at the point where $t = 0$.

Solution   (a)   Theorem 11.52 gives

$$\mathbf{r}'(t) = 3t^2\mathbf{i} + (1 - t)e^{-t}\mathbf{j} + 2\cos 2t\mathbf{k}$$

(b)   Since $\mathbf{r}(0) = \mathbf{i}$ and $\mathbf{r}'(0) = \mathbf{j} + 2\mathbf{k}$, the unit tangent vector at the point $(1, 0, 0)$ is

$$\mathbf{T}(0) = \frac{\mathbf{r}'(0)}{|\mathbf{r}'(0)|} = \frac{\mathbf{j} + 2\mathbf{k}}{\sqrt{1 + 4}} = \frac{1}{\sqrt{5}}\mathbf{j} + \frac{2}{\sqrt{5}}\mathbf{k} \qquad \bullet$$

● **Example 6**   Find parametric equations for the tangent line to the helix with parametric equations

$$x = 2\cos t \qquad y = \sin t \qquad z = t$$

at the point $(0, 1, \pi/2)$.

Solution   The vector equation of the helix is $\mathbf{r}(t) = \langle 2\cos t, \sin t, t \rangle$, so

$$\mathbf{r}'(t) = \langle -2\sin t, \cos t, 1 \rangle$$

The parameter value corresponding to the point $(0, 1, \pi/2)$ is $t = \pi/2$, so the tangent vector there is $\mathbf{r}'(\pi/2) = \langle -2, 0, 1 \rangle$. The tangent line is the line through $(0, 1, \pi/2)$ parallel to the vector $\langle -2, 0, 1 \rangle$, so by Equations 11.36 its parametric equations are

$$x = -2t \qquad y = 1 \qquad z = \frac{\pi}{2} + t \qquad\qquad \bullet$$

The next theorem shows that the differentiation formulas for real-valued functions have their counterparts for vector-valued functions.

Theorem (11.53)   Suppose $\mathbf{u}$ and $\mathbf{v}$ are differentiable vector functions, c is a scalar, and f is a real-valued function. Then

1. $\dfrac{d}{dt}\left[\mathbf{u}(t) + \mathbf{v}(t)\right] = \mathbf{u}'(t) + \mathbf{v}'(t)$

2. $\dfrac{d}{dt}\left[c\mathbf{u}(t)\right] = c\mathbf{u}'(t)$

3. $\dfrac{d}{dt}\left[f(t)\mathbf{u}(t)\right] = f'(t)\mathbf{u}(t) + f(t)\mathbf{u}'(t)$

4. $\dfrac{d}{dt}\left[\mathbf{u}(t) \cdot \mathbf{v}(t)\right] = \mathbf{u}'(t) \cdot \mathbf{v}(t) + \mathbf{u}(t) \cdot \mathbf{v}'(t)$

5. $\dfrac{d}{dt}\left[\mathbf{u}(t) \times \mathbf{v}(t)\right] = \mathbf{u}'(t) \times \mathbf{v}(t) + \mathbf{u}(t) \times \mathbf{v}'(t)$

6. $\dfrac{d}{dt}\left[\mathbf{u}(f(t))\right] = f'(t)\mathbf{u}'(f(t))$ *(Chain Rule)*

This theorem can be proved either directly from Definition 11.51 or by using Theorem 11.52 and the corresponding differentiation formulas for real-valued functions. The proofs of Formulas 1, 2, 3, 5, and 6 are left as exercises.

Proof of Formula 4   Let

$$\mathbf{u}(t) = \langle f_1(t), f_2(t), f_3(t) \rangle \qquad \mathbf{v}(t) = \langle g_1(t), g_2(t), g_3(t) \rangle$$

Then

$$\mathbf{u}(t) \cdot \mathbf{v}(t) = f_1(t)g_1(t) + f_2(t)g_2(t) + f_3(t)g_3(t) = \sum_{i=1}^{3} f_i(t)g_i(t)$$

so the ordinary product rule gives

$$\frac{d}{dt}\left[\mathbf{u}(t) \cdot \mathbf{v}(t)\right] = \frac{d}{dt}\sum_{i=1}^{3} f_i(t)g_i(t) = \sum_{i=1}^{3}\frac{d}{dt}\left[f_i(t)g_i(t)\right]$$

$$= \sum_{i=1}^{3}\left[f_i'(t)g_i(t) + f_i(t)g_i'(t)\right]$$

$$= \sum_{i=1}^{3} f_i'(t)g_i(t) + \sum_{i=1}^{3} f_i(t)g_i'(t)$$

$$= \mathbf{u}'(t) \cdot \mathbf{v}(t) + \mathbf{u}(t) \cdot \mathbf{v}'(t)$$

● **Example 7**   Show that if $|\mathbf{r}(t)| = c$ (a constant), then $\mathbf{r}'(t)$ is orthogonal to $\mathbf{r}(t)$ for all t.

Solution   Since

$$\mathbf{r}(t) \cdot \mathbf{r}(t) = |\mathbf{r}(t)|^2 = c^2$$

and c^2 is a constant, Formula 4 of Theorem 11.53 gives

$$0 = \frac{d}{dt}\left[\mathbf{r}(t) \cdot \mathbf{r}(t)\right] = \mathbf{r}'(t) \cdot \mathbf{r}(t) + \mathbf{r}(t) \cdot \mathbf{r}'(t) = 2\mathbf{r}'(t) \cdot \mathbf{r}(t)$$

Thus $\mathbf{r}'(t) \cdot \mathbf{r}(t) = 0$, which says that $\mathbf{r}'(t)$ is orthogonal to $\mathbf{r}(t)$.

Geometrically, this result says that if a curve lies on a sphere, then the tangent vector $\mathbf{r}'(t)$ is always perpendicular to the position vector $\mathbf{r}(t)$. ●

The **definite integral** of a continuous vector function $\mathbf{r}(t)$ can be defined in much the same way as for real-valued functions except that the integral is a vector. But then we can express the integral of $\mathbf{r}$ in terms of the integrals of its component functions f, g, and h as follows. (We use the notation of Chapter 5.)

$$\int_a^b \mathbf{r}(t)\,dt = \lim_{\|P\| \to 0} \sum_{i=1}^{n} \mathbf{r}(t_i^*)\,\Delta t_i$$

$$= \lim_{\|P\| \to 0}\left[\left(\sum_{i=1}^{n} f(t_i^*)\,\Delta t_i\right)\mathbf{i} + \left(\sum_{i=1}^{n} g(t_i^*)\,\Delta t_i\right)\mathbf{j} + \left(\sum_{i=1}^{n} h(t_i^*)\,\Delta t_i\right)\mathbf{k}\right]$$

and so

$$\int_a^b \mathbf{r}(t)\,dt = \left(\int_a^b f(t)\,dt\right)\mathbf{i} + \left(\int_a^b g(t)\,dt\right)\mathbf{j} + \left(\int_a^b h(t)\,dt\right)\mathbf{k}$$

The Fundamental Theorem of Calculus for vector functions says that

$$\int_a^b \mathbf{r}(t)\,dt = \mathbf{R}(t)\Big]_a^b = \mathbf{R}(b) - \mathbf{R}(a)$$

where $\mathbf{R}$ is an antiderivative of $\mathbf{r}$, that is, $\mathbf{R}'(t) = \mathbf{r}(t)$. We also use the notation $\int \mathbf{r}(t)\,dt$ for indefinite integrals (antiderivatives).

● **Example 8**   If $\mathbf{r}(t) = 2\cos t\mathbf{i} + \sin t\mathbf{j} + 2t\mathbf{k}$, then

$$\int \mathbf{r}(t)\,dt = \left(\int 2\cos t\,dt\right)\mathbf{i} + \left(\int \sin t\,dt\right)\mathbf{j} + \left(\int 2t\,dt\right)\mathbf{k}$$

$$= 2\sin t\mathbf{i} - \cos t\mathbf{j} + t^2\mathbf{k} + \mathbf{C}$$

where **C** is a vector constant of integration, and

$$\int_0^{\pi/2} \mathbf{r}(t)\,dt = [2\sin t\mathbf{i} - \cos t\mathbf{j} + t^2\mathbf{k}]_0^{\pi/2}$$

$$= 2\mathbf{i} + \mathbf{j} + \frac{\pi^2}{4}\mathbf{k}$$ ●

EXERCISES 11.7

In Exercises 1–6 sketch the curve with the given vector
equation.

1. $\mathbf{r}(t) = \langle t, -t, 2t \rangle$ **2.** $\mathbf{r}(t) = \langle t^2, t, 2 \rangle$

3. $\mathbf{r}(t) = \langle \sin t, 3, \cos t \rangle$ **4.** $\mathbf{r}(t) = \langle \sin t, t, \cos t \rangle$

5. $\mathbf{r}(t) = t\mathbf{i} + t^2\mathbf{j} + t^3\mathbf{k}$

6. $\mathbf{r}(t) = \sin t\mathbf{i} + \sin t\mathbf{j} + \sqrt{2}\cos t\mathbf{k}$

In Exercises 7–10 find the given limits.

7. $\lim_{t\to 0}\langle t, \cos t, 2 \rangle$

8. $\lim_{t\to 0}\left\langle \dfrac{1 - \cos t}{t}, t^3, e^{-1/t^2} \right\rangle$

9. $\lim_{t\to 1}\left(\sqrt{t+3}\,\mathbf{i} + \dfrac{t-1}{t^2-1}\mathbf{j} + \dfrac{\tan t}{t}\mathbf{k} \right)$

10. $\lim_{t\to\infty}\left(e^{-t}\mathbf{i} + \dfrac{t-1}{t+1}\mathbf{j} + \tan^{-1}t\mathbf{k} \right)$

In Exercises 11–18 find the domain and derivative of the given
vector function.

11. $\mathbf{r}(t) = \langle t, t^2, t^3 \rangle$

12. $\mathbf{r}(t) = \langle t^2 - 4, \sqrt{t-4}, \sqrt{6-t} \rangle$

13. $\mathbf{r}(t) = \mathbf{i} + \tan t\mathbf{j} + \sec t\mathbf{k}$

14. $\mathbf{r}(t) = te^{2t}\mathbf{i} + \dfrac{t-1}{t+1}\mathbf{j} + \tan^{-1}t\mathbf{k}$

15. $\mathbf{r}(t) = \ln(4 - t^2)\mathbf{i} + \sqrt{1+t}\,\mathbf{j} - 4e^{3t}\mathbf{k}$

16. $\mathbf{r}(t) = e^{-t}\cos t\mathbf{i} + e^{-t}\sin t\mathbf{j} + \ln|t|\,\mathbf{k}$

17. $\mathbf{r}(t) = \mathbf{a} + t\mathbf{b} + t^2\mathbf{c}$ **18.** $\mathbf{r}(t) = t\mathbf{a} \times (\mathbf{b} + t\mathbf{c})$

In Exercises 19–22 (a) sketch the plane curve with given
vector equation, (b) find $\mathbf{r}'(t)$, and (c) sketch the position
vector $\mathbf{r}(t)$ and the tangent vector $\mathbf{r}'(t)$ for the given value
of t.

19. $\mathbf{r}(t) = \langle \cos t, \sin t \rangle, t = \dfrac{\pi}{4}$

20. $\mathbf{r}(t) = \langle t^3, t^2 \rangle, t = 1$

21. $\mathbf{r}(t) = (1 + t)\mathbf{i} + t^2\mathbf{j}, t = 1$

22. $\mathbf{r}(t) = 2\sin t\mathbf{i} + 3\cos t\mathbf{j}, t = \dfrac{\pi}{3}$

In Exercises 23–26 find the unit tangent vector $\mathbf{T}(t)$ at the
point with the given value of the parameter t.

23. $\mathbf{r}(t) = \langle 2t, 3t^2, 4t^3 \rangle, t = 1$

24. $\mathbf{r}(t) = \langle e^{2t}, e^{-2t}, te^{2t} \rangle, t = 0$

25. $\mathbf{r}(t) = t\mathbf{i} + 2\sin t\mathbf{j} + 3\cos t\mathbf{k}, t = \dfrac{\pi}{6}$

26. $\mathbf{r}(t) = e^{2t}\cos t\mathbf{i} + e^{2t}\sin t\mathbf{j} + e^{2t}\mathbf{k}, t = \dfrac{\pi}{2}$

In Exercises 27–32 find parametric equations for the tangent line to the curve with given parametric equations at the given point.

27. $x = t, y = t^2, z = t^3$; $(1, 1, 1)$

28. $x = 1 + 2t, y = 1 + t - t^2, z = 1 - t + t^2 - t^3$; $(1, 1, 1)$

29. $x = t, y = \sqrt{2}\cos t, z = \sqrt{2}\sin t$; $\left(\frac{\pi}{4}, 1, 1\right)$

30. $x = \sin \pi t, y = \sqrt{t}, z = \cos \pi t$; $(0, 1, -1)$

31. $x = t \cos 2\pi t, y = t \sin 2\pi t, z = 4t$; $(0, \frac{1}{4}, 1)$

32. $x = \cos t, y = 3e^{2t}, z = 3e^{-2t}$; $(1, 3, 3)$

33. The curves $\mathbf{r}_1(t) = \langle t, t^2, t^3 \rangle$ and $\mathbf{r}_2(t) = \langle \sin t, \sin 2t, t \rangle$ intersect at the origin. Find their angle of intersection correct to the nearest degree.

34. At what point do the curves $\mathbf{r}_1(t) = \langle t, 1 - t, 3 + t^2 \rangle$ and $\mathbf{r}_2(s) = \langle 3 - s, s - 2, s^2 \rangle$ intersect? Find their angle of intersection correct to the nearest degree.

In Exercises 35–38 evaluate the given integral.

35. $\int_0^1 (t\mathbf{i} + t^2\mathbf{j} + t^3\mathbf{k}) \, dt$

36. $\int_1^2 [(1 + t^2)\mathbf{i} - 4t^4\mathbf{j} - (t^2 - 1)\mathbf{k}] \, dt$

37. $\int_0^{\pi/4} (\cos 2t\mathbf{i} + \sin 2t\mathbf{j} + t \sin t\mathbf{k}) \, dt$

38. $\int_1^4 \left(\sqrt{t}\mathbf{i} + te^{-t}\mathbf{j} + \frac{1}{t^2}\mathbf{k} \right) dt$

39. Find $\mathbf{r}(t)$ if $\mathbf{r}'(t) = t^2\mathbf{i} + 4t^3\mathbf{j} - t^2\mathbf{k}$ and $\mathbf{r}(0) = \mathbf{j}$.

40. Find $\mathbf{r}(t)$ if $\mathbf{r}'(t) = \sin t\mathbf{i} - \cos t\mathbf{j} + 2t\mathbf{k}$ and $\mathbf{r}(0) = \mathbf{i} + \mathbf{j} + 2\mathbf{k}$.

41. Suppose $\mathbf{u}$ and $\mathbf{v}$ are vector functions that possess limits as $t \to a$ and let c be a constant. Prove the following properties of limits.

(a) $\lim_{t \to a} [\mathbf{u}(t) + \mathbf{v}(t)] = \lim_{t \to a} \mathbf{u}(t) + \lim_{t \to a} \mathbf{v}(t)$

(b) $\lim_{t \to a} c\mathbf{u}(t) = c \lim_{t \to a} \mathbf{u}(t)$

(c) $\lim_{t \to a} [\mathbf{u}(t) \cdot \mathbf{v}(t)] = \lim_{t \to a} \mathbf{u}(t) \cdot \lim_{t \to a} \mathbf{v}(t)$

(d) $\lim_{t \to a} [\mathbf{u}(t) \times \mathbf{v}(t)] = \lim_{t \to a} \mathbf{u}(t) \times \lim_{t \to a} \mathbf{v}(t)$

42. Show that $\lim_{t \to a} \mathbf{r}(t) = \mathbf{b}$ if and only if for every $\varepsilon > 0$ there is a number $\delta > 0$ such that $|\mathbf{r}(t) - \mathbf{b}| < \varepsilon$ whenever $0 < |t - a| < \delta$.

43. Prove Formula 1 of Theorem 11.53.

44. Prove Formula 3 of Theorem 11.53.

45. Prove Formula 5 of Theorem 11.53.

46. Prove Formula 6 of Theorem 11.53.

47. If $\mathbf{u}(t) = \mathbf{i} - 2t^2\mathbf{j} + 3t^3\mathbf{k}$ and $\mathbf{v}(t) = t\mathbf{i} + \cos t\mathbf{j} + \sin t\mathbf{k}$, find $D_t[\mathbf{u}(t) \cdot \mathbf{v}(t)]$.

48. If $\mathbf{u}$ and $\mathbf{v}$ are the vector functions in Exercise 47, find $D_t[\mathbf{u}(t) \times \mathbf{v}(t)]$.

49. Show that if $\mathbf{r}$ is a vector function such that $\mathbf{r}'' = (\mathbf{r}')'$ exists, then

$$\frac{d}{dt}[\mathbf{r}(t) \times \mathbf{r}'(t)] = \mathbf{r}(t) \times \mathbf{r}''(t)$$

50. Find an expression for

$$\frac{d}{dt}[\mathbf{u}(t) \cdot (\mathbf{v}(t) \times \mathbf{w}(t))]$$

51. If $\mathbf{r}(t) \neq \mathbf{0}$, show that

$$\frac{d}{dt}|\mathbf{r}(t)| = \frac{1}{|\mathbf{r}(t)|}\mathbf{r}(t) \cdot \mathbf{r}'(t)$$

SECTION 11.8

Arc Length and Curvature

Recall that we defined the length of a plane curve $x = f(t)$ and $y = g(t)$, $a \leqslant t \leqslant b$, as the limit of lengths of inscribed polygons and, for the case where f' and g' are continuous, we arrived at the formula

(11.54) $$L = \int_a^b \sqrt{[f'(t)]^2 + [g'(t)]^2} \, dt = \int_a^b \sqrt{\left(\frac{dx}{dt}\right)^2 + \left(\frac{dy}{dt}\right)^2} \, dt$$

in Theorem 9.8.

The length of a space curve is defined in exactly the same way. If the curve has the vector equation $\mathbf{r}(t) = \langle f(t), g(t), h(t) \rangle$, $a \leqslant t \leqslant b$, or equivalently, the parametric equations $x = f(t)$, $y = g(t)$, and $z = h(t)$, where f', g', and h' are continuous, then it can be shown that its length is

(11.55)

$$L = \int_a^b \sqrt{[f'(t)]^2 + [g'(t)]^2 + [h'(t)]^2}\, dt$$

$$= \int_a^b \sqrt{\left(\frac{dx}{dt}\right)^2 + \left(\frac{dy}{dt}\right)^2 + \left(\frac{dz}{dt}\right)^2}\, dt$$

Notice that both of the arc length formulas (11.54) and (11.55) can be put into the more compact form

(11.56)

$$L = \int_a^b |\mathbf{r}'(t)|\, dt$$

because, for plane curves $\mathbf{r}(t) = f(t)\mathbf{i} + g(t)\mathbf{j}$,

$$|\mathbf{r}'(t)| = |f'(t)\mathbf{i} + g'(t)\mathbf{j}| = \sqrt{[f'(t)]^2 + [g'(t)]^2}$$

whereas, for space curves $\mathbf{r}(t) = f(t)\mathbf{i} + g(t)\mathbf{j} + h(t)\mathbf{k}$,

$$|\mathbf{r}'(t)| = |f'(t)\mathbf{i} + g'(t)\mathbf{j} + h'(t)\mathbf{k}| = \sqrt{[f'(t)]^2 + [g'(t)]^2 + [h'(t)]^2}$$

● **Example 1**   Find the length of the arc of the circular helix with vector equation $\mathbf{r}(t) = \cos t\,\mathbf{i} + \sin t\,\mathbf{j} + t\,\mathbf{k}$ from the point $(1, 0, 0)$ to the point $(1, 0, 2\pi)$.

Solution   Since $\mathbf{r}'(t) = -\sin t\,\mathbf{i} + \cos t\,\mathbf{j} + \mathbf{k}$, we have

$$|\mathbf{r}'(t)| = \sqrt{(-\sin t)^2 + \cos^2 t + 1} = \sqrt{2}$$

The arc from $(1, 0, 0)$ to $(1, 0, 2\pi)$ is described by the parameter interval $0 \leqslant t \leqslant 2\pi$ and so, from Formula 11.56, we have

$$L = \int_0^{2\pi} |\mathbf{r}'(t)|\, dt = \int_0^{2\pi} \sqrt{2}\, dt = 2\sqrt{2}\,\pi \qquad ●$$

A curve given by a vector function $\mathbf{r}(t)$ on an interval I is called **smooth** if $\mathbf{r}'$ is continuous and $\mathbf{r}'(t) \neq \mathbf{0}$ (except possibly at any endpoints of I). The significance of the condition $\mathbf{r}'(t) \neq \mathbf{0}$ is illustrated by the graph of the semi-cubical parabola $\mathbf{r}(t) = \langle 1 + t^3, t^2 \rangle$ in Figure 11.43. Since $\mathbf{r}'(t) = \langle 3t^2, 2t \rangle$, we have $\mathbf{r}'(0) = \langle 0, 0 \rangle = \mathbf{0}$. The point that corresponds to $t = 0$ is $(1, 0)$ and here there is a sharp corner that is called a **cusp**. We can think of smooth

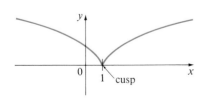

Figure 11.43

THE CURVE $\mathbf{r}(t) = \langle 1 + t^3, t^2 \rangle$

curves as curves with no such cusps. A curve, such as the semicubical parabola, that is made up of a finite number of smooth pieces is called **piecewise
smooth.** The arc length formula (11.56) holds for piecewise smooth curves.

A single curve C can be represented by more than one vector function.
For instance, the twisted cubic

$$(11.57) \qquad \mathbf{r}_1(t) = \langle t, t^2, t^3 \rangle \qquad 1 \leqslant t \leqslant 2$$

could also be represented by the function

$$(11.58) \qquad \mathbf{r}_2(u) = \langle e^u, e^{2u}, e^{3u} \rangle \qquad 0 \leqslant u \leqslant \ln 2$$

where the connection between the parameters t and u is given by $t = e^u$. We
say that Equations 11.57 and 11.58 are **parametrizations** of the curve C. If
you were to use Equation 11.56 to compute the length of C using Equations
11.57 and 11.58, you would get the same answer. In general it can be shown
(see Exercise 28) that when Equation 11.56 is used to compute the length of
any piecewise smooth curve, the arc length is independent of the parametrization that is used.

Now we suppose that C is a piecewise smooth curve given by
$\mathbf{r}(t) = f(t)\mathbf{i} + g(t)\mathbf{j} + h(t)\mathbf{k}$, $a \leqslant t \leqslant b$. As in Section 8.3, we define its **arc
length function** s by

$$(11.59) \qquad s(t) = \int_a^t |\mathbf{r}'(u)| \, du = \int_a^t \sqrt{\left(\frac{dx}{du}\right)^2 + \left(\frac{dy}{du}\right)^2 + \left(\frac{dz}{du}\right)^2} \, du$$

Thus $s(t)$ is the length of the part of C between $\mathbf{r}(a)$ and $\mathbf{r}(t)$ (see Figure 11.44).
If we differentiate both sides of Equation 11.59 using Part 1 of the Funda-
mental Theorem of Calculus, we obtain

$$(11.60) \qquad \frac{ds}{dt} = |\mathbf{r}'(t)|$$

Curvature

If C is a smooth curve defined by the vector function $\mathbf{r}$, then $\mathbf{r}'(t) \neq \mathbf{0}$. Recall
that the unit tangent vector $\mathbf{T}(t)$ is given by

$$\mathbf{T}(t) = \frac{\mathbf{r}'(t)}{|\mathbf{r}'(t)|}$$

and indicates the direction of the curve. From Figure 11.45 you can see that
$\mathbf{T}(t)$ changes direction very slowly when C is fairly straight but it changes
direction more quickly when C bends or twists more sharply.

The curvature of C at a given point is a measure of how quickly the
curve changes direction at that point. Specifically, we define it to be the magnitude of the rate of change of the unit tangent vector with respect to arc
length. (We use arc length so that the curvature will be independent of the
parametrization.)

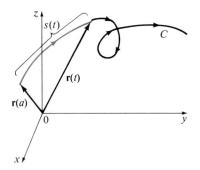

Figure 11.44

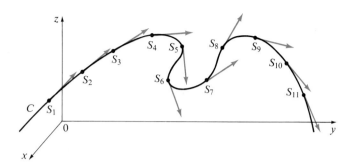

Figure 11.45

UNIT TANGENT VECTORS AT EQUALLY
SPACED POINTS ON C

Definition (11.61)

The **curvature** of a curve is

$$\kappa = \left| \frac{d\mathbf{T}}{ds} \right|$$

where $\mathbf{T}$ is the unit tangent vector.

The curvature is easier to compute if it is expressed in terms of the parameter t instead of s, so we use the Chain Rule (Theorem 11.53, Formula 6) to write

$$\frac{d\mathbf{T}}{dt} = \frac{d\mathbf{T}}{ds}\frac{ds}{dt} \qquad \text{and} \qquad \kappa = \left| \frac{d\mathbf{T}}{ds} \right| = \left| \frac{d\mathbf{T}/dt}{ds/dt} \right|$$

But $ds/dt = |\mathbf{r}'(t)|$ from Equation 11.60, so

(11.62)

$$\kappa(t) = \frac{|\mathbf{T}'(t)|}{|\mathbf{r}'(t)|}$$

● **Example 2**   Show that the curvature of a circle of radius a is $1/a$.

Solution   We can take the circle to have center the origin and then a parametrization is

$$\mathbf{r}(t) = a\cos t\,\mathbf{i} + a\sin t\,\mathbf{j}$$

Therefore $\mathbf{r}'(t) = -a\sin t\,\mathbf{i} + a\cos t\,\mathbf{j}$ and $|\mathbf{r}'(t)| = a$

so $$\mathbf{T}(t) = \frac{\mathbf{r}'(t)}{|\mathbf{r}'(t)|} = -\sin t\,\mathbf{i} + \cos t\,\mathbf{j}$$

and $$\mathbf{T}'(t) = -\cos t\,\mathbf{i} - \sin t\,\mathbf{j}$$

This gives $|\mathbf{T}'(t)| = 1$, so using Equation 11.62, we have

$$\kappa(t) = \frac{|\mathbf{T}'(t)|}{|\mathbf{r}'(t)|} = \frac{1}{a}$$

    The result of Example 2 shows that small circles have large curvature and large circles have small curvature, in accordance with our intuition. We can see directly from the definition of curvature that the curvature of a straight line is always 0 because the tangent vector is constant.

    Although Formula 11.62 can be used in all cases to compute the curvature, the formula given by the following theorem is often more convenient to apply.

Theorem (11.63)

The curvature of the curve given by the vector function $\mathbf{r}$ is

$$\kappa(t) = \frac{|\mathbf{r}'(t) \times \mathbf{r}''(t)|}{|\mathbf{r}'(t)|^3}$$

Proof   Since $\mathbf{T} = \mathbf{r}'/|\mathbf{r}'|$ and $|\mathbf{r}'| = ds/dt$, we have

$$\mathbf{r}' = |\mathbf{r}'|\mathbf{T} = \frac{ds}{dt}\mathbf{T}$$

so the Product Rule (Theorem 11.53, Formula 3) gives

$$\mathbf{r}'' = \frac{d^2s}{dt^2}\mathbf{T} + \frac{ds}{dt}\mathbf{T}'$$

Using the fact that $\mathbf{T} \times \mathbf{T} = \mathbf{0}$ (Example 2 in Section 11.4), we have

$$\mathbf{r}' \times \mathbf{r}'' = \left(\frac{ds}{dt}\right)^2 (\mathbf{T} \times \mathbf{T}')$$

Now $|\mathbf{T}(t)| = 1$ for all t, so $\mathbf{T}$ and $\mathbf{T}'$ are orthogonal by Example 7 in Section 11.7. Therefore, by Theorem 11.29,

$$|\mathbf{r}' \times \mathbf{r}''| = \left(\frac{ds}{dt}\right)^2 |\mathbf{T} \times \mathbf{T}'| = \left(\frac{ds}{dt}\right)^2 |\mathbf{T}||\mathbf{T}'| = \left(\frac{ds}{dt}\right)^2 |\mathbf{T}'|$$

Thus

$$|\mathbf{T}'| = \frac{|\mathbf{r}' \times \mathbf{r}''|}{\left(\dfrac{ds}{dt}\right)^2} = \frac{|\mathbf{r}' \times \mathbf{r}''|}{|\mathbf{r}'|^2}$$

and

$$\kappa = \frac{|\mathbf{T}'|}{|\mathbf{r}'|} = \frac{|\mathbf{r}' \times \mathbf{r}''|}{|\mathbf{r}'|^3}$$

● **Example 3**   Find the curvature of the twisted cubic $\mathbf{r}(t) = \langle t, t^2, t^3 \rangle$ at a general point and at $(0, 0, 0)$.

Solution   We first compute the required ingredients:

$$\mathbf{r}'(t) = \langle 1, 2t, 3t^2 \rangle \qquad \mathbf{r}''(t) = \langle 0, 2, 6t \rangle$$

$$|\mathbf{r}'(t)| = \sqrt{1 + 4t^2 + 9t^4}$$

$$\mathbf{r}'(t) \times \mathbf{r}''(t) = \begin{vmatrix} \mathbf{i} & \mathbf{j} & \mathbf{k} \\ 1 & 2t & 3t^2 \\ 0 & 2 & 6t \end{vmatrix} = 6t^2\mathbf{i} - 6t\mathbf{j} + 2\mathbf{k}$$

$$|\mathbf{r}'(t) \times \mathbf{r}''(t)| = \sqrt{36t^4 + 36t^2 + 4} = 2\sqrt{9t^4 + 9t^2 + 1}$$

Theorem 11.63 then gives

$$\kappa(t) = \frac{|\mathbf{r}'(t) \times \mathbf{r}''(t)|}{|\mathbf{r}'(t)|^3} = \frac{2\sqrt{1 + 9t^2 + 9t^4}}{(1 + 4t^2 + 9t^4)^{3/2}}$$

At the origin the curvature is $\kappa(0) = 2$.

For the special case of a plane curve with equation $y = f(x)$ we can choose x as the parameter and write $\mathbf{r}(x) = x\mathbf{i} + f(x)\mathbf{j}$. Then $\mathbf{r}'(x) = \mathbf{i} + f'(x)\mathbf{j}$, $\mathbf{r}''(x) = f''(x)\mathbf{j}$, and, since $\mathbf{i} \times \mathbf{j} = \mathbf{k}$ and $\mathbf{j} \times \mathbf{j} = \mathbf{0}$, we have $\mathbf{r}'(x) \times \mathbf{r}''(x) = f''(x)\mathbf{k}$. Also $|\mathbf{r}'(x)| = \sqrt{1 + [f'(x)]^2}$, so, by Theorem 11.63,

(11.64)
$$\kappa(x) = \frac{|f''(x)|}{[1 + (f'(x))^2]^{3/2}}$$

● **Example 4**   Find the curvature of the parabola $y = x^2$ at the points $(0, 0)$, $(1, 1)$, and $(2, 4)$.

Solution   Since $y' = 2x$ and $y'' = 2$, Formula 11.64 gives

$$\kappa(x) = \frac{|y''|}{[1 + (y')^2]^{3/2}} = \frac{2}{(1 + 4x^2)^{3/2}}$$

The curvature at $(0, 0)$ is $\kappa(0) = 2$. At $(1, 1)$ it is $\kappa(1) = 2/5^{3/2} \approx 0.18$. At $(2, 4)$ it is $\kappa(2) = 2/17^{3/2} \approx 0.03$. Observe that $\kappa(x) \to 0$ as $x \to \pm\infty$.

The Unit Normal

At a given point on a smooth space curve $\mathbf{r}(t)$, there are many vectors that are orthogonal to the unit tangent vector $\mathbf{T}(t)$ (see Figure 11.46). We single one out by observing that since $|\mathbf{T}(t)| = 1$ for all t, we have $\mathbf{T}(t) \cdot \mathbf{T}'(t) = 0$

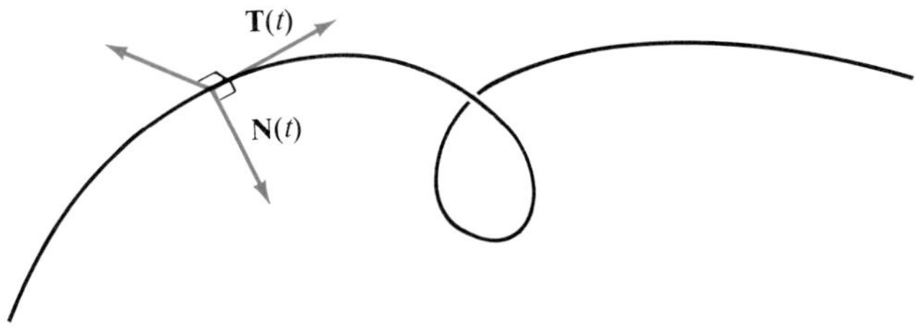

Figure 11.46

by Example 7 in Section 11.7, so $\mathbf{T}'(t)$ is orthogonal to $\mathbf{T}(t)$. If $\mathbf{r}'$ is also smooth, we can define the **principal unit normal vector** $\mathbf{N}(t)$ (or **unit normal** for short) as

$$\mathbf{N}(t) = \frac{\mathbf{T}'(t)}{|\mathbf{T}'(t)|}$$

● **Example 5**   Find the unit normal vector for the circular helix $\mathbf{r}(t) = \cos t\mathbf{i} + \sin t\mathbf{j} + t\mathbf{k}$.

Solution

$$\mathbf{r}'(t) = -\sin t\mathbf{i} + \cos t\mathbf{j} + \mathbf{k} \qquad |\mathbf{r}'(t)| = \sqrt{2}$$

$$\mathbf{T}(t) = \frac{\mathbf{r}'(t)}{|\mathbf{r}'(t)|} = \frac{1}{\sqrt{2}}(-\sin t\mathbf{i} + \cos t\mathbf{j} + \mathbf{k})$$

$$\mathbf{T}'(t) = \frac{1}{\sqrt{2}}(-\cos t\mathbf{i} - \sin t\mathbf{j}) \qquad |\mathbf{T}'(t)| = \frac{1}{\sqrt{2}}$$

$$\mathbf{N}(t) = \frac{\mathbf{T}'(t)}{|\mathbf{T}'(t)|} = -\cos t\mathbf{i} - \sin t\mathbf{j} = \langle -\cos t, -\sin t, 0 \rangle$$

This shows that the normal vector at a point on the helix is horizontal and points toward the z-axis.   ●

We summarize here the formulas for unit tangent and unit normal vectors and curvature.

$$\mathbf{T}(t) = \frac{\mathbf{r}'(t)}{|\mathbf{r}'(t)|} \qquad \mathbf{N}(t) = \frac{\mathbf{T}'(t)}{|\mathbf{T}'(t)|}$$

$$|\mathbf{T}| = |\mathbf{N}| = 1 \qquad \mathbf{T} \cdot \mathbf{N} = 0$$

$$\kappa = \left|\frac{d\mathbf{T}}{ds}\right| = \frac{|\mathbf{T}'(t)|}{|\mathbf{r}'(t)|} = \frac{|\mathbf{r}'(t) \times \mathbf{r}''(t)|}{|\mathbf{r}'(t)|^3}$$

EXERCISES 11.8

In Exercises 1–6 find the length of the given curve.

1. $\mathbf{r}(t) = \langle 2t, 3\sin t, 3\cos t \rangle, a \leqslant t \leqslant b$

2. $\mathbf{r}(t) = \langle e^t, e^t \sin t, e^t \cos t \rangle, 0 \leqslant t \leqslant 2\pi$

3. $\mathbf{r}(t) = 6t\mathbf{i} + 3\sqrt{2}t^2\mathbf{j} + 2t^3\mathbf{k}, 0 \leqslant t \leqslant 1$

4. $\mathbf{r}(t) = t^2\mathbf{i} + 2t\mathbf{j} + \ln t\mathbf{k}, 1 \leqslant t \leqslant e$

5. $x = 2t, y = t^2, z = t^2, 0 \leqslant t \leqslant 1$

6. $x = t\cos t, y = t\sin t, z = t, 0 \leqslant t \leqslant \dfrac{\pi}{2}$

In Exercises 7–12 (a) find the unit tangent and unit normal vectors $\mathbf{T}(t)$ and $\mathbf{N}(t)$, and (b) use Formula 11.62 to find the curvature.

7. $\mathbf{r}(t) = \langle \sin 4t, 3t, \cos 4t \rangle$

8. $\mathbf{r}(t) = \langle 6t, 3\sqrt{2}t^2, 2t^3 \rangle$

9. $\mathbf{r}(t) = \langle \sqrt{2}t, e^t, e^{-t} \rangle$

10. $\mathbf{r}(t) = \langle \sqrt{2}\cos t, \sin t, \sin t \rangle$

11. $\mathbf{r}(t) = \langle t^2, 2t^3/3, t \rangle$ **12.** $\mathbf{r}(t) = \langle t^2, 2t, \ln t \rangle$

In Exercises 13–18 use Theorem 11.63 to find the curvature.

13. $\mathbf{r}(t) = \mathbf{i} + t\mathbf{j} - t^2\mathbf{k}$

14. $\mathbf{r}(t) = (1 + t)\mathbf{i} + (1 - t)\mathbf{j} + 3t^2\mathbf{k}$

15. $\mathbf{r}(t) = 2t^3\mathbf{i} - 3t^2\mathbf{j} + 6t\mathbf{k}$

16. $\mathbf{r}(t) = (t^2 + 2)\mathbf{i} + (t^2 - 4t)\mathbf{j} + 2t\mathbf{k}$

17. $\mathbf{r}(t) = \sin t\mathbf{i} + \cos t\mathbf{j} + \sin t\mathbf{k}$

18. $\mathbf{r}(t) = t\mathbf{i} + 4t^{3/2}\mathbf{j} + t^2\mathbf{k}$

In Exercises 19–22 use Formula 11.64 to find the curvature.

19. $y = x^3$ **20.** $y = \sqrt{x}$

21. $y = \sin x$ **22.** $y = \ln x$

23. At what point does the curve $y = e^x$ have maximum curvature?

24. Use Theorem 11.63 to show that the curvature of a plane parametric curve $x = f(t)$, $y = g(t)$ is

$$\kappa = \frac{|\dot{x}\ddot{y} - \dot{y}\ddot{x}|}{[\dot{x}^2 + \dot{y}^2]^{3/2}}$$

where the dots indicate derivatives with respect to t.

In Exercises 25 and 26 use the formula in Exercise 24 to find the curvature.

25. $x = t^3$, $y = t^2$ **26.** $x = t\sin t$, $y = t\cos t$

27. Show that the curvature κ is related to the tangent and normal vectors by the equation

$$\frac{d\mathbf{T}}{ds} = \kappa\mathbf{N}$$

28. Show that arc length is independent of parametrization. [*Hint*: Suppose C is given by $\mathbf{r}_1(t)$, $a \leqslant t \leqslant b$, and also by $\mathbf{r}_2(u)$, $\alpha \leqslant u \leqslant \beta$, where $t = g(u)$ and $g'(u) > 0$. If $L_1 = \int_a^b |\mathbf{r}'_1(t)|\, dt$ and $L_2 = \int_\alpha^\beta |\mathbf{r}'_2(u)|\, du$, show that $L_1 = L_2$.]

SECTION 11.9

Motion in Space: Velocity and Acceleration

Suppose a particle moves through space so that its position vector at time t is $\mathbf{r}(t)$. Notice from Figure 11.42(b) that, for small values of t, the vector

(11.65)
$$\frac{\mathbf{r}(t + h) - \mathbf{r}(t)}{h}$$

approximates the direction of the particle moving along the curve $\mathbf{r}(t)$. Its magnitude measures the size of the displacement vector per unit time. The vector (11.65) gives the average velocity over a time interval of length h and its limit is the **velocity vector** $\mathbf{v}(t)$ at time t:

(11.66)
$$\mathbf{v}(t) = \lim_{h \to 0} \frac{\mathbf{r}(t + h) - \mathbf{r}(t)}{h} = \mathbf{r}'(t)$$

Thus the velocity vector points in the direction of the unit tangent vector.

The **speed** of the particle at time t is the magnitude of the velocity vector, that is, $|\mathbf{v}(t)|$. This is appropriate because, from Equations 11.66 and 11.60, we have

$$|\mathbf{v}(t)| = |\mathbf{r}'(t)| = \frac{ds}{dt} = \text{rate of change of distance with respect to time}$$

As in the case of one-dimensional motion, the **acceleration** of the particle is defined as the derivative of the velocity:

$$\mathbf{a}(t) = \mathbf{v}'(t) = \mathbf{r}''(t)$$

● **Example 1**   The position vector of an object moving in a plane is given by $\mathbf{r}(t) = t^3\mathbf{i} + t^2\mathbf{j}$, $t \geqslant 0$. Find its velocity, speed, and acceleration when $t = 1$ and illustrate geometrically.

Solution   The velocity and acceleration at time t are

$$\mathbf{v}(t) = \mathbf{r}'(t) = 3t^2\mathbf{i} + 2t\mathbf{j}$$

$$\mathbf{a}(t) = \mathbf{r}''(t) = 6t\mathbf{i} + 2\mathbf{j}$$

and the speed is

$$|\mathbf{v}(t)| = \sqrt{(3t^2)^2 + (2t)^2} = \sqrt{9t^4 + 4t^2}$$

When $t = 1$, we have

$$\mathbf{v}(1) = 3\mathbf{i} + 2\mathbf{j} \qquad \mathbf{a}(1) = 6\mathbf{i} + 2\mathbf{j} \qquad |\mathbf{v}(1)| = \sqrt{13}$$

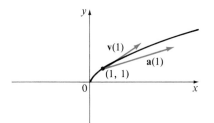

Figure 11.47

These velocity and acceleration vectors are shown in Figure 11.47.   ●

● **Example 2**   Find the velocity, acceleration, and speed of a particle with position vector $\mathbf{r}(t) = \langle t^2, e^t, te^t \rangle$.

Solution

$$\mathbf{v}(t) = \mathbf{r}'(t) = \langle 2t, e^t, (1 + t)e^t \rangle$$

$$\mathbf{a}(t) = \mathbf{v}'(t) = \langle 2, e^t, (2 + t)e^t \rangle$$

$$|\mathbf{v}(t)| = \sqrt{4t^2 + e^{2t} + (1 + t)^2 e^{2t}} \qquad ●$$

The vector integrals that were introduced in Section 11.7 can be used to find position vectors when velocity or acceleration vectors are known as in the following example.

● **Example 3**   A moving particle starts at an initial position $\mathbf{r}(0) = \langle 1, 0, 0 \rangle$ with initial velocity $\mathbf{v}(0) = \mathbf{i} - \mathbf{j} + \mathbf{k}$. Its acceleration is $\mathbf{a}(t) = 4t\mathbf{i} + 6t\mathbf{j} + \mathbf{k}$. Find its velocity and position at time t.

Solution   Since $\mathbf{a}(t) = \mathbf{v}'(t)$, we have

$$\mathbf{v}(t) = \int \mathbf{a}(t)\,dt = \int (4t\mathbf{i} + 6t\mathbf{j} + \mathbf{k})\,dt$$

$$= 2t^2\mathbf{i} + 3t^2\mathbf{j} + t\mathbf{k} + \mathbf{C}$$

To determine the value of the constant vector $\mathbf{C}$, we use the fact that $\mathbf{v}(0) = \mathbf{i} - \mathbf{j} + \mathbf{k}$. The preceding equation gives $\mathbf{v}(0) = \mathbf{C}$, so $\mathbf{C} = \mathbf{i} - \mathbf{j} + \mathbf{k}$ and

$$\mathbf{v}(t) = 2t^2\mathbf{i} + 3t^2\mathbf{j} + t\mathbf{k} + \mathbf{i} - \mathbf{j} + \mathbf{k}$$
$$= (2t^2 + 1)\mathbf{i} + (3t^2 - 1)\mathbf{j} + (t + 1)\mathbf{k}$$

Since $\mathbf{v}(t) = \mathbf{r}'(t)$, we have

$$\mathbf{r}(t) = \int \mathbf{v}(t)\, dt = \int [(2t^2 + 1)\mathbf{i} + (3t^2 - 1)\mathbf{j} + (t + 1)\mathbf{k}]\, dt$$
$$= \left(\frac{2t^3}{3} + t\right)\mathbf{i} + (t^3 - t)\mathbf{i} + \left(\frac{t^2}{2} + t\right)\mathbf{k} + \mathbf{D}$$

Putting $t = 0$, we find that $\mathbf{D} = \mathbf{r}(0) = \mathbf{i}$, so

$$\mathbf{r}(t) = \left(\frac{2t^3}{3} + t + 1\right)\mathbf{i} + (t^3 - t)\mathbf{j} + \left(\frac{t^2}{2} + t\right)\mathbf{k} \qquad \bullet$$

In general, vector integrals allow us to recover velocity when acceleration is known and position when velocity is known:

$$\mathbf{v}(t) = \mathbf{v}(t_0) + \int_{t_0}^{t} \mathbf{a}(u)\, du \qquad \mathbf{r}(t) = \mathbf{r}(t_0) + \int_{t_0}^{t} \mathbf{v}(u)\, du$$

If the force that acts on a particle is known, then the acceleration can be found from **Newton's Second Law of Motion.** The vector version of this law states that if, at any time t, a force $\mathbf{F}(t)$ acts on an object of mass m producing an acceleration $\mathbf{a}(t)$, then

$$\mathbf{F}(t) = m\mathbf{a}(t)$$

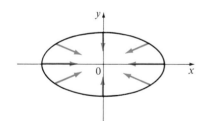

● **Example 4**   An object with mass m that moves in an elliptical path with constant angular speed ω has position vector $\mathbf{r}(t) = a\cos\omega t\,\mathbf{i} + b\sin\omega t\,\mathbf{j}$. Find the force acting on the object and show that it is directed toward the origin.

Solution $$\mathbf{v}(t) = \mathbf{r}'(t) = -a\omega\sin\omega t\,\mathbf{i} + b\omega\cos\omega t\,\mathbf{j}$$
$$\mathbf{a}(t) = \mathbf{v}'(t) = -a\omega^2\cos\omega t\,\mathbf{i} - b\omega^2\sin\omega t\,\mathbf{j}$$

Figure 11.48

Therefore Newton's Second Law gives the force as

$$\mathbf{F}(t) = m\mathbf{a}(t) = -m\omega^2(a\cos\omega t\,\mathbf{i} + b\sin\omega t\,\mathbf{j})$$

Notice that $\mathbf{F}(t) = -m\omega^2\mathbf{r}(t)$. This shows that the force acts in the direction opposite to the radius vector $\mathbf{r}(t)$ and therefore points toward the origin (see Figure 11.48). Such a force is called a *centripetal* (center-seeking) force.   ●

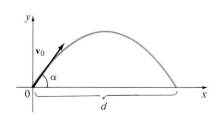

● **Example 5**   A projectile is fired with angle of elevation α and initial velocity $\mathbf{v}_0$ (see Figure 11.49). Assuming that air resistance is negligible and the only external force is due to gravity, find the position function $\mathbf{r}(t)$ of the

Figure 11.49

projectile. What value of α will maximize the range (the horizontal distance traveled)?

Solution   We set up the axes so that the projectile starts at the origin. Since the force due to gravity acts downward, we have

$$\mathbf{F} = m\mathbf{a} = -mg\mathbf{j}$$

where $g = |\mathbf{a}| \approx 9.8 \text{ m/s}^2$. Thus

$$\mathbf{a} = -g\mathbf{j}$$

Since $\mathbf{v}'(t) = \mathbf{a}$, we have

$$\mathbf{v}(t) = -gt\mathbf{j} + \mathbf{C}$$

where $\mathbf{C} = \mathbf{v}(0) = \mathbf{v}_0$. Therefore

$$\mathbf{r}'(t) = \mathbf{v}(t) = -gt\mathbf{j} + \mathbf{v}_0$$

Integrating again, we obtain

$$\mathbf{r}(t) = -\tfrac{1}{2}gt^2\mathbf{j} + t\mathbf{v}_0 + \mathbf{D}$$

But $\mathbf{D} = \mathbf{r}(0) = \mathbf{0}$, so the position vector of the projectile is given by

(11.67)
$$\mathbf{r}(t) = -\tfrac{1}{2}gt^2\mathbf{j} + t\mathbf{v}_0$$

If we write $|\mathbf{v}_0| = v_0$ (the initial speed of the projectile), then

$$\mathbf{v}_0 = v_0 \cos \alpha \mathbf{i} + v_0 \sin \alpha \mathbf{j}$$

and Equation 11.67 becomes

$$\mathbf{r}(t) = (v_0 \cos \alpha)t\mathbf{i} + [(v_0 \sin \alpha)t - \tfrac{1}{2}gt^2]\mathbf{j}$$

The parametric equations of the trajectory are therefore

(11.68)
$$x = (v_0 \cos \alpha)t \qquad y = (v_0 \sin \alpha)t - \tfrac{1}{2}gt^2$$

The horizontal distance d is the value of x when $y = 0$. Setting $y = 0$, we obtain $t = 0$ or $t = (2v_0 \sin \alpha)/g$. The latter value of t then gives

$$d = x = (v_0 \cos \alpha)\frac{2v_0 \sin \alpha}{g} = \frac{v_0^2 \sin 2\alpha}{g}$$

Clearly, d has its maximum value when $\sin 2\alpha = 1$, that is, $\alpha = \pi/4$.   ●

Tangential and Normal Components of Acceleration

When studying the motion of a particle, it is often of use to resolve the acceleration into two components, one in the direction of the tangent and the other in the direction of the normal. If we write $v = |\mathbf{v}|$ for the speed of the particle, then

$$\mathbf{T}(t) = \frac{\mathbf{r}'(t)}{|\mathbf{r}'(t)|} = \frac{\mathbf{v}(t)}{|\mathbf{v}(t)|} = \frac{\mathbf{v}}{v}$$

and so

$$\mathbf{v} = v\mathbf{T}$$

If we differentiate both sides of this equation with respect to t, we get

(11.69)
$$\mathbf{a} = \mathbf{v}' = v'\mathbf{T} + v\mathbf{T}'$$

If we use the expression for curvature given by Equation 11.62, then we have

(11.70)
$$\kappa = \frac{|\mathbf{T}'|}{|\mathbf{r}'|} = \frac{|\mathbf{T}'|}{v} \quad \text{so} \quad |\mathbf{T}'| = \kappa v$$

The unit normal vector was defined in the preceding section as $\mathbf{N} = \mathbf{T}'/|\mathbf{T}'|$, so (11.70) gives

$$\mathbf{T}' = |\mathbf{T}'|\mathbf{N} = \kappa v\mathbf{N}$$

and Equation 11.69 becomes

(11.71)
$$\boxed{\mathbf{a} = v'\mathbf{T} + \kappa v^2\mathbf{N}}$$

Writing a_T and a_N for the tangential and normal components of acceleration, we have

$$\mathbf{a} = a_T\mathbf{T} + a_N\mathbf{N}$$

(11.72) where $a_T = v' \quad \text{and} \quad a_N = \kappa v^2$

This resolution is illustrated in Figure 11.50.

Although we have expressions for the tangential and normal components of acceleration in Equations 11.72, it is desirable to have expressions that depend only on $\mathbf{r}$, $\mathbf{r}'$, and $\mathbf{r}''$. To this end we take the dot product of $\mathbf{v} = v\mathbf{T}$ with $\mathbf{a}$ as given by Equation 11.71:

$$\mathbf{v} \cdot \mathbf{a} = v\mathbf{T} \cdot (v'\mathbf{T} + \kappa v^2\mathbf{N})$$
$$= vv'\mathbf{T} \cdot \mathbf{T} + \kappa v^3\mathbf{T} \cdot \mathbf{N}$$
$$= vv' \quad (\text{since } \mathbf{T} \cdot \mathbf{T} = 1 \text{ and } \mathbf{T} \cdot \mathbf{N} = 0)$$

(11.73) Therefore $a_T = v' = \dfrac{\mathbf{v} \cdot \mathbf{a}}{v} = \dfrac{\mathbf{r}'(t) \cdot \mathbf{r}''(t)}{|\mathbf{r}'(t)|}$

Using the formula for curvature given by Theorem 11.63, we have

(11.74)
$$a_N = \kappa v^2 = \frac{|\mathbf{r}'(t) \times \mathbf{r}''(t)|}{|\mathbf{r}'(t)|^3} |\mathbf{r}'(t)|^2 = \frac{|\mathbf{r}'(t) \times \mathbf{r}''(t)|}{|\mathbf{r}'(t)|}$$

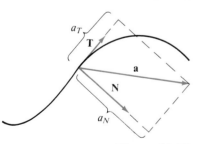

Figure 11.50

● **Example 6**   A particle moves with position function $r(t) = \langle t^2, t^2, t^3 \rangle$. Find the tangential and normal components of acceleration.

Solution

$$\mathbf{r}(t) = t^2\mathbf{i} + t^2\mathbf{j} + t^3\mathbf{k}$$

$$\mathbf{r}'(t) = 2t\mathbf{i} + 2t\mathbf{j} + 3t^2\mathbf{k}$$

$$\mathbf{r}''(t) = 2\mathbf{i} + 2\mathbf{j} + 6t\mathbf{k}$$

$$|\mathbf{r}'(t)| = \sqrt{8t^2 + 9t^4}$$

Therefore Equation 11.73 gives the tangential component as

$$a_T = \frac{\mathbf{r}'(t) \cdot \mathbf{r}''(t)}{|\mathbf{r}'(t)|} = \frac{8t + 18t^3}{\sqrt{8t^2 + 9t^4}}$$

Since

$$\mathbf{r}'(t) \times \mathbf{r}''(t) = \begin{vmatrix} \mathbf{i} & \mathbf{j} & \mathbf{k} \\ 2t & 2t & 3t^2 \\ 2 & 2 & 6t \end{vmatrix} = 6t^2\mathbf{i} - 6t^2\mathbf{j}$$

Equation 11.74 gives the normal component as

$$a_N = \frac{|\mathbf{r}'(t) \times \mathbf{r}''(t)|}{|\mathbf{r}'(t)|} = \frac{6\sqrt{2}t^2}{\sqrt{8t^2 + 9t^4}}$$ ●

EXERCISES 11.9

In Exercises 1–6 find the velocity, acceleration, and speed of a particle with the given position function. Sketch the path of the particle and draw the velocity and acceleration vectors for the given value of t.

1. $\mathbf{r}(t) = \langle t^2 - 1, t \rangle$, $t = 1$

2. $\mathbf{r}(t) = \langle \sqrt{t}, 1 - t \rangle$, $t = 1$

3. $\mathbf{r}(t) = e^t\mathbf{i} + e^{-t}\mathbf{j}$, $t = 0$

4. $\mathbf{r}(t) = \sin t\mathbf{i} + 2\cos t\mathbf{j}$, $t = \dfrac{\pi}{6}$

5. $\mathbf{r}(t) = \sin t\mathbf{i} + t\mathbf{j} + \cos t\mathbf{k}$, $t = 0$

6. $\mathbf{r}(t) = t\mathbf{i} + t^2\mathbf{j} + t^3\mathbf{k}$, $t = 1$

In Exercises 7–12 find the velocity, acceleration, and speed of a particle with the given position function.

7. $\mathbf{r}(t) = \langle t^3, t^2 + 1, t^3 - 1 \rangle$

8. $\mathbf{r}(t) = \langle \sqrt{t}, t, t\sqrt{t} \rangle$

9. $\mathbf{r}(t) = \dfrac{1}{t}\mathbf{i} + \mathbf{j} + t^2\mathbf{k}$

10. $\mathbf{r}(t) = e^t\mathbf{i} + 2t\mathbf{j} + e^{-t}\mathbf{k}$

11. $\mathbf{r}(t) = e^t(\cos t\mathbf{i} + \sin t\mathbf{j} + t\mathbf{k})$

12. $\mathbf{r}(t) = \cosh t\mathbf{i} + \sinh t\mathbf{j} + t\mathbf{k}$

In Exercises 13–16 find the velocity and position vectors of a particle that has the given acceleration and the given initial velocity and position.

13. $\mathbf{a}(t) = \mathbf{k}$, $\mathbf{v}(0) = \mathbf{i} - \mathbf{j}$, $\mathbf{r}(0) = \mathbf{0}$

14. $\mathbf{a}(t) = -10\mathbf{k}$, $\mathbf{v}(0) = \mathbf{i} + \mathbf{j} - \mathbf{k}$, $\mathbf{r}(0) = 2\mathbf{i} + 3\mathbf{j}$

15. $\mathbf{a}(t) = \mathbf{i} + 2\mathbf{j} + 2t\mathbf{k}$, $\mathbf{v}(0) = \mathbf{0}$, $\mathbf{r}(0) = \mathbf{i} + \mathbf{k}$

16. $\mathbf{a}(t) = t\mathbf{i} + t^2\mathbf{j} + \cos 2t\mathbf{k}$, $\mathbf{v}(0) = \mathbf{i} + \mathbf{k}$, $\mathbf{r}(0) = \mathbf{j}$

17. The position function of a particle is given by $\mathbf{r}(t) = \langle t^2, 5t, t^2 - 16t \rangle$. When is the speed a minimum?

18. What force is required so that a particle of mass m has the position function $\mathbf{r}(t) = t^3\mathbf{i} + t^2\mathbf{j} + t^3\mathbf{k}$?

19. A force with magnitude 20 N acts directly upward from the xy-plane on an object with mass 4 kg. The object starts at the origin with initial velocity $\mathbf{v}(0) = \mathbf{i} - \mathbf{j}$. Find its position function and its speed at time t.

20. Show that if a particle moves with constant speed, then the velocity and acceleration vectors are orthogonal.

21. A projectile is fired with an initial speed of 500 m/s and angle of elevation 30°. Find (a) the range of the projectile, (b) the maximum height reached, and (c) the speed at impact.

22. Do Exercise 21 if the projectile is fired from a position 200 m above the ground.

23. A ball is thrown at an angle of 45° to the ground. If the ball lands 90 m away, what was the initial speed of the ball?

24. A gun has muzzle speed 120 m/s. What angle of elevation should be used to hit an object 500 m away?

25. Show that the trajectory of a projectile is a parabola by eliminating the parameter from Equations 11.68.

In Exercises 26–31 find the tangential and normal components of the acceleration vector.

26. $\mathbf{r}(t) = (t^2 + 4)\mathbf{i} + (2t - 3)\mathbf{j}$

27. $\mathbf{r}(t) = (t - \sin t)\mathbf{i} + (1 - \cos t)\mathbf{j}$

28. $\mathbf{r}(t) = t\mathbf{i} + 4\sin t\mathbf{j} + 4\cos t\mathbf{k}$

29. $\mathbf{r}(t) = t^3\mathbf{i} + t^2\mathbf{j} + t\mathbf{k}$

30. $\mathbf{r}(t) = t\mathbf{i} + \cos^2 t\mathbf{j} + \sin^2 t\mathbf{k}$

31. $\mathbf{r}(t) = e^t\mathbf{i} + \sqrt{2}t\mathbf{j} + e^{-t}\mathbf{k}$

Exercises 32–35 are concerned with the derivation of Kepler's laws of planetary motion. Johannes Kepler (1571–1630) stated these laws after 20 years of studying the astronomical observations of Tycho Brahe. Later Newton was able to show that these three laws are consequences of two of his own laws. We use a coordinate system with the sun at the origin and let $\mathbf{r} = \mathbf{r}(t)$ be the position vector of a planet. (Equally well, $\mathbf{r}$ could be the position vector of the moon or a satellite moving around the earth or a comet moving around a star.) The velocity vector is $\mathbf{v} = \mathbf{r}'$ and the acceleration vector is $\mathbf{a} = \mathbf{r}''$. You may assume the following laws of Newton:

Second Law of Motion: $\mathbf{F} = m\mathbf{a}$

Law of Gravitation: $\mathbf{F} = -\dfrac{GMm}{r^3}\mathbf{r} = -\dfrac{GMm}{r^2}\mathbf{u}$

where $\mathbf{F}$ is the gravitational force on the planet, m and M are the masses of the planet and the sun, G is the universal gravitational constant, $r = |\mathbf{r}|$, and $\mathbf{u} = \mathbf{r}/r$ is the unit vector in the direction of $\mathbf{r}$.

32. Show that
$$\frac{d}{dt}(\mathbf{r} \times \mathbf{v}) = 0$$

It follows that
$$\mathbf{r} \times \mathbf{v} = \mathbf{h}$$

where $\mathbf{h}$ is a constant vector, and therefore the planet always lies in the plane through 0 perpendicular to $\mathbf{h}$. *The orbit of the planet is a plane curve.*

33. (a) Show that $\mathbf{h} = r^2\mathbf{u} \times \mathbf{u}'$.
    (b) Deduce that $\mathbf{a} \times \mathbf{h} = -GM\mathbf{u} \times (\mathbf{u} \times \mathbf{u}')$.
    (c) Deduce that $(\mathbf{v} \times \mathbf{h})' = GM\mathbf{u}'$. (*Hint:* Use Formula 6 of Theorem 11.31 and differentiate $\mathbf{u} \cdot \mathbf{u} = 1$ to get $\mathbf{u} \cdot \mathbf{u}' = 0$.)
    (d) Deduce that
$$\mathbf{v} \times \mathbf{h} = GM\mathbf{u} + \mathbf{c}$$

where $\mathbf{c}$ is a constant vector.
    (e) Deduce that
$$\mathbf{r} \cdot (\mathbf{v} \times \mathbf{h}) = GMr + cr\cos\theta$$

where $c = |\mathbf{c}|$ and θ is the angle between $\mathbf{c}$ and $\mathbf{r}$.
    (f) Deduce that
$$r = \frac{ed}{1 + e\cos\theta}$$

where $e = \dfrac{c}{GM}$ $d = \dfrac{h^2}{c}$ and $h = |\mathbf{h}|$

Comparing with Theorem 9.32, we see that this is the equation in polar coordinates (r, θ) of a conic section with focus at the origin and eccentricity e.
    We know that the orbit of a planet is a closed curve and so the conic must be an ellipse. This proves **Kepler's First Law: A planet revolves around the sun in an elliptical orbit with the sun at one focus.**

34. Choose the axes so that the standard basis vector $\mathbf{k}$ points in the direction of $\mathbf{h}$ and use polar coordinates with $\mathbf{i}$ in the direction of the polar axis so that $\mathbf{r} = (r\cos\theta)\mathbf{i} + (r\sin\theta)\mathbf{j}$.

    (a) Show that
$$\mathbf{h} = r^2\frac{d\theta}{dt}\mathbf{k}.$$

    (b) Deduce that
$$r^2\frac{d\theta}{dt} = h.$$

(c) If $A = A(t)$ is the area swept out by the radius vector $\mathbf{r} = \mathbf{r}(t)$ in the time interval $[t_0, t]$, show that

$$\frac{dA}{dt} = \frac{1}{2} r^2 \frac{d\theta}{dt}$$

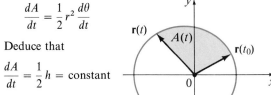

(d) Deduce that

$$\frac{dA}{dt} = \frac{1}{2} h = \text{constant}$$

This says that the rate at which A is swept out is constant and proves **Kepler's Second Law: The line joining the sun to a planet sweeps out equal areas in equal times.**

35. Let T be the period of a planet about the sun; that is, T is the time required for it to travel once around its elliptical orbit. Suppose that the lengths of the major and minor axes of the ellipse are $2a$ and $2b$.

(a) Use Exercise 34 part (d) to show that $T = 2\pi ab/h$.

(b) Show that

$$\frac{h^2}{GM} = ed = \frac{b^2}{a}$$

(c) Use parts (a) and (b) to show that

$$T^2 = \frac{4\pi^2}{GM} a^3$$

This proves **Kepler's Third Law: The square of the period of a planet is proportional to the cube of the semimajor axis of its orbit.** (Notice that the proportionality constant $4\pi^2/GM$ is independent of the planet.)

Cylindrical and Spherical Coordinates

Recall that in plane geometry we introduced the polar coordinate system in order to give a more convenient description of certain curves and regions. In three dimensions there are two coordinate systems that are similar to polar coordinates and give convenient descriptions of some commonly occurring surfaces and solids.

In the **cylindrical coordinate system**, a point P in three-dimensional space is represented by the ordered triple (r, θ, z) where r and θ are polar coordinates of the projection of P onto the xy-plane and z is the directed distance from the xy-plane to P (see Figure 11.51).

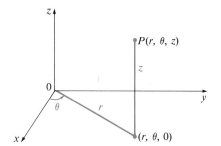

Figure 11.51

To convert from cylindrical to rectangular coordinates we use the equations

(11.75)
$$x = r \cos \theta \qquad y = r \sin \theta \qquad z = z$$

whereas to convert from rectangular to cylindrical coordinates we use

(11.76)
$$r^2 = x^2 + y^2 \qquad \tan \theta = \frac{y}{x} \qquad z = z$$

These equations follow from Equations 9.10 and 9.11.

Cylindrical coordinates are useful in problems that involve symmetry about the z-axis. For instance, the axis of the circular cylinder with Cartesian equation $x^2 + y^2 = c^2$ is the z-axis. In cylindrical coordinates this cylinder has the very simple equation $r = c$ (see Figure 11.52). This is the reason for the name "cylindrical" coordinates.

● **Example 1**   Describe the surface whose equation in cylindrical coordinates is $z = r$.

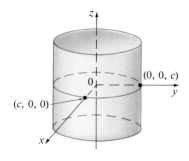

Figure 11.52

THE CYLINDER $r = c$

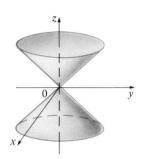

Figure 11.53

THE CONE $z = r$

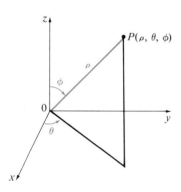

Figure 11.54

Solution   We first convert to an equation in rectangular coordinates. From the first equation in (11.76) we have

$$z^2 = r^2 = x^2 + y^2$$

We recognize the equation $z^2 = x^2 + y^2$ (by comparison with Equation 11.45) as being a circular cone whose axis is the z-axis (see Figure 11.53). ●

● **Example 2**   Find an equation in cylindrical coordinates for the ellipsoid $4x^2 + 4y^2 + z^2 = 1$.

Solution   Since $r^2 = x^2 + y^2$ from Equations 11.76, we have

$$z^2 = 1 - 4(x^2 + y^2) = 1 - 4r^2$$

So the equation of the ellipsoid in cylindrical coordinates is $z^2 = 1 - 4r^2$. ●

The **spherical coordinates** (ρ, θ, ϕ) of a point P in space are shown in Figure 11.54 where $\rho = |OP|$ is the distance from the origin to P, θ is the same angle as in cylindrical coordinates, and ϕ is the angle between the positive z-axis and the line segment OP. Note that

$$\rho \geqslant 0 \qquad 0 \leqslant \theta < 2\pi \qquad 0 \leqslant \phi \leqslant \pi$$

The spherical coordinate system is especially useful in problems where there is symmetry about the origin. For example, the sphere with center the origin and radius c has the simple equation $\rho = c$ (see Figure 11.55) and this is the reason for the name "spherical" coordinates. The graph of the equation $\theta = c$ is a vertical half-plane (see Figure 11.56), and the equation $\phi = c$ represents a half-cone with the z-axis as its axis (see Figure 11.57).

The relationship between rectangular and spherical coordinates can be seen from Figure 11.58. From triangles OPQ and OPP' we have

$$z = \rho \cos \phi \qquad r = \rho \sin \phi$$

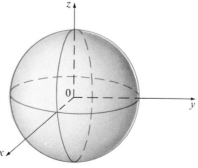

$\rho = c$, a sphere

Figure 11.55

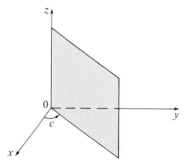

$\theta = c$, a half-plane

Figure 11.56

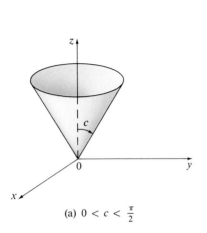

(a) $0 < c < \frac{\pi}{2}$

Figure 11.57

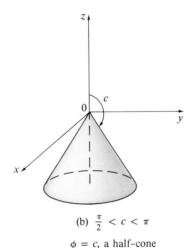

(b) $\frac{\pi}{2} < c < \pi$

$\phi = c$, a half–cone

Figure 11.58

But $x = r \cos \theta$ and $y = r \sin \theta$, so

(11.77)

$$x = \rho \sin \phi \cos \theta \qquad y = \rho \sin \phi \sin \theta \qquad z = \rho \cos \phi$$

Also, the distance formula shows that

(11.78)

$$\rho^2 = x^2 + y^2 + z^2$$

● **Example 3**   The point $(2, \pi/4, \pi/3)$ is given in spherical coordinates. Find its rectangular coordinates.

Solution   From Equations 11.77 we have

$$x = \rho \sin \phi \cos \theta = 2 \sin \frac{\pi}{3} \cos \frac{\pi}{4} = 2 \left(\frac{\sqrt{3}}{2} \right) \left(\frac{1}{\sqrt{2}} \right) = \sqrt{\frac{3}{2}}$$

$$y = \rho \sin \phi \sin \theta = 2 \sin \frac{\pi}{3} \sin \frac{\pi}{4} = 2 \left(\frac{\sqrt{3}}{2} \right) \left(\frac{1}{\sqrt{2}} \right) = \sqrt{\frac{3}{2}}$$

$$z = \rho \cos \phi = 2 \cos \frac{\pi}{3} = 2 \left(\frac{1}{2} \right) = 1$$

Thus the point $(2, \pi/4, \pi/3)$ is $\left(\sqrt{\frac{3}{2}}, \sqrt{\frac{3}{2}}, 1 \right)$ in rectangular coordinates.   ●

● **Example 4**   The point $(0, 2\sqrt{3}, -2)$ is given in rectangular coordinates. Find its spherical coordinates.

Solution   From Equation 11.78 we have

$$\rho = \sqrt{x^2 + y^2 + z^2} = \sqrt{0 + 12 + 4} = 4$$

and so Equations 11.77 give

$$\cos\phi = \frac{z}{\rho} = \frac{-2}{4} = -\frac{1}{2} \qquad \phi = \frac{2\pi}{3}$$

$$\cos\theta = \frac{x}{\rho\sin\phi} = 0 \qquad \theta = \frac{\pi}{2}$$

(Note that $\theta \neq 3\pi/2$ because $y = 2\sqrt{3} > 0$.) Therefore the spherical coordinates of the given point are $(4, \pi/2, 2\pi/3)$.   ●

● **Example 5**   Find an equation in spherical coordinates for the hyperboloid of two sheets with equation $x^2 - y^2 - z^2 = 1$.

Solution   Substituting the expressions in Equations 11.77 into the given equation, we have

$$\rho^2 \sin^2\phi\cos^2\theta - \rho^2\sin^2\phi\sin^2\theta - \rho^2\cos^2\phi = 1$$

$$\rho^2[\sin^2\phi(\cos^2\theta - \sin^2\theta) - \cos^2\phi] = 1$$

or $$\rho^2(\sin^2\phi\cos 2\theta - \cos^2\phi) = 1$$ ●

● **Example 6**   Find a rectangular equation for the surface whose spherical equation is $\rho = \sin\theta\sin\phi$.

Solution   From Equations 11.78 and 11.77 we have

$$x^2 + y^2 + z^2 = \rho^2 = \rho\sin\theta\sin\phi = y$$

or $$x^2 + (y - \tfrac{1}{2})^2 + z^2 = \tfrac{1}{4}$$

which is the equation of a sphere with center $(0, \tfrac{1}{2}, 0)$ and radius $\tfrac{1}{2}$.   ●

EXERCISES 11.10

In Exercises 1–4 change from cylindrical to rectangular coordinates.

1. $\left(3, \dfrac{\pi}{2}, 1\right)$

2. $\left(\sqrt{2}, \dfrac{\pi}{4}, \sqrt{2}\right)$

3. $\left(2, \dfrac{4\pi}{3}, 8\right)$

4. $\left(6, -\dfrac{\pi}{6}, 5\right)$

In Exercises 5–8 change from rectangular to cylindrical coordinates.

5. $(-1, 0, 0)$

6. $(1, 1, 1)$

7. $(\sqrt{3}, 1, 4)$

8. $(-\sqrt{2}, \sqrt{2}, 0)$

In Exercises 9–12 change from spherical to rectangular coordinates.

9. $(1, 0, 0)$

10. $(3, 0, \pi)$

11. $\left(1, \dfrac{\pi}{6}, \dfrac{\pi}{6}\right)$

12. $\left(2, \dfrac{\pi}{2}, \dfrac{3\pi}{4}\right)$

In Exercises 13–16 change from rectangular to spherical coordinates.

13. $(-3, 0, 0)$

14. $(1, 1, \sqrt{2})$

15. $(\sqrt{3}, 0, 1)$

16. $(-\sqrt{3}, -3, -2)$

In Exercises 17–32 identify the surface whose equation is given.

17. $r = 3$

18. $\rho = 3$

19. $\phi = \dfrac{\pi}{3}$

20. $\theta = \dfrac{\pi}{3}$

21. $z = r^2$

22. $r = 4 \sin \theta$

23. $\rho \cos \phi = 2$

24. $\rho \sin \phi = 2$

25. $\phi = 0$

26. $\phi = \dfrac{\pi}{2}$

27. $r^2 + z^2 = 25$

28. $r^2 - 2z^2 = 4$

29. $\rho^2(\sin^2 \phi \cos^2 \theta + \cos^2 \phi) = 4$

30. $\rho^2(\sin^2 \phi - 4 \cos^2 \phi) = 1$

31. $r^2 = r$

32. $\rho^2 - 6\rho + 8 = 0$

In Exercises 33–40 write the given equation (a) in cylindrical coordinates and (b) in spherical coordinates.

33. $x^2 + y^2 + z^2 = 16$

34. $x^2 + y^2 - z^2 = 16$

35. $x + 2y + 3z = 6$

36. $x^2 + y^2 = 2z$

37. $x^2 - y^2 - 2z^2 = 4$

38. $y^2 + z^2 = 1$

39. $x^2 + y^2 = 2y$

40. $z = x^2 - y^2$

REVIEW OF CHAPTER 11

Define, state, or discuss the following:

1. R^3

2. Distance formula in R^3

3. Equation of a sphere

4. Vectors in V_2 and V_3

5. Position vector

6. Length of a vector

7. Addition of vectors

8. Multiplication of a vector by a scalar

9. Unit vector

10. Dot product and its properties

11. Angle between two vectors

12. Orthogonal vectors

13. Parallel vectors

14. Direction cosines

15. Vector and scalar projections

16. Work done by a constant force

17. Cross product and its properties

18. Scalar triple product

19. Vector, parametric, and symmetric equations of a line

20. Vector and scalar equations of a plane

21. Distance from a point to a plane

22. Standard equations of ellipsoids, hyperboloids, paraboloids, and cones

23. Vector functions and component functions

24. Limit of a vector function

25. Continuity of a vector function

26. Derivative of a vector function

27. Space curve

28. Tangent vector and tangent line

29. Differentiation formulas for vector functions

30. Integrals of vector functions

31. Length of a curve

32. Smooth curve

33. Arc length function

34. Curvature

35. Velocity and acceleration along a curve

36. Tangential and normal components of acceleration

37. Cylindrical coordinates

38. Spherical coordinates

REVIEW EXERCISES FOR CHAPTER 11

1. Find the lengths of the sides of the triangle with vertices $A(2, 6, -4)$, $B(-1, 2, 8)$, and $C(0, 1, 2)$.

2. Find an equation of the sphere with center $(1, -1, 2)$ and radius 3.

3. Find the center and radius of the sphere $x^2 + y^2 + z^2 + 4x + 6y - 10z + 2 = 0$.

Calculate the quantities in Exercises 4–16 if

$$\mathbf{a} = \mathbf{i} + \mathbf{j} - 2\mathbf{k} \qquad \mathbf{b} = 3\mathbf{i} - 2\mathbf{j} + \mathbf{k} \qquad \mathbf{c} = \mathbf{j} - 5\mathbf{k}$$

4. $2\mathbf{a} + 3\mathbf{b}$

5. $6\mathbf{a} - 5\mathbf{c}$

6. $|\mathbf{b}|$

7. $\mathbf{a} \cdot \mathbf{b}$

8. $\mathbf{a} \times \mathbf{b}$

9. $|\mathbf{b} \times \mathbf{c}|$

10. $\mathbf{a} \cdot (\mathbf{b} \times \mathbf{c})$

11. $\mathbf{c} \times \mathbf{c}$

12. $\mathbf{a} \times (\mathbf{b} \times \mathbf{c})$

13. The angle between $\mathbf{a}$ and $\mathbf{b}$ (correct to the nearest degree)

14. The direction cosines of $\mathbf{b}$

15. The scalar projection of $\mathbf{b}$ onto $\mathbf{a}$

16. The vector projection of $\mathbf{b}$ onto $\mathbf{a}$

17. Find the value of x such that the vectors $\langle 2, x, 4 \rangle$ and $\langle 2x, 3, -7 \rangle$ are orthogonal.

18. Find two unit vectors orthogonal to both $\langle 1, 0, 1 \rangle$ and $\langle 2, 3, 4 \rangle$.

19. (a) Find a vector perpendicular to the plane through the points $A(1, 0, 0)$, $B(2, 0, -1)$, and $C(1, 4, 3)$.
    (b) Find the area of triangle ABC.

20. A constant force $\mathbf{F} = 3\mathbf{i} + 5\mathbf{j} + 10\mathbf{k}$ moves an object along the line segment from $(1, 0, 2)$ to $(5, 3, 8)$. Find the work done if the distance is measured in meters and the force in newtons.

In Exercises 21–23 find parametric equations for the line that satisfies the given conditions.

21. Passing through $(1, 2, 4)$ and in the direction of $\mathbf{v} = 2\mathbf{i} - \mathbf{j} + 3\mathbf{k}$

22. Passing through $(-6, -1, 0)$ and $(2, -3, 5)$

23. Passing through $(1, 0, 1)$ and parallel to the line with parametric equations $x = 4t$, $y = 1 - 3t$, $z = 2 + 5t$

In Exercises 24–26 find an equation of the plane that satisfies the given conditions.

24. Passing through $(4, -1, -1)$ and with normal vector $\langle 2, 6, -3 \rangle$

25. Passing through $(-4, 1, 2)$ and parallel to the plane $x + 2y + 5z = 3$

26. Passing through $(-1, 2, 0)$, $(2, 0, 1)$, and $(-5, 3, 1)$

27. Determine whether the lines given by the symmetric equations

$$\frac{x - 1}{2} = \frac{y - 2}{3} = \frac{z - 3}{4}$$

and

$$\frac{x + 1}{6} = \frac{y - 3}{-1} = \frac{z + 5}{2}$$

are parallel, skew, or intersect.

28. (a) Show that the planes $x + y - z = 1$ and $2x - 3y + 4z = 5$ are neither parallel nor perpendicular.
    (b) Find, correct to the nearest degree, the angle between these planes.

29. Find the distance from the point $(6, 2, -1)$ to the plane $3x + y - 4z = 2$.

In Exercises 30–38 identify and sketch the graph of the given surface.

30. $x = 6$

31. $y = z$

32. $6x + 4y + 3z = 12$

33. $y = x^2 + z^2$

34. $225x^2 + 9y^2 + 25z^2 = 225$

35. $x^2 = y^2 + 4z^2$

36. $y^2 + z^2 = 1 + x^2$

37. $-4x^2 + y^2 - 4z^2 = 4$

38. $y = z^2$

39. (a) Sketch the curve with vector function $\mathbf{r}(t) = 2\mathbf{i} + \sin t\,\mathbf{j} + \cos t\,\mathbf{k}$.
    (b) Find $\mathbf{r}'(t)$ and $\mathbf{r}''(t)$.

40. Find parametric equations for the tangent line to the curve given by $x = 2 + t^2$, $y = \sqrt{3 + t}$, $z = \cos \pi t$ at the point $(3, 2, -1)$.

41. If $\mathbf{r}(t) = (t + t^2)\mathbf{i} + (2 + t^3)\mathbf{j} + t^4\mathbf{k}$, evaluate $\int_0^1 \mathbf{r}(t)\,dt$.

42. Find the length of the curve $\mathbf{r}(t) = \langle 2t^{3/2}, \cos 2t, \sin 2t \rangle$, $0 \leqslant t \leqslant 1$.

43. For the curve given by $\mathbf{r}(t) = \langle t^3/3, t^2/2, t \rangle$, find (a) the unit tangent vector, (b) the unit normal vector, and (c) the curvature.

44. Find the curvature of the curve $y = x^4$ at the point $(1, 1)$.

45. Find the curvature of the ellipsoid $x = 3\cos t$, $y = 4\sin t$ at the points $(3,0)$ and $(0,4)$.

46. A particle moves with position function $\mathbf{r}(t) = 2\sqrt{2}t\mathbf{i} + e^{2t}\mathbf{j} + e^{-2t}\mathbf{k}$. Find the velocity, speed, and acceleration of the particle.

47. A particle starts at the origin with initial velocity $\mathbf{i} + 2\mathbf{j} + \mathbf{k}$. Its acceleration is $\mathbf{a}(t) = t\mathbf{i} + \mathbf{j} + t^2\mathbf{k}$. Find its position function.

48. Find the tangential and normal components of the acceleration vector of a particle with position function $\mathbf{r}(t) = t\mathbf{i} + 2t\mathbf{j} + t^2\mathbf{k}$.

49. The cylindrical coordinates of a point are $(2, \pi/6, 2)$. Find the rectangular and spherical coordinates of the point.

50. The rectangular coordinates of a point are $(2, 2, -1)$. Find the cylindrical and spherical coordinates of the point.

51. The spherical coordinates of a point are $(4, \pi/3, \pi/6)$. Find the rectangular and cylindrical coordinates of the point.

In Exercises 52–55 identify the surface whose equation is given.

52. $\phi = \dfrac{\pi}{4}$

53. $\theta = \dfrac{\pi}{4}$

54. $r = \cos\theta$

55. $\rho = 3\sec\phi$

In Exercises 56–58 write the given equation in cylindrical coordinates and in spherical coordinates.

56. $x^2 + y^2 = 4$

57. $x^2 + y^2 + z^2 = 4$

58. $x^2 + y^2 + z^2 = 2x$

12 Partial Derivatives

Strange as it may seem, the power
of mathematics rests on its evasion of
all unnecessary thought and on its
wonderful saving of mental operations.
Ernst Mach

So far we have dealt with the calculus of functions of a single variable. But, in the real world, physical quantities usually depend on two or more variables, so in this chapter we turn our attention to functions of several variables and extend the basic ideas of differential calculus to such functions.

Functions of Several Variables

The temperature T at a point on the surface of the earth depends on the longitude x and latitude y of the point. We can think of T as being a function of the two variables x and y, or as a function of the pair (x, y), and we indicate this functional dependence by writing $T = f(x, y)$.

The volume V of a circular cylinder depends on its radius r and height h. In fact, we know from Chapter 8 that $V = \pi r^2 h$. We say that V is a function of r and h and we write $V(r, h) = \pi r^2 h$.

Definition (12.1)

> Let $D \subset R^2$. A **function f of two variables** is a rule that assigns to each ordered pair (x, y) in D a unique real number denoted by $f(x, y)$. The set D is the **domain** of f and its **range** is the set of values that f takes on, that is, $\{f(x, y) \mid (x, y) \in D\}$.

We often write $z = f(x, y)$ to make explicit the value taken on by f at the general point (x, y). The variables x and y are **independent variables** and z is the **dependent variable.** [Compare this with the notation $y = f(x)$ for functions of a single variable.]

The situation described in Definition 12.1 is indicated by the notation $f: D \to R$. A function of two variables is just a special case of the general idea of a function $f: X \to Y$ (see Section 1.5) where $X = D \subset R^2$ and $Y \subset R$. One way of visualizing such a function is by means of an arrow diagram (see Figure 12.1) where the domain D is represented as a subset of the xy-plane.

Figure 12.1

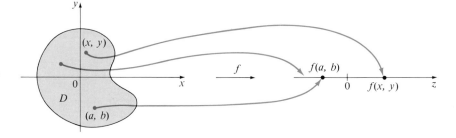

If a function f is given by a formula and no domain is specified, then the domain of f is understood to be the set of all pairs (x, y) for which the given expression is a well-defined real number.

● **Example 1**   Find the domains of the following functions and evaluate $f(3, 2)$.

(a) $f(x, y) = \dfrac{\sqrt{x + y + 1}}{x - 1}$ (b) $f(x, y) = x \ln(y^2 - x)$

Solution   (a) $f(3, 2) = \dfrac{\sqrt{3 + 2 + 1}}{3 - 1} = \dfrac{\sqrt{6}}{2}$

The expression for f will make sense if the denominator is not 0 and the quantity under the square root sign is nonnegative. So the domain of f is

$$D = \{(x, y) \mid x + y + 1 \geqslant 0, x \neq 1\}$$

The inequality $x + y + 1 \geqslant 0$, or $y \geqslant -x - 1$, describes the points that lie on or above the line $y = -x - 1$, while $x \neq 1$ means that the points on the line $x = 1$ must be excluded from the domain. [See Figure 12.2(a).]

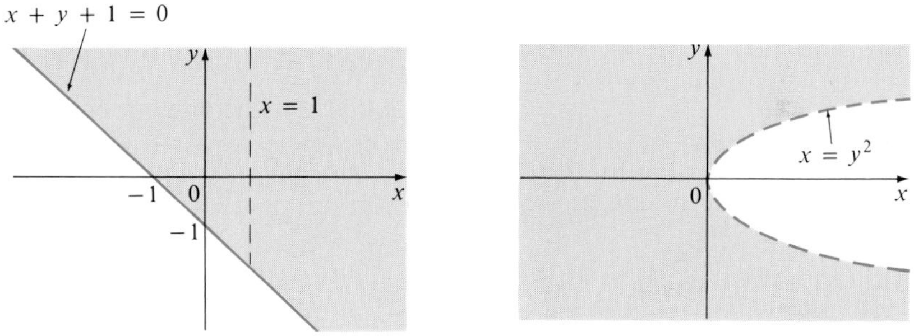

Figure 12.2

(a) Domain of $f(x, y) = \sqrt{x + y + 1}/(x - 1)$ 　　　　　 (b) Domain of $f(x, y) = x \ln(y^2 - x)$

(b) $f(3, 2) = 3 \ln(2^2 - 3) = 3 \ln 1 = 0$

Since $\ln(y^2 - x)$ is defined only when $y^2 - x > 0$, that is, $x < y^2$, the domain of f is $D = \{(x, y) \mid x < y^2\}$. This is the set of points to the left of the parabola $x = y^2$. [See Figure 12.2(b).]   ●

Definition (12.2)

> If f is a function of two variables with domain D, the **graph** of f is the set
>
> $$S = \{(x, y, z) \in R^3 \mid z = f(x, y), (x, y) \in D\}$$

Just as the graph of a function f of one variable is a curve C with equation $y = f(x)$, so the graph of a function f of two variables is a surface S with equation $z = f(x, y)$. We can visualize the graph S of f as lying directly above or below its domain D in the xy-plane (see Figure 12.3).

● **Example 2**   Sketch the graph of the function $f(x, y) = 6 - 3x - 2y$.

Solution   The graph of f has the equation $z = 6 - 3x - 2y$, or $3x + 2y + z = 6$, which is a plane. The portion of this graph that lies in the first octant is sketched in Figure 12.4.   ●

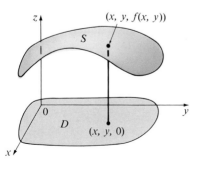

Figure 12.3

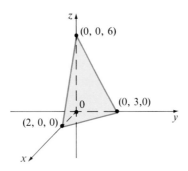

Figure 12.4

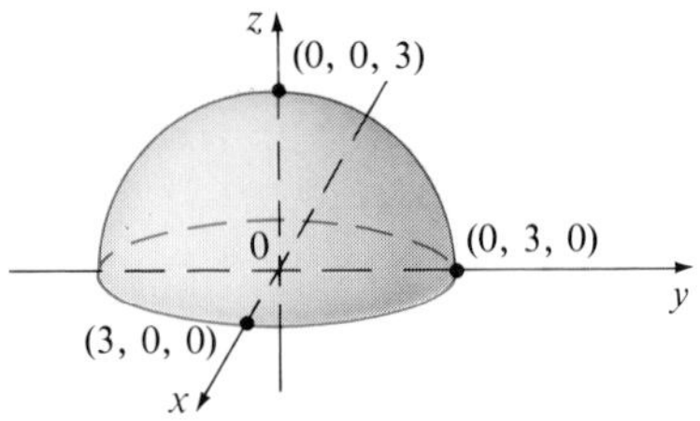

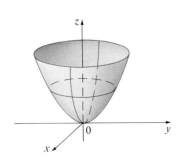

Figure 12.5

GRAPH OF $g(x, y) = \sqrt{9 - x^2 - y^2}$

Figure 12.6

GRAPH OF $h(x, y) = 4x^2 + y^2$

● **Example 3**   Find the domain and range and sketch the graph of

$$g(x, y) = \sqrt{9 - x^2 - y^2}$$

Solution   The domain of g is

$$D = \{(x, y) \,|\, 9 - x^2 - y^2 \geq 0\} = \{(x, y) \,|\, x^2 + y^2 \leq 9\}$$

which is the disk with center 0 and radius 3. The range of g is

$$\{z \,|\, z = \sqrt{9 - x^2 - y^2}, (x, y) \in D\}$$

Since z is a positive square root, $z \geq 0$. Also

$$9 - x^2 - y^2 \leq 9 \quad \Rightarrow \quad \sqrt{9 - x^2 - y^2} \leq 3$$

So the range is

$$\{z \,|\, 0 \leq z \leq 3\} = [0, 3]$$

The graph has equation $z = \sqrt{9 - x^2 - y^2}$. We square both sides of this equation to obtain $z^2 = 9 - x^2 - y^2$, or $x^2 + y^2 + z^2 = 9$, which we recognize as the equation of the sphere with center the origin and radius 3. But, since $z \geq 0$, the graph of g is just the top half of this sphere (see Figure 12.5). ●

● **Example 4**   Sketch the graph of the function $h(x, y) = 4x^2 + y^2$.

Solution   The graph of f has the equation $z = 4x^2 + y^2$, which is an elliptic paraboloid (see Equation 11.46). Traces in the horizontal planes $z = k$ are ellipses if $k > 0$. Vertical traces are parabolas. (See Figure 12.6.) ●

Computer programs are now readily available for graphing functions of two variables. In most such programs, traces in the vertical planes $x = k$ and $y = k$ are drawn for equally spaced values of k. Figure 12.7 shows computer-generated graphs of several functions. Notice that we get an especially good picture of a function when we view its graph from more than one point.

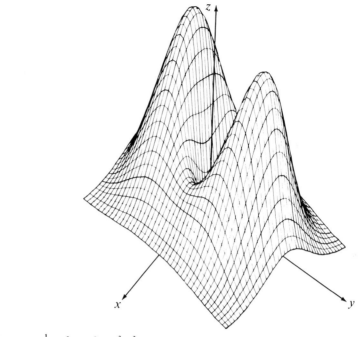

$f(x, y) = \dfrac{1}{8}(3x^2 + y^2)e^{-x^2 - y^2}$, $-2 \leqslant x \leqslant 2$, $-2 \leqslant y \leqslant 2$, viewed from $(1, 1, 1)$

Figure 12.7

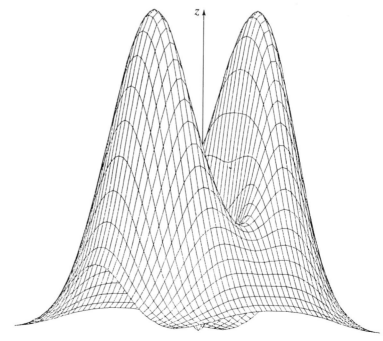

$f(x, y) = \dfrac{1}{8}(3x^2 + y^2)e^{-x^2 - y^2}$, $-2 \leqslant x \leqslant 2$, $-2 \leqslant y \leqslant 2$, viewed from $(1, -1, 0)$

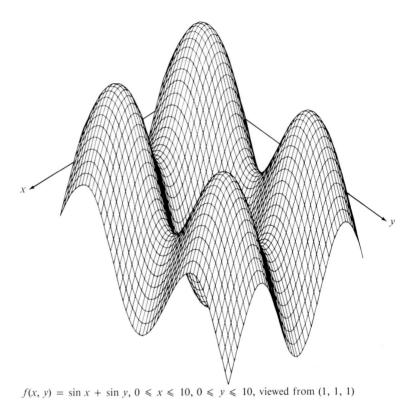

$f(x, y) = \sin x + \sin y,\ 0 \leqslant x \leqslant 10,\ 0 \leqslant y \leqslant 10$, viewed from $(1, 1, 1)$

Figure 12.7 (continued)

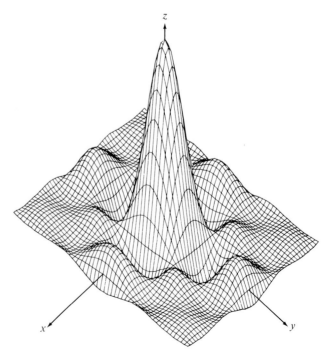

$f(x, y) = \dfrac{\sin x \sin y}{xy},\ -10 \leqslant x \leqslant 10,\ -10 \leqslant y \leqslant 10$, viewed from $(1, 1, 1)$

So far we have two methods for visualizing functions: arrow diagrams and graphs. A third method, borrowed from map makers, is a contour map on which points of constant elevation are joined to form *contour curves* or *level curves*.

Definition (12.3)

The **level curves** of a function f of two variables are the curves with equations $f(x, y) = k$, where k is a constant (in the range of f).

Figure 12.8

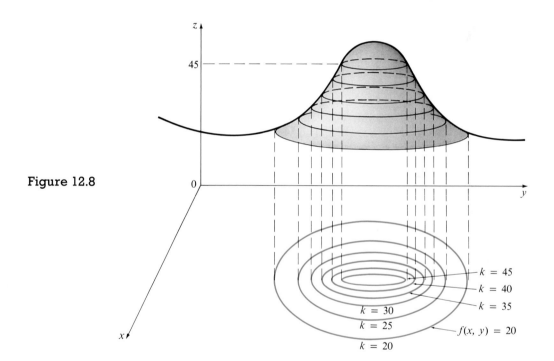

A level curve $f(x, y) = k$ is the locus of all points at which f takes on a given value k. In other words, it shows where the graph of f has height k.

You can see from Figure 12.8 the relation between level curves and horizontal traces. The level curves $f(x, y) = k$ are just the traces of the graph of f in the horizontal plane $z = k$ projected down to the xy-plane. So if you draw the level curves of a function and visualize them being lifted up to the surface at the indicated height, then you can mentally piece together a picture of the graph. The surface is steep where the level curves are close together. It is somewhat flatter where they are farther apart.

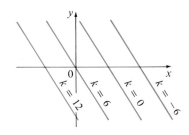

Figure 12.9

CONTOUR MAP FOR
$f(x, y) = 6 - 3x - 2y$

● **Example 5**   Sketch the level curves of the function $f(x, y) = 6 - 3x - 2y$ for $k = -6, 0, 6, 12$.

Solution   The level curves are

$$6 - 3x - 2y = k \quad \text{or} \quad 3x + 2y + (k - 6) = 0$$

This is a family of lines with slope $-\frac{3}{2}$. The four particular level curves with $k = -6, 0, 6,$ and 12 are $3x + 2y - 12 = 0, 3x + 2y - 6 = 0, 3x + 2y = 0,$ and $3x + 2y + 6 = 0$. They are sketched in Figure 12.9.   ●

● **Example 6**   Sketch the level curves of the function

$$g(x, y) = \sqrt{9 - x^2 - y^2} \text{ for } k = 0, 1, 2, 3$$

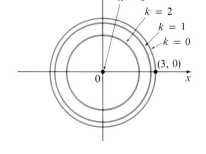

Solution   The level curves are

$$\sqrt{9 - x^2 - y^2} = k \quad \text{or} \quad x^2 + y^2 = 9 - k^2$$

This is a family of concentric circles with center $(0, 0)$ and radius $\sqrt{9 - k^2}$. The cases $k = 0, 1, 2, 3$ are shown in Figure 12.10. Try to visualize these level curves lifted up to form a surface and compare with the graph of g in Figure 12.5.   ●

Figure 12.10

CONTOUR MAP OF
$g(x, y) = \sqrt{9 - x^2 - y^2}$

● **Example 7**   Sketch some level curves of the function $h(x, y) = 4x^2 + y^2$.

Solution   The level curves are

$$4x^2 + y^2 = k \quad \text{or} \quad \frac{x^2}{\dfrac{k}{4}} + \frac{y^2}{k} = 1$$

which, for $k > 0$, describes a family of ellipses with semiaxes $\sqrt{k}/2$ and $\sqrt{k}$. Compare the sketches of these level curves in Figure 12.11 with the graph of h in Figure 12.6.   ●

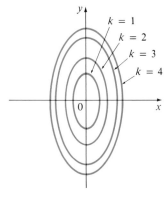

● **Example 8**   Sketch the level curves of $f(x, y) = x + y^2$ for $k = -1, 0, 1, 2, 3$.

Solution   The level curves are $x + y^2 = k$. These form a family of parabolas with x-intercept k. Although it is difficult to draw the graph $z = x + y^2$, we can visualize how this surface is put together from the level curves sketched in Figure 12.12.

Figure 12.11

CONTOUR MAP OF $h(x, y) = 4x^2 + y^2$

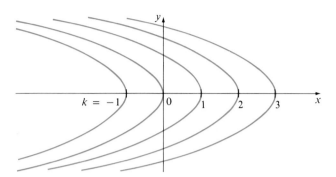

Figure 12.12

CONTOUR MAP OF $f(x, y) = x + y^2$

Figure 12.13 on page 722 shows some computer-generated level curves for the function $f(x, y) = xye^{-x^2-y^2}$ together with its computer-generated graph.

A **function of three variables,** f, is a rule that assigns to each ordered triple (x, y, z) in a domain $D \subset R^3$ a unique real number denoted by $f(x, y, z)$. For instance, in the illustration at the beginning of this section the temperature T depends on the time t as well as on the longitude x and latitude y, so we could write $T = f(x, y, t)$.

It is possible to visualize a function f of three variables by an arrow diagram but not by its graph since that would lie in a four-dimensional space. However we do gain some insight into f by examining its **level surfaces,** which are the surfaces with equations $f(x, y, z) = k$, where k is a constant. If the point (x, y, z) moves along a level surface, the value of $f(x, y, z)$ remains fixed.

● **Example 9**   Find the domain of f if $f(x, y, z) = \ln(z - y) + xy \sin z$.

Solution   The expression for $f(x, y, z)$ is defined as long as $z - y > 0$, so the domain of f is

$$D = \{(x, y, z) \in R^3 \,|\, z > y\}$$

This is a **half-space** consisting of all points that lie above the plane $z = y$.

● **Example 10**   Find the level surfaces of the function

$$f(x, y, z) = x^2 + y^2 + z^2$$

Solution   The level surfaces are $x^2 + y^2 + z^2 = k$, where $k \geqslant 0$. These form a family of concentric spheres with radius $\sqrt{k}$. Thus as (x, y, z) varies over any sphere with center O, the value of $f(x, y, z)$ remains fixed.

Functions of any number of variables can also be considered. A **function of n variables** is a rule that assigns a number $z = f(x_1, x_2, \ldots, x_n)$ to an n-tuple $(x_1, x_2, \ldots, x_n)$ of real numbers. For example, if a company uses n different ingredients in making a food product, c_i is the cost per unit of the ith ingredient, and x_i units of the ith ingredient are used, then the total cost C of the

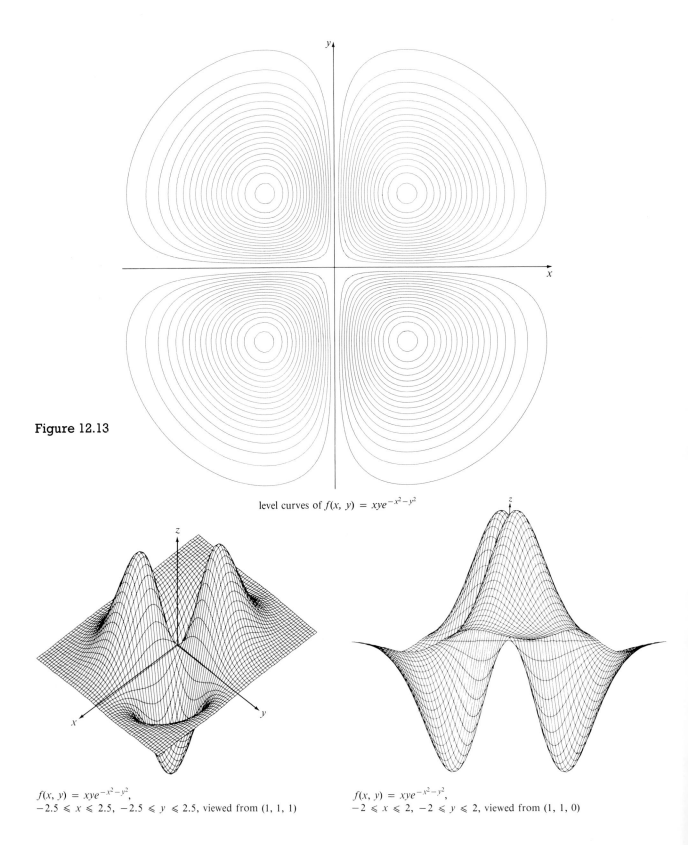

Figure 12.13

level curves of $f(x, y) = xye^{-x^2-y^2}$

$f(x, y) = xye^{-x^2-y^2}$,
$-2.5 \leqslant x \leqslant 2.5$, $-2.5 \leqslant y \leqslant 2.5$, viewed from $(1, 1, 1)$

$f(x, y) = xye^{-x^2-y^2}$,
$-2 \leqslant x \leqslant 2$, $-2 \leqslant y \leqslant 2$, viewed from $(1, 1, 0)$

ingredients is a function of the n variables $x_1, x_2, \ldots, x_n$:

(12.4)
$$C = f(x_1, x_2, \ldots, x_n) = c_1 x_1 + c_2 x_2 + \cdots + c_n x_n$$

The notation

$$f: D \subset R^n \to R$$

is used to signify that f is a real-valued function whose domain D is a subset of R^n. Sometimes we use vector notation to write functions more compactly: If $\mathbf{x} = \langle x_1, x_2, \ldots, x_n \rangle$, we often write $f(\mathbf{x})$ in place of $f(x_1, x_2, \ldots, x_n)$. With this notation we could rewrite the function defined in Equation 12.4 as

$$f(\mathbf{x}) = \mathbf{c} \cdot \mathbf{x}$$

where $\mathbf{c} = \langle c_1, c_2, \ldots, c_n \rangle$.

In view of the one-to-one correspondence between points $(x_1, x_2, \ldots, x_n)$ in R^n and their position vectors $\mathbf{x} = \langle x_1, x_2, \ldots, x_n \rangle$ in V_n, we have three ways of looking at a function $f: R^n \to R$:

1.   As a function of n real variables $x_1, x_2, \ldots, x_n$
2.   As a function of a single point variable $(x_1, x_2, \ldots, x_n)$
3.   As a function of a single vector variable $\mathbf{x} = \langle x_1, x_2, \ldots, x_n \rangle$

We shall see that all three points of view are useful.

EXERCISES 12.1

1. If $f(x, y) = x^2 - y^2 + 4xy - 7x + 10$, find (a) $f(2, 1)$, (b) $f(-3, 5)$, (c) $f(x + h, y)$, (d) $f(x, y + k)$, and (e) $f(x, x)$.

2. If $g(x, y) = \ln(xy + y - 1)$, find (a) $g(1, 1)$, (b) $g(e, 1)$, (c) $g(x, 1)$, (d) $g(x + h, y)$, and (e) $g(x, y + k)$.

3. If $F(x, y) = 3xy/(x^2 + 2y^2)$, find (a) $F(1, 1)$, (b) $F(-1, 2)$, (c) $F(t, 1)$, (d) $F(-1, y)$, and (e) $F(x, x^2)$.

4. If $G(x, y, z) = x \sin y \cos z$, find (a) $G(2, \pi/6, \pi/3)$, (b) $G(4, \pi/4, 0)$, (c) $G(t, t, t)$, (d) $G(u, v, 0)$, and (e) $G(x, x + y, x)$.

In Exercises 5–14 find the domain and range of the given function.

5. $f(x, y) = x + 2y - 5$

6. $f(x, y) = \sqrt{x - y}$

7. $f(x, y) = \dfrac{2}{x + y}$

8. $f(x, y) = \tan^{-1}\left(\dfrac{y}{x}\right)$

9. $f(x, y) = e^{x^2 - y}$

10. $f(x, y) = \sqrt{36 - 9x^2 - 4y^2}$

11. $f(x, y, z) = x^2 \ln(x - y + z)$

12. $f(x, y, z) = \dfrac{x}{yz}$

13. $f(x, y, z) = x \sin(y + z)$

14. $f(x, y, z) = \dfrac{1}{\sqrt{x^2 + y^2 + z^2 - 1}}$

In Exercises 15–30 find and sketch the domains of the given functions.

15. $f(x, y) = \sqrt[4]{y - 2x}$

16. $f(x, y) = \sqrt{x} + \sqrt{y}$

17. $f(x, y) = \dfrac{\sqrt{9 - x^2 - y^2}}{(x + 2y)}$

18. $f(x, y) = \dfrac{x^2 + y^2}{x^2 - y^2}$

19. $f(x, y) = xy\sqrt{x^2 + y}$ **20.** $f(x, y) = \tan(x - y)$

21. $f(x, y) = \ln(xy - 1)$ **22.** $f(x, y) = \ln(x^2 - y^2)$

23. $f(x, y) = x^2 \sec y$

24. $f(x, y) = \sqrt{x^2 + y^2 - 1} + \ln(4 - x^2 - y^2)$

25. $f(x, y) = \sin^{-1}(x + y)$

26. $f(x, y) = \sqrt{4 - 2x^2 - y^2}$

27. $f(x, y) = \ln x + \ln \sin y$

28. $f(x, y) = \sqrt{y - x} \ln(y + x)$

29. $f(x, y, z) = \sqrt{1 - x^2 - y^2 - z^2}$

30. $f(x, y, z) = \ln(16 - 4x^2 - 4y^2 - z^2)$

In Exercises 31–42 sketch the graphs of the given functions.

31. $f(x, y) = 3$ **32.** $f(x, y) = x$

33. $f(x, y) = 1 - x - y$ **34.** $f(x, y) = y^2$

35. $f(x, y) = x^2 + 9y^2$ **36.** $f(x, y) = 3 - x^2 - y^2$

37. $f(x, y) = \sqrt{x^2 + y^2}$

38. $f(x, y) = \sqrt{16 - x^2 - 16y^2}$

39. $f(x, y) = y^2 - x^2$

40. $f(x, y) = \sin y$

41. $f(x, y) = 1 - x^2$

42. $f(x, y) = x^2 + y^2 - 4x - 2y + 5$

In Exercises 43–52 draw a contour map of the given function showing several level curves.

43. $f(x, y) = xy$ **44.** $f(x, y) = x^2 - y^2$

45. $f(x, y) = x^2 + 9y^2$ **46.** $f(x, y) = e^{xy}$

47. $f(x, y) = \dfrac{x}{y}$ **48.** $f(x, y) = \dfrac{x + y}{x - y}$

49. $f(x, y) = \sqrt{x + y}$ **50.** $f(x, y) = y - \cos x$

51. $f(x, y) = x - y^2$ **52.** $f(x, y) = e^{1/(x^2 + y^2)}$

In Exercises 53–56 describe the level surfaces of the given functions.

53. $f(x, y, z) = x + 3y + 5z$

54. $f(x, y, z) = x^2 + 3y^2 + 5z^2$

55. $f(x, y, z) = x^2 - y^2 + z^2$

56. $f(x, y, z) = x^2 - y^2$

57. A thin metal plate, located in the xy-plane, has temperature $T(x, y)$ at the point (x, y). The level curves of T are called *isothermals* because at all points on an isothermal the temperature is the same. Sketch some isothermals if the temperature function is given by $T(x, y) = 100/(1 + x^2 + 2y^2)$.

58. If $V(x, y)$ is the electric potential at a point (x, y) in the xy-plane, then the level curves of V are called *equipotential curves* because at all points on such a curve the electric potential is the same. Sketch some equipotential curves if $V(x, y) = c/\sqrt{r^2 - x^2 - y^2}$, where c is a positive constant.

SECTION 12.2

Limits and Continuity

Consider the function $f(x, y) = \sqrt{9 - x^2 - y^2}$ whose domain is the closed disk $D = \{(x, y)|x^2 + y^2 \leq 9\}$ shown in Figure 12.14(a) and whose graph is the hemisphere shown in Figure 12.14(b).

If the point (x, y) is close to the origin, then x and y are both close to 0, and so $f(x, y)$ is close to 3. In fact, if (x, y) lies in a small open disk $x^2 + y^2 < \delta^2$, then

$$f(x, y) = \sqrt{9 - (x^2 + y^2)} > \sqrt{9 - \delta^2}$$

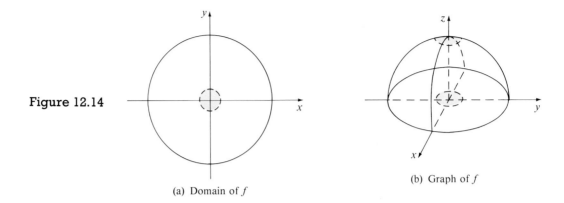

Figure 12.14

(a) Domain of f

(b) Graph of f

Thus we can make the values of $f(x, y)$ as close to 3 as we like by taking (x, y) in a small enough disk with center $(0, 0)$. We describe this situation by using the notation

$$\lim_{(x,y)\to(0,0)} \sqrt{9 - x^2 - y^2} = 3$$

In general, the notation

$$\lim_{(x,y)\to(a,b)} f(x, y) = L$$

means that the values of $f(x, y)$ can be made as close as we wish to the number L by taking the point (x, y) close enough to the point (a, b). A more precise definition follows.

Definition (12.5)

> Let f be a function of two variables defined on a disk with center (a, b), except possibly at (a, b). Then we say that the **limit of $f(x, y)$ as (x, y) approaches (a, b)** is L and we write
>
> $$\lim_{(x,y)\to(a,b)} f(x, y) = L$$
>
> if for every number $\varepsilon > 0$ there is a corresponding number $\delta > 0$ such that
>
> $$|f(x, y) - L| < \varepsilon \qquad \text{whenever} \qquad 0 < \sqrt{(x - a)^2 + (y - b)^2} < \delta$$

Other notations for the limit in Definition 12.5 are

$$\lim_{\substack{x\to a \\ y\to b}} f(x, y) = L \qquad \text{and} \qquad f(x, y) \to L \text{ as } (x, y) \to (a, b)$$

Since $|f(x, y) - L|$ is the distance between the numbers $f(x, y)$ and L, and $\sqrt{(x - a)^2 + (y - b)^2}$ is the distance between the point (x, y) and the point (a, b), Definition 12.5 says that the distance between $f(x, y)$ and L can

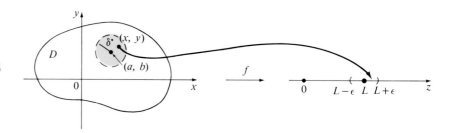

Figure 12.15

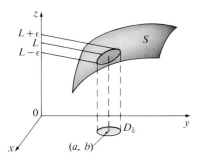

Figure 12.16

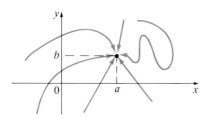

Figure 12.17

be made arbitrarily small by making the distance from (x, y) to (a, b) sufficiently small (but not 0). Figure 12.15 illustrates Definition 12.5 by means of an arrow diagram. If any small interval $(L - \varepsilon, L + \varepsilon)$ is given around L, then we can find a disk D_δ with center (a, b) and radius $\delta > 0$ such that f maps all the points in D_δ [except possibly (a, b)] into the interval $(L - \varepsilon, L + \varepsilon)$.

Another illustration of Definition 12.5 is given in Figure 12.16 where the surface S is the graph of f. If $\varepsilon > 0$ is given, we can find $\delta > 0$ such that if (x, y) is restricted to lie in the disk D_δ and $(x, y) \neq (a, b)$, then the corresponding part of S lies between the horizontal planes $z = L - \varepsilon$ and $z = L + \varepsilon$.

For functions of a single variable, when we let x approach a, there are only two possible directions of approach, from the left or right. Recall from Chapter 2 that if $\lim_{x \to a^-} f(x) \neq \lim_{x \to a^+} f(x)$, then $\lim_{x \to a} f(x)$ does not exist.

For functions of two variables the situation is not as simple because we can let (x, y) approach (a, b) from an infinite number of directions in any manner whatsoever (see Figure 12.17).

Definition 12.5 refers only to the *distance* between (x, y) and (a, b). It does not refer to the direction of approach. Therefore if the limit exists, then $f(x, y)$ must approach the same limit no matter how (x, y) approaches (a, b). Thus if we can find two different paths of approach along which $f(x, y)$ has different limits, then it follows that $\lim_{(x,y) \to (a,b)} f(x, y)$ does not exist.

If $f(x, y) \to L_1$ as $(x, y) \to (a, b)$ along a path C_1, and $f(x, y) \to L_2$ as $(x, y) \to (a, b)$ along a path C_2, where $L_1 \neq L_2$, then $\lim_{(x,y) \to (a,b)} f(x, y)$ does not exist.

● **Example 1**   Find $\displaystyle\lim_{(x,y) \to (0,0)} \frac{x^2 - y^2}{x^2 + y^2}$ if it exists.

Solution   Let $f(x, y) = (x^2 - y^2)/(x^2 + y^2)$. First let us approach $(0,0)$ along the x-axis. Then $y = 0$ gives $f(x, 0) = x^2/x^2 = 1$ for all $x \neq 0$, so

$$f(x, y) \to 1 \quad \text{as} \quad (x, y) \to (0, 0) \text{ along the } x\text{-axis}$$

We now approach along the y-axis by putting $x = 0$. Then $f(0, y) = -y^2/y^2 = -1$ for all $y \neq 0$, so

$$f(x, y) \to -1 \quad \text{as} \quad (x, y) \to (0, 0) \text{ along the } y\text{-axis}$$

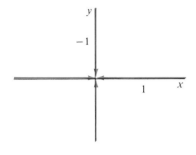

Figure 12.18

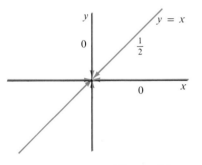

Figure 12.19

(See Figure 12.18.) Since f has two different limits along two different lines, the given limit does not exist. ●

● **Example 2**   If $f(x, y) = xy/(x^2 + y^2)$, does $\lim_{(x,y) \to (0,0)} f(x, y)$ exist?

Solution   If $y = 0$, then $f(x, 0) = 0/x^2 = 0$. Therefore

$$f(x, y) \to 0 \quad \text{as} \quad (x, y) \to (0, 0) \text{ along the } x\text{-axis}$$

If $x = 0$, then $f(0, y) = 0/y^2 = 0$, so

$$f(x, y) \to 0 \quad \text{as} \quad (x, y) \to (0, 0) \text{ along the } y\text{-axis}$$

Although we have obtained identical limits along the axes, that does not show that the given limit is 0. Let us now approach $(0, 0)$ along another line, say $y = x$. For all $x \neq 0$,

$$f(x, x) = \frac{x^2}{x^2 + x^2} = \frac{1}{2}$$

Therefore

$$f(x, y) \to \frac{1}{2} \quad \text{as} \quad (x, y) \to (0, 0) \text{ along } y = x$$

(See Figure 12.19.) Since we obtained different limits along different paths, the given limit does not exist. ●

● **Example 3**   If $f(x, y) = \dfrac{xy^2}{x^2 + y^4}$, does $\lim_{(x,y) \to (0,0)} f(x, y)$ exist?

Solution   With the solution of Example 2 in mind, let us try to save time by letting $(x, y) \to (0, 0)$ along any line through the origin. Then $y = mx$, where m is the slope, and if $m \neq 0$,

$$f(x, y) = f(x, mx) = \frac{x(mx)^2}{x^2 + (mx)^4} = \frac{m^2 x^3}{x^2 + m^4 x^4} = \frac{m^2 x}{1 + m^4 x^2}$$

So

$$f(x, y) \to 0 \quad \text{as} \quad (x, y) \to (0, 0) \text{ along } y = mx$$

Thus f has the same limiting value along every line through the origin. But that does not show that the given limit is 0, for if we now let $(x, y) \to (0, 0)$ along the parabola $x = y^2$ we have

$$f(x, y) = f(y^2, y) = \frac{y^2 \cdot y^2}{(y^2)^2 + y^4} = \frac{y^4}{2y^4} = \frac{1}{2}$$

so

$$f(x, y) \to \frac{1}{2} \quad \text{as} \quad (x, y) \to (0, 0) \text{ along } x = y^2$$

Since different paths lead to different limiting values, the given limit does not exist. ●

● **Example 4**   Find $\lim_{(x,y) \to (0,0)} \dfrac{3x^2 y}{x^2 + y^2}$ if it exists.

Solution   As in Example 3, one can show that the limit along any line through the origin is 0. This does not prove that the given limit is 0, but the limits along the parabolas $y = x^2$ and $x = y^2$ also turn out to be 0, so we begin to suspect that the limit does exist.

Let $\varepsilon > 0$. We want to find $\delta > 0$ such that

$$\left| \frac{3x^2 y}{x^2 + y^2} - 0 \right| < \varepsilon \qquad \text{whenever} \qquad 0 < \sqrt{x^2 + y^2} < \delta$$

that is, $\dfrac{3x^2 |y|}{x^2 + y^2} < \varepsilon$ whenever $0 < \sqrt{x^2 + y^2} < \delta$

But $x^2 \leqslant x^2 + y^2$ since $y^2 \geqslant 0$, so

$$\frac{3x^2 |y|}{x^2 + y^2} \leqslant 3|y| = 3\sqrt{y^2} \leqslant 3\sqrt{x^2 + y^2}$$

Thus if we choose $\delta = \varepsilon/3$ and let $0 < \sqrt{x^2 + y^2} < \delta$, then

$$\left| \frac{3x^2 y}{x^2 + y^2} - 0 \right| \leqslant 3\sqrt{x^2 + y^2} < 3\delta = 3\left(\frac{\varepsilon}{3}\right) = \varepsilon$$

Hence, by Definition 12.5,

$$\lim_{(x,y)\to(0,0)} \frac{3x^2 y}{x^2 + y^2} = 0 \qquad\qquad \bullet$$

Just as for functions of one variable, the calculation of limits can be greatly simplified by the use of properties of limits and by the use of continuity.

The properties of limits listed in Tables 2.14 and 2.15 can be extended to functions of two variables. The limit of a sum is the sum of the limits, the limit of a product is the product of the limits, and so on.

Recall that evaluating limits of *continuous* functions of a single variable is easy. It can be accomplished by direct substitution because the defining property of a continuous function is $\lim_{x\to a} f(x) = f(a)$. Continuous functions of two variables are also defined by the direct substitution property.

Definition (12.6)

> Let f be a function of two variables defined on a disk with center (a, b). Then f is called **continuous at** (a, b) if
>
> $$\lim_{(x,y)\to(a,b)} f(x, y) = f(a, b)$$

If the domain of f is a set $D \subset R^2$, then Definition 12.6 defines the continuity of f at an **interior point** (a, b) of D, that is, a point that is contained

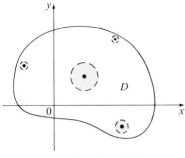

(a) Interior points of D

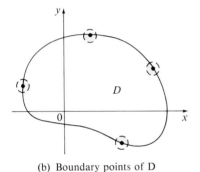

(b) Boundary points of D

Figure 12.20

in a disk $D_\delta \subset D$ [see Figure 12.20(a)]. But D may also contain a **boundary point,** that is, a point (a, b) such that every disk with center (a, b) contains points in D and also points not in D [see Figure 12.20(b)].

If (a, b) is a boundary point of D, then Definition 12.5 is modified so that the last line reads

$$|f(x, y) - L| < \varepsilon \quad \text{whenever } (x, y) \in D \text{ and } 0 < \sqrt{(x - a)^2 + (y - b)^2} < \delta$$

With this convention, Definition 12.6 also applies when f is defined at a boundary point (a, b) of D.

Finally, we say f is **continuous on** D if f is continuous at every point (a, b) in D.

The intuitive meaning of continuity is that if the point (x, y) changes by a small amount, then the value of $f(x, y)$ changes by a small amount. This means that a surface that is the graph of a continuous function has no holes or breaks.

Using the properties of limits, you can see that sums, differences, products, and quotients of continuous functions are continuous on their domains. Let us use this fact to give examples of continuous functions.

A **polynomial function of two variables** (or polynomial, for short) is a sum of terms of the form $cx^m y^n$, where c is a constant and m and n are non-negative integers. A **rational function** is a ratio of polynomials. For instance,

$$f(x, y) = x^4 + 5x^3 y^2 + 6xy^4 - 7y + 6$$

is a polynomial, whereas

$$g(x, y) = \frac{2xy + 1}{x^2 + y^2}$$

is a rational function.

From Definition 12.5 it can be shown (see Exercise 49) that

$$\lim_{(x,y) \to (a,b)} x = a \qquad \lim_{(x,y) \to (a,b)} y = b \qquad \lim_{(x,y) \to (a,b)} c = c$$

These limits show that the functions $f(x, y) = x$, $g(x, y) = y$, and $h(x, y) = c$ are continuous. Since any polynomial can be built up out of the simple functions f, g, and h by multiplication and addition, it follows that *all polynomials are continuous on* R^2. Likewise, any rational function is continuous on its domain since it is a quotient of continuous functions.

● **Example 5**   Evaluate $\lim\limits_{(x,y) \to (1,2)} (x^2 y^3 - x^3 y^2 + 3x + 2y)$.

Solution   Since $f(x, y) = x^2 y^3 - x^3 y^2 + 3x + 2y$ is a polynomial, it is continuous everywhere, so the limit can be found by direct substitution:

$$\lim_{(x,y) \to (1,2)} (x^2 y^3 - x^3 y^2 + 3x + 2y) = 1^2 \cdot 2^3 - 1^3 \cdot 2^2 + 3 \cdot 1 + 2 \cdot 2 = 11$$

●

● **Example 6**   Where is the function

$$f(x, y) = \frac{x^2 - y^2}{x^2 + y^2}$$

continuous?

Solution   The function f is discontinuous at $(0,0)$ because it is not defined there. Since f is a rational function it is continuous on its domain $D = \{(x, y) \mid (x, y) \neq (0,0)\}$.   ●

● **Example 7**   Let

$$g(x, y) = \begin{cases} \dfrac{x^2 - y^2}{x^2 + y^2} & \text{if } (x, y) \neq (0,0) \\ 0 & \text{if } (x, y) = (0,0) \end{cases}$$

Here g is defined at $(0,0)$ but g is still discontinuous at 0 because $\lim_{(x,y) \to (0,0)} g(x, y)$ does not exist (see Example 1).   ●

● **Example 8**   Let

$$f(x, y) = \begin{cases} \dfrac{3x^2 y}{x^2 + y^2} & \text{if } (x, y) \neq (0,0) \\ 0 & \text{if } (x, y) = (0,0) \end{cases}$$

We know f is continuous for $(x, y) \neq (0,0)$ since it is equal to a rational function there. Also, from Example 4, we have

$$\lim_{(x,y) \to (0,0)} f(x, y) = \lim_{(x,y) \to (0,0)} \frac{3x^2 y}{x^2 + y^2} = 0 = f(0,0)$$

Therefore f is continuous at $(0, 0)$, and so it is continuous on R^2.   ●

● **Example 9**   Let

$$h(x, y) = \begin{cases} \dfrac{3x^2 y}{x^2 + y^2} & \text{if } (x, y) \neq (0,0) \\ 17 & \text{if } (x, y) = (0,0) \end{cases}$$

Again from Example 4, we have

$$\lim_{(x,y) \to (0,0)} g(x, y) = \lim_{(x,y) \to (0,0)} \frac{3x^2 y}{x^2 + y^2} = 0 \neq 17 = g(0,0)$$

and so g is discontinuous at $(0, 0)$. But g is continuous on the set $S = \{(x, y) \mid (x, y) \neq (0,0)\}$ since it is equal to a rational function on S.   ●

Composition is another way of combining two continuous functions to get a third. The proof of the following theorem is similar to that of Theorem 2.27.

Theorem (12.7)

> If f is continuous at (a, b) and g is a function of a single variable that is continuous at $f(a, b)$, then the composite function $h = g \circ f$ defined by $h(x, y) = g(f(x, y))$ is continuous at (a, b).

● **Example 10**   On what set is the function $h(x, y) = \ln(x^2 + y^2 - 1)$ continuous?

Solution   Let $f(x, y) = x^2 + y^2 - 1$ and $g(t) = \ln t$. Then

$$g(f(x, y)) = \ln(x^2 + y^2 - 1) = h(x, y)$$

so $h = g \circ f$. Now f is continuous everywhere since it is a polynomial and g is continuous on its domain $\{t \mid t > 0\}$. Thus, by Theorem 12.7, h is continuous on its domain

$$D = \{(x, y) \mid x^2 + y^2 - 1 > 0\} = \{(x, y) \mid x^2 + y^2 > 1\}$$

which consists of all points outside the circle $x^2 + y^2 = 1$.   ●

Everything in this section can be extended to functions of three or more variables. The distance between two points (x, y, z) and (a, b, c) in R^3 is $\sqrt{(x - a)^2 + (y - b)^2 + (z - c)^2}$, so the definitions of limit and continuity of a function of three variables are as follows.

Definition (12.8)

> Let $f: D \subset R^3 \to R$.
>
> (a) $$\lim_{(x,y,z) \to (a,b,c)} f(x, y, z) = L$$
>
> means that for every number $\varepsilon > 0$ there is a corresponding number $\delta > 0$ such that
>
> $$|f(x, y, z) - L| < \varepsilon \qquad \text{whenever } (x, y, z) \in D \text{ and}$$
> $$0 < \sqrt{(x - a)^2 + (y - b)^2 + (z - c)^2} < \delta$$
>
> (b) f is **continuous** at (a, b, c) if
>
> $$\lim_{(x,y,z) \to (a,b,c)} f(x, y, z) = f(a, b, c)$$

If we use the vector notation introduced at the end of Section 12.1, then the definitions of a limit for functions of two or three variables can be written in a single compact form as follows.

(12.9)

> If $f: D \subset R^n \to R$, then $\lim_{\mathbf{x} \to \mathbf{a}} f(\mathbf{x}) = L$ means that for every number $\varepsilon > 0$ there is a corresponding number $\delta > 0$ such that
>
> $$|f(\mathbf{x}) - L| < \varepsilon \qquad \text{whenever} \qquad 0 < |\mathbf{x} - \mathbf{a}| < \delta$$

Notice that if $n = 1$, then $\mathbf{x} = x$ and $\mathbf{a} = a$, and (12.9) is just the definition of a limit for functions of a single variable. If $n = 2$, then $\mathbf{x} = \langle x, y \rangle$, $\mathbf{a} = \langle a, b \rangle$, and $|\mathbf{x} - \mathbf{a}| = \sqrt{(x - a)^2 + (y - b)^2}$, so (12.9) becomes Definition 12.5. If $n = 3$, then $\mathbf{x} = \langle x, y, z \rangle$, $\mathbf{a} = \langle a, b, c \rangle$, and (12.9) becomes part (a) of Definition 12.8. In each case the definition of continuity can be written as

$$\lim_{\mathbf{x} \to \mathbf{a}} f(\mathbf{x}) = f(\mathbf{a})$$

EXERCISES 12.2

In Exercises 1–30 find the limit, if it exists, or show that the limit does not exist.

1. $\displaystyle\lim_{(x,y) \to (2,3)} (x^2 y^2 - 2xy^5 + 3y)$

2. $\displaystyle\lim_{(x,y) \to (-3,4)} (x^3 + 3x^2 y^2 - 5y^3 + 1)$

3. $\displaystyle\lim_{(x,y) \to (0,0)} \frac{x^2 y^3 + x^3 y^2 - 5}{2 - xy}$

4. $\displaystyle\lim_{(x,y) \to (-2,1)} \frac{x^2 + xy + y^2}{x^2 - y^2}$

5. $\displaystyle\lim_{(x,y) \to (\pi,\pi)} x \sin\left(\frac{x + y}{4}\right)$

6. $\displaystyle\lim_{(x,y) \to (1,4)} e^{\sqrt{x + 2y}}$

7. $\displaystyle\lim_{(x,y) \to (0,0)} \frac{\sin(x + y)}{x + y}$

8. $\displaystyle\lim_{(x,y) \to (0,0)} \frac{x^2 - y^2}{x + y}$

9. $\displaystyle\lim_{(x,y) \to (0,0)} \frac{x - y}{x^2 + y^2}$

10. $\displaystyle\lim_{(x,y) \to (0,0)} \frac{x^2}{x^2 + y^2}$

11. $\displaystyle\lim_{(x,y) \to (0,0)} \frac{8x^2 y^2}{x^4 + y^4}$

12. $\displaystyle\lim_{(x,y) \to (0,0)} \frac{x^3 + xy^2}{x^2 + y^2}$

13. $\displaystyle\lim_{(x,y) \to (0,0)} \frac{2xy}{x^2 + 2y^2}$

14. $\displaystyle\lim_{(x,y) \to (0,0)} \frac{(x + y)^2}{x^2 + y^2}$

15. $\displaystyle\lim_{(x,y) \to (0,0)} \frac{xy}{\sqrt{x^2 + y^2}}$

16. $\displaystyle\lim_{(x,y) \to (0,0)} \frac{2x^2 + 3xy + 4y^2}{3x^2 + 5y^2}$

17. $\displaystyle\lim_{(x,y) \to (0,0)} \frac{xy + 1}{x^2 + y^2 + 1}$

18. $\displaystyle\lim_{(x,y) \to (0,0)} \frac{xy^3}{x^2 + y^6}$

19. $\displaystyle\lim_{(x,y) \to (0,0)} \frac{2x^2 y}{x^4 + y^2}$

20. $\displaystyle\lim_{(x,y) \to (0,0)} \frac{x^3 y^2}{x^2 + y^2}$

21. $\displaystyle\lim_{(x,y) \to (0,0)} \frac{x^2 + y^2}{\sqrt{x^2 + y^2 + 1} - 1}$

22. $\displaystyle\lim_{(x,y) \to (0,0)} \frac{\sqrt{x^2 y^2 + 1} - 1}{x^2 + y^2}$

23. $\displaystyle\lim_{(x,y) \to (0,1)} \frac{xy - x}{x^2 + y^2 - 2y + 1}$

24. $\displaystyle\lim_{(x,y) \to (1,-1)} \frac{x^2 + y^2 - 2x - 2y}{x^2 + y^2 - 2x + 2y + 2}$

25. $\displaystyle\lim_{(x,y,z) \to (1,2,3)} \frac{xz^2 - y^2 z}{xyz - 1}$

26. $\displaystyle\lim_{(x,y,z)\to(2,3,0)} \left[xe^z + \ln(2x - y) \right]$

27. $\displaystyle\lim_{(x,y,z)\to(0,0,0)} \frac{x^2 - y^2 - z^2}{x^2 + y^2 + z^2}$

28. $\displaystyle\lim_{(x,y,z)\to(0,0,0)} \frac{xy + yz + zx}{x^2 + y^2 + z^2}$

29. $\displaystyle\lim_{(x,y,z)\to(0,0,0)} \frac{xy + yz^2 + xz^2}{x^2 + y^2 + z^4}$

30. $\displaystyle\lim_{(x,y,z)\to(0,0,0)} \frac{x^2 y^2 z^2}{x^2 + y^2 + z^2}$

In Exercises 31–34 find $h(x, y) = g(f(x, y))$ and the set on which h is continuous.

31. $g(t) = e^{-t} \cos t, \ f(x, y) = x^4 + x^2 y^2 + y^4$

32. $g(t) = \dfrac{\sqrt{t} - 1}{\sqrt{t} + 1}, \ f(x, y) = x^2 - y$

33. $g(t) = t^2 + \sqrt{t}, \ f(x, y) = 2x + 3y - 6$

34. $g(z) = \sin z, \ f(x, y) = y \ln x$

In Exercises 35–48 determine the largest set on which the given function is continuous.

35. $F(x, y) = \dfrac{x^2 + y^2 + 1}{x^2 + y^2 - 1}$

36. $F(x, y) = \dfrac{x^6 + x^3 y^3 + y^6}{x^3 + y^3}$

37. $F(x, y) = \tan(x^4 - y^4)$

38. $F(x, y) = \ln(2x + 3y)$

39. $G(x, y) = e^{xy} \sin(x + y)$

40. $G(x, y) = \sin^{-1}(x^2 + y^2)$

41. $G(x, y) = \sqrt{x + y} - \sqrt{x - y}$

42. $G(x, y) = 2^{x \tan y}$

43. $f(x, y, z) = x \ln(yz)$

44. $f(x, y, z) = x + y\sqrt{x + z}$

45. $f(x, y) = \begin{cases} \dfrac{2x^2 - y^2}{2x^2 + y^2} & \text{if } (x, y) \neq (0, 0) \\ 0 & \text{if } (x, y) = (0, 0) \end{cases}$

46. $f(x, y) = \begin{cases} \dfrac{x^2 y^3}{2x^2 + y^2} & \text{if } (x, y) \neq (0, 0) \\ 0 & \text{if } (x, y) = (0, 0) \end{cases}$

47. $f(x, y) = \begin{cases} \dfrac{x^2 y^3}{2x^2 + y^2} & \text{if } (x, y) \neq (0, 0) \\ 1 & \text{if } (x, y) = (0, 0) \end{cases}$

48. $f(x, y) = \begin{cases} \dfrac{xy}{x^2 + xy + y^2} & \text{if } (x, y) \neq (0, 0) \\ 0 & \text{if } (x, y) = (0, 0) \end{cases}$

49. Prove, using Definition 12.5, that

(a) $\displaystyle\lim_{(x,y)\to(a,b)} x = a$ (b) $\displaystyle\lim_{(x,y)\to(a,b)} y = b$

(c) $\displaystyle\lim_{(x,y)\to(a,b)} c = c$

50. For what values of the number r is the function

$$f(x, y, z) = \begin{cases} \dfrac{(x + y + z)^r}{x^2 + y^2 + z^2} & \text{if } (x, y, z) \neq (0, 0, 0) \\ 0 & \text{if } (x, y, z) = (0, 0, 0) \end{cases}$$

continuous on R^3?

51. Show that the function $f: R^n \to R$ given by $f(\mathbf{x}) = |\mathbf{x}|$ is continuous on R^n.
[*Hint:* Consider $|\mathbf{x} - \mathbf{a}|^2 = (\mathbf{x} - \mathbf{a}) \cdot (\mathbf{x} - \mathbf{a})$.]

52. If $\mathbf{c} \in V_n$, show that the function $f: R^n \to R$ given by $f(\mathbf{x}) = \mathbf{c} \cdot \mathbf{x}$ is continuous on R^n.

SECTION 12.3

Partial Derivatives

If f is a function of two variables x and y, suppose we let only x vary while keeping y fixed, say $y = b$, where b is a constant. Then we are really considering a function of a single variable x, namely, $g(x) = f(x, b)$. If g has a derivative at a, then we call it the **partial derivative of f with respect to x at** (**a**, **b**) and denote it by $f_x(a, b)$. Thus

(12.10)

$$f_x(a, b) = g'(a) \qquad \text{where } g(x) = f(x, b)$$

By Definition 3.2 we know that

$$g'(a) = \lim_{h \to 0} \frac{g(a + h) - g(a)}{h}$$

and so Equation 12.10 becomes

(12.11)

$$f_x(a, b) = \lim_{h \to 0} \frac{f(a + h, b) - f(a, b)}{h}$$

Similarly the **partial derivative of f with respect to y at (a, b),** denoted by $f_y(a, b)$, is obtained by keeping x fixed ($x = a$) and finding the ordinary derivative at b of the function $G(y) = f(a, y)$:

(12.12)

$$f_y(a, b) = \lim_{h \to 0} \frac{f(a, b + h) - f(a, b)}{h}$$

If we now let the point (a, b) vary, f_x and f_y become functions of two variables.

(12.13)

If f is a function of two variables, its **partial derivatives** are the functions f_x and f_y defined by

$$f_x(x, y) = \lim_{h \to 0} \frac{f(x + h, y) - f(x, y)}{h}$$

$$f_y(x, y) = \lim_{h \to 0} \frac{f(x, y + h) - f(x, y)}{h}$$

There are many alternative notations for partial derivatives. For instance, instead of f_x we can write f_1 or $D_1 f$ (to indicate differentiation with respect to the *first* variable) or $\partial f/\partial x$. But here $\partial f/\partial x$ cannot be interpreted as a ratio of differentials.

Notations for Partial Derivatives

If $z = f(x, y)$ we write

$$f_x(x, y) = f_x = \frac{\partial f}{\partial x} = \frac{\partial}{\partial x} f(x, y) = \frac{\partial z}{\partial x} = f_1 = D_1 f = D_x f$$

$$f_y(x, y) = f_y = \frac{\partial f}{\partial y} = \frac{\partial}{\partial y} f(x, y) = \frac{\partial z}{\partial y} = f_2 = D_2 f = D_y f$$

To compute partial derivatives, all you have to do is remember from Equation 12.10 that the partial derivative with respect to x is just the *ordinary* derivative of the function g of a single variable that you get by keeping y fixed. Thus we have the following rule:

Rule for Finding Partial Derivatives of $z = f(x, y)$

1. To find f_x, regard y as a constant and differentiate $f(x, y)$ with respect to x.

2. To find f_y, regard x as a constant and differentiate $f(x, y)$ with respect to y.

● **Example 1**   If $f(x, y) = x^3 + x^2 y^3 - 2y^2$, find $f_x(2, 1)$, $f_y(2, 1)$.

Solution   Holding y constant and differentiating with respect to x, we get

$$f_x(x, y) = 3x^2 + 2xy^3$$

and so
$$f_x(2, 1) = 3 \cdot 2^2 + 2 \cdot 2 \cdot 1^3 = 16$$

Holding x constant and differentiating with respect to y, we get

$$f_y(x, y) = 3x^2 y^2 - 4y$$
$$f_y(2, 1) = 3 \cdot 2^2 \cdot 1^2 - 4 \cdot 1 = 8$$

● **Example 2**   If $f(x, y) = \sin\left(\dfrac{x}{1 + y}\right)$, calculate $\dfrac{\partial f}{\partial x}, \dfrac{\partial f}{\partial y}$.

Solution   Using the Chain Rule for functions of one variable, we have

$$\frac{\partial f}{\partial x} = \cos\left(\frac{x}{1 + y}\right) \cdot \frac{\partial}{\partial x}\left(\frac{x}{1 + y}\right) = \cos\left(\frac{x}{1 + y}\right) \cdot \frac{1}{1 + y}$$

$$\frac{\partial f}{\partial y} = \cos\left(\frac{x}{1 + y}\right) \cdot \frac{\partial}{\partial y}\left(\frac{x}{1 + y}\right) = -\cos\left(\frac{x}{1 + y}\right) \cdot \frac{x}{(1 + y)^2}$$

● **Example 3**   Find $\partial z/\partial x$ and $\partial z/\partial y$ if z is defined implicitly as a function of x and y by the equation

$$x^3 + y^3 + z^3 + 6xyz = 1$$

Solution   To find $\partial z/\partial x$ we differentiate implicitly with respect to x, being careful to treat y as a constant:

$$3x^2 + 3z^2\,\frac{\partial z}{\partial x} + 6yz + 6xy\,\frac{\partial z}{\partial x} = 0$$

Solving this equation for $\partial z/\partial x$, we obtain

$$\frac{\partial z}{\partial x} = -\frac{x^2 + 2yz}{z^2 + 2xy}$$

Similarly, implicit differentiation with respect to y gives

$$\frac{\partial z}{\partial y} = -\frac{y^2 + 2xz}{z^2 + 2xy}$$

●

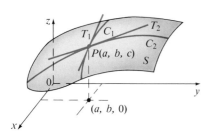

Figure 12.21

To give a geometric interpretation of partial derivatives we recall that the equation $z = f(x, y)$ represents a surface S (the graph of f). If $f(a, b) = c$, then the point $P(a, b, c)$ lies on S. The vertical plane $y = b$ intersects S in a curve C_1. (In other words, C_1 is the trace of S in the plane $y = b$.) Likewise the vertical plane $x = a$ intersects S in a curve C_2. Both of the curves C_1, C_2 pass through the point P (see Figure 12.21).

Notice that the curve C_1 is the graph of the function $g(x) = f(x, b)$, so the slope of its tangent T_1 at P is $g'(a) = f_x(a, b)$. The curve C_2 is the graph of the function $G(y) = f(a, y)$, so the slope of its tangent T_2 at P is $G'(b) = f_y(a, b)$.

Thus the partial derivatives $f_x(a, b)$ and $f_y(a, b)$ can be interpreted geometrically as the slopes of the tangent lines at $P(a, b, c)$ to the traces C_1 and C_2 of S in the planes $y = b$ and $x = a$.

Partial derivatives can also be interpreted as rates of change. If $z = f(x, y)$, then $\partial z/\partial x$ represents the rate of change of z with respect to x when y is fixed. Similarly $\partial z/\partial y$ represents the rate of change of z with respect to y when x is fixed. For instance, referring to the temperature function T at the beginning of Section 12.1, we see that $\partial T/\partial x$ is the rate at which the temperature changes in the east–west direction and $\partial T/\partial y$ is the rate at which it changes in the north–south direction.

Functions of More Than Two Variables

Partial derivatives can also be defined for functions of three or more variables. For example, if f is a function of three variables x, y, and z, then its partial derivative with respect to x is defined as

$$f_x(x, y, z) = \lim_{h \to 0} \frac{f(x + h, y, z) - f(x, y, z)}{h}$$

and it is found by regarding y and z as constants and differentiating $f(x, y, z)$ with respect to x.

In general if u is a function of n variables, $u = f(x_1, x_2, \ldots, x_n)$, its partial derivative with respect to the ith variable x_i is

$$\frac{\partial u}{\partial x_i} = \lim_{h \to 0} \frac{f(x_1, \ldots, x_{i-1}, x_i + h, x_{i+1}, \ldots, x_n) - f(x_1, \ldots, x_i, \ldots, x_n)}{h}$$

and we also write

$$\frac{\partial u}{\partial x_i} = \frac{\partial f}{\partial x_i} = f_{x_i} = f_i = D_i f$$

● **Example 4**   Find f_x, f_y, f_z if $f(x, y, z) = e^{xy} \ln z$.

Solution   Holding y and z constant and differentiating with respect to x, we have

$$f_x = y e^{xy} \ln z$$

Similarly,

$$f_y = x e^{xy} \ln z \qquad \text{and} \qquad f_z = \frac{e^{xy}}{z} \qquad\qquad ●$$

Higher Derivatives

If f is a function of two variables, then its partial derivatives f_x and f_y are also functions of two variables, so we can consider their partial derivatives $(f_x)_x$, $(f_x)_y$, $(f_y)_x$, $(f_y)_y$, which are called the **second partial derivatives** of f. If $z = f(x, y)$, we use the following notation:

$$(f_x)_x = f_{xx} = f_{11} = \frac{\partial}{\partial x}\left(\frac{\partial f}{\partial x}\right) = \frac{\partial^2 f}{\partial x^2} = \frac{\partial^2 z}{\partial x^2}$$

$$(f_x)_y = f_{xy} = f_{12} = \frac{\partial}{\partial y}\left(\frac{\partial f}{\partial x}\right) = \frac{\partial^2 f}{\partial y\, \partial x} = \frac{\partial^2 z}{\partial y\, \partial x}$$

$$(f_y)_x = f_{yx} = f_{21} = \frac{\partial}{\partial x}\left(\frac{\partial f}{\partial y}\right) = \frac{\partial^2 f}{\partial x\, \partial y} = \frac{\partial^2 z}{\partial x\, \partial y}$$

$$(f_y)_y = f_{yy} = f_{22} = \frac{\partial}{\partial y}\left(\frac{\partial f}{\partial y}\right) = \frac{\partial^2 f}{\partial y^2} = \frac{\partial^2 z}{\partial y^2}$$

Thus the notation f_{xy} (or $\partial^2 f / \partial y\, \partial x$) means that we first differentiate with respect to x and then with respect to y, whereas in computing f_{yx} the order is reversed.

● **Example 5**   Find the second partial derivatives of

$$f(x, y) = x^3 + x^2 y^3 - 2y^2$$

Solution   In Example 1 we found that

$$f_x(x, y) = 3x^2 + 2xy^3 \qquad f_y(x, y) = 3x^2y^2 - 4y$$

Therefore

$$f_{xx} = \frac{\partial}{\partial x}(3x^2 + 2xy^3) = 6x + 2y^3$$

$$f_{xy} = \frac{\partial}{\partial y}(3x^2 + 2xy^3) = 6xy^2$$

$$f_{yx} = \frac{\partial}{\partial x}(3x^2y^2 - 4y) = 6xy^2$$

$$f_{yy} = \frac{\partial}{\partial y}(3x^2y^2 - 4y) = 6x^2y - 4$$

Notice that $f_{xy} = f_{yx}$ in Example 5. This is not just a coincidence. It turns out that the mixed partial derivatives f_{xy} and f_{yx} are equal for most functions that one meets in practice. The following theorem, discovered by the French mathematician Alexis Clairaut (1713–1765), gives conditions under which we can assert that $f_{xy} = f_{yx}$. The proof is in Appendix C.

Clairaut's Theorem (12.14)

> Suppose f is defined on a disk D that contains the point (a, b). If the functions f_{xy} and f_{yx} are both continuous on D, then
>
> $$f_{xy}(a, b) = f_{yx}(a, b)$$

● **Example 6**   Show that the function $u(x, y) = e^x \sin y$ is a solution of the partial differential equation $u_{xx} + u_{yy} = 0$. [This is called Laplace's equation after Pierre Laplace (1749–1827). Solutions of this equation are called harmonic functions and play a role in problems of heat conduction and electric potential.]

Solution
$$u_x = e^x \sin y \qquad u_y = e^x \cos y$$
$$u_{xx} = e^x \sin y \qquad u_{yy} = -e^x \sin y$$
$$u_{xx} + u_{yy} = e^x \sin y - e^x \sin y = 0$$

Therefore u satisfies Laplace's equation. ●

Partial derivatives of order 3 or higher can also be defined. For instance,

$$f_{xyy} = (f_{xy})_y = \frac{\partial}{\partial y}\left(\frac{\partial^2 z}{\partial y\, \partial x}\right) = \frac{\partial^3 z}{\partial y^2\, \partial x}$$

and using Clairaut's Theorem it can be shown that $f_{xyy} = f_{yxy} = f_{yyx}$ if these functions are continuous.

● **Example 7**   Calculate f_{xxyz} if $f(x, y, z) = \sin(3x + yz)$.

Solution
$$f_x = 3\cos(3x + yz)$$
$$f_{xx} = -9\sin(3x + yz)$$
$$f_{xxy} = -9z\cos(3x + yz)$$
$$f_{xxyz} = -9\cos(3x + yz) + 9yz\sin(3x + yz)$$

EXERCISES 12.3

In Exercises 1–18 find the indicated partial derivatives.

1. $f(x, y) = x^3 y^5$; $f_x(3, -1)$

2. $f(x, y) = \sqrt{2x + 3y}$; $f_y(2, 4)$

3. $f(x, y) = xe^{-y} + 3y$; $\dfrac{\partial f}{\partial y}(1, 0)$

4. $f(x, y) = \sin(y - x)$; $\dfrac{\partial f}{\partial x}(3, 3)$

5. $z = \dfrac{x^3 + y^3}{x^2 + y^2}$; $\dfrac{\partial z}{\partial x}, \dfrac{\partial z}{\partial y}$

6. $z = x\sqrt{y} - \dfrac{y}{\sqrt{x}}$; $\dfrac{\partial z}{\partial x}, \dfrac{\partial z}{\partial y}$

7. $z = \dfrac{x}{y} + \dfrac{y}{x}$; $\dfrac{\partial z}{\partial x}$

8. $z = (3xy^2 - x^4 + 1)^4$; $\dfrac{\partial z}{\partial x}, \dfrac{\partial z}{\partial y}$

9. $xy + yz = xz$; $\dfrac{\partial z}{\partial x}, \dfrac{\partial z}{\partial y}$

10. $xyz = \cos(x + y + z)$; $\dfrac{\partial z}{\partial x}, \dfrac{\partial z}{\partial y}$

11. $x^2 + y^2 - z^2 = 2x(y + z)$; $\dfrac{\partial z}{\partial x}, \dfrac{\partial z}{\partial y}$

12. $xy^2z^3 + x^3y^2z = x + y + z$; $\dfrac{\partial z}{\partial x}, \dfrac{\partial z}{\partial y}$

13. $u = xy\sec(xy)$; $\dfrac{\partial u}{\partial x}$

14. $u = \dfrac{x}{x + t}$; $\dfrac{\partial u}{\partial x}, \dfrac{\partial u}{\partial t}$

15. $f(x, y, z) = xyz$; $f_y(0, 1, 2)$

16. $f(x, y, z) = \sqrt{x^2 + y^2 + z^2}$; $f_z(0, 3, 4)$

17. $u = xy + yz + zx$; u_x, u_y, u_z

18. $u = x^2 y^3 t^4$; u_x, u_y, u_t

In Exercises 19–50 find the first partial derivatives of the given functions.

19. $f(x, y) = x^3 y^5 - 2x^2 y + x$

20. $f(x, y) = x^2 y^2 (x^4 + y^4)$

21. $f(x, y) = x^4 + x^2 y^2 + y^4$

22. $f(x, y) = \ln(x^2 + y^2)$

23. $f(x, y) = \dfrac{x - y}{x + y}$ **24.** $f(x, y) = x^y$

25. $f(x, y) = e^x \tan(x - y)$ **26.** $f(x, y) = e^{xy} \cos x \sin y$

27. $f(s, t) = \sqrt{2 - 3s^2 - 5t^2}$

28. $f(s, t) = \dfrac{s}{\sqrt{s^2 + t^2}}$

29. $f(u, v) = \tan^{-1}\left(\dfrac{u}{v}\right)$ **30.** $f(x, t) = e^{\sin(t/x)}$

31. $g(x, y) = y\tan(x^2 y^3)$ **32.** $g(x, y) = \ln(x + \ln y)$

33. $z = \ln(x + \sqrt{x^2 + y^2})$ **34.** $z = x^{xy}$

35. $z = \sinh\sqrt{3x + 4y}$ **36.** $z = \log_x y$

37. $f(x, y) = \displaystyle\int_x^y e^{t^2}\, dt$ **38.** $f(x, y) = \displaystyle\int_y^x \dfrac{e^t}{t}\, dt$

39. $f(x, y, z) = x^2 yz^3 + xy - z$

40. $f(x, y, z) = x\sqrt{yz}$ **41.** $f(x, y, z) = x^{yz}$

42. $f(x, y, z) = xe^y + ye^z + ze^x$

43. $u = z \sin \dfrac{y}{x + z}$

44. $u = x^{y/z}$

45. $u = xy^2z^3 \ln(x + 2y + 3z)$

46. $u = x^{y^z}$

47. $f(x, y, z, t) = \dfrac{x - y}{z - t}$

48. $f(x, y, z, t) = xy^2z^3t^4$

49. $u = \sqrt{x_1^2 + x_2^2 + \cdots + x_n^2}$

50. $u = \sin(x_1 + 2x_2 + \cdots + nx_n)$

In Exercises 51–56 find $\partial z/\partial x$ and $\partial z/\partial y$.

51. $z = f(x) + g(y)$

52. $z = f(x)g(y)$

53. $z = f(x + y)$

54. $z = f(xy)$

55. $z = f\left(\dfrac{x}{y}\right)$

56. $z = f(ax + by)$

In Exercises 57–62 find all the second partial derivatives.

57. $f(x, y) = x^2y + x\sqrt{y}$

58. $f(x, y) = \sin(x + y) + \cos(x - y)$

59. $z = (x^2 + y^2)^{3/2}$

60. $z = \cos^2(5x + 2y)$

61. $z = t \sin^{-1}\sqrt{x}$

62. $z = x^{\ln t}$

In Exercises 63–66 verify that the conclusion of Clairaut's Theorem holds, that is, $u_{xy} = u_{yx}$.

63. $u = x^5y^4 - 3x^2y^3 + 2x^2$

64. $u = \sin^2 x \cos y$

65. $u = \sin^{-1}(xy^2)$

66. $u = x^2y^3z^4$

In Exercises 67–74 find the indicated partial derivative.

67. $f(x, y) = x^2y^3 - 2x^4y; \ f_{xxx}$

68. $f(x, y) = e^{xy^2}; \ f_{xxy}$

69. $f(x, y, z) = x^5 + x^4y^4z^3 + yz^2; \ f_{xyz}$

70. $f(x, y, z) = e^{xyz}; \ f_{yzy}$

71. $z = x \sin y; \ \dfrac{\partial^3 z}{\partial y^2 \, \partial x}$

72. $z = \ln \sin(x - y); \ \dfrac{\partial^3 z}{\partial y \, \partial x^2}$

73. $u = \ln(x + 2y^2 + 3z^3); \ \dfrac{\partial^3 u}{\partial x \, \partial y \, \partial z}$

74. $u = x^ay^bz^c; \ \dfrac{\partial^6 u}{\partial x \, \partial y^2 \, \partial z^3}$

75. Verify that the function $u = e^{-\alpha^2 k^2 t} \sin kx$ is a solution of the heat conduction equation $u_t = \alpha^2 u_{xx}$.

76. Which of the following functions are solutions of Laplace's equation $u_{xx} + u_{yy} = 0$?
(a) $u = x^2 + y^2$
(b) $u = x^2 - y^2$
(c) $u = x^3 + 3xy^2$
(d) $u = \ln\sqrt{x^2 + y^2}$
(e) $u = \sin x \cosh y + \cos x \sinh y$
(f) $u = e^{-x}\cos y - e^{-y}\cos x$

77. Verify that the function $u = 1/\sqrt{x^2 + y^2 + z^2}$ is a solution of the three-dimensional Laplace equation $u_{xx} + u_{yy} + u_{zz} = 0$.

78. Show that the following functions are solutions of the wave equation $u_{tt} = a^2 u_{xx}$.
(a) $u = \sin(kx)\sin(akt)$
(b) $u = t/(a^2t^2 - x^2)$
(c) $u = (x - at)^6 + (x + at)^6$
(d) $u = \sin(x - at) + \ln(x + at)$

79. If f and g are twice differentiable functions of a single variable, show that the function

$$u(x, t) = f(x + at) + g(x - at)$$

is a solution of the wave equation given in Exercise 78.

80. If f and g are twice differentiable functions of a single variable, show that the function

$$u(x, y) = xf(x + y) + yg(x + y)$$

satisfies the equation $u_{xx} - 2u_{xy} + u_{yy} = 0$.

81. Show that the function $z = xe^y + ye^x$ is a solution of the equation

$$\frac{\partial^3 z}{\partial x^3} + \frac{\partial^3 z}{\partial y^3} = x\frac{\partial^3 z}{\partial x \, \partial y^2} + y\frac{\partial^3 z}{\partial x^2 \, \partial y}$$

82. The temperature at a point (x, y) on a flat metal plate is given by $T(x, y) = 60/(1 + x^2 + y^2)$ where T is measured in °C and x, y in meters. Find the rate of change of temperature with respect to distance at the point $(2, 1)$ in (a) the x-direction and (b) the y-direction.

83. The total resistance R produced by three conductors with resistances R_1, R_2, R_3 connected in a parallel electrical circuit is given by the formula

$$\frac{1}{R} = \frac{1}{R_1} + \frac{1}{R_2} + \frac{1}{R_3}$$

Find $\partial R/\partial R_1$.

84. The gas law for a fixed mass m of an ideal gas at absolute temperature T, pressure P, and volume V is $PV = mRT$, where R is the gas constant. Show that

$$\frac{\partial P}{\partial V} \frac{\partial V}{\partial T} \frac{\partial T}{\partial P} = -1$$

85. The kinetic energy of a body with mass m and velocity v is $K = \frac{1}{2}mv^2$. Show that

$$\frac{\partial K}{\partial m} \frac{\partial^2 K}{\partial v^2} = K$$

86. If a, b, c are the sides of a triangle and A, B, C are the opposite angles, find $\partial A/\partial a, \partial A/\partial b, \partial A/\partial c$ by implicit differentiation of the Law of Cosines.

87. Use Clairaut's Theorem to show that if the third order partial derivatives of f are continuous, then

$$f_{xyy} = f_{yxy} = f_{yyx}$$

88. (a) How many nth order partial derivatives does a function of two variables have?

(b) If these partial derivatives are all continuous, how many of them can be distinct?

(c) Answer the question in part (a) for a function of three variables.

89. Let

$$f(x, y) = \begin{cases} \dfrac{x^3 y - xy^3}{x^2 + y^2} & \text{if } (x, y) \neq (0,0) \\ 0 & \text{if } (x, y) = (0,0) \end{cases}$$

(a) Find $f_x(x, y)$ and $f_y(x, y)$ when $(x, y) \neq (0,0)$.

(b) Find $f_x(0, 0)$ and $f_y(0, 0)$ using Equations 12.11 and 12.12.

(c) Show that $f_{xy}(0, 0) = -1$ and $f_{yx}(0, 0) = 1$.

(d) Does the result of part (c) contradict Clairaut's Theorem?

90. If $u = e^{a_1 x_1 + a_2 x_2 + \cdots + a_n x_n}$, where $a_1^2 + a_2^2 + \cdots + a_n^2 = 1$, show that

$$\frac{\partial^2 u}{\partial x_1^2} + \frac{\partial^2 u}{\partial x_2^2} + \cdots + \frac{\partial^2 u}{\partial x_n^2} = u$$

Tangent Planes and Differentials

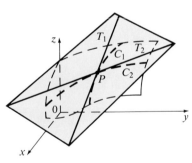

Figure 12.22

Suppose a surface S has equation $z = f(x, y)$, where f has continuous first partial derivatives, and let $P(x_0, y_0, z_0)$ be a point on S. As in the preceding section, let C_1 and C_2 be the curves obtained by intersecting the vertical planes $y = y_0$ and $x = x_0$ with the surface S. Then the point P lies on both C_1 and C_2. Let T_1 and T_2 be the tangent lines to the curves C_1 and C_2 at the point P. Then the **tangent plane** to the surface S at the point P is defined to be the plane that contains both of the tangent lines T_1 and T_2 (see Figure 12.22).

We shall see in Section 12.6 that if C is any other curve that lies on the surface S and passes through P, then its tangent line at P also lies in the tangent plane. Therefore you can think of the tangent plane to S at P as consisting of all possible tangent lines at P to curves that lie on S and pass through P. The tangent plane at P is the plane that most closely approximates the surface S near the point P.

We know from Equation 11.40 that any plane passing through $P(x_0, y_0, z_0)$ has an equation of the form

$$A(x - x_0) + B(y - y_0) + C(z - z_0) = 0$$

By dividing this equation by C and letting $a = -A/C$ and $b = -B/C$, we can write it in the form

(12.15)
$$z - z_0 = a(x - x_0) + b(y - y_0)$$

If Equation 12.15 represents the tangent plane at P, then its intersection with the plane $y = y_0$ must be the tangent line T_1. Setting $y = y_0$ in Equation 12.15 gives

$$z - z_0 = a(x - x_0) \qquad y = y_0$$

and we recognize these as the equations (in slope-point form) of a line with slope a. But from Section 12.3 we know that the slope of T_1 is $f_x(x_0, y_0)$. Therefore $a = f_x(x_0, y_0)$.

Similarly putting $x = x_0$ in Equation 12.15 we get $z - z_0 = b(y - y_0)$, which must represent the tangent line T_2, so $b = f_y(x_0, y_0)$.

(12.16)

The equation of the tangent plane to the surface $z = f(x, y)$ at the point $P(x_0, y_0, z_0)$ is

$$z - z_0 = f_x(x_0, y_0)(x - x_0) + f_y(x_0, y_0)(y - y_0)$$

Note the similarity between the equation (12.16) of a tangent plane and the equation (3.4) of a tangent line.

● **Example 1**   Find the tangent plane to the elliptic paraboloid $z = 2x^2 + y^2$ at the point $(1, 1, 3)$.

Solution   Let $f(x, y) = 2x^2 + y^2$. Then

$$f_x(x, y) = 4x \qquad f_y(x, y) = 2y$$
$$f_x(1, 1) = 4 \qquad f_y(1, 1) = 2$$

Then (12.16) gives the equation of the tangent plane at $(1, 1, 3)$ as

$$z - 3 = 4(x - 1) + 2(y - 1)$$

or $\qquad\qquad 4x + 2y - z = 3$ ●

Differentials

Recall that for a function of one variable, $y = f(x)$, we defined the increment of y as

$$\Delta y = f(x + \Delta x) - f(x)$$

and the differential of y as

(12.17)

$$dy = f'(x)\,dx$$

(See Section 3.9.) Figure 12.23 shows the relationship between Δy and dy: Δy represents the change in height of the curve $y = f(x)$ and dy represents

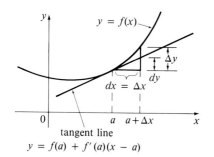

Figure 12.23

the change in height of the tangent line when x changes by an amount $dx = \Delta x$.

We note that $\Delta y - dy$ approaches 0 faster than Δx because if we let ε denote the ratio of these quantities, then

$$\varepsilon = \frac{\Delta y - dy}{\Delta x} = \frac{f(x + \Delta x) - f(x) - f'(x)\Delta x}{\Delta x}$$

$$= \frac{f(x + \Delta x) - f(x)}{\Delta x} - f'(x)$$

$$\to f'(x) - f'(x) = 0 \quad \text{as} \quad \Delta x \to 0$$

Therefore we have

(12.18)
$$\Delta y = dy + \varepsilon \Delta x \qquad \text{where } \varepsilon \to 0 \text{ as } \Delta x \to 0$$

Now consider a function of two variables, $z = f(x, y)$. If x and y are given increments Δx and Δy, then the corresponding **increment** of z is

(12.19)
$$\Delta z = f(x + \Delta x, y + \Delta y) - f(x, y)$$

Thus Δz represents the change in the value of f when (x, y) changes to $(x + \Delta x, y + \Delta y)$.

The **differentials** dx and dy are independent variables; that is, they can be given any values. Then the **differential** dz, also called the **total differential,** is defined by

(12.20)
$$dz = f_x(x, y)\, dx + f_y(x, y)\, dy = \frac{\partial z}{\partial x}\, dx + \frac{\partial z}{\partial y}\, dy$$

(Compare with Equation 12.17.) Sometimes the notation df is used in place of dz.

If we take

$$dx = \Delta x = x - a \qquad dy = \Delta y = y - b$$

in Equation 12.20, then the differential of z is

(12.21)
$$dz = f_x(a, b)(x - a) + f_y(a, b)(y - b)$$

On the other hand, if f_x and f_y are continuous, we see from (12.16) that the equation of the tangent plane to the surface $z = f(x, y)$ at the point $(a, b, f(a, b))$ is

(12.22)
$$z - f(a, b) = f_x(a, b)(x - a) + f_y(a, b)(y - b)$$

Comparing Equations 12.21 and 12.22, we see that dz represents the change in height of the tangent plane, whereas Δz represents the change in height of the surface $z = f(x, y)$ when (x, y) changes from (a, b) to $(a + \Delta x, b + \Delta y)$. (See Figure 12.24 and compare it with Figure 12.23.)

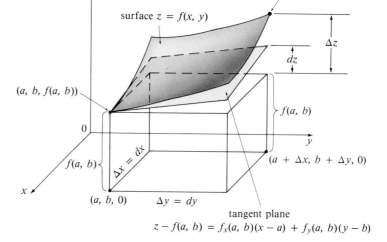

Figure 12.24

surface $z = f(x, y)$

$(a + \Delta x, b + \Delta y, f(a + \Delta x, b + \Delta y))$

Δz

dz

$(a, b, f(a, b))$

$f(a, b)$

0

y

$f(a, b)$

$\Delta x = dx$

$(a + \Delta x, b + \Delta y, 0)$

x

$(a, b, 0)$

$\Delta y = dy$

tangent plane

$z - f(a, b) = f_x(a, b)(x - a) + f_y(a, b)(y - b)$

It will be proved later in this section (Theorem 12.24) that if f_x and f_y are continuous, then

$$\Delta z - dz = \varepsilon_1 \Delta x + \varepsilon_2 \Delta y$$

where ε_1 and ε_2 are functions of Δx and Δy, which approach 0 as Δx and Δy approach 0. This means that $\Delta z - dz \approx 0$ and so

$$\Delta z \approx dz$$

which says that the actual change in z is approximately equal to the differential dz when Δx and Δy are small. This allows us to estimate the value of $f(a + \Delta x, b + \Delta y)$ when $f(a, b)$ is known:

(12.23) $$f(a + \Delta x, b + \Delta y) \approx f(a, b) + dz$$

When we use the approximation in (12.23) we are using the tangent plane at $P(a, b, f(a, b))$ as an approximation to the surface $z = f(x, y)$ when (x, y) is close to (a, b).

● **Example 2**   (a) If $z = f(x, y) = x^2 + 3xy - y^2$, find the differential dz. (b) If x changes from 2 to 2.05 and y changes from 3 to 2.96, compare the values of Δz and dz.

Solution   (a)   Definition 12.20 gives

$$dz = \frac{\partial z}{\partial x} dx + \frac{\partial z}{\partial y} dy$$

$$= (2x + 3y) dx + (3x - 2y) dy$$

(b)   Putting $x = 2$, $dx = \Delta x = 0.05$, $y = 3$, and $dy = \Delta y = -0.04$, we get

$$dz = [2(2) + 3(3)]0.05 + [3(2) - 2(3)](-0.04)$$

$$= 0.65$$

The increment of z is

$$\Delta z = f(2.05, 2.96) - f(2, 3)$$
$$= [(2.05)^2 + 3(2.05)(2.96) - (2.96)^2] - [2^2 + 3(2)(3) - 3^2]$$
$$= 0.6449$$

Notice that $\Delta z \approx dz$ but dz is easier to compute.   ●

● **Example 3**   Use differentials to find an approximate value for $\sqrt{9(1.95)^2 + (8.1)^2}$.

Solution   Consider the function $z = f(x, y) = \sqrt{9x^2 + y^2}$ and observe that we can easily calculate $f(2, 8) = 10$. Therefore we take $a = 2$, $b = 8$, $dx = \Delta x = -0.05$, $dy = \Delta y = 0.1$ in (12.23). Since

$$f_x(x, y) = \frac{9x}{\sqrt{9x^2 + y^2}} \quad \text{and} \quad f_y(x, y) = \frac{y}{\sqrt{9x^2 + y^2}}$$

we have $\sqrt{9(1.95)^2 + (8.1)^2} = f(1.95, 8.1)$

$$\approx f(2, 8) + dz$$
$$= f(2, 8) + f_x(2, 8)\,dx + f_y(2, 8)\,dy$$
$$= 10 + \tfrac{18}{10}(-0.05) + \tfrac{8}{10}(0.1)$$
$$= 9.99$$

You can check with a calculator that this approximation is accurate to two decimal places.   ●

● **Example 4**   The base radius and height of a right circular cone are measured as 10 cm and 25 cm, respectively, with a possible error in measurement of as much as 0.1 cm in each. Use differentials to estimate the maximum error in the calculated volume of the cone.

Solution   The volume V of a cone with base radius r and height h is $V = \pi r^2 h/3$. So the differential of V is

$$dV = \frac{\partial V}{\partial r}\,dr + \frac{\partial V}{\partial h}\,dh = \frac{2\pi rh}{3}\,dr + \frac{\pi r^2}{3}\,dh$$

Since the errors are at most 0.1 cm, we have $|\Delta x| \leqslant 0.1, |\Delta y| \leqslant 0.1$. To find the largest error in the volume we take the largest error in the measurement of r and h. Therefore we take $dr = 0.1$ and $dh = 0.1$ along with $r = 10$, $h = 25$. This gives

$$dV = \frac{500\pi}{3}(0.1) + \frac{100\pi}{3}(0.1) = 20\pi$$

Thus the maximum error in the calculated volume is about 20π cm^3 ≈ 63 cm^3.   ●

The following theorem says that dz is a good approximation to Δz when Δx and Δy are small, provided that f_x and f_y are both continuous.

Theorem (12.24)

Suppose that f_x and f_y exist on a rectangular region R with sides parallel to the axes and containing the points (a, b) and $(a + \Delta x, b + \Delta y)$. Suppose that f_x and f_y are continuous at the point (a, b) and let

$$\Delta z = f(a + \Delta x, b + \Delta y) - f(a, b)$$

Then $\qquad \Delta z = f_x(a, b)\, \Delta x + f_y(a, b)\, \Delta y + \varepsilon_1\, \Delta x + \varepsilon_2\, \Delta y$

where ε_1 and ε_2 are functions of Δx and Δy that approach 0 as $(\Delta x, \Delta y) \to (0, 0)$.

Proof   Referring to Figure 12.25 we write

(12.25) $\qquad \Delta z = [f(a + \Delta x, b + \Delta y) - f(a, b + \Delta y)] + [f(a, b + \Delta y) - f(a, b)]$

Figure 12.25

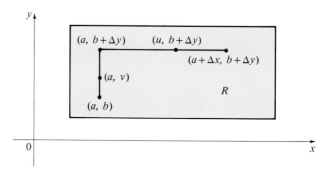

Observe that the function of a single variable

$$g(x) = f(x, b + \Delta y)$$

is defined on the interval $[a, a + \Delta x]$ and $g'(x) = f_x(x, b + \Delta y)$. If we apply the Mean Value Theorem to g, we get

$$g(a + \Delta x) - g(a) = g'(u)\, \Delta x$$

where u is some number between a and $a + \Delta x$. In terms of f, this equation becomes

$$f(a + \Delta x, b + \Delta y) - f(a, b + \Delta y) = f_x(u, b + \Delta y)\, \Delta x$$

This gives us an expression for the first part of the right side of Equation 12.25. For the second part we let $h(y) = f(a, y)$. Then h is a function of a single variable defined on the interval $[b, b + \Delta y]$ and $h'(y) = f_y(a, y)$. A second application of the Mean Value Theorem then gives

$$h(b + \Delta y) - h(b) = h'(v)\, \Delta y$$

where v is some number between b and $b + \Delta y$. In terms of f, this becomes

$$f(a, b + \Delta y) - f(a, b) = f_y(a, v) \Delta y$$

We now substitute these expressions into Equation 12.25 and obtain

$$\begin{aligned}
\Delta z &= f_x(u, b + \Delta y) \Delta x + f_y(a, v) \Delta y \\
&= f_x(a, b) \Delta x + [f_x(u, b + \Delta y) - f_x(a, b)] \Delta x + f_y(a, b) \Delta y \\
&\quad + [f_y(a, v) - f_y(a, b)] \Delta y \\
&= f_x(a, b) \Delta x + f_y(a, b) \Delta y + \varepsilon_1 \Delta x + \varepsilon_2 \Delta y
\end{aligned}$$

where $\qquad \varepsilon_1 = f_x(u, b + \Delta y) - f_x(a, b) \qquad \varepsilon_2 = f_y(a, v) - f_y(a, b)$

Since $(u, b + \Delta y) \to (a, b)$ and $(a, v) \to (a, b)$ as $(\Delta x, \Delta y) \to (0, 0)$ and since f_x and f_y are continuous at (a, b), we see that $\varepsilon_1 \to 0$ and $\varepsilon_2 \to 0$ as $(\Delta x, \Delta y) \to (0, 0)$. $\qquad\qquad\qquad\qquad\qquad\qquad\qquad\qquad\qquad\bullet$

The conclusion of Theorem 12.24 can be written as

$$\Delta z = dz + \varepsilon_1 \Delta x + \varepsilon_2 \Delta y$$

where ε_1 and $\varepsilon_2 \to 0$ as $(\Delta x, \Delta y) \to (0, 0)$. This is the two-dimensional version of Equation 12.18, which is equivalent to the differentiability of a function of a single variable. Therefore we use the following definition for the differentiability of a function of two variables.

Definition (12.26)

If $z = f(x, y)$, then f is **differentiable** at (a, b) if Δz can be expressed in the form

$$\Delta z = f_x(a, b) \Delta x + f_y(a, b) \Delta y + \varepsilon_1 \Delta x + \varepsilon_2 \Delta y$$

where ε_1 and $\varepsilon_2 \to 0$ as $(\Delta x, \Delta y) \to (0, 0)$.

Thus Theorem 12.24 says that if f_x and f_y exist near (a, b) and are continuous at (a, b), then f is differentiable at (a, b). In particular, polynomials and rational functions are differentiable on their domains because their partial derivatives are continuous.

It can be proved that, just as in single-variable calculus, all differentiable functions are continuous (see Exercise 37). But, unlike the situation in one-dimensional calculus, Exercise 38 gives an example of a function whose partial derivatives exist but which is not differentiable.

Functions of Three or More Variables

Differentials and differentiability can be defined in a similar manner for functions of more than two variables. For instance, if $w = f(x, y, z)$, then the **increment** of w is

$$\Delta w = f(x + \Delta x, y + \Delta y, z + \Delta z) - f(x, y, z)$$

The **differential** dw is defined in terms of the differentials dx, dy, and dz of the independent variables by

$$dw = \frac{\partial w}{\partial x} \, dx + \frac{\partial w}{\partial y} \, dy + \frac{\partial w}{\partial z} \, dz$$

If $dx = \Delta x$, $dy = \Delta y$, and $dz = \Delta z$ are all small and f has continuous partial derivatives, then dw can be used to approximate Δw.

Differentiability can be defined by an expression similar to the one in Definition 12.26.

● **Example 5**   The dimensions of a rectangular box are measured to be 75 cm, 60 cm, and 40 cm, and the measurements are each correct to within 0.2 cm. Use differentials to estimate the largest possible error when the volume of the box is calculated from these measurements.

Solution   If the dimensions of the box are x, y, and z, its volume is $V = xyz$ and so

$$dV = \frac{\partial V}{\partial x} \, dx + \frac{\partial V}{\partial y} \, dy + \frac{\partial V}{\partial z} \, dz = yz \, dx + xz \, dy + xy \, dz$$

We are given that $|\Delta x| \leqslant 0.2$, $|\Delta y| \leqslant 0.2$, and $|\Delta z| \leqslant 0.2$. To find the largest error in the volume we therefore use $dx = 0.2$, $dy = 0.2$, and $dz = 0.2$ together with $x = 75$, $y = 60$, and $z = 40$:

$$\Delta V \approx dV = (60)(40)(0.2) + (75)(40)(0.2) + (75)(60)(0.2)$$

$$= 1980$$

Thus errors of only 0.2 cm in measuring each dimension could lead to an error of as much as 1980 cm^3 in the calculated volume!   ●

EXERCISES 12.4

In Exercises 1–8 find an equation of the tangent plane to the given surface at the given point.

1. $z = x^2 + 4y^2$, $(2, 1, 8)$ **2.** $z = x^2 - y^2$, $(3, -2, 5)$

3. $z = 5 + (x - 1)^2 + (y + 2)^2$, $(2, 0, 10)$

4. $z = \sin(x + y)$, $(1, -1, 0)$

5. $z = \ln(2x + y)$, $(-1, 3, 0)$ **6.** $z = e^x \ln y$, $(3, 1, 0)$

7. $z = xy$, $(-1, 2, -2)$ **8.** $z = \sqrt{x - y}$, $(5, 1, 2)$

In Exercises 9–18 find the differential of the given function.

9. $z = x^2 y^3$

10. $z = x^4 - 5x^2 y + 6xy^3 + 10$

11. $z = \dfrac{1}{x^2 + y^2}$ **12.** $z = ye^{xy}$

13. $u = e^x \cos xy$ **14.** $v = \ln(2x - 3y)$

15. $w = x^2 y + y^2 z$ **16.** $w = x \sin yz$

17. $w = \ln\sqrt{x^2 + y^2 + z^2}$ **18.** $w = \dfrac{x + y}{y + z}$

19. If $z = 5x^2 + y^2$ and (x, y) changes from $(1, 2)$ to $(1.05, 2.1)$, compare the values of Δz and dz.

20. If $z = x^2 - xy + 3y^2$ and (x, y) changes from $(3, -1)$ to $(2.96, -0.95)$, compare the values of Δz and dz.

In Exercises 21–26 use differentials to approximate the value of f at the given point.

21. $f(x, y) = \sqrt{x^2 - y^2}$, $(5.01, 4.02)$

22. $f(x, y) = \sqrt{20 - x^2 - 7y^2}$, $(1.95, 1.08)$

23. $f(x, y) = \ln(x - 3y)$, $(6.9, 2.06)$

24. $f(x, y) = xe^{xy}$, $(5.9, 0.01)$

25. $f(x, y, z) = x^2 y^3 z^4$, $(1.05, 0.9, 3.01)$

26. $f(x, y, z) = xy^2 \sin \pi z$, $(3.99, 4.98, 4.03)$

In Exercises 27–30 use differentials to approximate the given number.

27. $8.94\sqrt{9.99 - (1.01)^3}$ **28.** $(\sqrt{99} + \sqrt[3]{124})^4$

29. $\sqrt{0.99}\, e^{0.02}$

30. $\sqrt{(3.02)^2 + (1.97)^2 + (5.99)^2}$

31. The length and width of a rectangle are measured as 30 cm and 24 cm, respectively, with an error in measurement of at most 0.1 cm in each. Use differentials to estimate the maximum error in the calculated area of the rectangle.

32. The dimensions of a closed rectangular box are measured as 80 cm, 60 cm, and 50 cm, respectively, with a possible error of 0.2 cm in each dimension. Use differentials to estimate the maximum error in calculating the surface area of the box.

33. Use differentials to estimate the amount of tin in a closed tin can with diameter 8 cm and height 12 cm if the tin is 0.04 cm thick.

34. The pressure, volume, and temperature of a mole of an ideal gas are related by the equation $PV = 8.31T$, where P is measured in kilopascals, V in liters, and T in $^\circ$K $(=\,^\circ$C $+ 273)$. Use differentials to find the approximate change in the pressure if the volume increases from 12 L to 12.3 L and the temperature decreases from 310°K to 305°K.

35. If R is the total resistance of three resistors, connected in parallel, with resistances R_1, R_2, R_3, then

$$\frac{1}{R} = \frac{1}{R_1} + \frac{1}{R_2} + \frac{1}{R_3}$$

If the resistances are measured as $R_1 = 25$ ohms, $R_2 = 40$ ohms, and $R_3 = 50$ ohms, with possible errors of 0.5% in each case, estimate the maximum error in the calculated value of R.

36. Four positive numbers, each less than 50, are rounded to the first decimal place and then multiplied together. Use differentials to estimate the maximum possible error in the computed product that might result from the rounding.

37. Prove that if f is a function of two variables that is differentiable at (a, b), then f is continuous at (a, b).

38. If

$$f(x, y) = \begin{cases} \dfrac{xy}{x^2 + y^2} & \text{if } (x, y) \neq (0, 0) \\ 0 & \text{if } (x, y) = (0, 0) \end{cases}$$

show that $f_x(0, 0)$ and $f_y(0, 0)$ both exist but f is not differentiable at $(0, 0)$.

SECTION 12.5

The Chain Rule

We first recall that the Chain Rule for functions of a single variable gives the rule for differentiating a composite function: If $y = f(x)$ and $x = g(t)$, where f and g are differentiable functions, then y is indirectly a differentiable function of t and

(12.27)

$$\frac{dy}{dt} = \frac{dy}{dx}\frac{dx}{dt}$$

There are several versions of the Chain Rule for functions of more than one variable, each of them giving a rule for differentiating a composite function. The first version (Theorem 12.28) deals with the case where $z = f(x, y)$ and each of the variables x and y is in turn a function of a variable t. This means that z is indirectly a function of t $[z = f(g(t), h(t))]$ and the Chain Rule gives a formula for differentiating z as a function of t. We shall be assuming that f is differentiable (Definition 12.26). Recall from Theorem 12.24 that this will be the case if f_x and f_y are continuous.

The Chain Rule (Case 1) (12.28)

Suppose that $z = f(x, y)$ is a differentiable function of x and y, where $x = g(t)$ and $y = h(t)$ are both differentiable functions of t. Then z is a differentiable function of t and

$$\frac{dz}{dt} = \frac{\partial f}{\partial x}\frac{dx}{dt} + \frac{\partial f}{\partial y}\frac{dy}{dt}$$

Proof   A change of Δt in t produces changes of Δx in x and Δy in y. These, in turn, produce a change of Δz in z, and from Definition 12.26 we have

$$\Delta z = \frac{\partial f}{\partial x}\Delta x + \frac{\partial f}{\partial y}\Delta y + \varepsilon_1 \Delta x + \varepsilon_2 \Delta y$$

where $\varepsilon_1 \to 0$ and $\varepsilon_2 \to 0$ as $(\Delta x, \Delta y) \to (0, 0)$. [If the functions ε_1 and ε_2 are not defined at $(0, 0)$, we can define them to be 0 there.] Dividing both sides of this equation by Δt, we have

$$\frac{\Delta z}{\Delta t} = \frac{\partial f}{\partial x}\frac{\Delta x}{\Delta t} + \frac{\partial f}{\partial y}\frac{\Delta y}{\Delta t} + \varepsilon_1 \frac{\Delta x}{\Delta t} + \varepsilon_2 \frac{\Delta y}{\Delta t}$$

If we now let $\Delta t \to 0$, then $\Delta x = g(t + \Delta t) - g(t) \to 0$ because g is differentiable and therefore continuous. Similarly, $\Delta y \to 0$. This, in turn, means that $\varepsilon_1 \to 0$ and $\varepsilon_2 \to 0$, so

$$\frac{dz}{dt} = \lim_{\Delta t \to 0} \frac{\Delta z}{\Delta t}$$

$$= \frac{\partial f}{\partial x} \lim_{\Delta t \to 0} \frac{\Delta x}{\Delta t} + \frac{\partial f}{\partial y} \lim_{\Delta t \to 0} \frac{\Delta y}{\Delta t} + \lim_{\Delta t \to 0} \varepsilon_1 \lim_{\Delta t \to 0} \frac{\Delta x}{\Delta t} + \lim_{\Delta t \to 0} \varepsilon_2 \lim_{\Delta t \to 0} \frac{\Delta y}{\Delta t}$$

$$= \frac{\partial f}{\partial x}\frac{dx}{dt} + \frac{\partial f}{\partial y}\frac{dy}{dt} + 0 \cdot \frac{dx}{dt} + 0 \cdot \frac{dy}{dt}$$

$$= \frac{\partial f}{\partial x}\frac{dx}{dt} + \frac{\partial f}{\partial y}\frac{dy}{dt}$$

Since we often write $\partial z/\partial x$ in place of $\partial f/\partial x$, we can rewrite the Chain Rule in the form

$$\frac{dz}{dt} = \frac{\partial z}{\partial x}\frac{dx}{dt} + \frac{\partial z}{\partial y}\frac{dy}{dt}$$

● **Example 1**   If $z = x^2y + 3xy^4$, where $x = e^t$ and $y = \sin t$, find dz/dt.

Solution   The Chain Rule gives

$$\frac{dz}{dt} = \frac{\partial z}{\partial x}\frac{dx}{dt} + \frac{\partial z}{\partial y}\frac{dy}{dt}$$

$$= (2xy + 3y^4)e^t + (x^2 + 12xy^3)\cos t$$

$$= (2e^t \sin t + 3\sin^4 t)e^t + (e^{2t} + 12e^t \sin^3 t)\cos t \qquad ●$$

Note 1:   Although we have expressed the answer to Example 1 totally in terms of t, the expression in terms of x, y, and t is adequate for some purposes. For instance, if we had been asked to find the value of dz/dt when $t = 0$, we could simply observe that $x = 1$ and $y = 0$ when $t = 0$ and so

$$\frac{dz}{dt}\bigg]_{t=1} = 0e^0 + 1\cos 0 = 1$$

Note 2:   If $T(x, y)$ represents the temperature at a point (x, y) and $x = f(t)$ and $y = g(t)$ are the parametric equations of a curve C, then the composite function $z = T(f(t), g(t))$ represents the temperature at points on C and the derivative dz/dt represents the rate of change of temperature along the curve.

● **Example 2**   The pressure P (in kilopascals), volume V (in liters), and temperature T (in $°$K) of a mole of an ideal gas are related by the equation $PV = 8.31T$. Find the rate at which the pressure is changing when the temperature is $300°$K and increasing at a rate of $0.1°$K/s and the volume is 100 L and increasing at a rate of 0.2 L/s.

Solution   If t represents the time elapsed in seconds, then at the given instant we have $T = 300$, $dT/dt = 0.1$, $V = 100$, $dV/dt = 0.2$. Since

$$P = 8.31\frac{T}{V}$$

the Chain Rule gives

$$\frac{dP}{dt} = \frac{\partial P}{\partial T}\frac{dT}{dt} + \frac{\partial P}{\partial V}\frac{dV}{dt} = \frac{8.31}{V}\frac{dT}{dt} - \frac{8.31T}{V^2}\frac{dV}{dt}$$

$$= \frac{8.31}{100}(0.1) - \frac{8.31(300)}{100^2}(0.2)$$

$$= -0.04155$$

The pressure is decreasing at a rate of about 0.042 kPa/s.   ●

We now consider the situation where $z = f(x, y)$ but each of x and y is a function of two variables s and t: $x = g(s, t)$, $y = h(s, t)$. Then z is indirectly a function of s and t and we wish to find $\partial z/\partial s$ and $\partial z/\partial t$. Recall that in computing $\partial z/\partial t$ we hold s fixed and compute the ordinary derivative of z with respect to t. Therefore we can apply Theorem 12.28 to obtain

$$\frac{\partial z}{\partial t} = \frac{\partial z}{\partial x}\frac{\partial x}{\partial t} + \frac{\partial z}{\partial y}\frac{\partial y}{\partial t}$$

A similar argument holds for $\partial z/\partial s$ and so we have proved the following version of the Chain Rule.

The Chain Rule (Case 2) (12.29)

Suppose that $z = f(x, y)$ is a differentiable function of x and y, where $x = g(s, t)$, $y = h(s, t)$, and the partial derivatives g_s, g_t, h_s, and h_t exist. Then

$$\frac{\partial z}{\partial s} = \frac{\partial z}{\partial x}\frac{\partial x}{\partial s} + \frac{\partial z}{\partial y}\frac{\partial y}{\partial s}$$

$$\frac{\partial z}{\partial t} = \frac{\partial z}{\partial x}\frac{\partial x}{\partial t} + \frac{\partial z}{\partial y}\frac{\partial y}{\partial t}$$

● **Example 3**   If $z = e^x \sin y$ where $x = st^2$ and $y = s^2t$, find $\partial z/\partial s$ and $\partial z/\partial t$.

Solution   Applying Case 2 of the Chain Rule, we get

$$\frac{\partial z}{\partial s} = \frac{\partial z}{\partial x}\frac{\partial x}{\partial s} + \frac{\partial z}{\partial y}\frac{\partial y}{\partial s} = (e^x \sin y)(t^2) + (e^x \cos y)(2st)$$

$$= t^2 e^{st^2} \sin(s^2t) + 2st e^{st^2} \cos(s^2t)$$

$$\frac{\partial z}{\partial t} = \frac{\partial z}{\partial x}\frac{\partial x}{\partial t} + \frac{\partial z}{\partial y}\frac{\partial y}{\partial t} = (e^x \sin y)(2st) + (e^x \cos y)(s^2)$$

$$= 2st e^{st^2} \sin(s^2t) + s^2 e^{st^2} \cos(s^2t)$$ ●

In Case 2 of the Chain Rule there are three types of variables: s and t are **independent** variables, x and y are called **intermediate** variables, and z is the **dependent** variable. Notice that in Theorem 12.29 there is one term for each intermediate variable and each of these terms resembles the one-dimensional Chain Rule in Equation 12.27.

Now we consider the general situation where a dependent variable u is a function of n intermediate variables $x_1, \ldots, x_n$, each of which is, in turn, a function of m independent variables $t_1, \ldots, t_m$. Notice that there are n terms, one for each intermediate variable. The proof is similar to that of Case 1.

The Chain Rule (General Version)
(12.30)

Suppose that u is a differentiable function of the n variables $x_1, x_2, \ldots, x_n$ and each x_j is a function of the m variables $t_1, t_2, \ldots, t_m$ such that all the partial derivatives $\partial x_j/\partial t_i$ exist ($j = 1, \ldots, n$; $i = 1, \ldots, m$). Then u is a function of $t_1, \ldots, t_m$ and

$$\frac{\partial u}{\partial t_i} = \frac{\partial u}{\partial x_1}\frac{\partial x_1}{\partial t_i} + \frac{\partial u}{\partial x_2}\frac{\partial x_2}{\partial t_i} + \cdots + \frac{\partial u}{\partial x_n}\frac{\partial x_n}{\partial t_i}$$

for each $i = 1, 2, \ldots, m$.

● **Example 4** Write out the Chain Rule for the case where $w = f(x, y, z, t)$ and $x = x(u, v)$, $y = y(u, v)$, $z = z(u, v)$, and $t = t(u, v)$.

Solution Applying Theorem 12.30 with $n = 4$ and $m = 2$, we have

$$\frac{\partial w}{\partial u} = \frac{\partial w}{\partial x}\frac{\partial x}{\partial u} + \frac{\partial w}{\partial y}\frac{\partial y}{\partial u} + \frac{\partial w}{\partial z}\frac{\partial z}{\partial u} + \frac{\partial w}{\partial t}\frac{\partial t}{\partial u}$$

$$\frac{\partial w}{\partial v} = \frac{\partial w}{\partial x}\frac{\partial x}{\partial v} + \frac{\partial w}{\partial y}\frac{\partial y}{\partial v} + \frac{\partial w}{\partial z}\frac{\partial z}{\partial v} + \frac{\partial w}{\partial t}\frac{\partial t}{\partial v}$$

●

● **Example 5** If $u = x^4 y + y^2 z^3$ where $x = rse^t$, $y = rs^2 e^{-t}$, and $z = r^2 s \sin t$, find the value of $\partial u/\partial s$ when $r = 2$, $s = 1$, $t = 0$.

Solution Using the Chain Rule with $n = m = 3$, we have

$$\frac{\partial u}{\partial s} = \frac{\partial u}{\partial x}\frac{\partial x}{\partial s} + \frac{\partial u}{\partial y}\frac{\partial y}{\partial s} + \frac{\partial u}{\partial z}\frac{\partial z}{\partial s}$$

$$= (4x^3 y)(re^t) + (x^4 + 2yz^3)(2rse^{-t}) + (3y^2 z^2)(r^2 \sin t)$$

When $r = 2$, $s = 1$, and $t = 0$, we have $x = 2$, $y = 2$, and $z = 0$, so

$$\frac{\partial u}{\partial s} = (64)(2) + (16)(4) + (0)(0) = 192$$

●

● **Example 6** If $g(s, t) = f(s^2 - t^2, t^2 - s^2)$ and f is differentiable, show that g satisfies the equation

$$t\frac{\partial g}{\partial s} + s\frac{\partial g}{\partial t} = 0$$

Solution Let $x = s^2 - t^2$ and $y = t^2 - s^2$. Then $g(s, t) = f(x, y)$ and the Chain Rule gives

$$\frac{\partial g}{\partial s} = \frac{\partial f}{\partial x}\frac{\partial x}{\partial s} + \frac{\partial f}{\partial y}\frac{\partial y}{\partial s} = \frac{\partial f}{\partial x}(2s) + \frac{\partial f}{\partial y}(-2s)$$

$$\frac{\partial g}{\partial t} = \frac{\partial f}{\partial x}\frac{\partial x}{\partial t} + \frac{\partial f}{\partial y}\frac{\partial y}{\partial t} = \frac{\partial f}{\partial x}(-2t) + \frac{\partial f}{\partial y}(2t)$$

Therefore

$$t\frac{\partial g}{\partial s} + s\frac{\partial g}{\partial t} = \left(2st\frac{\partial f}{\partial x} - 2st\frac{\partial f}{\partial y}\right) + \left(-2st\frac{\partial f}{\partial x} + 2st\frac{\partial f}{\partial y}\right) = 0 \qquad \bullet$$

● **Example 7**   If $z = f(x, y)$ has continuous second order partial derivatives and $x = r^2 + s^2$ and $y = 2rs$, find   (a) $\partial z/\partial r$ and   (b) $\partial^2 z/\partial r^2$.

Solution   (a)   The Chain Rule gives

$$\frac{\partial z}{\partial r} = \frac{\partial z}{\partial x}\frac{\partial x}{\partial r} + \frac{\partial z}{\partial y}\frac{\partial y}{\partial r} = \frac{\partial z}{\partial x}(2r) + \frac{\partial z}{\partial y}(2s)$$

(b)   Applying the Product Rule to the expression in part (a), we get

$$\frac{\partial^2 z}{\partial r^2} = \frac{\partial}{\partial r}\left(2r\frac{\partial z}{\partial x} + 2s\frac{\partial z}{\partial y}\right)$$

(12.31)

$$= 2\frac{\partial z}{\partial x} + 2r\frac{\partial}{\partial r}\left(\frac{\partial z}{\partial x}\right) + 2s\frac{\partial}{\partial r}\left(\frac{\partial z}{\partial y}\right)$$

But, using the Chain Rule again, we have

$$\frac{\partial}{\partial r}\left(\frac{\partial z}{\partial x}\right) = \frac{\partial}{\partial x}\left(\frac{\partial z}{\partial x}\right)\frac{\partial x}{\partial r} + \frac{\partial}{\partial y}\left(\frac{\partial z}{\partial x}\right)\frac{\partial y}{\partial r}$$

$$= \frac{\partial^2 z}{\partial x^2}(2r) + \frac{\partial^2 z}{\partial y\,\partial x}(2s)$$

$$\frac{\partial}{\partial r}\left(\frac{\partial z}{\partial y}\right) = \frac{\partial}{\partial x}\left(\frac{\partial z}{\partial y}\right)\frac{\partial x}{\partial r} + \frac{\partial}{\partial y}\left(\frac{\partial z}{\partial y}\right)\frac{\partial y}{\partial r}$$

$$= \frac{\partial^2 z}{\partial x\,\partial y}(2r) + \frac{\partial^2 z}{\partial y^2}(2s)$$

Putting these expressions into Equation 12.31 and using the equality of the mixed second order derivatives, we obtain

$$\frac{\partial^2 z}{\partial r^2} = 2\frac{\partial z}{\partial x} + 2r\left(2r\frac{\partial^2 z}{\partial x^2} + 2s\frac{\partial^2 z}{\partial y\,\partial x}\right) + 2s\left(2r\frac{\partial^2 z}{\partial x\,\partial y} + 2s\frac{\partial^2 z}{\partial y^2}\right)$$

$$= 2\frac{\partial z}{\partial x} + 4r^2\frac{\partial^2 z}{\partial x^2} + 8rs\frac{\partial^2 z}{\partial x\,\partial y} + 4s^2\frac{\partial^2 z}{\partial y^2} \qquad \bullet$$

Implicit Differentiation

The Chain Rule can be used to give a more complete description of the process of implicit differentiation that was introduced in Sections 3.5 and 12.3. We suppose that an equation of the form $F(x, y) = 0$ defines y implicitly as a differentiable function of x, that is, $y = f(x)$, where $F(x, f(x)) = 0$ for all x in the domain of f. If F is differentiable, we can apply Case 1 of the Chain Rule to differentiate both sides of the equation $F(x, y) = 0$ with respect to x. Since both x and y are functions of x, we obtain

$$\frac{\partial F}{\partial x}\frac{dx}{dx} + \frac{\partial F}{\partial y}\frac{dy}{dx} = 0$$

But $dx/dx = 1$, so if $\partial F/\partial y \neq 0$ we solve for dy/dx and obtain

(12.32)

$$\frac{dy}{dx} = -\frac{\dfrac{\partial F}{\partial x}}{\dfrac{\partial F}{\partial y}} = -\frac{F_x}{F_y}$$

● **Example 8**   Find y' if $x^3 + y^3 = 6xy$.

Solution   The given equation can be written as

$$F(x, y) = x^3 + y^3 - 6xy = 0$$

so Equation 12.32 gives

$$\frac{dy}{dx} = -\frac{F_x}{F_y} = -\frac{3x^2 - 6y}{3y^2 - 6x} = -\frac{x^2 - 2y}{y^2 - 2x}$$

This solution should be compared with the one given in Example 2 in Section 3.5.   ●

Now we suppose that z is given implicitly as a function $z = f(x, y)$ by an equation of the form $F(x, y, z) = 0$. This means that $F(x, y, f(x, y)) = 0$ for all (x, y) in the domain of f. If F is differentiable and f_x, f_y exist, then we can use the Chain Rule to differentiate the equation $F(x, y, z) = 0$ as follows:

$$\frac{\partial F}{\partial x}\frac{\partial x}{\partial x} + \frac{\partial F}{\partial y}\frac{\partial y}{\partial x} + \frac{\partial F}{\partial z}\frac{\partial z}{\partial x} = 0$$

But $\dfrac{\partial}{\partial x}(x) = 1$ and $\dfrac{\partial}{\partial x}(y) = 0$

so this equation becomes

$$\frac{\partial F}{\partial x} + \frac{\partial F}{\partial z}\frac{\partial z}{\partial x} = 0$$

If $\partial F/\partial z \neq 0$, we solve for $\partial z/\partial x$ and obtain the first formula in Equations 12.33. The formula for $\partial z/\partial y$ is obtained in a similar manner.

(12.33)

$$\frac{\partial z}{\partial x} = -\frac{\dfrac{\partial F}{\partial x}}{\dfrac{\partial F}{\partial z}} \qquad \frac{\partial z}{\partial y} = -\frac{\dfrac{\partial F}{\partial y}}{\dfrac{\partial F}{\partial z}}$$

● **Example 9**   Find $\dfrac{\partial z}{\partial x}$ and $\dfrac{\partial z}{\partial y}$ if $x^3 + y^3 + z^3 + 6xyz = 1$.

Solution   Let $F(x, y, z) = x^3 + y^3 + z^3 + 6xyz - 1$. Then, from Equations 12.33, we have

$$\frac{\partial z}{\partial x} = -\frac{F_x}{F_z} = -\frac{3x^2 + 6yz}{3z^2 + 6xy} = -\frac{x^2 + 2yz}{z^2 + 2xy}$$

$$\frac{\partial z}{\partial y} = -\frac{F_y}{F_z} = -\frac{3y^2 + 6xz}{3z^2 + 6xy} = -\frac{y^2 + 2xz}{z^2 + 2xy}$$

This solution should be compared with Example 3 in Section 12.3.   ●

EXERCISES 12.5

In Exercises 1–8 use the Chain Rule to find dz/dt or dw/dt.

1. $z = x^2 + y^2$, $x = t^3$, $y = 1 + t^2$

2. $z = x^2 y^3$, $x = 1 + \sqrt{t}$, $y = 1 - \sqrt{t}$

3. $z = 6x^3 - 3xy + 2y^2$, $x = e^t$, $y = \cos t$

4. $z = x\sqrt{1 + y^2}$, $x = te^{2t}$, $y = e^{-t}$

5. $z = \ln(x + y^2)$, $x = \sqrt{1 + t}$, $y = 1 + \sqrt{t}$

6. $z = xe^{x/y}$, $x = \cos t$, $y = e^{2t}$

7. $w = xy^2 z^3$, $x = \sin t$, $y = \cos t$, $z = 1 + e^{2t}$

8. $w = \dfrac{x}{y} + \dfrac{y}{z}$, $x = \sqrt{t}$, $y = \cos 2t$, $z = e^{-3t}$

In Exercises 9–14 use the Chain Rule to find $\partial z/\partial s$ and $\partial z/\partial t$.

9. $z = x^2 \sin y$, $x = s^2 + t^2$, $y = 2st$

10. $z = \sin x \cos y$, $x = (s - t)^2$, $y = s^2 - t^2$

11. $z = x^2 - 3x^2 y^3$, $x = se^t$, $y = se^{-t}$

12. $z = x \tan^{-1}(xy)$, $x = t^2$, $y = se^t$

13. $z = 2^{x - 3y}$, $x = s^2 t$, $y = st^2$

14. $z = xe^y + ye^{-x}$, $x = e^t$, $y = st^2$

In Exercises 15–18 write out the Chain Rule for the given case (as in Example 4).

15. $u = f(x, y)$, where $x = x(r, s, t)$, $y = y(r, s, t)$

16. $w = f(x, y, z)$, where $x = x(t, u)$, $y = y(t, u)$, $z = z(t, u)$

17. $v = f(p, q, r)$, where $p = p(x, y, z)$, $q = q(x, y, z)$, $r = r(x, y, z)$

18. $u = f(s, t)$, where $s = s(w, x, y, z)$, $t = t(w, x, y, z)$

In Exercises 19–26 use the Chain Rule to find the indicated partial derivatives.

19. $w = x^2 + y^2 + z^2$, $x = st$, $y = s \cos t$, $z = s \sin t$; $\dfrac{\partial w}{\partial s}$, $\dfrac{\partial w}{\partial t}$ when $s = 1$, $t = 0$

20. $u = xy + yz + zx$, $x = st$, $y = e^{st}$, $z = t^2$; $\dfrac{\partial u}{\partial s}$, $\dfrac{\partial u}{\partial t}$ when $s = 0$, $t = 1$

21. $z = y^2 \tan x$, $x = t^2 uv$, $y = u + tv^2$; $\dfrac{\partial z}{\partial t}$, $\dfrac{\partial z}{\partial u}$, $\dfrac{\partial z}{\partial v}$ when $t = 2$, $u = 1$, $v = 0$

22. $z = \dfrac{x}{y}$, $x = re^{st}$, $y = rse^t$; $\dfrac{\partial z}{\partial r}$, $\dfrac{\partial z}{\partial s}$, $\dfrac{\partial z}{\partial t}$ when $r = 1$, $s = 2$, $t = 0$

23. $u = \dfrac{x + y}{y + z}$, $x = p + r + t$, $y = p - r + t$, $z = p + r - t$; $\dfrac{\partial u}{\partial p}$, $\dfrac{\partial u}{\partial r}$, $\dfrac{\partial u}{\partial t}$

24. $t = z \sec(xy)$, $x = uv$, $y = vw$, $z = wu$; $\dfrac{\partial t}{\partial u}, \dfrac{\partial t}{\partial v}, \dfrac{\partial t}{\partial w}$

25. $w = \cos(x - y)$, $x = rs^2 t^3 \sin \theta$, $y = r^2 st \cos \theta$; $\dfrac{\partial w}{\partial r}, \dfrac{\partial w}{\partial s}$, $\dfrac{\partial w}{\partial t}, \dfrac{\partial w}{\partial \theta}$

26. $u = pq - p^2 r^2 s$, $p = x + 2y$, $q = x - 2y$, $r = \dfrac{x}{y^4}$, $s = 2xy^{3/2}$; $\dfrac{\partial u}{\partial x}, \dfrac{\partial u}{\partial y}$

27. The radius of a right circular cylinder is decreasing at a rate of 1.2 cm/s while its height is increasing at a rate of 3 cm/s. At what rate is the volume of the cylinder changing when the radius is 80 cm and the height is 150 cm?

28. The radius of a right circular cone is increasing at a rate of 1.8 in./s while its height is decreasing at a rate of 2.5 in./s. At what rate is the volume of the cone changing when the radius is 120 in. and the height is 140 in.?

29. The pressure of one mole of an ideal gas is increasing at a rate of 0.05 kPa/s and the temperature is increasing at a rate of $0.15°$K/s. Use the equation in Example 2 to find the rate of change of the volume when the pressure is 20 kPa and the temperature is $320°$K.

30. Car A is traveling north on Highway 16 at 90 km/h. Car B is traveling west on Highway 83 at 80 km/h. Each car is approaching the intersection of these highways. How fast is the distance between the cars changing when car A is 0.3 km from the intersection and car B is 0.4 km from the intersection?

In Exercises 31–34 assume that all the given functions are differentiable.

31. If $z = f(x, y)$, where $x = r \cos \theta$ and $y = r \sin \theta$, (a) find $\partial z/\partial r$ and $\partial z/\partial \theta$ and (b) show that

$$\left(\frac{\partial z}{\partial x}\right)^2 + \left(\frac{\partial z}{\partial y}\right)^2 = \left(\frac{\partial z}{\partial r}\right)^2 + \frac{1}{r^2}\left(\frac{\partial z}{\partial \theta}\right)^2$$

32. If $u = f(x, y)$, where $x = e^s \cos t$ and $y = e^s \sin t$, show that

$$\left(\frac{\partial u}{\partial x}\right)^2 + \left(\frac{\partial u}{\partial y}\right)^2 = e^{-2s}\left[\left(\frac{\partial u}{\partial s}\right)^2 + \left(\frac{\partial u}{\partial t}\right)^2\right]$$

33. If $z = f(x - y)$, show that

$$\frac{\partial z}{\partial x} + \frac{\partial z}{\partial y} = 0$$

34. If $z = f(x, y)$, where $x = s + t$ and $y = s - t$, show that

$$\left(\frac{\partial z}{\partial x}\right)^2 - \left(\frac{\partial z}{\partial y}\right)^2 = \frac{\partial z}{\partial s}\frac{\partial z}{\partial t}$$

In Exercises 35–40 assume that all the given functions have continuous second order partial derivatives.

35. Show that any function of the form $z = f(x + at) + g(x - at)$ is a solution of the wave equation

$$\frac{\partial^2 z}{\partial t^2} = a^2 \frac{\partial^2 z}{\partial x^2}$$

(*Hint:* Let $u = x + at$, $v = x - at$.)

36. If $u = f(x, y)$, where $x = e^s \cos t$ and $y = e^s \sin t$, show that

$$\frac{\partial^2 u}{\partial x^2} + \frac{\partial^2 u}{\partial y^2} = e^{-2s}\left[\frac{\partial^2 u}{\partial s^2} + \frac{\partial^2 u}{\partial t^2}\right]$$

37. If $z = f(x, y)$, where $x = r^2 + s^2$, $y = 2rs$, find $\partial^2 z/\partial r\,\partial s$. (Compare with Example 7.)

38. If $z = f(x, y)$, where $x = r \cos \theta$, $y = r \sin \theta$, find (a) $\partial z/\partial r$, (b) $\partial z/\partial \theta$, and (c) $\partial^2 z/\partial r\,\partial \theta$.

39. If $z = f(x, y)$, where $x = r \cos \theta$, $y = r \sin \theta$, show that

$$\frac{\partial^2 z}{\partial x^2} + \frac{\partial^2 z}{\partial y^2} = \frac{\partial^2 z}{\partial r^2} + \frac{1}{r^2}\frac{\partial^2 z}{\partial \theta^2} + \frac{1}{r}\frac{\partial z}{\partial r}$$

40. Suppose $z = f(x, y)$, where $x = g(s, t)$ and $y = h(s, t)$.
(a) Show that

$$\frac{\partial^2 z}{\partial t^2} = \frac{\partial^2 z}{\partial x^2}\left(\frac{\partial x}{\partial t}\right)^2 + 2\frac{\partial^2 z}{\partial x\,\partial y}\frac{\partial x}{\partial t}\frac{\partial y}{\partial t} + \frac{\partial^2 z}{\partial y^2}\left(\frac{\partial y}{\partial t}\right)^2$$

$$+ \frac{\partial z}{\partial x}\frac{\partial^2 x}{\partial t^2} + \frac{\partial z}{\partial y}\frac{\partial^2 y}{\partial t^2}$$

(b) Find a similar formula for $\partial^2 z/\partial s\,\partial t$.

In Exercises 41–44 use Equation 12.32 to find dy/dx.

41. $x^2 - xy + y^3 = 8$ **42.** $y^5 + 3x^2 y^2 + 5x^4 = 12$

43. $2y^2 + \sqrt[3]{xy} = 3x^2 + 17$ **44.** $x \cos y + y \cos x = 1$

In Exercises 45–50 use Equations 12.33 to find $\partial z/\partial x$ and $\partial z/\partial y$.

45. $xy + yz - xz = 0$ **46.** $xyz = \cos(x + y + z)$

47. $x^2 + y^2 - z^2 = 2x(y + z)$

48. $xy^2 z^3 + x^3 y^2 z = x + y + z$

49. $xe^y + yz + ze^x = 0$ **50.** $y^2 z e^{x+y} - \sin(xyz) = 0$

SECTION 12.6

Directional Derivatives and the Gradient Vector

Recall that if $z = f(x, y)$, then the partial derivatives f_x and f_y are defined as

(12.34)

$$f_x(x_0, y_0) = \lim_{h \to 0} \frac{f(x_0 + h, y_0) - f(x_0, y_0)}{h}$$

$$f_y(x_0, y_0) = \lim_{h \to 0} \frac{f(x_0, y_0 + h) - f(x_0, y_0)}{h}$$

and represent the rates of change of z in the x- and y-directions, that is, in the directions of the unit vectors **i** and **j**.

Suppose that we now wish to find the rate of change of z at (x_0, y_0) in the direction of an arbitrary unit vector $\mathbf{u} = \langle a, b \rangle$ (see Figure 12.26). To do this we consider the surface S with equation $z = f(x, y)$ (the graph of f) and we let $z_0 = f(x_0, y_0)$. Then the point $P(x_0, y_0, z_0)$ lies on S. The vertical plane that passes through P in the direction **u** intersects S in a curve C (see Figure 12.27). The slope of the tangent line T to C at P will be the rate of change of z in the direction of **u**.

If $Q(x, y, z)$ is another point on C and P', Q' are the projections of P, Q on the xy-plane, then the vector $\overrightarrow{P'Q'}$ is parallel to **u** and so

$$\overrightarrow{P'Q'} = h\mathbf{u} = \langle ha, hb \rangle$$

for some scalar h. Therefore $x - x_0 = ha$, $y - y_0 = hb$, so $x = x_0 + ha$, $y = y_0 + hb$, and

$$\frac{\Delta z}{h} = \frac{z - z_0}{h} = \frac{f(x_0 + ha, y_0 + hb) - f(x_0, y_0)}{h}$$

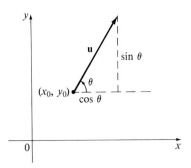

Figure 12.26

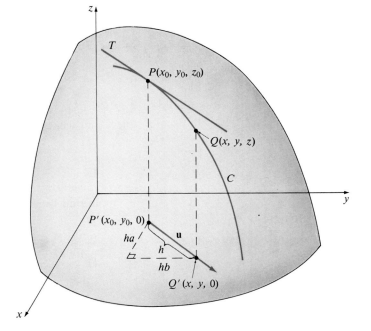

Figure 12.27

If we take the limit as $h \to 0$, we obtain the rate of change of z (with respect to distance) in the direction of $\mathbf{u}$, which is called the directional derivative of f in the direction of $\mathbf{u}$.

Definition (12.35)

The **directional derivative** of f at (x_0, y_0) in the direction of a unit vector $\mathbf{u} = \langle a, b \rangle$ is

$$D_{\mathbf{u}} f(x_0, y_0) = \lim_{h \to 0} \frac{f(x_0 + ha, y_0 + hb) - f(x_0, y_0)}{h}$$

if this limit exists.

By comparing Definition 12.35 with Equations 12.34, we see that if $\mathbf{u} = \mathbf{i} = \langle 1, 0 \rangle$, then $D_{\mathbf{i}} f = f_x$ and if $\mathbf{u} = \mathbf{j} = \langle 0, 1 \rangle$, then $D_{\mathbf{j}} f = f_y$. In other words, the partial derivatives of f with respect to x and y are just special cases of the directional derivative.

For computational purposes we generally use the formula given by the following theorem.

Theorem (12.36)

If f is a differentiable function of x and y, then f has a directional derivative in the direction of any unit vector $\mathbf{u} = \langle a, b \rangle$ and

$$D_{\mathbf{u}} f(x, y) = f_x(x, y)a + f_y(x, y)b$$

Proof   If we define a function g of the single variable h by

$$g(h) = f(x_0 + ha, y_0 + hb)$$

then by the definition of a derivative we have

(12.37)
$$g'(0) = \lim_{h \to 0} \frac{g(h) - g(0)}{h} = \lim_{h \to 0} \frac{f(x_0 + ha, y_0 + hb) - f(x_0, y_0)}{h}$$

$$= D_{\mathbf{u}} f(x_0, y_0)$$

On the other hand, we can write $g(h) = f(x, y)$, where $x = x_0 + ha$, $y = y_0 + hb$, so the Chain Rule (Theorem 12.28) gives

$$g'(h) = \frac{\partial f}{\partial x} \frac{dx}{dh} + \frac{\partial f}{\partial y} \frac{dy}{dh} = f_x(x, y)a + f_y(x, y)b$$

If we now put $h = 0$, then $x = x_0$, $y = y_0$, and

(12.38)
$$g'(0) = f_x(x_0, y_0)a + f_y(x_0, y_0)b$$

Comparing Equations 12.37 and 12.38, we see that

$$D_{\mathbf{u}}f(x_0, y_0) = f_x(x_0, y_0)a + f_y(x_0, y_0)b \qquad \bullet$$

If the unit vector $\mathbf{u}$ makes an angle θ with the positive x-axis (as in Figure 12.26), then we can write $\mathbf{u} = \langle \cos\theta, \sin\theta \rangle$ and the formula in Theorem 12.36 becomes

(12.39)
$$D_{\mathbf{u}}f(x, y) = f_x(x, y)\cos\theta + f_y(x, y)\sin\theta$$

● **Example 1**   Find the directional derivative $D_{\mathbf{u}}f(x, y)$ if

$$f(x, y) = x^3 - 3xy + 4y^2$$

and $\mathbf{u}$ is the unit vector given by angle $\theta = \pi/6$. What is $D_{\mathbf{u}}f(1, 2)$?

Solution   Formula 12.39 gives

$$D_{\mathbf{u}}f(x, y) = f_x(x, y)\cos\frac{\pi}{6} + f_y(x, y)\sin\frac{\pi}{6}$$

$$= (3x^2 - 3y)\frac{\sqrt{3}}{2} + (-3x + 8y)\frac{1}{2}$$

$$= \frac{1}{2}[3\sqrt{3}x^2 - 3x + (8 - 3\sqrt{3})y]$$

Therefore

$$D_{\mathbf{u}}f(1, 2) = \frac{1}{2}[3\sqrt{3}(1)^2 - 3(1) + (8 - 3\sqrt{3})(2)] = \frac{13 - 3\sqrt{3}}{2} \qquad \bullet$$

Notice from Theorem 12.36 that the directional derivative can be written as the dot product of two vectors:

(12.40)
$$D_{\mathbf{u}}f(x, y) = f_x(x, y)a + f_y(x, y)b$$

$$= \langle f_x(x, y), f_y(x, y) \rangle \cdot \langle a, b \rangle$$

$$= \langle f_x(x, y), f_y(x, y) \rangle \cdot \mathbf{u}$$

The first vector in this dot product occurs not only in computing directional derivatives but in many other contexts as well. So we give it a special name (the gradient of f) and a special notation (grad f or ∇f, which is read "del f").

Definition (12.41)

> If f is a function of two variables x and y, then the **gradient** of f is the vector function ∇f defined by
>
> $$\nabla f(x, y) = \langle f_x(x, y), f_y(x, y) \rangle = \frac{\partial f}{\partial x}\mathbf{i} + \frac{\partial f}{\partial y}\mathbf{j}$$

● **Example 2**   If $f(x, y) = \sin x + e^{xy}$, then

$$\nabla f(x, y) = \langle \cos x + ye^{xy}, xe^{xy} \rangle$$

and $\nabla f(0, 1) = \langle 2, 0 \rangle$ ●

With this notation for the gradient vector, we can rewrite the expression (12.40) for the directional derivative as

(12.42)

$$D_{\mathbf{u}} f(x, y) = \nabla f(x, y) \cdot \mathbf{u}$$

This expresses the directional derivative in the direction of $\mathbf{u}$ as the scalar projection of the gradient vector onto $\mathbf{u}$.

● **Example 3**   Find the directional derivation of the function $f(x, y) = x^3 y^4$ at the point $(6, -1)$ in the direction of the vector $\mathbf{v} = 2\mathbf{i} + 5\mathbf{j}$.

Solution   We first compute the gradient vector at $(6, -1)$:

$$\nabla f(x, y) = 3x^2 y^4 \mathbf{i} + 4x^3 y^3 \mathbf{j}$$

$$\nabla f(6, -1) = 108\mathbf{i} - 864\mathbf{j}$$

Note that $\mathbf{v}$ is not a unit vector, but, since $|\mathbf{v}| = \sqrt{29}$, the unit vector in the direction of $\mathbf{v}$ is

$$\mathbf{u} = \frac{\mathbf{v}}{|\mathbf{v}|} = \frac{2}{\sqrt{29}} \mathbf{i} + \frac{5}{\sqrt{29}} \mathbf{j}$$

Therefore, by Equation 12.42, we have

$$D_{\mathbf{u}} f(6, -1) = \nabla f(6, -1) \cdot \mathbf{u} = (108\mathbf{i} - 864\mathbf{j}) \cdot \left(\frac{2}{\sqrt{29}} \mathbf{i} + \frac{5}{\sqrt{29}} \mathbf{j} \right)$$

$$= \frac{2(108) - 5(864)}{\sqrt{29}} = -\frac{4104}{\sqrt{29}}$$ ●

For functions of three variables we can define directional derivatives in a similar manner. Again $D_{\mathbf{u}} f(x, y, z)$ can be interpreted as the rate of change of the function in the direction of a unit vector $\mathbf{u}$.

Definition (12.43)

The **directional derivative** of f at (x_0, y_0, z_0) in the direction of a unit vector $\mathbf{u} = \langle a, b, c \rangle$ is

$$D_{\mathbf{u}} f(x_0, y_0, z_0) = \lim_{h \to 0} \frac{f(x_0 + ha, y_0 + hb, z_0 + hc) - f(x_0, y_0, z_0)}{h}$$

if this limit exists.

If we use vector notation, then we can write both definitions (12.35 and 12.43) of the directional derivative in the compact form

(12.44)
$$D_{\mathbf{u}}f(\mathbf{x}_0) = \lim_{h \to 0} \frac{f(\mathbf{x}_0 + h\mathbf{u}) - f(\mathbf{x}_0)}{h}$$

where $\mathbf{x}_0 = \langle x_0, y_0 \rangle$ if $n = 2$ and $\mathbf{x}_0 = \langle x_0, y_0, z_0 \rangle$ if $n = 3$. This is reasonable because the vector equation of the line through $\mathbf{x}_0$ in the direction of the vector $\mathbf{u}$ is $\mathbf{x} = \mathbf{x}_0 + t\mathbf{u}$ (Equation 11.35) and so $f(\mathbf{x}_0 + h\mathbf{u})$ represents the value of f at a point on this line.

If $f(x, y, z)$ is differentiable and $\mathbf{u} = \langle a, b, c \rangle$, then the same method that was used to prove Theorem 12.36 can be used to show that

(12.45)
$$D_{\mathbf{u}}f(x, y, z) = f_x(x, y, z)a + f_y(x, y, z)b + f_z(x, y, z)c$$

For a function f of three variables, the **gradient vector,** denoted by ∇f or grad f, is

$$\nabla f(x, y, z) = \langle f_x(x, y, z), f_y(x, y, z), f_z(x, y, z) \rangle$$

or, for short,

(12.46)
$$\nabla f = \langle f_x, f_y, f_z \rangle = \frac{\partial f}{\partial x}\mathbf{i} + \frac{\partial f}{\partial y}\mathbf{j} + \frac{\partial f}{\partial z}\mathbf{k}$$

Then, just as with functions of two variables, Formula 12.45 for the directional derivative can be rewritten as

(12.47)
$$D_{\mathbf{u}}f(x, y, z) = \nabla f(x, y, z) \cdot \mathbf{u}$$

● **Example 4**   If $f(x, y, z) = x \sin yz$, (a) find the gradient of f and (b) find the directional derivative of f at $(1, 3, 0)$ in the direction of $\mathbf{v} = \mathbf{i} + 2\mathbf{j} - \mathbf{k}$.

Solution   (a)   The gradient of f is

$$\nabla f(x, y, z) = \langle f_x(x, y, z), f_y(x, y, z), f_z(x, y, z) \rangle$$
$$= \langle \sin yz, xz \cos yz, xy \cos yz \rangle$$

(b)   At $(1, 3, 0)$ we have $\nabla f(1, 3, 0) = \langle 0, 0, 3 \rangle$. The unit vector in the direction of $\mathbf{v} = \mathbf{i} + 2\mathbf{j} - \mathbf{k}$ is

$$\mathbf{u} = \frac{1}{\sqrt{6}}\mathbf{i} + \frac{2}{\sqrt{6}}\mathbf{j} - \frac{1}{\sqrt{6}}\mathbf{k}$$

Therefore Equation 12.47 gives

$$D_{\mathbf{u}}f(1,3,0) = \nabla f(1,3,0) \cdot \mathbf{u}$$

$$= 3\mathbf{k} \cdot \left(\frac{1}{\sqrt{6}}\mathbf{i} + \frac{2}{\sqrt{6}}\mathbf{j} - \frac{1}{\sqrt{6}}\mathbf{k} \right)$$

$$= 3\left(-\frac{1}{\sqrt{6}} \right) = -\sqrt{\frac{3}{2}}$$

Suppose we have a function f of two or three variables and we consider all possible directional derivatives of f at a given point. These give the rates of change of f in all possible directions. We can then ask the questions: In which of these directions does f change fastest and what is the maximum rate of change? The answers are provided by the following theorem.

Theorem (12.48)

> Suppose f is a differentiable function of two or three variables. The maximum value of the directional derivative $D_{\mathbf{u}}f(\mathbf{x})$ is $|\nabla f(\mathbf{x})|$ and it occurs when $\mathbf{u}$ has the same direction as the gradient vector $\nabla f(\mathbf{x})$.

Proof   From Equation 12.42 or 12.47 we have

$$D_{\mathbf{u}}f = \nabla f \cdot \mathbf{u}$$

$$= |\nabla f|\,|\mathbf{u}|\cos\theta \qquad \textit{(from Theorem 11.14)}$$

$$= |\nabla f|\cos\theta$$

where θ is the angle between ∇f and $\mathbf{u}$. The maximum value of $\cos\theta$ is 1 and this occurs when $\theta = 0$. Therefore the maximum value of $D_{\mathbf{u}}f$ is $|\nabla f|$ and it occurs when $\theta = 0$, that is, when $\mathbf{u}$ has the same direction as ∇f.   ●

● **Example 5**   (a) If $f(x, y) = xe^{y}$, find the rate of change of f at the point $P(2, 0)$ in the direction from P to $Q(5, 4)$. (b) In what direction does f have the maximum rate of change? What is this maximum rate of change?

Solution   (a)   We first compute the gradient vector:

$$\nabla f(x, y) = \langle f_x, f_y \rangle = \langle e^{y}, xe^{y} \rangle$$

$$\nabla f(2, 0) = \langle 1, 2 \rangle$$

The unit vector in the direction of $\overrightarrow{PQ} = \langle 3, 4 \rangle$ is $\mathbf{u} = \langle \frac{3}{5}, \frac{4}{5} \rangle$, so the rate of change of f in the direction from P to Q is

$$D_{\mathbf{u}}f(2, 0) = \nabla f(2, 0) \cdot \mathbf{u} = \langle 1, 2 \rangle \cdot \langle \tfrac{3}{5}, \tfrac{4}{5} \rangle$$

$$= 1(\tfrac{3}{5}) + 2(\tfrac{4}{5}) = \tfrac{11}{5}$$

(b)   According to Theorem 12.48, f increases fastest in the direction of the gradient vector $\nabla f(2,0) = \langle 1,2 \rangle$. The maximum rate of change is

$$|\nabla f(2,0)| = |\langle 1,2 \rangle| = \sqrt{5}$$

● **Example 6**   Suppose that the temperature at a point (x,y,z) in space is given by $T(x,y,z) = 80/(1 + x^2 + 2y^2 + 3z^2)$, where T is measured in °C and x,y,z in meters. In which direction does the temperature increase fastest at the point $(1,1,-2)$? What is the maximum rate of increase?

Solution   The gradient of T is

$$\nabla T = \frac{\partial T}{\partial x}\mathbf{i} + \frac{\partial T}{\partial y}\mathbf{j} + \frac{\partial T}{\partial z}\mathbf{k}$$

$$= -\frac{160x}{(1 + x^2 + 2y^2 + 3z^2)^2}\mathbf{i} - \frac{320y}{(1 + x^2 + 2y^2 + 3z^2)^2}\mathbf{j}$$

$$- \frac{480z}{(1 + x^2 + 2y^2 + 3z^2)^2}\mathbf{k}$$

$$= -\frac{160}{(1 + x^2 + 2y^2 + 3z^2)^2}(-x\mathbf{i} - 2y\mathbf{j} - 3z\mathbf{k})$$

At the point $(1,1,-2)$ the gradient vector is

$$\nabla T(1,2,-1) = \tfrac{160}{256}(-\mathbf{i} - 2\mathbf{j} + 6\mathbf{k}) = \tfrac{5}{8}(-\mathbf{i} - 2\mathbf{j} + 6\mathbf{k})$$

By Theorem 12.48 the temperature increases fastest in the direction of the gradient vector $\nabla T(1,2,-1) = \tfrac{5}{8}(-\mathbf{i} - 2\mathbf{j} + 6\mathbf{k})$ or, equivalently, in the direction of $-\mathbf{i} - 2\mathbf{j} + 6\mathbf{k}$ or the unit vector $(-\mathbf{i} - 2\mathbf{j} + 6\mathbf{k})/\sqrt{41}$. The maximum rate of increase is the length of the gradient vector:

$$|\nabla T(1,1,-2)| = \frac{5}{8}|-\mathbf{i} - 2\mathbf{j} + 6\mathbf{k}| = \frac{5\sqrt{41}}{8}$$

Therefore the maximum rate of increase of temperature is $5\sqrt{41}/8 \approx 4°/\text{m}$.

●

Tangent Planes to Level Surfaces

Suppose S is a surface with equation $F(x,y,z) = k$, that is, it is a level surface of a function F of three variables, and let $P(x_0, y_0, z_0)$ be a point on S. Let C be any curve that lies on the surface S and passes through the point P. Recall from Section 11.7 that C is described by a continuous vector function $\mathbf{r}(t) = \langle x(t), y(t), z(t) \rangle$. Let t_0 be the parameter value corresponding to P, that is, $\mathbf{r}(t_0) = \langle x_0, y_0, z_0 \rangle$. Since C lies on S, any point $(x(t), y(t), z(t))$ must satisfy the equation of S, that is,

(12.49) $$F(x(t), y(t), z(t)) = k$$

If x, y, and z are differentiable functions of t and F is also differentiable, then we can use the Chain Rule to differentiate both sides of Equation 12.49 as follows:

$$\textbf{(12.50)} \qquad \frac{\partial F}{\partial x}\frac{dx}{dt} + \frac{\partial F}{\partial y}\frac{dy}{dt} + \frac{\partial F}{\partial z}\frac{dz}{dt} = 0$$

But, since $\nabla F = \langle F_x, F_y, F_z \rangle$ and $\mathbf{r}'(t) = \langle x'(t), y'(t), z'(t) \rangle$, Equation 12.50 can be written in terms of a dot product as

$$\nabla F \cdot \mathbf{r}'(t) = 0$$

In particular, when $t = t_0$ we have $\mathbf{r}(t_0) = \langle x_0, y_0, z_0 \rangle$, so

$$\textbf{(12.51)} \qquad \nabla F(x_0, y_0, z_0) \cdot \mathbf{r}'(t_0) = 0$$

Equation 12.51 says that *the gradient vector at* P, $\nabla F(x_0, y_0, z_0)$, *is perpendicular to the tangent vector* $\mathbf{r}'(t_0)$ *to any curve* C *on* S *that passes through* P (see Figure 12.28). If $\nabla F(x_0, y_0, z_0) \neq \mathbf{0}$, it is therefore natural to define the **tangent plane to the level surface** $F(x, y, z) = k$ **at** $P(x_0, y_0, z_0)$ as the plane that passes through P and has normal vector $\nabla F(x_0, y_0, z_0)$. Using the standard equation of a plane (11.40), we can write the equation of this tangent plane as

$$\textbf{(12.52)} \qquad \boxed{\begin{aligned} F_x(x_0, y_0, z_0)(x - x_0) &+ F_y(x_0, y_0, z_0)(y - y_0) \\ &+ F_z(x_0, y_0, z_0)(z - z_0) = 0 \end{aligned}}$$

The **normal line** to S at P is the line passing through P and perpendicular to the tangent plane. Its direction is therefore given by the gradient vector $\nabla F(x_0, y_0, z_0)$ and so, by Equation 11.37, its symmetric equations are

$$\textbf{(12.53)} \qquad \frac{x - x_0}{F_x(x_0, y_0, z_0)} = \frac{y - y_0}{F_y(x_0, y_0, z_0)} = \frac{z - z_0}{F_z(x_0, y_0, z_0)}$$

In the special case where the equation of a surface S is of the form $z = f(x, y)$ (that is, S is the graph of a function f of two variables), then we can rewrite it as

$$F(x, y, z) = f(x, y) - z = 0$$

and regard S as a level surface (with $k = 0$) of F. Then

$$F_x(x_0, y_0, z_0) = f_x(x_0, y_0) \qquad F_y(x_0, y_0, z_0) = f_y(x_0, y_0) \qquad F_z(x_0, y_0, z_0) = -1$$

so Equation 12.52 becomes

$$f_x(x_0, y_0)(x - x_0) + f_y(x_0, y_0)(y - y_0) - (z - z_0) = 0$$

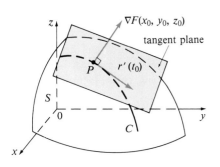

Figure 12.28

which is equivalent to (12.16). Thus our new, more general definition of a tangent plane is consistent with the definition that was given for the special case of Section 12.4.

● **Example 7**   Find the equations of the tangent plane and normal line at the point $(-2, 1, -3)$ to the ellipsoid

$$\frac{x^2}{4} + y^2 + \frac{z^2}{9} = 3$$

Solution   The ellipsoid is the level surface (with $k = 3$) of the function

$$F(x, y, z) = \frac{x^2}{4} + y^2 + \frac{z^2}{9}$$

Therefore we have

$$F_x(x, y, z) = \frac{x}{2} \qquad F_y(x, y, z) = 2y \qquad F_z(x, y, z) = \frac{2z}{9}$$

$$F_x(-2, 1, -3) = -1 \qquad F_y(-2, 1, -3) = 2 \qquad F_z(-2, 1, -3) = -\frac{2}{3}$$

Then Equation 12.52 gives the equation of the tangent plane at $(-2, 1, -3)$ as

$$-1(x + 2) + 2(y - 1) - \tfrac{2}{3}(z + 3) = 0$$

which simplifies to $3x - 6y + 2z + 18 = 0$.

By Equation 12.53, the symmetric equations of the normal line are

$$\frac{x + 2}{-1} = \frac{y - 1}{2} = \frac{z + 3}{-2/3}$$ ●

Significance of the Gradient Vector

Let us summarize the ways in which the gradient vector is significant. We first consider a function f of three variables and a point $P(x_0, y_0, z_0)$ in its domain. On the one hand, we know from Theorem 12.48 that the gradient vector $\nabla f(x_0, y_0, z_0)$ gives the direction of fastest increase of f. On the other hand, we know that $\nabla f(x_0, y_0, z_0)$ is orthogonal to the level surface S of f through P. (Refer to Figure 12.28.) These two properties are quite compatible intuitively because as we move away from P on the level surface S, the value of f does not change at all. So it seems reasonable that if we move in the perpendicular direction we get the maximum increase.

In like manner we consider a function f of two variables and a point $P(x_0, y_0)$ in its domain. Again the gradient vector $\nabla f(x_0, y_0)$ gives the direction of fastest increase of f. Also, by considerations similar to our discussion of tangent planes, it can be shown that $\nabla f(x_0, y_0)$ is perpendicular to the level curve $f(x, y) = k$ that passes through P. Again this is intuitively plausible because the values of f remain constant as we move along the curve. (See Figure 12.29.)

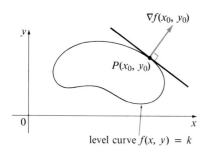

level curve $f(x, y) = k$

Figure 12.29

EXERCISES 12.6

In Exercises 1 and 2 find the directional derivative of f at the given point in the direction indicated by the given angle θ.

1. $f(x, y) = x^2y^3 + 2x^4y, \ (1, -2), \ \theta = \dfrac{\pi}{3}$

2. $f(x, y) = (x^2 - y)^3, \ (3, 1), \ \theta = \dfrac{3\pi}{4}$

In Exercises 3–6 (a) find the gradient of f, (b) evaluate the gradient at the given point P, and (c) find the rate of change of f at P in the direction of the given vector $\mathbf{u}$.

3. $f(x, y) = x^3 - 4x^2y + y^2, \ P(0, -1), \ \mathbf{u} = \langle \tfrac{3}{5}, \tfrac{4}{5} \rangle$

4. $f(x, y) = e^x \sin y, \ P\left(1, \dfrac{\pi}{4}\right), \ \mathbf{u} = \left\langle \dfrac{-1}{\sqrt{5}}, \dfrac{2}{\sqrt{5}} \right\rangle$

5. $f(x, y, z) = xy^2z^3, \ P(1, -2, 1), \ \mathbf{u} = \left\langle \dfrac{1}{\sqrt{3}}, \dfrac{-1}{\sqrt{3}}, \dfrac{1}{\sqrt{3}} \right\rangle$

6. $f(x, y, z) = xy + yz^2 + xz^3, \ P(2, 0, 3), \ \mathbf{u} = \langle -\tfrac{2}{3}, -\tfrac{1}{3}, \tfrac{2}{3} \rangle$

In Exercises 7–14 find the directional derivative of the given function at the given point in the direction of the given vector $\mathbf{v}$.

7. $f(x, y) = \sqrt{x - y}, \ (5, 1), \ \mathbf{v} = \langle 12, 5 \rangle$

8. $f(x, y) = \dfrac{x}{y}, \ (6, -2), \ \mathbf{v} = \langle -1, 3 \rangle$

9. $g(x, y) = xe^{xy}, \ (-3, 0), \ \mathbf{v} = 2\mathbf{i} + 3\mathbf{j}$

10. $g(x, y) = e^x \cos y, \ \left(1, \dfrac{\pi}{6}\right), \ \mathbf{v} = \mathbf{i} - \mathbf{j}$

11. $f(x, y, z) = \sqrt{xyz}, \ (2, 4, 2), \ \mathbf{v} = \langle 4, 2, -4 \rangle$

12. $f(x, y, z) = z^3 - x^2y, \ (1, 6, 2), \ \mathbf{v} = \langle 3, 4, 12 \rangle$

13. $g(x, y, z) = x \tan^{-1}\left(\dfrac{y}{z}\right), \ (1, 2, -2), \ \mathbf{v} = \mathbf{i} + \mathbf{j} - \mathbf{k}$

14. $g(x, y, z) = xe^{yz} + xye^z, \ (-2, 1, 1), \ \mathbf{v} = \mathbf{i} - 2\mathbf{j} + 3\mathbf{k}$

In Exercises 15–20 find the maximum rate of change of f at the given point and the direction in which it occurs.

15. $f(x, y) = xe^{-y} + 3y, \ (1, 0)$

16. $f(x, y) = \ln(x^2 + y^2), \ (1, 2)$

17. $f(x, y) = \sqrt{x^2 + 2y}, \ (4, 10)$

18. $f(x, y, z) = x + \dfrac{y}{z}, \ (4, 3, -1)$

19. $f(x, y) = \cos(3x + 2y), \ \left(\dfrac{\pi}{6}, -\dfrac{\pi}{8}\right)$

20. $f(x, y, z) = \dfrac{x}{y} + \dfrac{y}{z}, \ (4, 2, 1)$

21. Show that a differentiable function f decreases most rapidly at $\mathbf{x}$ in the direction opposite to the gradient vector, that is, in the direction of $-\nabla f(\mathbf{x})$.

22. Use the result of Exercise 21 to find the direction in which the function $f(x, y) = x^4y - x^2y^3$ decreases fastest at the point $(2, -3)$.

23. The temperature T in a metal ball is inversely proportional to the distance from the center of the ball, which we take to be the origin. The temperature at the point $(1, 2, 2)$ is $120°$.
(a) Find the rate of change of T at $(1, 2, 2)$ in the direction toward the point $(2, 1, 3)$.
(b) Show that at any point in the ball the direction of greatest increase in temperature is given by a vector that points toward the origin.

24. The temperature at a point (x, y, z) is given by

$$T(x, y, z) = 200e^{-x^2 - 3y^2 - 9z^2}$$

where T is measured in °C and x, y, z in meters.
(a) Find the rate of change of temperature at the point $P(2, -1, 2)$ in the direction toward the point $(3, -3, 3)$.
(b) In which direction does the temperature increase fastest at P?
(c) Find the maximum rate of increase at P.

25. Suppose that over a certain region of space the electrical potential V is given by
$V(x, y, z) = 5x^2 - 3xy + xyz$.
(a) Find the rate of change of the potential at $P(3, 4, 5)$ in the direction of the vector $\mathbf{v} = \mathbf{i} + \mathbf{j} - \mathbf{k}$.
(b) In which direction does V change most rapidly at P?
(c) What is the maximum rate of change at P?

26. Suppose that you are climbing a hill whose shape is given by the equation $z = 1000 - 0.01x^2 - 0.02y^2$ and you are standing at a point with coordinates $(60, 100, 764)$.
(a) In which direction should you proceed initially in order to reach the top of the hill fastest?
(b) If you climb in that direction, at what angle above the horizontal will you be climbing initially?

In Exercises 27–30 show that the operation of taking the gradient of a function has the given properties. Assume that u and v are differentiable functions of x and y and a, b are constants.

27. $\nabla(au+bv)=a\nabla u+b\nabla v$ **28.** $\nabla(uv) = u\nabla v + v\nabla u$

29. $\nabla\left(\dfrac{u}{v}\right) = \dfrac{v\nabla u - u\nabla v}{v^2}$ **30.** $\nabla u^n = nu^{n-1}\nabla u$

In Exercises 31–38 find equations of (a) the tangent plane and (b) the normal line to the given surface at the given point.

31. $4x^2 + y^2 + z^2 = 24,\ (2,2,2)$

32. $x^2 - 2y^2 + z^2 = 3,\ (-1,1,-2)$

33. $xy + yz + zx = 3,\ (1,1,1)$

34. $x^2 + y^2 - z^2 - 2xy + 4xz = 4,\ (1,0,1)$

35. $xyz = 6,\ (1,2,3)$

36. $x^2 - 2y^2 - 3z^2 + xyz = 4,\ (3,-2,-1)$

37. $z + 1 = xe^y\cos z,\ (1,0,0)$

38. $xe^{yz} = 1,\ (1,0,5)$

39. Show that the equation of the tangent plane to the ellipsoid $x^2/a^2 + y^2/b^2 + z^2/c^2 = 1$ at the point (x_0, y_0, z_0) can be written as
$$\frac{xx_0}{a^2} + \frac{yy_0}{b^2} + \frac{zz_0}{c^2} = 1$$

40. Find the equation of the tangent plane to the hyperboloid $x^2/a^2 + y^2/b^2 - z^2/c^2 = 1$ at (x_0, y_0, z_0) and express it in a form similar to the one in Exercise 39.

41. Show that the equation of the tangent plane to the elliptic paraboloid $z/c = x^2/a^2 + y^2/b^2$ at the point

(x_0, y_0, z_0) can be written as
$$\frac{2xx_0}{a^2} + \frac{2yy_0}{b^2} = \frac{z + z_0}{c}$$

42. Find the points on the ellipsoid $x^2 + 2y^2 + 3z^2 = 1$ where the tangent plane is parallel to the plane $3x - y + 3z = 1$.

43. Find the points on the hyperboloid $x^2 - y^2 + 2z^2 = 1$ where the normal line is parallel to the line that joins the points $(3, -1, 0)$ and $(5, 3, 6)$.

44. Show that the ellipsoid $3x^2 + 2y^2 + z^2 = 9$ and the sphere $x^2 + y^2 + z^2 - 8x - 6y - 8z + 24 = 0$ are tangent to each other at the point $(1, 1, 2)$. (This means that they have a common tangent plane at the point.)

45. Show that every plane that is tangent to the cone $x^2 + y^2 = z^2$ passes through the origin.

46. Show that every normal line to the sphere $x^2 + y^2 + z^2 = r^2$ passes through the center of the sphere.

47. Show that the sum of the x-, y-, and z-intercepts of any tangent plane to the surface $\sqrt{x} + \sqrt{y} + \sqrt{z} = \sqrt{c}$ is a constant.

48. Show that the product of the x-, y-, and z-intercepts of any tangent plane to the surface $xyz = c^3$ is a constant.

49. Show that if $z = f(x, y)$ is differentiable at $\mathbf{x}_0 = (x_0, y_0)$, then
$$\lim_{\mathbf{x}\to\mathbf{x}_0} \frac{f(\mathbf{x}) - f(\mathbf{x}_0) - \nabla f(\mathbf{x}_0)\cdot(\mathbf{x} - \mathbf{x}_0)}{|\mathbf{x} - \mathbf{x}_0|} = 0$$
(*Hint*: Use Definition 12.26 directly.)

SECTION 12.7

Maximum and Minimum Values

As we saw in Chapter 4, one of the main uses of ordinary derivatives is finding maximum and minimum values. In this section we shall see how to use partial derivatives to locate maxima and minima of functions of two variables.

Definition (12.54)

> If f is a function of two variables, $f(a,b)$ is a **local maximum** of f if $f(x, y) \leqslant f(a,b)$ for all points (x, y) in some disk with center (a,b). If $f(x, y) \geqslant f(a,b)$ for all (x, y) in such a disk, $f(a,b)$ is a **local minimum** of f.

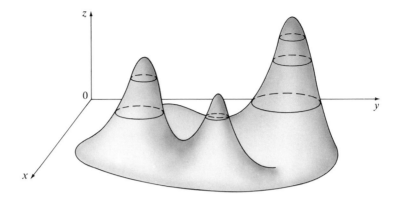

Figure 12.30

If the inequalities in Definition 12.54 hold for *all* points (x, y) in the domain of f, then $f(a, b)$ is the **absolute maximum** (or **absolute minimum**) of f.

The graph of a function with several maxima and minima is sketched in Figure 12.30. You can think of the local maxima as mountain peaks and the local minima as valley bottoms.

Theorem (12.55)

> If f has a local extremum (that is, a local maximum or minimum) at (a, b) and the first order partial derivatives of f exist there, then $f_x(a, b) = 0$ and $f_y(a, b) = 0$.

Proof   Let $g(x) = f(x, b)$. If f has a local extremum at (a, b), then g has a local extremum at a, so $g'(a) = 0$ by Fermat's Theorem (4.4). But, by Equation 12.10, $g'(a) = f_x(a, b)$, so $f_x(a, b) = 0$. Similarly by applying Fermat's Theorem to $G(y) = f(a, y)$ we obtain $f_y(a, b) = 0$.   ●

If we put $f_x(a, b) = 0$ and $f_y(a, b) = 0$ in the equation of a tangent plane (12.16) we get $z = z_0$. Thus the geometric interpretation of Theorem 12.55 is that if the graph of f has a tangent plane at a local extremum, then the tangent plane must be horizontal.

A point (a, b) such that $f_x(a, b) = 0$ and $f_y(a, b) = 0$, or one of these partial derivatives does not exist, is called a **critical point** (or *stationary point*) of f. Theorem 12.55 says that if f has a local extremum at (a, b), then (a, b) is a critical point of f. However, as in single-variable calculus, not all critical points give rise to extrema. At a critical point, a function could have a local maximum or a local minimum or neither.

● **Example 1**   Let $f(x, y) = x^2 + y^2 - 2x - 6y + 14$. Then

$$f_x(x, y) = 2x - 2 \qquad f_y(x, y) = 2y - 6$$

These partial derivatives are equal to 0 when $x = 1$ and $y = 3$, so the only critical point is $(1, 3)$. By completing the square we find that

$$f(x, y) = 4 + (x - 1)^2 + (y - 3)^2$$

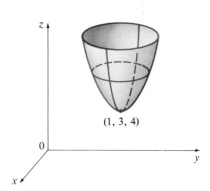

(1, 3, 4)

Figure 12.31

Since $(x - 1)^2 \geqslant 0$ and $(y - 3)^2 \geqslant 0$, we have $f(x, y) \geqslant 4$ for all values of x and y. Therefore $f(1, 3) = 4$ is a local minimum, and in fact it is the absolute minimum of f. This can be confirmed geometrically from the graph of f, which is the elliptic paraboloid with vertex $(1, 3, 4)$ shown in Figure 12.31.

● **Example 2**   Find the extreme values of $f(x, y) = y^2 - x^2$.

Solution   Since $f_x = -2x$ and $f_y = 2y$, the only critical point is $(0, 0)$. Notice that for points on the x-axis we have $y = 0$, so $f(x, y) = -x^2 < 0$ (if $x \neq 0$). However for points on the y-axis we have $x = 0$, so $f(x, y) = y^2 > 0$ (if $y \neq 0$). Thus every disk with center $(0, 0)$ contains points where f takes positive values as well as points where f takes negative values. Therefore $f(0, 0) = 0$ cannot be an extreme value for f, so f has no extreme values.

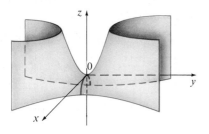

Figure 12.32

GRAPH OF $f(x, y) = y^2 - x^2$

Example 2 illustrates the fact that a function need not have a maximum or minimum value at a critical point. Figure 12.32 shows how this is possible. The graph of f is the hyperbolic paraboloid $z = y^2 - x^2$, which has a horizontal tangent plane ($z = 0$) at the origin. You can see that $f(0, 0) = 0$ is a maximum in the direction of the x-axis but a minimum in the direction of the y-axis. Near the origin the graph has the shape of a saddle and so $(0, 0)$ is called a *saddle point* of f.

We need to be able to determine whether or not a function has an extreme value at a critical point. The following test, which is proved at the end of this section, is analogous to the Second Derivatives Test for functions of one variable.

Second Derivatives Test (12.56)

Suppose the second partial derivatives of f are continuous in a disk with center (a, b), and suppose that $f_x(a, b) = 0$ and $f_y(a, b) = 0$ [that is, (a, b) is a critical point of f]. Let

$$D = D(a, b) = f_{xx}(a, b)f_{yy}(a, b) - [f_{xy}(a, b)]^2$$

(a)   If $D > 0$ and $f_{xx}(a, b) > 0$, then $f(a, b)$ is a local minimum.
(b)   If $D > 0$ and $f_{xx}(a, b) < 0$, then $f(a, b)$ is a local maximum.
(c)   If $D < 0$, then $f(a, b)$ is not a local extremum.

Note 1:   In case (c) the point (a, b) is called a **saddle point** of f and the graph of f crosses its tangent plane at (a, b).

Note 2:   If $D = 0$, the test gives no information: f could have a local maximum or local minimum at (a, b) or (a, b) could be a saddle point of f.

● **Example 3**   Find the local extrema of $f(x, y) = x^4 + y^4 - 4xy + 1$.

Solution   We first locate the critical points:

$$f_x = 4x^3 - 4y \qquad f_y = 4y^3 - 4x$$

Setting these partial derivatives equal to 0, we obtain the equations

$$x^3 - y = 0 \quad \text{and} \quad y^3 - x = 0$$

To solve these equations we substitute $y = x^3$ from the first equation into the second one. This gives

$$0 = x^9 - x = x(x^8 - 1) = x(x^4 - 1)(x^4 + 1) = x(x^2 - 1)(x^2 + 1)(x^4 + 1)$$

so there are three real roots: $x = 0, 1, -1$. The three critical points are $(0, 0)$, $(1, 1)$, and $(-1, -1)$.

Next we calculate the second partial derivatives and $D(x, y)$:

$$f_{xx} = 12x^2 \qquad f_{xy} = -4 \qquad f_{yy} = 12y^2$$
$$D(x, y) = f_{xx}f_{yy} - (f_{xy})^2 = 144x^2y^2 - 16$$

Since $D(0, 0) = -16 < 0$, it follows from case (c) of the Second Derivatives Test that $(0, 0)$ is a saddle point; that is, f has no local extremum at $(0, 0)$. Since $D(1, 1) = 128 > 0$ and $f_{xx}(1, 1) = 12 > 0$, case (a) of the test says that $f(1, 1) = -1$ is a local minimum. Similarly $D(-1, -1) = 128 > 0$ and $f_{xx}(-1, -1) = 12 > 0$, so $f(-1, -1) = -1$ is also a local minimum. ●

● **Example 4**   Find the shortest distance from the point $(1, 0, -2)$ to the plane $x + 2y + z = 4$.

Solution   The distance from any point (x, y, z) to the point $(1, 0, -2)$ is

$$d = \sqrt{(x - 1)^2 + y^2 + (z + 2)^2}$$

but if (x, y, z) lies on the plane $x + 2y + z = 4$, then $z = 4 - x - 2y$, so $d = \sqrt{(x - 1)^2 + y^2 + (6 - x - 2y)^2}$. We can minimize d by minimizing the simpler expression

$$d^2 = f(x, y) = (x - 1)^2 + y^2 + (6 - x - 2y)^2$$

By solving the equations

$$f_x = 2(x - 1) - 2(6 - x - 2y) = 4x + 4y - 14 = 0$$
$$f_y = 2y - 4(6 - x - 2y) = 4x + 10y - 24 = 0$$

we find that the only critical point is $(\frac{11}{6}, \frac{5}{3})$. Since $f_{xx} = 4$, $f_{xy} = 4$, $f_{yy} = 10$, we have $D(x, y) = f_{xx}f_{yy} - (f_{xy})^2 = 24 > 0$ and $f_{xx} > 0$, so by the Second Derivatives Test f has a local minimum at $(\frac{11}{6}, \frac{5}{3})$. Intuitively we can see that this local minimum is actually an absolute minimum because there must be a point on the given plane that is closest to $(1, 0, -2)$. If $x = \frac{11}{6}$ and $y = \frac{5}{3}$, then

$$d = \sqrt{(x - 1)^2 + y^2 + (6 - x - 2y)^2} = \sqrt{(\tfrac{5}{6})^2 + (\tfrac{5}{3})^2 + (\tfrac{5}{6})^2} = 5\sqrt{6}/6$$

The shortest distance from $(1, 0, -2)$ to the plane $x + 2y + z = 4$ is $5\sqrt{6}/6$.
                                                                              ●

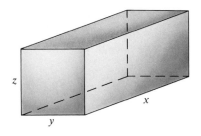

Figure 12.33

● **Example 5**   A rectangular box without a lid is to be made from 12 m² of cardboard. Find the maximum volume of such a box.

Solution   Let the length, width, and height of the box (in meters) be x, y, and z, as shown in Figure 12.33. Then the volume of the box is

$$V = xyz$$

We can express V as a function of just two variables x and y by using the fact that the area of the four sides and the bottom of the box is

$$2xz + 2yz + xy = 12$$

Solving this equation for z we get $z = (12 - xy)/2(x + y)$ so the expression for V becomes

$$V = xy \frac{12 - xy}{2(x + y)} = \frac{12xy - x^2y^2}{2(x + y)}$$

We compute the partial derivatives:

$$\frac{\partial V}{\partial x} = \frac{y^2(12 - 2xy - x^2)}{2(x + y)^2} \qquad \frac{\partial V}{\partial y} = \frac{x^2(12 - 2xy - y^2)}{2(x + y)^2}$$

If V is a maximum, then $\partial V/\partial x = \partial V/\partial y = 0$, but $x = 0$ or $y = 0$ will give $V = 0$, so we must solve the equations

$$12 - 2xy - x^2 = 0 \qquad 12 - 2xy - y^2 = 0$$

These imply that $x^2 = y^2$ and so $x = y$. (Note that x and y must both be positive in this problem.) If we put $x = y$ in either equation we get $12 - 3x^2 = 0$, which gives $x = 2$, $y = 2$, and $z = (12 - 2 \cdot 2)/2(2 + 2) = 1$.
    We could use the Second Derivatives Test to show that this gives a local maximum of V, or we could simply argue from the physical nature of this problem that there must be an absolute maximum volume that has to occur at a critical point of V, so it must occur when $x = 2$, $y = 2$, $z = 1$. Then $V = 2 \cdot 2 \cdot 1 = 4$, so the maximum volume of the box is 4 m³.   ●

    For a function f of one variable the Extreme Value Theorem says that if f is continuous on a closed interval $[a, b]$, then f has an absolute minimum value and an absolute maximum value. According to the procedure in (4.8) we found these by evaluating f not only at the critical numbers but also at the endpoints a and b.
    There is a similar situation for functions of two variables. Just as a closed interval contains its endpoints, a **closed set** in R^2 is one that contains all its boundary points. (See the discussion following Definition 12.6.) For instance, the disk

$$D = \{(x, y) \,|\, x^2 + y^2 \leqslant 1\}$$

consisting of all points on and inside the circle $x^2 + y^2 = 1$ is a closed set because it contains all of its boundary points (which are the points on the circle $x^2 + y^2 = 1$). But if even one point on the boundary curve were omitted, the set would not be closed. (See Figure 12.34.)

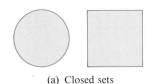

(a) Closed sets

(b) Sets that are not closed

Figure 12.34

A **bounded set** in R^2 is one that is contained within some disk. In other words, it is finite in extent. Then, in terms of closed and bounded sets, we can state the following analogue of the Extreme Value Theorem in two dimensions.

Extreme Value Theorem for Functions of Two Variables (12.57)

> If f is continuous on a closed, bounded set D in R^2, then f attains an absolute maximum value $f(x_1, y_1)$ and an absolute minimum value $f(x_2, y_2)$ at some points (x_1, y_1) and (x_2, y_2) in D.

To find the extreme values guaranteed by Theorem 12.57, we note that, by Theorem 12.55, if f has an extreme value at (x_1, y_1), then (x_1, y_1) is either a critical point of f or a boundary point of D. Thus we have the following analogue of (4.8):

(12.58)

> To find the absolute maximum and minimum values of a continuous function f on a closed bounded set D:
>
> 1. Find the values of f at the critical points of f in D.
> 2. Find the extreme values of f on the boundary of D.
> 3. The largest of the values from steps 1 and 2 is the absolute maximum value; the smallest of these values is the absolute minimum value.

● **Example 6**   Find the absolute maximum and minimum values of $f(x, y) = x^2 - 2xy + 2y$ on the rectangle $D = \{(x, y) | 0 \leqslant x \leqslant 3, 0 \leqslant y \leqslant 2\}$.

Solution   Since f is a polynomial it is continuous on the closed bounded rectangle D, so Theorem 12.57 tells us there will be both an absolute maximum and an absolute minimum. According to step 1 we first find the critical points. These occur when

$$f_x = 2x - 2y = 0 \qquad f_y = -2x + 2 = 0$$

so the only critical point is $(1, 1)$ and the value of f there is $f(1, 1) = 1$.

In step 2 we look at the values of f on the boundary of D, which consists of the four line segments L_1, L_2, L_3, L_4 shown in Figure 12.35. On L_1 we have $y = 0$ and

$$f(x, 0) = x^2 \qquad 0 \leqslant x \leqslant 3$$

This is an increasing function of x, so its minimum value is $f(0, 0) = 0$ and its maximum value is $f(3, 0) = 9$. On L_2 we have $x = 3$ and

$$f(3, y) = 9 - 4y \qquad 0 \leqslant y \leqslant 2$$

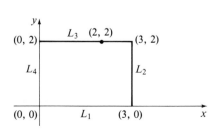

Figure 12.35

This is a decreasing function of y, so its maximum value is $f(3,0) = 9$ and its minimum value is $f(3,2) = 1$. On L_3 we have $y = 2$ and

$$f(x, 2) = x^2 - 4x + 4 \qquad 0 \leqslant x \leqslant 3$$

By the methods of Chapter 4, or simply by observing that $f(x, 2) = (x - 2)^2$, we see that the minimum value of this function is $f(2, 2) = 0$ and the maximum value is $f(0, 2) = 4$. Finally, on L_4 we have $x = 0$ and

$$f(0, y) = 2y \qquad 0 \leqslant y \leqslant 2$$

with maximum value $f(0, 2) = 4$ and minimum value $f(0, 0)$. Thus, on the boundary, the minimum value of f is 0 and the maximum is 9.

In step 3 we compare these values with the value $f(1, 1) = 1$ at the critical point and conclude that the absolute maximum value of f on D is $f(3, 0) = 9$ and the absolute minimum value is $f(0, 0) = f(2, 2) = 0$. ●

We close this section by giving a proof of the first part of the Second Derivatives Test. Notice the similarity to the proof of the Second Derivatives Test for functions of one variable. Parts (b) and (c) have similar proofs.

Proof of Theorem 12.56 part (a)   We compute the second order directional derivative of f in the direction of $\mathbf{u} = \langle h, k \rangle$. The first order derivative is given by Theorem 12.36:

$$D_{\mathbf{u}} f = f_x h + f_y k$$

Applying this theorem a second time, we have

$$D_{\mathbf{u}}^2 f = D_{\mathbf{u}}(D_{\mathbf{u}} f) = \frac{\partial}{\partial x}(D_{\mathbf{u}} f) h + \frac{\partial}{\partial y}(D_{\mathbf{u}} f) k$$

$$= (f_{xx} h + f_{yx} k) h + (f_{xy} h + f_{yy} k) k$$

$$= f_{xx} h^2 + 2 f_{xy} h k + f_{yy} k^2 \qquad (by\ Theorem\ 12.14)$$

If we complete the square in this expression, we obtain

(12.59)
$$D_{\mathbf{u}}^2 f = f_{xx}\left(h + \frac{f_{xy}}{f_{xx}} k\right)^2 + \frac{k^2}{f_{xx}}(f_{xx} f_{yy} - f_{xy}^2)$$

We are given that $f_{xx}(a, b) > 0$ and $D(a, b) > 0$. But f_{xx} and $D = f_{xx} f_{yy} - f_{xy}^2$ are continuous functions, so there is a disk B with center (a, b) and radius $\delta > 0$ such that $f_{xx}(x, y) > 0$ and $D(x, y) > 0$ whenever (x, y) is in B. Therefore, by looking at Equation 12.59, we see that $D_{\mathbf{u}}^2 f(x, y) > 0$ whenever (x, y) is in B. This means that if C is the curve obtained by intersecting the graph of f with the vertical plane through $P(a, b, f(a, b))$ in the direction of $\mathbf{u}$, then C is concave upward on an interval of length 2δ. This is true in the direction of every vector $\mathbf{u}$, so if we restrict (x, y) to lie in B, the graph of f lies above its horizontal tangent plane at P. Thus $f(x, y) \geqslant f(a, b)$ whenever (x, y) is in B. This shows that $f(a, b)$ is a local minimum. ●

EXERCISES 12.7

In Exercises 1–20 find the local maximum and minimum values and saddle points of the given function.

1. $f(x, y) = x^2 + y^2 + 4x - 6y$

2. $f(x, y) = 4x^2 + y^2 - 4x + 2y$

3. $f(x, y) = 2x^2 + y^2 + 2xy + 2x + 2y$

4. $f(x, y) = 1 + 2xy - x^2 - y^2$

5. $f(x, y) = x^2 + y^2 + x^2y + 4$

6. $f(x, y) = 2x^3 + xy^2 + 5x^2 + y^2$

7. $f(x, y) = x^3 - 3xy + y^3$

8. $f(x, y) = y\sqrt{x} - y^2 - x + 6y$

9. $f(x, y) = xy - 2x - y$

10. $f(x, y) = xy(1 - x - y)$

11. $f(x, y) = \dfrac{x^2y^2 - 8x + y}{xy}$

12. $f(x, y) = x^2 + y^2 + \dfrac{1}{x^2y^2}$

13. $f(x, y) = e^x \cos y$

14. $f(x, y) = (2x - x^2)(2y - y^2)$

15. $f(x, y) = 3x^2y + y^3 - 3x^2 - 3y^2 + 2$

16. $f(x, y) = xye^{-x^2-y^2}$

17. $f(x, y) = x \sin y$

18. $f(x, y) = \dfrac{(x + y + 1)^2}{x^2 + y^2 + 1}$

19. $f(x, y) = \sin x + \sin y + \sin(x + y), 0 \leqslant x \leqslant 2\pi,$
$0 \leqslant y \leqslant 2\pi$

20. $f(x, y) = \sin x + \sin y + \cos(x + y), 0 \leqslant x \leqslant \dfrac{\pi}{4},$

$0 \leqslant y \leqslant \dfrac{\pi}{4}$

In Exercises 21–24 find the absolute maximum and minimum values of f on the set D.

21. $f(x, y) = 5 - 3x + 4y$, D is the closed triangular region with vertices $(0, 0)$, $(4, 0)$, and $(4, 5)$

22. $f(x, y) = y\sqrt{x} - y^2 - x + 6y,$
$D = \{(x, y) | 0 \leqslant x \leqslant 9, 0 \leqslant y \leqslant 5\}$

23. $f(x, y) = x^2 + y^2 + x^2y + 4,$
$D = \{(x, y) | |x| \leqslant 1, |y| \leqslant 1\}$

24. $f(x, y) = 2x^2 + x + y^2 - 2, D = \{(x, y) | x^2 + y^2 \leqslant 4\}$

25. Find the point on the plane $x + 2y + 3z = 4$ that is closest to the origin.

26. Find the point on the plane $2x - y + z = 1$ that is closest to the point $(-4, 1, 3)$.

27. Find the shortest distance from the point $(2, -2, 3)$ to the plane $6x + 4y - 3z = 2$.

28. Find the shortest distance from the point (x_0, y_0, z_0) to the plane $Ax + By + Cz + D = 0$.

29. Find the points on the surface $z^2 = xy + 1$ that are closest to the origin.

30. Find the points on the surface $x^2y^2z = 1$ that are closest to the origin.

31. Find three positive numbers whose sum is 100 and whose product is a maximum.

32. Find three positive numbers x, y, and z whose sum is 100 such that $x^ay^bz^c$ is a maximum.

33. Find the volume of the largest rectangular box with edges parallel to the axes that can be inscribed in the ellipsoid $9x^2 + 36y^2 + 4z^2 = 36$.

34. Solve the problem in Exercise 33 for a general ellipsoid $x^2/a^2 + y^2/b^2 + z^2/c^2 = 1$.

35. Find the volume of the largest rectangular box in the first octant with three faces in the coordinate planes and one vertex in the plane $x + 2y + 3z = 6$.

36. Solve the problem in Exercise 35 for a general plane $x/a + y/b + z/c = 1$ where a, b, c are positive numbers.

37. Find the dimensions of a rectangular box of maximum volume such that the sum of the lengths of its 12 edges is a constant c.

38. Find the dimensions of the rectangular box with largest volume if the total surface area is given as 64 cm².

39. A cardboard box without a lid is to have a volume of 32,000 cm³. Find the dimensions that will minimize the amount of cardboard used.

40. The base of an aquarium with given volume V is made of slate and the sides are made of glass. If slate costs five times as much (per unit area) as glass, find the dimensions of the aquarium that will minimize the cost of the materials.

41. Suppose that a scientist has reason to believe that two quantities x and y are related linearly, that is $y = mx + b$, at least approximately, for some values of m and b. The scientist performs an experiment and collects data in the form of points $(x_1, y_1), (x_2, y_2), \ldots, (x_n, y_n)$, which he then plots. These points will not lie exactly on a straight line so he wants to find constants m and b so that the line $y = mx + b$ "fits" the points as well as possible. (See Figure 12.36.) Let $d_i = y_i - (mx_i + b)$ be the vertical deviation of the point (x_i, y_i) from the line. The **method of least squares** determines m and b so as to minimize $\sum_{i=1}^{n} d_i^2$, the sum of the squares of these deviations. Show that, according to this method, the line of best fit is obtained when

$$m \sum_{i=1}^{n} x_i + bn = \sum_{i=1}^{n} y_i$$

$$m \sum_{i=1}^{n} x_i^2 + b \sum_{i=1}^{n} x_i = \sum x_i y_i$$

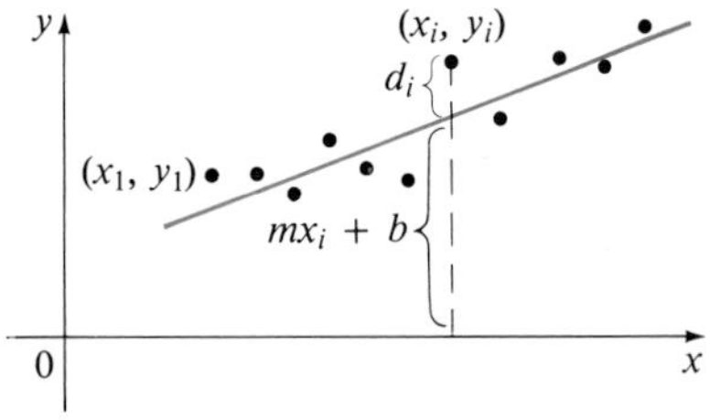

Figure 12.36

Thus the line is found by solving these two equations in the two unknowns m and b.

42. (a) Use the method of least squares (Exercise 41) to find the straight line $y = mx + b$ that best fits the data points $(1, 5.1)$, $(2, 6.8)$, $(3, 9.4)$, and $(4, 10.5)$.
    (b) As a check on your calculations in part (a), plot the four points and graph the line.

43. The following data give the height x (in inches) and weight y (in pounds) of ten 18-year-old boys. Use the method of least squares (Exercise 41) to fit these data to a straight line. Then use it to predict the weight of an 18-year-old boy who is 6 ft tall.

x	69	65	71	73	68	63	70	67	69	70
y	138	127	178	185	141	122	158	135	145	162

SECTION 12.8

Lagrange Multipliers

In Example 5 in Section 12.7 we maximized a volume function $V = xyz$ subject to the constraint $2xz + 2yz + xy = 12$, which expressed the side condition that the surface area was 12 m^2. In this section we present Lagrange's method for maximizing or minimizing a general function $f(x, y, z)$ subject to a constraint (or side condition) of the form $g(x, y, z) = k$.

Geometrically speaking, this amounts to finding the extreme values of $f(x, y, z)$ when the point (x, y, z) is restricted to lie on the surface S with equation $g(x, y, z) = k$ (a level surface of g). Suppose that f has an extreme value at $P(x_0, y_0, z_0)$ on S and let C be a curve with vector equation $\mathbf{r}(t) = \langle x(t), y(t), z(t) \rangle$ that lies on S and passes through P. If t_0 is the parameter value corresponding to P, then $\mathbf{r}(t_0) = \langle x_0, y_0, z_0 \rangle$. The composite function $h(t) = f(x(t), y(t), z(t))$ represents the values that f takes on the curve C. Since f has an extreme value at (x_0, y_0, z_0) it follows that h has an extreme value at t_0, so $h'(t_0) = 0$. But if f is differentiable, we can use the Chain Rule to write

$$0 = h'(t_0) = f_x(x_0, y_0, z_0)x'(t_0) + f_y(x_0, y_0, z_0)y'(t_0) + f_z(x_0, y_0, z_0)z'(t_0)$$

$$= \nabla f(x_0, y_0, z_0) \cdot \mathbf{r}'(t_0)$$

This shows that the gradient vector $\nabla f(x_0, y_0, z_0)$ is orthogonal to the tangent vector $\mathbf{r}'(t_0)$ to every such curve C. But we already know from Section 12.6 that the gradient vector of g, $\nabla g(x_0, y_0, z_0)$, is also orthogonal to $\mathbf{r}'(t_0)$ (see Equation 12.51). This means that the gradient vectors $\nabla f(x_0, y_0, z_0)$ and $\nabla g(x_0, y_0, z_0)$ must be parallel. Therefore, if $\nabla g(x_0, y_0, z_0) \neq \mathbf{0}$, there is a number λ such that

(12.60)

$$\nabla f(x_0, y_0, z_0) = \lambda \nabla g(x_0, y_0, z_0)$$

The number λ in Equation 12.60 is called a **Lagrange multiplier.** The procedure based on Equation 12.60 is called the **method of Lagrange multipliers** and is as follows:

To find the maximum and minimum values of $f(x, y, z)$ subject to the constraint $g(x, y, z) = k$ (assuming that these extreme values exist):

(a)   Find all values of x, y, z, and λ such that

$$\nabla f(x, y, z) = \lambda \nabla g(x, y, z)$$

and $\qquad\qquad g(x, y, z) = k$

(b)   Evaluate f at all the points (x, y, z) that arise from step (a). The largest of these values is the maximum value of f; the smallest is the minimum value of f.

If we write the vector equation $\nabla f = \lambda \nabla g$ in terms of its components, then the equations in step (a) become

$$f_x = \lambda g_x \qquad f_y = \lambda g_y \qquad f_z = \lambda g_z \qquad g(x, y, z) = k$$

This is a system of four equations in the four unknowns x, y, z, and λ, but it is not necessary to find explicit values for λ.

For functions of two variables the method of Lagrange multipliers is similar to the method just described. To find the extreme values of $f(x, y)$ subject to the constraint $g(x, y) = k$ (a level curve of g) we look for values of x, y, and λ such that

$$\nabla f(x, y) = \lambda \nabla g(x, y) \quad \text{and} \quad g(x, y) = k$$

This amounts to solving three equations in three unknowns:

$$f_x = \lambda g_x \qquad f_y = \lambda g_y \qquad g(x, y) = k$$

Our first illustration of Lagrange's method is to reconsider the problem given in Example 5 in Section 12.7.

● **Example 1**   A rectangular box without a lid is to be made from 12 m^2 of cardboard. Find the maximum volume of such a box.

Solution   As in Example 5 in Section 12.7 we let x, y, and z be the length, width, and height, respectively, of the box in meters. Then we wish to maximize

$$V = xyz$$

subject to the constraint

$$g(x, y, z) = 2xz + 2yz + xy = 12$$

Using the method of Lagrange multipliers, we look for values of x, y, z, and λ such that $\nabla V = \lambda \nabla g$ and $g(x, y, z) = 12$. This gives the equations

$$V_x = \lambda g_x \qquad V_y = \lambda g_y \qquad V_z = \lambda g_z \qquad 2xz + 2yz + xy = 12$$

which become
$$yz = \lambda(2z + y) \qquad (1)$$
$$xz = \lambda(2z + x) \qquad (2)$$
$$xy = \lambda(2x + 2y) \qquad (3)$$
$$2xz + 2yz + xy = 12 \qquad (4)$$

There are no general rules for solving systems of equations. Sometimes some ingenuity is required. In the present example you might notice that if we multiply (1) by x, (2) by y, and (3) by z, then the left sides of these equations will be identical. Doing this, we have

$$xyz = \lambda(2xz + xy) \qquad (5)$$
$$xyz = \lambda(2yz + xy) \qquad (6)$$
$$xyz = \lambda(2xz + 2yz) \qquad (7)$$

We observe that $\lambda \neq 0$ because $\lambda = 0$ would imply $yz = xz = xy = 0$ from (1), (2), and (3) and this would contradict (4). Therefore from (5) and (6), we have

$$2xz + xy = 2yz + xy$$

which gives $xz = yz$. But $z \neq 0$ (since $z = 0$ would give $V = 0$) so $x = y$. From (6) and (7), we have

$$2yz + xy = 2xz + 2yz$$

which gives $2xz = xy$ and so (since $x \neq 0$) $y = 2z$. If we now put $x = y = 2z$ in (4) we get

$$4z^2 + 4z^2 + 4z^2 = 12$$

Since x, y, z are all positive, we therefore have $z = 1$, $x = 2$, and $y = 2$ as before.   ●

● **Example 2**   Find the extreme values of the function $f(x, y) = x^2 + y$ on the circle $x^2 + y^2 = 1$.

Solution   We are asked for the extreme values of f subject to the constraint $g(x, y) = x^2 + y^2 = 1$. Using Lagrange multipliers, we solve the equations $\nabla f = \lambda \nabla g$, $g(x, y) = 1$, which can be written as

$$f_x = \lambda g_x \qquad f_y = \lambda g_y \qquad g(x, y) = 1$$

or

$$2x = 2x\lambda \qquad (1)$$

$$1 = 2y\lambda \qquad (2)$$

$$x^2 + y^2 = 1 \qquad (3)$$

From (1) we have $x = 0$ or $\lambda = 1$. If $x = 0$, then (3) gives $y = \pm 1$. If $\lambda = 1$, then $y = \frac{1}{2}$ from (2), so then (3) gives $x = \pm\sqrt{3}/2$. Therefore f has possible extreme values at the points $(0, 1), (0, -1), (\sqrt{3}/2, \frac{1}{2}), (-\sqrt{3}/2, \frac{1}{2})$. Evaluating f at these four points, we find that

$$f(0, 1) = 1 \qquad f(0, -1) = -1 \qquad f\left(\pm\frac{\sqrt{3}}{2}, \frac{1}{2}\right) = \frac{5}{4}$$

Therefore the maximum value of f on the circle $x^2 + y^2 = 1$ is $f(\pm\sqrt{3}/2, \frac{1}{2}) = \frac{5}{4}$ and the minimum value is $f(0, -1) = -1$.   ●

● **Example 3**   Find the points on the sphere $x^2 + y^2 + z^2 = 4$ that are closest to and farthest from the point $(3, 1, -1)$.

Solution   The distance from a point (x, y, z) to the point $(3, 1, -1)$ is

$$d = \sqrt{(x - 3)^2 + (y - 1)^2 + (z + 1)^2}$$

but the algebra is simpler if we instead maximize and minimize the square of the distance:

$$d^2 = f(x, y, z) = (x - 3)^2 + (y - 1)^2 + (z + 1)^2$$

The constraint is that the point (x, y, z) lies on the sphere, that is,

$$g(x, y, z) = x^2 + y^2 + z^2 = 4$$

According to the method of Lagrange multipliers, we solve $\nabla f = \lambda \nabla g$, $g = 4$. This gives

$$2(x - 3) = 2x\lambda \qquad (1)$$

$$2(y - 1) = 2y\lambda \qquad (2)$$

$$2(z + 1) = 2z\lambda \qquad (3)$$

$$x^2 + y^2 + z^2 = 4 \qquad (4)$$

The simplest way to solve these equations is to solve for x, y, and z in terms of λ from (1), (2), and (3) and then substitute these values into (4). From (1)

we have

$$x - 3 = x\lambda \quad \text{or} \quad x(1 - \lambda) = 3 \quad \text{or} \quad x = \frac{3}{1 - \lambda}$$

[Note that $1 - \lambda \neq 0$ because $\lambda = 1$ is impossible from (1).] Similarly, (2) and (3) yield

$$y = \frac{1}{1 - \lambda} \qquad z = -\frac{1}{1 - \lambda}$$

Therefore, from (4),

$$\frac{3^2}{(1 - \lambda)^2} + \frac{1^2}{(1 - \lambda)^2} + \frac{(-1)^2}{(1 - \lambda)^2} = 4$$

which gives $(1 - \lambda)^2 = \frac{11}{4}$, $1 - \lambda = \pm\sqrt{11}/2$, so

$$\lambda = 1 \pm \frac{\sqrt{11}}{2}$$

These values of λ then give the corresponding points (x, y, z):

$$\left(\frac{6}{\sqrt{11}}, \frac{2}{\sqrt{11}}, -\frac{2}{\sqrt{11}}\right) \quad \text{and} \quad \left(-\frac{6}{\sqrt{11}}, -\frac{2}{\sqrt{11}}, \frac{2}{\sqrt{11}}\right)$$

It is easy to see that f has a smaller value at the first of these points, so the closest point is $(6/\sqrt{11}, 2/\sqrt{11}, -2/\sqrt{11})$ and the farthest point is $(-6/\sqrt{11}, -2/\sqrt{11}, 2/\sqrt{11})$.   ●

Two Constraints

Suppose now that we want to find the maximum and minimum values of $f(x, y, z)$ subject to two constraints (side conditions) of the form $g(x, y, z) = k$ and $h(x, y, z) = c$. Geometrically, this means that we are looking for the extreme values of f when (x, y, z) is restricted to lie on the curve of intersection of the level surfaces $g(x, y, z) = k$ and $h(x, y, z) = c$. It can be shown that if an extreme value occurs at (x_0, y_0, z_0), then $\nabla f(x_0, y_0, z_0)$ is in the plane determined by $\nabla g(x_0, y_0, z_0)$ and $\nabla h(x_0, y_0, z_0)$ and so there are numbers λ and μ (called Lagrange multipliers) such that

(12.61)
$$\nabla f(x_0, y_0, z_0) = \lambda \nabla g(x_0, y_0, z_0) + \mu \nabla h(x_0, y_0, z_0)$$

In this case Lagrange's method is to look for extreme values by solving five equations in the five unknowns x, y, z, λ, and μ. These equations are obtained by writing Equation 12.61 in terms of its components and using the constraint equations:

$$f_x = \lambda g_x + \mu h_x$$

$$f_y = \lambda g_y + \mu h_y$$

$$f_z = \lambda g_z + \mu h_z$$

$$g(x, y, z) = k$$

$$h(x, y, z) = c$$

● **Example 4**   Find the maximum value of the function $f(x, y, z) = x + 2y + 3z$ on the curve of intersection of the plane $x - y + z = 1$ and the cylinder $x^2 + y^2 = 1$.

Solution   We maximize $f(x, y, z) = x + 2y + 3z$ subject to the constraints $g(x, y, z) = x - y + z = 1$ and $h(x, y, z) = x^2 + y^2 = 1$. The Lagrange condition is $\nabla f = \lambda \nabla g + \mu \nabla h$, so we solve the equations

$$1 = \lambda + 2x\mu \qquad (1)$$

$$2 = -\lambda + 2y\mu \qquad (2)$$

$$3 = \lambda \qquad (3)$$

$$x - y + z = 1 \qquad (4)$$

$$x^2 + y^2 = 1 \qquad (5)$$

Putting $\lambda = 3$ [from (3)] in (1), we get $2x\mu = -2$, so $x = -1/\mu$. Similarly, (2) gives $y = 5/2\mu$. Substitution in (5) then yields

$$\frac{1}{\mu^2} + \frac{25}{4\mu^2} = 1$$

and so $\mu^2 = \frac{29}{4}$, $\mu = \pm\sqrt{29}/2$. Then $x = \mp 2/\sqrt{29}$, $y = \pm 5/\sqrt{29}$, and, from (4), $z = 1 - x + y = 1 \pm 7/\sqrt{29}$. The corresponding values of f are

$$\mp \frac{2}{\sqrt{29}} + 2\left(\pm \frac{5}{\sqrt{29}}\right) + 3\left(1 \pm \frac{7}{\sqrt{29}}\right) = 3 \pm \sqrt{29}$$

Therefore the maximum value of f on the given curve is $3 + \sqrt{29}$.   ●

EXERCISES 12.8

In Exercises 1–15 use Lagrange multipliers to find the maximum and minimum values of the given function subject to the given constraint or constraints.

1. $f(x, y) = x^2 - y^2;\ x^2 + y^2 = 1$

2. $f(x, y) = 2x + y;\ x^2 + 4y^2 = 1$

3. $f(x, y) = xy;\ 9x^2 + y^2 = 4$

4. $f(x, y) = x^2 + y^2;\ x^4 + y^4 = 1$

5. $f(x, y, z) = x + 3y + 5z; x^2 + y^2 + z^2 = 1$

6. $f(x, y, z) = x - y + 3z; x^2 + y^2 + 4z^2 = 4$

7. $f(x, y, z) = xyz; x^2 + 2y^2 + 3z^2 = 6$

8. $f(x, y, z) = x^2 y^2 z^2; x^2 + y^2 + z^2 = 1$

9. $f(x, y, z) = x^2 + y^2 + z^2; x^4 + y^4 + z^4 = 1$

10. $f(x, y, z) = x^4 + y^4 + z^4; x^2 + y^2 + z^2 = 1$

11. $f(x, y, z, t) = x + y + z + t; x^2 + y^2 + z^2 + t^2 = 1$

12. $f(x_1, x_2, \ldots, x_n) = x_1 + x_2 + \cdots + x_n;$ $x_1^2 + x_2^2 + \cdots + x_n^2 = 1$

13. $f(x, y, z) = x + 2y; x + y + z = 1, y^2 + z^2 = 4$

14. $f(x, y, z) = 3x - y - 3z; x + y - z = 0,$ $x^2 + 2z^2 = 1$

15. $f(x, y, z) = yz + xy; xy = 1, y^2 + z^2 = 1$

16. Find the maximum and minimum volumes of a rectangular box whose surface area is 1500 cm^2 and whose total edge length is 200 cm.

17. A manufacturer produces a quantity Q of a certain product and Q depends on the amount x of labor used and the amount y of capital. A simple model that is sometimes used to express Q explicitly as a function of x and y is the Cobb-Douglas production function: $Q = Kx^\alpha y^{1-\alpha}$ where K and α are constants with $K > 0$ and $0 < \alpha < 1$. If the cost of a unit of labor is m and the cost of a unit of capital is n, and the company can spend only p dollars as its total budget, then it is required to maximize the production Q subject to the constraint that $mx + ny = p$. Show that the maximum production occurs when

$$x = \frac{\alpha p}{m} \quad \text{and} \quad y = \frac{(1 - \alpha)p}{n}$$

18. Referring to Exercise 17, we now suppose that the production is fixed at $Kx^\alpha y^{1-\alpha} = Q$, where Q is a constant. What values of x and y will minimize the cost function $C(x, y) = mx + ny$?

19. Use Lagrange multipliers to prove that the rectangle with maximum area that has a given perimeter p is a square.

20. Use Lagrange multipliers to prove that the triangle with maximum area that has a given perimeter p is equilateral. [*Hint:* Use Heron's formula for the area: $A = \sqrt{s(s - x)(s - y)(s - z)}$ where $s = p/2$ and x, y, z are the lengths of the sides.]

21–36. Use Lagrange multipliers to give alternate solutions to Exercises 25–40 of Section 12.7.

REVIEW OF CHAPTER 12

Define, state, or discuss the following:

1. Functions of two variables and their graphs
2. Level curves
3. Functions of three variables
4. Level surfaces
5. Limit of a function of two or three variables
6. Continuity of such a function
7. Partial derivatives
8. Clairaut's Theorem
9. Equation of a tangent plane to the surface $z = f(x, y)$
10. Equation of a tangent plane to the surface $F(x, y, z) = 0$
11. Differential
12. Differentiable function
13. The Chain Rule
14. Implicit differentiation
15. Directional derivative
16. Gradient and its significance
17. Local and absolute maximum and minimum values
18. Critical point
19. Saddle point
20. Second Derivatives Test
21. Extreme Value Theorem for functions of two variables
22. Lagrange multipliers

REVIEW EXERCISES FOR CHAPTER 12

In Exercises 1–4 find and sketch the domain of the given function.

1. $f(x, y) = \dfrac{\ln(x + y + 1)}{x - 1}$

2. $f(x, y) = \sqrt{\sin \pi(x^2 + y^2)}$

3. $f(x, y) = \cos^{-1} x + \tan^{-1} y$

4. $f(x, y, z) = \sqrt{z - x^2 - y^2}$

In Exercises 5 and 6 sketch the graph of the given function.

5. $f(x, y) = 1 - x^2 - y^2$

6. $f(x, y) = \sqrt{x^2 + y^2 - 1}$

In Exercises 7 and 8 sketch several level curves of the given function.

7. $f(x, y) = e^{-(x^2 + y^2)}$ **8.** $f(x, y) = x^2 + 4y$

In Exercises 9 and 10 evaluate the limit or show that it does not exist.

9. $\lim\limits_{(x,y) \to (0,0)} \dfrac{x^2 y^2}{x^2 + 2y^2}$ **10.** $\lim\limits_{(x,y) \to (0,0)} \dfrac{2xy}{x^2 + 2y^2}$

In Exercises 11–16 find the first partial derivatives.

11. $f(x, y) = 3x^4 - x\sqrt{y}$ **12.** $g(x, y) = \dfrac{x}{\sqrt{x + 2y}}$

13. $f(s, t) = e^{2s} \cos \pi t$ **14.** $g(r, s) = r \sin \sqrt{r^2 + s^2}$

15. $f(x, y, z) = xy^z$ **16.** $g(u, v, w) = w^2 e^{u/v}$

In Exercises 17–20 find all second partial derivatives of f.

17. $f(x, y) = x^2 y^3 - 2x^4 + y^2$

18. $f(x, y) = x^3 \ln(x - y)$

19. $f(x, y, z) = xy^2 z^3$ **20.** $f(x, y, z) = xe^y \cos z$

21. If $u = x^y$, show that

$$\frac{x}{y} \frac{\partial u}{\partial x} + \frac{1}{\ln x} \frac{\partial u}{\partial y} = 2u$$

22. If $\rho = \sqrt{x^2 + y^2 + z^2}$, show that

$$\frac{\partial^2 \rho}{\partial x^2} + \frac{\partial^2 \rho}{\partial y^2} + \frac{\partial^2 \rho}{\partial z^2} = \frac{2}{\rho}$$

In Exercises 23–28 find an equation of the tangent plane to the given surface at the given point.

23. $z = x^2 + y^2 + 4y$, $(0, 1, 5)$

24. $z = x^3 + 2xy$, $(1, 2, 5)$

25. $z = xe^y$, $(1, 0, 1)$

26. $x^2 + y^2 + z^2 = 29$, $(2, 3, 4)$

27. $x^2 + 2y^2 - 3z^2 = 14$, $(3, 2, -1)$

28. $xy^2 z^3 = 12$, $(3, 2, 1)$

29. Find the points on the sphere $x^2 + y^2 + z^2 = 1$ where the tangent plane is parallel to the plane $2x + y - 3z = 2$.

30. Find dz if $z = x^2 \tan^{-1} y$.

31. Use differentials to approximate the number $(1.98)^3 \sqrt{(3.01)^2 + (3.97)^2}$.

32. The two legs of a right triangle are measured as 5 m and 12 m, respectively, with a possible error in measurement of at most 0.2 cm in each. Use differentials to estimate the maximum error in the calculated value of (a) the area of the triangle and (b) the length of the hypotenuse.

33. If $w = \sqrt{x} + y^2/z$, where $x = e^{2t}$, $y = t^3 + 4t$, $z = t^2 - 4$, use the Chain Rule to find dw/dt.

34. If $z = \cos xy + y \cos x$, where $x = u^2 + v$, $y = u - v^2$, use the Chain Rule to find $\partial z/\partial u$ and $\partial z/\partial v$.

35. If $z = y + f(x^2 - y^2)$, where f is differentiable, show that

$$y \frac{\partial z}{\partial x} + x \frac{\partial z}{\partial y} = x$$

36. The length x of a side of a triangle is increasing at a rate of 3 in./s, the length y of another side is decreasing at a rate of 2 in./s, and the contained angle θ is increasing at a rate of 0.05 radian/s. How fast is the area of the triangle changing when $x = 40$ in., $y = 50$ in., and $\theta = \pi/6$?

37. If $z = f(u, v)$, where $u = xy$, $v = y/x$, and f has continuous second partial derivatives, show that

$$x^2 \frac{\partial^2 z}{\partial x^2} - y^2 \frac{\partial^2 z}{\partial y^2} = -4uv \frac{\partial^2 z}{\partial u \, \partial v} + 2v \frac{\partial z}{\partial v}$$

38. If $yz^4 + x^2 z^3 = e^{xyz}$, find $\dfrac{\partial z}{\partial x}$ and $\dfrac{\partial z}{\partial y}$.

39. Find the gradient of the function $f(x, y, z) = z^2 e^{x\sqrt{y}}$.

In Exercises 40–42 find the directional derivative of f at the given point in the indicated direction.

40. $f(x, y) = x^2 y^3 + 4xy^5$, $(1, -1)$, in the direction of $\mathbf{v} = \langle 3, -4 \rangle$

41. $f(x, y) = 2\sqrt{x} - y^2$, $(1, 5)$, in the direction toward the point $(4, 1)$

42. $f(x, y, z) = x^2 y + x\sqrt{1 + z}$, $(1, 2, 3)$, in the direction of $\mathbf{v} = 2\mathbf{i} + \mathbf{j} - 2\mathbf{k}$

43. Find the maximum rate of change of $f(x, y) = x^2 y + \sqrt{y}$ at the point $(2, 1)$. In which direction does it occur?

44. Find the direction in which $f(x, y, z) = ze^{xy}$ increases most rapidly at the point $(0, 1, 2)$. What is the maximum rate of increase?

In Exercises 45–48 find the local maximum and minimum values and saddle points of the given function.

45. $f(x, y) = x^2 - xy + y^2 + 9x - 6y + 10$

46. $f(x, y) = x^3 - 6xy + 8y^3$

47. $f(x, y) = 3xy - x^2y - xy^2$

48. $f(x, y) = (x^2 + y)e^{y/2}$

In Exercises 49 and 50 find the absolute maximum and minimum values of f on the set D.

49. $f(x, y) = 4xy^2 - x^2y^2 - xy^3$; D is the closed triangular region in the xy-plane with vertices $(0, 0)$, $(0, 6)$, and $(6, 0)$

50. $f(x, y) = e^{-x^2 - y^2}(x^2 + 2y^2)$; D is the disk $x^2 + y^2 \leqslant 4$

In Exercises 51–54 use Lagrange multipliers to find the maximum and minimum values of f subject to the given constraints.

51. $f(x, y) = x^2y$; $x^2 + y^2 = 1$

52. $f(x, y) = \dfrac{1}{x} + \dfrac{1}{y}$; $\dfrac{1}{x^2} + \dfrac{1}{y^2} = 1$

53. $f(x, y) = x + y + z$; $\dfrac{1}{x} + \dfrac{1}{y} + \dfrac{1}{z} = 1$

54. $f(x, y, z) = x^2 + 2y^2 + 3z^2$; $x + y + z = 1$, $x - y + 2z = 2$

55. Find the points on the surface $xy^2z^3 = 2$ that are closest to the origin.

56. A package in the shape of a rectangular box can be mailed parcel post if the sum of its length and girth (the perimeter of a cross-section perpendicular to the length) is at most 84 in. Find the dimensions of the package with largest volume that can be mailed parcel post.

13 Multiple Integrals

Mathematics is the tool specially
suited to dealing with abstract
concepts of any kind and there is no
limit to its power in this field. For
this reason a book on the new
physics, if not purely descriptive
of experimental work, must be
essentially mathematical.

P. A. M. Dirac

In this chapter we extend the idea of a definite integral to double and triple integrals of functions of two or three variables. These ideas are then used to compute volumes, masses, and centroids of more general regions than we were able to consider in Chapter 9.

Double Integrals over Rectangles

As a guide to defining the double integral of a function of two variables, let us first recall the basic facts concerning definite integrals of functions of a single variable. If f is defined on a closed interval $[a, b]$, we start by taking a partition P of $[a, b]$ into subintervals $[x_{i-1}, x_i]$ where

$$a = x_0 < x_1 < \cdots < x_{n-1} < x_n = b$$

We choose points x_i^* in $[x_{i-1}, x_i]$ and let $\Delta x_i = x_i - x_{i-1}$ and $\|P\| = \max\{\Delta x_i\}$. Then we form the Riemann sum

(13.1)
$$\sum_{i=1}^{n} f(x_i^*) \, \Delta x_i$$

and take the limit of such sums as $\|P\| \to 0$ to obtain the definite integral of f from a to b:

(13.2)
$$\int_a^b f(x) \, dx = \lim_{\|P\| \to 0} \sum_{i=1}^{n} f(x_i^*) \, \Delta x_i$$

In the special case where $f(x) \geqslant 0$, the Riemann sum can be interpreted as the sum of the areas of the approximating rectangles in Figure 13.1 and $\int_a^b f(x) \, dx$ represents the area under the curve $y = f(x)$ from a to b.

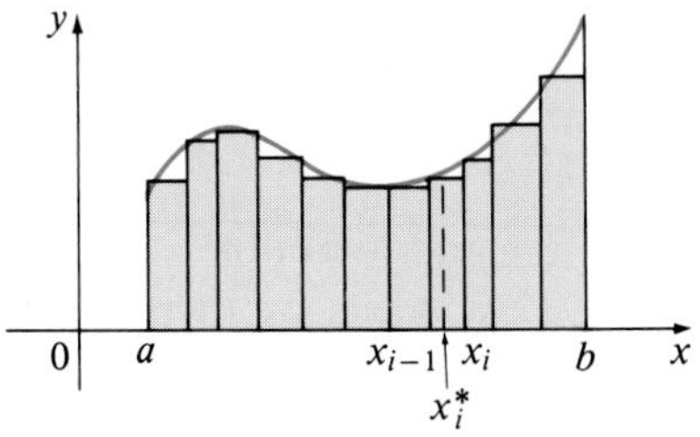

In a similar manner we define the double integral of a function f of two variables that is defined on a closed rectangle

$$R = [a, b] \times [c, d] = \{(x, y) \in R^2 \mid a \leqslant x \leqslant b, c \leqslant y \leqslant d\}$$

Figure 13.1

The first step is to take a partition P of R into subrectangles. This is accomplished by partitioning the intervals $[a, b]$ and $[c, d]$ as follows:

$$a = x_0 < x_1 < \cdots < x_{i-1} < x_i < \cdots < x_m = b$$
$$c = y_0 < y_1 < \cdots < y_{j-1} < y_j < \cdots < y_n = d$$

By drawing lines parallel to the coordinate axes through these partition points as in Figure 13.2, we form the subrectangles

$$R_{ij} = [x_{i-1}, x_i] \times [y_{j-1}, y_j] = \{(x, y) \mid x_{i-1} \leqslant x \leqslant x_i, \, y_{j-1} \leqslant y \leqslant y_j\}$$

for $i = 1, \ldots, m; j = 1, \ldots, n$. There are mn of these subrectangles which cover R. If we let

$$\Delta x_i = x_i - x_{i-1} \qquad \Delta y_j = y_j - y_{j-1}$$

then the area of R_{ij} is

$$\Delta A_{ij} = \Delta x_i \, \Delta y_j$$

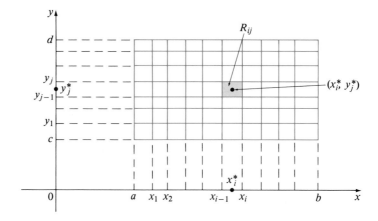

Figure 13.2

PARTITION OF A RECTANGLE

Next we choose a point (x_i^*, y_j^*) in R_{ij} and, by analogy with the Riemann sum (13.1), we form the **double Riemann sum**

(13.3)
$$\sum_{i=1}^{m} \sum_{j=1}^{n} f(x_i^*, y_j^*) \Delta A_{ij}$$

This double sum means that for each subrectangle we evaluate f at the chosen point and multiply by the area of the subrectangle, and then we add the results. When written out in full, the double Riemann sum (13.3) becomes

$$f(x_1^*, y_1^*) \Delta A_{11} + f(x_1^*, y_2^*) \Delta A_{12} + \cdots + f(x_1^*, y_n^*) \Delta A_{1n}$$
$$+ f(x_2^*, y_1^*) \Delta A_{21} + f(x_2^*, y_2^*) \Delta A_{22} + \cdots + f(x_2^*, y_n^*) \Delta A_{2n}$$
$$+ \cdots$$
$$+ f(x_m^*, y_1^*) \Delta A_{m1} + f(x_m^*, y_2^*) \Delta A_{m2} + \cdots + f(x_m^*, y_n^*) \Delta A_{mn}$$

The final ingredient that we need for the definition of a double integral is the **norm** of the partition, which is the length of the longest diagonal of all the subrectangles R_{ij} and is denoted by $\|P\|$. Note that if we let $\|P\| \to 0$, then the partition becomes finer. By analogy with the defining equation for a single integral (13.2), we make the following definition.

Definition (13.4)

The **double integral** of f over the rectangle R is

$$\iint\limits_{R} f(x, y)\, dA = \lim_{\|P\| \to 0} \sum_{i=1}^{m} \sum_{j=1}^{n} f(x_i^*, y_j^*) \Delta A_{ij}$$

if this limit exists.

Note 1:   The precise meaning of the limit that defines the double integral is as follows:

$$\iint\limits_{R} f(x, y)\, dA = I$$

means that for every $\varepsilon > 0$ there is a corresponding number $\delta > 0$ such that

$$\left| I - \sum_{i=1}^{m} \sum_{j=1}^{n} f(x_i^*, y_j^*) \right| < \varepsilon$$

for all partitions P of R with $\|P\| < \delta$ and for all possible choices of (x_i^*, y_j^*) in R_{ij}.

In other words, all Riemann sums can be made arbitrarily close to I by taking sufficiently fine partitions.

Note 2:   In view of the fact that $\Delta A_{ij} = \Delta x_i \Delta y_j$, another notation that is sometimes used for the double integral is

$$\iint_R f(x, y)\, dA = \iint_R f(x, y)\, dx\, dy$$

Note 3:   A function f is called **integrable** if the limit in Definition 13.4 exists. It is shown in courses on advanced calculus that all continuous functions are integrable. In fact, the double integral of f will exist provided that f is "not too discontinuous."

● **Example 1**   Find an approximate value for the integral $\iint_R (x - 3y^2)\, dA$, where $R = \{(x, y) \mid 0 \leq x \leq 2, 1 \leq y \leq 2\}$, by computing the double Riemann sum with partition lines $x = 1$ and $y = \frac{3}{2}$ and taking (x_i^*, y_j^*) to be the center of each rectangle.

Solution   The partition is shown in Figure 13.3. The area of each subrectangle is $\Delta A_{ij} = \frac{1}{2}$, (x_i^*, y_j^*) is the center of R_{ij}, and $f(x, y) = x - 3y^2$. So the corresponding Riemann sum is

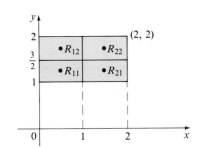

Figure 13.3

$$\sum_{i=1}^{2} \sum_{j=1}^{2} f(x_i^*, y_j^*)\, \Delta A_{ij}$$

$$= f(x_1^*, y_1^*)\, \Delta A_{11} + f(x_1^*, y_2^*)\, \Delta A_{12} + f(x_2^*, y_1^*)\, \Delta A_{21} + f(x_2^*, y_2^*)\, \Delta A_{22}$$

$$= f\left(\tfrac{1}{2}, \tfrac{5}{4}\right) \Delta A_{11} + f\left(\tfrac{1}{2}, \tfrac{7}{4}\right) \Delta A_{12} + f\left(\tfrac{3}{2}, \tfrac{5}{4}\right) \Delta A_{21} + f\left(\tfrac{3}{2}, \tfrac{7}{4}\right) \Delta A_{22}$$

$$= \left(-\tfrac{67}{16}\right)\tfrac{1}{2} + \left(-\tfrac{139}{16}\right)\tfrac{1}{2} + \left(-\tfrac{51}{16}\right)\tfrac{1}{2} + \left(-\tfrac{123}{16}\right)\tfrac{1}{2}$$

$$= -\tfrac{95}{8} = -11.875$$

Thus we have

$$\iint_R (x - 3y^2)\, dA \approx -11.875$$

●

Note: It is very difficult to calculate double integrals directly from the definition. In the next section we shall develop an efficient method for computing double integrals and then we shall see that the exact value of the double integral in Example 1 is -12.

Interpretation of Double Integrals as Volumes

Just as single integrals of positive functions can be interpreted as areas, double integrals of positive functions can be interpreted as volumes in the following way. Suppose that $f(x, y) \geq 0$ and f is defined on the rectangle $R = [a, b] \times [c, d]$. The graph of f is a surface with equation $z = f(x, y)$.

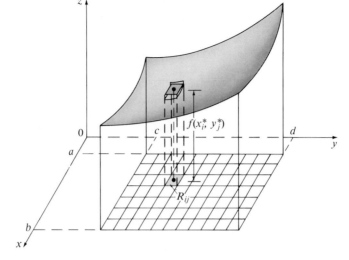

Figure 13.4

Let S be the solid that lies above R and under the graph of f, that is,

$$S = \{(x, y, z) \in R^3 \,|\, 0 \leqslant z \leqslant f(x, y), (x, y) \in R\}$$

If we partition R into subrectangles R_{ij} and choose (x_i^*, y_j^*) in R_{ij}, then we can approximate the part of S that lies above R_{ij} by a thin rectangular box (or "column") with base R_{ij} and height $f(x_i^*, y_j^*)$ as shown in Figure 13.4. (Compare with Figure 13.1.) The volume of this box is

$$V_{ij} = f(x_i^*, y_j^*) \Delta A_{ij}$$

If we follow this procedure for all the rectangles and add the volumes of the corresponding boxes, we get an approximation to the total volume of S:

(13.5)
$$V \approx \sum_{i=1}^{m} \sum_{j=1}^{n} f(x_i^*, y_j^*) \Delta A_{ij}$$

Notice that this double sum is just the Riemann sum (13.3). Our intuition tells us that the approximation given in (13.5) becomes better if we use a finer partition P and so we would expect that

$$V = \lim_{||P|| \to 0} \sum_{i=1}^{m} \sum_{j=1}^{n} f(x_i^*, y_j^*) \Delta A_{ij} = \iint_R f(x, y) \, dA$$

(the latter equality being true by the definition of a double integral). It can be shown, using our previous definition of volume (8.1), that this expectation is correct.

Theorem (13.6)

> If $f(x, y) \geqslant 0$ and f is integrable over the rectangle R, then the volume V of the solid that lies above R and under the surface $z = f(x, y)$ is
>
> $$V = \iint_R f(x, y) \, dA$$

It should be remembered that the interpretation of a double integral as a volume is valid only when the integrand f is a *positive* function. The integrand in Example 1 is not a positive function, so its integral is not a volume.

● **Example 2**   Estimate the volume of the solid that lies above the square $R = [0, 2] \times [0, 2]$ and below the elliptic paraboloid $z = 16 - x^2 - 2y^2$. Use the partition of R into four equal squares and choose (x_i^*, y_j^*) to be the upper right corner of R_{ij}. Sketch the solid and the approximating rectangular boxes.

Solution   The partition is shown in Figure 13.5(a). The paraboloid is the graph of $f(x, y) = 16 - x^2 - 2y^2$ and the area of each square is 1. Approximating the volume by the Riemann sum, we have

$$V \approx f(1, 1)\Delta A_{11} + f(1, 2)\Delta A_{12} + f(2, 1)\Delta A_{21} + f(2, 2)\Delta A_{22}$$

$$= 13(1) + 7(1) + 10(1) + 4(1) = 34$$

Figure 13.5

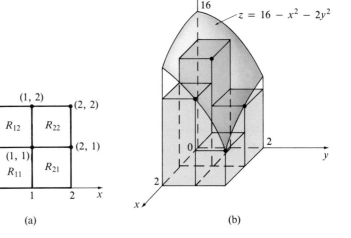

(a)   (b)

This is the volume of the approximating rectangular boxes shown in Figure 13.5(b).   ●

We list here some properties of double integrals that can be proved in the same manner as in Section 6.4. We assume that all of the integrals exist.

(13.7)
$$\iint\limits_{R} [f(x, y) + g(x, y)]\, dA = \iint\limits_{R} f(x, y)\, dA + \iint\limits_{R} g(x, y)\, dA$$

(13.8)
$$\iint\limits_{R} cf(x, y)\, dA = c \iint\limits_{R} f(x, y)\, dA \qquad \text{where } c \text{ is a constant}$$

(13.9)   If $f(x, y) \geqslant g(x, y)$ for all (x, y) in R, then

$$\iint\limits_{R} f(x, y)\, dA \geqslant \iint\limits_{R} g(x, y)\, dA$$

EXERCISES 13.1

1. Find approximations to $\iint_R (x - 3y^2)\,dA$ using the same partition as in Example 1 but choosing (x_i^*, y_j^*) to be the (a) upper left corner, (b) upper right corner, (c) lower left corner, (d) lower right corner of R_{ij}.

2. Find the approximation to the volume in Example 2 if (x_i^*, y_j^*) is chosen to be the center of each square.

In Exercises 3–8 calculate the double Riemann sum of f for the partition of R given by the indicated lines and the given choice of (x_i^*, y_j^*). Also calculate the norm of the partition.

3. $f(x, y) = x^2 + 4y$, $R = \{(x, y) | 0 \leqslant x \leqslant 2, 0 \leqslant y \leqslant 3\}$, $x = 1$, $y = 1$, $y = 2$; $(x_i^*, y_j^*) =$ upper right corner of R_{ij}

4. $f(x, y) = x^2 + 4y$, $R = \{(x, y) | 0 \leqslant x \leqslant 2, 0 \leqslant y \leqslant 3\}$, $x = 1$, $y = 1$, $y = 2$; $(x_i^*, y_j^*) =$ center of R_{ij}

5. $f(x, y) = xy - y^2$, $R = \{(x, y) | 0 \leqslant x \leqslant 5, 0 \leqslant y \leqslant 4\}$, $x = 1$, $x = 2$, $x = 3$, $x = 4$, $y = 2$; $(x_i^*, y_j^*) =$ center of R_{ij}

6. $f(x, y) = 2x + x^2y$, $R = \{(x, y) | -2 \leqslant x \leqslant 2, -1 \leqslant y \leqslant 1\}$, $x = -1$, $x = 0$, $x = 1$, $y = -\frac{1}{2}$, $y = 0$, $y = \frac{1}{2}$; $(x_i^*, y_j^*) =$ lower left corner of R_{ij}

7. $f(x, y) = x^2 - y^2$, $R = [0, 5] \times [0, 2]$, $x = 1$, $x = 3$, $x = 4$, $y = \frac{1}{2}$, $y = 1$; $(x_i^*, y_j^*) =$ upper left corner of R_{ij}

8. $f(x, y) = 5xy^2$, $R = [1, 3] \times [1, 4]$, $x = 1.8$, $x = 2.5$, $y = 2$, $y = 3$; $(x_i^*, y_j^*) =$ lower right corner of R_{ij}

9. If f is a constant function, $f(x, y) = k$, and $R = [a, b] \times [c, d]$, show that

$$\iint_R k\,dA = k(b - a)(d - c)$$

10. If $R = [0, 1] \times [0, 1]$, show that

$$0 \leqslant \iint_R \sin(x + y)\,dA \leqslant 1$$

SECTION 13.2

Iterated Integrals

Recall that it is usually difficult to evaluate single integrals from first principles but the Fundamental Theorem of Calculus provides a much easier method. The evaluation of double integrals from first principles is even more difficult, but in this section we shall see how to express a double integral as an iterated integral which can then be evaluated by calculating two single integrals.

Suppose f is a function of two variables that is integrable over the rectangle $R = [a, b] \times [c, d]$. We use the notation $\int_c^d f(x, y)\,dy$ to mean that x is held fixed and $f(x, y)$ is integrated with respect to y from $y = c$ to $y = d$. This procedure is called *partial integration with respect to y*. (Notice its similarity to partial differentiation.) Now $\int_c^d f(x, y)\,dy$ is a number that depends on the value of x, so it defines a function of x:

$$A(x) = \int_c^d f(x, y)\,dy$$

If we now integrate the function A with respect to x from $x = a$ to $x = b$, we get

(13.10)
$$\int_a^b A(x)\,dx = \int_a^b \left[\int_c^d f(x, y)\,dy \right] dx$$

The integral on the right side of Equation 13.10 is called an **iterated integral.**

Usually the brackets are omitted. Thus

(13.11)
$$\int_a^b \int_c^d f(x, y)\, dy\, dx = \int_a^b \left[\int_c^d f(x, y)\, dy \right] dx$$

means that we first integrate with respect to y from c to d and then with respect to x from a to b.

Similarly, the iterated integral

(13.12)
$$\int_c^d \int_a^b f(x, y)\, dx\, dy = \int_c^d \left[\int_a^b f(x, y)\, dx \right] dy$$

means that we first integrate with respect to x (holding y fixed) from $x = a$ to $x = b$ and then we integrate the resulting function of y with respect to y from $y = c$ to $y = d$. Notice that in both Equations 13.11 and 13.12 we work from the inside out.

● **Example 1**   Evaluate the iterated integrals: (a) $\int_0^3 \int_1^2 x^2 y\, dy\, dx$ and (b) $\int_1^2 \int_0^3 x^2 y\, dx\, dy$.

Solution   (a)   Regarding x as a constant, we obtain

$$\int_1^2 x^2 y\, dy = \left[x^2 \frac{y^2}{2} \right]_{y=1}^{y=2}$$

$$= x^2 \left(\frac{2^2}{2} \right) - x^2 \left(\frac{1^2}{2} \right) = \frac{3}{2} x^2$$

Thus the function A in the preceding discussion is given by $A(x) = 3x^2/2$ in this example. We now integrate this function of x from 0 to 3:

$$\int_0^3 \int_1^2 x^2 y\, dy\, dx = \int_0^3 \left[\int_1^2 x^2 y\, dy \right] dx$$

$$= \int_0^3 \frac{3}{2} x^2\, dx = \frac{x^3}{2} \Big]_0^3 = \frac{27}{2}$$

(b)   Here we first integrate with respect to x:

$$\int_1^2 \int_0^3 x^2 y\, dx\, dy = \int_1^2 \left[\int_0^3 x^2 y\, dx \right] dy$$

$$= \int_1^2 \left[\frac{x^3}{3} y \right]_{x=0}^{x=3} dy$$

$$= \int_1^2 9y\, dy = 9 \frac{y^2}{2} \Big]_1^2 = \frac{27}{2} \qquad ●$$

Notice that in Example 1 we obtained the same answer whether we integrated with respect to y or x first. In general it turns out (see Theorem 13.13) that the two iterated integrals in Equations 13.11 and 13.12 are always equal;

that is, the order of integration does not matter. (This is similar to Theorem 12.14 on the equality of the mixed partial derivatives.)

The following theorem gives a practical method for evaluating a double integral by expressing it as an iterated integral (in either order). It is named after the Italian mathematician Guido Fubini (1879–1943).

Fubini's Theorem (13.13)

If f is integrable over the rectangle $R = \{(x, y) \mid a \leqslant x \leqslant b, c \leqslant y \leqslant d\}$, then

$$\iint\limits_{R} f(x, y)\, dA = \int_a^b \int_c^d f(x, y)\, dy\, dx = \int_c^d \int_a^b f(x, y)\, dx\, dy$$

The proof of Fubini's Theorem is too difficult to include in this book, but we can at least give an intuitive indication of why it is true for the case where $f(x, y) \geqslant 0$. Recall from Theorem 13.6 that if f is positive, then we can interpret the double integral $\iint_R f(x, y)\, dA$ as the volume V of the solid S that lies above R and under the surface $z = f(x, y)$. But we have another formula that we used for volume in Chapter 8, namely,

$$V = \int_a^b A(x)\, dx$$

where $A(x)$ is the area of a cross-section of S in the plane through x perpendicular to the x-axis. From Figure 13.6(a) you can see that $A(x)$ is the area under the curve C whose equation is $z = f(x, y)$ where x is held constant and $c \leqslant y \leqslant d$. Therefore

$$A(x) = \int_c^d f(x, y)\, dy$$

and we have

$$\iint\limits_{R} f(x, y)\, dA = V = \int_a^b A(x)\, dx = \int_a^b \int_c^d f(x, y)\, dy\, dx$$

Figure 13.6

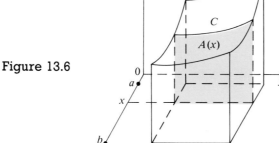

(a)

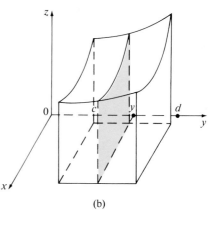

(b)

A similar argument, using cross-sections perpendicular to the y-axis as in Figure 13.6(b), shows that

$$\iint\limits_R f(x, y)\, dA = \int_c^d \int_a^b f(x, y)\, dx\, dy$$

● **Example 2**   Evaluate $\iint_R (x - 3y^2)\, dA$ where $R = \{(x, y) \mid 0 \leqslant x \leqslant 2,\ 1 \leqslant y \leqslant 2\}$. (Compare with Example 1 in Section 13.1.)

Solution 1   Fubini's Theorem gives

$$\iint\limits_R (x - 3y^2)\, dA = \int_0^2 \int_1^2 (x - 3y^2)\, dy\, dx$$

$$= \int_0^2 \Big[xy - y^3 \Big]_{y=1}^{y=2} dx$$

$$= \int_0^2 (x - 7)\, dx = \frac{x^2}{2} - 7x \Big]_0^2 = -12$$

Solution 2   Again applying Fubini's Theorem, but this time integrating with respect to x first, we have

$$\iint\limits_R (x - 3y^2)\, dA = \int_1^2 \int_0^2 (x - 3y^2)\, dx\, dy$$

$$= \int_1^2 \left[\frac{x^2}{2} - 3xy^2 \right]_{x=0}^{x=2} dy$$

$$= \int_1^2 (2 - 6y^2)\, dy = 2y - 2y^3 \Big]_1^2 = -12 \qquad ●$$

● **Example 3**   Evaluate $\iint_R y \sin(xy)\, dA$ where $R = [1, 2] \times [0, \pi]$.

Solution 1   If we first integrate with respect to x, we get

$$\iint\limits_R y \sin(xy)\, dA = \int_0^\pi \int_1^2 y \sin(xy)\, dx\, dy$$

$$= \int_0^\pi \Big[-\cos(xy) \Big]_{x=1}^{x=2} dy$$

$$= \int_0^\pi (-\cos 2y + \cos y)\, dy$$

$$= -\tfrac{1}{2} \sin 2y + \sin y \Big]_0^\pi = 0$$

Solution 2   If we reverse the order of integration, we get

$$\iint\limits_R y \sin(xy)\, dA = \int_1^2 \int_0^\pi y \sin(xy)\, dy\, dx$$

To evaluate the inner integral we use integration by parts with

$$u = y \qquad dv = \sin(xy)\,dy$$

$$du = dy \qquad v = -\frac{\cos(xy)}{x}$$

and so
$$\int_0^\pi y \sin(xy)\,dy = -\frac{y\cos(xy)}{x}\Bigg]_{y=0}^{y=\pi} + \frac{1}{x}\int_0^\pi \cos(xy)\,dy$$

$$= -\frac{\pi\cos\pi x}{x} + \frac{1}{x^2}\Bigg[\sin(xy)\Bigg]_{y=0}^{y=\pi}$$

$$= -\frac{\pi\cos\pi x}{x} + \frac{\sin\pi x}{x^2}$$

If we now integrate the first term by parts with $u = -1/x$ and $dv = \pi\cos\pi x\,dx$, we get $du = dx/x^2$, $v = \sin\pi x$, and

$$\int\left(-\frac{\pi\cos\pi x}{x}\right)dx = -\frac{\sin\pi x}{x} - \int\frac{\sin\pi x}{x^2}\,dx$$

Therefore
$$\int\left(-\frac{\pi\cos\pi x}{x} + \frac{\sin\pi x}{x^2}\right)dx = -\frac{\sin\pi x}{x}$$

and so
$$\int_1^2\int_0^\pi y\sin(xy)\,dy\,dx = \left[-\frac{\sin\pi x}{x}\right]_1^2$$

$$= -\frac{\sin 2\pi}{2} + \sin\pi = 0 \qquad\bullet$$

Note: In Example 2, Solutions (1) and (2) are equally straightforward, but in Example 3 the first solution is much easier than the second one. Therefore in evaluating double integrals it is wise to choose the order of integration that gives easier integrals.

● **Example 4**   Find the volume of the solid S that is bounded by the elliptic paraboloid $x^2 + 2y^2 + z = 16$, the planes $x = 2$ and $y = 2$, and the three coordinate planes.

Solution   We first observe that S is the solid that lies under the surface $z = 16 - x^2 - 2y^2$ and above the square $R = [0,2] \times [0,2]$. This solid was considered in Exercise 2 in Section 13.1 (see the sketch in Figure 13.5) but we are now in a position to evaluate the double integral using Fubini's Theorem. Therefore

$$V = \iint_R (16 - x^2 - 2y^2)\,dA = \int_0^2\int_0^2 (16 - x^2 - 2y^2)\,dx\,dy$$

$$= \int_0^2\left[16x - \frac{x^3}{3} - 2y^2x\right]_{x=0}^{x=2}\,dy$$

$$= \int_0^2\left(\frac{88}{3} - 4y^2\right)dy = \left[\frac{88}{3}y - \frac{4}{3}y^3\right]_0^2 = 48 \qquad\bullet$$

In the special case where $f(x, y)$ can be factored as the product of a function of x only and a function of y only, the double integral of f can be written in a particularly simple form. To be specific, suppose that $f(x, y) = g(x)h(y)$ and $R = [a, b] \times [c, d]$. Then Fubini's Theorem gives

$$\iint\limits_{R} f(x, y)\, dA = \int_{c}^{d} \int_{a}^{b} g(x)h(y)\, dx\, dy = \int_{c}^{d} \left[\int_{a}^{b} g(x)h(y)\, dx \right] dy$$

In the inner integral y is a constant, so $h(y)$ is a constant and we can write

$$\int_{c}^{d} \left[\int_{a}^{b} g(x)h(y)\, dx \right] dy = \int_{c}^{d} \left[h(y) \left(\int_{a}^{b} g(x)\, dx \right) \right] dy$$
$$= \int_{a}^{b} g(x)\, dx \int_{c}^{d} h(y)\, dy$$

since $\int_{a}^{b} g(x)\, dx$ is a constant. Therefore, in this case, the double integral of f can be written as the product of two single integrals:

$$\boxed{\iint\limits_{R} g(x)h(y)\, dA = \int_{a}^{b} g(x)\, dx \int_{c}^{d} h(y)\, dy \qquad \text{where } R = [a, b] \times [c, d]}$$

● **Example 5**   If $R = \left[0, \dfrac{\pi}{2} \right] \times \left[0, \dfrac{\pi}{2} \right]$, then

$$\iint\limits_{R} \sin x \cos y\, dA = \int_{0}^{\pi/2} \sin x\, dx \int_{0}^{\pi/2} \cos y\, dy$$
$$= \left[-\cos x \right]_{0}^{\pi/2} \left[\sin y \right]_{0}^{\pi/2}$$
$$= 1 \cdot 1 = 1$$

●

EXERCISES 13.2

In Exercises 1–4 find $\int_{0}^{2} f(x, y)\, dy$ and $\int_{0}^{1} f(x, y)\, dx$.

1. $f(x, y) = x^2 y^3$

2. $f(x, y) = 2xy - 3x^2$

3. $f(x, y) = xe^{x+y}$

4. $f(x, y) = \dfrac{x}{y^2 + 1}$

In Exercises 5–12 calculate the iterated integral.

5. $\int_{0}^{4} \int_{0}^{2} x\sqrt{y}\, dx\, dy$

6. $\int_{0}^{2} \int_{0}^{3} e^{x-y}\, dy\, dx$

7. $\int_{-1}^{1} \int_{0}^{1} (x^3 y^3 + 3xy^2)\, dy\, dx$

8. $\int_0^1 \int_1^2 (x^4 - y^2)\,dx\,dy$

9. $\int_0^3 \int_0^1 \sqrt{x + y}\,dx\,dy$

10. $\int_0^{\pi/2} \int_0^{\pi/2} \sin(x + y)\,dy\,dx$

11. $\int_0^{\pi/4} \int_0^3 \sin x\,dy\,dx$

12. $\int_1^2 \int_0^1 (x + y)^{-2}\,dx\,dy$

In Exercises 13–20 calculate the double integral.

13. $\iint_R (2y^2 - 3xy^3)\,dA,\ R = \{(x, y)\,|\,1 \leqslant x \leqslant 2, 0 \leqslant y \leqslant 3\}$

14. $\iint_R \left(xy^2 + \dfrac{y}{x}\right)dA,\ R = \{(x, y)\,|\,2 \leqslant x \leqslant 3, -1 \leqslant y \leqslant 0\}$

15. $\iint_R x \sin y\,dA,\ R = \left\{(x, y)\,|\,1 \leqslant x \leqslant 4, 0 \leqslant y \leqslant \dfrac{\pi}{6}\right\}$

16. $\iint_R \dfrac{1 + x}{1 + y}\,dA,\ R = \{(x, y)\,|\,-1 \leqslant x \leqslant 2, 0 \leqslant y \leqslant 1\}$

17. $\iint_R x \sin(x + y)\,dA,\ R = \left[0, \dfrac{\pi}{6}\right] \times \left[0, \dfrac{\pi}{3}\right]$

18. $\iint_R xe^{xy}\,dA,\ R = [0, 1] \times [0, 1]$

19. $\iint_R \dfrac{1}{x + y}\,dA,\ R = [1, 2] \times [0, 1]$

20. $\iint_R xye^{xy^2}\,dA,\ R = [0, 1] \times [0, 1]$

21. Find the volume of the solid lying under the circular paraboloid $z = x^2 + y^2$ and above the rectangle $R = [-2, 2] \times [-3, 3]$.

22. Find the volume of the solid lying under the elliptic paraboloid $x^2/4 + y^2/9 + z = 1$ and above the square $R = [-1, 1] \times [-2, 2]$.

23. Find the volume of the solid lying under the hyperbolic paraboloid $z = y^2 - x^2$ and above the square $R = [-1, 1] \times [1, 3]$.

24. Find the volume of the solid bounded by the elliptic paraboloid $z = 1 + (x - 1)^2 + 4y^2$, the planes $x = 3$ and $y = 2$, and the coordinate planes.

25. Find the volume of the solid bounded by the surface $z = 6 - xy$ and the planes $x = 2$, $x = -2$, $y = 0$, $y = 3$, and $z = 0$.

26. Find the volume of the solid bounded by the surface $z = x\sqrt{x^2 + y}$ and the planes $x = 0$, $x = 1$, $y = 0$, $y = 1$, and $z = 0$.

SECTION 13.3

Double Integrals over General Regions

For single integrals, the region over which we integrate is always an interval. But, for double integrals, we want to be able to integrate not just over rectangles but also over regions D of more general shape, such as the one illustrated in Figure 13.7(a). We suppose that D is a bounded region, which means that D can be enclosed in a rectangular region R as in Figure 13.7(b).

Figure 13.7

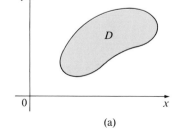

(a)

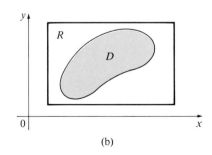

(b)

Then we define a new function F with domain R by

(13.14)
$$F(x, y) = \begin{cases} f(x, y) & \text{if } (x, y) \text{ is in } D \\ 0 & \text{if } (x, y) \text{ is in } R \text{ but not in } D \end{cases}$$

If F is integrable over R, then we say f is **integrable** over D and we define the **double integral of f over D** by

(13.15)
$$\iint\limits_{D} f(x, y)\, dA = \iint\limits_{R} F(x, y)\, dA$$

where F is given by Equation 13.14.

Definition 13.15 makes sense because R is a rectangle and so $\iint_R F(x, y)\, dA$ has been previously defined in Section 13.1. The procedure that we have used is reasonable because the values of $F(x, y)$ are 0 when (x, y) lies outside D and so they contribute nothing to the integral. This means that it does not matter what rectangle R we use as long as it contains D.

In the case where $f(x, y) \geqslant 0$ we can still interpret $\iint_D f(x, y)\, dA$ as the volume of the solid that lies above D and under the surface $z = f(x, y)$ (the graph of f). You can see that this is reasonable by comparing the graphs of f and F in Figure 13.8 and remembering that $\iint_R F(x, y)\, dA$ is the volume under the graph of F.

Figure 13.8(b) also shows that F is likely to have discontinuities at the boundary points of D. Nonetheless if f is continuous on D and the boundary curve of D is "well-behaved" (in a sense outside the scope of this book), then it can be shown that F is integrable over R and therefore f is integrable over D. In particular, this will be the case for the following types of regions.

A plane region D is said to be of **type I** if it lies between the graphs of two continuous functions of x, that is,

$$D = \{(x, y) \mid a \leqslant x \leqslant b, g_1(x) \leqslant y \leqslant g_2(x)\}$$

Figure 13.8

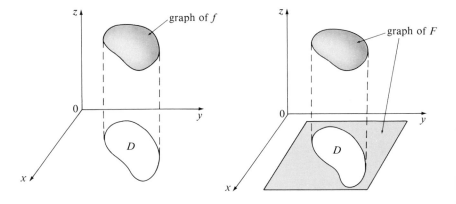

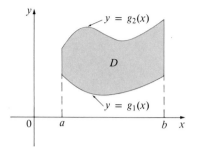

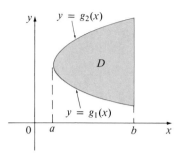

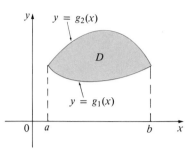

Figure 13.9

SOME TYPE I REGIONS

where g_1 and g_2 are continuous on $[a, b]$. Some examples of type I regions are shown in Figure 13.9.

In order to evaluate $\iint_D f(x, y)\, dA$ when D is a region of type I we choose a rectangle $R = [a, b] \times [c, d]$ that contains D as in Figure 13.10 and we let F be the function given by Equation 13.14; that is, F agrees with f on D and F is 0 outside D. Then, by Fubini's Theorem,

$$\iint_D f(x, y)\, dA = \iint_R F(x, y)\, dA = \int_a^b \int_c^d F(x, y)\, dy\, dx$$

Observe that $F(x, y) = 0$ if $y < g_1(x)$ or $y > g_2(x)$ since then (x, y) lies outside D. Therefore

$$\int_c^d F(x, y)\, dy = \int_{g_1(x)}^{g_2(x)} F(x, y)\, dy = \int_{g_1(x)}^{g_2(x)} f(x, y)\, dy$$

because $F(x, y) = f(x, y)$ when $g_1(x) \leqslant y \leqslant g_2(x)$. Thus we have the following formula that enables us to evaluate the double integral as an iterated integral.

(13.16)

> If f is continuous on a type I region
> $D = \{(x, y) \mid a \leqslant x \leqslant b,\ g_1(x) \leqslant y \leqslant g_2(x)\}$, then
>
> $$\iint_D f(x, y)\, dA = \int_a^b \int_{g_1(x)}^{g_2(x)} f(x, y)\, dy\, dx$$

Figure 13.10

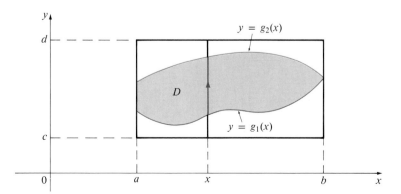

The integral on the right side of (13.16) is an iterated integral that is similar to the ones we considered in the preceding section except that in the inner integral we regard x as being constant not only in $f(x, y)$ but also in the limits of integration, $g_1(x)$ and $g_2(x)$.

We also consider plane regions of **type II** which can be expressed as

(13.17)
$$D = \{(x, y) | c \leqslant y \leqslant d, h_1(y) \leqslant x \leqslant h_2(y)\}$$

where h_1 and h_2 are continuous. Two such regions are illustrated in Figure 13.11.

Figure 13.11

SOME TYPE II REGIONS

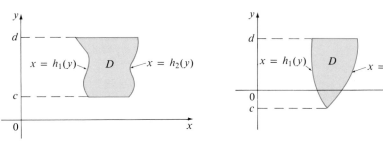

Using the same methods that were used in establishing (13.16) it can be shown that

(13.18)
$$\iint\limits_{D} f(x, y) \, dA = \int_{c}^{d} \int_{h_1(y)}^{h_2(y)} f(x, y) \, dx \, dy$$

where D is a type II region given by Equation 13.17.

● **Example 1**   Evaluate $\iint_D (x + 2y) \, dA$ where D is the region bounded by the parabolas $y = 2x^2$ and $y = 1 + x^2$.

Solution   The parabolas intersect when $2x^2 = 1 + x^2$, that is, $x^2 = 1$, so $x = \pm 1$. We note that the region D, sketched in Figure 13.12, is a type I region but not a type II region and we can write

$$D = \{(x, y) | -1 \leqslant x \leqslant 1, 2x^2 \leqslant y \leqslant 1 + x^2\}$$

Since the lower boundary is $y = 2x^2$ and the upper boundary is $y = 1 + x^2$, (13.16) gives

$$\iint\limits_{D} (x + 2y) \, dA = \int_{-1}^{1} \int_{2x^2}^{1+x^2} (x + 2y) \, dy \, dx$$

$$= \int_{-1}^{1} \left[xy + y^2 \right]_{y=2x^2}^{y=1+x^2} dx$$

$$= \int_{-1}^{1} \left[x(1 + x^2) + (1 + x^2)^2 - x(2x^2) - (2x^2)^2 \right] dx$$

$$= \int_{-1}^{1} (-3x^4 - x^3 + 2x^2 + x + 1) \, dx$$

$$= -3 \frac{x^5}{5} - \frac{x^4}{4} + 2 \frac{x^3}{3} + \frac{x^2}{2} + x \Big]_{-1}^{1} = \frac{32}{15}$$

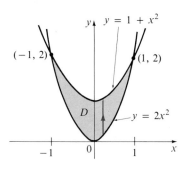

Figure 13.12

Note: When setting up a double integral as in Example 1 it is essential to draw a diagram. Often it is helpful to draw a vertical arrow as in Figure 13.12. Then the limits of integration for the *inner* integral can be read from the diagram as follows: The arrow starts at the lower boundary $y = g_1(x)$ which gives the lower limit in the integral, and the arrow ends at the upper boundary $y = g_2(x)$ which gives the upper limit of integration. For a type II region the arrow is horizontal and goes from the left boundary to the right boundary.

● **Example 2** Find the volume of the solid that lies under the paraboloid $z = x^2 + y^2$ and above the region D in the xy-plane bounded by the line $y = 2x$ and the parabola $y = x^2$.

Solution 1 From Figure 13.13(a) we see that D is a type I region and

$$D = \{(x, y) \mid 0 \leqslant x \leqslant 2, x^2 \leqslant y \leqslant 2x\}$$

Therefore the volume under $z = x^2 + y^2$ and above D is

$$V = \iint_D (x^2 + y^2)\, dA = \int_0^2 \int_{x^2}^{2x} (x^2 + y^2)\, dy\, dx$$

$$= \int_0^2 \left[x^2 y + \frac{y^3}{3} \right]_{y=x^2}^{y=2x} dx = \int_0^2 \left[x^2(2x) + \frac{(2x)^3}{3} - x^2 x^2 - \frac{(x^2)^3}{3} \right] dx$$

$$= \int_0^2 \left(-\frac{x^6}{3} - x^4 + \frac{14x^3}{3} \right) dx = -\frac{x^7}{21} - \frac{x^5}{5} + \frac{7x^4}{6} \Big]_0^2 = \frac{216}{35}$$

Solution 2 From Figure 13.13(b) we see that D can also be written as a type II region:

$$D = \left\{ (x, y) \mid 0 \leqslant y \leqslant 4, \frac{y}{2} \leqslant x \leqslant \sqrt{y} \right\}$$

Therefore another expression for V is

$$V = \iint_D (x^2 + y^2)\, dA = \int_0^4 \int_{y/2}^{\sqrt{y}} (x^2 + y^2)\, dx\, dy$$

$$= \int_0^4 \left[\frac{x^3}{3} + y^2 x \right]_{x=y/2}^{x=\sqrt{y}} dy = \int_0^4 \left(\frac{y^{3/2}}{3} + y^{5/2} - \frac{y^3}{24} - \frac{y^3}{2} \right) dy$$

$$= \frac{2}{15} y^{5/2} + \frac{2}{7} y^{7/2} - \frac{13}{96} y^4 \Big]_0^4 = \frac{216}{35} \qquad ●$$

● **Example 3** Evaluate $\iint_D xy\, dA$ where D is the region bounded by the line $y = x - 1$ and the parabola $y^2 = 2x + 6$.

Solution The region D is shown in Figure 13.14. Again D is both type I and type II but the description of D as a type I region is more complicated because the lower boundary consists of two parts. Therefore we prefer to express D as a type II region:

$$D = \left\{ (x, y) \mid -2 \leqslant y \leqslant 4, \frac{y^2}{2} - 3 \leqslant x \leqslant y + 1 \right\}$$

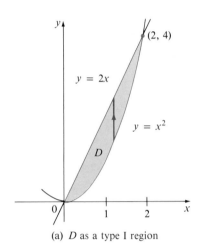

(a) D as a type I region

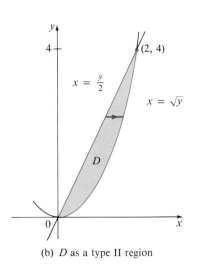

(b) D as a type II region

Figure 13.13

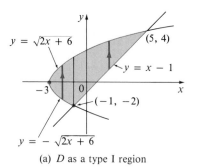

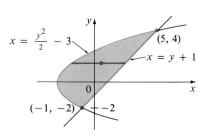

(a) D as a type I region

(b) D as a type II region

Figure 13.14

Then (13.18) gives

$$\iint_D xy\, dA = \int_{-2}^4 \int_{y^2/2-3}^{y+1} xy\, dx\, dy = \int_{-2}^4 \left[\frac{x^2}{2}y\right]_{x=y^2/2-3}^{x=y+1} dy$$

$$= \frac{1}{2}\int_{-2}^4 y\left[(y+1)^2 - \left(\frac{y^2}{2}-3\right)^2\right]dy$$

$$= \frac{1}{2}\int_{-2}^4 \left(-\frac{y^5}{4} + 4y^3 + 2y^2 - 8y\right)dy$$

$$= \frac{1}{2}\left[-\frac{y^6}{24} + y^4 + 2\frac{y^3}{3} - 4y^2\right]_{-2}^4 = 36$$

If we had expressed D as a type I region using Figure 13.14(a), then we would have obtained

$$\iint_D xy\, dA = \int_{-3}^{-1} \int_{-\sqrt{2x+6}}^{\sqrt{2x+6}} xy\, dy\, dx + \int_{-1}^5 \int_{x-1}^{\sqrt{2x+6}} xy\, dy\, dx$$

but this would have involved more work than the above method.   ●

● **Example 4**   Find the volume of the tetrahedron bounded by the planes $x + 2y + z = 2$, $x = 2y$, $x = 0$, and $z = 0$.

Solution   In a question such as this, it is wise to draw two diagrams: one of the three-dimensional solid and another of the plane region D over which it lies. Figure 13.15(a) shows the tetrahedron T bounded by the coordinate planes $x = 0$, $z = 0$, the vertical plane $x = 2y$, and the plane $x + 2y + z = 2$. Since the plane $x + 2y + z = 2$ intersects the xy-plane (whose equation is $z = 0$) in the line $x + 2y = 2$, we see that T lies above the triangular region D in the xy-plane bounded by the lines $x = 2y$, $x + 2y = 2$, $x = 0$ [see Figure 13.15(b)].

The plane $x + 2y + z = 2$ can be written as $z = 2 - x - 2y$, so the required volume lies under the graph of the function $z = 2 - x - 2y$ and above

$$D = \left\{(x,y)\,\middle|\,0 \leqslant x \leqslant 1, \frac{x}{2} \leqslant y \leqslant 1 - \frac{x}{2}\right\}$$

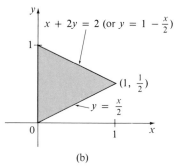

(a)

(b)

Figure 13.15

Therefore

$$V = \iint_D (2 - x - 2y)\, dA = \int_0^1 \int_{x/2}^{1-x/2} (2 - x - 2y)\, dy\, dx$$

$$= \int_0^1 \left[2y - xy - y^2 \right]_{y=x/2}^{y=1-x/2} dy$$

$$= \int_0^1 \left[2 - x - x\left(1 - \frac{x}{2}\right) - \left(1 - \frac{x}{2}\right)^2 - x + \frac{x^2}{2} + \frac{x^2}{4} \right] dx$$

$$= \int_0^1 (x^2 - 2x + 1)\, dx = \frac{x^3}{3} - x^2 + x \Big]_0^1 = \frac{1}{3}$$

● **Example 5**   Evaluate the iterated integral $\int_0^1 \int_x^1 \sin(y^2)\, dy\, dx$.

Solution   If we try to evaluate the integral as it stands, we are faced with the task of first evaluating $\int \sin(y^2)\, dy$. But it is impossible to do so in finite terms because $\int \sin(y^2)\, dy$ is not an elementary function. (See the end of Section 7.6.) So we must change the order of integration. This is accomplished by first expressing the given iterated integral as a double integral. Using (13.16) backward, we have

$$\int_0^1 \int_x^1 \sin(y^2)\, dy\, dx = \iint_D \sin(y^2)\, dA$$

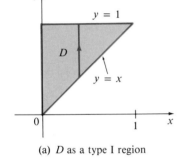

(a) D as a type I region

where $D = \{(x, y) \,|\, 0 \leqslant x \leqslant 1, x \leqslant y \leqslant 1\}$

We sketch this region D in Figure 13.16(a). Then from Figure 13.16(b) we see that an alternative description of D is

$$D = \{(x, y) \,|\, 0 \leqslant y \leqslant 1, 0 \leqslant x \leqslant y\}$$

This enables us to use (13.18) to express the double integral as an iterated integral in the reverse order:

$$\int_0^1 \int_x^1 \sin(y^2)\, dy\, dx = \iint_D \sin(y^2)\, dA$$

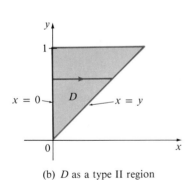

(b) D as a type II region

Figure 13.16

$$= \int_0^1 \int_0^y \sin(y^2)\, dx\, dy = \int_0^1 \left[x \sin(y^2) \right]_{x=0}^{x=y} dy$$

$$= \int_0^1 y \sin(y^2)\, dy = -\frac{1}{2} \cos(y^2) \Big]_0^1$$

$$= \frac{1}{2}(1 - \cos 1)$$ ●

Properties of Double Integrals

We assume that all of the following integrals exist. The first three properties of double integrals over a region D follow immediately from Definition 13.15 and Properties 13.7, 13.8, and 13.9.

(13.19)
$$\iint_D [f(x,y) + g(x,y)]\, dA = \iint_D f(x,y)\, dA + \iint_D g(x,y)\, dA$$

(13.20)
$$\iint_D cf(x,y)\, dA = c \iint_D f(x,y)\, dA$$

(13.21)   If $f(x,y) \geqslant g(x,y)$ for all (x,y) in D, then

$$\iint_D f(x,y)\, dA \geqslant \iint_D g(x,y)\, dA$$

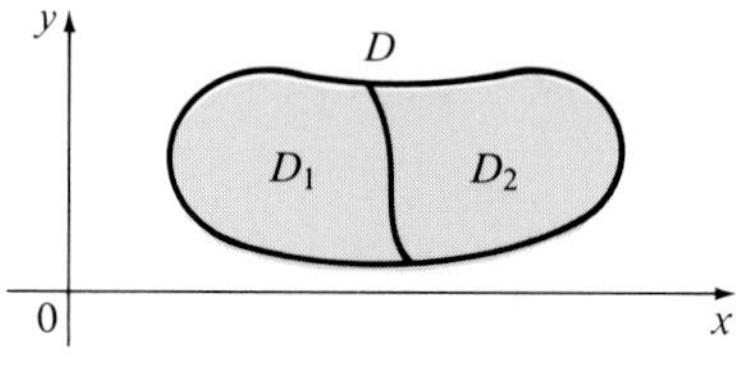

The next property of double integrals is the analogue of the property of single integrals given by the equation $\int_a^b f = \int_a^c f + \int_c^b f$.

If $D = D_1 \cup D_2$, where D_1 and D_2 do not overlap except perhaps on their boundaries (see Figure 13.17), then

Figure 13.17

(13.22)
$$\iint_D f(x,y)\, dA = \iint_{D_1} f(x,y)\, dA + \iint_{D_2} f(x,y)\, dA$$

Property 13.22 can be used to evaluate double integrals over regions D that are neither type I nor type II but can be expressed as a union of type I or II regions. Figure 13.18 gives an illustration of this procedure. (See Exercises 43 and 44.)

Figure 13.18

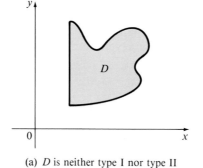

(a) D is neither type I nor type II

(b) $D = D_1 \cup D_2$, D_1 is type I, D_2 is type II

The next property of integrals says that if we integrate the constant function $f(x,y) = 1$ over a region D, we get the area of D:

(13.23)
$$\iint_D 1\, dA = A(D)$$

For instance, if D is a type I region and we put $f(x, y) = 1$ in Formula 13.14, we get

$$\iint_D 1 \, dA = \int_a^b \int_{g_1(x)}^{g_2(x)} 1 \, dy \, dx = \int_a^b [g_2(x) - g_1(x)] \, dx = A(D)$$

by Equation 5.37.

Finally we obtain an analogue of Property 8 of single integrals (Theorem 5.14) by combining Properties 13.20, 13.21, and 13.23. (See Exercise 47.)

(13.24)

> If $m \le f(x, y) \le M$ for all (x, y) in D, then
>
> $$mA(D) \le \iint_D f(x, y) \, dA \le MA(D)$$

● **Example 6**   Use Property 13.24 to estimate the integral $\iint_D e^{\sin x \cos y} \, dA$ where D is the disk with center the origin and radius 2.

Solution   Since $-1 \le \sin x \le 1$ and $-1 \le \cos y \le 1$, we have $-1 \le \sin x \cos y \le 1$ and therefore

$$e^{-1} \le e^{\sin x \cos y} \le e^1 = e$$

Thus, using $m = e^{-1} = 1/e$, $M = e$, and $A(D) = \pi(2)^2$ in Property 13.24, we obtain

$$\frac{4\pi}{e} \le \iint_D e^{\sin x \cos y} \, dA \le 4\pi e$$ ●

EXERCISES 13.3

In Exercises 1–6 evaluate the iterated integral.

1. $\int_0^1 \int_0^y x \, dx \, dy$

2. $\int_0^1 \int_0^y y \, dx \, dy$

3. $\int_0^2 \int_{\sqrt{x}}^3 (x^2 + y) \, dy \, dx$

4. $\int_0^1 \int_{1-x}^{1+x} (2x - 3y^2) \, dy \, dx$

5. $\int_0^1 \int_0^x \sin(x^2) \, dy \, dx$

6. $\int_0^1 \int_{x-1}^0 \frac{2y}{x + 1} \, dy \, dx$

In Exercises 7–22 evaluate the double integral.

7. $\iint_D xy \, dA$, $D = \{(x, y) | 0 \le x \le 1, x^2 \le y \le \sqrt{x}\}$

8. $\iint_D (x - 2y) \, dA$,
$D = \{(x, y) | 1 \le x \le 3, 1 + x \le y \le 2x\}$

9. $\iint_D (3x + y) \, dA$,
$D = \left\{(x, y) \left| \frac{\pi}{6} \le x \le \frac{\pi}{4}, \sin x \le y \le \cos x \right.\right\}$

10. $\iint_D (x^2 - 2xy)\, dA$,
$D = \{(x, y)\,|\, 0 \leqslant x \leqslant 1, \sqrt{x} \leqslant y \leqslant 2 - x\}$

11. $\iint_D (y - xy^2)\, dA$,
$D = \{(x, y)\,|\, 0 \leqslant y \leqslant 1, -y \leqslant x \leqslant 1 + y\}$

12. $\iint_D x \sin y\, dA$, $D = \left\{(x, y)\,\middle|\, 0 \leqslant y \leqslant \dfrac{\pi}{2}, 0 \leqslant x \leqslant \cos y\right\}$

13. $\iint_D e^{x/y}\, dA$, $D = \{(x, y)\,|\, 1 \leqslant y \leqslant 2, y \leqslant x \leqslant y^3\}$

14. $\iint_D \dfrac{1}{x}\, dA$, $D = \{(x, y)\,|\, 1 \leqslant y \leqslant e, y^2 \leqslant x \leqslant y^4\}$

15. $\iint_D x \cos y\, dA$, D is bounded by $y = 0$, $y = x^2$, $x = 1$

16. $\iint_D e^{x+y}\, dA$, D is bounded by $y = 0$, $y = x$, $x = 1$

17. $\iint_D (x^2 + y)\, dA$, D is bounded by $y = x^2$, $y = 2 - x^2$

18. $\iint_D 3xy\, dA$, D is bounded by $y = x$, $y = x^2 - 4x + 4$

19. $\iint_D 4y^3\, dA$, D is bounded by $y = x - 6$, $y^2 = x$

20. $\iint_D (y^2 - x)\, dA$, D is bounded by $x = y^2$, $x = 3 - 2y^2$

21. $\iint_D xy\, dA$, D is the first quadrant part of the disk with center $(0, 0)$ and radius 1

22. $\iint_D ye^x\, dA$, D is the triangular region with vertices $(0, 0)$, $(2, 4)$, and $(6, 0)$

In Exercises 23–34 find the volume of the given solid.

23. Under the paraboloid $z = x^2 + y^2$ and above the region bounded by $y = x^2$ and $x = y^2$

24. Under the paraboloid $z = 3x^2 + y^2$ and above the region bounded by $y = x$ and $x = y^2 - y$

25. Under the surface $z = xy$ and above the triangle with vertices $(1, 1)$, $(4, 1)$, and $(1, 2)$

26. Under the surface $z = 1 + xy$ and above the triangle with vertices $(1, 1)$, $(4, 1)$, and $(3, 2)$

27. Bounded by the paraboloid $z = x^2 + y^2 + 4$ and the planes $x = 0$, $y = 0$, $z = 0$, $x + y = 1$

28. Bounded by the cylinder $x^2 + z^2 = 9$ and the planes $x = 0$, $y = 0$, $z = 0$, $x + 2y = 2$ in the first octant

29. Bounded by the cylinder $y^2 + z^2 = 4$ and the planes $x = 2y$, $x = 0$, $z = 0$ in the first octant

30. Bounded by the cylinder $y^2 + z^2 = 9$ and the planes $y = 3x$, $y = 0$, $z = 0$ in the first octant

31. Bounded by the planes $x = 0$, $y = 0$, $z = 0$, $x + y + z = 1$

32. Bounded by the planes $y = 0$, $z = 0$, $y = x$, $6x + 2y + 3z = 6$

33. Bounded by the cylinder $x^2 + y^2 = 1$ and the planes $y = z$, $x = 0$, $z = 0$ in the first octant

34. Bounded by the cylinders $x^2 + y^2 = r^2$, $y^2 + z^2 = r^2$

In Exercises 35–38 sketch the region of integration and change the order of integration.

35. $\displaystyle\int_0^1 \int_0^x f(x, y)\, dy\, dx$

36. $\displaystyle\int_0^{\pi/2} \int_0^{\sin x} f(x, y)\, dy\, dx$

37. $\displaystyle\int_1^2 \int_0^{\ln x} f(x, y)\, dy\, dx$

38. $\displaystyle\int_0^1 \int_{y^2}^{2-y} f(x, y)\, dx\, dy$

In Exercises 39–42 evaluate the given integral by reversing the order of integration.

39. $\displaystyle\int_0^1 \int_{3y}^3 e^{x^2}\, dx\, dy$

40. $\displaystyle\int_0^1 \int_{\sqrt{y}}^1 \sqrt{x^3 + 1}\, dx\, dy$

41. $\displaystyle\int_0^1 \int_{e^x}^e \frac{1}{\ln y}\, dy\, dx$

42. $\displaystyle\int_0^1 \int_{x^2}^1 x^3 \sin(y^3)\, dy\, dx$

In Exercises 43 and 44 express D as a union of regions of type I or type II and evaluate the integral.

43. $\displaystyle\iint_D x^2\, dA$

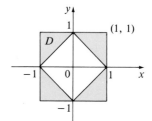

44. $\displaystyle\iint_D xy\, dA$

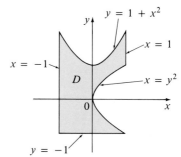

In Exercises 45 and 46 use Property 13.24 to estimate the value of the given integral.

45. $\displaystyle\iint_D \sqrt{x^3 + y^3}\, dA$, $D = [0, 1] \times [0, 1]$

46. $\iint_D e^{x^2 + y^2}\, dA$, D is the disk with center the origin and radius $\frac{1}{2}$

47. Prove Property 13.24.

Double Integrals in Polar Coordinates

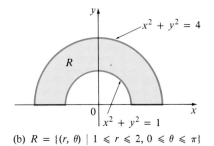

(a) $R = \{(r, \theta) \mid 0 \leqslant r \leqslant 1, 0 \leqslant \theta \leqslant 2\pi\}$

Suppose that we want to evaluate a double integral $\iint_R f(x, y)\, dA$ where the region R is one of the regions shown in Figure 13.19. In both cases the description of R in terms of rectangular coordinates is rather complicated but R is easily described using polar coordinates.

These regions are special cases of a **polar rectangle**

$$R = \{(r, \theta) \mid a \leqslant r \leqslant b, \alpha \leqslant \theta \leqslant \beta\}$$

which is shown in Figure 13.20. In order to compute $\iint_R f(x, y)\, dA$ where R is a polar rectangle, we start with a partition of $[a, b]$ into m subintervals:

$$a = r_0 < r_1 < r_2 < \cdots < r_{i-1} < r_i < \cdots < r_m = b$$

and a partition of $[\alpha, \beta]$ into n subintervals:

$$\alpha = \theta_0 < \theta_1 < \theta_2 < \cdots < \theta_{j-1} < \theta_j < \cdots < \theta_n = \beta$$

(b) $R = \{(r, \theta) \mid 1 \leqslant r \leqslant 2, 0 \leqslant \theta \leqslant \pi\}$

Figure 13.19

Then the circles $r = r_i$ and the rays $\theta = \theta_j$ determine a **polar partition** P of R into the small polar rectangles shown in Figure 13.21. The norm $\|P\|$ of the polar partition is the length of the longest diagonal of all the polar subrectangles.

The "center" of the polar subrectangle

$$R_{ij} = \{(r, \theta) \mid r_{i-1} \leqslant r \leqslant r_i, \theta_{j-1} \leqslant \theta \leqslant \theta_j\}$$

has polar coordinates

$$r_i^* = \tfrac{1}{2}(r_{i-1} + r_i) \qquad \theta_j^* = \tfrac{1}{2}(\theta_{j-1} + \theta_j)$$

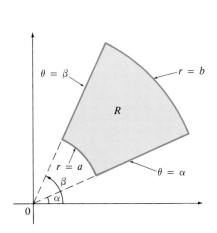

Figure 13.20

POLAR RECTANGLE

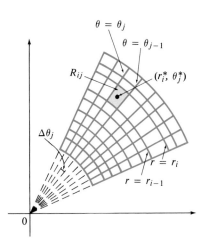

Figure 13.21

POLAR PARTITION

We compute the area of R_{ij} using the fact that the area of a sector of a circle with radius r and central angle θ is $\frac{1}{2}r^2\theta$. Subtracting the areas of two such sectors, each of which has central angle $\Delta\theta_j = \theta_j - \theta_{j-1}$, we find that the area of R_{ij} is

$$
\begin{aligned}
\Delta A_{ij} &= \tfrac{1}{2}r_i^2 \,\Delta\theta_j - \tfrac{1}{2}r_{i-1}^2 \,\Delta\theta_j \\
&= \tfrac{1}{2}(r_i^2 - r_{i-1}^2)\,\Delta\theta_j \\
&= \tfrac{1}{2}(r_i + r_{i-1})(r_i - r_{i-1})\,\Delta\theta_j \\
&= r_i^* \,\Delta r_i \,\Delta\theta_j
\end{aligned}
$$

where $\Delta r_i = r_i - r_{i-1}$.

Although we have defined the double integral $\iint_R f(x, y)\,dA$ in terms of rectangular partitions, it can be shown that, for continuous functions f, we always get the same answer using polar partitions. The rectangular coordinates of the center of R_{ij} are $(r_i^* \cos\theta_j^*, r_i^* \sin\theta_j^*)$, so a typical Riemann sum is

$$(13.25) \qquad \sum_{i=1}^{m}\sum_{j=1}^{n} f(r_i^*\cos\theta_j^*, r_i^*\sin\theta_j^*)\,\Delta A_{ij} = \sum_{i=1}^{m}\sum_{j=1}^{n} f(r_i^*\cos\theta_j^*, r_i^*\sin\theta_j^*)r_i^*\,\Delta r_i\,\Delta\theta_j$$

If we write $g(r, \theta) = rf(r\cos\theta, r\sin\theta)$, then the Riemann sum in Equation 13.25 can be written as

$$\sum_{i=1}^{m}\sum_{j=1}^{n} g(r_i^*, \theta_j^*)\,\Delta r_i\,\Delta\theta_j$$

which is a Riemann sum for the double integral

$$\int_\alpha^\beta \int_a^b g(r, \theta)\,dr\,d\theta$$

Therefore we have

$$
\begin{aligned}
\iint_R f(x, y)\,dA &= \lim_{\|P\|\to 0} \sum_{i=1}^{m}\sum_{j=1}^{n} f(r_i^*\cos\theta_j^*, r_i^*\sin\theta_j^*)\,\Delta A_{ij} \\
&= \lim_{\|P\|\to 0} \sum_{i=1}^{m}\sum_{j=1}^{n} g(r_i^*, \theta_j^*)\,\Delta r_i\,\Delta\theta_j \\
&= \int_\alpha^\beta \int_a^b g(r, \theta)\,dr\,d\theta = \int_\alpha^\beta \int_a^b f(r\cos\theta, r\sin\theta)r\,dr\,d\theta
\end{aligned}
$$

Change to Polar Coordinates in a Double Integral (13.26)

If f is continuous on a polar rectangle R given by $0 \leqslant a \leqslant r \leqslant b$, $\alpha \leqslant \theta \leqslant \beta$, where $0 \leqslant \beta - \alpha \leqslant 2\pi$, then

$$\iint_R f(x, y)\,dA = \int_\alpha^\beta \int_a^b f(r\cos\theta, r\sin\theta)r\,dr\,d\theta$$

The formula in (13.26) says that we convert from rectangular to polar coordinates in a double integral by writing $x = r\cos\theta$ and $y = r\sin\theta$, using the appropriate limits of integration for r and θ, and replacing dA by $r\,dr\,d\theta$. Be careful not to forget the additional factor r on the right side of Formula 13.26. Some people like to remember it by means of Figure 13.22 where the "infinitesimal" polar rectangle could be thought of as an ordinary rectangle with dimensions $r\,d\theta$ and dr and therefore has "area" $dA = r\,dr\,d\theta$.

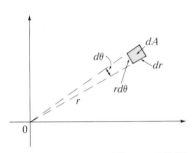

Figure 13.22

● **Example 1**   Evaluate $\iint_R (3x + 4y^2)\,dA$ where R is the region in the upper half-plane bounded by the circles $x^2 + y^2 = 1$ and $x^2 + y^2 = 4$.

Solution   The region R can be described as

$$R = \{(x, y)\,|\,y \geqslant 0,\ 1 \leqslant x^2 + y^2 \leqslant 4\}$$

It is the half-ring shown in Figure 13.19(b), and in polar coordinates it is given by $1 \leqslant r \leqslant 2,\ 0 \leqslant \theta \leqslant \pi$. Therefore, by Formula 13.26,

$$\iint_R (3x + 4y^2)\,dA = \int_0^\pi \int_1^2 (3r\cos\theta + 4r^2\sin^2\theta)r\,dr\,d\theta$$

$$= \int_0^\pi \int_1^2 (3r^2\cos\theta + 4r^3\sin^2\theta)\,dr\,d\theta$$

$$= \int_0^\pi \Big[r^3\cos\theta + r^4\sin^2\theta\Big]_{r=1}^{r=2} d\theta$$

$$= \int_0^\pi (7\cos\theta + 15\sin^2\theta)\,d\theta$$

$$= \int_0^\pi \Big[7\cos\theta + \frac{15}{2}(1 - \cos 2\theta)\Big]d\theta$$

$$= 7\sin\theta + \frac{15\theta}{2} - \frac{15}{4}\sin 2\theta\Big]_0^\pi = \frac{15\pi}{2} \qquad ●$$

● **Example 2**   Find the volume of the solid bounded by the paraboloid $z = 1 - x^2 - y^2$ and the plane $z = 0$.

Solution   If we put $z = 0$ in the equation of the paraboloid, we get $x^2 + y^2 = 1$. This means that the plane intersects the paraboloid in the circle $x^2 + y^2 = 1$, so the solid lies under the paraboloid and above the circular disk D given by $x^2 + y^2 \leqslant 1$ [see Figures 13.23 and 13.19(a)]. In polar coordinates D is given by $0 \leqslant r \leqslant 1,\ 0 \leqslant \theta \leqslant 2\pi$. Since $1 - x^2 - y^2 = 1 - r^2$, the volume is

$$V = \iint_D (1 - x^2 - y^2)\,dA = \int_0^{2\pi} \int_0^1 (1 - r^2)r\,dr\,d\theta$$

$$= \int_0^{2\pi} d\theta \int_0^1 (r - r^3)\,dr = 2\pi \Big[\frac{r^2}{2} - \frac{r^4}{4}\Big]_0^1$$

$$= \frac{\pi}{2}$$

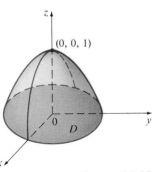

Figure 13.23

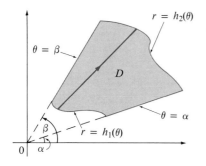

Figure 13.24

TYPE II POLAR REGION
$D = \{(r, \theta) \,|\, \alpha \leq \theta \leq \beta, \; h_1(\theta) \leq r \leq h_2(\theta)\}$

If we had used rectangular coordinates instead of polar coordinates, then we would have obtained

$$V = \iint_D (1 - x^2 - y^2)\, dA = \int_{-1}^{1} \int_{-\sqrt{1-x^2}}^{\sqrt{1-x^2}} (1 - x^2 - y^2)\, dy\, dx$$

which is not easy to evaluate because it involves finding the following integrals:

$$\int \sqrt{1 - x^2}\, dx \qquad \int x^2 \sqrt{1 - x^2}\, dx \qquad \int (1 - x^2)^{3/2}\, dx \qquad \bullet$$

What we have done so far can be extended to the more complicated type of region shown in Figure 13.24. We could call it a type II polar region because it is analogous to the type II rectangular regions considered in Section 13.3. In fact, by combining Formulas 13.26 and 13.18, we obtain the following formula:

(13.27)

> If f is continuous on a polar region of the form
>
> $$D = \{(r, \theta) \,|\, \alpha \leq \theta \leq \beta, \, h_1(\theta) \leq r \leq h_2(\theta)\}$$
>
> then $\qquad \displaystyle\iint_D f(x, y)\, dA = \int_{\alpha}^{\beta} \int_{h_1(\theta)}^{h_2(\theta)} f(r\cos\theta, r\sin\theta)\, r\, dr\, d\theta$

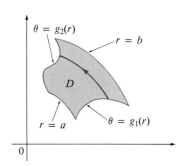

Figure 13.25

TYPE I POLAR REGION
$D = \{(r, \theta) \,|\, a \leq r \leq b, \; g_1(r) \leq \theta \leq g_2(r)\}$

In particular, taking $f(x, y) = 1$, $h_1(\theta) = 0$, and $h_2(\theta) = h(\theta)$ in this formula, we see that the area of the region D bounded by $\theta = \alpha$, $\theta = \beta$, and $r = h(\theta)$ is

$$A(D) = \iint_D 1\, dA = \int_{\alpha}^{\beta} \int_0^{h(\theta)} r\, dr\, d\theta$$

$$= \int_{\alpha}^{\beta} \left[\frac{r^2}{2} \right]_0^{h(\theta)} d\theta = \int_{\alpha}^{\beta} \frac{1}{2} [h(\theta)]^2\, d\theta$$

and this agrees with Formula 9.14.

Similarly, the analogue of a type I rectangular region is a polar region of the form

$$D = \{(r, \theta) \,|\, a \leq r \leq b, \, g_1(r) \leq \theta \leq g_2(r)\}$$

(see Figure 13.25) and Formula 13.16, together with Formula 13.26, shows that

(13.28)

$$\iint_D f(x, y)\, dA = \int_a^b \int_{g_1(r)}^{g_2(r)} f(r\cos\theta, r\sin\theta)\, r\, d\theta\, dr$$

● **Example 3**   Find the volume of the solid that lies under the paraboloid $z = x^2 + y^2$, above the xy-plane, and inside the cylinder $x^2 + y^2 = 2x$.

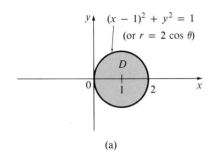

(a)

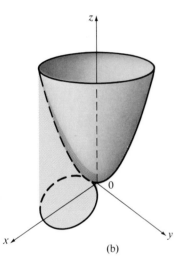

(b)

Figure 13.26

Solution   The solid lies above the disk D whose boundary circle has equation $x^2 + y^2 = 2x$ or

$$(x - 1)^2 + y^2 = 1$$

(see Figure 13.26). In polar coordinates we have $x^2 + y^2 = r^2$ and $x = r\cos\theta$, so the boundary circle becomes $r^2 = 2r\cos\theta$ or $r = 2\cos\theta$. Thus the disk D is given by

$$D = \left\{ (r, \theta) \,\middle|\, -\frac{\pi}{2} \leqslant \theta \leqslant \frac{\pi}{2},\, 0 \leqslant r \leqslant 2\cos\theta \right\}$$

and, by Formula 13.27, we have

$$V = \iint_D (x^2 + y^2)\,dA = \int_{-\pi/2}^{\pi/2} \int_0^{2\cos\theta} r^2 r\,dr\,d\theta$$

$$= \int_{-\pi/2}^{\pi/2} \left[\frac{r^4}{4} \right]_0^{2\cos\theta} d\theta = 4 \int_{-\pi/2}^{\pi/2} \cos^4\theta\,d\theta$$

$$= 8 \int_0^{\pi/2} \cos^4\theta\,d\theta = 8 \int_0^{\pi/2} \left(\frac{1 + \cos 2\theta}{2} \right)^2 d\theta$$

$$= 2 \int_0^{\pi/2} \left[1 + 2\cos 2\theta + \frac{1}{2}(1 + \cos 4\theta) \right] d\theta$$

$$= 2 \left[\frac{3}{2}\theta + \sin 2\theta + \frac{1}{8}\sin 4\theta \right]_0^{\pi/2}$$

$$= 2 \left(\frac{3}{2} \right) \left(\frac{\pi}{2} \right) = \frac{3\pi}{2} \qquad \bullet$$

EXERCISES 13.4

In Exercises 1–8 evaluate the given integral by changing to polar coordinates.

1. $\iint_R x\,dA$ where R is the disk with center the origin and radius 5

2. $\iint_R y\,dA$ where R is the region in the first quadrant bounded by the circle $x^2 + y^2 = 9$ and the lines $y = x$ and $y = 0$

3. $\iint_R xy\,dA$ where R is the region in the first quadrant that lies between the circles $x^2 + y^2 = 4$ and $x^2 + y^2 = 25$

4. $\iint_R \sin(x^2 + y^2)\,dA$ where R is the annular region $1 \leqslant x^2 + y^2 \leqslant 16$

5. $\iint_D 1/\sqrt{x^2 + y^2}\,dA$ where D is the region bounded by the cardioid $r = 1 + \sin\theta$

6. $\iint_D \sqrt{x^2 + y^2}\,dA$ where D is the region bounded by the cardioid $r = 1 + \cos\theta$

7. $\iint_D (x^2 + y^2)\,dA$ where D is the region bounded by the spirals $r = \theta$ and $r = 2\theta$ for $0 \leqslant \theta \leqslant 2\pi$

8. $\iint_D x\,dA$ where D is the region in the first quadrant that lies between the circles $x^2 + y^2 = 4$ and $x^2 + y^2 = 2x$

In Exercises 9–17 use polar coordinates to find the volume of the given solid.

9. Under the paraboloid $z = x^2 + y^2$ and above the disk $x^2 + y^2 \leqslant 9$

10. Under the cone $z = \sqrt{x^2 + y^2}$ and above the ring $4 \leqslant x^2 + y^2 \leqslant 25$

11. Under the plane $6x + 4y + z = 12$ and above the disk with boundary circle $x^2 + y^2 = y$

12. Bounded by the paraboloid $z = 10 - 3x^2 - 3y^2$ and the plane $z = 4$

13. Above the cone $z = \sqrt{x^2 + y^2}$ and below the sphere $x^2 + y^2 + z^2 = 1$

14. Bounded by the paraboloids $z = 3x^2 + 3y^2$ and $z = 4 - x^2 - y^2$

15. Inside both the cylinder $x^2 + y^2 = 4$ and the ellipsoid $4x^2 + 4y^2 + z^2 = 64$

16. Inside the sphere $x^2 + y^2 + z^2 = 4a^2$ and outside the cylinder $x^2 + y^2 = 2ax$

17. A sphere of radius a

18. A cylindrical drill with radius r_1 is used to bore a hole through the center of a sphere of radius r_2. Find the volume of the remaining solid.

In Exercises 19–22 evaluate the given iterated integral by converting to polar coordinates.

19. $\int_0^1 \int_0^{\sqrt{1-x^2}} e^{x^2+y^2} \, dy \, dx$

20. $\int_{-a}^{a} \int_0^{\sqrt{a^2-y^2}} (x^2 + y^2)^{3/2} \, dx \, dy$

21. $\int_0^2 \int_{-\sqrt{4-y^2}}^{\sqrt{4-y^2}} x^2 y^2 \, dx \, dy$

22. $\int_0^2 \int_0^{\sqrt{2x-x^2}} \sqrt{x^2 + y^2} \, dy \, dx$

23. (a) We define the improper integral (over the entire plane R^2)
$$I = \iint_{R^2} e^{-(x^2+y^2)} \, dA = \int_{-\infty}^{\infty} \int_{-\infty}^{\infty} e^{-(x^2+y^2)} \, dy \, dx$$
$$= \lim_{a \to \infty} \iint_{D_a} e^{-(x^2+y^2)} \, dA$$
where D_a is the disk with radius a and center the origin. Show that
$$\int_{-\infty}^{\infty} \int_{-\infty}^{\infty} e^{-(x^2+y^2)} \, dA = \pi$$

(b) An equivalent definition of the improper integral in part (a) is
$$\iint_{R^2} e^{-(x^2+y^2)} \, dA = \lim_{a \to \infty} \iint_{S_a} e^{-(x^2+y^2)} \, dA$$
where S_a is the square with vertices $(\pm a, \pm a)$. Use this to show that
$$\int_{-\infty}^{\infty} e^{-x^2} \, dx \int_{-\infty}^{\infty} e^{-y^2} \, dy = \pi$$

(c) Deduce that
$$\int_{-\infty}^{\infty} e^{-x^2} \, dx = \sqrt{\pi}$$

(d) By making the change of variable $t = \sqrt{2}x$, show that
$$\int_{-\infty}^{\infty} e^{-x^2/2} \, dx = \sqrt{2\pi}$$
(This is a fundamental result for probability and statistics.)

SECTION 13.5

Applications of Double Integrals

We have already seen one application of double integrals: computing volumes. Another geometric application is finding areas of surfaces and this will be done in the next chapter. In this section we explore physical applications such as computing mass, electric charge, center of mass, and moment of inertia.

In Chapter 8 we were able to use single integrals to compute moments and the center of mass of a thin plate or lamina with constant density. But now, equipped with the double integral, we can consider a lamina with variable density. Suppose the lamina occupies a region D of the xy-plane and its density (in units of mass per unit area) at a point (x, y) in D is given

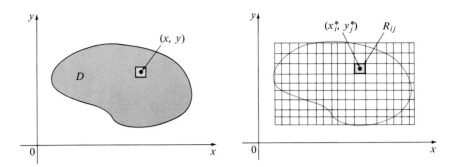

Figure 13.27   **Figure 13.28**

by $\rho(x, y)$ where ρ is a continuous function on D. This means that

$$\rho(x, y) = \lim \frac{\Delta m}{\Delta A}$$

where Δm and ΔA are the mass and area of a small rectangle that contains (x, y) and the limit is taken as the dimensions of the rectangle approach 0 (see Figure 13.27).

To find the total mass m of the lamina we partition a rectangle R containing D into subrectangles R_{ij} (as in Figure 13.28) and consider $\rho(x, y)$ to be 0 outside D. If we choose a point (x_i^*, y_j^*) in R_{ij}, then the mass of the part of the lamina that occupies R_{ij} is approximately $\rho(x_i^*, y_j^*) \Delta A_{ij}$ where ΔA_{ij} is the area of R_{ij}. If we add all such masses, we get an approximation to the total mass:

$$m \approx \sum_{i=1}^{m} \sum_{j=1}^{n} \rho(x_i^*, y_j^*) \Delta A_{ij}$$

If we now take finer partitions, we obtain the total mass m of the lamina as the limiting value of the approximations:

(13.29)
$$m = \lim_{\|P\| \to 0} \sum_{i=1}^{m} \sum_{j=1}^{n} \rho(x_i^*, y_j^*) \Delta A_{ij} = \iint\limits_{D} \rho(x, y)\, dA$$

Physicists also consider other types of density that can be treated in the same manner. For example, if an electric charge is distributed over a region D and the charge density (in units of charge per unit area) is given by $\sigma(x, y)$ at a point (x, y) in D, then the total charge Q is given by

(13.30)
$$Q = \iint\limits_{D} \sigma(x, y)\, dA$$

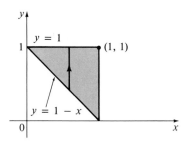

Figure 13.29

● **Example 1**   Charge is distributed over the triangular region D in Figure 13.29 so that the charge density at (x, y) is $\sigma(x, y) = xy$, measured in coulombs per square meter (C/m^2). Find the total charge.

Solution   From Equation 13.30 and Figure 13.29, we have

$$Q = \iint_D \sigma(x, y)\, dA = \int_0^1 \int_{1-x}^1 xy\, dy\, dx$$

$$= \int_0^1 \left[x\, \frac{y^2}{2} \right]_{y=1-x}^{y=1} dx = \int_0^1 \frac{x}{2}\, [1^2 - (1 - x)^2]\, dx$$

$$= \frac{1}{2} \int_0^1 (2x^2 - x^3)\, dx = \frac{1}{2} \left[\frac{2x^3}{3} - \frac{x^4}{4} \right]_0^1 = \frac{5}{24}$$

Thus the total charge is $\frac{5}{24}$ C.   ●

Now let us find the center of mass of a lamina with density function $\rho(x, y)$ that occupies a region D. Recall that in Section 8.7 we defined the moment of a particle about an axis as the product of its mass and its directed distance from the axis. We partition D into small rectangles as in Figure 13.28. Then the mass of R_{ij} is approximately $\rho(x_i^*, y_j^*)\Delta A_{ij}$, so we can approximate the moment of R_{ij} with respect to the x-axis by

$$[\rho(x_i^*, y_j^*)\Delta A_{ij}]y_j^*$$

If we now add these quantities and take the limit as the norm $\|P\|$ approaches 0, we obtain the **moment** of the entire lamina **about the x-axis:**

(13.31)
$$M_x = \lim_{\|P\| \to 0} \sum_{i=1}^m \sum_{j=1}^n y_j^* \rho(x_i^*, y_j^*)\Delta A_{ij} = \iint_D y\rho(x, y)\, dA$$

Similarly the **moment about the y-axis** is

(13.32)
$$M_y = \lim_{\|P\| \to 0} \sum_{i=1}^m \sum_{j=1}^n x_i^* \rho(x_i^*, y_j^*)\Delta A_{ij} = \iint_D x\rho(x, y)\, dA$$

As before, we define the center of mass $(\bar{x}, \bar{y})$ so that $m\bar{x} = M_y$ and $m\bar{y} = M_x$. The physical significance is that the lamina behaves as if its entire mass is concentrated at its center of mass. Thus the lamina will balance horizontally when supported at its center of mass (see Figure 8.48).

(13.33)

The coordinates $(\bar{x}, \bar{y})$ of the center of mass of a lamina occupying the region D and having density function $\rho(x, y)$ are

$$\bar{x} = \frac{M_y}{m} = \frac{1}{m} \iint_D x\rho(x, y) \, dA \qquad \bar{y} = \frac{M_x}{m} = \frac{1}{m} \iint_D y\rho(x, y) \, dA$$

where the mass m is given by

$$m = \iint_D \rho(x, y) \, dA$$

● **Example 2**   Find the mass and center of mass of a triangular lamina with vertices $(0, 0)$, $(1, 0)$, and $(0, 2)$ if the density function is $\rho(x, y) = 1 + 3x + y$.

Solution   The triangle is shown in Figure 13.30. (Note that the equation of the upper boundary is $y = 2 - 2x$.) The mass of the lamina is

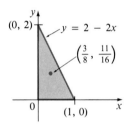

Figure 13.30

$$m = \iint_D \rho(x, y) \, dA = \int_0^1 \int_0^{2-2x} (1 + 3x + y) \, dy \, dx$$

$$= \int_0^1 \left[y + 3xy + \frac{y^2}{2} \right]_{y=0}^{y=2-2x} dx$$

$$= 4 \int_0^1 (1 - x^2) \, dx = 4 \left[x - \frac{x^3}{3} \right]_0^1 = \frac{8}{3}$$

Then the formulas in (13.33) give

$$\bar{x} = \frac{1}{m} \iint_D x\rho(x, y) \, dA = \frac{3}{8} \int_0^1 \int_0^{2-2x} (x + 3x^2 + xy) \, dy \, dx$$

$$= \frac{3}{8} \int_0^1 \left[xy + 3x^2 y + x\frac{y^2}{2} \right]_{y=0}^{y=2-2x} dx$$

$$= \frac{3}{2} \int_0^1 (x - x^3) \, dx = \left[\frac{x^2}{2} - \frac{x^4}{4} \right]_0^1 = \frac{3}{8}$$

$$\bar{y} = \frac{1}{m} \iint_D y\rho(x, y) \, dA = \frac{3}{8} \int_0^1 \int_0^{2-2x} (y + 3xy + y^2) \, dy \, dx$$

$$= \frac{3}{8} \int_0^1 \left[\frac{y^2}{2} + 3x\frac{y^2}{2} + \frac{y^3}{3} \right]_{y=0}^{y=2-2x} dx$$

$$= \frac{9}{4} \int_0^1 (7 - 9x - 3x^2 + 5x^3) \, dx$$

$$= \frac{1}{4} \left[7x - 9\frac{x^2}{2} - x^3 + 5\frac{x^4}{4} \right]_0^1 = \frac{11}{16}$$

The center of mass is at the point $(\frac{3}{8}, \frac{11}{16})$.

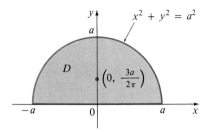

Figure 13.31

● **Example 3**   The density at any point on a semicircular lamina is proportional to the distance from the center of the circle. Find the center of mass of the lamina.

Solution   Let us place the lamina as the upper half of the circle $x^2 + y^2 = a^2$ (Figure 13.31). Then the distance from a point (x, y) to the center of the circle (the origin) is $\sqrt{x^2 + y^2}$. Therefore the density function is

$$\rho(x, y) = K\sqrt{x^2 + y^2}$$

where K is some constant. Both the density function and the shape of the lamina suggest that we convert to polar coordinates. Then $\sqrt{x^2 + y^2} = r$ and the region D is given by $0 \leqslant r \leqslant a, 0 \leqslant \theta \leqslant \pi$. Thus the mass of the lamina is

$$m = \iint_D \rho(x, y)\, dA = \iint_D K\sqrt{x^2 + y^2}\, dA$$

$$= \int_0^\pi \int_0^a (Kr)r\, dr\, d\theta = K\int_0^\pi d\theta \int_0^a r^2\, dr$$

$$= K\pi \frac{r^3}{3}\Big]_0^a = \frac{K\pi a^3}{3}$$

Both the lamina and the density function are symmetric with respect to the y-axis, so the center of mass must lie on the y-axis, that is, $\bar{x} = 0$. The y-coordinate is given by

$$\bar{y} = \frac{1}{m} \iint_D y\rho(x, y)\, dA$$

$$= \frac{3}{K\pi a^3} \int_0^\pi \int_0^a r\sin\theta(Kr)r\, dr\, d\theta$$

$$= \frac{3}{\pi a^3} \int_0^\pi \sin\theta\, d\theta \int_0^a r^3\, dr$$

$$= \frac{3}{\pi a^3} \Big[-\cos\theta\Big]_0^\pi \left[\frac{r^4}{4}\right]_0^a$$

$$= \frac{3}{\pi a^3} \frac{2a^4}{4} = \frac{3a}{2\pi}$$

Therefore the center of mass is at the point $(0, 3a/2\pi)$. (Compare this with Example 3 in Section 8.7 where we found that the center of mass of a lamina with the same shape but uniform density is at the point $(0, 4a/3\pi)$.)   ●

Moment of Inertia

The **moment of inertia** (also called the **second moment**) of a particle of mass m about an axis is defined to be mr^2 where r is the distance from the particle to the axis. We extend this concept to a lamina with density function $\rho(x, y)$ and occupying a region D by proceeding as we did for ordinary moments.

We partition D into small rectangles, approximate the moment of inertia of each subrectangle about the x-axis, and take the limit of the sum as $\|P\|$ approaches 0. The result is the **moment of inertia** of the lamina **about the x-axis:**

(13.34)

$$I_x = \lim_{\|P\| \to 0} \sum_{i=1}^{m} \sum_{j=1}^{n} (y_j^*)^2 \rho(x_i^*, y_j^*) \Delta A_{ij} = \iint_D y^2 \rho(x, y) \, dA$$

Similarly, the **moment of inertia about the y-axis** is

(13.35)

$$I_y = \lim_{\|P\| \to 0} \sum_{i=1}^{m} \sum_{j=1}^{n} (x_i^*)^2 \rho(x_i^*, y_j^*) \Delta A_{ij} = \iint_D x^2 \rho(x, y) \, dA$$

It is also of interest to consider the **moment of inertia about the origin,** also called the **polar moment of inertia:**

(13.36)

$$I_0 = \lim_{\|P\| \to 0} \sum_{i=1}^{m} \sum_{j=1}^{n} [(x_i^*)^2 + (y_j^*)^2] \rho(x_i^*, y_j^*) \Delta A_{ij} = \iint_D (x^2 + y^2) \rho(x, y) \, dA$$

Note that $I_0 = I_x + I_y$.

● **Example 4**   Find the moments of inertia I_x, I_y, and I_0 of a homogeneous disk D with density $\rho(x, y) = \rho$, center the origin, and radius a.

Solution   The boundary of D is the circle $x^2 + y^2 = a^2$ and in polar coordinates D is described by $0 \leqslant \theta \leqslant 2\pi, 0 \leqslant r \leqslant a$. Let us compute I_0 first:

$$I_0 = \iint_D (x^2 + y^2) \rho \, dA = \rho \int_0^{2\pi} \int_0^a r^2 r \, dr \, d\theta$$

$$= \rho \int_0^{2\pi} d\theta \int_0^a r^3 \, dr = 2\pi\rho \left[\frac{r^4}{4} \right]_0^a = \frac{\pi\rho a^4}{2}$$

Instead of computing I_x and I_y directly, we use the facts that $I_x + I_y = I_0$ and $I_x = I_y$ (from the symmetry of the problem). Thus

$$I_x = I_y = \frac{I_0}{2} = \frac{\pi\rho a^4}{4}$$

●

In Example 4, notice that the mass of the disk is

$$m = \text{density} \times \text{area} = \rho(\pi a^2)$$

so the moment of inertia of the disk about the origin (like a wheel about its axle) can be written as

$$I_0 = \tfrac{1}{2}ma^2$$

Thus if we increase the mass or the radius of the disk, we thereby increase the moment of inertia. In general, the moment of inertia plays much the same role in rotational motion that mass plays in linear motion. The moment of inertia of a wheel is what makes it difficult to start or stop the rotation of the wheel, just as the mass of a car is what makes it difficult to start or stop the motion of the car.

The **radius of gyration of a lamina about an axis** is the number R such that

(13.37)
$$mR^2 = I$$

where m is the mass of the lamina and I is the moment of inertia about the given axis. Equation 13.37 says that if the mass of the lamina were concentrated at a distance R from the axis, then the moment of inertia of this "point mass" would be the same as the moment of inertia of the lamina.

In particular, the radius of gyration $\bar{\bar{y}}$ with respect to the x-axis and the radius of gyration $\bar{\bar{x}}$ with respect to the y-axis are given by the equations

(13.38)
$$m\bar{\bar{y}}^2 = I_x \qquad m\bar{\bar{x}}^2 = I_y$$

Thus $(\bar{\bar{x}}, \bar{\bar{y}})$ is the point at which the mass of the lamina can be concentrated without changing the moments of inertia with respect to the coordinate axes. (Note the analogy with the center of mass.)

● **Example 5**   Find the radius of gyration about the x-axis of the disk in Example 4.

Solution   As noted, the mass of the disk is $m = \rho\pi a^2$, so from Equations 13.38 we have

$$\bar{\bar{y}}^2 = \frac{I_x}{m} = \frac{\dfrac{\pi\rho a^4}{4}}{\rho\pi a^2} = \frac{a^2}{4}$$

Therefore the radius of gyration about the x-axis is

$$\bar{\bar{y}} = \frac{a}{2}$$

which is half the radius of the disk.   ●

EXERCISES 13.5

1. Electric charge is distributed over the rectangle $0 \leqslant x \leqslant 2, 1 \leqslant y \leqslant 2$ so that the charge density at (x, y) is $\sigma(x, y) = x^2 + 3y^2$ measured in coulombs per square meter. Find the total charge on the rectangle.

2. Electric charge is distributed over the unit disk $x^2 + y^2 \leqslant 1$ so that the charge density at (x, y) is $\sigma(x, y) = 1 + x^2 + y^2$ measured in coulombs per square meter. Find the total charge on the disk.

In Exercises 3–12 find the mass and center of mass of the lamina that occupies the given region D and has the given density function ρ.

3. $D = \{(x, y) \mid -1 \leqslant x \leqslant 1, 0 \leqslant y \leqslant 1\}$; $\rho(x, y) = x^2$

4. $D = \{(x, y) \mid 0 \leqslant x \leqslant 2, 0 \leqslant y \leqslant 3\}$; $\rho(x, y) = y$

5. D is the triangular region with vertices $(0, 0), (2, 1),$ $(0, 3)$; $\rho(x, y) = x + y$.

6. D is the triangular region with vertices $(0, 0), (1, 1),$ $(4, 0)$; $\rho(x, y) = x$.

7. D is the region in the first quadrant bounded by the parabola $y = x^2$ and the line $y = 1$; $\rho(x, y) = xy$.

8. D is bounded by the parabola $y = 9 - x^2$ and the x-axis; $\rho(x, y) = y$.

9. D is bounded by the parabola $x = y^2$ and the line $y = x - 2$; $\rho(x, y) = 3$.

10. D is bounded by the cardioid $r = 1 + \sin \theta$; $\rho(x, y) = 2$.

11. $D = \{(x, y) \mid 0 \leqslant y \leqslant \sin x, 0 \leqslant x \leqslant \pi\}$; $\rho(x, y) = y$

12. $D = \left\{(x, y) \mid 0 \leqslant y \leqslant \cos x, 0 \leqslant x \leqslant \dfrac{\pi}{2}\right\}$; $\rho(x, y) = x$

13. A lamina occupies the part of the disk $x^2 + y^2 \leqslant 1$ in the first quadrant. Find its center of mass if the density

at any point is proportional to its distance from the x-axis.

14. Find the center of mass of the lamina in Exercise 13 if the density at any point is proportional to the square of its distance from the origin.

15. Find the center of mass of a lamina in the shape of an isosceles right triangle with equal sides of length a if the density at any point is proportional to the square of the distance from the vertex opposite the hypotenuse.

16. A lamina occupies the region inside the circle $x^2 + y^2 = 2y$ but outside the circle $x^2 + y^2 = 1$. Find the center of mass if the density at any point is inversely proportional to its distance from the origin.

17. Find the moments of inertia I_x, I_y, I_0 for the lamina of Exercise 7.

18. Find the moments of inertia I_x, I_y, I_0 for the lamina of Exercise 4.

19. Find the moments of inertia I_x, I_y, I_0 for the lamina of Exercise 9.

20. Find the moments of inertia I_x, I_y, I_0 for the lamina of Exercise 14.

21. A lamina with constant density $\rho(x, y) = \rho$ occupies a square with vertices $(0, 0), (a, 0), (a, a),$ and $(0, a)$. Find the moments of inertia I_x and I_y and the radii of gyration $\bar{\bar{x}}$ and $\bar{\bar{y}}$.

22. A lamina with constant density $\rho(x, y) = \rho$ occupies the region under the curve $y = \sin x$ from $x = 0$ to $x = \pi$. Find the moments of inertia I_x, I_y and the radii of gyration $\bar{\bar{x}}$ and $\bar{\bar{y}}$.

23. Show that the formulas for $\bar{x}$ and $\bar{y}$ in (13.33) become Equations 8.40 when $\rho(x, y) = \rho$ (the density is constant).

SECTION 13.6

Triple Integrals

Just as we defined single integrals for functions of one variable and double integrals for functions of two variables, so we can define triple integrals for functions of three variables. Let us first deal with the simplest case where f is defined on a rectangular box:

(13.39)
$$B = \{(x, y, z) \mid a \leqslant x \leqslant b, c \leqslant y \leqslant d, r \leqslant z \leqslant s\}$$

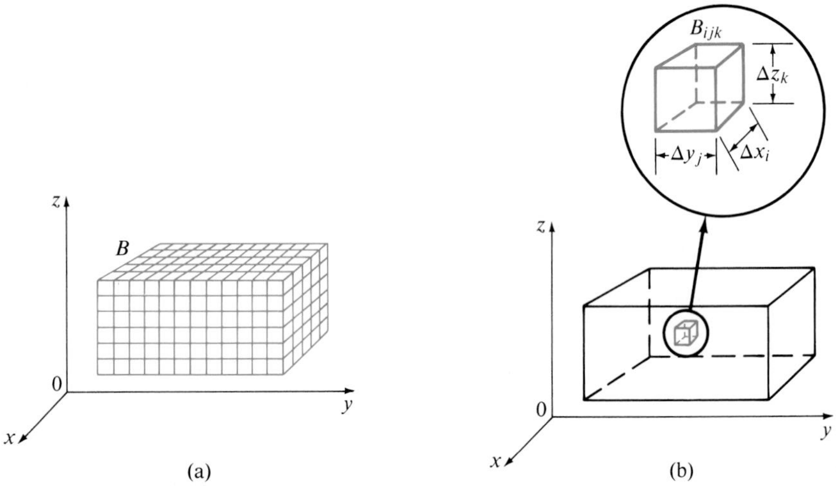

Figure 13.32

(a)   (b)

The first step is to partition the intervals $[a, b]$, $[c, d]$, and $[r, s]$ as follows:

$$a = x_0 < x_1 < \cdots < x_{i-1} < x_i < \cdots < x_l = b$$

$$c = y_0 < y_1 < \cdots < y_{j-1} < y_j < \cdots < y_m = d$$

$$r = z_0 < z_1 < \cdots < z_{k-1} < z_k < \cdots < z_n = s$$

The planes through these partition points parallel to the coordinate planes will divide the box B into lmn sub-boxes

$$B_{ijk} = [x_{i-1}, x_i] \times [y_{j-1}, y_j] \times [z_{k-1}, z_k]$$

which are shown in Figure 13.32.

The volume of B_{ijk} is

$$\Delta V_{ijk} = \Delta x_i \, \Delta y_j \, \Delta z_k$$

where $\Delta x_i = x_i - x_{i-1}$ $\Delta y_j = y_j - y_{j-1}$ $\Delta z_k = z_k - z_{k-1}$

Then we form the **triple Riemann sum**

(13.40)
$$\sum_{i=1}^{l} \sum_{j=1}^{m} \sum_{k=1}^{n} f(x_i^*, y_j^*, z_k^*) \, \Delta V_{ijk}$$

where (x_i^*, y_j^*, z_k^*) is in B_{ijk}. We define the **norm** of the partition P to be the length of the longest diagonal of all the boxes B_{ijk} and we denote the norm by $\|P\|$. Then, by analogy with the definition of a double integral (13.4), we define the triple integral as the limit of the triple Riemann sums in (13.40).

Definition (13.41)

> The **triple integral** of f over the box B is
>
> $$\iiint\limits_{B} f(x, y, z) \, dV = \lim_{\|P\| \to 0} \sum_{i=1}^{l} \sum_{j=1}^{m} \sum_{k=1}^{n} f(x_i^*, y_j^*, z_k^*) \, \Delta V_{ijk}$$
>
> if this limit exists.

The precise meaning of the limit in Definition 13.41 is similar to that given in Note 1 after Definition 13.4. Again, the triple integral will always exist if f is continuous. An alternative notation is

$$\iiint_B f(x, y, z)\, dV = \iiint_B f(x, y, z)\, dx\, dy\, dz$$

Just as for double integrals, the practical method for evaluating triple integrals is to express them as iterated integrals as follows.

Fubini's Theorem for Triple Integrals (13.42)

If f is integrable over the rectangular box $B = [a, b] \times [c, d] \times [r, s]$, then

$$\iiint_B f(x, y, z)\, dV = \int_r^s \int_c^d \int_a^b f(x, y, z)\, dx\, dy\, dz$$

The iterated integral on the right side of Fubini's Theorem means that we integrate first with respect to x (keeping y and z fixed), then we integrate with respect to y (keeping z fixed), and finally we integrate with respect to z. There are five other possible orders in which we can integrate. For instance, if we integrate with respect to y, then z, and then x, we have

$$\iiint_B f(x, y, z)\, dV = \int_a^b \int_r^s \int_c^d f(x, y, z)\, dy\, dz\, dx$$

● **Example 1**   Evaluate the triple integral $\iiint_B xyz^2\, dV$ where B is the rectangular box given by

$$B = \{(x, y, z)\,|\,0 \leqslant x \leqslant 1, -1 \leqslant y \leqslant 2, 0 \leqslant z \leqslant 3\}$$

Solution   We could use any of the six possible orders of integration. If we choose to integrate with respect to x, then y, and then z, we obtain

$$\iiint_B xyz^2\, dV = \int_0^3 \int_{-1}^2 \int_0^1 xyz^2\, dx\, dy\, dz$$

$$= \int_0^3 \int_{-1}^2 \left[\frac{x^2 yz^2}{2}\right]_{x=0}^{x=1} dy\, dz$$

$$= \int_0^3 \int_{-1}^2 \frac{yz^2}{2}\, dy\, dz = \int_0^3 \left[\frac{y^2 z^2}{4}\right]_{y=-1}^{y=2} dz$$

$$= \int_0^3 \frac{3z^2}{4}\, dz = \frac{z^3}{4}\bigg]_0^3 = \frac{27}{4}$$

●

Now we define the **triple integral over a general bounded region E** in three-dimensional space (a solid) by the same procedure that we used for double integrals (13.15). We enclose E in a box B of the type given by Equation 13.39.

Then we define F so that it agrees with f on E but is 0 for points in B that are outside E. By definition,

$$\iiint_E f(x, y, z)\, dV = \iiint_B F(x, y, z)\, dV$$

This integral will exist if f is continuous and the boundary of E is "reasonably smooth." The triple integral has essentially the same properties as the double integral (13.19–13.22).

We restrict our attention to continuous functions f and to certain simple types of regions. A solid region E is said to be of **type 1** if it lies between the graphs of two continuous functions of x and y, that is,

(13.43)
$$E = \{(x, y, z) \,|\, (x, y) \in D,\ \phi_1(x, y) \leqslant z \leqslant \phi_2(x, y)\}$$

where D is the projection of E onto the xy-plane as shown in Figure 13.33. Notice that the upper boundary of the solid E is the surface with equation $z = \phi_2(x, y)$, while the lower boundary is the surface $z = \phi_1(x, y)$.

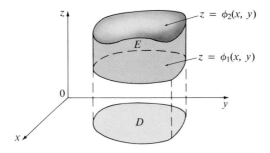

Figure 13.33

A TYPE 1 SOLID REGION

By the same sort of argument that led to (13.16), it can be shown that, if E is a type 1 region given by Equation 13.43, then

(13.44)
$$\iiint_E f(x, y, z)\, dV = \iint_D \left[\int_{\phi_1(x,y)}^{\phi_2(x,y)} f(x, y, z)\, dz \right] dA$$

The meaning of the inner integral on the right side of Equation 13.44 is that x and y are held fixed, and therefore $\phi_1(x, y)$ and $\phi_2(x, y)$ are regarded as constants, while $f(x, y, z)$ is integrated with respect to z.

In particular, if the projection D of E onto the xy-plane is a type I plane region [as in Figure 13.34(a)], then

$$E = \{(x, y, z) \,|\, a \leqslant x \leqslant b, g_1(x) \leqslant y \leqslant g_2(x), \phi_1(x, y) \leqslant z \leqslant \phi_2(x, y)\}$$

and Equation 13.44 becomes

(13.45)
$$\iiint_E f(x, y, z)\, dV = \int_a^b \int_{g_1(x)}^{g_2(x)} \int_{\phi_1(x,y)}^{\phi_2(x,y)} f(x, y, z)\, dz\, dy\, dx$$

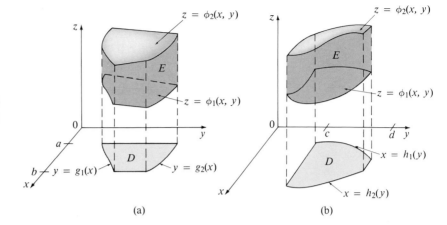

Figure 13.34
SOME TYPE 1 SOLID REGIONS

(a)

(b)

If, on the other hand, D is a type II plane region [see Figure 13.34(b)], then

$$E = \{(x, y, z) \,|\, c \leqslant y \leqslant d, h_1(y) \leqslant x \leqslant h_2(y), \phi_1(x, y) \leqslant z \leqslant \phi_2(x, y)\}$$

and Equation 13.44 becomes

(13.46)

$$\iiint_E f(x, y, z) \, dV = \int_c^d \int_{h_1(y)}^{h_2(y)} \int_{\phi_1(x,y)}^{\phi_2(x,y)} f(x, y, z) \, dz \, dx \, dy$$

● **Example 2**   Evaluate $\iiint_E z \, dV$ where E is the solid tetrahedron bounded by the four planes $x = 0$, $y = 0$, $z = 0$, and $x + y + z = 1$.

Solution   When setting up a triple integral it is wise to draw *two* diagrams: one of the solid region E and one of its projection D on the xy-plane (see Figure 13.35). The lower boundary of the tetrahedron is the plane $z = 0$ and the upper boundary is the plane $x + y + z = 1$ (or $z = 1 - x - y$), so we use $\phi_1(x, y) = 0$ and $\phi_2(x, y) = 1 - x - y$ in Formula 13.45. Notice that the planes $x + y + z = 1$ and $z = 0$ intersect in the line $x + y = 1$ (or

Figure 13.35

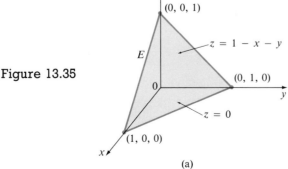

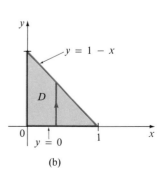

(a)

(b)

$y = 1 - x$) in the xy-plane. So the projection of E is the triangular region shown in Figure 13.35(b) and we have

(13.47) $$E = \{(x, y, z) \mid 0 \leqslant x \leqslant 1, 0 \leqslant y \leqslant 1 - x, 0 \leqslant z \leqslant 1 - x - y\}$$

This description of E as a type 1 region enables us to evaluate the integral as follows:

$$\iiint_E z \, dV = \int_0^1 \int_0^{1-x} \int_0^{1-x-y} z \, dz \, dy \, dx$$

$$= \int_0^1 \int_0^{1-x} \left[\frac{z^2}{2} \right]_{z=0}^{z=1-x-y} dy \, dx$$

$$= \frac{1}{2} \int_0^1 \int_0^{1-x} (1 - x - y)^2 \, dy \, dx$$

$$= \frac{1}{2} \int_0^1 \left[-\frac{(1 - x - y)^3}{3} \right]_{y=0}^{y=1-x} dx$$

$$= \frac{1}{6} \int_0^1 (1 - x)^3 \, dx = \frac{1}{6} \left[-\frac{(1 - x)^4}{4} \right]_0^1$$

$$= \frac{1}{24}$$ ●

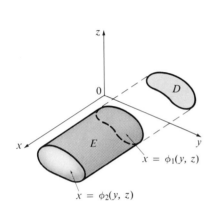

Figure 13.36

A TYPE 2 REGION

A solid region E is of **type 2** if it is of the form

$$E = \{(x, y, z) \mid (y, z) \in D, \phi_1(y, z) \leqslant x \leqslant \phi_2(y, z)\}$$

where, this time, D is the projection of E onto the yz-plane (see Figure 13.36). The back surface is $x = \phi_1(y, z)$, the front surface is $x = \phi_2(y, z)$, and we have

(13.48) $$\iiint_E f(x, y, z) \, dV = \iint_D \left[\int_{\phi_1(y,z)}^{\phi_2(y,z)} f(x, y, z) \, dx \right] dA$$

Finally, a **type 3** region is of the form

$$E = \{(x, y, z) \mid (x, z) \in D, \phi_1(x, z) \leqslant y \leqslant \phi_2(x, z)\}$$

where D is the projection of E onto the xz-plane, $y = \phi_1(x, z)$ is the left surface, and $y = \phi_2(x, z)$ is the right surface (see Figure 13.37). For this type of region we have

(13.49) $$\iiint_E f(x, y, z) \, dV = \iint_D \left[\int_{\phi_1(x,z)}^{\phi_2(x,z)} f(x, y, z) \, dy \right] dA$$

In each of Equations 13.48 and 13.49 there may be two possible expressions for the integral depending on whether D is a type I or II plane region (and corresponding to Equations 13.45 and 13.46).

● **Example 3**   Evaluate $\iiint_E \sqrt{x^2 + z^2} \, dV$, where E is the region bounded by the paraboloid $y = x^2 + z^2$ and the plane $y = 4$.

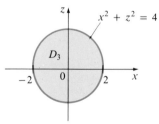

Figure 13.37

A TYPE 3 REGION

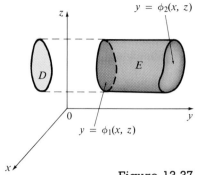

(a) Region of integration

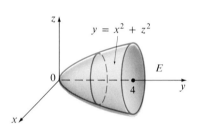

(b) Projection on xy-plane

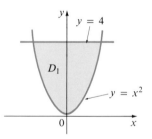

(c) Projection on xz-plane

Figure 13.38

Solution   The solid E is shown in Figure 13.38(a). If we regard it as a type 1 region, then we need to consider its projection D_1 onto the xy-plane, which is the parabolic region in Figure 13.38(b). (The trace of $y = x^2 + z^2$ in the plane $z = 0$ is the parabola $y = x^2$.) From $y = x^2 + z^2$ we obtain $z = \pm \sqrt{y - x^2}$, so the lower boundary surface of E is $z = -\sqrt{y - x^2}$ and the upper surface is $z = \sqrt{y - x^2}$. Therefore the description of E as a type 1 region is

$$E = \{(x, y, z)\,|-2 \leqslant x \leqslant 2, x^2 \leqslant y \leqslant 4, -\sqrt{y - x^2} \leqslant z \leqslant \sqrt{y - x^2}\}$$

and so we obtain

$$\iiint_E \sqrt{x^2 + z^2}\, dV = \int_{-2}^{2} \int_{x^2}^{4} \int_{-\sqrt{y-x^2}}^{\sqrt{y-x^2}} \sqrt{x^2 + z^2}\, dz\, dy\, dx$$

Although this expression is correct, it is extremely difficult to evaluate. So let us instead consider E as a type 3 region. As such, its projection D_3 onto the xz-plane is the disk $x^2 + z^2 \leqslant 4$ shown in Figure 13.38(c).

Then the left boundary of E is the paraboloid $y = x^2 + z^2$ and the right boundary is the plane $y = 4$, so taking $\phi_1(x, z) = x^2 + z^2$ and $\phi_2(x, z) = 4$ in Equation 13.49, we have

$$\iiint_E \sqrt{x^2 + z^2}\, dV = \iint_{D_3} \left[\int_{x^2+z^2}^{4} \sqrt{x^2 + z^2}\, dy \right] dA$$

$$= \iint_{D_3} (4 - x^2 - z^2)\sqrt{x^2 + z^2}\, dA$$

Although this integral could be written as

$$\int_{-2}^{2} \int_{-\sqrt{4-x^2}}^{\sqrt{4-x^2}} (4 - x^2 - z^2)\sqrt{x^2 + z^2}\, dz\, dx$$

it is easier to convert to polar coordinates in the xz-plane: $x = r\cos\theta$, $z = r\sin\theta$. This gives

$$\iiint_E \sqrt{x^2 + z^2}\, dV = \iint_{D_3} (4 - x^2 - z^2)\sqrt{x^2 + z^2}\, dA$$

$$= \int_0^{2\pi} \int_0^2 (4 - r^2) r\, r\, dr\, d\theta$$

$$= \int_0^{2\pi} d\theta \int_0^2 (4r^2 - r^4)\, dr = 2\pi \left[\frac{4r^3}{3} - \frac{r^5}{5} \right]_0^2$$

$$= \frac{128\pi}{15} \qquad\qquad\bullet$$

Note: The most difficult step in evaluating a triple integral is setting up an expression for the region of integration (such as Equation 13.47 in Example 2). Remember that the limits of integration in the inner integral contain at most two variables, the limits of integration in the middle integral contain at most one variable, and the limits of integration in the outer integral must be constants.

Within Figure 13.37 labels: $y = \phi_2(x, z)$, E, D, 0, y, $y = \phi_1(x, z)$, x, z

Applications of Triple Integrals

Recall that if $f(x) \geq 0$, then the single integral $\int_a^b f(x)\,dx$ represents the area under the curve $y = f(x)$ from a to b, and if $f(x, y) \geq 0$, then the double integral $\iint_D f(x, y)\,dA$ represents the volume under the surface $z = f(x, y)$ and above D. The corresponding interpretation of a triple integral $\iiint_E f(x, y, z)\,dV$, where $f(x, y, z) \geq 0$, is not very useful because it would be the "hypervolume" of a four-dimensional object and, of course, that is impossible to visualize. (Remember that E is just the *domain* of f; the graph of f lies in four-dimensional space.) Nonetheless, the triple integral $\iiint_E f(x, y, z)\,dV$ can be interpreted in different ways in different physical situations, depending on the physical interpretations of x, y, z, and $f(x, y, z)$.

Let us begin with the special case where $f(x, y, z) = 1$ for all points in E. Then the triple integral does represent the volume of E:

(13.50)
$$V(E) = \iiint_E dV$$

For example, you can see this in the case of a type 1 region simply by putting $f(x, y, z) = 1$ in Formula 13.44:

$$\iiint_E 1\,dV = \iint_D \left[\int_{\phi_1(x,y)}^{\phi_2(x,y)} dz \right] dA = \iint_D [\phi_2(x, y) - \phi_1(x, y)]\,dA$$

and from Section 13.3 we know this represents the volume that lies between the surfaces $z = \phi_1(x, y)$ and $z = \phi_2(x, y)$.

● **Example 4**   Use a triple integral to find the volume of the tetrahedron T bounded by the planes $x + 2y + z = 2$, $x = 2y$, $x = 0$, and $z = 0$.

Solution   The tetrahedron T and its projection D on the xy-plane are shown in Figure 13.15. The lower boundary of T is the plane $z = 0$ and the upper boundary is the plane $x + 2y + z = 2$, that is, $z = 2 - x - 2y$. Therefore we have

$$V(T) = \iiint_T dV = \int_0^1 \int_{x/2}^{1-x/2} \int_0^{2-x-2y} dz\,dy\,dx$$

$$= \int_0^1 \int_{x/2}^{1-x/2} (2 - x - 2y)\,dy\,dx = \frac{1}{3}$$

by the same calculation as in Example 4 in Section 13.3.

(Notice that it is not necessary to use triple integrals to compute volumes. They simply give an alternative method for setting up the calculation.)
●

All of the applications of double integrals in Section 13.5 can be immediately extended to triple integrals. For example, if the density function of a solid object that occupies the region E is $\rho(x, y, z)$, in units of mass per unit volume, at any given point (x, y, z), then its **mass** is

(13.51)
$$m = \iiint_E \rho(x, y, z) \, dV$$

and its **moments** about the three coordinate planes are

(13.52)
$$M_{yz} = \iiint_E x\rho(x, y, z) \, dV \qquad M_{xz} = \iiint_E y\rho(x, y, z) \, dV$$

$$M_{xy} = \iiint_E z\rho(x, y, z) \, dV$$

The **center of mass** is located at the point $(\bar{x}, \bar{y}, \bar{z})$ where

(13.53)
$$\bar{x} = \frac{M_{yz}}{m} \qquad \bar{y} = \frac{M_{xz}}{m} \qquad \bar{z} = \frac{M_{xy}}{m}$$

If the density is constant, the center of mass of the solid is called the **centroid** of E. The **moments of inertia** about the three coordinate axes are

(13.54)
$$I_x = \iiint_E (y^2 + z^2)\rho(x, y, z) \, dV \qquad I_y = \iiint_E (x^2 + z^2)\rho(x, y, z) \, dV$$

$$I_z = \iiint_E (x^2 + y^2)\rho(x, y, z) \, dV$$

Also, as in Section 13.5, the total *electric charge* on a solid object occupying a region E and having charge density $\sigma(x, y, z)$ is

$$Q = \iiint_E \sigma(x, y, z) \, dV$$

● **Example 5**   Find the center of mass of a solid of constant density that is bounded by the parabolic cylinder $x = y^2$ and the planes $x = z$, $z = 0$, and $x = 1$.

Solution   The solid E and its projection onto the xy-plane are shown in Figure 13.39. The lower and upper surfaces of E are the planes $z = 0$ and $z = x$, so we describe E as a type 1 region:

$$E = \{(x, y, z) \,|\, -1 \leqslant y \leqslant 1, \, y^2 \leqslant x \leqslant 1, \, 0 \leqslant z \leqslant x\}$$

Then, if the density is $\rho(x, y, z) = \rho$, the mass is

$$m = \iiint_E \rho \, dV = \int_{-1}^{1} \int_{y^2}^{1} \int_{0}^{x} \rho \, dz \, dx \, dy$$

$$= \rho \int_{-1}^{1} \int_{y^2}^{1} x \, dx \, dy = \rho \int_{-1}^{1} \left[\frac{x^2}{2}\right]_{x=y^2}^{x=1} dy$$

$$= \frac{\rho}{2} \int_{-1}^{1} (1 - y^4) \, dy = \rho \int_{0}^{1} (1 - y^4) \, dy$$

$$= \rho \left[y - \frac{y^5}{5}\right]_{0}^{1} = \frac{4\rho}{5}$$

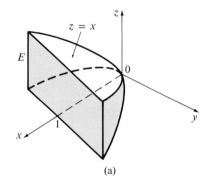

(a)

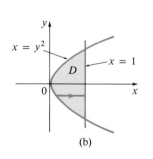

(b)

Figure 13.39

Because of the symmetry of E and ρ about the xz-plane, we can immediately say that $M_{xz} = 0$ and therefore $\bar{y} = 0$. The other moments are

$$
M_{yz} = \iiint_E x\rho \, dV = \int_{-1}^{1} \int_{y^2}^{1} \int_0^x x\rho \, dz \, dx \, dy
$$

$$
= \rho \int_{-1}^{1} \int_{y^2}^{1} x^2 \, dx \, dy = \rho \int_{-1}^{1} \left[\frac{x^3}{3} \right]_{x=y^2}^{x=1} dy
$$

$$
= \frac{2\rho}{3} \int_0^1 (1 - y^6) \, dy = \frac{2\rho}{3} \left[y - \frac{y^7}{7} \right]_0^1 = \frac{4\rho}{7}
$$

$$
M_{xy} = \iiint_E z\rho \, dV = \int_{-1}^{1} \int_{y^2}^{1} \int_0^x z\rho \, dz \, dx \, dy
$$

$$
= \rho \int_{-1}^{1} \int_{y^2}^{1} \left[\frac{z^2}{2} \right]_{z=0}^{z=x} dx \, dy = \frac{\rho}{2} \int_{-1}^{1} \int_{y^2}^{1} x^2 \, dx \, dy
$$

$$
= \frac{\rho}{3} \int_0^1 (1 - y^6) \, dy = \frac{2\rho}{7}
$$

Therefore the center of mass is

$$
(\bar{x}, \bar{y}, \bar{z}) = \left(\frac{M_{yz}}{m}, \frac{M_{xz}}{m}, \frac{M_{xy}}{m} \right) = \left(\frac{5}{7}, 0, \frac{5}{14} \right)
$$

EXERCISES 13.6

1. Evaluate the integral in Example 1 integrating first with respect to z, then x, and then y.

2. Evaluate the integral $\iiint_E (x^2 + yz) \, dV$, where

$$
E = \{(x, y, z) \mid 0 \leqslant x \leqslant 2, -3 \leqslant y \leqslant 0, -1 \leqslant z \leqslant 1\}
$$

using six different orders of integration.

In Exercises 3–6 evaluate the iterated integral.

3. $\int_0^1 \int_0^z \int_0^y xyz \, dx \, dy \, dz$

4. $\int_0^1 \int_x^{2x} \int_0^{x+y} 2xy \, dz \, dy \, dx$

5. $\int_0^{\pi} \int_0^2 \int_0^{\sqrt{4 - z^2}} z \sin y \, dx \, dz \, dy$

6. $\int_0^3 \int_0^{\sqrt{9 - x^2}} \int_0^x yz \, dy \, dz \, dx$

In Exercises 7–16 evaluate the triple integral.

7. $\iiint_E yz \, dV$, where
   $E = \{(x, y, z) \mid 0 \leqslant z \leqslant 1, 0 \leqslant y \leqslant 2z, 0 \leqslant x \leqslant z + 2\}$

8. $\iiint_E e^x \, dV$, where
   $E = \{(x, y, z) \mid 0 \leqslant y \leqslant 1, 0 \leqslant x \leqslant y, 0 \leqslant z \leqslant x + y\}$

9. $\iiint_E y \, dV$, where E lies under the plane $z = x + 2y$ and above the region in the xy-plane bounded by the curves $y = x^2$, $y = 0$, and $x = 1$

10. $\iiint_E x \, dV$, where E is bounded by the planes $x = 0$, $y = 0$, $z = 0$, and $3x + 2y + z = 6$

11. $\iiint_E xy \, dV$, where E is the solid tetrahedron with vertices $(0, 0, 0)$, $(1, 0, 0)$, $(0, 2, 0)$, and $(0, 0, 3)$

12. $\iiint_E xz \, dV$, where E is the solid tetrahedron with vertices $(0, 0, 0)$, $(0, 1, 0)$, $(1, 1, 0)$, and $(0, 1, 1)$

13. $\iiint_E z\,dV$, where E is bounded by the planes $x = 0$, $y = 0$, $z = 0$, $y + z = 1$, and $x + z = 1$

14. $\iiint_E (x + 2y)\,dV$, where E is bounded by the parabolic cylinder $y = x^2$ and the planes $x = z$, $x = y$, and $z = 0$

15. $\iiint_E z\,dV$, where E is bounded by the paraboloid $x = 4y^2 + 4z^2$ and the plane $x = 4$

16. $\iiint_E z\,dV$, where E is bounded by the cylinder $y^2 + z^2 = 9$ and the planes $y = 0$, $x = 1$, and $z = 0$ in the first octant

In Exercises 17–20 express the integral $\iiint_E f(x, y, z)\,dV$ as an iterated integral in six different ways, where E is the solid bounded by the given surfaces.

17. $x^2 + z^2 = 4$, $y = 0$, $y = 6$

18. $z = 0$, $x = 0$, $y = 2$, $z = y - 2x$

19. $z = 0$, $z = y$, $x^2 = 1 - y$

20. $9x^2 + 4y^2 + z^2 = 1$

In Exercises 21 and 22 give five other iterated integrals that are equal to the given iterated integral.

21. $\int_0^1 \int_y^1 \int_0^y f(x, y, z)\,dz\,dx\,dy$

22. $\int_0^1 \int_0^{x^2} \int_0^y f(x, y, z)\,dz\,dy\,dx$

In Exercises 23 and 24 use a triple integral to find the volume of the given solid.

23. The wedge in the first octant that is cut from the cylinder $y^2 + z^2 = 1$ by the planes $y = x$ and $x = 1$

24. The solid bounded by the elliptic cylinder $4x^2 + z^2 = 4$ and the planes $y = 0$ and $y = z + 2$

In Exercises 25–28 find the mass and center of mass of the given solid E with the given density function ρ.

25. E is the solid of Exercise 9; $\rho(x, y, z) = 2$.

26. E is bounded by the parabolic cylinder $z = 1 - y^2$ and the planes $x + z = 1$, $x = 0$, and $z = 0$; $\rho(x, y, z) = 4$.

27. E is the cube given by $0 \leqslant x \leqslant a$, $0 \leqslant y \leqslant a$, $0 \leqslant z \leqslant a$; $\rho(x, y, z) = x^2 + y^2 + z^2$.

28. E is the tetrahedron bounded by the planes $x = 0$, $y = 0$, $z = 0$, $x + y + z = 1$; $\rho(x, y, z) = y$.

In Exercises 29–32 set up, but do not evaluate, integral expressions for (a) the mass, (b) the center of mass, (c) the moment of inertia about the z-axis.

29. The solid in the first octant bounded by the cylinder $x^2 + y^2 = 1$ and the planes $y = z$, $x = 0$, and $z = 0$; $\rho(x, y, z) = 1 + x + y + z$

30. The hemisphere $x^2 + y^2 + z^2 = 1$, $z \geqslant 0$; $\rho(x, y, z) = \sqrt{x^2 + y^2 + z^2}$

31. The solid of Exercise 15; $\rho(x, y, z) = x^2 + y^2 + z^2$

32. The solid of Exercise 16; $\rho(x, y, z) = x^2 + y^2$

SECTION 13.7

Triple Integrals in Cylindrical and Spherical Coordinates

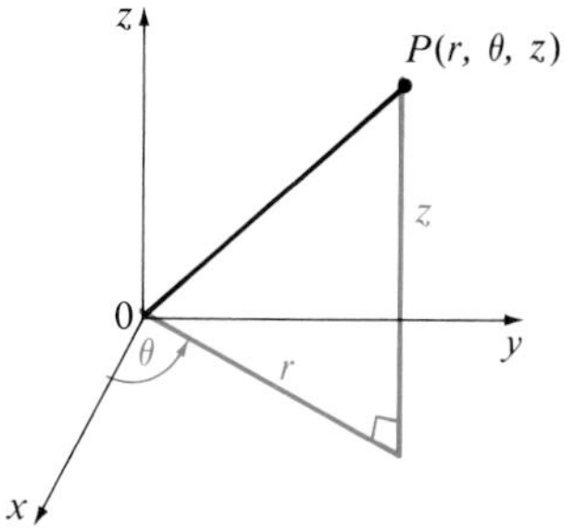

Figure 13.40

We saw in Section 13.4 that some double integrals are easier to evaluate using polar coordinates. In this section we shall see that some triple integrals are easier to evaluate using cylindrical or spherical coordinates.

Recall from Section 11.10 that the cylindrical coordinates of a point P are (r, θ, z) where r, θ, and z are shown in Figure 13.40. Suppose that E is a type 1 region whose projection D on the xy-plane is conveniently described in polar coordinates (see Figure 13.41). In particular, suppose that f is continuous and

$$E = \{(x, y, z)\,|\,(x, y) \in D,\ \phi_1(x, y) \leqslant z \leqslant \phi_2(x, y)\}$$

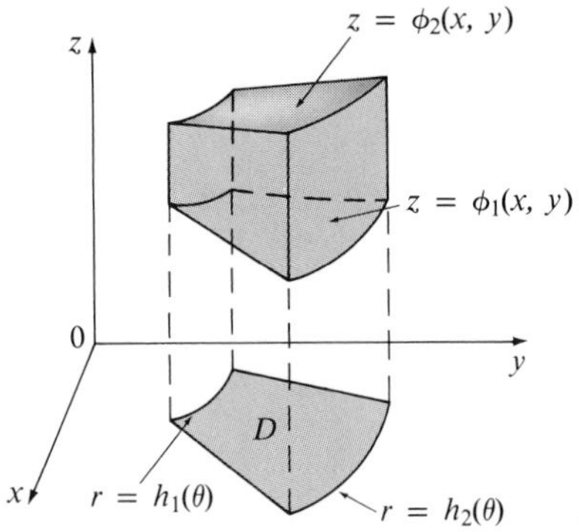

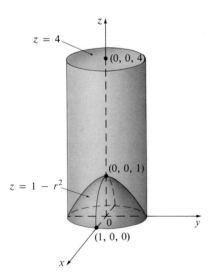

$$z = \phi_2(x, y)$$
$$z = \phi_1(x, y)$$
$$r = h_1(\theta)$$
$$r = h_2(\theta)$$

Figure 13.41

where D is given in polar coordinates by

$$D = \{(r, \theta) \mid \alpha \leqslant \theta \leqslant \beta, h_1(\theta) \leqslant r \leqslant h_2(\theta)\}$$

We know from Equation 13.44 that

(13.55)
$$\iiint\limits_{E} f(x, y, z)\, dV = \iint\limits_{D} \left[\int_{\phi_1(x,y)}^{\phi_2(x,y)} f(x, y, z)\, dz \right] dA$$

But we also know how to evaluate double integrals in polar coordinates. In fact, combining Equation 13.55 with (13.27), we obtain

(13.56)
$$\iiint\limits_{E} f(x, y, z)\, dV = \int_{\alpha}^{\beta} \int_{h_1(\theta)}^{h_2(\theta)} \int_{\phi_1(r\cos\theta,\, r\sin\theta)}^{\phi_2(r\cos\theta,\, r\sin\theta)} f(r\cos\theta, r\sin\theta, z)\, r\, dz\, dr\, d\theta$$

Formula 13.56 is the **formula for triple integration in cylindrical coordinates.** It says that we convert a triple integral from rectangular to cylindrical coordinates by writing $x = r\cos\theta$, $y = r\sin\theta$, leaving z as it is, using the appropriate limits of integration for z, r, and θ, and replacing dV by $r\, dz\, dr\, d\theta$. It is worthwhile to use this formula when E is a solid region easily described in cylindrical coordinates, and especially when the function $f(x, y, z)$ involves the expression $x^2 + y^2$.

● **Example 1**   A solid E lies within the cylinder $x^2 + y^2 = 1$, below the plane $z = 4$, and above the paraboloid $z = 1 - x^2 - y^2$ (see Figure 13.42). The density at any point is proportional to its distance from the axis of the cylinder. Find the mass of E.

Solution   In cylindrical coordinates the cylinder is $r = 1$ and the paraboloid is $z = 1 - r^2$, so we can write

$$E = \{(r, \theta, z) \mid 0 \leqslant \theta \leqslant 2\pi, 0 \leqslant r \leqslant 1, 1 - r^2 \leqslant z \leqslant 4\}$$

Since the density at (x, y, z) is proportional to the distance from the z-axis, the density function is

$$f(x, y, z) = K\sqrt{x^2 + y^2} = Kr$$

Figure 13.42

where K is the proportionality constant. Therefore, from Formula 13.51, the mass of E is

$$m = \iiint_E K\sqrt{x^2 + y^2}\, dV = \int_0^{2\pi} \int_0^1 \int_{1-r^2}^4 (Kr)r\, dz\, dr\, d\theta$$

$$= \int_0^{2\pi} d\theta \int_0^1 Kr^2[4 - (1 - r^2)]\, dr$$

$$= 2\pi K \int_0^1 (3r^2 + r^4)\, dr$$

$$= 2\pi K \left[r^3 + \frac{r^5}{5} \right]_0^1 = \frac{12\pi K}{5} \qquad \bullet$$

Figure 13.43

SPHERICAL COORDINATES OF A POINT P

Spherical Coordinates

In Section 11.10 we defined the spherical coordinates (ρ, θ, ϕ) of a point (see Figure 13.43) and we demonstrated the following relationships between rectangular coordinates and spherical coordinates:

(13.57) $x = \rho \sin \phi \cos \theta \qquad y = \rho \sin \phi \sin \theta \qquad z = \rho \cos \phi$

In this coordinate system the analogue of a rectangular box is a **spherical wedge**

$$E = \{(\rho, \theta, \phi) \,|\, a \leqslant \rho \leqslant b, \alpha \leqslant \theta \leqslant \beta, c \leqslant \phi \leqslant d\}$$

where $a \geqslant 0$, $\beta - \alpha \leqslant 2\pi$, and $d - c \leqslant \pi$. To integrate over such a region we consider a spherical partition P of E into smaller spherical wedges E_{ijk} by means of spheres $\rho = \rho_i$, half-planes $\theta = \theta_j$, and half-cones $\phi = \phi_k$. The norm of P, $\|P\|$, is the length of the longest diagonal of these wedges. If $\|P\|$ is small, then E_{ijk} is approximately a rectangular box with dimensions $\Delta\rho_i$, $\rho_i \Delta\phi_k$, and $\rho_i \sin \phi_k \Delta\theta_j$ (see Figure 13.44). So an approximation to the volume

Figure 13.44

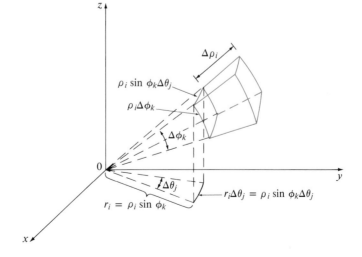

of E_{ijk} is given by

$$\Delta V_{ijk} \approx \rho_i^2 \sin \phi_k \, \Delta \rho_i \, \Delta \theta_j \, \Delta \phi_k$$

In fact, it can be shown, with the aid of the Mean Value Theorem (Exercise 29), that the volume of E_{ijk} is given exactly by

$$\Delta V_{ijk} = \tilde{\rho}_i^2 \sin \tilde{\phi}_k \, \Delta \rho_i \, \Delta \theta_j \, \Delta \phi_k$$

where $(\tilde{\rho}_i, \tilde{\theta}_j, \tilde{\phi}_k)$ is some point in E_{ijk}. Let (x_i^*, y_j^*, z_k^*) be the rectangular coordinates of this point. Although triple integrals were defined using rectangular partitions, it can be shown that spherical partitions could be used instead. Therefore

$$\iiint_E f(x, y, z) \, dV = \lim_{\|P\| \to 0} \sum_{i=1}^{l} \sum_{j=1}^{m} \sum_{k=1}^{n} f(x_i^*, y_j^*, z_k^*) \, \Delta V_{ijk}$$

$$= \lim_{\|P\| \to 0} \sum_{i=1}^{l} \sum_{j=1}^{m} \sum_{k=1}^{n} f(\tilde{\rho}_i \sin \tilde{\phi}_k \cos \tilde{\theta}_j, \, \tilde{\rho}_i \sin \tilde{\phi}_k \sin \tilde{\theta}_j, \, \tilde{\rho}_i \cos \tilde{\phi}_k) \tilde{\rho}_i^2 \sin \tilde{\phi}_k \, \Delta \rho_i \, \Delta \theta_j \, \Delta \phi_k$$

But this sum is a Riemann sum for the function

$$F(\rho, \theta, \phi) = \rho^2 \sin \phi \, f(\rho \sin \phi \cos \theta, \rho \sin \phi \sin \theta, \rho \cos \phi)$$

Consequently we have the following **formula for triple integration in spherical coordinates:**

(13.58)

$$\iiint_E f(x, y, z) \, dV$$

$$= \int_c^d \int_\alpha^\beta \int_a^b f(\rho \sin \phi \cos \theta, \rho \sin \phi \sin \theta, \rho \cos \phi) \rho^2 \sin \phi \, d\rho \, d\theta \, d\phi$$

where E is a spherical wedge given by

$$E = \{(\rho, \theta, \phi) \mid a \leqslant \rho \leqslant b, \alpha \leqslant \theta \leqslant \beta, c \leqslant \phi \leqslant d\}$$

Formula 13.58 says that we convert a triple integral from rectangular coordinates to spherical coordinates by writing $x = \rho \sin \phi \cos \theta$, $y = \rho \sin \phi \sin \theta$, $z = \rho \cos \phi$, using the appropriate limits of integration, and replacing dV by $\rho^2 \sin \phi \, d\rho \, d\theta \, d\phi$.

This formula can be extended to include more general spherical regions such as

$$E = \{(\rho, \theta, \phi) \mid \alpha \leqslant \theta \leqslant \beta, c \leqslant \phi \leqslant d, g_1(\theta, \phi) \leqslant \rho \leqslant g_2(\theta, \phi)\}$$

In this case the formula is the same as in (13.58) except that the limits of integration for ρ are $g_1(\theta, \phi)$ and $g_2(\theta, \phi)$.

Usually, spherical coordinates are used in triple integrals when surfaces such as cones and spheres form the boundary of the region of integration.

● **Example 2**   Evaluate $\iiint_B e^{(x^2+y^2+z^2)^{3/2}}\, dV$, where B is the unit ball:

$$B = \{(x, y, z)\,|\,x^2 + y^2 + z^2 \leqslant 1\}$$

Solution   Since the boundary of B is a sphere, we shall use spherical coordinates:

$$B = \{(\rho, \theta, \phi)\,|\,0 \leqslant \rho \leqslant 1, 0 \leqslant \theta \leqslant 2\pi, 0 \leqslant \phi \leqslant \pi\}$$

In addition, spherical coordinates are appropriate because

$$x^2 + y^2 + z^2 = \rho^2$$

Thus (13.58) gives

$$\iiint_B e^{(x^2+y^2+z^2)^{3/2}}\, dV = \int_0^\pi \int_0^{2\pi} \int_0^1 e^{(\rho^2)^{3/2}} \rho^2 \sin\phi\, d\rho\, d\theta\, d\phi$$

$$= \int_0^\pi \sin\phi\, d\phi \int_0^{2\pi} d\theta \int_0^1 \rho^2 e^{\rho^3}\, d\rho$$

$$= \Big[-\cos\phi\Big]_0^\pi (2\pi)\Big[\tfrac{1}{3} e^{\rho^3}\Big]_0^1$$

$$= \frac{4\pi}{3}(e - 1)$$ ●

Note: It would have been extremely awkward to evaluate the integral in Example 2 without spherical coordinates. In rectangular coordinates the iterated integral would have been

$$\int_{-1}^1 \int_{-\sqrt{1-x^2}}^{\sqrt{1-x^2}} \int_{-\sqrt{1-x^2-y^2}}^{\sqrt{1-x^2-y^2}} e^{(x^2+y^2+z^2)^{3/2}}\, dz\, dy\, dx$$

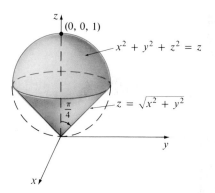

Figure 13.45

● **Example 3**   Use spherical coordinates to find the volume of the solid that lies above the cone $z = \sqrt{x^2 + y^2}$ and below the sphere $x^2 + y^2 + z^2 = z$ (see Figure 13.45).

Solution   Notice that the sphere passes through the origin and has center $(0, 0, \tfrac{1}{2})$. We write the equation of the sphere in spherical coordinates as

$$\rho^2 = \rho\cos\phi \qquad \text{or} \qquad \rho = \cos\phi$$

The cone can be written as

$$\rho\cos\phi = \sqrt{\rho^2 \sin^2\phi \cos^2\theta + \rho^2 \sin^2\phi \sin^2\theta} = \rho\sin\phi$$

This gives $\sin\phi = \cos\phi$ or $\phi = \pi/4$. Therefore the description of the solid E in spherical coordinates is

$$E = \{(\rho, \theta, \phi)\,|\,0 \leqslant \theta \leqslant 2\pi, 0 \leqslant \phi \leqslant \frac{\pi}{4}, 0 \leqslant \rho \leqslant \cos\phi\}$$

and its volume is

$$
V(E) = \iiint_E dV = \int_0^{2\pi} \int_0^{\pi/4} \int_0^{\cos\phi} \rho^2 \sin\phi \, d\rho \, d\phi \, d\theta
$$

$$
= \int_0^{2\pi} d\theta \int_0^{\pi/4} \sin\phi \left[\frac{\rho^3}{3} \right]_{\rho=0}^{\rho=\cos\phi} d\phi
$$

$$
= \frac{2\pi}{3} \int_0^{\pi/4} \sin\phi \cos^3\phi \, d\phi = \frac{2\pi}{3} \left[-\frac{\cos^4\phi}{4} \right]_0^{\pi/4}
$$

$$
= \frac{\pi}{8}
$$

EXERCISES 13.7

In Exercises 1–8 use cylindrical coordinates.

1. Evaluate $\iiint_E(x^2 + y^2)\,dV$, where E is the region bounded by the cylinder $x^2 + y^2 = 4$ and the planes $z = -1$ and $z = 2$.

2. Evaluate $\iiint_E \sqrt{x^2 + y^2}\,dV$, where E is the solid bounded by the paraboloid $z = 9 - x^2 - y^2$ and · the xy-plane.

3. Evaluate $\iiint_E y\,dV$, where E is the solid that lies between the cylinders $x^2 + y^2 = 1$ and $x^2 + y^2 = 4$, above the xy-plane, and below the plane $z = x + 2$.

4. Evaluate $\iiint_E xz\,dV$, where E is bounded by the planes $z = 0$, $z = y$, and the cylinder $x^2 + y^2 = 1$ in the half-space $y \geq 0$.

5. Find the volume of the region E bounded by the paraboloids $z = x^2 + y^2$ and $z = 36 - 3x^2 - 3y^2$.

6. Find the centroid of the region of Exercise 5.

7. Find the mass and center of mass of the solid S bounded by the paraboloid $z = 4x^2 + 4y^2$ and the plane $z = a$ $(a > 0)$ if S has constant density K.

8. Find the mass of a ball B given by $x^2 + y^2 + z^2 \leq a^2$ if the density at any point is proportional to its distance from the z-axis.

In Exercises 9–22 use spherical coordinates.

9. Evaluate $\iiint_B(x^2 + y^2 + z^2)\,dV$, where B is the unit ball $x^2 + y^2 + z^2 \leq 1$.

10. Evaluate $\iiint_H(x^2 + y^2)\,dV$, where H is the hemispherical region that lies above the xy-plane and below the sphere $x^2 + y^2 + z^2 = 1$.

11. Evaluate $\iiint_E y^2\,dV$, where E is the part of the unit ball $x^2 + y^2 + z^2 \leq 1$ that lies in the first octant.

12. Evaluate $\iiint_E xe^{(x^2+y^2+z^2)^2}\,dV$, where E is the solid that lies between the spheres $x^2 + y^2 + z^2 = 1$ and $x^2 + y^2 + z^2 = 4$ in the first octant.

13. Evaluate $\iiint_E \sqrt{x^2 + y^2 + z^2}\,dV$, where E is bounded below by the cone $\phi = \pi/6$ and above by the sphere $\rho = 2$.

14. Evaluate $\iiint_E x^2\,dV$, where E lies between the spheres $\rho = 1$ and $\rho = 3$ and above the cone $\phi = \pi/4$.

15. Find the volume of the solid that lies above the cone $\phi = \pi/3$ and below the sphere $\rho = 4\cos\phi$.

16. Find the centroid of the solid in Exercise 15.

17. Find the mass of a solid hemisphere H of radius a if the density at any point is proportional to its distance from the center of the base.

18. Find the center of mass of the solid H in Exercise 17.

19. Find the moment of inertia of the solid H in Exercise 17 about its axis.

20. Find the centroid of a solid homogeneous hemisphere of radius a.

21. Find the moment of inertia about a diameter of the base of a solid homogeneous hemisphere of radius a.

22. Find the mass and center of mass of a solid hemisphere of radius a if the density at any point is proportional to its distance from the base.

In Exercises 23–26 use cylindrical or spherical coordinates, whichever seems more appropriate.

23. Find the volume and centroid of the solid E that lies above the cone $z = \sqrt{x^2 + y^2}$ and below the sphere $x^2 + y^2 + z^2 = 1$.

24. Find the volume of the smaller wedge cut from a sphere of radius a by two planes that intersect along a diameter at an angle of $\pi/6$.

25. Evaluate $\iiint_E z \, dV$ where E lies above the paraboloid $z = x^2 + y^2$ and below the plane $z = 2y$.

26. Find the volume enclosed by the torus $\rho = \sin \phi$.

In Exercises 27 and 28 evaluate the integral by changing to spherical coordinates.

27. $\displaystyle\int_{-3}^{3} \int_{-\sqrt{9-x^2}}^{\sqrt{9-x^2}} \int_{0}^{\sqrt{9-x^2-y^2}} z\sqrt{x^2 + y^2 + z^2} \, dz \, dy \, dx$

28. $\displaystyle\int_{0}^{3} \int_{0}^{\sqrt{9-y^2}} \int_{\sqrt{x^2+y^2}}^{\sqrt{18-x^2-y^2}} (x^2 + y^2 + z^2) \, dz \, dx \, dy$

29. (a) Use cylindrical coordinates to show that the volume of the solid bounded above by the sphere $r^2 + z^2 = a^2$ and below by the cone $z = r \cot \phi_0$ (or $\phi = \phi_0$), where $0 < \phi_0 < \pi/2$, is

$$V = \frac{2\pi a^3}{3}(1 - \cos \phi_0)$$

(b) Deduce that the volume of the spherical wedge given by $\rho_1 \leqslant \rho \leqslant \rho_2$, $\theta_1 \leqslant \theta \leqslant \theta_2$, $\phi_1 \leqslant \phi \leqslant \phi_2$ is

$$\Delta V = \frac{\rho_2^3 - \rho_1^3}{3}(\cos \phi_1 - \cos \phi_2)(\theta_2 - \theta_1)$$

(c) Use the Mean Value Theorem to show that the volume in part (b) can be written as

$$\Delta V = \tilde{\rho}^2 \sin \tilde{\phi} \, \Delta\rho \, \Delta\theta \, \Delta\phi$$

where $\tilde{\rho}$ lies between ρ_1 and ρ_2, $\tilde{\phi}$ lies between ϕ_1 and ϕ_2, $\Delta\rho = \rho_2 - \rho_1$, $\Delta\theta = \theta_2 - \theta_1$, and $\Delta\phi = \phi_2 - \phi_1$.

SECTION 13.8

Change of Variables in Multiple Integrals

In one-dimensional calculus we often use a change of variable (a substitution) to simplify an integral. By reversing the roles of x and u, the Substitution Rule (5.32) can be written as

(13.59)
$$\int_a^b f(x) \, dx = \int_c^d f(g(u))g'(u) \, du$$

where $x = g(u)$ and $a = g(c)$, $b = g(d)$. Another way of writing Formula 13.59 is as follows:

(13.60)
$$\int_a^b f(x) \, dx = \int_c^d f(x(u)) \frac{dx}{du} \, du$$

A change of variables can also be useful in double integrals. We have already seen one example of this: conversion to polar coordinates. The new variables r and θ are related to the old variables x and y by the equations

$$x = r \cos \theta \qquad y = r \sin \theta$$

and the change of variables formula (13.26) could be written as

$$\iint_R f(x, y) \, dA = \iint_S f(r \cos \theta, r \sin \theta) r \, dA$$

where S is the region in the $r\theta$-plane that corresponds to the region R in the xy-plane.

More generally we consider a change of variables that is given by a **transformation** T from the uv-plane to the xy-plane:

$$T(u, v) = (x, y)$$

where x and y are related to u and v by the equations

(13.61)
$$x = g(u, v) \qquad y = h(u, v)$$

or, as we sometimes write,

$$x = x(u, v) \qquad y = y(u, v)$$

We shall usually assume that T is a C^1 **transformation,** which means that g and h have continuous first order partial derivatives.

A transformation T is really just a function whose domain and range are both subsets of R^2. If $T(u_1, v_1) = (x_1, y_1)$, then the point (x_1, y_1) is called the **image** of the point (u_1, v_1). If no two points have the same image, T is called **one-to-one.** Figure 13.46 shows the effect of a transformation T on a region S in the uv-plane. T transforms S into a region R in the xy-plane called the **image of S,** consisting of the images of all points in S.

Figure 13.46

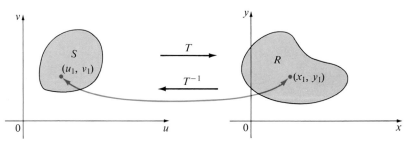

If T is a one-to-one transformation, then it has an **inverse transformation** T^{-1} from the xy-plane to the uv-plane and it may be possible to solve Equations 13.61 for u and v in terms of x and y:

$$u = G(x, y) \qquad v = H(x, y)$$

● **Example 1**   A transformation is defined by the equations

$$x = u^2 - v^2 \qquad y = 2uv$$

Find the image of the square $S = \{(u, v) \mid 0 \le u \le 1, 0 \le v \le 1\}$.

Solution   First let us find the images of the sides of S. The first side, S_1, is given by $v = 0$ $(0 \le u \le 1)$ (see Figure 13.47). From the given equations we have $x = u^2$, $y = 0$, and so $0 \le x \le 1$. Thus S_1 is mapped into the line segment from $(0, 0)$ to $(1, 0)$ in the xy-plane. The second side, S_2, is $u = 1$ $(0 \le v \le 1)$ and, putting $u = 1$ in the given equations, we get

$$x = 1 - v^2 \qquad y = 2v$$

Figure 13.47

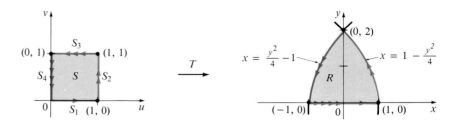

Eliminating v, we obtain

(13.62)
$$x = 1 - \frac{y^2}{4} \qquad 0 \leqslant x \leqslant 1$$

which is part of a parabola. Similarly, S_3 is given by $v = 1$ $(0 \leqslant u \leqslant 1)$ whose image is the parabolic arc

(13.63)
$$x = \frac{y^2}{4} - 1 \qquad -1 \leqslant x \leqslant 0$$

Finally, S_4 is given by $u = 0$ $(0 \leqslant v \leqslant 1)$ whose image is $x = -v^2$, $y = 0$, that is, $-1 \leqslant x \leqslant 0$. (Notice that as we move around the square in the counterclockwise direction, we also move around the parabolic region in the counterclockwise direction.) The image of S is the region R (shown in Figure 13.47) bounded by the x-axis and the parabolas given by Equations 13.62 and 13.63.   ●

Now let us see how a change of variables affects a double integral. We start with a small rectangle S in the uv-plane whose lower left corner is the point (u_0, v_0) and whose dimensions are Δu and Δv (see Figure 13.48). Its image is a region R in the xy-plane one of whose boundary points is $(x_0, y_0) = T(u_0, v_0)$. The equation of the lower side of S is $v = v_0$ whose image curve is given by $x = g(u, v_0)$, $y = h(u, v_0)$, or, in vector form,

$$g(u, v_0)\mathbf{i} + h(u, v_0)\mathbf{j}$$

The tangent vector at (x_0, y_0) to this image curve is

$$\mathbf{T}_u = g_u(u_0, v_0)\mathbf{i} + h_u(u_0, v_0)\mathbf{j} = \frac{\partial x}{\partial u}\mathbf{i} + \frac{\partial y}{\partial u}\mathbf{j}$$

Similarly, the tangent vector at (x_0, y_0) to the image curve of the left side of S (namely, $u = u_0$) is

$$\mathbf{T}_v = g_v(u_0, v_0)\mathbf{i} + h_v(u_0, v_0)\mathbf{j} = \frac{\partial x}{\partial v}\mathbf{i} + \frac{\partial y}{\partial v}\mathbf{j}$$

This means that we can approximate the image region $R = T(S)$ by a parallelogram determined by the vectors $\Delta u \mathbf{T}_u$ and $\Delta v \mathbf{T}_v$ (see Figure 13.48). Therefore we can approximate the area of R by the area of this parallelogram,

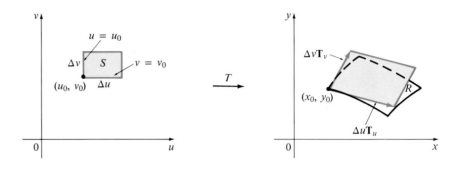

Figure 13.48

which, from Section 11.4, is

(13.64)
$$\|(\Delta u \mathbf{T}_u) \times (\Delta v \mathbf{T}_v)\| = \|\mathbf{T}_u \times \mathbf{T}_v\| \, \Delta u \, \Delta v$$

Computing the cross product, we obtain

$$\mathbf{T}_u \times \mathbf{T}_v = \begin{vmatrix} \mathbf{i} & \mathbf{j} & \mathbf{k} \\ \dfrac{\partial x}{\partial u} & \dfrac{\partial y}{\partial u} & 0 \\ \dfrac{\partial x}{\partial v} & \dfrac{\partial y}{\partial v} & 0 \end{vmatrix} = \begin{vmatrix} \dfrac{\partial x}{\partial u} & \dfrac{\partial y}{\partial u} \\ \dfrac{\partial x}{\partial v} & \dfrac{\partial y}{\partial v} \end{vmatrix} \mathbf{k} = \begin{vmatrix} \dfrac{\partial x}{\partial u} & \dfrac{\partial x}{\partial v} \\ \dfrac{\partial y}{\partial u} & \dfrac{\partial y}{\partial v} \end{vmatrix} \mathbf{k}$$

The determinant that arises in this calculation is called the *Jacobian* of the transformation, after Carl Jacobi (1804–1851), and is given a special notation.

Definition (13.65)

> The **Jacobian** of the transformation T given by $x = g(u, v)$ and $y = h(u, v)$ is
>
> $$\frac{\partial(x, y)}{\partial(u, v)} = \begin{vmatrix} \dfrac{\partial x}{\partial u} & \dfrac{\partial x}{\partial v} \\ \dfrac{\partial y}{\partial u} & \dfrac{\partial y}{\partial v} \end{vmatrix} = \frac{\partial x}{\partial u} \frac{\partial y}{\partial v} - \frac{\partial x}{\partial v} \frac{\partial y}{\partial u}$$

With this notation we can use Equation 13.64 to give an approximation to the area ΔA of R:

(13.66)
$$\Delta A \approx \left| \frac{\partial(x, y)}{\partial(u, v)} \right| \Delta u \, \Delta v$$

where the Jacobian is evaluated at (u_0, v_0).

Next we partition a region S in the uv-plane into rectangles S_{ij} and call their images in the xy-plane R_{ij} (see Figure 13.49). Applying the approxi-

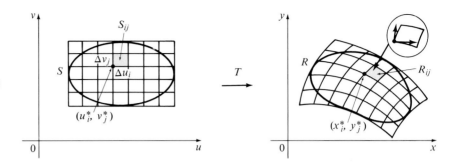

Figure 13.49

mation in (13.66) to each R_{ij}, we approximate the double integral of f over R as follows:

$$\iint_R f(x, y)\, dA \approx \sum_{i=1}^{m} \sum_{j=1}^{n} f(x_i^*, y_j^*)\, \Delta A_{ij}$$

$$\approx \sum_{i=1}^{m} \sum_{j=1}^{n} f(g(u_i^*, v_j^*),\, h(u_i^*, v_j^*)) \left| \frac{\partial(x, y)}{\partial(u, v)} \right| \Delta u_i\, \Delta v_j$$

where the Jacobian is evaluated at (u_i^*, v_j^*). Notice that this double sum is a Riemann sum for the integral

$$\iint_S f(g(u, v),\, h(u, v)) \left| \frac{\partial(x, y)}{\partial(u, v)} \right| du\, dv$$

The foregoing argument suggests that the following theorem is true. (A full proof is given in books on advanced calculus.)

Change of Variables in a Double Integral (13.67)

Suppose that T is a one-to-one C^1 transformation whose Jacobian is nonzero and that maps a region S in the uv-plane onto a region R in the xy-plane. Suppose that f is continuous on R and that R and S are type I or II plane regions. Then

$$\iint_R f(x, y)\, dx\, dy = \iint_S f(x(u, v),\, y(u, v)) \left| \frac{\partial(x, y)}{\partial(u, v)} \right| du\, dv$$

Theorem 13.67 says that we change from an integral in x and y to an integral in u and v by expressing x and y in terms of u and v and writing

$$dA = \left| \frac{\partial(x, y)}{\partial(u, v)} \right| du\, dv$$

Notice the similarity between Theorem 13.67 and the one-dimensional formula in Equation 13.60. Instead of the derivative dx/du, we have the absolute value of the Jacobian: $|\partial(x, y)/\partial(u, v)|$.

Figure 13.50

THE POLAR COORDINATE
TRANSFORMATION

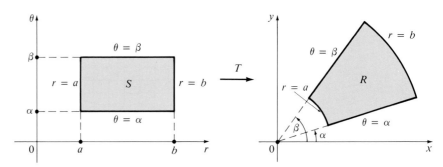

As a first illustration of Theorem 13.67, let us show that the formula for integration in polar coordinates is just a special case. Here the transformation T from the $r\theta$-plane to the xy-plane is given by

$$x = g(r, \theta) = r \cos \theta \qquad y = h(r, \theta) = r \sin \theta$$

and the geometry of the transformation is shown in Figure 13.50. T maps an ordinary rectangle in the $r\theta$-plane to a polar rectangle in the xy-plane. The Jacobian of T is

$$\frac{\partial(x, y)}{\partial(r, \theta)} = \begin{vmatrix} \dfrac{\partial x}{\partial r} & \dfrac{\partial x}{\partial \theta} \\[2mm] \dfrac{\partial y}{\partial r} & \dfrac{\partial y}{\partial \theta} \end{vmatrix} = \begin{vmatrix} \cos \theta & -r \sin \theta \\ \sin \theta & r \cos \theta \end{vmatrix} = r \cos^2 \theta + r \sin^2 \theta = r > 0$$

Thus Theorem 13.67 gives

$$\iint_R f(x, y)\, dx\, dy = \iint_S f(r \cos \theta, r \sin \theta) \left| \frac{\partial(x, y)}{\partial(r, \theta)} \right| dr\, d\theta$$

$$= \int_\alpha^\beta \int_a^b f(r \cos \theta, r \sin \theta) r\, dr\, d\theta$$

which is the same as (13.26).

● **Example 2**   Use the change of variables $x = u^2 - v^2$, $y = 2uv$ to evaluate the integral $\iint_R y\, dA$, where R is the region bounded by the x-axis and the parabolas $y^2 = 4 - 4x$ and $y^2 = 4 + 4x$.

Solution   The region R is pictured in Figure 13.47. In Example 1 we discovered that $T(S) = R$, where S is the square $[0, 1] \times [0, 1]$. Indeed, the reason for making the change of variables to evaluate the integral is that S is a much simpler region than R. First we need to compute the Jacobian:

$$\frac{\partial(x, y)}{\partial(u, v)} = \begin{vmatrix} \dfrac{\partial x}{\partial u} & \dfrac{\partial x}{\partial v} \\[2mm] \dfrac{\partial y}{\partial u} & \dfrac{\partial y}{\partial v} \end{vmatrix} = \begin{vmatrix} 2u & -2v \\ 2v & 2u \end{vmatrix} = 4u^2 + 4v^2 > 0$$

Therefore, by Theorem 13.67,

$$
\iint_R y \, dA = \iint_S 2uv \left| \frac{\partial(x, y)}{\partial(u, v)} \right| dA
$$

$$
= \int_0^1 \int_0^1 (2uv) 4(u^2 + v^2) \, du \, dv
$$

$$
= 8 \int_0^1 \int_0^1 (u^3 v + uv^3) \, du \, dv
$$

$$
= 8 \int_0^1 \left[\frac{u^4}{4} v + \frac{u^2}{2} v^3 \right]_{u=0}^{u=1} dv
$$

$$
= \int_0^1 (2v + 4v^3) \, dv
$$

$$
= \left[v^2 + v^4 \right]_0^1 = 2
$$

Note: Example 2 was not a very difficult problem to solve because we were given a suitable change of variables. If we are not supplied with a transformation, then the first step will be to think of an appropriate change of variables. If $f(x, y)$ is difficult to integrate, then the form of $f(x, y)$ may suggest a transformation. If the region of integration R is awkward, then the transformation should be chosen so that the corresponding region S in the uv-plane has a convenient description.

● **Example 3**   Evaluate the integral $\iint_R e^{(x+y)/(x-y)} \, dA$, where R is the triangular region with vertices $(0, 0)$, $(1, 0)$, and $(0, -1)$.

Solution   Since it is not easy to integrate $e^{(x+y)/(x-y)}$, we make a change of variables suggested by the form of this function:

(13.68) $u = x + y \qquad v = x - y$

These equations define a transformation T^{-1} from the xy-plane to the uv-plane. Theorem 13.67 talks about a transformation T from the uv-plane to the xy-plane. It is obtained by solving Equations 13.68 for x and y:

(13.69) $x = \tfrac{1}{2}(u + v) \qquad y = \tfrac{1}{2}(u - v)$

The Jacobian of T is

$$
\frac{\partial(x, y)}{\partial(u, v)} =
\begin{vmatrix}
\dfrac{\partial x}{\partial u} & \dfrac{\partial x}{\partial v} \\[2mm]
\dfrac{\partial y}{\partial u} & \dfrac{\partial y}{\partial v}
\end{vmatrix}
=
\begin{vmatrix}
\dfrac{1}{2} & \dfrac{1}{2} \\[2mm]
\dfrac{1}{2} & -\dfrac{1}{2}
\end{vmatrix}
= -\frac{1}{2}
$$

To find the region S in the uv-plane corresponding to R, we note that the three sides of R lie on the lines

$$
y = 0 \qquad x - y = 1 \qquad x = 0
$$

and, from either Equations 13.68 or Equations 13.69, the image lines in the uv-plane are

$$u = v \qquad v = 1 \qquad u = -v$$

Thus, the region S is the triangular region with vertices $(0,0)$, $(1,1)$, and $(-1,1)$ shown in Figure 13.51. Since

$$S = \{(u, v) | 0 \leqslant v \leqslant 1, \ -v \leqslant u \leqslant v\}$$

Figure 13.51

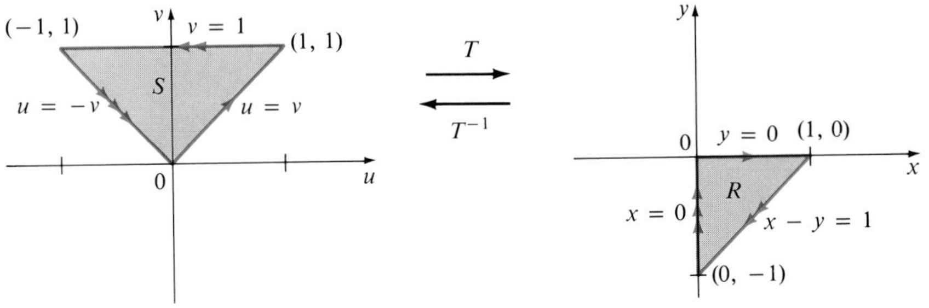

Theorem 13.67 gives

$$\iint_R e^{(x+y)/(x-y)} \, dA = \iint_S e^{u/v} \left| \frac{\partial(x, y)}{\partial(u, v)} \right| \, du \, dv$$

$$= \int_0^1 \int_{-v}^{v} e^{u/v} \left(\frac{1}{2} \right) du \, dv$$

$$= \frac{1}{2} \int_0^1 \left[v e^{u/v} \right]_{u=-v}^{u=v} dv$$

$$= \frac{1}{2} \int_0^1 (e - e^{-1}) v \, dv = \frac{e - e^{-1}}{4} \qquad \bullet$$

There is a similar change of variables formula for triple integrals. Let T be a transformation that maps a region S in uvw-space onto a region R in xyz-space by means of the equations

$$x = g(u, v, w) \qquad y = h(u, v, w) \qquad z = k(u, v, w)$$

The **Jacobian** of T is the following 3×3 determinant:

(13.70)

$$\frac{\partial(x, y, z)}{\partial(u, v, w)} = \begin{vmatrix} \dfrac{\partial x}{\partial u} & \dfrac{\partial x}{\partial v} & \dfrac{\partial x}{\partial w} \\[2mm] \dfrac{\partial y}{\partial u} & \dfrac{\partial y}{\partial v} & \dfrac{\partial y}{\partial w} \\[2mm] \dfrac{\partial z}{\partial u} & \dfrac{\partial z}{\partial v} & \dfrac{\partial z}{\partial w} \end{vmatrix}$$

Under hypotheses similar to those in Theorem 13.67, we have the following formula for triple integrals:

(13.71)

$$\iiint\limits_{R} f(x, y, z)\, dx\, dy\, dz$$

$$= \iiint\limits_{S} f(x(u, v, w),\, y(u, v, w),\, z(u, v, w)) \left| \frac{\partial(x, y, z)}{\partial(u, v, w)} \right| du\, dv\, dw$$

● **Example 4**   Use Formula 13.71 to derive the formula for triple integration in spherical coordinates.

Solution   Here the change of variables is given by

$$x = \rho \sin \phi \cos \theta \qquad y = \rho \sin \phi \sin \theta \qquad z = \rho \cos \phi$$

We compute the Jacobian as follows:

$$\frac{\partial(x, y, z)}{\partial(\rho, \theta, \phi)} = \begin{vmatrix} \sin \phi \cos \theta & -\rho \sin \phi \sin \theta & \rho \cos \phi \cos \theta \\ \sin \phi \sin \theta & \rho \sin \phi \cos \theta & \rho \cos \phi \sin \theta \\ \cos \phi & 0 & -\rho \sin \phi \end{vmatrix}$$

$$= \cos \phi \begin{vmatrix} -\rho \sin \phi \sin \theta & \rho \cos \phi \cos \theta \\ \rho \sin \phi \cos \theta & \rho \cos \phi \sin \theta \end{vmatrix}$$

$$\quad - \rho \sin \phi \begin{vmatrix} \sin \phi \cos \theta & -\rho \sin \phi \sin \theta \\ \sin \phi \sin \theta & \rho \sin \phi \cos \theta \end{vmatrix}$$

$$= \cos \phi (-\rho^2 \sin \phi \cos \phi \sin^2 \theta - \rho^2 \sin \phi \cos \phi \cos^2 \theta)$$

$$\quad - \rho \sin \phi (\rho \sin^2 \phi \cos^2 \theta + \rho \sin^2 \phi \sin^2 \theta)$$

$$= -\rho^2 \sin \phi \cos^2 \phi - \rho^2 \sin \phi \sin^2 \phi = -\rho^2 \sin \phi$$

Since $0 \leqslant \phi \leqslant \pi$, we have $\sin \phi \geqslant 0$. Therefore

$$\left| \frac{\partial(x, y, z)}{\partial(\rho, \theta, \phi)} \right| = |-\rho^2 \sin \phi| = \rho^2 \sin \phi$$

·and Formula 13.71 gives

$$\iiint\limits_{R} f(x, y, z)\, dx\, dy\, dz$$

$$= \iiint\limits_{S} f(\rho \sin \phi \cos \theta,\, \rho \sin \phi \sin \theta,\, \rho \cos \phi) \rho^2 \sin \phi\, d\rho\, d\theta\, d\phi$$

which is equivalent to Formula 13.58.   ●

EXERCISES 13.8

In Exercises 1–6 find the Jacobian of the transformation.

1. $x = u - 2v, y = 2u - v$

2. $x = u - v^2, y = u + v^2$

3. $x = e^{2u} \cos v, y = e^{2u} \sin v$

4. $x = se^t, y = se^{-t}$

5. $x = u + v + w, y = u + v - w, z = u - v + w$

6. $x = 2u, y = 3v^2, z = 4w^3$

In Exercises 7 and 8 find the image of the given set S under the given transformation.

7. $S = \{(u, v) \mid 0 \le u \le 2, 0 \le v \le 1\}, x = u - 2v,$
$y = 2u - v$

8. $S = \{(u, v) \mid 0 \le u \le 1, u \le v \le 1\}, x = u^2, y = v$

In Exercises 9–16 use the given transformation to evaluate the given integral.

9. $\iint_R (3x + 4y) \, dA$, where R is the region bounded by the lines $y = x, y = x - 2, y = -2x$, and $y = 3 - 2x$; $x = \frac{1}{3}(u + v), y = \frac{1}{3}(v - 2u)$

10. $\iint_R (x + y) \, dA$, where R is the square with vertices $(0, 0)$, $(2, 3), (5, 1)$, and $(3, -2)$; $x = 2u + 3v, y = 3u - 2v$

11. $\iint_R x^2 \, dA$, where R is the region bounded by the ellipse $9x^2 + 4y^2 = 36$; $x = 2u, y = 3v$

12. $\iint_R (x^2 - xy + y^2) \, dA$, where R is the region bounded by the ellipse $x^2 - xy + y^2 = 2$; $x = \sqrt{2}u - \sqrt{2/3}v$, $y = \sqrt{2}u + \sqrt{2/3}v$

13. $\iint_R xy \, dA$, where R is the region in the first quadrant bounded by the lines $y = x$ and $y = 3x$ and the hyperbolas $xy = 1, xy = 3$; $x = u/v, y = v$

14. $\iint_R y^2 \, dA$, where R is the region bounded by the curves $xy = 1, xy = 2, xy^2 = 1, xy^2 = 2$; $u = xy, v = xy^2$

15. $\iiint_E dV$, where E is the solid enclosed by the ellipsoid $x^2/a^2 + y^2/b^2 + z^2/c^2 = 1$; $x = au, y = bv, z = cw$
(This integral gives the volume of the ellipsoid.)

16. $\iiint_E x^2 y \, dV$, where E is the solid of Exercise 15

In Exercises 17–20 evaluate the integral by making an appropriate change of variables.

17. $\iint_R xy \, dA$, where R is the region bounded by the lines $2x - y = 1, 2x - y = -3, 3x + y = 1$, and $3x + y = -2$

18. $\iint_R \dfrac{x + 2y}{\cos(x - y)} \, dA$, where R is the parallelogram bounded by the lines $y = x, y = x - 1, x + 2y = 0$, and $x + 2y = 2$

19. $\iint_R \cos\left(\dfrac{y - x}{y + x}\right) \, dA$, where R is the trapezoidal region with vertices $(1, 0), (2, 0), (0, 2)$, and $(0, 1)$

20. $\iint_R \sin(9x^2 + 4y^2) \, dA$, where R is the region in the first quadrant bounded by the ellipse $9x^2 + 4y^2 = 1$

REVIEW OF CHAPTER 13

Define, state, or discuss the following:

1. Double Riemann sum

2. Double integral over a rectangle

3. Iterated integral

4. Fubini's Theorem

5. Double integral over a general region

6. Regions of type I and II

7. Evaluating double integrals

8. Volume

9. Properties of double integrals

10. Conversion from rectangular to polar coordinates in a double integral

11. Electric charge

12. Mass and center of mass of a lamina

13. Moment of inertia

14. Radius of gyration

15. Triple integral

16. Evaluation of triple integrals

17. Mass, center of mass, and moments of inertia of a solid region

18. Triple integrals in cylindrical coordinates

19. Triple integrals in spherical coordinates

20. Jacobian of a transformation

21. Formula for change of variables in a double integral

22. Formula for change of variables in a triple integral

REVIEW EXERCISES FOR CHAPTER 13

In Exercises 1–6 calculate the iterated integral.

1. $\int_{-2}^{2} \int_{0}^{4} (4x^3 + 3xy^2)\,dx\,dy$

2. $\int_{0}^{\pi} \int_{0}^{1} x\cos(xy)\,dy\,dx$

3. $\int_{1}^{2} \int_{0}^{x^2} \frac{1}{x+y}\,dy\,dx$

4. $\int_{0}^{1} \int_{0}^{y^2} e^{y^3}\,dx\,dy$

5. $\int_{0}^{1} \int_{0}^{x^2} \int_{0}^{y} y^2 z\,dz\,dy\,dx$

6. $\int_{0}^{1} \int_{\sqrt{y}}^{1} \int_{0}^{y} xy\,dz\,dx\,dy$

In Exercises 7 and 8 calculate the iterated integral by first reversing the order of integration.

7. $\int_{0}^{1} \int_{x}^{1} e^{x/y}\,dy\,dx$

8. $\int_{0}^{1} \int_{y^2}^{1} y\sin(x^2)\,dx\,dy$

In Exercises 9–22 calculate the value of the multiple integral.

9. $\iint_{R} \frac{1}{(x-y)^2}\,dA,\ R = \{(x,y)\,|\,0 \leqslant x \leqslant 1,\, 2 \leqslant y \leqslant 4\}$

10. $\iint_{D} x^3\,dA,\ D = \{(x,y)\,|\,-1 \leqslant x \leqslant 1,\, x^2 - 1 \leqslant y \leqslant x + 1\}$

11. $\iint_{D} xy\,dA,\ D$ is bounded by $y^2 = x^3$ and $y = x$

12. $\iint_{D} xe^y\,dA,\ D$ is bounded by $y = 0$, $y = x^2$, $x = 1$

13. $\iint_{D} (xy + 2x + 3y)\,dA,\ D$ is the region in the first quadrant bounded by $x = 1 - y^2$, $y = 0$, $x = 0$

14. $\iint_{D} y\,dA,\ D$ is the region in the first quadrant bounded by the hyperbola $xy = 1$ and the lines $y = x$, $y = 2$

15. $\iint_{D} (x^2 + y^2)^{3/2}\,dA,\ D$ is the region in the first quadrant bounded by the lines $y = 0$, $y = \sqrt{3}x$, and the circle $x^2 + y^2 = 9$

16. $\iint_{D} \sqrt{x^2 + y^2}\,dA,\ D$ is the closed disk with radius 1 and center $(0,1)$

17. $\iiint_{E} x^2 z\,dV$,
$E = \{(x,y,z)\,|\,0 \leqslant x \leqslant 2,\, 0 \leqslant y \leqslant 2x,\, 0 \leqslant z \leqslant x\}$

18. $\iiint_{T} y\,dV$, T is the tetrahedron bounded by the planes $x = 0$, $y = 0$, $z = 0$, and $2x + y + z = 2$

19. $\iiint_{E} y^2 z^2\,dV$, E is bounded by the paraboloid $x = 1 - y^2 - z^2$ and the plane $x = 0$

20. $\iiint_{E} z\,dV$, E is bounded by the planes $y = 0$, $z = 0$, $x + y = 2$ and the cylinder $y^2 + z^2 = 1$ in the first octant

21. $\iiint_{E} yz\,dV$, E lies above the plane $z = 0$, below the plane $z = y$, and inside the cylinder $x^2 + y^2 = 4$

22. $\iiint_{H} z^3\sqrt{x^2 + y^2 + z^2}\,dV$, H is the solid hemisphere with center the origin, radius 1, that lies above the xy-plane

In Exercises 23–28 find the volume of the given solid.

23. Under the paraboloid $z = x^2 + 4y^2$ and above the rectangle $R = [0,2] \times [1,4]$

24. Under the surface $z = x^2 y$ and above the triangle in the xy-plane with vertices $(1,0)$, $(2,1)$, and $(4,0)$

25. The solid tetrahedron with vertices $(0,0,0)$, $(0,0,1)$, $(0,2,0)$, and $(2,2,0)$

26. Bounded by the cylinder $x^2 + y^2 = 4$ and the planes $z = 0$ and $y + z = 3$

27. One of the wedges cut from the cylinder $x^2 + 9y^2 = a^2$ by the planes $z = 0$ and $z = mx$

28. Above the paraboloid $z = x^2 + y^2$ and below the half-cone $z = \sqrt{x^2 + y^2}$

29. Find the mass and center of mass of a lamina that occupies the region D bounded by the parabola $x = 1 - y^2$ and the coordinate axes in the first quadrant if the density function is $\rho(x,y) = y$.

30. Find the moments of inertia and radii of gyration about the x- and y-axes for the lamina of Exercise 29.

31. A lamina occupies the part of the disk $x^2 + y^2 \leqslant a^2$ that lies in the first quadrant. Find the centroid of the lamina.

32. Find the center of mass of the lamina in Exercise 31 if the density function is $\rho(x, y) = xy^2$.

33. Find the centroid of a right circular cone with height h and base radius a. (Place the cone so that its base is in the xy-plane with center the origin and its axis along the positive z-axis.)

34. Find the moment of inertia of the cone of Exercise 33 about its axis (the z-axis).

35. Use polar coordinates to evaluate

$$\int_0^{\sqrt{2}} \int_y^{\sqrt{4-y^2}} \frac{1}{1 + x^2 + y^2}\, dx\, dy$$

36. Give five other iterated integrals that are equal to

$$\int_0^2 \int_0^{y^3} \int_0^{y^2} f(x, y, z)\, dz\, dx\, dy$$

37. Use the transformation $u = x - y$, $u = x + y$ to evaluate $\iint_R (x - y)/(x + y)\, dA$, where R is the square with vertices $(0, 2)$, $(1, 1)$, $(2, 2)$, and $(1, 3)$.

38. Use the transformation $x = u^2$, $y = v^2$, $z = w^2$ to find the volume of the region bounded by the surface $\sqrt{x} + \sqrt{y} + \sqrt{z} = 1$ and the coordinate planes.

14 Vector Calculus

It is mathematics that offers the exact
mathematical sciences a certain measure
of security which, without mathematics,
they could not obtain.

Albert Einstein

In this chapter we study the calculus of vector fields. (These are functions that assign vectors to points in space.) In particular we define line integrals (which can be used to find the work done by a force field in moving an object along a curve). Then we define surface integrals (which can be used to find the rate of fluid flow across a surface). The connections between these new types of integrals and the single, double, and triple integrals that we have already met are given by the higher-dimensional analogues of the Fundamental Theorem of Calculus: Green's Theorem, Stokes' Theorem, and the Divergence Theorem.

Vector Fields

In Section 11.7 we studied vector functions whose domains were sets of real numbers and whose ranges were sets of vectors. Now we look at a type of function, called a vector field, whose domain is a set of points in R^2 (or R^3) and whose range is a set of vectors in V_2 (or V_3).

Definition (14.1)

> Let D be a set in R^2 (a plane region). A **vector field on R^2** is a function **F** that assigns to each point (x, y) in D a two-dimensional vector $\mathbf{F}(x, y)$.

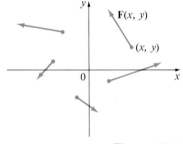

Figure 14.1
VECTOR FIELD ON R^2

The best way to picture a vector field is to draw the arrow representing the vector $\mathbf{F}(x, y)$ starting at the point (x, y). Of course, it is impossible to do this for all points (x, y) but we can gain a reasonable impression of **F** by doing it for a few representative points in D as in Figure 14.1. Since $\mathbf{F}(x, y)$ is a two-dimensional vector we can write it in terms of its **component functions** P and Q as follows:

$$\mathbf{F}(x, y) = P(x, y)\mathbf{i} + Q(x, y)\mathbf{j} = \langle P(x, y), Q(x, y)\rangle$$

or, for short,

$$\mathbf{F} = P\mathbf{i} + Q\mathbf{j}$$

Notice that P and Q are scalar functions of two variables and are sometimes called **scalar fields** to distinguish them from vector fields.

Definition (14.2)

> Let E be a subset of R^3. A **vector field on R^3** is a function **F** that assigns to each point (x, y, z) in E a three-dimensional vector $\mathbf{F}(x, y, z)$.

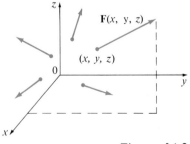

Figure 14.2
VECTOR FIELD ON R^3

A vector field **F** on R^3 is pictured in Figure 14.2. We can express it in terms of its component functions P, Q, and R as

$$\mathbf{F}(x, y, z) = P(x, y, z)\mathbf{i} + Q(x, y, z)\mathbf{j} + R(x, y, z)\mathbf{k}$$

As with the vector functions in Section 11.7, we can define continuity of vector fields and show that **F** is continuous if and only if its component functions P, Q, and R are continuous.

We sometimes identify a point (x, y, z) with its position vector $\mathbf{x} = \langle x, y, z\rangle$ and write $\mathbf{F}(\mathbf{x})$ instead of $\mathbf{F}(x, y, z)$. Then **F** becomes a function that assigns a vector $\mathbf{F}(\mathbf{x})$ to a vector $\mathbf{x}$.

● **Example 1**   A vector field on R^2 is defined by

$$\mathbf{F}(x, y) = -y\mathbf{i} + x\mathbf{j}$$

Describe **F** by sketching some of the vectors $\mathbf{F}(x, y)$ as in Figure 14.1.

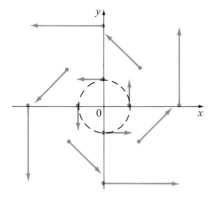

Figure 14.3

$\mathbf{F}(x, y) = -y\mathbf{i} + x\mathbf{j}$

Solution   The required sketch is given in Figure 14.3. It appears that each arrow is tangent to a circle with center the origin. To confirm this, we take the dot product of the position vector $\mathbf{x} = x\mathbf{i} + y\mathbf{j}$ with the vector $\mathbf{F}(\mathbf{x}) = \mathbf{F}(x, y)$:

$$\mathbf{x} \cdot \mathbf{F}(\mathbf{x}) = (x\mathbf{i} + y\mathbf{j}) \cdot (-y\mathbf{i} + x\mathbf{j})$$

$$= -xy + yx = 0$$

This shows that $\mathbf{F}(x, y)$ is perpendicular to the position vector $\langle x, y \rangle$ and is therefore tangent to a circle with center the origin and radius $|\mathbf{x}| = \sqrt{x^2 + y^2}$. Notice also that

$$|\mathbf{F}(x, y)| = \sqrt{(-y)^2 + x^2} = \sqrt{x^2 + y^2} = |\mathbf{x}|$$

so the magnitude of the vector $\mathbf{F}(x, y)$ is equal to the radius of the circle.   ●

● **Example 2**   Sketch the vector field on R^3 given by $\mathbf{F}(x, y, z) = z\mathbf{k}$.

Solution   The sketch is shown in Figure 14.4. Notice that all vectors are vertical and point upward above the xy-plane and downward below it. The magnitude increases with the distance from the xy-plane.   ●

● **Example 3**   Imagine a fluid flowing steadily along a pipe and let $\mathbf{V}(x, y, z)$ be the velocity vector at a point (x, y, z). Then $\mathbf{V}$ assigns a vector to each point (x, y, z) in a certain domain E (the interior of the pipe) and so $\mathbf{V}$ is a vector field on R^3 called a **velocity field.** A possible velocity field is illustrated in Figure 14.5. The speed at any given point is indicated by the length of the arrow.

Velocity fields also occur in other areas of physics. For instance, the vector field in Example 1 could be used as the velocity field describing the counterclockwise rotation of a wheel.   ●

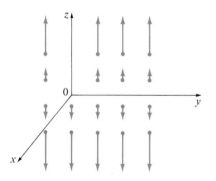

Figure 14.4

$\mathbf{F}(x, y, z) = z\mathbf{k}$

● **Example 4**   Newton's Law of Gravitation states that the magnitude of the gravitational force between two objects with masses m and M is

$$|\mathbf{F}| = \frac{mMG}{r^2}$$

where r is the distance between the objects and G is the gravitational constant. (This is an example of an inverse square law.) Let us assume that the object with mass M is located at the origin in R^3. (For instance, M could be the mass of the earth and the origin would be at the center of the earth.) Let the position vector of the object with mass m be $\mathbf{x} = \langle x, y, z \rangle$. Then $r = |\mathbf{x}|$, so $r^2 = |\mathbf{x}|^2$. The gravitational force exerted on this second object acts toward the origin, and the unit vector in this direction is

$$-\frac{\mathbf{x}}{|\mathbf{x}|}$$

Therefore the gravitational force acting on the object at $\mathbf{x} = \langle x, y, z \rangle$ is

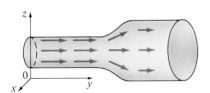

Figure 14.5

VELOCITY FIELD IN FLUID FLOW

(14.3)

$$\mathbf{F}(\mathbf{x}) = -\frac{mMG}{|\mathbf{x}|^3}\mathbf{x}$$

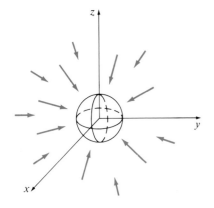

Figure 14.6

GRAVITATIONAL FORCE FIELD

[Physicists often use the notation **r** instead of **x** for the position vector, so you may see Formula 14.3 written in the form $\mathbf{F} = -(mMG/r^3)\mathbf{r}$.] The function given by Equation 14.3 is an example of a vector field, called the **gravitational field,** because it associates a vector [the force $\mathbf{F}(\mathbf{x})$] with every point **x** in space.

Formula 14.3 is a compact way of writing the gravitational field but we can also write it in terms of its component functions by using the facts that $\mathbf{x} = x\mathbf{i} + y\mathbf{j} + z\mathbf{k}$ and $|\mathbf{x}| = \sqrt{x^2 + y^2 + z^2}$:

$$\mathbf{F}(x, y, z) = \frac{-mMGx}{(x^2 + y^2 + z^2)^{3/2}}\,\mathbf{i} + \frac{-mMGy}{(x^2 + y^2 + z^2)^{3/2}}\,\mathbf{j}$$

$$+ \frac{-mMGz}{(x^2 + y^2 + z^2)^{3/2}}\,\mathbf{k}$$

The gravitational field **F** is pictured in Figure 14.6.   ●

● **Example 5**   Suppose an electric charge Q is located at the origin. According to Coulomb's Law, the electric force $\mathbf{F}(\mathbf{x})$ exerted by this charge on a charge q located at a point (x, y, z) with position vector $\mathbf{x} = \langle x, y, z \rangle$ is

(14.4)
$$\mathbf{F}(\mathbf{x}) = \frac{\varepsilon qQ}{|\mathbf{x}|^3}\,\mathbf{x}$$

where ε is a constant (which depends on the units that are used). For like charges we have $qQ > 0$ and the force is repulsive; for unlike charges we have $qQ < 0$ and the force is attractive. Notice the similarity between Formulas 14.3 and 14.4. Both vector fields are examples of **force fields.**

Instead of considering the electric force **F**, physicists often consider the force per unit charge:

$$\mathbf{E}(\mathbf{x}) = \frac{1}{q}\,\mathbf{F}(\mathbf{x}) = \frac{\varepsilon Q}{|\mathbf{x}|^3}\,\mathbf{x}$$

Then **E** is a vector field on R^3 called the **electric field** of Q.   ●

● **Example 6**   If f is a scalar function of two variables, recall from Section 12.6 that its gradient ∇f (or grad f) is defined by

$$\nabla f(x, y) = f_x(x, y)\mathbf{i} + f_y(x, y)\mathbf{j}$$

Therefore ∇f is really a vector field on R^2 and is called a **gradient vector field.** For instance, if

$$f(x, y) = x^2y - y^3$$

then its gradient vector field is given by

$$\nabla f(x, y) = 2xy\mathbf{i} + (x^2 - 3y^2)\mathbf{j}$$

Likewise if f is a scalar function of three variables, its gradient is a vector field on R^3 given by

$$\nabla f(x, y, z) = f_x(x, y, z)\mathbf{i} + f_y(x, y, z)\mathbf{j} + f_z(x, y, z)\mathbf{k}$$ ●

A vector field $\mathbf{F}$ is called a **conservative vector field** if it is the gradient of some scalar function, that is, if there exists a function f such that $\mathbf{F} = \nabla f$. In this situation f is called a **potential function** for $\mathbf{F}$.

Not all vector fields are conservative but such fields do arise frequently in physics. For example, the gravitational field $\mathbf{F}$ in Example 4 is conservative because if we define

$$f(x, y, z) = \frac{mMG}{\sqrt{x^2 + y^2 + z^2}}$$

then

$$\nabla f(x, y, z) = \frac{\partial f}{\partial x}\mathbf{i} + \frac{\partial f}{\partial y}\mathbf{j} + \frac{\partial f}{\partial z}\mathbf{k}$$

$$= \frac{-mMGx}{(x^2 + y^2 + z^2)^{3/2}}\mathbf{i} + \frac{-mMGy}{(x^2 + y^2 + z^2)^{3/2}}\mathbf{j}$$

$$+ \frac{-mMGz}{(x^2 + y^2 + z^2)^{3/2}}\mathbf{k}$$

$$= \mathbf{F}(x, y, z)$$

Later in this chapter (Theorems 14.23 and 14.35) we shall learn how to tell whether or not a given vector field is conservative.

EXERCISES 14.1

In Exercises 1–8 sketch the vector field $\mathbf{F}$ by drawing a diagram like Figure 14.3 or 14.4.

1. $\mathbf{F}(x, y) = x\mathbf{i} + y\mathbf{j}$

2. $\mathbf{F}(x, y) = x\mathbf{i} - y\mathbf{j}$

3. $\mathbf{F}(x, y) = y\mathbf{i} + \mathbf{j}$

4. $\mathbf{F}(x, y) = \dfrac{y\mathbf{i} - x\mathbf{j}}{\sqrt{x^2 + y^2}}$

5. $\mathbf{F}(x, y, z) = \mathbf{j}$

6. $\mathbf{F}(x, y, z) = z\mathbf{j}$

7. $\mathbf{F}(x, y, z) = y\mathbf{j}$

8. $\mathbf{F}(x, y, z) = \mathbf{j} + \mathbf{k}$

In Exercises 9–14 find the gradient vector field of f.

9. $f(x, y) = x^5 - 4x^2y^3$

10. $f(x, y) = \sin(2x + 3y)$

11. $f(x, y) = e^{3x}\cos 4y$

12. $f(x, y, z) = xyz$

13. $f(x, y, z) = xy^2 - yz^3$

14. $f(x, y, z) = x\ln(y - z)$

SECTION 14.2

Line Integrals

In this section we define an integral that is similar to a single integral except that instead of integrating over an interval $[a, b]$, we integrate over a curve C. Such integrals are called *line integrals*, although "curve integrals" would be better terminology.

We start with a plane curve C given by parametric equations

(14.5)
$$x = x(t) \qquad y = y(t) \qquad a \leqslant t \leqslant b$$

or, equivalently, by the vector equation $\mathbf{r}(t) = x(t)\mathbf{i} + y(t)\mathbf{j}$, and we assume that C is a smooth curve. [This means that $\mathbf{r}'$ is continuous and $\mathbf{r}'(t) \neq \mathbf{0}$. See Section 11.8.] A partition of the parameter interval $[a, b]$ by points t_i with

$$a = t_0 < t_1 < t_2 < \cdots < t_n = b$$

will determine a partition P of the curve by points $P_i(x_i, y_i)$, where $x_i = x(t_i)$ and $y_i = y(t_i)$ (see Figure 14.7). These points P_i divide C into n subarcs with lengths $\Delta s_1, \Delta s_2, \ldots, \Delta s_n$. The **norm** $\|P\|$ of the partition is the longest of these lengths. We choose any point $P_i^*(x_i^*, y_i^*)$ in the ith subarc. (This will correspond to a point t_i^* in $[t_{i-1}, t_i]$.) Now if f is any function of two variables whose domain includes the curve C, we evaluate f at the point (x_i^*, y_i^*), multiply by the length Δs_i of the subarc, and form the sum

$$\sum_{i=1}^{n} f(x_i^*, y_i^*) \Delta s_i$$

which is similar to a Riemann sum. Then we take the limit of these sums and make the following definition by analogy with a single integral.

Definition (14.6)

If f is defined on a smooth curve C given by Equations 14.5, then the **line integral of f along C** is

$$\int_C f(x, y)\, ds = \lim_{\|P\| \to 0} \sum_{i=1}^{n} f(x_i^*, y_i^*) \Delta s_i$$

if this limit exists.

Figure 14.7

A PARTITION OF $[a, b]$
DETERMINES A PARTITION OF C

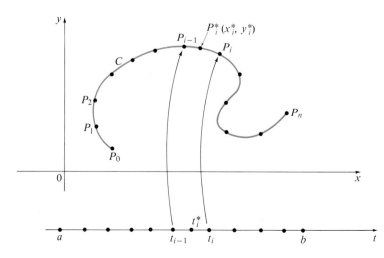

It can be shown that if f is a continuous function, then the limit in Definition 14.6 will always exist and the following formula can be used to evaluate the line integral:

(14.7)

$$\int_C f(x, y)\, ds = \int_a^b f(x(t), y(t)) \sqrt{\left(\frac{dx}{dt}\right)^2 + \left(\frac{dy}{dt}\right)^2}\, dt$$

The way to remember Formula 14.7 is to express everything in terms of the parameter t. Use the parametric equations to express x and y in terms of t and recall the arc length formula from Section 9.3:

$$ds = \sqrt{\left(\frac{dx}{dt}\right)^2 + \left(\frac{dy}{dt}\right)^2}\, dt$$

In the special case where C is the line segment that joins $(a, 0)$ to $(b, 0)$, then, using x as the parameter, we can write the parametric equations of C as follows: $x = x$, $y = 0$, $a \leqslant x \leqslant b$. Formula 14.7 then becomes

$$\int_C f(x, y)\, ds = \int_a^b f(x, 0)\, dx$$

and so the line integral reduces to an ordinary single integral in this case.

● **Example 1**   Evaluate $\int_C (2 + x^2 y)\, ds$, where C is the upper half of the unit circle $x^2 + y^2 = 1$.

Solution   Recall that the unit circle can be parametrized by means of the equations

$$x = \cos t \qquad y = \sin t$$

and the upper half of the circle is described by the parameter interval $0 \leqslant t \leqslant \pi$ (see Figure 14.8). Therefore Formula 14.7 gives

$$\int_C (2 + x^2 y)\, ds = \int_0^\pi (2 + \cos^2 t \sin t) \sqrt{\left(\frac{dx}{dt}\right)^2 + \left(\frac{dy}{dt}\right)^2}\, dt$$

$$= \int_0^\pi (2 + \cos^2 t \sin t) \sqrt{\sin^2 t + \cos^2 t}\, dt$$

$$= \int_0^\pi (2 + \cos^2 t \sin t)\, dt = \left[2t - \frac{\cos^3 t}{3} \right]_0^\pi$$

$$= 2\pi + \frac{2}{3}$$

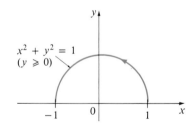

Figure 14.8

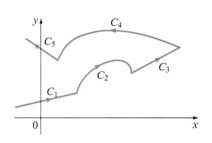

Figure 14.9

A PIECEWISE SMOOTH CURVE

Suppose now that C is a **piecewise smooth curve**; that is, C is a union of a finite number of smooth curves $C_1, C_2, \ldots, C_n$, where, as illustrated in Figure 14.9, the initial point of C_{i+1} is the terminal point of C_i. Then we define the integral of f along C as the sum of the integrals of f along each of the smooth pieces of C:

$$\int_C f(x, y)\, ds = \int_{C_1} f(x, y)\, ds + \int_{C_2} f(x, y)\, ds + \cdots + \int_{C_n} f(x, y)\, ds$$

● **Example 2**   Evaluate $\int_C 2x\, ds$, where C consists of the arc C_1 of the parabola $y = x^2$ from $(0,0)$ to $(1,1)$ followed by the vertical line segment C_2 from $(1,1)$ to $(1,2)$.

Solution   The curve C is shown in Figure 14.10. C_1 is the graph of a function of x, so we can choose x as the parameter and the equations for C_1 become

$$x = x \qquad y = x^2 \qquad 0 \leqslant x \leqslant 1$$

Therefore

$$\int_{C_1} 2x\, ds = \int_0^1 2x \sqrt{\left(\frac{dx}{dx}\right)^2 + \left(\frac{dy}{dx}\right)^2}\, dx$$

$$= \int_0^1 2x\sqrt{1 + 4x^2}\, dx = \frac{1}{4} \cdot \frac{2}{3}(1 + 4x^2)^{3/2}\Big]_0^1 = \frac{5\sqrt{5} - 1}{6}$$

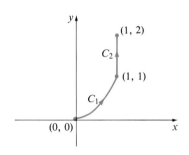

Figure 14.10
$C = C_1 \cup C_2$

On C_2 we choose y as the parameter, so the equations of C_2 are

$$x = 1 \qquad y = y \qquad 1 \leqslant y \leqslant 2$$

and

$$\int_{C_2} 2x\, ds = \int_1^2 2(1) \sqrt{\left(\frac{dx}{dy}\right)^2 + \left(\frac{dy}{dy}\right)^2}\, dy = \int_1^2 2\, dy = 2$$

Thus

$$\int_C 2x\, ds = \int_{C_1} 2x\, ds + \int_{C_2} 2x\, ds = \frac{5\sqrt{5} - 1}{6} + 2 \qquad ●$$

Any physical interpretation of a line integral $\int_C f(x, y)\, ds$ will depend on the physical interpretation of the function f. Suppose that $\rho(x, y)$ represents the linear density at a point (x, y) of a thin wire shaped like a curve C. Then the mass of the part of the wire from P_{i-1} to P_i in Figure 14.7 is approximately $\rho(x_i^*, y_i^*)\Delta s_i$ and so the total mass of the wire is approximately $\sum \rho(x_i^*, y_i^*)\Delta s_i$. By taking finer partitions of the curve, we obtain the mass m of the wire as the limiting value of these approximations:

$$m = \lim_{\|P\| \to 0} \sum_{i=1}^{n} \rho(x_i^*, y_i^*)\Delta s_i = \int_C \rho(x, y)\, ds$$

[For example, if $f(x, y) = 2 + x^2 y$ represents the density of a semicircular wire, then the integral in Example 1 would represent the mass of the wire.] The center of mass of the wire with density function ρ is at the point $(\bar{x}, \bar{y})$, where

(14.8)
$$\bar{x} = \frac{1}{m} \int_C x\rho(x, y)\, ds \qquad \bar{y} = \frac{1}{m} \int_C y\rho(x, y)\, ds$$

Other physical interpretations of line integrals will be discussed later in this chapter.

There are two other line integrals that are obtained by replacing Δs_i by $\Delta x_i = x_i - x_{i-1}$ or $\Delta y_i = y_i - y_{i-1}$ in Definition 14.6. They are called the **line integrals of f along C with respect to x and y:**

(14.9)
$$\int_C f(x, y)\, dx = \lim_{\|P\| \to 0} \sum_{i=1}^{n} f(x_i^*, y_i^*)\, \Delta x_i$$

(14.10)
$$\int_C f(x, y)\, dy = \lim_{\|P\| \to 0} \sum_{i=1}^{n} f(x_i^*, y_i^*)\, \Delta y_i$$

When we want to distinguish the original line integral $\int_C f(x, y)\, ds$ from those in Equations 14.9 and 14.10 we call it the **line integral with respect to arc length.**

The following formulas say that line integrals with respect to x and y can also be evaluated by expressing everything in terms of t: $x = x(t)$, $y = y(t)$, $dx = x'(t)\, dt$, $dy = y'(t)\, dt$.

(14.11)
$$\int_C f(x, y)\, dx = \int_a^b f(x(t), y(t)) x'(t)\, dt$$

$$\int_C f(x, y)\, dy = \int_a^b f(x(t), y(t)) y'(t)\, dt$$

It frequently happens that line integrals with respect to x and y occur together. When this happens it is customary to abbreviate by writing

$$\int_C P(x, y)\, dx + \int_C Q(x, y)\, dy = \int_C P(x, y)\, dx + Q(x, y)\, dy$$

● **Example 3**   Evaluate $\int_C y^2\, dx + x\, dy$, where (a) $C = C_1$ is the line segment from $(-5, -3)$ to $(0, 2)$ and (b) $C = C_2$ is the arc of the parabola $x = 4 - y^2$ from $(-5, -3)$ to $(0, 2)$ (see Figure 14.11).

Solution   (a)   A parametric representation for the line segment is

$$x = 5t - 5 \qquad y = 5t - 3 \qquad 0 \leqslant t \leqslant 1$$

(Use Equation 14.14 with $\mathbf{r}_0 = \langle -5, -3 \rangle$ and $\mathbf{r}_1 = \langle 0, 2 \rangle$.) Then $dx = 5\, dt$, $dy = 5\, dt$, and Formula 14.11 gives

$$\int_{C_1} y^2\, dx + x\, dy = \int_0^1 (5t - 3)^2 (5\, dt) + (5t - 5)(5\, dt)$$

$$= 5 \int_0^1 (25t^2 - 25t + 4)\, dt$$

$$= 5 \left[\frac{25t^3}{3} - \frac{25t^2}{2} + 4t \right]_0^1 = -\frac{5}{6}$$

(b)   Since the parabola is given as a function of y, let us take y as the parameter and write C as

$$x = 4 - y^2 \qquad y = y \qquad -3 \leqslant y \leqslant 2$$

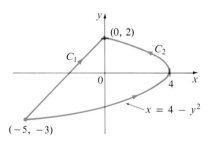

Figure 14.11

Then $dx = -2y\,dy$ and by Formula 14.11 we have

$$\int_{C_2} y^2\,dx + x\,dy = \int_{-3}^{2} y^2(-2y)\,dy + (4 - y^2)\,dy$$

$$= \int_{-3}^{2} (-2y^3 - y^2 + 4)\,dy$$

$$= \left[-\frac{y^4}{2} - \frac{y^3}{3} + 4y \right]_{-3}^{2} = 40\tfrac{5}{6} \qquad \bullet$$

Notice that we got different answers in parts (a) and (b) of Example 3 even though the two curves had the same endpoints. Thus, in general, the value of a line integral depends not just on the endpoints of the curve but also on the path. (But see Section 14.3 for conditions under which the integral will be independent of the path.)

Notice also that the answers in Example 3 depend on the direction, or orientation, of the curve. If $-C_1$ denotes the line segment from $(0, 2)$ to $(-5, -3)$, you can verify, using the parametrization

$$x = -5t \qquad y = 2 - 5t \qquad 0 \leqslant t \leqslant 1$$

that

$$\int_{-C_1} y^2\,dx + x\,dy = \frac{5}{6}$$

In general, a given parametrization $x = x(t)$, $y = y(t)$, $a \leqslant t \leqslant b$, determines an **orientation** of a curve C, with the positive direction corresponding to increasing values of the parameter t. (See Figure 14.12 where the initial point A corresponds to the parameter value a and the terminal point B corresponds to $t = b$.)

If $-C$ denotes the curve consisting of the same points as C but with the opposite orientation (from initial point B to terminal point A in Figure 14.12), then we have

$$\int_{-C} f(x, y)\,dx = -\int_{C} f(x, y)\,dx \qquad \int_{-C} f(x, y)\,dy = -\int_{C} f(x, y)\,dy$$

But if we integrate with respect to arc length, the value of the line integral does *not* change when we reverse the orientation of the curve:

$$\int_{-C} f(x, y)\,ds = \int_{C} f(x, y)\,ds$$

(This is because Δs_i is always positive whereas Δx_i and Δy_i change sign when we reverse the orientation of C.)

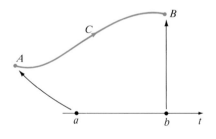

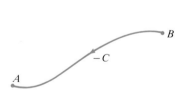

Figure 14.12

Line Integrals in Space

We now suppose that C is a smooth space curve given by parametric equations

$$x = x(t) \qquad y = y(t) \qquad z = z(t) \qquad a \leqslant t \leqslant b$$

or by a vector equation $\mathbf{r}(t) = x(t)\mathbf{i} + y(t)\mathbf{j} + z(t)\mathbf{k}$. If f is a function of three variables that is continuous on some region containing C, then we define the

line integral of f along C (with respect to arc length) in a manner similar to that for plane curves:

$$\int_C f(x, y, z)\, ds = \lim_{\|P\| \to 0} \sum_{i=1}^n f(x_i^*, y_i^*, z_i^*)\, \Delta s_i$$

We evaluate it using a formula similar to Formula 14.7:

(14.12)
$$\int_C f(x, y, z)\, ds = \int_a^b f(x(t), y(t), z(t)) \sqrt{\left(\frac{dx}{dt}\right)^2 + \left(\frac{dy}{dt}\right)^2 + \left(\frac{dz}{dt}\right)^2}\, dt$$

Observe that the integrals in both Formulas 14.7 and 14.12 can be written in the more compact vector notation

$$\int_a^b f(\mathbf{r}(t)) |\mathbf{r}'(t)|\, dt$$

In the special case where $f(x, y, z) \equiv 1$, we get

$$\int_C ds = \int_a^b |\mathbf{r}'(t)|\, dt = L$$

where L is the length of the curve C (see Formula 11.56).

Line integrals along C with respect to x, y, and z can also be defined. For example,

$$\int_C f(x, y, z)\, dz = \lim_{\|P\| \to 0} \sum_{i=1}^n f(x_i^*, y_i^*, z_i^*)\, \Delta z_i$$
$$= \int_a^b f(x(t), y(t), z(t)) z'(t)\, dt$$

Therefore, as with line integrals in the plane, we evaluate integrals of the form

(14.13)
$$\int_C P(x, y, z)\, dx + Q(x, y, z)\, dy + R(x, y, z)\, dz$$

by expressing everything (x, y, z, dx, dy, dz) in terms of the parameter t.

● **Example 4**   Evaluate $\int_C y \sin z\, ds$, where C is the circular helix given by $x = \cos t$, $y = \sin t$, $z = t$, $0 \leqslant t \leqslant 2\pi$.

Solution   Formula 14.12 gives

$$\int_C y \sin z\, ds = \int_0^{2\pi} (\sin t) \sin t \sqrt{\left(\frac{dx}{dt}\right)^2 + \left(\frac{dy}{dt}\right)^2 + \left(\frac{dz}{dt}\right)^2}\, dt$$
$$= \int_0^{2\pi} \sin^2 t \sqrt{\sin^2 t + \cos^2 t + 1}\, dt$$
$$= \sqrt{2} \int_0^{2\pi} \frac{1}{2}(1 - \cos 2t)\, dt = \frac{\sqrt{2}}{2}\left[t - \frac{1}{2}\sin 2t \right]_0^{2\pi} = \sqrt{2}\pi \quad ●$$

Note: In setting up a line integral, sometimes the most difficult thing is to think of a parametric representation for a curve whose geometric description is given. It is useful to remember that the vector representation of the line segment that starts at $\mathbf{r}_0$ and ends at $\mathbf{r}_1$ is given by

(14.14)

$$\mathbf{r}(t) = (1 - t)\mathbf{r}_0 + t\mathbf{r}_1 \qquad 0 \leqslant t \leqslant 1$$

(See Figure 11.26 and Equation 11.35 with $\mathbf{v} = \mathbf{r}_1 - \mathbf{r}_0$.)

● **Example 5**   Evaluate $\int_C y\,dx + z\,dy + x\,dz$, where C consists of the line segment C_1 from $(2, 0, 0)$ to $(3, 4, 5)$ followed by the vertical line segment C_2 from $(3, 4, 5)$ to $(3, 4, 0)$.

Solution   The curve C is shown in Figure 14.13. Using Equation 14.14 we write C_1 as

$$\mathbf{r}(t) = (1 - t)\langle 2, 0, 0 \rangle + t\langle 3, 4, 5 \rangle = \langle 2 + t, 4t, 5t \rangle$$

or, in parametric form, as

$$x = 2 + t \qquad y = 4t \qquad z = 5t \qquad 0 \leqslant t \leqslant 1$$

Thus

$$\int_{C_1} y\,dx + z\,dy + x\,dz = \int_0^1 (4t)\,dt + (5t)4\,dt + (2 + t)5\,dt$$

$$= \int_0^1 (10 + 29t)\,dt = 10t + 29\frac{t^2}{2}\bigg]_0^1 = 24.5$$

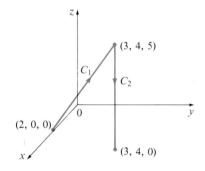

Figure 14.13

Likewise C_2 can be written in the form

$$\mathbf{r}(t) = (1 - t)\langle 3, 4, 5 \rangle + t\langle 3, 4, 0 \rangle = \langle 3, 4, 5 - 5t \rangle$$

or $x = 3 \qquad y = 4 \qquad z = 5 - 5t \qquad 0 \leqslant t \leqslant 1$

Then $dx = 0 = dy$, so

$$\int_{C_2} y\,dx + z\,dy + x\,dz = \int_0^1 3(-5)\,dt = -15$$

Adding the values of these integrals together, we obtain

$$\int_{C_2} y\,dx + z\,dy + x\,dz = 24.5 - 15 = 9.5 \qquad\qquad ●$$

Line Integrals of Vector Fields

Recall from Section 8.5 that the work done by a variable force $f(x)$ in moving a particle from a to b along the x-axis is $W = \int_a^b f(x)\,dx$. Then in Section 11.3 we found that the work done by a constant force $\mathbf{F}$ in moving an object from a point P to another point Q in space is $W = \mathbf{F} \cdot \mathbf{D}$, where $\mathbf{D} = \overrightarrow{PQ}$ is the displacement vector.

Now suppose that $\mathbf{F} = P\mathbf{i} + Q\mathbf{j} + R\mathbf{k}$ is a continuous force field on R^3, such as the gravitational field of Example 4 in Section 14.1 or the electric force field of Example 5 in Section 14.1. (A force field on R^2 could be regarded as a special case where $R = 0$ and P and Q depend only on x and y.) We

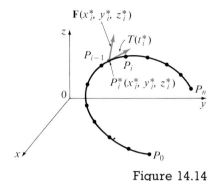

Figure 14.14

wish to compute the work done by this force in moving a particle along a smooth curve C.

We partition C into subarcs $P_{i-1}P_i$ with lengths Δs_i by means of a partition of the parameter interval $[a,b]$. (See Figure 14.7 for the two-dimensional case or Figure 14.14 for the three-dimensional case.) Choose a point $P_i^*(x_i^*, y_i^*, z_i^*)$ on the ith subarc corresponding to the parameter value t_i^*. If Δs_i is small, then as the particle moves from P_{i-1} to P_i along the curve it proceeds approximately in the direction of $\mathbf{T}(t_i^*)$, the unit tangent vector at P_i^*. Therefore the work done by the force $\mathbf{F}$ in moving the particle from P_{i-1} to P_i is approximately

$$\mathbf{F}(x_i^*, y_i^*, z_i^*) \cdot [\Delta s_i \mathbf{T}(t_i^*)] = [\mathbf{F}(x_i^*, y_i^*, z_i^*) \cdot \mathbf{T}(t_i^*)]\Delta s_i$$

and the total work done in moving the particle along C is approximately

(14.15)
$$\sum_{i=1}^{n} [\mathbf{F}(x_i^*, y_i^*, z_i^*) \cdot \mathbf{T}(x_i^*, y_i^*, z_i^*)]\Delta s_i$$

where $\mathbf{T}(x, y, z)$ is the unit tangent vector at the point (x, y, z) on C. Intuitively, we see that these approximations ought to become better as $\|P\|$ becomes smaller. Therefore we define the **work** W done by the force field $\mathbf{F}$ as the limit of the Riemann sums in (14.15), namely,

(14.16)
$$W = \int_C F(x, y, z) \cdot \mathbf{T}(x, y, z)\, ds = \int_C \mathbf{F} \cdot \mathbf{T}\, ds$$

Equation 14.16 says that *work is the integral with respect to arc length of the tangential component of the force.*

If the curve C is given by the vector equation $\mathbf{r}(t) = x(t)\mathbf{i} + y(t)\mathbf{j} + z(t)\mathbf{k}$, then $\mathbf{T}(t) = \mathbf{r}'(t)/|\mathbf{r}'(t)|$, so using Equation 14.12 we can rewrite Equation 14.16 in the form

$$W = \int_a^b \left[\mathbf{F}(\mathbf{r}(t)) \cdot \frac{\mathbf{r}'(t)}{|\mathbf{r}'(t)|} \right] |\mathbf{r}'(t)|\, dt = \int_a^b \mathbf{F}(\mathbf{r}(t)) \cdot \mathbf{r}'(t)\, dt$$

This latter integral is often abbreviated as $\int_C \mathbf{F} \cdot d\mathbf{r}$ and occurs in other areas of physics as well. Therefore we make the following definition for the line integral of *any* continuous vector field.

Definition (14.17)

Let $\mathbf{F}$ be a continuous vector field defined on a smooth curve C given by a vector function $\mathbf{r}(t)$, $a \leqslant t \leqslant b$. Then the **line integral of F along C** is

$$\int_C \mathbf{F} \cdot d\mathbf{r} = \int_a^b \mathbf{F}(\mathbf{r}(t)) \cdot \mathbf{r}'(t)\, dt = \int_C \mathbf{F} \cdot \mathbf{T}\, ds$$

When using Definition 14.17 remember that $\mathbf{F}(\mathbf{r}(t))$ is just short for $\mathbf{F}(x(t), y(t), z(t))$, so we simply evaluate $\mathbf{F}(\mathbf{r}(t))$ by putting $x = x(t)$, $y = y(t)$, and $z = z(t)$ in the expression for $\mathbf{F}(x, y, z)$. Notice also that we can formally write $d\mathbf{r} = \mathbf{r}'(t)dt$.

● **Example 6**   Find the work done by the force field

$$\mathbf{F}(x, y) = -y^2\mathbf{i} + xy\mathbf{j}$$

in moving a particle along the semicircle $\mathbf{r}(t) = \cos t\mathbf{i} + \sin t\mathbf{j}, 0 \leqslant t \leqslant \pi$.

Solution   Since $x = \cos t$ and $y = \sin t$, we have

$$\mathbf{F}(\mathbf{r}(t)) = -\sin^2 t\mathbf{i} + \cos t \sin t\mathbf{j}$$

and

$$\mathbf{r}'(t) = -\sin t\mathbf{i} + \cos t\mathbf{j}$$

Therefore the work done is

$$\int_C \mathbf{F} \cdot d\mathbf{r} = \int_0^\pi \mathbf{F}(\mathbf{r}(t)) \cdot \mathbf{r}'(t)\, dt = \int_0^\pi (\sin^3 t + \cos^2 t \sin t)\, dt$$

$$= \int_0^\pi \sin t\, dt = -\cos t\Big]_0^\pi = 2$$ ●

Note: Even though $\int_C \mathbf{F} \cdot d\mathbf{r} = \int_C \mathbf{F} \cdot \mathbf{T}\, ds$ and integrals with respect to arc length are unchanged when orientation is reversed, it is still true that

$$\int_{-C} \mathbf{F} \cdot d\mathbf{r} = -\int_C \mathbf{F} \cdot d\mathbf{r}$$

because the unit tangent vector $\mathbf{T}$ is replaced by its negative when C is replaced by $-C$.

● **Example 7**   Evaluate $\int_C \mathbf{F} \cdot d\mathbf{r}$, where $\mathbf{F}(x, y, z) = xy\mathbf{i} + yz\mathbf{j} + zx\mathbf{k}$ and C is the twisted cubic given by

$$x = t \qquad y = t^2 \qquad z = t^3 \qquad 0 \leqslant t \leqslant 1$$

Solution   We have

$$\mathbf{r}(t) = t\mathbf{i} + t^2\mathbf{j} + t^3\mathbf{k}$$

$$\mathbf{r}'(t) = \mathbf{i} + 2t\mathbf{j} + 3t^2\mathbf{k}$$

$$\mathbf{F}(\mathbf{r}(t)) = t^3\mathbf{i} + t^5\mathbf{j} + t^4\mathbf{k}$$

Thus

$$\int_C \mathbf{F} \cdot d\mathbf{r} = \int_0^1 \mathbf{F}(\mathbf{r}(t)) \cdot \mathbf{r}'(t)\, dt$$

$$= \int_0^1 (t^3 + 5t^6)\, dt = \frac{t^4}{4} + \frac{5t^7}{7}\Big]_0^1 = \frac{27}{28}$$ ●

Finally we note the connection between line integrals of vector fields and line integrals of scalar fields. Suppose the vector field $\mathbf{F}$ on R^3 is given in component form by $\mathbf{F} = P\mathbf{i} + Q\mathbf{j} + R\mathbf{k}$. We use Definition 14.17 to compute its line integral along C:

$$\int_C \mathbf{F} \cdot d\mathbf{r} = \int_a^b \mathbf{F}(\mathbf{r}(t)) \cdot \mathbf{r}'(t)\, dt$$

$$= \int_a^b (P\mathbf{i} + Q\mathbf{j} + R\mathbf{k}) \cdot (x'(t)\mathbf{i} + y'(t)\mathbf{j} + z'(t)\mathbf{k})\, dt$$

$$= \int_a^b [P(x(t), y(t), z(t))x'(t) + Q(x(t), y(t), z(t))y'(t)$$

$$+ R(x(t), y(t), z(t))z'(t)]\, dt$$

But this last integral is precisely the line integral in (14.13). Therefore we have

$$\int_C \mathbf{F} \cdot d\mathbf{r} = \int_C P\,dx + Q\,dy + R\,dz$$

where

$$\mathbf{F} = P\mathbf{i} + Q\mathbf{j} + R\mathbf{k}$$

For example, the integral $\int_C y\,dx + z\,dy + x\,dz$ in Example 5 could be expressed as $\int_C \mathbf{F} \cdot d\mathbf{r}$ where

$$\mathbf{F}(x, y, z) = y\mathbf{i} + z\mathbf{j} + x\mathbf{k}$$

EXERCISES 14.2

In Exercises 1–16 evaluate the line integral, where C is the given curve.

1. $\int_C x\,ds$, C: $x = t^3$, $y = t$, $0 \leqslant t \leqslant 1$

2. $\int_C y\,ds$, C: $x = t^3$, $y = t^2$, $0 \leqslant t \leqslant 1$

3. $\int_C xy^4\,ds$, C is the right half of the circle $x^2 + y^2 = 16$

4. $\int_C xy\,ds$, C is the line segment joining $(-1, 1)$ to $(2, 3)$

5. $\int_C (x - 2y^2)\,dy$, C is the arc of the parabola $y = x^2$ from $(-2, 4)$ to $(1, 1)$

6. $\int_C \sin x\,dx$, C is the arc of the curve $x = y^4$ from $(1, -1)$ to $(1, 1)$

7. $\int_C xy\,dx + (x - y)\,dy$, C consists of line segments from $(0, 0)$ to $(2, 0)$ and from $(2, 0)$ to $(3, 2)$

8. $\int_C x\sqrt{y}\,dx + 2y\sqrt{x}\,dy$, C consists of the arc of the circle $x^2 + y^2 = 1$ from $(1, 0)$ to $(0, 1)$ and the line segment from $(0, 1)$ to $(4, 3)$

9. $\int_C xyz\,ds$, C: $x = 2t$, $y = 3\sin t$, $z = 3\cos t$, $0 \leqslant t \leqslant \pi/2$

10. $\int_C x^2 z\,ds$, C: $x = \sin 2t$, $y = 3t$, $z = \cos 2t$, $0 \leqslant t \leqslant \pi/4$

11. $\int_C xy^2 z\,ds$, C is the line segment from $(1, 0, 1)$ to $(0, 3, 6)$

12. $\int_C xz\,ds$, C: $x = 6t$, $y = 3\sqrt{2}t^2$, $z = 2t^3$, $0 \leqslant t \leqslant 1$

13. $\int_C x^3 y^2 z\,dz$, C: $x = 2t$, $y = t^2$, $z = t^2$, $0 \leqslant t \leqslant 1$

14. $\int_C yz\,dy + xy\,dz$, C: $x = \sqrt{t}$, $y = t$, $z = t^2$, $0 \leqslant t \leqslant 1$

15. $\int_C z^2\,dx - z\,dy + 2y\,dz$, C consists of line segments from $(0, 0, 0)$ to $(0, 1, 1)$, from $(0, 1, 1)$ to $(1, 2, 3)$, and from $(1, 2, 3)$ to $(1, 2, 4)$

16. $\int_C yz\,dx + xz\,dy + xy\,dz$, C consists of line segments from $(0, 0, 0)$ to $(2, 0, 0)$, from $(2, 0, 0)$ to $(1, 3, -1)$, and from $(1, 3, -1)$ to $(1, 3, 0)$

In Exercises 17–22 evaluate the line integral $\int_C \mathbf{F} \cdot d\mathbf{r}$, where C is given by the vector function $\mathbf{r}(t)$.

17. $\mathbf{F}(x, y) = x^2 y\mathbf{i} - xy\mathbf{j}$, $\mathbf{r}(t) = t^3\mathbf{i} + t^4\mathbf{j}$, $0 \leqslant t \leqslant 1$

18. $\mathbf{F}(x, y) = e^x\mathbf{i} + xy\mathbf{j}$, $\mathbf{r}(t) = t^2\mathbf{i} + t^3\mathbf{j}$, $0 \leqslant t \leqslant 1$

19. $\mathbf{F}(x, y, z) = x\mathbf{i} - z\mathbf{j} + y\mathbf{k}$, $\mathbf{r}(t) = 2t\mathbf{i} + 3t\mathbf{j} - t^2\mathbf{k}$, $-1 \leqslant t \leqslant 1$

20. $\mathbf{F}(x, y, z) = (y + z)\mathbf{i} - x^2\mathbf{j} - 4y^2\mathbf{k}$, $\mathbf{r}(t) = t\mathbf{i} + t^2\mathbf{j} + t^4\mathbf{k}$, $0 \leqslant t \leqslant 1$

21. $\mathbf{F}(x, y, z) = \sin x\mathbf{i} + \cos y\mathbf{j} + xz\mathbf{k}$, $\mathbf{r}(t) = t^3\mathbf{i} - t^2\mathbf{j} + t\mathbf{k}$, $0 \leqslant t \leqslant 1$

22. $\mathbf{F}(x, y, z) = x^2\mathbf{i} + xy\mathbf{j} + z^2\mathbf{k}$, $\mathbf{r}(t) = \sin t\mathbf{i} + \cos t\mathbf{j} + t^2\mathbf{k}$, $0 \leqslant t \leqslant \pi/2$

23. A thin wire is bent into the shape of a semicircle $x^2 + y^2 = 4$, $x \geqslant 0$. If the linear density is a constant k, find the mass and center of mass of the wire.

24. Find the mass and center of mass of a thin wire in the shape of a quarter-circle $x^2 + y^2 = r^2$, $x \geqslant 0$, $y \geqslant 0$, if the density function is $\rho(x, y) = x + y$.

25. (a) Write the formulas analogous to Equations 14.8 for the center of mass $(\bar{x}, \bar{y}, \bar{z})$ of a thin wire with density function $\rho(x, y, z)$ in the shape of a space curve C.
   (b) Find the center of mass of a wire in the shape of the helix $x = 2 \sin t$, $y = 2 \cos t$, $z = 3t$, $0 \leqslant t \leqslant 2\pi$, if the density is a constant k.

26. Find the mass and center of mass of a wire in the shape of the helix $x = t$, $y = \cos t$, $z = \sin t$, $0 \leqslant t \leqslant 2\pi$, if the density at any point is equal to the square of the distance from the origin.

27. Find the work done by the force field $\mathbf{F}(x, y) = x\mathbf{i} + (y + 2)\mathbf{j}$ in moving an object along an arch of the cycloid $\mathbf{r}(t) = (t - \sin t)\mathbf{i} + (1 - \cos t)\mathbf{j}$, $0 \leqslant t \leqslant 2\pi$.

28. Find the work done by the force field $\mathbf{F}(x, y) = x \sin y\mathbf{i} + y\mathbf{j}$ on a particle that moves along the parabola $y = x^2$ from $(-1, 1)$ to $(2, 4)$.

29. Find the work done by the force field $\mathbf{F}(x, y, z) = xz\mathbf{i} + yx\mathbf{j} + zy\mathbf{k}$ on a particle that moves along the curve $\mathbf{r}(t) = t^2\mathbf{i} - t^3\mathbf{j} + t^4\mathbf{k}$, $0 \leqslant t \leqslant 1$.

30. The force exerted by an electric charge at the origin on a charged particle at a point (x, y, z) with position vector $\mathbf{r} = \langle x, y, z \rangle$ is $\mathbf{F}(\mathbf{r}) = K\mathbf{r}/|\mathbf{r}|^3$ where K is a constant. (See Example 5 in Section 14.1.) Find the work done as the particle moves along a straight line from $(2, 0, 0)$ to $(2, 1, 5)$.

SECTION 14.3

The Fundamental Theorem for Line Integrals

Recall from Section 5.5 that Part 2 of the Fundamental Theorem of Calculus can be written as

(14.18)

$$\int_a^b F'(x)\, dx = F(b) - F(a)$$

where F' is continuous on $[a, b]$. If we think of the gradient vector ∇f of a function f of two or three variables as a sort of derivative of f, then the following theorem can be regarded as a version of the fundamental theorem for line integrals.

Theorem (14.19)

> Let C be a smooth curve given by the vector function $\mathbf{r}(t)$, $a \leqslant t \leqslant b$. Let f be a differentiable function of two or three variables whose gradient vector ∇f is continuous on C. Then
>
> $$\int_C \nabla f \cdot d\mathbf{r} = f(\mathbf{r}(b)) - f(\mathbf{r}(a))$$

Note: Theorem 14.19 says that we can evaluate the line integral of a conservative vector field (the gradient vector field of the potential function f) simply by knowing the value of f at the endpoints of C. If f is a function of two variables and C is a plane curve with initial point $A(x_1, y_1)$ and ter-

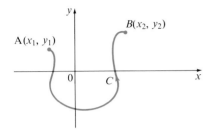

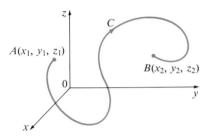

Figure 14.15

minal point $B(x_2, y_2)$ as in Figure 14.15, then Theorem 14.19 becomes

$$\int_C \nabla f \cdot d\mathbf{r} = f(x_2, y_2) - f(x_1, y_1)$$

If f is a function of three variables and C is a space curve joining $A(x_1, y_1, z_1)$ to $B(x_2, y_2, z_2)$, then we have

$$\int_C \nabla f \cdot d\mathbf{r} = f(x_2, y_2, z_2) - f(x_1, y_1, z_1)$$

Let us prove Theorem 14.19 for the latter case.

Proof of Theorem 14.19   Using Definition 14.17, we have

$$
\begin{aligned}
\int_C \nabla f \cdot d\mathbf{r} &= \int_a^b \nabla f(\mathbf{r}(t)) \cdot \mathbf{r}'(t)\, dt \\
&= \int_a^b \left(\frac{\partial f}{\partial x}\frac{dx}{dt} + \frac{\partial f}{\partial y}\frac{dy}{dt} + \frac{\partial f}{\partial z}\frac{dz}{dt} \right) dt \\
&= \int_a^b \frac{d}{dt} f(\mathbf{r}(t))\, dt \qquad \text{(by the Chain Rule)} \\
&= f(\mathbf{r}(b)) - f(\mathbf{r}(a))
\end{aligned}
$$

The last step follows from the Fundamental Theorem of Calculus (Equation 14.18). ●

Although we have proved Theorem 14.19 for smooth curves, it is also true for piecewise smooth curves. This can be seen by subdividing C into a finite number of smooth curves and adding the resulting integrals.

● **Example 1**   Find the work done by the gravitational field

$$\mathbf{F}(\mathbf{x}) = -\frac{mMG}{|\mathbf{x}|^3}\, \mathbf{x}$$

in moving a particle with mass m from the point $(3, 4, 12)$ to the point $(2, 2, 0)$ along a piecewise smooth curve C. (See Example 4 in Section 14.1.)

Solution   From Section 14.1 we know that $\mathbf{F}$ is a conservative vector field and, in fact, $\mathbf{F} = \nabla f$ where

$$f(x, y, z) = \frac{mMG}{\sqrt{x^2 + y^2 + z^2}}$$

Therefore, by Theorem 14.19, the work done is

$$
\begin{aligned}
W &= \int_C \mathbf{F} \cdot d\mathbf{r} = \int_C \nabla f \cdot d\mathbf{r} \\
&= f(2, 2, 0) - f(3, 4, 12) \\
&= \frac{mMG}{\sqrt{2^2 + 2^2}} - \frac{mMG}{\sqrt{3^2 + 4^2 + 12^2}} = mMG\left(\frac{1}{2\sqrt{2}} - \frac{1}{13} \right)
\end{aligned}
$$

●

Independence of Path

Suppose C_1 and C_2 are two piecewise smooth curves (which are called **paths**) that have the same initial point A and terminal point B. We know from Example 3 in Section 14.2 that, in general, $\int_{C_1} \mathbf{F} \cdot d\mathbf{r} \neq \int_{C_2} \mathbf{F} \cdot d\mathbf{r}$. But one implication of Theorem 14.19 is that $\int_{C_1} \nabla f \cdot d\mathbf{r} = \int_{C_2} \nabla f \cdot d\mathbf{r}$ whenever ∇f is continuous. In other words, the line integral of a conservative vector field depends on only the initial point and terminal point of a curve.

In general, if $\mathbf{F}$ is a continuous vector field with domain D, we say that the line integral $\int_C \mathbf{F} \cdot d\mathbf{r}$ is **independent of path** if $\int_{C_1} \mathbf{F} \cdot d\mathbf{r} = \int_{C_2} \mathbf{F} \cdot d\mathbf{r}$ for any two paths C_1 and C_2 in D that have the same initial and terminal points. With this terminology we can say that **line integrals of conservative vector fields are independent of path.**

Figure 14.16

A CLOSED CURVE

A curve is called **closed** if its terminal point coincides with its initial point, that is, $\mathbf{r}(b) = \mathbf{r}(a)$ (see Figure 14.16). If $\int_C \mathbf{F} \cdot d\mathbf{r}$ is independent of path in D and C is any closed path in D, we can choose any two points A and B on C and regard C as being composed of the path C_1 from A to B followed by the path C_2 from B to A (see Figure 14.17). Then

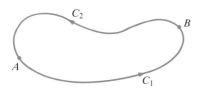

Figure 14.17

$$\int_C \mathbf{F} \cdot d\mathbf{r} = \int_{C_1} \mathbf{F} \cdot d\mathbf{r} + \int_{C_2} \mathbf{F} \cdot d\mathbf{r} = \int_{C_1} \mathbf{F} \cdot d\mathbf{r} - \int_{-C_2} \mathbf{F} \cdot d\mathbf{r} = 0$$

since C_1 and $-C_2$ have the same initial and terminal points.

Conversely, if it is true that $\int_C \mathbf{F} \cdot d\mathbf{r} = 0$ whenever C is a closed path in D, then we demonstrate independence of path as follows. Take any two paths C_1 and C_2 from A to B in D and define C to be the curve consisting of C_1 followed by $-C_2$. Then

$$0 = \int_C \mathbf{F} \cdot d\mathbf{r} = \int_{C_1} \mathbf{F} \cdot d\mathbf{r} + \int_{-C_2} \mathbf{F} \cdot d\mathbf{r} = \int_{C_1} \mathbf{F} \cdot d\mathbf{r} - \int_{C_2} \mathbf{F} \cdot d\mathbf{r}$$

and so $\int_{C_1} \mathbf{F} \cdot d\mathbf{r} = \int_{C_2} \mathbf{F} \cdot d\mathbf{r}$. Thus we have proved the following:

Theorem (14.20)

> $\int_C \mathbf{F} \cdot d\mathbf{r}$ is independent of path in D if and only if $\int_C \mathbf{F} \cdot d\mathbf{r} = 0$ for every closed path C in D.

Since we know that the line integral of any conservative vector field $\mathbf{F}$ is independent of path, it follows that $\int_C \mathbf{F} \cdot d\mathbf{r} = 0$ for closed paths. The physical interpretation is that the work done by a conservative force field (such as the gravitational or electric field in Section 14.1) as it moves an object around a closed path is 0.

The following theorem says that the *only* vector fields that are independent of path are conservative. It is stated and proved for plane curves but there is a similar version for space curves. We assume that D is **open**, which means that for every point P in D there is a disk with center P that lies entirely in D. (Thus every point in D is an interior point in the sense of Section 12.2.) In addition, we assume that D is **connected**. This means that any two points in D can be joined by a path that lies in D.

Theorem (14.21)

Suppose $\mathbf{F}$ is a vector field that is continuous on an open connected region D. If $\int_C \mathbf{F} \cdot d\mathbf{r}$ is independent of path in D, then $\mathbf{F}$ is a conservative vector field on D; that is, there exists a function f such that $\nabla f = \mathbf{F}$.

Proof   Let $A(a, b)$ be a fixed point in D. We construct the desired potential function f by defining

$$f(x, y) = \int_{(a,b)}^{(x,y)} \mathbf{F} \cdot d\mathbf{r}$$

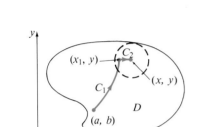

(x_1, y) — C_2

C_1 (x, y)

D

(a, b)

(a)

for any point (x, y) in D. Since $\int_C \mathbf{F} \cdot d\mathbf{r}$ is independent of path, it does not matter which path C from (a, b) to (x, y) is used to evaluate $f(x, y)$. Since D is open, there exists a disk contained in D with center (x, y). Choose any point (x_1, y) in the disk with $x_1 < x$ and let C consist of any path C_1 from (a, b) to (x_1, y) followed by the horizontal line segment C_2 from (x_1, y) to (x, y) [see Figure 14.18(a)]. Then

$$f(x, y) = \int_{C_1} \mathbf{F} \cdot d\mathbf{r} + \int_{C_2} \mathbf{F} \cdot d\mathbf{r} = \int_{(a,b)}^{(x_1,y)} \mathbf{F} \cdot d\mathbf{r} + \int_{C_2} \mathbf{F} \cdot d\mathbf{r}$$

Notice that the first of these integrals does not depend on x, so

$$\frac{\partial}{\partial x} f(x, y) = 0 + \frac{\partial}{\partial x} \int_{C_2} \mathbf{F} \cdot d\mathbf{r}$$

If we write $\mathbf{F} = P\mathbf{i} + Q\mathbf{j}$, then

$$\int_{C_2} \mathbf{F} \cdot d\mathbf{r} = \int_{C_2} P \, dx + Q \, dy$$

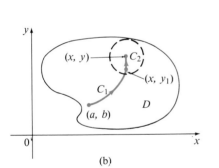

(x, y) — C_2

C_1 (x, y_1)

D

(a, b)

(b)

Figure 14.18

On C_2, y is constant, so $dy = 0$. Using t as the parameter, where $x_1 \leqslant t \leqslant x$, we have

$$\frac{\partial}{\partial x} f(x, y) = \frac{\partial}{\partial x} \int_{C_2} P \, dx + Q \, dy$$

$$= \frac{\partial}{\partial x} \int_{x_1}^{x} P(t, y) \, dt = P(x, y)$$

by Part 1 of the Fundamental Theorem of Calculus (5.18). A similar argument, using a vertical line segment [see Figure 14.18(b)], shows that

$$\frac{\partial}{\partial y} f(x, y) = \frac{\partial}{\partial y} \int_{C_2} P \, dx + Q \, dy$$

$$= \frac{\partial}{\partial y} \int_{y_1}^{y} Q(x, t) \, dt = Q(x, y)$$

Thus

$$\mathbf{F} = P\mathbf{i} + Q\mathbf{j} = \frac{\partial f}{\partial x} \mathbf{i} + \frac{\partial f}{\partial y} \mathbf{j} = \nabla f$$

which says that $\mathbf{F}$ is conservative.

The question remains: How is it possible to determine whether or not a vector field $\mathbf{F}$ is conservative? Suppose it is known that $\mathbf{F} = P\mathbf{i} + Q\mathbf{j}$ is conservative, where P and Q have continuous first order partial derivatives. Then there is a function f such that $\mathbf{F} = \nabla f$, that is,

$$P = \frac{\partial f}{\partial x} \quad \text{and} \quad Q = \frac{\partial f}{\partial y}$$

Therefore, by Clairaut's Theorem (12.14),

$$\frac{\partial P}{\partial y} = \frac{\partial^2 f}{\partial y\, \partial x} = \frac{\partial^2 f}{\partial x\, \partial y} = \frac{\partial Q}{\partial x}$$

Theorem (14.22)

> If $\mathbf{F}(x, y) = P(x, y)\mathbf{i} + Q(x, y)\mathbf{j}$ is a conservative vector field, where P and Q have continuous first order partial derivatives on a domain D, then throughout D we have
>
> $$\frac{\partial P}{\partial y} = \frac{\partial Q}{\partial x}$$

The converse of Theorem 14.22 is true only for a special type of region. To explain this we first need the concept of a **simple curve,** which is a curve that does not intersect itself anywhere between its endpoints. [See Figure 14.19; $\mathbf{r}(a) = \mathbf{r}(b)$ for a simple closed curve, but $\mathbf{r}(t_1) \neq \mathbf{r}(t_2)$ when $a < t_1 < t_2 < b$.]

In Theorem 14.21 we needed an open connected region. For the next theorem we need a stronger condition. A **simply-connected region** in the plane is a connected region D such that every simple closed curve in D encloses only points that are in D. Notice from Figure 14.20 that, intuitively speaking,

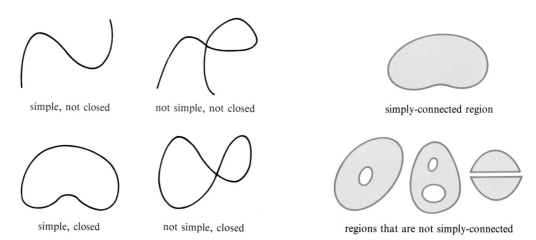

| simple, not closed | not simple, not closed | simply-connected region |
| simple, closed | not simple, closed | regions that are not simply-connected |

Figure 14.19
TYPES OF CURVES

Figure 14.20

a simply-connected region contains no holes and cannot consist of two separate pieces.

In terms of simply-connected regions we can now state a partial converse to Theorem 14.22 that gives a convenient method for verifying that a vector field on R^2 is conservative. The proof will be given in the next section as a consequence of Green's Theorem.

Theorem (14.23)

Let $\mathbf{F} = P\mathbf{i} + Q\mathbf{j}$ be a vector field on an open simply-connected region D. Suppose that P and Q have continuous first order derivatives and

$$\frac{\partial P}{\partial y} = \frac{\partial Q}{\partial x} \qquad \text{throughout } D$$

Then $\mathbf{F}$ is conservative.

● **Example 2**   Determine whether or not the vector field

$$\mathbf{F}(x, y) = 2xy\mathbf{i} + xy^3\mathbf{j}$$

is conservative.

Solution   Let $P(x, y) = 2xy$ and $Q(x, y) = xy^3$. Then

$$\frac{\partial P}{\partial y} = 2x \qquad \frac{\partial Q}{\partial x} = y^3$$

Since $\partial P/\partial y \neq \partial Q/\partial x$, $\mathbf{F}$ is not conservative by Theorem 14.22.   ●

● **Example 3**   Determine whether or not the vector field

$$\mathbf{F}(x, y) = (3 + 2xy)\mathbf{i} + (x^2 - 3y^2)\mathbf{j}$$

is conservative.

Solution   Let $P(x, y) = 3 + 2xy$ and $Q(x, y) = x^2 - 3y^2$. Then

$$\frac{\partial P}{\partial y} = 2x = \frac{\partial Q}{\partial x}$$

Also the domain of $\mathbf{F}$ is the entire plane ($D = R^2$) which is open and simply-connected. Therefore we can apply Theorem 14.23 and conclude that $\mathbf{F}$ is conservative.   ●

In Example 3, Theorem 14.23 told us that $\mathbf{F}$ is conservative but it did not tell us how to find the (potential) function f such that $\mathbf{F} = \nabla f$. The proof of Theorem 14.21 gives us a clue as to how to find f. We use "partial integration" as in the following example.

● **Example 4**   (a) If $\mathbf{F}(x, y) = (3 + 2xy)\mathbf{i} + (x^2 - 3y^2)\mathbf{j}$, find a function f such that $\mathbf{F} = \nabla f$. (b) Evaluate the line integral $\int_C \mathbf{F} \cdot d\mathbf{r}$, where C is the curve given by $\mathbf{r}(t) = e^t \sin t\mathbf{i} + e^t \cos t\mathbf{j}$, $0 \leqslant t \leqslant \pi$.

Solution   (a)   From Example 3 we know $\mathbf{F}$ is conservative and so there exists a function f with $\nabla f = \mathbf{F}$, that is,

$$f_x(x, y) = 3 + 2xy \qquad (1)$$

$$f_y(x, y) = x^2 - 3y^2 \qquad (2)$$

Integrating (1) with respect to x, we obtain

$$f(x, y) = 3x + x^2y + g(y) \qquad (3)$$

Notice that the constant of integration is a constant with respect to x, that is, a function of y that we have called $g(y)$. Next we differentiate both sides of (3) with respect to y:

$$f_y(x, y) = x^2 + g'(y) \qquad (4)$$

Comparing (2) and (4) we see that

$$g'(y) = -3y^2$$

Integrating with respect to y, we have

$$g(y) = -y^3 + K$$

where K is a constant. Putting this in (3), we have

$$f(x, y) = 3x + x^2y - y^3 + K$$

as the desired potential function.

(b)   Using Theorem 14.19, all we have to know are the initial and terminal points of C, namely, $\mathbf{r}(0) = (0, 1)$ and $\mathbf{r}(\pi) = (0, -e^{\pi})$. In the expression for $f(x, y)$ in part (a), any value of the constant K will do, so let us choose $K = 0$. Then we have

$$\int_C \mathbf{F} \cdot d\mathbf{r} = \int_C \nabla f \cdot d\mathbf{r} = f(0, -e^{\pi}) - f(0, 1)$$

$$= e^{3\pi} - (-1) = e^{3\pi} + 1$$

This method is much shorter than the straightforward method for evaluating line integrals that we learned in Section 14.2.   ●

A criterion for determining whether or not a vector field $\mathbf{F}$ on R^3 is conservative will be given in Section 14.5. Meanwhile, the next example shows that the technique for finding the potential function is much the same as for vector fields on R^2.

● **Example 5**   If $\mathbf{F}(x, y, z) = y^2\mathbf{i} + (2xy + e^{3z})\mathbf{j} + 3ye^{3z}\mathbf{k}$, find a function f such that $\nabla f = \mathbf{F}$.

Solution   If there is such a function f, then

$$f_x(x, y, z) = y^2 \qquad (1)$$

$$f_y(x, y, z) = 2xy + e^{3z} \qquad (2)$$

$$f_z(x, y, z) = 3ye^{3z} \qquad (3)$$

Integrating (1) with respect to x, we get

$$f(x, y, z) = xy^2 + g(y, z) \qquad (4)$$

where $g(y, z)$ is a constant with respect to x. Then differentiating (4) with respect to y, we have

$$f_y(x, y, z) = 2xy + g_y(y, z)$$

and comparison with (2) gives

$$g_y(y, z) = e^{3z}$$

Thus $g(y, z) = ye^{3z} + h(z)$ and we rewrite (4) as

$$f(x, y, z) = xy^2 + ye^{3z} + h(z)$$

Finally, differentiating with respect to z and comparing with (3), we obtain $h'(z) = 0$ and therefore $h(z) = K$, a constant. The desired function is

$$f(x, y, z) = xy^2 + ye^{3z} + K$$

It is easily verified that $\nabla f = \mathbf{F}$.   ●

Conservation of Energy

Let us apply the ideas of this chapter to a continuous force field $\mathbf{F}$ that moves an object along a path C given by $\mathbf{r}(t)$, $a \leqslant t \leqslant b$, where $\mathbf{r}(a) = A$ is the initial point and $\mathbf{r}(b) = B$ is the terminal point of C. According to Newton's Second Law of Motion (see Section 11.9), the force $\mathbf{F}(\mathbf{r}(t))$ at a point on C is related to the acceleration $\mathbf{a}(t) = \mathbf{r}''(t)$ by the equation

$$\mathbf{F}(\mathbf{r}(t)) = m\mathbf{r}''(t)$$

So the work done by the force on the object is

$$
\begin{aligned}
W &= \int_C \mathbf{F} \cdot d\mathbf{r} = \int_a^b \mathbf{F}(\mathbf{r}(t)) \cdot \mathbf{r}'(t)\, dt \\
&= \int_a^b m\mathbf{r}''(t) \cdot \mathbf{r}'(t)\, dt \\
&= \frac{m}{2} \int_a^b \frac{d}{dt}\left[\mathbf{r}'(t) \cdot \mathbf{r}'(t)\right] dt \qquad \text{(Theorem 11.53, Formula 4)} \\
&= \frac{m}{2} \int_a^b \frac{d}{dt}\, |\mathbf{r}'(t)|^2\, dt \\
&= \frac{m}{2}\left[|\mathbf{r}'(t)|^2\right]_a^b \qquad \text{(Fundamental Theorem of Calculus)} \\
&= \frac{m}{2}(|\mathbf{r}'(b)|^2 - |\mathbf{r}'(a)|^2)
\end{aligned}
$$

Therefore

(14.24)
$$W = \tfrac{1}{2}m|\mathbf{v}(b)|^2 - \tfrac{1}{2}m|\mathbf{v}(a)|^2$$

where $\mathbf{v} = \mathbf{r}'$ is the velocity.

The quantity $\tfrac{1}{2}m|\mathbf{v}(t)|^2$, that is, half the mass times the square of the speed, is called the **kinetic energy** of the object. Therefore we can rewrite Equation 14.24 as

(14.25)
$$W = K(B) - K(A)$$

which says that the work done by the force field along C is equal to the change in kinetic energy at the endpoints of C.

Now let us further assume that $\mathbf{F}$ is a conservative force field; that is, we can write $\mathbf{F} = \nabla f$. In physics, the **potential energy** of an object at the point (x, y, z) is defined as $P(x, y, z) = -f(x, y, z)$, so we have $\mathbf{F} = -\nabla P$. Then by Theorem 14.19 we have

$$W = \int_C \mathbf{F} \cdot d\mathbf{r} = -\int_C \nabla P \cdot d\mathbf{r}$$
$$= -[P(\mathbf{r}(b)) - P(\mathbf{r}(a))] = P(A) - P(B)$$

Comparing this equation with Equation 14.25, we see that

$$P(A) + K(A) = P(B) + K(B)$$

which says that if an object moves from one point A to another point B under the influence of a conservative force field, then the sum of its potential energy and its kinetic energy remains constant. This is called the **Law of Conservation of Energy** and it is the reason that the vector field is called **conservative.**

EXERCISES 14.3

In Exercises 1–10 determine whether or not $\mathbf{F}$ is a conservative vector field. If it is, find a function f such that $\mathbf{F} = \nabla f$.

1. $\mathbf{F}(x, y) = (2x - 3y)\mathbf{i} + (2y - 3x)\mathbf{j}$

2. $\mathbf{F}(x, y) = (3x^2 - 4y)\mathbf{i} + (4y^2 - 2x)\mathbf{j}$

3. $\mathbf{F}(x, y) = (x^2 + y)\mathbf{i} + x^2\mathbf{j}$

4. $\mathbf{F}(x, y) = (x^2 + y)\mathbf{i} + (y^2 + x)\mathbf{j}$

5. $\mathbf{F}(x, y) = (1 + 4x^3y^3)\mathbf{i} + 3x^4y^2\mathbf{j}$

6. $\mathbf{F}(x, y) = (y\cos x - \cos y)\mathbf{i} + (\sin x + x\sin y)\mathbf{j}$

7. $\mathbf{F}(x, y) = (e^{2x} + x\sin y)\mathbf{i} + x^2\cos y\mathbf{j}$

8. $\mathbf{F}(x, y) = (ye^{xy} + 4x^3y)\mathbf{i} + (xe^{xy} + x^4)\mathbf{j}$

9. $\mathbf{F}(x, y) = (ye^x + \sin y)\mathbf{i} + (e^x + x\cos y)\mathbf{j}$

10. $\mathbf{F}(x, y) = (x + y^2)\mathbf{i} + (2xy + y^2)\mathbf{j}$

In Exercises 11–18 (a) find a function f such that $\mathbf{F} = \nabla f$ and (b) use part (a) to evaluate $\int_C \mathbf{F} \cdot d\mathbf{r}$ along the given curve C.

11. $\mathbf{F}(x, y) = x\mathbf{i} + y\mathbf{j}$, C is the arc of the parabola $y = x^2$ from $(-1, 1)$ to $(3, 9)$

12. $\mathbf{F}(x, y) = y\mathbf{i} + x\mathbf{j}$, C is the arc of the curve $y = x^4 - x^3$ from $(1, 0)$ to $(2, 8)$

13. $\mathbf{F}(x, y) = 2xy^3\mathbf{i} + 3x^2y^2\mathbf{j}$, C: $\mathbf{r}(t) = \sin t\mathbf{i} + (t^2 + 1)\mathbf{j}$, $0 \le t \le \pi/2$

14. $\mathbf{F}(x, y) = e^{2y}\mathbf{i} + (1 + 2xe^{2y})\mathbf{j}$, C: $\mathbf{r}(t) = te^t\mathbf{i} + (1 + t)\mathbf{j}$, $0 \le t \le 1$

15. $\mathbf{F}(x, y, z) = y\mathbf{i} + (x + z)\mathbf{j} + y\mathbf{k}$, C is the line segment from $(2, 1, 4)$ to $(8, 3, -1)$

16. $\mathbf{F}(x, y, z) = 2xy^3z^4\mathbf{i} + 3x^2y^2z^4\mathbf{j} + 4x^2y^3z^3\mathbf{k}$, C: $x = t$, $y = t^2$, $z = t^3$, $0 \le t \le 2$

17. $\mathbf{F}(x, y, z) = (2xz + \sin y)\mathbf{i} + x\cos y\mathbf{j} + x^2\mathbf{k}$, C: $\mathbf{r}(t) = \cos t\mathbf{i} + \sin t\mathbf{j} + t\mathbf{k}$, $0 \le t \le 2\pi$

18. $\mathbf{F}(x, y, z) = 4xe^z\mathbf{i} + \cos y\mathbf{j} + 2x^2e^z\mathbf{k}$, C: $\mathbf{r}(t) = t\mathbf{i} + t^2\mathbf{j} + t^4\mathbf{k}$, $0 \le t \le 1$

In Exercises 19 and 20 show that the line integral is independent of path and evaluate the integral.

19. $\int_C 2x\sin y\, dx + (x^2\cos y - 3y^2)\, dy$, C is any path from $(-1, 0)$ to $(5, 1)$

20. $\int_C (2y^2 - 12x^3y^3)\, dx + (4xy - 9x^4y^2)\, dy$, C is any path from $(1, 1)$ to $(3, 2)$

21. Show that if the vector field $\mathbf{F} = P\mathbf{i} + Q\mathbf{j} + R\mathbf{k}$ is conservative and P, Q, R have continuous first order partial derivatives, then

$$\frac{\partial P}{\partial y} = \frac{\partial Q}{\partial x} \qquad \frac{\partial P}{\partial z} = \frac{\partial R}{\partial x} \qquad \text{and} \qquad \frac{\partial Q}{\partial z} = \frac{\partial R}{\partial y}$$

22. Use Exercise 21 to show that the line integral $\int_C y\, dx + x\, dy + xyz\, dz$ is not independent of path.

23. Let $\mathbf{F}(x, y) = \dfrac{-y\mathbf{i} + x\mathbf{j}}{x^2 + y^2}$.

(a) Show that $\partial P/\partial y = \partial Q/\partial x$.

(b) Show that $\int_C \mathbf{F} \cdot d\mathbf{r}$ is not independent of path. [*Hint:* Compute $\int_{C_1} \mathbf{F} \cdot d\mathbf{r}$ and $\int_{C_2} \mathbf{F} \cdot d\mathbf{r}$, where C_1 and C_2 are the upper and lower halves of the circle $x^2 + y^2 = 1$ from $(1, 0)$ to $(-1, 0)$.] Does this contradict Theorem 14.23?

SECTION 14.4

Green's Theorem

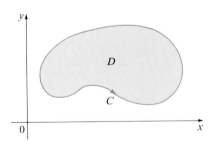

Figure 14.21

Green's Theorem, named after the English mathematical physicist Sir George Green (1793–1841), gives the relationship between a line integral around a simple closed curve C and a double integral over the plane region D bounded by C. (See Figure 14.21. We assume that D consists of all points inside C as well as all points on C.) In stating Green's Theorem we use the convention that the **positive orientation** of a simple closed curve C refers to a single *counterclockwise* traversal of C. Thus if C is given by the vector function $\mathbf{r}(t)$, $a \le t \le b$, then the region D is always on the left as the point $\mathbf{r}(t)$ traverses C (see Figure 14.22).

Figure 14.22

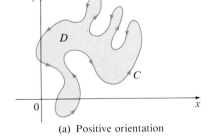

(a) Positive orientation   (b) Negative orientation

Green's Theorem (14.26)

Let C be a positively oriented, piecewise-smooth, simple closed curve in the plane and let D be the region bounded by C. If P and Q have continuous partial derivatives on an open region that contains D, then

$$\int_C P\,dx + Q\,dy = \iint\limits_D \left(\frac{\partial Q}{\partial x} - \frac{\partial P}{\partial y} \right) dA$$

Note: The notation

$$\oint_C P\,dx + Q\,dy \qquad \text{or} \qquad \oint_C P\,dx + Q\,dy$$

is sometimes used to indicate that the line integral is calculated using the positive orientation of the closed curve C. Another notation for the positively oriented boundary curve of D is ∂D, so the equation in Green's Theorem can be written as

(14.27)
$$\iint\limits_D \left(\frac{\partial Q}{\partial x} - \frac{\partial P}{\partial y} \right) dA = \int_{\partial D} P\,dx + Q\,dy$$

Green's Theorem should be regarded as the analogue of the Fundamental Theorem of Calculus for double integrals. Compare Equation 14.27 with the statement of the Fundamental Theorem of Calculus, Part 2, in the following equation:

$$\int_a^b F'(x)\,dx = F(b) - F(a)$$

In both cases there is an integral involving derivatives (F', $\partial Q/\partial x$, and $\partial P/\partial y$) on the left side of the equation. And in both cases the right side involves the values of the original functions (F, Q, and P) only on the *boundary* of the domain. (In the one-dimensional case the domain is an interval $[a, b]$ whose boundary consists of just two points, a and b.)

Green's Theorem is not easy to prove in the generality stated in Theorem 14.26, but we can give a proof for the special case where the region is both of type I and of type II (see Section 13.3). Let us call such regions **simple regions.**

Proof of Green's Theorem for the Case Where D Is a Simple Region   Notice that Green's Theorem will be proved if we can show that

(14.28)
$$\int_C P\,dx = -\iint\limits_D \frac{\partial P}{\partial y}\,dA$$

and

(14.29)
$$\int_C Q\, dy = \iint_D \frac{\partial Q}{\partial x}\, dA$$

We prove Equation 14.28 by expressing D as a type I region:

$$D = \{(x, y)\,|\, a \leqslant x \leqslant b,\, g_1(x) \leqslant y \leqslant g_2(x)\}$$

where g_1 and g_2 are continuous functions. This enables us to compute the double integral on the right side of Equation 14.28 as follows:

(14.30)
$$\iint_D \frac{\partial P}{\partial y}\, dA = \int_a^b \int_{g_1(x)}^{g_2(x)} \frac{\partial P}{\partial y}(x, y)\, dy\, dx$$

$$= \int_a^b [P(x, g_2(x)) - P(x, g_1(x))]\, dx$$

where the last step follows from the Fundamental Theorem of Calculus.

Now we compute the left side of Equation 14.28 by breaking up C as the union of the four curves C_1, C_2, C_3, and C_4 shown in Figure 14.23. On C_1 we take x as the parameter and write the parametric equations as $x = x$, $y = g_1(x)$, $a \leqslant x \leqslant b$. Thus

$$\int_{C_1} P(x, y)\, dx = \int_a^b P(x, g_1(x))\, dx$$

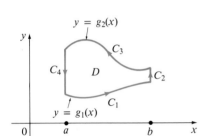

Figure 14.23

Observe that C_3 goes from right to left but $-C_3$ goes from left to right, so we can write the parametric equations of $-C_3$ as $x = x$, $y = g_2(x)$, $a \leqslant x \leqslant b$. Therefore

$$\int_{C_3} P(x, y)\, dx = -\int_{-C_3} P(x, y)\, dx = -\int_a^b P(x, g_2(x))\, dx$$

On C_2 or C_4 (either of which might reduce to just a single point), x is constant, so $dx = 0$ and

$$\int_{C_2} P(x, y)\, dx = 0 = \int_{C_4} P(x, y)\, dx$$

Hence

$$\int_C P(x, y)\, dx = \int_{C_1} P(x, y)\, dx + \int_{C_2} P(x, y)\, dx$$

$$+ \int_{C_3} P(x, y)\, dx + \int_{C_4} P(x, y)\, dx$$

$$= \int_a^b P(x, g_1(x))\, dx - \int_a^b P(x, g_2(x))\, dx$$

Comparing this expression with the one in Equation 14.30, we see that

$$\int_C P(x, y)\, dx = -\iint_D \frac{\partial P}{\partial y}\, dA$$

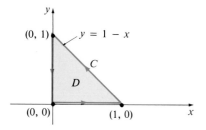

Figure 14.24

Equation 14.29 can be proved in much the same way by expressing D as a type II region (see Exercise 24). Then, by adding Equations 14.28 and 14.29, we obtain Green's Theorem. ●

● **Example 1**　　Evaluate $\int_C x^4\,dx + xy\,dy$ where C is the triangular curve consisting of the line segments from $(0,0)$ to $(1,0)$, from $(1,0)$ to $(0,1)$, and from $(0,1)$ to $(0,0)$.

Solution　　Although the given line integral could be evaluated as usual by the methods of Section 14.2, that would involve setting up three separate integrals along the three sides of the triangle, so let us instead use Green's Theorem. Notice that the region D enclosed by C is simple and C has positive orientation (see Figure 14.24). If we let $P(x, y) = x^4$ and $Q(x, y) = xy$, then we have

$$\int_C x^4\,dx + xy\,dy = \iint_D \left(\frac{\partial Q}{\partial x} - \frac{\partial P}{\partial y}\right) dA$$

$$= \int_0^1 \int_0^{1-x} (y - 0)\,dy\,dx$$

$$= \int_0^1 \left[\frac{y^2}{2}\right]_{y=0}^{y=1-x} dx = \frac{1}{2}\int_0^1 (1 - x)^2\,dx$$

$$= -\frac{1}{6}(1 - x)^3\Big]_0^1 = \frac{1}{6}$$

●

● **Example 2**　　Evaluate $\oint_C (3y - e^{\sin x})\,dx + (7x + \sqrt{y^4 + 1})\,dy$, where C is the circle $x^2 + y^2 = 9$.

Solution　　The region D bounded by C is the disk $x^2 + y^2 \leqslant 9$, so let us change to polar coordinates after applying Green's Theorem:

$$\oint_C (3y - e^{\sin x})\,dx + (7x + \sqrt{y^4 + 1})\,dy$$

$$= \iint_D \left[\frac{\partial}{\partial x}(7x + \sqrt{y^4 + 1}) - \frac{\partial}{\partial y}(3y - e^{\sin x})\right] dA$$

$$= \int_0^{2\pi} \int_0^3 (7 - 3)r\,dr\,d\theta$$

$$= 4\int_0^{2\pi} d\theta \int_0^3 r\,dr = 36\pi$$

●

In Examples 1 and 2 we found that the double integral was easier to evaluate than the line integral. (Try setting up the line integral in Example 2 and you will be convinced!) But sometimes it is easier to evaluate the line integral and Green's Theorem is used in the reverse direction. For instance, if it is known that $P(x, y) = Q(x, y) = 0$ on the curve C, then Green's Theorem gives

$$\iint_D \left(\frac{\partial Q}{\partial x} - \frac{\partial P}{\partial y}\right) dA = \int_C P\,dx + Q\,dy = 0$$

no matter what values P and Q assume in the region D.

Another application of the reverse direction of Green's Theorem is in computing areas. Since the area of D is $\iint_D 1 \, dA$, we wish to choose P and Q so that

$$\frac{\partial Q}{\partial x} - \frac{\partial P}{\partial y} = 1$$

There are several possibilities: $P(x, y) = 0$ and $Q(x, y) = x$; $P(x, y) = -y$ and $Q(x, y) = 0$; or $P(x, y) = -y/2$ and $Q(x, y) = x/2$. Then Green's Theorem gives the following formulas for the area of D:

(14.31)

$$A = \oint_C x \, dy = -\oint_C y \, dx = \frac{1}{2} \oint_C x \, dy - y \, dx$$

● **Example 3**   Find the area enclosed by the ellipse $\dfrac{x^2}{a^2} + \dfrac{y^2}{b^2} = 1$.

Solution   The ellipse has parametric equations $x = a \cos t$ and $y = b \sin t$, $0 \leqslant t \leqslant 2\pi$. Using the third formula in Equation 14.31, we have

$$A = \frac{1}{2} \int_C x \, dy - y \, dx$$

$$= \frac{1}{2} \int_0^{2\pi} (a \cos t)(b \cos t) \, dt - (b \sin t)(-a \sin t) \, dt$$

$$= \frac{ab}{2} \int_0^{2\pi} dt = \pi ab$$

●

Although we have proved Green's Theorem only for the case where D is simple, we can now extend it to the case where D is a finite union of simple regions. For example, if D is the region shown in Figure 14.25, then we can write $D = D_1 \cup D_2$ where D_1 and D_2 are both simple. The boundary of D_1 is $C_1 \cup C_3$ and the boundary of D_2 is $C_2 \cup (-C_3)$ so, applying Green's Theorem to D_1 and D_2 separately, we get

$$\int_{C_1 \cup C_3} P \, dx + Q \, dy = \iint_{D_1} \left(\frac{\partial Q}{\partial x} - \frac{\partial P}{\partial y} \right) dA$$

$$\int_{C_2 \cup (-C_3)} P \, dx + Q \, dy = \iint_{D_2} \left(\frac{\partial Q}{\partial x} - \frac{\partial P}{\partial y} \right) dA$$

If we add these two equations, the line integrals along C_3 and $-C_3$ will cancel, so we get

$$\int_{C_1 \cup C_2} P \, dx + Q \, dy = \iint_D \left(\frac{\partial Q}{\partial x} - \frac{\partial P}{\partial y} \right) dA$$

which is Green's Theorem for $D = D_1 \cup D_2$ since its boundary is $C = C_1 \cup C_2$.

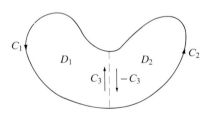

Figure 14.25

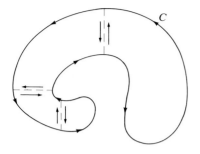

Figure 14.26

The same sort of argument allows us to establish Green's Theorem for any finite union of simple regions (see Figure 14.26).

● **Example 4**   Evaluate $\oint_C y^2\,dx + 3xy\,dy$, where C is the boundary of the semiannular region D in the upper half-plane between the circles $x^2 + y^2 = 1$ and $x^2 + y^2 = 4$.

Solution   Notice that although D is not simple, the y-axis divides it into two simple regions (see Figure 13.19). In polar coordinates we can write

$$D = \{(r,\theta)\,|\,1 \leq r \leq 2, 0 \leq \theta \leq \pi\}$$

Therefore Green's Theorem gives

$$\int_C y^2\,dx + 3xy\,dy = \iint_D \left[\frac{\partial}{\partial x}(3xy) - \frac{\partial}{\partial y}(y^2)\right]dA$$

$$= \iint_D y\,dA = \int_0^\pi \int_1^2 (r\sin\theta)r\,dr\,d\theta$$

$$= \int_0^\pi \sin\theta\,d\theta \int_1^2 r^2\,dr = \left[-\cos\theta\right]_0^\pi\left[\frac{r^3}{3}\right]_1^2$$

$$= \frac{14}{3}$$ ●

Green's Theorem can be extended to apply to regions with holes, that is, regions that are not simply-connected. Observe that the boundary C of the region D in Figure 14.27(a) consists of two simple closed curves C_1 and C_2. We assume that these boundary curves are oriented so that the region D is always on the left as the curve C is traversed. Thus the positive direction is counterclockwise for the outer curve C_1 but clockwise for the inner curve C_2. If we divide D into two regions D' and D'' by means of the lines shown in Figure 14.27(b) and then apply Green's Theorem to each of D' and D'', we get

$$\iint_D \left(\frac{\partial Q}{\partial x} - \frac{\partial P}{\partial y}\right)dA = \iint_{D'} \left(\frac{\partial Q}{\partial x} - \frac{\partial P}{\partial y}\right)dA + \iint_{D''} \left(\frac{\partial Q}{\partial x} - \frac{\partial P}{\partial y}\right)dA$$

$$= \int_{\partial D'} P\,dx + Q\,dy + \int_{\partial D''} P\,dx + Q\,dy$$

Figure 14.27

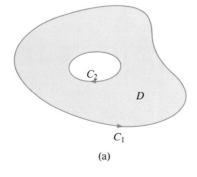

(a)

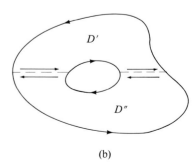

(b)

Since the line integrals along the common boundary lines are in opposite directions, they cancel and we get

$$\iint_D \left(\frac{\partial Q}{\partial x} - \frac{\partial P}{\partial y}\right) dA = \int_{C_1} P\,dx + Q\,dy + \int_{C_2} P\,dx + Q\,dy = \int_C P\,dx + Q\,dy$$

which is Green's Theorem for the region D.

● **Example 5**   If $\mathbf{F}(x, y) = (-y\mathbf{i} + x\mathbf{j})/(x^2 + y^2)$, show that $\int \mathbf{F} \cdot d\mathbf{r} = 2\pi$ for every simple closed path that encloses the origin.

Solution   Since C is an *arbitrary* closed path that encloses the origin, it is difficult to compute the given integral directly. So let us consider a counterclockwise-oriented circle C' with center the origin and radius a, where a is chosen to be small enough that C' lies inside C (see Figure 14.28). Let D be the region bounded by C and C'. Then its positively-oriented boundary is $C \cup (-C')$ and so the general version of Green's Theorem discussed above gives

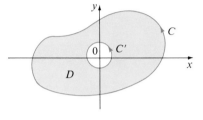

Figure 14.28

$$\int_C P\,dx + Q\,dy + \int_{-C'} P\,dx + Q\,dy = \iint_D \left(\frac{\partial Q}{\partial x} - \frac{\partial P}{\partial y}\right) dA$$

$$= \iint_D \left[\frac{y^2 - x^2}{(x^2 + y^2)^2} - \frac{y^2 - x^2}{(x^2 + y^2)^2}\right] dA$$

$$= 0$$

Therefore

$$\int_C P\,dx + Q\,dy = \int_{C'} P\,dx + Q\,dy$$

that is,

$$\int_C \mathbf{F} \cdot d\mathbf{r} = \int_{C'} \mathbf{F} \cdot d\mathbf{r}$$

We now easily compute the latter integral using the parametrization $\mathbf{r}(t) = a\cos t\,\mathbf{i} + a\sin t\,\mathbf{j}$, $0 \le t \le 2\pi$. Thus

$$\int_C \mathbf{F} \cdot d\mathbf{r} = \int_{C'} \mathbf{F} \cdot d\mathbf{r} = \int_0^{2\pi} \mathbf{F}(\mathbf{r}(t)) \cdot \mathbf{r}'(t)\,dt$$

$$= \int_0^{2\pi} \frac{(-a\sin t)(-a\sin t) + (a\cos t)(a\cos t)}{a^2\cos^2 t + a^2\sin^2 t}\,dt$$

$$= \int_0^{2\pi} dt = 2\pi$$ ●

We end this section by using Green's Theorem to prove a result that was stated in the preceding section.

Proof of Theorem 14.23   We are assuming that $\mathbf{F} = P\mathbf{i} + Q\mathbf{j}$ is a vector field on an open simply-connected region D, P and Q have continuous first order partial derivatives, and

$$\frac{\partial P}{\partial y} = \frac{\partial Q}{\partial x} \qquad \text{throughout } D$$

If C is any simple closed path in D, and R is the region that D encloses, then Green's Theorem gives

$$\oint_C \mathbf{F} \cdot d\mathbf{r} = \oint_C P\,dx + Q\,dy = \iint_R \left(\frac{\partial Q}{\partial x} - \frac{\partial P}{\partial y} \right) dA = \iint_R 0\,dA = 0$$

Therefore $\int_C \mathbf{F} \cdot d\mathbf{r}$ is independent of path in D by Theorem 14.20. It follows that $\mathbf{F}$ is a conservative vector field by Theorem 14.21.   ●

EXERCISES 14.4

In Exercises 1–4 evaluate the line integral by two methods: (a) directly and (b) using Green's Theorem.

1. $\oint_C x^2 y\,dx + xy^3\,dy$, where C is the square with vertices $(0,0)$, $(1,0)$, $(1,1)$, and $(0,1)$

2. $\oint_C x\,dx - x^2 y^2\,dy$, where C is the triangle with vertices $(0,0)$, $(1,1)$, and $(0,1)$

3. $\oint_C (x + 2y)\,dx + (x - 2y)\,dy$, where C consists of the arc of the parabola $y = x^2$ from $(0,0)$ to $(1,1)$ followed by the line segment from $(1,1)$ to $(0,0)$

4. $\oint_C (x^2 + y^2)\,dx + 2xy\,dy$, where C consists of the arc of the parabola $y = x^2$ from $(0,0)$ to $(2,4)$ and the line segments from $(2,4)$ to $(0,4)$ and from $(0,4)$ to $(0,0)$

In Exercises 5–16 use Green's Theorem to evaluate the line integral along the given positively-oriented curve.

5. $\int_C xy\,dx + y^5\,dy$, where C is the triangle with vertices $(0,0)$, $(2,0)$, and $(2,1)$

6. $\int_C x^2 y\,dx + xy^5\,dy$, where C is the square with vertices $(\pm 1, \pm 1)$

7. $\int_C (y + e^{\sqrt{x}})\,dx + (2x + \cos y^2)\,dy$, where C is the boundary of the region enclosed by the parabolas $y = x^2$ and $x = y^2$

8. $\int_C (y^2 - \tan^{-1} x)\,dx + (3x + \sin y)\,dy$, where C is the boundary of the region enclosed by the parabola $y = x^2$ and the line $y = 4$

9. $\int_C x^2\,dx + y^2\,dy$, where C is the curve $x^6 + y^6 = 1$

10. $\int_C x^2 y\,dx - 3y^2\,dy$, where C is the circle $x^2 + y^2 = 1$

11. $\int_C xy\,dx + 2x^2\,dy$, where C consists of the line segment from $(-2,0)$ to $(2,0)$ and the top half of the circle $x^2 + y^2 = 4$

12. $\int_C 2xy\,dx + x^2\,dy$, where C is the cardioid $r = 1 + \cos\theta$

13. $\int_C (xy + e^{x^2})\,dx + (x^2 - \ln(1 + y))\,dy$, where C consists of the line segment from $(0,0)$ to $(\pi,0)$ and the curve $y = \sin x$, $0 \leqslant x \leqslant \pi$

14. $\int_C (x^3 - y^3)\,dx + (x^3 + y^3)\,dy$, where C is the boundary of the region between the circles $x^2 + y^2 = 1$ and $x^2 + y^2 = 9$

15. $\int_C \mathbf{F} \cdot d\mathbf{r}$, where $\mathbf{F}(x, y) = (y^2 - x^2 y)\mathbf{i} + xy^2\mathbf{j}$ and C consists of the circle $x^2 + y^2 = 4$ from $(2,0)$ to $(\sqrt{2}, \sqrt{2})$ and the line segments from $(\sqrt{2}, \sqrt{2})$ to $(0,0)$ and from $(0,0)$ to $(2,0)$

16. $\int_C \mathbf{F} \cdot d\mathbf{r}$, where $\mathbf{F}(x, y) = x^3 y\mathbf{i} + x^4\mathbf{j}$ and C is the curve $x^4 + y^4 = 1$

In Exercises 17 and 18 find the area of the given region using one of the formulas in Equation 14.31.

17. The region bounded by the hypocycloid with vector equation $\mathbf{r}(t) = \cos^3 t\,\mathbf{i} + \sin^3 t\,\mathbf{j}$, $0 \leqslant t \leqslant 2\pi$

18. The region bounded by the curve with vector equation $\mathbf{r}(t) = \cos t\,\mathbf{i} + \sin^3 t\,\mathbf{j}$, $0 \leqslant t \leqslant 2\pi$

19. Let D be a region bounded by a simple closed path C in the xy-plane. Use Green's Theorem to prove that the coordinates of the centroid $(\bar{x}, \bar{y})$ of D are

$$\bar{x} = \frac{1}{2A} \oint_C x^2\,dy \qquad \bar{y} = -\frac{1}{2A} \oint_C y^2\,dx$$

where A is the area of D.

20. Use Exercise 19 to find the centroid of a semicircular region of radius a.

21. A plane lamina with constant density $\rho(x, y) = \rho$ occupies a region in the xy-plane bounded by a simple closed path C. Show that its moments of inertia about the axes are

$$I_x = -\frac{\rho}{3} \oint_C y^3 \, dx \qquad I_y = \frac{\rho}{3} \oint_C x^3 \, dy$$

22. Use Exercise 21 to find the moment of inertia of a circular disk of radius a with constant density ρ about a diameter. (Compare with Example 4 in Section 13.5.)

23. If $\mathbf{F}$ is the vector field of Example 5, show that $\int_C \mathbf{F} \cdot d\mathbf{r} = 0$ for every simple closed path that does not pass through or enclose the origin.

24. Complete the proof of the special case of Green's Theorem by proving Equation 14.29.

SECTION 14.5

Curl and Divergence

In this section we define two operations that can be performed on vector fields. Each operation resembles differentiation but one produces a vector field whereas the other produces a scalar field.

If $\mathbf{F} = P\mathbf{i} + Q\mathbf{j} + R\mathbf{k}$ is a vector field on R^3 and the partial derivatives of P, Q, and R all exist, then the **curl** of $\mathbf{F}$ is the vector field on R^3 defined by

(14.32)

$$\text{curl } \mathbf{F} = \left(\frac{\partial R}{\partial y} - \frac{\partial Q}{\partial z}\right)\mathbf{i} + \left(\frac{\partial P}{\partial z} - \frac{\partial R}{\partial x}\right)\mathbf{j} + \left(\frac{\partial Q}{\partial x} - \frac{\partial P}{\partial y}\right)\mathbf{k}$$

As an aid to the memory, let us rewrite Equation 14.32 using operator notation. We introduce the vector differential operator ∇ ("del") as

$$\nabla = \mathbf{i}\frac{\partial}{\partial x} + \mathbf{j}\frac{\partial}{\partial y} + \mathbf{k}\frac{\partial}{\partial z}$$

It has meaning when it operates on a scalar function to produce the gradient of f:

$$\nabla f = \mathbf{i}\frac{\partial f}{\partial x} + \mathbf{j}\frac{\partial f}{\partial y} + \mathbf{k}\frac{\partial f}{\partial z} = \frac{\partial f}{\partial x}\mathbf{i} + \frac{\partial f}{\partial y}\mathbf{j} + \frac{\partial f}{\partial z}\mathbf{k}$$

If we think of ∇ as a vector with components $\partial/\partial x$, $\partial/\partial y$, and $\partial/\partial z$, we can also consider the formal cross product of ∇ with the vector field $\mathbf{F}$ as follows:

$$\nabla \times \mathbf{F} = \begin{vmatrix} \mathbf{i} & \mathbf{j} & \mathbf{k} \\ \dfrac{\partial}{\partial x} & \dfrac{\partial}{\partial y} & \dfrac{\partial}{\partial z} \\ P & Q & R \end{vmatrix}$$

$$= \left(\frac{\partial R}{\partial y} - \frac{\partial Q}{\partial z}\right)\mathbf{i} + \left(\frac{\partial P}{\partial z} - \frac{\partial R}{\partial x}\right)\mathbf{j} + \left(\frac{\partial Q}{\partial x} - \frac{\partial P}{\partial y}\right)\mathbf{k}$$

$$= \text{curl } \mathbf{F}$$

Thus the easiest way to remember Definition 14.32 is by means of the symbolic expression

(14.33)

$$\text{curl } \mathbf{F} = \nabla \times \mathbf{F}$$

● **Example 1**   If $\mathbf{F}(x, y, z) = xz\mathbf{i} + xyz\mathbf{j} - y^2\mathbf{k}$, find curl $\mathbf{F}$.

Solution   Using Equation 14.33, we have

$$\text{curl } \mathbf{F} = \nabla \times \mathbf{F} = \begin{vmatrix} \mathbf{i} & \mathbf{j} & \mathbf{k} \\ \dfrac{\partial}{\partial x} & \dfrac{\partial}{\partial y} & \dfrac{\partial}{\partial z} \\ xz & xyz & -y^2 \end{vmatrix}$$

$$= \left[\frac{\partial}{\partial y}(-y^2) - \frac{\partial}{\partial z}(xyz) \right]\mathbf{i} - \left[\frac{\partial}{\partial x}(-y^2) - \frac{\partial}{\partial z}(xz) \right]\mathbf{j}$$

$$+ \left[\frac{\partial}{\partial x}(xyz) - \frac{\partial}{\partial y}(xz) \right]\mathbf{k}$$

$$= (-2y - xy)\mathbf{i} - (0 - x)\mathbf{j} + (yz - 0)\mathbf{k}$$

$$= -y(2 + x)\mathbf{i} + x\mathbf{j} + yz\mathbf{k}$$

Recall that the gradient of a function f of three variables is a vector field on R^3 and so we can compute its curl. The following theorem says that the curl of a gradient vector field is **0**. Notice the similarity to Example 2 in Section 11.4: $\mathbf{a} \times \mathbf{a} = \mathbf{0}$ for every three-dimensional vector $\mathbf{a}$.

Theorem (14.34)

If f is a function of three variables that has continuous second order partial derivatives, then

$$\text{curl}(\nabla f) = \mathbf{0}$$

Proof   We have

$$\text{curl}(\nabla f) = \nabla \times (\nabla f) = \begin{vmatrix} \mathbf{i} & \mathbf{j} & \mathbf{k} \\ \dfrac{\partial}{\partial x} & \dfrac{\partial}{\partial y} & \dfrac{\partial}{\partial z} \\ \dfrac{\partial f}{\partial x} & \dfrac{\partial f}{\partial y} & \dfrac{\partial f}{\partial z} \end{vmatrix}$$

$$= \left(\frac{\partial^2 f}{\partial y\,\partial z} - \frac{\partial^2 f}{\partial z\,\partial y} \right)\mathbf{i} + \left(\frac{\partial^2 f}{\partial z\,\partial x} - \frac{\partial^2 f}{\partial x\,\partial z} \right)\mathbf{j} + \left(\frac{\partial^2 f}{\partial x\,\partial y} - \frac{\partial^2 f}{\partial y\,\partial x} \right)\mathbf{k}$$

$$= 0\mathbf{i} + 0\mathbf{j} + 0\mathbf{k} = \mathbf{0}$$

by Clairaut's Theorem (12.14).

Since a conservative vector field is one for which $\mathbf{F} = \nabla f$, Theorem 14.34 can be rephrased as saying that if $\mathbf{F}$ is conservative, then curl $\mathbf{F} = \mathbf{0}$. (Compare this with Exercise 21 in Section 14.3.) This gives us a way of verifying that a vector field is not conservative.

● **Example 2**   Show that the vector field $\mathbf{F}(x, y, z) = xz\mathbf{i} + xyz\mathbf{j} - y^2\mathbf{k}$ is not conservative.

Solution   In Example 1 we showed that

$$\text{curl } \mathbf{F} = -y(2 + x)\mathbf{i} + x\mathbf{j} + yz\mathbf{k}$$

This shows that curl $\mathbf{F} \neq \mathbf{0}$ and so, by Theorem 14.34, $\mathbf{F}$ is not conservative.

●

The converse of Theorem 14.34 is not true in general but the following theorem says the converse is true if $\mathbf{F}$ is defined everywhere. (More generally it will be true if the domain "has no holes.") Theorem 14.35 is the three-dimensional analogue of Theorem 14.23. Its proof is omitted because it requires Stokes' Theorem (see Section 14.8).

Theorem (14.35)

> If $\mathbf{F}$ is a vector field defined on all of R^3 whose component functions have continuous partial derivatives and curl $\mathbf{F} = \mathbf{0}$, then $\mathbf{F}$ is a conservative vector field.

● **Example 3**   (a) Show that $\mathbf{F}(x, y, z) = y^2z^3\mathbf{i} + 2xyz^3\mathbf{j} + 3xy^2z^2\mathbf{k}$ is a conservative vector field. (b) Find a function f such that $\mathbf{F} = \nabla f$.

Solution   (a)   We compute the curl of $\mathbf{F}$:

$$\text{curl } \mathbf{F} = \nabla \times \mathbf{F} = \begin{vmatrix} \mathbf{i} & \mathbf{j} & \mathbf{k} \\ \dfrac{\partial}{\partial x} & \dfrac{\partial}{\partial y} & \dfrac{\partial}{\partial z} \\ y^2z^3 & 2xyz^3 & 3xy^2z^2 \end{vmatrix}$$

$$= (6xyz^2 - 6xyz^2)\mathbf{i} - (3y^2z^2 - 3y^2z^2)\mathbf{j} + (2yz^3 - 2yz^3)\mathbf{k}$$

$$= \mathbf{0}$$

Since curl $\mathbf{F} = \mathbf{0}$ and the domain of $\mathbf{F}$ is R^3, $\mathbf{F}$ is a conservative vector field by Theorem 14.35.

(b)   The technique for finding f was given in Section 14.3. We have

$$f_x(x, y, z) = y^2z^3 \qquad\qquad (1)$$

$$f_y(x, y, z) = 2xyz^3 \qquad\qquad (2)$$

$$f_z(x, y, z) = 3xy^2z^2 \qquad\qquad (3)$$

Integrating (1) with respect to x, we obtain

$$f(x, y, z) = xy^2z^3 + g(y, z) \qquad (4)$$

Differentiating (4) with respect to y, we get $f_y(x, y, z) = 2xyz^3 + g_y(y, z)$, so comparison with (2) gives $g_y(y, z) = 0$. Thus $g(y, z) = h(z)$ and

$$f_z(x, y, z) = 3xy^2z^2 + h'(z)$$

Then (3) gives $h'(z) = 0$. Therefore

$$f(x, y, z) = xy^2z^3 + K \qquad \bullet$$

The reason for the name *curl* can be understood in the case where **F** represents the velocity field in fluid flow (see Example 3 in Section 14.1). There curl **F** describes how the fluid rotates, or curls, about an axis. [The axis points in the direction of the vector curl $\mathbf{F}(x, y, z)$ and the length of the vector indicates the tendency of the fluid to rotate.] In particular, if curl $\mathbf{F} = \mathbf{0}$, then there is no tendency to rotate and **F** is called **irrotational.** In other words, there are no whirlpools or eddies in the moving fluid. Justification of these ideas requires the use of Stokes' Theorem (see Section 14.8).

Divergence

If $\mathbf{F} = P\mathbf{i} + Q\mathbf{j} + R\mathbf{k}$ is a vector field on R^3 and $\partial P/\partial x$, $\partial Q/\partial y$, and $\partial R/\partial z$ exist, then the **divergence** of **F** is the function of three variables defined by

(14.36)
$$\boxed{\text{div } \mathbf{F} = \frac{\partial P}{\partial x} + \frac{\partial Q}{\partial y} + \frac{\partial R}{\partial z}}$$

Observe that curl **F** is a vector field but div **F** is a scalar field. In terms of the gradient operator $\nabla = (\partial/\partial x)\mathbf{i} + (\partial/\partial y)\mathbf{j} + (\partial/\partial z)\mathbf{k}$, the divergence of **F** can be written symbolically as the dot product of ∇ and **F**:

(14.37)
$$\boxed{\text{div } \mathbf{F} = \nabla \cdot \mathbf{F}}$$

● **Example 4**   If $\mathbf{F}(x, y, z) = xz\mathbf{i} + xyz\mathbf{j} - y^2\mathbf{k}$, find div **F**.

Solution   By definition of divergence (Equation 14.36 or 14.37) we have

$$\text{div } \mathbf{F} = \nabla \cdot \mathbf{F} = \frac{\partial}{\partial x}(xz) + \frac{\partial}{\partial y}(xyz) + \frac{\partial}{\partial z}(-y^2)$$

$$= z + xz \qquad \bullet$$

If $\mathbf{F}$ is a vector field on R^3, then curl $\mathbf{F}$ is also a vector field on R^3. As such, we can compute its divergence. The next theorem shows that the result is 0. [Note the analogy with the scalar triple product: $\mathbf{a} \cdot (\mathbf{a} \times \mathbf{b}) = 0$.]

Theorem (14.38)

If $\mathbf{F} = P\mathbf{i} + Q\mathbf{j} + R\mathbf{k}$ is a vector field on R^3 and P, Q, and R have continuous second order partial derivatives, then

$$\text{div curl } \mathbf{F} = 0$$

Proof   Using the definitions of divergence and curl, we have

$$\text{div curl } \mathbf{F} = \nabla \cdot (\nabla \times F)$$

$$= \frac{\partial}{\partial x}\left(\frac{\partial R}{\partial y} - \frac{\partial Q}{\partial z}\right) + \frac{\partial}{\partial y}\left(\frac{\partial P}{\partial z} - \frac{\partial R}{\partial x}\right) + \frac{\partial}{\partial z}\left(\frac{\partial Q}{\partial x} - \frac{\partial P}{\partial y}\right)$$

$$= \frac{\partial^2 R}{\partial x\, \partial y} - \frac{\partial^2 Q}{\partial x\, \partial z} + \frac{\partial^2 P}{\partial y\, \partial z} - \frac{\partial^2 R}{\partial y\, \partial x} + \frac{\partial^2 Q}{\partial z\, \partial x} - \frac{\partial^2 P}{\partial z\, \partial y}$$

$$= 0$$

because the terms cancel in pairs by Clairaut's Theorem (12.14).   ●

● **Example 5**   Show that the vector field $\mathbf{F}(x, y, z) = xz\mathbf{i} + xyz\mathbf{j} - y^2\mathbf{k}$ cannot be written as the curl of another vector field, that is, $\mathbf{F} \neq \text{curl } \mathbf{G}$.

Solution   In Example 4 we showed that

$$\text{div } \mathbf{F} = z + xz$$

and therefore div $\mathbf{F} \neq 0$. If it were true that $\mathbf{F} = \text{curl } \mathbf{G}$, then Theorem 14.38 would give

$$\text{div } \mathbf{F} = \text{div curl } \mathbf{G} = 0$$

which contradicts div $\mathbf{F} \neq 0$. Therefore $\mathbf{F}$ is not the curl of another vector field.   ●

Again the reason for the name *divergence* can be understood in the context of fluid flow. If $\mathbf{F}(x, y, z)$ is the velocity of a fluid (or gas), then div $\mathbf{F}(x, y, z)$ represents the rate of change (with respect to time) of the mass of fluid (or gas) flowing from the point (x, y, z) per unit volume. In other words, div $\mathbf{F}(x, y, z)$ measures the tendency of the fluid to diverge from the point (x, y, z). If div $\mathbf{F} = 0$, then $\mathbf{F}$ is called **incompressible.**

Another differential operator occurs when we compute the divergence of a gradient vector field ∇f. If f is a function of three variables, we have

$$\text{div}(\nabla f) = \nabla \cdot (\nabla f) = \frac{\partial^2 f}{\partial x^2} + \frac{\partial^2 f}{\partial y^2} + \frac{\partial^2 f}{\partial z^2}$$

and this expression occurs so often that we abbreviate it as $\nabla^2 f$. The operator

$$\nabla^2 = \nabla \cdot \nabla$$

is called the **Laplace operator** because of its relation to **Laplace's equation**

$$\nabla^2 f = \frac{\partial^2 f}{\partial x^2} + \frac{\partial^2 f}{\partial y^2} + \frac{\partial^2 f}{\partial z^2} = 0$$

We can also apply the Laplace operator ∇^2 to a vector field

$$\mathbf{F} = P\mathbf{i} + Q\mathbf{j} + R\mathbf{k}$$

in terms of its components:

$$\nabla^2 \mathbf{F} = \nabla^2 P\mathbf{i} + \nabla^2 Q\mathbf{j} + \nabla^2 R\mathbf{k}$$

Vector Forms of Green's Theorem

The curl and divergence operators allow us to rewrite Green's Theorem in versions that will be useful in our later work. We suppose that the plane region D, its boundary curve C, and the functions P and Q satisfy the hypotheses of Green's Theorem (14.26). Then we consider the vector field $\mathbf{F} = P\mathbf{i} + Q\mathbf{j}$. Its line integral is

$$\oint_C \mathbf{F} \cdot d\mathbf{r} = \oint_C P\,dx + Q\,dy$$

and its curl is

$$\text{curl } \mathbf{F} = \begin{vmatrix} \mathbf{i} & \mathbf{j} & \mathbf{k} \\ \dfrac{\partial}{\partial x} & \dfrac{\partial}{\partial y} & \dfrac{\partial}{\partial z} \\ P(x,y) & Q(x,y) & 0 \end{vmatrix} = \left(\frac{\partial Q}{\partial x} - \frac{\partial P}{\partial y} \right)\mathbf{k}$$

Therefore $\quad (\text{curl } \mathbf{F}) \cdot \mathbf{k} = \left(\dfrac{\partial Q}{\partial x} - \dfrac{\partial P}{\partial y} \right)\mathbf{k} \cdot \mathbf{k} = \dfrac{\partial Q}{\partial x} - \dfrac{\partial P}{\partial y}$

and we can now rewrite the equation in Green's Theorem in the vector form

(14.39)

$$\oint_C \mathbf{F} \cdot d\mathbf{r} = \iint_D (\text{curl } \mathbf{F}) \cdot \mathbf{k}\,dA$$

Equation 14.39 expresses the line integral of the tangential component of $\mathbf{F}$ along C as the double integral of $(\text{curl } \mathbf{F}) \cdot \mathbf{k}$ over the region D enclosed by C. We now derive a similar formula involving the *normal* component of $\mathbf{F}$.

If C is given by the vector equation

$$\mathbf{r}(t) = x(t)\mathbf{i} + y(t)\mathbf{j} \qquad a \leqslant t \leqslant b$$

then the unit tangent vector (see Section 11.7) is

$$\mathbf{T}(t) = \frac{x'(t)}{|\mathbf{r}'(t)|}\mathbf{i} + \frac{y'(t)}{|\mathbf{r}'(t)|}\mathbf{j}$$

You can verify that the outward unit normal vector to C is given by

$$\mathbf{n}(t) = \frac{y'(t)}{|\mathbf{r}'(t)|}\mathbf{i} - \frac{x'(t)}{|\mathbf{r}'(t)|}\mathbf{j}$$

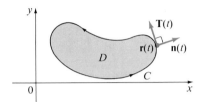

(See Figure 14.29.) Then, from Equation 14.7, we have

$$\int_C \mathbf{F} \cdot \mathbf{n}\, ds = \int_a^b (\mathbf{F} \cdot \mathbf{n})(t)|\mathbf{r}'(t)|\, dt$$

$$= \int_a^b \left[\frac{P(x(t), y(t))y'(t)}{|\mathbf{r}'(t)|} - \frac{Q(x(t), y(t))x'(t)}{|\mathbf{r}'(t)|} \right]|\mathbf{r}'(t)|\, dt$$

$$= \int_a^b P(x(t), y(t))y'(t)\, dt - Q(x(t), y(t))x'(t)\, dt$$

$$= \int_C P\, dy - Q\, dx$$

$$= \iint_D \left(\frac{\partial P}{\partial x} + \frac{\partial Q}{\partial y} \right) dA$$

Figure 14.29

by Green's Theorem. But the integrand in this double integral is just the divergence of $\mathbf{F}$. So we have a second vector form of Green's Theorem:

(14.40)

$$\oint_C \mathbf{F} \cdot \mathbf{n}\, ds = \iint_D \operatorname{div} \mathbf{F}(x, y)\, dA$$

This version says that the line integral of the normal component of $\mathbf{F}$ along C is equal to the double integral of the divergence of $\mathbf{F}$ over the region D enclosed by C.

EXERCISES 14.5

In Exercises 1–10 find (a) the curl and (b) the divergence of the vector field.

1. $\mathbf{F}(x, y, z) = x\mathbf{i} + y\mathbf{j} + z\mathbf{k}$

2. $\mathbf{F}(x, y, z) = x^2y\mathbf{i} + yz^2\mathbf{j} + zx^2\mathbf{k}$

3. $\mathbf{F}(x, y, z) = yz\mathbf{i} + xz\mathbf{j} + xy\mathbf{k}$

4. $\mathbf{F}(x, y, z) = y^2z\mathbf{i} - x^2yz\mathbf{k}$

5. $\mathbf{F}(x, y, z) = xy\mathbf{j} + xyz\mathbf{k}$

6. $F(x, y, z) = \sin x\mathbf{i} + \cos x\mathbf{j} + z^2\mathbf{k}$

7. $F(x, y, z) = e^{xz}\mathbf{i} - 2e^{yz}\mathbf{j} + 3xe^y\mathbf{k}$

8. $F(x, y, z) = (x + 3y - 5z)\mathbf{i} + (z - 3y)\mathbf{j}$
$+ (5x + 6y - z)\mathbf{k}$

9. $F(x, y, z) = xe^y\mathbf{i} - ze^{-y}\mathbf{j} + y\ln z\mathbf{k}$

10. $F(x, y, z) = e^{xyz}\mathbf{i} + \sin(x - y)\mathbf{j} - \dfrac{xy}{z}\mathbf{k}$

In Exercises 11–18 determine whether or not the given vector field is conservative. If it is conservative, find a function f such that $F = \nabla f$.

11. $F(x, y, z) = y\mathbf{i} + x\mathbf{j} + \mathbf{k}$

12. $F(x, y, z) = x\mathbf{i} + y\mathbf{j} + x\mathbf{k}$

13. $F(x, y, z) = yz\mathbf{i} - z^2\mathbf{j} + x^2\mathbf{k}$

14. $F(x, y, z) = z\mathbf{i} + 2yz\mathbf{j} + (x + y^2)\mathbf{k}$

15. $F(x, y, z) = \cos y\mathbf{i} + \sin x\mathbf{j} + \tan z\mathbf{k}$

16. $F(x, y, z) = x\mathbf{i} + e^y\sin z\mathbf{j} + e^y\cos z\mathbf{k}$

17. $F(x, y, z) = yz\mathbf{i} + (y^2 + xz)\mathbf{j} + xy\mathbf{k}$

18. $F(x, y, z) = zx\mathbf{i} + xy\mathbf{j} + yz\mathbf{k}$

19. Is there a vector field $\mathbf{G}$ on R^3 such that curl $\mathbf{G} = xy^2\mathbf{i} + yz^2\mathbf{j} + zx^2\mathbf{k}$? Explain.

20. Is there a vector field $\mathbf{G}$ on R^3 such that curl $\mathbf{G} = yz\mathbf{i} + xyz\mathbf{j} + xy\mathbf{k}$? Explain.

21. Show that any vector field of the form
$F(x, y, z) = f(x)\mathbf{i} + g(y)\mathbf{j} + h(z)\mathbf{k}$, where f, g, h are differentiable functions, is irrotational.

22. Show that any vector field of the form
$F(x, y, z) = f(y, z)\mathbf{i} + g(x, z)\mathbf{j} + h(x, y)\mathbf{k}$ is incompressible.

Prove the identities in Exercises 23–30 assuming that the appropriate partial derivatives exist and are continuous.

If f is a scalar field and $\mathbf{F}, \mathbf{G}$ are vector fields, then $f\mathbf{F}$, $\mathbf{F} \cdot \mathbf{G}$, and $\mathbf{F} \times \mathbf{G}$ are vector fields defined by

$$(f\mathbf{F})(x, y, z) = f(x, y, z)\mathbf{F}(x, y, z)$$

$$(\mathbf{F} \cdot \mathbf{G})(x, y, z) = \mathbf{F}(x, y, z) \cdot \mathbf{G}(x, y, z)$$

$$(\mathbf{F} \times \mathbf{G})(x, y, z) = \mathbf{F}(x, y, z) \times \mathbf{G}(x, y, z)$$

23. $\operatorname{div}(\mathbf{F} + \mathbf{G}) = \operatorname{div}\mathbf{F} + \operatorname{div}\mathbf{G}$

24. $\operatorname{curl}(\mathbf{F} + \mathbf{G}) = \operatorname{curl}\mathbf{F} + \operatorname{curl}\mathbf{G}$

25. $\operatorname{div}(f\mathbf{F}) = f\operatorname{div}\mathbf{F} + \mathbf{F} \cdot \nabla f$

26. $\operatorname{curl}(f\mathbf{F}) = f\operatorname{curl}\mathbf{F} + (\nabla f) \times \mathbf{F}$

27. $\operatorname{div}(\mathbf{F} \times \mathbf{G}) = \mathbf{G} \cdot \operatorname{curl}\mathbf{F} - \mathbf{F} \cdot \operatorname{curl}\mathbf{G}$

28. $\operatorname{div}(\nabla f \times \nabla g) = 0$

29. $\operatorname{curl}\operatorname{curl}\mathbf{F} = \operatorname{grad}\operatorname{div}\mathbf{F} - \nabla^2\mathbf{F}$

30. $\nabla(\mathbf{F} \cdot \mathbf{G}) = (\mathbf{F} \cdot \nabla)\mathbf{G} + (\mathbf{G} \cdot \nabla)\mathbf{F}$
$+ \mathbf{F} \times \operatorname{curl}\mathbf{G} + \mathbf{G} \times \operatorname{curl}\mathbf{F}$

31. Use Green's Theorem in the form of Equation 14.40 to prove **Green's first identity:**

$$\iint_D f\nabla^2 g\, dA = \oint_C f(\nabla g) \cdot \mathbf{n}\, ds - \iint_D \nabla f \cdot \nabla g\, dA$$

where D and C satisfy the hypotheses of Green's Theorem and the appropriate partial derivatives of f and g exist and are continuous. (The quantity $\nabla g \cdot \mathbf{n} = D_\mathbf{n}g$ occurs in the line integral. This is the directional derivative in the direction of the normal vector $\mathbf{n}$ and is called the **normal derivative** of g.)

32. Use Green's first identity (Exercise 31) to prove **Green's second identity:**

$$\iint_D (f\nabla^2 g - g\nabla^2 f)\, dA = \oint_C (f\nabla g - g\nabla f) \cdot \mathbf{n}\, ds$$

where D and C satisfy the hypotheses of Green's Theorem and the appropriate partial derivatives of f and g exist and are continuous.

SECTION 14.6

The Area of a Surface

In Section 8.4 we found the area of a very special type of surface—a surface of revolution—using the methods of single-variable calculus. Since then we have studied more general types of surfaces given by equations of the form $z = f(x, y)$ (the graph of a function f of two variables) or $F(x, y, z) = k$ (the level surface of a function F of three variables). In this section we shall try to find the areas of such surfaces and still more general surfaces called parametric surfaces.

In much the same way that we describe a space curve by a vector function $\mathbf{r}(t)$ of a single parameter t, we can describe a surface by a vector function $\mathbf{r}(u, v)$ of two parameters u and v. We suppose that

(14.41)
$$\mathbf{r}(u, v) = x(u, v)\mathbf{i} + y(u, v)\mathbf{j} + z(u, v)\mathbf{k}$$

is a vector-valued function defined on a region D in the uv-plane and the partial derivatives of x, y, and z with respect to u and v are all continuous. The set of all points (x, y, z) in R^3, such that

(14.42)
$$x = x(u, v) \qquad y = y(u, v) \qquad z = z(u, v)$$

and (u, v) varies throughout D, is called a **parametric surface** S and Equations 14.42 are called the **parametric equations** of S. In other words, the surface S is traced out by the position vector $\mathbf{r}(u, v)$ as (u, v) moves throughout the region D (see Figure 14.30).

Figure 14.30

A PARAMETRIC SURFACE

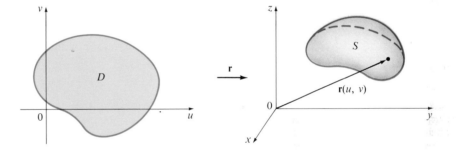

● **Example 1**   Find a parametric representation of the sphere
$$x^2 + y^2 + z^2 = a^2$$

Solution   The sphere has a simple representation $\rho = a$ in spherical coordinates, so let us choose the angles ϕ and θ in spherical coordinates as the parameters (see Section 11.10). Then, putting $\rho = a$ in the equations for conversion from spherical to rectangular coordinates (Equations 11.77), we obtain

$$x = a \sin \phi \cos \theta \qquad y = a \sin \phi \sin \theta \qquad z = a \cos \phi$$

as the parametric equations of the sphere. The corresponding vector equation is

$$\mathbf{r}(\phi, \theta) = a \sin \phi \cos \theta \mathbf{i} + a \sin \phi \sin \theta \mathbf{j} + a \cos \phi \mathbf{k}$$

Since $0 \leqslant \phi \leqslant \pi$ and $0 \leqslant \theta \leqslant 2\pi$, the parameter domain is the rectangle $D = [0, \pi] \times [0, 2\pi]$.   ●

● **Example 2**   Find a parametric representation for the cylinder
$$x^2 + y^2 = 4, \qquad 0 \leqslant z \leqslant 1$$

Solution   The cylinder has a simple representation $r = 2$ in cylindrical coordinates, so we choose as parameters θ and z in cylindrical coordinates. Then the parametric equations of the cylinder are

$$x = 2\cos\theta \qquad y = 2\sin\theta \qquad z = z$$

where $0 \leqslant \theta \leqslant 2\pi$ and $0 \leqslant z \leqslant 1$.   ●

● **Example 3**   Find a parametric representation for the elliptic paraboloid $z = x^2 + 2y^2$.

Solution   If we regard x and y as parameters, then the parametric equations are simply

$$x = x \qquad y = y \qquad z = x^2 + 2y^2$$

and the vector equation is

$$\mathbf{r}(x, y) = x\mathbf{i} + y\mathbf{j} + (x^2 + 2y^2)\mathbf{k}$$ ●

In general, a surface given as the graph of a function of x and y, that is, with an equation of the form $z = f(x, y)$, can always be regarded as a parametric surface by taking x and y as parameters and writing the parametric equations as

$$x = x \qquad y = y \qquad z = f(x, y)$$

Let us now find the tangent plane to a parametric surface S given by a vector function $\mathbf{r}(u, v)$ at a point P_0 with position vector $\mathbf{r}(u_0, v_0)$. If we keep u constant by putting $u = u_0$, then $\mathbf{r}(u_0, v)$ becomes a vector function of the single parameter v and defines a curve C_1 lying on S (see Figure 14.31). The tangent vector to C_1 at P_0 is

(14.43)
$$\mathbf{T}_v = \frac{\partial x}{\partial v}(u_0, v_0)\mathbf{i} + \frac{\partial y}{\partial v}(u_0, v_0)\mathbf{j} + \frac{\partial z}{\partial v}(u_0, v_0)\mathbf{k}$$

Similarly if we keep v constant by putting $v = v_0$, we get a curve C_2 given by $\mathbf{r}(u, v_0)$ that lies on S and its tangent vector at P_0 is

(14.44)
$$\mathbf{T}_u = \frac{\partial x}{\partial u}(u_0, v_0)\mathbf{i} + \frac{\partial y}{\partial u}(u_0, v_0)\mathbf{j} + \frac{\partial z}{\partial u}(u_0, v_0)\mathbf{k}$$

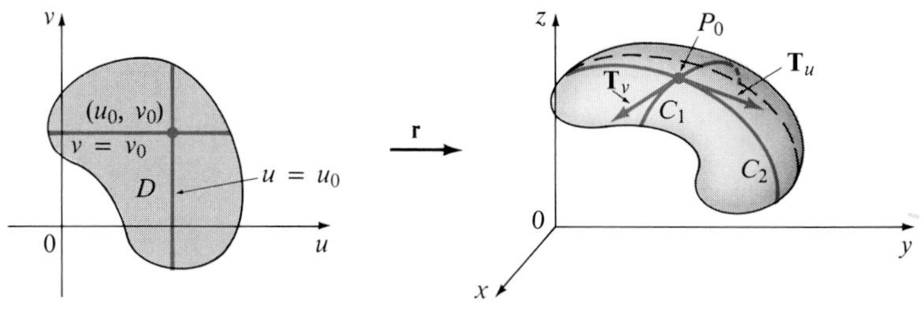

Figure 14.31

If the **normal vector** $\mathbf{T}_u \times \mathbf{T}_v$ is not **0**, then the surface S is called **smooth.** (It has no "corners.") In this case the tangent plane to S at P_0 exists and can be found as usual using the normal vector.

● **Example 4**   Find the tangent plane to the surface with parametric equations $x = u^2$, $y = v^2$, $z = u + 2v$ at the point $(1, 1, 3)$.

Solution   We first compute the tangent vectors:

$$\mathbf{T}_u = \frac{\partial x}{\partial u}\mathbf{i} + \frac{\partial y}{\partial u}\mathbf{j} + \frac{\partial z}{\partial u}\mathbf{k} = 2u\mathbf{i} + \mathbf{k}$$

$$\mathbf{T}_v = \frac{\partial x}{\partial v}\mathbf{i} + \frac{\partial y}{\partial v}\mathbf{j} + \frac{\partial z}{\partial v}\mathbf{k} = 2v\mathbf{j} + 2\mathbf{k}$$

Thus the normal vector is

$$\mathbf{T}_u \times \mathbf{T}_v = \begin{vmatrix} \mathbf{i} & \mathbf{j} & \mathbf{k} \\ 2u & 0 & 1 \\ 0 & 2v & 2 \end{vmatrix} = -2v\mathbf{i} - 4u\mathbf{j} + 4uv\mathbf{k}$$

Notice that the point $(1, 1, 3)$ corresponds to the parameter values $u = 1$ and $v = 1$, so the normal vector there is

$$-2\mathbf{i} - 4\mathbf{j} + 4\mathbf{k}$$

Therefore the tangent plane at $(1, 1, 3)$ is

$$-2(x - 1) - 4(y - 1) + 4(z - 3) = 0$$

or
$$x + 2y - 2z + 3 = 0 \qquad\qquad ●$$

Now we define the surface area of a general parametric surface given by Equation 14.41. For simplicity we start by considering a surface whose parameter domain D is a rectangle and we partition it into subrectangles R_{ij}. Let us choose (u_i^*, v_j^*) to be the lower left corner of R_{ij} (see Figure 14.32). The part S_{ij} of the surface S that corresponds to R_{ij} has the point P_{ij} with position vector $\mathbf{r}(u_i^*, v_j^*)$ as one of its corners. Let

$$\mathbf{T}_{u_i} = \mathbf{T}_u(u_i^*, v_j^*) \quad \text{and} \quad \mathbf{T}_{v_j} = \mathbf{T}_v(u_i^*, v_j^*)$$

Figure 14.32

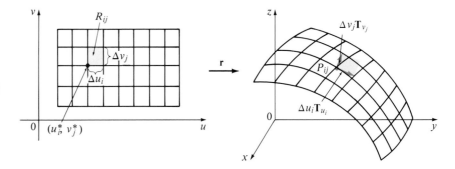

be the tangent vectors at P_{ij} as given by Equations 14.44 and 14.43. We approximate S_{ij} by the parallelogram determined by the vectors $\Delta u_i \mathbf{T}_{u_i}$ and $\Delta v_j \mathbf{T}_{v_j}$. (This parallelogram is shown in Figure 14.32 and lies in the tangent plane to S at P_{ij}.) The area of this parallelogram is

$$|(\Delta u_i \mathbf{T}_{u_i}) \times (\Delta v_j \mathbf{T}_{v_j})| = |\mathbf{T}_{u_i} \times \mathbf{T}_{v_j}| \Delta u_i \Delta v_j$$

and so an approximation to the area of S is

$$\sum_{i=1}^{m} \sum_{j=1}^{n} |\mathbf{T}_{u_i} \times \mathbf{T}_{v_j}| \Delta u_i \Delta v_j$$

Our intuition tells us that this approximation gets better as $\|P\| \to 0$, and we recognize the double sum as a Riemann sum for the double integral $\iint_D |\mathbf{T}_u \times \mathbf{T}_v| \, du \, dv$. This motivates the following definition:

Definition (14.45)

> The **surface area** of the parametric surface S given by Equation 14.41 is
>
> $$A(S) = \iint_D |\mathbf{T}_u \times \mathbf{T}_v| \, dA$$
>
> where
>
> $$\mathbf{T}_u = \frac{\partial x}{\partial u} \mathbf{i} + \frac{\partial y}{\partial u} \mathbf{j} + \frac{\partial z}{\partial u} \mathbf{k} \qquad \mathbf{T}_v = \frac{\partial x}{\partial v} \mathbf{i} + \frac{\partial y}{\partial v} \mathbf{j} + \frac{\partial z}{\partial v} \mathbf{k}$$

Computing the cross product of the tangent vectors, we get

$$\mathbf{T}_u \times \mathbf{T}_v = \begin{vmatrix} \mathbf{i} & \mathbf{j} & \mathbf{k} \\ \dfrac{\partial x}{\partial u} & \dfrac{\partial y}{\partial u} & \dfrac{\partial z}{\partial u} \\ \dfrac{\partial x}{\partial v} & \dfrac{\partial y}{\partial v} & \dfrac{\partial z}{\partial v} \end{vmatrix}$$

$$= \left(\frac{\partial y}{\partial u} \frac{\partial z}{\partial v} - \frac{\partial z}{\partial u} \frac{\partial y}{\partial v} \right) \mathbf{i} + \left(\frac{\partial z}{\partial u} \frac{\partial x}{\partial v} - \frac{\partial x}{\partial u} \frac{\partial z}{\partial v} \right) \mathbf{j} + \left(\frac{\partial x}{\partial u} \frac{\partial y}{\partial v} - \frac{\partial y}{\partial u} \frac{\partial x}{\partial v} \right) \mathbf{k}$$

$$= \frac{\partial(y, z)}{\partial(u, v)} \mathbf{i} + \frac{\partial(z, x)}{\partial(u, v)} \mathbf{j} + \frac{\partial(x, y)}{\partial(u, v)} \mathbf{k}$$

where we have used the notation for Jacobians given in Definition 13.65. Thus the formula for surface area in Definition 14.45 becomes

(14.46)

$$A(S) = \iint_D \sqrt{\left[\frac{\partial(x, y)}{\partial(u, v)} \right]^2 + \left[\frac{\partial(y, z)}{\partial(u, v)} \right]^2 + \left[\frac{\partial(x, z)}{\partial(u, v)} \right]^2} \, dA$$

Notice the similarity between the surface area formula in Equation 14.46 and the arc length formula in Equation 11.55.

● **Example 5**   Find the surface area of a sphere of radius a.

Solution   In Example 1 we found the parametric representation

$$x = a \sin \phi \cos \theta \qquad y = a \sin \phi \sin \theta \qquad z = a \cos \phi$$

where the parameter domain is

$$D = \{(\phi, \theta) | 0 \leqslant \phi \leqslant \pi, 0 \leqslant \theta \leqslant 2\pi\}$$

We first compute the cross product of the tangent vectors:

$$\mathbf{T}_\phi \times \mathbf{T}_\theta = \begin{vmatrix} \mathbf{i} & \mathbf{j} & \mathbf{k} \\ \dfrac{\partial x}{\partial \phi} & \dfrac{\partial y}{\partial \phi} & \dfrac{\partial z}{\partial \phi} \\ \dfrac{\partial x}{\partial \theta} & \dfrac{\partial y}{\partial \theta} & \dfrac{\partial z}{\partial \theta} \end{vmatrix} = \begin{vmatrix} \mathbf{i} & \mathbf{j} & \mathbf{k} \\ a \cos \phi \cos \theta & a \cos \phi \sin \theta & -a \sin \phi \\ -a \sin \phi \sin \theta & a \sin \phi \cos \theta & 0 \end{vmatrix}$$

$$= a^2 \sin^2 \phi \cos \theta \mathbf{i} + a^2 \sin^2 \phi \sin \theta \mathbf{j} + a^2 \sin \phi \cos \phi \mathbf{k}$$

Thus

$$\begin{aligned} |\mathbf{T}_\phi \times \mathbf{T}_\theta| &= \sqrt{a^4 \sin^4 \phi \cos^2 \theta + a^4 \sin^4 \phi \sin^2 \theta + a^4 \sin^2 \phi \cos^2 \phi} \\ &= \sqrt{a^4 \sin^4 \phi + a^4 \sin^2 \phi \cos^2 \phi} \\ &= a^2 \sqrt{\sin^2 \phi} = a^2 \sin \phi \end{aligned}$$

since $\sin \phi \geqslant 0$ for $0 \leqslant \phi \leqslant \pi$. Therefore the area of the sphere is

$$\begin{aligned} A &= \iint\limits_D |\mathbf{T}_\phi \times \mathbf{T}_\theta| dA = \int_0^{2\pi} \int_0^{\pi} a^2 \sin \phi \, d\phi \, d\theta \\ &= a^2 \int_0^{2\pi} d\theta \int_0^{\pi} \sin \phi \, d\phi = a^2 (2\pi) 2 = 4\pi a^2 \end{aligned}$$

●

For the special case of a surface S with equation $z = f(x, y)$, where (x, y) lies in D and f has continuous partial derivatives, we take x and y as parameters. The parametric equations are

$$x = x \qquad y = y \qquad z = f(x, y)$$

so

$$\mathbf{T}_x = \mathbf{i} + \left(\frac{\partial f}{\partial x}\right) \mathbf{k} \qquad \mathbf{T}_y = \mathbf{j} + \left(\frac{\partial f}{\partial y}\right) \mathbf{k}$$

and

(14.47)

$$\mathbf{T}_x \times \mathbf{T}_y = \begin{vmatrix} \mathbf{i} & \mathbf{j} & \mathbf{k} \\ 1 & 0 & \dfrac{\partial f}{\partial x} \\ 0 & 1 & \dfrac{\partial f}{\partial y} \end{vmatrix} = -\frac{\partial f}{\partial x} \mathbf{i} - \frac{\partial f}{\partial y} \mathbf{j} + \mathbf{k}$$

Thus the surface area formula in Definition 14.45 becomes

(14.48)
$$A(S) = \iint_D \sqrt{1 + \left(\frac{\partial z}{\partial x}\right)^2 + \left(\frac{\partial z}{\partial y}\right)^2}\, dA$$

This should be compared with the arc length formula

$$L = \int_a^b \sqrt{1 + \left(\frac{dy}{dx}\right)^2}\, dx$$

● **Example 6**   Find the area of the part of the paraboloid $z = x^2 + y^2$ that lies under the plane $z = 9$.

Solution   The plane intersects the paraboloid in the circle $x^2 + y^2 = 9$, $z = 9$. Therefore the given surface lies above the disk D with center the origin and radius 3. Using Formula 14.48, we have

$$A = \iint_D \sqrt{1 + \left(\frac{\partial z}{\partial x}\right)^2 + \left(\frac{\partial z}{\partial y}\right)^2}\, dA = \iint_D \sqrt{1 + (2x)^2 + (2y)^2}\, dA$$

$$= \iint_D \sqrt{1 + 4(x^2 + y^2)}\, dA$$

Converting to polar coordinates, we obtain

$$A = \int_0^{2\pi} \int_0^3 \sqrt{1 + 4r^2}\, r\, dr\, d\theta = \int_0^{2\pi} d\theta \int_0^3 r\sqrt{1 + 4r^2}\, dr$$

$$= 2\pi \left(\frac{1}{8}\right) \frac{2}{3} (1 + 4r^2)^{3/2}\Big]_0^3 = \frac{\pi}{6} (37\sqrt{37} - 1)$$ ●

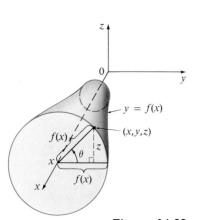

Figure 14.33

The question remains whether our definition of surface area (14.45) is consistent with the surface area formula from single-variable calculus (8.20). Let us consider the surface S obtained by rotating the curve $y = f(x)$, $a \leqslant x \leqslant b$, about the x-axis, where $f(x) \geqslant 0$ and f' is continuous. Let θ be the angle of rotation as shown in Figure 14.33. If (x, y, z) is a point on S, then

(14.49)
$$x = x \qquad y = f(x)\cos\theta \qquad z = f(x)\sin\theta$$

Therefore we take x and θ as parameters and regard Equations 14.49 as the parametric equations of S. The parameter domain is given by $a \leqslant x \leqslant b$, $0 \leqslant \theta \leqslant 2\pi$. To compute the surface area of S we need the tangent vectors

$$\mathbf{T}_x = \mathbf{i} + f'(x)\cos\theta\,\mathbf{j} + f'(x)\sin\theta\,\mathbf{k}$$
$$\mathbf{T}_\theta = -f(x)\sin\theta\,\mathbf{j} + f(x)\cos\theta\,\mathbf{k}$$

Thus
$$\mathbf{T}_x \times \mathbf{T}_\theta = \begin{vmatrix} \mathbf{i} & \mathbf{j} & \mathbf{k} \\ 1 & f'(x)\cos\theta & f'(x)\sin\theta \\ 0 & -f(x)\sin\theta & f(x)\cos\theta \end{vmatrix}$$

$$= f(x)f'(x)\mathbf{i} - f(x)\cos\theta\mathbf{j} - f(x)\sin\theta\mathbf{k}$$

and so

$$|\mathbf{T}_x \times \mathbf{T}_\theta| = \sqrt{[f(x)]^2[f'(x)]^2 + [f(x)]^2\cos^2\theta + [f(x)]^2\sin^2\theta}$$
$$= \sqrt{[f(x)]^2[1 + (f'(x))^2]} = f(x)\sqrt{1 + [f'(x)]^2}$$

because $f(x) \geqslant 0$. Therefore the area of S is

$$A = \iint_D |\mathbf{T}_x \times \mathbf{T}_\theta|\, dA = \int_0^{2\pi}\int_a^b f(x)\sqrt{1 + [f'(x)]^2}\, dx\, d\theta$$

$$= 2\pi \int_a^b f(x)\sqrt{1 + [f'(x)]^2}\, dx$$

This is precisely the formula that was used to define the area of a surface of revolution in single-variable calculus (8.20).

EXERCISES 14.6

In Exercises 1–8 find a parametric representation for the surface.

1. The upper half of the ellipsoid $3x^2 + 2y^2 + z^2 = 1$

2. The part of the hyperboloid $-x^2 - y^2 + z^2 = 1$ that lies below the rectangle $[-1,1] \times [-3,3]$

3. The part of the elliptic paraboloid $y = 6 - 3x^2 - 2z^2$ that lies to the right of the xz-plane

4. The part of the elliptic paraboloid $x + y^2 + 2z^2 = 4$ that lies in front of the plane $x = 0$

5. The part of the sphere $x^2 + y^2 + z^2 = 4$ that lies above the cone $z = \sqrt{x^2 + y^2}$

6. The part of the cylinder $x^2 + z^2 = 1$ that lies between the planes $y = -1$ and $y = 3$

7. The part of the plane $z = 5$ that lies inside the cylinder $x^2 + y^2 = 16$

8. The part of the plane $z = x + 3$ that lies inside the cylinder $x^2 + y^2 = 1$

In Exercises 9–12 find an equation of the tangent plane to the given parametric surface at the given point.

9. $x = u + v$, $y = 3u^2$, $z = u - v$; $(2,3,0)$

10. $x = u^2$, $y = u - v^2$, $z = v^2$; $(1,0,1)$

11. $\mathbf{r}(u,v) = uv\mathbf{i} + ue^v\mathbf{j} + ve^u\mathbf{k}$; $(0,0,0)$

12. $\mathbf{r}(u,v) = (u + v)\mathbf{i} + u\cos v\mathbf{j} + v\sin u\mathbf{k}$; $(1,1,0)$

In Exercises 13–22 find the area of the given surface.

13. The part of the plane $x + 2y + z = 4$ that lies inside the cylinder $x^2 + y^2 = 4$

14. The part of the surface $z = x + y^2$ that lies above the triangle with vertices $(0,0)$, $(1,1)$, and $(0,1)$

15. The part of the hyperbolic paraboloid $z = y^2 - x^2$ that lies between the cylinders $x^2 + y^2 = 1$ and $x^2 + y^2 = 4$

16. The part of the paraboloid $x = y^2 + z^2$ that lies inside the cylinder $y^2 + z^2 = 9$

17. The part of the surface $y = 4x + z^2$ that lies between the planes $x = 0$, $x = 1$, $z = 0$, and $z = 1$

18. The part of the cylinder $x^2 + z^2 = a^2$ that lies inside the cylinder $x^2 + y^2 = a^2$

19. The part of the sphere $x^2 + y^2 + z^2 = a^2$ that lies inside the cylinder $x^2 + y^2 = ax$

20. The ellipsoid $x^2/a^2 + y^2/b^2 + z^2/c^2 = 1$ (Set up the integral but do not evaluate it.)

21. The surface with parametric equations $x = uv$, $y = u + v$, $z = u - v$, $u^2 + v^2 \leqslant 1$

22. The helicoid (or spiral ramp) with vector equation $\mathbf{r}(u, v) = u \cos v \mathbf{i} + u \sin v \mathbf{j} + v \mathbf{k}$, $\quad 0 \leqslant u \leqslant 1$, $0 \leqslant v \leqslant \pi$. Also make a sketch of the helicoid.

23. Find a parametric representation for the torus obtained by rotating about the z-axis the circle in the

xz-plane with center $(b, 0, 0)$ and radius $a < b$. (*Hint:* Take as parameters the angles θ and α shown in Figure 14.34.)

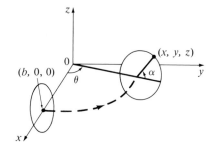

Figure 14.34

24. Use the parametric representation from Exercise 23 to find the surface area of the torus.

Surface Integrals

The relationship between surface integrals and surface area is much the same as the relationship between line integrals and arc length. Suppose that f is a function of three variables whose domain includes the smooth surface with vector equation

$$\mathbf{r}(u, v) = x(u, v)\mathbf{i} + y(u, v)\mathbf{j} + z(u, v)\mathbf{k} \qquad (u, v) \in D$$

We first assume that the parameter domain D is a rectangle. A partition of D into subrectangles R_{ij} with dimensions Δu_i and Δv_j will determine a partition of S into curvilinear regions S_{ij} that lie on S (see Figure 14.35). A point (u_i^*, v_j^*) in R_{ij} determines a point on S_{ij} with position vector $\mathbf{r}(u_i^*, v_j^*)$. By analogy with the line integral with respect to arc length (see Definition 14.6),

Figure 14.35

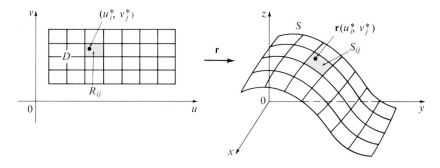

we evaluate f at $\mathbf{r}(u_i^*, v_j^*)$, multiply by the area $A(S_{ij})$, and form the sum

$$\sum_{i=1}^{m} \sum_{j=1}^{n} f(\mathbf{r}(u_i^*, v_j^*)) A(S_{ij})$$

Then we take the limit as $\|P\| \to 0$ and define the **surface integral of f over the surface S** as

(14.50)
$$\iint_S f(x, y, z)\, dS = \lim_{\|P\| \to 0} \sum_{i=1}^{m} \sum_{j=1}^{n} f(\mathbf{r}(u_i^*, v_j^*)) A(S_{ij})$$

Notice not only the analogy with the definition of a line integral (14.6) but also the analogy with the definition of a double integral (13.4).

By approximating S_{ij} by part of a tangent plane as we did in defining surface area (see Figure 14.32) it can be shown that the limit on the right side of Equation 14.50 is equal to the double integral

$$\iint_D f(\mathbf{r}(u, v))|\mathbf{T}_u \times \mathbf{T}_v|\, dA$$

where $\qquad \mathbf{T}_u = \dfrac{\partial x}{\partial u}\mathbf{i} + \dfrac{\partial y}{\partial u}\mathbf{j} + \dfrac{\partial z}{\partial u}\mathbf{k} \qquad \mathbf{T}_v = \dfrac{\partial x}{\partial v}\mathbf{i} + \dfrac{\partial y}{\partial v}\mathbf{j} + \dfrac{\partial z}{\partial v}\mathbf{k}$

This procedure can be carried out even when D is not a rectangle but is a type I or II region. Thus we have the following formula for a surface integral over S:

(14.51)
$$\iint_S f(x, y, z)\, dS = \iint_D f(\mathbf{r}(u, v))|\mathbf{T}_u \times \mathbf{T}_v|\, dA$$

This should be compared with the formula for a line integral:

$$\int_C f(x, y, z)\, ds = \int_a^b f(\mathbf{r}(t))|\mathbf{r}'(t)|\, dt$$

Formula 14.51 allows us to compute a surface integral by converting it into a double integral over the parameter domain D. When using this formula, remember that $f(\mathbf{r}(u, v))$ is evaluated by writing $x = x(u, v)$, $y = y(u, v)$, and $z = z(u, v)$ in the formula for $f(x, y, z)$. In fact, by using the same calculation that led to Equation 14.46, we can write the right side of Equation 14.51 as

$$\iint_D f(x(u, v),\, y(u, v),\, z(u, v)) \sqrt{\left[\frac{\partial(x, y)}{\partial(u, v)}\right]^2 + \left[\frac{\partial(y, z)}{\partial(u, v)}\right]^2 + \left[\frac{\partial(x, z)}{\partial(u, v)}\right]^2}\, dA$$

which is the analogue of Equation 14.12. Observe that $\iint_S 1\, dS = A(S)$.

Surface integrals have applications similar to those for the integrals we have previously considered. For example, if a thin sheet (say, of aluminum foil) has the shape of a surface S and the density (mass per unit area) at the point (x, y, z) is $\rho(x, y, z)$, then the total mass of the sheet is

$$m = \iint_S \rho(x, y, z)\, dS$$

and the center of mass is $(\bar{x}, \bar{y}, \bar{z})$, where

$$\bar{x} = \frac{1}{m} \iint_S x\rho(x, y, z)\, dS \quad \bar{y} = \frac{1}{m} \iint_S y\rho(x, y, z)\, dS \quad \bar{z} = \frac{1}{m} \iint_S z\rho(x, y, z)\, dS$$

Moments of inertia can also be defined as before.

● **Example 1**   Compute the surface integral $\iint_S x^2\, dS$, where S is the unit sphere $x^2 + y^2 + z^2 = 1$.

Solution   As in Example 1 in Section 14.6 we use the parametric representation

$$x = \sin\phi\cos\theta \quad y = \sin\phi\sin\theta \quad z = \cos\phi \quad 0 \leqslant \phi \leqslant \pi, \quad 0 \leqslant \theta \leqslant 2\pi$$

that is,

$$\mathbf{r}(\phi, \theta) = \sin\phi\cos\theta\,\mathbf{i} + \sin\phi\sin\theta\,\mathbf{j} + \cos\phi\,\mathbf{k}$$

As in Example 5 in Section 14.6, we can compute that

$$|\mathbf{T}_\phi \times \mathbf{T}_\theta| = \sin\phi$$

Therefore, by Formula 14.51,

$$\iint_S x^2\, dS = \iint_D (\sin\phi\cos\theta)^2 |\mathbf{T}_\phi \times \mathbf{T}_\theta|\, dA$$

$$= \int_0^{2\pi} \int_0^\pi \sin^2\phi\cos^2\theta\,\sin\phi\, d\phi\, d\theta$$

$$= \int_0^{2\pi} \cos^2\theta\, d\theta \int_0^\pi \sin^3\phi\, d\phi$$

$$= \int_0^{2\pi} \frac{1}{2}(1 + \cos 2\theta)\, d\theta \int_0^\pi (\sin\phi - \sin\phi\cos^2\phi)\, d\phi$$

$$= \frac{1}{2}\left[\theta + \frac{1}{2}\sin 2\theta\right]_0^{2\pi} \left[-\cos\phi + \frac{1}{3}\cos^3\phi\right]_0^\pi$$

$$= \frac{4\pi}{3} \qquad ●$$

In the preceding section we mentioned that any surface that is the graph of a function of x and y can be regarded as a parametric surface with parameters x and y. Suppose S has equation $z = g(x, y)$, $(x, y) \in D$, where g has continuous partial derivatives. Then the parametric equations of S are

$$x = x \qquad y = y \qquad z = g(x, y)$$

and from Equation 14.47 we have

$$|\mathbf{T}_x \times \mathbf{T}_y| = \sqrt{1 + \left(\frac{\partial z}{\partial x}\right)^2 + \left(\frac{\partial z}{\partial y}\right)^2}$$

Therefore, in this case, Formula 14.51 becomes

(14.52)
$$\iint_S f(x, y, z)\, dS = \iint_D f(x, y, g(x, y)) \sqrt{1 + \left(\frac{\partial z}{\partial x}\right)^2 + \left(\frac{\partial z}{\partial y}\right)^2}\, dA$$

● **Example 2**   Evaluate $\iint_S y\, dS$, where S is the surface $z = x + y^2$, $0 \leqslant x \leqslant 1, 0 \leqslant y \leqslant 2$.

Solution   Since

$$\frac{\partial z}{\partial x} = 1 \qquad \text{and} \qquad \frac{\partial z}{\partial y} = 2y$$

Formula 14.52 gives

$$\iint_S y\, dS = \iint_D y \sqrt{1 + \left(\frac{\partial z}{\partial x}\right)^2 + \left(\frac{\partial z}{\partial y}\right)^2}\, dA$$

$$= \int_0^1 \int_0^2 y\sqrt{1 + 1 + 4y^2}\, dy\, dx$$

$$= \int_0^1 dx \sqrt{2} \int_0^2 y\sqrt{1 + 2y^2}\, dy$$

$$= \sqrt{2}\left(\frac{1}{4}\right)\frac{2}{3}(1 + 2y^2)^{3/2}\Big]_0^2 = \frac{13\sqrt{2}}{3}$$

●

If S is a piecewise smooth surface, that is, a finite union of smooth surfaces $S_1, S_2, \ldots, S_n$ that intersect only along their boundaries, then the surface integral of f over S is defined by

$$\iint_S f(x, y, z)\, dS = \iint_{S_1} f(x, y, z)\, dS + \cdots + \iint_{S_n} f(x, y, z)\, dS$$

● **Example 3**   Evaluate $\iint_S z\, dS$ where S is the surface whose sides S_1 are given by the cylinder $x^2 + y^2 = 1$, whose bottom S_2 is the disk $x^2 + y^2 \leqslant 1$ in the plane $z = 0$, and whose top S_3 is the part of the plane $z = x + 1$ that lies above S_2.

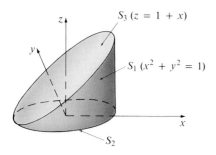

Figure 14.36

Solution   The surface S is shown in Figure 14.36. (We have changed the usual position of the axes to get a better look at S.) For S_1 we use θ and z as parameters (see Example 2 in Section 14.6) and write its parametric equations as

$$x = \cos\theta \qquad y = \sin\theta \qquad z = z$$

where $0 \leqslant \theta \leqslant 2\pi$ and $0 \leqslant z \leqslant 1 + x = 1 + \cos\theta$

Therefore

$$\mathbf{T}_\theta \times \mathbf{T}_z = \begin{vmatrix} \mathbf{i} & \mathbf{j} & \mathbf{k} \\ -\sin\theta & \cos\theta & 0 \\ 0 & 0 & 1 \end{vmatrix} = \cos\theta\,\mathbf{i} + \sin\theta\,\mathbf{j}$$

and $|\mathbf{T}_\theta \times \mathbf{T}_z| = \sqrt{\cos^2\theta + \sin^2\theta} = 1$

Thus the surface integral over S_1 is

$$\iint_{S_1} z\,dS = \iint_D z|\mathbf{T}_\theta \times \mathbf{T}_z|\,dA$$

$$= \int_0^{2\pi}\int_0^{1+\cos\theta} z\,dz\,d\theta = \int_0^{2\pi}\frac{1}{2}(1 + \cos\theta)^2\,d\theta$$

$$= \frac{1}{2}\int_0^{2\pi}\left[1 + 2\cos\theta + \frac{1}{2}(1 + \cos 2\theta)\right]d\theta$$

$$= \frac{1}{2}\left[\frac{3}{2}\theta + 2\sin\theta + \frac{1}{4}\sin 2\theta\right]_0^{2\pi} = \frac{3\pi}{2}$$

Since S_2 lies in the plane $z = 0$, we have

$$\iint_{S_2} z\,dS = \iint_{S_2} 0\,dS = 0$$

The top surface S_3 lies above the unit disk D and is part of the plane $z = 1 + x$. So, taking $g(x, y) = 1 + x$ in Formula 14.52 and converting to polar coordinates, we have

$$\iint_{S_3} z\,dS = \iint_D (1 + x)\sqrt{1 + \left(\frac{\partial z}{\partial x}\right)^2 + \left(\frac{\partial z}{\partial y}\right)^2}\,dA$$

$$= \int_0^{2\pi}\int_0^1 (1 + r\cos\theta)\sqrt{1 + 1 + 0}\,r\,dr\,d\theta$$

$$= \sqrt{2}\int_0^{2\pi}\int_0^1 (r + r^2\cos\theta)\,dr\,d\theta$$

$$= \sqrt{2}\int_0^{2\pi}\left(\frac{1}{2} + \frac{1}{3}\cos\theta\right)d\theta = \sqrt{2}\left[\frac{\theta}{2} + \frac{\sin\theta}{3}\right]_0^{2\pi} = \sqrt{2}\,\pi$$

Therefore
$$\iint\limits_{S} z \, dS = \iint\limits_{S_1} z \, dS + \iint\limits_{S_2} z \, dS + \iint\limits_{S_3} z \, dS$$

$$= \frac{3\pi}{2} + 0 + \sqrt{2}\pi = \left(\frac{3}{2} + \sqrt{2}\right)\pi \qquad \bullet$$

Oriented Surfaces

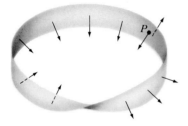

Figure 14.37

A MÖBIUS STRIP

In order to define surface integrals of vector fields we need to rule out non-orientable surfaces such as the Möbius strip shown in Figure 14.37. [It is named after the German geometer August Möbius (1790–1868).] You can construct one for yourself by taking a long rectangular strip of paper, giving it a half-twist, and taping the short edges together as in Figure 14.38. If an ant were to crawl along the Möbius strip starting at a point P, it would end up on the "other side" of the strip (that is, with its upper side pointing in the opposite direction). Then, if the ant continues to crawl in the same direction, it will end up back at the same point P without ever having crossed an edge. (If you have constructed a Möbius strip, try drawing a pencil line down the middle.) Therefore a Möbius strip really has only one side.

From now on we shall consider only orientable surfaces. We start with a two-sided surface S that has a tangent plane at every point (x, y, z) on S (except at any boundary points). There are two unit normal vectors $\mathbf{n}_1$ and $\mathbf{n}_2 = -\mathbf{n}_1$ at (x, y, z) (see Figure 14.39). If it is possible to choose a unit normal vector $\mathbf{n}$ at every such point (x, y, z) so that $\mathbf{n}$ varies continuously over S, then S is called an **oriented surface** and the given choice of $\mathbf{n}$ provides S with an **orientation.** There are two possible orientations for any orientable surface (see Figure 14.40).

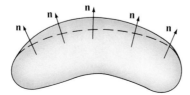

Figure 14.38

CONSTRUCTING A MÖBIUS STRIP

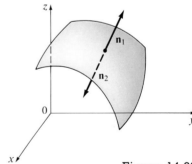

Figure 14.39

Figure 14.40

THE TWO ORIENTATIONS OF AN
ORIENTABLE SURFACE

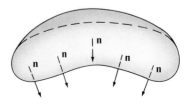

If S is a smooth orientable surface given in parametric form by a vector function $\mathbf{r}(u, v)$, then it is automatically supplied with the orientation of the unit normal vector

(14.53)
$$\mathbf{n} = \frac{\mathbf{T}_u \times \mathbf{T}_v}{|\mathbf{T}_u \times \mathbf{T}_v|}$$

and the opposite orientation is given by $-\mathbf{n}$. For instance, in Example 1 in Section 14.6 we found the parametric representation

$$\mathbf{r}(\phi, \theta) = a \sin \phi \cos \theta \mathbf{i} + a \sin \phi \sin \theta \mathbf{j} + a \cos \phi \mathbf{k}$$

for the sphere $x^2 + y^2 + z^2 = a^2$. Then in Example 5 in Section 14.6 we found that

$$\mathbf{T}_\phi \times \mathbf{T}_\theta = a^2 \sin^2 \phi \cos \theta \mathbf{i} + a^2 \sin^2 \phi \sin \theta \mathbf{j} + a^2 \sin \phi \cos \phi \mathbf{k}$$

and
$$|\mathbf{T}_\phi \times \mathbf{T}_\theta| = a^2 \sin \phi$$

So the orientation induced by $\mathbf{r}(\phi, \theta)$ is defined by the unit normal vector

$$\mathbf{n} = \frac{\mathbf{T}_\phi \times \mathbf{T}_\theta}{|\mathbf{T}_\phi \times \mathbf{T}_\theta|} = \sin \phi \cos \theta \mathbf{i} + \sin \phi \sin \theta \mathbf{j} + \cos \phi \mathbf{k} = \frac{1}{a} \mathbf{r}(\phi, \theta)$$

Observe that $\mathbf{n}$ points in the same direction as the position vector, that is, outward from the sphere (see Figure 14.41). The opposite (inward) orientation would have been obtained [see Figure 14.41(b)] if we had reversed the order of the parameters since $\mathbf{T}_\theta \times \mathbf{T}_\phi = -\mathbf{T}_\phi \times \mathbf{T}_\theta$.

For a **closed surface,** that is, a surface that is the boundary of a solid region E, the convention is that the **positive orientation** is the one for which the normal vectors point *outward* from E and inward-pointing normals give the negative orientation (see Figure 14.41).

For a surface $z = g(x, y)$ given as the graph of g, we use Equation 14.47 and see that the induced orientation is given by the unit normal vector

(14.54)
$$\mathbf{n} = \frac{-\dfrac{\partial g}{\partial x} \mathbf{i} - \dfrac{\partial g}{\partial y} \mathbf{j} + \mathbf{k}}{\sqrt{1 + \left(\dfrac{\partial g}{\partial x}\right)^2 + \left(\dfrac{\partial g}{\partial y}\right)^2}}$$

Figure 14.41

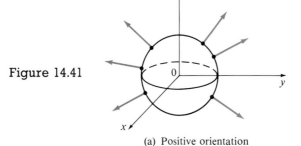

(a) Positive orientation

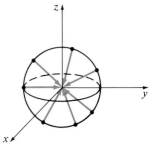

(b) Negative orientation

Since the **k**-component is positive, this gives the *upward* orientation of the surface.

Surface Integrals of Vector Fields

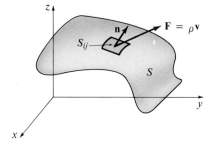

Figure 14.42

Suppose that S is an oriented surface with unit normal vector **n** and imagine a fluid with density $\rho(x, y, z)$ and velocity field $\mathbf{v}(x, y, z)$ flowing through S. (Think of S as an imaginary surface so that it does not impede the fluid flow.) Then the rate of flow (mass per unit time) per unit area is $\rho\mathbf{v}$. If we partition S into small parts S_{ij} as in Figure 14.42 (compare with Figure 14.35), then S_{ij} is nearly planar and so we can approximate the mass of fluid crossing S_{ij} in the direction of the normal **n** per unit time by the quantity

$$\rho\mathbf{v} \cdot \mathbf{n} A(S_{ij})$$

where ρ, **v**, and **n** are evaluated at some point on S_{ij}. (Recall that the component of the vector $\rho\mathbf{v}$ in the direction of the unit vector **n** is $\rho\mathbf{v} \cdot \mathbf{n}$.) By summing these quantities and taking the limit we get, according to Definition 14.50, the surface integral of the function $\rho\mathbf{v} \cdot \mathbf{n}$ over S:

(14.55)
$$\iint_S \rho\mathbf{v} \cdot \mathbf{n}\, dS = \iint_S \rho(x, y, z)\mathbf{v}(x, y, z) \cdot \mathbf{n}(x, y, z)\, dS$$

and this is interpreted physically as the rate of flow through S.

If we write $\mathbf{F} = \rho\mathbf{v}$, then **F** is also a vector field on R^3 and the integral in Equation 14.55 becomes

$$\iint_S \mathbf{F} \cdot \mathbf{n}\, dS$$

A surface integral of this form occurs frequently in physics, even when **F** is not $\rho\mathbf{v}$, and is called the surface integral of **F** over S. (In physics it is also called the **flux** of **F** over S.)

Definition (14.56)

If **F** is a continuous vector field defined on an oriented surface S with unit normal vector **n**, then the **surface integral of F over S** is

$$\iint_S \mathbf{F} \cdot d\mathbf{S} = \iint_S \mathbf{F} \cdot \mathbf{n}\, dS$$

In words, Definition 14.56 says that the surface integral of a vector field over S is equal to the surface integral of its normal component over S (as previously defined).

If S is given by a vector function $\mathbf{r}(u, v)$, then $\mathbf{n}$ is given by Equation 14.53 and from Definition 14.56 and Equation 14.51 we have

$$\iint_S \mathbf{F} \cdot d\mathbf{S} = \iint_S \mathbf{F} \cdot \frac{\mathbf{T}_u \times \mathbf{T}_v}{|\mathbf{T}_u \times \mathbf{T}_v|} \, dS$$

$$= \iint_D \left[\mathbf{F}(\mathbf{r}(u, v)) \cdot \frac{\mathbf{T}_u \times \mathbf{T}_v}{|\mathbf{T}_u \times \mathbf{T}_v|} \right] |\mathbf{T}_u \times \mathbf{T}_v| \, dA$$

where D is the parameter domain. Thus we have

(14.57)
$$\boxed{\iint_S \mathbf{F} \cdot d\mathbf{S} = \iint_D \mathbf{F} \cdot (\mathbf{T}_u \times \mathbf{T}_v) \, dA}$$

which is comparable to the expression for evaluating line integrals of vector fields in Definition 14.17.

● **Example 4**   Find the flux of the vector field $\mathbf{F}(x, y, z) = z\mathbf{i} + y\mathbf{j} + x\mathbf{k}$ over the unit sphere $x^2 + y^2 + z^2 = 1$.

Solution   Using the parametric representation

$$\mathbf{r}(\phi, \theta) = \sin\phi \cos\theta \mathbf{i} + \sin\phi \sin\theta \mathbf{j} + \cos\phi \mathbf{k} \qquad 0 \leqslant \phi \leqslant \pi, \ \ 0 \leqslant \theta \leqslant 2\pi$$

we have $\mathbf{F}(\mathbf{r}(\phi, \theta)) = \cos\phi \mathbf{i} + \sin\phi \sin\theta \mathbf{j} + \sin\phi \cos\theta \mathbf{k}$

and, from Example 5 in Section 14.6,

$$\mathbf{T}_\phi \times \mathbf{T}_\theta = \sin^2\phi \cos\theta \mathbf{i} + \sin^2\phi \sin\theta \mathbf{j} + \sin\phi \cos\phi \mathbf{k}$$

Therefore

$$\mathbf{F}(\mathbf{r}(\phi, \theta)) \cdot (\mathbf{T}_\phi \times \mathbf{T}_\theta) = \cos\phi \sin^2\phi \cos\theta + \sin^3\phi \sin^2\theta + \sin^2\phi \cos\phi \cos\theta$$

and, by Formula 14.57, the flux is

$$\iint_S \mathbf{F} \cdot d\mathbf{S} = \iint_D \mathbf{F} \cdot (\mathbf{T}_\phi \times \mathbf{T}_\theta) \, dA$$

$$= \int_0^{2\pi} \int_0^\pi (2\sin^2\phi \cos\phi \cos\theta + \sin^3\phi \sin^2\theta) \, d\phi \, d\theta$$

$$= 2\int_0^\pi \sin^2\phi \cos\phi \, d\phi \int_0^{2\pi} \cos\theta \, d\theta + \int_0^\pi \sin^3\phi \, d\phi \int_0^{2\pi} \sin^2\theta \, d\theta$$

$$= 0 + \int_0^\pi \sin^3\phi \, d\phi \int_0^{2\pi} \sin^2\theta \, d\theta \qquad \left(since \int_0^{2\pi} \cos\theta \, d\theta = 0 \right)$$

$$= \frac{4\pi}{3}$$

by the same calculation as in Example 1.   ●

If, for instance, the vector field in Example 1 is a velocity field describing the flow of a fluid with density 1, then the answer, $4\pi/3$, represents the rate of flow through the unit sphere in units of mass per unit time.

In the case of a surface S given by a graph $z = g(x, y)$ and oriented upward, Equation 14.47 gives

$$\mathbf{T}_x \times \mathbf{T}_y = -\frac{\partial g}{\partial x}\,\mathbf{i} - \frac{\partial g}{\partial y}\,\mathbf{j} + \mathbf{k}$$

and so, using

$$\iint_S \mathbf{F} \cdot d\mathbf{S} = \iint_D \mathbf{F} \cdot (\mathbf{T}_x \times \mathbf{T}_y)\, dA$$

with $\mathbf{F} = P\mathbf{i} + Q\mathbf{j} + R\mathbf{k}$, we obtain

(14.58)
$$\boxed{\iint_S \mathbf{F} \cdot d\mathbf{S} = \iint_D \left(-P\frac{\partial g}{\partial x} - Q\frac{\partial g}{\partial y} + R \right) dA}$$

● **Example 5**   Evaluate $\iint_S \mathbf{F} \cdot d\mathbf{S}$, where $\mathbf{F}(x, y, z) = y\mathbf{i} + x\mathbf{j} + z\mathbf{k}$ and S is the boundary of the solid region E enclosed by the paraboloid $z = 1 - x^2 - y^2$ and the plane $z = 0$.

Solution S consists of a parabolic top surface S_1 and a circular bottom surface S_2 (see Figure 14.43). Since S is a closed surface we use the convention of positive (outward) orientation. This means that S_1 is oriented upward and we can use Equation 14.58 with D being the projection of S_1 on the xy-plane, namely, the disk $x^2 + y^2 \le 1$. Since

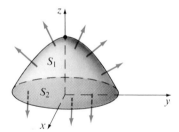

$$P(x, y, z) = y \qquad Q(x, y, z) = x \qquad R(x, y, z) = z = 1 - x^2 - y^2$$

on S_1 and

$$\frac{\partial g}{\partial x} = -2x \qquad \frac{\partial g}{\partial y} = -2y$$

Figure 14.43

we have

$$\iint_{S_1} \mathbf{F} \cdot d\mathbf{S} = \iint_D \left(-P\frac{\partial g}{\partial x} - Q\frac{\partial g}{\partial y} + R \right) dA$$

$$= \iint_D \left[-y(-2x) - x(-2y) + 1 - x^2 - y^2 \right] dA$$

$$= \iint_D (1 + 4xy - x^2 - y^2)\, dA$$

$$= \int_0^{2\pi} \int_0^1 (1 + 4r^2 \cos\theta \sin\theta - r^2)\, r\, dr\, d\theta$$

$$= \int_0^{2\pi} \int_0^1 (r - r^3 + 4r^3 \cos\theta \sin\theta)\, dr\, d\theta$$

$$= \int_0^{2\pi} \left(\frac{1}{4} + \cos\theta \sin\theta \right) d\theta$$

$$= \frac{1}{4}(2\pi) + 0 = \frac{\pi}{2}$$

The disk S_2 is oriented downward, so its unit normal vector is $\mathbf{n} = -\mathbf{k}$ and we have

$$\iint_{S_2} \mathbf{F} \cdot d\mathbf{S} = \iint_{S_2} \mathbf{F} \cdot (-\mathbf{k})\, dS = \iint_D (-z)\, dA$$
$$= \iint_D 0\, dA = 0$$

since $z = 0$ on S_2. Finally we compute, by definition, $\iint_S \mathbf{F} \cdot d\mathbf{S}$ as the sum of the surface integrals of $\mathbf{F}$ over the pieces S_1 and S_2:

$$\iint_S \mathbf{F} \cdot d\mathbf{S} = \iint_{S_1} \mathbf{F} \cdot d\mathbf{S} + \iint_{S_2} \mathbf{F} \cdot d\mathbf{S} = \frac{\pi}{2} + 0 = \frac{\pi}{2} \qquad \bullet$$

Although we motivated the surface integral of a vector field using the example of fluid flow, this concept also arises in other physical situations. For instance, if $\mathbf{E}$ is an electric field (see Example 5 in Section 14.1), then the surface integral

$$\iint_S \mathbf{E} \cdot d\mathbf{S}$$

is called the **electric flux** of $\mathbf{E}$ through the surface S. One of the important laws of electrostatics is **Gauss's Law,** which says that the net charge enclosed by a closed surface S is

(14.59)
$$Q = \varepsilon_0 \iint_S \mathbf{E} \cdot d\mathbf{S}$$

where ε_0 is a constant (called the permittivity of free space) that depends on the units used. (In the SI system, $\varepsilon_0 \approx 8.8542 \times 10^{-12}\ \mathrm{C}^2/\mathrm{N\text{-}m}^2$.) Thus if the vector field $\mathbf{F}$ in Example 4 represents an electric field, we can conclude that the charge enclosed by S is $Q = 4\pi\varepsilon_0/3$.

Another application of surface integrals occurs in the study of heat flow. Suppose the temperature at a point (x, y, z) in a body is $u(x, y, z)$. Then the **heat flow** is defined as the vector field

$$\mathbf{F} = -K\,\nabla u$$

where K is an experimentally determined constant called the **conductivity** of the substance. The rate of heat flow across the surface S in the body is then given by the surface integral

$$\iint_S \mathbf{F} \cdot d\mathbf{S} = -K \iint_S \nabla u \cdot d\mathbf{S}$$

● **Example 6**   The temperature u in a metal ball is proportional to the square of the distance from the center of the ball. Find the rate of heat flow across a sphere S of radius a with center at the center of the ball.

Solution   Taking the center of the ball to be at the origin, we have

$$u(x, y, z) = C(x^2 + y^2 + z^2)$$

where C is the proportionality constant. Then the heat flow is

$$\mathbf{F}(x, y, z) = -K \nabla u = -KC(2x\mathbf{i} + 2y\mathbf{j} + 2z\mathbf{k})$$

where K is the conductivity of the metal. Instead of using the usual parametrization of the sphere as in Example 4, we observe that the outward unit normal to the sphere $x^2 + y^2 + z^2 = a^2$ at the point (x, y, z) is

$$\mathbf{n} = \frac{1}{a}(x\mathbf{i} + y\mathbf{j} + z\mathbf{k})$$

and so
$$\mathbf{F} \cdot \mathbf{n} = -\frac{2KC}{a}(x^2 + y^2 + z^2)$$

But on S we have $x^2 + y^2 + z^2 = a^2$, so $\mathbf{F} \cdot \mathbf{n} = -2aKC$. Therefore the rate of heat flow across S is

$$\iint_S \mathbf{F} \cdot d\mathbf{S} = \iint_S \mathbf{F} \cdot \mathbf{n} \, dS$$

$$= -2aKC \iint_S dS = -2aKCA(S)$$

$$= -2aKC(4\pi a^2) = -8KC\pi a^3 \qquad \bullet$$

EXERCISES 14.7

In Exercises 1–12 evaluate the given surface integral.

1. $\iint_S (x^2z + y^2z) \, dS$, where S is the hemisphere $x^2 + y^2 + z^2 = 4, z \geqslant 0$

2. $\iint_S xyz \, dS$, where S is the part of the sphere $x^2 + y^2 + z^2 = 1$ that lies above the cone $z = \sqrt{x^2 + y^2}$

3. $\iint_S (x^2y + z^2) \, dS$, where S is the part of the cylinder $x^2 + y^2 = 9$ between the planes $z = 0$ and $z = 2$

4. $\iint_S (x^2 + y^2 + z^2) \, dS$, where S consists of the cylinder in Exercise 3 together with its top and bottom disks

5. $\iint_S y \, dS$, where S is the part of the plane $3x + 2y + z = 6$ that lies in the first octant

6. $\iint_S xz \, dS$, where S is the triangle with vertices $(1, 0, 0)$, $(0, 1, 0)$, and $(0, 0, 1)$

7. $\iint_S x \, dS$, where S is the surface $y = x^2 + 4z, 0 \leqslant x \leqslant 2$, $0 \leqslant z \leqslant 2$

8. $\iint_S (y^2 + z^2) \, dS$, where S is the part of the paraboloid $x = 4 - y^2 - z^2$ that lies in front of the plane $x = 0$

9. $\iint_S yz \, dS$, where S is the part of the plane $z = y + 3$ that lies inside the cylinder $x^2 + y^2 = 1$

10. $\iint_S xy \, dS$, where S is the boundary of the region enclosed by the cylinder $x^2 + z^2 = 1$ and the planes $y = 0$ and $x + y = 2$

11. $\iint_S yz \, dS$, where S is the surface with parametric equations $x = uv, y = u + v, z = u - v$, $u^2 + v^2 \leqslant 1$

12. $\iint_S \sqrt{1 + x^2 + y^2} \, dS$, where S is the helicoid with vector equation $\mathbf{r}(u, v) = u \cos v \mathbf{i} + u \sin v \mathbf{j} + v\mathbf{k}$
$0 \leqslant u \leqslant 1, \quad 0 \leqslant v \leqslant \pi$

In Exercises 13–22 evaluate the surface integral $\iint_S \mathbf{F} \cdot d\mathbf{S}$ for the given vector field $\mathbf{F}$ and the given oriented surface S. For closed surfaces, use the positive (outward) orientation.

13. $\mathbf{F}(x, y, z) = x\mathbf{i} + y\mathbf{j} + z\mathbf{k}$, S is the sphere $x^2 + y^2 + z^2 = 9$

14. $\mathbf{F}(x, y, z) = -y\mathbf{i} + x\mathbf{j} + 3z\mathbf{k}$, S is the hemisphere $z = \sqrt{16 - x^2 - y^2}$ with upward orientation

15. $\mathbf{F}(x, y, z) = e^y\mathbf{i} + ye^x\mathbf{j} + x^2y\mathbf{k}$, S is the part of the paraboloid $z = x^2 + y^2$ that lies above the square $0 \leqslant x \leqslant 1, 0 \leqslant y \leqslant 1$ and has upward orientation

16. $\mathbf{F}(x, y, z) = x^2y\mathbf{i} - 3xy^2\mathbf{j} + 4y^3\mathbf{k}$, S is the part of the elliptic paraboloid $z = x^2 + y^2 - 9$ that lies below the square $0 \leqslant x \leqslant 2, 0 \leqslant y \leqslant 1$ and has downward orientation

17. $\mathbf{F}(x, y, z) = x\mathbf{i} + xy\mathbf{j} + xz\mathbf{k}$, S is the surface of Exercise 5 with upward orientation

18. $\mathbf{F}(x, y, z) = -x\mathbf{i} - y\mathbf{j} + z^2\mathbf{k}$, S is the part of the cone $z = \sqrt{x^2 + y^2}$ between the planes $z = 1$ and $z = 2$ with upward orientation

19. $\mathbf{F}(x, y, z) = y\mathbf{j} - z\mathbf{k}$, S consists of the paraboloid $y = x^2 + z^2, 0 \leqslant y \leqslant 1$, and the disk $x^2 + z^2 \leqslant 1$, $y = 1$

20. $\mathbf{F}(x, y, z) = x\mathbf{i} + y\mathbf{j} + 5\mathbf{k}$, S is the surface of Exercise 10

21. $\mathbf{F}(x, y, z) = x\mathbf{i} + 2y\mathbf{j} + 3z\mathbf{k}$, S is the cube with vertices $(\pm 1, \pm 1, \pm 1)$

22. $\mathbf{F}(x, y, z) = y\mathbf{i} + x\mathbf{j} + z^2\mathbf{k}$, S is the helicoid of Exercise 12

23. Find the center of mass of the hemisphere $x^2 + y^2 + z^2 = a^2, z \geqslant 0$, if it has constant density.

24. Find the mass of a thin funnel in the shape of a cone $z = \sqrt{x^2 + y^2}, 1 \leqslant z \leqslant 4$, if its density function is $\rho(x, y, z) = 10 - z$.

25. (a) Give an integral expression for the moment of inertia I_z about the z-axis of a thin sheet in the shape of a surface S if the density function is ρ.
    (b) Find the moment of inertia about the z-axis of the funnel in Exercise 24.

26. The conical surface $z^2 = x^2 + y^2, 0 \leqslant z \leqslant a$, has constant density k. Find (a) the center of mass, (b) the moment of inertia about the z-axis.

27. A fluid with density 1200 flows with velocity $\mathbf{v} = y\mathbf{i} + \mathbf{j} + z\mathbf{k}$. Find the rate of flow upward through the paraboloid $z = 9 - (x^2 + y^2)/4, x^2 + y^2 \leqslant 36$.

28. A fluid has density 1500 and velocity field $\mathbf{v} = -y\mathbf{i} + x\mathbf{j} + 2z\mathbf{k}$. Find the rate of flow outward through the sphere $x^2 + y^2 + z^2 = 25$.

29. Use Gauss's Law to find the charge contained in the solid hemisphere $x^2 + y^2 + z^2 \leqslant a^2, z \geqslant 0$, if the electric field is $\mathbf{E}(x, y, z) = x\mathbf{i} + y\mathbf{j} + 2z\mathbf{k}$.

30. Use Gauss's Law to find the charge enclosed by the cube with vertices $(\pm 1, \pm 1, \pm 1)$ if the electric field is $\mathbf{E}(x, y, z) = x\mathbf{i} + y\mathbf{j} + z\mathbf{k}$.

31. The temperature at the point (x, y, z) in a substance with conductivity $K = 6.5$ is $u(x, y, z) = 2y^2 + 2z^2$. Find the rate of heat flow inward across the cylindrical surface $y^2 + z^2 = 6, 0 \leqslant x \leqslant 4$.

32. The temperature at a point in a ball with conductivity K is inversely proportional to the distance from the center of the ball. Find the rate of heat flow across a sphere S of radius a with center at the center of the ball.

SECTION 14.8

Stokes' Theorem

Stokes' Theorem, named after the Irish mathematical physicist Sir George Stokes (1819–1903), can be regarded as a higher-dimensional version of Green's Theorem. Whereas Green's Theorem relates a double integral over a plane region D to a line integral around its plane boundary curve, Stokes' Theorem relates a surface integral over a surface S to a line integral around the boundary curve of S (which will be a space curve). Figure 14.44 shows an oriented surface with unit normal vector $\mathbf{n}$. The orientation of S will induce the **positive orientation of the boundary curve** C shown in the figure.

This means that if you walk in the positive direction around C with your head pointing in the direction of **n**, then the surface will always be on your left.

Stokes' Theorem (14.60)

Let S be an oriented piecewise-smooth surface that is bounded by a simple, closed, piecewise-smooth boundary curve C with positive orientation. Let **F** be a vector field whose components have continuous partial derivatives on an open region in R^3 that contains S. Then

$$\int_C \mathbf{F} \cdot d\mathbf{r} = \iint_S \operatorname{curl} \mathbf{F} \cdot d\mathbf{S}$$

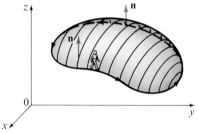

Figure 14.44

Since

$$\int_C \mathbf{F} \cdot d\mathbf{r} = \int_C \mathbf{F} \cdot \mathbf{T}\, ds \quad \text{and} \quad \iint_S \operatorname{curl} \mathbf{F} \cdot d\mathbf{S} = \iint_S \operatorname{curl} \mathbf{F} \cdot \mathbf{n}\, dS$$

Stokes' Theorem says that the line integral around the boundary curve of S of the tangential component of **F** is equal to the surface integral of the normal component of the curl of **F**.

The positively oriented boundary curve of the oriented surface S is often written as ∂S, so Stokes' Theorem can be expressed as

(14.61)

$$\iint_S \operatorname{curl} \mathbf{F} \cdot d\mathbf{S} = \int_{\partial S} \mathbf{F} \cdot d\mathbf{r}$$

There is an analogy among Stokes' Theorem, Green's Theorem, and the Fundamental Theorem of Calculus. As before, there is an integral involving derivatives on the left side of Equation 14.61 (recall that curl **F** is a sort of derivative of **F**) and the right side involves the values of **F** only on the *boundary* of S.

In fact, in the special case where the surface S is flat and lies in the xy-plane with upward orientation, then the unit normal is **k**, the surface integral becomes a double integral, and Stokes' Theorem becomes

$$\int_C \mathbf{F} \cdot d\mathbf{r} = \iint_S \operatorname{curl} \mathbf{F} \cdot d\mathbf{S} = \iint_S (\operatorname{curl} \mathbf{F}) \cdot \mathbf{k}\, dA$$

This is precisely the vector form of Green's Theorem given in Equation 14.39. Thus we see that Green's Theorem is really a special case of Stokes' Theorem.

Although Stokes' Theorem is too difficult for us to prove in its full generality, we can give a proof when S is a graph and **F**, S, and C are well-behaved.

Proof of a Special Case of Stokes' Theorem   We assume that the equation of S is $z = g(x, y), (x, y) \in D$, where g has continuous second order partial derivatives and D is a simple plane region whose boundary curve C_1 corresponds to C. If the orientation of S is upward, then the positive orientation of C corresponds to the positive orientation of C_1 (see Figure 14.45). We are also assuming that $\mathbf{F} = P\mathbf{i} + Q\mathbf{j} + R\mathbf{k}$, where the partial derivatives of P, Q, and R are continuous.

Since S is a graph we can apply Formula 14.58 with $\mathbf{F}$ replaced by curl $\mathbf{F}$. The result is

$$(14.62) \qquad \iint_S \operatorname{curl} \mathbf{F} \cdot d\mathbf{S} = \iint_D \left[-\left(\frac{\partial R}{\partial y} - \frac{\partial Q}{\partial z}\right)\frac{\partial z}{\partial x} - \left(\frac{\partial P}{\partial z} - \frac{\partial R}{\partial x}\right)\frac{\partial z}{\partial y} + \left(\frac{\partial Q}{\partial x} - \frac{\partial P}{\partial y}\right) \right] dA$$

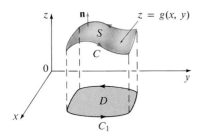

where the partial derivatives of P, Q, and R are evaluated at $(x, y, f(x, y))$. If

$$x = x(t) \qquad y = y(t) \qquad a \leqslant t \leqslant b$$

is a parametric representation of C_1, then a parametric representation of C is

$$x = x(t) \qquad y = y(t) \qquad z = g(x(t), y(t)) \qquad a \leqslant t \leqslant b$$

Figure 14.45

This allows us, with the aid of the Chain Rule, to evaluate the line integral as follows:

$$\int_C \mathbf{F} \cdot d\mathbf{r} = \int_a^b \left(P\frac{dx}{dt} + Q\frac{dy}{dt} + R\frac{dz}{dt} \right) dt$$

$$= \int_a^b \left[P\frac{dx}{dt} + Q\frac{dy}{dt} + R\left(\frac{\partial z}{\partial x}\frac{dx}{dt} + \frac{\partial z}{\partial y}\frac{dy}{dt}\right) \right] dt$$

$$= \int_a^b \left[\left(P + R\frac{\partial z}{\partial x}\right)\frac{dx}{dt} + \left(Q + R\frac{\partial z}{\partial y}\right)\frac{dy}{dt} \right] dt$$

$$= \int_{C_1} \left(P + R\frac{\partial z}{\partial x}\right)dx + \left(Q + R\frac{\partial z}{\partial y}\right)dy$$

$$= \iint_D \left[\frac{\partial}{\partial x}\left(Q + R\frac{\partial z}{\partial y}\right) - \frac{\partial}{\partial y}\left(P + R\frac{\partial z}{\partial x}\right) \right] dA$$

where we have used Green's Theorem in the last step. Then, using the Chain Rule again and remembering that P, Q, and R are functions of x, y, and z and that z is itself a function of x and y, we get

$$\int_C \mathbf{F} \cdot d\mathbf{r} = \iint_D \left[\left(\frac{\partial Q}{\partial x} + \frac{\partial Q}{\partial z}\frac{\partial z}{\partial x} + \frac{\partial R}{\partial x}\frac{\partial z}{\partial y} + \frac{\partial R}{\partial z}\frac{\partial z}{\partial x}\frac{\partial z}{\partial y} + R\frac{\partial^2 z}{\partial x\,\partial y}\right) \right.$$

$$\left. - \left(\frac{\partial P}{\partial y} + \frac{\partial P}{\partial z}\frac{\partial z}{\partial y} + \frac{\partial R}{\partial y}\frac{\partial z}{\partial x} + \frac{\partial R}{\partial z}\frac{\partial z}{\partial y}\frac{\partial z}{\partial x} + R\frac{\partial^2 z}{\partial y\,\partial x}\right) \right] dA$$

Four of the terms in this double integral cancel and the remaining six terms can be arranged to coincide with the right side of Equation 14.62. Therefore

$$\int_C \mathbf{F} \cdot d\mathbf{r} = \iint_S \text{curl } \mathbf{F} \cdot d\mathbf{S} \qquad \bullet$$

● **Example 1**   Evaluate $\int_C \mathbf{F} \cdot d\mathbf{r}$, where $\mathbf{F}(x, y, z) = -y^2\mathbf{i} + x\mathbf{j} + z^2\mathbf{k}$ and C is the curve of intersection of the plane $y + z = 2$ and the cylinder $x^2 + y^2 = 1$. (Orient C to be counterclockwise when viewed from above.)

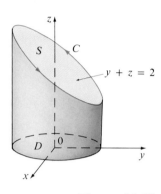

Figure 14.46

Solution   The curve C (an ellipse) is shown in Figure 14.46. Although $\int_C \mathbf{F} \cdot d\mathbf{r}$ could be evaluated directly, it is not easy to parametrize C so let us use Stokes' Theorem instead. We first compute

$$\text{curl } \mathbf{F} = \begin{vmatrix} \mathbf{i} & \mathbf{j} & \mathbf{k} \\ \dfrac{\partial}{\partial x} & \dfrac{\partial}{\partial y} & \dfrac{\partial}{\partial z} \\ -y^2 & x & z^2 \end{vmatrix} = (1 + 2y)\mathbf{k}$$

Let us choose S to be the elliptical region in the plane $y + z = 2$ that is bounded by C. If we orient S upward, then C has the induced positive orientation. The projection D of S on the xy-plane is the disk $x^2 + y^2 \leqslant 1$, so using Equation 14.58 with $z = g(x, y) = 2 - y$, we have

$$\int_C \mathbf{F} \cdot d\mathbf{r} = \iint_S \text{curl } \mathbf{F} \cdot d\mathbf{S} = \iint_D (1 + 2y)\, dA$$

$$= \int_0^{2\pi} \int_0^1 (1 + 2r\sin\theta) r\, dr\, d\theta$$

$$= \int_0^{2\pi} \left[\frac{r^2}{2} + 2\frac{r^3}{3}\sin\theta \right]_0^1 d\theta = \int_0^{2\pi} \left(\frac{1}{2} + \frac{2}{3}\sin\theta \right) d\theta$$

$$= \frac{1}{2}(2\pi) + 0 = \pi \qquad \bullet$$

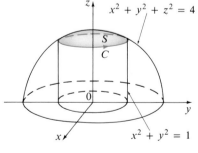

Figure 14.47

● **Example 2**   Use Stokes' Theorem to compute $\iint_S \text{curl } \mathbf{F} \cdot d\mathbf{S}$, where $\mathbf{F}(x, y, z) = yz\mathbf{i} + xz\mathbf{j} + xy\mathbf{k}$ and S is the part of the sphere $x^2 + y^2 + z^2 = 4$ that lies inside the cylinder $x^2 + y^2 = 1$ and above the xy-plane (see Figure 14.47).

Solution   To find the boundary curve C we solve the equations $x^2 + y^2 + z^2 = 4$ and $x^2 + y^2 = 1$. Subtracting, we get $z^2 = 3$ and so $z = \sqrt{3}$ (since $z > 0$). Thus C is the circle given by the equations $x^2 + y^2 = 1$, $z = \sqrt{3}$. The vector equation of C is

$$\mathbf{r}(t) = \cos t\,\mathbf{i} + \sin t\,\mathbf{j} + \sqrt{3}\,\mathbf{k} \qquad 0 \leqslant t \leqslant 2\pi$$

so

$$\mathbf{r}'(t) = -\sin t\,\mathbf{i} + \cos t\,\mathbf{j}$$

Also we have

$$\mathbf{F}(r(t)) = \sqrt{3}\sin t\,\mathbf{i} + \sqrt{3}\cos t\,\mathbf{j} + \cos t \sin t\,\mathbf{k}$$

Therefore, by Stokes' Theorem,

$$\iint_S \text{curl } \mathbf{F} \cdot d\mathbf{S} = \int_C \mathbf{F} \cdot d\mathbf{r} = \int_0^{2\pi} \mathbf{F}(\mathbf{r}(t)) \cdot \mathbf{r}'(t)\, dt$$

$$= \int_0^{2\pi}(-\sqrt{3}\sin^2 t + \sqrt{3}\cos^2 t)\, dt$$

$$= \sqrt{3}\int_0^{2\pi} \cos 2t\, dt = 0 \qquad \bullet$$

Note that in Example 2 we computed a surface integral simply by knowing the values of $\mathbf{F}$ on the boundary curve C. This means that if we have a different oriented surface with the same boundary curve C, then we would get exactly the same value for the surface integral!

In general, if S_1 and S_2 are oriented surfaces with the same oriented boundary curve C and both satisfy the hypotheses of Stokes' Theorem, then

(14.63)
$$\iint_{S_1} \text{curl } \mathbf{F} \cdot d\mathbf{S} = \int_C \mathbf{F} \cdot d\mathbf{r} = \iint_{S_2} \text{curl } \mathbf{F} \cdot d\mathbf{S}$$

This fact is useful when it is difficult to integrate over one surface but easy to integrate over the other.

Let us now use Stokes' Theorem to throw some light on the meaning of the curl vector. Suppose that C is an oriented closed curve and $\mathbf{v}$ represents the velocity field in fluid flow. Consider the line integral

$$\int_C \mathbf{v} \cdot d\mathbf{r} = \int_C \mathbf{v} \cdot \mathbf{T}\, ds$$

(a) $\int_C \mathbf{v} \cdot d\mathbf{r} > 0$, positive circulation

and recall that $\mathbf{v} \cdot \mathbf{T}$ is the component of $\mathbf{v}$ in the direction of the unit tangent vector $\mathbf{T}$. This means that the closer the direction of $\mathbf{v}$ is to the direction of $\mathbf{T}$, the larger the value of $\mathbf{v} \cdot \mathbf{T}$. Thus $\int_C \mathbf{v} \cdot d\mathbf{r}$ is a measure of the tendency of the fluid to move around C and is called the **circulation** of $\mathbf{v}$ around C (see Figure 14.48).

Now let $P_0(x_0, y_0, z_0)$ be a point in the fluid and let S_a be a small disk with radius a and center P_0. Then $(\text{curl } \mathbf{F})(P) \approx (\text{curl } \mathbf{F})(P_0)$ for all points P on S_a because curl $\mathbf{F}$ is continuous. Thus, by Stokes' Theorem, we get the following approximation to the circulation around the boundary circle C_a:

(b) $\int_C \mathbf{v} \cdot d\mathbf{r} < 0$, negative circulation

Figure 14.48

$$\int_{C_a} \mathbf{v} \cdot d\mathbf{r} = \iint_{S_a} \text{curl } \mathbf{v} \cdot d\mathbf{S} = \iint_{S_a} \text{curl } \mathbf{v} \cdot \mathbf{n}\, dS$$

$$\approx \iint_{S_a} \text{curl } \mathbf{v}(P_0) \cdot \mathbf{n}(P_0)\, dS$$

$$= \text{curl } \mathbf{v}(P_0) \cdot \mathbf{n}(P_0)\pi a^2$$

This approximation becomes better as $a \to 0$ and we have

(14.64)
$$\text{curl } \mathbf{v}(P_0) \cdot \mathbf{n}(P_0) = \lim_{a \to 0} \frac{1}{\pi a^2} \int_{C_a} \mathbf{v} \cdot d\mathbf{r}$$

Equation 14.64 gives the relationship between the curl and the circulation. It shows that curl $\mathbf{v} \cdot \mathbf{n}$ is a measure of the rotating effect of the fluid about the axis $\mathbf{n}$. The curling effect is greatest about the axis parallel to curl $\mathbf{v}$.

Finally we mention that Stokes' Theorem can be used to prove Theorem 14.35 (which states that if curl $\mathbf{F} = \mathbf{0}$ on all of R^3, then $\mathbf{F}$ is conservative). From our previous work (Theorems 14.20 and 14.21) we know that $\mathbf{F}$ will be conservative if $\int_C \mathbf{F} \cdot d\mathbf{r} = 0$ for every closed path C. Given C, suppose we can find a surface S whose boundary is C. (This can be done but the proof requires advanced techniques.) Then Stokes' Theorem gives

$$\int_C \mathbf{F} \cdot d\mathbf{r} = \iint_S \operatorname{curl} \mathbf{F} \cdot d\mathbf{S} = \iint_S \mathbf{0} \cdot d\mathbf{S} = 0$$

EXERCISES 14.8

In Exercises 1–6 use Stokes' Theorem to evaluate $\iint_S \operatorname{curl} \mathbf{F} \cdot d\mathbf{S}$.

1. $\mathbf{F}(x, y, z) = xyz\mathbf{i} + x\mathbf{j} + e^{xy}\cos z\mathbf{k}$, S is the hemisphere $x^2 + y^2 + z^2 = 1$, $z \geqslant 0$, oriented upward

2. $\mathbf{F}(x, y, z) = y^2z\mathbf{i} + xz\mathbf{j} + x^2y^2\mathbf{k}$, S is the part of the paraboloid $z = x^2 + y^2$ that lies inside the cylinder $x^2 + y^2 = 1$, oriented upward

3. $\mathbf{F}(x, y, z) = yz^3\mathbf{i} + \sin(xyz)\mathbf{j} + x^3\mathbf{k}$, S is the part of the paraboloid $y = 1 - x^2 - z^2$ that lies to the right of the xz-plane, oriented toward the xz-plane

4. $\mathbf{F}(x, y, z) = (x + \tan^{-1} yz)\mathbf{i} + y^2z\mathbf{j} + z\mathbf{k}$, S is the part of the hemisphere $x = \sqrt{9 - y^2 - z^2}$ that lies inside the cylinder $y^2 + z^2 = 4$, oriented in the direction of the positive x-axis

5. $\mathbf{F}(x, y, z) = xyz\mathbf{i} + xy\mathbf{j} + x^2yz\mathbf{k}$, S consists of the top and the four sides (but not the bottom) of the cube with vertices $(\pm 1, \pm 1, \pm 1)$, oriented outward. (*Hint:* Use Equation 14.63.)

6. $\mathbf{F}(x, y, z) = xy\mathbf{i} + e^z\mathbf{j} + xy^2\mathbf{k}$, S consists of the four sides of the pyramid with vertices $(0, 0, 0)$, $(1, 0, 0)$, $(0, 0, 1)$, $(1, 0, 1)$, and $(0, 1, 0)$ that lie to the right of the xz-plane, oriented in the direction of the positive y-axis (*Hint:* Use Equation 14.63.)

In Exercises 7–10 use Stokes' Theorem to evaluate $\int_C \mathbf{F} \cdot d\mathbf{r}$. In each case C is oriented counterclockwise as viewed from above.

7. $\mathbf{F}(x, y, z) = xz\mathbf{i} + 2xy\mathbf{j} + 3xy\mathbf{k}$, C is the boundary of the part of the plane $3x + y + z = 3$ in the first octant

8. $\mathbf{F}(x, y, z) = z^2\mathbf{i} + y^2\mathbf{j} + xy\mathbf{k}$, C is the triangle with vertices $(1, 0, 0)$, $(0, 1, 0)$, and $(0, 0, 2)$

9. $\mathbf{F}(x, y, z) = 2z\mathbf{i} + 4x\mathbf{j} + 5y\mathbf{k}$, C is the curve of intersection of the plane $z = x + 4$ and the cylinder $x^2 + y^2 = 4$

10. $\mathbf{F}(x, y, z) = x^2z\mathbf{i} + xy^2\mathbf{j} + z^2\mathbf{k}$, C is the curve of intersection of the plane $x + y + z = 1$ and the cylinder $x^2 + y^2 = 9$

In Exercises 11–14 verify that Stokes' Theorem is true for the given vector field $\mathbf{F}$ and surface S.

11. $\mathbf{F}(x, y, z) = 3y\mathbf{i} + 4z\mathbf{j} - 6x\mathbf{k}$, S is the part of the paraboloid $z = 9 - x^2 - y^2$ that lies above the xy-plane, oriented upward

12. $\mathbf{F}(x, y, z) = xy\mathbf{i} + yz\mathbf{j} + xz\mathbf{k}$, S is the hemisphere $z = \sqrt{a^2 - x^2 - y^2}$, oriented upward

13. $\mathbf{F}(x, y, z) = y\mathbf{i} + z\mathbf{j} + x\mathbf{k}$, S is the part of the plane $x + y + z = 1$ that lies in the first octant, oriented upward

14. $\mathbf{F}(x, y, z) = y\mathbf{i} + z\mathbf{j} + x\mathbf{k}$, S is the helicoid of Exercise 22 in Section 14.6

15. If S is a sphere and $\mathbf{F}$ satisfies the hypotheses of Stokes' Theorem, show that $\iint_S \operatorname{curl} \mathbf{F} \cdot d\mathbf{S} = 0$.

16. If S and C satisfy the hypotheses of Stokes' Theorem and f, g have continuous second order partial derivatives, show that

(a) $\int_C (f \nabla g) \cdot d\mathbf{r} = \iint_S (\nabla f \times \nabla g) \cdot d\mathbf{S}$

(b) $\int_C (f \nabla f) \cdot d\mathbf{r} = 0$

(c) $\int_C (f \nabla g + g \nabla f) \cdot d\mathbf{r} = 0$

The Divergence Theorem

In Section 14.5 we rewrote Green's Theorem in a vector version as

$$\int_C \mathbf{F} \cdot \mathbf{n}\, ds = \iint_D \text{div } \mathbf{F}(x, y)\, dA$$

where C is the positively oriented boundary curve of the plane region D. If we were seeking to extend this theorem to vector fields on R^3, we might make the guess that

(14.65)
$$\iint_S \mathbf{F} \cdot \mathbf{n}\, dS = \iiint_E \text{div } \mathbf{F}(x, y, z)\, dV$$

where S is the boundary surface of the solid region E. It turns out that Equation 14.65 is true, under appropriate hypotheses, and is called the Divergence Theorem or Gauss's Theorem after the German mathematician Karl Friedrich Gauss (1777–1855). Notice its similarity to Green's Theorem and Stokes' Theorem in that it relates the integral of a derivative of a function (div $\mathbf{F}$ in this case) over a region to the integral of the original function $\mathbf{F}$ over the boundary of the region.

At this stage you may wish to review the various types of regions over which we were able to evaluate triple integrals in Section 13.6. We shall state and prove the Divergence Theorem for regions E that are simultaneously of types 1, 2, and 3 and we shall call such regions **simple solid regions.** (For instance, regions bounded by ellipsoids or rectangular boxes are simple solid regions.) The boundary of E will be a closed surface and we use the convention, introduced in Section 14.7, that the positive orientation is outward; that is, the unit normal vector $\mathbf{n}$ is directed outward from E.

The Divergence Theorem (14.66)

> Let E be a simple solid region whose boundary surface S has positive (outward) orientation. Let $\mathbf{F}$ be a vector field whose component functions have continuous partial derivatives on an open region that contains E. Then
>
> $$\iint_S \mathbf{F} \cdot d\mathbf{S} = \iiint_E \text{div } \mathbf{F}\, dV$$

Proof   Let $\mathbf{F} = P\mathbf{i} + Q\mathbf{j} + R\mathbf{k}$. Then

$$\text{div } \mathbf{F} = \frac{\partial P}{\partial x} + \frac{\partial Q}{\partial y} + \frac{\partial R}{\partial z}$$

so
$$\iiint_E \text{div } \mathbf{F}\, dV = \iiint_E \frac{\partial P}{\partial x}\, dV + \iiint_E \frac{\partial Q}{\partial y}\, dV + \iiint_E \frac{\partial R}{\partial z}\, dV$$

If **n** is the unit outward normal of S, then the surface integral on the left side of the Divergence Theorem is

$$\iint_S \mathbf{F} \cdot d\mathbf{S} = \iint_S \mathbf{F} \cdot \mathbf{n}\, dS = \iint_S (P\mathbf{i} + Q\mathbf{j} + R\mathbf{k}) \cdot \mathbf{n}\, dS$$

$$= \iint_S P\mathbf{i} \cdot \mathbf{n}\, dS + \iint_S Q\mathbf{j} \cdot \mathbf{n}\, dS + \iint_S R\mathbf{k} \cdot \mathbf{n}\, dS$$

Therefore, to prove the Divergence Theorem, it suffices to prove the three equations:

(14.67)
$$\iint_S P\mathbf{i} \cdot \mathbf{n}\, dS = \iiint_E \frac{\partial P}{\partial x}\, dV$$

(14.68)
$$\iint_S Q\mathbf{j} \cdot \mathbf{n}\, dS = \iiint_E \frac{\partial Q}{\partial y}\, dV$$

(14.69)
$$\iint_S R\mathbf{k} \cdot \mathbf{n}\, dS = \iiint_E \frac{\partial R}{\partial z}\, dV$$

To prove Equation 14.69 we use the fact that E is a type 1 region:

$$E = \{(x, y, z) \,|\, (x, y) \in D, \phi_1(x, y) \leqslant z \leqslant \phi_2(x, y)\}$$

where D is the projection of E onto the xy-plane. By Equation 13.44, we have

$$\iiint_E \frac{\partial R}{\partial z}\, dV = \iint_D \left[\int_{\phi_1(x,y)}^{\phi_2(x,y)} \frac{\partial R}{\partial z}(x, y, z)\, dz \right] dA$$

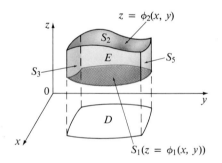

Figure 14.49

and therefore, by the Fundamental Theorem of Calculus,

(14.70)
$$\iiint_E \frac{\partial R}{\partial z}\, dV = \iint_D [R(x, y, \phi_2(x, y)) - R(x, y, \phi_1(x, y))]\, dA$$

The boundary surface S consists of six pieces: the bottom surface S_1, the top surface S_2, and the four vertical sides S_3, S_4, S_5, and S_6 (see Figure 14.49). Notice that on each of the vertical sides we have $\mathbf{k} \cdot \mathbf{n} = 0$ and so

$$\iint_{S_i} R\mathbf{k} \cdot \mathbf{n}\, dS = \iint_{S_i} 0\, dS = 0 \qquad i = 3, 4, 5, 6$$

This gives

(14.71)
$$\iint_S R\mathbf{k} \cdot \mathbf{n}\, dS = \iint_{S_1} R\mathbf{k} \cdot \mathbf{n}\, dS + \iint_{S_2} R\mathbf{k} \cdot \mathbf{n}\, dS$$

(It may happen that some of the vertical sides do not appear, as in the case of an ellipsoid, but in any case Equation 14.71 is still true.)

The equation of S_2 is $z = \phi_2(x, y)$, $(x, y) \in D$, and the outward normal $\mathbf{n}$ points upward, so from Equation 14.58 (with $\mathbf{F}$ replaced by $R\mathbf{k}$) we have

$$\iint\limits_{S_2} R\mathbf{k} \cdot \mathbf{n}\, dS = \iint\limits_{D} R(x, y, \phi_2(x, y))\, dA$$

On S_1 we have $z = \phi_1(x, y)$, but here the outward normal $\mathbf{n}$ points downward, so

$$\mathbf{n} = \frac{1}{\sqrt{\left(\dfrac{\partial \phi_1}{\partial x}\right)^2 + \left(\dfrac{\partial \phi_1}{\partial y}\right)^2 + 1}}\left(\frac{\partial \phi_1}{\partial x}\mathbf{i} + \frac{\partial \phi_1}{\partial y}\mathbf{j} - \mathbf{k}\right)$$

and

$$\iint\limits_{S_1} R\mathbf{k} \cdot \mathbf{n}\, dS$$

$$= \iint\limits_{D} R(x, y, \phi_1(x, y)) \frac{(-1)}{\sqrt{\left(\dfrac{\partial \phi_1}{\partial x}\right)^2 + \left(\dfrac{\partial \phi_1}{\partial y}\right)^2 + 1}} \sqrt{1 + \left(\frac{\partial \phi_1}{\partial x}\right)^2 + \left(\frac{\partial \phi_1}{\partial y}\right)^2}\, dA$$

$$= -\iint\limits_{D} R(x, y, \phi_1(x, y))\, dA$$

Therefore Equation 14.71 gives

$$\iint\limits_{S} R\mathbf{k} \cdot \mathbf{n}\, dS = \iint\limits_{D} [R(x, y, \phi_2(x, y)) - R(x, y, \phi_1(x, y))]\, dA$$

Comparison with Equation 14.70 gives

$$\iint\limits_{S} R\mathbf{k} \cdot \mathbf{n}\, dS = \iiint\limits_{E} \frac{\partial R}{\partial z}\, dV$$

Equations 14.67 and 14.68 are proved in a similar manner using the expressions for E as a type 2 or 3 region. ●

Notice that the method of proof of the Divergence Theorem is very similar to that of Green's Theorem.

● **Example 1**   Find the flux of the vector field $\mathbf{F}(x, y, z) = z\mathbf{i} + y\mathbf{j} + x\mathbf{k}$ over the unit sphere $x^2 + y^2 + z^2 = 1$.

Solution   First we compute the divergence of $\mathbf{F}$:

$$\text{div } \mathbf{F} = \frac{\partial}{\partial x}(z) + \frac{\partial}{\partial y}(y) + \frac{\partial}{\partial z}(x) = 1$$

The unit sphere S is the boundary of the unit ball B given by $x^2 + y^2 + z^2 \leqslant 1$.

Thus the Divergence Theorem gives the flux as

$$\iint_S \mathbf{F} \cdot d\mathbf{S} = \iiint_B \operatorname{div} \mathbf{F} \, dV = \iiint_B 1 \, dV$$

$$= V(B) = \frac{4}{3} \pi (1)^3 = \frac{4\pi}{3}$$

This solution should be compared with the solution in Example 4 in Section 14.7.

● **Example 2**   Evaluate $\iint_S \mathbf{F} \cdot d\mathbf{S}$, where

$$\mathbf{F}(x, y, z) = xy\mathbf{i} + (y^2 + e^{xz^2})\mathbf{j} + \sin(xy)\mathbf{k}$$

and S is the surface of the region E bounded by the parabolic cylinder $z = 1 - x^2$ and the planes $z = 0$, $y = 0$, and $y + z = 2$ (see Figure 14.50).

Solution   It would be extremely difficult to evaluate the given surface integral directly. (We would have to evaluate four surface integrals corresponding to the four pieces of S.) Furthermore, the divergence of $\mathbf{F}$ is much less complicated than $\mathbf{F}$ itself:

$$\operatorname{div} \mathbf{F} = \frac{\partial}{\partial x}(xy) + \frac{\partial}{\partial y}(y^2 + e^{xz^2}) + \frac{\partial}{\partial z}(\sin xy)$$

$$= y + 2y = 3y$$

Therefore we shall use the Divergence Theorem to transform the given surface integral into a triple integral. The easiest way to evaluate the triple integral is to express E as a type 3 region:

$$E = \{(x, y, z) \mid -1 \leqslant x \leqslant 1, 0 \leqslant z \leqslant 1 - x^2, 0 \leqslant y \leqslant 2 - z\}$$

Then we have

$$\iint_S \mathbf{F} \cdot d\mathbf{S} = \iiint_E \operatorname{div} \mathbf{F} \, dV = \iiint_E 3y \, dV$$

$$= 3 \int_{-1}^1 \int_0^{1-x^2} \int_0^{2-z} y \, dy \, dz \, dx$$

$$= 3 \int_{-1}^1 \int_0^{1-x^2} \frac{(2 - z)^2}{2} \, dz \, dx$$

$$= \frac{3}{2} \int_{-1}^1 \left[-\frac{(2 - z)^3}{3} \right]_0^{1-x^2} dx$$

$$= -\frac{1}{2} \int_{-1}^1 [(x^2 + 1)^3 - 8] \, dx$$

$$= -\int_0^1 (x^6 + 3x^4 + 3x^2 - 7) \, dx = \frac{184}{35}$$

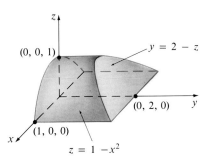

$(0, 0, 1)$ $y = 2 - z$

$(0, 2, 0)$ y

$(1, 0, 0)$

$z = 1 - x^2$

Figure 14.50

Although we have proved the Divergence Theorem only for simple solid regions, it can be proved for regions that are finite unions of simple solid regions. (The procedure is analogous to the one we used in Section 14.4 to extend Green's Theorem.)

For example, let us consider the region E that lies between the closed surfaces S_1 and S_2, where S_1 lies inside S_2. Let n_1 and n_2 be outward normals of S_1 and S_2. Then the boundary surface of E is $S = S_1 \cup S_2$ and its normal $\mathbf{n}$ is given by $\mathbf{n} = -\mathbf{n}_1$ on S_1 and $\mathbf{n} = \mathbf{n}_2$ on S_2 (see Figure 14.51). Applying the Divergence Theorem to S, we get

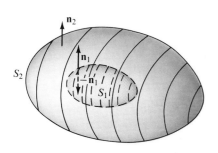

(14.72)

$$\iiint_E \operatorname{div} \mathbf{F}\, dV = \iint_S \mathbf{F} \cdot d\mathbf{S} = \iint_S \mathbf{F} \cdot \mathbf{n}\, dS$$

$$= \iint_{S_1} \mathbf{F} \cdot (-\mathbf{n}_1)\, dS + \iint_{S_2} \mathbf{F} \cdot \mathbf{n}_2\, dS$$

$$= -\iint_{S_1} \mathbf{F} \cdot d\mathbf{S} + \iint_{S_2} \mathbf{F} \cdot d\mathbf{S}$$

Figure 14.51

Let us apply this to the electric field (see Example 5 in Section 14.1):

$$\mathbf{E}(\mathbf{x}) = \frac{\varepsilon Q}{|\mathbf{x}|^3}\, \mathbf{x}$$

where S_1 is a small sphere with radius a and center the origin. You can verify that $\operatorname{div} \mathbf{E} = 0$ (see Exercise 21). Therefore Equation 14.72 gives

$$\iint_{S_2} \mathbf{E} \cdot d\mathbf{S} = \iint_{S_1} \mathbf{E} \cdot d\mathbf{S} + \iiint_E \operatorname{div} \mathbf{E}\, dV$$

$$= \iint_{S_1} \mathbf{E} \cdot d\mathbf{S} = \iint_{S_1} \mathbf{E} \cdot \mathbf{n}\, dS$$

The point of this calculation is that we can compute the surface integral over S_1 because S_1 is a sphere. The normal vector at $\mathbf{x}$ is $\mathbf{x}/|\mathbf{x}|$. Therefore

$$\mathbf{E} \cdot \mathbf{n} = \frac{\varepsilon Q}{|\mathbf{x}|^3}\, \mathbf{x} \cdot \left(\frac{\mathbf{x}}{|\mathbf{x}|}\right) = \frac{\varepsilon Q}{|\mathbf{x}|^4}\, \mathbf{x} \cdot \mathbf{x}$$

$$= \frac{\varepsilon Q}{|\mathbf{x}|^2} = \frac{\varepsilon Q}{a^2}$$

since the equation of S_1 is $|\mathbf{x}| = a$. Thus we have

$$\iint_{S_2} \mathbf{E} \cdot d\mathbf{S} = \iint_{S_1} \mathbf{E} \cdot \mathbf{n}\, dS = \frac{\varepsilon Q}{a^2} \iint_{S_1} dS$$

$$= \frac{\varepsilon Q}{a^2}\, A(S_1) = \frac{\varepsilon Q}{a^2}\, 4\pi a^2 = 4\pi\varepsilon Q$$

This shows that the electric flux of $\mathbf{E}$ is $4\pi\varepsilon Q$ through *any* closed surface S_2 that contains the origin. [This is a special case of Gauss's Law (Equation 14.59) for a single charge. The relationship between ε and ε_0 is $\varepsilon = 1/4\pi\varepsilon_0$.]

Another application of the Divergence Theorem occurs in fluid flow. Let $\mathbf{v}(x, y, z)$ be the velocity field of a fluid with constant density ρ. Then $\mathbf{F} = \rho\mathbf{v}$ is the rate of flow per unit area. If $P_0(x_0, y_0, z_0)$ is a point in the fluid and

B_a is a ball with center P_0 and very small radius a, then div $\mathbf{F}(P) \approx$ div $\mathbf{F}(P_0)$ for all points in B_a since div $\mathbf{F}$ is continuous. We approximate the flux over the boundary sphere S_a as follows:

$$\iint_{S_a} \mathbf{F} \cdot d\mathbf{S} = \iiint_{B_a} \operatorname{div} \mathbf{F}\, dV \approx \iiint_{B_a} \operatorname{div} \mathbf{F}(P_0)\, dV$$

$$= \operatorname{div} \mathbf{F}(P_0) V(B_a)$$

This approximation becomes better as $a \to 0$ and suggests that

(14.73)
$$\operatorname{div} \mathbf{F}(P_0) = \lim_{a \to 0} \frac{1}{V(B_a)} \iint_{S_a} \mathbf{F} \cdot d\mathbf{S}$$

Equation 14.73 says that div $\mathbf{F}(P_0)$ is the net rate of outward flux per unit volume at P_0. (This is the reason for the name *divergence*.) If div $\mathbf{F}(P) > 0$, the net flow is outward near P and P is called a **source.** If div $\mathbf{F}(P) < 0$, the net flow is inward near P and P is called a **sink.**

EXERCISES 14.9

In Exercises 1 and 2 verify that the Divergence Theorem is true for the given vector field $\mathbf{F}$ on the given region E.

1. $\mathbf{F}(x, y, z) = 3x\mathbf{i} + xy\mathbf{j} + 2xz\mathbf{k}$, E is the cube bounded by the planes $x = 0$, $x = 1$, $y = 0$, $y = 1$, $z = 0$, $z = 1$

2. $\mathbf{F}(x, y, z) = xz\mathbf{i} + yz\mathbf{j} + 3z^2\mathbf{k}$, E is the solid bounded by the paraboloid $z = x^2 + y^2$ and the plane $z = 1$

In Exercises 3–14 use the Divergence Theorem to calculate the surface integral $\iint_S \mathbf{F} \cdot d\mathbf{S}$.

3. $\mathbf{F}(x, y, z) = 3y^2z^3\mathbf{i} + 9x^2yz^2\mathbf{j} - 4xy^2\mathbf{k}$, S is the surface of the cube with vertices $(\pm 1, \pm 1, \pm 1)$.

4. $\mathbf{F}(x, y, z) = x^2y\mathbf{i} - x^2z\mathbf{j} + z^2y\mathbf{k}$, S is the surface of the rectangular box bounded by the planes $x = 0$, $x = 3$, $y = 0$, $y = 2$, $z = 0$, $z = 1$

5. $\mathbf{F}(x, y, z) = -xz\mathbf{i} - yz\mathbf{j} + z^2\mathbf{k}$, S is the ellipsoid $x^2/a^2 + y^2/b^2 + z^2/c^2 = 1$

6. $\mathbf{F}(x, y, z) = 3xy\mathbf{i} + y^2\mathbf{j} - x^2y^4\mathbf{k}$, S is the surface of the tetrahedron with vertices $(0, 0, 0)$, $(1, 0, 0)$, $(0, 1, 0)$, and $(0, 0, 1)$

7. $\mathbf{F}(x, y, z) = z \cos y\mathbf{i} + x \sin z\mathbf{j} + xz\mathbf{k}$, S is the surface of the tetrahedron bounded by the planes $x = 0$, $y = 0$, $z = 0$, and $2x + y + z = 2$

8. $\mathbf{F}(x, y, z) = (x + e^{y \tan z})\mathbf{i} + 3xe^{xz}\mathbf{j} + (\cos y - z)\mathbf{k}$, S is the surface with equation $x^4 + y^4 + z^4 = 1$

9. $\mathbf{F}(x, y, z) = x^3\mathbf{i} + y^3\mathbf{j} + z^3\mathbf{k}$, S is the sphere $x^2 + y^2 + z^2 = 1$

10. $\mathbf{F}(x, y, z) = x^3\mathbf{i} + 2xz^2\mathbf{j} + 3y^2z\mathbf{k}$, S is the surface of the solid bounded by the paraboloid $z = 4 - x^2 - y^2$ and the xy-plane

11. $\mathbf{F}(x, y, z) = ye^{z^2}\mathbf{i} + y^2\mathbf{j} + e^{xy}\mathbf{k}$, S is the surface of the solid bounded by the cylinder $x^2 + y^2 = 9$ and the planes $z = 0$ and $z = y - 3$

12. $\mathbf{F}(x, y, z) = (x^3 + y \sin z)\mathbf{i} + (y^3 + z \sin x)\mathbf{j} + 3z\mathbf{k}$, S is the surface of the solid bounded by the hemispheres $z = \sqrt{4 - x^2 - y^2}$, $z = \sqrt{1 - x^2 - y^2}$ and the plane $z = 0$

13. $\mathbf{F}(x, y, z) = xy^2\mathbf{i} + yz\mathbf{j} + zx^2\mathbf{k}$, S is the surface of the solid that lies between the cylinders $x^2 + y^2 = 1$ and $x^2 + y^2 = 4$ and between the planes $z = 1$ and $z = 3$

14. $\mathbf{F}(x, y, z) = (x^3 + yz)\mathbf{i} + x^2y\mathbf{j} + xy^2\mathbf{k}$, S is the surface of the solid bounded by the spheres $x^2 + y^2 + z^2 = 4$ and $x^2 + y^2 + z^2 = 9$

Prove the identities in Exercises 15–20 assuming that
S and E satisfy the conditions of the Divergence Theorem
and the scalar functions and components of the vector
fields have continuous second order partial derivatives.

15. $\iint_S \mathbf{a} \cdot \mathbf{n}\, dS = 0$, where $\mathbf{a}$ is a constant vector

16. $V(E) = \frac{1}{3}\iint_S \mathbf{F} \cdot d\mathbf{S}$, where $\mathbf{F}(x, y, z) = x\mathbf{i} + y\mathbf{j} + z\mathbf{k}$

17. $\iint_S \text{curl}\,\mathbf{F} \cdot d\mathbf{S} = 0$

18. $\iint_S D_n f\, dS = \iiint_E \nabla^2 f\, dV$

19. $\iint_S (f\,\nabla g) \cdot \mathbf{n}\, dS = \iiint_E (f\,\nabla^2 g + \nabla f \cdot \nabla g)\, dV$

20. $\iint_S (f\,\nabla g - g\,\nabla f) \cdot \mathbf{n}\, dS = \iiint_E (f\,\nabla^2 g - g\,\nabla^2 f)\, dV$

21. Verify that div $\mathbf{E} = 0$ for the electric field
$\mathbf{E}(\mathbf{x}) = \varepsilon Q \mathbf{x}/|\mathbf{x}|^3$.

SECTION 14.10

Summary

The main results of this chapter are all higher-dimensional analogues of the Fundamental Theorem of Calculus. To help you remember them, we collect them together here (without hypotheses) so that you can see more easily their essential similarity. Notice that in each case we have an integral of a "derivative" over a region on the left side, and the right side involves the values of the original function only on the *boundary* of the region.

Fundamental Theorem of Calculus	$\displaystyle \int_a^b F'(x)\, dx = F(b) - F(a)$	
Fundamental Theorem for Line Integrals	$\displaystyle \int_C \nabla f \cdot d\mathbf{r} = f(\mathbf{r}(b)) - f(\mathbf{r}(a))$	
Green's Theorem	$\displaystyle \iint_D \left(\frac{\partial Q}{\partial x} - \frac{\partial P}{\partial y}\right) dA = \int_C P\, dx + Q\, dy$	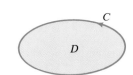
Stokes' Theorem	$\displaystyle \iint_S \text{curl}\,\mathbf{F} \cdot d\mathbf{S} = \int_C \mathbf{F} \cdot d\mathbf{r}$	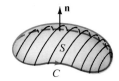
Divergence Theorem	$\displaystyle \iiint_E \text{div}\,\mathbf{F}\, dV = \iint_S \mathbf{F} \cdot d\mathbf{S}$	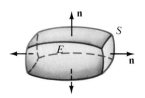

REVIEW OF CHAPTER 14

Define, state, or discuss the following:

1. Vector field
2. Conservative vector field
3. Potential function
4. Line integral of a scalar function with respect to arc length
5. Line integral of a scalar function with respect to x, y, and z
6. Line integral of a vector field
7. Fundamental theorem for line integrals
8. Independence of path
9. Green's Theorem
10. Curl
11. Divergence
12. Parametric surface
13. Normal vector to a parametric surface
14. Surface area of a parametric surface
15. Formula for the surface area of a graph $z = f(x, y)$
16. Surface integral of a scalar function
17. Oriented surface
18. Surface integral of a vector field
19. Stokes' Theorem
20. The Divergence Theorem

REVIEW EXERCISES FOR CHAPTER 14

In Exercises 1–9 evaluate the given line integral.

1. $\int_C y \, ds$, where C is the arc of the parabola $y^2 = 2x$ from $(0, 0)$ to $(2, 2)$

2. $\int_C yz^2 \, ds$, where C is the line segment from $(-1, 1, 3)$ to $(0, 3, 5)$

3. $\int_C x^3 z \, ds$, where C is given by $x = 2 \sin t$, $y = t$, $z = 2 \cos t$, $0 \leqslant t \leqslant \pi/2$

4. $\int_C xy \, dx + y \, dy$, where C is the sine curve $y = \sin x$, $0 \leqslant x \leqslant \pi/2$

5. $\int_C x^3 y \, dx - x \, dy$, where C is the circle $x^2 + y^2 = 1$ with counterclockwise orientation

6. $\int_C x \sin y \, dx + xyz \, dz$, where C is given by $\mathbf{r}(t) = t\mathbf{i} + t^2\mathbf{j} + t^3\mathbf{k}$, $0 \leqslant t \leqslant 1$

7. $\int_C y \, dx + z \, dy + x \, dz$, where C consists of the line segments from $(0, 0, 0)$ to $(1, 1, 2)$ and from $(1, 1, 2)$ to $(3, 1, 4)$

8. $\int_C \mathbf{F} \cdot d\mathbf{r}$, where $\mathbf{F}(x, y) = x^2 y \mathbf{i} + e^y \mathbf{j}$ and C is given by $\mathbf{r}(t) = t^2 \mathbf{i} - t^3 \mathbf{j}$, $0 \leqslant t \leqslant 1$

9. $\int_C \mathbf{F} \cdot d\mathbf{r}$ where $\mathbf{F}(x, y, z) = (x + y)\mathbf{i} + z\mathbf{j} + x^2 y \mathbf{k}$ and C is given by $\mathbf{r}(t) = 2t\mathbf{i} + t^2\mathbf{j} + t^4\mathbf{k}$, $0 \leqslant t \leqslant 1$

10. Find the work done by the force field $\mathbf{F}(x, y, z) = z\mathbf{i} + x\mathbf{j} + y\mathbf{k}$ in moving a particle from the point $(3, 0, 0)$ to the point $(0, \pi/2, 3)$ (a) along a straight line, (b) along the helix $x = 3 \cos t$, $y = t$, $z = 3 \sin t$.

In Exercises 11 and 12 show that $\mathbf{F}$ is a conservative vector field. Then find a function f such that $\mathbf{F} = \nabla f$.

11. $\mathbf{F}(x, y) = \sin y \mathbf{i} + (x \cos y + \sin y)\mathbf{j}$

12. $\mathbf{F}(x, y, z) = (2xy^3 + z^2)\mathbf{i} + (3x^2 y^2 + 2yz)\mathbf{j} + (y^2 + 2xz)\mathbf{k}$

In Exercises 13 and 14 show that $\mathbf{F}$ is conservative and use this fact to evaluate $\int_C \mathbf{F} \cdot d\mathbf{r}$ along the given curve.

13. $\mathbf{F}(x, y) = (2x + y^2 + 3x^2 y)\mathbf{i} + (2xy + x^3 + 3y^2)\mathbf{j}$, C is the arc of the curve $y = x \sin x$ from $(0, 0)$ to $(\pi, 0)$

14. $\mathbf{F}(x, y, z) = yz(2x + y)\mathbf{i} + xz(x + 2y)\mathbf{j} + xy(x + y)\mathbf{k}$, C is given by $\mathbf{r}(t) = (1 + t)\mathbf{i} + (1 + 2t^2)\mathbf{j} + (1 + 3t^3)\mathbf{k}$, $0 \leqslant t \leqslant 1$

15. Verify that Green's Theorem is true for the line integral $\int_C xy \, dx + x^2 \, dy$ where C is the triangle with vertices $(0, 0)$, $(1, 0)$, and $(1, 2)$.

16. Use Green's theorem to evaluate $\int_C (1 + \tan y) \, dx + (x^2 + e^y) \, dy$ where C is the positively oriented boundary of the region enclosed by the curves $y = \sqrt{x}$, $x = 1$, and $y = 0$.

17. Use Green's Theorem to evaluate $\int_C x^2 y \, dx - xy^2 \, dy$ where C is the circle $x^2 + y^2 = 4$ with counterclockwise orientation.

18. Find curl $\mathbf{F}$ and div $\mathbf{F}$ if $\mathbf{F}(x, y, z) = x^2 z \mathbf{i} + 2x \sin y \mathbf{j} + 2z \cos y \mathbf{k}$.

19. Show that there is no vector field $\mathbf{G}$ such that curl $\mathbf{G} = 2x\mathbf{i} + 3yz\mathbf{j} - xz^2\mathbf{k}$.

20. Show that, under conditions to be stated on the vector fields **F** and **G**,

$$\text{curl}(\mathbf{F} \times \mathbf{G}) = \mathbf{F} \, \text{div} \, \mathbf{G} - \mathbf{G} \, \text{div} \, \mathbf{F} + (\mathbf{G} \cdot \nabla)\mathbf{F}$$
$$- (\mathbf{F} \cdot \nabla)\mathbf{G}$$

21. Find an equation of the tangent plane at the point $(4, -2, 1)$ to the parametric surface given by $\mathbf{r}(u, v) = v^2\mathbf{i} - uv\mathbf{j} + u^2\mathbf{k}$.

22. Find the area of the part of the surface $z = x^2 + 2y$ that lies above the triangle with vertices $(0, 0)$, $(1, 0)$, and $(1, 2)$.

In Exercises 23–26 evaluate the surface integral.

23. $\iint_S z \, dS$, where S is the part of the paraboloid $z = x^2 + y^2$ that lies under the plane $z = 4$

24. $\iint_S (x^2 z + y^2 z) \, dS$, where S is the part of the plane $z = 4 + x + y$ that lies inside the cylinder $x^2 + y^2 = 4$

25. $\iint_S \mathbf{F} \cdot d\mathbf{S}$, where $\mathbf{F}(x, y, z) = xz\mathbf{i} - 2y\mathbf{j} + 3x\mathbf{k}$ and S is the sphere $x^2 + y^2 + z^2 = 4$ with outward orientation

26. $\iint_S \mathbf{F} \cdot d\mathbf{S}$, where $\mathbf{F}(x, y, z) = x^2\mathbf{i} + xy\mathbf{j} + z\mathbf{k}$ and S is the part of the paraboloid $z = x^2 + y^2$ below the plane $z = 1$ with upward orientation

27. Verify that Stokes' Theorem is true for the vector field $\mathbf{F}(x, y, z) = x^2\mathbf{i} + y^2\mathbf{j} + z^2\mathbf{k}$, where S is the part of the paraboloid $z = 1 - x^2 - y^2$ that lies above the xy-plane and S has upward orientation.

28. Use Stokes' Theorem to evaluate $\iint_S \text{curl} \, \mathbf{F} \cdot d\mathbf{S}$, where $\mathbf{F}(x, y, z) = x^2yz\mathbf{i} + yz^2\mathbf{j} + z^3 e^{xy}\mathbf{k}$, S is the part of the sphere $x^2 + y^2 + z^2 = 5$ that lies above the plane $z = 1$, and S is oriented upward.

29. Use Stokes' Theorem to evaluate $\int_C \mathbf{F} \cdot d\mathbf{r}$, where $\mathbf{F}(x, y, z) = xy\mathbf{i} + yz\mathbf{j} + zx\mathbf{k}$ and C is the triangle with vertices $(1, 0, 0)$, $(0, 1, 0)$, and $(0, 0, 1)$, oriented counterclockwise as viewed from above.

30. Use the Divergence Theorem to calculate the surface integral $\iint_S \mathbf{F} \cdot d\mathbf{S}$, where $\mathbf{F}(x, y, z) = x^3\mathbf{i} + y^3\mathbf{j} + z^3\mathbf{k}$ and S is the surface of the solid bounded by the cylinder $x^2 + y^2 = 1$ and the planes $z = 0$ and $z = 2$.

31. Verify that the Divergence Theorem is true for the vector field $\mathbf{F}(x, y, z) = x\mathbf{i} + y\mathbf{j} + z\mathbf{k}$, where E is the unit ball $x^2 + y^2 + z^2 \leqslant 1$.

32. Compute the outward flux of

$$\mathbf{F}(x, y, z) = (x\mathbf{i} + y\mathbf{j} + z\mathbf{k})/(x^2 + y^2 + z^2)^{3/2}$$

through the ellipsoid $4x^2 + 9y^2 + 6z^2 = 36$.

33. Let $\mathbf{F}(x, y, z) = (3x^2yz - 3y)\mathbf{i} + (x^3z - 3x)\mathbf{j} + (x^3y + 2z)\mathbf{k}$. Evaluate $\int_C \mathbf{F} \cdot d\mathbf{r}$, where C is the curve

with initial point $(0, 0, 2)$ and terminal point $(0, 3, 0)$ shown in the figure.

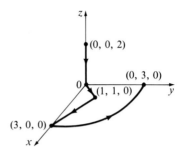

Figure for Exercise 33

34. Let

$$\mathbf{F}(x, y) = \frac{(2x^3 + 2xy^2 - 2y)\mathbf{i} + (2y^3 + 2x^2y + 2x)\mathbf{j}}{x^2 + y^2}$$

Evaluate $\oint_C \mathbf{F} \cdot d\mathbf{r}$ where C is shown in the figure.

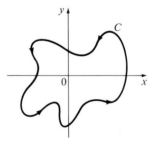

Figure for Exercise 34

35. Find $\iint_S \mathbf{F} \cdot \mathbf{n} \, dS$, where $\mathbf{F}(x, y, z) = x\mathbf{i} + y\mathbf{j} + z\mathbf{k}$ and S is the outwardly oriented surface shown in the figure (the boundary surface of a cube with a unit corner cube removed).

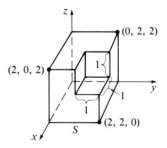

Figure for Exercise 35

36. If the components of **F** have continuous second partial derivatives and S is the boundary surface of a simple solid region, show that $\iint_S \text{curl} \, \mathbf{F} \cdot d\mathbf{S} = 0$.

15 Differential Equations

Among all the mathematical
disciplines the theory of differential
equations is the most important. It
furnishes the explanation of all those
elementary manifestations of nature
which involve time.

Sophus Lie

Throughout this book we have come into contact now and then with differential equations. For instance, in Section 6.7 we saw that the only solutions of the equation $y' = ky$ are of the form $y = Ce^{kx}$ and in Section 8.8 we learned how to solve separable differential equations. We encountered Laplace's equation and other partial differential equations in Section 12.3. In this chapter we give an introduction to the general study of differential equations by showing how to solve several basic types of equations and how to apply them.

Basic Concepts

There are two kinds of differential equations. An **ordinary differential equation** involves an unknown function of a single variable and some of its derivatives (ordinary derivatives). A **partial differential equation** involves an unknown function of two or more variables and some of its partial derivatives. The **order** of a differential equation is the order of the highest derivative that appears in the equation. Thus

$$x \frac{dy}{dx} = e^{xy}$$

is an ordinary differential equation of order 1 and

$$y''' - 2xy' + y = \sin x$$

is a third order ordinary differential equation, whereas

$$\frac{\partial^2 u}{\partial x^2} + 2xy \frac{\partial^2 u}{\partial x \, \partial y} - \frac{\partial^2 u}{\partial y^2} = 0$$

is a second order partial differential equation. We shall be studying only ordinary differential equations in this chapter, so when we talk about a differential equation we mean an ordinary differential equation.

A first order differential equation has the form

(15.1)
$$\frac{dy}{dx} = F(x, y)$$

where F is some function of the two variables x and y. The special case where F can be factored as a function of x times a function of y, that is, $F(x, y) = g(x)f(y)$, is called a **separable equation** and we learned how to solve this type of equation in Section 8.8. (That section should be read or reviewed at this time.)

An **initial-value problem** for a first order differential equation consists of finding a solution of Equation 15.1 that also satisfies an **initial condition** of the form

(15.2)
$$y(x_0) = y_0$$

Such a solution is a function ϕ that satisfies both

$$\phi'(x) = F(x, \phi(x)) \qquad \text{for all } x \text{ in some interval}$$

and
$$\phi(x_0) = y_0$$

It is proved in advanced books on differential equations that if F and $\partial F/\partial y$ are continuous on an open region of the xy-plane and $(x_0, y_0) \in D$, then there exists a unique solution to the initial-value problem given by Equations 15.1 and 15.2.

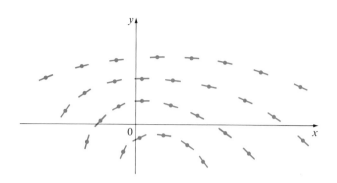

Figure 15.1

DIRECTION FIELD FOR $y' = F(x, y)$

Even if it is not possible to solve explicitly a differential equation of the form of Equation 15.1, we can still get a rough picture of what the graphs of the solutions look like. (These are called **solution curves.**) At any point (x, y) in the domain of F, we know that the number $F(x, y) = dy/dx$ represents the slope of the solution curve. Thus, at each point (x, y), the differential equation tells us the direction in which the curve is proceeding. If we draw a short line segment with slope $F(x, y)$ at a large number of points (x, y), the result is called a **direction field** and allows us to visualize the general shape of the solution curves. The more line segments we draw, the clearer the picture becomes (see Figure 15.1).

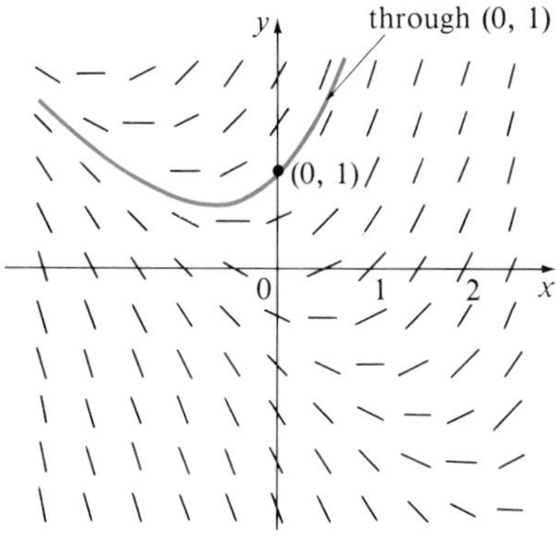

solution curve through (0, 1)

Figure 15.2

DIRECTION FIELD FOR $y' = x + y$

● **Example 1**   (a) Sketch the direction field for the differential equation $y' = x + y$.   (b) Use the direction field to sketch the solution curve that passes through the point $(0, 1)$.

Solution   (a)   We first compute the slope at several points in the following chart.

x	0	0	0	0	0	0.5	0.5	0.5	0.5	0.5	1	1	1	$\cdots$
y	0	0.5	1	-0.5	-1	0	0.5	1	-0.5	-1	0	1	-1	$\cdots$
$y' = x + y$	0	0.5	1	-0.5	-1	0.5	1	1.5	0	-0.5	1	2	0	$\cdots$

Then we draw line segments with these slopes at these points in Figure 15.2.
    (b)   We sketch the solution curve through $(0, 1)$ in Figure 15.2 by following the direction field.   ●

● **Example 2**   (a) Sketch the direction field for the differential equation $y' = -x/4y$.   (b) Solve the differential equation and sketch the solution curves.

Solution   (a)   By computing slopes as in Example 1, we sketch the direction field in Figure 15.3(a). Notice that the slope is constant on lines through the origin.

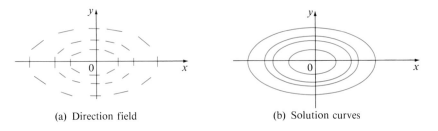

(a) Direction field   (b) Solution curves

Figure 15.3
$y' = -x/4y$

(b)   Since the differential equation is separable we can solve it by the method of Section 8.8:

$$\frac{dy}{dx} = -\frac{x}{4y}$$

$$4y\,dy = -x\,dx$$

$$\int 4y\,dy = -\int x\,dx$$

$$2y^2 = -\frac{x^2}{2} + C$$

$$\frac{x^2}{4} + y^2 = C_1 \qquad \left(C_1 = \frac{C}{2}\right)$$

We recognize these solutions as being a family of concentric ellipses, which are sketched in Figure 15.3(b). Notice the similarity to the direction field in Figure 15.3(a).   ●

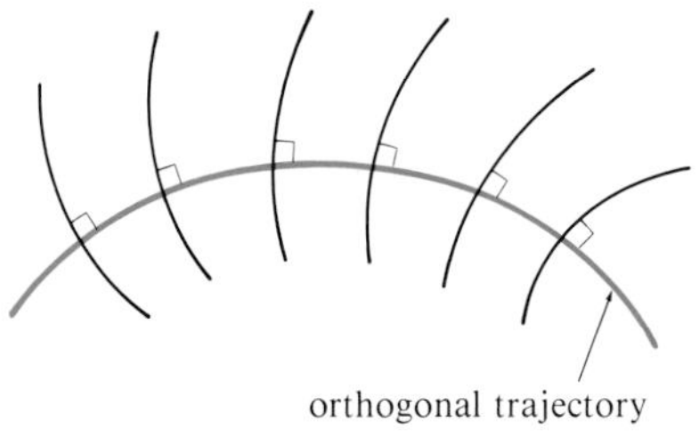

Orthogonal Trajectories

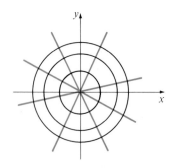

orthogonal trajectory

Figure 15.4

An **orthogonal trajectory** of a family of curves is a curve that intersects each curve of the family orthogonally, that is, at right angles (see Figure 15.4). For instance, each member of the family $y = mx$ of straight lines through the origin is an orthogonal trajectory of the family $x^2 + y^2 = r^2$ of concentric circles with center the origin (see Figure 15.5). We say that the two families are orthogonal trajectories of each other.

● **Example 3**   Find the orthogonal trajectories of the family of curves $x = ky^2$, where k is an arbitrary constant.

Solution   The curves $x = ky^2$ form a family of parabolas whose axis of symmetry is the x-axis. The first step is to find a single differential equation that is satisfied by all members of the family. If we differentiate $x = ky^2$, we get

$$1 = 2ky\frac{dy}{dx} \qquad \text{or} \qquad \frac{dy}{dx} = \frac{1}{2ky}$$

This is a differential equation but it depends on k. To eliminate k we note that, from the equation of the general parabola $x = ky^2$, we have $k = x/y^2$

Figure 15.5

and so the differential equation can be written as

$$\frac{dy}{dx} = \frac{1}{2ky} = \frac{1}{2\dfrac{x}{y^2}\, y}$$

or
$$\frac{dy}{dx} = \frac{y}{2x}$$

This means that the slope of the tangent line at any point (x, y) on one of the parabolas is $y' = y/2x$. On an orthogonal trajectory the slope of the tangent line must be the negative reciprocal of this slope. Therefore the orthogonal trajectories must satisfy the differential equation

$$\frac{dy}{dx} = -\frac{2x}{y}$$

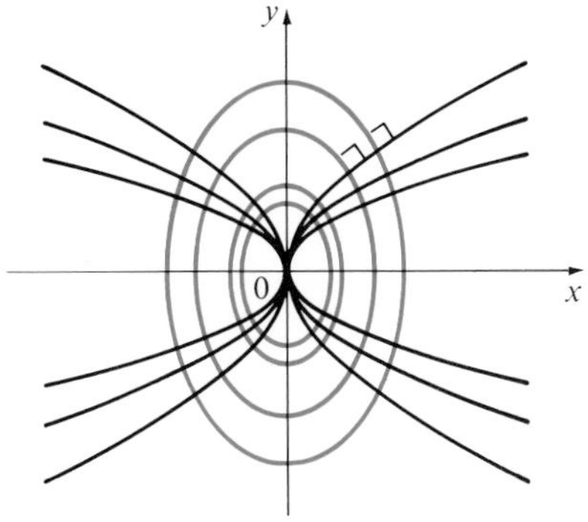

This differential equation is separable and we solve it as follows:

$$\int y\, dy = -\int 2x\, dx$$

$$\frac{y^2}{2} = -x^2 + C$$

(15.3)
$$x^2 + \frac{y^2}{2} = C$$

Figure 15.6

where C is an arbitrary positive constant. Thus the orthogonal trajectories are the family of ellipses given by Equation 15.3 and sketched in Figure 15.6.

●

Orthogonal trajectories occur in various branches of physics. For example, in an electrostatic field the lines of force are orthogonal to the lines of constant potential. And the streamlines in aerodynamics are orthogonal trajectories of the velocity-equipotential curves.

EXERCISES 15.1

In Exercises 1–4 sketch the direction field of the given differential equation. Then use it to sketch a solution curve that passes through the given point.

1. $y' = y^2, (0, 1)$ **2.** $y' = x^2 + y, (1, 1)$

3. $y' = x^2 + y^2, (0, 0)$ **4.** $y' = y(4 - y), (0, 1)$

In Exercises 5–8 (a) sketch the direction field, (b) sketch some solution curves without solving the differential

equation, (c) solve the differential equation, and (d) sketch the solutions obtained in part (c) and compare with those from part (b).

5. $y' = \dfrac{1}{y}$ **6.** $y' = xy$

7. $y' = -\dfrac{y}{x}$ **8.** $y' = \dfrac{x^2}{y}$

In Exercises 9–14 find the orthogonal trajectories of the given family of curves. Sketch several members of each family.

9. $y = kx^2$

10. $x^2 - y^2 = k$

11. $y = (x + k)^{-1}$

12. $y = kx^3$

13. $x^2 - 2y^2 = k$

14. $y = ke^{-x}$

SECTION 15.2

Homogeneous Equations

A first order differential equation $y' = f(x, y)$ is called **homogeneous** if $f(x, y)$ can be written as $g(y/x)$ where g is a function of a single variable. For instance, the differential equation

$$\frac{dy}{dx} = \frac{x^2 - xy + y^2}{x^2 - y^2} + \ln x - \ln y + \frac{x + y}{x + 2y}$$

is homogeneous because it can be written as

$$\frac{dy}{dx} = \frac{1 - \left(\dfrac{y}{x}\right) + \left(\dfrac{y}{x}\right)^2}{1 - \left(\dfrac{y}{x}\right)^2} - \ln\left(\frac{y}{x}\right) + \frac{1 + \dfrac{y}{x}}{1 + 2\left(\dfrac{y}{x}\right)}$$

However the differential equation

$$\frac{dy}{dx} = \frac{x^2 - xy^2 + y^2}{x^2 - y^2}$$

is *not* homogeneous because the right side cannot be written as a function of y/x.

A homogeneous differential equation $y' = g(y/x)$ can always be transformed into a separable equation by making the change of variable

$$v = \frac{y}{x}$$

Then

$$y = xv$$

and so

$$y' = v + xv'$$

Thus the differential equation $y' = g(y/x)$ becomes

$$v + xv' = g(v)$$

or

$$v' = \frac{g(v) - v}{x}$$

After we solve this separable differential equation for v as a function of x, we have the solution $y = xv(x)$ of the original differential equation.

● **Example 1**   Solve the differential equation

$$y' = \frac{xy + y^2}{x^2}$$

Solution   First we notice that this equation is homogeneous since we can rewrite it as

$$y' = \left(\frac{y}{x}\right) + \left(\frac{y}{x}\right)^2$$

Therefore we make the substitution $v = y/x$, which gives

$$y = xv \qquad y' = v + xv'$$

So the differential equation becomes

$$v + xv' = v + v^2$$

or
$$v' = \frac{v^2}{x}$$

We solve this separable differential equation in the usual way:

$$\frac{dv}{dx} = \frac{v^2}{x}$$

$$\int \frac{dv}{v^2} = \int \frac{dx}{x} \qquad v \neq 0$$

$$-\frac{1}{v} = \ln|x| + C$$

$$v = -\frac{1}{\ln|x| + C}$$

Then the solution to the original differential equation is

$$y(x) = xv(x) = -\frac{x}{\ln|x| + C}$$

Notice that in solving the differential equation we had to rule out $v = 0$. But if $v = 0$, then $y = 0$, and we can verify directly that $y = 0$ gives another solution of the differential equation.   ●

EXERCISES 15.2

Solve the differential equations in Exercises 1–8.

1. $y' = \dfrac{x - y}{x}$

2. $y' = \dfrac{x + y}{x - y}$

5. $(x^2 - y^2)y' = 2xy$

6. $(y - 2x)y' = x - 2y$

3. $\dfrac{dy}{dx} = \dfrac{x^2 + y^2}{xy}$

4. $\dfrac{dy}{dx} = \dfrac{y^2 - x^2}{2xy}$

7. $xy' = y + xe^{y/x}$

8. $xy' \sin\dfrac{y}{x} = y \sin\dfrac{y}{x} - x$

Solve the initial-value problems in Exercises 9–12.

9. $x^2 + xy + y^2 - x^2y' = 0$, $y(1) = 0$

10. $xy' - y = \sqrt{x^2 - y^2}$, $x > 0$, $y(1) = 0$

11. $x^3 + y^3 = xy^2 \dfrac{dy}{dx}$, $y(1) = 2$

12. $xy' = y \ln y - y \ln x$, $y(1) = 1$

SECTION 15.3

First Order Linear Equations

A first order linear differential equation is of the form

(15.4)
$$\frac{dy}{dx} + P(x)y = Q(x)$$

where P and Q are continuous functions. This type of equation occurs frequently in various sciences, as we shall see.

The standard method for solving Equation 15.4 is to multiply both sides of the equation by a suitable function $I(x)$ called an integrating factor. We try to find I so that the left side of Equation 15.4, when multiplied by $I(x)$, becomes the derivative of the product $I(x)y$:

(15.5)
$$I(x)(y' + P(x)y) = (I(x)y)'$$

If we can find such a function I, then Equation 15.4 becomes

$$(I(x)y)' = I(x)Q(x)$$

Integrating both sides, we would have

$$I(x)y = \int I(x)Q(x)\,dx + C$$

so the solution would be

(15.6)
$$y(x) = \frac{1}{I(x)}\left[\int I(x)Q(x)\,dx + C\right]$$

To find such an I we expand Equation 15.5 and cancel terms:

$$I(x)y' + I(x)P(x)y = (I(x)y)' = I'(x)y + I(x)y'$$

$$I(x)P(x) = I'(x)$$

This is a separable differential equation for I, which we solve as in Section 8.8:

$$\int \frac{dI}{I} = \int P(x)\,dx$$

$$\ln|I| = \int P(x)\,dx$$

$$I = Ae^{\int P(x)\,dx}$$

where $A = \pm e^C$. We are looking for a particular integrating factor, not the most general one, so we take $A = 1$ and use

(15.7)
$$I(x) = e^{\int P(x)\,dx}$$

Thus a formula for the general solution to Equation 15.4 is provided by Equation 15.6 where I is given by Equation 15.7. Instead of memorizing this formula, however, we just remember the form of the integrating factor.

To solve the linear differential equation $y' + P(x)y = Q(x)$, multiply both sides by the function $I(x) = e^{\int P(x)\,dx}$.

● **Example 1**   Solve the differential equation $\dfrac{dy}{dx} + 3x^2 y = 6x^2$.

Solution   The given equation is linear since it has the form of Equation 15.4 with $P(x) = 3x^2$ and $Q(x) = 6x^2$. An integrating factor is

$$I(x) = e^{\int 3x^2\,dx} = e^{x^3}$$

Multiplying both sides of the differential equation by e^{x^3}, we get

$$e^{x^3}\frac{dy}{dx} + 3x^2 e^{x^3} y = 6x^2 e^{x^3}$$

or

$$\frac{d}{dx}(e^{x^3} y) = 6x^2 e^{x^3}$$

Integrating both sides, we have

$$e^{x^3} y = \int 6x^2 e^{x^3}\,dx = 2e^{x^3} + C$$

$$y = 2 + Ce^{-x^3} \qquad\qquad\qquad ●$$

● **Example 2**   Find the solution of the initial-value problem

$$x^2 y' + xy = 1 \qquad x > 0 \qquad y(1) = 2$$

Solution   We must first divide both sides by the coefficient of y' to put the differential equation into standard form:

(15.8)
$$y' + \frac{1}{x}y = \frac{1}{x^2} \qquad x > 0$$

The integrating factor is

$$I(x) = e^{\int(1/x)\,dx} = e^{\ln x} = x$$

Multiplication of Equation 15.8 by x gives

$$xy' + y = \frac{1}{x}$$

or

$$(xy)' = \frac{1}{x}$$

Then

$$xy = \int \frac{1}{x} \, dx = \ln x + C$$

and so

$$y = \frac{\ln x + C}{x}$$

Since $y(1) = 2$, we have

$$2 = \frac{\ln 1 + C}{1} = C$$

Therefore the solution to the initial-value problem is

$$y = \frac{\ln x + 2}{x}$$

● **Example 3**   Solve $y' + 2xy = 1$.

Solution   The given equation is in the standard form for a linear equation. Multiplying by the integrating factor

$$e^{\int 2x \, dx} = e^{x^2}$$

we get

$$e^{x^2}y' + 2xe^{x^2}y = e^{x^2}$$

or

$$(e^{x^2}y)' = e^{x^2}$$

Therefore

$$e^{x^2}y = \int e^{x^2} \, dx + C$$

Recall from Section 7.6 that $\int e^{x^2} \, dx$ cannot be expressed in terms of elementary functions. Nonetheless it is a perfectly good function and we can leave the answer as

$$y = e^{-x^2} \int e^{x^2} \, dx + Ce^{-x^2}$$

Another way of writing the solution is

$$y = e^{-x^2} \int_0^x e^{t^2} \, dt + Ce^{-x^2}$$

(Any number can be chosen for the lower limit of integration.)   ●

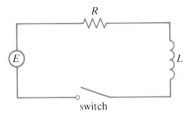

Figure 15.7

Application to Electric Circuits

The simple electric circuit shown in Figure 15.7 contains an electromotive force (usually a battery or generator) that produces a voltage of $E(t)$ volts (V) and a current of $I(t)$ amperes (A) at time t. The circuit also contains a resistor with a resistance of R ohms (Ω) and an inductor with an inductance of L henries (H).

Ohm's Law gives the drop in voltage due to the resistor as RI. The voltage drop due to the inductor is $L(dI/dt)$. One of Kirchhoff's laws says that the sum of the voltage drops is equal to the supplied voltage $E(t)$. Thus we have

$$(15.9) \qquad L\frac{dI}{dt} + RI = E(t)$$

which is a first order linear differential equation. The solution gives the current I at time t.

● **Example 4**   Suppose that in the simple circuit of Figure 15.7 the resistance is 12 Ω and the inductance is 4 H. If a battery gives a constant voltage of 60 V and the switch is closed when $t = 0$ so the current starts with $I(0) = 0$, find (a) $I(t)$, (b) the current after 1 s, and (c) the limiting value of the current.

Solution   (a)   If we put $L = 4$, $R = 12$, and $E(t) = 60$ in Equation 15.9, we obtain the initial-value problem

$$4\frac{dI}{dt} + 12I = 60 \qquad I(0) = 0$$

or

$$\frac{dI}{dt} + 3I = 15 \qquad I(0) = 0$$

Multiplying by the integrating factor $e^{\int 3dt} = e^{3t}$, we get

$$e^{3t}\frac{dI}{dt} + 3e^{3t}I = 15e^{3t}$$

$$\frac{d}{dt}(e^{3t}I) = 15e^{3t}$$

$$e^{3t}I = \int 15e^{3t}\,dt = 5e^{3t} + C$$

$$I(t) = 5 + Ce^{-3t}$$

Since $I(0) = 0$ we have $5 + C = 0$, so $C = -5$ and

$$I(t) = 5(1 - e^{-3t})$$

(b)   After 1 s the current is

$$I(1) = 5(1 - e^{-3}) \approx 4.75 \text{ A}$$

(c) $\lim\limits_{t \to \infty} I(t) = \lim\limits_{t \to \infty} 5(1 - e^{-3t})$

$$= 5 - 5 \lim_{t \to \infty} e^{-3t}$$

$$= 5 - 0 = 5$$

● **Example 5**   Suppose that the resistance and inductance remain as in Example 4 but, instead of the battery, we use a generator that produces a variable voltage of $E(t) = 60 \sin 30t$ volts. Find $I(t)$.

Solution   This time the differential equation becomes

$$4 \frac{dI}{dt} + 12I = 60 \sin 30t \qquad \text{or} \qquad \frac{dI}{dt} + 3I = 15 \sin 30t$$

The same integrating factor e^{3t} gives

$$\frac{d}{dt}(e^{3t}I) = e^{3t}\frac{dI}{dt} + 3e^{3t}I = 15e^{3t}\sin 30t$$

Using Formula 98 in the Table of Integrals, we have

$$e^{3t}I = \int 15e^{3t}\sin 30t\, dt = 15 \frac{e^{3t}}{909}(3\sin 30t - 30\cos 30t) + C$$

$$I = \frac{5}{101}(\sin 30t - 10\cos 30t) + Ce^{-3t}$$

Since $I(0) = 0$, we get

$$-\tfrac{50}{101} + C = 0$$

so $I(t) = \tfrac{5}{101}(\sin 30t - 10\cos 30t) + \tfrac{50}{101}e^{-3t}$ ●

EXERCISES 15.3

Solve the differential equations in Exercises 1–10.

1. $y' - 3y = e^x$

2. $y' + 4y = x$

3. $y' - 2xy = x$

4. $y' + (\cos x)y = \cos x$

5. $xy' + y = x \cos x$

6. $xy' + 2y = e^{x^2}$

7. $y' \cos x = y \sin x + \sin 2x$

8. $1 + xy = xy'$

9. $\dfrac{dy}{dx} + 2xy = x^2$

10. $\dfrac{dy}{dx} = x \sin 2x + y \tan x, \quad -\dfrac{\pi}{2} < x < \dfrac{\pi}{2}$

In Exercises 11–16 solve the initial-value problem.

11. $y' + y = x + e^x, \; y(0) = 0$

12. $xy' - 3y = x^2, \; x > 0, \; y(1) = 0$

13. $y' - 2xy = 2xe^{x^2}, \; y(0) = 3$

14. $(1 + x^2)y' + 2xy = 3\sqrt{x}, \; y(0) = 2$

15. $x^2 \dfrac{dy}{dx} + 2xy = \cos x$, $y(\pi) = 0$

16. $x \dfrac{dy}{dx} - \dfrac{y}{x+1} = x$, $y(1) = 0$

17. A **Bernoulli differential equation** [named after James Bernoulli (1654–1705)] is of the form

$$\frac{dy}{dx} + P(x)y = Q(x)y^n$$

Observe that, if $n = 0$ or 1, the Bernoulli equation is linear. For other values of n, show that the substitution $u = y^{1-n}$ transforms the Bernoulli equation into the linear equation

$$\frac{du}{dx} + (1 - n)P(x)u = (1 - n)Q(x)$$

In Exercises 18–20 use the method of Exercise 17 to solve the given differential equation.

18. $xy' + y = -xy^2$ **19.** $y' + \dfrac{2}{x}y = \dfrac{y^3}{x^2}$

20. $y' + y = xy^3$

21. In the circuit shown in Figure 15.7 a battery supplies a constant voltage of 40 V, the inductance is 2 H, the resistance is 10 Ω, and $I(0) = 0$.
  (a) Find $I(t)$.
  (b) Find the current after 0.1 s.

22. In the circuit shown in Figure 15.7 a generator supplies a voltage of $E(t) = 40 \sin 60t$ volts, the inductance is 1 H, the resistance is 20 Ω, and $I(0) = 1$ A.
  (a) Find $I(t)$.
  (b) Find the current after 0.1 s.

23. Figure 15.8 shows a circuit containing an electromotive force, a capacitor with a capacitance of C farads (F), and a resistor with a resistance of R ohms (Ω). The voltage drop across the capacitor is Q/C, where Q is the charge (in coulombs), so in this case Kirchhoff's Law gives

$$RI + \frac{Q}{C} = E(t)$$

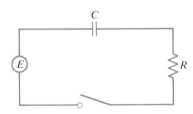

Figure 15.8

But $I = dQ/dt$ (see Example 2 in Section 3.7) so we have

$$R \frac{dQ}{dt} + \frac{1}{C}Q = E(t)$$

Suppose the resistance is 5 Ω, the capacitance is 0.05 F, a battery gives a constant voltage of 60 V, and the initial charge is $Q(0) = 0$ C. Find the charge and the current at time t.

24. In the circuit of Exercise 23, $R = 2$ Ω, $C = 0.01$ F, $Q(0) = 0$, and $E(t) = 10 \sin 60t$. Find the charge and the current at time t.

25. Psychologists interested in learning theory study **learning curves.** A learning curve is the graph of a function $P(t)$, the performance of someone learning a skill as a function of the training time t. The derivative dP/dt represents the rate at which performance improves. If M is the maximum level of performance of which the learner is capable, it is reasonable to assume that dP/dt is proportional to $M - P(t)$. (At first, learning is rapid. Then, as performance increases and approaches its maximal value, the rate of learning decreases.) Thus

$$\frac{dP}{dt} = k(M - P(t))$$

where k is a positive constant. Solve this linear differential equation and sketch the learning curve.

26. Two new workers were hired for an assembly line. Jim processed 25 units during the first hour and 45 units the second hour. Mark processed 35 units during the first hour and 50 units the second hour. Using the model of Exercise 25, estimate the maximum number of units per hour that each worker is capable of processing.

27. An object with mass m is dropped from rest and we assume that the air resistance is proportional to the speed of the object. If $s(t)$ is the distance dropped after t seconds, then the speed is $v = s'(t)$ and the acceleration is $a = v'(t)$. If g is the acceleration due to gravity, then the downward force on the object is $mg - cv$, where c is a positive constant, and Newton's Second Law gives

$$m \frac{dv}{dt} = mg - cv$$

  (a) Solve this differential equation to find the velocity at time t.
  (b) What is the limiting velocity?
  (c) Find the distance the object has fallen after t seconds.

Exact Equations

Suppose the equation

(15.10)
$$f(x, y) = C$$

defines y implicitly as a differentiable function of x. Then $y = y(x)$ satisfies a first order differential equation obtained by using the Chain Rule (12.28) to differentiate both sides of Equation 15.10 with respect to x:

(15.11)
$$f_x(x, y) + f_y(x, y)y' = 0$$

A differential equation of the form of Equation 15.11 is called *exact*.

Definition (15.12)

A first order differential equation of the form

$$P(x, y) + Q(x, y)\frac{dy}{dx} = 0$$

is called **exact** if there is a function $f(x, y)$ such that

$$f_x(x, y) = P(x, y) \qquad \text{and} \qquad f_y(x, y) = Q(x, y)$$

If the function f in Definition 15.12 is known, then the exact differential equation $P(x, y) + Q(x, y)y' = 0$ is easy to solve because

$$\frac{d}{dx} f(x, y(x)) = f_x(x, y) + f_y(x, y)y'$$

$$= P(x, y) + Q(x, y)y' = 0$$

Thus the solution is given implicitly by

(15.13)
$$f(x, y) = C$$

We may be able to solve Equation 15.13 for y as an explicit function of x.

How can we tell whether or not a differential equation of the form $P(x, y) + Q(x, y)y' = 0$ is exact? If we consider the vector field $\mathbf{F}(x, y) = \langle P(x, y), Q(x, y) \rangle$, then the condition for exactness in Definition 15.12 can be written as $\mathbf{F}(x, y) = \nabla f(x, y)$; that is, $\mathbf{F}$ is a conservative vector field. But we know from Theorems 14.22 and 14.23 that a vector field $\mathbf{F} = P\mathbf{i} + Q\mathbf{j}$ is conservative if and only if $\partial P/\partial y = \partial Q/\partial x$ (assuming these partial derivatives are continuous and the domain is simply-connected). Therefore we have the following convenient method for testing the exactness of a differential equation.

Theorem (15.14)

Suppose P and Q have continuous partial derivatives on a simply-connected domain. Then the differential equation

$$P(x, y) + Q(x, y)\frac{dy}{dx} = 0$$

is exact if and only if

$$\frac{\partial P}{\partial y} = \frac{\partial Q}{\partial x}$$

Furthermore, the procedures that we used in Section 14.3 for determining the (potential) function f such that $\nabla f = \mathbf{F}$ can also be used here to solve an exact differential equation.

● **Example 1**   Solve the differential equation

$$4x + 3y + 3(x + y^2)y' = 0$$

Solution   Here

$$P(x, y) = 4x + 3y \quad \text{and} \quad Q(x, y) = 3x + 3y^2$$

have continuous partial derivatives on R^2. Also

$$\frac{\partial P}{\partial y} = 3 = \frac{\partial Q}{\partial x}$$

so the differential equation is exact by Theorem 15.14. Thus there exists a function f such that

$$f_x(x, y) = 4x + 3y \qquad (1)$$

and
$$f_y(x, y) = 3x + 3y^2 \qquad (2)$$

To determine f we first integrate (1) with respect to x:

$$f(x, y) = 2x^2 + 3xy + g(y) \qquad (3)$$

(The constant of integration is a function of y.) Now we differentiate (3) with respect to y:

$$f_y(x, y) = 3x + g'(y) \qquad (4)$$

Comparing (2) and (4), we see that

$$g'(y) = 3y^2$$

and so
$$g(y) = y^3$$

(We do not need the arbitrary constant here.) Thus

$$f(x, y) = 2x^2 + 3xy + y^3$$

and by Equation 15.13 the solution is given implicitly by

$$2x^2 + 3xy + y^3 = C$$

As a check on our work we can verify this as follows:

$$\frac{d}{dx}(2x^2 + 3xy + y^3) = 4x + 3y + 3xy' + 3y^2y'$$

$$= (4x + 3y) + 3(x + y^2)y' = 0$$

from the differential equation. Thus

$$2x^2 + 3xy + y^3 = C \qquad \bullet$$

Integrating Factors

If the differential equation

$$P(x, y) + Q(x, y)\frac{dy}{dx} = 0$$

is not exact, it may be possible to find an integrating factor $I(x, y)$ such that, after multiplication by $I(x, y)$, the resulting equation

(15.15) $$I(x, y)P(x, y) + I(x, y)Q(x, y)\frac{dy}{dx} = 0$$

is exact.

To find such an integrating factor we use Theorem 15.14. Equation 15.15 will be exact if

$$\frac{\partial}{\partial y}(IP) = \frac{\partial}{\partial x}(IQ)$$

that is $$I_yP + IP_y = I_xQ + IQ_x$$

(15.16) or $$PI_y - QI_x = I(Q_x - P_y)$$

In general it is harder to solve this partial differential equation than the original differential equation. But it is sometimes possible to find I that is a function of x or y alone. For instance, suppose I is a function of x alone. Then $I_y = 0$, so Equation 15.16 becomes

(15.17) $$\frac{dI}{dx} = \frac{P_y - Q_x}{Q}I$$

If $(P_y - Q_x)/Q$ is a function of x alone, then Equation 15.17 is a first order linear (and separable) ordinary differential equation that can be solved for $I(x)$. Then Equation 15.15 will be exact and can be solved as in Example 1.

● **Example 2**   Solve the equation $2x + y^2 + xyy' = 0$.

Solution   Here

$$P(x, y) = 2x + y^2 \qquad Q(x, y) = xy$$

$$\frac{\partial P}{\partial y} = 2y \qquad\qquad \frac{\partial Q}{\partial x} = y$$

Since $\partial P/\partial y \neq \partial Q/\partial x$, the given equation is not exact. But

$$\frac{P_y - Q_x}{Q} = \frac{2y - y}{xy} = \frac{1}{x}$$

is a function of x alone, so by Equation 15.17 there is an integrating factor I that satisfies

$$\frac{dI}{dx} = \frac{I}{x}$$

Solving this equation as a separable differential equation, we get $I(x) = x$. (We do not need the most general integrating factor, just a particular one.) Multiplying the original equation by x, we get

(15.18)
$$2x^2 + xy^2 + x^2yy' = 0$$

If we let $p(x, y) = 2x^2 + xy^2 \qquad q(x, y) = x^2y$

then $$\frac{\partial p}{\partial y} = 2xy = \frac{\partial q}{\partial x}$$

so Equation 15.18 is now exact. Thus there is a function f such that

$$f_x(x, y) = 2x^2 + xy^2 \qquad f_y(x, y) = x^2y$$

Integrating the first of these equations we get

$$f(x, y) = \tfrac{2}{3}x^3 + \tfrac{1}{2}x^2y^2 + g(y)$$

so $$f_y(x, y) = x^2y + g'(y)$$

Comparison then gives $g'(y) = 0$, so g is a constant (which we can take to be 0). Therefore

$$f(x, y) = \tfrac{2}{3}x^3 + \tfrac{1}{2}x^2y^2$$

and the solution is given by

$$\tfrac{2}{3}x^3 + \tfrac{1}{2}x^2y^2 = C \qquad\qquad ●$$

EXERCISES 15.4

In Exercises 1–12 determine whether the given differential equation is exact. If so, solve it.

1. $2x + y + (x + 2y)y' = 0$

2. $2x - y + (x - 2y)y' = 0$

3. $3xy - 2 + (3y^2 - x^2)y' = 0$

4. $3x^2 - 2x + 3y + (3x - 2y)y' = 0$

5. $\sin y + (1 + x \cos y)y' = 0$

6. $y - e^x \cos y + (x + e^x \sin y)y' = 0$

7. $(x + y)e^{x/y} + \left(x - \dfrac{x^2}{y} \right)e^{x/y}y' = 0$

8. $e^x + y \cos x + (e^y - y \sin x)y' = 0$

9. $x \ln y\, dx - (x + y \ln x)\, dy = 0$

10. $(2x^3 y^2 - \tfrac{1}{2}e^{2y})\, dx + (x^4 y - xe^{2y})\, dy = 0$

11. $\dfrac{1}{y} + \dfrac{2y}{x^3} = \left(\dfrac{x}{y^2} + \dfrac{1}{x^2} \right)\dfrac{dy}{dx}$

12. $\dfrac{dy}{dx} = \dfrac{\cos y + y \cos x}{x \sin y - \sin x}$

Solve the initial-value problems in Exercises 13–16.

13. $3x^2 + 2xy + 3y^2 + (x^2 + 6xy)y' = 0$, $y = 2$ when $x = 1$

14. $3x^2 y^2 + 8xy^5 + (2x^3 y + 20x^2 y^4)y' = 0$, $y = 1$ when $x = 2$

15. $1 + y \cos xy + (x \cos xy)y' = 0$, $y = 0$ when $x = 1$

16. $\ln y + 3y^2 + \left(\dfrac{x}{y} + 6xy \right)y' = 0$, $y > 0$, $y = 1$ when $x = 1$

In Exercises 17–20 show that the given equation is not exact but becomes exact when multiplied by the given integrating factor. Then solve the equation.

17. $y^2 + (1 + xy)y' = 0$, $I(x, y) = e^{xy}$

18. $y(x + y) - x^2 y' = 0$, $I(x, y) = \dfrac{1}{xy^2}$

19. $y + y^3 + (x + x^3)y' = 0$, $I(x, y) = \dfrac{1}{(1 + x^2 + y^2)^{3/2}}$

20. $2y \cos x - xy \sin x + (2x \cos x)y' = 0$, $I(x, y) = xy$

In Exercises 21–23 find an integrating factor and thus solve the given equation.

21. $3xy + 2y^2 + (x^2 + 2xy)y' = 0$

22. $1 - xy + x(y - x)y' = 0$, $x > 0$

23. $2xy + 3x^2 y + 3y^2 + (x^2 + 2y)y' = 0$

24. Show that if $(Q_x - P_y)/P$ is a function of y alone then Equation 15.16 enables us to find an integrating factor for $P + Qy' = 0$. Use this method to solve the equation
$$y - 6x^2 y^3 + (2x - 8x^3 y^2)y' = 0$$

25. Show that every separable differential equation is exact.

Strategy for Solving First Order Equations

In solving first order differential equations we used the technique for separable equations in Section 8.8 and the method for linear equations in Section 15.3. We also developed methods for solving homogeneous equations in Section 15.2 and exact equations in Section 15.4. In this section we present a miscellaneous collection of first order differential equations and part of the problem will be to recognize which technique should be used on which equation.

As with the strategy of integration (Section 7.6) and the strategy of testing series (Section 10.7), the main idea is to classify the equation according to its *form*. Here, however, the important thing is not so much the form of the functions involved as it is the form of the equation itself.

Recall that a **separable** equation can be written in the form

(15.19)
$$\frac{dy}{dx} = g(x)f(y)$$

that is, the expression for dy/dx can be factored as a product of a function of x and a function of y.

A **linear** equation can be put into the form

(15.20)
$$\frac{dy}{dx} + P(x)y = Q(x)$$

A **homogeneous** equation can be expressed in the form

(15.21)
$$\frac{dy}{dx} = g\left(\frac{y}{x}\right)$$

An **exact** equation has the form

(15.22)
$$P(x, y) + Q(x, y)\frac{dy}{dx} = 0$$

where $\partial P/\partial y = \partial Q/\partial x$.

If an equation has none of these forms, we can, as a last resort, attempt to find an **integrating factor** and thus make the equation exact.

In each of these cases, some preliminary algebra may be required in order to put a given equation into one of the preceding forms. (This step is analogous to Step 1 of the strategy for integration: algebraic simplification.)

It may happen that a given equation is of more than one type. For instance, the equation

$$xy' = y$$

is separable because dy/dx can be written as

$$\frac{dy}{dx} = \left(\frac{1}{x}\right)y \qquad \textit{(Compare with Equation 15.19.)}$$

It is also linear since we can write the equation as

$$\frac{dy}{dx} + \left(-\frac{1}{x}\right)y = 0 \qquad \textit{(Compare with Equation 15.20.)}$$

Furthermore it is also homogeneous because we can write it as

$$\frac{dy}{dx} = \frac{y}{x} \qquad \textit{(Compare with Equation 15.21.)}$$

In such a case we could solve the equation using any one of the corresponding methods, although one of the methods might be easier than the others.

In the following examples we identify the type of each equation without working out the details of the solutions.

● **Example 1** $y' = 1 - x - y + xy$

Initially this equation may not appear to be in any of the forms of Equations 15.19, 15.20, 15.21, or 15.22, but observe that we can factor the right side and therefore write the equation as

$$y' = (1 - x)(1 - y)$$

We now recognize the equation as being separable and we can solve it using the methods of Section 8.8.   ●

● **Example 2** $x^2 - y^2 + 2xyy' = 0$

The equation is clearly not separable, nor is it linear. Since $P_y = -2y$ and $Q_x = 2y$, it is not exact. But if we solve for y' we get

$$y' = \frac{y^2 - x^2}{2xy} = \frac{1}{2}\left[\frac{y}{x} - \frac{1}{y/x}\right]$$

which shows that y' is a function of y/x and the equation is homogeneous (see Equation 15.21). (We could have anticipated this because the expressions x^2, y^2, and $2xy$ are all of degree 2.) The change of variable $v = y/x$ will convert the equation into a separable equation.   ●

● **Example 3** $y' = -\dfrac{3x^2 + 2xy^3}{3x^2 y^2}$

This equation is not separable, linear, or homogeneous. We suspect it might be exact, so we write it in the form

$$(3x^2 + 2xy^3) + (3x^2 y^2)y' = 0$$

If $P(x, y) = 3x^2 + 2xy^3$ and $Q(x, y) = 3x^2 y^2$, then $P_y = 6xy^2 = Q_x$. Therefore the equation is indeed exact and can be solved by the methods of Section 15.4.   ●

● **Example 4** $(x + 1)y - xy' = x^3 - x^2$

If we put the equation in the form

$$y' - \left(\frac{x + 1}{x}\right)y = x - x^2$$

we recognize it as having the form of Equation 15.20. It is therefore linear and can be solved using the integrating factor

$$e^{-\int(1 + 1/x)\,dx}$$ ●

EXERCISES 15.5

Classify and solve the following differential equations.

1. $y' = y + \sin x$

2. $y' = y \sin x$

3. $x - e^x + 4y^3 y' = 0$

4. $3x^2 y^2 + 2x^3 yy' = 0$

5. $(x^2 - 2xy - y^2)y' = x^2 + 2xy - y^2$

6. $\dfrac{dy}{dx} = \dfrac{e^x - y}{x}$

7. $(2xy - 3\tan x)y' = 3y\sec^2 x - y^2$

8. $xy' + y^2 = 1$

9. $x - (x^2y + y)y' = 0$

10. $y' = \dfrac{\cos y}{x\sin y - 1}$

11. $xy' = \sqrt{1 + x^2} + 2y$

12. $x(x + y)^2 y' = y(x^2 + xy + y^2)$

13. $y^2\sqrt{1 + x^3} + y' = 0$

14. $xy' = 2 - x^2 + 2y^2 - x^2y^2$

15. $2(x + yy') + e^y(1 + xy') = 0$

16. $xy' = x^2 + x^3 + y$

17. $[\sin(xy) + xy\cos(xy)]y' = 1 - y^2\cos(xy)$

18. $2x^2y + y^2 + (x^3 + 2xy\ln x)y' = 0$

19. $xy' = 2\sqrt{xy} - y$ **20.** $y' = \dfrac{x^2 + xy}{x^2 - 1}$

SECTION 15.6

Second Order Linear Equations

A **second order linear differential equation** has the form

(15.23)
$$P(x)\frac{d^2y}{dx^2} + Q(x)\frac{dy}{dx} + R(x)y = G(x)$$

where P, Q, R, and G are continuous functions. We shall see in Section 15.9 that equations of this type arise in the study of the motion of a spring as well as in electric circuits.

In this section we study the case where $G(x) = 0$, for all x, in Equation 15.23. Such equations are called **homogeneous** linear equations. (This use of the word *homogeneous* has nothing to do with the meaning given in Section 15.2.) Thus the form of a second order linear homogeneous differential equation is

(15.24)
$$P(x)\frac{d^2y}{dx^2} + Q(x)\frac{dy}{dx} + R(x)y = 0$$

If $G(x) \neq 0$ for some x, Equation 15.23 is **nonhomogeneous** and will be discussed in Section 15.8.

There are two basic facts that enable us to solve homogeneous linear equations. The first of these says that if we know two solutions y_1 and y_2 of such an equation, then the **linear combination** $y = c_1y_1 + c_2y_2$ is also a solution.

Theorem (15.25)

> If $y_1(x)$ and $y_2(x)$ are both solutions of the linear homogeneous equation (15.24) and c_1 and c_2 are any constants, then the function
>
> $$y(x) = c_1y_1(x) + c_2y_2(x)$$
>
> is also a solution of Equation 15.24.

Proof   Since y_1 and y_2 are solutions of Equation 15.24, we have

$$P(x)y_1'' + Q(x)y_1' + R(x)y_1 = 0$$

and $$P(x)y_2'' + Q(x)y_2' + R(x)y_2 = 0$$

Therefore, using the basic rules for differentiation, we have

$$P(x)y'' + Q(x)y' + R(x)y$$

$$= P(x)(c_1y_1 + c_2y_2)'' + Q(x)(c_1y_1 + c_2y_2)' + R(x)(c_1y_1 + c_2y_2)$$

$$= P(x)(c_1y_1'' + c_2y_2'') + Q(x)(c_1y_1' + c_2y_2') + R(x)(c_1y_1 + c_2y_2)$$

$$= c_1(P(x)y_1'' + Q(x)y_1' + R(x)y_1) + c_2(P(x)y_2'' + Q(x)y_2' + R(x)y_2)$$

$$= c_1(0) + c_2(0) = 0$$

Thus $y = c_1y_1 + c_2y_2$ is a solution of Equation 15.24.   ●

The other fact we need is given by the following theorem, which is proved in more advanced courses. It says that the general solution is a linear combination of two **linearly independent** solutions y_1 and y_2. This means that neither y_1 nor y_2 is a constant multiple of the other.

Theorem (15.26)

> If y_1 and y_2 are linearly independent solutions of Equation 15.24, then the general solution is given by
>
> $$y(x) = c_1y_1(x) + c_2y_2(x)$$
>
> where c_1 and c_2 are arbitrary constants.

Theorem 15.26 is very useful because it says that if we know *two* particular linearly independent solutions, then we know *every* solution.

In general it is not an easy matter to discover particular solutions to a second order linear equation. But it is always possible to do so if the coefficient functions P, Q, and R are constant functions, that is, if the differential equation has the form

(15.27)

> $$ay'' + by' + cy = 0$$

where a, b, and c are constants.

It is not hard to think of some likely candidates for particular solutions of Equation 15.27 if we state the equation verbally. We are looking for a function y such that a constant times its second derivative y'' plus another constant times y' plus a third constant times y is equal to 0. We know that the exponential function $y = e^{rx}$ (where r is a constant) has the property that its derivative is a constant multiple of itself: $y' = re^{rx}$. Furthermore, $y'' = r^2 e^{rx}$. If we substitute these expressions into Equation 15.27, we see that $y = e^{rx}$ will be a solution if

$$ar^2 e^{rx} + bre^{rx} + ce^{rx} = 0$$

or

$$(ar^2 + br + c)e^{rx} = 0$$

But e^{rx} is never 0. Therefore $y = e^{rx}$ will be a solution of Equation 15.27 if r is a root of the equation

(15.28)

$$\boxed{ar^2 + br + c = 0}$$

Equation 15.28 is called the **auxiliary equation** (or **characteristic equation**) of the differential equation $ay'' + by' + cy = 0$. Notice that it is an algebraic equation that is obtained from the differential equation by replacing y'' by r^2, y' by r, and y by 1.

Sometimes the roots r_1 and r_2 of the auxiliary equation can be found by factoring. In other cases they are found by using the quadratic formula:

(15.29)

$$r_1 = \frac{-b + \sqrt{b^2 - 4ac}}{2a} \qquad r_2 = \frac{-b - \sqrt{b^2 - 4ac}}{2a}$$

We distinguish three cases according to the sign of the discriminant $b^2 - 4ac$. The third case ($b^2 - 4ac < 0$) is left to the next section.

Case I: $b^2 - 4ac > 0$.   In this case the roots r_1 and r_2 of the auxiliary equation are real and distinct, so $y_1 = e^{r_1 x}$ and $y_2 = e^{r_2 x}$ are two linearly independent solutions of Equation 15.27. (Note that $e^{r_2 x}$ is not a constant multiple of $e^{r_1 x}$.) Thus, by Theorem 15.26, we have the following:

(15.30)

If the roots r_1 and r_2 of the auxiliary equation $ar^2 + br + c = 0$ are real and unequal, then the general solution of $ay'' + by' + cy = 0$ is

$$y = c_1 e^{r_1 x} + c_2 e^{r_2 x}$$

● **Example 1**   Solve the equation $y'' + y' - 6y = 0$.

Solution   The auxiliary equation is

$$r^2 + r - 6 = (r - 2)(r + 3) = 0$$

whose roots are $r = 2, -3$. Therefore, by (15.30) the general solution of the given differential equation is

$$y = c_1 e^{2x} + c_2 e^{-3x}$$

●

● **Example 2**   Solve $3\dfrac{d^2y}{dx^2} + \dfrac{dy}{dx} - y = 0$.

Solution   To solve the auxiliary equation $3r^2 + r - 1 = 0$ we use the quadratic formula:

$$r = \frac{-1 \pm \sqrt{13}}{6}$$

Since the roots are real and distinct, the general solution is

$$y = c_1 e^{(-1+\sqrt{13})x/6} + c_2 e^{(-1-\sqrt{13})x/6}$$

●

Case II: $b^2 - 4ac = 0$.   In this case $r_1 = r_2$; that is, the roots of the auxiliary equation are real and equal. Let us denote by r the common value of r_1 and r_2. Then, from Equations 15.29, we have

(15.31)
$$r = -\frac{b}{2a} \qquad \text{so} \qquad 2ar + b = 0$$

We know that $y_1 = e^{rx}$ is one solution of Equation 15.27. We now verify that $y_2 = xe^{rx}$ is also a solution:

$$ay_2'' + by_2' + cy_2 = a(2re^{rx} + r^2xe^{rx}) + b(e^{rx} + rxe^{rx}) + cxe^{rx}$$
$$= (2ar + b)e^{rx} + (ar^2 + br + c)xe^{rx}$$
$$= 0(e^{rx}) + 0(xe^{rx}) = 0$$

The first term is 0 by Equations 15.31; the second term is 0 because r is a root of the auxiliary equation. Since $y_1 = e^{rx}$ and $y_2 = xe^{rx}$ are linearly independent solutions, Theorem 15.26 provides us with the general solution.

(15.32)

> If the auxiliary equation $ar^2 + br + c = 0$ has only one real root r, then the general solution of $ay'' + by' + cy = 0$ is
>
> $$y = c_1 e^{rx} + c_2 xe^{rx}$$

● **Example 3**   Solve the equation $4y'' + 12y' + 9y = 0$.

Solution   The auxiliary equation $4r^2 + 12r + 9 = 0$ can be factored as

$$(2r + 3)^2 = 0$$

so the only root is $r = -\frac{3}{2}$. By (15.32) the general solution is

$$y = c_1 e^{-3x/2} + c_2 x e^{-3x/2}$$

Initial- and Boundary-Value Problems

An **initial-value problem** for the second order Equation 15.23 or 15.24 consists of finding a solution y of the differential equation that also satisfies initial conditions of the form

$$y(x_0) = y_0 \qquad y'(x_0) = y_1$$

where y_0 and y_1 are given constants. If P, Q, R, and G are continuous on an interval and $P(x) \neq 0$ there, then a theorem found in more advanced books guarantees the existence and uniqueness of a solution to this initial-value problem. Example 4 illustrates the technique for solving such a problem.

A **boundary-value problem** for Equation 15.23 consists of finding a solution y of the differential equation that also satisfies boundary conditions of the form

$$y(x_0) = y_0 \qquad y(x_1) = y_1$$

In contrast with the situation for initial-value problems, a boundary-value problem does not always have a solution. The method is illustrated in Example 5.

● **Example 4**   Solve the initial-value problem

$$y'' + y' - 6y = 0 \qquad y(0) = 1 \qquad y'(0) = 0$$

Solution   From Example 1 we know that the general solution of the differential equation is

$$y(x) = c_1 e^{2x} + c_2 e^{-3x}$$

Differentiating this solution, we get

$$y'(x) = 2c_1 e^{2x} - 3c_2 e^{-3x}$$

To satisfy the initial conditions we require that

$$y(0) = c_1 + c_2 = 1 \qquad (1)$$
$$y'(0) = 2c_1 - 3c_2 = 0 \qquad (2)$$

From (2) we have $c_2 = \frac{2}{3}c_1$ and so (1) gives

$$c_1 + \tfrac{2}{3}c_1 = 1 \qquad c_1 = \tfrac{3}{5} \qquad c_2 = \tfrac{2}{5}$$

Thus the required solution of the initial-value problem is

$$y = \tfrac{3}{5}e^{2x} + \tfrac{2}{5}e^{-3x}$$

● **Example 5** Solve the boundary-value problem

$$y'' + 2y' + y = 0 \qquad y(0) = 1 \qquad y(1) = 3$$

Solution The auxiliary equation is

$$r^2 + 2r + 1 = 0 \qquad \text{or} \qquad (r + 1)^2 = 0$$

whose only root is $r = -1$. Therefore the general solution is

$$y(x) = c_1 e^{-x} + c_2 x e^{-x}$$

The boundary conditions will be satisfied if

$$y(0) = c_1 = 1$$
$$y(1) = c_1 e^{-1} + c_2 e^{-1} = 3$$

The first condition gives $c_1 = 1$, so the second condition becomes

$$e^{-1} + c_2 e^{-1} = 3$$

Solving this equation for c_2 by first multiplying through by e, we get

$$1 + c_2 = 3e \qquad c_2 = 3e - 1$$

Thus the solution of the boundary-value problem is

$$y = e^{-x} + (3e - 1)x e^{-x}$$

●

EXERCISES 15.6

Solve the differential equations in Exercises 1–12.

1. $y'' - 3y' + 2y = 0$

2. $y'' - y' = 0$

3. $3y'' - 8y' - 3y = 0$

4. $y'' + 9y' + 20y = 0$

5. $y'' = y$

6. $9y'' - 30y' + 25y = 0$

7. $2y'' + y' = 0$

8. $y'' - 2y' - 4y = 0$

9. $\dfrac{d^2y}{dx^2} + 2\dfrac{dy}{dx} - y = 0$

10. $6\dfrac{d^2y}{dx^2} - \dfrac{dy}{dx} - 2y = 0$

11. $\dfrac{d^2y}{dx^2} - 8\dfrac{dy}{dx} + 16y = 0$

12. $2\dfrac{d^2y}{dx^2} + 5\dfrac{dy}{dx} + y = 0$

Solve the initial-value problems in Exercises 13–16.

13. $y'' + 3y' - 4y = 0$, $y(0) = 2$, $y'(0) = -3$

14. $y'' - 4y = 0$, $y(0) = 1$, $y'(0) = 0$

15. $y'' - 2y' - 3y = 0$, $y(1) = 3$, $y'(1) = 1$

16. $y'' - 2y' + y = 0$, $y(2) = 0$, $y'(2) = 1$

Solve the boundary-value problems in Exercises 17–20.

17. $y'' + 4y' + 4y = 0$, $y(0) = 0$, $y(1) = 3$

18. $y'' + 5y' - 6y = 0$, $y(0) = 0$, $y(2) = 1$

19. $y'' - y' - 2y = 0$, $y(-1) = 1$, $y(1) = 0$

20. $y'' + 4y' + 3y = 0$, $y(1) = 0$, $y(3) = 2$

SECTION 15.7

Complex Roots

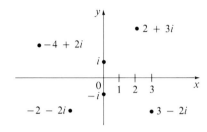

Figure 15.9

COMPLEX NUMBERS AS POINTS IN THE
ARGAND PLANE

In solving the linear homogeneous second order differential equation $ay'' + by' + cy = 0$ in the preceding section, we found that there are solutions of the form $y = e^{rx}$, where r is a root of the auxiliary equation $ar^2 + br + c = 0$. We have already investigated the case where there are two distinct real roots ($b^2 - 4ac > 0$) and the case where the roots coincide ($b^2 - 4ac = 0$). The remaining case ($b^2 - 4ac < 0$) involves complex roots, so we first give a brief discussion of complex numbers.

A **complex number** can be represented by an expression of the form $a + bi$, where a and b are real numbers and i is a symbol with the property that $i^2 = -1$. The complex number $a + bi$ can also be represented by the ordered pair (a, b) and plotted as a point in a plane (called the Argand plane) as in Figure 15.9. Thus the complex number $i = 0 + 1 \cdot i$ is identified with the point $(0, 1)$. Addition and multiplication of complex numbers are defined as follows:

$$(a + bi) + (c + di) = (a + c) + (b + d)i$$
(15.33)
$$(a + bi)(c + di) = (ac - bd) + (ad + bc)i$$

A real number a can be regarded as a special type of complex number by writing $a = a + 0i$. Then, if we put $a = c = 0$, $b = d = 1$ in Equation 15.33, we get

$$i^2 = (0 + i)(0 + i) = -1 + 0i = -1$$

So the equation $i^2 = -1$ is now literally true. Equation 15.33 can then be remembered as saying that multiplication is performed just as for real numbers but using the fact that $i^2 = -1$.

It is now reasonable to write $\sqrt{-1} = i$ and with this convention we can express the roots of quadratic equations with negative discriminant in terms of complex numbers.

● **Example 1**   Find the roots of the equation $x^2 + x + 1 = 0$.

Solution   Using the quadratic formula, we have

$$x = \frac{-1 \pm \sqrt{1^2 - 4 \cdot 1}}{2} = \frac{-1 \pm \sqrt{-3}}{2} = \frac{-1 \pm \sqrt{3}i}{2}$$
●

We also need to give a meaning to the expression e^z when $z = x + iy$ is a complex number. The theory of infinite series as developed in Chapter 10 can be extended to the case where the terms are complex numbers. Using the Taylor series for e^x (10.47) as our guide, we define

(15.34)
$$e^z = \sum_{n=0}^{\infty} \frac{z^n}{n!} = 1 + z + \frac{z^2}{2!} + \frac{z^3}{3!} + \cdots$$

and it turns out that this complex exponential function has the same properties as the real exponential function. In particular, it is true that

(15.35)
$$e^{z_1 + z_2} = e^{z_1} e^{z_2}$$

If we put $z = iy$, where y is a real number, in Equation 15.34, and use the facts that

$$i^2 = -1, \quad i^3 = i^2 i = -i, \quad i^4 = 1, \quad i^5 = i, \quad \ldots$$

we get
$$e^{iy} = 1 + iy + \frac{(iy)^2}{2!} + \frac{(iy)^3}{3!} + \frac{(iy)^4}{4!} + \frac{(iy)^5}{5!} + \cdots$$

$$= 1 + iy - \frac{y^2}{2!} - i\frac{y^3}{3!} + \frac{y^4}{4!} + i\frac{y^5}{5!} + \cdots$$

$$= \left(1 - \frac{y^2}{2!} + \frac{y^4}{4!} - \frac{y^6}{6!} + \cdots\right) + i\left(y - \frac{y^3}{3!} + \frac{y^5}{5!} - \cdots\right)$$

$$= \cos y + i \sin y$$

Here we have used the Taylor series for $\cos y$ and $\sin y$ (Equations 10.50 and 10.49). The result is a famous formula called **Euler's formula:**

(15.36)
$$\boxed{e^{iy} = \cos y + i \sin y}$$

Combining Euler's formula with Equation 15.35, we get

(15.37)
$$e^{x + iy} = e^x e^{iy} = e^x(\cos y + i \sin y)$$

The rule for differentiating complex exponential functions turns out to be the same as for differentiating real exponential functions. In particular,

$$\frac{d}{dx} e^{rx} = re^{rx}$$

even when r is a complex number. (See Exercise 20.) Therefore the same analysis as in Section 15.6 shows that if r_1 and r_2 are complex roots of the auxiliary equation $ar^2 + br + c = 0$, then the general solution of $ay'' + by' + cy = 0$ is

(15.38)
$$y = c_1 e^{r_1 x} + c_2 e^{r_2 x}$$

We can write

$$r_1 = \alpha + i\beta \qquad r_2 = \alpha - i\beta$$

where α and β are real numbers. (In fact, $\alpha = -b/2a$, $\beta = \sqrt{4ac - b^2}/2a$.) Then, using Equation 15.37, we rewrite the solution (Equation 15.38) as

$$
\begin{aligned}
y &= C_1 e^{(\alpha + i\beta)x} + C_2 e^{(\alpha - i\beta)x} \\
&= C_1 e^{\alpha x}(\cos \beta x + i \sin \beta x) + C_2 e^{\alpha x}(\cos \beta x - i \sin \beta x) \\
&= e^{\alpha x}[(C_1 + C_2)\cos \beta x + i(C_1 - C_2)\sin \beta x] \\
&= e^{\alpha x}(c_1 \cos \beta x + c_2 \sin \beta x)
\end{aligned}
$$

where $c_1 = C_1 + C_2$, $c_2 = i(C_1 - C_2)$. This will give all solutions (real or complex) of the differential equation. The solutions will be real when the constants c_1 and c_2 are real. We summarize the discussion as follows:

(15.39)

> If the roots of the auxiliary equation $ar^2 + br + c = 0$ are the complex numbers $r_1 = \alpha + i\beta$, $r_2 = \alpha - i\beta$, then the general solution of $ay'' + by' + cy = 0$ is
>
> $$y = e^{\alpha x}(c_1 \cos \beta x + c_2 \sin \beta x)$$

● **Example 2**   Solve the equation $y'' - 6y' + 13y = 0$.

Solution   The auxiliary equation is $r^2 - 6r + 13 = 0$. By the quadratic formula, the roots are

$$r = \frac{6 \pm \sqrt{36 - 52}}{2} = \frac{6 \pm \sqrt{-16}}{2} = 3 \pm 2i$$

By (15.39) the general solution of the differential equation is

$$y = e^{3x}(c_1 \cos 2x + c_2 \sin 2x) \qquad ●$$

● **Example 3**   Solve the initial-value problem

$$y'' + y = 0 \qquad y(0) = 2 \qquad y'(0) = 3$$

Solution   The auxiliary equation is $r^2 + 1 = 0$, or $r^2 = -1$, whose roots are $\pm i$. Thus $\alpha = 0$, $\beta = 1$, and since $e^{0x} = 1$, the general solution is

$$y(x) = c_1 \cos x + c_2 \sin x$$

Since

$$y'(x) = -c_1 \sin x + c_2 \cos x$$

the initial conditions become

$$y(0) = c_1 = 2$$

$$y'(0) = c_2 = 3$$

Therefore the solution of the initial-value problem is

$$y(x) = 2\cos x + 3\sin x \qquad ●$$

EXERCISES 15.7

Solve the differential equations in Exercises 1–8.

1. $y'' + 2y' + 10y = 0$

2. $y'' + 10y' + 41y = 0$

3. $y'' + 25y = 0$

4. $y'' - 4y' + 13y = 0$

5. $y'' - y' + 2y = 0$

6. $y'' = -5y$

7. $\dfrac{d^2y}{dx^2} - 2\dfrac{dy}{dx} + 5y = 0$

8. $2\dfrac{d^2y}{dx^2} + \dfrac{dy}{dx} + 3y = 0$

Solve the initial-value problems in Exercises 9–12.

9. $y'' - 2y' + 2y = 0$, $y(0) = 1$, $y'(0) = 2$

10. $y'' + 4y' + 6y = 0$, $y(0) = 2$, $y'(0) = 4$

11. $y'' + 9y = 0$, $y(\pi/3) = 0$, $y'(\pi/3) = 1$

12. $y'' + 4y = 0$, $y(\pi/6) = 1$, $y'(\pi/6) = 0$

In Exercises 13–16 solve the boundary-value problems if possible.

13. $y'' + y = 0$, $y(0) = 1$, $y(\pi) = 0$

14. $y'' + 9y = 0$, $y(0) = 1$, $y(\pi/2) = 0$

15. $y'' + 4y' + 13y = 0$, $y(0) = 2$, $y(\pi/2) = 1$

16. $y'' + 2y' + 5y = 0$, $y(0) = 1$, $y(\pi) = 2$

17. If a, b, and c are all positive constants and $y(x)$ is a solution of the differential equation $ay'' + by' + cy = 0$, show that $\lim_{x \to \infty} y(x) = 0$.

18. Use Euler's formula (15.36) to prove De Moivre's formula:

$$(\cos x + i\sin x)^n = \cos nx + i\sin nx$$

19. Use Euler's formula to prove the following formulas for $\cos x$ and $\sin x$:

$$\cos x = \frac{e^{ix} + e^{-ix}}{2} \qquad \sin x = \frac{e^{ix} - e^{-ix}}{2i}$$

20. Use Formula 15.37 to prove that $D_x e^{rx} = re^{rx}$ when $r = a + ib$ is a complex number.

SECTION 15.8

Nonhomogeneous Linear Equations

In this section we learn how to solve second order nonhomogeneous linear differential equations with constant coefficients, that is, equations of the form

(15.40)
$$ay'' + by' + cy = G(x)$$

where a, b, c are constants and G is a continuous function. The related homogeneous equation

(15.41)
$$ay'' + by' + cy = 0$$

is called the **complementary equation** and plays an important role in the solution of the original nonhomogeneous equation (15.40).

Theorem (15.42)

> The general solution of the nonhomogeneous differential equation (15.40) can be written as
>
> $$y(x) = y_p(x) + y_c(x)$$
>
> where y_p is a particular solution of Equation 15.40 and y_c is the general solution of the complementary Equation 15.41.

Proof All we have to do is verify that if y is any solution of Equation 15.40, then $y - y_p$ is a solution of the complementary Equation 15.41. Indeed

$$a(y - y_p)'' + b(y - y_p)' + c(y - y_p) = ay'' - ay_p'' + by' - by_p' + cy - cy_p$$
$$= (ay'' + by' + cy) - (ay_p'' + by_p' + cy_p)$$
$$= g(x) - g(x) = 0 \qquad \bullet$$

We know from Sections 15.6 and 15.7 how to solve the complementary equation. (Recall that the solution is $y_c = c_1 y_1 + c_2 y_2$ where y_1 and y_2 are linearly independent solutions of Equation 15.41.) Therefore Theorem 15.42 says that we know the general solution of the nonhomogeneous equation as soon as we know a particular solution y_p. There are two methods for finding a particular solution. The method of undetermined coefficients is straightforward but works only for a restricted class of functions G. The method of variation of parameters works for every function G but is usually more difficult to apply in practice.

The Method of Undetermined Coefficients

We first illustrate the method of undetermined coefficients for the equation

$$ay'' + by' + cy = G(x)$$

where $G(x)$ is a polynomial. It is reasonable to guess that there is a particular solution y_p that is a polynomial of the same degree as G because if y is a polynomial then $ay'' + by' + cy$ is also a polynomial. We therefore substitute $y_p(x) =$ a polynomial (of the same degree as G) into the differential equation and determine the coefficients.

● **Example 1** Solve the equation $y'' + y' - 2y = x^2$.

Solution The auxiliary equation of $y'' + y' - 2y = 0$ is

$$r^2 + r - 2 = (r - 1)(r + 2) = 0$$

with roots $r = 1, -2$. So the solution of the complementary equation is

$$y_c = c_1 e^x + c_2 e^{-2x}$$

Since $G(x) = x^2$ is a polynomial of degree 2, we seek a particular solution of the form

$$y_p(x) = Ax^2 + Bx + C$$

Then $y_p' = 2Ax + B$ and $y_p'' = 2A$ so, substituting into the given differential equation, we have

$$(2A) + (2Ax + B) - 2(Ax^2 + Bx + C) = x^2$$

or $\qquad -2Ax^2 + (2A - 2B)x + (2A + B - 2C) = x^2$

Polynomials are equal when their coefficients are equal. Thus

$$-2A = 1 \qquad 2A - 2B = 0 \qquad 2A + B - 2C = 0$$

The solution of this system of equations is

$$A = -\tfrac{1}{2} \qquad B = -\tfrac{1}{2} \qquad C = -\tfrac{3}{4}$$

The particular solution is therefore

$$y_p(x) = -\tfrac{1}{2}x^2 - \tfrac{1}{2}x - \tfrac{3}{4}$$

and, by Theorem 15.42, the general solution is

$$y = y_c + y_p = c_1 e^x + c_2 e^{-2x} - \tfrac{1}{2}x^2 - \tfrac{1}{2}x - \tfrac{3}{4} \qquad \bullet$$

If $G(x)$, the right side of Equation 15.40, is of the form Ce^{kx}, where C and k are constants, then we take as trial solution a function of the same form, $y_p(x) = Ae^{kx}$, because the derivatives of e^{kx} are constant multiples of e^{kx}.

● **Example 2**   Solve $y'' + 4y = e^{3x}$.

Solution   The auxiliary equation is $r^2 + 4 = 0$ with roots $\pm 2i$, so the solution of the complementary equation is

$$y_c(x) = c_1 \cos 2x + c_2 \sin 2x$$

For the particular solution we try $y_p(x) = Ae^{3x}$. Then $y_p' = 3Ae^{3x}$ and $y_p'' = 9Ae^{3x}$. Substituting into the differential equation, we have

$$9Ae^{3x} + 4(Ae^{3x}) = e^{3x}$$

so $13Ae^{3x} = e^{3x}$ and $A = \tfrac{1}{13}$. Thus the particular solution is

$$y_p(x) = \tfrac{1}{13}e^{3x}$$

and the general solution is

$$y(x) = c_1 \cos 2x + c_2 \sin 2x + \tfrac{1}{13}e^{3x} \qquad \bullet$$

If $G(x)$ is either $C \cos kx$ or $C \sin kx$, then, because of the rules for differentiating the sine and cosine functions, we take as trial particular solution a function of the form

$$y_p(x) = A \cos kx + B \sin kx$$

● **Example 3**   Solve $y'' + y' - 2y = \sin x$.

Solution   We try the particular solution

$$y_p(x) = A \cos x + B \sin x$$

Then

$$y_p' = -A \sin x + B \cos x \qquad y_p'' = -A \cos x - B \sin x$$

so substitution in the differential equation gives

$$(-A\cos x - B\sin x) + (-A\sin x + B\cos x) - 2(A\cos x + B\sin x) = \sin x$$

or
$$(-3A + B)\cos x + (-A - 3B)\sin x = \sin x$$

This will be true if

$$-3A + B = 0 \quad \text{and} \quad -A - 3B = 1$$

The solution of this system is

$$A = -\tfrac{1}{10} \qquad B = -\tfrac{3}{10}$$

so the particular solution is

$$y_p(x) = -\tfrac{1}{10}\cos x - \tfrac{3}{10}\sin x$$

In Example 1 we found that the solution of the complementary equation is $y_c = c_1 e^x + c_2 e^{-2x}$. Thus the general solution of the given equation is

$$y_p(x) = c_1 e^x + c_2 e^{-2x} - \tfrac{1}{10}(\cos x + 3\sin x) \qquad \bullet$$

If $G(x)$ is a product of functions of the preceding types, then we take the trial solution to be a product of functions of the same type. For instance, in solving the differential equation

$$y'' + 2y' + 4y = x\cos 3x$$

we would try

$$y_p(x) = (Ax + B)(C\cos 3x + D\sin 3x)$$

If $G(x)$ is a sum of functions of the above types, we use the easily verified principle of superposition, which says that if y_{p_1} and y_{p_2} are solutions of

$$ay'' + by' + cy = G_1(x) \qquad ay'' + by' + cy = G_2(x)$$

respectively, then $y_{p_1} + y_{p_2}$ is a solution of

$$ay'' + by' + cy = G_1(x) + G_2(x)$$

● **Example 4**   Solve $y'' - 4y = xe^x + \cos 2x$.

Solution   The auxiliary equation is $r^2 - 4 = 0$ with roots ± 2, so the solution of the complementary equation is $y_c(x) = c_1 e^{2x} + c_2 e^{-2x}$. For the equation $y'' - 4y = xe^x$ we try

$$y_{p_1}(x) = (Ax + B)e^x$$

Then $y_p' = (Ax + A + B)e^x$, $y_p'' = (Ax + 2A + B)e^x$, so substitution in the equation gives

$$(Ax + 2A + B)e^x - 4(Ax + B)e^x = xe^x$$

or
$$(-3Ax + 2A - 3B)e^x = xe^x$$

Thus $-3A = 1$ and $2A - 3B = 0$, so $A = -\frac{1}{3}$, $B = -\frac{2}{9}$, and

$$y_{p_1}(x) = (-\tfrac{1}{3}x - \tfrac{2}{9})e^x$$

For the equation $y'' - 4y = \cos 2x$, we try

$$y_{p_2}(x) = C \cos 2x + D \sin 2x$$

Substitution gives

$$-4C \cos 2x - 4D \sin 2x - 4(C \cos 2x + D \sin 2x) = \cos 2x$$

or $-8C \cos 2x - 8D \sin 2x = \cos 2x$

Therefore $-8C = 1$, $-8D = 0$, and

$$y_{p_2}(x) = -\tfrac{1}{8} \cos 2x$$

By the superposition principle, the general solution is

$$y = y_c + y_{p_1} + y_{p_2} = c_1 e^{2x} + c_2 e^{-2x} - (\tfrac{1}{3}x + \tfrac{2}{9})e^x - \tfrac{1}{8} \cos 2x \qquad \bullet$$

Finally we note that the recommended trial solution y_p sometimes turns out to be a solution of the complementary equation and therefore cannot be a solution of the nonhomogeneous equation. In such cases we multiply the recommended trial solution by x (or x^2 if need be) so that no term in $y_p(x)$ is a solution of the complementary equation.

● **Example 5**   Solve $y'' + y = \sin x$.

Solution   The auxiliary equation is $r^2 + 1 = 0$ with roots $\pm i$, so the solution of the complementary equation is

$$y_c(x) = c_1 \cos x + c_2 \sin x$$

Ordinarily we would use the trial solution

$$y_p(x) = A \cos x + B \sin x$$

but we observe that it is a solution of the complementary equation, so instead we try

$$y_p(x) = Ax \cos x + Bx \sin x$$

Then $y_p'(x) = A \cos x - Ax \sin x + B \sin x + Bx \cos x$

$$y_p''(x) = -2A \sin x - Ax \cos x + 2B \cos x - Bx \sin x$$

Substitution in the differential equation gives

$$y_p'' + y_p = -2A \sin x + 2B \cos x = \sin x$$

so $A = -\frac{1}{2}$, $B = 0$, and

$$y_p(x) = -\tfrac{1}{2}x \cos x$$

The general solution is

$$y(x) = c_1 \cos x + c_2 \sin x - \tfrac{1}{2}x \cos x \qquad \bullet$$

The Method of Variation of Parameters

Suppose we have already solved the homogeneous equation $ay'' + by' + cy = 0$ and written the solution as

(15.43)
$$y(x) = c_1 y_1(x) + c_2 y_2(x)$$

where y_1 and y_2 are linearly independent solutions. Let us replace the constants (or parameters) c_1 and c_2 in Equation 15.43 by arbitrary functions $u_1(x)$ and $u_2(x)$. We look for a particular solution of the nonhomogeneous equation $ay'' + by' + cy = G(x)$ of the form

(15.44)
$$y_p(x) = u_1(x)y_1(x) + u_2(x)y_2(x)$$

(This method is called **variation of parameters** because we have varied the parameters c_1 and c_2 to make them functions.) Differentiating Equation 15.44, we get

(15.45)
$$y_p' = (u_1'y_1 + u_2'y_2) + (u_1 y_1' + u_2 y_2')$$

Since u_1 and u_2 are arbitrary functions, we can impose two conditions on them. One condition will be that y_p is a solution of the differential equation; we can choose the other condition so as to simplify our calculations. In view of the expression in Equation 15.45, let us impose the condition that

(15.46)
$$u_1'y_1 + u_2'y_2 = 0$$

Then
$$y_p'' = u_1'y_1' + u_2'y_2' + u_1 y_1'' + u_2 y_2''$$

Substituting in the differential equation, we get

$$a(u_1'y_1' + u_2'y_2' + u_1 y_1'' + u_2 y_2'') + b(u_1 y_1' + u_2 y_2') + c(u_1 y_1 + u_2 y_2) = G$$

or

(15.47)
$$u_1(ay_1'' + by_1' + cy_1) + u_2(ay_2'' + by_2' + cy_2) + a(u_1'y_1' + u_2'y_2') = G$$

But y_1 and y_2 are solutions of the complementary equation, so

$$ay_1'' + by_1' + cy_1 = 0 \quad \text{and} \quad ay_2'' + by_2' + cy_2 = 0$$

and Equation 15.47 simplifies to

(15.48)
$$a(u_1'y_1' + u_2'y_2') = G$$

Equations 15.46 and 15.48 form a system of two equations in the unknown functions u_1' and u_2'. After solving this system we may be able to integrate to find u_1 and u_2 and then the particular solution is given by Equation 15.44.

● **Example 6**   Solve the equation $y'' + y = \tan x$, $0 < x < \dfrac{\pi}{2}$.

Solution   The auxiliary equation is $r^2 + 1 = 0$ with roots $\pm i$, so the solution of $y'' + y = 0$ is $c_1 \sin x + c_2 \cos x$. Using variation of parameters we seek a solution of the form

$$y_p(x) = u_1(x)\sin x + u_2(x)\cos x$$

Then $\qquad y_p' = (u_1'\sin x + u_2'\cos x) + (u_1\cos x - u_2\sin x)$

Set

(15.49) $\qquad\qquad\qquad\qquad u_1'\sin x + u_2'\cos x = 0$

Then $\qquad y_p'' = u_1'\cos x - u_2'\sin x - u_1\sin x - u_2\cos x$

For y_p to be a solution we must have

(15.50) $\qquad\qquad y_p'' + y_p = u_1'\cos x - u_2'\sin x = \tan x$

Solving Equations 15.49 and 15.50, we get

$$u_1'(\sin^2 x + \cos^2 x) = \cos x \tan x$$

$$u_1' = \sin x \qquad u_1(x) = -\cos x$$

(We seek a particular solution, so we do not need a constant of integration here.) Then, from Equation 15.49, we obtain

$$u_2' = -\frac{\sin x}{\cos x}u_1' = -\frac{\sin^2 x}{\cos x} = \frac{\cos^2 x - 1}{\cos x} = \cos x - \sec x$$

So $\qquad u_2(x) = \sin x - \ln(\sec x + \tan x)$

(Note that $\sec x + \tan x > 0$ for $0 < x < \pi/2$.) Therefore

$$y_p(x) = -\cos x \sin x + (\sin x - \ln(\sec x + \tan x))\cos x$$

$$= -\cos x \ln(\sec x + \tan x)$$

and the general solution is

$$y(x) = c_1\sin x + c_2\cos x - \cos x \ln(\sec x + \tan x)$$

●

EXERCISES 15.8

Solve the differential equations and initial-value problems in Exercises 1–12 using the method of undetermined coefficients.

1. $y'' - y' - 6y = \cos 3x$

2. $y'' + 2y' + 2y = x^3 - 1$

3. $y'' - 4y' + 4y = e^{-x}$

4. $y'' - 7y' + 12y = \sin x - \cos x$

5. $y'' + 36y = 2x^2 - x$

6. $y'' + 2y' + y = xe^{-x}$

7. $4y'' + 5y' + y = e^x$

8. $2y'' + 3y' + y = 1 + x \cos 2x$

9. $y'' - 2y' + 5y = x + \sin 3x,\ y(0) = 1,\ y'(0) = 2$

10. $y'' + 2y = e^x \sin x,\ y(0) = 1,\ y'(0) = 1$

11. $y'' - y = xe^{3x},\ y(0) = 0,\ y'(0) = 1$

12. $y'' - 3y' + 2y = 2x - e^{-2x},\ y(0) = 1,\ y'(0) = 0$

In Exercises 13–16 write a trial solution for the method of undetermined coefficients. Do not determine the coefficients.

13. $y'' + 2y' + 6y = x^4 e^{2x}$

14. $y'' + 6y' + 2y = x^3 + e^x \sin 2x$

15. $y'' - 2y' + 2y = e^x \cos x$

16. $y'' + 3y' = 1 + xe^{-3x}$

Solve the differential equations in Exercises 17–22 using the method of variation of parameters.

17. $y'' + y = \sec x,\ 0 < x < \dfrac{\pi}{2}$

18. $y'' + y = \cot x,\ 0 < x < \dfrac{\pi}{2}$

19. $y'' - 3y' + 2y = \dfrac{1}{1 + e^{-x}}$

20. $y'' + 3y' + 2y = \sin(e^x)$

21. $y'' - y = \dfrac{1}{x}$

22. $y'' + 4y' + 4y = \dfrac{e^{-2x}}{x^3}$

SECTION 15.9

Applications of Second Order Differential Equations

Second order linear differential equations have a variety of applications in science and engineering. In this section we explore two of them: the vibration of springs and electric circuits.

Vibrating Springs

We consider the motion of an object with mass m at the end of a spring that is either vertical (as in Figure 15.10) or horizontal on a level surface (as in Figure 15.11). In Section 8.5 we discussed Hooke's Law, which says that if the spring is stretched (or compressed) x units from its natural length, then it exerts a force that is proportional to x:

$$\text{restoring force} = -kx$$

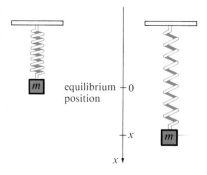

Figure 15.10

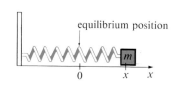

Figure 15.11

where k is a positive constant (called the *spring constant*). If we ignore any external resisting forces (due to air resistance or friction) then, by Newton's Second Law (force equals mass times acceleration), we have

(15.51)
$$m\frac{d^2x}{dt^2} = -kx \qquad \text{or} \qquad m\frac{d^2x}{dt^2} + kx = 0$$

This is a second order linear differential equation. Its auxiliary equation is $mr^2 + k = 0$ with roots $r = \pm\omega i$, where $\omega = \sqrt{k/m}$. Thus the general solution is

$$x(t) = c_1 \cos \omega t + c_2 \sin \omega t$$

which can also be written as

$$x(t) = A \cos(\omega t + \delta)$$

where
$$\omega = \sqrt{\frac{k}{m}} \qquad \text{(frequency)}$$

$$A = \sqrt{c_1^2 + c_2^2} \qquad \text{(amplitude)}$$

$$\cos \delta = \frac{c_1}{A}, \quad \sin \delta = -\frac{c_2}{A} \qquad \text{(}\delta\text{ is the phase angle)}$$

This type of motion is called **simple harmonic motion.**

● **Example 1**   A spring with a mass of 2 kg has natural length 0.5 m. A force of 25.6 N is required to stretch it to a length of 0.7 m. If the spring is stretched to a length of 0.7 m and then released with initial velocity 0, find the position of the mass at any time t.

Solution   From Hooke's Law, the force required to stretch the spring is

$$k(0.2) = 25.6$$

so $k = 25.6/0.2 = 128$. Using this value of the spring constant k, together with $m = 2$, in Equation 15.51, we have

$$2\frac{d^2x}{dt^2} + 128x = 0$$

As in the earlier general discussion, the solution of this equation is

(15.52)
$$x(t) = c_1 \cos 8t + c_2 \sin 8t$$

We are given the initial condition that $x(0) = 0.2$. But, from Equation 15.52, $x(0) = c_1$. Therefore $c_1 = 0.2$. Differentiating Equation 15.52, we get

$$x'(t) = -8c_1 \sin 8t + 8c_2 \cos 8t$$

Since the initial velocity is given as $x'(0) = 0$, we have $c_2 = 0$ and so the solution is

$$x(t) = \tfrac{1}{5}\cos 8t$$

Damped Vibrations

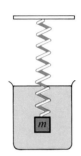

Figure 15.12

We next consider the motion of a spring that is subject to a frictional force (in the case of the horizontal spring of Figure 15.11) or a damping force (in the case where a vertical spring moves through a fluid as in Figure 15.12). An example is the damping force supplied by a shock absorber in a car.

We assume that the damping force is proportional to the velocity of the mass and acts in the direction opposite to the motion. (This has been confirmed, at least approximately, by some physical experiments.) Thus

$$\text{damping force} = -c\frac{dx}{dt}$$

where c is a positive constant, called the **damping constant.** Thus, in this case, Newton's Second Law gives

$$m\frac{d^2x}{dt^2} = \text{restoring force} + \text{damping force}$$

$$= -kx - c\frac{dx}{dt}$$

or

(15.53)

$$m\frac{d^2x}{dt^2} + c\frac{dx}{dt} + kx = 0$$

This is a second order linear differential equation with auxiliary equation $mr^2 + cr + k = 0$. The roots are

(15.54)

$$r_1 = \frac{-c + \sqrt{c^2 - 4mk}}{2m} \qquad r_2 = \frac{-c - \sqrt{c^2 - 4mk}}{2m}$$

According to Sections 15.6 and 15.7, we need to discuss three cases.

Case I: $c^2 - 4mk > 0$ *(overdamping)*.   In this case r_1 and r_2 are distinct real roots and

$$x = c_1 e^{r_1 t} + c_2 e^{r_2 t}$$

Since c, m, and k are all positive we have $\sqrt{c^2 - 4mk} < c$, so the roots r_1 and r_2 given by Equations 15.54 must both be negative. This shows that $x \to 0$ as $t \to \infty$. The graph of x as a function of t is shown in Figure 15.13. Notice that oscillations do not occur. This is because $c^2 > 4mk$ means that there is a strong damping force (high-viscosity oil or grease) compared with a weak spring or small mass.

Case II: $c^2 - 4mk = 0$ *(critical damping)*.   This case corresponds to equal roots

$$r_1 = r_2 = -\frac{c}{2m}$$

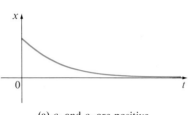

(a) c_1 and c_2 are positive   (b) c_1 and c_2 have opposite signs

Figure 15.13
OVERDAMPING

The solution is given by

$$x = (c_1 + c_2 t)e^{-(c/2m)t}$$

and its graph is shown in Figure 15.14. It is similar to Case I but the damping is just sufficient to suppress vibrations. Any decrease in the viscosity of the fluid will lead to the vibrations of the following case.

Figure 15.14
CRITICAL DAMPING

Case III: $c^2 - 4mk < 0$ (underdamping).   Here the roots are complex:

$$\left.\begin{matrix} r_1 \\ r_2 \end{matrix}\right\} = -\frac{c}{2m} \pm \omega i \qquad \text{where } \omega = \frac{\sqrt{4mk - c^2}}{2m}$$

The solution is given by

$$x = e^{-(c/2m)t}(c_1 \cos \omega t + c_2 \sin \omega t)$$

We see that there are oscillations that are damped by the factor $e^{-(c/2m)t}$. Since $c > 0$ and $m > 0$, we have $-(c/2) < 0$ so $e^{-(c/2m)t} \to 0$ as $t \to \infty$. This implies that $x \to 0$ as $t \to \infty$; that is, the motion decays to 0 as time increases. The graph is shown in Figure 15.15.

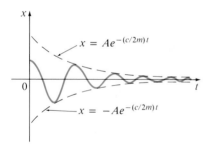

Figure 15.15
UNDERDAMPING

● **Example 2**   Suppose that the spring of Example 1 is immersed in a fluid with damping constant $c = 40$. Find the position of the mass at any time t if it starts from the equilibrium position and is given a push to start it with an initial velocity of 0.6 m/s.

Solution   From Example 1 the mass is $m = 2$ and the spring constant is $k = 128$, so the differential equation (15.53) becomes

$$2\frac{d^2x}{dt^2} + 40\frac{dx}{dt} + 128x = 0$$

or $$\frac{d^2x}{dt^2} + 20\frac{dx}{dt} + 64x = 0$$

The auxiliary equation is $r^2 + 20r + 64 = (r + 4)(r + 16) = 0$ with roots $-4, -16$, so the motion is overdamped and the solution is

$$x(t) = c_1 e^{-4t} + c_2 e^{-16t}$$

We are given that $x(0) = 0$, so $c_1 + c_2 = 0$. Differentiating, we get

$$x'(t) = -4c_1 e^{-4t} - 16c_2 e^{-16t}$$

so $$x'(0) = -4c_1 - 16c_2 = 0.6$$

Since $c_2 = -c_1$, this gives $12c_1 = 0.6$ or $c_1 = 0.05$. Therefore

$$x = 0.05(e^{-4t} - e^{-16t})$$ ●

Forced Vibrations

Suppose that, in addition to the restoring force and the damping force, the motion of the spring is also affected by an external force $F(t)$. Then Newton's Second Law gives

$$m\frac{d^2x}{dt^2} = \text{restoring force} + \text{damping force} + \text{external force}$$

$$= -kx - c\frac{dx}{dt} + F(t)$$

Thus, instead of the homogeneous equation (15.53), the motion of the spring is now governed by the following nonhomogeneous differential equation:

(15.55) $$m\frac{d^2x}{dt^2} + c\frac{dx}{dt} + kx = F(t)$$

The motion of the spring can be determined by the methods of Section 15.8. A commonly occurring type of external force is a periodic force function

$$F(t) = F_0 \cos \omega_0 t \qquad \text{where } \omega_0 \neq \omega = \sqrt{\frac{k}{m}}$$

In this case, and in the absence of a damping force ($c = 0$), you are asked in Exercise 7 to use the method of undetermined coefficients to show that

(15.56) $$x(t) = c_1 \cos \omega t + c_2 \sin \omega t + \frac{F_0}{m(\omega^2 - \omega_0^2)} \cos \omega_0 t$$

If $\omega_0 = \omega$, then the applied frequency reinforces the natural frequency and the result is vibrations of large amplitude. This is the phenomenon of **resonance** (see Exercise 8).

Electric Circuits

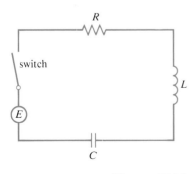

Figure 15.16

In Section 15.3 we were able to use first order linear equations to analyze electric circuits that contain a resistor and inductor (see Figure 15.7) or a resistor and capacitor (see Figure 15.8). Now that we know how to solve second order linear equations, we are in a position to analyze the circuit shown in Figure 15.16. It contains an electromotive force E (supplied by a battery or generator), a resistor R, an inductor L, and a capacitor C, in series. If the charge on the capacitor at time t is $Q = Q(t)$, then the current is the rate of change of Q with respect to t: $I = dQ/dt$. As in Section 15.3, it is known from physics that the voltage drops across the resistor, inductor, and capacitor are

$$RI \qquad L\frac{dI}{dt} \qquad \frac{Q}{C}$$

respectively. Kirchhoff's voltage law says that the sum of these voltage drops is equal to the supplied voltage:

$$L\frac{dI}{dt} + RI + \frac{Q}{C} = E(t)$$

Since $I = dQ/dt$, this equation becomes

(15.57)

$$L\frac{d^2Q}{dt^2} + R\frac{dQ}{dt} + \frac{1}{C}Q = E(t)$$

which is a second order linear differential equation with constant coefficients. If the charge Q_0 and the current I_0 are known at time 0, then we have the initial conditions

$$Q(0) = Q_0 \qquad Q'(0) = I(0) = I_0$$

and the initial-value problem can be solved by the methods of Section 15.8.

A differential equation for the current can be obtained by differentiating Equation 15.57 with respect to t and remembering that $I = dQ/dt$:

$$L\frac{d^2I}{dt^2} + R\frac{dI}{dt} + \frac{1}{C}I = E'(t)$$

● **Example 3**   Find the charge and current at time t in the circuit of Figure 15.16 if $R = 40\ \Omega$, $L = 1H$, $C = 16 \times 10^{-4}$ F, $E(t) = 100\cos 10t$, and the initial charge and current are both 0.

Solution   With the given values of L, R, C, and $E(t)$, Equation 15.57 becomes

(15.58)

$$\frac{d^2Q}{dt^2} + 40\frac{dQ}{dt} + 625Q = 100\cos 10t$$

The auxiliary equation is $r^2 + 40r + 625 = 0$ with roots

$$r = \frac{-40 \pm \sqrt{-900}}{2} = -20 \pm 15i$$

so the solution of the complementary equation is

$$Q_c(t) = e^{-20t}(c_1\cos 15t + c_2\sin 15t)$$

For the method of undetermined coefficients we try the particular solution

$$Q_p(t) = A \cos 10t + B \sin 10t$$

Then
$$Q_p'(t) = -10A \sin 10t + 10B \cos 10t$$

$$Q_p''(t) = -100A \cos 10t - 100B \sin 10t$$

Substituting into Equation 15.58, we have

$$(-100A \cos 10t - 100B \sin 10t) + 40(-10A \sin 10t + 10B \cos 10t)$$

$$+ 625(A \cos 10t + B \sin 10t) = 100 \cos 10t$$

or

$$(525A + 400B) \cos 10t + (-400A + 525B) \sin 10t = 100 \cos 10t$$

Equating coefficients, we have

$$525A + 400B = 100 \qquad 21A + 16B = 4$$
$$\text{or}$$
$$-400A + 525B = 0 \qquad -16A + 21B = 0$$

The solution of this system is $A = \frac{84}{697}$, $B = \frac{64}{697}$, so the particular solution is

$$Q_p(t) = \frac{1}{697}(84 \cos 10t + 64 \sin 10t)$$

and the general solution is

$$Q(t) = Q_c(t) + Q_p(t)$$

$$= e^{-20t}(c_1 \cos 15t + c_2 \sin 15t) + \frac{4}{697}(21 \cos 10t + 16 \sin 10t)$$

Imposing the initial condition $Q(0) = 0$, we get

$$Q(0) = c_1 + \frac{84}{697} = 0 \qquad c_1 = -\frac{84}{697}$$

To impose the other initial condition we first differentiate to find the current:

$$I = \frac{dQ}{dt} = e^{-20t}[(-20c_1 + 15c_2) \cos 15t + (-15c_1 - 20c_2) \sin 15t]$$

$$+ \frac{40}{697}(-21 \sin 10t + 16 \cos 10t)$$

$$I(0) = -20c_1 + 15c_2 + \frac{640}{697} = 0 \qquad c_2 = -\frac{464}{2091}$$

Thus the formula for the charge is

$$Q(t) = \frac{4}{697}\left[\frac{e^{-20t}}{3}(-63 \cos 15t - 116 \sin 15t) + (21 \cos 10t + 16 \sin 10t)\right]$$

and the expression for the current is

$$I(t) = \frac{1}{2091}[e^{-20t}(-1920 \cos 15t + 13{,}060 \sin 15t)$$

$$+ 120(-21 \sin 10t + 16 \cos 10t)]$$

Note 1:   In Example 3 the solution for $Q(t)$ consists of two parts. Since $e^{-20t} \to 0$ as $t \to \infty$ and both $\cos 15t$ and $\sin 15t$ are bounded functions,

$$Q_c(t) = \tfrac{4}{2091}e^{-20t}(-63\cos 15t - 116\sin 15t) \to 0 \qquad \text{as } t \to \infty$$

So, for large values of t,

$$Q(t) \approx Q_p(t) = \tfrac{4}{697}(21\cos 10t + 16\sin 10t)$$

and, for this reason, $Q_p(t)$ is called the **steady state solution.**

Note 2:   Comparing Equations 15.55 and 15.57, we see that mathematically they are identical. This suggests the analogies given in the following chart between physical situations that, at first glance, are very different.

Spring system		Electric circuit	
x	displacement	Q	charge
dx/dt	velocity	$I = dQ/dt$	current
m	mass	L	inductance
c	damping constant	R	resistance
k	spring constant	$1/C$	elastance
$F(t)$	external force	$E(t)$	electromotive force

We can also transfer other ideas from one situation to the other. For instance, the steady state solution discussed in Note 1 makes sense in the spring system. And the phenomenon of resonance in the spring system can be usefully carried over to electric circuits as electrical resonance.

EXERCISES 15.9

1. A spring with a 3-kg mass is stretched 0.6 m by a force of 20 N. If the spring begins at its equilibrium position but a push gives it an initial velocity of 1.2 m/s, find the position of the mass after t seconds.

2. A spring with a 4-kg mass has natural length 1 m and is stretched to a length of 1.3 m by a force of 24.3 N. If the spring is compressed to a length of 0.8 m and then released with zero velocity, find the position of the mass at any time t.

3. A spring with a mass of 2 kg has damping constant 14, and a force of 6 N is required to stretch the spring 0.5 m beyond its natural length. The spring is stretched 1 m beyond its natural length and then released with zero velocity. Find the position of the mass at any t.

4. A spring with a mass of 3 kg has damping constant 30 and spring constant 123. Find the position of the mass at time t if it starts at the equilibrium position with a velocity of 2 m/s.

5. For the spring in Exercise 3, find the mass that would produce critical damping.

6. For the spring in Exercise 4, find the damping constant that would produce critical damping.

7. Suppose a spring has mass m and spring constant k and let $\omega = \sqrt{k/m}$. Suppose that the damping constant is so small that the damping force is negligible. If an external force $F(t) = F_0 \cos \omega_0 t$ is applied, where $\omega_0 \neq \omega$, use the method of undetermined coefficients to show that the motion of the mass is described by Equation 15.56.

8. As in Exercise 7, consider a spring with mass m, spring constant k, and damping constant $c = 0$, and let $\omega = \sqrt{k/m}$. If an external force $F(t) = F_0 \cos \omega t$ is applied (the applied frequency equals the natural frequency), use the method of undetermined coefficients to show that the motion of the mass is given by

$$x(t) = c_1 \cos \omega t + c_2 \sin \omega t + \frac{F_0}{2m\omega} t \sin \omega t$$

9. A series circuit consists of a resistor with $R = 20 \, \Omega$, an inductor with $L = 1$ H, a capacitor with $C = 0.002$ F, and a 12-volt battery. If the initial charge and current are both 0, find the charge and current at time t.

10. A series circuit contains a resistor with $R = 24 \, \Omega$, an inductor with $L = 2$ H, a capacitor with $C = 0.005$ F, and a 12-volt battery. The initial charge is $Q = 0.001$ C and the initial current is 0. Find the charge and current at time t.

11. The battery in Exercise 9 is replaced by a generator producing a voltage of $E(t) = 12 \sin 10t$. Find the charge at time t.

12. The battery in Exercise 10 is replaced by a generator producing a voltage of $E(t) = 12 \sin 10t$. Find the charge at time t.

SECTION 15.10

Series Solutions

So far, the only second order differential equations that we have been able to solve are linear equations with constant coefficients. Even a simple-looking equation like

(15.59)
$$y'' - 2xy' + y = 0$$

is not easy to solve. But it is important to be able to solve equations such as Equation 15.59 since they arise from physical problems and, in particular, in connection with the Schrödinger equation in quantum mechanics. In such a case we use the method of power series; that is, we look for a solution of the form

$$y = f(x) = \sum_{n=0}^{\infty} a_n x^n = a_0 + a_1 x + a_2 x^2 + a_3 x^3 + \cdots$$

The method is to substitute this expression into the differential equation and determine the values of the coefficients $a_0, a_1, a_2, \ldots$. This technique resembles the method of undetermined coefficients discussed in Section 15.8.

Before using power series to solve Equation 15.59, we first illustrate the method on the simpler equation $y'' + y = 0$. It is true that we already know how to solve this equation by the techniques of Section 15.7, but it is easier to understand the power series method when it is applied to this simpler equation.

● **Example 1** Use power series to solve the equation $y'' + y = 0$.

Solution We assume there is a solution of the form

(15.60)
$$y = a_0 + a_1 x + a_2 x^2 + a_3 x^3 + \cdots = \sum_{n=0}^{\infty} a_n x^n$$

By Theorem 10.39 we can differentiate power series term by term, so

$$y' = a_1 + 2a_2x + 3a_3x^2 + \cdots = \sum_{n=1}^{\infty} na_n x^{n-1}$$

(15.61)
$$y'' = 2a_2 + 2 \cdot 3a_3x + \cdots = \sum_{n=2}^{\infty} n(n-1)a_n x^{n-2}$$

In order to compare the expressions for y and y'' more easily, we rewrite y'' as follows:

(15.62)
$$y'' = \sum_{n=0}^{\infty} (n+2)(n+1)a_{n+2} x^n$$

By writing out the first few terms of Equation 15.62 you can see that it is the same as Equation 15.61. To obtain Equation 15.62 we replaced n by $n+2$ and began the summation at 0 instead of 2. Substituting these expressions into the differential equation, we obtain

$$\sum_{n=0}^{\infty} (n+2)(n+1)a_{n+2} x^n + \sum_{n=0}^{\infty} a_n x^n = 0$$

or

(15.63)
$$\sum_{n=0}^{\infty} [(n+2)(n+1)a_{n+2} + a_n]x^n = 0$$

If two power series are equal, then the corresponding coefficients must be equal (from Section 10.9). Therefore the coefficients of x^n in Equation 15.63 must be 0:

$$(n+2)(n+1)a_{n+2} + a_n = 0$$

(15.64)
$$a_{n+2} = -\frac{a_n}{(n+1)(n+2)} \qquad n = 0, 1, 2, 3, \ldots$$

Equation 15.64 is called a recursion relation. If a_0 and a_1 are known, it allows us to determine the remaining coefficients recursively by putting $n = 0, 1, 2, 3, \ldots$ in succession.

Put $n = 0$: $a_2 = -\dfrac{a_0}{1 \cdot 2}$

Put $n = 1$: $a_3 = -\dfrac{a_1}{2 \cdot 3}$

Put $n = 2$: $a_4 = -\dfrac{a_2}{3 \cdot 4} = \dfrac{a_0}{1 \cdot 2 \cdot 3 \cdot 4} = \dfrac{a_0}{4!}$

Put $n = 3$: $a_5 = -\dfrac{a_3}{4 \cdot 5} = \dfrac{a_1}{2 \cdot 3 \cdot 4 \cdot 5} = \dfrac{a_1}{5!}$

Put $n = 4$: $a_6 = -\dfrac{a_4}{5 \cdot 6} = -\dfrac{a_0}{4!5 \cdot 6} = -\dfrac{a_0}{6!}$

Put $n = 5$: $a_7 = -\dfrac{a_5}{6 \cdot 7} = -\dfrac{a_1}{5! \, 6 \cdot 7} = -\dfrac{a_1}{7!}$

By now we see the pattern:

For the even coefficients, $a_{2n} = (-1)^n \dfrac{a_0}{(2n)!}$

For the odd coefficients, $a_{2n+1} = (-1)^n \dfrac{a_1}{(2n+1)!}$

Putting these values back into Equation 15.60, we write the solution as

$$y = a_0 + a_1 x + a_2 x^2 + a_3 x^3 + a_4 x^4 + a_5 x^5 + \cdots$$

$$= a_0 \left(1 - \frac{x^2}{2!} + \frac{x^4}{4!} - \frac{x^6}{6!} + \cdots + (-1)^n \frac{x^{2n}}{(2n)!} + \cdots \right)$$

$$+ a_1 \left(x - \frac{x^3}{3!} + \frac{x^5}{5!} - \frac{x^7}{7!} + \cdots + (-1)^n \frac{x^{2n+1}}{(2n+1)!} + \cdots \right)$$

$$= a_0 \sum_{n=0}^{\infty} (-1)^n \frac{x^{2n}}{(2n)!} + a_1 \sum_{n=0}^{\infty} (-1)^n \frac{x^{2n+1}}{(2n+1)!} \qquad \bullet$$

Note 1:   We recognize the series obtained in Example 1 as being the Maclaurin series for $\cos x$ and $\sin x$. (See Equations 10.50 and 10.49.) Therefore we could write the solution as

$$y(x) = a_0 \cos x + a_1 \sin x$$

But we will not usually be able to express power series solutions of differential equations in terms of known functions.

● **Example 2**   Solve $y'' - 2xy' + y = 0$.

Solution   We assume there is a solution of the form

$$y = \sum_{n=0}^{\infty} a_n x^n$$

Then $y' = \displaystyle\sum_{n=1}^{\infty} n a_n x^{n-1}$

and $y'' = \displaystyle\sum_{n=2}^{\infty} n(n-1)a_n x^{n-2} = \sum_{n=0}^{\infty} (n+2)(n+1)a_{n+2} x^n$

as in Example 1. Substituting in the differential equation, we get

$$\sum_{n=0}^{\infty} (n+2)(n+1)a_{n+2} x^n - 2x \sum_{n=1}^{\infty} n a_n x^{n-1} + \sum_{n=0}^{\infty} a_n x^n = 0$$

$$\sum_{n=0}^{\infty} (n+2)(n+1)a_{n+2} x^n - \sum_{n=1}^{\infty} 2n a_n x^n + \sum_{n=0}^{\infty} a_n x^n = 0$$

$$\sum_{n=0}^{\infty} \left[(n+2)(n+1)a_{n+2} - (2n-1)a_n \right] x^n = 0$$

This will be true if the coefficient of x^n is 0:

$$(n + 2)(n + 1)a_{n+2} - (2n - 1)a_n = 0$$

(15.65)
$$a_{n+2} = \frac{2n - 1}{(n + 1)(n + 2)} a_n \qquad n = 0, 1, 2, 3, \dots$$

We solve this recursion relation by putting $n = 0, 1, 2, 3, \dots$ successively in Equation 15.65:

Put $n = 0$: $\quad a_2 = \dfrac{-1}{1 \cdot 2} a_0$

Put $n = 1$: $\quad a_3 = \dfrac{1}{2 \cdot 3} a_1$

Put $n = 2$: $\quad a_4 = \dfrac{3}{3 \cdot 4} a_2 = -\dfrac{3}{1 \cdot 2 \cdot 3 \cdot 4} a_0 = -\dfrac{3}{4!} a_0$

Put $n = 3$: $\quad a_5 = \dfrac{5}{4 \cdot 5} a_3 = \dfrac{1 \cdot 5}{2 \cdot 3 \cdot 4 \cdot 5} a_1 = \dfrac{1 \cdot 5}{5!} a_1$

Put $n = 4$: $\quad a_6 = \dfrac{7}{5 \cdot 6} a_4 = -\dfrac{3 \cdot 7}{4! \, 5 \cdot 6} a_0 = -\dfrac{3 \cdot 7}{6!} a_0$

Put $n = 5$: $\quad a_7 = \dfrac{9}{6 \cdot 7} a_5 = \dfrac{1 \cdot 5 \cdot 9}{5! \, 6 \cdot 7} a_1 = \dfrac{1 \cdot 5 \cdot 9}{7!} a_1$

Put $n = 6$: $\quad a_8 = \dfrac{11}{7 \cdot 8} a_6 = -\dfrac{3 \cdot 7 \cdot 11}{8!} a_0$

Put $n = 7$: $\quad a_9 = \dfrac{13}{8 \cdot 9} a_7 = \dfrac{1 \cdot 5 \cdot 9 \cdot 13}{9!} a_1$

In general the even coefficients are given by

$$a_{2n} = -\frac{3 \cdot 7 \cdot 11 \cdot \, \cdots \, \cdot (4n - 5)}{(2n)!} a_0$$

and the odd coefficients are given by

$$a_{2n+1} = \frac{1 \cdot 5 \cdot 9 \cdot \, \cdots \, \cdot (4n - 3)}{(2n + 1)!} a_1$$

The solution is

$$y = a_0 + a_1 x + a_2 x^2 + a_3 x^3 + a_4 x^4 + \cdots$$

$$= a_0 \left(1 - \frac{1}{2!} x^2 - \frac{3}{4!} x^4 - \frac{3 \cdot 7}{6!} x^6 - \frac{3 \cdot 7 \cdot 11}{8!} x^8 - \cdots \right)$$

$$+ a_1 \left(x + \frac{1}{3!} x^3 + \frac{1 \cdot 5}{5!} x^5 + \frac{1 \cdot 5 \cdot 9}{7!} x^7 + \frac{1 \cdot 5 \cdot 9 \cdot 13}{9!} x^9 + \cdots \right)$$

or

$$(15.66) \qquad y = a_0 \left(1 - \sum_{n=1}^{\infty} \frac{3 \cdot 7 \cdot \cdots \cdot (4n - 5)}{(2n)!} x^{2n} \right)$$

$$+ a_1 \sum_{n=0}^{\infty} \frac{1 \cdot 5 \cdot 9 \cdot \cdots \cdot (4n - 3)}{(2n + 1)!} x^{2n + 1} \qquad \bullet$$

Note 2:   In Example 2 we had to assume that the differential equation had a series solution. But now we could verify directly that the function given by Equation 15.66 is indeed a solution.

Note 3:   Unlike the situation of Example 1, the power series that arise in the solution of Example 2 do not define elementary functions. The functions

$$y_1(x) = 1 - \sum_{n=1}^{\infty} \frac{3 \cdot 7 \cdot \cdots \cdot (4n - 5)}{(2n)!} x^{2n}$$

and

$$y_2(x) = \sum_{n=0}^{\infty} \frac{1 \cdot 5 \cdot 9 \cdot \cdots \cdot (4n - 3)}{(2n + 1)!} x^{2n + 1}$$

are perfectly good functions but they cannot be expressed in terms of familiar functions.

Note 4:   If we were asked to solve the initial-value problem

$$y'' - 2xy' + y = 0 \qquad y(0) = 0 \qquad y'(0) = 1$$

we would observe from Theorem 10.44 that

$$a_0 = y(0) = 0 \qquad a_1 = y'(0) = 1$$

This would simplify the calculations in Example 2 since all of the even co-efficients would be 0. The solution to the initial-value problem is

$$y(x) = \sum_{n=0}^{\infty} \frac{1 \cdot 5 \cdot 9 \cdot \cdots \cdot (4n - 3)}{(2n + 1)!} x^{2n + 1}$$

EXERCISES 15.10

Use power series to solve the following differential equations.

1. $y' = 6y$

2. $y' = xy$

3. $y' = x^2 y$

4. $y'' = y$

5. $y'' + 3xy' + 3y = 0$

6. $y'' = xy$

7. $(x^2 + 1)y'' + xy' - y = 0$

8. $y'' - xy' + 2y = 0$

9. $y'' - xy' - y = 0$, $y(0) = 1$, $y'(0) = 0$

10. $y'' + x^2 y = 0$, $y(0) = 1$, $y'(0) = 0$

11. $y'' + x^2 y' + xy = 0$, $y(0) = 0$, $y'(0) = 1$

12. $x^2 y'' + xy' + x^2 y = 0$, $y(0) = 1$, $y'(0) = 0$

(The solution of this initial-value problem is called a Bessel function of order 0.)

REVIEW OF CHAPTER 15

Write the general form of the following types of differential equation. Then discuss the method of solution in each case.

1. First order: (a) separable, (b) homogeneous, (c) linear, (d) exact, (e) integrating factor

2. Second order linear with constant coefficients: (a) homogeneous, (b) nonhomogeneous

REVIEW EXERCISES FOR CHAPTER 15

Solve the differential equations in Exercises 1–22.

1. $1 + 2xy^2 + 2x^2yy' = 0$

2. $y' + 2xy = 2x^3$

3. $xy' - 2y = x^3$

4. $xy' = y + x\cos^2\left(\dfrac{y}{x}\right)$

5. $(2y - 3y^2)y' = x\cos x$

6. $y' = 2 + 2x^2 + y + x^2y$

7. $(x^2 + xy)y' = x^2 + y^2$

8. $(2x\cos y - 3y^2)y' = 2(x - \sin y)$

9. $1 + y^2 - \sqrt{1 - x^2}\,y' = 0$

10. $yy' = x\sqrt{1 + x^2}\,\sqrt{1 + y^2}$

11. $y' - y = e^{2x}$ **12.** $y'' - y = e^{2x}$

13. $y'' - 6y' + 34y = 0$ **14.** $4y'' - 20y' + 25y = 0$

15. $2y'' + y' = y$ **16.** $3y'' + 2y' + y = 0$

17. $\dfrac{d^2y}{dx^2} + 2\dfrac{dy}{dx} + y = \sin 3x$

18. $\dfrac{d^2y}{dx^2} - 2\dfrac{dy}{dx} - 3y = \cos 4x$

19. $4\dfrac{d^2y}{dx^2} + 9y = 2x^2 - 3$

20. $\dfrac{d^2y}{dx^2} - 4\dfrac{dy}{dx} + 20y = xe^x$

21. $\dfrac{d^2y}{dx^2} - 3\dfrac{dy}{dx} + 2y = e^{2x}$

22. $\dfrac{d^2y}{dx^2} + y = \csc x,\ 0 < x < \dfrac{\pi}{2}$

Solve the initial-value problems in Exercises 23–28.

23. $y' + y = \sqrt{x}\,e^{-x},\ y(0) = 3$

24. $1 + x = 2xyy',\ x > 0,\ y(1) = -2$

25. $y'' + 6y' = 0,\ y(1) = 3,\ y'(1) = 12$

26. $y'' - 6y' + 25y = 0,\ y(0) = 2,\ y'(0) = 1$

27. $y'' - 5y' + 4y = 0,\ y(0) = 0,\ y'(0) = 1$

28. $9y'' + y = 3x + e^{-x},\ y(0) = 1,\ y'(0) = 2$

In Exercises 29 and 30 find the orthogonal trajectories of the given family of curves.

29. $kx^2 + y^2 = 1$ **30.** $y = \dfrac{k}{1 + x^2}$

31. Use power series to solve the initial-value problem $y'' + xy' + y = 0,\ y(0) = 0,\ y'(0) = 1$.

32. Use power series to solve the equation $y'' - xy' - 2y = 0$.

33. A series circuit contains a resistor with $R = 40\ \Omega$, an inductor with $L = 2$ H, a capacitor with $C = 0.0025$ F, and a 12-V battery. The initial charge is $Q = 0.01$ C and the initial current is 0. Find the charge at time t.

34. A spring with a mass of 2 kg has damping constant 16, and a force of 12.8 N stretches the spring 0.2 m beyond its natural length. Find the position of the mass at time t if it starts at the equilibrium position with a velocity of 2.4 m/s.

A Review of Algebra

Arithmetic Operations

The real numbers have the following properties:

$$
\begin{array}{lll}
a + b = b + a & ab = ba & \text{(Commutative Law)} \\
(a + b) + c = a + (b + c) & (ab)c = a(bc) & \text{(Associative Law)} \\
a(b + c) = ab + ac & & \text{(Distributive Law)}
\end{array}
$$

In particular, putting $a = -1$ in the Distributive Law, we get

$$-(b + c) = (-1)(b + c) = (-1)b + (-1)c$$

and so

$$-(b + c) = -b - c$$

● **Example 1**

(a) $(3xy)(-4x) = 3(-4)x^2y = -12x^2y$

(b) $2t(7x + 2tx - 11) = 14tx + 4t^2x - 22t$

(c) $4 - 3(x - 2) = 4 - 3x + 6 = 10 - 3x$ ●

If we use the Distributive Law three times, we get

$$(a + b)(c + d) = (a + b)c + (a + b)d = ac + bc + ad + bd$$

This says that we multiply two factors by multiplying each term in one factor by each term in the other factor and adding the products. Schematically, we have

$$(a + b)(c + d)$$

In the case where $c = a$ and $d = b$, we have

$$(a + b)^2 = a^2 + ba + ab + b^2$$

or

(1)
$$(a + b)^2 = a^2 + 2ab + b^2$$

Similarly, we obtain

(2)
$$(a - b)^2 = a^2 - 2ab + b^2$$

● **Example 2**

(a) $(2x + 1)(3x - 5) = 6x^2 + 3x - 10x - 5 = 6x^2 - 7x - 5$

(b) $(x + 6)^2 = x^2 + 12x + 36$

(c) $3(x - 1)(4x + 3) - 2(x + 6) = 3(4x^2 - x - 3) - 2x - 12$

$$= 12x^2 - 3x - 9 - 2x - 12$$
$$= 12x^2 - 5x - 21$$ ●

Fractions

To add two fractions with the same denominator, we use the Distributive Law:

$$\frac{a}{b} + \frac{c}{b} = \frac{1}{b} \times a + \frac{1}{b} \times c = \frac{1}{b}(a + c) = \frac{a + c}{b}$$

Thus it is true that

$$\frac{a + c}{b} = \frac{a}{b} + \frac{c}{b}$$

But remember to avoid the following common error:

$$\frac{a}{b+c} \ne \frac{a}{b} + \frac{a}{c}$$

(For instance, take $a = b = c = 1$ to see the error.)

To add two fractions with different denominators, we use a common denominator:

$$\frac{a}{b} + \frac{c}{d} = \frac{ad + bc}{bd}$$

We multiply them as follows:

$$\frac{a}{b} \cdot \frac{c}{d} = \frac{ac}{bd}$$

In particular, it is true that

$$\frac{-a}{b} = -\frac{a}{b} = \frac{a}{-b}$$

To divide two fractions, we invert and multiply:

$$\frac{\dfrac{a}{b}}{\dfrac{c}{d}} = \frac{a}{b} \times \frac{d}{c} = \frac{ad}{bc}$$

● **Example 3**

(a) $\dfrac{x + 3}{x} = \dfrac{x}{x} + \dfrac{3}{x} = 1 + \dfrac{3}{x}$

(b) $\dfrac{3}{x - 1} + \dfrac{x}{x + 2} = \dfrac{3(x + 2) + x(x - 1)}{(x - 1)(x + 2)} = \dfrac{3x + 6 + x^2 - x}{x^2 + x - 2}$

$$= \dfrac{x^2 + 2x + 6}{x^2 + x - 2}$$

(c) $\dfrac{s^2 t}{u} \cdot \dfrac{ut}{-2} = \dfrac{s^2 t^2 u}{-2u} = -\dfrac{s^2 t^2}{2}$

(d) $\dfrac{\dfrac{x}{y}+1}{1-\dfrac{y}{x}} = \dfrac{\dfrac{x+y}{y}}{\dfrac{x-y}{x}} = \dfrac{x+y}{y} \times \dfrac{x}{x-y} = \dfrac{x(x+y)}{y(x-y)} = \dfrac{x^2+xy}{xy-y^2}$ ●

Factoring

We have used the Distributive Law to expand certain algebraic expressions. We sometimes need to reverse this process (again using the Distributive Law) by factoring an expression as a product of simpler ones. The easiest situation occurs when there is a common factor as follows:

$$\xrightarrow{\quad\text{Expanding}\quad}$$

$$3x(x-2) = 3x^2 - 6x$$

$$\xleftarrow{\quad\text{Factoring}\quad}$$

To factor a quadratic of the form $x^2 + bx + c$ we note that

$$(x+r)(x+s) = x^2 + (r+s)x + rs$$

so we need to choose numbers r and s so that $r + s = b$ and $rs = c$.

● **Example 4**   Factor $x^2 + 5x - 24$.

Solution   The two integers that add to give 5 and multiply to give -24 are -3 and 8. Therefore

$$x^2 + 5x - 24 = (x-3)(x+8)$$ ●

● **Example 5**   Factor $2x^2 - 7x - 4$.

Solution   Even though the coefficient of x^2 is not 1, we can still look for factors of the form $2x + r$ and $x + s$, where $rs = -4$. Experimentation reveals that

$$2x^2 - 7x - 4 = (2x+1)(x-4)$$ ●

Some special quadratics can be factored by using Equations 1 or 2 (from right to left) or by using the formula for a difference of squares:

(3)
$$a^2 - b^2 = (a-b)(a+b)$$

The analogous formula for a difference of cubes is

(4)
$$a^3 - b^3 = (a-b)(a^2 + ab + b^2)$$

which you can verify by expanding the right side. For a sum of cubes we have

(5)
$$a^3 + b^3 = (a + b)(a^2 - ab + b^2)$$

● **Example 6**
(a) $x^2 - 6x + 9 = (x - 3)^2$ *(Equation 2; a = x, b = 3)*
(b) $4x^2 - 25 = (2x - 5)(2x + 5)$ *(Equation 3; a = 2x, b = 5)*
(c) $x^3 + 8 = (x + 2)(x^2 - 2x + 4)$ *(Equation 5; a = x, b = 2)*   ●

● **Example 7**   Simplify $\dfrac{x^2 - 16}{x^2 - 2x - 8}$.

Solution   Factoring numerator and denominator, we have

$$\frac{x^2 - 16}{x^2 - 2x - 8} = \frac{(x - 4)(x + 4)}{(x - 4)(x + 2)} = \frac{x + 4}{x + 2}$$ ●

To factor polynomials of degree 3 or more, we sometimes use the following fact.

The Factor Theorem (6)   If P is a polynomial and $P(b) = 0$, then $x - b$ is a factor of $P(x)$.

● **Example 8**   Factor $x^3 - 3x^2 - 10x + 24$.

Solution   Let $P(x) = x^3 - 3x^2 - 10x + 24$. If $P(b) = 0$, where b is an integer, then b is a factor of 24. Thus the possibilities for b are $\pm 1, \pm 2, \pm 3,$ $\pm 4, \pm 6, \pm 8, \pm 12, \pm 24$. We find that $P(1) = 12$, $P(-1) = 30$, $P(2) = 0$. By the Factor Theorem, $x - 2$ is a factor. Instead of substituting further, we use long division as follows:

$$
\begin{array}{r}
x^2 - x - 12 \\
x - 2 \overline{\smash{\big)}\, x^3 - 3x^2 - 10x + 24} \\
\underline{x^3 - 2x^2} \\
-x^2 - 10x \\
\underline{-x^2 + 2x} \\
-12x + 24 \\
\underline{-12x + 24}
\end{array}
$$

Therefore

$$x^3 - 3x^2 - 10x + 24 = (x - 2)(x^2 - x - 12) = (x - 2)(x + 3)(x - 4)$$ ●

Completing the Square

For purposes of graphing parabolas (as in Example 2 in Section 1.7) or integrating rational functions (as in Example 6 in Section 7.4) the technique of completing the square is useful. This means rewriting a quadratic $ax^2 + bx + c$ in the form $a(x + p)^2 + q$ and can be accomplished by:

1. Factoring the number a from the terms involving x
2. Adding and subtracting the square of half the coefficient of x

In general, we have

$$ax^2 + bx + c = a\left[x^2 + \frac{b}{a}x\right] + c$$

$$= a\left[x^2 + \frac{b}{a}x + \left(\frac{b}{2a}\right)^2 - \left(\frac{b}{2a}\right)^2\right] + c$$

$$= a\left(x + \frac{b}{2a}\right)^2 + \left(c - \frac{b^2}{4a}\right)$$

● **Example 9**   Rewrite $x^2 + x + 1$ by completing the square.

Solution   The square of half the coefficient of x is $\frac{1}{4}$. Thus

$$x^2 + x + 1 = x^2 + x + \tfrac{1}{4} - \tfrac{1}{4} + 1 = (x + \tfrac{1}{2})^2 + \tfrac{3}{4} \qquad ●$$

● **Example 10**

$$2x^2 - 12x + 11 = 2[x^2 - 6x] + 11 = 2[x^2 - 6x + 9 - 9] + 11$$

$$= 2[(x - 3)^2 - 9] + 11 = 2(x - 3)^2 - 7 \qquad ●$$

Quadratic Formula

By completing the square as above we can obtain the following formula for the roots of a quadratic equation.

The Quadratic Formula (7)

The roots of the quadratic equation $ax^2 + bx + c = 0$ are

$$x = \frac{-b \pm \sqrt{b^2 - 4ac}}{2a}$$

● **Example 11**   Solve the equation $5x^2 + 3x - 3 = 0$.

Solution   With $a = 5$, $b = 3$, $c = -3$, the quadratic formula gives the solutions

$$x = \frac{-3 \pm \sqrt{3^2 - 4(5)(-3)}}{2(5)} = \frac{-3 \pm \sqrt{69}}{10}$$

●

The quantity $b^2 - 4ac$ that appears in the quadratic formula is called the **discriminant.** There are three possibilities:

1. If $b^2 - 4ac > 0$, the equation has two real roots.
2. If $b^2 - 4ac = 0$, the roots are equal.
3. If $b^2 - 4ac < 0$, there is no real root. (The roots are complex.)

These three cases correspond to the fact that the number of times the parabola $y = ax^2 + bx + c$ crosses the x-axis is 2, 1, or 0 (see Figure 1). In case 3 the quadratic $ax^2 + bx + c$ cannot be factored and is called **irreducible.**

Figure 1
POSSIBLE GRAPHS OF
$y = ax^2 + bx + c$

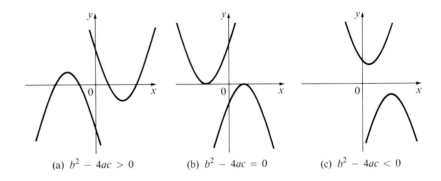

(a) $b^2 - 4ac > 0$ (b) $b^2 - 4ac = 0$ (c) $b^2 - 4ac < 0$

● **Example 12**   The quadratic $x^2 + x + 2$ is irreducible because its discriminant is negative:

$$b^2 - 4ac = 1^2 - 4(1)(2) = -7 < 0$$

Therefore it is impossible to factor $x^2 + x + 2$.   ●

The Binomial Theorem

Recall the binomial expansion from Equation 1:

$$(a + b)^2 = a^2 + 2ab + b^2$$

If we multiply both sides by $(a + b)$ and simplify, we get the binomial expansion

(8)

$$(a + b)^3 = a^3 + 3a^2b + 3ab^2 + b^3$$

Repeating this procedure, we get

$$(a + b)^4 = a^4 + 4a^3b + 6a^2b^2 + 4ab^3 + b^4$$

In general, we have the following formula (which is a special case of the Binomial Theorem of Section 10.10):

The Binomial Theorem (9)

If k is a positive integer, then

$$(a + b)^k = a^k + ka^{k-1}b + \frac{k(k - 1)}{1 \cdot 2} a^{k-2}b^2$$

$$+ \frac{k(k - 1)(k - 2)}{1 \cdot 2 \cdot 3} a^{k-3}b^3$$

$$+ \cdots + \frac{k(k - 1) \cdots (k - n + 1)}{1 \cdot 2 \cdot 3 \cdot \cdots \cdot n} a^{k-n}b^n$$

$$+ \cdots + kab^{k-1} + b^k$$

● **Example 13**   Expand $(x - 2)^5$.

Solution   Using the Binomial Theorem with $a = x$, $b = -2$, $k = 5$, we have

$$(x - 2)^5 = x^5 + 5x^4(-2) + \frac{5 \cdot 4}{1 \cdot 2} x^3(-2)^2 + \frac{5 \cdot 4 \cdot 3}{1 \cdot 2 \cdot 3} x^2(-2)^3$$

$$+ 5x(-2)^4 + (-2)^5$$

$$= x^5 - 10x^4 + 40x^3 - 80x^2 + 80x - 32$$ ●

Radicals

The most commonly occurring radicals are square roots. The symbol $\sqrt{}$ means "the positive square root of." Thus

$$x = \sqrt{a} \quad \text{means} \quad x^2 = a \text{ and } x \geqslant 0$$

Since $a = x^2 \geqslant 0$, the symbol $\sqrt{a}$ makes sense only when $a \geqslant 0$. Here are two rules for working with square roots:

(10)

$$\sqrt{ab} = \sqrt{a}\sqrt{b} \qquad \sqrt{\frac{a}{b}} = \frac{\sqrt{a}}{\sqrt{b}}$$

However there is no similar rule for the square root of a sum. In fact, you should remember to avoid the following common error:

$$\sqrt{a + b} \not= \sqrt{a} + \sqrt{b}$$

(For instance, take $a = 9$ and $b = 16$ to see the error.)

● **Example 14**

(a) $\dfrac{\sqrt{18}}{\sqrt{2}} = \sqrt{\dfrac{18}{2}} = \sqrt{9} = 3$

(b) $\sqrt{x^2 y} = \sqrt{x^2}\sqrt{y} = |x|\sqrt{y}$

Notice that $\sqrt{x^2} = |x|$ because $\sqrt{}$ indicates the positive square root. (See Section 1.1.) ●

In general, if n is a positive integer,

$$x = \sqrt[n]{a} \quad \text{means} \quad x^n = a$$

If n is even, then $a \geqslant 0$ and $x \geqslant 0$.

Thus $\sqrt[3]{-8} = -2$ because $(-2)^3 = -8$, but $\sqrt[4]{-8}$ and $\sqrt[6]{-8}$ are not defined The following rules are valid:

$$\sqrt[n]{ab} = \sqrt[n]{a}\sqrt[n]{b} \qquad \sqrt[n]{\dfrac{a}{b}} = \dfrac{\sqrt[n]{a}}{\sqrt[n]{b}}$$

● **Example 15** $\sqrt[3]{x^4} = \sqrt[3]{x^3 x} = \sqrt[3]{x^3}\sqrt[3]{x} = x\sqrt[3]{x}$ ●

Exponents

Let a be any positive number and let n be a positive integer. Then, by definition,

1. $a^n = \underbrace{a \cdot a \cdot \cdots \cdot a}_{n \text{ factors}}$

2. $a^0 = 1$

3. $a^{-n} = \dfrac{1}{a^n}$

4. $a^{1/n} = \sqrt[n]{a}$
   $a^{m/n} = \sqrt[n]{a^m} = (\sqrt[n]{a})^m$ m is any integer

Laws of Exponents (11)

Let a and b be positive numbers and let r and s be any rational numbers (that is, ratios of integers). Then

1. $a^r \times a^s = a^{r+s}$ 2. $\dfrac{a^r}{a^s} = a^{r-s}$

3. $(a^r)^s = a^{rs}$

4. $(ab)^r = a^r b^r$ 5. $\left(\dfrac{a}{b}\right)^r = \dfrac{a^r}{b^r}$ $b \neq 0$

In words, these five laws can be stated as follows:

1. To multiply two powers of the same number, we add the exponents.
2. To divide two powers of the same number, we subtract the exponents.
3. To raise a power to a new power, we multiply the exponents.
4. To raise a product to a power, we raise each factor to the power.
5. To raise a quotient to a power, we raise both numerator and denominator to the power.

● **Example 16**

(a) $2^8 \times 8^2 = 2^8 \times (2^3)^2 = 2^8 \times 2^6 = 2^{14}$

(b) $\dfrac{x^{-2} - y^{-2}}{x^{-1} + y^{-1}} = \dfrac{\dfrac{1}{x^2} - \dfrac{1}{y^2}}{\dfrac{1}{x} + \dfrac{1}{y}} = \dfrac{\dfrac{y^2 - x^2}{x^2 y^2}}{\dfrac{y + x}{xy}} = \dfrac{y^2 - x^2}{x^2 y^2} \cdot \dfrac{xy}{y + x}$

$\qquad\qquad = \dfrac{(y - x)(y + x)}{xy(y + x)} = \dfrac{y - x}{xy}$

(c) $4^{3/2} = \sqrt{4^3} = \sqrt{64} = 8$ Alternative solution: $4^{3/2} = (\sqrt{4})^3 = 2^3 = 8$

(d) $\dfrac{1}{\sqrt[3]{x^4}} = \dfrac{1}{x^{4/3}} = x^{-4/3}$

(e) $\left(\dfrac{x}{y}\right)^3 \left(\dfrac{y^2 x}{z}\right)^4 = \dfrac{x^3}{y^3} \cdot \dfrac{y^8 x^4}{z^4} = x^7 y^5 z^{-4}$ ●

Mathematical Induction

The principle of mathematical induction is useful when proving a statement S_n about the positive integer n. For instance, if S_n is the statement

$$(ab)^n = a^n b^n$$

then

S_1 says that $ab = ab$
S_2 says that $(ab)^2 = a^2 b^2$

and so on.

Let S_n be a statement about the positive integer n. Suppose that
1. S_1 is true
2. S_{k+1} is true whenever S_k is true
Then S_n is true for all positive integers n.

This is reasonable because, since S_1 is true, it follows from condition 2 (with $k = 1$) that S_2 is true. Then, using condition 2 with $k = 2$, we see that S_3 is true. Again using condition 2, this time with $k = 3$, we have that S_4 is true. This procedure can be followed indefinitely.

In using the principle of mathematical induction, there are three steps.

Step 1: Prove that S_n is true when $n = 1$.

Step 2: Assume that S_n is true when $n = k$ and deduce that S_n is true when $n = k + 1$.

Step 3: Conclude that S_n is true for all n by the principle of mathematical induction.

● **Example 17** If a and b are real numbers, prove that $(ab)^n = a^n b^n$ for every positive integer n.

Solution Let S_n be the given statement.
1. S_1 is true because $(ab)^1 = ab = a^1 b^1$.
2. Assume that S_k is true, that is, $(ab)^k = a^k b^k$. Then

$$(ab)^{k+1} = (ab)^k(ab) = a^k b^k ab$$
$$= (a^k a)(b^k b) = a^{k+1} b^{k+1}$$

This says that S_{k+1} is true.
3. Therefore, by the principle of mathematical induction, S_n is true for all n; that is, $(ab)^n = a^n b^n$ for every positive integer n. ●

● **Example 18** Prove that, for every positive integer n,

$$1 + 2 + 3 + \cdots + n = \frac{n(n + 1)}{2}$$

Solution Let S_n be the given statement.
1. S_1 is true because

$$1 = \frac{1(1 + 1)}{2}$$

2. Assume that S_k is true, that is,

$$1 + 2 + \cdots + k = \frac{k(k + 1)}{2}$$

Then

$$1 + 2 + \cdots + (k + 1) = (1 + 2 + \cdots + k) + (k + 1)$$

$$= \frac{k(k + 1)}{2} + k + 1$$

$$= \frac{k(k + 1) + 2(k + 1)}{2}$$

$$= \frac{(k + 1)(k + 2)}{2}$$

Thus

$$1 + 2 + \cdots + (k + 1) = \frac{(k + 1)[(k + 1) + 1]}{2}$$

which shows that S_{k+1} is true.

3.   Therefore S_n is true for all n by mathematical induction, that is,

$$1 + 2 + \cdots + (n + 1) = \frac{n(n + 1)}{2}$$

for every positive integer n.   ●

EXERCISES A

In Exercises 1–16 expand and simplify.

1. $(-6ab)(0.5ac)$

2. $-(2x^2y)(-xy^4)$

3. $2x(x - 5)$

4. $(4 - 3x)x$

5. $-2(4 - 3a)$

6. $8 - (4 + x)$

7. $4(x^2 - x + 2) - 5(x^2 - 2x + 1)$

8. $5(3t - 4) - (t^2 + 2) - 2t(t - 3)$

9. $(4x - 1)(3x + 7)$

10. $x(x - 1)(x + 2)$

11. $(2x - 1)^2$

12. $(2 + 3x)^2$

13. $y^4(6 - y)(5 + y)$

14. $(t - 5)^2 - 2(t + 3)(8t - 1)$

15. $(1 + 2x)(x^2 - 3x + 1)$

16. $(1 + x - x^2)^2$

In Exercises 17–28 perform the indicated operations and simplify.

17. $\dfrac{2 + 8x}{2}$

18. $\dfrac{9b - 6}{3b}$

19. $\dfrac{1}{x + 5} + \dfrac{2}{x - 3}$

20. $\dfrac{1}{x + 1} + \dfrac{1}{x - 1}$

21. $u + 1 + \dfrac{u}{u + 1}$

22. $\dfrac{2}{a^2} - \dfrac{3}{ab} + \dfrac{4}{b^2}$

23. $\dfrac{x/y}{z}$

24. $\dfrac{x}{y/z}$

25. $\left(\dfrac{-2r}{s}\right)\left(\dfrac{s^2}{-6t}\right)$

26. $\dfrac{a}{bc} \div \dfrac{b}{ac}$

27. $\dfrac{1 + \dfrac{1}{c-1}}{1 - \dfrac{1}{c-1}}$

28. $1 + \dfrac{1}{1 + \dfrac{1}{1+x}}$

In Exercises 29–48 factor the given expression.

29. $2x + 12x^3$

30. $5ab - 8abc$

31. $x^2 + 7x + 6$

32. $x^2 - x - 6$

33. $x^2 - 2x - 8$

34. $2x^2 + 7x - 4$

35. $9x^2 - 36$

36. $8x^2 + 10x + 3$

37. $6x^2 - 5x - 6$

38. $x^2 + 10x + 25$

39. $t^3 + 1$

40. $4t^2 - 9s^2$

41. $4t^2 - 12t + 9$

42. $x^3 - 27$

43. $x^3 + 2x^2 + x$

44. $x^3 - 4x^2 + 5x - 2$

45. $x^3 + 3x^2 - x - 3$

46. $x^3 - 2x^2 - 23x + 60$

47. $x^3 + 5x^2 - 2x - 24$

48. $x^3 - 3x^2 - 4x + 12$

In Exercises 49–54 simplify the given expression.

49. $\dfrac{x^2 + x - 2}{x^2 - 3x + 2}$

50. $\dfrac{2x^2 - 3x - 2}{x^2 - 4}$

51. $\dfrac{x^2 - 1}{x^2 - 9x + 8}$

52. $\dfrac{x^3 + 5x^2 + 6x}{x^2 - x - 12}$

53. $\dfrac{1}{x+3} + \dfrac{1}{x^2 - 9}$

54. $\dfrac{x}{x^2 + x - 2} - \dfrac{2}{x^2 - 5x + 4}$

In Exercises 55–60 complete the square.

55. $x^2 + 2x + 5$

56. $x^2 - 16x + 80$

57. $x^2 - 5x + 10$

58. $x^2 + 3x + 1$

59. $4x^2 + 4x - 2$

60. $3x^2 - 24x + 50$

In Exercises 61–68 solve the given equation.

61. $x^2 + 9x - 10 = 0$

62. $x^2 - 2x - 8 = 0$

63. $x^2 + 9x - 1 = 0$

64. $x^2 - 2x - 7 = 0$

65. $3x^2 + 5x + 1 = 0$

66. $2x^2 + 7x + 2 = 0$

67. $x^3 - 2x + 1 = 0$

68. $x^3 + 3x^2 + x - 1 = 0$

Which of the quadratics in Exercises 69–72 are irreducible?

69. $2x^2 + 3x + 4$

70. $2x^2 + 9x + 4$

71. $3x^2 + x - 6$

72. $x^2 + 3x + 6$

In Exercises 73–76 use the Binomial Theorem to expand the given expression.

73. $(a + b)^6$

74. $(a + b)^7$

75. $(x^2 - 1)^4$

76. $(3 + x^2)^5$

In Exercises 77–82 simplify the given radicals.

77. $\sqrt{32}\sqrt{2}$

78. $\dfrac{\sqrt[3]{-2}}{\sqrt[3]{54}}$

79. $\dfrac{\sqrt[4]{32x^4}}{\sqrt[4]{2}}$

80. $\sqrt{xy}\sqrt{x^3 y}$

81. $\sqrt{16a^4 b^3}$

82. $\dfrac{\sqrt[5]{96a^6}}{\sqrt[5]{3a}}$

In Exercises 83–100 use the Laws of Exponents to rewrite and simplify the given expression.

83. $3^{10} \times 9^8$

84. $2^{16} \times 4^{10} \times 16^6$

85. $\dfrac{x^9 (2x)^4}{x^3}$

86. $\dfrac{a^n \times a^{2n+1}}{a^{n-2}}$

87. $\dfrac{a^{-3} b^4}{a^{-5} b^5}$

88. $\dfrac{x^{-1} + y^{-1}}{(x + y)^{-1}}$

89. $3^{-1/2}$

90. $96^{1/5}$

91. $125^{2/3}$

92. $64^{-4/3}$

93. $(2x^2 y^4)^{3/2}$

94. $(x^{-5} y^3 z^{10})^{-3/5}$

95. $\sqrt[5]{y^6}$

96. $(\sqrt[4]{a})^3$

97. $\dfrac{1}{(\sqrt{t})^5}$

98. $\dfrac{\sqrt[8]{x^5}}{\sqrt[4]{x^3}}$

99. $\sqrt[4]{\dfrac{t^{1/2}\sqrt{st}}{s^{2/3}}}$

100. $\sqrt[4]{r^{2n+1}} \times \sqrt[4]{r^{-1}}$

In Exercises 101–108 state whether or not the given equation is true for all values of the variable.

101. $\sqrt{x^2} = x$

102. $\sqrt{x^2 + 4} = |x| + 2$

103. $\dfrac{16 + a}{16} = 1 + \dfrac{a}{16}$

104. $\dfrac{1}{x^{-1} + y^{-1}} = x + y$

105. $\dfrac{x}{x+y} = \dfrac{1}{1+y}$ **106.** $\dfrac{2}{4+x} = \dfrac{1}{2} + \dfrac{2}{x}$

107. $(x^3)^4 = x^7$

108. $6 - 4(x+a) = 6 - 4x - 4a$

In Exercises 109–118 n represents a positive integer. Use mathematical induction to prove the given statement.

109. $2^n > n$

110. $3^n > 2n$

111. $(1+x)^n > 1 + nx$ (where $x \geqslant -1$)

112. If $0 \leqslant a < b$, then $a^n < b^n$.

113. $7^n - 1$ is divisible by 6.

114. $\left(\dfrac{a}{b}\right)^n = \dfrac{a^n}{b^n}$

115. $1 + 3 + 5 + \cdots + (2n - 1) = n^2$

116. $2 + 6 + 12 + \cdots + n(n+1) = \dfrac{n(n+1)(n+2)}{3}$

117. $\dfrac{1}{2} + \dfrac{1}{6} + \dfrac{1}{12} + \cdots + \dfrac{1}{n(n+1)} = \dfrac{n}{n+1}$

118. $a + ar + ar^2 + \cdots + ar^{n-1} = \dfrac{a(1 - r^n)}{1 - r}$ $(r \neq -1)$

B Review of Trigonometry

Angles

Angles can be measured in degrees or radians (abbreviated as rad). The angle given by a complete revolution contains $360°$, which is the same as 2π rad. Therefore

(1)
$$\pi \text{ rad} = 180°$$

and

(2)
$$1 \text{ rad} = \left(\frac{180}{\pi}\right)° \approx 57.3° \qquad 1° = \frac{\pi}{180} \text{ rad} \approx 0.017 \text{ rad}$$

● **Example 1** (a) Find the radian measure of $60°$. (b) Express $5\pi/4$ rad in degrees.

Solution (a) From Equation 1 or 2 we see that to convert from degrees to radians we multiply by $\pi/180$. Therefore

$$60° = 60\left(\frac{\pi}{180}\right) = \frac{\pi}{3} \text{ rad}$$

(b) To convert from radians to degrees we multiply by $180/\pi$. Thus

$$\frac{5\pi}{4} \text{ rad} = \frac{5\pi}{4}\left(\frac{180}{\pi}\right) = 225°$$

In calculus we use radians to measure angles except when otherwise indicated. The following table gives the correspondence between degree and radian measures of some common angles.

Degrees	0°	30°	45°	60°	90°	120°	135°	150°	180°	270°	360°
Radians	0	$\dfrac{\pi}{6}$	$\dfrac{\pi}{4}$	$\dfrac{\pi}{3}$	$\dfrac{\pi}{2}$	$\dfrac{2\pi}{3}$	$\dfrac{3\pi}{4}$	$\dfrac{5\pi}{6}$	π	$\dfrac{3\pi}{2}$	2π

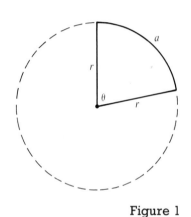

Figure 1

Figure 1 shows a sector of a circle with central angle θ and radius r subtending an arc with length a. Since the length of the arc is proportional to the size of the angle, and since the entire circle has circumference $2\pi r$ and central angle 2π, we have

$$\frac{\theta}{2\pi} = \frac{a}{2\pi r}$$

Solving this equation for θ and for a, we obtain

(3)
$$\boxed{\theta = \frac{a}{r}} \qquad \boxed{a = r\theta}$$

Remember that Equations 3 are valid only when θ is measured in radians.

In particular, putting $a = r$ in Equation 3, we see that an angle of 1 rad is the angle subtended at the center of a circle by an arc equal in length to the radius of the circle (see Figure 2).

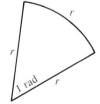

Figure 2

● **Example 2**   (a) If the radius of a circle is 5 cm, what angle is subtended by an arc of 6 cm? (b) If a circle has radius 3 cm, what is the length of an arc subtended by a central angle of $3\pi/8$ rad?

Solution   (a)   Using Equation 3 with $a = 6$ and $r = 5$, we see that the angle is

$$\theta = \frac{6}{5} = 1.2 \text{ rad}$$

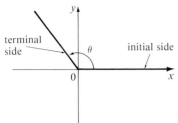

Figure 3

$\theta \geqslant 0$

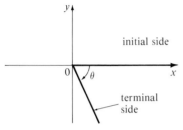

Figure 4

$\theta < 0$

(b)   With $r = 3$ cm and $\theta = 3\pi/8$ rad, the arc length is

$$a = r\theta = 3\left(\frac{3\pi}{8}\right) = \frac{9\pi}{8} \text{ cm}$$

The **standard position** of an angle occurs when we place its vertex at the origin of a coordinate system and its initial side on the positive x-axis as in Figure 3. A **positive** angle is obtained by rotating the initial side counterclockwise until it coincides with the terminal side. Likewise, **negative** angles are obtained by clockwise rotation as in Figure 4. Figure 5 shows several examples of angles in standard position. Notice that different angles can have the same terminal side. For instance, the angles $3\pi/4$, $-5\pi/4$, and $11\pi/4$ have the same initial and terminal sides because

$$\frac{3\pi}{4} - 2\pi = -\frac{5\pi}{4} \qquad \frac{3\pi}{4} + 2\pi = \frac{11\pi}{4}$$

and 2π rad represents a complete revolution.

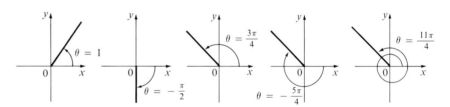

Figure 5

ANGLES IN STANDARD POSITION

The Trigonometric Functions

For an acute angle θ the six trigonometric functions are defined as ratios of lengths of sides of a right triangle as follows (see Figure 6):

(4)

$$\sin\theta = \frac{\text{opp}}{\text{hyp}} \qquad \csc\theta = \frac{\text{hyp}}{\text{opp}}$$

$$\cos\theta = \frac{\text{adj}}{\text{hyp}} \qquad \sec\theta = \frac{\text{hyp}}{\text{adj}}$$

$$\tan\theta = \frac{\text{opp}}{\text{adj}} \qquad \cot\theta = \frac{\text{adj}}{\text{opp}}$$

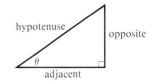

Figure 6

This procedure does not work for obtuse or negative angles, so for a general angle θ in standard position we let $P(x, y)$ be any point on the terminal side of θ and we let r be the distance $|OP|$ as in Figure 7. Then we define

(5)
$$\sin \theta = \frac{y}{r} \qquad \csc \theta = \frac{r}{y}$$

$$\cos \theta = \frac{x}{r} \qquad \sec \theta = \frac{r}{x}$$

$$\tan \theta = \frac{y}{x} \qquad \cot \theta = \frac{x}{y}$$

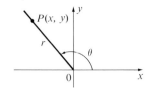

Figure 7

Since division by 0 is not defined, $\tan \theta$ and $\sec \theta$ are undefined when $x = 0$ and $\csc \theta$ and $\cot \theta$ are undefined when $y = 0$. Notice that the definitions in (4) and (5) are consistent when θ is an acute angle.

If θ is a number, the convention is that $\sin \theta$ means the sine of the angle whose *radian* measure is θ. For example, the expression $\sin 3$ implies that we are dealing with an angle of 3 rad. When finding a calculator approximation to this number we must remember to set our calculator in radian mode and then we obtain

$$\sin 3 \approx 0.14112$$

If we want to know the sine of the angle $3°$ we would write $\sin 3°$ and, with our calculator in degree mode, we find that

$$\sin 3° \approx 0.05234$$

The exact trigonometric ratios for certain angles can be read from the triangles in Figure 8. For instance,

$$\sin \frac{\pi}{4} = \frac{1}{\sqrt{2}} \qquad \sin \frac{\pi}{6} = \frac{1}{2} \qquad \sin \frac{\pi}{3} = \frac{\sqrt{3}}{2}$$

$$\cos \frac{\pi}{4} = \frac{1}{\sqrt{2}} \qquad \cos \frac{\pi}{6} = \frac{\sqrt{3}}{2} \qquad \cos \frac{\pi}{3} = \frac{1}{2}$$

$$\tan \frac{\pi}{4} = 1 \qquad \tan \frac{\pi}{6} = \frac{1}{\sqrt{3}} \qquad \tan \frac{\pi}{3} = \sqrt{3}$$

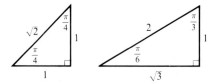

Figure 8

The signs of the trigonometric functions for angles in each of the four quadrants can be remembered by means of the CAST rule as follows:

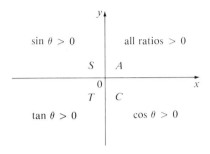

- **Example 3**   Find the exact trigonometric ratios for $\theta = 2\pi/3$.

Solution   From Figure 9 we see that a point on the terminal line for $\theta = 2\pi/3$ is $P(-1, \sqrt{3})$. Therefore, taking

$$x = -1 \qquad y = \sqrt{3} \qquad r = 2$$

in the definitions of the trigonometric ratios, we have

$$\sin\frac{2\pi}{3} = \frac{\sqrt{3}}{2} \qquad \cos\frac{2\pi}{3} = -\frac{1}{2} \qquad \tan\frac{2\pi}{3} = -\sqrt{3}$$

$$\csc\frac{2\pi}{3} = \frac{2}{\sqrt{3}} \qquad \sec\frac{2\pi}{3} = -2 \qquad \cot\frac{2\pi}{3} = -\frac{1}{\sqrt{3}}$$

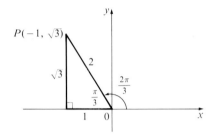

Figure 9

The following table gives some values of $\sin\theta$ and $\cos\theta$ found by the method of Example 3.

θ	0	$\dfrac{\pi}{6}$	$\dfrac{\pi}{4}$	$\dfrac{\pi}{3}$	$\dfrac{\pi}{2}$	$\dfrac{2\pi}{3}$	$\dfrac{3\pi}{4}$	$\dfrac{5\pi}{6}$	π	$\dfrac{3\pi}{2}$	2π
$\sin\theta$	0	$\dfrac{1}{2}$	$\dfrac{1}{\sqrt{2}}$	$\dfrac{\sqrt{3}}{2}$	1	$\dfrac{\sqrt{3}}{2}$	$\dfrac{1}{\sqrt{2}}$	$\dfrac{1}{2}$	0	-1	0
$\cos\theta$	1	$\dfrac{\sqrt{3}}{2}$	$\dfrac{1}{\sqrt{2}}$	$\dfrac{1}{2}$	0	$-\dfrac{1}{2}$	$-\dfrac{1}{\sqrt{2}}$	$-\dfrac{\sqrt{3}}{2}$	-1	0	1

- **Example 4**   If $\cos\theta = \frac{2}{5}$ and $0 < \theta < \pi/2$, find the other five trigonometric functions of θ.

Solution   Since $\cos\theta = \frac{2}{5}$, we can label the hypotenuse as having length 5 and the adjacent side as having length 2 in Figure 10. If the opposite side has length x, then the Pythagorean Theorem gives $x^2 + 4 = 25$ and so

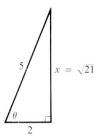

Figure 10

$x^2 = 21$, $x = \sqrt{21}$. We can now use the diagram to write the other five trigonometric functions:

$$\sin\theta = \frac{\sqrt{21}}{5} \qquad \tan\theta = \frac{\sqrt{21}}{2}$$

$$\csc\theta = \frac{5}{\sqrt{21}} \qquad \sec\theta = \frac{5}{2} \qquad \cot\theta = \frac{2}{\sqrt{21}} \qquad \bullet$$

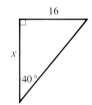

Figure 11

● **Example 5**   Use a calculator to approximate the value of x in Figure 11.

Solution   From the diagram we see that

$$\tan 40° = \frac{16}{x}$$

Therefore

$$x = \frac{16}{\tan 40°} \approx 19.07 \qquad \bullet$$

Trigonometric Identities

A trigonometric identity is a relationship among the trigonometric functions. The most elementary are the following, which are immediate consequences of the definitions of the trigonometric functions.

(6)

$$\csc\theta = \frac{1}{\sin\theta} \qquad \sec\theta = \frac{1}{\cos\theta} \qquad \cot\theta = \frac{1}{\tan\theta}$$

$$\tan\theta = \frac{\sin\theta}{\cos\theta} \qquad \cot\theta = \frac{\cos\theta}{\sin\theta}$$

For the next identity we refer to Figure 7. The distance formula (or, equivalently, the Pythagorean Theorem) tells us that $x^2 + y^2 = r^2$. Therefore

$$\sin^2\theta + \cos^2\theta = \frac{y^2}{r^2} + \frac{x^2}{r^2} = \frac{x^2 + y^2}{r^2} = \frac{r^2}{r^2} = 1$$

We have therefore proved one of the most useful of all trigonometric identities:

(7)

$$\sin^2\theta + \cos^2\theta = 1$$

If we now divide both sides of Equation 7 by $\cos^2\theta$ and use Equations 6, we get

(8)
$$\tan^2 \theta + 1 = \sec^2 \theta$$

Similarly, if we divide both sides of Equation 7 by $\sin^2 \theta$, we get

(9)
$$1 + \cot^2 \theta = \csc^2 \theta$$

The identities

(10a)
(10b)
$$\sin(-\theta) = -\sin \theta$$
$$\cos(-\theta) = \cos \theta$$

show that sin is an odd function and cos is an even function. (See Section 1.5.) They are easily proved by drawing a diagram showing θ and $-\theta$ in standard position (see Exercise 39).

Since the angles θ and $\theta + 2\pi$ have the same terminal side, we have

(11)
$$\sin(\theta + 2\pi) = \sin \theta \qquad \cos(\theta + 2\pi) = \cos \theta$$

These identities show that the sine and cosine functions are periodic with period 2π.

The remaining trigonometric identities are all consequences of two basic identities called the **addition formulas:**

(12a)
(12b)
$$\sin(x + y) = \sin x \cos y + \cos x \sin y$$
$$\cos(x + y) = \cos x \cos y - \sin x \sin y$$

The proofs of these addition formulas are indicated in Exercises 85, 86, and 87.

By substituting $-y$ for y in Equations 12a and 12b and using Equations 10a and 10b we obtain the following **subtraction formulas:**

(13a)
(13b)
$$\sin(x - y) = \sin x \cos y - \cos x \sin y$$
$$\cos(x - y) = \cos x \cos y + \sin x \sin y$$

Then, by dividing the formulas in Equations 12 or Equations 13, we obtain the corresponding formulas for $\tan(x \pm y)$:

(14a)
$$\tan(x + y) = \frac{\tan x + \tan y}{1 - \tan x \tan y}$$

(14b)
$$\tan(x - y) = \frac{\tan x - \tan y}{1 + \tan x \tan y}$$

If we put $y = x$ in the addition formulas (12) we get the **double angle formulas:**

(15a)
$$\sin 2x = 2 \sin x \cos x$$

(15b)
$$\cos 2x = \cos^2 x - \sin^2 x$$

Then, by using the identity $\sin^2 x + \cos^2 x = 1$, we obtain the following alternate forms of the double angle formulas for $\cos 2x$:

(16a)
$$\cos 2x = 2 \cos^2 x - 1$$

(16b)
$$\cos 2x = 1 - 2 \sin^2 x$$

If we now solve these equations for $\cos^2 x$ and $\sin^2 x$ we get the following **half-angle formulas,** which are useful in integral calculus:

(17a)
$$\cos^2 x = \frac{1 + \cos 2x}{2}$$

(17b)
$$\sin^2 x = \frac{1 - \cos 2x}{2}$$

Finally we state the **product formulas,** which can be deduced from Equations 12 and 13.

(18a)
$$\sin x \cos y = \tfrac{1}{2}[\sin(x + y) + \sin(x - y)]$$

(18b)
$$\cos x \cos y = \tfrac{1}{2}[\cos(x + y) + \cos(x - y)]$$

(18c)
$$\sin x \sin y = \tfrac{1}{2}[\cos(x - y) - \cos(x + y)]$$

There are many other trigonometric identities but the ones we have stated are the ones used most often in calculus. If you forget any of them, remember that they can all be deduced from Equations 13a and 13b.

● **Example 6**   Find all values of x in the interval $[0, 2\pi]$ such that $\sin x = \sin 2x$.

Solution   Using the double angle formula (15a), we rewrite the given equation as

$$\sin x = 2 \sin x \cos x$$

or $$\sin x(1 - 2 \cos x) = 0$$

Therefore there are two possibilities:

$$\sin x = 0 \qquad \text{or} \qquad 1 - 2 \cos x = 0$$

$$x = 0, \pi, 2\pi \qquad\qquad \cos x = \frac{1}{2}$$

$$x = \frac{\pi}{3}, \frac{5\pi}{3}$$

There are five solutions to the given equation: $0, \pi/3, \pi, 5\pi/3$, and 2π.   ●

Graphs of the Trigonometric Functions

The graph of the function $f(x) = \sin x$, shown in Figure 12(a), is obtained by plotting points for $0 \leqslant x \leqslant 2\pi$ and then using the periodic nature (from Equation 11) to complete the graph. Notice that the zeros of the sine function occur at the integer multiples of π, that is,

$$\sin x = 0 \qquad \text{when } x = n\pi \qquad n \text{ an integer}$$

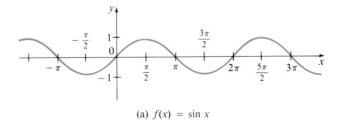

(a) $f(x) = \sin x$

Figure 12

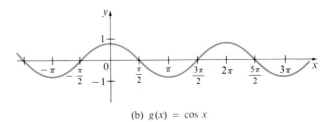

(b) $g(x) = \cos x$

Because of the identity

$$\cos x = \sin\left(x + \frac{\pi}{2}\right)$$

(which can be verified using Equation 12a) the graph of cosine is obtained by shifting the graph of sine by an amount $\pi/2$ to the left [see Figure 12(b)]. Note that for both the sine and cosine functions the domain is $(-\infty, \infty)$ and the range is the closed interval $[-1, 1]$. Thus, for all values of x, we have

$$-1 \leqslant \sin x \leqslant 1 \qquad -1 \leqslant \cos x \leqslant 1$$

The graphs of the remaining four trigonometric functions are shown in Figure 13 and their domains are indicated there. Notice that tangent and cotangent have range $(-\infty, \infty)$, whereas cosecant and secant have range $(-\infty, -1] \cup [1, \infty)$. All four functions are periodic: tangent and cotangent have period π, whereas cosecant and secant have period 2π.

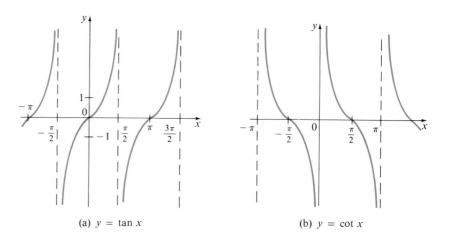

(a) $y = \tan x$ (b) $y = \cot x$

Figure 13

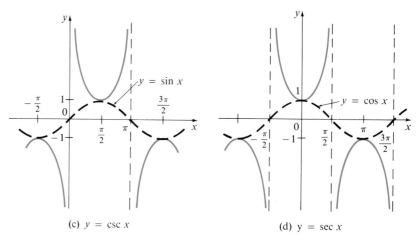

(c) $y = \csc x$ (d) $y = \sec x$

EXERCISES B

In Exercises 1–6 convert from degrees to radians.

1. 210° **2.** 300° **3.** 9°

4. −315° **5.** 900° **6.** 36°

In Exercises 7–12 convert from radians to degrees.

7. 4π **8.** $-\dfrac{7\pi}{2}$ **9.** $\dfrac{5\pi}{12}$

10. $\dfrac{8\pi}{3}$ **11.** $-\dfrac{3\pi}{8}$ **12.** 5

13. Find the length of a circular arc subtended by an angle of $\pi/12$ rad if the radius of the circle is 36 cm.

14. If a circle has radius 10 cm, what is the length of the arc subtended by a central angle of 72°?

15. A circle has radius 1.5 m. What angle is subtended at the center of the circle by an arc 1 m long?

16. Find the radius of a circular sector with angle $3\pi/4$ and arc length 6 cm.

In Exercises 17–22 draw, in standard position, the angle whose measure is given.

17. 315° **18.** −150° **19.** $-\dfrac{3\pi}{4}$ rad

20. $\dfrac{7\pi}{3}$ rad **21.** 2 rad **22.** −3 rad

In Exercises 23–28 find the exact trigonometric ratios for the angles whose radian measures are given.

23. $\dfrac{3\pi}{4}$ **24.** $\dfrac{4\pi}{3}$ **25.** $\dfrac{9\pi}{2}$

26. -5π **27.** $\dfrac{5\pi}{6}$ **28.** $\dfrac{11\pi}{4}$

In Exercises 29–34 find the remaining trigonometric ratios.

29. $\sin\theta = \dfrac{3}{5}, 0 < \theta < \dfrac{\pi}{2}$

30. $\tan\alpha = 2, 0 < \alpha < \dfrac{\pi}{2}$

31. $\sec\phi = -1.5, \dfrac{\pi}{2} < \phi < \pi$

32. $\cos x = -\dfrac{1}{3}, \pi < x < \dfrac{3\pi}{2}$

33. $\cot\beta = 3, \pi < \beta < 2\pi$

34. $\csc\theta = -\dfrac{4}{3}, \dfrac{3\pi}{2} < \theta < 2\pi$

In Exercises 35–38 find, correct to five decimal places, the length of the side labeled x.

35.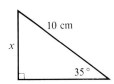

36.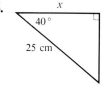

37.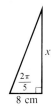

38.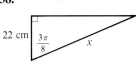

39. (a) Prove Equation 10a.
     (b) Prove Equation 10b.

40. (a) Prove Equation 14a.
     (b) Prove Equation 14b.

41. (a) Prove Equation 18a.
     (b) Prove Equation 18b.
     (c) Prove Equation 18c.

Prove the identities in Exercises 42–58.

42. $\cos\left(\dfrac{\pi}{2} - x\right) = \sin x$ **43.** $\sin\left(\dfrac{\pi}{2} + x\right) = \cos x$

44. $\sin(\pi - x) = \sin x$ **45.** $\sin\theta\cot\theta = \cos\theta$

46. $(\sin x + \cos x)^2 = 1 + \sin 2x$

47. $\sec y - \cos y = \tan y \sin y$

48. $\tan^2\alpha - \sin^2\alpha = \tan^2\alpha\sin^2\alpha$

49. $\cot^2\theta + \sec^2\theta = \tan^2\theta + \csc^2\theta$

50. $2\csc 2t = \sec t\csc t$

51. $\tan 2\theta = \dfrac{2\tan\theta}{1 - \tan^2\theta}$

52. $\dfrac{1}{1 - \sin\theta} + \dfrac{1}{1 + \sin\theta} = 2\sec^2\theta$

53. $\sin x\sin 2x + \cos x\cos 2x = \cos x$

54. $\sin^2 x - \sin^2 y = \sin(x + y)\sin(x - y)$

55. $\dfrac{\sin\phi}{1 - \cos\phi} = \csc\phi + \cot\phi$

56. $\tan x + \tan y = \dfrac{\sin(x + y)}{\cos x \cos y}$

57. $\sin 3\theta + \sin\theta = 2\sin 2\theta \cos\theta$

58. $\cos 3\theta = 4\cos^3\theta - 3\cos\theta$

If $\sin x = \frac{1}{3}$ and $\sec y = \frac{5}{4}$, where x and y lie between 0 and $\pi/2$, evaluate the expressions in Exercises 59–64.

59. $\sin(x + y)$ **60.** $\cos(x + y)$ **61.** $\cos(x - y)$

62. $\sin(x - y)$ **63.** $\sin 2y$ **64.** $\cos 2y$

In Exercises 65–72 find all values of x in the interval $[0, 2\pi]$ that satisfy the given equation.

65. $2\cos x - 1 = 0$ **66.** $3\cot^2 x = 1$

67. $2\sin^2 x = 1$ **68.** $|\tan x| = 1$

69. $\sin 2x = \cos x$ **70.** $2\cos x + \sin 2x = 0$

71. $\sin x = \tan x$ **72.** $2 + \cos 2x = 3\cos x$

In Exercises 73–76 find all values of x in the interval $[0, 2\pi]$ that satisfy the given inequality.

73. $\sin x \leqslant \frac{1}{2}$ **74.** $2\cos x + 1 > 0$

75. $-1 < \tan x < 1$ **76.** $\sin x > \cos x$

In Exercises 77–82 graph the given functions by starting with the graphs in Figures 12 and 13 and applying the transformations of Section 1.7 where appropriate.

77. $y = \cos\left(x - \dfrac{\pi}{3}\right)$ **78.** $y = \tan 2x$

79. $y = \dfrac{1}{3}\tan\left(x - \dfrac{\pi}{2}\right)$ **80.** $y = 1 + \sec x$

81. $y = |\sin x|$ **82.** $y = 2 + \sin\left(x + \dfrac{\pi}{4}\right)$

83. Prove the **Law of Cosines:** If a triangle has sides with lengths a, b, and c, and θ is the angle between the sides with lengths a and b, then

$$c^2 = a^2 + b^2 - 2ab\cos\theta$$

(*Hint:* Introduce a coordinate system so that θ is in standard position as in Figure 14. Express x and y in terms of θ and then use the distance formula to compute c.)

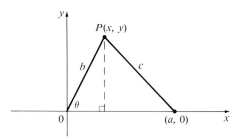

Figure 14

84. In order to find the distance $|AB|$ across a small inlet, a point C is located as in Figure 15 and the following measurements were recorded: $\angle C = 103°$, $|AC| = 820$ m, $|BC| = 910$ m. Use the Law of Cosines to find the required distance.

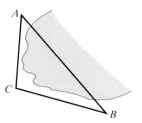

Figure 15

85. Use Figure 16 to prove the subtraction formula

$$\cos(\alpha - \beta) = \cos\alpha\cos\beta + \sin\alpha\sin\beta$$

[*Hint:* Compute c^2 in two ways (using the Law of Cosines and also using the distance formula) and compare the two expressions.]

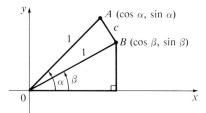

Figure 16

86. Use the formula in Exercise 85 to prove the addition formula for cosine (12b).

87. Use the subtraction formula for cosine and the identities

$$\cos\left(\dfrac{\pi}{2} - \theta\right) = \sin\theta \qquad \sin\left(\dfrac{\pi}{2} - \theta\right) = \cos\theta$$

to prove the subtraction formula for the sine function.

C Proofs of Theorems

In this appendix we present proofs of several theorems that were stated in the main body of the text. These theorems are numbered according to their original statement.

Properties of Limits (2.14)

Suppose that c is a constant and the limits

$$\lim_{x \to a} f(x) = L \quad \text{and} \quad \lim_{x \to a} g(x) = M$$

exist. Then

1. $\lim_{x \to a}[f(x) + g(x)] = L + M$

2. $\lim_{x \to a}[f(x) - g(x)] = L - M$

3. $\lim_{x \to a}[cf(x)] = cL$

4. $\lim_{x \to a}[f(x)g(x)] = LM$

5. $\lim_{x \to a} \dfrac{f(x)}{g(x)} = \dfrac{L}{M} \quad$ if $M \neq 0$

Proof of Property 1   Let $\varepsilon > 0$ be given. We must find $\delta > 0$ such that

$$\left|f(x) + g(x) - (L + M)\right| < \varepsilon \qquad \text{whenever } 0 < |x - a| < \delta$$

Using the Triangle Inequality [(1.5) in Section 1.1] we can write

(*)
$$\left|f(x) + g(x) - (L + M)\right| = \left|(f(x) - L) + (g(x) - M)\right|$$
$$\leqslant \left|f(x) - L\right| + \left|g(x) - M\right|$$

We shall make $\left|f(x) + g(x) - (L + M)\right|$ less than ε by making each of the terms $\left|f(x) - L\right|$ and $\left|g(x) - M\right|$ less than $\varepsilon/2$.

Since $\varepsilon/2 > 0$ and $\lim_{x \to a} f(x) = L$, there exists a number $\delta_1 > 0$ such that

$$\left|f(x) - L\right| < \frac{\varepsilon}{2} \qquad \text{whenever } 0 < |x - a| < \delta_1$$

Similarly, since $\lim_{x \to a} g(x) = M$, there exists a number $\delta_2 > 0$ such that

$$\left|g(x) - M\right| < \frac{\varepsilon}{2} \qquad \text{whenever } 0 < |x - a| < \delta_2$$

Let $\delta = \min\{\delta_1, \delta_2\}$. (This symbol means the smaller of the two numbers δ_1 and δ_2.) Notice that

$$\text{if } 0 < |x - a| < \delta, \quad \text{then} \quad 0 < |x - a| < \delta_1 \text{ and } 0 < |x - a| < \delta_2$$

and so
$$\left|f(x) - L\right| < \frac{\varepsilon}{2} \text{ and } \left|g(x) - M\right| < \frac{\varepsilon}{2}$$

Therefore, by (*)

$$\left|f(x) + g(x) - (L + M)\right| \leqslant \left|f(x) - L\right| + \left|g(x) - M\right|$$
$$< \frac{\varepsilon}{2} + \frac{\varepsilon}{2} = \varepsilon$$

To summarize,

$$\left|f(x) + g(x) - (L + M)\right| < \varepsilon \qquad \text{whenever } 0 < |x - a| < \delta$$

Thus, by definition of a limit,

$$\lim_{x \to a}\left[f(x) + g(x)\right] = L + M \qquad \bullet$$

Proof of Property 4   Let $\varepsilon > 0$ be given. We want to find $\delta > 0$ such that

$$\left|f(x)g(x) - LM\right| < \varepsilon \qquad \text{whenever } 0 < |x - a| < \delta$$

In order to get terms that contain $|f(x) - L|$ and $|g(x) - M|$ we add and subtract $Lg(x)$ as follows:

$$
\begin{aligned}
|f(x)g(x) - LM| &= |f(x)g(x) - Lg(x) + Lg(x) - LM| \\
&= |(f(x) - L)g(x) + L(g(x) - M)| \\
&\leq |(f(x) - L)g(x)| + |L(g(x) - M)| \qquad (\textit{Triangle Inequality}) \\
&= |f(x) - L||g(x)| + |L||g(x) - M|
\end{aligned}
$$

We want to make each of these terms less than $\varepsilon/2$.

Since $\lim_{x \to a} g(x) = M$, there is a number $\delta_1 > 0$ such that

$$
|g(x) - M| < \frac{\varepsilon}{2(1 + |L|)} \qquad \text{whenever } 0 < |x - a| < \delta_1
$$

Also, there is a number $\delta_2 > 0$ such that if $0 < |x - a| < \delta_2$, then

$$
|g(x) - M| < 1
$$

and therefore

$$
|g(x)| = |g(x) - M + M| \leq |g(x) - M| + |M| < 1 + |M|
$$

Since $\lim_{x \to a} f(x) = L$, there is a number $\delta_3 > 0$ such that

$$
|f(x) - L| < \frac{\varepsilon}{2(1 + |M|)} \qquad \text{whenever } 0 < |x - a| < \delta_3
$$

Let $\delta = \min\{\delta_1, \delta_2, \delta_3\}$. If $0 < |x - a| < \delta$, then we have $0 < |x - a| < \delta_1$, $0 < |x - a| < \delta_2$, and $0 < |x - a| < \delta_3$, so we can combine the inequalities to obtain

$$
\begin{aligned}
|f(x)g(x) - LM| &\leq |f(x) - L||g(x)| + |L||g(x) - M| \\
&< \frac{\varepsilon}{2(1 + |M|)}(1 + |M|) + |L|\frac{\varepsilon}{2(1 + |L|)} \\
&< \frac{\varepsilon}{2} + \frac{\varepsilon}{2} = \varepsilon
\end{aligned}
$$

This shows that $\lim_{x \to a} f(x)g(x) = LM$.   ●

Proof of Property 3   If we take $g(x) = c$ in Property 4, we get

$$
\begin{aligned}
\lim_{x \to a}[cf(x)] &= \lim_{x \to a}[g(x)f(x)] \\
&= \lim_{x \to a} g(x) \cdot \lim_{x \to a} f(x) \\
&= \lim_{x \to a} c \cdot \lim_{x \to a} f(x) \\
&= c \lim_{x \to a} f(x) \qquad (\textit{by Property 7})
\end{aligned}
$$
●

Proof of Property 2   Using Property 1 and Property 3 with $c = -1$, we have

$$\lim_{x \to a}[f(x) - g(x)] = \lim_{x \to a}[f(x) + (-1)g(x)]$$

$$= \lim_{x \to a} f(x) + \lim_{x \to a}(-1)g(x)$$

$$= \lim_{x \to a} f(x) + (-1)\lim_{x \to a} g(x)$$

$$= \lim_{x \to a} f(x) - \lim_{x \to a} g(x) \qquad \bullet$$

Proof of Property 5   First let us show that

$$\lim_{x \to a} \frac{1}{g(x)} = \frac{1}{M}$$

To do this we must show that, given $\varepsilon > 0$, there exists $\delta > 0$ such that

$$\left| \frac{1}{g(x)} - \frac{1}{M} \right| < \varepsilon \qquad \text{whenever } 0 < |x - a| < \delta$$

Observe that

$$\left| \frac{1}{g(x)} - \frac{1}{M} \right| = \frac{|M - g(x)|}{|Mg(x)|}$$

We know that we can make the numerator small. But we also need to know that the denominator is not small when x is near a. Since $\lim_{x \to a} g(x) = M$, there is a number $\delta_1 > 0$ such that, whenever $0 < |x - a| < \delta_1$, we have

$$|g(x) - M| < \frac{|M|}{2}$$

and therefore

$$|M| = |M - g(x) + g(x)| \leqslant |M - g(x)| + |g(x)|$$

$$< \frac{|M|}{2} + |g(x)|$$

This shows that

$$|g(x)| > \frac{M}{2} \qquad \text{whenever } 0 < |x - a| < \delta_1$$

and so, for these values of x,

$$\frac{1}{|Mg(x)|} = \frac{1}{|M||g(x)|} < \frac{1}{|M|} \cdot \frac{2}{|M|} = \frac{2}{M^2}$$

Also, there exists $\delta_2 > 0$ such that

$$|g(x) - M| < \frac{M^2}{2}\varepsilon \qquad \text{whenever } 0 < |x - a| < \delta_2$$

Let $\delta = \min\{\delta_1, \delta_2\}$. Then, for $0 < |x - a| < \delta$, we have

$$\left|\frac{1}{g(x)} - \frac{1}{M}\right| = \frac{|M - g(x)|}{|Mg(x)|} < \frac{2}{M^2}\frac{M^2}{2}\varepsilon = \varepsilon$$

It follows that $\lim_{x \to a} 1/g(x) = 1/M$. Finally, using Property 4, we obtain

$$\lim_{x \to a}\frac{f(x)}{g(x)} = \lim_{x \to a} f(x)\left(\frac{1}{g(x)}\right)$$

$$= \lim_{x \to a} f(x) \lim_{x \to a}\frac{1}{g(x)}$$

$$= L \cdot \frac{1}{M} = \frac{L}{M} \qquad \bullet$$

Theorem (2.16)

> If $f(x) \leqslant g(x)$ for all x in an open interval that contains a (except possibly at a) and
>
> $$\lim_{x \to a} f(x) = L \quad \text{and} \quad \lim_{x \to a} g(x) = M$$
>
> then $L \leqslant M$.

Proof We use the method of proof by contradiction. Suppose, if possible, that $L > M$. Property 2 of limits says that

$$\lim_{x \to a}[g(x) - f(x)] = M - L$$

Therefore, for any $\varepsilon > 0$, there exists $\delta > 0$ such that

$$\left|[g(x) - f(x)] - (M - L)\right| < \varepsilon \qquad \text{whenever } 0 < |x - a| < \delta$$

In particular, taking $\varepsilon = L - M$ (note that $L - M > 0$ by hypothesis), we have a number $\delta > 0$ such that

$$\left|[g(x) - f(x)] - (M - L)\right| < L - M \qquad \text{whenever } 0 < |x - a| < \delta$$

Since $a \leqslant |a|$ for any number a, we have

$$[g(x) - f(x)] - (M - L) < L - M \qquad \text{whenever } 0 < |x - a| < \delta$$

which simplifies to

$$g(x) < f(x) \qquad \text{whenever } 0 < |x - a| < \delta$$

But this contradicts $f(x) \leqslant g(x)$. Thus the inequality $L > M$ must be false. Therefore $L \leqslant M$. ●

The Squeeze Theorem (2.17)

> If $f(x) \leqslant g(x) \leqslant h(x)$ for all x in an open interval that contains a (except possibly at a) and
>
> $$\lim_{x \to a} f(x) = \lim_{x \to a} h(x) = L$$
>
> then $$\lim_{x \to a} g(x) = L$$

Proof   Let $\varepsilon > 0$ be given. Since $\lim_{x \to a} f(x) = L$, there is a number $\delta_1 > 0$ such that

$$|f(x) - L| < \varepsilon \qquad \text{whenever } 0 < |x - a| < \delta_1$$

that is, $L - \varepsilon < f(x) < L + \varepsilon \qquad \text{whenever } 0 < |x - a| < \delta_1$

Since $\lim_{x \to a} h(x) = L$, there is a number $\delta_2 > 0$ such that

$$|h(x) - L| < \varepsilon \qquad \text{whenever } 0 < |x - a| < \delta_2$$

that is, $L - \varepsilon < h(x) < L + \varepsilon \qquad \text{whenever } 0 < |x - a| < \delta_2$

Let $\delta = \min\{\delta_1, \delta_2\}$. If $0 < |x - a| < \delta$, then $0 < |x - a| < \delta_1$ and $0 < |x - a| < \delta_2$, so

$$L - \varepsilon < f(x) \leqslant g(x) \leqslant h(x) < L + \varepsilon$$

In particular,

$$L - \varepsilon < g(x) < L + \varepsilon$$

and so $|g(x) - L| < \varepsilon$. Therefore $\lim_{x \to a} g(x) = L$. ●

Theorem (2.26)

> If f is continuous at b and $\lim_{x \to a} g(x) = b$, then
>
> $$\lim_{x \to a} f(g(x)) = f(b)$$

Proof   Let $\varepsilon > 0$ be given. We want to find a number $\delta > 0$ such that

$$|f(g(x)) - f(b)| < \varepsilon \qquad \text{whenever } 0 < |x - a| < \delta$$

Since f is continuous at b, we have

$$\lim_{y \to b} f(y) = f(b)$$

and so there exists $\delta_1 > 0$ such that

$$|f(y) - f(b)| < \varepsilon \qquad \text{whenever } |y - b| < \delta_1$$

Since $\lim_{x \to a} g(x) = b$, there exists $\delta > 0$ such that

$$\left| g(x) - b \right| < \delta_1 \qquad \text{whenever } 0 < \left| x - a \right| < \delta$$

Combining these two statements, we see that whenever $0 < \left| x - a \right| < \delta$ we have $\left| g(x) - b \right| < \delta_1$, which implies that $\left| f(g(x)) - f(b) \right| < \varepsilon$. Therefore $\lim_{x \to a} f(g(x)) = f(b)$. ●

The proof of the following result was promised in the proof of Theorem 3.21.

Theorem

If $0 < \theta < \dfrac{\pi}{2}$, then $\theta \leqslant \tan \theta$.

Proof Figure 1 shows a sector of a circle with center O, central angle θ, and radius 1. Then

$$\left| AD \right| = \left| OA \right| \tan \theta = \tan \theta$$

We approximate the arc AB by an inscribed polygon consisting of n equal line segments and we look at a typical segment PQ. We extend the lines OP and OQ to meet AD in the points R and S. Then we draw $RT \parallel PQ$ as in Figure 1. Observe that

$$\angle RTO = \angle PQO < 90°$$

Figure 1

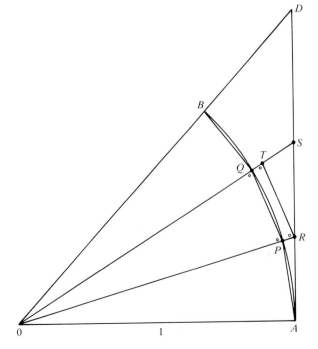

and so $\angle RTS > 90°$. Therefore we have

$$|PQ| < |RT| < |RS|$$

If we add n such inequalities, we get

$$L_n < |AD| = \tan\theta$$

where L_n is the length of the inscribed polygon. Thus, by Theorem 2.16, we have

$$\lim_{n\to\infty} L_n \leqslant \tan\theta$$

But the arc length was defined in Equation 8.9 as the limit of the lengths of inscribed polygons, so

$$\theta = \lim_{n\to\infty} L_n \leqslant \tan\theta \qquad \bullet$$

Property 5 of Integrals
(5.13)

$$\int_a^b f(x)\, dx = \int_a^c f(x)\, dx + \int_c^b f(x)\, dx$$

if all of these integrals exist.

Proof   We first assume that $a < c < b$. Since we are assuming that $\int_a^b f(x)\, dx$ exists, we can compute it as a limit of Riemann sums using only partitions P that include c as one of the partition points. If P is such a partition, let P_1 be the corresponding partition of $[a, c]$ determined by those partition points of P that lie in $[a, c]$. Similarly, P_2 will denote the corresponding partition of $[c, b]$. Note that $\|P_1\| \leqslant \|P\|$ and $\|P_2\| \leqslant \|P\|$. Thus, if $\|P\| \to 0$, it follows that $\|P_1\| \to 0$ and $\|P_2\| \to 0$. If $\{x_i | 1 \leqslant i \leqslant n\}$ is the set of partition points for P and $n = k + m$, where k is the number of subintervals in $[a, c]$ and m is the number of subintervals in $[c, b]$, then $\{x_i | 1 \leqslant i \leqslant k\}$ is the set of partition points for P_1. If we write $t_j = x_{k+j}$ for the partition points to the right of c, then $\{t_j | 1 \leqslant j \leqslant m\}$ is the set of partition points for P_2. Thus we have

$$a = x_0 < x_1 < \cdots < x_k < x_{k+1} < \cdots < x_n = b$$

$$c < t_1 < \cdots < t_m = b$$

Choosing $x_i^* = x_i$ and letting $\Delta t_j = t_j - t_{j-1}$, we compute $\int_a^b f(x)\, dx$ as follows:

$$\int_a^b f(x)\, dx = \lim_{\|P\|\to 0} \sum_{i=1}^n f(x_i)\, \Delta x_i$$

$$= \lim_{\|P\|\to 0} \left[\sum_{i=1}^k f(x_i)\, \Delta x_i + \sum_{i=k+1}^n f(x_i)\, \Delta x_i \right]$$

$$= \lim_{\|P\| \to 0} \left[\sum_{i=1}^{k} f(x_i) \Delta x_i + \sum_{j=1}^{m} f(t_j) \Delta t_j \right]$$

$$= \lim_{\|P_1\| \to 0} \sum_{i=1}^{k} f(x_i) \Delta x_i + \lim_{\|P_2\| \to 0} \sum_{j=1}^{m} f(t_j) \Delta t_j$$

$$= \int_{a}^{c} f(x)\,dx + \int_{c}^{b} f(t)\,dt$$

Now suppose that $c < a < b$. By what we have already proved, we have

$$\int_{c}^{b} f(x)\,dx = \int_{c}^{a} f(x)\,dx + \int_{a}^{b} f(x)\,dx$$

Therefore
$$\int_{a}^{b} f(x)\,dx = -\int_{c}^{a} f(x)\,dx + \int_{c}^{b} f(x)\,dx$$

$$= \int_{a}^{c} f(x)\,dx + \int_{c}^{b} f(x)\,dx$$

(See Note 6 in Section 5.3.) The proofs are similar for the remaining four orderings of a, b, and c. ●

Theorem (6.11)

> If f is a one-to-one continuous function defined on an interval (a,b), then its inverse function f^{-1} is also continuous.

Proof   First we show that if f is both one-to-one and continuous on (a, b), then it must be either increasing or decreasing on (a, b). If it were neither increasing nor decreasing, then there would exist numbers x_1, x_2, and x_3 in (a, b) with $x_1 < x_2 < x_3$ such that $f(x_2)$ does not lie between $f(x_1)$ and $f(x_3)$. There are two possibilities: either (1) $f(x_3)$ lies between $f(x_1)$ and $f(x_2)$ or (2) $f(x_1)$ lies between $f(x_2)$ and $f(x_3)$. (Draw a picture.) In case (1) we apply the Intermediate Value Theorem to the continuous function f to get a number c between x_1 and x_2 such that $f(c) = f(x_3)$. In case (2) the Intermediate Value Theorem gives a number c between x_2 and x_3 such that $f(c) = f(x_1)$. In either case we have contradicted the fact that f is one-to-one.

Let us assume, for the sake of definiteness, that f is increasing on (a, b). We take any number y_0 in the domain of f^{-1} and we let $f^{-1}(y_0) = x_0$; that is, x_0 is the number in (a, b) such that $f(x_0) = y_0$. To show that f^{-1} is continuous at y_0 we take any $\varepsilon > 0$ such that the interval $(x_0 - \varepsilon, x_0 + \varepsilon)$ is contained in (a, b). Since f is increasing, it maps the numbers in the interval $(x_0 - \varepsilon, x_0 + \varepsilon)$ onto the numbers in the interval $(f(x_0 - \varepsilon), f(x_0 + \varepsilon))$ and f^{-1} reverses the correspondence. If we let δ denote the smaller of the numbers $\delta_1 = y_0 - f(x_0 - \varepsilon)$ and $\delta_2 = f(x_0 + \varepsilon) - y_0$, then the interval $(y_0 - \delta, y_0 + \delta)$ is contained in the interval $(f(x_0 - \varepsilon), f(x_0 + \varepsilon))$ and so is mapped into the interval $(x_0 - \varepsilon, x_0 + \varepsilon)$ by f^{-1}. (See the arrow diagram in Figure 2.) We have therefore found a number $\delta > 0$ such that

$$\left| f^{-1}(y) - f^{-1}(y_0) \right| < \varepsilon \qquad \text{whenever } |y - y_0| < \delta$$

This shows that $\lim_{y \to y_0} f^{-1}(y) = f^{-1}(y_0)$ and so f^{-1} is continuous at any number y_0 in its domain.

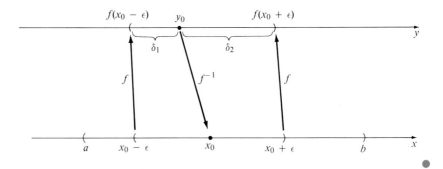

Figure 2

In order to prove Theorem 10.38 we first need the following results:

Theorem

1. If a power series $\sum a_n x^n$ converges when $x = b$ (where $b \neq 0$), then it converges whenever $|x| < |b|$.
2. If a power series $\sum a_n x^n$ diverges when $x = d$ (where $d \neq 0$), then it diverges whenever $|x| > |d|$.

Proof of 1   Suppose that $\sum a_n b^n$ converges. Then, by Theorem 10.17, $\lim_{n \to \infty} a_n b^n = 0$. According to Definition 10.1 with $\varepsilon = 1$, there is a positive integer N such that $|a_n b^n| < 1$ whenever $n \geqslant N$. Thus, for $n \geqslant N$, we have

$$|a_n x^n| = \left| \frac{a_n b^n x^n}{b^n} \right| = |a_n b^n| \left| \frac{x}{b} \right|^n < \left| \frac{x}{b} \right|^n$$

If $|x| < |b|$, then $|x/b| < 1$, so $\sum |x/b|^n$ is a convergent geometric series. Therefore, by the Comparison Test, the series $\sum_{n=N}^{\infty} |a_n x^n|$ is convergent. Thus the series $\sum a_n x^n$ is absolutely convergent and therefore convergent.

Proof of 2   Suppose that $\sum a_n d^n$ diverges. If x is any number such that $|x| > |d|$, then $\sum a_n x^n$ cannot converge because, by part 1, the convergence of $\sum a_n x^n$ would imply the convergence of $\sum a_n d^n$. Therefore $\sum a_n x^n$ diverges whenever $|x| > |d|$.

Theorem

For a power series $\sum a_n x^n$ there are only three possibilities:
1. The series converges only when $x = 0$.
2. The series converges for all x.
3. There is a positive number R such that the series converges if $|x| < R$ and diverges if $|x| > R$.

Proof   Suppose that neither 1 nor 3 is true. Then there are nonzero numbers b and d such that $\sum a_n x^n$ converges for $x = b$ and diverges for

$x = d$. Thus the set $S = \{x \mid \sum a_n x^n \text{ converges}\}$ is not empty. By the preceding theorem, the series diverges if $|x| > |d|$, so $|x| \leq |d|$ for all $x \in S$. This says that $|d|$ is an upper bound for the set S. Thus, by the Completeness Axiom (see Section 10.1), S has a least upper bound R. If $|x| > R$, then $x \notin S$, so $\sum a_n x^n$ diverges. If $|x| < R$, then $|x|$ is not an upper bound for S and so there exists $b \in S$ such that $b > |x|$. Since $b \in S$, $\sum a_n b^n$ converges, so by the preceding theorem $\sum a_n x^n$ converges. ●

Theorem (10.38)

For a power series $\sum a_n (x - c)^n$ there are only three possibilities:
1. The series converges only when $x = c$.
2. The series converges for all x.
3. There is a positive number R such that the series converges if $|x - c| < R$ and diverges if $|x - c| > R$.

Proof   If we make the change of variable $u = x - c$, then the power series becomes $\sum a_n u^n$ and we can apply the preceding theorem to this series. In case 3 we have convergence for $|u| < R$ and divergence for $|u| > R$. Thus we have convergence for $|x - c| < R$ and divergence for $|x - c| > R$. ●

Clairaut's Theorem (12.14)

Suppose f is defined on a disk D containing the point (a, b). If the functions f_{xy} and f_{yx} are both continuous on D, then $f_{xy}(a, b) = f_{yx}(a, b)$.

Proof   For small values of h, $h \neq 0$, consider the difference

$$\Delta(h) = [f(a + h, b + h) - f(a + h, b)] - [f(a, b + h) - f(a, b)]$$

Notice that if we let $g(x) = f(x, b + h) - f(x, b)$, then

$$\Delta(h) = g(a + h) - g(a)$$

By the Mean Value Theorem, there is a number c between a and $a + h$ such that

$$g(a + h) - g(a) = g'(c)h = h[f_x(c, b + h) - f_x(c, b)]$$

Applying the Mean Value Theorem again, this time to f_x, we get a number d between b and $b + h$ such that

$$f_x(c, b + h) - f_x(c, b) = f_{xy}(c, d)h$$

Combining these equations, we obtain

$$\Delta(h) = h^2 f_{xy}(c, d)$$

If $h \to 0$, then $(c, d) \to (a, b)$, so the continuity of f_{xy} at (a, b) gives

$$\lim_{h \to 0} \frac{\Delta(h)}{h^2} = \lim_{(c,d) \to (a,b)} f_{xy}(c, d) = f_{xy}(a, b)$$

Similarly, by writing

$$\Delta(h) = [f(a + h, b + h) - f(a, b + h)] - [f(a + h, b) - f(a, b)]$$

and using the Mean Value Theorem twice and the continuity of f_{yx} at (a, b), we obtain

$$\lim_{h \to 0} \frac{\Delta(h)}{h^2} = f_{yx}(a, b)$$

It follows that $f_{xy}(a, b) = f_{yx}(a, b)$.   ●

D Lies My Calculator Told Me

There is a wide variety of pocket-size calculating devices on the market. All of them have certain limitations in common: a limited range of magnitude (usually less than 10^{100}) and a bound on accuracy (typically eight or more digits).

A calculator usually comes with an owner's manual. Read it! The manual will tell you about further limitations (e.g., for angles when entering trigonometric functions) and how to overcome them.

Preliminary Experiments with Your Calculator

To have a first look at the limitations and quality of your calculator, make it compute $2 \div 3$. Of course, the answer is not a terminating decimal so it cannot be represented exactly on your calculator. If the last displayed digit is 6 rather than 7, then your calculator approximates $\frac{2}{3}$ by truncating instead of rounding so be prepared for loss of accuracy in longer calculations.

Now multiply the result by 3; that is, calculate $(2 \div 3) \times 3$. If the answer is 2, then subtract 2 from the result, thereby calculating $(2 \div 3) \times 3 - 2$. Instead of obtaining 0 as the answer, you might obtain a small negative number, which depends on the construction of the circuits. (The calculator keeps, in this case, a few "spare" digits that are remembered but not shown.)

This is all right because, as previously mentioned, the finite number of digits makes it impossible to represent $2 \div 3$ exactly.

A similar situation occurs when you calculate $(\sqrt{0.6})^2 - 0.6$. If you do not obtain 0, the order of magnitude of the result will tell you how many digits the calculator uses internally.

Next, try to compute $(-1)^5$ using the y^x button. Many calculators will indicate an error because they are built to attempt $e^{5\ln(-1)}$. One way to overcome this is to use the fact that $(-1)^k = \cos k\pi$ when k is an integer.

The Perils of Subtraction

You might have observed that subtraction of two numbers that are close to each other is a tricky operation. The difficulty is similar to this thought exercise: Imagine that you walk blindfolded 100 steps forward and then turn around and walk 99 steps. Are you sure that you end up exactly one step from where you started?

The name of this phenomenon is "loss of significant digits." To illustrate it on an example let us calculate

$$8721\sqrt{3} - 10{,}681\sqrt{2}$$

The approximations from my calculator are

$$8721\sqrt{3} \approx 15105.21509 \qquad \text{and} \qquad 10{,}681\sqrt{2} \approx 15105.21506$$

and so we get $8721\sqrt{3} - 10{,}681\sqrt{2} \approx 0.00003$. Even with three spare digits exposed, the difference comes out as 0.00003306. As you can see, the two ten-digit numbers agree in nine digits that, after subtraction, become zeros before the first nonzero digit. To make things worse, the formerly small errors in the square roots become more visible. In this particular example we can use rationalization to write

$$8721\sqrt{3} - 10{,}681\sqrt{2} = \frac{1}{8721\sqrt{3} + 10{,}681\sqrt{2}}$$

(work out the details!) and now the loss of significant digits does not occur:

$$\frac{1}{8721\sqrt{3} + 10{,}681\sqrt{2}} \approx 0.00003310115 \qquad \text{to seven digits}$$

(It would take too much space to explain why all seven digits are reliable; a subject called numerical analysis deals with these and similar situations.)

Now you can see why in Exercise 48 in Section 2.2 the guess at the limit was bound to go wrong: $\tan x$ becomes so close to x that the values will eventually agree in all digits that the calculator is capable of carrying. Similarly, if you start with just about any continuous function f and try to guess the value of

$$f'(x) = \lim_{h \to 0} \frac{f(x + h) - f(x)}{h}$$

long enough using a calculator, you will end up with a zero, despite all the rules in Section 3.2!

Where Calculus Is More Powerful Than Calculators

One of the secrets of success of calculus in overcoming the difficulties connected with subtraction is symbolic manipulation. For instance, $(a + b) - a$ is always b, although the calculated value may be different. Try it with $a = 10^7$ and $b = \sqrt{2} \times 10^{-5}$. Another powerful tool is the use of inequalities; a good example is the Squeeze Theorem (2.17) as demonstrated in Example 4 in Section 2.5. Yet another method for avoiding computational difficulties is provided by the Mean Value Theorem (4.10) (see Exercise 5) and its consequences, such as l'Hospital's Rule (5.56) (which helps solve the aforementioned Exercise 48 in Section 2.3 and others) and Taylor's Formula (10.54).

The limitations of calculators are further illustrated by infinite series. It is a common misconception that a series can be summed by adding terms until there is "practically nothing to add" and "the error is less than the first neglected term." The latter statement is true for most alternating series (Theorem 10.28) but not in general. As an example to refute these misconceptions, let us consider the series

$$\sum_{n=1}^{\infty} \frac{1}{n^{1.001}}$$

which is a convergent p-series ($p = 1.001 > 1$). Suppose we were to try to sum this series correct to eight decimal places by adding terms until they are less than 5 in the ninth decimal place. In other words, we stop when

$$\frac{1}{n^{1.001}} < 0.000000005$$

that is, when $n = N = 196{,}217{,}284$. (This would require a high-speed computer and increased precision.) After going to all this trouble, we would end up with the approximating partial sum

$$S_N = \sum_{n=1}^{N} \frac{1}{n^{1.001}} < 19.5$$

But from the proof of the Integral Test [see Figure 10.8(b)], we have

$$\sum_{n=1}^{\infty} \frac{1}{n^{1.001}} > \int_{1}^{\infty} \frac{dx}{x^{1.001}} = 1000$$

Thus the machine result represents less than 2% of the correct answer!

Suppose that we then wanted to add a huge number of terms of this series, say 10^{100} terms, in order to approximate the infinite sum more closely. (This number 10^{100}, called a googol, is outside the range of pocket calculators and is much larger than the number of elementary particles in our solar system.) If we were to add 10^{100} terms of the above series (only in theory; a million years is less than 10^{26} microseconds) we would still obtain a sum of less than 207 compared with the true sum of more than 1000. This estimate of 207 was obtained by using a more precise form of the Integral Test, known as the Euler-Maclaurin Formula, and only then using a calculator.

There are, and always will be, mathematical problems untreatable by a calculator or computer, regardless of its size or speed. A calculator does stretch the human capability to handle numbers, but there is still considerable scope for "thinking before doing."

EXERCISES D

1. Guess the value of

$$\lim_{x\to 0}\left(\frac{1}{\sin^2 x} - \frac{1}{x^2}\right)$$

and determine when to stop guessing before the loss of significant digits destroys your results. (The answer will depend on the calculator.) Then find the precise answer using an appropriate calculus method.

2. Guess the value of

$$\lim_{h\to 0}\frac{\ln(1 + h)}{h}$$

and determine when to stop guessing before the loss of significant digits destroys your results. This time the detrimental subtraction takes place inside the machine. Explain how (assuming that the Taylor series with $c = 1$ is used to approximate $\ln x$). Then find the precise answer using an appropriate calculus method.

3. Use your calculator to evaluate

$$2\sqrt{13}\cos\left(\frac{1}{3}\arctan\frac{18\sqrt{3}}{35}\right)$$

The answer looks conspicuously simple. Can you prove that the calculated result is correct?

4. Even innocent-looking calculus problems can lead to numbers beyond the calculator range. Show that the maximum value of the function

$$f(x) = \frac{x^{25}}{(1.0001)^x}$$

is greater than 10^{124}. (*Hint:* Use logarithms.) What is the limit of $f(x)$ as $x \to \infty$?

5. Try to evaluate

$$D = \ln\ln(10^9 + 1) - \ln\ln(10^9)$$

on your calculator. These numbers are so close together that you will likely obtain 0 or just a few digits of

accuracy. However we can use the Mean Value Theorem to achieve much greater accuracy.

(a) Let $f(x) = \ln\ln x$, $a = 10^9$, and $b = 10^9 + 1$. Then the Mean Value Theorem gives

$$f(b) - f(a) = f'(c)(b - a) = f'(c)$$

where $a < c < b$. Since f' is decreasing, we have $f'(a) > f'(c) > f'(b)$. Use this to estimate the value of D.

(b) Use the Mean Value Theorem a second time to discover why the quantities $f'(a)$ and $f'(b)$ in part (a) are so close to each other.

6. The positive numbers

$$a_n = \int_0^1 e^{1-x}x^n\,dx$$

can, in theory, be calculated from a reduction formula obtained by integration by parts: $a_0 = e - 1$, $a_n = na_{n-1} - 1$. Prove, using $1 \leqslant e^{1-x} \leqslant e$ and the Squeeze Theorem, that $\lim_{n\to\infty} a_n = 0$. Then try to calculate a_{20} from the reduction formula using your calculator. What went wrong?

The initial term $a_0 = e - 1$ cannot be represented exactly in a calculator. Let us call c the approximation of $e - 1$ that we can enter. Verify from the reduction formula (by observing the pattern after a few steps) that

$$a_n = \left[c - \left(\frac{1}{1!} + \frac{1}{2!} + \cdots + \frac{1}{n!}\right)\right]n!$$

and recall from Equation 10.48 that

$$\frac{1}{1!} + \frac{1}{2!} + \cdots + \frac{1}{n!}$$

converges to $e - 1$ as $n \to \infty$. The expression in square brackets converges to $c - (e - 1)$, a nonzero number, which gets multiplied by a fast-growing factor $n!$. We conclude that even if all further calculations (after entering a_0) were performed without errors, the initial accuracy would cause the computed sequence $\{a_n\}$ to diverge.

Answers to
E Odd-Numbered Exercises

Chapter 1

Exercises 1.1

1. $(-2, \infty)$ **3.** $[-1, \infty)$ **5.** $(3, \infty)$

7. $(2, 6)$ **9.** $(0, 1]$ **11.** $[-1, \frac{1}{2})$

13. $[2, 3]$ **15.** $(-\infty, 1) \cup (2, \infty)$ **17.** $[-1, \frac{1}{2}]$

19. $(-\infty, \infty)$ **21.** $(-\sqrt{3}, \sqrt{3})$

23. $(-1, 0) \cup (1, \infty)$ **25.** $(-\infty, 0) \cup (\frac{1}{4}, \infty)$

27. $(-2, 0) \cup (2, \infty)$ **29.** $(-\infty, 5) \cup (16, \infty)$

31. $(-\infty, -1] \cup [1, \infty)$ **33.** $\pm \frac{3}{2}$ **35.** $2, -\frac{4}{3}$ **37.** $(-3, 3)$ **39.** $(3, 5)$

41. $(-\infty, -7] \cup [-3, \infty)$ **43.** $[1.3, 1.7]$ **45.** $[-4, -1] \cup [1, 4]$ **47.** $(\frac{1}{2}, \infty)$ **49.** $(-1, \infty)$ **51.** $(-\frac{1}{2}, \frac{5}{2})$
53. $x \geqslant (a + b)c/ab$ **55.** $x > (c - b)/a$

Exercises 1.2

1. 5   **3.** $\sqrt{74}$ **5.** $2\sqrt{37}$ **11.** $(0, -4)$ **13.** (a) $(4, 9)$ (b) $(3.5, -3)$

15.   **17.**   **19.**   **21.**

23.   **25.**   **27.** 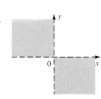   **29.**

31.   **33.** $(x - 3)^2 + (y + 1)^2 = 25$ **35.** $x^2 + y^2 = 65$ **37.** $(2, -5), 4$ **39.** $(-\frac{1}{2}, 0), \frac{1}{2}$
41. $(\frac{1}{4}, -\frac{1}{4}), \sqrt{5}/2\sqrt{2}$

Exercises 1.3

1. 2   **3.** $-\frac{9}{2}$ **5.** $6x - y - 15 = 0$ **7.** $2x - 3y + 19 = 0$ **9.** $5x + y = 11$ **11.** $y = 3x - 2$
13. $3x - y - 3 = 0$ **15.** $x - \sqrt{3}y = 2 + 4\sqrt{3}$ **17.** $y = 5$ **19.** $x + 2y + 11 = 0$ **21.** $5x - 2y + 1 = 0$
27. $x - y - 3 = 0$ **29.** $m = -\frac{1}{3}, b = 0$ **31.** $m = 0, b = -2$ **33.** $m = \frac{3}{4}, b = -3$ **35.** $45°$

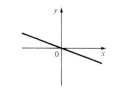   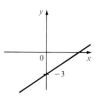

37. $146°$ **39.** $(1, 1), 72°$ (or $108°$)   **41.** $(3, -1), 8°$ (or $172°$)   **45.**   **47.**

Exercises 1.4

1. Parabola   **3.** Ellipse   **5.** Hyperbola   **7.** Ellipse   **9.** Parabola

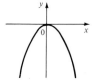   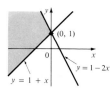   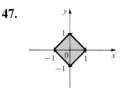

11. Hyperbola

13. Hyperbola

15. Ellipse

17. Parabola

19. Parabola

21. Ellipse

23.

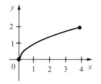

25. $y = x^2 - 2x$

27.

Exercises 1.5

1. $0, 12, \frac{3}{4}, 7 - 3\sqrt{5}, a^2 - 3a + 2, a^2 + 3a + 2$ **3.** $-\frac{1}{3}, -3, (1 - \pi)/(1 + \pi), (1 - a)/(1 + a), (2 - a)/a, (1 + a)/(1 - a)$

5. $x \rightarrow \boxed{\text{square root}} \rightarrow \sqrt{x}$

$4 \rightarrow \boxed{\text{square root}} \rightarrow 2$

$2 \rightarrow \boxed{\text{square root}} \rightarrow \sqrt{2}$

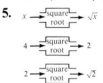

7. $\{0, 1, 2, 4\}$

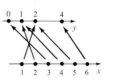

9. $[-2, 3], [-6, 14]$ **11.** $\{x \mid x \neq \frac{5}{3}\} = (-\infty, \frac{5}{3}) \cup (\frac{5}{3}, \infty), \{y \mid y \neq 0\} = (-\infty, 0) \cup (0, \infty)$ **13.** $[\frac{5}{2}, \infty), [0, \infty)$

15. $\{x \mid |x| \leq 1\} = [-1, 1], [0, 1]$ **17.** $\{x \mid x \neq \pm 1\} = (-\infty, -1) \cup (-1, 1) \cup (1, \infty)$

19. $\{x \mid x \leq 0 \text{ or } x \geq 6\} = (-\infty, 0] \cup [6, \infty)$ **21.** $[0, \pi)$ **23.** $(-\infty, \infty)$

25. $(-\infty, \infty)$ **27.** $(-\infty, \infty)$ **29.** $(-\infty, \infty)$ **31.** $(-\infty, \infty)$

33. $(-\infty, \infty)$ **35.** $(-\infty, 0]$ **37.** $[-2, 2]$ **39.** $\{x \mid x \neq 0\}$

41. $(-\infty, \infty)$ **43.** $(-\infty, \infty)$ **45.** $(-\infty, 0) \cup (0, \infty)$ **47.** $(-\infty, 1) \cup (1, \infty)$

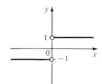

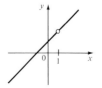

49. $(-\infty, \infty)$ **51.** $(-\infty, \infty)$ **53.** $(-\infty, \infty)$ **55.** $(-\infty, \infty)$

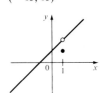

57. $(-\infty, \infty)$ **59.** $(-\infty, \infty)$ **61.** Yes, domain $[-3, 2]$, range $[-2, 2]$ **63.** No

65. $f(x) = -\frac{7}{6}x - \frac{4}{3}, -2 \leqslant x \leqslant 4$ **67.** $f(x) = 1 - \sqrt{-x}$

69. $A = 10x - x^2, 0 < x < 10$ **71.** $A = (\sqrt{3}/4)x^2, x > 0$

73. $S = x^2 + (8/x), x > 0$ **75.** Even   **77.** Neither   **79.** Odd

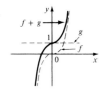

Exercises 1.6

1. $(f + g)(x) = x^2 + 5, (-\infty, \infty); (f - g)(x) = x^2 - 2x - 5, (-\infty, \infty); (fg)(x) = x^3 + 4x^2 - 5x, (-\infty, \infty);$
$(f/g)(x) = (x^2 - x)/(x + 5), (-\infty, -5) \cup (-5, \infty)$

3. $(f + g)(x) = \sqrt{1 + x} + \sqrt{1 - x}, [-1, 1]; (f - g)(x) = \sqrt{1 + x} - \sqrt{1 - x}, [-1, 1]; (fg)(x) = \sqrt{1 - x^2}, [-1, 1];$
$(f/g)(x) = \sqrt{(1 + x)/(1 - x)}, [-1, 1)$

5. $(f + g)(x) = \sqrt{x} + \sqrt[3]{x}, [0, \infty); (f - g)(x) = \sqrt{x} - \sqrt[3]{x}, [0, \infty); (fg)(x) = x^{5/6}, [0, \infty); (f/g)(x) = \sqrt[6]{x}, (0, \infty)$

7. $\{x | -3 \leqslant x \leqslant 4, x \neq \pm\sqrt{2}\} = [-3, -\sqrt{2}) \cup (-\sqrt{2}, \sqrt{2}) \cup (\sqrt{2}, 4]$

9.   **11.**

13. $(f \circ g)(x) = 8x + 1, (-\infty, \infty); (g \circ f)(x) = 8x + 11, (-\infty, \infty); (f \circ f)(x) = 4x + 9, (-\infty, \infty);$
$(g \circ g)(x) = 16x - 5, (-\infty, \infty)$

15. $(f \circ g)(x) = 3(6x^2 + 7x + 2), (-\infty, \infty); (g \circ f)(x) = 6x^2 - 3x + 2, (-\infty, \infty);$
$(f \circ f)(x) = 8x^4 - 8x^3 + x, (-\infty, \infty); (g \circ g)(x) = 9x + 8, (-\infty, \infty)$

17. $(f \circ g)(x) = 1/(x^3 + 2x), x \neq 0; (g \circ f)(x) = 1/x^3 + 2/x, x \neq 0; (f \circ f)(x) = x, x \neq 0;$
$(g \circ g)(x) = x^9 + 6x^7 + 12x^5 + 10x^3 + 4x, (-\infty, \infty)$

19. $(f \circ g)(x) = \sqrt[3]{1 - \sqrt{x}}, [0, \infty); (g \circ f)(x) = 1 - \sqrt[6]{x}, [0, \infty); (f \circ f)(x) = \sqrt[9]{x}, (-\infty, \infty); (g \circ g)(x) = 1 - \sqrt{1 - \sqrt{x}}, [0, 1]$

21. $(f \circ g)(x) = (3x - 4)/(3x - 2), x \neq 2, x \neq \frac{2}{3}; (g \circ f)(x) = -(x + 2)/3x, x \neq -\frac{1}{2}, x \neq 0;$
$(f \circ f)(x) = (5x + 4)/(4x + 5), x \neq -\frac{1}{2}, x \neq -\frac{5}{4}; (g \circ g)(x) = x/(4 - x), x \neq 2, x \neq 4$

23. $(f \circ g \circ h)(x) = \sqrt{x - 1} - 1$

25. $(f \circ g \circ h)(x) = (\sqrt{x} - 5)^4 + 1$ **27.** $g(x) = x - 9, f(x) = x^5$ **29.** $g(x) = x^2, f(x) = x/(x + 4)$

31. $h(x) = x^2, g(x) = x + 1, f(x) = 1/x$ **33.** $g(x) = x^2 + x - 1$

Exercises 1.7

1.   **3.**   **5.** 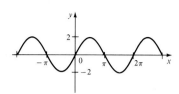 **7.**

9.   **11.** 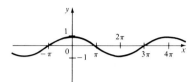 **13.**

15. 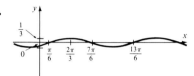 **17.**   **19.**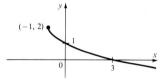

Review Exercises for Chapter 1

1. $(-\infty, 1]$ **3.** $(-10, 4)$ **5.** $(-\infty, 2) \cup (3, \infty)$ **7.** 10 **9.** $(x - 2)^2 + (y - 1)^2 = 9$
11. Center $(-1, 4)$, radius 3   **13.** $5x - 3y = 13$ **15.** $x + 2y = 8$ **17.** Slope $-\frac{3}{5}$, y-intercept 2
19. Parabola   **21.** Hyperbola   **23.** Ellipse

25. $3, 1 + 2\sqrt{2}, 1 + \sqrt{-x - 1}, 1 + \sqrt{x^2 - 1}, x + 2\sqrt{x - 1}$ **27.** $\{x \mid x \neq (-1 \pm \sqrt{5})/2\}$
29. 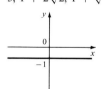   **31.**   **33.**   **35.**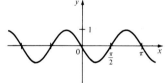

37. (a) $(f + g)(x) = x^2 + x + 2, (-\infty, \infty)$ (b) $(f/g)(x) = x^2/(x + 2), (-\infty, -2) \cup (-2, \infty)$
(c) $(f \circ g)(x) = (x + 2)^2, (-\infty, \infty)$ (d) $(g \circ f)(x) = x^2 + 2, (-\infty, \infty)$ **39.** $f(x) = \sqrt{x}, g(x) = x^2 + x + 9$

Chapter 2

Exercises 2.1

1. (a) $(x^2 + x - 2)/(x - 1)$ (b) $3.5, 2.5, 3.1, 2.9, 3.01, 2.99, 3.001, 2.999$ (c) 3 (d) $y = 3x$
3. (a) $(x^3 - 8)/(x - 2)$ (b) $15.25, 9.25, 12.61, 11.41, 12.0601, 11.9401, 12.006001, 11.994001$ (c) 12
(d) $y = 12x - 16$
5. (a) $(4x^2 - x)/x$ (b) $1, -3, -0.6, -1.4, -0.96, -1.04, -0.996, -1.004$ (c) -1 (d) $x + y = 1$
7. (a) $[(1/x) - 1]/(x - 1)$
(b) $-\frac{2}{3}, -2, -0.909091, -1.111111, -0.990099, -1.010101, -0.999001, -1.001001$ (c) -1 (d) $x + y = 2$

9. (a) $(\sqrt{x} - 2)/(x - 4)$ (b) 0.242641, 0.258343, 0.248457, 0.251582, 0.249844, 0.250156, 0.249984, 0.250016
    (c) $\frac{1}{4}$ (d) $x - 4y + 4 = 0$

11. (a) (i) -32 ft/s   (ii) -25.6 ft/s   (iii) -24.8 ft/s   (iv) -24.16 ft/s   (v) -24 ft/s

13. (a) (i) 3 m/s   (ii) 2 m/s   (iii) 1.5 m/s   (iv) 1.1 m/s   (b) 1 m/s

    (c), (d)

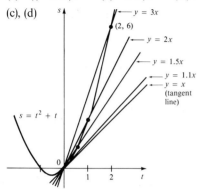

15. (a) (i) $-1.2°$/h   (ii) $-1.25°$/h   (iii) $-1.3°$/h   (b) $-1.6°$/h

17. (a) (i) \$20.25/unit   (ii) \$20.05/unit   (b) \$20.00/unit

Exercises 2.2

1. 2   **3.** 1   **5.** 2   **7.** 0   **9.** 1   **11.** 1   **13.** 7   **15.** 3   **17.** 2   **19.** 1   **21.** -1

23. Does not exist   **25.** 1   **27.** 12   **29.** 4   **31.** $-\frac{1}{5}$ **33.** $\frac{1}{16}$ **35.** Does not exist   **37.** 0

39. (a) -1 (b) 1   (c) Does not exist   **41.** (a) 3   (b) 0   (c) Does not exist   **43.** $-\frac{1}{4}$

45. (a) 0.459698, 0.489670, 0.493369, 0.496261, 0.498336, 0.499583, 0.499896, 0.499996   (b) $\frac{1}{2}$

47. 2, 2.59374246, 2.70481383, 2.71692393, 2.71814593, 2.71826824, 2.71828047, 2.71828169, 2.71828182, 2.71828183; 2.71828

Exercises 2.3

1. (a) $|x - 3| < 0.1/6$ (b) $|x - 3| < 0.01/6$

Exercises 2.4

1. 75   **3.** 60   **5.** $\frac{1}{2}$ **7.** 2   **9.** -3 **11.** -2 **13.** $\frac{2}{5}$ **15.** (a) 5   (b) 9   (c) 2   (d) $-\frac{1}{3}$

    (e) $-\frac{3}{8}$ (f) 0   (g) Does not exist   (h) $-\frac{6}{11}$ **17.** Does not exist   **19.** 6   **21.** Does not exist

23. -2 **25.** 12   **27.** 108   **29.** $-\frac{1}{2}$ **31.** $\frac{2}{3}$ **33.** 3   **37.** 5   **39.** -1.5 **41.** -1

43. Does not exist   **45.** 9   **47.** Does not exist   **49.** 9   **51.** 0   **53.** 0   **55.** 0

59. (a) 1, 1   (b) Yes, 1   (c)   **61.** (a) (i) $n - 1$ (ii) n (b) a is not an integer

63. (a) (i) 2   (ii) -2 (b) No   (c)

67. $a = 0$, $f(x) = H(x)$, $g(x) = 1 - H(x)$, where H is defined in Example 9 in Section 2.2   **69.** $\frac{2}{3}$

Exercises 2.5

1. (a) $-5, -3, -1, 3, 5, 8, 10$　　(b) Left, left, neither, neither, neither, right, neither
11. $f(2)$ undefined　　**13.** $f(1)$ undefined　　**15.** $\lim_{x \to 1} f(x)$ does not exist　　**17.** $\lim_{x \to -3} f(x) \neq f(-3)$

19. R　　**21.** $x \neq -1$　　**23.** $(-1, \infty)$　　**25.** R　　**27.** R　　**29.** $(-\infty, -5) \cup [2, \infty)$
33. 0, continuous from the right　　**35.** Continuous at all points　　**37.** $0, 1$, continuous from right at both

　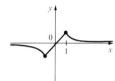　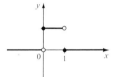

39. $\{n/2 \mid n \text{ is an integer}\}$, continuous from the right　　**41.** $\frac{1}{3}$　　**53.** $(-1.33, -1.32)$　　**55.** Never continuous

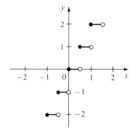

Review Exercises for Chapter 2

1. (a) -8　　(b) $8x + y = 17$　　**3.** $\sqrt{6}$　　**5.** $\frac{1}{2}$　　**7.** 2　　**9.** 0　　**11.** $-\frac{1}{8}$　　**13.** -1
17. (a) (i) 3　　(ii) 0　　(iii) Does not exist　　(iv) (0)　　(v) 0　　(vi) 0　　(b) At 0 and 3
(c) 　　　　　　　　　　　　　　**19.** $(-\infty, \infty)$

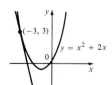

Chapter 3

Exercises 3.1

1. $-4, 4x + y + 9 = 0$　　**3.** $\frac{1}{6}, x - 6y + 16 = 0$　　　　**5.** $10x - y + 13 = 0$　　**7.** $x + 8y - 5 = 0$

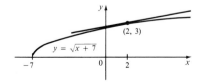

9. 7 m/s, 6.1 m/s, 6.01 m/s; 6 m/s　　**11.** 3.31 m/s, 3.0301 m/s, 3.003001 m/s, $(3 + 3h + h^2)$ m/s; 3 m/s
13. (a) 9.8 m/s　　(b) 19.6 m/s　　(c) 49 m/s　　(d) $2\sqrt{2695} \approx 104$ m/s

15. (a) $(2a - 6)$ ft/s, -4 ft/s, -2 ft/s
  (b) $t = 3$ (c) $t > 3$
  (d) 10 ft
  (e)
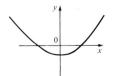

17. (a) $(6a^2 - 18a + 12)$ ft/s, 0 ft/s, 0 ft/s
  (b) $t = 1, t = 2$ (c) $t < 1, t > 2$
  (d) 34 ft
  (e)
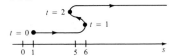

19. $-1/(2a - 1)^2$ **21.** $1/(3 - a)^{3/2}$ **23.** $1/3a^{2/3}$ **25.** $f(x) = \sqrt{x}, a = 1$ **27.** $f(x) = x^9, a = 1$
29. $f(t) = \sin t, a = \pi/2$ **31.** $f'(x) = 5, R, R$ **33.** $f'(x) = 3x^2 - 2x + 2, R, R$
35. $f'(x) = 1 + 2/x^2, \{x \mid x \neq 0\}, \{x \mid x \neq 0\}$ **37.** $g'(x) = 1/\sqrt{1 + 2x}, [-\frac{1}{2}, \infty), (-\frac{1}{2}, \infty)$
39. $G'(x) = -10/(2 + x)^2, \{x \mid x \neq -2\}, \{x \mid x \neq -2\}$ **41.** $f'(t) = 3(t + 1)^2, R, R$ **43.** $f'(x) = 2ax + b, R, R$
45. $1, 2x, 3x^2, nx^{n-1}, 5x^4$

47.
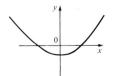
49.
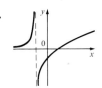
51.
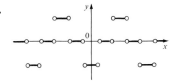

53. $f'(x) = \begin{cases} 1 & \text{if } x > 6 \\ -1 & \text{if } x < 6 \end{cases}$ or $f'(x) = \dfrac{x - 6}{|x - 6|}$
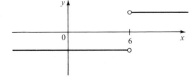

55. (a)

  (b) All values of x (c) $f'(x) = 2|x|$ **57.** Does not exist

Exercises 3.2

1. $f'(x) = 2x - 10$ **3.** $V'(r) = 4\pi r^2$ **5.** $F'(x) = 12{,}288x^2$ **7.** $Y'(t) = -54t^{-10}$ **9.** $g'(x) = 2x - 2/x^3$
11. $h'(x) = -3/(x - 1)^2$ **13.** $G'(s) = (2s + 1)(s^2 + 2) + (s^2 + s + 1)(2s)$ $[= 4s^3 + 3s^2 + 6s + 2]$
15. $y' = \frac{3}{2}\sqrt{x} + 2/\sqrt{x} - 3/2x\sqrt{x}$ **17.** $y' = \sqrt{5}/2\sqrt{x}$ **19.** $y' = -(4x^3 + 2x)/(x^4 + x^2 + 1)^2$
21. $y' = 2ax + b$ **23.** $y' = (-3t^2 + 14t + 23)/(t^2 + 5t - 4)^2$ **25.** $y' = 1 + 2/5\sqrt[5]{x^3}$ **27.** $u' = \sqrt{2}x^{\sqrt{2}-1}$
29. $v' = \frac{3}{2}\sqrt{x} - 5/2x^3\sqrt{x}$ **31.** $f'(x) = 2cx/(x^2 + c)^2$ **33.** $f'(x) = 2x^4(x^3 - 5)/(x^3 - 2)^2$
35. $s' = (1 + 2t - 2t^2)/t^2(t + 1)^2$ **37.** $P'(x) = na_n x^{n-1} + (n - 1)a_{n-1}x^{n-2} + \cdots + 2a_2 x + a_1$ **39.** $y = 4$
41. $x - 2y + 2 = 0$ **43.** $2, (-2 \pm \sqrt{3}, (1 \mp \sqrt{3})/2)$ **45.** $(1, 0), (-\frac{1}{3}, \frac{32}{27})$ **47.** $(4, 8)$
49. $x - 4y - 14 = 0$ **51.** $12x + y + 98 = 0$ **53.** $(-\frac{1}{4}, \frac{1}{256})$

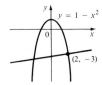

55. (a) $v(t) = 12t^3 - 48t^2 + 36t$ (b) $t = 0, 1, 3$ (c) $0 < t < 1, t > 3$
57. (a) $v(t) = (1 - t^2)/(t^2 + 1)^2$ (b) $t = 1$ (c) $0 \leqslant t < 1$

59. No

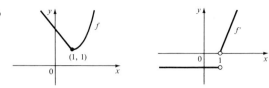

61. (a) -16 (b) $-\frac{20}{9}$ (c) 20 **65.** $y' = (x^4 + x + 1)(2x - 3)/2\sqrt{x} + \sqrt{x}\,[(4x^3 + 1)(2x - 3) + 2(x^4 + x + 1)]$

Exercises 3.3

1. 1 **3.** $(\sqrt{3} - 1)/2$ **5.** $2\sqrt{2}/3\pi$ **7.** 0 **9.** 5 **11.** $\sin 1$ **13.** $1/\pi$ **15.** $1/2$ **17.** 9
23. $\cos x - \sin x$ **25.** $-\csc x \cot^2 x - \csc^3 x$ **27.** $(x \sec^2 x - \tan x)/x^2$
29. $(\sin x + \cos x + x \sin x - x \cos x)/(1 + \sin 2x)$ **31.** $x^{-4} \sin x (x + x \sec^2 x - 3 \tan x)$
33. $(x^2 + 2x \tan x)/\sec x$ **35.** $4x - 2y = \pi - 2$ **37.** $(2n + 1)\pi \pm \pi/3$, n an integer   **39.** $\frac{1}{2}$ **41.** $\frac{1}{2}$
43. $\frac{1}{2}$ **45.** 1 **47.** (a) $\sec^2 x = 1/\cos^2 x$ (b) $\sec x \tan x = \sin x/\cos^2 x$ (c) $\cos x - \sin x = (\cot x - 1)/\csc x$

Exercises 3.4

1. $4u(x + 1), 48$ **3.** $3u^2(1 - 1/x^2), 0$ **5.** $F'(x) = 10(x^2 + 4x + 6)^4(x + 2)$
7. $G'(x) = 6(3x - 2)^9(5x^2 - x + 1)^{11}(85x^2 - 51x + 9)$ **9.** $f'(t) = -16(2t^2 - 6t + 1)^{-9}(2t - 3)$
11. $g'(x) = (2x - 7)/2\sqrt{x^2 - 7x}$ **13.** $h'(t) = \frac{3}{2}(t - 1/t)^{1/2}(1 + 1/t^2)$ **15.** $F'(y) = 39(y - 6)^2/(y + 7)^4$
17. $f'(z) = -\frac{2}{5}(2z - 1)^{-6/5}$ **19.** $y' = 8(2x - 5)^3(8x^2 - 5)^{-4}(-4x^2 + 30x - 5)$ **21.** $y' = 3 \sec^2 3x$
23. $y' = -3x^2 \sin(x^3)$ **25.** $y' = -12 \cos x \sin x(1 + \cos^2 x)^5$ **27.** $y' = -\sin(\tan x) \sec^2 x$ **29.** $y' = 0$
31. $y' = -(1/3)\csc(x/3)\cot(x/3)$ **33.** $y' = 3 \sin x \cos x(\sin x - \cos x)$ **35.** $y' = -\cos(1/x)/x^2$
37. $y' = 4 \cos 2x/(1 - \sin 2x)^2$ **39.** $y' = 6x^2 \tan(x^3) \sec^2(x^3)$ **41.** $y' = \sin[2(1 - \sqrt{x})/(1 + \sqrt{x})]/\sqrt{x}(1 + \sqrt{x})^2$
43. $y' = 0$ **45.** $y' = (1 + 1/2\sqrt{x})/2\sqrt{x + \sqrt{x}}$ **47.** $f'(x) = 9[x^3 + (2x - 1)^3]^2(9x^2 - 8x + 2)$
49. $p'(t) = -2[(1 + 2/t)^{-1} + 3t]^{-3}[2(t + 2)^{-2} + 3]$ **51.** $y' = \cos(\tan\sqrt{\sin x})(\sec^2\sqrt{\sin x})(\cos x)/2\sqrt{\sin x}$
53. $y = 0$ **55.** $3x + 16y = 44$ **57.** $4x + y = \pi + 1$ **59.** 28
61. $f'(x) = 2x \sec^2 3x(1 + 3x \tan 3x)$, $\mathrm{dom}(f) = \mathrm{dom}(f') = \{x \,|\, x \neq (2n - 1)\pi/6$, n an integer$\}$
63. $f'(x) = -(\sin\sqrt{x})/4\sqrt{x}\sqrt{\cos\sqrt{x}}$, $\mathrm{dom}(f) = \{x \,|\, 0 \leqslant x \leqslant \pi^2/4$ or $[(4n - 1)\pi/2]^2 \leqslant x \leqslant [(4n + 1)\pi/2]^2$ for
   some $n = 1, 2, \ldots\}$, $\mathrm{dom}(f') = \{x \,|\, 0 < x < \pi^2/4$ or $[(4n - 1)\pi/2]^2 < x < [(4n + 1)\pi/2]^2$ for some $n = 1, 2, \ldots\}$
65. $(5\pi/2)\cos 10\pi t$ cm/s   **67.** (a) On $(0, \infty)$ (b) $G'(x) = h'(\sqrt{x})/2\sqrt{x}$
69. (a) $F'(x) = -\sin x f'(\cos x)$ (b) $G'(x) = -\sin(f(x))f'(x)$ **73.** $f'(x) = x/|x|$
75. $h'(x) = |2x - 1| + 2x(2x - 1)/|2x - 1|$

Exercises 3.5

1. (a) $y' = -(2x + y + 3)/x$ (b) $y = 5/x - x - 3$, $y' = -5/x^2 - 1$
3. (a) $y' = -y^2/x^2$ (b) $y = x/(3x - 1)$, $y' = -1/(3x - 1)^2$
5. (a) $y' = (2x - y)/(x + 4y)$ (b) $y = (-x \pm \sqrt{9x^2 + 24})/4$, $y' = -\frac{1}{4} \pm 9x/4\sqrt{9x^2 + 24}$
7. $y' = (y - 2x)/(3y^2 - x)$ **9.** $y' = (18x - x^{-2/3}y^{1/3})/(12y + x^{1/3}y^{-2/3})$ **11.** $y' = -x^3/y^3$
13. $y' = (3x\sqrt{x^2 + y^2} - 2y)/(2x - 3y\sqrt{x^2 + y^2})$ **15.** $y' = 2(x - y)^2 + y/x$ [or $(3x^2 + 1 - 2xy)/(x^2 + 2)$]
17. $y' = [y \cos x + \sin(x - y)]/[\sin(x - y) - \sin x]$ **19.** $y' = -y/x$
21. $dx/dy = (1 - 4y^3 - 2x^2y - x^4)/(2xy^2 + 4x^3y)$ **23.** $-\frac{1}{6}$ **25.** $5x + 4y + 16 = 0$ **27.** $y = x$
29. $9x + 13y - 40 = 0$ **31.** $(\pm 5\sqrt{3}/4, \pm 5/4)$ **33.** $x_0 x/a^2 - y_0 y/b^2 = 1$ **37.** $45°$ (or $135°$)
39. $0°$ at $(0, 0)$, $8°$ (or $172°$) at $(1, 1)$ **41.** $72°$ (or $108°$)   **43.** $63°$ (or $117°$)   **45.** $63°$ (or $117°$)   **47.** $18°$ (or $162°$)
49. $71°$ (or $109°$)   **51.** $(0, 0)$ **59.** $\frac{3}{4}$

Exercises 3.6

1. $f'(x) = 4x^3 - 9x^2 + 16$, $f''(x) = 12x^2 - 18x$ **3.** $h'(x) = x/\sqrt{x^2 + 1}$, $h''(x) = 1/(x^2 + 1)^{3/2}$
5. $F'(s) = 24(3s + 5)^7$, $F''(s) = 504(3s + 5)^6$ **7.** $y' = 1/(1 - x)^2$, $y'' = 2/(1 - x)^3$
9. $y' = -\frac{3}{2}x(1 - x^2)^{-1/4}$, $y'' = \frac{3}{4}(x^2 - 2)(1 - x^2)^{-5/4}$

11. $H'(t) = 6\tan^2(2t - 1)\sec^2(2t - 1), H''(t) = 24\tan(2t - 1)\sec^4(2t - 1) + 24\tan^3(2t - 1)\sec^2(2t - 1)$

13. $F'(r) = (\sec\sqrt{r}\tan\sqrt{r})/2\sqrt{r}, F''(r) = (\sec^3\sqrt{r}\tan^2\sqrt{r} + \sec^3\sqrt{r} - r^{-1/2}\sec\sqrt{r}\tan\sqrt{r})/4r$ **15.** 0

17. $\frac{375}{8}(5t - 1)^{-5/2}$ **19.** $1/\sqrt{2}, 3/4\sqrt{2}, 27/16\sqrt{2}, 405/64\sqrt{2}$ **21.** -80 **23.** $-2x/y^5$ **25.** $64/(3x + y)^3$

27. $f'(x) = 1 - 2x + 3x^2 - 4x^3 + 5x^4 - 6x^5, f''(x) = -2 + 6x - 12x^2 + 20x^3 - 30x^4,$
$f'''(x) = 6 - 24x + 60x^2 - 120x^3, f^{(4)}(x) = -24 + 120x - 360x^2, f^{(5)}(x) = 120 - 720x,$
$f^{(6)}(x) = -720, f^{(n)}(x) = 0$ for $7 \leqslant n \leqslant 73$ **29.** $n!$ **31.** $(-1)^n(n + 2)!/6x^{n+3}$ **33.** $-2^{50}\cos 2x$

35. (a) $v(t) = 3t^2 - 3, a(t) = 6t$ (b) 6 m/s^2 (c) $a(1) = 6$ m/s^2

37. (a) $v(t) = 2At + B, a(t) = 2A$ (b) $2A$ m/s^2 (c) $2A$ m/s^2

39. (a) $t = 0, 2$ (b) $s(0) = 2$ m, $v(0) = 0$ m/s, $s(2) = -14$ m, $v(2) = -16$ m/s

41. (a) $y'(t) = A\omega\cos\omega t, y''(t) = -A\omega^2\sin\omega t$ **43.** $P(x) = x^2 - x + 3$

Exercises 3.7

1. (a) (i) 91   (ii) 76.51   (iii) 75.1501   (b) 75

3. (a) 7200π cm^2/s   (b) $21,600\pi$ cm^2/s   (c) $36,000\pi$ cm^2/s   **5.** (a) 8π ft^2/ft   (b) 16π ft^2/ft   (c) 24π ft^2/ft

7. (a) 6 kg/m   (b) 12 kg/m   (c) 18 kg/m   **9.** (a) 4.75 A   (b) 5 A   **11.** $dV/dP = -C/P^2$

13. -0.04 **15.** ≈ -92.6 (cm/s)/cm   **17.** $C'(x) = 1.5 + 0.004x$; $1.90, 1.902$

19. $C'(x) = 3 + 0.02x + 0.0006x^2$; $11, $11.07

Exercises 3.8

1. $dV/dt = 3x^2\,dx/dt$ **3.** -1 **5.** $1/50\pi$ cm/min   **7.** (a) $\frac{25}{3}$ ft/s   (b) $\frac{10}{3}$ ft/s   **9.** $250\sqrt{3}$ mi/h

11. 65 mi/h   **13.** $\frac{720}{13} \approx 55.4$ km/h   **15.** -1.6 cm/min   **17.** $(10,000 + 800,000\pi/9) \approx 2.89 \times 10^5$ cm^3/min

19. $\frac{10}{3}$ cm/min   **21.** 80 cm^3/min   **23.** $1650/\sqrt{31} \approx 296$ km/h   **25.** $\sqrt{2/5}$ rad/s

Exercises 3.9

1. $dy = 5x^4\,dx$ **3.** $dy = [(2x^3 + x)/\sqrt{x^4 + x^2 + 1}]\,dx$ **5.** $dy = 7\,dx/(2x + 3)^2$ **7.** $dy = 2\cos 2x\,dx$

9. (a) $dy = -2x\,dx$ (b) -5 **11.** (a) $dy = 6x(x^2 + 5)^2\,dx$ (b) 10.8

13. (a) $dy = -(3x + 2)^{-4/3}\,dx$ (b) 0.0025 **15.** (a) $dy = -\sin x\,dx$ (b) -0.025

17. $\Delta y = 1.25, dy = 1$ **19.** $\Delta y = 1.44, dy = 1.6$

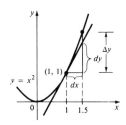 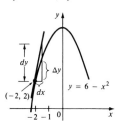

21. $\Delta y = 77, 33.25, 5.882, 0.571802$; $dy = 57, 28.5, 5.7, 0.57$; $\Delta y - dy = 20, 4.75, 0.182, 0.001802$ **23.** $6 + \frac{1}{120} \approx 6.0083$

25. $6 + \frac{1}{54} \approx 6.0185$ **27.** 0.099   **29.** 0.857   **31.** $1 + \pi/45 \approx 1.07$ **33.** (a) 270 cm^3 (b) 36 cm^2

35. (a) $84/\pi \approx 27$ cm^2 (b) $\frac{1}{84} \approx 0.012$ **37.** (a) $V \approx 2\pi rh\,\Delta r$ (b) $\pi h(\Delta r)^2$

Exercises 3.10

1. -0.6860 **3.** 1.5850   **5.** 2.165737   **7.** 1.618034   **9.** 3.992020   **11.** 1.895494

13. $-2.114908, 0.254102, 1.860806$ **15.** $1, -0.569840$ **17.** $0, 1.109144, 3.698154$ **19.** (b) 31.622777

Review Exercises for Chapter 3

1. $f'(x) = 3x^2 + 5$ **3.** $f'(x) = -5/2\sqrt{3 - 5x}$ **5.** $y' = 2(7x + 18)(x + 2)^7(x + 3)^5$ **7.** $y' = (9 - 2x)/(9 - 4x)^{3/2}$

9. $y' = (1 - 2xy^3)/(3x^2y^2 + 6y + 4)$ **11.** $y' = 7/8\sqrt[8]{x}$ **13.** $y' = 8/(8 - 3x)^2$

15. $y' = \frac{1}{5}(x\tan x)^{-4/5}(\tan x + x\sec^2 x)$ **17.** $y' = 2x/(1 + 2y)$ **19.** $y' = 2(2x - 5)/(x - 2)^2(x - 3)^2$

21. $y' = -\sec^2(\sqrt{1 - x})/2\sqrt{1 - x}$ **23.** $y' = \cos(\tan\sqrt{1 + x^3})\sec^2(\sqrt{1 + x^3})\frac{3}{2}x^2/\sqrt{1 + x^3}$

25. $y' = -6x\csc^2(3x^2 + 5)$ **27.** $y' = (mx\cos mx - \sin mx)/x^2$ **29.** $y' = -\sin(2\tan x)\sec^2 x$

31. $y' = 15x^2(x^3 + 7)^4\sqrt{7 - x^2} - x(x^3 + 7)^5/\sqrt{7 - x^2}$ **33.** -120 **35.** $-5x^4/y^{11}$ **37.** $3x + 2y - 8 = 0$
39. $4x - y + \sqrt{3} - 4\pi/3 = 0$ **41.** $(\pi/4, \sqrt{2}), (5\pi/4, -\sqrt{2})$
43. (a) $v(t) = 3t^2 - 12, a(t) = 6t$ (b) Upward when $t > 2$, downward when $0 \leqslant t < 2$ (c) 23
45. $f'(x) = 2xg(x) + x^2g'(x)$ **47.** $f'(x) = 2g(x)g'(x)$ **49.** $f'(x) = g'(g(x))g'(x)$ **51.** 4 kg/m   **53.** $\frac{4}{3}$ cm²/min
55. 13 ft/s   **57.** $-3x(4 - x^2)^{1/2} dx$ **59.** $20 - 0.1/6 \approx 19.983$ **61.** 0.724492   **63.** $(0.9340, -2.0634)$

Chapter 4

Exercises 4.1

1. Absolute maximum at e, local maximum at b and e, absolute minimum at d, local minimum at d and s

3.

No maximum, absolute minimum
$f(-1) = -1$

5.

Absolute maximum $f(-2) = 2$,
local and absolute minimum $f(0) = 0$

7.

No maximum or minimum

9.

Absolute maximum $f(0) = 1$,
no minimum

11.

Local and absolute maximum
$f(0) = 1$, absolute minimum
$f(-2) = -3$

13.

No maximum or minimum

15.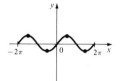

Local and absolute maximum
$f(\pi/2) = f(-3\pi/2) = 1$, local and
absolute minimum
$f(3\pi/2) = f(-\pi/2) = -1$

17.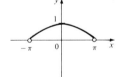

Local and absolute maximum
$f(0) = 1$, no minimum

19.

No maximum or minimum

21.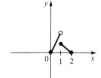

No maximum, absolute minimum
$f(0) = f(2) = 0$

23. $\frac{1}{3}$ **25.** ± 1 **27.** None   **29.** $(-1 \pm \sqrt{5})/2$ **31.** 0   **33.** $-2, 0$ **35.** ± 1 **37.** $0, 4, \frac{8}{7}$
39. 2   **41.** $n\pi/4$ (n any integer)   **43.** $f(3) = 5, f(1) = 1$ **45.** $f(5) = 66, f(2) = -15$

47. $f(1) = 9, f(-2) = 0$ **49.** $f(-3) = 47, f(\pm\sqrt{2}) = -2$ **51.** $f(2) = 5, f(1) = 3$ **53.** $f(-32) = 16, f(0) = 0$
55. $f(-1) = 1, f(1) = -1$ **57.** $f(\pi/4) = \sqrt{2}, f(0) = 1$ **63.**

Exercises 4.2

1. $\pm 1/\sqrt{3}$ **3.** $\pi/2$ **7.** $\frac{3}{2}$ **9.** $\pm\sqrt{7/3}$ **11.** $\sqrt{2}$ **13.** f is not differentiable at 1   **21.** No   **25.** No

Exercises 4.3

1. (a) Increasing on $(-\infty, -\frac{1}{2}]$, decreasing on $[-\frac{1}{2}, \infty)$
(b) Local maximum $f(-\frac{1}{2}) = 20.25$
(c)

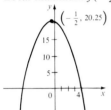

3. (a) Increasing on $(-\infty, \infty)$
(b) No maximum or minimum
(c)

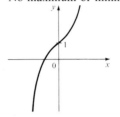

5. (a) Increasing on $(-\infty, \frac{1}{3}]$, decreasing on $[\frac{1}{3}, 1]$, increasing on $[1, \infty)$
(b) Local maximum $f(\frac{1}{3}) = \frac{4}{27}$, minimum $f(1) = 0$
(c)

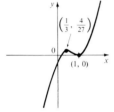

7. (a) Decreasing on $(-\infty, \frac{1}{3}]$, increasing on $[\frac{1}{3}, 3]$, decreasing on $[3, \infty)$
(b) Local minimum $f(\frac{1}{3}) = \frac{14}{27}$, maximum $f(3) = 10$
(c)

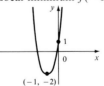

9. (a) Increasing on $(-\infty, -1]$ and $[0, 1]$, decreasing on $[-1, 0]$ and $[1, \infty)$
(b) Local maximum $f(\pm 1) = 1$, local minimum $f(0) = 0$
(c)

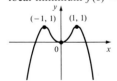

11. (a) Decreasing on $(-\infty, -1]$, increasing on $[-1, \infty)$
(b) Local minimum $f(-1) = -2$
(c)

13. (a) Increasing on $(-\infty, \frac{12}{7}]$ and $[4, \infty)$, decreasing on $[\frac{12}{7}, 4]$
(b) Local maximum $f(\frac{12}{7}) = 12^3 \cdot 16^4/7^7$, local minimum $f(4) = 0$
(c)

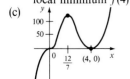

15. (a) Increasing on $(-\infty, 4]$, decreasing on $[4, 6]$
(b) Local maximum $f(4) = 4\sqrt{2}$
(c)

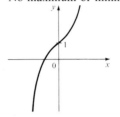

17. (a) Decreasing on $(-\infty, -\frac{1}{6}]$, increasing on $[-\frac{1}{6}, \infty)$
(b) Local minimum $f(-\frac{1}{6}) = -5/6^{1.2}$
(c)

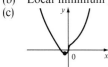

19. (a) Increasing on $[0, \frac{3}{4}]$, decreasing on $[\frac{3}{4}, 1]$
(b) Local maximum $f(\frac{3}{4}) = 3\sqrt{3}/16$
(c)

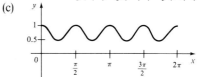

21. (a) Decreasing on $[0, \pi/3]$ and $[5\pi/3, 2\pi]$, increasing on $[\pi/3, 5\pi/3]$
(b) Local maximum $f(5\pi/3) = 5\pi/3 + \sqrt{3}$, minimum $f(\pi/3) = \pi/3 - \sqrt{3}$
(c)

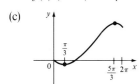

23. (a) Decreasing on $[0, \pi/4]$, $[\pi/2, 3\pi/4]$, $[\pi, 5\pi/4]$, $[3\pi/2, 7\pi/4]$, increasing on $[\pi/4, \pi/2]$, $[3\pi/4, \pi]$, $[5\pi/4, 3\pi/2]$, $[7\pi/4, 2\pi]$
(b) Local maximum $f(\pi/2) = f(\pi) = f(3\pi/2) = 1$, minimum $f(\pi/4) = f(3\pi/4) = f(5\pi/4) = f(7\pi/4) = \frac{1}{2}$
(c)

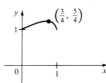

25. Increasing on $(-\infty, (-2-\sqrt{7})/3]$ and $[(-2+\sqrt{7})/3, \infty)$, decreasing on $[(-2-\sqrt{7})/3, (-2+\sqrt{7})/3]$
27. Decreasing on $(-\infty, -2]$, increasing on $[-2, \infty)$
29. Absolute maximum $f(1) = 2$, absolute minimum $f(-1) = -12$, no local extremum

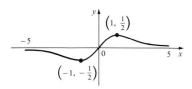

31. Local and absolute maximum $f(\frac{3}{4}) = \frac{5}{4}$, absolute minimum $f(0) = f(1) = 1$, no local minimum

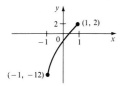

33. Local and absolute maximum $g(1) = \frac{1}{2}$, local and absolute minimum $g(-1) = -\frac{1}{2}$

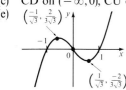

41. $f(x) = \frac{1}{9}(2x^3 + 3x^2 - 12x + 7)$

Exercises 4.4

1. (a) Increasing on $(-\infty, -1/\sqrt{3}]$ and $[1/\sqrt{3}, \infty)$, decreasing on $[-1/\sqrt{3}, 1/\sqrt{3}]$
(b) Local maximum $f(-1/\sqrt{3}) = 2/3\sqrt{3}$, local minimum $f(1/\sqrt{3}) = -2/3\sqrt{3}$
(c) CD on $(-\infty, 0)$, CU on $(0, \infty)$ (d) 0
(e) $\left(\frac{-1}{\sqrt{3}}, \frac{2}{3\sqrt{3}}\right)$

3. (a) Increasing on $(-\infty, -\frac{1}{3}]$ and $[1, \infty)$, decreasing on $[-\frac{1}{3}, 1]$
(b) Local maximum $f(-\frac{1}{3}) = \frac{32}{27}$, local minimum $f(1) = 0$
(c) CD on $(-\infty, \frac{1}{3})$, CU on $(\frac{1}{3}, \infty)$ (d) $\frac{1}{3}$
(e) $\left(-\frac{1}{3}, \frac{32}{27}\right)$

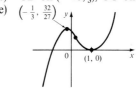

5. (a) Decreasing on $(-\infty, \frac{1}{4}]$, increasing on $[\frac{1}{4}, \infty)$
(b) Local minimum $g(\frac{1}{4}) = -\frac{27}{256}$
(c) CU on $(-\infty, \frac{1}{2})$ and $(1, \infty)$,
　　CD on $(\frac{1}{2}, 1)$　　(d) $\frac{1}{2}, 1$
(e)

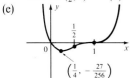

$\left(\frac{1}{4}, -\frac{27}{256}\right)$

7. (a) Increasing on $(-\infty, -1]$ and $[1, \infty)$,
　　decreasing on $[-1, 1]$
(b) Local maximum $h(-1) = 5$, minimum $h(1) = 1$
(c) CD on $(-\infty, -1/\sqrt{2})$ and $(0, 1/\sqrt{2})$,
　　CU on $(-1/\sqrt{2}, 0)$ and $(1/\sqrt{2}, \infty)$　　(d) $0, \pm 1/\sqrt{2}$

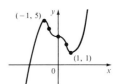

$(-1, 5)$
$(1, 1)$

9. (a) Decreasing on $(-\infty, \frac{1}{2}]$, increasing on $[\frac{1}{2}, \infty)$
(b) Local minimum $F(\frac{1}{2}) = -\frac{1}{64}$
(c) CU on $(-\infty, 0)$, $((5 - \sqrt{5})/10, (5 + \sqrt{5})/10)$ and
　　$(1, \infty)$, CD on $(0, (5 - \sqrt{5})/10)$ and $((5 + \sqrt{5})/10, 1)$
(d) $0, 1, (5 \pm \sqrt{5})/10$
(e)

$\left(\frac{1}{2}, -\frac{1}{64}\right)$

11. (a) Decreasing on $(-\infty, \infty)$
(b) No maximum or minimum
(c) CD on $(-\infty, 0)$, CU on $(0, \infty)$　　(d) 0
(e)

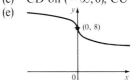

$(0, 8)$

13. (a) Increasing on $(-\infty, \infty)$
(b) No maximum or minimum
(c) CD on $(-\infty, 0)$, CU on $(0, \infty)$　　(d) 0
(e)

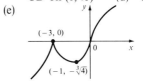

15. (a) Increasing on $(-\infty, -3]$ and $[-1, \infty)$,
　　decreasing on $[-3, -1]$
(b) Local maximum $Q(-3) = 0$,
　　minimum $Q(-1) = -\sqrt[3]{4}$
(c) CU on $(-\infty, -3)$ and $(-3, 0)$,
　　CD on $(0, \infty)$　　(d) 0
(e)

$(-3, 0)$
$(-1, -\sqrt[3]{4})$

17. (a) Increasing on $[n\pi, (2n + 1)\pi/2]$,
　　decreasing on remaining intervals
(b) Maximum $f((2n + 1)\pi/2) = 1$,
　　minimum $f(n\pi) = 0$
(c) CD on $(n\pi + \pi/4, n\pi + 3\pi/4)$, CU on
　　$(n\pi - \pi/4, n\pi + \pi/4)$　　(d) $n\pi \pm \pi/4$
(e)

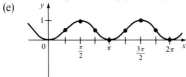

19. (a) Increasing on $(-\infty, \infty)$
(b) No maximum or minimum
(c) CD on $(2n\pi, (2n + 1)\pi)$, CU
　　on $((2n + 1)\pi, (2n + 2)\pi)$　　(d) $n\pi$
(e)

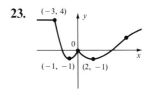

21.

$(-1, 4)$

$(1, 0)$

23.

$(-3, 4)$

$(-1, -1)$　$(2, -1)$

Exercises 4.5

1. 0   **3.** −2   **5.** $\frac{1}{2}$ **7.** $\frac{1}{6}$ **9.** 0   **11.** 0   **13.** 2   **15.** 0   **17.** −1   **19.** 0   **21.** 0
23. $-\frac{1}{2}$ **25.** Does not exist   **27.** $y = 2$ **29.** $y = 1, y = -1$
31. Horizontal asymptote $y = 0$, increasing on $[-1, 1]$, decreasing $(-\infty, -1]$ and $[1, \infty)$, CD on $(-\infty, -\sqrt{3})$ and $(0, \sqrt{3})$, CU on $(-\sqrt{3}, 0)$ and $(\sqrt{3}, \infty)$

33. Horizontal asymptote $y = 1$, decreasing on $(-\infty, 0]$, increasing on $[0, \infty)$, CU on $(-1/\sqrt{2}, 1/\sqrt{2})$, CD on $(-\infty, -1/\sqrt{2})$ and $(1/\sqrt{2}, \infty)$

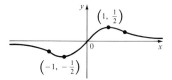

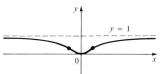

35. (a) 1   (b)

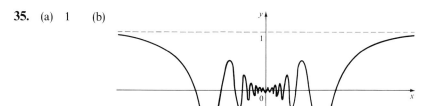

37.

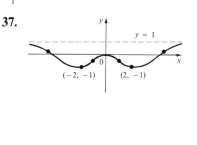

39. 0, 0.5, 1.0, 1.125, 1.0, 0.78125, 0.5625, 0.3828125, 0.25, 0.158203125, 0.09765625, 0.00038147, 2.204×10^{-12}, 7.8886×10^{-27}; 0

Exercises 4.6

1. $x = -9, x = -4, x = 3, x = 7, y = -2, y = 2$ **3.** $-\infty$ **5.** ∞ **7.** $-\infty$ **9.** $-\infty$ **11.** ∞
13. $-\infty$ **15.** ∞ **17.** $-\infty$ **19.** $-\infty$ **21.** ∞ **23.** ∞ **25.** $-\infty$ **27.** ∞ **29.** $-\infty$
31. 0   **33.** ∞ **35.** $-\infty$ **37.** $x = 1; x = -1, y = 1$ **39.** $x = 0, y = 1$
41. $x = 4, y = 0$ **43.** $x = 1, x = -1, y = 0$ **45.** (a) 0   (b) ∞ or $-\infty$ **47.** $|x - (-3)| < 0.1$

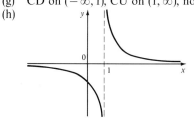

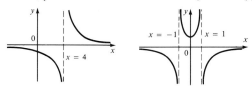

Exercises 4.7 (*abbreviations:* VA, vertical asymptote; HA, horizontal asymptote; IP, inflection point)

1. (a) $\{x \mid x \neq 1\}$ (b) y-intercept -1
(c) None   (d) VA $x = 1$, HA $y = 0$
(e) Decreasing on $(-\infty, 1)$ and $(1, \infty)$
(f) No maximum or minimum
(g) CD on $(-\infty, 1)$, CU on $(1, \infty)$, no IP
(h)

3. (a) $\{x \mid x \neq \pm 3\}$ (b) y-intercept $-\frac{1}{9}$
(c) About y-axis   (d) VA $x = \pm 3$, HA $y = 0$
(e) Increasing on $(-\infty, -3)$ and $(-3, 0]$, decreasing on $[0, 3)$ and $(3, \infty)$
(f) Local maximum $f(0) = -\frac{1}{9}$
(g) CU on $(-\infty, -3)$ and $(3, \infty)$, CD on $(-3, 3)$, no IP
(h)

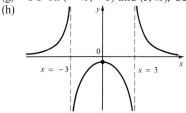

5. (a) $\{x \mid x \neq \pm 1\}$ (b) x-intercept 0, y-intercept 0
   (c) About the origin
   (d) VA $x = \pm 1$, slant asymptote $y = x$
   (e) Increasing on $(-\infty, -\sqrt{3}]$ and $[\sqrt{3}, \infty)$,
       decreasing on $[-\sqrt{3}, -1)$, $(-1, 1)$, and $(1, \sqrt{3}]$
   (f) Local maximum $f(-\sqrt{3}) = -3\sqrt{3}/2$,
       minimum $f(\sqrt{3}) = 3\sqrt{3}/2$
   (g) CD on $(-\infty, -1)$ and $(0, 1)$, CU on $(-1, 0)$ and
       $(1, \infty)$, IP$(0, 0)$
   (h)

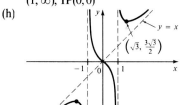

7. (a) $\{x \mid x \neq \frac{3}{2}\}$ (b) x-intercept 0, y-intercept 0
   (c) None   (d)   HA $y = 0$, VA $x = \frac{3}{2}$
   (e) Decreasing on $(-\infty, -\frac{3}{2}]$ and $(\frac{3}{2}, \infty)$,
       increasing on $[-\frac{3}{2}, \frac{3}{2})$
   (f) Local minimum $f(-\frac{3}{2}) = -\frac{1}{24}$
   (g) CD on $(-\infty, -3)$, CU on $(-3, \frac{3}{2})$ and $(\frac{3}{2}, \infty)$,
       IP$(-3, -\frac{1}{27})$
   (h)

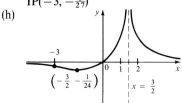

9. (a) $\{x \mid x \neq 0\}$ (b)   No intercepts
   (c) About $(0, 0)$
   (d) VA $x = 0$, slant asymptote $y = x$
   (e) Increasing on $(-\infty, -2]$ and $[2, \infty)$,
       decreasing on $[-2, 0)$ and $(0, 2]$
   (f) Local maximum $f(-2) = -4$, minimum $f(2) = 4$
   (g) CD on $(-\infty, 0)$, CU on $(0, \infty)$, no IP
   (h)

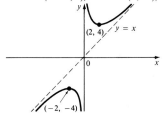

11. (a) $\{x \mid x \neq -3\}$
   (b) x-intercept 3, y-intercept -1 (c) None
   (d) VA $x = -3$, HA $y = 1$
   (e) Increasing on $(-\infty, -3)$ and $(-3, \infty)$
   (f) No maximum or minimum
   (g) CU on $(-\infty, -3)$, CD on $(-3, \infty)$, no IP
   (h)

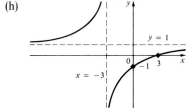

13. (a) $\{x \mid x \neq 1, -2\}$ (b) y-intercept $-\frac{1}{2}$
   (c) None   (d)   VA $x = 1$, $x = -2$, HA $y = 0$
   (e) Increasing on $(-\infty, -2)$ and $(-2, -\frac{1}{2}]$,
       decreasing on $[-\frac{1}{2}, 1)$ and $(1, \infty)$
   (f) Local maximum $f(-\frac{1}{2}) = -\frac{4}{9}$
   (g) CU on $(-\infty, -2)$ and $(1, \infty)$, CD on $(-2, 1)$,
       no IP
   (h)

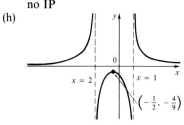

15. (a) $\{x \mid x \neq \pm 1\}$ (b) y-intercept 1
   (c) About y-axis   (d)   VA $x = \pm 1$, HA $y = -1$
   (e) Decreasing on $(-\infty, -1)$ and $(-1, 0]$,
       increasing on $[0, 1)$ and $(1, \infty)$
   (f) Local minimum $f(0) = 1$
   (g) CD on $(-\infty, -1)$ and $(1, \infty)$, CU on $(-1, 1)$, no IP
   (h)

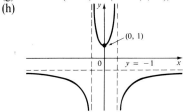

17. (a) $\{x \mid x \neq -1\}$
(b) x-intercept 1, y-intercept -1 (c) None
(d) VA $x = -1$, HA $y = 1$
(e) Increasing on $(-\infty, -1)$ and $(-1, \infty)$
(f) No maximum or minimum
(g) CU on $(-\infty, -1)$ and $(0, 1/\sqrt[3]{2})$,
    CD on $(-1, 0)$ and $(1/\sqrt[3]{2}, \infty)$,
    IP$(0, -1)$, $(1/\sqrt[3]{2}, -\frac{1}{3})$
(h)

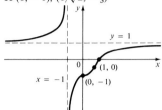

19. (a) $\{x \mid x \neq 1\}$
(b) y-intercept -1, x-intercepts $(1 \pm \sqrt{5})/2$
(c) None
(d) VA $x = 1$, slant asymptote $y = -x$
(e) Decreasing on $(-\infty, 1)$ and $(1, \infty)$
(f) No maximum or minimum
(g) CD on $(-\infty, 1)$, CU on $(1, \infty)$, no IP
(h)

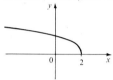

21. (a) $\{x \mid x \neq 0, \pm 1\}$ (b) None
(c) About $(0, 0)$
(d) VA $x = -1$, $x = 0$, $x = 1$; HA $y = 0$
(e) Decreasing on $(-\infty, -1)$, $(-1, -1/\sqrt{3}]$, $[1/\sqrt{3}, 1)$,
    and $(1, \infty)$, increasing on $[-1/\sqrt{3}, 0)$ and $(0, 1/\sqrt{3}]$
(f) Local minimum $f(-1/\sqrt{3}) = 3\sqrt{3}/2$,
    maximum $f(1/\sqrt{3}) = -3\sqrt{3}/2$
(g) CD on $(-\infty, -1)$, $(0, 1)$; CU on $(-1, 0)$, $(1, \infty)$
(h)

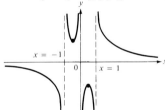

23. (a) $(-\infty, 2]$ (b) x-intercept 2, y-intercept $\sqrt{2}$
(c) None   (d) None
(e) Decreasing on $(-\infty, 2]$
(f) No local maximum or minimum
(g) CD on $(-\infty, 2)$, no IP
(h)

25. (a) $[0, \infty)$ (b) x-intercept 0, y-intercept 0
(c) None   (d) None   (e) Increasing on $[0, \infty)$
(f) No local maximum or minimum
(g) CD on $(0, \infty)$, no IP
(h)

27. (a) R (b) y-intercept 1   (c) None
(d) HA $y = 0$ (e) Decreasing on $(-\infty, \infty)$
(f) No local maximum or minimum
(g) CU on $(-\infty, \infty)$, no IP
(h)

29. (a) $(0, \infty)$ (b) x-intercept 1   (c) None
(d) VA $x = 0$ (e) Decreasing on $(0, \infty)$
(f) No local maximum or minimum
(g) CU on $(0, \infty)$, no IP
(h)

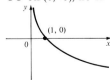

31. (a) $\{x \mid |x| \geq 5\} = (-\infty, -5] \cup [5, \infty)$
(b) x-intercepts ± 5 (c) About y-axis
(d) None
(e) Decreasing on $(-\infty, -5]$, increasing on $[5, \infty)$
(f) No local maximum or minimum
(g) CD on $(-\infty, -5)$ and $(5, \infty)$, no IP
(h)

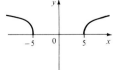

33. (a) $\{x \mid |x| \geqslant 3\} = (-\infty, -3] \cup [3, \infty)$
(b) x-intercepts ± 3 (c) About $(0,0)$ (d) None
(e) Increasing on $(-\infty, -3]$ and $[3, \infty)$
(f) No local maximum or minimum
(g) CU on $(-3\sqrt{3/2}, -3)$ and $(3\sqrt{3/2}, \infty)$,
CD on $(-\infty, -3\sqrt{3/2})$ and $(3, 3\sqrt{3/2})$,
IP when $x = \pm 3\sqrt{3/2}$
(h)

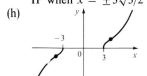

35. (a) $\{x \mid |x| \leqslant 1, x \neq 0\} = [-1, 0) \cup (0, 1]$
(b) x-intercepts ± 1 (c) About $(0,0)$
(d) VA $x = 0$
(e) Decreasing on $[-1, 0)$ and $(0, 1]$
(f) No local maximum or minimum
(g) CU on $(-1, -\sqrt{2/3})$ and $(0, \sqrt{2/3})$,
CD on $(-\sqrt{2/3}, 0)$ and $(\sqrt{2/3}, 1)$,
IP $(\pm\sqrt{2/3}, \pm 1/\sqrt{2})$
(h)

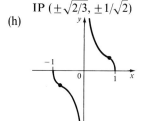

37. (a) $(-\infty, \infty)$ (b) x-intercepts 0, -27, y-intercept 0
(c) None   (d) None
(e) Increasing on $(-\infty, -8]$ and $[0, \infty)$,
decreasing on $[-8, 0]$
(f) Local minimum $f(0) = 0$,
local maximum $f(-8) = 4$
(g) CD on $(-\infty, 0)$ and $(0, \infty)$
(h)

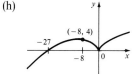

39. (a) R (b) x-intercepts $-1, 0$; y-intercept 0
(c) None   (d) None
(e) Increasing on $(-\infty, -\frac{1}{4}]$ and $[0, \infty)$,
decreasing on $[-\frac{1}{4}, 0]$
(f) Local maximum $f(-\frac{1}{4}) = \frac{1}{4}$, minimum $f(0) = 0$
(g) CD on $(-\infty, 0)$ and $(0, \infty)$, no IP
(h)

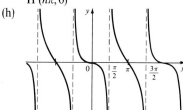

41. (a) R
(b) y-intercept 1, x-intercepts $n\pi + \pi/4$ (n an integer)
(c) Period 2π (d) None
(e) Decreasing on $[2n\pi - \pi/4, 2n\pi + 3\pi/4]$,
increasing on $[2n\pi + 3\pi/4, 2n\pi + 7\pi/4]$
(f) Local minimum $f(2n\pi + 3\pi/4) = -\sqrt{2}$,
local maximum $f(2n\pi - \pi/4) = \sqrt{2}$
(g) CU on $(2n\pi + \pi/4, 2n\pi + 5\pi/4)$,
CD on $(2n\pi - 3\pi/4, 2n\pi + \pi/4)$, IP$(n\pi + \pi/4, 0)$
(h)

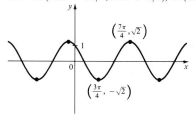

43. (a) $\{x \mid x \neq (2n + 1)\pi/2\}$
(b) y-intercept 0, x-intercept $n\pi$
(c) About $(0,0)$ and period 2π
(d) VA $x = (2n + 1)\pi/2$
(e) Decreasing on $((2n - 1)\pi/2, (2n + 1)\pi/2)$
(f) No maximum or minimum
(g) CU on $((2n - 1)\pi/2, n\pi)$, CD on $(n\pi, (2n + 1)\pi/2)$,
IP$(n\pi, 0)$
(h)

45. (a) $(-\pi/2, \pi/2)$ (b) x-intercept 0, y-intercept 0
(c) About y-axis   (d) VA $x = \pm\pi/2$
(e) Decreasing on $(-\pi/2, 0]$, increasing on $[0, \pi/2)$
(f) Local minimum $f(0) = 0$
(g) CU on $(-\pi/2, \pi/2)$, no IP
(h)

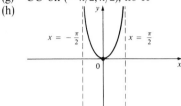

47. (a) $(0, 3\pi)$ (c) None   (d) None
(e) Decreasing on $(0, \pi/3)$ and $[5\pi/3, 7\pi/3]$,
increasing on $[\pi/3, 5\pi/3]$ and $[7\pi/3, 3\pi)$
(f) Local minimum $f(\pi/3) = \pi/6 - \sqrt{3}/2$,
$f(7\pi/3) = 7\pi/6 - \sqrt{3}/2$,
maximum $f(5\pi/3) = 5\pi/6 + \sqrt{3}/2$
(g) CU on $(0, \pi)$ and $(2\pi, 3\pi)$, CD on $(\pi, 2\pi)$,
IP$(\pi, \pi/2)$, $(2\pi, \pi)$
(h)

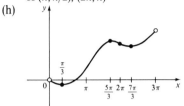

49. (a) $(-\infty, \infty)$ (b) y-intercept 2
(c) Period 2π (d) None
(e) Increasing on $[(2n - 1)\pi, 2n\pi]$,
decreasing on $[2n\pi, (2n + 1)\pi]$
(f) Maximum $f(2n\pi) = 2$,
minimum $f((2n + 1)\pi) = -2$
(g) CD on $(2n\pi - 2\pi/3, 2n\pi + 2\pi/3)$,
CU on remaining intervals,
IP when $x = 2n\pi \pm 2\pi/3$
(h)

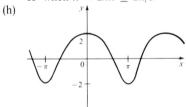

51. (a) R (b) y-intercept 0, x-intercept $n\pi$
(c) About $(0, 0)$, period 2π (d) None
(e) Increasing on $[-\pi, -2\pi/3]$ and $[2\pi/3, \pi]$,
decreasing on $[-2\pi/3, 2\pi/3]$
(f) Local maximum $f(-2\pi/3) = 3\sqrt{3}/2$,
local minimum $f(2\pi/3) = -3\sqrt{3}/2$
(g) CD on $(-\pi, -\alpha)$ and $(0, \alpha)$, where $\cos\alpha = \frac{1}{4}$,
CU on $(-\alpha, 0)$ and (α, π), IP when $x = 0, \pm\alpha, \pi$
(h)

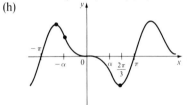

Exercises 4.8

1. 50, 50   **3.** 10, 10   **9.** 14,062.5 ft^2 **11.** 4000 cm^3 **13.** \$191.28   **15.** $(1.2, -0.6)$ **17.** $(1, \pm\sqrt{5})$
19. Square, side $\sqrt{2}r$ **21.** $L/2, \sqrt{3}L/4$ **23.** Base $\sqrt{3}r$, height $3r/2$ **25.** $4\pi r^3/3\sqrt{3}$ **27.** $\pi r^2(1 + \sqrt{5})$
29. 24 cm, 36 cm   **31.** (a) All of the wire for the square   (b) $40\sqrt{3}/(9 + 4\sqrt{3})$ m for the square
33. Height $=$ radius $= \sqrt[3]{V/\pi}$ cm   **35.** $2\pi l^3/9\sqrt{3}$ **37.** 30   **39.** $10\sqrt[3]{3}/(1 + \sqrt[3]{3})$ ft from the stronger source
43. $x = 6$ in.   **45.** \$11.50   **47.** $(L + W)^2/2$

Exercises 4.9

1. (a) \$1,035,000, \$1035, \$2025/unit   (b) 100   (c) \$225
3. (a) \$2330.71, \$2.33, \$4.07/unit   (b) 159   (c) \$1.07
5. (a) \$188.25, \$0.19, \$0.28/unit   (b) 400   (c) \$0.15   **7.** 400   **9.** 16,667   **11.** 672   **13.** 100
15. (a) $p(x) = 19 - x/3000$ (b) \$9.50   **17.** (a) $p(x) = 550 - x/10$ (b) \$175   (c) \$100

Exercises 4.10

1. $4x^3 + 3x^2 - 5x + C$ **3.** $3x^{10}/5 - x^8/2 + x^3 + x + C$ **5.** $2x^{3/2}/3 + 3x^{4/3}/4 + C$
7. $-3/2x^4 + C_1$ if $x > 0$, $-3/2x^4 + C_2$ if $x < 0$ **9.** $2x\sqrt{x}/3 + 2\sqrt{x} + C$ **11.** $2t^{7/2}/7 + 4t^{5/2}/5 + C$
13. $-\cos x - 2\sin x + C$ **15.** $\tan t + t^3/3 + C_n$, $(2n - 1)\pi/2 < t < (2n + 1)\pi/2$

17. $x^5/5 - 2x^3/3 + x^2/2 - x + C$ **19.** $x^4/12 + x^5/20 + Cx + D$ **21.** $x^2/2 + Cx + D$
23. $x^4 + Cx^2/2 + Dx + E$ **25.** $2x^2 + 3x - 9$ **27.** $2x^{3/2} - 2\sqrt{x} + 2$ **29.** $3\sin x - 5\cos x + 9$
31. $2x + 5x^{8/5}/8 + 3/8$ **33.** $-4x^2 + 5x + 6$ **35.** $x^5 - 5x^2 + 5$ **37.** $x^4/12 - 3\cos x + 3x + 5$
39. $x^3 + 3x^2 - 5x + 4$ **41.** $f(x) = 1/2x + x/4 - 3/4$ **43.** $s(t) = 3t - t^2 + 4$ **45.** $s(t) = t^3/2 + 4t^2 - 2t + 1$
47. $s(t) = t^4/12 - t^3/6 - 10t$ **49.** (a) $s(t) = 450 - 4.9t^2$ (b) $\sqrt{450/4.9} \approx 9.58$ s
  (c) $-9.8\sqrt{450/4.9} \approx -93.9$ m/s
51. (a) $s(t) = 450 + 5t - 4.9t^2$ (b) $(5 + \sqrt{8845})/9.8 \approx 10.1$ s   (c) ≈ -94.0 m/s   **55.** $\frac{88}{15}$ ft/s²   **57.** 225 ft

Review Exercises for Chapter 4

1. Absolute and local maximum $f(-2) = 21$, local minimum $f(2) = -11$, absolute minimum $f(-5) = -60$
3. Absolute maximum $f(4) = \frac{1}{3}$, absolute minimum $f(0) = -1$
5. Local and absolute minimum $f(\frac{\pi}{4}) = \frac{\pi}{4} - 1$, absolute maximum $f(\pi) = \pi$ **7.** ∞ **9.** ∞ **11.** 0
13. $\frac{1}{2}$ **15.** 2   **17.** $-\infty$ **19.** ∞
21. (a) R (b) y-intercept 1   (c) None   (d) None   **23.** (a) $\{x \mid x \neq 0, 3\} = (-\infty, 0) \cup (0, 3) \cup (3, \infty)$
  (e) Increasing on $(-\infty, \infty)$ (f) None   (b) None   (c) None
  (g) CD on $(-\infty, 0)$, CU on $(0, \infty)$, IP$(0, 1)$ (d) HA $y = 0$, VA $x = 0$, $x = 3$
  (h)   (e) Decreasing on $(-\infty, 0)$, $(0, 1]$, $(3, \infty)$;
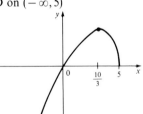   increasing on $[1, 3)$
                                                                              (f) Local minimum $f(1) = \frac{1}{4}$
                                                                              (g) CU on $(0, 3)$, $(3, \infty)$, CD on $(-\infty, 0)$
                                                                              (h)

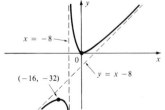

25. (a) $(-\infty, 5]$ (b) x-intercepts 0, 5, y-intercept 0   **27.** (a) $\{x \mid x \neq -8\}$
  (c) None   (d) None   (b) x-intercept 0, y-intercept 0   (c) None
  (e) Increasing on $(-\infty, \frac{10}{3}]$, decreasing on $[\frac{10}{3}, 5]$ (d) VA $x = -8$, slant asymptote $y = x - 8$
  (f) Local maximum $f(\frac{10}{3}) = 10\sqrt{5}/3\sqrt{3}$ (e) Increasing on $(-\infty, -16]$ and $[0, \infty)$,
  (g) CD on $(-\infty, 5)$ decreasing on $[-16, -8)$ and $(-8, 0]$
  (h)   (f) Local minimum $f(-16) = -32$,
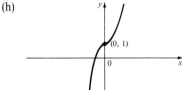   local minimum $f(0) = 0$
                                                                              (g) CD on $(-\infty, -8)$, CU on $(-8, \infty)$, no IP
                                                                              (h)

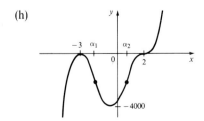

29. (a) R (b) x-intercepts $-3, 2$;   (h)
       y-intercept -2592 (c) None
  (d) None   (e) Increasing on $(-\infty, -3]$ and
       $[-\frac{7}{9}, \infty)$, decreasing on $[-3, -\frac{7}{9}]$
  (f) Local maximum $f(-3) = 0$, minimum
       $f(-\frac{7}{9}) = -20^4 \cdot 25^5 \cdot 9^{-9} \approx -4033$
  (g) CD on $(-\infty, \alpha_1)$ and $(\alpha_2, 2)$, CU on (α_1, α_2)
       and $(2, \infty)$ where $\alpha_1, \alpha_2 = (-14 \pm 5\sqrt{10})/18$,
       IP when $x = \alpha_1, \alpha_2, 2$

31. (a) R (b) x-intercept 0, y-intercept 0
(c) About $(0,0)$ (d) None
(e) Increasing on $(-\infty, \infty)$ (f) None
(g) CD on $(-\infty, 0)$, CU on $(0, \infty)$, IP$(0, 0)$
(h)

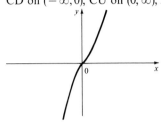

33. (a) $(-\pi/2, \pi/2)$ (b) y-intercept 0
(c) About $(0,0)$ (d) VA $x = \pm \pi/2$
(e) Increasing on $[-\pi/3, \pi/3]$,
decreasing on $(-\pi/2, -\pi/3]$ and $[\pi/3, \pi/2)$
(f) Local maximum $f(\pi/3) = 4\pi/3 - \sqrt{3}$,
minimum $f(-\pi/3) = -4\pi/3 + \sqrt{3}$
(g) CU on $(-\pi/2, 0)$, CD on $(0, \pi/2)$, IP$(0, 0)$
(h)

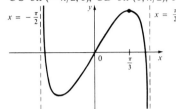

37.

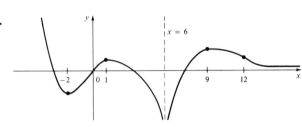

41. $3\sqrt{3}r^2$ **43.** $4/\sqrt{3}$ cm from D
45. Width $2R/\sqrt{3}$, depth $2\sqrt{2}R/\sqrt{3}$
47. $x^2/2 - 4x^{5/4}/5 + C$ **49.** $2\sqrt{x} + 2x^{3/2}/3 - 8/3$
51. $x^5/20 + x^3/6 + x - 1$ **53.** $s(t) = 4t^{5/2}/15 + 2t$

Chapter 5

Exercises 5.1

1. $\sqrt{1} + \sqrt{2} + \sqrt{3} + \sqrt{4} + \sqrt{5}$ **3.** $3^4 + 3^5 + 3^6$ **5.** $-1 + \frac{1}{3} + \frac{3}{5} + \frac{5}{7} + \frac{7}{9}$ **7.** $1^{10} + 2^{10} + 3^{10} + \cdots + n^{10}$

9. $1 - 1 + 1 - 1 + \cdots + (-1)^{n-1}$ **11.** $\sum_{i=1}^{10} i$ **13.** $\sum_{i=1}^{19} \frac{i}{i+1}$ **15.** $\sum_{i=1}^{n} 2i$ **17.** $\sum_{i=0}^{5} 2^i$ **19.** $\sum_{i=1}^{n} x^i$

21. 80 **23.** 3276 **25.** 0 **27.** 61 **29.** $n/(n+1)$ **31.** $n(n+1)$ **33.** $n(n^2 + 6n + 17)/3$
35. $n(n^2 + 6n + 11)/3$ **37.** $n(n^3 + 2n^2 - n - 10)/4$ **45.** $\frac{1}{3}$ **47.** 14

Exercises 5.2

1. (a) 1 (b) 50 **3.** (a) 1 (b) 43 **5.** (a) 1 (b) 35 **7.** (a) 0.5 (b) 12.1875
(c)

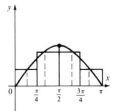

(c)

(c)

(c)

9. (a) $\pi/4$ (b) $\pi(1 + \sqrt{3})/2$ **13.** $\frac{256}{3}$ **15.** 20 **17.** 37.5
(c)

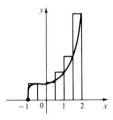

19. 8   **21.** $\frac{16}{15}$ 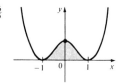   **23.** 2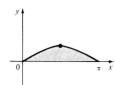

Exercises 5.3

1. (a) 1.2   (b) 4   **3.** (a) 1   (b) 2.336   **5.** (a) 0.5   (b) -0.028 **7.** $c(b - a)$ **9.** 15
11. 19.5   **13.** $\frac{4}{3}$ **15.** $b^4/4 + 2b^2$ **17.** $\frac{32}{5}$ **21.** $\int_0^1 (2x^2 - 5x)\,dx$ **23.** $\int_0^\pi \cos x\,dx$ **25.** $\int_0^1 x^4\,dx$
27. $\int_1^3 (3x^5 - 6)\,dx$ **29.** $-\frac{38}{3}$ **33.** $\frac{1}{2}$

Exercises 5.4

1. 12   **3.** $3\sqrt{3}$ **5.** 16   **7.** 22.5   **9.** $\frac{8}{3}$ **11.** $\pi/3 - \sqrt{3}$ **13.** 5   **15.** 7   **17.** 2   **19.** $\frac{20}{3}$
21. $\frac{1000}{3}$ **23.** $\int_1^{12} f(x)\,dx$ **25.** $\int_7^{10} f(x)\,dx$ **39.** $2 \le \int_1^3 x^3\,dx \le 54$ **41.** $\frac{1}{2} \le \int_1^2 (1/x)\,dx \le 1$
43. $-3 \le \int_{-3}^0 (x^2 + 2x)\,dx \le 9$ **45.** $2 \le \int_{-1}^1 \sqrt{1 + x^4}\,dx \le 2\sqrt{2}$

Exercises 5.5

1. $g'(x) = (x^2 - 1)^{20}$ **3.** $g'(u) = 1/(1 + u^4)$ **5.** $F'(x) = -\cos(x^2)$ **7.** $h'(x) = -\sin^4(1/x)/x^2$
9. $y' = -\sin(\tan^4 x)\sec^2 x$ **11.** $y' = 5/(25x^2 + 10x - 4)$ **13.** $g'(x) = 3[(3x - 1)/(3x + 1)] - 2[(2x - 1)/(2x + 1)]$
15. $y' = 3x^{7/2}\sin(x^3) - (\sin\sqrt{x})/2x^{1/4}$ **17.** 54   **19.** -1 **21.** 231   **23.** $\frac{28}{81}$ **25.** $(18\sqrt{2} - 12)/5$
27. $\frac{29}{35}$ **29.** $\frac{28}{3}$ **31.** 0   **33.** Does not exist   **35.** $\frac{2}{3}$ **37.** $\frac{1}{4}$ **39.** $(\sqrt{2} - 1)/2$ **41.** Does not exist
43. $2\sqrt{3}/3$ **45.** 1   **47.** -3.5 **49.** 10.7   **51.** $\frac{26}{3}$ **53.** $\frac{243}{4}$ **55.** 2   **57.** $(1 + 2^{1.8})/1.8$
63. $(2/5)x^{5/2} + C$ **65.** $4x - (8/3)x^{3/2} + x^2/2 + C$ **67.** $x^2 + \sec x + C$ **71.** $\ln 2$ **73.** $2^8/\ln 2$ **75.** $\pi/2$
77. $e^2/2 + e - \frac{1}{2}$ **79.** $x^3/3 + x + \tan^{-1}x + C$

Exercises 5.6

1. $(x^2 - 1)^{100}/200 + C$ **3.** $-\cos 4x/4 + C$ **5.** $-1/2(x^2 + 6x) + C$ **7.** $(x^2 + x + 1)^4/4 + C$
9. $\frac{2}{3}(x - 1)^{3/2} + C$ **11.** $(2 + x^4)^{3/2}/6 + C$ **13.** $-1/6(3x^2 - 2x + 1)^3 + C$ **15.** $-2/5(t + 1)^5 + C$
17. $-(1 - 2y)^{2.3}/4.6 + C$ **19.** $\frac{1}{2}\sin 2\theta + C$ **21.** $\frac{4}{7}(x + 2)^{7/4} - \frac{8}{3}(x + 2)^{3/4} + C$ **23.** $-\frac{1}{2}\cos(t^2) + C$
25. $(1 - x^2)^{7/2}/7 - (1 - x^2)^{5/2}/5 + C$ **27.** $\frac{1}{4}\sin^4 x + C$ **29.** $\frac{2}{3}(1 + \sec x)^{3/2} + C$ **31.** $\sqrt{ax^2 + 2bx + c} + C$
33. $-\frac{1}{2}\cos(2x + 3) + C$ **35.** $(\sin 3\alpha)x + \frac{1}{3}\cos 3x + C$ **37.** $2(b + cx^{a+1})^{3/2}/3c(a + 1) + C$ **39.** $\frac{1}{101}$ **41.** $\frac{32}{3}$
43. $\frac{16}{15}$ **45.** 0   **47.** $4\sqrt{2}/3 - 5\sqrt{5}/12$ **49.** 1   **51.** 3   **53.** $a^3/3$ **55.** 0   **57.** $\frac{14}{3}$ **59.** $2 - \sqrt{3}$
61. $\frac{10}{11}$ **65.** $\frac{1}{2}\ln|2x - 1| + C$ **67.** $(\ln x)^3/3 + C$ **69.** $(1 + e^x)^{11}/11 + C$ **71.** $\ln|\ln x| + C$
73. $x - e^{-x} + C$ **75.** $\frac{1}{2}\ln|x^2 + 2x| + C$ **77.** $\tan^{-1}(e^x) + C$ **79.** $\tan^{-1}x + \frac{1}{2}\ln(1 + x^2) + C$
81. $\frac{1}{2}\tan^{-1}(x^2) + C$ **83.** $\frac{1}{2}\ln 3$ **85.** 2   **87.** $1 - e^{-2}$

Exercises 5.7

1. $\frac{20}{3}$ 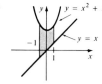   **3.** 16.5 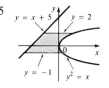   **5.** 36   **7.** $\frac{1}{6}$

9. $\frac{1}{3}$ **11.** 4 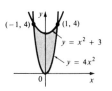   **13.** 36 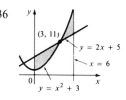   **15.** 8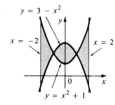

17. $\frac{32}{3}$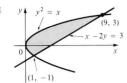

19. $\frac{8}{3}$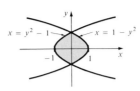

21. $\frac{37}{12}$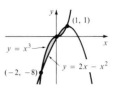

23. $5\pi^2/32 + 1/\sqrt{2} - 2$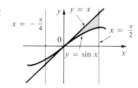

25. $\frac{1}{2}$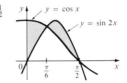

27. $\frac{1}{2}$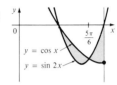

29. 34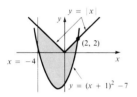

31. 12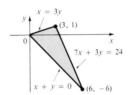

33. 9   **35.** 14.5

37. 1.5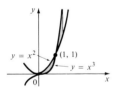

39. 3.75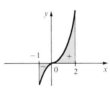

41. 0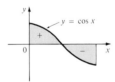

43. $\ln 2 - \frac{1}{2}$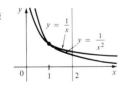

45. $\pi - \frac{2}{3}$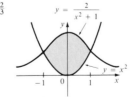

47. $e^3/3 - e + \frac{2}{3}$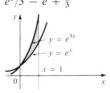

Exercises 5.8

1. (a) $\frac{74}{3}$ m   (b) $\frac{74}{3}$ m   **3.** (a) $-\frac{10}{3}$ m   (b) $\frac{98}{3}$ m   **5.** (a) $(10\sqrt{5} - 29)/3$ m   (b) $(10\sqrt{5} - 7)/3$ m

7. (a) $v(t) = (t^2/2) + 4t + 5$ (b) $\frac{1250}{3}$ m   **9.** (a) $v(t) = t^3 - t^2 - 2t$ (b) 6.5 m

11. $45 + \frac{4}{3}(7\sqrt{7} - 2\sqrt{2})$ kg   **13.** \$388,280,000   **15.** $16(2\sqrt{2} - 1)/3$ **17.** $(5/4\pi)[1 - \cos(2\pi t/5)]$ L

Exercises 5.9

1. -2 **3.** $-\frac{11}{3}$ **5.** $\frac{1}{5}$ **7.** $(4/9\pi)(1 + 1/2\sqrt{2})$ **9.** (a) 3   (b) $\frac{3}{2}$ (c)

11. (a) 8/3   (b) $2/\sqrt{3}$ (c) 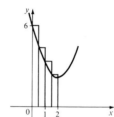   **13.** $(50 + 28/\pi)°$ F $\approx 59°$ F   **15.** 6 kg/m   **17.** $5/4\pi \approx 0.4$I

Review Exercises for Chapter 5

1. 0.00909   **3.** $2n + n^2(n + 1)^2/4$ **5.** 7.75   **7.** -18 **9.** $\frac{875}{12}$ **11.** $\frac{875}{12}$ **13.** $\frac{9}{10}$

15. $\frac{1209}{28}$ **17.** 3480   **19.** 2   **21.** Does not exist   **23.** $-1/25(2 + x^5)^5 + C$ **25.** $-(\cos \pi x)/\pi + C$
27. $-\sin(1/t) + C$ **29.** $-\tan(\cos x) + C$ **31.** 4   **33.** $F'(x) = \sqrt{1 + x^2 + x^4}$ **35.** $g'(x) = 3x^5/\sqrt{1 + x^9}$
37. $y' = (2\cos x - \cos\sqrt{x})/2x$ **39.** $\frac{4}{3}$ **41.** 108   **43.** 1   **45.** $2\sqrt{2}$ **47.** $4 \leqslant \int_1^3 \sqrt{x^2 + 3}\,dx \leqslant 4\sqrt{3}$
53. 30   **57.** $2e^{\sqrt{x}} + C$ **59.** $-\frac{1}{2}[\ln(\cos x)]^2 + C$ **61.** $\frac{1}{4}\ln(1 + x^4) + C$

Chapter 6

Exercises 6.1

1.

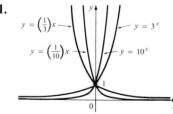

3. 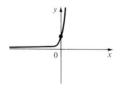   **5.** 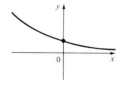   **7.**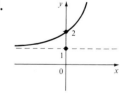

9. 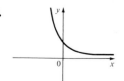   **11.** 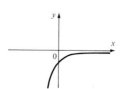   **13.** 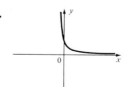   **15.**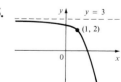

17. ∞ **19.** 0   **21.** ∞ **23.** 0   **25.** ∞ **27.** ∞ **29.** 1   **31.** ∞ **33.** 1   **35.** ∞

Exercises 6.2

1. No   **3.** Yes   **5.** No   **7.** Yes   **9.** Yes   **11.** No   **13.** $f^{-1}(x) = (x - 7)/4$

15. $f^{-1}(x) = (5x - 1)/(2x + 3)$ **17.** $f^{-1}(x) = (x^2 - 2)/5$, $x \geqslant 0$

19. (b) $\frac{1}{2}$ **21.** (b) $\frac{1}{12}$
    (c) $g(x) = (x - 1)/2$, domain $= R =$ range   (c) $g(x) = \sqrt[3]{x}$, domain $= R =$ range
    (e)   (e)

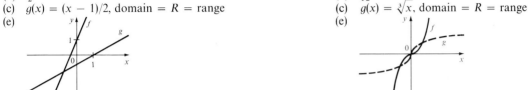

23. (b) $-\frac{1}{2}$
    (c) $g(x) = \sqrt{9 - x}$, domain $= [0, 9]$, range $= [0, 3]$
    (e)

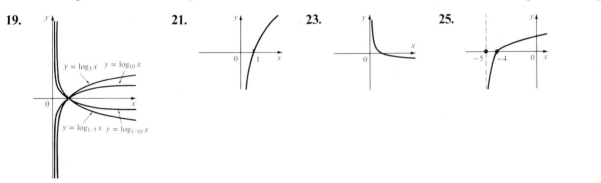

Exercises 6.3

1. 6   **3.** $\frac{1}{3}$ **5.** -3 **7.** $\sqrt{2}$ **9.** 2   **11.** 1   **13.** 15   **15.** $\log_5(ab/c)$ **17.** $\ln[\sqrt[3]{x}/(2x + 3)^4]$

19.   **21.**   **23.**   **25.**

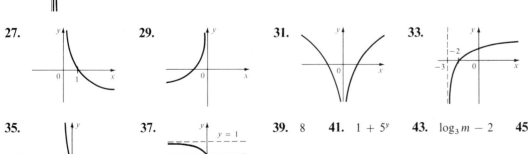

27.   **29.**   **31.**   **33.**

35.   **37.**   **39.** 8   **41.** $1 + 5^y$ **43.** $\log_3 m - 2$ **45.** e^2

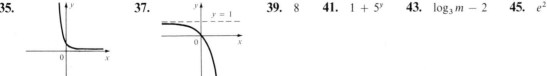

47. 40   **49.** 3   **51.** e^{ey} **57.** 8.9   **59.** $-\infty$ **61.** ∞ **63.** 1   **65.** $-\infty$ **67.** 0   **69.** 0
71. 0   **73.** 0   **75.** $(-\infty, 1), (-\infty, \infty)$ **77.** $(1, \infty), (-\infty, \infty)$ **79.** $y = e^x - 3$ **81.** $y = (\ln x)^2$
83. $y = \log_{10}[x/(1 - x)]$ **87.** e^5

Exercises 6.4

1. $f'(x) = 1/(x + 1)$, $(-1, \infty)$, $(-1, \infty)$ **3.** $f'(x) = (\log_3 e)2x/(x^2 - 4)$, $|x| > 2$, $|x| > 2$

5. $f'(x) = -\tan x$, $\{x \mid (4n - 1)\pi/2 < x < (4n + 1)\pi/2, n = 0, \pm 1, \pm 2, \ldots\}$, same

7. $f'(x) = (-1 - 2x)/(2 - x - x^2)$, $(-2, 1)$, $(-2, 1)$ **9.** $f'(x) = 2x \ln(1 - x^2) - 2x^3/(1 - x^2)$, $(-1, 1)$, $(-1, 1)$

11. $y' = 1/x \ln 10$, $y'' = -1/x^2 \ln 10$ **13.** $y' = 1 + \ln x$, $y'' = 1/x$ **15.** $f'(x) = (2 + \ln x)/2\sqrt{x}$

17. $g'(x) = -2a/(a^2 - x^2)$ **19.** $F'(x) = 1/2x$ **21.** $f'(t) = (\log_2 e)(4t^3 - 2t)/(t^4 - t^2 + 1)$

23. $h'(y) = 3/y + \cot y$ **25.** $g'(u) = -2/u(1 + \ln u)^2$ **27.** $y' = 3 \cot x(\ln \sin x)^2$

29. $y' = [1 + x^2(1 - 2\ln x)]/x(1 + x^2)^2$ **31.** $y' = -6/5(x^2 - 1)$ **33.** $y' = (3x - 2)/x(x - 1)$

35. $y' = (1 + 1/x)/(x + \ln x)$ **37.** $\ln 2$ **39.** $e^2/2 + e - \frac{1}{2}$ **41.** $\frac{1}{2}\ln|2x - 1| + C$ **43.** $\frac{1}{2}\ln|x^2 + 2x| + C$

45. $(\ln x)^3/3 + C$ **47.** $-\ln(1 + \cos x) + C$ **51.** $y' = (3x - 7)^4(8x^2 - 1)^3[12/(3x - 7) + 48x/(8x^2 - 1)]$

53. $y' = (x + 1)^4(x - 5)^3(x - 3)^{-8}[4/(x + 1) + 3/(x - 5) - 8/(x - 3)]$

55. $y' = e^x\sqrt{x^5 + 2}(x + 1)^{-4}(x^2 + 3)^{-2}[1 + 5x^4/2(x^5 + 2) - 4/(x + 1) - 4x/(x^2 + 3)]$

57. $y' = 2x/(x^2 + y^2 - 2y)$ **59.** 0 **61.** $x - ey = e$ **63.** $f^{(n)}(x) = (-1)^{n-1}(n - 1)!/(x - 1)^n$ **65.** $\frac{1}{3}$

67. CD on $(0, e^{8/3})$, CU on $(e^{8/3}, \infty)$, IP$(e^{8/3}, 8e^{-4/3}/3)$

69. (a) $\bigcup_{n=-\infty}^{\infty}(2n\pi - \pi/2, 2n\pi + \pi/2)$

     (b) x-intercept $2n\pi$, y-intercept 0

     (c)   About y-axis and period 2π

     (d)   Vertical asymptote $x = (2n + 1)\pi/2$

     (e)   Increasing on $(2n\pi - \pi/2, 2n\pi]$, decreasing on $[2n\pi, 2n\pi + \pi/2)$

     (f)   Local maximum $f(2n\pi) = 0$

     (g)   CD on $(2n\pi - \pi/2, 2n\pi + \pi/2)$, no IP

     (h)

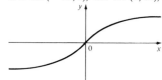

71. (a) $(0, \infty)$ (c)   None

     (d)   Vertical asymptote $x = 0$

     (e)   Increasing on $(0, \infty)$

     (f)   No maximum or minimum

     (g)   CD on $(0, \infty)$, no IP

     (h)

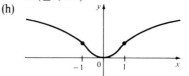

73. (a) R (b) x-intercept 0, y-intercept 0

     (c)   About $(0, 0)$ (d)   None

     (e)   Increasing on $(-\infty, \infty)$ (f)   None

     (g)   CU on $(-\infty, 0)$, CD on $(0, \infty)$, IP$(0, 0)$

     (h)

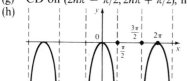

75. (a) R (b) x-intercept 0, y-intercept 0

     (c)   About y-axis   (d)   None

     (e)   Decreasing on $(-\infty, 0]$, increasing on $[0, \infty)$

     (f)   Local minimum $f(0) = 0$

     (g)   CD on $(-\infty, -1)$ and $(1, \infty)$, CU on $(-1, 1)$, IP$(\pm 1, \ln 2)$

     (h)

77. $\ln x + C_1$ if $x > 0$, $\ln|x| + C_2$ if $x < 0$

Exercises 6.5

1. $f'(x) = e^{\sqrt{x}}/2\sqrt{x}$, $[0, \infty)$, $(0, \infty)$ **3.** $f'(x) = (4\ln 7)x^3 7^{x^4}$, R, R

5. $f'(x) = \sqrt{2}^{\sin x}(\ln \sqrt{2})\cos x$, R, R **7.** $f'(x) = \sqrt{5}(\ln x)^{\sqrt{5}-1}/x$, $(1, \infty)$, $(1, \infty)$ **9.** $y' = -e^{-x}$, $y'' = e^{-x}$

11. $y' = \sec^2 x\, e^{\tan x}$, $y'' = \sec^2 x\, e^{\tan x}(\sec^2 x + 2\tan x)$ **13.** $f'(x) = (\ln 2)x \cdot 2^{x^2+1}$ **15.** $F'(x) = e^x(\ln x + 1/x)$

17. $f'(t) = -\pi^{-t}\ln \pi$ **19.** $h'(t) = 3t^2 - 3^t \ln 3$ **21.** $y' = e^{2x}(2x + 1)$ **23.** $y' = (\ln 2)(\ln 3)2^{3^x}3^x$

25. $y' = \cos(3^x) \cdot 3^x \ln 3$ **27.** $y' = 2e^{2x}/(e^{2x} - 2)$ **29.** $y' = e^{-1/x}/x^2$ **31.** $y' = e^{x+e^x}(1 + e^x)$

33. $y' = -x/(1 + x)$ **35.** $y' = x^{\sin x}(\cos x \ln x + \sin x/x)$ **37.** $y' = e^x x^{e^x}(\ln x + 1/x)$

39. $y' = (\ln x)^x(\ln \ln x + 1/\ln x)$ **41.** $y' = 0$ **43.** $y' = x^{1/x}(1 - \ln x)/x^2$ **45.** $y' = x^{x^2+1}(1 + 2\ln x)$

47. $y' = 1 + (x + 1)e^x/\sin(x - y)$ **49.** -2 **51.** $x + e^\pi y = \pi$ **53.** $x - 2y + \ln 4 = 0$

57. $256e^{-2x}$ **59.** (b) $c = $ initial velocity   (c) $t = (\ln 2)/k$

61. (a) $2500 \times 2^{t/3}$ (b) $(2500 \ln 2)2^{t/3}/3$ (c) $3 \ln 6/\ln 2 \approx 7.75$ h

63. (b) $K(be^{-bt} - ae^{-at})$ (c) $t = \ln(b/a)/(b - a)$ **67.** $500/\ln 5$ **69.** $-e^{-6x}/6 + C$ **71.** $(1 + e^x)^{11}/11 + C$

73. $x - e^{-x} + C$ **75.** $-\frac{1}{2}e^{-x^2} + C$ **77.** $\frac{1}{2}e^{x^2 - 4x - 3} + C$ **79.** $1 - e^{-2} \approx 0.865$

81. $e^3/3 - e + 2/3 \approx 4.644$

83. (a) Decreasing on $(-\infty, -1]$, increasing on $[-1, \infty)$ (b) CD on $(-\infty, -2)$, CU on $(-2, \infty)$
   (c) $(-2, -2e^{-2})$

85. (a) $(-\infty, \infty)$ **87.** (a) $\{x \mid x \neq -1\}$ (b) y-intercept $f(0) = 1/e$
   (b) x-intercept $(\ln 2)/2$, y-intercept -1 (c) None   (d) Horizontal asymptote $y = 1$,
   (c) None   (d) None   Vertical asymptote $x = -1$
   (e) Increasing on $(-\infty, \infty)$ (e) Increasing on $(-\infty, -1)$ and $(-1, \infty)$
   (f) None   (f) No maximum, or minimum
   (g) CD on $(-\infty, \frac{1}{2}\ln 2)$, CU on $(\frac{1}{2}\ln 2, \infty)$, IP $(\frac{1}{2}\ln 2, 0)$ (g) CU on $(-\infty, -1)$ and $(-1, -\frac{1}{2})$,
   (h)   CD on $(-\frac{1}{2}, \infty)$, IP $(-\frac{1}{2}, 1/e^2)$
      (h)

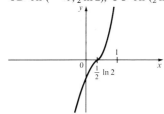 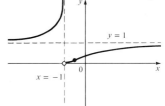

89. (a) R (b) y-intercept 1   (c) About y-axis   **91.** (a) R (b) x-intercept 0, y-intercept 0
   (d) Horizontal asymptote $y = 0$ (c) About $(0, 0)$ (d) None
   (e) Increasing on $(-\infty, 0]$, decreasing on $[0, \infty)$ (e) Increasing on $(-\infty, \infty)$ (f) None
   (f) Local maximum $f(0) = 1$ (g) CD on $(-\infty, 0)$, CU on $(0, \infty)$, IP $(0, 0)$
   (g) CU on $(-\infty, -1/\sqrt{2})$ and $(1/\sqrt{2}, \infty)$,   (h)
   CD on $(-1/\sqrt{2}, 1/\sqrt{2})$, IP $(\pm 1/\sqrt{2}, 1/\sqrt{e})$
   (h)

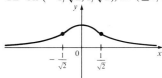

93. (a) $(0, \infty)$ (b) None   (c) None   **95.** (a) $\{x \mid x \neq 0\}$ (b) None   (c) About y-axis
   (d) Vertical asymptote $x = 0$ (d) Horizontal asymptote $y = 1$
   (e) Decreasing on $(0, 1]$, increasing on $[1, \infty)$ (e) Increasing on $(-\infty, 0)$, decreasing on $(0, \infty)$
   (f) Local minimum $f(1) = 1$ (f) None   (g) CU on $(-\infty, 0)$ and $(0, \infty)$, no IP
   (g) CU on $(0, \infty)$, no IP   (h)
   (h)

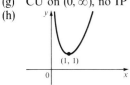 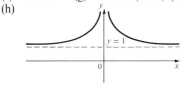

97. $2e^x + 2x^{3/2} + C$ **99.** $\frac{1}{2}$

Exercises 6.7

1. (a) 100×2^{3t} (b) $\approx 1.07 \times 10^{11}$ (c) $\ln 100/3 \ln 2 \approx 2.2$ h

3. (a) $500e^{(\ln 16)t/3}$ (b) $\approx 20{,}159$ (c) $3 \ln 60/\ln 16 \approx 4.4$ h   **5.** (a) $506{,}157$ (b) $683{,}241$

7. (a) $Ce^{-0.0005t}$ (b) $-2000 \ln 0.9 \approx 211$ s

9. (a) $50 \times 2^{-t/0.00014}$ (b) $\approx 1.57 \times 10^{-20}$ mg   (c) $\approx 4.5 \times 10^{-5}$ s   **11.** $21 + 12e^{-t/10}$

13. (a) ≈ 64.5 kPa   (b) ≈ 39.9 kPa

15. (a) \$4615.87   (b) \$4658.91   (c) \$4697.04   (d) \$4703.11   (e) \$4704.67   (f) \$4704.94

17. (a) $450e^{-0.4} \approx 302$ kg   (b) ≈ 30.4 min

Exercises 6.8

1. π **3.** $\pi/3$ **5.** $\pi/4$ **7.** $5\pi/6$ **9.** $-\pi/4$ **11.** $-\pi/6$ **13.** 0.7 **15.** 10 **17.** $1/2$ **19.** $3/5$
21. $-\pi/4$ **23.** $119/169$ **25.** $(\sqrt{5}+4\sqrt{2})/9$ **29.** $x/\sqrt{x^2+1}$ **37.** $g'(x)=3x^2/(1+x^6)$
39. $y'=2x/\sqrt{1-x^4}$ **41.** $F'(x)=a/(a^2+x^2)$ **43.** $H'(x)=1+2x\arctan x$
45. $g'(t)=-4/\sqrt{t^4-16t^2}$ **47.** $G'(t)=-1/\sqrt{2(-2t^2+3t-1)}$ **49.** $y'=x/|x|(1+x^2)$
51. $y'=\cos x/(1+\sin^2 x)$ **53.** $y'=\pi/2\sqrt{1-x^2}\,(\cos^{-1}x)^2$ **55.** $y'=-1/(1+x^2)(\tan^{-1}x)^2$
57. $y'=2x\cot^{-1}3x-3x^2/(1+9x^2)$ **59.** $y'=-\sqrt{a^2-b^2}\sin x/|(a+b\cos x)\sin x|$
61. $g'(x)=3/\sqrt{-9x^2-6x},\ \left[-\tfrac{2}{3},0\right],\ (-\tfrac{2}{3},0)$
63. $S'(x)=1/(1+x^2)\sqrt{1-(\tan^{-1}x)^2},\ [-\tan 1,\tan 1],\ (-\tan 1,\tan 1)$
65. $G'(x)=-1/2x\sqrt{(x^2-1)}\csc^{-1}x,\ |x|\geq 1,\ |x|>1$ **67.** $U'(t)=2^{\arctan t}(\ln 2)/(1+t^2),\ (-\infty,\infty),\ (-\infty,\infty)$ **69.** $\pi/6$
71. $-\pi/2$ **73.** π **75.** π **77.** $\pi^2/4$ **79.** 0 **81.** At a distance $\tfrac{5}{3}(4-\sqrt{10})$ from A
83. (a) $\left[-\tfrac{1}{2},\infty\right)$ (b) x-intercept 0, y-intercept 0   **85.** (a) $\{0\}$
    (c) None   (d) Horizontal asymptote $y=\pi/2$ (h)
    (e) Increasing on $\left[-\tfrac{1}{2},\infty\right)$
    (f) No maximum or minimum
    (g) CD on $(-\tfrac{1}{2},\infty)$, no IP
    (h)

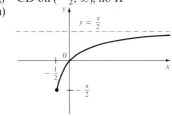

87. (a) $\{x\mid x\neq 0\}$ (b) None   **89.** (a) $(-\infty,\infty)$ (b) x-intercept 0, y-intercept 0
    (c) About $(0,0)$ (d) $y=\pm\pi/2$ (c) Odd   (d) Slant asymptotes $y=x\pm\pi/2$
    (e) Constant on $(-\infty,0)$ and on $(0,\infty)$ (e) Increasing on $(-\infty,\infty)$ (f) None
    (f) Local maximum and minimum at every point   (g) CD on $(-\infty,0)$, CU on $(0,\infty)$, IP $(0,0)$
    (h)   (h)

91. $x^2+5\sin^{-1}x+C$ **93.** $\pi/2$ **95.** $\tfrac{1}{3}\sin^{-1}(x^3)+C$ **97.** $\tfrac{1}{2}\ln(x^2+9)+3\tan^{-1}(x/3)+C$
99. $\tfrac{1}{3}\tan^{-1}(3x)+C$ **101.** $\tan^{-1}(e^x)+C$ **103.** $\pi^2/72$

Exercises 6.9

17. $\coth x=\tfrac{5}{4}$, $\operatorname{sech} x=\tfrac{3}{5}$, $\cosh x=\tfrac{5}{3}$, $\sinh x=\tfrac{4}{3}$, $\operatorname{csch} x=\tfrac{3}{4}$ **19.** (a) 1   (b) -1 (c) ∞ (d) $-\infty$
    (e) 0   (f) 1   (g) ∞ (h) $-\infty$ (i) 0   **27.** $f'(x)=3\operatorname{sech}^2 3x$ **29.** $h'(x)=4x^3\sinh(x^4)$
31. $G'(x)=2x\operatorname{sech} x-x^2\operatorname{sech} x\tanh x$ **33.** $H'(t)=e^t\operatorname{sech}^2(e^t)$ **35.** $y'=x^{\cosh x}(\sinh x\ln x+\cosh x/x)$
37. $y'=2x/\sqrt{x^4-1}$ **39.** $y'=\ln(\operatorname{sech} 4x)-4x\tanh 4x$ **41.** $y'=a/(a^2-x^2)$ **43.** $y'=-4/x\sqrt{x^8+1}$
45. $y'=-1/x\sqrt{x^2+1}$ **47.** $\tfrac{1}{2}\cosh 2x+C$ **49.** $\ln|\sinh x|+C$ **51.** $\sinh^{-1}(x/2)+C$ **53.** $\tfrac{1}{2}\ln 3$
55. $(\ln(1+\sqrt{2}),\sqrt{2})$

Exercises 6.10

1. $\tfrac{1}{4}$ **3.** $\tfrac{3}{2}$ **5.** 1   **7.** ∞ **9.** $\tfrac{1}{2}$ **11.** 0   **13.** ∞ **15.** $1/3a^{2/3}$ **17.** $\tfrac{1}{2}$ **19.** 0   **21.** $\tfrac{1}{2}$
23. ∞ **25.** 0   **27.** $\tfrac{2}{3}$ **29.** α **31.** $\tfrac{2}{3}$ **33.** -2 **35.** 0   **37.** $\tfrac{1}{2}$ **39.** 0   **41.** 0   **43.** 0
45. 1   **47.** ∞ **49.** 0   **51.** 0   **53.** 0   **55.** 1   **57.** e **59.** e^{-2} **61.** e^3 **63.** 1   **65.** 1
67. $1/e$ **69.** 0   **71.** $\tfrac{1}{2}$ **73.** $-\infty$ **75.** 4   **77.** $\tfrac{1}{24}$

79. (a) R (b) x-intercept 0, y-intercept 0
(c) None   (d) Horizontal asymptote $y = 0$
(e) Increasing on $(-\infty, 1]$, decreasing on $[1, \infty)$
(f) Local maximum $f(1) = 1/e$
(g) CD on $(-\infty, 2)$, CU on $(2, \infty)$, IP$(2, 2/e^2)$
(h)

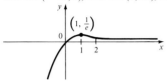

81. (a) $(0, \infty)$ (b) x-intercept 1
(c) None   (d) None
(e) Decreasing on $(0, 1/e)$, increasing on $[1/e, \infty)$
(f) Local minimum $(1/e, -1/e)$
(g) CU on $(0, \infty)$, no IP
(h)

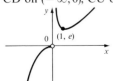

83. (a) $(0, \infty)$ (b) x-intercept 1
(c) None   (d) None
(e) Decreasing on $(0, 1/\sqrt{e}\,]$, increasing on $[1/\sqrt{e}, \infty)$
(f) Local minimum $f(1/\sqrt{e}) = -1/2e$
(g) CD on $(0, e^{-3/2})$, CU on $(e^{-3/2}, \infty)$,
IP$(e^{-3/2}, -3/2e^3)$
(h)

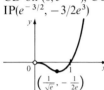

85. (a) R (b) x-intercept 0, y-intercept 0
(c) About $(0, 0)$ (d) Horizontal asymptote $y = 0$
(e) Decreasing on $(-\infty, -1/\sqrt{2}\,]$ and $[1/\sqrt{2}, \infty)$,
increasing on $[-1/\sqrt{2}, 1/\sqrt{2}\,]$
(f) Local minimum $f(-1/\sqrt{2}) = -1/\sqrt{2e}$, local
maximum $f(1/\sqrt{2}) = 1/\sqrt{2e}$
(g) CD on $(-\infty, -\sqrt{3/2})$ and $(0, \sqrt{3/2})$, CU on
$(-\sqrt{3/2}, 0)$ and $(\sqrt{3/2}, \infty)$,
IP$(\pm\sqrt{3/2}, \pm\sqrt{3/2}\,e^{-3/2})$ and $(0, 0)$
(h)

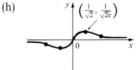

87. (a) $\{x \mid x \neq 0\}$ (b) None   (c) None
(d) Horizontal asymptote $y = 0$, vertical asymptote
$x = 0$
(e) Decreasing on $(-\infty, 0)$ and $(0, 1]$, increasing on
$[1, \infty)$ (f) Local minimum $f(1) = e$
(g) CD on $(-\infty, 0)$, CU on $(0, \infty)$, no IP
(h)

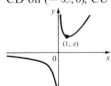

89. (a) $\{x \mid x \neq 0\}$ (b) None   (c) None
(d) Horizontal asymptote $y = 0$, vertical asymptote
$x = 0$
(e) Decreasing on $(-\infty, 0)$ and $(0, 3]$, increasing on
$[3, \infty)$ (f) Local minimum $f(3) = e^3/27$
(g) CD on $(-\infty, 0)$, CU on $(0, \infty)$, no IP
(h)

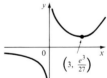

91. (a) $\{x \mid x \neq 0\}$ (b) None
(c) None   (d) Vertical asymptote $x = 0$
(e) Increasing on $(-\infty, 0)$ and $[1, \infty)$, decreasing on $(0, 1]$
(f) Local minimum $f(1) = e$
(g) CD on $(-\infty, 0)$, CU on $(0, \infty)$, no IP
(h)

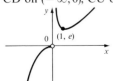

93. (a) R (b) y-intercept 1
(c) None   (d) Slant asymptote $y = x$
(e) Increasing on $(-\infty, \infty)$ (f) None
(g) CU on $(-\infty, \infty)$
(h)

95. (a) $(-1, \infty)$ (b) x-intercept 0, y-intercept 0
(c)   None   (d)   Vertical asymptote $x = -1$
(e)   Decreasing on $(-1, 0]$, increasing on $[0, \infty)$
(f)   Local minimum $f(0) = 0$
(g)   CU on $(-1, \infty)$, no IP
(h)

97. (a) $\{x \mid x > 0, x \neq 1\} = (0, 1) \cup (1, \infty)$ (b)   None
(c)   None   (d)   Horizontal asymptote $y = e$
(e)   Constant on $(0, 1) \cup (1, \infty)$
(f)   Local maximum and minimum at every point
(h)

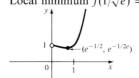

99. (a) $(0, \infty)$ (b)   None
(c)   None   (d)   None
(e)   Decreasing on $(0, e^{-2}]$, increasing on $[e^{-2}, \infty)$
(f)   Local minimum $f(e^{-2}) = e^{-2/e}$
(h)

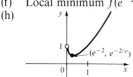

101. (a) $(0, \infty)$ (b)   None
(c)   None   (d)   None
(e)   Decreasing on $(0, 1/\sqrt{e}]$, increasing on $[1/\sqrt{e}, \infty)$
(f)   Local minimum $f(1/\sqrt{e}) = e^{-1/2e}$
(h)

103. (a) $\{x \mid x > -1, x \neq 0\} = (-1, 0) \cup (0, \infty)$
(b)   None   (c)   None
(d)   Horizontal asymptote $y = 1$, vertical asymptote $x = -1$
(e)   Decreasing on $(-1, 0) \cup (0, \infty)$ (f)   None
(h)

Review Exercises for Chapter 6

1.    **3.** 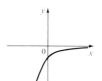   **5.** 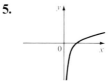   **7.**

9. $\ln 5$ **11.** $\ln 10$ **13.** $e^{2/\pi}$ **15.** $\tan^{-1} 4 + n\pi$, n an integer   **17.** $(\log_{10} e)(2x - 1)/(x^2 - 1)$
19. $\sqrt{x + 1}(2 - x)^5(x + 3)^{-7}[1/2(x + 1) - 5/(2 - x) - 7/(x + 3)]$ **21.** $(1 + c^2)e^{cx} \sin x$ **23.** $2 \tan x$
25. $e^{-1/x}(1 + 1/x)$ **27.** $(\cos^{-1} x)^{\sin^{-1} x}[\cos^{-1} x \ln(\cos^{-1} x) - \sin^{-1} x]/(\cos^{-1} x)\sqrt{1 - x^2}$ **29.** $e^{x + e^x}$
31. $-1/x - 1/x(\ln x)^2$ **33.** $7^{\sqrt{2x}} \ln 7/\sqrt{2x}$ **35.** $e^{\sec^{-1} x}(1 + 1/\sqrt{x^2 - 1})$ **37.** $3 \tanh 3x$
39. $\cosh x/\sqrt{\sinh^2 x - 1}$ **41.** $\cot x - \sin x \cos x$ **43.** $y = 1/\sqrt{x}(x + 1)$ **45.** $1/(x^4 + x^2 + 1)$ **47.** $2^x(\ln 2)^n$
49. $3x - 2y + \ln 4 = 0$ **51.** $(-3, 0)$ **53.** ∞ **55.** $-\infty$ **57.** ∞ **59.** $\pi/3$ **61.** $-1/2\pi$ **63.** 0
65. $-\frac{1}{3}$ **67.** 0 **69.** 0 **71.** 1

73. (a) $\{x \mid x \neq 0\}$ (b) None   (c) About $(0,0)$
(d) Horizontal asymptote $y = 0$
(e) Decreasing on $(-\infty, 0)$ and $(0, \infty)$
(f) None
(g) CD on $(-\infty, 0)$, CU on $(0, \infty)$, no IP
(h)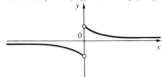

75. (a) $\{x \mid x \neq 1\}$ (b) y-intercept $\frac{1}{2}$ (c) None
(d) Horizontal asymptote $y = 1$, vertical asymptote $x = 1$
(e) Decreasing on $(-\infty, 1)$ and $(1, \infty)$ (f) None
(g) CD on $(-\infty, 1 - \ln\sqrt{2})$, CU on $(1 - \ln\sqrt{2}, 1)$ and $(1, \infty)$, IP when $x = 1 - \ln\sqrt{2}$
(h)

77. (a) R (b) y-intercept 2
(c) None   (d) None
(e) Decreasing on $(-\infty, \frac{1}{4}\ln 3]$, increasing on $[\frac{1}{4}\ln 3, \infty)$
(f) Local minimum $f(\frac{1}{4}\ln 3) = 3^{1/4} + 3^{-3/4}$
(g) CU on $(-\infty, \infty)$, no IP
(h)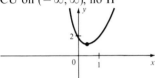

79. (a) $(0, \infty)$ (b) None
(c) None   (d) Vertical asymptote $x = 0$
(e) Decreasing on $(0, \frac{1}{2}]$, increasing on $[\frac{1}{2}, \infty)$
(f) Local minimum $f(\frac{1}{2}) = \frac{1}{2} + \ln 2$
(g) CU on $(0, \infty)$, no IP
(h)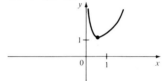

81. (a) $1000e^{(\ln 9)t/2} = 1000 \times 3^t$ (b) 27,000   (c) $\ln 2/\ln 3$ h
83. $v(t) = -Ae^{-ct}[c\cos(\omega t + \delta) + \omega\sin(\omega t + \delta)]$, $a(t) = Ae^{-ct}[(c^2 - \omega^2)\cos(\omega t + \delta) + 2c\omega\sin(\omega t + \delta)]$
85. $f'(x) = e^{g(x)}g'(x)$ **87.** $f'(x) = g'(x)/g(x)$ **89.** $\frac{\pi}{6}$ **91.** $\ln 2 - \frac{7}{4}$ **93.** $2e^{\sqrt{x}} + C$
95. $-\frac{1}{2}[\ln(\cos x)]^2 + C$ **97.** $\frac{1}{4}\ln(1 + x^4) + C$ **99.** $\ln|1 + \sec\theta| + C$ **101.** $\frac{1}{3}\sinh 3t + C$ **105.** $e^{\sqrt{x}}/2x$
107. $\frac{1}{3}\ln 4$ **109.** $\frac{2}{3}$

Chapter 7

Exercises 7.1

1. $xe^{2x}/2 - e^{2x}/4 + C$ **3.** $-\frac{1}{4}x\cos 4x + \frac{1}{16}\sin 4x + C$ **5.** $\frac{1}{3}x^2\sin 3x + \frac{2}{9}x\cos 3x - \frac{2}{27}\sin 3x + C$
7. $x(\ln x)^2 - 2x\ln x + 2x + C$ **9.** $\frac{1}{8}(\sin 2\theta - 2\theta\cos 2\theta) + C$ **11.** $t^3(3\ln t - 1)/9 + C$
13. $e^{2\theta}(2\sin 3\theta - 3\cos 3\theta)/13 + C$ **15.** $y\cosh y - \sinh y + C$ **17.** $1 - 2/e$ **19.** $-\frac{1}{2}$ **21.** 1
23. $(5\sin 3x\sin 5x + 3\cos 3x\cos 5x)/16 + C$ **25.** $\sin x(\ln\sin x - 1) + C$ **27.** $(2x + 1)e^x + C$
29. $x(\cos\ln x + \sin\ln x)/2 + C$ **31.** $4\ln 2 - \frac{3}{2}$ **33.** $2(\sin\sqrt{x} - \sqrt{x}\cos\sqrt{x}) + C$ **35.** $e^{x^2}(x^4/2 - x^2 + 1) + C$
37. $-\frac{1}{4}\cos x\sin^3 x + 3x/8 - \frac{3}{16}\sin 2x + C$ **39.** (b) $\frac{2}{3}, \frac{8}{15}$ **45.** $x[(\ln x)^3 - 3(\ln x)^2 + 6\ln x - 6] + C$
47. $\pi/2 - 1$ **49.** 1   **51.** $2 - e^{-t}(t^2 + 2t + 2)$ m   **55.** 1

Exercises 7.2

1. $\pi/4$ **3.** $\frac{3}{8}x + \frac{1}{4}\sin 2x + \frac{1}{32}\sin 4x + C$ **5.** $\frac{1}{7}\cos^7 x - \frac{1}{5}\cos^5 x + C$ **7.** $(3\pi - 4)/192$
9. $3x/2 + \cos 2x - \frac{1}{8}\sin 4x + C$ **11.** $\frac{1}{6}\sin^6 x - \frac{1}{4}\sin^8 x + \frac{1}{10}\sin^{10} x + C$ or $-\frac{1}{6}\cos^6 x + \frac{1}{4}\cos^8 x - \frac{1}{10}\cos^{10} x + C_1$
13. $\frac{1}{8}[5x/2 + 2\sin 2x + \frac{3}{8}\sin 4x - \frac{1}{3}\sin^3 2x] + C$ **15.** $-\frac{1}{5}\cos^5 x + \frac{2}{3}\cos^3 x - \cos x + C$
17. $[\frac{2}{7}\cos^3 x - \frac{2}{3}\cos x]\sqrt{\cos x} + C$ **19.** $\sqrt{x} + \frac{1}{2}\sin 2\sqrt{x} + C$ **21.** $\frac{1}{2}\cos^2 x - \ln|\cos x| + C$
23. $\ln(1 + \sin x) + C$ **25.** $\tan x - x + C$ **27.** $\tan x + \frac{1}{3}\tan^3 x + C$ **29.** $\frac{1}{5}$ **31.** $\frac{1}{3}\sec^3 x + C$
33. $\frac{1}{4}\sec^4 x - \tan^2 x + \ln|\sec x| + C$ **35.** $\frac{38}{15}$ **37.** $\frac{1}{6}\sec^6 x + C$ [or $\frac{1}{2}\tan^2 x + \frac{1}{2}\tan^4 x + \frac{1}{6}\tan^6 x + C_1$]
39. $\frac{1}{2}\tan^2 x + C$ **41.** $\sqrt{3} - \pi/3$ **43.** $-\frac{1}{5}\cot^5 x - \frac{1}{7}\cot^7 x + C$ **45.** $\ln|\csc x - \cot x| + C$
47. $\ln|\csc x - \cot x| + \cos x + C$ **49.** $\frac{1}{2}[\frac{1}{3}\sin 3x - \frac{1}{7}\sin 7x] + C$ **51.** $\frac{1}{2}[\sin x + \frac{1}{7}\sin 7x] + C$
53. $\frac{1}{2}[\frac{1}{4}\cos 4x - \frac{1}{6}\cos 6x] + C$ **55.** $\frac{1}{2}\sin 2x + C$ **57.** 0   **59.** $\frac{1}{3}$ **61.** $s = (1 - \cos^3\omega t)/3\omega$

Exercises 7.3

1. $2/\sqrt{3}$ **3.** $-\sqrt{1 - x^2} + C$ **5.** $\frac{1}{4}\sin^{-1}(2x) + \frac{1}{2}x\sqrt{1 - 4x^2} + C$ **7.** $\ln(1 + \sqrt{2})$

9. $\sqrt{x^2 - 16}/32x^2 + \frac{1}{128}\sec^{-1}(x/4) + C$ **11.** $\sqrt{9x^2 - 4} - 2\sec^{-1}(3x/2) + C$

13. $x/\sqrt{a^2 - x^2} - \sin^{-1}(x/a) + C$ **15.** $(1/\sqrt{3})\ln|(\sqrt{x^2 + 3} - \sqrt{3})/x| + C$ **17.** $\frac{64}{1215}$ **19.** $\frac{5}{3}(1 + x^2)^{3/2} + C$

21. $\frac{1}{4}[\sqrt{x^2 - 2}/x - (x^2 - 2)^{3/2}/3x^3] + C$ **23.** $\frac{1}{2}[\sin^{-1}(x - 1) + (x - 1)\sqrt{2x - x^2}] + C$

25. $\frac{1}{3}\ln|3x + 1 + \sqrt{9x^2 + 6x - 8}| + C$ **27.** $\frac{1}{2}[\tan^{-1}(x + 1) + (x + 1)/(x^2 + 2x + 2)] + C$

29. $\frac{1}{2}[e^t\sqrt{9 - e^{2t}} + 9\sin^{-1}(e^t/3)] + C$

Exercises 7.4

1. $\dfrac{A}{x - 1} + \dfrac{B}{x + 2}$ **3.** $\dfrac{A}{x} + \dfrac{B}{x + 2}$ **5.** $\dfrac{A}{2x - 1} + \dfrac{B}{(2x - 1)^2} + \dfrac{C}{2x + 3}$ **7.** $\dfrac{A}{x} + \dfrac{B}{x^2} + \dfrac{C}{x^3} + \dfrac{D}{x - 1}$

9. $1 + \dfrac{A}{x - 1} + \dfrac{B}{x + 1}$ **11.** $\dfrac{A}{x} + \dfrac{Bx + C}{x^2 + 2}$ **13.** $\dfrac{Ax + B}{x^2 + 1} + \dfrac{Cx + D}{x^2 + 4} + \dfrac{Ex + F}{(x^2 + 4)^2}$

15. $\dfrac{Ax + B}{x^2 + 9} + \dfrac{Cx + D}{(x^2 + 9)^2} + \dfrac{Ex + F}{(x^2 + 9)^3}$ **17.** $\dfrac{A}{x} + \dfrac{B}{x^2} + \dfrac{Cx + D}{x^2 + x + 2}$ **19.** $\dfrac{x^2}{2} - x + \ln|x + 1| + C$

21. $\ln 3 + 3\ln 6 - 3\ln 4 = \ln\frac{81}{8}$ **23.** $3x - 7\ln|2x + 3| + C$

25. $x - \ln|x| + 2\ln|x - 1| + C = x + \ln((x - 1)^2/|x|) + C$ **27.** $2\ln 2 + \frac{1}{2}$

29. $\frac{1}{3}\ln|x| - \ln|x + 1| + \frac{2}{3}\ln|2x + 3| + C$ **31.** $4\ln 6 - 3\ln 5$ **33.** $-1/5(x - 1) + \frac{1}{25}\ln|(x + 4)/(x - 1)| + C$

35. $2\ln|x| + 3\ln|x + 2| + 1/x + C$ **37.** $2x + 3\ln|x + 2| + \ln|x - 1| + 1/(x - 1) + C$

39. $\frac{1}{3}\ln|x^3 + 3x^2 + 4| + C$ **41.** $\ln|1 - 1/x| + 1/(x - 1) - 1/2(x - 1)^2 + C$

43. $\ln|x + 1| + 2/(x + 1) - 1/2(x + 1)^2 + C$ **45.** $\frac{3}{16}\ln|(x + 1)/(x - 1)| - 1/8(x - 1) - 1/4(x + 1) - 1/8(x + 1)^2 + C$

47. $1/x + \frac{1}{2}\ln|(x - 1)/(x + 1)| + C$ **49.** $(1 - \ln 2)/2$ **51.** $\ln\sqrt{3} - \pi/6\sqrt{3}$

53. $\ln(x - 1)^2 + \ln\sqrt{x^2 + 1} - 3\tan^{-1}x + C$ **55.** $-\ln|x + 1| + \ln(x^2 - 4x + 7) + (5/\sqrt{3})\tan^{-1}((x - 2)/\sqrt{3}) + C$

57. $\frac{1}{3}\ln|x - 1| - \frac{1}{6}\ln(x^2 + x + 1) - (1/\sqrt{3})\tan^{-1}((2x + 1)/\sqrt{3}) + C$

59. $\ln|x - 1| - \ln\sqrt{x^2 + 1} + 1/(x - 1) + \tan^{-1}x + C$ **61.** $\frac{3}{2}\ln(x^2 + 1) - 3\tan^{-1}x + \sqrt{2}\tan^{-1}(x/\sqrt{2}) + C$

63. $\ln|x + 1| - \frac{1}{2}\ln|x + 2| - \frac{1}{4}\ln(x^2 + 2x + 2) - \frac{1}{2}\tan^{-1}(x + 1) + C$

65. $-1/2(x^2 + 2x + 4) - (2/3\sqrt{3})\tan^{-1}((x + 1)/\sqrt{3}) - 2(x + 1)/3(x^2 + 2x + 4) + C$ **67.** $\ln|x| + 1/(x^2 + 1) + C$

69. $\frac{1}{2}\ln(x^2 + 2x + 2) - \frac{7}{2}\tan^{-1}(x + 1) - (3x + 5)/2(x^2 + 2x + 2) + C$ **71.** $-2/(x^2 + 4)^2 + C$

73. $\ln|\sin^2 x - 3\sin x + 2| + C$ **77.** $\frac{1}{2}\ln|(x - 2)/x| + C$

79. $\frac{1}{2}\ln|x^2 + x - 1| - (1/2\sqrt{5})\ln|(2x + 1 - \sqrt{5})/(2x + 1 + \sqrt{5})| + C$ **81.** $1 + 2\ln 2$

Exercises 7.5

1. $2(1 - \ln 2)$ **3.** $2(\sqrt{x} - \tan^{-1}\sqrt{x}) + C$ **5.** $\frac{3}{2}\ln|x^{2/3} - 1| + C$ **7.** $\frac{1676}{15}$

9. $4[(\sqrt{1 + \sqrt{x}})^3/3 - \sqrt{1 + \sqrt{x}}] + C$ **11.** $x + 4\sqrt{x} + 4\ln|\sqrt{x} - 1| + C$

13. $\frac{3}{10}(x^2 + 1)^{5/3} - \frac{3}{4}(x^2 + 1)^{2/3} + C$ **15.** $2\sqrt{x} - 3\sqrt[3]{x} + 6\sqrt[6]{x} - 6\ln(\sqrt[6]{x} + 1) + C$

17. $2\sqrt{x} - 4\sqrt[4]{x} + 4\ln(1 + \sqrt[4]{x}) + C$ **19.** $(2/\sqrt{bc})\tan^{-1}\sqrt{bx/c} + C$

21. $3\sqrt[3]{x - 1} - \ln|1 + \sqrt[3]{x - 1}| + \frac{1}{2}\ln[(x - 1)^{2/3} - (x - 1)^{1/3} + 1] - \sqrt{3}\tan^{-1}[(2\sqrt[3]{x - 1} - 1)/\sqrt{3}] + C$

23. $\sqrt{x - x^2} - \tan^{-1}\sqrt{(1 - x)/x} + C$ **25.** $\ln|\sin x/(1 + \sin x)| + C$ **27.** $\ln[(e^x + 2)^2/(e^x + 1)] + C$

29. $2\sqrt{1 - e^x} + \ln[(1 - \sqrt{1 - e^x})/(1 + \sqrt{1 - e^x})] + C$ **31.** $C - \cot(x/2)$ or $-\cot x - \csc x + C$

33. $-\sqrt{2}\ln(\sqrt{2} - 1) = (1/\sqrt{2})\ln(3 + 2\sqrt{2})$ **35.** $\frac{1}{5}\ln|(2\tan(x/2) + 1)/(\tan(x/2) - 2)| + C$

37. $\frac{1}{4}\ln|\tan(x/2)| + \frac{1}{8}\tan^2(x/2) + C$ **39.** $(1/\sqrt{a^2 + b^2})\ln|(b\tan(x/2) - a + \sqrt{a^2 + b^2})/(b\tan(x/2) - a - \sqrt{a^2 + b^2})| + C$

Exercises 7.6

1. $\frac{1}{3}\sin^3 x - \frac{1}{5}\sin^5 x + C$ **3.** $1 - \sqrt{3}/2$ **5.** $2\sqrt{x - 2} - 4\tan^{-1}(\sqrt{x - 2}/2) + C$

7. $x\ln(1 + x^2) - 2x + 2\tan^{-1}x + C$ **9.** $\frac{4097}{45}$ **11.** $\frac{1}{2}\ln(x^2 - 2x + 2) + \tan^{-1}(x - 1) + C$

13. $3\ln|(3 - \sqrt{9 - x^2})/x| + \sqrt{9 - x^2} + C$ **15.** $(x^2 + 2)\sinh x - 2x\cosh x + C$ **17.** $\tan^{-1}(\sin x) + C$
19. $2/\pi$ **21.** $e^{3x}(5\sin 5x + 3\cos 5x)/34 + C$ **23.** $\frac{1}{2}[\ln|x + 1| - \frac{1}{2}\ln(x^2 + 1) + \tan^{-1}x] + C$
25. $-(x^3 + 1)e^{-x^3}/3 + C$ **27.** $\frac{1}{3}\ln|3x + 2 + \sqrt{9x^2 + 12x - 5}| + C$ **29.** $\frac{3}{4}x^{4/3} - \frac{6}{11}x^{11/6} + C$
31. $2x + 11\ln|x - 3| + C$ **33.** $\frac{1}{16}(x - \frac{1}{4}\sin 4x + \frac{1}{3}\sin^3 2x) + C$ **35.** $-\ln(1 + \sqrt{1 - x^2}) + C$
37. $\frac{1}{2}\ln|(e^x - 1)/(e^x + 1)| + C$ **39.** 0 **41.** $\frac{86}{3}$ **43.** $\frac{1}{2}(\ln\sin x)^2 + C$ **45.** $\frac{1}{6}\ln((x^2 + 1)/(x^2 + 4)) + C$
47. $3((x + c)^{7/3}/7 - c(x + c)^{4/3}/4) + C$ **49.** $3\ln(\sqrt{x + 1} + 3) - \ln(\sqrt{x + 1} + 1) + C$
51. $\frac{1}{2}(x + 2)\sin 2x - \frac{1}{4}(2x^2 + 8x - 7)\cos 2x + C$ **53.** $\frac{1}{2}\sin^{-1}(x^2/4) + C$ **55.** $\frac{1}{6}\csc^3 2x - \frac{1}{10}\csc^5 2x + C$
57. $e^{\arctan x} + C$ **59.** $-e^{-2t}(4t^3 + 6t^2 + 6t + 3)/8 + C$ **61.** $\frac{1}{24}\cos 6x - \frac{1}{16}\cos 4x - \frac{1}{8}\cos 2x + C$
63. $\sin^{-1}x - \sqrt{1 - x^2} + C$ **65.** $\ln\sqrt{x^2 + a^2} + \tan^{-1}(x/a) + C$ **67.** $\frac{1}{20}\tan^{-1}(x^5/4) + C$
69. $x + 2/[1 + \tan(x/2)] + C$ (or $x + \sec x - \tan x + C$)   **71.** $x\sec x - \ln|\sec x + \tan x| + C$
73. $\frac{2}{3}[(x + 1)^{3/2} - x^{3/2}] + C$ **75.** $2\sqrt{x}\tan^{-1}\sqrt{x} - \ln(1 + x) + C$ **77.** $\frac{1}{2}\ln(2x + 1 + \sqrt{4x^2 + 4x + 5}) + C$
79. $e^{-x} + \frac{1}{2}\ln|(e^x - 1)/(e^x + 1)| + C$

Exercises 7.7

1. (a) 0.265625   (b) 0.250000; 0.25   **3.** (a) 4.661488   (b) 4.666563; $\frac{14}{3} \approx 4.666667$
5. (a) 0.612115   (b) 0.611887; $\frac{1}{2}\ln\frac{17}{5} \approx 0.611888$ **7.** (a) 1.913972   (b) 1.934766
9. (a) 0.746211   (b) 0.746825   **11.** (a) 0.481672   (b) 0.481172   **13.** (a) 0.132465   (b) 0.132727
15. (a) 0.409140   (b) 0.395802   **17.** (a) 0.0017   (b) 0.000016 (or 0.000007)
19. (a) 0.0703   (c) 0.00076   **21.** 130   **23.** 0.310   **25.** 15.4   **27.** 8.6 mi

Exercises 7.8

1. Divergent   **3.** $\frac{1}{2}$ **5.** Divergent   **7.** 1   **9.** 0   **11.** $-\ln\frac{2}{3}$ **13.** Divergent   **15.** Divergent
17. $e^2/4$ **19.** Divergent   **21.** Divergent   **23.** 1   **25.** $1/\ln 2$ **27.** $2\sqrt{3}$ **29.** Divergent
31. Divergent   **33.** $\frac{5}{3}$ **35.** Divergent   **37.** 2   **39.** Divergent   **41.** Divergent   **43.** $\frac{4}{3}$ **45.** $-\frac{1}{4}$

47. e 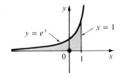   **49.** $\pi/2$ 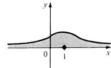   **51.** Infinite

53. π **55.** $1/(1 - p), p < 1$ **57.** $-1/(p + 1)^2, p > -1$ **61.** Convergent   **63.** Convergent

Exercises 7.9

1. $\frac{1}{25}e^{-3x}(-3\cos 4x + 4\sin 4x) + C$ **3.** $-\sqrt{9x^2 - 1}/x + 3\ln|3x + \sqrt{9x^2 - 1}| + C$ **5.** $e^{3x}(9x^2 - 6x + 2)/27 + C$
7. $\frac{1}{2}[x^2\sin^{-1}(x^2) + \sqrt{1 - x^4}] + C$ **9.** $\tan^{-1}(\sinh e^x) + C$ **11.** $\frac{1}{2}(x + 2)\sqrt{5 - 4x - x^2} + \frac{9}{2}\sin^{-1}[(x + 2)/3] + C$
13. $\frac{1}{4}\tan x\sec^3 x + \frac{3}{8}\tan x\sec x + \frac{3}{8}\ln|\sec x + \tan x| + C$ **15.** $\frac{1}{9}\sin^3 x[3\ln(\sin x) - 1] + C$
17. $-2\sqrt{2 + 3\cos x} - \sqrt{2}\ln|(\sqrt{2 + 3\cos x} - \sqrt{2})/(\sqrt{2 + 3\cos x} + \sqrt{2})| + C$ **19.** $\frac{8}{15}$

Review Exercises for Chapter 7

1. $x - 2\ln|x + 1| + C$ **3.** $(\arctan x)^6/6 + C$ **5.** $-1/e^{\sin x} + C$ **7.** $\frac{1}{25}x^5(5\ln x - 1) + C$
9. $-\frac{1}{2}\cos(x^2) + C$ **11.** $\frac{1}{3}\ln|(x - 2)/(2x - 1)| + C$ **13.** $\frac{1}{9}\sec^9 x - \frac{3}{7}\sec^7 x + \frac{3}{5}\sec^5 x - \frac{1}{3}\sec^3 x + C$
15. $\sqrt{1 + 2x} - 3\ln(\sqrt{1 + 2x} + 3) + C$ **17.** $2e^{\sqrt{x}} + C$ **19.** $(x + 1)\ln(x^2 + 2x + 2) - 2x + 2\tan^{-1}(x + 1) + C$
21. $\ln|x| - \frac{1}{2}\ln(x^2 + 1) + C$ **23.** $-\cot x - x + C$ **25.** $-x/\sqrt{x^2 - 1} + C$
27. $2\ln|x - 1| - \frac{1}{2}\tan^{-1}[(x + 1)/2] + C$ **29.** $e^{e^x} + C$ **31.** $\ln x[\ln(\ln x) - 1] + C$ **33.** $\frac{2}{5}$ **35.** Divergent
37. $1 - \pi/2$ **39.** $\frac{1}{24}$ **41.** 2   **43.** $8\ln\frac{4}{3} - 1$ **45.** $\pi/4$ **47.** $\sqrt{3} - \pi/3$
49. $-a/(a^2 + b^2)$ if $a < 0$, divergent if $a \geqslant 0$ **51.** $\frac{1}{2}[e^x\sqrt{1 - e^{2x}} + \sin^{-1}(e^x)] + C$
53. $\frac{1}{4}(2x + 1)\sqrt{x^2 + x + 1} + \frac{3}{8}\ln|x + \frac{1}{2} + \sqrt{x^2 + x + 1}| + C$ **55.** (a) 1.090608   (b) 1.089429
57. 0.0067   **59.** 17.74   **61.** 2   **63.** No

Chapter 8

Exercises 8.1

1. $\pi/5$

3. $\pi/3$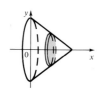

5. 8π

7. $3\pi/10$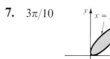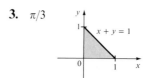

9. $64\pi/15$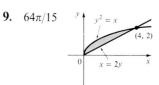

11. $208\pi/45$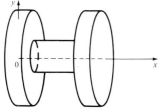

13. $32\pi/3$ **15.** $128\pi/3$ **17.** $128\pi/15$ **19.** $112\pi/15$ **21.** $64\pi/5$ **23.** $16\pi/5$ **25.** $46\pi/15$
27. $\pi(e^2 - 1)/2$ **29.** $2\pi(\tan 1 - 1)$ **31.** $\pi^2/4$ **33.** $\pi^2/2 - \pi$ **35.** $198\pi/5$ **37.** $2\pi + \pi^2/4$ **39.** 3π
41. 21π 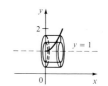   **43.** $\pi r^2 h/3$ **45.** $\pi h^2(r - h/3)$ **47.** $2b^2 h/3$ **49.** $2\pi^2 R r^2$

51. $16r^3/3$ **53.** 24 **55.** 2 **57.** $\frac{1}{3}$ **61.** $4\pi abc/3$ **63.** $8\int_0^r \sqrt{r^2 - x^2}\sqrt{R^2 - x^2}\, dx$

Exercises 8.2

1. $15\pi/2$ **3.** $(16\pi/3)(5\sqrt{5} - 1)$ **5.** 8π **7.** $\pi/10$ **9.** $10\pi^2$ **11.** 16π **13.** $\pi(e^2 + 1)/2$ **15.** $4096\pi/9$
17. $1944\pi/5$ **19.** 216π **21.** $5\pi/6$

23. $124\pi/5$ 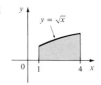    **25.** $17\pi/6$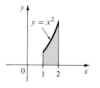

27. $2\pi e$ **29.** $\pi/2$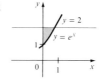

31. $81\pi/10$ **33.** $4\pi^2$ **35.** $16\pi/15$ **37.** $4\pi/3$ **39.** $\frac{4}{3}\pi r^3$ **41.** $\frac{1}{3}\pi r^2 h$

Exercises 8.3

1. $(577^{3/2} - 145^{3/2})/216$ **3.** $(13\sqrt{13} - 8)/27$ **5.** $\frac{59}{24}$ **7.** $\frac{4}{3}$ **9.** $\frac{181}{9}$ **11.** $\ln(1 + \sqrt{2})$
13. $\ln 3 - \frac{1}{2}$ **15.** $\ln(\sqrt{e^2 + 1} - 1) + \sqrt{1 + e^2} - \ln(\sqrt{2} - 1) - 1 - \sqrt{2}$ **17.** $\sinh 1$ **19.** $\int_0^1 \sqrt{1 + 9x^4}\, dx$
21. $\int_0^\pi \sqrt{1 + \cos^2 x}\, dx$ **23.** $\int_0^{\pi/2} \sqrt{1 + e^{2x}(1 - \sin 2x)}\, dx$ **25.** $s(x) = \frac{2}{27}[(1 + 9x)^{3/2} - 10\sqrt{10}]$ **27.** 7798 m
29. 6   **31.** 1.548   **33.** 3.820

Exercises 8.4

1. $\pi(37\sqrt{37} - 17\sqrt{17})/6$ **3.** $5\sqrt{5}\pi$ **5.** $12{,}289\pi/192$ **7.** $2\pi[\sqrt{2} + \ln(1 + \sqrt{2})]$ **9.** $\pi(e^2 + 4 - e^{-2})/4$
11. $5813\pi/30$ **13.** $\pi(65\sqrt{65} - 17\sqrt{17})/24$ **15.** $2\pi(\sqrt{7} - \frac{1}{2}) + (\sqrt{2}\pi/2)[\ln(1 + 1/\sqrt{2}) - \ln(2 + \sqrt{\frac{7}{2}})]$
17. $\pi(145\sqrt{145} - 10\sqrt{10})/27$ **19.** $\pi(3200\sqrt{10} - 494\sqrt{13}/5)/243$
21. $(\pi/2)[e\sqrt{4e^2 + 1} - \sqrt{5} + \frac{1}{2}\ln((2e + \sqrt{4e^2 + 1})/(2 + \sqrt{5}))]$ **23.** $(\pi/8)[21 - 8\ln 2 - (\ln 2)^2]$ **25.** ≈ 3.44
27. $\pi/4$ **29.** $3\pi a^2$ **33.** $2\pi[b^2 + a^2 b \sin^{-1}(\sqrt{a^2 - b^2}/a)/\sqrt{a^2 - b^2}]$ **37.** $4\pi^2 r^2$

Exercises 8.5

1. 7200 J   **3.** $\frac{5030}{3}$ ft-lb   **5.** $\frac{15}{4}$ ft-lb   **7.** $\frac{25}{24} \approx 1.04$ J   **9.** 10.8 cm   **11.** 625 ft-lb
13. 650,000 ft-lb   **15.** $\approx 2.45 \times 10^3$ J   **17.** $\approx 1.06 \times 10^6$ J   **19.** $\approx 5.8 \times 10^3$ ft-lb   **21.** $\approx 7.20 \times 10^5$ J
25. $Gm_1 m_2(1/a - 1/b)$

Exercises 8.6

1. (a) 9.8 kPa   (b) 1.96×10^4 N   (c) 4.9×10^3 N   **3.** 6.5×10^6 N   **5.** $1000 g\pi r^3$ N   **7.** 3.00×10^3 lb
9. 1.56×10^3 lb   **11.** 3.47×10^4 lb   **13.** 5.27×10^5 N   **15.** (a) 314 N   (b) 353 N   **17.** 8.32×10^4 N
19. (a) 5.63×10^3 lb   (b) 5.06×10^4 lb   (c) 4.88×10^4 lb   (d) 3.03×10^5 lb

Exercises 8.7

1. $40, 12, (1, \frac{10}{3})$ **3.** $-15, 17, (\frac{17}{11}, -\frac{15}{11})$ **5.** $(1.5, 1.2)$ **7.** $(\frac{11}{13}, \frac{49}{13})$
9. $(2(2\sqrt{2} - 1)/3(\sqrt{2} + \ln(1 + \sqrt{2})), 4/3(\sqrt{2} + \ln(1 + \sqrt{2})))$
11. $(0, \pi/8)$ **13.** $((e - 1)^{-1}, (e + 1)/4)$ **15.** $(0.4, 0.5)$
17. $((\pi\sqrt{2} - 4)/4(\sqrt{2} - 1), 1/4(\sqrt{2} - 1))$ **19.** $(0, 0)$ **21.** $\frac{4}{3}, 0, (0, \frac{2}{3})$ **23.** $-\frac{44}{3}, 0, (0, -\frac{11}{15})$ **27.** $\pi r^2 h/3$

Exercises 8.8

1. $y = -1/(x + C)$ or $y = 0$ **3.** $x^2 - y^2 = C$ **5.** $y = Ce^{1/x}$ **7.** $(y - 1)e^y = \frac{1}{3}(x^2 + 1)^{3/2} + C$
9. $u = -\ln(C - e^{2t}/2)$ **11.** $y = \tan(x - 1)$ **13.** $ye^y = x^3 - 8$ **15.** $\cos y = \cos x - 1$
17. $x = \sqrt{3 + 2e^t(t - 1)}$ **19.** $u = 1 - \sqrt{t^2 + t + 4}$ **21.** $f(x) = e^{x^4/4}$
23. (a) $15e^{-t/100}$ (b) $15e^{-0.2} \approx 12.3$ kg   **25.** $x = ab(e^{(b - a)kt} - 1)/(be^{(b - a)kt} - a)$
27. $5 \times 10^9 e^{0.02(t - 1986)}$ (a) 6.6 billion   (b) 49 billion   (c) 146 trillion   (a) 56   (b) 7.6   (c) 0.0025
29. (a) $dy/dt = ky(1 - y)$ (b) $y = y_0/[y_0 + (1 - y_0)e^{-kt}]$ (c) 3:36 P.M.

Review Exercises for Chapter 8

1. 2π **3.** $16\pi/3$ **5.** $(4\pi/3)(2ah + h^2)^{3/2}$ **7.** $23\pi/210$ **9.** 36   **11.** $\frac{2}{3}(5\sqrt{5} - 2\sqrt{2})$ **13.** $47\pi/16$
15. 1.297   **17.** 3.2 J   **19.** $8000\pi/3$ ft-lb   **23.** $(-\frac{1}{2}, \frac{12}{5})$ **25.** $y = \sqrt[3]{\frac{3}{2}x^2 - 3\cos x + C}$
27. $y = 2 \pm \sqrt{2\tan^{-1} x + C}$ **29.** $50 + 10e^{-3t/2000}$, 50 kg

Chapter 9

Exercises 9.1

1. (a)

(b) $3x + y = 5$

3. (a)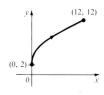

(b) $x = \frac{3}{25}(y - 2)^2,$
$2 \leqslant y \leqslant 12$

5. (a)

(b) $y = 1 - x^2, x \geqslant 0$

7. (a)

(b) $x^2 + y^2 = 1, x \geqslant 0$

9. (a)

(b) $x + y = 1, 0 \leqslant x \leqslant 1$

11. (a)

(b) $y = x, x > 0$

13. (a)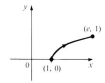

(b) $y = x^2, 0 \leqslant x \leqslant 1$

15. (a)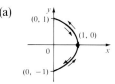

(b) $x = 1 - y^2, -1 \leqslant y \leqslant 1$

17. (a)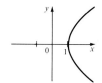

(b) $y = \sqrt{\ln x}, 1 \leqslant x \leqslant e$, or $x = e^{y^2}, 0 \leqslant y \leqslant 1$

19. (a)

(b) $x^2 - y^2 = 1, x \geqslant 1$

21. Moves counterclockwise along the circle $x^2 + y^2 = 1$ from $(-1, 0)$ to $(1, 0)$
23. Moves along the line $x + 8y = 13$ from $(-3, 2)$ to $(5, 1)$
25. Moves once clockwise around the ellipse $x^2/4 + y^2/9 = 1$, starting and ending at $(0, 3)$

27.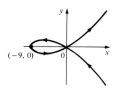

29.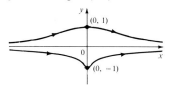

Exercises 9.2

1. $x + y = 0$ **3.** $x - 36y + 52 = 0$ **5.** $3x + 2y = 6\sqrt{2}$ **7.** $y = 1$ **9.** $3x + 4y = 25$
11. $2t/(2t + 1), 2/(2t + 1)^3$ **13.** $2(2t - 3)\sqrt{t + 1}, 2(6t + 1)$ **15.** $-\tan \pi t, -\sec^3 \pi t$
17. $-e^{3t}(2t + 1), e^{4t}(6t + 5)$

19. Horizontal at $(1, 5)$, $(1, -1)$; vertical at $(-1, 2)$, $(3, 2)$

21. Horizontal at $(0, -9)$; vertical at $(\pm 2, -6)$

23. Horizontal at $(0, 0)$, $(2^{1/3}, 2^{2/3})$; vertical at $(0, 0)$, $(2^{2/3}, 2^{1/3})$

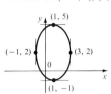

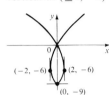

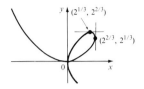

25. $y = x$, $y = -x$

27. $(0, 0)$, $y = \pm\sqrt{3}\,x$

29. (a) $d\sin\theta/(r - d\cos\theta)$

31. $(-5, 6)$, $(-\frac{208}{27}, \frac{32}{3})$

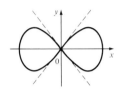

33. πab **35.** $(e^{\pi/2} - 1)/2$ **37.** $2\pi r^2 + \pi d^2$

Exercises 9.3

1. 2π **3.** $4\sqrt{34}$ **5.** $\sqrt{2}(e^{\pi} - 1)$ **7.** $-\ln(\sqrt{2} - 1)$ **9.** $\sqrt{13}$ **11.** $6\sqrt{2}$, $\sqrt{2}$ **15.** $47,104\pi/15$
17. $2\sqrt{2}\pi(2e^{\pi} + 1)/5$ **19.** $128\pi/5$ **21.** $24\pi(949\sqrt{26} + 1)/5$
23. (a) $2\pi[b^2 + (ab/e)\sin^{-1}e]$, $e = \sqrt{a^2 - b^2}/a = $ eccentricity   (b) $2\pi[a^2 + (b^2/2e)\ln((1 + e)/(1 - e))]$

Exercises 9.4

1.

$(1, 5\pi/2)$, $(-1, 3\pi/2)$

3.

$(1, 6\pi/5)$, $(-1, 11\pi/5)$

5.

$(4, 4\pi/3)$, $(-4, \pi/3)$

7.

$(1, 0)$, $(-1, 3\pi)$

9.

$(1, 1)$

11.

$(0, -1.5)$

13.

$(-2\sqrt{3}, 2)$

15.

$(-\frac{1}{2}, -\sqrt{3}/2)$

17. $(\sqrt{2}, 3\pi/4)$ **19.** $(4, 11\pi/6)$ **21.**

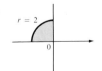

23.

25.

27. $\frac{1}{2}\sqrt{40 + 6\sqrt{6} - 6\sqrt{2}}$ **29.** $y = 2$ **31.** $y^2 = 2x + 1$ **33.** $(x^2 + y^2)^2 = 2xy$ **35.** $r \sin \theta = 5$

37. $r = 5$ **39.** $r^2 = \csc 2\theta$

41.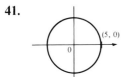

43.

45.

47.

49.

51.

53.

55.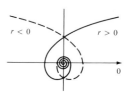

57.

59.

61.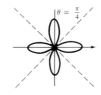

63.

65.

67.

69.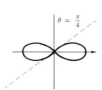

71.

73.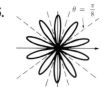

75. $1/\sqrt{3}$ **77.** $-2/\pi$ **79.** -1 **81.** $\sqrt{3}$

83. Horizontal at $(3/\sqrt{2}, \pi/4)$, $(-3/\sqrt{2}, 3\pi/4)$; vertical at $(3, 0)$, $(0, \pi/2)$

85. Horizontal at $(1, 3\pi/2)$, $(1, \pi/2)$, $(\frac{2}{3}, \alpha)$, $(\frac{2}{3}, \pi - \alpha)$, $(\frac{2}{3}, \pi + \alpha)$, $(\frac{2}{3}, 2\pi - \alpha)$, where $\alpha = \sin^{-1}(1/\sqrt{6})$; vertical at $(1, 0)$, $(1, \pi)$, $(\frac{2}{3}, 3\pi/2 - \alpha)$, $(\frac{2}{3}, 3\pi/2 + \alpha)$, $(\frac{2}{3}, \pi/2 - \alpha)$, $(\frac{2}{3}, \pi/2 + \alpha)$

87. Horizontal at $(\frac{3}{2}, \pi/3)$, $(\frac{3}{2}, 5\pi/3)$, and the pole; vertical at $(2, 0)$, $(\frac{1}{2}, 2\pi/3)$, $(\frac{1}{2}, 4\pi/3)$

89. Center $(b/2, a/2)$, radius $\sqrt{a^2 + b^2}/2$

Exercises 9.5

1. $\pi^3/6$ **3.** $\pi/6 + \sqrt{3}/4$ **5.** $121\pi^5/160$ **7.** $(4\pi - 3\sqrt{3})/96$

9. $25\pi/4$ 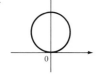   **11.** $3\pi/2$ **13.** 4 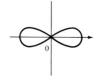   **15.** $33\pi/2$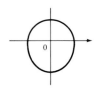

17. $\pi/2$

19. $\pi/12$ **21.** $\pi/20$ **23.** $\pi - 3\sqrt{3}/2$ **25.** $9\sqrt{3}/8 - \pi/4$ **27.** $4\pi/3 + 2\sqrt{3}$

29. π **31.** $(\pi - 2)/8$ **33.** $\pi/2 - 1$ **35.** $19\pi/3 - 11\sqrt{3}/2$ **37.** $(\pi + 3\sqrt{3})/4$ **39.** $(1/\sqrt{2}, \pi/4)$ and the pole
41. $(\frac{1}{2}, \pi/3)$, $(\frac{1}{2}, 5\pi/3)$ and the pole **43.** $(\sqrt{3}/2, \pi/3)$, $(\sqrt{3}/2, 2\pi/3)$, and the pole **45.** $15\pi/4$
47. $\sqrt{1 + (\ln 2)^2} \, (4^\pi - 1)/\ln 2$ **49.** $\frac{8}{3}[(\pi^2 + 1)^{3/2} - 1]$ **51.** $\frac{16}{3}$ **53.** (b) $2\pi(2 - \sqrt{2})$

Exercises 9.6

1. $(0,0)$, $(0, -2)$, $y = 2$ **3.** $(0,0)$, $(\frac{1}{4}, 0)$, $x = -\frac{1}{4}$ **5.** $(-1, 3)$, $(-\frac{7}{8}, 3)$, $x = -\frac{9}{8}$ **7.** $(2, 4)$, $(\frac{3}{2}, 4)$, $x = \frac{5}{2}$

9. $(\pm 4, 0)$, $(\pm 2\sqrt{3}, 0)$ **11.** $(0, \pm 5)$, $(0, \pm 4)$ **13.** $(\pm 12, 0)$, $(\pm 13, 0)$, $y = \pm\frac{5}{12}x$ **15.** $(0, \pm 1)$, $(0, \pm\sqrt{10})$, $y = \pm x/3$

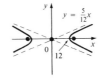
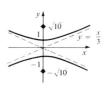

17. $(1, \pm 3)$, $(1, \pm\sqrt{5})$ **19.** $(2 \pm \sqrt{6}, 1)$, $(2 \pm \sqrt{15}, 1)$, $y - 1 = \pm(\sqrt{6}/2)(x - 2)$ **21.** $x^2 = 12y$

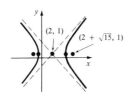

23. $y^2 = 4(x - 2)$ **25.** $y^2 = 16x$ **27.** $x^2/4 + y^2/3 = 1$ **29.** $(x - 3)^2/8 + y^2/9 = 1$
31. $(x - 2)^2/9 + (y - 2)^2/5 = 1$ **33.** $y^2 - x^2/8 = 1$ **35.** $(x - 4)^2/4 - (y - 3)^2/5 = 1$ **37.** $x^2/9 - y^2/36 = 1$
39. $x^2/3,763,600 + y^2/3,753,196 = 1$ **41.** (a) $121x^2/1,500,625 - 121y^2/3,339,375 = 1$ (b) ≈ 248 mi
45. (a) Ellipse (b) Hyperbola (c) No curve **47.** 9.69

Exercises 9.7

1. $r = 6/(3 + 2\cos\theta)$ **3.** $r = 2/(1 + \sin\theta)$ **5.** $r = 20/(1 + 4\cos\theta)$ **7.** $r = 10/(1 + \sin\theta)$
9. (a) 3 (b) Hyperbola (c) $x = \frac{4}{3}$ **11.** (a) 1 (b) Parabola (c) $x = -2$
(d)
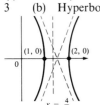
(d)
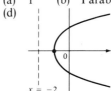

13. (a) $\frac{1}{2}$ (b) Ellipse   (c) $y = 6$ **15.** (a) $\frac{3}{4}$ (b) Ellipse   (c) $x = -\frac{1}{3}$
(d) 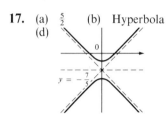   (d)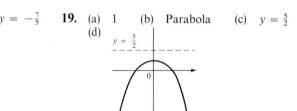

17. (a) $\frac{5}{2}$ (b) Hyperbola   (c) $y = -\frac{7}{5}$ **19.** (a) 1   (b) Parabola   (c) $y = \frac{5}{2}$
(d) 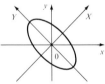   (d)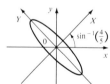

Exercises 9.8

1. $((\sqrt{3} + 4)/2, (4\sqrt{3} - 1)/2)$ **3.** $(2\sqrt{3} - 1, \sqrt{3} + 2)$ **5.** $X = \sqrt{2}\,Y^2$, parabola

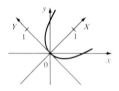

7. $3X^2 + Y^2 = 2$, ellipse   **9.** $X^2 + Y^2/9 = 1$, ellipse   **11.** $(X - 1)^2 - 3Y^2 = 1$, hyperbola

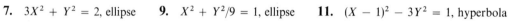

13. (a) $Y - 1 = 4X^2$ (b) $(0, \frac{17}{16}), (-\frac{17}{20}, \frac{51}{80})$ (c) $64x - 48y + 75 = 0$

Review Exercises for Chapter 9

1. 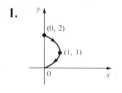   **3.** 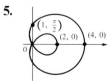   **5.**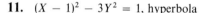

$x = 2y - y^2, 0 \leqslant y \leqslant 2$ $(x - 1)^2 + (y - 2)^2 = 1$

7.   **9.**   **11.**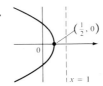

13. $r = 4\cos\theta$ **15.** $\frac{1}{2}$ **17.** $(4 + \pi)/(4 - \pi)$ **19.** $(\sin t + t\cos t)/(\cos t - t\sin t)$, $(t^2 + 2)/(\cos t - t\sin t)^3$
21. Vertical tangent at $(3a/2, \pm\sqrt{3}a/2)$, $(-3a, 0)$; horizontal tangent at $(a, 0)$, $(-a/2, \pm 3\sqrt{3}a/2)$

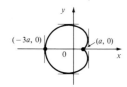

23. 9   **25.** $(2, \pm\pi/3)$ **27.** $(\pi - 1)/2$ **29.** $2(5\sqrt{5} - 1)$
31. $(2\sqrt{\pi^2 + 1} - \sqrt{4\pi^2 + 1})/2\pi + \ln[(2\pi + \sqrt{4\pi^2 + 1})/(\pi + \sqrt{\pi^2 + 1})]$ **33.** $471{,}295\pi/1024$
35. $(0, \pm 1)$, $(0, \pm 3)$ **37.** $(-\frac{25}{24}, 3)$, $(-1, 3)$ **39.** $x^2 = 8(y - 4)$ **41.** $5x^2 - 20y^2 = 36$

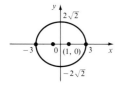   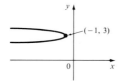

43. $3X^2 - Y^2 = 1$, hyperbola   **45.** $r = 4/(3 + \cos\theta)$

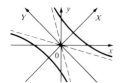

Chapter 10

Exercises 10.1

1. $\{\frac{1}{3}, \frac{2}{5}, \frac{3}{7}, \frac{4}{9}, \frac{5}{11}, \ldots\}$ **3.** $\{\frac{1}{2}, -\frac{2}{4}, \frac{3}{8}, -\frac{4}{16}, \frac{5}{32}, \ldots\}$ **5.** $\{1, \frac{3}{2}, \frac{5}{2}, \frac{35}{8}, \frac{63}{8}, \ldots\}$ **7.** $\{1, 0, -1, 0, 1, \ldots\}$
9. $\{1, \frac{1}{2}, \frac{2}{3}, \frac{3}{5}, \frac{5}{8}, \ldots\}$ **11.** $a_n = 1/2^n$ **13.** $a_n = 3n - 2$ **15.** $a_n = (-1)^n n!$ **17.** $a_n = (-1)^{n+1}(n + 1)/(2n + 1)$
19. 0   **21.** 1   **23.** Diverges (to ∞)   **25.** 0   **27.** 0   **29.** Diverges   **31.** Diverges (to ∞)   **33.** $\pi/2$
35. 0   **37.** 0   **39.** 0   **41.** 0   **43.** 1   **45.** 0   **47.** $\frac{1}{2}$ **49.** Diverges (to ∞)   **51.** 0   **53.** 0
55. $-1 < r < 1$ **57.** Decreasing   **59.** Increasing   **61.** Not monotonic   **63.** Decreasing   **65.** 2
67. (b) $(3 + \sqrt{5})/2$ **69.** (b) $(1 + \sqrt{5})/2$

Exercises 10.2

1. $\frac{20}{3}$ **3.** $\frac{1}{2}$ **5.** Divergent   **7.** 8   **9.** $5e/(3 - e)$ **11.** $\frac{8}{3}$ **13.** Divergent   **15.** $\frac{400}{7}$
17. Divergent   **19.** Divergent   **21.** $\frac{1}{3}$ **23.** $\frac{17}{36}$ **25.** Divergent   **27.** $\frac{3}{4}$ **29.** $\frac{3}{2}$ **31.** $\sin 1$
33. Divergent   **35.** $\frac{5}{9}$ **37.** $307/999$ **39.** $41{,}111/333{,}000$ **41.** $2 < x < 4$, $1/(4 - x)$
43. $-5 < x < 5$, $x^2/5(5 - x)$ **45.** $|x - n\pi| < \pi/6$, n any integer, $1/(1 - 2\sin x)$ **47.** 12 m
49. The series is divergent.

Exercises 10.3 (*abbreviations*: C, convergent; D, divergent)

1. D　**3.** C　**5.** C　**7.** D　**9.** C　**11.** D　**13.** D　**15.** C　**17.** C　**19.** C　**21.** C
23. $p > 1$　**25.** $p < -1$　**27.** $(1, \infty)$

Exercises 10.4

1. C　**3.** C　**5.** D　**7.** C　**9.** C　**11.** D　**13.** C　**15.** C　**17.** C　**19.** C　**21.** C
23. D　**25.** C　**27.** D　**29.** C　**31.** C　**33.** C　**35.** D　**37.** D

Exercises 10.5

1. C　**3.** D　**5.** C　**7.** D　**9.** C　**11.** C　**13.** D　**15.** C　**17.** C　**19.** C　**21.** D
23. $\{a_n\}$ is not decreasing.　**25.** p is not a negative integer.　**27.** 0.82　**29.** 0.13 (or 0.137)　**31.** 0.8415
33. 0.6065

Exercises 10.6 (*abbreviations*: AC, absolutely convergent; CC, conditionally convergent)

1. AC　**3.** D　**5.** CC　**7.** AC　**9.** CC　**11.** D　**13.** AC　**15.** AC　**17.** AC　**19.** CC
21. D　**23.** AC　**25.** D　**27.** AC　**29.** AC　**31.** D　**33.** AC　**35.** AC

Exercises 10.7

1. C　**3.** C　**5.** C　**7.** C　**9.** C　**11.** D　**13.** C　**15.** C　**17.** C　**19.** C　**21.** D
23. D　**25.** C　**27.** D　**29.** C　**31.** C　**33.** C　**35.** C　**37.** D　**39.** C

Exercises 10.8

1. $[-1, 1)$　**3.** $(-1, 1)$　**5.** $(-\infty, \infty)$　**7.** $(-2, 2]$　**9.** $\left[-\frac{1}{3}, \frac{1}{3}\right]$　**11.** $[-1, 1)$　**13.** $(0, 2]$
15. $(-\infty, \infty)$　**17.** $[2.5, 3.5)$　**19.** $[-14, -6)$　**21.** $\{-6\}$　**23.** $[0, 1]$　**25.** $(e - 1, e + 1)$
27. $(-\infty, \infty)$　**29.** $(-\infty, \infty)$

Exercises 10.9

1. $\displaystyle\sum_{n=0}^{\infty} (-1)^n \frac{x^{2n}}{(2n)!}, R = \infty$　**3.** $\displaystyle\sum_{n=0}^{\infty} \frac{(-1)^{n(n-1)/2}}{\sqrt{2}} \frac{(x - \pi/4)^n}{n!}, R = \infty$　**5.** $\displaystyle\sum_{n=0}^{\infty} (-1)^n (n + 1) x^n, R = 1$

7. $\displaystyle\sum_{n=0}^{\infty} (-1)^n (x - 1)^n, R = 1$　**9.** $\displaystyle\sum_{n=0}^{\infty} \frac{e^3}{n!} (x - 3)^n, R = \infty$　**11.** $\displaystyle\sum_{n=0}^{\infty} \frac{x^{2n+1}}{(2n + 1)!}, R = \infty$　**13.** $\displaystyle\sum_{n=0}^{\infty} (-1)^n x^n, R = 1$

15. $\displaystyle\sum_{n=0}^{\infty} (-1)^n (n + 1) x^n, R = 1$　**17.** $\displaystyle\sum_{n=0}^{\infty} (-1)^n 4^n x^{2n}, R = \frac{1}{2}$　**19.** $\displaystyle\sum_{n=0}^{\infty} \frac{(-1)^n}{4^{n+1}} x^{2n}, R = 2$　**21.** $\displaystyle\sum_{n=0}^{\infty} x^{2n}, R = 1$

23. $\displaystyle\sum_{n=0}^{\infty} \frac{2x^{2n+1}}{2n + 1}, R = 1$　**25.** $\displaystyle\sum_{n=0}^{\infty} \frac{3^n x^n}{n!}, R = \infty$　**27.** $\displaystyle\sum_{n=0}^{\infty} \frac{(-1)^n x^{2n+2}}{(2n)!}, R = \infty$　**29.** $\displaystyle\sum_{n=0}^{\infty} \frac{(-1)^n x^{2n+2}}{2^{2n+1}(2n + 1)!}, R = \infty$

31. $\displaystyle\sum_{n=1}^{\infty} \frac{(-1)^{n+1} 2^{2n-1} x^{2n}}{(2n)!}, R = \infty$　**33.** $\displaystyle\sum_{n=0}^{\infty} \frac{(-1)^n x^{2n}}{(2n + 1)!}, R = \infty$

35. $1 + \dfrac{x}{2} + \displaystyle\sum_{n=2}^{\infty} (-1)^{n-1} \frac{1 \cdot 3 \cdot 5 \cdot \,\cdots\, \cdot (2n - 3)}{2^n n!} x^n, R = 1$　**37.** $1 + \displaystyle\sum_{n=1}^{\infty} \frac{1 \cdot 4 \cdot 7 \cdot \,\cdots\, \cdot (3n - 2)}{3^n n!} x^n, R = 1$

39. $\displaystyle\sum_{n=0}^{\infty} \frac{(-1)^n}{2} (n + 1)(n + 2) x^n, R = 1$　**41.** $\ln 5 + \displaystyle\sum_{n=1}^{\infty} \frac{(-1)^{n-1} x^n}{5^n n}, R = 5$　**43.** $\displaystyle\sum_{n=1}^{\infty} (-1)^{n-1} \frac{x^n}{n}, 0.09531$

45. $C + \displaystyle\sum_{n=0}^{\infty} \frac{(-1)^n x^{4n+3}}{(4n + 3)(2n + 1)!}$　**47.** $C + \displaystyle\sum_{n=0}^{\infty} \frac{(-1)^n x^{4n+1}}{4n + 1}$　**49.** $C + x + \dfrac{x^4}{8} + \displaystyle\sum_{n=2}^{\infty} (-1)^{n-1} \frac{1 \cdot 3 \cdot 5 \cdot \,\cdots\, \cdot (2n - 3)}{2^n n!(3n + 1)} x^{3n+1}$

51. 0.310　**53.** 0.4989　**55.** 0.0354　**59.** $[-1, 1], [-1, 1), (-1, 1)$

Exercises 10.10

1. $1 + \dfrac{x}{2} + \displaystyle\sum_{n=2}^{\infty} (-1)^{n-1} \dfrac{1 \cdot 3 \cdot 5 \cdot \cdots \cdot (2n-3)}{2^n n!} x^n, R = 1$ **3.** $\displaystyle\sum_{n=0}^{\infty} (-1)^n \dfrac{(n+1)(n+2)(n+3)2^n}{6} x^n, R = \dfrac{1}{2}$

5. $x + \displaystyle\sum_{n=1}^{\infty} \dfrac{1 \cdot 3 \cdot 5 \cdot \cdots \cdot (2n-1)}{2^n n!} x^{n+1}, R = 1$ **7.** $\dfrac{1}{2} + \dfrac{1}{2} \displaystyle\sum_{n=1}^{\infty} \dfrac{(-1)^n 1 \cdot 4 \cdot 7 \cdot \cdots \cdot (3n-2)}{24^n n!} x^n, R = 8$

9. $1 - \dfrac{x^4}{4} - \displaystyle\sum_{n=2}^{\infty} \dfrac{3 \cdot 7 \cdot 11 \cdot \cdots \cdot (4n-5)}{4^n n!} x^{4n}, R = 1$ **11.** $\displaystyle\sum_{n=0}^{\infty} \dfrac{(n+1)(n+2)(n+3)(n+4)}{24} x^{n+5}, R = 1$

13. (a) $1 + \displaystyle\sum_{n=1}^{\infty} \dfrac{1 \cdot 3 \cdot 5 \cdot \cdots \cdot (2n-1)}{2^n n!} x^{2n}$ (b) $x + \displaystyle\sum_{n=1}^{\infty} \dfrac{1 \cdot 3 \cdot 5 \cdot \cdots \cdot (2n-1)}{2^n n!} \dfrac{x^{2n+1}}{2n+1}$

15. (a) $1 + \displaystyle\sum_{n=1}^{\infty} (-1)^n \dfrac{1 \cdot 3 \cdot 5 \cdot \cdots \cdot (2n-1)}{2^n n!} x^n$ (b) 0.953 **17.** (a) $\displaystyle\sum_{n=1}^{\infty} nx^n$ (b) 2

19. (a) $1 + \dfrac{x^2}{2} + \displaystyle\sum_{n=2}^{\infty} (-1)^{n-1} \dfrac{1 \cdot 3 \cdot 5 \cdot \cdots \cdot (2n-3)}{2^n n!} x^{2n}$ (b) $99{,}225$

Exercises 10.11

1. $-2 + 8(x+1) - 9(x+1)^2 + 4(x+1)^3$ **3.** $\dfrac{1}{2} + \dfrac{\sqrt{3}}{2}\left(x - \dfrac{\pi}{6}\right) - \dfrac{1}{4}\left(x - \dfrac{\pi}{6}\right)^2 - \dfrac{\sqrt{3}}{12}\left(x - \dfrac{\pi}{6}\right)^3$ **5.** $x + \tfrac{1}{3}x^3$

7. $x + x^2 + \tfrac{1}{3}x^3$ **9.** $3 + \tfrac{1}{6}(x-9) - \tfrac{1}{216}(x-9)^2 + \tfrac{1}{3888}(x-9)^3$ **11.** $-\dfrac{1}{2}\left(x - \dfrac{\pi}{2}\right)^2$

13. $T_1(x) = 1, T_2(x) = 1 - \dfrac{x^2}{2}, T_3(x) = 1 - \dfrac{x^2}{2}, T_4(x) = 1 - \dfrac{x^2}{2} + \dfrac{x^4}{24}$

15. (a) $\sqrt{1+x} = 1 + \dfrac{x}{2} - \dfrac{1}{8(1+z)^{3/2}} x^2$ (b) 0.00125

17. (a) $\dfrac{1}{\sqrt{2}} + \dfrac{1}{\sqrt{2}}\left(x - \dfrac{\pi}{4}\right) - \dfrac{1}{\sqrt{2}}\left(x - \dfrac{\pi}{4}\right)^2 - \dfrac{1}{6\sqrt{2}}\left(x - \dfrac{\pi}{4}\right)^3 + \dfrac{1}{24\sqrt{2}}\left(x - \dfrac{\pi}{4}\right)^4 + \dfrac{1}{120\sqrt{2}}\left(x - \dfrac{\pi}{4}\right)^5 - \dfrac{\sin z}{720}\left(x - \dfrac{\pi}{4}\right)^6$
(b) 0.00033

19. (a) $\dfrac{1}{(1+2x)^4} = 1 - 8x + 40x^2 - 160x^3 + \dfrac{560}{(1+2z)^8} x^4$ (b) 0.34

21. (a) $\tan x = x + \dfrac{x^3}{3} + \dfrac{(\sec^2 z \tan^3 z + 2\sec^4 z \tan z)}{3} x^4$ (b) 0.06

23. (a) $e^{x^2} = 1 + x^2 + \dfrac{e^{z^2}(3 + 12z^2 + 4z^4)}{6} x^4$ (b) 0.00006

25. (a) $x^{3/4} = 8 + \dfrac{3}{8}(x-16) - \dfrac{3}{1024}(x-16)^2 + \dfrac{15}{196{,}608}(x-16)^3 - \dfrac{135(x-16)^4}{6144 z^{13/4}}$ (b) 0.0000034

27. 1.10517 **29.** 1.0192 **31.** 0.336 **33.** 0.4794 **35.** 0.98481 **37.** 0.57358 **39.** $|x| \leqslant 1$

Review Exercises for Chapter 10

1. Convergent, $\tfrac{1}{2}$ **3.** Divergent   **5.** Divergent   **7.** Convergent, e^{12} **9.** D   **11.** C   **13.** C   **15.** C
17. C   **19.** D   **21.** 8   **23.** $\tfrac{4111}{3330}$ **25.** 0.9721 **29.** $3, [-3, 3]$ **31.** $[2.5, 3.5)$

33. $\dfrac{1}{2} + \dfrac{\sqrt{3}}{2}\left(x - \dfrac{\pi}{6}\right) - \dfrac{1}{2}\dfrac{1}{2!}\left(x - \dfrac{\pi}{6}\right)^2 - \dfrac{\sqrt{3}}{2}\dfrac{1}{3!}\left(x - \dfrac{\pi}{6}\right)^3 + \cdots = \dfrac{1}{2}\displaystyle\sum_{n=0}^{\infty} (-1)^n \left[\dfrac{1}{(2n)!}\left(x - \dfrac{\pi}{6}\right)^{2n} + \dfrac{\sqrt{3}}{(2n+1)!}\left(x - \dfrac{\pi}{6}\right)^{2n+1}\right]$

35. $\displaystyle\sum_{n=0}^{\infty}(-1)^n x^{n+2}, 1$ **37.** $\displaystyle-\sum_{n=1}^{\infty}\frac{x^n}{n}, 1$ **39.** $\displaystyle\sum_{n=0}^{\infty}(-1)^n\frac{x^{8n+4}}{(2n+1)!}, \infty$

41. $\dfrac{1}{2}\left[1+\dfrac{x}{2^6}+\dfrac{1\cdot 5}{2!}\dfrac{x^2}{2^{12}}+\dfrac{1\cdot 5\cdot 9}{3!}\dfrac{x^3}{2^{18}}+\cdots+\dfrac{1\cdot 5\cdot 9\cdots\cdot(4n-3)}{n!\,2^{6n}}x^n+\cdots\right], 16$ **43.** $\ln|x|+\displaystyle\sum_{n=1}^{\infty}\dfrac{x^n}{n\cdot n!}+C$

45. 0.7788   **47.** $\sqrt{x}=1+\dfrac{1}{2}(x-1)-\dfrac{1}{8}(x-1)^2+\dfrac{1}{16}(x-1)^3-\dfrac{5}{128z^{7/2}}(x-1)^4, 0.000006$

Chapter 11

Exercises 11.1

1. (a)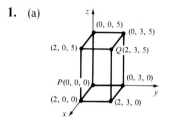

3. (a)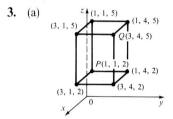

5. (a)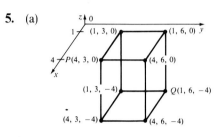

(b) $|PQ|=\sqrt{38}$ (b) $|PQ|=\sqrt{22}$ (b) $|PQ|=\sqrt{34}$

7. $|AB|=\sqrt{21}, |BC|=\sqrt{6}, |CA|=3\sqrt{3}$; right triangle
9. $|AB|=\sqrt{69}, |BC|=\sqrt{158}, |CA|=13$; neither   **11.** Not collinear
13. $x^2+(y-1)^2+(z+1)^2=16$ **15.** $(x+6)^2+(y+1)^2+(z-2)^2=12$ **17.** $(-1,-4,2), 7$
19. $(-\frac{1}{2},1,-3), \frac{7}{2}$ **21.** $(\frac{1}{2},0,0), \frac{1}{2}$ **23.** $14x-6y-10z=9$, a plane perpendicular to AB
27. $\sqrt{85}/2, \frac{5}{2}, \sqrt{94}/2$ **29.** A plane parallel to the yz-plane and nine units in front of it
31. A half-space consisting of all points to the right of the plane $y=2$
33. A vertical plane through 0 intersecting the xy-plane in the line $y=x$
35. Circular cylinder, radius 1, axis the z-axis
37. All points outside the sphere with radius 1 and center 0
39. All points inside the sphere with radius 2 and center $(0,0,1)$
41. Hyperbolic cylinder   **43.** All points on and between the horizontal planes $z=2$ and $z=-2$

Exercises 11.2

1. $\mathbf{a}=\langle 3,1\rangle$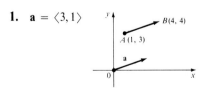

3. $\mathbf{a}=\langle 0,-2\rangle$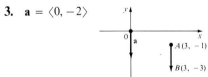

5. $\mathbf{a}=\langle 2,0,-2\rangle$ **7.** $\langle 5,-1\rangle$ **9.** $\langle 1,0,2\rangle$

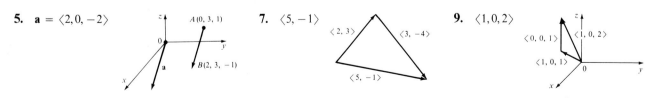

11. $13, \langle 3,-4\rangle, \langle 7,-20\rangle, \langle 10,-24\rangle, \langle 7,-4\rangle$ **13.** $7, \langle 3,-2,10\rangle, \langle 1,-4,2\rangle, \langle 4,-6,12\rangle, \langle 10,-5,34\rangle$
15. $\sqrt{2}, 2\mathbf{i}, -2\mathbf{j}, 2\mathbf{i}-2\mathbf{j}, 7\mathbf{i}+\mathbf{j}$ **17.** $\sqrt{3}, 3\mathbf{i}+4\mathbf{k}, -\mathbf{i}+2\mathbf{j}-2\mathbf{k}, 2\mathbf{i}+2\mathbf{j}+2\mathbf{k}, 11\mathbf{i}-\mathbf{j}+15\mathbf{k}$ **19.** $\langle 1/\sqrt{5}, 2/\sqrt{5}\rangle$
21. $\langle -2/\sqrt{29}, 4/\sqrt{29}, 3/\sqrt{29}\rangle$ **23.** $(\mathbf{i}+\mathbf{j})/\sqrt{2}$

Exercises 11.3

1. -1 **3.** -5 **5.** 2π **7.** -11 **11.** $\cos^{-1}\frac{11}{15} \approx 43°$ **13.** $\cos^{-1}(2/13\sqrt{5}) \approx 86°$
15. $\cos^{-1}(1/7\sqrt{3}) \approx 85°$ **17.** $114°, 33°, 33°$ **19.** Parallel **21.** Neither **23.** Orthogonal **25.** $-\frac{4}{5}$
27. $(\mathbf{i} - \mathbf{j} - \mathbf{k})/\sqrt{3}$ [or $(-\mathbf{i} + \mathbf{j} + \mathbf{k})/\sqrt{3}$] **29.** $\frac{1}{3}, \frac{2}{3}, \frac{2}{3}$; $71°, 48°, 48°$ **31.** $-8/\sqrt{77}, 3/\sqrt{77}, 2/\sqrt{77}$; $156°, 70°, 77°$
33. $5/\sqrt{38}, 3/\sqrt{38}, 2/\sqrt{38}$; $36°, 61°, 71°$ **35.** $3/\sqrt{5}, \langle 6/5, 3/5, 0 \rangle$ **37.** $1/\sqrt{2}, (\mathbf{i} + \mathbf{k})/2$ **39.** 38 J

Exercises 11.4

1. $\langle -1, 0, 1 \rangle$ **3.** $\langle 3, 14, -9 \rangle$ **5.** $-2\mathbf{i} + 2\mathbf{j}$ **7.** $2\mathbf{i} - \mathbf{j} + 4\mathbf{k}$ **9.** $\langle -2, 6, -3 \rangle, \langle 2, -6, 3 \rangle$
11. $\langle -2/\sqrt{6}, -1/\sqrt{6}, 1/\sqrt{6} \rangle, \langle 2/\sqrt{6}, 1/\sqrt{6}, -1/\sqrt{6} \rangle$ **19.** (a) $\langle 6, 3, 2 \rangle$ (b) $\frac{7}{2}$
21. (a) $\langle -10, -3, 7 \rangle$ (b) $\sqrt{158}/2$ **23.** 226 **25.** 21 **29.** (b) $\sqrt{\frac{97}{3}}$

Exercises 11.5

1. $\mathbf{r} = \langle 3, -1, 8 \rangle + t\langle 2, 3, 5 \rangle, x = 3 + 2t, y = -1 + 3t, z = 8 + 5t$
3. $\mathbf{r} = (\mathbf{j} + 2\mathbf{k}) + t(6\mathbf{i} + 3\mathbf{j} + 2\mathbf{k}), x = 6t, y = 1 + 3t, z = 2 + 2t$
5. $x = 2 + 4t, y = 1 - t, z = 8 - 5t; (x - 2)/4 = (y - 1)/(-1) = (z - 8)/(-5)$
7. $x = 3, y = 1 + t, z = -1 - 5t; x = 3, y - 1 = (z + 1)/(-5)$
9. $x = (-1 + t)/3, y = 1 + 4t, z = 1 - 9t; (x + \frac{1}{3})/(\frac{1}{3}) = (y - 1)/4 = (z - 1)/(-9)$
13. (a) $x/2 = (y - 2)/3 = (z + 1)/(-7)$ (b) $(-\frac{2}{7}, \frac{11}{7}, 0), (-\frac{4}{3}, 0, \frac{11}{3}), (0, 2, -1)$ **15.** Skew **17.** Parallel
19. $7x + y + 4z = 31$ **21.** $5x + 3y - 4z + 1 = 0$ **23.** $x + y - z = 13$ **25.** $3x - 4y - 6z = 33$
27. $x - 2y + z = 0$ **29.** $17x - 6y - 5z = 32$ **31.** $25x + 14y + 8z = 77$ **33.** Neither, $60°$
35. Perpendicular **37.** $x - 2 = y/(-8) = z/(-7)$ **39.** $x/a + y/b + z/c = 1$ **41.** $\frac{25}{3}$ **43.** 6 **45.** 2

Exercises 11.6

1. $x = k, z^2 - y^2 = 1 - k^2$, hyperbola
$y = k, x^2 + z^2 = 1 + k^2$, circle
$z = k, x^2 - y^2 = 1 - k^2$, hyperbola
Hyperboloid of one sheet with axis the y-axis

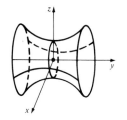

3. $x = k, y^2 + 4z^2 = 4 - 4k^2/9$, ellipse ($|k| < 3$)
$y = k, x^2 + 9z^2 = 9 - 9k^2/4$, ellipse ($|k| < 2$)
$z = k, 4x^2 + 9y^2 = 36(1 - k^2)$, ellipse ($|k| < 1$)
Ellipsoid

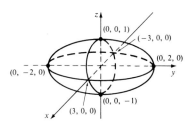

5. $x = k, 4z^2 - y^2 = 1 + k^2$, hyperbola
$y = k, 4z^2 - x^2 = 1 + k^2$, hyperbola
$z = k, x^2 + y^2 = 4k^2 - 1$, circle ($|k| > \frac{1}{2}$)
Hyperboloid of two sheets

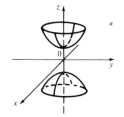

7. $x = k, z = y^2$, parabola
$y = k, z = k^2$, line
$z = k, y \pm \sqrt{k}$, lines ($k > 0$)
Parabolic cylinder

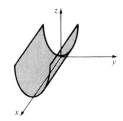

9. $x = k$, $y^2 - z^2 = k^2$, hyperbola $(k \neq 0)$
     $y = k$, $x^2 + z^2 = k^2$, circle $(k \neq 0)$
     $z = k$, $y^2 - x^2 = k^2$, hyperbola $(k \neq 0)$
     $x = 0$, $y = \pm z$, lines
     Cone

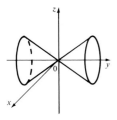

11. $x = k$, $y = 4z^2 + k^2$, parabola
     $y = k$, $x^2 + 4z^2 = k$, ellipse $(k > 0)$
     $z = k$, $y = x^2 + 4k^2$, parabola
     Elliptic paraboloid with axis the y-axis

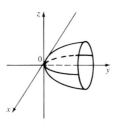

13. $x^2/4 + y^2/3 - z^2/12 = 1$
     Hyperboloid of one sheet with axis the z-axis

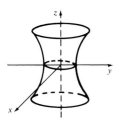

15. $z - 1 = x^2 + y^2$
     Circular paraboloid with vertex $(0, 0, 1)$ and axis the z-axis

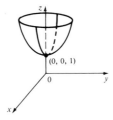

17. $(x + 2)^2 + (y - 3)^2 - 4(z + 1)^2 = 22$
     Hyperboloid of one sheet with center $(-2, 3, -1)$
     and axis parallel to the z-axis

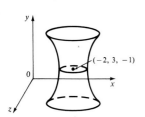

19. $x^2/100 + y^2/25 = 1$
     Elliptic cylinder with axis the z-axis

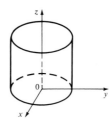

21. $z = (y - 2)^2 - x^2$
     Hyperbolic paraboloid with center $(0, 2, 0)$

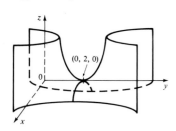

23. $(x - 2)^2 = (y + 2)^2 + z^2$
     Cone with vertex at $(2, -2, 0)$

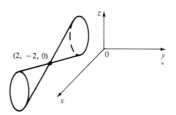

25. $y = x^2 + z^2$

27. $-4x = y^2 + z^2$, paraboloid

Exercises 11.7

1.   **3.**   **5.** 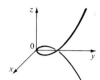   **7.** $\langle 0, 1, 2 \rangle$ **9.** $2\mathbf{i} + \tfrac{1}{2}\mathbf{j} + (\tan 1)\mathbf{k}$

11. $R, \mathbf{r}'(t) = \langle 1, 2t, 3t^2 \rangle$ **13.** $\{t \,|\, t \neq (2n + 1)\pi/2, n \text{ an integer}\}, \mathbf{r}'(t) = \sec^2 t\mathbf{j} + \sec t \tan t\,\mathbf{k}$
15. $[-1, 2), \mathbf{r}'(t) = -2t/(4 - t^2)\mathbf{i} + 1/2\sqrt{1 + t}\,\mathbf{j} - 12e^{3t}\mathbf{k}$ **17.** $R, \mathbf{r}(t) = \mathbf{b} + 2t\mathbf{c}$
19. (a), (c)

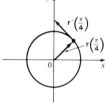

(b) $r'(t) = \langle -\sin t, \cos t \rangle$

21. (a), (c)

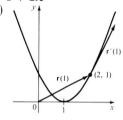

(b) $r'(t) = \mathbf{i} + 2t\mathbf{j}$

23. $\langle 1/\sqrt{46}, 3/\sqrt{46}, 6/\sqrt{46} \rangle$ **25.** $\tfrac{2}{5}\mathbf{i} + (2\sqrt{3}/5)\mathbf{j} - \tfrac{3}{5}\mathbf{k}$ **27.** $x = 1 + t, y = 1 + 2t, z = 1 + 3t$
29. $x = (\pi/4) + t, y = 1 - t, z = 1 + t$ **31.** $x = -(\pi/2)t, y = \tfrac{1}{4} + t, z = 1 + 4t$ **33.** $66°$ **35.** $\tfrac{1}{2}\mathbf{i} + \tfrac{1}{3}\mathbf{j} + \tfrac{1}{4}\mathbf{k}$
37. $\tfrac{1}{2}\mathbf{i} + \tfrac{1}{2}\mathbf{j} + (4 - \pi)/4\sqrt{2}\,\mathbf{k}$ **39.** $(t^3/3)\mathbf{i} + (t^4 + 1)\mathbf{j} - (t^3/3)\mathbf{k}$ **47.** $1 - 4t \cos t + 11t^2 \sin t + 3t^3 \cos t$

Exercises 11.8

1. $\sqrt{13}(b - a)$ **3.** 8 **5.** $\sqrt{3} + (1/\sqrt{2}) \ln(\sqrt{2} + \sqrt{3})$
7. (a) $\langle \tfrac{4}{5}\cos 4t, \tfrac{3}{5}, -\tfrac{4}{5}\sin 4t \rangle, \langle -\sin 4t, 0, -\cos 4t \rangle$ (b) $\tfrac{16}{25}$
9. (a) $\langle \sqrt{2}e^t, e^{2t}, -1 \rangle/(e^{2t} + 1), \langle 1 - e^{2t}, \sqrt{2}e^t, \sqrt{2}e^t \rangle/(e^{2t} + 1)$ (b) $\sqrt{2}e^{2t}/(e^{2t} + 1)^2$
11. (a) $\langle 2t/(2t^2 + 1), 2t^2/(2t^2 + 1), 1/(2t^2 + 1) \rangle, \langle (1 - 2t^2)/(2t^2 + 1), 2t/(2t^2 + 1), -2t/(2t^2 + 1) \rangle$ (b) $2/(2t^2 + 1)^2$
13. $2/(4t^2 + 1)^{3/2}$ **15.** $\sqrt{t^4 + 4t^2 + 1}/6(t^4 + t^2 + 1)^{3/2}$ **17.** $\sqrt{2}/(1 + \cos^2 t)^{3/2}$ **19.** $6|x|/(1 + 9x^4)^{3/2}$
21. $|\sin x|/(1 + \cos^2 x)^{3/2}$ **23.** $(-\tfrac{1}{2}\ln 2, 1/\sqrt{2})$ **25.** $6/|t|(9t^2 + 4)^{3/2}$

Exercises 11.9

1. $\mathbf{v}(t) = \langle 2t, 1 \rangle$
$\mathbf{a}(t) = \langle 2, 0 \rangle$
$|\mathbf{v}(t)| = \sqrt{4t^2 + 1}$

3. $\mathbf{v}(t) = e^t\mathbf{i} - e^{-t}\mathbf{j}$
$\mathbf{a}(t) = e^t\mathbf{i} + e^{-t}\mathbf{j}$
$|\mathbf{v}(t)| = \sqrt{e^{2t} + e^{-2t}}$

5. $\mathbf{v}(t) = \cos t\mathbf{i} + \mathbf{j} - \sin t\mathbf{k}$
$\mathbf{a}(t) = -\sin t\mathbf{i} - \cos t\mathbf{k}$
$|\mathbf{v}(t)| = \sqrt{2}$

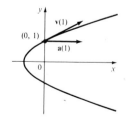   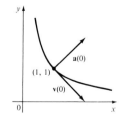   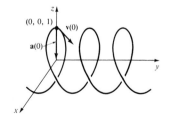

7. $\langle 3t^2, 2t, 3t^2 \rangle, \langle 6t, 2, 6t \rangle, \sqrt{18t^4 + 4t^2}$ **9.** $-t^{-2}\mathbf{i} + 2t\mathbf{k}, 2t^{-3}\mathbf{i} + 2\mathbf{k}, \sqrt{t^{-4} + 4t^2}$
11. $e^t((\cos t - \sin t)\mathbf{i} + (\sin t + \cos t)\mathbf{j} + (t + 1)\mathbf{k}), e^t(-2\sin t\mathbf{i} + 2\cos t\mathbf{j} + (t + 2)\mathbf{k}), e^t\sqrt{t^2 + 2t + 3}$
13. $\mathbf{v}(t) = \mathbf{i} - \mathbf{j} + t\mathbf{k}, \mathbf{r}(t) = t\mathbf{i} - t\mathbf{j} + (t^2/2)\mathbf{k}$ **15.** $\mathbf{v}(t) = t\mathbf{i} + 2t\mathbf{j} + t^2\mathbf{k}, \mathbf{r}(t) = (1 + t^2/2)\mathbf{i} + t^2\mathbf{j} + (1 + t^3/3)\mathbf{k}$
17. $t = 4$ **19.** $\mathbf{r}(t) = t\mathbf{i} - t\mathbf{j} + \tfrac{5}{2}t^2\mathbf{k}, |\mathbf{v}(t)| = \sqrt{25t^2 + 2}$ **21.** (a) 22 km   (b) 3.2 km   (c) 500 m/s

23. 30 m/s   **27.** $\sin t/\sqrt{2(1 - \cos t)}, \sqrt{(1 - \cos t)/2}$
29. $(18t^3 + 4t)/\sqrt{9t^4 + 4t^2 + 1}, 2\sqrt{9t^4 + 9t^2 + 1}/\sqrt{9t^4 + 4t^2 + 1}$ **31.** $e^t - e^{-t}, \sqrt{2}$

Exercises 11.10

1. $(0, 3, 1)$ **3.** $(-1, -\sqrt{3}, 8)$ **5.** $(1, \pi, 0)$ **7.** $(2, \pi/6, 4)$ **9.** $(0, 0, 1)$ **11.** $(\sqrt{3}/4, \frac{1}{4}, \sqrt{3}/2)$ **13.** $(3, \pi, \pi/2)$
15. $(2, 0, \pi/3)$ **17.** Cylinder with radius 3   **19.** Half-cone   **21.** Circular paraboloid   **23.** Horizontal plane
25. Positive z-axis   **27.** Sphere, radius 5, center the origin   **29.** Circular cylinder, radius 2, axis the y-axis
31. Cylinder, radius 1, together with the z-axis   **33.** (a) $r^2 + z^2 = 16$ (b) $\rho = 4$
35. (a) $r \cos \theta + 2r \sin \theta + 3z = 6$ (b) $\rho (\sin \phi \cos \theta + 2 \sin \phi \sin \theta + 3 \cos \phi) = 6$
37. (a) $2z^2 = r^2 \cos 2\theta - 4$ (b) $\rho^2(\sin^2 \phi \cos 2\theta - 2\cos^2 \phi) = 4$ **39.** (a) $r = 2 \sin \theta$ (b) $\rho \sin \phi = 2 \sin \theta$

Review Exercises for Chapter 11

1. $|AB| = 13, |BC| = \sqrt{38}, |AC| = \sqrt{65}$ **3.** $(-2, -3, 5), 6$ **5.** $6\mathbf{i} + \mathbf{j} + 13\mathbf{k}$ **7.** -1 **9.** $3\sqrt{35}$ **11.** $\mathbf{0}$
13. $96°$ **15.** $-1/\sqrt{6}$ **17.** 4 **19.** (a) $\langle 4, -3, 4 \rangle$ (b) $\sqrt{41}/2$ **21.** $x = 1 + 2t, y = 2 - t, z = 4 + 3t$
23. $x = 1 + 4t, y = -3t, z = 1 + 5t$ **25.** $x + 2y + 5z = 8$ **27.** Skew   **29.** $22/\sqrt{26}$
31. Plane   **33.** Circular paraboloid   **35.** Cone

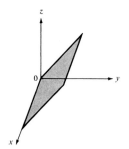

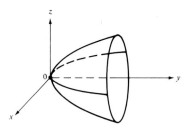

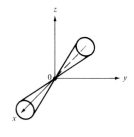

37. Hyperboloid of two sheets   **39.** (a)   **41.** $\frac{5}{6}\mathbf{i} + \frac{9}{4}\mathbf{j} + \frac{1}{5}\mathbf{k}$

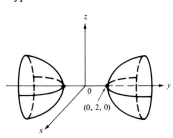

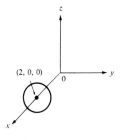

(b) $\cos t\mathbf{j} - \sin t\mathbf{k},$
$-\sin t\mathbf{j} - \cos t\mathbf{k}$

43. (a) $\langle t^2, t, 1 \rangle/\sqrt{t^4 + t^2 + 1}$ (b) $\langle 2t, 1 - t^4, -2t^3 - t \rangle/\sqrt{t^8 + 4t^6 + 2t^4 + 5t^2}$
(c) $\sqrt{t^8 + 4t^6 + 2t^4 + 5t^2}/(t^4 + t^2 + 1)^2$
45. $\frac{3}{16}, \frac{4}{9}$ **47.** $\langle t + t^3/6, 2t + t^2/2, t + t^4/12 \rangle$ **49.** $(\sqrt{3}, 1, 2), (2\sqrt{2}, \pi/6, \pi/4)$ **51.** $(1, \sqrt{3}, 2\sqrt{3}), (2, \pi/3, 2\sqrt{3})$
53. A half-plane   **55.** The horizontal plane $z = 3$ **57.** $r^2 + z^2 = 4, \rho = 2$

Chapter 12

Exercises 12.1

1. (a) 7 (b) -45 (c) $x^2 + 2xh + h^2 - y^2 + 4xy + 4hy - 7x - 7h + 10$
(d) $x^2 - y^2 - 2yk - k^2 + 4xy + 4xk - 7x + 10$ (e) $4x^2 - 7x + 10$
3. (a) 1 (b) $-\frac{2}{3}$ (c) $3t/(t^2 + 2)$ (d) $-3y/(1 + 2y^2)$ (e) $3x/(1 + 2x^2)$ **5.** $R^2; R$
7. $\{(x, y) \mid x + y \neq 0\}; \{z \mid z \neq 0\}$ **9.** $R^2; \{z \mid z > 0\}$ **11.** $\{(x, y, z) \mid x - y + z > 0\}; R$ **13.** $R^3; R$

15. $\{(x, y) \mid y \geqslant 2x\}$

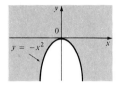

17. $\{(x, y) \mid x^2 + y^2 \leqslant 9, x \neq -2y\}$

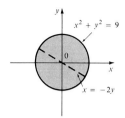

19. $\{(x, y) \mid y \geqslant -x^2\}$

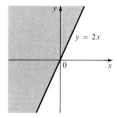

21. $\{(x, y) \mid xy > 1\}$

23. $\{(x, y) \mid y \neq (2n + 1)\pi/2, n \text{ an integer}\}$

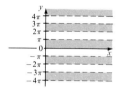

25. $\{(x, y) \mid -1 \leqslant x + y \leqslant 1\}$

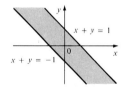

27. $\{(x, y) \mid x > 0, 2n\pi < y < (2n + 1)\pi, n \text{ an integer}\}$

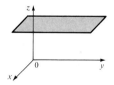

29. $\{(x, y, z) \mid x^2 + y^2 + z^2 \leqslant 1\}$

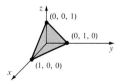

31. $z = 3$, horizontal plane

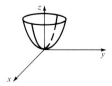

33. $x + y + z = 1$, plane

35. $z = x^2 + 9y^2$, elliptic paraboloid

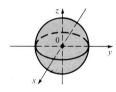

37. $z = \sqrt{x^2 + y^2}$, top half of cone

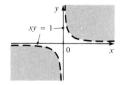

39. $z = y^2 - x^2$, hyperbolic paraboloid. See Figure 11.34(b)   **41.** $z = 1 - x^2$, parabolic cylinder

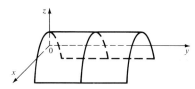

43. $xy = k$ **45.** $x^2 + 9y^2 = k$ **47.** $x/y = k$

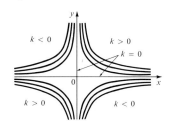

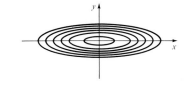

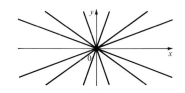

49. $\sqrt{x + y} = k$ **51.** $x = y^2 + k$ **53.** Family of parallel planes

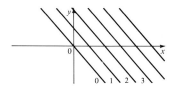

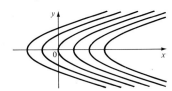

55. Family of hyperboloids of one or two sheets with axis the y-axis   **57.**

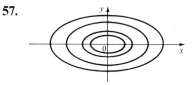

Exercises 12.2

1. -927 **3.** $-\frac{5}{2}$ **5.** π **7.** 1   **9.** Does not exist   **11.** Does not exist   **13.** Does not exist
15. 0   **17.** 1   **19.** Does not exist   **21.** 2   **23.** Does not exist   **25.** $-\frac{3}{5}$ **27.** Does not exist
29. Does not exist   **31.** $h(x, y) = e^{-(x^4 + x^2 y^2 + y^4)} \cos(x^4 + x^2 y^2 + y^4);\ R^2$
33. $h(x, y) = 4x^2 + 9y^2 + 12xy - 24x - 36y + 36 + \sqrt{2x + 3y - 6};\ \{(x, y) | 2x + 3y \geqslant 6\}$ **35.** $\{(x, y) | x^2 + y^2 \neq 1\}$
37. $\{(x, y) | x^4 - y^4 \neq (2n + 1)\pi/2, n \text{ an integer}\}$ **39.** R^2 **41.** $\{(x, y) | x \geqslant |y|\}$ **43.** $\{(x, y, z) | yz > 0\}$
45. $\{(x, y) | (x, y) \neq (0, 0)\}$ **47.** $\{(x, y) | (x, y) \neq (0, 0)\}$

Exercises 12.3

1. -27 **3.** 2   **5.** $(x^4 + 3x^2 y^2 - 2xy^3)/(x^2 + y^2)^2, (y^4 + 3x^2 y^2 - 2x^3 y)/(x^2 + y^2)^2$ **7.** $1/y - y/x^2$
9. $(y - z)/(x - y), (x + z)/(x - y)$ **11.** $(x - y - z)/(x + z), (y - x)/(x + z)$ **13.** $y \sec(xy) + xy^2 \sec(xy) \tan(xy)$
15. 0   **17.** $y + z, x + z, x + y$ **19.** $f_x(x, y) = 3x^2 y^5 - 4xy + 1, f_y(x, y) = 5x^3 y^4 - 2x^2$
21. $f_x(x, y) = 4x^3 + 2xy^2, f_y(x, y) = 2x^2 y + 4y^3$ **23.** $f_x(x, y) = 2y/(x + y)^2, f_y(x, y) = -2x/(x + y)^2$
25. $f_x = e^x[\tan(x - y) + \sec^2(x - y)], f_y = -e^x \sec^2(x - y)$ **27.** $f_s = -3s/\sqrt{2 - 3s^2 - 5t^2}, f_t = -5t/\sqrt{2 - 3s^2 - 5t^2}$
29. $f_u = v/(u^2 + v^2), f_v = -u/(u^2 + v^2)$ **31.** $g_x = 2xy^4 \sec^2(x^2 y^3), g_y = \tan(x^2 y^3) + 3x^2 y^3 \sec^2(x^2 y^3)$
33. $\partial z/\partial x = 1/\sqrt{x^2 + y^2}, \partial z/\partial y = y/(x^2 + y^2 + x\sqrt{x^2 + y^2})$
35. $\partial z/\partial x = 3 \cosh(\sqrt{3x + 4y})/2\sqrt{3x + 4y}, \partial z/\partial y = 2 \cosh(\sqrt{3x + 4y})/\sqrt{3x + 4y}$ **37.** $f_x = -e^{x^2}, f_y = e^{y^2}$
39. $f_x = 2xyz^3 + y, f_y = x^2 z^3 + x, f_z = 3x^2 yz^2 - 1$ **41.** $f_x = yzx^{yz-1}, f_y = zx^{yz} \ln x, f_z = yx^{yz} \ln x$

43. $u_x = -yz\cos(y/(x+z))/(x+z)^2$, $u_y = z\cos(y/(x+z))/(x+z)$, $u_z = \sin(y/(x+z)) - yz\cos(y/(x+z))/(x+z)^2$

45. $u_x = y^2z^3[\ln(x+2y+3z) + x/(x+2y+3z)]$, $u_y = 2xyz^3[\ln(x+2y+3z) + y/(x+2y+3z)]$,
$u_z = 3xy^2z^2[\ln(x+2y+3z) + z/(x+2y+3z)]$

47. $f_x = 1/(z-t)$, $f_y = 1/(t-z)$, $f_z = (y-x)/(z-t)^2$, $f_t = (x-y)/(z-t)^2$ **49.** $\partial u/\partial x_i = x_i/\sqrt{x_1^2 + x_2^2 + \cdots + x_n^2}$

51. $f'(x), g'(y)$ **53.** $f'(x+y), f'(x+y)$ **55.** $f'(x/y)/y, -xf'(x/y)/y^2$

57. $f_{xx} = 2y$, $f_{xy} = 2x + 1/2\sqrt{y} = f_{yx}$, $f_{yy} = -x/4y\sqrt{y}$

59. $z_{xx} = 3(2x^2 + y^2)/\sqrt{x^2 + y^2}$, $z_{xy} = 3xy/\sqrt{x^2 + y^2} = z_{yx}$, $z_{yy} = 3(x^2 + 2y^2)/\sqrt{x^2 + y^2}$

61. $z_{tt} = 0$, $z_{tx} = 1/2\sqrt{x - x^2} = z_{xt}$, $z_{xx} = t(2x-1)/4(x-x^2)^{3/2}$ **67.** $-48xy$ **69.** $48x^3y^3z^2$ **71.** $-\sin y$

73. $72yz^2/(x + 2y + 3z^3)^3$ **83.** R^2/R_1^2

89. (a) $f_x(x,y) = (x^4y + 4x^2y^3 - y^5)/(x^2 + y^2)^2$, $f_y(x,y) = (x^5 - 4x^3y^2 - xy^4)/(x^2 + y^2)^2$ (b) $0, 0$
  (d) No, since f_{xy} and f_{yx} are not continuous

Exercises 12.4

1. $4x + 8y - z = 8$ **3.** $2x + 4y - z + 6 = 0$ **5.** $2x + y - z = 1$ **7.** $2x - y - z + 2 = 0$

9. $dz = 2xy^3\,dx + 3x^2y^2\,dy$ **11.** $dz = -2x(x^2 + y^2)^{-2}\,dx - 2y(x^2 + y^2)^{-2}\,dy$

13. $du = e^x(\cos xy - y\sin xy)\,dx - xe^x\sin xy\,dy$ **15.** $dw = 2xy\,dx + (x^2 + 2yz)\,dy + y^2\,dz$

17. $dw = (x^2 + y^2 + z^2)^{-1}(x\,dx + y\,dy + z\,dz)$ **19.** $\Delta z = 0.9225$, $dz = 0.9$ **21.** 2.99 **23.** -0.28

25. 65.88 **27.** 26.76 **29.** 1.015 **31.** 5.4 cm^2 **33.** 16 cm^3 **35.** $\frac{1}{17} \approx 0.059$ ohm

Exercises 12.5

1. $6t^5 + 4t^3 + 4t$ **3.** $(18e^{2t} - 3\cos t)e^t + (3e^t - 4\cos t)\sin t$ **5.** $(x + y^2)^{-1}(1/2\sqrt{1 + t} + y/\sqrt{t})$

7. $y^2z^3\cos t - 2xyz^3\sin t + 6xy^2z^2e^{2t}$ **9.** $\partial z/\partial s = 4sx\sin y + 2tx^2\cos y$, $\partial z/\partial t = 4tx\sin y + 2sx^2\cos y$

11. $\partial z/\partial s = 2x(1 - 3y^3)e^t - 9x^2y^2e^{-t}$, $\partial z/\partial t = 2x(1 - 3y^3)se^t + 9x^2y^2se^{-t}$

13. $\partial z/\partial s = 2^{x-3y}(\ln 2)(2st - 3t^2)$, $\partial z/\partial t = 2^{x-3y}(\ln 2)(s^2 - 6st)$

15.
$$\frac{\partial u}{\partial r} = \frac{\partial u}{\partial x}\frac{\partial x}{\partial r} + \frac{\partial u}{\partial y}\frac{\partial y}{\partial r}$$
$$\frac{\partial u}{\partial s} = \frac{\partial u}{\partial x}\frac{\partial x}{\partial s} + \frac{\partial u}{\partial y}\frac{\partial y}{\partial s}$$
$$\frac{\partial u}{\partial t} = \frac{\partial u}{\partial x}\frac{\partial x}{\partial t} + \frac{\partial u}{\partial y}\frac{\partial y}{\partial t}$$

17.
$$\frac{\partial v}{\partial x} = \frac{\partial v}{\partial p}\frac{\partial p}{\partial x} + \frac{\partial v}{\partial q}\frac{\partial q}{\partial x} + \frac{\partial v}{\partial r}\frac{\partial r}{\partial x}$$
$$\frac{\partial v}{\partial y} = \frac{\partial v}{\partial p}\frac{\partial p}{\partial y} + \frac{\partial v}{\partial q}\frac{\partial q}{\partial y} + \frac{\partial v}{\partial r}\frac{\partial r}{\partial y}$$
$$\frac{\partial v}{\partial z} = \frac{\partial v}{\partial p}\frac{\partial p}{\partial z} + \frac{\partial v}{\partial q}\frac{\partial q}{\partial z} + \frac{\partial v}{\partial r}\frac{\partial r}{\partial z}$$

19. $2, 0$ **21.** $0, 0, 4$ **23.** $\partial u/\partial p = 2(z - x)/(y + z)^2 = -t/p^2$, $\partial u/\partial r = 0$, $\partial u/\partial t = 2/(y + z) = 1/p$

25. $\partial w/\partial r = \sin(x - y)(2rst\cos\theta - s^2t^3\sin\theta)$, $\partial w/\partial s = \sin(x - y)(r^2t\cos\theta - 2rst^3\sin\theta)$,
  $\partial w/\partial t = \sin(x - y)(r^2s\cos\theta - 3rs^2t^2\sin\theta)$, $\partial w/\partial\theta = -\sin(x - y)(r^2st\sin\theta + rs^2t^3\cos\theta)$ **27.** $-9600\pi\text{ cm}^3/\text{s}$

29. -0.27 L/s   **31.** (a) $\partial z/\partial r = (\partial z/\partial x)\cos\theta + (\partial z/\partial y)\sin\theta$, $\partial z/\partial\theta = -(\partial z/\partial x)r\sin\theta + (\partial z/\partial y)r\cos\theta$

37. $4rs\,\partial^2z/\partial x^2 + (4r^2 + 4s^2)\,\partial^2z/\partial x\,\partial y + 4rs\,\partial^2z/\partial y^2 + 2\,\partial z/\partial y$ **41.** $(y - 2x)/(3y^2 - x)$

43. $(18x - x^{-2/3}y^{1/3})/(12y + x^{1/3}y^{-2/3})$ **45.** $(y - z)/(x - y)$, $(x + z)/(x - y)$

47. $(x - y - z)/(x + z)$, $(y - x)/(x + z)$ **49.** $-(e^y + ze^x)/(y + e^x)$, $-(xe^y + z)/(y + e^x)$

Exercises 12.6

1. $7\sqrt{3} - 16$ **3.** (a) $\nabla f(x,y) = \langle 3x^2 - 8xy, -4x^2 + 2y\rangle$ (b) $\langle 0, -2\rangle$ (c) $-\frac{8}{5}$

5. (a) $\nabla f(x,y,z) = \langle y^2z^3, 2xyz^3, 3xy^2z^2\rangle$ (b) $\langle 4, -4, 12\rangle$ (c) $20/\sqrt{3}$ **7.** $\frac{7}{52}$ **9.** $29/\sqrt{13}$ **11.** $\frac{1}{6}$

13. $-\pi/4\sqrt{3}$ **15.** $\sqrt{5}, \langle 1, 2\rangle$ **17.** $\sqrt{17}/6, \langle 4, 1\rangle$ **19.** $\sqrt{\frac{13}{2}}, \langle -3, -2\rangle$ **23.** (a) $-40/3\sqrt{3}$

25. (a) $32/\sqrt{3}$ (b) $\langle 38, 6, 12\rangle$ (c) $2\sqrt{406}$ **31.** (a) $4x + y + z = 12$ (b) $(x - 2)/4 = y - 2 = z - 2$

33. (a) $x + y + z = 3$ (b) $x - 1 = y - 1 = z - 1$ (or $x = y = z$)

35. (a) $6x + 3y + 2z = 18$ (b) $(x - 1)/6 = (y - 2)/3 = (z - 3)/2$

37. (a) $x + y - z = 1$ (b) $x - 1 = y = -z$ **43.** $(\pm 6\sqrt{3}, \mp 2\sqrt{6}/3, \pm\sqrt{6}/2)$

Exercises 12.7

1. Minimum $f(-2, 3) = -13$ **3.** Minimum $f(0, -1) = -1$ **5.** Minimum $f(0, 0) = 4$, saddle points $(\pm\sqrt{2}, -1)$

7. Minimum $f(1, 1) = -1$, saddle point $(0, 0)$ **9.** Saddle point $(1, 2)$ **11.** Maximum $f(-\frac{1}{2}, 4) = -6$

13. None   **15.** Maximum $f(0,0) = 2$, minimum $f(0,2) = -2$, saddle points $(\pm 1, 1)$
17. Saddle points $(0, n\pi)$, n an integer   **19.** Maximum $f(\pi/3, \pi/3) = 3\sqrt{3}/2$, minimum $f(5\pi/3, 5\pi/3) = -3\sqrt{3}/2$
21. Minimum $f(4,0) = -7$, maximum $f(4,5) = 13$ **23.** Maximum $f(\pm 1, 1) = 7$, minimum $f(0,0) = 4$
25. $(\frac{2}{7}, \frac{4}{7}, \frac{6}{7})$ **27.** $7/\sqrt{61}$ **29.** $(0,0,1)$, $(0,0,-1)$ **31.** $\frac{100}{3}, \frac{100}{3}, \frac{100}{3}$ **33.** $16/\sqrt{3}$ **35.** $\frac{4}{3}$
37. Cube, edge length $c/12$ **39.** Square base of side 40 cm, height 20 cm   **43.** $y = 6.67x - 308$, 172.2 lb

Exercises 12.8

1. Maximum $f(\pm 1, 0) = 1$, minimum $f(0, \pm 1) = -1$
3. Maximum $f(\pm\sqrt{2/3}, \pm\sqrt{2}) = \frac{2}{3}$, minimum $f(\pm\sqrt{2/3}, \mp\sqrt{2}) = -\frac{2}{3}$
5. Maximum $f(1/\sqrt{35}, 3/\sqrt{35}, 5/\sqrt{35}) = \sqrt{35}$, minimum $f(-1/\sqrt{35}, -3/\sqrt{35}, -5/\sqrt{35}) = -\sqrt{35}$
7. Maximum $2/\sqrt{3}$, minimum $-2/\sqrt{3}$ **9.** Maximum $\sqrt{3}$, minimum 1
11. Maximum $f(\frac{1}{2}, \frac{1}{2}, \frac{1}{2}, \frac{1}{2}) = 2$, minimum $f(-\frac{1}{2}, -\frac{1}{2}, -\frac{1}{2}, -\frac{1}{2}) = -2$
13. Maximum $f(1, \sqrt{2}, -\sqrt{2}) = 1 + 2\sqrt{2}$, minimum $f(1, -\sqrt{2}, \sqrt{2}) = 1 - 2\sqrt{2}$ **15.** Maximum $\frac{3}{2}$, minimum $\frac{1}{2}$
21-35. See Exercises 25-39 in Section 12.7.

Review Exercises for Chapter 12

1. $\{(x, y) \,|\, y > -x - 1, x \neq 1\}$ **3.** $\{(x, y) \,|\, -1 \leqslant x \leqslant 1\}$ **5.**   **7.**

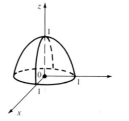

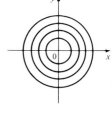

9. 0   **11.** $f_x = 12x^3 - \sqrt{y}$, $f_y = -\frac{1}{2}x/\sqrt{y}$ **13.** $f_s = 2e^{2s}\cos \pi t$, $f_t = -\pi e^{2s}\sin \pi t$
15. $f_x = y^z$, $f_y = xzy^{z-1}$, $f_z = xy^z \ln y$ **17.** $f_{xx} = 2y^3 - 24x^2$, $f_{xy} = 6xy^2$, $f_{yy} = 6x^2y + 2$
19. $f_{xx} = 0$, $f_{yy} = 2xz^3$, $f_{zz} = 6xy^2z$, $f_{xy} = 2yz^3$, $f_{xz} = 3y^2z^2$, $f_{yz} = 6xyz^2$ **23.** $6y - z = 1$ **25.** $x + y - z = 0$
27. $3x + 4y + 3z = 14$ **29.** $(\pm\sqrt{\frac{2}{7}}, \pm 1/\sqrt{14}, \mp 3/\sqrt{14})$ **31.** 38.656   **33.** $e^t + 2(y/z)(3t^2 + 4) - 2t(y^2/z^2)$
39. $ze^{x\sqrt{y}}\langle z\sqrt{y}, xz/2\sqrt{y}, 2\rangle$ **41.** $\frac{43}{5}$ **43.** $\sqrt{145}/2$, $\langle 4, 9/2\rangle$ **45.** Minimum $f(-4, 1) = -9$
47. Maximum $f(1, 1) = 1$, saddle points $(0, 0)$, $(0, 3)$, $(3, 0)$ **49.** Maximum $f(1, 2) = 4$, minimum $f(2, 4) = -64$
51. Maximum $f(\pm\sqrt{\frac{2}{3}}, 1/\sqrt{3}) = 2/3\sqrt{3}$, minimum $f(\pm\sqrt{\frac{2}{3}}, -1/\sqrt{3}) = -2/3\sqrt{3}$
53. Maximum $f(3, 3, 3) = 9$, minimum $f(1, 1, -1) = f(1, -1, 1) = f(-1, 1, 1) = 1$
55. $(\pm 3^{-1/4}, 3^{-1/4}\sqrt{2}, \pm 3^{1/4})$, $(\pm 3^{-1/4}, -3^{-1/4}\sqrt{2}, \pm 3^{1/4})$

Chapter 13

Exercises 13.1

1. (a) -17.75 (b) -15.75 (c) -8.75 (d) -6.75 **3.** $63, \sqrt{2}$ **5.** $0, \sqrt{5}$ **7.** $\frac{247}{8}, \sqrt{5}$

Exercises 13.2

1. $4x^2, y^3/3$ **3.** $(e^2 - 1)xe^x, e^y$ **5.** $\frac{32}{3}$ **7.** 0   **9.** $\frac{4}{15}(31 - 9\sqrt{3})$ **11.** $3(1 - 1/\sqrt{2})$ **13.** $-\frac{585}{8}$
15. $\frac{15}{4}(2 - \sqrt{3})$ **17.** $(\sqrt{3} - 1)/2 - \pi/12$ **19.** $\ln\frac{27}{16}$ **21.** 104   **23.** 16   **25.** 72

Exercises 13.3

1. $\frac{1}{6}$ **3.** $16(1 - \sqrt{2}/7)$ **5.** $(1 - \cos 1)/2$ **7.** $\frac{1}{12}$ **9.** $\pi(3\sqrt{2} - 1 - \sqrt{3})/4 + (14 - 13\sqrt{3})/8$ **11.** $\frac{3}{4}$
13. $\frac{1}{2}e^4 - 2e$ **15.** $(1 - \cos 1)/2$ **17.** $\frac{16}{5}$ **19.** $\frac{500}{3}$ **21.** $\frac{1}{8}$ **23.** $\frac{6}{35}$ **25.** $\frac{31}{8}$ **27.** $\frac{13}{6}$ **29.** $\frac{16}{3}$

31. $\frac{1}{6}$ **33.** $\frac{1}{3}$ **35.** $\int_0^1 \int_y^1 f(x, y) \, dx \, dy$ **37.** $\int_0^{\ln 2} \int_{e^y}^2 f(x, y) \, dx \, dy$ **39.** $(e^9 - 1)/6$ **41.** $e - 1$

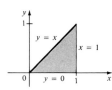

43. 1   **45.** $0 \leq \iint_D \sqrt{x^3 + y^3} \, dA \leq \sqrt{2}$

Exercises 13.4

1. 0   **3.** $\frac{609}{8}$ **5.** 2π **7.** $24\pi^5$ **9.** $81\pi/2$ **11.** $5\pi/2$ **13.** $(2\pi/3)(1 - 1/\sqrt{2})$ **15.** $(8\pi/3)(64 - 24\sqrt{3})$
17. $\frac{4}{3}\pi a^3$ **19.** $(\pi/4)(e - 1)$ **21.** $4\pi/3$

Exercises 13.5

1. $\frac{50}{3} C$ **3.** $\frac{2}{3}, (0, 1/2)$ **5.** $6, (\frac{3}{4}, \frac{3}{2})$ **7.** $\frac{1}{6}, (\frac{4}{7}, \frac{3}{4})$ **9.** $\frac{27}{2}, (\frac{8}{5}, \frac{1}{2})$ **11.** $\pi/4, (\pi/2, 16/9\pi)$
13. $(\frac{3}{8}, 3\pi/16)$ **15.** $(2a/5, 2a/5)$ if vertex is $(0, 0)$ and sides are along positive axes   **17.** $\frac{1}{10}, \frac{1}{16}, \frac{13}{80}$
19. $\frac{189}{20}, \frac{1269}{28}, \frac{1917}{35}$ **21.** $\rho a^4/3, \rho a^4/3; a/\sqrt{3}, a/\sqrt{3}$

Exercises 13.6

3. $\frac{1}{48}$ **5.** $\frac{16}{3}$ **7.** $\frac{7}{5}$ **9.** $\frac{5}{28}$ **11.** $\frac{1}{10}$ **13.** $\frac{1}{12}$ **15.** 0

17. $\displaystyle \int_{-2}^{2} \int_0^6 \int_{-\sqrt{4-x^2}}^{\sqrt{4-x^2}} f(x, y, z) \, dz \, dy \, dx = \int_0^6 \int_{-2}^2 \int_{-\sqrt{4-x^2}}^{\sqrt{4-x^2}} f(x, y, z) \, dz \, dx \, dy = \int_{-2}^2 \int_0^6 \int_{-\sqrt{4-z^2}}^{\sqrt{4-z^2}} f(x, y, z) \, dx \, dy \, dz$

$\displaystyle = \int_0^6 \int_{-2}^2 \int_{-\sqrt{4-z^2}}^{\sqrt{4-z^2}} f(x, y, z) \, dx \, dz \, dy = \int_{-2}^2 \int_{-\sqrt{4-x^2}}^{\sqrt{4-x^2}} \int_0^6 f(x, y, z) \, dy \, dz \, dx = \int_{-2}^2 \int_{-\sqrt{4-z^2}}^{\sqrt{4-z^2}} \int_0^6 f(x, y, z) \, dy \, dx \, dz$

19. $\displaystyle \int_{-1}^1 \int_0^{1-x^2} \int_0^y f(x, y, z) \, dz \, dy \, dx = \int_0^1 \int_{-\sqrt{1-y}}^{\sqrt{1-y}} \int_0^y f(x, y, z) \, dz \, dx \, dy = \int_0^1 \int_z^1 \int_{-\sqrt{1-y}}^{\sqrt{1-y}} f(x, y, z) \, dx \, dy \, dz$

$\displaystyle = \int_0^1 \int_0^y \int_{-\sqrt{1-y}}^{\sqrt{1-y}} f(x, y, z) \, dx \, dz \, dy = \int_{-1}^1 \int_0^{1-x^2} \int_z^{1-x^2} f(x, y, z) \, dy \, dz \, dx = \int_0^1 \int_{-\sqrt{1-z}}^{\sqrt{1-z}} \int_z^{1-x^2} f(x, y, z) \, dy \, dx \, dz$

21. $\displaystyle \int_0^1 \int_0^x \int_0^y f(x, y, z) \, dz \, dy \, dx = \int_0^1 \int_z^1 \int_y^1 f(x, y, z) \, dx \, dy \, dz = \int_0^1 \int_0^y \int_y^1 f(x, y, z) \, dx \, dz \, dy = \int_0^1 \int_0^x \int_z^x f(x, y, z) \, dy \, dz \, dx$

$\displaystyle = \int_0^1 \int_z^1 \int_z^x f(x, y, z) \, dy \, dx \, dz$ **23.** $\frac{\pi}{4} - \frac{1}{3}$ **25.** $\frac{9}{10}, (\frac{22}{27}, \frac{25}{63}, \frac{152}{189})$ **27.** $a^5, \left(\frac{7a}{12}, \frac{7a}{12}, \frac{7a}{12} \right)$

29. (a) $m = \displaystyle \int_0^1 \int_0^{\sqrt{1-x^2}} \int_0^y (1 + x + y + z) \, dz \, dy \, dx$ (b) $(\bar{x}, \bar{y}, \bar{z})$, where $\bar{x} = \dfrac{1}{m} \displaystyle \int_0^1 \int_0^{\sqrt{1-x^2}} \int_0^y x(1 + x + y + z) \, dz \, dy \, dx$,

$\bar{y} = \dfrac{1}{m} \displaystyle \int_0^1 \int_0^{\sqrt{1-x^2}} \int_0^y y(1 + x + y + z) \, dz \, dy \, dx$, $\bar{z} = \dfrac{1}{m} \displaystyle \int_0^1 \int_0^{\sqrt{1-x^2}} \int_0^y z(1 + x + y + z) \, dz \, dy \, dx$

(c) $\displaystyle \int_0^1 \int_0^{\sqrt{1-x^2}} \int_0^y (x^2 + y^2)(1 + x + y + z) \, dz \, dy \, dx$

31. (a) $m = \displaystyle \int_{-1}^1 \int_{-\sqrt{1-y^2}}^{\sqrt{1-y^2}} \int_{4y^2 + 4z^2}^4 (x^2 + y^2 + z^2) \, dx \, dz \, dy$

(b) $(\bar{x}, \bar{y}, \bar{z})$, where $\bar{x} = \dfrac{1}{m} \displaystyle \int_{-1}^1 \int_{-\sqrt{1-y^2}}^{\sqrt{1-y^2}} \int_{4y^2 + 4z^2}^4 x(x^2 + y^2 + z^2) \, dx \, dz \, dy$,

$\bar{y} = \dfrac{1}{m} \displaystyle \int_{-1}^1 \int_{-\sqrt{1-y^2}}^{\sqrt{1-y^2}} \int_{4y^2 + 4z^2}^4 y(x^2 + y^2 + z^2) \, dx \, dz \, dy$, $\bar{z} = \dfrac{1}{m} \displaystyle \int_{-1}^1 \int_{-\sqrt{1-y^2}}^{\sqrt{1-y^2}} \int_{4y^2 + 4z^2}^4 z(x^2 + y^2 + z^2) \, dx \, dz \, dy$

(c) $\displaystyle \int_{-1}^1 \int_{-\sqrt{1-y^2}}^{\sqrt{1-y^2}} \int_{4y^2 + 4z^2}^4 (x^2 + y^2)(x^2 + y^2 + z^2) \, dx \, dz \, dy$

Exercises 13.7

1. 24π **3.** 0 **5.** 162π **7.** $\pi Ka^2/8,\ (0,0,2a/3)$ **9.** $4\pi/5$ **11.** $\pi/30$ **13.** $4\pi(2-\sqrt{3})$ **15.** 10π
17. $K\pi a^4/2$, where K is the proportionality constant **19.** $2\pi Ka^6/9$ **21.** $4K\pi a^5/15$
23. $(2\pi/3)(1-1/\sqrt{2}),\ (0,0,3/8(2-\sqrt{2}))$ **25.** $5\pi/6$ **27.** $243\pi/5$

Exercises 13.8

1. 3 **3.** $2e^{4u}$ **5.** -4 **7.** The parallelogram bounded by the lines $y=2x,\ y=2x+3,\ x=2y,\ x=2y-6$
9. $\frac{11}{3}$ **11.** 6π **13.** $2\ln 3$ **15.** $\frac{4}{3}\pi abc$ **17.** $-\frac{66}{125}$ **19.** $\frac{3}{2}\sin 1$

Review Exercises for Chapter 13

1. 1152 **3.** $\ln(\frac{27}{4})-1$ **5.** $\frac{1}{110}$ **7.** $(e-1)/2$ **9.** $\ln(3/2)$ **11.** $\frac{1}{40}$ **13.** $\frac{41}{30}$ **15.** $81\pi/5$
17. $\frac{32}{3}$ **19.** $\pi/96$ **21.** $\frac{64}{15}$ **23.** 176 **25.** $\frac{2}{3}$ **27.** $2ma^3/9$ **29.** $\frac{1}{12},\ (\frac{1}{3},\frac{8}{15})$
31. $(4a/3\pi,4a/3\pi)$ **33.** $(0,0,h/4)$ **35.** $(\pi/8)\ln 5$ **37.** $-\ln 2$

Chapter 14

Exercises 14.1

1.

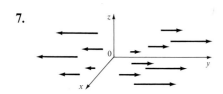

3.

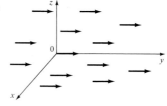

5.

7.

9. $\nabla f(x,y)=(5x^4-8xy^3)\mathbf{i}-12x^2y^2\mathbf{j}$

11. $\nabla f(x,y)=3e^{3x}\cos 4y\,\mathbf{i}-4e^{3x}\sin 4y\,\mathbf{j}$ **13.** $\nabla f(x,y,z)=y^2\mathbf{i}+(2xy-z^3)\mathbf{j}-3yz^2\mathbf{k}$

Exercises 14.2

1. $(10\sqrt{10}-1)/54$ **3.** 1638.4 **5.** 48 **7.** $\frac{17}{3}$ **9.** $9\sqrt{13}\pi/4$ **11.** $3\sqrt{35}$ **13.** $\frac{16}{11}$ **15.** $\frac{77}{6}$
17. $-\frac{19}{143}$ **19.** -2 **21.** $\frac{6}{5}-\cos 1-\sin 1$ **23.** $2\pi k,\ (4/\pi,0)$
25. (a) $\bar{x}=(1/m)\int_C x\rho(x,y,z)\,ds,\ \bar{y}=(1/m)\int_C y\rho(x,y,z)\,ds,\ \bar{z}=(1/m)\int_C z\rho(x,y,z)\,ds$, where $m=\int_C \rho(x,y,z)\,ds$
    (b) $2\sqrt{13}k\pi,\ (0,0,3\pi)$
27. $2\pi^2$ **29.** $\frac{23}{88}$

Exercises 14.3

1. $f(x,y)=x^2-3xy+y^2+K$ **3.** Not conservative **5.** $f(x,y)=x^4y^3+x+K$ **7.** Not conservative
9. $f(x,y)=ye^x+x\sin y+K$ **11.** (a) $f(x,y)=(x^2+y^2)/2$ (b) 44 **13.** (a) $f(x,y)=x^2y^3$ (b) $(1+\pi^2/4)^3$
15. (a) $f(x,y,z)=xy+yz$ (b) 15 **17.** (a) $f(x,y,z)=x^2z+x\sin y$ (b) 2π **19.** $25\sin 1-1$

Exercises 14.4

1. $-\frac{1}{12}$ **3.** $-\frac{1}{6}$ **5.** $-\frac{4}{3}$ **7.** $\frac{1}{3}$ **9.** 0 **11.** 0 **13.** π **15.** $\pi+\frac{16}{3}(1/\sqrt{2}-1)$ **17.** $3\pi/8$

Exercises 14.5

1. (a) $\mathbf{0}$ (b) 3 **3.** (a) $\mathbf{0}$ (b) 0 **5.** (a) $xz\mathbf{i} - yz\mathbf{j} + y\mathbf{k}$ (b) $x + xy$
7. (a) $(3xe^y + 2ye^{yz})\mathbf{i} + (xe^{xz} - 3e^y)\mathbf{j}$ (b) $ze^{xz} - 2ze^{yz}$ **9.** (a) $(\ln z + e^{-y})\mathbf{i} - xe^y\mathbf{k}$ (b) $e^y + ze^{-y} + y/z$
11. $f(x, y, z) = xy + z + K$ **13.** Not conservative **15.** Not conservative **17.** $f(x, y, z) = xyz + y^3/3 + K$
19. No

Exercises 14.6

1. $x = x, y = y, z = \sqrt{1 - 3x^2 - 2y^2}$ [or $x = (1/\sqrt{3})\sin\phi\cos\theta, y = (1/\sqrt{2})\sin\phi\sin\theta,$
$z = \cos\phi, 0 \le \phi \le \pi/2, 0 \le \theta \le 2\pi$] **3.** $x = x, y = 6 - 3x^2 - 2z^2, z = z, 3x^2 + 2z^2 \le 6$
5. $x = 2\sin\phi\cos\theta, y = 2\sin\phi\sin\theta, z = 2\cos\phi, 0 \le \phi \le \pi/4, 0 \le \theta \le 2\pi$
7. $x = r\cos\theta, y = r\sin\theta, z = 5, 0 \le r \le 4, 0 \le \theta \le 2\pi$ (or $x = x, y = y, z = 5, x^2 + y^2 \le 16$) **9.** $3x - y + 3z = 3$
11. $x = 0$ **13.** $4\sqrt{6}\pi$ **15.** $(\pi/6)(17\sqrt{17} - 5\sqrt{5})$ **17.** $(\sqrt{21}/2) + \frac{17}{4}[\ln(2 + \sqrt{21}) - \ln\sqrt{17}]$ **19.** $2a^2(\pi - 2)$
21. $\pi(2\sqrt{6} - \frac{8}{3})$ **23.** $x = (b + a\cos\alpha)\cos\theta, y = (b + a\cos\alpha)\sin\theta, z = a\sin\alpha, 0 \le \theta \le 2\pi, 0 \le \alpha \le 2\pi$

Exercises 14.7

1. 16π **3.** 16π **5.** $3\sqrt{14}$ **7.** $(33\sqrt{33} - 17\sqrt{17})/6$ **9.** $\pi\sqrt{2}/4$ **11.** 0 **13.** 108π
15. $(11 - 10e)/6$ **17.** 12 **19.** 0 **21.** 48 **23.** $(0, 0, a/2)$
25. (a) $I_z = \iint_S (x^2 + y^2)\rho(x, y, z)\, dS$ (b) $4329\sqrt{2}\pi/5$ **27.** $194{,}400\pi$ **29.** $8\pi a^3 \varepsilon_0/3$ **31.** 1248π

Exercises 14.8

1. π **3.** $3\pi/4$ **5.** 0 **7.** 3.5 **9.** -4π

Exercises 14.9

3. 8 **5.** 0 **7.** $\frac{1}{6}$ **9.** $12\pi/5$ **11.** $-81\pi/2$ **13.** 27π

Review Exercises for Chapter 14

1. $(5\sqrt{5} - 1)/3$ **3.** $4\sqrt{5}$ **5.** $-\pi$ **7.** $\frac{17}{2}$ **9.** 5 **11.** $f(x, y) = x\sin y - \cos y$ **13.** π^2 **17.** -8π
21. $x + 4y + 4z = 0$ **23.** $\pi(391\sqrt{17} + 1)/60$ **25.** $-64\pi/3$ **29.** $-\frac{1}{2}$ **33.** -4 **35.** 21

Chapter 15

Exercises 15.1

1.

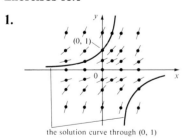

the solution curve through (0, 1)

3.

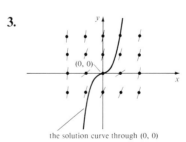

the solution curve through (0, 0)

5. (a), (b)

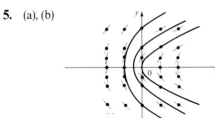

(c) $y = \pm\sqrt{2(x + c)}$ (d)

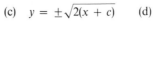

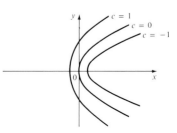

7. (a), (b) 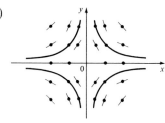   (c) $y = c/x$ (d)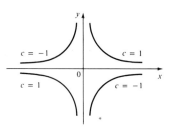

9. $x^2 + 2y^2 = C$

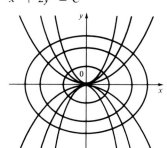

11. $y^3 = 3(x + c)$

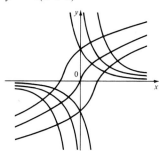

13. $y = c/2x$

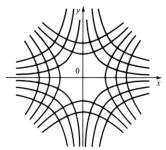

Exercises 15.2

1. $y = \frac{1}{2}(x - k/x)$ **3.** $y = x\sqrt{C + \ln x^2}$ **5.** $y = (1 \pm \sqrt{1 - 4k^2x^2})/2k$ **7.** $y = -x\ln(C - \ln|x|)$
9. $y = x\tan(\ln|x|)$ **11.** $y = x(8 + \ln|x^3|)^{1/3}$

Exercises 15.3

1. $y = Ce^{3x} - e^x/2$ **3.** $y = Ce^{x^2} - \frac{1}{2}$ **5.** $y = \sin x + (\cos x)/x + C/x$ **7.** $y = \frac{1}{2}\sec x - \cos x + C\sec x$
9. $y = x/2 + Ce^{-x^2} + \frac{1}{2}e^{-x^2}\int e^{x^2}\,dx$ **11.** $y = x - 1 + \frac{1}{2}(e^x + e^{-x}) = x - 1 + \cosh x$ **13.** $y = (x^2 + 3)e^{x^2}$
15. $y = (\sin x)/x^2$ **19.** $y = \pm 1/\sqrt{Cx^4 + 1}$ **21.** (a) $I(t) = 4 - 4e^{-5t}$ (b) $4 - 4e^{-1/2} \approx 1.57$ A
23. $Q(t) = 3(1 - e^{-4t})$, $I(t) = 12e^{-4t}$ **25.** $P(t) = M + Ce^{-kt}$

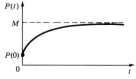

27. (a) $(mg/c)(1 - e^{-ct/m})$ (b) mg/c (c) $(mg/c)[t + (m/c)e^{-ct/m}] - m^2g/c^2$

Exercises 15.4

1. $y = \frac{1}{2}(-x \pm \sqrt{4c - 3x^2})$ **3.** Not exact   **5.** $x\sin y + y = C$ **7.** $xye^{x/y} = C$ **9.** Not exact
11. $y = \pm\sqrt{C - x^3}$ **13.** $y = (-x^2 + \sqrt{180x - 11x^4})/6x$ **15.** $y = [\sin^{-1}(1 - x)]/x$ **17.** $ye^{xy} = C$
19. $xy/\sqrt{1 + x^2 + y^2} = C$ **21.** $x^3y + x^2y^2 = C$ **23.** $e^{3x}(x^2y + y^2) = C$

Exercises 15.5

1. $y = Ce^x - \frac{1}{2}(\sin x + \cos x)$ **3.** $y = \pm\sqrt[4]{e^x - \frac{1}{2}x^2 + C}$ **5.** $x + y = k(x^2 + y^2)$ **7.** $xy^2 - 3y\tan x = C$
9. $y^2 = \ln(x^2 + 1) + k$ **11.** $y = Cx^2 - \frac{1}{2}[\sqrt{1 + x^2} - x^2\ln|(\sqrt{x^2 + 1} + 1)/x|]$ **13.** $y = 1/(k + \int\sqrt{1 + x^3}\,dx)$
15. $x^2 + y^2 + xe^y = C$ **17.** $y\sin(xy) - x = C$ **19.** $y = x[1 - (C/x)]^2$

Exercises 15.6

1. $y = c_1e^x + c_2e^{2x}$ **3.** $y = c_1e^{-x/3} + c_2e^{3x}$ **5.** $y = c_1e^x + c_2e^{-x}$ **7.** $y = c_1 + c_2e^{-x/2}$
9. $y = c_1e^{(-1+\sqrt{2})x} + c_2e^{(-1-\sqrt{2})x}$ **11.** $y = c_1e^{4x} + c_2xe^{4x}$ **13.** $y = e^x + e^{-4x}$ **15.** $y = 2e^{1-x} + e^{3(x-1)}$
17. $y = 3xe^{-2x+2}$ **19.** $y = [e^{5-x} - e^{2(1+x)}]/(e^6 - 1)$

Exercises 15.7

1. $y = e^{-x}(c_1 \cos 3x + c_2 \sin 3x)$ **3.** $y = c_1 \cos 5x + c_2 \sin 5x$ **5.** $y = e^{x/2}[c_1 \cos(\sqrt{7}x/2) + c_2 \sin(\sqrt{7}x/2)]$
7. $y = e^x(c_1 \cos 2x + c_2 \sin 2x)$ **9.** $y = e^x(\cos x + \sin x)$ **11.** $y = -\frac{1}{3}\sin 3x$ **13.** No solution
15. $y = e^{-2x}(2 \cos 3x - e^\pi \sin 3x)$

Exercises 15.8

1. $y = c_1 e^{3x} + c_2 e^{-2x} - \frac{5}{78}\cos 3x - \frac{1}{78}\sin 3x$ **3.** $y = e^{2x}(c_1 x + c_2) + e^{-x}/9$
5. $y = c_1 \cos 6x + c_2 \sin 6x + x^2/18 - x/36 - \frac{1}{324}$ **7.** $y = c_1 e^{-x/4} + c_2 e^{-x} + e^x/10$
9. $y = e^x[\frac{63}{130}\cos 2x + \frac{201}{260}\sin 2x] + \frac{1}{5}(x + 2) + \frac{3}{26}\cos 3x - \frac{1}{13}\sin 3x$ **11.** $y = \frac{5}{8}e^x - \frac{17}{32}e^{-x} + e^{3x}[\frac{1}{8}x - \frac{3}{32}]$
13. $y_p = (Ax^4 + Bx^3 + Cx^2 + Dx + E)e^{2x}$ **15.** $y_p = xe^x(A \cos x + B \sin x)$
17. $y = (c_1 + x)\sin x + (c_2 + \ln \cos x)\cos x$ **19.** $y = [c_1 + \ln(1 + e^{-x})]e^x + [c_2 - e^{-x} + \ln(1 + e^{-x})]e^{2x}$
21. $y = [c_1 - \int(e^x/2x)\,dx]e^{-x} + [c_2 + \int(e^{-x}/2x)\,dx]e^x$

Exercises 15.9

1. $x = 0.36 \sin(10t/3)$ **3.** $x = -\frac{1}{5}e^{-6t} + \frac{6}{5}e^{-t}$ **5.** $\frac{49}{12}$ kg
9. $Q(t) = (-e^{-10t}/250)(6 \cos 20t + 3 \sin 20t) + \frac{3}{125}, I(t) = \frac{3}{5}e^{-10t}\sin 20t$
11. $Q(t) = e^{-10t}[\frac{3}{250}\cos 20t - \frac{3}{500}\sin 20t] - \frac{3}{250}\cos 10t + \frac{3}{125}\sin 10t$

Exercises 15.10

1. $\displaystyle\sum_{n=0}^{\infty} a_0 \frac{6^n}{n!} x^n = a_0 e^{6x}$ **3.** $\displaystyle\sum_{n=0}^{\infty} \frac{a_0}{3^n n!} x^{3n} = a_0 e^{x^3/3}$ **5.** $a_0 \displaystyle\sum_{n=0}^{\infty} \left(-\frac{3}{2}\right)^n \frac{1}{n!} x^{2n} + a_1 \displaystyle\sum_{n=0}^{\infty} \frac{(-6)^n n!}{(2n + 1)!} x^{2n+1}$

7. $a_0 + a_1 x + a_0 \dfrac{x^2}{2} + a_0 \displaystyle\sum_{n=2}^{\infty} \frac{(-1)^{n-1}(2n - 3)!}{2^{2n-2}n!(n - 2)!} x^{2n}$ **9.** $\displaystyle\sum_{n=0}^{\infty} \frac{x^{2n}}{2^n n!} = e^{x^2/2}$

11. $x + \displaystyle\sum_{n=0}^{\infty} \frac{(-1)^n 2^2 5^2 \cdots (3n - 1)^2}{(3n + 1)!} x^{3n+1}$

Review Exercises for Chapter 15

1. $x + x^2 y^2 = C$ **3.** $y = Cx^2 + x^4/2$ **5.** $y^2 - y^3 = x \sin x + \cos x + C$ **7.** $y/x + \ln(x - y)^2 = \ln|x| + C$
9. $y = \tan(\sin^{-1} x + C)$ **11.** $y = e^{2x} + Ce^x$ **13.** $y = e^{3x}(c_1 \cos 5x + c_2 \sin 5x)$ **15.** $y = c_1 e^{-x} + c_2 e^{x/2}$
17. $y = e^{-x}(c_1 + c_2 x) - \frac{3}{50}\cos 3x - \frac{2}{25}\sin 3x$ **19.** $y = c_1 \cos(3x/2) + c_2 \sin(3x/2) + \frac{2}{9}x^2 - \frac{43}{81}$
21. $y = c_1 e^x + c_2 e^{2x} + xe^{2x}$ **23.** $y = e^{-x}[\frac{2}{3}x^{3/2} + 3]$ **25.** $y = 5 - 2e^{-6(x-1)}$ **27.** $y = (e^{4x} - e^x)/3$

29. $y^2 - 2\ln|y| + x^2 = K$ **31.** $\displaystyle\sum_{n=0}^{\infty} \frac{(-2)^n n!}{(2n + 1)!} x^{2n+1}$ **33.** $Q(t) = -0.02e^{-10t}(\cos 10t + \sin 10t) + 0.03$

Appendix A Exercises

1. $-3a^2 bc$ **3.** $2x^2 - 10x$ **5.** $-8 + 6a$ **7.** $-x^2 + 6x + 3$ **9.** $12x^2 + 25x - 7$ **11.** $4x^2 - 4x + 1$
13. $30y^4 + y^5 - y^6$ **15.** $2x^3 - 5x^2 - x + 1$ **17.** $1 + 4x$ **19.** $(3x + 7)/(x^2 + 2x - 15)$
21. $(u^2 + 3u + 1)/(u + 1)$ **23.** $x/(yz)$ **25.** $rs/(3t)$ **27.** $c/(c - 2)$ **29.** $2x(1 + 6x^2)$ **31.** $(x + 1)(x + 6)$
33. $(x + 2)(x - 4)$ **35.** $9(x - 2)(x + 2)$ **37.** $(3x + 2)(2x - 3)$ **39.** $(t + 1)(t^2 - t + 1)$ **41.** $(2t - 3)^2$
43. $x(x + 1)^2$ **45.** $(x - 1)(x + 1)(x + 3)$ **47.** $(x - 2)(x + 3)(x + 4)$ **49.** $(x + 2)/(x - 2)$
51. $(x + 1)/(x - 8)$ **53.** $(x - 2)/(x^2 - 9)$ **55.** $(x + 1)^2 + 4$ **57.** $(x - \frac{5}{2})^2 + \frac{15}{4}$ **59.** $(2x + 1)^2 - 3$
61. $1, -10$ **63.** $(-9 \pm \sqrt{85})/2$ **65.** $(-5 \pm \sqrt{13})/6$ **67.** $1, (-1 \pm \sqrt{5})/2$ **69.** Irreducible
71. Not irreducible (two real roots)   **73.** $a^6 + 6a^5 b + 15a^4 b^2 + 20a^3 b^3 + 15a^2 b^4 + 6ab^5 + b^6$
75. $x^8 - 4x^6 + 6x^4 - 4x^2 + 1$ **77.** 8   **79.** $2|x|$ **81.** $4a^2 b\sqrt{b}$ **83.** 3^{26} **85.** $16x^{10}$ **87.** $a^2 b^{-1}$
89. $1/\sqrt{3}$ **91.** 25   **93.** $2\sqrt{2}|x|^3 y^6$ **95.** $y^{6/5}$ **97.** $t^{-5/2}$ **99.** $t^{1/4}/s^{1/24}$ **101.** False   **103.** True
105. False   **107.** False

Appendix B Exercises

1. $7\pi/6$ **3.** $\pi/20$ **5.** 5π **7.** $720°$ **9.** $75°$ **11.** $67.5°$ **13.** 3π cm   **15.** $\frac{2}{3}$ rad $= (\frac{120}{\pi})°$

17.   **19.**   **21.**

23. $\sin(3\pi/4) = 1/\sqrt{2}$, $\cos(3\pi/4) = -1/\sqrt{2}$, $\tan(3\pi/4) = -1$, $\csc(3\pi/4) = \sqrt{2}$, $\sec(3\pi/4) = -\sqrt{2}$, $\cot(3\pi/4) = -1$
25. $\sin(9\pi/2) = 1$, $\cos(9\pi/2) = 0$, $\csc(9\pi/2) = 1$, $\cot(9\pi/2) = 0$, $\tan(9\pi/2)$ and $\sec(9\pi/2)$ undefined
27. $\frac{1}{2}$, $-\sqrt{3}/2$, $-1/\sqrt{3}$, 2, $-2/\sqrt{3}$, $-\sqrt{3}$ **29.** $\cos\theta = \frac{4}{5}$, $\tan\theta = \frac{3}{4}$, $\csc\theta = \frac{5}{3}$, $\sec\theta = \frac{5}{4}$, $\cot\theta = \frac{4}{3}$
31. $\sin\phi = \sqrt{5}/3$, $\cos\phi = -\frac{2}{3}$, $\tan\phi = -\sqrt{5}/2$, $\csc\phi = 3/\sqrt{5}$, $\cot\phi = -2/\sqrt{5}$
33. $\sin\beta = -1/\sqrt{10}$, $\cos\beta = -3/\sqrt{10}$, $\tan\beta = \frac{1}{3}$, $\csc\beta = -\sqrt{10}$, $\sec\beta = -\sqrt{10}/3$ **35.** 5.73576 cm
37. 24.62147 cm   **59.** $(4 + 6\sqrt{2})/15$ **61.** $(3 + 8\sqrt{2})/15$ **63.** $\frac{24}{25}$ **65.** $\pi/3$, $5\pi/3$ **67.** $\pi/4$, $3\pi/4$, $5\pi/4$, $7\pi/4$
69. $\pi/6$, $\pi/2$, $5\pi/6$, $3\pi/2$ **71.** 0, π, 2π **73.** $0 \leqslant x \leqslant \pi/6$ and $5\pi/6 \leqslant x \leqslant 2\pi$
75. $0 \leqslant x < \pi/4$, $3\pi/4 < x < 5\pi/4$, $7\pi/4 < x \leqslant 2\pi$

77.   **79.**

   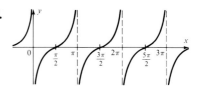

81.

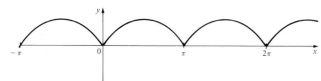

Index

Abel, Niels, 136
Abscissa, 10
Absolutely convergent series, 601
Absolute maximum and minimum, 174, 769
Absolute value, 7
Absolute value function, 33
Acceleration, 143, 698
  normal component of, 701
  tangential component of, 701
Achilles and the tortoise, xxvi
Addition formulas for sine and cosine, A21
Addition of vectors, 651
Algebra, A1
Algebraic function, 42
Alternating harmonic series, 598, 601
Alternating series, 596
Alternating Series Test, 596
Analytic function, 619
Analytic geometry, 11
Angle, A15
  between curves, 137
  direction, 658
  of inclination, 17
  between lines, 18
  between vectors, 656
Antiderivative, 243
Approximation, 2
  by differentials, 161
  to e, 330
  by Newton's method, 164
  by Riemann sums, 270
  by Simpson's Rule, 445
  tangent line, 161
  tangent plane, 744
  by Taylor's Formula, 631
  by the Trapezoidal Rule, 441
Arc length, 481, 482, 531, 546, 691
Area, xxii, 259
  under a curve, 263, 270
  between curves, 298
  by exhaustion, xxii
  by Green's Theorem, 879
  under a parametric curve, 528
  in polar coordinates, 544
  of a sector of a circle, 419, 543

Area (*continued*)
  of a surface, 894, 896
  of a surface of revolution, 488, 533
Area problem, xxii, 259
Arrow diagram, 28, 714
Astroid, 140, 525, 534
Asymptote, 223
  horizontal, 202, 223
  of a hyperbola, 24
  slant, 224
  vertical, 212, 223
Auxiliary equation, 949
Average cost, 239
Average rate of change, 56, 146
Average value of a function, 308
Average velocity, xxv, 55
Axes:
  coordinate, 11, 645
  of ellipse, 552
  rotation of, 562

Bacterial growth, 360
Barrow, Isaac, xxiv, 284
Base of cylinder, 464
Base of logarithm, 327
  change of, 334
Bernoulli, James, 939
Bernoulli, John, 384, 524
Bernoulli equation, 939
Bessel function, 611, 616, 975
Binomial series, 627
Binomial Theorem, 625, A8
Blood flow, 150, 238, 305
Bound, greatest lower, 577
Bound, least upper, 577
Boundary condition, 951
Boundary point, 729
Boundary value problem, 951
Bounded function, 271
Bounded sequence, 576
Bounded set, 773
Boyle's Law, 154
Brachistochrone problem, 524
Brahe, Tycho, 703
Branches of hyperbola, 24

C^1 transformation, 838
Cable (hanging), 377
Calculators, A39
Calculus, xxix
Cancellation equations, 322
Cardioid, 525, 538, 545, 547
Cartesian coordinate system, 11
Cartesian plane, 10
Catenary, 377
Cauchy, Augustin, 390
Cauchy-Schwarz inequality, 661
Cauchy's Mean Value Theorem, 390
Cavalieri's Principle, 474
Center of gravity, 501
Center of mass, 501, 504, 817, 829, 900
Centroid:
  of a plane region, 503, 505
  of a solid, 829
Chain Rule, 125
  for several variables, 750, 752, 753
Change of base, 334
Change of variables, 290
  in multiple integrals, 810, 832, 834, 841, 845
Characteristic equation, 949
Charge, 815, 829
Charge density, 815, 829
Chemical reaction, 147
Circle, 12
Circular cylinder, 464
Circulation of a vector field, 914
Cissoid, 543
Clairaut, Alexis, 738
Clairaut's Theorem, 738
Closed curve, 868
Closed interval, 3
Closed set, 772
Closed surface, 904
Combinations of functions, 36
Comparison Test, 592
Comparison Theorem for integrals, 457
Complementary equation, 956
Completeness Axiom, 577
Completing the square, A6
Complex number, 953
Component function, 683, 852

Components of a vector, 649
Composition of functions, 38, 731
    continuity of, 88, 731
    derivative of, 125
Compound interest, 363
Compressibility, 148
Computer-generated graphs, 717, 718, 721
Concavity, 194
Concentration, 148
Conchoid, 140, 543
Conditionally convergent series, 602
Conductivity, 908
Cone, 473, 679
Conic section, 21, 549, 557
    directrix, 558
    eccentricity, 558
    focus, 558
    polar equation, 560
Connected region, 868
Conservation of energy, 873
Conservative vector field, 855
Constant function, 40, 106
Constrained extrema, 777
Constraint, 776
Continuity:
    of a function, 82, 728, 731
    on an interval, 84
    from the left, 84
    piecewise, 271
    from the right, 84
Convergence:
    absolute, 601
    conditional, 602
    of an improper integral, 449, 453
    interval of, 612
    radius of, 612
    of a sequence, 571
    of a series, 580
Convergent improper integral, 449, 453
Convergent sequence, 571
Convergent series, 580
Coordinate(s), 3, 645
    Cartesian, 11
    cylindrical, 704, 831
    polar, 535
    rectangular, 11
    spherical, 705, 833
Coordinate axes, 11, 644
Coordinate planes, 644
Cosine function, A17
    derivative, 122
    graph, A23
    power series, 621
Cost function, 152, 239
Critically damped vibration, 965
Critical number, 179
Critical point, 769
Cross product, 662

Cross-section, 464
Curl of a vector field, 883
Current, 147, 967
Curvature, 693
Curve(s):
    closed, 868
    length of, 481, 531, 691
    level, 719
    orientation of, 860
    orthogonal, 137
    parametric, 522, 684
    piecewise smooth, 692, 857
    polar, 537
    simple, 870
    smooth, 691
    space, 684
Curve-sketching procedure, 222
Cusp, 691
Cycloid, 523, 528, 532
Cylinder, 464, 680
Cylindrical coordinates, 704, 831
Cylindrical shell, 475

Damped vibrations, 965
Damping constant, 965
Damping force, 965
Decay:
    law of natural, 359
    radioactive, 361
Decibel, 334
Decreasing function, 188
Decreasing sequence, 575
Definite integral, 268
    of a vector-valued function, 688
Definite integration:
    by parts, 403
    by substitution, 293
Degree of a polynomial, 42
Del (∇), 760
Delta (Δ) notation, 56, 145
Demand function, 241
Density:
    area, 814
    linear, 146
    liquid, 497
    of a solid, 828
Dependent variable, 30, 714, 752
Derivative(s), 96
    of composite functions, 125
    of a constant function, 106
    directional, 759, 761
    domain of, 100
    of exponential functions, 343, 344
    as a function, 100
    higher, 141
    of hyperbolic functions, 378
    of an integral, 280
    of an inverse function, 324

Derivative(s) (*continued*)
    of inverse trigonometric functions, 372
    of logarithmic functions, 335
    notation, 101, 735
    partial, 733, 736
    of a power function, 107, 114, 345
    of a power series, 616
    of a product, 110
    of a quotient, 111
    as a rate of change, 98
    as the slope of a tangent, 97
    of trigonometric functions, 120, 122
    of a vector-valued function, 685
Descartes, René, 11
Determinant, 662
Deviation, 776
Differentiable function, 102, 747
Differential, 160, 743, 748
    total, 743
Differential equation, 245, 359, 508
    Bernoulli, 939
    exact, 940
    first-order, 928
    general solution, 509
    homogeneous, 932, 947
    integrating factor of, 942
    linear, 934, 947
    logistic, 514
    nonhomogeneous, 947, 956
    order of, 508, 928
    ordinary, 928
    partial, 928
    particular solution of, 956
    separable, 509
    series solution of, 971
    solution of, 509
    strategy for solving, 944
Differential operator, 101
Differentiation, 101
    formulas for, 115
    implicit, 134, 754
    logarithmic, 340
    partial, 735
    of power series, 616
    of vector functions, 687
Directional derivative, 759, 761
Direction angle, 658
Direction cosine, 658
Direction field, 929
Direction numbers, 670
Directrix:
    of a conic section, 558
    of a parabola, 549
Discontinuity, 82
Discontinuous function, 82, 83
Discriminant, A7
Disk method, 468
Displacement, 55, 303, 660

Distance:
  between point and line, 668, 675
  between point and plane, 668
  between points in a plane, 11
  between points in space, 646
  between real numbers, 8
Distance formula, 8, 646
Divergence:
  of an improper integral, 449, 453
  of an infinite series, 580
  test for, 584
  of a vector field, 886
Divergence Theorem, 916
Divergent improper integral, 449, 453
Divergent sequence, 571
Divergent series, 580
Domain of a function, 27, 714
Dot product, 655
Double-angle formulas, A22
Double integral, 789, 800
  in polar coordinates, 810
Downward concavity, 194
Dummy variable, 269

e (the number), 329, 354, 620
Earthquakes, 334
Eccentricity, 558
Electric circuit, 937, 967
Electric force field, 854
Elementary functions, 438
Ellipse, 23, 551
  area, 414
  directrix, 558
  eccentricity, 558
  foci, 551
  major axis, 552
  polar equation, 560
  reflecting property, 557
  vertices, 552
Ellipsoid, 677
Elliptic cylinder, 681
Elliptic paraboloid, 679
Endpoint extrema, 179
Energy:
  conservation of, 873
  kinetic, 741, 874
  potential, 874
Epicycloid, 525
Equation(s):
  of a circle, 13
  differential (see Differential equation)
  of a graph, 12
  heat, 740
  of a line, 15, 16
  of a line in space, 669, 670
  Laplace's, 738
  linear, 17, 673
  logistic, 514

Equation(s) (continued)
  parametric, 522
  of a plane, 672
  point-slope, 15
  polar, 537
  second-degree, 20
  slope-intercept, 14
  of sphere, 647
  wave, 740
Equilateral hyperbola, 24
Equipotential curves, 724
Error, 162
  percentage, 163
  relative, 163
  in Taylor approximation, 631
Error estimate:
  for alternating series, 599
  for Simpson's Rule, 445
  for the Trapezoidal Rule, 442
Eudoxus, A2
Euler's constant, 591
Euler's formula, 954
Even function, 33, 222, 294
Exact differential equation, 940
Exponential decay, 359
Exponential function, 316, 386
  derivative of, 343, 344
  limits of, 318
  power series for, 619
Exponential growth, 359
Exponents, laws of, 317, 355, A10
Extreme value, 174, 768, 769
Extreme Value Theorem, 175, 773

Factorial, 142
Factoring, 980
Factor Theorem, A5
Fermat, Pierre, xxiv, 11, 177
Fermat's Principle, 237
Fermat's Theorem, 177
Fibonacci, 571
Fibonacci sequence, 570, 579
Field:
  conservative, 855
  electric, 854
  force, 854
  gravitational, 854
  scalar, 852
  vector, 852
  velocity, 853
First Derivative Test, 190
  for absolute extrema, 233
First octant, 644
Fluid flow, 853
Flux, 905
Focus, of a conic section, 558
Focus, of a parabola, 549
Folium of Descartes, 133

Force, 493
  exerted by liquid, 498
Force field, 854
Fourier, Joseph, 154
Four-leaved rose, 539, 545
Fractions (partial), 420
Frustum:
  of a cone, 473, 487
  of a pyramid, 473
Fubini, Guido, 795
Fubini's Theorem, 795, 823
Function(s), 27
  algebraic, 42
  analytic, 619
  arc length, 484, 692
  arrow diagram of, 28, 714
  average value of, 308
  bounded, 271
  composite, 38, 731
  constant, 40, 106
  continuous, 82, 728, 731
  cost, 152, 239
  cubic, 42
  decreasing, 188
  demand, 241
  derivative of, 96
  difference of, 36
  differentiable, 102, 747
  discontinuous, 82
  domain of, 27, 714
  elementary, 438
  even, 33, 222
  exponential, 316, 356
  extreme values of, 174
  gradient of, 760, 762
  graph of, 30, 715
  greatest integer, 79
  harmonic, 738
  hyperbolic, 376
  implicit, 134
  increasing, 188
  integrable, 268, 790
  inverse, 321
  inverse hyperbolic, 379
  inverse trigonometric, 365
  limit of, 60, 69, 725, 731
  linear, 42
  logarithmic, 327
  machine diagram of, 27
  marginal cost, 152, 239
  marginal profit, 241
  marginal revenue, 241
  maximum value of, 174
  minimum value of, 174
  of n variables, 722
  natural logarithmic, 330, 351
  odd, 33, 222
  one-to-one, 320

Function(s) (*continued*)
  periodic, 223
  piecewise continuous, 271
  polynomial, 42, 729
  position, 55
  potential, 855
  power, 40, 107
  product of, 36
  profit, 241
  quadratic, 42
  quotient of, 36
  range of, 27
  rational, 42, 729
  root, 41
  sum of, 36
  of three variables, 720
  transcendental, 42
  trigonometric, A17
  of two variables, 714
  value of, 27
  vector-valued, 683
Fundamental Theorem for line integrals, 866
Fundamental Theorem of Calculus, 280,
    283, 287

g, 493
G, 497
Galileo, 524, 529
Galois, Evariste, 136
Gas law, 741
Gauss, Karl Friedrich, 255
Gauss's Law, 908
Gauss's Theorem, 916
Geometric series, 581
Gompertz function, 516
Gradient, 760, 762
Gradient vector field, 854
Graph:
  of an equation, 12
  of a function, 30, 714
  of a parametric curve, 522
  polar, 537
Graphical addition, 37
Gravitational acceleration, 493
Gravitational field, 854
Gravitation law, 497
Greater than, 3
Greatest integer function, 79
Green, Sir George, 875
Green's identities, 890
Green's Theorem, 876
  vector forms, 888, 889
Growth, law of natural, 359
Growth rate, 149
Gyration, radius of, 821

Half-angle formulas, A22
Half-life, 361

Harmonic function, 738
Harmonic series, 583, 591
Heat conductivity, 908
Heat equation, 740
Heat flow, 908
Heaviside, function, 65
Heaviside, Oliver, 65
Helix, 685
Higher derivatives, 141, 737
Homogeneous differential equation, 932, 947
Hooke's Law, 494
Horizontal asymptote, 202
Horizontal line, 16
Horizontal Line Test, 320
Horizontal plane, 645
Huygens, 524
Hydrostatic pressure and force, 497
Hyperbola, 23, 553
  asymptotes, 24, 553
  branches, 24
  directrix, 558
  eccentricity, 550
  equation, 23, 554
  equilateral, 24
  foci, 553
  polar equation, 560
  reflecting property, 567
  vertices, 553
Hyperbolic functions, 376
  derivatives, 378
  inverses, 379
Hyperbolic identities, 377
Hyperbolic paraboloid, 680
Hyperbolic substitution, 416
Hyperboloid:
  of one sheet, 678
  of two sheets, 679
Hypocycloid, 525

i, 953
i, 653
Image, 27, 838
Implicit differentiation, 134, 754
Implicit function, 134
Improper integrals, 449, 453
Inclination, angle of, 17
Incompressible velocity field, 887
Increasing function, 188
Increasing sequence, 575
Increment, 56, 145, 743, 747
Indefinite integral, 285
Independence of path, 868
Independent variable, 30, 714, 752
Indeterminate form, 383, 384, 387, 388
Index of summation, 254
Induction principle, A11
Inequalities, 3
  Cauchy-Schwarz, 661

Inequalities (*continued*)
  rules for, 4
  triangle, 8, 661
Inertia (moment of), 818, 829
Infinite discontinuity, 83
Infinite interval, 4
Infinite limit, 211, 217
Infinite sequence (*see* Sequence)
Infinite series (*see* Series)
Inflection point, 196
Initial condition, 511, 928
Initial-value problem, 511, 928, 951
Inner product, 655
Instantaneous rate of change, 56, 98, 146
Instantaneous velocity, 55, 98
Integer, 2
Integrable function, 268
Integral(s):
  approximations to, 440
  change of variables in, 290, 293
  definite, 268
  derivative of, 280
  double, 789, 800
  indefinite, 285
  iterated, 793, 801
  line, 856, 859, 861, 863
  multiple, 786
  surface, 899, 901, 905
  table of, 434, endpapers
  triple, 822
  of vector function, 688
Integral Test, 588
Integrand, 269
Integrating factor, 942
Integration, 269
  formulas, 434, endpapers
  indefinite, 285
  limits of, 269
  numerical, 440
  partial, 793
  by partial fractions, 419
  by parts, 400, 403
  of power series, 616
  reversing the order of, 805
  strategy for, 433
  by substitution, 290, 293
  tables, use of, 457
  by trigonometric substitution, 412
Intercepts, 23, 222
Interest compounded continuously, 363
Interior point, 728
Intermediate Value Theorem, 89
Intermediate variable, 756
Intersection of polar graphs, 546
Intersection of sets, 3
Interval, 3
Interval of convergence, 612
Inverse function, 321

Inverse hyperbolic functions, 379
Inverse trigonometric functions, 365
   derivatives, 372
Involute, 526
Irrational number, 2
Irreducible quadratic, A7
Irrotational vector field, 886
Isothermal compressibility, 149
Isothermals, 724
Iterated integral, 793, 801

j, 653
Jacobi, Carl, 840
Jacobian, 840, 844
Joule, 493
Jump discontinuity, 83

k, 653
Kepler, Johann, 703
Kepler's Laws, 703, 704
Kinetic energy, 741, 874
Kirchhoff's Laws, 937, 968

Lagrange, Joseph, 183
Lagrange form of remainder term, 630
Lagrange multiplier, 777, 780
Lamina, 503, 814
Laplace, Pierre, 738
Laplace operator, 888
Laplace's equation, 738, 888
Law of Conservation of Energy, 874
Law of Cosines, A26
Law of laminar flow, 151
Law of natural decay, 359
Law of natural growth, 359
Laws of exponents, 317, 355, A10
Laws of logarithms, 328, 352
Learning curve, 153, 939
Least squares method, 776
Least upper bound, 577
Left-hand limit, 65, 72
Leibniz, Gottfried Wilhelm, xxiv, 284
Leibniz notation, 101
Lemniscate, 140, 543
Length:
   of a curve, 481–483
   of a line segment, 8, 646
   of a parametric curve, 531
   of a polar curve, 546
   of a space curve, 691
   of a vector, 650
Less than, 3
Level curve, 719
Level surface, 720
l'Hospital, Marquis de, 384
l'Hospital's Rule, 384
Limaçon, 542
Limit Comparison Test, 594

Limits, xxii–xxix
   of a function, 60, 69, 725
   infinite, 211, 217
   at infinity, 200, 207
   of integration, 269
   left-hand, 65, 72
   one-sided, 65, 72
   properties of, 74, 75
   right-hand, 65
   of a sequence, 571
   of a sum, 74
   of a vector-valued function, 683
Linear combination, 947
Linear differential equation, 934, 947
Linear equation, 17
Linear function, 42
Linearly independent solutions, 948
Line(s) in the plane, 14
   equations of, 15, 16
   horizontal, 16
   normal, 116, 765
   parallel, 17
   perpendicular, 18
   secant, xxiii, 56
   slope of, 14
   tangent, 51, 97
Line(s) in space:
   parametric equations of, 669
   symmetric equations of, 670
   vector equation of, 669
Line integral, 856, 859, 861, 863
Liquid force, 498
Lituus, 543
Local maximum, 174, 768
Local minimum, 174, 768
Logarithm(s), 327
   laws of, 328
   natural, 330, 351
Logarithmic differentiation, 339
Logarithmic function, 327, 357
   derivative of 335, 336, 358
   limits of, 329
Logistic differential equation, 514

Machine diagram of a function, 27
Maclaurin, Colin, 619
Maclaurin series, 619
Magnitude of a vector, 650
Major axis of ellipse, 552
Marginal cost function, 152, 239
Marginal profit function, 241
Marginal revenue function, 241
Mass, 304, 815, 829, 900
   center of, 501, 504, 817, 829, 900
Mathematical induction, A10
Maximum value, 174, 768, 769
Mean Value Theorem, 183
Mean Value Theorem for integrals, 309

Method of cylindrical shells, 475
Method of exhaustion, xxii
Method of Lagrange multipliers, 777, 780
Method of least squares, 776
Method of undetermined coefficients, 957
Method of variation of parameters, 961
Midpoint formula, 14
Minimum value, 174, 768, 769
Möbius strip, 903
Moment:
   about axis, 502, 816
   of inertia, 818, 829
   of a lamina, 504
   of a particle, 502
   about a plane, 829
   polar, 819
   of a solid, 829
   of a system of particles, 502
Monotonic function, 188
Monotonic sequence, 575
Multiple integrals, 786
Multiplier (Lagrange), 777, 780

Natural exponential function, 354
Natural growth law, 359
Natural logarithm function, 330, 351
Newton, Sir Isaac, xxiv, xxix, 164, 284, 535, 626
Newton (unit of force), 493
Newton's Law of Cooling, 362
Newton's Law of Gravitation, 497, 703, 853
Newton's method, 164
Newton's Second Law, 493, 699
Nonhomogeneous differential equation, 947, 956
Normal component of acceleration, 701
Normal derivative, 890
Normal line, 116, 765
Normal vector, 671, 893
Norm of a partition, 261, 789
Number:
   complex, 953
   irrational, 2
   rational, 2
   real, 2
Numerical integration, 440

Octant, 644
Odd function, 33, 222, 294
One-sided limits, 65, 72
One-to-one function, 320
One-to-one transformation, 838
Open interval, 3
Open region, 868
Orbit, 703
Ordered pair, 10
Ordered triple, 645
Order of a differential equation, 508, 928

Order of integration (reversing), 805
Ordinary differential equation, 928
Ordinate, 10
Oriented surface, 903
Orientation of a curve, 860
Orientation of surface, 903
Origin, 2, 10
Orthogonal curves, 137
Orthogonal trajectories, 139, 930
Orthogonal vectors, 657
Overdamped vibration, 965

p-series, 590
Pappus's Theorem, 506
Parabola, 21, 549
   axis, 549
   directrix, 549
   equation, 21, 550
   focus, 549
   polar equation, 560
   reflection property, 557
   vertex, 21, 549
Parabolic cylinder, 681
Paraboloid, 679
Paradoxes of Zeno, xxvi, xxviii
Parallelepiped, 667
Parallel lines, 17
Parallelogram law, 651
Parallel planes, 673
Parallel vectors, 652
Parameter, 522, 684
Parametric curve, 522, 684
Parametric equations, 522, 684, 891
Parametric surface, 891
Parametrization of a curve, 692
Partial derivative, 733, 736
Partial fractions, 419
Partial integration, 793
Partial sum of a series, 580
Partition:
   of an interval, 261
   polar, 809
   of a rectangle, 788
   of a rectangular box, 822
   regular, 271
Parts, integration by, 400, 403
Pascal, Blaise, 534
Path, 868
Percentage error, 163
Period, 223
Periodic function, 223
Perpendicular lines, 18
Perpendicular vectors, 657
Piecewise continuous function, 271
Piecewise smooth curve, 857
Piriform, 140
Plane:
   equation of, 672

Plane (*continued*)
   parallel, 673
   tangent to a surface, 741, 765
Planetary motion, 703
Point of inflection, 196
Point-slope equation of a line, 15
Poiseuille's Law, 151, 238, 306
Polar axis, 535
Polar coordinates, 535
Polar equations, 537
   of conics, 560
Polar partition, 809
Polar rectangle, 809
Pole, 535
Polynomial, 42, 729
Population, 149
   of bacteria, 360
   of world, 516
Position function, 55
Position vector, 649
Positive orientation:
   of a curve, 875
   of a surface, 904
Potential energy, 874
Potential function, 855
Pound, 493
Power function, 40
Power Rule, 107, 113, 114, 140, 345
Power series, 610, 611
   differentiation of, 616
   integration of, 616
   interval of convergence, 612
   radius of convergence, 612
   representation of functions, 615
   solution of differential equations, 971
Pressure exerted by a liquid, 497
Principle of mathematical induction, A11
Principle of superposition, 959
Product:
   cross, 662
   dot, 655
   of functions, 36
   scalar triple, 666
Product Rule, 110
Profit function, 241
Projectile, 525, 699
Projection, 645, 659

Quadrant, 11
Quadratic formula, A6
Quadratic function, 42
Quadric cylinder, 680
Quadric surface, 677
Quotient Rule, 111

Radian measure, A15
Radioactive decay, 361
Radius of convergence, 612

Radius of gyration, 820
Range of a function, 27, 714
Rate of change:
   average, 56, 146
   derivative as, 98
   instantaneous, 56, 98, 146
Rate of growth, 149
Rate of reaction, 148
Rates, related, 155
Rational function, 42, 729
Rationalizing substitutions, 430
Rational number, 2
Ratio Test, 603
Real line, 3
Real number, 2
Rearrangement of a series, 606
Rectangular coordinate system, 11
Reduction formula, 404
Reflecting functions, 44
Reflection property:
   of ellipse, 557
   of hyperbola, 567
   of parabola, 557
Region:
   closed, 772
   connected, 868
   under a graph, 263, 270
   open, 868
   simple, 876
   simply-connected, 870
   between two graphs, 297
   of Type I, 800
   of Type II, 802
Regular partition, 271
Related rates, 155
Relative error, 163
Relative maximum and minimum, 174
Remainder in Taylor's Formula, 630
Removable discontinuity, 83
Resonance, 967
Revenue function, 241
Reversing order of integration, 805
Revolution:
   ellipsoid of, 677
   paraboloid of, 680
   solid of, 469
   surface of, 486
Richter scale, 334
Riemann, Georg Bernhard, 269, 607
Riemann integral, 269
Riemann sum, 269, 789, 822
Right-hand limit, 65, 72
Right-hand rule, 644, 664
Roberval, Gilles, 284, 529
Rolle, Michel, 182
Rolle's Theorem, 182
Root function, 41
Root Test, 606

Rose:
  eight-leaved, 542
  five-leaved, 543
  four-leaved, 539, 545
  three-leaved, 542
Rotation of axes, 562

Saddle point, 770
Scalar, 651
Scalar field, 852
Scalar multiple of a vector, 651
Scalar product, 655
Scalar projection, 659
Scalar triple product, 666
Secant function, A17
  derivative, 122
  graph, A24
Secant line, xxiii, 56
Second derivative, 141, 737
Second Derivative Test, 197
Second Derivatives Test, 770
Sector of a circle, 419
Separable differential equation, 509
Sequence, xxvi, 570
  bounded, 576
  convergent, 571
  decreasing, 575
  divergent, 571
  increasing, 575
  limit of, xxvi, 571
  monotonic, 575
  of partial sums, 580
  term of, 570
Series, 579
  absolutely convergent, 601
  alternating, 596
  alternating harmonic, 598, 601
  binomial, 627
  conditionally convergent, 602
  convergent, 580
  divergent, 580
  geometric, 581
  harmonic, 583, 591
  Maclaurin, 619
  p-, 590
  partial sum of, 580
  power, 610
  strategy for testing, 608
  sum of, xxviii, 580
  Taylor, 619
  term of, 579
Series solution of differential equation, 971
Set, 3
Shell method, 476
Shifts of functions, 43
Sigma notation, 254
Simple curve, 868
Simple harmonic motion, 964

Simple region, 876
Simple solid region, 916
Simply-connected region, 870
Simpson, Thomas, 445
Simpson's Rule, 445
  error in, 445
Sine function, A17
  derivative, 121
  graph, A23
  power series, 620
Skew lines, 671
Slant asymptote, 224
Slope, 94
  of tangent line, 51
Slope-intercept equation of a line, 14
Smooth curve, 691
Snell's Law, 237
Solid, 464
  volume of, 465, 791
Solid of revolution, 467
Solution curve, 929
Source, 921
Space, three-dimensional, 644
Space curve, 684
Speed, 98, 698
Sphere, 647
Spherical coordinates, 705, 833
Spiral, 542
Spring constant, 494, 964
Squeeze Theorem, 77, 573
Standard basis vectors, 653
Steady state solution, 970
Stokes, Sir George, 910
Stokes' Theorem, 911
Strategy:
  for integration, 433
  for solving differential equations, 944
  for testing series, 608
Stretching functions, 44
Substitution Rule, 290, 293
Sum:
  of functions, 36
  of a geometric series, 581
  of an infinite series, 580
  Riemann, 269
  for vectors, 651
Summation notation, 254
Surface, 645
  closed, 904
  computer-drawn, 717, 718, 721
  level, 720
  oriented, 903
  parametric, 891
  quadric, 677
  of revolution, 486
  smooth, 893
Surface area:
  of a parametric surface, 894

Surface area (*continued*)
  of a sphere, 895
  of a surface of revolution, 488, 533
  of a surface $z = f(x, y)$, 896
Surface integral, 899, 901, 905
Symmetric equations of a line, 670
Symmetry, 21, 33, 222
  in polar graphs, 540
Symmetry principle, 503

Tables of integrals, 434, endpapers
  use of, 457
Tangent function, A17
  derivative, 122
  graph, A24
Tangential component of acceleration, 701
Tangent line:
  to a curve, xxiii, 51, 97
  to a parametric curve, 526
  to a polar curve, 540
  slope of, 51, 97
  to a space curve, 685
  vertical, 220
Tangent line approximation, 161
Tangent plane:
  to a parametric surface, 892
  to a surface $f(x, y, z) = k$, 765
  to a surface $z = f(x, y)$, 741
Tangent plane approximation, 744
Tangent problem, xxiii, 50
Tangent vector, 685
Tautochrone problem, 524
Taylor, Brook, 619
Taylor polynomial, 629
Taylor series, 619
Taylor's Formula, 630
Techniques of integration, summary, 433
Telescoping sum, 258
Term:
  of a sequence, 570
  of a series, 579
Test for Concavity, 195
Test for Monotonic Functions, 189
Tests for convergence and divergence of series:
  Alternating Series Test, 596
  Comparison Test, 592
  Integral Test, 588
  Limit Comparison Test, 594
  Ratio Test, 603
  Root Test, 606
  summary of tests, 608
  Test for Divergence, 584
Third derivative, 141
Three-leaved rose, 542
Torque, 667
Torricelli, Evangelista, 529
Torricelli's Law, 59
Torus, 474

Total differential, 743
Trace, 677
Trajectory, 700
Transcendental function, 43
Transcendental number, 330
Transformation, 838
Transformed functions, 43
Trapezoidal Rule, 441
   error in, 442
Triangle inequality, 8, 661
Triangle law, 651
Trigonometric functions, A17
   derivatives of, 120
   graphs of, A23
   integrals of, 406
   inverse, 365
   limits of, 117
Trigonometric identities, A20
Trigonometric integrals, 406
Trigonometric substitution, 412
Triple integrals, 822
   in cylindrical coordinates, 832
   in spherical coordinates, 834
Triple scalar product, 660
Trochoid, 525
Type I plane region, 800
Type II plane region, 802
Type 1 solid region, 824
Type 2 solid region, 826
Type 3 solid region, 826

Underdamping, 966
Undetermined coefficients, 958
Union of sets, 3
Unit normal vector, 696
Unit tangent vector, 685
Unit vector, 654
Upward concavity, 194

Value of a function, 27
Variable:
   change of, 290
   dependent, 30, 714, 752
   dummy, 269
   independent, 30, 714, 752
   intermediate, 752
Variation of parameters, 961
Vascular branching, 238
Vector(s), 648
   acceleration, 698

Vector(s) (*continued*)
   addition of, 651
   angle between, 656
basis, 653
   components of, 649
cross product of, 662
   difference of, 652
   dot product of, 655
   force, 854
   gradient, 760, 762
   i, **j**, and **k**, 653
   length of, 650
   linear combination of, 947
   magnitude of, 650
   normal, 671, 893
   orthogonal, 657
   parallel, 652
   perpendicular, 657
   position, 649
   projection, 659
   representation of, 649
   scalar multiple of, 651
   subtraction of, 652
   sum of, 651
   tangent, 685
   three-dimensional, 649
   two-dimensional, 649
   unit, 654
   unit normal, 696
   unit tangent, 685
   velocity, 697
   zero, 650
Vector equation of a line, 669
Vector equation of a plane, 672
Vector field, 852
   conservative, 855
   curl of, 893
   divergence of, 886
   flux of, 905
   irrotational, 886
   velocity, 853
Vector product, 662
Vector space, 653
Vector triple product, 668
Vector-valued function, 683
   continuous, 684
   derivative of, 685
   integral of, 688
   limit of, 683

Velocity, xxiv, 52
   average, xxv, 55
   instantaneous, 55
Velocity field, 853
Velocity vector, 697
Verhulst, 514
Vertex:
   of ellipse, 552
   of hyperbola, 553
   of parabola, 21, 549
   of paraboloid, 679
Vertical Line Test, 31
Vertical tangent line, 220
Vibrating spring, 963
Volume, 465
   by cross-sections, 465
   by cylindrical shells, 475
   by disks, 468
   by double integrals, 791
   of a solid of revolution, 467
   by triple integrals, 828
   by washers, 469

Wallis, John, xxiv
Washer method, 469
Wave equation, 740
Weierstrass, Karl, 431
Weierstrass substitution, 431
Weight, 493
Whispering gallery, 557
Work, 493, 660, 863
Wren, Sir Christopher, 533

x-axis, 10, 644
x-coordinate, 10, 645
x-intercept, 23, 222
xy-plane, 644
xz-plane, 644

y-axis, 10, 644
y-coordinate, 10, 645
y-intercept, 23, 222
yz-plane, 644

z-axis, 644
z-coordinate, 645
Zeno, xxvi
Zeno's paradoxes, xxvi, xxviii
Zero vector, 650
Zone of a sphere, 492

Trigonometric Forms

63 $\displaystyle\int \sin^2 u\, du = \frac{1}{2}u - \frac{1}{4}\sin 2u + C$

64 $\displaystyle\int \cos^2 u\, du = \frac{1}{2}u + \frac{1}{4}\sin 2u + C$

65 $\displaystyle\int \tan^2 u\, du = \tan u - u + C$

66 $\displaystyle\int \cot^2 u\, du = -\cot u - u + C$

67 $\displaystyle\int \sin^3 u\, du = -\frac{1}{3}(2 + \sin^2 u)\cos u + C$

68 $\displaystyle\int \cos^3 u\, du = \frac{1}{3}(2 + \cos^2 u)\sin u + C$

69 $\displaystyle\int \tan^3 u\, du = \frac{1}{2}\tan^2 u + \ln|\cos u| + C$

70 $\displaystyle\int \cot^3 u\, du = -\frac{1}{2}\cot^2 u - \ln|\sin u| + C$

71 $\displaystyle\int \sec^3 u\, du = \frac{1}{2}\sec u \tan u + \frac{1}{2}\ln|\sec u + \tan u| + C$

72 $\displaystyle\int \csc^3 u\, du = -\frac{1}{2}\csc u \cot u + \frac{1}{2}\ln|\csc u - \cot u| + C$

73 $\displaystyle\int \sin^n u\, du = -\frac{1}{n}\sin^{n-1} u \cos u + \frac{n-1}{n}\int \sin^{n-2} u\, du$

74 $\displaystyle\int \cos^n u\, du = \frac{1}{n}\cos^{n-1} u \sin u + \frac{n-1}{n}\int \cos^{n-2} u\, du$

75 $\displaystyle\int \tan^n u\, du = \frac{1}{n-1}\tan^{n-1} u - \int \tan^{n-2} u\, du$

76 $\displaystyle\int \cot^n u\, du = \frac{-1}{n-1}\cot^{n-1} u - \int \cot^{n-2} u\, du$

77 $\displaystyle\int \sec^n u\, du = \frac{1}{n-1}\tan u \sec^{n-2} u + \frac{n-2}{n-1}\int \sec^{n-2} u\, du$

78 $\displaystyle\int \csc^n u\, du = \frac{-1}{n-1}\cot u \csc^{n-2} u + \frac{n-2}{n-1}\int \csc^{n-2} u\, du$

79 $\displaystyle\int \sin au \sin bu\, du = \frac{\sin(a-b)u}{2(a-b)} - \frac{\sin(a+b)u}{2(a+b)} + C$

80 $\displaystyle\int \cos au \cos bu\, du = \frac{\sin(a-b)u}{2(a-b)} + \frac{\sin(a+b)u}{2(a+b)} + C$

$\displaystyle\int \sin au \cos bu\, du = -\frac{\cos(a-b)u}{2(a-b)} - \frac{\cos(a+b)u}{2(a+b)} + C$

82 $\displaystyle\int u \sin u\, du = \sin u - u\cos u + C$

83 $\displaystyle\int u \cos u\, du = \cos u + u\sin u + C$

84 $\displaystyle\int u^n \sin u\, du = -u^n \cos u + n\int u^{n-1}\cos u\, du$

85 $\displaystyle\int u^n \cos u\, du = u^n \sin u - n\int u^{n-1}\sin u\, du$

86 $\displaystyle\int \sin^n u \cos^m u\, du = -\frac{\sin^{n-1} u \cos^{m+1} u}{n+m} + \frac{n-1}{n+m}\int \sin^{n-2} u \cos^m u\, du$

$\displaystyle\quad\quad = \frac{\sin^{n+1} u \cos^{m-1} u}{n+m} + \frac{m-1}{n+m}\int \sin^n u \cos^{m-2} u\, du$

Inverse Trigonometric Forms

87 $\displaystyle\int \sin^{-1} u\, du = u\sin^{-1} u + \sqrt{1-u^2} + C$

88 $\displaystyle\int \cos^{-1} u\, du = u\cos^{-1} u - \sqrt{1-u^2} + C$

89 $\displaystyle\int \tan^{-1} u\, du = u\tan^{-1} u - \frac{1}{2}\ln(1+u^2) + C$

90 $\displaystyle\int u \sin^{-1} u\, du = \frac{2u^2 - 1}{4}\sin^{-1} u + \frac{u\sqrt{1-u^2}}{4} + C$

91 $\displaystyle\int u \cos^{-1} u\, du = \frac{2u^2 - 1}{4}\cos^{-1} u - \frac{u\sqrt{1-u^2}}{4} + C$

92 $\displaystyle\int u \tan^{-1} u\, du = \frac{u^2 + 1}{2}\tan^{-1} u - \frac{u}{2} + C$

93 $\displaystyle\int u^n \sin^{-1} u\, du = \frac{1}{n+1}\left[u^{n+1}\sin^{-1} u - \int \frac{u^{n+1}\, du}{\sqrt{1-u^2}} \right], \quad n \neq -1$

94 $\displaystyle\int u^n \cos^{-1} u\, du = \frac{1}{n+1}\left[u^{n+1}\cos^{-1} u + \int \frac{u^{n+1}\, du}{\sqrt{1-u^2}} \right], \quad n \neq -1$

95 $\displaystyle\int u^n \tan^{-1} u\, du = \frac{1}{n+1}\left[u^{n+1}\tan^{-1} u - \int \frac{u^{n+1}\, du}{1+u^2} \right], \quad n \neq -1$